The Palgrave Handbook of Critical Menstruation Studies

Chris Bobel · Inga T. Winkler ·
Breanne Fahs · Katie Ann Hasson ·
Elizabeth Arveda Kissling · Tomi-Ann Roberts
Editors

The Palgrave Handbook of Critical Menstruation Studies

Editors
Chris Bobel
Department of Women's, Gender, and
Sexuality Studies
University of Massachusetts Boston
Boston, MA, USA

Breanne Fahs
Women and Gender Studies & Social
and Cultural Analysis
Arizona State University
Glendale, AZ, USA

Elizabeth Arveda Kissling
Women's & Gender Studies
Eastern Washington University
Cheney, WA, USA

Inga T. Winkler
Institute for the Study of Human Rights
Columbia University
New York, NY, USA

Katie Ann Hasson
Center for Genetics and Society
Berkeley, CA, USA

Tomi-Ann Roberts
Department of Psychology
Colorado College
Colorado Springs, CO, USA

ISBN 978-981-15-0613-0   ISBN 978-981-15-0614-7   (eBook)
https://doi.org/10.1007/978-981-15-0614-7

Cover image: © Jen Lewis

This Palgrave Macmillan imprint is published by the registered company Springer Nature
Singapore Pte Ltd.
The registered company address is: 152 Beach Road, #21-01/04 Gateway East, Singapore
189721, Singapore

ACKNOWLEDGMENTS

A book project of this scope and scale requires the creativity, grit, tenacity, and goodwill of legions—more than can be properly acknowledged here. Our exhaustive outreach depended on many intersecting networks of countless scholars, advocates, and others who helped connect us with the right person to write the right piece at the right time. We know that every chapter in this book is possible because of the labor of many and we regret that we cannot list each of these behind-the-scenes helpers.

But we will take a moment to explicitly name a few people and organizations whose support of this project was invaluable.

Sharra Vostral helped conceive the rationale and framework for this handbook. Her visionary work crafting the proposal for this Handbook set the project in motion, and now, several years later, we remain in her debt. Our thanks also go to the anonymous peer reviewers who provided incisive feedback [and encouragement] at both proposal and clearance review stages. They, too, helped shape this Handbook.

We leaned heavily on several editors and editorial assistants along the way. In the early days, Michelle Chouinard managed the communication and organization of our call for proposals. Trisha Maharaj, Victoria Miller, Laura Charney, and Sydney Amoakoh provided invaluable support for many chapters. During the final and all-important stage of preparing the book for production, Sydney Amoakoh also single-handedly managed the abstracts, bios, images, figures and tables, and various consent forms plus more for more than 130 contributors. Her calm efficiency and capacity to track detail is a marvel. We also benefited from the hand of Dakota Porter, who stepped in to help with myriad administrative tasks in the last phase of manuscript preparation. Many thanks also to Virginia Roaf who provided editorial support and *special appreciation* to the peerless Perri Schenker whose invaluable editorial skills were essential to producing this resource. Others who stepped in at key moments include Adrian Jjuuko, Marcy Darnovsky, the Center for Genetics and Society, Radu Dondera, Dawn Dow, and Anna Krakus. We thank them each.

We also note with gratitude the team at Palgrave Macmillan/Springer Nature, especially Holly Tyler who first pitched the idea of a handbook to Chris with irresistible enthusiasm, and Joshua Pitt who succeeded her and walked with us throughout the subsequent years of this project. He and editorial assistant Sophie Li responded to every query—the trivial, the profound, and the anxious–with equanimity and unflagging support for our vision for this book. "Thank you" is too small a phrase.

Finally, we appreciatively acknowledge those who donated resources to support the book. First, we thank artist Jen Lewis, self-described 'menstrual designer' whose arresting 2015 macrophotograph "The Crimson Wave" (2015) graces our cover. Second, we express our gratitude to our generous funders—the Center for the Study of Social Difference at Columbia University through its Working Group on Menstrual Health and Gender Justice and the University of Massachusetts Boston Periodic Multi-Year Review Fund. Without their support, we would not have been able to meet our ambitious goal of publishing this robust and richly diverse body of work. And above all, we express our sincerest gratitude to the Water Supply and Sanitation Collaborative Council whose abiding belief in the value of this book enabled us to not only engage crucial editorial help, but also covered the fees necessary to make the digital edition permanently open access worldwide. From the very beginning, our fervent hope for this book was that it function as a reliable and accessible 'go to' resource for the widest possible audience, and WSSCC's generosity makes this truly possible. Thank you!

About the Cover: Beauty in Blood— A Macrophotographic Lens on Menstruation, Body Politics, and Visual Art

"The Crimson Wave" (2015) exemplifies the *Beauty in Blood* collection, my feminist, bioartography project that seeks to confront social taboos pertaining to menstruation and the female body through macrophotography of menstrual fluid. I challenge the notion that menstruation is "gross," "vulgar," or "unrefined" through candid, real-life photos of my menstrual blood which force viewers to see and think about menstruation in an entirely new way. There is an abstract artistic quality when blood meets water that warrants a closer look not only by women but also by society as a whole. Capturing the artful quality of this natural occurrence is my way of progressing society's view and conversation around menstruation as well as redefining some traditional fine art aesthetics.

In my opinion, society's squeamishness about menstruation is completely ridiculous considering its graphic consumption of bloodshed through violence in pop culture entertainment, that is, blood sports like boxing, hockey, and wrestling; video games like *Call of Duty*; shows and movies like *Dexter* and *Twilight*; and even the news media. Pacifying social taboos only serves to give more power to society than to the self, and as women we have done that for far too long. My work quashes this taboo, reclaims feminine power, and puts menstruation in the context it so rightly deserves.

Creating each piece of work is a four-step process: media (aka blood/menstrual fluid) collection, design layout (aka pouring), photoshoot, and finally photo selection. The images of menstrual fluid are obtained in two different manners. During the early stages, we captured images by mounting a camera on a tripod and strategically angling it over the toilet bowl, so Rob, my husband, artistic collaborator and project photographer, could snap photos as soon as I poured the freshly collected menstrual fluid from my cup. After several shoots and a desire to capture more dynamic imagery, we began shooting in a small aquarium (about 15 gallons). Rob discovered a fluid photography technique that greatly improved our final designs. Both Rob and myself approach each shoot with an experimental spirit and love to play with variables to see how it will effect the menstrual fluid's movement in the water, for example, salt density, ratio of freshwater to saltwater, and tools to distribute the blood. The clarity of the final images can be credited to the use of saltwater, which slows menstrual fluid movement, and macro lenses, which show us more than the naked eye can see.

If I have learned anything over the past few years of producing *Beauty in Blood* it is that menstruation matters more than most people in society are willing to recognize; it is deeply embedded in our global body politics and is a major contributor to the vast gender inequity between men and women today. Institutionalized hierarchies maintain and support the outdated patriarchal belief that menstruation makes the female body inferior to the male body. Billions of dollars are spent annually trying to make women's bodies conform to male "norms" by suppressing the natural menstrual cycle through hormonal birth control. The feminine "hygiene" industry perpetuates taboo thinking by suggesting the monthly cycle is dirty and socially impolite; it should be concealed in frilly pink wrappers like candy and only very loosely referenced with blue liquid in product commercials. In my experience, women and men are hungry for an authentic dialogue about menstruation and all that encompasses.

It is clear the time is now to stand up and speak out on behalf of menstruation. It is a natural, messy but beautiful part of life, and just because it is not a shared experience doesn't mean it needs to be a divisive topic that aids gender inequity. *Beauty in Blood* asserts that menstruation needs to be seen to help normalize the menstruating body and to acknowledge this part of the menstruator's life experience by inviting the viewer to take a closer look and reflect on their personal gut reactions to the subject of "menstruation."

Jen Lewis
Menstrual Designer

CONTENTS

NOTES ON CONTRIBUTORS

Jane Hartman Adamé is a customer engagement and user research professional and former hairdresser living with hypermobile Ehlers-Danlos Syndrome, a connective tissue disorder. Jane is a co-creator of FLEX Cup, an inclusively designed menstrual cup made in collaboration with Andy Miller, a medical device designer. Jane turns customers into co-designers from her home in Oakland, CA.

Rockaya Aidara is a gender, equity, and human rights policy specialist with over 10 years' experience in development and international cooperation. At the Water Supply and Sanitation Collaborative Council she designed and implemented the Joint Programme on Gender, Hygiene and Sanitation, which used menstrual hygiene management as an entry point to address gender inequalities in WASH. Prior to joining WSSCC, Rockaya worked with the UN Agency on Gender Equality and Women's Empowerment, the UN Intergovernmental Panel on Climate Change, and the European Foundation FEDRE. She supported programs on women's political participation, as well as advocacy campaigns on violence against women, peace and security, and climate change.

Lise Ulrik Andreasen is a Ph.D. fellow at The Danish School of Education at Aarhus University, Denmark. Based on fieldwork, her Ph.D. project examines lived and embodied experiences of young menstruators in Denmark. Lise's research on menstruation and youth intertwines with her interests in feminist theories of gender, sexuality, science, methodologies, materiality, affect, politics, utopia, care, and ethics. She holds an M.A. in women's studies from University College Cork, Ireland, and an M.A. in educational anthropology from Aarhus University. Lise is a member of the Society for Menstrual Cycle Research and lives in Copenhagen, Denmark with her partner and two children.

Robert Bain is a statistics specialist at UNICEF and has been a member of the WHO/UNICEF Joint Monitoring Programme for Water Supply, Sanitation and Hygiene since 2014. Prior to joining UNICEF, he worked as a researcher for the Water Institute at the University of North Carolina at Chapel Hill and the University of Bristol with a primary research focus on monitoring drinking water quality. Robert received his Master of Engineering from the University of Cambridge and MIT in 2008.

Jessica L. Barnack-Tavlaris is an associate professor in the Department of Psychology at The College of New Jersey; she teaches classes in health and social psychology, and research methods. Her research interests include attitudes toward menstruation, stigma toward women's reproductive health, and the transition from infertility to motherhood. She is the book and media review editor for *Women's Reproductive Health*, the official journal of the *Society for Menstrual Cycle Research*.

Anja Benshaul-Tolonen has been an assistant professor of economics at Columbia University's Barnard College since 2015, working on economic development and applied economics. One strand of her research focuses on health and gender, including menstruation and school absenteeism, stigma around menstruation, and household health investment and knowledge. Another strand focuses on natural resource extraction and how the sector interacts with local economic development, health, and gender. Her research methods include quasi-experimental analysis and randomized control trials, and large datasets. She also teaches econometrics and development economics.

Mayuri Bhattacharjee is a menstrual health educator and trainer who has reached more than 8000 menstruators through her Menstrual Health Workshops in Assam and West Bengal. As a changemaker of change.org's She Creates Change Fellowship, she runs a digital campaign called *Dignity in Floods* (www.change.org/dignityinfloods) to build women-friendly flood relief shelters in Assam. She is a climate reality leader at The Climate Reality Project and a World Economic Forum Global Shaper. She won the 2019 Ton Schouten Award for WASH storytelling from IRC WASH.

Ingrid Goldbloom Bloch is a self-taught artist who sees beauty in common objects. She is known for creating art that is humorous and thought-provoking, and transforms everyday objects into something entirely different from their intended purpose with the goal of creating conversations. Composed of hardware store finds, street debris, and stumbled upon items, her mixed-media sculptures draw upon the traditions of contemporary fiber arts and assemblage. Ingrid's work has been collected by museums as far-reaching as Germany's The Bikini Museum, Azerbaijan's The Waste to Art Museum, and Ripley's Believe It or Not! Museums in Orlando and Los Angeles. She lives in Needham, Massachusetts with her husband, two teenage sons, and a parakeet.

Chris Bobel is professor and chair of women's, gender and sexuality studies at the University of Massachusetts Boston. Chris is the author of *The Managed Body: Developing Girls and Menstrual Health in the Global South* (Palgrave Macmillan), *New Blood: Third Wave Feminism and the Politics of Menstruation* (Rutgers University Press), *The Paradox of Natural Mothering* (Temple University Press), the co-edited collections (with Samantha Kwan) *Embodied Resistance: Breaking the Rules, Challenging the Norms,* and *Body Battlegrounds: Transgressions, Tensions and Transformations* (both with Vanderbilt University Press). Chris is the past president of the Society for Menstrual Cycle Research and a fellow of the Working Group on Menstrual Health & Gender Justice at Columbia University. She is often consulted by the mainstream media about the rapidly growing menstrual activist movement. She is at work on a new ethnographic project exploring contemporary activism inspired by grief and trauma.

Lacey Bobier is a sociology Ph.D. candidate at the University of Michigan. Her research focuses on adolescent girls, sexual subjectivity, and their roles in the gender power structure. Her previous publications examine early childhood sexuality education, while her current work considers the construction and regulation of girls' bodies through such mediums as magazines and school policies.

Danielle Boodoo-Fortunè is a poet and visual artist from Trinidad and Tobago. Her first collection of poems, *Doe Songs* (Peepal Tree Press) was awarded the 2019 Bocas OCM Price in Poetry. Her paintings have been featured in numerous arts publications and exhibitions in the Caribbean and abroad.

Gabriella Boros has shown her prints, paintings, and multimedia works nationally and internationally. Currently focusing on woodblock prints and handmade books, she also does nature photography, acrylic on wood panel, drawings, sculptures, and found object cheese boxes. Gabriella's narratives reflect her European parentage, Israeli birth, and American childhood. Her latest works include a solo show at Stockholm's Ze Zig Zag Zone and a print in the "Spinoza: Marrano of Reason" show in Amsterdam. In 2020, she will complete a residency at the Bernheim Arboretum and Research Forest, where she will create a series of installations commemorating Kentucky women and the native plants that represent them.

Chandra Bozelko was the first incarcerated person to have a regular byline in a publication outside of prison. Her newspaper column, "Prison Diaries," became an award-winning blog. She has won many awards and fellowships for her writing and criticism of the United States criminal justice system. Bozelko is now a syndicated columnist with Creators Syndicate and serves as the vice president of the National Society of Newspaper Columnists.

Janelle Chambers is a mother of three children, two sons and a daughter. She identifies as a lesbian woman and is in a long-term relationship with a very loving wife.

Venkatraman Chandra-Mouli leads the work on adolescent sexual and reproductive health (ASRH) in the World Health Organization's Department of Reproductive Health and Research. His work includes building the evidence base on ASRH, and supporting countries to translate evidence into action through well-conceived and well-managed policies and programs. His experience is global in scope and spans over 30 years, during which he has contributed substantially to a number of WHO publications and the work of numerous national-level bodies and front-line organizations around the world. Dr. Chandra-Mouli has presented in global, regional, and national conferences, and (co)authored books, book chapters, articles, blog pieces, and around 90 peer-reviewed journal articles.

Jieun Choi is a freelance journalist and videographer currently based in Seoul, South Korea. She finds beauty in telling stories of the unheard. With relentless curiosity, Jieun dives into various realms of the society in which she lives. Previously, she worked at a media startup, Korea Exposé, covering mainly society, culture, and gender issues of the Korean Peninsula. She has experience working in arts and media scene in Seoul, Hong Kong, and Melbourne. Choi holds a B.A. in fine arts from the University of Hong Kong.

Joan C. Chrisler is the Class of 1943 Professor Emerita of Psychology at Connecticut College, where she taught courses in gender, social, and health psychology. She is internationally known for her research and writing on the psychology of women and women's health issues, including menstruation, PMS, body image, and aging. She is editor of the *Women's Reproductive Health* journal. Her most recent books are *The Routledge Handbook of Women's Sexual and Reproductive Health* (forthcoming), *Woman's Embodied Self: Feminist Perspectives on Identity and Image,* and *Lectures on the Psychology of Women.*

Ilana Cohen is an independent researcher. She holds an M.A. in anthropology and women, gender, and sexuality studies from Brandeis University, where she studied the menstrual hygiene management sector and menarche ceremonies in Tamil Nadu, India. She earned B.A.s in cultural anthropology and Jewish women and gender studies from the School of General Studies at Columbia University and List College at the Jewish Theological Seminary. She is a research associate at Verité, a nonprofit organization dedicated to ensuring safe, fair, and legal working conditions worldwide.

Berkley D. Conner is a doctoral student in communication studies with a concentration in rhetoric and public advocacy at the University of Iowa. Her scholarship broadly examines health and medicine from a humanistic perspective, particularly around cultural rhetorics of menstrual health. She is especially interested in how menstruators' subjectivities are negotiated between their capacity as regulated spaces and their capacity to weaponize their bodies for resistive purposes. Her current research explores medical and public discourses about various modes of menstrual management.

Laura Crane is an associate professor at the Centre for Research in Autism and Education (CRAE) at UCL Institute of Education, where she is also deputy director. Laura's work focuses on two main areas: the education and healthcare experiences of autistic people, their families and the professionals who work with them; and developing an evidence base to promote access to justice for witnesses on the autism spectrum (in both the criminal and family justice systems). Laura is a strong advocate of public engagement and community outreach; ensuring that research is accessible to the public, to policymakers, and—importantly—to the autistic community and their allies.

Amina Darwish is the Muslim chaplain at Columbia University. She has received ijaza in Islamic studies and in the 10 Qira'at. She also studied individually with various Islamic scholars. She earned a B.S. in chemical engineering from Kuwait University, an M.S. in industrial engineering and a Ph.D. in chemical engineering from the University of Cincinnati. She previously served as an adjunct professor in Islamic studies at Northern Kentucky University, the Muslim chaplain at the University of Cincinnati, and as the content development coordinator at the Muslim Youth of North America. She is the founder and CEO of Mercy in Action.

Catarina de Albuquerque is chief executive officer for the global multi-stakeholder partnership, Sanitation and Water for All. From 2008 to 2014, she was the first UN Special Rapporteur on the right to safe drinking water and sanitation. Between 2004 and 2008 she presided over the negotiations of the Optional Protocol to the International Covenant on Economic, Social and Cultural Rights, which the UN General Assembly approved by consensus on December 10, 2008. Ms. de Albuquerque was awarded the Human Rights Golden Medal by the Portuguese Parliament (December 10, 2009) for outstanding work in the area of human rights.

Jac Dellaria is a queer, trans illustrator, and cartoonist currently based in Chicago, IL. His work focuses on his personal experiences with transitioning and managing the balance between one's sexual orientation and gender identity. Jac studied comics at the University of Wisconsin–Madison, and also creates work under the name Wrigley. Jac's art can be found at www.jacdellaria.com.

Radhika Desai has a Ph.D. in sociology from Indiana University, Bloomington. Her work spans women's work, early childhood development, financial inclusion, livelihood promotion, microfinance, and entrepreneurship. Radhika has brought together knowledge and practice as a program manager, social impact and gender evaluation specialist, researcher, and teacher for postgraduate students of women's studies. Her writings include *Women's Work Counts: Feminist Arguments for Human Rights at Work* (PWESCR, 2015) and *"Livelihoods of the Poor" in the 2011 State of India's Livelihoods Report* (SAGE Publications, 2011).

Heather Dillaway is a professor of sociology, director of the Bachelor of Science in Public Health Program, and associate dean in the College of Liberal Arts and Sciences at Wayne State University in Detroit, Michigan. Dillaway's research focuses on women's menopause experiences and the reproductive health experiences of women with physical disabilities.

Anna Druet is a researcher, science and education manager at Clue, as well as a public health and femtech advocate. She aims to help people know more about their bodies and raise awareness of the central importance of reproductive and biological autonomy to global welfare.

Noémie Elhadad is an associate professor of biomedical informatics, affiliated with Columbia University's Computer Science and Data Science Institute. Her research interests lie at the intersection of machine learning, natural language processing, and medicine. She investigates ways in which observational clinical data and patient-generated data can enhance access to relevant information for clinicians, patients, and health researchers alike, and the ultimate potential of such access to impact healthcare and the health of patients.

Daniel A. Epstein is an assistant professor in the Department of Informatics at the University of California, Irvine. His research is in the area of human–computer interaction (HCI), where he studies how personal tracking technology can acknowledge the realities of everyday life. He leverages this understanding to develop and evaluate new apps and interfaces which better account for those realities. He holds a Ph.D. in computer science and engineering from the University of Washington.

Mindy J. Erchull is a professor of psychological science and a member of the Women's and Gender Studies Program at the University of Mary Washington. She has a Ph.D. in social psychology from Arizona State University. Her research focuses on the objectification and sexualization of women, feminist identity, division of labor and parenting, and women's reproductive health. Her menstrual cycle research has largely focused on education about and attitudes toward menstruation and menstruators.

Breanne Fahs is a professor of women and gender studies at Arizona State University, where she specializes in studying women's sexuality, critical embodiment studies, radical feminism, and political activism. She has authored five books and three edited collections: *Performing Sex, Valerie Solanas, Out for Blood, Firebrand Feminism, Women, Sex, and Madness, The Moral Panics of Sexuality, Transforming Contagion*, and *Burn it Down*. She also works as a clinical psychologist in a private practice, where she specializes in sexuality, couples work, and trauma recovery.

Johanna Falzone attributes her creative roots to growing up in the 90s under the influence of punk rock music, feminism, Nickelodeon cartoons, and Barbie. These forces have incited her attraction to pretty imagery ranging from the grotesque to the overtly feminine with whimsical nods to childhood in her paintings, illustrations, poetry, films, short stories, and screenplays. Johanna is also classically trained in ballet and modern dance. She attended Suzanne Farrell's Young Dancer's Workshop in 2007 and 2008; as well as Canada's Royal Winnipeg Ballet School's 2008 Summer Program. Currently based out of Florida, she remains a fierce Winnipeg Jets fan and Tim Hortons iced coffee and donut lover.

Alex Farley has worked with WoMena as a research and project management officer. She holds an M.Sc. in African development from the London School of Economics, with a specialty in gender-sensitive humanitarian policy and programming.

Sarah Fox is a President's Postdoctoral Fellow at Carnegie Mellon University's Human Computer Interaction Institute. Her research focuses on how technological artifacts challenge or propagate social exclusions by examining existing systems and building alternatives. Her work has earned awards in leading computing venues, including ACM, CSCW, CHI, and DIS, and has been featured in the *Journal of Peer Production*, *Design Issues*, and *New Media and Society*. She holds a Ph.D. in human-centered design and engineering from the University of Washington.

S. E. Frank is a Ph.D. candidate at the University of Wisconsin–Madison in the Department of Sociology. She currently studies menstruation in United States institutions, including law and the military. Sarah lectures for Madison's Department of Legal Studies and Sociology and leads graduate teaching trainings across the university. The present research on queering menstruation won the Alpha Kappa Delta Sociology Honors Society Graduate Student Paper Award at the American Sociological Association in 2019 and the 3-Minute Thesis Competition at the Midwest Sociological Society in 2019. Follow her work at https://teachingfrankly.com.

Rosa Freedman is the inaugural professor of law, conflict, and global development at the University of Reading. She received her LLB, LLM, and Ph.D. from the University of London, and is a member of Gray's Inn, the UN Secretary-General's Civil Society Advisory Board, and the UK Foreign Office's Women Peace and Security Steering Committee. Freedman's research and publications focus on the UN, particularly human rights bodies and systems, peacekeeping, and accountability for human rights abuses committed during such operations. Her publications include two monographs, two co-edited collections, and articles in the *American Journal of International Law, European Journal of International Law, Leiden Journal of International Law,* and *Human Rights Quarterly,* among others.

Saniya Lee Ghanoui is a Ph.D. candidate in history at the University of Illinois at Urbana-Champaign. Her dissertation is a transnational cultural history that investigates the development of the movements for sex education in the United States and Sweden from 1910 through 1962, the interactions between these two countries, and their signature method of education: the sex education film.

Krystal Nandini Ghisyawan is an independent Indo-Caribbean, queer feminist scholar working in the areas of female same-sex desire, LGBTQI advocacy, and women in Hinduism. She holds a Ph.D. in sociology from the University of the West Indies and is a former postdoctoral associate at Rutgers University. She is a director on the board of the Silver Lining Foundation and has guided the organization's research and development agenda since 2014. She is currently completing her manuscript, *Erotic Cartographies: Mapping Caribbean Subjectivities, Spaces, and Queer Decolonial Praxis*, which explores the space-making practices of same-sex loving women in Trinidad.

Carla Giacummo has channeled her passion for promoting open discussion on menstruation and elevating it as a vital sign into building Eco-Ser in 2012. She has also been a Menstrupedia co-publisher for Spanish since 2015. She regards the platform as the perfect tool for girls around 9 to learn about periods, and as an invaluable community of nonprofits, health institutions, teachers, doctors, and others who promote menstrual literacy in Latin America, the United States, Spain, and other countries worldwide. Driven by her love for the art of connection, Giacummo has also worked as an executive secretary, piloted her own clay atelier for children 10 and older, and is the mother of two boys.

Sarah Goddard is a global health and international development professional. Her work has focused on governance, health, water, and sanitation, and sustainable urban development in low- and middle-income countries. Sarah has a Master of Public Health and Master of Arts in international affairs from Columbia University and an undergraduate degree from Brown University.

Beth Goldblatt is an associate professor in the Faculty of Law at the University of Technology Sydney, Australia, and an honorary associate professor in the School of Law at the University of the Witwatersrand, South Africa. She works on equality, human rights, comparative constitutional law and feminist legal theory, focusing on gender and poverty. She is the author of *Developing the Right to Social Security—A Gender Perspective* and co-editor of two collections on women's social and economic rights. Beth is a member of the UTS Law Health Justice Research Centre and a co-convener of the UTS Feminist Legal Research Group. She previously worked as a researcher on disability issues.

Alma Gottlieb is the (co)author/(co)editor of nine books. Gottlieb began her publishing career with *Blood Magic: The Anthropology of Menstruation*, an award-winning collection that helped inaugurate a modern, feminist approach to menstruation cross-culturally. Gottlieb has held fellowships and grants from the Guggenheim Foundation, National Endowment for the Humanities, and Social Science Research Council, and has held teaching/research appointments at Princeton University, École des Hautes Études (Paris), Catholic University of Leuven, and elsewhere. A Professor Emerita at the University of Illinois at Urbana-Champaign, Gottlieb is currently a Visiting Scholar in Anthropology at Brown University. She holds a Ph.D. in anthropology from the University of Virginia.

Ruth Green-Cole is a mother, artist, curator, academic, curriculum developer, educationalist, and the former director of the Whangarei Art Museum, a regional art gallery in Northland, New Zealand. Her research interests include the leaky and maternal body in contemporary art, the sacred feminine, gender studies, feminist theory, and contemporary and modern New Zealand art. She received a Master of Art with first-class honors in art history from the University of Auckland in 2014 for her thesis, "Visualising Menstruation: Gendered Blood in Contemporary Art." Green-Cole posts about menstruation and visual art on her blog at http://hyperheterotopia.com/.

Heather C. Guidone is the program director of the Center for Endometriosis Care. For more than 25 years, she has focused on endometriosis education, research facilitation, policy reform, patient-centered care, health literacy, engagement and adherence, and more. A board-certified patient advocate and health educator, she serves on many councils, committees, and special interest groups on endometriosis, pelvic pain, gynepathologies, and women's health issues, and has contributed to countless books, articles, and publications on these topics. She is active in several professional health organizations, including as a PCORI ambassador and contributing member of the Society for Menstrual Cycle Research.

Miren Guilló-Arakistain is a professor in the social anthropology program in the Department of Philosophy of Values and Social Anthropology at the University of the Basque Country (UPV/EHU). She is a graduate of UPV-EHU in social and cultural anthropology and pedagogy and holds a master's in feminist and gender studies. Her research interests are anthropology of medicine and health, social theory of the body, feminist epistemologies, agency, and social change. Her forthcoming doctoral thesis examines the politics of menstruation, gender relations, identities, and corporalities. She is part of AFIT Feminist Anthropology Research Group at UPV-EHU.

Katie Ann Hasson writes, speaks, researches, and teaches about the social and political aspects of human genetic and reproductive technologies. She is currently program director on genetic justice at the Center for Genetics and Society. Katie earned her Ph.D. in sociology with a designated emphasis in women, gender, and sexuality from the University of California, Berkeley, and was previously an assistant professor of sociology and gender studies at the University of Southern California.

Alexandra J. Hawkey is a postdoctoral researcher at the Translational Health Research Institute at Western Sydney University, Australia. Her research interest is women's sexual and reproductive health, including women's fertility and contraception choices, cancer screening and survivorship, sexuality, sexual health, and menstruation and menopause. Alex also has a special interest in working alongside marginalized communities, such as migrant and refugee women.

Lubabah Helwani currently works in bioethics at the University of Southern California. Her educational background includes an M.S. in medical and cultural anthropology from Harvard University, with a focus on women's menstrual health from the Ash-Sham region of Syria.

Julie Hennegan is a research associate at the Johns Hopkins Bloomberg School of Public Health. Her research focuses on menstrual health and hygiene, and the design and evaluation of complex social and behavioral interventions for women's health. Julie holds a D.Phil. from the Centre for Evidence Based Intervention at the University of Oxford, an M.Sc. in evidence-based social intervention from the University of Oxford (UK), as well as a B.Psy.Sc. (Hons I) from the University of Queensland (Australia).

Lauren C. Houghton first became interested in women's health as an anthropologist when she learned women in the Global South menstruate three times less across their lifetimes than women in the Global North. In the Department of Epidemiology at Columbia University's School of Public Health, she now uses mixed-methods to understand how culture gets beneath the skin through hormones, specifically regarding puberty, the menstrual cycle, breast cancer risk, and women's broader reproductive lives. She is currently exploring the use of digital menstrual health in studying the causes of breast cancer, and in the dissemination and implementation of the latest breast cancer science.

Andisheh Jahangir currently works with the World Health Organization country office in Iran and volunteers with WoMena. She holds a Master of International Public Health from the University of Sydney.

Ingrid Johnston is an experimental psychologist with expertise in social and health psychology. She taught full-time at SUNY, Fredonia for 12 years in the Psychology Department, focusing on women's health, psychology of women and health psychology. She is currently a professor of psychology and associate dean for Lesley University's College of Liberal Arts and Sciences. She has published extensively in the area of women's reproductive health, with particular emphasis on psychosocial aspects of embodiment. She has served on the board for the Society for Menstrual Cycle Research for almost 20 years. Johnston has served this organization as program chair, president, and past president.

Ina Jurga is an engineer, educator, networker, and advocate with more than 15 years' experience in the WASH sector. Working for the Berlin-based NGO, WASH United, she co-initiated and coordinates the international Menstrual Hygiene Day (28 May). Each year this day is dedicated globally to breaking the silence around menstruation and menstrual hygiene management.

Kalvikarasi Karunanithy has a B.A. in commerce from Pondicherry University and an M.A. in business administration from Sathyabama University. She works at Eco Femme in sales and marketing and is a menstrual educator in the organization's Pad for Pad Program. She feels a strong connection to nature and the environment and currently resides in Tamil Nadu, India.

Danielle Keiser has been a vivid and integral player in the menstrual health community since 2013, when she helped launch and grow 28 May, Menstrual Hygiene Day. Danielle is the CEO and executive director of the Menstrual Health Hub (MH Hub), a female health impact organization focused on ecosystem building, knowledge sharing, and high-level advocacy around menstrual health worldwide. Using women-centered design and a human rights approach, the MH Hub consults various entities on female health innovation, investment, communications, and business strategy.

Sally King is the founder of Menstrual Matters (www.menstrual-matters.com), a freely accessible and evidence-based website about how to identify and manage menstrual cycle-related symptoms. She also writes a popular blog about the way in which menstrual health relates to gender inequalities. Sally has over a decade's experience in research quality assurance roles within human rights organizations and programs. She has an M.A. in research methods (qualitative & quantitative) and is currently doing a Ph.D. on the topic of premenstrual syndrome at King's College London.

Elizabeth Arveda Kissling is professor of women's and gender studies at Eastern Washington University. Her research focuses on women's health, bodies, and feminism, and especially how these issues are represented in media. Her newest book about abortion activism and social media, *From a Whisper to a Shout*, was published in 2018 by Repeater Books. As the author of *Capitalizing on the Curse* and related articles, she is best known for her research on media representations of menstruation. Her pronouns are *she* and *her*.

Kuntala Lahiri-Dutt is a professor at Australian National University, and teaches gender and development in the university's Masters in Applied Anthropology and Participatory Development Program. She has written extensively on women and gender in relation to the environment, focusing on water, agriculture, and extractive resources. More information can be gleaned from her staff page: https://crawford.anu.edu.au/people/academic/kuntala-lahiri-dutt.

Gerda Larsson is co-founder and managing director of The Case for Her, an innovative funding collaborative that invests in early-stage markets within women's and girls' health. Driven by a passion for women's rights and gendered development, Gerda has built a career scaling CSR efforts, organizations, and philanthropic foundations. She is also the chair of the Mitt Alby Foundation, chair of the 1325 Policy Group, a board member of the East African e-commerce company, Kasha, and a jurist for the feminist film price, The Anna Award. Gerda has a B.A. in urban planning and a master's in development practice from Stockholm University.

Rachel B. Levitt is a master's student in clinical mental health counseling student at Monmouth University. Her research interests include sexuality and gender identity, attitudes towards menstruation, the mental health effects of internalizing the male gaze, and feminist counseling.

Jen Lewis is the conceptual artist and menstrual designer behind *Beauty in Blood*, a transformative macrophotography and video art project that confronts the social taboos pertaining to menstruation and the female body. She received her B.A. in the history of art from the University of Michigan (Ann Arbor) in 2001. Her work has been displayed in group exhibitions internationally, such as *Period Pieces* at the Urban Artroom (Sweden) and the *9th Annual Juried Art Show* at The Kinsey Institute (USA). Jen also curated a special theme exhibit, "Widening the Cycle: A Menstrual Cycle and Reproductive Justice Art Show" for the joint 2015 conference of the Society for Menstrual Cycle Research and the Center for Women's Health and Human Rights.

Libbet Loughnan is a data and monitoring specialist. She has worked in international development, including the World Bank, UNICEF, and WHO since 2003. She works across the full data cycle, particularly in the monitoring and analysis of progress on WASH and gender-related SDG indicators, program monitoring, the methodological development of equality measures and indicators, surveys, and in supporting data partnerships. Libbet has a Master of Public Health with the LSHTM, and an undergraduate degree from the University of Melbourne.

Trisha Maharaj is an independent researcher focusing on cultural and religious practices related to menstruation and women's experiences and attitudes in the Hindu diaspora of Trinidad. She recently graduated from Columbia University with an M.A. in human rights studies. She also holds a B.A. in international studies with a regional focus in Africa from American University.

Thérèse Mahon is WaterAid's global lead on menstrual hygiene management and has been working on the issue since 2006. Thérèse works with WaterAid's country programs to develop and implement MHM programming; and to generate evidence on MHM to influence policy and practice globally. She is the co-author of the book, *Menstrual Hygiene Matters* and led a regional situation analysis of MHM in schools in South Asia. She also contributed to the development of the Joint Monitoring Programme (JMP) global guidance for monitoring MHM-related indicators for SDG4 and 6 in schools.

Phoebe Man is a multimedia artist, independent curator and associate professor at the School of Creative Media, City University of Hong Kong. Her socially engaged animations, videos, and installations call for active engagement from her audiences, and have been featured in over 180 exhibitions and festivals worldwide. In 2017, Man was selected as one of four international artists to join the Wapping Project Berlin Residency program. Her most recent work, *Free Coloring: If I Were* centers on sexual assault, inviting audiences to engage in discussions and create artwork from one of three perspectives: "if I were a victim," "if I were a perpetrator," and "if I were a bystander."

Swatija Manorama has been active in the campaign group, the Forum Against Oppression of Women, Mumbai, since the mid-1980s. She holds a bachelor's degree in microbiology, a master's in anthropology, and a postgraduate diploma in gerontology. She has authored and co-authored various books and papers addressing issues such as women and religion, science, health and reproductive health, including *Coping with Plural Identities* (Red Globe Press, 2002) and *Introduction to Fertile Futures: Grounding Feminist Science Studies Across Communities* (Routledge, 2001) with co-author J. Elaine Walters.

Lina Acca Mathew has twelve years' experience teaching undergraduate and postgraduate law courses in India. She is an assistant professor at the Government Law College Kozhikode and has taught in various law colleges in Kerala. She was awarded her Ph.D. from the Faculty of Law at Queensland University of Technology, Australia in 2017 on legislative models for prosecuting child sexual abuse in India. She completed her LLM at the National Law School of India University and her LLB at the Government Law College Thiruvananthapuram. She has publications and conference presentations concerning laws on women and children, cyber law, and legal education.

Mbarou Gassama Mbaye holds an Education Doctorate in international education from UMass, Amherst. She has been working for the last twenty years on gender issues in West and Central Africa. She has also coordinated programs at UN Women on gender, public policies, and budget, mainly in the sectors of health, education, environment, and water and sanitation.

Annie McCarthy is an anthropologist interested in the ways marginalized children negotiate and challenge institutions that seek to preserve, foster or establish "childhood." McCarthy's doctoral research explored the ways a group of slum children in Delhi, India, navigate the complexities and contradictions of development through their participation in NGO programs. McCarthy has also explored missionary efforts to "rescue" girls in early twentieth century south India and is currently developing a project to ethnographically engage with ideas of children's growth beyond the biomedical paradigm of stunting. She currently works at the University of Canberra as an assistant professor of global studies.

Maureen C. McHugh is a Distinguished University Professor at Indiana University of Pennsylvania (IUP), where she teaches graduate and undergraduate courses in gender, sexuality, and diversity. She has published journal articles and chapters in many *Psychology of Women Quarterly* anthologies and in handbooks addressing gender (differences), feminist methods, sexuality, violence against women, size bias, and women and aging. She co-edited *The Wrong Prescription for Women: How Medicine and Media Create a "Need" for Treatments, Drugs, and Surgery*. Her current research interests include women and shame, slut shaming, genital shaming, menstrual shame, fat shaming, femininity, and sexual violence.

Ginny Mendis works for MAS Holdings (Pvt) Ltd., a multinational manufacturer of intimate apparel, sports swim and performance wear headquartered in Sri Lanka. She is currently leading the menstruation and incontinence space for the FemTeach team at MAS. Their vision is to be the go-to innovator and manufacturer in the FemTech apparel space, addressing women's health needs. MAS FemTech is focused on innovating "functional, lifestyle and wellness-oriented solutions for the female reproductive cycle from menarche to menopause."

Sheryl E. Mendlinger received her B.A. in English literature and linguistics, as well as her M.A. and Ph.D. in education, from Israel's Ben-Gurion University. Sheryl's academic expertise and publications focus on intergenerational transmission of knowledge and health behaviors in mother–daughter dyads from multicultural populations in Israel, with a focus on menstruation. Her other research interests include knowledge acquisition in agriculture in Tanzania, the economic development of the Massai tribe in a remote area of Tanzania, and educational success in the women's prison educational program. She recently co-authored *Schlopping: Developing Relationships, Self-Image & Memories* with her daughter Yael Magen, Esq.

Victoria Miller is a recent graduate of Columbia University, where she received her master's in human rights studies. She focused on the sanitization and narrative of menstruation. Previously, she worked for Penguin Random House and holds a B.A. from New York University in English and American literature and journalism.

Tova Mirvis is the author of the memoir *The Book of Separation*, which was a *New York Times* Book Review Editor's Choice and was excerpted in the *New York Times* "Modern Love" column. She has also written three novels: *Visible City*, *The Outside World*, and *The Ladies Auxiliary*, which was a national bestseller. Her essays have appeared in *The Boston Globe Magazine*, *The Washington Post*, *Real Simple*, and *Psychology Today*, and her fiction has been broadcast on National Public Radio.

Vinod Mishra carries more than 18 years' experience in the water and sanitation sector, with a focus on providing WASH programs across district, state, and national levels in India with support for project planning, management, capacity building, and implementation. As India Coordinator for the WSSCC, he has developed a strategy to support the Swachh Bharat Mission in conducting policy advocacy for collective behavior change regarding equity, inclusion, capacity building, MHM, and rapid learning in order to make India open defecation free. He holds a master's degree in political science and international relations from the University of Allahabad, and an M.B.A. from Delhi's Indira Gandhi National Open University.

Alfred Muli is a public health, monitoring, evaluation, and learning specialist with close to a decade of hands-on experience in the water, sanitation and hygiene (WASH) and reproductive health spaces. He is the regional program manager for East Africa at Ruby Cup, an award-winning social business and pioneer in providing menstrual cups, as well as education on reproductive health and menstrual care, to girls and women in 11 countries of the Global South. Alfred has previously worked with, among others, WASH United as National Coordinator for Menstrual Hygiene Day (Kenya) and ZanaAfrica, where he managed an RCT.

Shardi Nahavandi has expertise in many different sectors from business development and urban design, to programming and neuroplasticity. She holds five degrees, four of which are from University College London. She founded the UK-based medical technology startup, Pexxi, which uses genetic testing and AI to help people find hormonal contraceptives that complement their unique biological profiles. Shardi also advises various startups at Cambridge University and the Royal Society of Engineering across health and tech verticals.

Shamirah Nakalema is the training coordinator and project manager at WoMena Uganda. She holds a bachelor's degree in adult and community education.

Diana Nalunga is a trainer with WoMena Uganda and acts as the organization's administrative assistant. She holds a Bachelor's degree in industrial and organizational psychology and has trained in menstrual health management and sexual and reproductive health rights.

Dalitso Ndaferankhande is a Malawian girls' and women's rights advocate with a background in education, SRHR and Violence Against Women and Girls (VAWG). In 2016, she established the Mizuyathu Foundation to reduce grade repetition and high drop-out rates among primary school girls. Before that, she coordinated projects to combat VAWG. Currently, Dalitso leads efforts to establish a pan-African alliance for maternal mental health. She has advocated for changing harmful cultural practices, promoted safe mother-hood and completed community-level research on reusable sanitary pads for tackling menstruation-related absenteeism. Dalitso is a published author of short stories and a proud mother and wife.

Jocelyne Alice Ngo Njiki is a Cameroonian rural engineer. She has been working in the water sector for nearly ten years and is very devoted to her profession. Since 2015 she has been trained on the issues of gender, water, and sanitation, and has spoken publicly about such issues, raising aware-ness and advocating for changes in policies and practices. Improving the liv-ing conditions of populations is her main goal. Ngo also enjoys reading and traveling.

Stella Nyanzi is a Ugandan scholar, writer, and activist. Her work focuses on social anthropology, sexuality, gender, marginalized groups, and freedom of expression, including through the #Pads4GirlsUg campaign. She is known for her provocative poetry using expletives and vulgarity to upset notions of what is deemed acceptable—employing 'radical rudeness.' Nyanzi worked as a researcher at the Makerere Institute of Social Research until 2016, from which she was dismissed after staging a naked protest. She holds a Ph.D. in anthropology from the London School of Hygiene and Tropical Medicine.

Neville Okwaro is a water, sanitation and hygiene specialist and men-strual hygiene management trainer for East and Southern Africa with the Water Supply and Sanitation Collaborative Council. He coordinates national WASH actors in Kenya, the National Environmental Sanitation and Hygiene Inter-agency Coordinating Committee and the seven technical working groups at the Kenyan Ministry of Health's Division of Environmental Health. Neville has helped steer the development of Kenya's Menstrual Hygiene Management Policy and Strategy, MHM in Schools Teachers' Handbook, and MHM Monitoring and Evaluation Indicators. He sees MHM as a cross-cutting rights issue concerning equality, nondiscrimination, and inclu-sion in the various sectors related to daily life.

Julitta Onabanjo is the regional director for UNFPA East and Southern Africa. Dr. Onabanjo joined UNFPA in 1995 as a national program officer and thereafter as a program specialist in Swaziland and Kenya. She subse-quently served at UNFPA headquarters as technical advisor, HIV/AIDS, and later as special assistant to the Executive Director. More recently, she has held UNFPA representative posts in Tanzania and South Africa.

Sheila Vipul Patel is a public health analyst in the Health Coverage for Low-Income and Uninsured Populations Program at RTI International. She applies managerial and technical research skills to supporting projects that inform evidence-based health care decision-making and projects that evaluate health care transformation. Ms. Patel is also a Ph.D. Candidate in health policy and management at the University of North Carolina, Chapel Hill, where she focuses on the implementation of effective behavioral, sexual, and reproductive health services. Prior to joining RTI in 2014, Ms. Patel worked with the Harvard School of Public Health and the World Health Organization on issues related to youth.

Archana Patkar is an independent advisor on gender equality, health, participation and inclusion; committed to research, policy, and practices to advance dignity and rights across the human life course. As head of Policy Advocacy and Operations at the UN Water Supply and Sanitation Collaborative Council, she introduced menstrual hygiene management to the WSSCC and to the sector in several countries globally, particularly within Africa and Asia. Her current focus is on eliminating cervical cancer, HIV, the integration of sexual and reproductive health rights into universal health care, and the essential links between people and the planet.

Radha Paudel is a nurse, humanist activist, author, and entrepreneur. As a researcher, trainer, author, advocate, and producer of biodegradable pads, she has pioneered the dialogue on "dignified menstruation," linking women's participation in peace and politics, human rights, empowerment, and the sustainable development goals, including health, education, WASH, and environment. She has been a speaker at universities, forums, and conferences all around the world since 2008. She has published four books: *Khalangama Hamala* (best literary award winner, 2013), *Shantika Pailaharu* (2018), *Dignified Menstruation is an Everyone's Business* (2018), and *Apabitra Ragat* (*An Impure Blood*, 2019).

Elizabeth (Liz) Pellicano is a developmental cognitive scientist committed to understanding the distinctive opportunities and challenges often faced by autistic people and tracing their impact on everyday life. She trained as an educational psychologist and completed a Ph.D. on the cognitive profile of children with autism in Perth, Australia before becoming a Research Fellow in Psychiatry at the University of Oxford. She was the director of the Centre for Research in Autism and Education (CRAE) at London's Institute of Education from 2013 to 2017. She is now a professor at Macquarie University's Department of Educational Studies.

Milena Bacalja Perianes is a gender researcher and feminist entrepreneur specializing in female health. With an M.Phil. in multidisciplinary gender studies from the University of Cambridge and a master's in international development from RMIT, she works with public and private sector actors to advance gender equality. Milena's career started in international development, working across Asia and Africa for the UN. Now, as the co-founder of the Menstrual Health Hub and a gender specialist for the Criterion Institute, she promotes women-centered design in products, services, research, and programming.

Josefin Persdotter is a Ph.D. candidate in sociology and science and technology studies at the Department of Sociology and Work Science at the University of Gothenburg, Sweden. Her dissertation deals with everyday aspects of menstrual life, and she specifically studies menstrual practices in the bathroom in an effort to illuminate obscurities surrounding menstruality, namely practices that are typically non-spoken, non-worded, and even non-thought. Persdotter is also a leading menstrual activist in Sweden, and is a known menstrual artist and co-founder of the organization MENSEN—forum for menstruation.

Janette Perz is a professor of health psychology and director of the Translational Health Research Institute at Western Sydney University, Australia. She is co-editor of *The Routledge Handbook of Women's Sexual and Reproductive Health* with Jane Ussher and Joan Chrisler. She has undertaken a significant research program in sexual and reproductive health, including the experience of premenstrual syndrome (PMS) in heterosexual and lesbian relationships; the development and evaluation of a couple-based psychological intervention for PMS; sexual well-being and reproductive needs in CALD populations; and LGBTI cancer care.

Penelope A. Phillips-Howard, a reader at the Liverpool School of Tropical Medicine, has been a public health epidemiologist for the past 30 years. Her interests have broadened to determining the harms associated with poor menstrual hygiene management (MHM) and what interventions can mitigate risk. She has been a principal investigator on studies in Kenya and India, including a current cluster randomized controlled trial evaluating the effect of menstrual cups or cash transfer to reduce sexual and reproductive harms, and school-related indices in schoolgirls. She provides technical support on MHM-related committees and working groups, and has authored some 20 papers on this topic.

Niva Piran is a clinical psychologist, academic researcher, writer on embodiment amongst girls and women, and is Professor Emerita at the University of Toronto. A Fellow of the American Psychological Association and the Academy of Eating Disorders, Dr. Piran has received a 2018 Association for Women in Psychology Distinguished Publication Award for her book, *Journeys of Embodiment at the Intersection of Body and Culture: The Developmental Theory of Embodiment*. She is the co-editor of four books on body image and eating disorders and a former body image consultant to the National Ballet School of Canada.

Priti Shrestha Piya studies gender and development, reproductive health and women's economic status in the Global South. She has conducted fieldwork through her home country, Nepal. Her work has been featured in research publications, grassroots organization reports and policy briefs. She holds a master's degree in sociology and an M.A. in social change and development studies.

Jennifer Poole is the founder of Medical Services Pacific, and served as the organization's executive director from 2010 to 2019. She is currently serving as Permanent Secretary in Fiji's Ministry of Women, Children and Poverty Alleviation. For years, she has worked in multiple capacities for various humanitarian agencies such as World Vision International, CAFOD UK and Caritas Pakistan. Her consulting expertise includes project design and evaluation, risk assessment and conflict management, gender assessment, training, and capacity building.

Kamini Prakash was a member of WSSCC's India Support Unit in New Delhi from 2015 to 2019, and was in charge of the menstrual hygiene management interventions in India. She also led the development of MHM training materials for persons with visual and hearing impairments and trained special educators in the use of these materials. She is a member of the Menstrual Health Alliance India.

Ela Przybylo is an assistant professor at Illinois State University's Department of English. Her forthcoming book *Asexual Erotics: Intimate Readings of Compulsory Sexuality* seeks to rethink the role of compulsory sexuality in feminist and queer thought and practice. Ela's work on asexuality has appeared in *GLQ, Sexualities, Feminism and Psychology* and *Asexualities: Feminist and Queer Perspectives*, and *Introducing the New Sexuality Studies* (3rd ed.); and her work on crippling menstrual pain has appeared in *Feminist Formations*. She is co-editor of *On the Politics of Ugliness* and of special issues of *Ada: A Journal of Gender, New Media, and Technology* and *English Studies in Canada*. She is also a founding editor of *Feral Feminisms*. You can find her online at https://przybyloela.wordpress.com/.

Maria Carmen Punzi started her research on social enterprises in the menstrual health space in 2016 for her M.Sc. thesis in global business and sustainability at the Rotterdam School of Management. After graduating, she joined the Business Society Management Department as a research assistant and became the innovation advisor for the Menstrual Health Hub. In 2018, Maria Carmen joined PSI-Europe, working as the menstrual health focal point for the network members of Population Services International. In 2019, she started her Ph.D. studies on menstrual health and societal change under the Rotterdam School of Management's Dynamics of Inclusive Prosperity initiative.

Isabella Mema Rasch is a co-founder of MANA Care Products, a social enterprise which produces environmentally friendly and reusable sanitary products and which provides extension services for menstrual health management to local rural communities. With a working background in environmental management, she and Angelica Salele-Sefo founded MANA Care to eliminate the environmental impact caused by traditional disposable products, promote proper menstrual health practices, and to break period poverty and the challenges surrounding menstruation that exist in the Pacific.

Nancy King Reame is the Mary Dickey Lindsay Professor Emerita of Health Promotion and Risk Reduction in Columbia University's School of Nursing. At the University of Michigan, she is the Rhetaugh G. Dumas Professor Emerita of Nursing and Research Scientist Emerita, Department of Obstetrics and Gynecology, School of Medicine.

Anna Remington conducts research on the superior abilities that we so often see in autistic people, specifically with respect to attention and perception. She is interested in how and why these superiorities develop, and ways in which we might capitalize on these strengths. Among others, her current research projects include working with autistic people in the family justice system, investigating autistic people's greater capacity to detect sound and ways to promote autistic employment. Anna is also co-founder and director of MiniManuscript.com, and she became the director of the Centre for Research in Autism and Education in 2017.

Virginia Roaf is an independent development consultant working in the area of human rights and water and sanitation. From 2010 to 2015 she was senior advisor to the UN Special Rapporteur on the human rights to water and sanitation. She is currently involved with various programs that promote the human rights to water and sanitation globally and locally.

Tomi-Ann Roberts is professor of psychology at Colorado College. Her work focuses on the psychological consequences of the sexualization and objectification of girls and women. The first paper she co-authored on this topic, *Objectification Theory*, is the most cited article in the history of the journal, *Psychology of Women Quarterly*. In addition to her scholarly publications, she served on the American Psychological Association's Task Force on the Sexualization of Girls, the Task Force on Educating Through Feminist Research and as President of the Society for Menstrual Cycle Research from 2017 to 2019. She leverages her feminist psychological science as an expert witness and consultant in cases involving objectification as a form of sexism and gender discrimination.

Jennifer Rothchild is associate professor of sociology and coordinator of the gender, women, and sexuality studies program at the University of Minnesota, Morris. For more than twenty years, she has conducted community-based research in South Asia and the United States and is considered one of the leading scholars on gender and development in Nepal. She is the author of the book *Gender Trouble Makers: Education and Empowerment in Nepal* (Routledge, 2006), as well as book chapters, essays, and policy reports.

Eilish Mairi Roy is an applied psychology undergraduate student at the University of Kent. As part of her degree course, she undertook a one-year placement at the Centre for Research in Autism and Education (CRAE) at UCL Institute for Education.

Klara Rydström holds an M.A. degree in Women's and Gender Studies from the University of Hull, England and Universidad de Oviedo, Spain. Her research interests are related to sex and gender, bodies, feminist theory, and menstruation. She also is an activist within the Swedish menstrual movement. Since 2018, Rydström has served as project manager for the organization MENSEN—forum för menstruation, where she is developing a menstrual certification for workplaces.

Angelica Salele is a co-founder of the startup MANA Care Products, a Samoan-based social enterprise that provides women and girls with affordable, safe, and environmentally friendly menstrual products including reusable menstrual cups. She aims to not only address period poverty and the stigma around menstruation but to fight plastic pollution from single-use plastic pads. In 2018, Salele was one of 12 contestants to win the UN Environment Asia-Pacific Low-Carbon Lifestyles Challenge.

Ursula Maschette Santos is the Brazilian America Coordinator of the Menstrual Health Hub. She holds a degree in psychology from the Mackenzie Presbyterian University, a master's in education, health promotion, and international development from University College London and has over five years' experience in planning, implementation, and evaluation of educational and community-based projects in countries like Brazil, England, and Italy. As a menstrual health activist, she has dedicated the past few years to studies regarding gender equality, sexuality, sexual and reproductive rights, as well as sexual and reproductive education, especially regarding menstrual health.

Musu Bakoto Sawo is the national coordinator for Think Young Women and the deputy executive secretary of the Truth, Reconciliation and Reparations Commission of The Gambia. She has gained in-depth knowledge of human rights through more than 17 years of activism in children and women's rights, and her membership in different community-based organizations. She has translated this knowledge into capacity building, research, networking, program development, and practical engagement with human rights mechanisms, as well as with grassroots, national, and international organizations and platforms. She holds an LLM in human rights and democratization in Africa.

Vanita Singh is pursuing a Ph.D. in public systems at the Indian Institute of Management, Ahmedabad. She has completed a master's in health administration from the Tata Institute of Social Sciences, Mumbai. Her interest areas include health policy, women and child health, universal health coverage, and equity in health care.

M. Sivakami is a professor at the School of Health Systems Studies, Tata Institute of Social Sciences (TISS) in Mumbai, India. She is passionate about issues of women and children. She broadly works in the area of demography, gender, and health. She has widely published in national and international journals on health and public health.

Marni Sommer has worked in global health and development on issues ranging from improving access to essential medicines to humanitarian relief in conflict settings. Dr. Sommer's particular areas of expertise include conducting participatory research with adolescents, understanding and promoting healthy transitions to adulthood, the intersection of public health and education, gender and sexual health, and the implementation and evaluation of adolescent-focused interventions. Dr. Sommer presently leads the Gender, Adolescent Transitions and Environment (GATE) Program, based in the Department of Sociomedical Sciences in the Mailman School of Public Health at Columbia University.

Nichole Speciale is an artist whose work is primarily two-dimensional, often problematizing the illusory canvas surface of a painting, often pushing it into the object realm. She sources imagery from popularized science and Americana, and uses traditionally feminine material to create works that point to a larger scale understanding of our universal context. Nichole received her MFA from the University of California, San Diego, and now practices in the Boston area.

Swetha Sridhar is an independent researcher, with experience working across sectors such as sexual and reproductive health, menstrual hygiene management, and WASH. Her work has involved tracking and analyzing the changing policy fields in these sectors and designing strategies to generate evidence-based impact. She attended the University of Cambridge on a Lady Meherbai D Tata Scholarship, where she read for an M.Phil. in gender studies. She also holds an M.A. in development studies from the Indian Institute of Technology, Madras. She has previously been awarded the Mitacs Globalink Fellowship and the DAAD Scholarship.

Linda Steele is a socio-legal researcher exploring intersections of disability, gender, law, and justice. She has completed a monograph for Routledge's Social Justice series, *Disability, Criminal Justice and Law: Reconsidering Court Diversion*, and is co-editing *The Legacies of Institutionalisation: Disability, Law and Policy in the 'Deinstitutionalised' Community*. Linda is a senior lecturer in the UTS Faculty of Law, a member of the UTS Law Health Justice Research Centre and a co-convener of the UTS Feminist Legal Research Group. She is also a Senior Visiting Research Fellow at the Faculty of Law, Humanities and the Arts, University of Wollongong, Australia. She was formerly a disability rights lawyer at the Intellectual Disability Rights Service.

Gloria Steinem is a writer, political activist, and feminist organizer. She established *New York* and *Ms.* magazines and co-founded the National Women's Political Caucus, the Ms. Foundation for Women, the Free to Be Foundation and the Women's Media Center in the United States. She has received multiple awards for her journalism, the Society of Writers Award from the United Nations, as well as the Presidential Medal of Freedom from President Barack Obama in 2013. In 2016, she and Amy Richards co-produced a series of eight documentaries on violence against women around the world for VICELAND.

Evelina W. Sterling is the director of research and strategic initiatives and assistant professor of sociology at Kennesaw State University. Evelina worked as a public health researcher and medical sociologist focusing on women's health issues for over 25 years. She is the author of six consumer health books, including books addressing polycystic ovary syndrome (PCOS) and primary ovarian insufficiency (POI), and a board member of the Society for Menstrual Cycle Research (SMCR).

Robyn Steward is autistic and has spent over 15 years raising awareness internationally about the experiences of those with autism. This topic is the main focus of her research with University College London and the Wellcome Trust, and two books: *The Independent Woman's Handbook for Super Safe Living on the Autistic Spectrum* and *The Autism Friendly Guide to Periods*. Steward also (co)hosts a BBC Podcast called *1800 Seconds on Autism*, and *The Autism Journal podcast*. Steward was a joint awardee of the 2015 National Autistic Society (NAS) Professional Award, was on power100's 2018 list of Most Influential Disabled People in the UK, and is an NAS ambassador.

Margaret L. (Peggy) Stubbs is Professor Emerita of Psychology at Chatham University in Pittsburgh, PA, where she previously directed the undergraduate Psychology and Women's Studies programs. As a social and developmental psychologist interested in well-being across the lifespan, her research includes a special focus on girls' and women's development, psychosocial aspects of the menstrual experience, and menstrual education. A longtime member of the Society for Menstrual Cycle Research, she has served in a variety of roles involving organizational leadership and the publication of menstrual cycle research in mainstream journals.

Deepthi Sukumar is a Dalit woman and an activist. She has been working for the liberation and rehabilitation of women engaged in manual scavenging for more than two decades. Her parents migrated to the city of Chennai from a remote area in Andhra Pradesh for education and employment. She travels widely to villages and small towns to meet Dalit women living in difficult circumstances and who have become victims of human rights violations.

Vicci Tallis is a feminist who has been working on gender, HIV and AIDS, sexual and reproductive rights, and LGBTI issues in South and southern Africa for over 30 years. She is currently working as an independent consultant. Previously, she was the director of programs at the SRHR Africa Trust; and prior to that she was the program manager for the HIV and AIDS Unit at the Open Society Initiative for Southern Africa. Vicca has a Ph.D. in development studies from the University of KwaZulu Natal, South Africa. She is the author of *Feminisms, HIV and AIDS. Subverting Power, Reducing Vulnerability*.

Eugenia Tarzibachi is a bilingual psychologist and Ph.D. in social sciences specialized in fostering diversity and inclusion strategies for internationally underrepresented groups, particularly women and girls. She authored *Women's Thing. Menstruation, Gender and Power* (Penguin Random House Argentina) and *Gender Mainstreaming in Health: Progress and Challenges for a New Road Map Toward the 2030 Agenda for the Americas* (Panamerican Health Organization/World Health Organization). In 2016, Tarzibachi was distinguished by the US Library of Congress for her doctoral research and has received several awards and academic merit scholarships from institutions in Spain and Argentina. She is also a member of the Health Equity Network of the Americas and the Board of Directors of the Society for Menstrual Cycle Research.

Marianne Tellier is a public health professional with over 10 years' experience and the co-founder and Board Chair of WoMena. She holds an B.Sc./M.Sc. in public health from the University of Copenhagen and an M.Sc. in health policy, planning and financing from the London School of Hygiene & Tropical Medicine and the London School of Economics.

Siri Tellier is senior reproductive health advisor for WoMena. She is the course leader for Health in Emergencies and Refugee Health at the University of Copenhagen. She holds an M.Sc./HYG from Harvard School of Public Health.

Jane M. Ussher is professor of Women's Health Psychology in the Translational Health Research Institute at Western Sydney University, Australia. Her research focuses on examining subjectivity in relation to the reproductive body and sexuality, and the gendered experience of cancer and cancer care. She is the author of over 250 papers and chapters, and 11 books, including *The Madness of Women: Myth and Experience* and *Managing the Monstrous Feminine: Regulating the Reproductive Body*. Jane is also editor of the Routledge *Women and Psychology* book series. Her current research focuses on older women's sexual embodiment and LGBTI experiences of cancer.

Shailini Vora has a breadth of experience working within the charity sector for causes such as criminal justice, women's empowerment, and sustainable economies. Her main project has been researching the effects of menstruation on homeless women with the social enterprise, No More Taboo. She developed long-lasting solutions to the issue of period poverty, including working directly with vulnerable women, delivering training to organizations, and lobbying for improved menstrual education. She is one of the authors of the groundbreaking *Break the Barriers* report on menstrual education, published by Plan International UK in 2018. She now works with St Mungo's, tackling the root causes of homelessness.

Sharra L. Vostral is an associate professor of history at Purdue University. She is the author of Under Wraps: A History of Menstrual Hygiene Technology (2008) and Toxic Shock: A Social History (2018) for which she earned a National Science Foundation grant to complete its research. She has been interviewed and quoted in *The New York Times Magazine*, *The Atlantic*, *Wired*, *NPR*, *CNN*, and *Newsweek*.

Steve Nganga Wambui is a graduate of the University of Nairobi and holds a Bachelor's degree in social work and sociology. As a social worker, he currently consults on topics related to reproductive and menstrual health with The Cup organization in Kenya. He is also a project officer in the Kipepo Mentorship Program based in the urban slums of Nairobi.

Jennifer Weiss-Wolf is vice president and Women and Democracy Fellow at the Brennan Center for Justice at NYU School of Law. A leading advocate for issues of gender, politics, and menstruation, she was dubbed the "architect of the U.S. campaign to squash the tampon tax" by *Newsweek*. Her 2017 book *Periods Gone Public: Taking a Stand for Menstrual Equity* was lauded by Gloria Steinem as "the beginning of liberation for us all." Weiss-Wolf's writing and policy work have been featured by *Ms.* Magazine, *The New York Times*, *The Washington Post*, *TIME*, *Cosmopolitan*, *Harper's Bazaar*, *Teen Vogue*, *Marie Claire*, *Vox*, *Vice*, and *NPR*, among others. She is also a contributor to the 2018 Young Adult anthology, *Period.: Twelve Voices Tell the Bloody Truth*. Weiss-Wolf received her J.D. from the Benjamin N. Cardozo School of Law.

Mirjam Werner is an assistant professor in the Business-Society Management Department at Rotterdam School of Management, Erasmus University (RSM). She holds an M.Sc. in cultural anthropology from the University of Amsterdam (2005) and a Ph.D. in political science from the University of Leeds (2011). Her research interests include social movements and political activism, framing and sensemaking, organizational change, organizational identity and culture, and emotions. Her current research projects concern social movements as motors of bottom-up change within organizational contexts and exploring the performative nature of emotions in social interaction (i.e., what do emotions do?).

Inga T. Winkler is a lecturer at the Institute for the Study of Human Rights and the Director of the Working Group on Menstrual Health & Gender Justice at Columbia University. She is particularly interested in the intersections of menstruation, human rights, and culture and focuses on questions of inequalities, marginalization, and representation. Another strand of her research builds on her policy and consulting experience and engages directly with policy-makers on menstrual health. Her books include the first comprehensive monograph on the human right to water and an edited volume on the Sustainable Development Goals. She is affiliated faculty at the Columbia Water Center in the Earth Institute, the Economic and Social Rights Working Group at the Human Rights Institute at the University of Connecticut, and the Center on Law and Social Transformation at the University of Bergen, Norway. She is the former legal adviser to the UN Special Rapporteur on the human rights to water and sanitation.

Camilla Wirseen grew up in Sweden and studied architecture in Italy. She has been working as a photographer, a curator at major cultural institutions and a university lecturer. In 2005, her career changed direction when she became the co-founder of Peepoople, and again in 2012 when she started The Cup Foundation to help underprivileged girls access sustainable menstrual cups and comprehensive education on sexuality and reproductive rights. Since its launch, her program has reached more than 20,000 girls and 10,000 boys in Kenya. Today she also provides trainer-to-trainer workshops and is creating awareness of girls' challenges by blogging, running a podcast and managing a unique gift shop in Kibera, an informal settlement in Nairobi.

Jill M. Wood is a teaching professor in the Department of Women's, Gender, and Sexuality Studies at Penn State University. As a feminist teacher and researcher, she specializes in women's health (specifically menstruation and childbirth) and women's sexualities (particularly sexual response during the menopausal transition). Professor Wood also writes and works on topics in feminist pedagogies, as she believes that education is a potentially transformative and empowering experience, particularly for marginalized students. Jill is a self-proclaimed foodie and gardener, a budding yoga teacher, and the proud mama of 3 fantastic kiddos, "Mister" their dog, and 6 backyard chickens.

Garazi Zulaika is a public health epidemiologist who has worked in global health research on the issues of adolescent sexual and reproductive health (SRH) and menstruation. Ms. Zulaika currently works with the Liverpool School of Tropical Medicine, Department of Clinical Sciences as a technical officer and studies the public health epidemiology of menstrual cup and cash transfer interventions on girls schooling in western Kenya. There, she is also pursuing her doctoral research assessing these interventions' effects on girls' SRH outcomes and risk behaviors.

List of Figures

LIST OF TABLES

REPRINT CREDITS

Several chapters in this Handbook were previously published. They include:

"Prisons that Withhold Menstrual Pads Humiliate Women and Violate Basic Rights" by Chandra Bozelko was first published on June 12, 2015 in *The Guardian*. Reprinted with permission. No further reproduction or distribution of the material is allowed without permission from the publisher.

"Navigating the Binary: A Visual Narrative of Trans and Genderqueer Menstruation" S.E. Frank and Jac Dellaria was published in an extended form as Frank S. E. (2020) "Queering Menstruation: Trans and Non-Binary Identity and Body Politics." *Sociological Inquiry*. 90 (2). Reprinted with permission. [OA CC-BY 4.0].

"Out of the Mikvah, into the World" by Tova Mirvas was first published on September 19, 2017 in *The Lenny Letter*. It is excerpted from the 2018 memoir *The Book of Separation*. Boston: Houghton Mifflin Harcourt. Reprinted with permission. No further reproduction or distribution of the material is allowed without permission from the publisher.

"The Menstrual Mark: Menstruation as Social Stigma" by Ingrid Johnston-Robledo and Joan C. Chrisler was first published in 2013 in *Sex Roles*. 68 (1–2): 9–18. Reprinted with permission. No further reproduction or distribution of the material is allowed without permission from the publisher.

"If Men Could Menstruate" by Gloria Steinem was first published in October 1978 in *Ms.* Magazine. Reprinted with permission. No further reproduction or distribution of the material is allowed without permission from the publisher.

"Mapping the Knowledge and Understanding of Menarche, Menstrual Hygiene and Menstrual Health among Adolescent Girls in Low- and Middle-Income Countries" by Venkatraman Chandra-Mouli and Sheila Vipul Patel

CHAPTER 1

Introduction: Menstruation as Lens—Menstruation as Opportunity

Chris Bobel

The field of critical menstruation studies is burgeoning. And so this Handbook arrives just in time to capture a robust and carefully curated view of where we are now and where we might go next.

But it is 2020, and menstruation is as old as humanity itself. Why is this the first handbook to bring together this body of knowledge?

To state the obvious, menstruation and more broadly, the menstrual cycle are often dismissed and derided. The same goes for menopause, at the further end of the reproductive life span. It is transgressive to resist the norm of menstrual (and menopausal) concealment. With notable exceptions, across cultures and historical eras, we socialize this biological process—including serious inquiry into its form, function, and meaning—into hiding. This is shortsighted and at the same time deeply revealing, as it shines a bright spotlight on the need for change. After all, a dearth of attention to a fundamental reality and indeed a vital sign is not only a profound knowledge gap, it is an exposure of the power of misogyny and stigma to suppress knowledge production. And when we lack knowledge, we cannot effectively act to effect change.

Menstruation as lens

Of course, there *has* been relevant scholarship, but until recently menstruation as a subject of research and advocacy has been relegated to the fringes. There have been moments when menstruation broke through, such as when feminist artist Judy Chicago created her iconic lithograph "Red Flag" in 1971, a depiction of a hand removing a tampon, shocking viewers into engaging the everyday reality of menstruation. In 1977 The Society for Menstrual Cycle Research (SMCR) was founded by a multidisciplinary group of scholars who were feminist pathbreakers in understanding the centrality of

© The Author(s) 2020
C. Bobel et al. (eds.), *The Palgrave Handbook of Critical Menstruation Studies*, https://doi.org/10.1007/978-981-15-0614-7_1

menstrual cycle research to women's health. In 1978 Gloria Steinem penned her classic satirical essay "If Men Could Menstruate"—a piece, included in this Handbook because it continues to slyly expose the sexism that shapes our menstrual culture. And in the early 1980s, an outbreak of Toxic Shock Syndrome, a rare but severe illness, was linked to the use of super-absorbent tampons. These breakthroughs—artistic, scholarly, popular, and tragic—laid important ground now being built upon in the early twenty-first century, a time in which the menstrual cycle moves from margin to center as a subject of urgent concern and enthusiastic exploration.

Menstruation as opportunity

It has been said so often it is now cliché—"menstruation is having its moment!"

November's issue of *Cosmopolitan* dubbed 2015 "the year the period went public," and indeed, the half decade since has brought us a tremendous diversity of menstrual -positive expressions—from the artistic to the practical, the serious and the playful, the local and the global. Instagram made the news when Rupi Kaur's photo of her period-stained pajama pants was (twice) removed and outcry across social media was loud and persistent. The unique menstrual challenges of women and girls living on the streets and schoolgirls in low- and middle-income countries inspired a raft of grassroots campaigns. Efforts to de-tax menstrual products succeeded in multiple countries—first in Kenya in 2004. Canada dropped the tax in 2015, and Malaysia, India, and Australia followed in 2018. In 2019, a short documentary about the birth of a menstrual-pad-making microbusiness won the Academy Award for best short documentary just a year after a biopic about the inventor of the machine making those pads, Arunachalam Muruganantham, enjoyed Bollywood success—and beyond. We finally have a period emoji—no small thing, given the centrality of phones and social media in everyday life—and menstrual apps abound. National, state, and municipal programs in countries from Kenya to Scotland provide free menstrual supplies to menstruators in schools, prisons, shelters, and other public facilities. Considered together, these events constitute a shift. Since these watershed moments, attention to menstruation has intensified and diversified. To those of us working in this area, we find ourselves breathless, trying to keep up.

But we are not complaining!

Menstruation *is* having its moment—no doubt. And we aim to seize it in the shape of the *Palgrave Handbook of Critical Menstruation Studies*. We built this book—numbering 72 chapters, written by a total 134 contributors from 23 countries—to provide an unmatched resource for scholars, activists, policy makers, and practitioners, both those new to and already familiar with the field.

At its core, the Handbook is animated by two intertwined central questions: What new lines of inquiry, including research questions and social justice engagements, are possible when we center our attention on menstrual health and politics across the lifespan? And what knowledge is gained when menstruation emerges as a dynamic category of analysis? The answers to these questions take shape in this collection of empirical research and theoretical essays that are supplemented with first-person narratives; practice notes from those working in the field; poetry, and visual art. We conclude each of our six sections with what we call "transnational engagements"—rich conversations across diverse spaces, experiences, and identities which appear as actual dialogues in some sections and as distinct voices responding to a shared set of questions in others. What unites these different forms of knowledge is a shared commitment to advancing menstruation as a way to make sense of political, social, medical, and/or biological processes, and the recursive work embedded in the menstrual cycle's myriad social constructions.

Our choices here deviate from those made in most conventional academic handbooks. In this rapidly growing field of inquiry and advocacy, a diversity of voices and approaches shape what we know—and this Handbook aims to capture those many articulations. We take very seriously feminist critiques of epistemological rigidity that reflect a very narrow (and privileged) idea of what counts as knowledge. Indeed, expertise comes in many forms. The broad range of the content is its strength, but it also stretches the limits of what some readers might consider a more uniform set of readings. Through our editorial processes, we chose to preserve the unique writing styles of our contributors, pushing against the usual impulses in edited collections to standardize content so that each chapter aligns nicely with the next. In our section titled "Menstruation as Structural," for example, several practice notes written by policy makers are peppered among more scholarly chapters penned by academics who review the extant literature and/or offer new insights based on their original research. The Handbook also includes personal narratives that explore cultural and religious practices related to menstruation, menstruating while in detention, and the relationship between child marriage and menstruation. These chapters bring together different ways of peering inside what's at stake when menstruation is regarded as a structural issue, one ripe for policy interventions, with real-life implications for human beings. Throughout the book, our editorial choices should make clear that we value the implied dialogue and symbiosis between those living the issues, those conducting research, and those putting it to work.

With this in mind, the chapters in this collection reflect different forms of knowledge that are shaping critical menstruation studies—a field that, from its beginning, has been a site where activists, artists, journalists, clinicians, and researchers have each contributed to its articulation and application. A field that, until recently, went largely unnamed. Similar to critical race studies or critical gender studies, *critical menstruation studies* is premised upon menstruation as a category of analysis: asking how systems of power and knowledge are built upon its understanding and, furthermore, who benefits from these social constructions. Critical menstruation studies—which some argue might be more aptly named critical *menstrual* studies, to capture the menstrual cycle across the life course, including, but not limited to, menstruation itself—is a coherent and multidimensional transdisciplinary subject of inquiry and advocacy, one that enables an exciting epistemological clarity that holds significant potential for knowledge production and social transformation. This Handbook is the first to coin the term—with thanks to Sharra Vostral who suggested it and conceptualized the Handbook structure with me at the advent of this project. The Handbook's purpose, then, is to represent a particular landscape of knowledge that highlights its current diversity and promise as the field rapidly develops and expands. We seek to explore this landscape in all its diversity with lively intent.

But this is not an intellectual playground where ideas are vetted out of reach of the lived experiences of real people in real time. The stakes in this emerging field are high. Between 1970 and 1980, 941 American women were diagnosed with Toxic Shock Syndrome, 73 of which died (Vostral 2018). The story of tampon-related TSS is an object lesson at the intersection of capitalism, gendered consumption, and faulty techno-science, a tragic illustration of both the literal and figurative costs of stigma. Stigma's impact can be quite insidious and expansive in ways that capture far less media attention than the TSS crisis. To wit, we include a chapter about endometriosis by patient advocate and health educator Heather Guidone who describes the damage done—physically, psychologically, emotionally, and socially—through a combination of lengthy diagnostic delays and disease illiteracy, causing both patients and practitioners to dismiss the disease's wide-ranging symptoms as routine. In short, and quite literally, menstrual stigma harms.

And that is why we are unequivocal. Attention to menstrual issues across the life span surfaces broader societal issues and tensions, including gender inequality, practices and discourses of embodiment, processes of racialization and commodification, and emergent technologies as read through various disciplines and interdisciplines (for example, history, psychology, communication studies, sociology, anthropology, art, nursing, gender studies, public health, law policy analysis—the list goes on). Put differently, menstruation-as-unit-of-analysis serves as a gateway—both conceptually and symbolically—to reveal, unpack, and complicate inequalities across biological, social, cultural, religious, political, and historical dimensions. Yes. Menstruation matters.

Menstruation as lens: menstruation as opportunity

The members of the editorial team share a commitment to produce a collection that is purposely interdisciplinary and transnational. It draws on fields in the humanities and social sciences, intentionally stopping at the boundary of basic biomedical research about menstruation. Here, we used "menstruation as lens, menstruation as opportunity" to think beyond anatomy and biology. We chose to dig into the *meanings* of menstruation. As such, we opted to organize the Handbook outside the more normative life course approach from menarche to menopause. While we acknowledge this linear process, we also recognize its limitations. Here, we are doing something different. We have organized the book thematically into six overlapping sections, each edited by an associate editor: *Menstruation as Fundamental, Menstruation as Embodied, Menstruation as Rationale, Menstruation as Structural, Menstruation as Material* and *Menstruation as Narrative*. Each of these sections is introduced by a short framing essay, authored by its editor. We acknowledge that the Handbook is hardly comprehensive. For one, we failed—in spite of our best efforts—to produce a collection that adequately decentered Western voices by engaging *more* scholars from the Global South. We hope that subsequent editions of the Handbook will more successfully meet this crucial goal. And of course, there are many topics left unaddressed. We need more work that explores the measured impacts of menstrual stigma, for example, especially for marginalized menstruators. There is a need for research that bridges menstrual and menopausal realities in the Global South and Global North, to substantively and responsibly explore not only the differences, but also the similarities in these spaces. No doubt, readers will see, and we fervently hope, respond to innumerable opportunities for further study. Because, done right, critical menstruation studies not only sheds light on diverse experiences across the menstrual life course, it also brings fresh fodder to persistent questions: What is the relationship between embodiment and identity? What constitutes a health crisis? How do we navigate the tensions between tradition and modernity? How do we create a world where all bodies thrive? Through these sections and the Handbook as a whole, we aim to demonstrate the richness that is the field of critical menstruation studies, a polyvocal constellation of scholarship and advocacy that is finally coming into its own.

Reference

Vostral, Sharra. 2018. *Toxic Shock: A Social History*. New York, NY: New York University Press.

PART I

Menstruation as Fundamental

CHAPTER 2

Introduction: Menstruation as Fundamental

Inga T. Winkler

Most articles on menstruation start by pointing out that menstruation is a normal biological process. This, of course, is true. But at the same time, menstruation is so much more for many people; in fact, it is *fundamental*. Menstruation unites the personal and the political, the intimate and the public, and the physiological and the socio-cultural.

Menstruation is fundamental because it either facilitates or impedes the realization of a whole range of human rights. In 2019, a group of United Nations human rights experts acknowledged that

> The stigma and shame generated by stereotypes around menstruation have severe impacts on all aspects of women's and girls' human rights, including their human rights to equality, health, housing, water, sanitation, education, freedom of religion or belief, safe and healthy working conditions, and to take part in cultural life and public life without discrimination. (United Nations 2019)

Because these dimensions of menstruation in different spheres of life shape lived experiences, we begin this Handbook with a series of chapters that focus on the voices and lived experiences of menstruators in different contexts. They all menstruate, but their unique socio-cultural, religious, and political contexts differentially shape and provide meaning to their experiences.

The coherence of this set of chapters lies in its deliberate diversity—in content, experiences, formats, and authors. Regarding the latter, while I aimed to include an even more diverse representation, I am keen to acknowledge the range of backgrounds of those who wrote for this section across geography, culture, religion, race, ethnicity, caste, age, sexual orientation, and gender identity. Sometimes more explicitly, sometimes more implicitly, they all bring their own lived experiences into their research and writing.

© The Author(s) 2020
C. Bobel et al. (eds.), *The Palgrave Handbook of Critical Menstruation Studies*, https://doi.org/10.1007/978-981-15-0614-7_2

Not only do the chapters highlight the uniqueness of different experiences, but they also present them in diverse ways. Some chapters are traditional research chapters, others are personal narratives, and yet another takes the form of a conversation between different contributors, which we call "Transnational Engagements" that punctuate each section. As editor, I value these different forms of knowledge and how they contribute to a better understanding of menstruation. I consider this all the more important in such a rapidly emerging field as Critical Menstruation Studies, in which many questions are still underexplored.

To begin to address these questions, contributors in this section address menstruation in different conditions, including informal settlements, homelessness, detention, disability, child marriage, and migration. The latter chapters in this section complement these perspectives by adding the layers of religion, culture, and caste. Annie McCarthy and Kuntala Lahiri-Dutt focus on the experiences of women in Delhi's informal settlements to describe the ways women manage the structural deficits they face, reconfigure notions of privacy, and navigate changing gender relations. The theme of navigating the need for privacy in public space is also central to Shailini Vora's exploration of homelessness for women in Bristol, UK who are conscious of their doubly stigmatized status as 'homeless' menstruators—a precarious reality that forces strategic management and concealment of their menstrual status.

The unwanted publicness of menstrual experiences is further put into stark relief by Chandra Bozelko speaking about her experience living in detention, most poignantly, the shame and humiliation of staining her clothes and having to ask male guards for tampons and pads. Tomi-Ann Roberts complements this perspective by detailing the experiences of women deprived of liberty who undergo a degrading strip search *en masse*. She argues that this is a uniquely misogynist form of punishment which women experience against the background of the objectification and self-objectification of their bodies. Trans and genderqueer menstruators also deal with unwanted publicness as well as social expectations, standards of femininity, and a range of constraints in social and physical spheres. S. E. Frank and Jac Dellaria present these in a visual narrative that focuses on everyday experiences.

The power over women's bodies is a central theme in Linda Steele's and Beth Goldblatt's chapter. The authors powerfully demonstrate that women and girls with disabilities are perceived as mentally and physically incapable of meeting gendered norms of menstrual concealment which leads to coercive interventions by parents, carers, medical professionals, and judges, particularly through sterilization. In a different context, Musu Bakoto Sawo presents a narrative of coercion and her journey into forced womanhood. Her story powerfully describes how she turned from a survivor of child marriage into a children's and women's rights activist.

All these chapters show how experiences of menstruation are shaped by gendered expectations about women's bodies in social context. Alex Hawkey's, Jane Ussher's, and Janette Perz's contribution is a potent demonstration of these forces as it emphasizes the shifting constructions and experiences of menarche and menstruation from the perspective of migrant and refugee women resettled in Australia and Canada.

As the section proceeds, the frame broadens to offer religious and cultural perspectives on menstruation. Most often, when discussing religion and menstruation, the language is one of restriction and oppression. Ilana Cohen's exploration of the menstrual traditions in both Judaism and Hinduism succeeds in adding complexity to that frame. She examines how menstrual practices contribute to a better understanding of the ways a religious community defines and (re)produces itself. This overview is complemented by two personal narratives that provide additional perspectives on menstruation and religion. Tova Mirvis offers a personal reflection on the Jewish practice of *mikvah*, or ritual bath after the completion of her menstrual period. She shares her growing doubts about her religious beliefs and laws which required the *mikvah*, eventually leading her to leave the religious world of which she was a part. Deepthi Sukumar's narrative addresses the intersection of Hinduism, menstruation, and caste. She compellingly details that menstrual restrictions often associated with Hinduism have never affected her, a Dalit. She argues that "caste is her period:" whether menstruating or not, Dalit women are considered 'impure' and 'polluting.'

Alma Gottlieb contributes a chapter that reflects on menstrual taboos. We often hear that menstruation is shrouded in taboos, myths, and silence. But what do we really mean by this claim? In response, Gottlieb disentangles the idea of taboos, taking the reader to the origin of the Polynesian word *tapu*, which is neither negative nor positive but invokes the notion of a state of being that is too powerful to act on. From there, she discusses a diverse range of encounters with menstruation in various cultures and inserts greater nuance into the discussion on taboos. To conclude, the individuals participating in this section's "Transnational Engagements" on cultural and religious menstrual practices edited by Trisha Maharaj and Inga T. Winkler further the diversity of perspectives. The contributors demonstrate varying perceptions of menstrual practices including how they exercise their agency when deciding if or how to engage in these practices and/or their transformation. This conversation thus productively complicates the too-common depiction of all menstrual practices as restrictions necessarily forced upon women.

The chapters in this section demonstrate the importance—and indeed urgency—of considering the lived experiences of all menstruators. These vary widely and are shaped by a range of different factors including religion, culture, political systems, socialization, caste, disability, place of residence, among many others. In many cases, an intersection of factors such as gender *and* disability; or gender, religion, *and* caste determine menstrual experiences. This material offers insights into some individuals' menstrual experiences many of whom are marginalized on different grounds. The chapters in this section are complemented by additional perspectives in other sections of the Handbook, such as the experiences of women and girls in refugee camps described by Siri Tellier et al., autistic experiences of menstruation described by Robyn Steward et al., and trans menstruation addressed by Klara Rydström.

What stands out throughout the section are the tensions between 'the public' and 'the private.' Many individuals shared feelings of embarrassment when publicly disclosing their menstrual status or shame when having to request menstrual products. To avoid such discomfort, individuals often seek out privacy when menstruating. This forces us to ask: Why do we think that menstruation should be kept invisible and private? Why is this natural biological process considered embarrassing? And how do gendered social norms and perceptions of modesty inform our understanding of what menstruation 'should be?' In answering these questions, we must be very careful not to impose the burden of transforming societal norms on individuals alone who are often in the most marginalized or vulnerable situations. Such transformation requires us all to contribute to broader societal change.

Menstruation is fundamental because it is ultimately about power relations—the power of the guard in the prison or staff in a homeless shelter to dispense or withhold menstrual products, the power of judges to authorize sterilizations, the power of parents and relatives to force young girls to marry, and the power of religious authorities to expect unflinching conformity with religious norms. Adopting a human rights perspective to addressing menstruation forces us to rethink and shift these power relations. At the core of human rights is the dignity and agency of every individual, and the voices included in this section powerfully demonstrate that such agency can take many different forms: turning from a survivor of child marriage or someone formerly living in detention into advocates whose voices are heard widely; transforming socio-cultural norms; and finding meaning in religious menstrual norms. Considering menstruation as fundamental means to enable women and girls and anyone who menstruates to exercise their agency.

Reference

United Nations. 2019. "International Women's Day—8 March 2019 Women's Menstrual Health Should No Longer Be a Taboo." Accessed July 26, 2019. https://www.ohchr.org/EN/NewsEvents/Pages/DisplayNews.aspx?NewsID=24258&LangID=E.

CHAPTER 3

Bleeding in Public? Rethinking Narratives of Menstrual Management from Delhi's Slums

Annie McCarthy and Kuntala Lahiri-Dutt

Flowing from inside the body to out, menstrual blood is experienced in both public and private realms—where hierarchies of gender, knowledge, and power position menstruators[1] as responsible for, but not always in control of, the meanings attached to their own bodies. Menstruating bodies are, thus, both objects and agents, where agency is at once the agency *of* the body as an independent actor that is not always or easily controlled and agency *over* the body (Fingerson 2006, 23). As "both the objects and subjects of their bodies, of menarche and menstruation" (Puri 1999, 43), menstruators are positioned betwixt and between the public and the private (see Vora [Chapter 4] in this volume). Experiences of menstruation, while deeply personal and embodied, also have an external biomedical framing; menstrual blood is a private secret that is expected to be concealed (see Wood [Chapter 25] in this volume), yet menarche can have radical implications for a girl's lifestyle and mobility (Jewitt and Ryley 2014; Puri 1999). Hygiene is considered an individual pursuit—yet, in the absence of sanitation infrastructure, it becomes a public issue. The shifting meanings and values assigned to 'public' and 'private' across regimes of knowledge, culture, and environment are, thus, the key to understanding experiences of menstruation in any context. To illustrate the salience of these ideas for Menstrual Hygiene Management (MHM) in the Indian context, this chapter analyzes the gendered challenges of everyday life in informal settlements in Delhi. We explore the experiences of women and girls who manage menstruation in conditions of extreme congestion to argue that the way privacy is conceived of in MHM initiatives—as self-evidently material—erases the complex ways privacy is socially constructed, gendered, and layered with power dynamics. To draw out these points, this paper will

© The Author(s) 2020
C. Bobel et al. (eds.), *The Palgrave Handbook of Critical Menstruation Studies*, https://doi.org/10.1007/978-981-15-0614-7_3

first introduce the deficit framing of MHM in India and then move through a discussion of key themes: privacy, space, and knowledge before introducing the water and sanitation issues faced in Delhi. Subsequently, it will more explicitly introduce the context of informal settlements in Delhi, before lastly focusing on a particular settlement and the story of one woman: Champa.

This chapter brings together the findings and insights of a number of research projects. The first author draws on her experience of fieldwork with children from four slum communities in Delhi, in which she documented the ways these children and their communities were framed by sanitation and hygiene promotion campaigns (see McCarthy 2015). The second author contributes specific data on women's experiences of menstruation, collected through long interviews, focus group discussions, and participant observation carried out in one particular slum cluster in the New Okhla Industrial Development Authority (NOIDA). The women in this latter study group were selected on the basis of familiarity developed from earlier research.

INDICATING DEFICITS

When engaging with the MHM literature on India, one is invariably and immediately bombarded with a range of 'alarming' statistics that starkly highlight the 'deficiencies' in Indian women's and girls' menstrual knowledge and practices. These studies report statistics such as 88% of menstruating women in India use fabric, rags, ash, straw or wood shavings to absorb their menstrual flow; "70% of mothers consider menstruation 'dirty,' perpetuating a culture of shame and ignorance. . . . Girls are typically absent for 20% of the school year due to menstruation"; and poor menstruation hygiene practices cause a "70% increase in incidence of reproductive tract infections" (USAID, Kiawah Trust, and Dasra 2014, 2). In these studies, individual experiences of menstruation, studied in a specific context, are transformed into numerical data sets that circulate globally to justify interventions largely focused on poor and marginalized women in the Global South. Key indicators[2] informing these data sets are knowledge of menstruation at menarche; use of menstrual products; days absent from school or work as a result of menstruation; rates of reproductive tract infections (RTIs); access to clean water and toilet facilities; experience of menstrual taboos; and methods of disposing absorbents (see Rajagopal and Mathur 2017; Mahon and Fernandes 2010; Kumar and Srivastava 2011 for Indian examples). These indicators, taken together, transform menstruation into a series of milestones and practices that can be assessed against global rubrics of health, dignity, education, and productivity (for one example see Sommer 2010). Yet, the benevolence and utilitarianism of this language obscures the fact that these indicators are embedded in specific contexts, and that a variety of structural, religious, cultural, and gendered practices—that both construct and obstruct the 'management' of menstruation—are involved in determining an individual's menstrual management practices.

Rather than any specific engagement with these local meanings of menstruation, contemporary MHM initiatives emerging out of transnational human rights and development discourses, assume a universal human subject with rights to 'dignity,' 'privacy,' 'hygiene,' 'health,' and 'productivity.' While claiming universality, each of these terms has a unique history that cannot be separated from the violent construction of colonized, feminized, 'dirty,' 'lazy' bodies as the 'other' against which first colonial, and later developmentalist projects were created and sustained. Today, alongside 'health' it is the paradigm of dignity that is the key discursive tool mobilized by contemporary MHM initiatives (for examples see Mahon and Fernandes 2010; Phillips-Howard et al. 2016 and for a critique see Bobel 2019). Yet the discursive production of 'indignity' in studies on MHM in India necessitates critiques similar to those mounted by Bhaskar Mukhopadhyay (2006) in his attack on Arjun Appadurai's (2004) 'apologia' of World Bank programs targeting open defecation. Mukhopadhyay (2006, 227) accuses Appadurai of "maintaining the moral purity of categories—dignity, humiliation, purity, pollution, right, wrong" at the expense of excluding "a more porous field of responses." Mukhopadhyay (2006) suggests that in abandoning these pure categories, we should not ignore the 'problem' but, instead, engage "with popular or subaltern practices as ethico-political responses and [reflect] on their sources of authority rather than simply denigrating them from the vantage point of some absolute wisdom (227)." This kind of commitment informs our approach, particularly our representation of women's and girls' practices of managing menstruation in the latter half of this paper.

Managing Privacy and Knowledge

The translation of 'individual' practices into indicators used to promote standardized education and MHM campaigns forces us to re-examine the language of 'management' and ask "who is doing the managing?" and "what exactly is it that they are managing?" Lahiri-Dutt's (2015) critique of MHM projects suggests that these programs aim to 'empower' menstruators to manage their periods as individual, private concerns free from social taboos or stigma. Yet underlying these efforts are a set of assumptions about the positive relationship between privacy, bodily autonomy and empowerment, and negative connotations of 'public' as the space where stigma and restriction are imposed. This binary—and the responsibility it places on women to 'manage' their own bodies rather than on society to 'manage' its expectations—have long been challenged by feminist and radical menstruation activists both in the West (see Bobel 2010) and more recently in India where there has been a recent spate of menstruation-related activism (Prasanna 2016; Fadnis 2017). These recent campaigns in India have confronted restrictions on entering places of worship while menstruating and government taxes on menstrual products. The latter have involved challenging public stigma around menstruation by brandishing bloody pads and mobilizing hashtags like #DontTaxMyPeriod (Fadnis 2017) to oppose the very real nexus of gendered violence, capitalism, and medicalization that makes bleeding bodies a 'problem.' Yet outside of these moments that intentionally orchestrate the private 'bleeding over' into the public, everyday experiences of menstruation in India are similarly saturated with complex violences that are neither entirely public nor completely private, entirely biological or cultural, familial, or individual. Jyoti Puri's (1999) work among middle-class Indian women speaks to this tension. Bitter tales of the rules and restrictions imposed post menarche intertwine with individual impulses to conceal and individually manage menstruation. These stories point to the impossibility of containing menstruation within the bounds of an individual body, and force us when considering experiences of menstruation in informal settlements to interrogate global development funders' MHM interventions that largely seek to render menstruation a 'technical' problem with technical solutions (Li 2007).[3]

Congestion is a defining feature of life in informal settlements and plays a powerful role in shaping the lives of women and girls affecting their personal care, physical and mental well-being (see Reddy and Snehalatha 2011; Joshi, Fawcett, and Mannan 2011). These cramped conditions also contribute to the reconfiguring of the very notion of privacy itself. Thus, in informal settlements, privacy is not necessarily a material space but a technique of modulating, or even countering forces of social control. Indeed, Gan (2009, 3) observes that "in a rigid social environment, privacy preserves a small breathing space, providing privacy from others while also enabling self-expression, the privacy to do something else or to be someone else." Following Moore's (1984, 6) definition of privacy as "a desire for socially approved protection against painful social obligations," we are reminded of the ways that privacy for women in informal settlements produces spaces for an alternative identity that is different from their traditional domestic role or their role as wage earners. Women in informal settlements in Delhi are typically migrants from rural areas, who, on entering the sprawling urban metropolis, find themselves surrounded by rapid-paced social change. Thus, following Gan's (2009, 3) argument—that privacy is a way to shield oneself from the grasping hand of social convention—we argue that rural women moving into the congested metropolis utilize techniques for the production of privacy to protect themselves from the impact of change and to pause and reconsider their place in the social mosaic. Here privacy is not the ability to be alone with or have full control over one's body, but is enabled by a series of techniques and practices that assert a claim to autonomy and identity that is neither entirely spatial nor social. Ayona Datta's (2008) reflections on her fieldwork in informal settlements in Delhi reminds us that gender is key to these practices, forcing us to recognize how "intimately the materialities of bodies and bodily performances are connected to masculine or feminine marking of places, and how these are regulated and given form through narratives and discourses (202)."

Yet these local narratives and discourses are framed as at best irrelevant and at worst outright dangerous by MHM initiatives in India that highlight deficiencies in girls' knowledge of menstruation prior to menarche. Studies of Indian girls have found that anywhere between 35 and 81% of girls surveyed are unaware of menstruation prior to menarche (see Van Eijk et al. 2016; USAID, Kiawah Trust, and Dasra 2014; Rajagopal and Mathur 2017; Zaidi, Sivakami, and Ramasamy 2015; Bhattacherjee et al. 2013). Noting this lack of knowledge, many of these studies go on to comment on the source and 'quality' of the knowledge these women and girls do possess (see Van Eijk et al. 2016; Dasgupta and Sarkar 2008). These studies emphasize how the lack or 'poor quality' of girls' knowledge leads to distress, typically by describing how menarche triggers anxiety, panic, fear, and worries about imminent death. One study in Ranchi notes that among the slum girls surveyed, 54.5% were frightened and cried (Kumar and Srivastava 2011, 596). Yet in these studies this distress is not linked to bodily changes but an 'ignorance' of biological process that produce them. This discourse of ignorance-as-inevitably-causing-distress leads to the double victimization of women and girls as subject not only to their own bodies, but also to the ignorance of their own families and communities. Local ways of knowing about menstruation are further stigmatized by being labeled cultural and religious, and are framed almost entirely through the twin lenses of taboo and restriction. Following Mohanty's (1984) classic text on the production of 'third world difference,' we argue that such studies reinscribe the object status of these girls and women and in the process affirm teachers, public health officials, and development workers as the only true 'subjects' of MHM interventions (for examples of this see Rajagopal and Mathur 2017; Kumar and Srivastava 2011). While we acknowledge that efforts to improve menstrual literacy are important, we suggest they must be framed in ways that acknowledge local knowledges and stop short of privileging outsider expertise as inherently superior.

WASHing in Delhi[4]

Although constantly welcoming new inhabitants, many of Delhi's slums are decades old. From the beginning, "the building of planned Delhi was mirrored in the simultaneously mushrooming of the unplanned Delhi" (Baviskar 2003, 91). Today in Delhi, estimates of the population of slum settlements, officially called *jhuggī jhoprī* (JJ) clusters,[5] range from 15% to over 50% of the urban population (Ghertner 2015, 6). This numerical uncertainty is both produced by and, in turn, produces spaces of infrastructural neglect, where informality is equated with illegality and slum dwellers are stigmatized as 'migrants' deemed ineligible to share civic rights to the modern city (Ghertner 2015; Baviskar 2003; Datta 2012). Growing voices of discontent, powerfully mobilizing through Resident Welfare Associations (RWA), argue that slums are unsightly stains on the city, the last remnants of which have to be erased for Delhi to take its rightful place on the global stage (see Baviskar 2003; Bhan 2009; Datta 2012; Ghertner 2015). But such narratives of slums—as sites of filth and disease, as obstacles to development and modernity—are far older than the post-independence mushrooming of Delhi, and far more widespread. These have their origins in the colonial period when "doctors and surgeons helped to form and give seemingly scientific precision to abiding impressions of India as a land of dirt and disease, of lethargy and superstition, of backwardness and barbarity" (Arnold 1993, 292). When in 1888 Viceroy Lord Dufferin requested a general inquiry into the hygiene habits of India, colonial officials in their report wrote "to the masses of the people . . . sanitation is foolishness" (Prashad 2002, 47–48). But sanitation became "foolishness" as Prashad (2001) persuasively argues for colonial Delhi, precisely because the British pathologized Indians as impossibly dirty and declined to spend money on modern sanitation infrastructure.

Contemporary WASH initiatives targeting informal settlements in Delhi articulate many of the same issues and doubts as Delhi's colonial administrators. Here questions about the attitudes and dispositions of the 'unclean' combine with the practical difficulties of making infrastructural changes in dense settlements to produce both a systematic disinclination, as well as financial and bureaucratic barriers that prevent many of the innumerable small NGOs who have mushroomed in Delhi in recent years from engaging in costly infrastructure projects (one example of an organization encountered by McCarthy during her fieldwork is WASH United).[6] This lack of engagement is stark considering that the urban poor in Delhi have "particularly vulnerable water access . . . millions lack official connections or even rights to public water supplies" (Truelove 2011, 147). Yaffa Truelove (2011) notes that even when available the "water supply is marked by such dramatic unreliability that the majority of residents engage in informal and supplemental water sources and practices" (147). Despite these gaps, many interventions into WASH by small NGOs witnessed by McCarthy during her fieldwork in Delhi in 2013 occurred entirely at the level of education and behavior change. These programs create a trap, in which slum dwellers without access to adequate sanitation infrastructure can never be clean enough to shatter ideas of their innate proclivity toward filth, nor adequately demonstrate their desire and worthiness to be given the infrastructure that would allow them to be 'properly clean.' Seemingly stuck in this endless cycle of education campaigns, slum dwellers drift in and out of WASH initiatives that, just like the handwashing promotion campaign studied by McCarthy (2015) do not provide soap nor address issues of water accessibility.

In the context of MHM, these education programs are mostly silent about one of the key biomedical facts of menarche: fertility. This is rather ironic given that, as Emily Martin (1999) has shown, the biomedical paradigm of menstruation is one of failed reproduction. This becomes even more paradoxical in the context of informal settlements in which a girl's newly attained fertility is the driving force behind efforts to control or constrain her movements in ways that ensure the preservation of her family's *izzat* (honor/integrity), preserve the possibilities for arranging a 'good' marriage, and ensure that both the daughter and her family do not become the subjects of cruel community gossip (see Chakraborty 2010). A training manual developed by the Water Supply and Sanitation Collaborative Council (WSSCC) in collaboration with the Government of India (2013, 31) explicitly advises trainers to de-link the "teaching of sex education from training in menstrual hygiene practices, to avoid causing ethical or religious offense." But this focus on menstruation as solely a "hygiene crisis" actively ignores many of the local social and moral meanings attached to menstruating bodies (Bobel 2019, 295).

In the context of informal settlements, moral and social meanings are equally inscribed in local infrastructure and resources, often in deeply gendered ways. In the communities in which McCarthy worked in 2013, water was the greatest source of community tension; neighbors continually evaluated each other's 'private' water use, misuse, or overuse in moral terms. In one slum community, girls vehemently denounced their neighbors' water use, framing whole families as selfish and only able to think of themselves. An individual's or family's, use or perceived misuse of shared resources—water pumps and public toilets—becomes thickly layered with moral significance; people consider others 'good' or 'bad' based on their usage of resources. Yet the necessity of using these 'public' resources, means that sanitation, in particular, the care of the menstruating body, in this context cannot be reduced to an individual 'private' project. Specifically, this moral quality means that 'privacy,' if conceived of spatially, is not just a question of being able to conceal one's body and its processes, such as menstruation, but also the ability to shield oneself from the nexus of community observation and gossip. Yet returning to Moore's (1984) definition of privacy as a "socially approved protection" (6) understanding privacy simply in terms of spatiality does not do justice to the extent to which community relations and gendered forms of participation in networks of sharing and speaking construct meaningful personhood in these communities. To explore this further let us now explore the space of one informal settlement in Delhi.

NOIDA

In and around NOIDA, the popular acronym for New Okhla Industrial Development Authority, informal JJ clusters number between 200 and 310, with an additional 150–180 in the areas around Greater NOIDA. Each cluster may have 20–50 *jhuggīs* or dwellings, although often, there are several *jhuggīs* on a single plot of land. Living in these various communities are about 8000–10,000 women, most of whom work as domestic helpers in middle-class residences. Typically working between 5 a.m. and 9 p.m., these women earn between INR 15,000 and INR 18,000 (USD 300–USD 360) a month. Having migrated from the Indian states of Uttar Pradesh, Bihar, and West Bengal—and also from Bangladesh—these predominantly low-caste women live in these slum settlements to work for the period of time it takes to save up to build a *pakka* (concrete) house in their home village or to save for their children's marriages. In slum communities in NOIDA 20 or 30 people share a common latrine—made of bamboo sticks and tarpaulin with holes in the bottom to carry the excrement to the nearby field. In these communities, each shanty home has three sections: a front, a middle, and a back. The small front section, made of bricks on the mud, is used for washing dishes and clothes; some women take their bath here—in their *sari*. This affords them some privacy—only their husband and children can come here, or see them. Then there is a small, canopied area where the family sleeps during the hot, humid summer, or entertains house guests, particularly men. At the very back is the family's private area—possibly a wooden bed, with a mosquito net; and a shelf, or a box, that contains the family's personal possessions. This rudimentary division of space is one technique of producing privacy that demonstrates both its spatial, social, and gendered characteristics.

CHAMPA

Champa, a woman of around 32 years, lives with her husband and two children in such one settlement. Champa is from West Bengal. There is no government school for children living in this informal settlement, so her children do not go to school. Champa and her husband go out to work in the day, and the children cannot stay at home by themselves, so Champa takes them along to work. Sometimes they also help her with household chores or help their father collect and recycle garbage. Champa's son, 11, is good at sweeping and scrubbing. She proudly recounted how her son often mocks her cleaning: "How do you clean rich people's houses? How could they take your work and not complain?" Cleanliness and hygiene is, thus, a matter of pride for Champa and her son, who asserts his own claims to hygienic standards higher than his mothers.[7]

While menstruating, Champa, like other women in the cluster, uses homemade napkins made from rolled up sections of old cloth. She says old cloth is soft. Champa uses stronger fabrics, such as *sarī* borders, to make a string to hold the roll in place. She throws away the cloth after one use; she never reuses the cloth because, unlike in her village, it is impossible to wash the used cloth in the cramped conditions of the shanty colony or in its crowded bathing place. When we asked how she ensured a steady supply of old saris, Champa said that she received old saris as gifts. She added that in the harsh weather of Delhi, the saris do not last long, soon becoming ready for use as rags. There are several waste bins that the slum community has set up and usually the garbage is burned by the community when the bins start to overflow. When she puts on the cloth in her shanty, her husband stands guard outside to ensure her privacy. She puts on one cloth bundle before starting work early in the morning. She is often busy during the better part of the day, and is only able to change the pack after her late afternoon shower.

The role of husbands in assisting women with managing their menstruation was commonly reported during our focus groups. This points to the way privacy in this context is something that is relationally constructed, both along and across gendered lines. Yet that the couple collaborates to manage menstruation dismisses many beliefs of gender-based separation of male–female domains that are prevalent in feminist discourse.

Champa cannot reuse her cloths, as many Indian women do, because she lacks privacy and access to resources to wash and dry cloths. MHM experts deplore the practice of reusing homemade 'sanitary napkins' particularly when they cannot, as is typically the case in informal settlements, be dried properly in full sunlight (for an example of this rhetoric see Mahon and Fernandes 2010). While public health experts stress that sunlight removes all traces of dampness and has a sterilizing effect, in cramped slum communities exposing one's cloths to the sun would also mean exposing the fact of one's menstruation to the community. Reuse of improperly clean and dried cloths is in much MHM literature cited as a key factor of recurrent and dangerous RTI, yet this is an assumption that has been questioned in a systematic review (see Sumpter and Torondel 2013 for both examples and refutation). Despite this recent research, many MHM initiatives would likely celebrate Champa's adoption of single-use absorbents. Yet such an evaluation would fail to take into account the broader ways the overcrowding in Champa's settlement, its stretched water resources, the role of her husband in securing her privacy, and her labor to continually source and make new cloths to absorb her menses, structure her life. By ignoring these social, infrastructural, and interpersonal conditions actually required to 'manage' menstruation, any simple celebration of the use of single-use absorbents in this context would fail to recognize the extent to which local menstrual management practices cut across public and private domains, configuring Champa as both an agent and something that is acted upon. If Champa had greater access to privacy, no doubt she would reuse her cloths, as many other Indian women do. This, in itself, should provoke us to think further about the politics of defining experiences of 'lack,' and to look further into the ways value is assigned to particular practices in particular contexts. Additionally, it forces us to ask, with Barbara Penner (2010): "When is provision good enough, dignified enough? And who decides?"

Changing Meanings of Menstruation Over the Life Course

The simultaneously public and private nature of menstruation was further borne out during a focus group discussion with eight, initially reluctant, women living in the JJ cluster. To attempt to alleviate this discomfort we started asking these women about experiences of menarche of girls in their community. Women in our focus group said that girls in this community get their period between the ages of 10 to 12. When they lived in the village, simple *stree achar* (feminine rituals) were performed to mark a girl's puberty and menarche.

While these rituals have been altered or simply omitted in the new context of the slum (see Hawkey, Ussher, and Perz [Chapter 10] in this volume), the onset of menstruation nonetheless sends out a 'warning' signal to a girl's parents. In the village, as soon as the girl reaches puberty, the elders in the family began to alert the parents to start looking for a groom, to protect her virginity. Within a year or two of her first period, the girl will be married (see Sawo [Chapter 9] in this volume). In the slum, keeping a menstruating girl unmarried and at home for long is considered neither safe for the girl nor the family who must bear the burden of an unexpected pregnancy from a love affair or rape. While marriage is still considered the 'only' option for many families, growing discourses of girls' education and awareness of laws prohibiting child marriage place many slum-dwelling families in precarious positions in relation to their teenage daughters who themselves are increasingly educated about their rights in local NGO programs teaching girls' empowerment. For this generation of girls who have lived more of their life in the slum than the village and have often spent years in NGO programs cultivating educational and career aspirations, menstruation can signal much larger contestations about the meanings of childhood and adulthood.

As a woman matures however, the significance of her menstruation changes. Women told us that a housewife in a village experiencing menstruation is seen as a body that needs to rest: she is not meant to enter the *mandir* (temple; or, at home, the corner where the idols are housed) or perform *puja* (worship) (see Cohen [Chapter 11] as well as Sukumar [Chapter 13] in this volume). She is not expected to go outdoors or mingle with others, and she is expected to rest in a room alone during those four days. Married women abstained from cooking for their families on those four days, and other women had to take up these tasks. In a large joint, or extended, family, this does not pose a problem, as extra hands are always around. In a village, menstruation tends to bind a woman's body and her activities to a private space. Rather than seeing this entirely through the paradigm of 'restriction' and thus, negative gendered constructions of 'cleanliness' and 'purity,' we must realize the ways these practices allowed women to rest completely for four days and offered their bodies a break from relentless labor. This other side of 'confinement' is that the necessity of contributing labor and its subsequent income to their smaller family units in Delhi, meant most women in slums avoided taking time off during menstruation. Yet even though they did not take this time off, women continued to engage in practices of cleansing the body post menstruation, that would have previously initiated their return to everyday life. Thus, in spite of difficulties accessing water, and as testament to the importance of this ritual, most women wash their hair and clean their body once their period is over.

Older women in the focus group noted that menstruation necessitated non-participation in devotional activities but also allowed them to avoid having sex with their husbands. Younger women agreed: "When we are in the village", Champa said, "a husband does not come near the woman who is having her period, sex is completely forbidden." However, once couples are in the slum away from the joint family social controls are loosened, and husbands expect their wives to have sex on demand. From our discussions, this seemed to be one of the primary reasons of quarrels and ensuing fights, leading to violence between couples in the slum. Here competing demands made on the female body highlight tensions between reproductive labor and productive labor that women in informal settlements had to continually renegotiate in new settings far from their broader kin networks and the gendered norms that defined them.

These selected stories, drawn from across spaces of the rural and urban and of adolescence and wifehood illuminate the ways menstruation—its broad socio-cultural meaning and its management at the bodily level—cannot be regarded as a singular or static process, but rather one that is mediated across the life course, through relations with others, by configuring and reconfiguring space and by navigating expectations of productive and reproductive labor.

In these communities, young girls look up to their mothers to teach them about self-care, and to their mother-in-law after marriage. In other words, knowledge about menstruation management is transmitted generationally. While their current location—far from their home village, state, or country—can and does disrupt traditional celebrations, cultural, familial, and generational knowledge of menstruation and related practices continue to be important even as they are challenged. Consequently, these resources—such as the stories shared in our focus groups documenting collaboration, nostalgia, and emerging spaces of contestation—remain major keys to understanding women's narratives of menstrual management practices. Largely deemed insufficient when examined through biomedical and hygienic lenses, this knowledge and the stories which encode it are fact deeply valued by women, and constitutes the first and primary site of information, meaning-making and support for most women. Returning to Mukhopadhyay (2006, 227) we must ask ourselves what it would mean to stop "denigrating" this knowledge "from the vantage point of some absolute wisdom" and engage with it as a form of agency imbued with clues to what women and their communities value and seek.

Conclusion

As we have shown, women's own voices have for a long time been ignored in debates about and interventions into women's MHM practices. In highlighting interventions that denigrate women's experiences and knowledge, we seek to challenge contemporary MHM initiatives and the extent to which they preserve and produce powerful discourses of 'third world difference' (Mohanty 1984). We argue that in Delhi's informal settlements, women who are marginalized across multiple axes of class, caste, and gender face daily struggles to claim recognition and access to the city's resources. We suggest that we can learn important lessons by drawing on the conversations with Champa and other women who constantly manage these structural deficits alongside the discourses of deficiency that are used to frame their lives. These women who were nostalgic for the menstrual management practices of village life and the corresponding period of rest force us to recognize that "women's participation in gender-traditional religions" may not signify passivity (Talukdar 2014, 141). Thus, by paying attention to forms of knowledge transmission and meaning-making of the kind expressed by these women we can better understand women's lived experiences of menstrual 'management.'

These meanings are shaped by environmental factors—rural *vs* urban settings, joint families *vs* nuclear, private *vs* public access water—and how notions of public and private are reconfigured and remade in informal settlements. As women in the focus group noted, once in the city, the special significance attributed to menstruation's monthly occurrence is remarkably reduced, and other things, such as the commitment to be at the place of work on time, begin to dictate the daily rhythm of life. The absence of a clearer distinction between private and public spaces within *jhuggīs*, and the JJ clusters more broadly, means that new meanings are attributed to spaces, new ways of performing everyday practices are imagined and invented, and menstruation begins to assume and convey new connotations. Yet, these practices are not fixed, they change constantly, as women and men—living in extremely congested conditions that often provide no spatial privacy at all—assiduously and constantly recreate new ways of being. The tools used to 'manage' menstruation promoted by MHM initiatives—biomedical knowledge, single-use absorbents, access to toilets and water sources—are important, but seem to occupy a world apart from Champa's, whose management techniques rely on kinship and relationality. By rendering menstruation a technical, hygienic crisis these initiatives at best ignore and at worst stigmatize the very things that make menstruation meaningful in the lives of women and girls in informal settlements.

Notes

1. We will endeavor to use the gender-neutral language of menstruators throughout this paper, to avoid gender essentialism and acknowledge what Chris Bobel (2010, 164) refers to as the "inclusion fundamental to third-wave feminism." But also, and perhaps more significantly, given the subject matter of this paper, we use this language to acknowledge the extent to which global inequalities of calorie intake, access to medical care, housing, and working conditions mean that menstruation can by no means be assumed or presumed to flow naturally and regularly from all bodies sexed female of reproductive age.
2. We borrow the language of indicators from Sally Engle Merry's work, specifically *The Seductions of Quantification: Measuring Human Rights, Gender Violence, and Sex Trafficking* (2016).
3. Tania Li (2007, 7) reminds us, "the practice of 'rendering technical' confirms expertise and constitutes the boundary between those who are positioned as trustees, with the capacity to diagnose deficiencies in others, and those who are subject to expert direction."

4. WASH is a widely used development acronym that stands for water, sanitation and hygiene.
5. Both the words *jhuggī* and *jhoprī* are used to describe an individual dwelling within a slum settlement with the phrase *jhuggī jhoprī* cluster or JJ cluster being used to describe the settlement itself.
6. In 2015 the Central Bureau of Investigation (CBI) in an affidavit to the supreme court stated that there were 29.9 lakh or almost 3 million NGOs registered in India under the Society Registration Act (Rajagopal 2015). 76,566 of these NGOs were registered in Delhi—constituting roughly one registered NGO for every 248 people.
7. Inter-generational conflicts and claims to expertise around hygiene, dominated many children's responses to the handwashing campaign observed by McCarthy (2015) during her fieldwork.

References

Appadurai, Arjun. 2004. "The Capacity to Aspire: Culture and the Terms of Recognition." In *Culture and Public Action*, edited by V. Rao and M. Walton. Stanford: Stanford University Press.

Arnold, David. 1993. *Colonising the Body: State Medicine and Epidemic Disease in Nineteenth-Century India*. Berkeley: University of California Press.

Baviskar, Amita. 2003. "Between Violence and Desire: Space, Power, Identity in the Making of Metropolitan Delhi." *International Social Science Journal* 55 (175): 89–98.

Bhan, G. 2009. "'This Is No Longer the City I Once Knew': Evictions, the Urban Poor and the Right to the City in Millennial Delhi." *Environment and Urbanisation* 21 (1): 127–42.

Bhattacherjee, Sharmistha, Kuntala Ray, Romy Biswas, and Manasi Chakraborty. 2013. "Menstruation: Experiences of Adolescent Slum Dwelling Girls of Siliguri City, West Bengal, India." *Journal of Basic and Clinical Reproductive Sciences* 2 (2): 85–91.

Bobel, Chris. 2010. *New Blood: Third-Wave Feminism and the Politics of Menstruation*. New Brunswick: Rutgers University Press.

———. 2019. *The Managed Body: Developing Girls & Menstrual Health in the Global South*. Cham, Switzerland: Palgrave Macmillan.

Chakraborty, Kabita. 2010. "The Sexual Lives of Muslim Girls in the Bustees of Kolkatta, India." *Sex Education: Sexuality, Society and Learning* 10 (1): 1–21.

Dasgupta, A., and M. Sarkar. 2008 "Menstrual Hygiene: How Hygienic Is the Adolescent Girl?" *Indian Journal of Community Medicine* 33 (2): 77–80.

Datta, Ayona. 2008. "Spatialising Performance: Masculinities and Femininities in a 'Fragmented' Field." *Gender, Place and Culture* 15 (2): 189–204.

———. 2012 *The Illegal City: Space, Law and Gender in a Delhi Squatter Settlement*. Surrey: Ashgate.

Van Eijk, Anna Maria, M. Sivakami, Mahmita Bora Thakkar, Ashley Bauman, Kayla F. Laserson, Susanne Coates, and Penelope A. Phillios-Howard. 2016. "Menstrual Hygiene Management among Adolescent Girls in India: A Systematic Review and Meta-analysis." *BMJ Open* 6. https://doi.org/10.1136/bmjopen-2015-010290.

Fadnis, Deepa. 2017. "Feminist Activists Protest Tax on Sanitary Pads: Attempts to Normalize Conversations about Menstruation in India Using Hashtag Activism." *Feminist Media Studies* 17 (41): 1–4.

Fingerson, Laura. 2006. *Girls in Power: Gender, Body and Menstruation in Adolescence.* New York: State University of New York Press.

Gan, Wendy. 2009. *Women, Privacy and Modernity in Early Twentieth-Century British Writing.* London: Palgrave Macmillan.

Ghertner, Asher. 2015. *Rule by Aesthetics: World-Class City Making in Delhi.* New York: Oxford University Press.

Jewitt, Sarah, and Harriet Ryley. 2014. "It's a Girl Thing: Menstruation, School Attendance, Spatial Mobility and Wider Gender Inequalities in Keynya." *Geoforum* 56: 137–47.

Joshi, Deepa, Ben Fawcett, and Fouzia Mannan. 2011. "Health, Hygiene and Appropriate Sanitation: Experiences and Perceptions of the Urban Poor." *Environment and Urbanization* 23 (1): 91–111.

Kumar, Anant, and Kamiya Srivastava. 2011. "Cultural and Social Practices Regarding Menstruation among Adolescent Girls." *Social Work in Public Health* 26 (6): 594–604.

Lahiri-Dutt, Kuntala. 2015. "Medicalising Menstruation: A Feminist Political Economic Critique of Menstrual Hygiene Management in South Asia." *Gender, Place and Culture: A Journal of Feminist Geography* 22 (8): 1158–776.

Li, Tania. 2007. *The Will to Improve: Governmentality, Development, and the Practice of Politics.* Durham: Duke University Press.

Mahon, T., and M. Fernandes. 2010. "Menstrual Hygiene in South Asia: A Neglected Issue for WASH (Water, Sanitation and Hygiene) Programs." *Gender and Development* 18 (1): 99–113.

Martin, Emily. 1999. "The Woman in the Flexible Body." In *Revision Women, Health and Healing: Feminist, Cultural and Technoscience Perspectives*, edited by A. Clarke and V. L. Olesen. New York: Routledge.

McCarthy, Annie. 2015. "Telling Stories, Washing Hands: Exploring the Role of Narrative in Development Programs Targeting Children." *South Asian History and Culture* 6 (3): 401–16.

Merry, Sally Engle. 2016. *The Seductions of Quantification: Measuring Human Rights, Gender Violence, and Sex Trafficking.* Chicago: University of Chicago Press.

Mohanty, Chandra Talpade. 1984. "Under Western Eyes: Feminist Scholarship and Colonial Discourses." *boundary 2* 12 (3): 333–58.

Moore, B., Jr. 1984. *Privacy: Studies in Social and Cultural History.* London: Pantheon Books.

Mukhopadhyay, Bhaskar. 2006. "Crossing the Howrah Bridge: Calcutta, Filth, and Dwelling—Forms, Fragments, Phantasms." *Theory, Culture and Society* 23: 221–41.

Penner, Barbara. 2010. "Flush with Inequality: Sanitation in South Africa." *Places Journal*, November. https://placesjournal.org/article/flush-with-inequality-sanitation-in-south-africa/.

Phillips-Howard, Penelope A., Bethany Caruso, Belen Torondel, Garazi Zulaika, Murat Sahin, and Marni Sommer. 2016. "Menstrual Hygiene Management among Adolescent Schoolgirls in Low- and Middle-Income Countries: Research Priorities." *Global Health Action* 9 (1): 33032.

Prasanna, Chitra Karunakaran. 2016. "Claiming the Public Sphere: Menstrual Taboos and the Rising Dissent in India." *Agenda: Empowering Women for Gender Equity* 30 (3): 91–95.

Prashad, Vijay. 2001 "The Technology of Sanitation in Colonial Delhi." *Modern Asian Studies* 35 (1): 113–15.

———. 2002. *Untouchable Freedom: A Social History of a Dalit Community.* New Delhi: Oxford University Press.

Puri, Jyoti. 1999. *Woman, Body, Desire in Post-Colonial India: Narratives of Gender and Sexuality.* New York: Routledge.

Rajagopal, K. 2015. "Only 10% of NGOs Have Filed Financial Details with Govt: CBI to SC." *The Hindu*, September 18. https://www.thehindu.com/news/national/only-10-ngos-have-filed-financial-details-with-govtcbi-to-sc/article7665565.ece.

Rajagopal, Shobita, and Kanchan Mathur. 2017. "Breaking the Silence around Menstruation: Experiences of Adolescent Girls in an Urban Setting in India." *Gender and Development* 25 (2): 303–17.

Reddy, B. S., and M. Snehalatha. 2011. "Sanitation and Personal Hygiene What Does It Mean to Poor and Vulnerable Women?" *Indian Journal of Gender Studies* 18 (3): 381–404.

Sommer, Marni. 2010. "Where the Education System and Women's Bodies Collide: The Social and Health Impact of Girls' Experiences of Menstruation and Schooling in Tanzania." *Journal of Adolescence* 33 (4): 521–29.

Sumpter, C., and B. Torondel. 2013. "A Systematic Review of the Health and Social Effects of Menstrual Hygiene Management." *PLoS One* 8 (4): e62004.

Talukdar, Jaitar. 2014. "Rituals and Embodiment: Class Differences in Religious Fasting Practices of Bengali Hindu Women." *Sociological Focus* 47: 141–62.

Truelove, Yaffa. 2011. "(Re-)Conceptualizing Water Inequality in Delhi, India through a Feminist Political Ecology Framework." *Geoforum* 42: 143–52.

USAID, Kiawah Trust and Dasra. 2014. *Spot On! Improving Menstrual Hygiene in India.* https://www.dasra.org/cause/improving-menstrual-health-and-hygiene.

Water Supply and Sanitation Collaborative Council and Governmnet of India. 2013. *WASH and Health for Menstrual Hygiene Mangagement: Traing of Trainers Maunal, v1.0.* https://www.wsscc.org/resources-feed/training-of-trainers-manual-wash-and-health-for-menstrual-hygiene-management/.

Zaidi, Syed Hasan Nawaz, A. Sivakami, and D. Jegadeesh Ramasamy. 2015. "Menstrual Hygiene and Sanitation Practices among Adolescent School Going Girls: A Study from a South Indian Town." *International Journal of Community Medicine and Public Health* 2 (2): 189–94.

CHAPTER 4

The Realities of Period Poverty: How Homelessness Shapes Women's Lived Experiences of Menstruation

Shailini Vora

INTRODUCTION

Being on your period is the worst time for a woman to be homeless – it gives you that extra blow. —Simran

While estimates vary, in the US, 553,000 people are experiencing homelessness on any single night (US Department of Housing and Urban Development 2018, 1), while this figure in the UK is roughly 320,000 (Shelter 2018). Single women make up over a quarter of the users of homeless services in the UK (Homeless Link 2017, 23), and this percentage is similar in the US, with 28% of people experiencing homelessness being single women (US Department of Housing and Urban Development 2018, 11). These numbers are likely to be significant underestimates given the number of women experiencing "hidden homelessness" (Watson with Austerberry 1986), who do not access homeless services but stay in other temporary forms of accommodation such as the houses of relatives, friends, hostels, or bed and breakfasts. There are millions of people living in makeshift, precarious housing situations who lack complete and reliable access to private, safe and clean water, and sanitation facilities (see also McCarthy and Lahiri-Dutt [Chapter 3] in this volume).

The experiences of menstruation by people who are homeless, however, has been historically overlooked by the public, civil society actors, policy makers, and academics. Much progress has been made in recent years within academic literature to deepen our understanding of the multifaceted issue of women's homelessness, through studies of health, abuse, trauma, and specific policy responses

C. Bobel et al. (eds.), *The Palgrave Handbook of Critical Menstruation Studies*, https://doi.org/10.1007/978-981-15-0614-7_4

and intervention (for example, Padgett et al. 2006; Vijayaraghavan et al. 2012; Schutt and Garrett 2013). Taking into account the specific exclusions and traumas that are faced by women who are homeless, this literature is very productive and useful in order to work toward alleviating these challenges and providing gender-specific support for recovery. The topic of menstruation, however, is expressed only as a factor of the reproductive health of women who are homeless, as outlined by medical narratives (for example Ensign 2001). To my knowledge, there has been no explicit study conducted on the ways in which women in precarious housing situations understand menstruation in relation to their own corporealities and subjectivities (see Sebert Kuhlmann et al. 2019).

The growing presence of a class-aware menstrual activist movement, and the subsequent practical initiatives that have emerged to alleviate the challenges faced by marginalized women are seeking to reverse this inattention. The energy and persistence of grassroots campaigns have prompted responses at local government levels to the issue of period poverty. Period poverty, a term used mostly in the UK, refers to the state in which people who menstruate find themselves without the financial resources to access suitable menstrual products.

Despite this burgeoning movement within activist spaces, academic scholarship on menstruation has been largely inattentive to the socioeconomic diversity of women, failing to take into account how their experiences and identities transform the ways in which they relate to their menstruating bodies. Existing literature within the social sciences about menstruation has been focused on the issues of stigma, commodification, menstrual health, and medicalization (Kissling 2006; Johnston-Robledo and Stubbs 2013; Lahiri-Dutt 2015). These have been extremely productive for the understanding of the politics of menstruation, however many Anglo-American texts fail to address intersectionality. Menstrual literature in the Global North has been written about, and for, white,[1] middle-class, cisgender women (Johnston-Robledo and Stubbs 2013, 4), or in a developmental context, addressing the exclusions of women living in poverty in the Global South (for example Dhingra, Kumar, and Kour 2009; Boosey, Prestwich, and Deave 2014; Smiles, Short, and Sommer 2017). The situated-ness of these debates, it seems, is polarized: either addressing the privileged middle classes in the Global North or the socioeconomically marginalized in the Global South. However, disenfranchised women within societies in the West have been neglected: those who may not have financial or material resources to manage menstruation in a way that meets societal expectations.[2] It is imperative, therefore, in order to work toward a truly emancipatory and revolutionary feminism, that the politics of difference is included within gendered debates. How do socioeconomic disparities among women affect the way that they relate to themselves as menstruators?

This chapter therefore attempts to bring about a 'class consciousness' (bell hooks 2000) in the mapping of the lived experience of menstruation, through an increased understanding of the experiences of women who are homeless. It offers an insight to the ways in which women experiencing homelessness understand and negotiate their menstrual bodies within contexts of limited financial and material resources. This study explores the scale of the personal, offering a phenomenological insight into their experiences. This contrasts with atheoretical texts within academic literature that seek to homogenize homeless populations (DeVerteuil, May, and Von Mahs 2009, 658). It also attempts to critically analyze the current policy and third sector initiatives across the Global North that hope to minimize the effects of period poverty upon marginalized menstruators through practical interventions.

PHENOMENOLOGY, EMOTIONS, AND THE BODY

Phenomenology is the theoretical starting point for my exploration of the rich depths of the menstrual experiences of women who are homeless. I give authority to lived experience to 'capture life as it is lived' (Moran 2000, 5). Phenomenology attends to a complex interrelatedness between the material flesh, the body, and the consciousness of the human subject (Merleau-Ponty 2002, 5), and it is in the footsteps of this theory that I explore the experiential, affective and contingent nature of the lives of marginalized women throughout menstruation.

I attempt to generate this understanding by prioritizing the feelings and emotions that tint perceptions and embodied experience. Not only is emotional experience important to paint a rich portrayal of the modalities of menstruators who are homeless, but as a feminist narrative, it frames 'the personal as political' (Pile 2010, 7). The privileging of emotional experience and understandings of the self (Bondi 2005, 6) allows feminist geographers to unpick the felt complexities of gendered experience (Pile 2010, 7).

However, by focusing on the realm of the immaterial, I do not want to neglect the fleshy ontology of the body (Grosz 1994, ix). While phenomenology and emotional geography foreground the body as the site of unique experience, scholars "still often fail to talk about a body that breaks its boundaries – urinates, bleeds, vomits, farts, [and] engulfs tampons" (Longhurst 2001, 23). A dismissal of fleshy corporeality denies the agency of the material body itself. If it is true that a person lives through their body, then it is through the freedoms and restrictions of their bodily capacities and mobility that this living occurs (Young 2005, 16).

Bodies on the Borderline

The normal body is not a bleeding body. Encounters with bodily fluids provoke averse responses such as nausea, disgust, and horror (Kristeva 1982, 3). A body threatening to burst its boundaries and give birth (Longhurst 2000, 455), leak milk from her breasts (Boyer 2012, 553) and spill blood from between her legs (Young 2005, 97) is viewed with horror and fascination (Kristeva 1982, 3). It is this cultural representation of a sticky, messy femininity that places menstruating women at the borders of social legitimacy. Any manifestation, therefore, of menstruation, whether it be a bloody stain or an emotional expression, suggest that "women are not men, cannot be men, and as so cannot exist in the world as men do" (MacDonald 2007, 348). It is through the concealment of this process that menstruators claim normalcy.

The abject is also embodied by the homeless woman. She represents a seepage beyond the boundaries of 'acceptable' social life (Butler 1993, xiii). Her body is subject to stigmatization and marginalization; in the public imaginary she is cast as 'dirty,' 'deviant,' and 'transgressive' (Radley, Hodgetts, and Cullen 2006, 438). Rough sleepers, in the public imagination, are "lives who are not considered to be 'lives', and whose materiality is understood not to 'matter'" (Meijer and Prins 1998, 281).

What happens when the abject fluid comes to plague the abject body? A stark paradox appears for a woman sleeping in the streets, constantly within the public gaze, attempting to uphold the privacy of her flowing, leaky body. Against this background, my research is guided by three questions: (1) How do women experiencing homelessness negotiate the emotional and affective experiences of menstruation? (2) How is menstruation materially managed? and (3) To what extent are third sector initiatives in the US and UK effective in addressing the challenges marginalized menstruators face?

METHODOLOGY

For the case study, primary research interviews were undertaken with 40 women in the city of Bristol, UK, who were accessing a range of services that support vulnerable people in precarious housing situations for their various needs: shelters, drug support groups, day centers, and food banks. Bristol, located in the south west of the UK, has a significant homeless population, with statutory homelessness being over twice the national average (Gouk 2017). Rough sleeping has increased consistently over the last five years (Yong 2017) due to pernicious austerity measures by the UK government and an on-going housing crisis. Private rental costs in Bristol are the highest in the UK outside of London, and the area has 9% less social housing than the national average (D'Arcy 2017, 32).

In order to analyze the effectiveness of charitable initiatives in tackling period poverty, I use two methods. I firstly compare the needs highlighted by women in the interviews to the needs addressed by activities undertaken by charitable groups, and secondly attempt to unpick the extent to which these initiatives seek to resolve underlying, long-term issues such as breaking down stigmas around menstruation. This research has been undertaken through my first-hand experience as working for No More Taboo, a social enterprise seeking to alleviate the issues around period poverty in the UK, and through secondary research using online sources.

An important caveat must be highlighted, however, before the trajectory of this study continues. Menstruation is not a uniquely female experience. "Not all women menstruate, and not only women menstruate" (Bobel 2010, 11).

Menopause and external factors that can alter the menstrual cycle, such as contraception, body weight, and stress (Stöppler 2015) create a large group of women who do not menstruate. In contrast, transgender men and intersex people are not female, yet can still menstruate (Bobel 2010, 12). For the purpose of this study, however, the experience of cisgender, menstruating women will be explored as all the women interviewed identify as such.

EMOTIONAL EXPERIENCES OF MENSTRUATION

For me, my period means problems. Mental and physical problems. (Safiya)

The majority of the women interviewed, when asked to describe their experience of menstruation, framed their monthly bleed as an emotional and painful period, rife with negative sensations, such as irritability, stress, vulnerability, and symptoms of low mood as well as anxiety and depression.

Cheryl expresses the need for rest and privacy, despite finding herself in a state of constant flux due to her insecure housing situation:

> It's quite tough and it's embarrassing when you think you're smelly. I feel that people know that I'm on, even if I know they don't know, I think they do! [Menstruation] makes me irritable, it makes me tired and it gives me back problems, and I can't move, and obviously in the situation that I'm in in the moment, it's quite difficult.

She accounts experiences of 'felt stigma' (Scambler 2009, 445), constructing her body as malodourous and deviant. The heightened awareness of her menstrual self as potentially disgusting within the social sphere creates internalized sensations of shame and guilt. She anticipates and imagines people's adverse reactions to her menstruating body, and this threat causes her discomfort and embarrassment. She describes herself as more emotionally sensitive and drained, while acknowledging her precarious housing situation—sleeping on the sofas in the houses of friends and family. This highlights the tension between her homeless body and menstrual body. While she is menstruating she "can't move," while simultaneously having to be constantly mobile, changing from one house to another. This mobility undermines her ability to self-care and fully manage the pains and the stresses that she documents as part of her menstruating experience. Her corporeal vulnerabilities are intensified through her homeless situation.

Mary-Ann extends this notion of 'felt stigma' and applies it to her status as a woman experiencing homelessness. "You want to be having a wash, but you can't. When you're homeless, you're embarrassed about your situation anyway." The embarrassment Mary-Ann feels of her leaky, menstruating body is exacerbated by her lack of stable accommodation. Mary-Ann rearticulates common notions of both the homeless (Gerrard and Farrugia 2015, 2220) and the menstrual body (Lee and Sasser-Coen 2015, 10) as messy, dirty and impure. Her embodiment of both abject categories places her in a doubly stigmatized position, her body marginalized by a culture that eschews and rejects manifestations of poverty and leaky corporeality.

Discursive debates seem far removed from the everyday realities of life on the streets, however they manifest in the material ways that the bodies of menstruating, marginalized women are perceived and controlled, not only by external disciplinary forces but also by their internal, self-regulatory actions. Both Cheryl and Mary-Ann attempt to uphold culturally dominant rituals of self-purification and self-regulation, attempting to conceal their menstruating status and remain within the margins of acceptable cleanliness. To a woman restricted by her mobility and financial resources, a socially legitimate, clean, concealed period is difficult to attain.

Affective Management Strategies

The emotional-corporeal experience of menstruation influences the everyday routines of the women and spaces in which they inhabit. They use "tactical rationalities" (Cloke, May, and Johnsen 2008, 243) to actively negotiate their menstrual state while simultaneously accommodating for their marginalized position as women experiencing homelessness.

Warmth, Comfort, and Safety

Many of the interviewees highlighted warmth and comfort as an important factor to alleviate negative symptoms of menstruation. Naomi emphasizes the significance of café spaces; when feeling dispirited, with painful menstrual cramps, she would "practically live in cafés. They're warm and they've got comfy sofas." The participants considered spaces such as multi-functional day centers extremely useful in this respect, considering that "if you stay too long [in a café], they ask you to buy something or make you leave" (Naomi).

Kim, having slept on the streets for the past month, says that "places like this [a Christian day center] give people comfort. So they know they're safe an' that. 'Cause on the streets it's not safe." She echoes the narratives of many female rough sleepers in the cartographies of fear within urban spaces (Radley, Hodgetts, and Cullen 2006, 441), and finds refuge and solace in the spaces of the center, where she can socialize and receive free tea and warm food. This highlights the importance of homelessness organizations and shelters as "spaces of care" (Johnsen, Cloke, and May 2005, 790), not only offering nutritional sustenance, but an opportunity to alleviate emotional isolation and provide physical safety.

However, this 'space of care' within Mary-Ann's life-world forms only a fraction of her affective resources. While she was menstruating, she also had to rely on existing social networks and friendships. "I've been sleeping in the town centre, but 'cause I'm on[3] now, I slept at my mate's last night 'cause I couldn't take no more" (Mary-Ann). This is not a unique occurrence. Adesola, has nowhere to go in the daytime, as the night shelter in which she is temporarily residing closes its doors throughout the day. She employs social and affective networks of housed friends to stay comfortable and warm throughout the day while she is menstruating: "I have got some good people on my side, I'll turn up and say 'right, I need a couple of hours.' They'll let me go round and sit there with a hot water bottle." The participants who sought the help of friends and acquaintances did not express any qualms for doing so. For the disconnected woman, without such affective networks, she must take rest and refuge within the public spaces of the city: "When it was hurting a lot, I just had to sit down for a bit, just on the bench. I had nowhere else I could go" (Rhian).

Yearning for Privacy

Women experiencing homelessness reside in spaces that are not theirs. Shopping centers, libraries, and parks are common public spaces that are used by women who are homeless to take respite (Reeve, Casey, and Goudie 2006, 7). Shelters and couches in friends' homes barely offer true privacy. A stable, private space is required for the intimate act of menstrual management and negotiation of subsequent emotional fluctuations and fatigue. The majority of women interviewed yearned for privacy while menstruating, especially for the act of managing their menstrual flows, as public toilets were deemed "disgusting" (Mia) and "terrible" (Erin). The constant upheaval and mobility of Cheryl's life does not allow her to fully relax, as she would if she had her own, private space. "I'm stopping here and there and everywhere, and I have very severe, heavy periods, so it's very uncomfortable." This sentiment is echoed by Simran: "I would just love to be somewhere . . . and not feel like an intruder." The extreme visibility of homelessness contrasts sharply with the intimacy needed to manage the physical and emotional aspects of menstruation.

MATERIAL MENSTRUAL MANAGEMENT

I haven't used pads, 'cause I haven't had no money. I've had to use toilet paper. (Jennifer)

Menstruators negotiate their periods through its management (Young 2005, 103), as socially produced scripts of purity and pollution (Douglas 1966, 35) require its secrecy and concealment. The practical considerations of menstruation were a concern raised by the interviewees. The commodification of sanitary products within Western society means that they come at a cost, a cost that many financially marginalized women, struggling to pay for basic amenities, may not be able to bear. Jennifer documents her socially irregular use of toilet paper, gleaned from public toilets, to ensure that her menstrual blood is absorbed. In her situation, she cannot afford to purchase commodified sanitary products. One participant also spoke of having to occasionally shoplift such items, in desperation, echoing a worrying trend in "survival shoplifting" for those in such financial destitution (Hall 2017, 23).

Some women explained that, in "desperate situations" (Jenelle), they would ask for menstrual products in the institutional locations of the spaces in which the interviews took place. Their experiences, however, are varied. In a short-term shelter, Jenelle affirms that "we're pretty lucky here, 'cause they've always got some in stock." She relies on the shelter to provide her the material resources for managing her menstruation. However, Kate, who uses the services of a Christian day center, highlights the inattention to service users' menstrual needs: "Sometimes they have some . . . sometimes they don't. I don't really understand why that is, they should have a lot more considering that it's an obvious priority." One of the functions of the center, to Kate, is the provision of basic amenities that are unobtainable given the financial situation of their female service users. She problematizes the inconsistency of the supply of sanitary products, framing them as an "obvious priority:" the management of an essential, unpredictable and uncontrollable bodily process that has been overlooked by the staff.

Some women, however, were reluctant to ask the staff as they felt it was a shameful subject. "I would never ask, I'd be too embarrassed" (Michelle). Such embarrassment, as described by Michelle, is the result of a persistent socio-cultural stigma surrounding menstruation (Laws 1990). The inability to access products at the shelter also highlights an inadequate system of communication about the availability of menstrual products. Cynthia comments that she does not know "if they give them out here," illustrating a gap between the supply of products to shelters via donations and the actual receipt and usage of the products by service users.[4]

However, in some situations, this embarrassment is present not only due to the shroud of socio-cultural stigma that surrounds menstruation, but also the institutional power relations between the women and those working at the shelters. Emily, a resident at a longer-term shelter, felt uncomfortable discussing her intimate, bodily needs with those employed to regulate and govern her behavior within her living space. "I wouldn't want to say this but it's kind of like a prison here, they come and do their checks to see if you're all right which means coming into the flat, and if they don't hear you they'll come into your bedroom. It's like having a warden, and you don't really want to be like 'please sir, can I have something for my bleeding.'" Using the analogy of a prison, Emily echoes geographical analyses of shelters as disciplinary institutions through which homeless bodies are governed and controlled (Williams 1996, 85). Nearly every aspect of their lives is subject to scrutiny by the wardens, their finances, their health, and their living space. She resists the total scrutiny of her body by refusing to disclose information about her intimate bodily processes to the staff. The need to ask for products is highlighted as unfair by some participants, due to this regime of bodily control exerted by staff: "I feel like they should be in all toilets, and you can help yourself" (Samantha).

Embarrassment and the resistance of power relationships within the shelter are not the only rationales for refusing to ask for sanitary products. Frankie states: "It's not their responsibility really. You're supposed to be getting used to providing for yourself again." She believes that menstruation is a personal process that should be managed individually, and views the ability to afford menstrual products as a progression toward being able to support herself economically. The ability to provide for herself again shows the symbolic value of embracing a self-sufficient, economically responsible subjectivity (Farrugia 2011, 82). It lifts her from the marginal zone of the abject (Kristeva 1982, 3), and allows her to break away from common perceptions of people who are homeless as 'irresponsible' and 'morally deviant.'

'Alternative' Practices

The consumption of menstrual hygiene products has become naturalized in Western society, with disposable tampons and pads being the most common method used to manage menstruation. The awareness and use of reusable sanitary products such as washable pads and menstrual cups are now growing as a form of 'alternative' menstrual hygiene management. Despite their popularity and prevalence being trivial in the market compared to disposables (Atkin 2018), demand is rising due to an increasingly environmentally conscious consumer market.

However, the use of such products was not viewed as possible nor desirable for the interviewees. The participants communicated the difficulty in the cleaning of reusable sanitary products due to their constant spatial mobility (for the 'rough sleepers' and the 'hidden homeless' who are constantly moving around the houses of friends and hostels) and the lack of privacy of shared cleaning facilities (for those in shelters). The initial costs of the products were also too steep for the women, who possess limited financial resources. The extent to which the participants can partake in an environmental menstrual feminism, therefore, is extremely limited, due to its complete reliance on women's consumption patterns and privacy of sanitation facilities.

The subjective experiences of the participants are inextricably bound to contradictory tensions that arise between their homeless and menstrual status. The participants embody sensations of moral judgment that arise from the double stigma of menstruation and homelessness. They attempt to find solutions to rid themselves of visceral feelings of impurity which ties them to an abject condition. However, their visibility as women experiencing homelessness within the public gaze and constant spatial mobility diminishes their ability to manage menstruation in the ways they would like to: in privacy, and with a stable, safe place to rest. They improvise and find makeshift ways to ease these tensions through informal networks and homeless institutions, despite framing difficulties in accessing the products. Against this background, the next section of this chapter will move to asking: What are third sector organizations doing to minimize the hurdles faced by the participants?

Initiatives to Combat Period Poverty

Civil society has been quick to galvanize attention and support for addressing period poverty, since the issue came into public consciousness (Gharib 2015). Activity has flourished in both the US and the UK within the voluntary sector, working to alleviate the situations of millions of menstruators living in poverty. The majority of such initiatives in both countries adhere to a donation-based approach. Rallying up support from the public, they collect menstrual products (disposable pads and tampons), and donate these unconditionally to soup runs, homeless shelters, domestic violence refuges, schools and foster care agencies. Examples of organizations following this approach include I Support the Girls and #HappyPeriod in the US, and #TheHomelessPeriod and Freedom4Girls in the UK.

The provisioning of menstrual products to "spaces of care" (Johnsen, Cloke, and May 2005, 790) for children and adults in vulnerable situations can be lauded for attempting to alleviate the issue in its immediacy. They seek to ameliorate the lives of people experiencing social and financial marginalization, operating through a philosophy of inclusion. There are no conditions in exchange for the receipt of menstrual products. They are supplied without judgment, without questioning the 'deservedness' of the clients (Johnsen, Cloke, and May 2005, 805). This non-interventionist approach (Watts, Fitzpatrick, and Johnsen 2018, 237) provides a valuable safety net for the vulnerable menstruator. These initiatives respond to the situation expressed by Erin: "I mean, what can you do when you've got no money?"

Critiques of the Donation-Based Approach

Does the free provision of products, however, address the complex, multivalent issues that the participants in this study have expressed? The narratives outlined in this paper indicate that there may need to be a more holistic approach to period poverty if it is to be tackled in a productive and adequate manner. I critique the unconditional donation of products based on three main themes. Firstly, I address issues of communication and agency, and then discuss the importance of breaking taboos surrounding menstruation. I then go on to discuss considerations around the environment and sustainability.

The supply of menstrual products to service users depends wholly on how the homelessness service, whether it be a shelter, day center, or soup run, choose to distribute the products. As seen above, many of the women who participated in the study did not know who to approach to talk about menstruation, and many confessed to feeling embarrassed to speak to members of staff about accessing menstrual products. Others were simply not informed that menstrual products were available. Homelessness services desperately need to improve communication around the availability of menstrual products. In a day center, where interviews for this study were conducted, the staff disclosed that they had received large donations of menstrual products, and possessed years' worth of supplies. However, the women interviewed did not know that such products were available to them at this particular service.

How can organizations focused on the distribution of menstrual products ensure that these products reach the hands of those who really need them? As a start, more focus should be put on how many people receive menstrual products, instead of celebrating the number of donations made to shelters and refuges. Supplies within homelessness services should be accessible without needing to ask for them, diminishing the power staff have over service users' bodies. In one emergency night shelter, a range of menstrual products were supplied in every bathroom, and users therefore had the agency to choose which products were suitable for them, and they were not needlessly submitted to admitting their menstrual status to those working in the shelter. Increased efforts should be made among third sector practitioners to deepen understanding of the issues surrounding menstruating while homeless, and to break down the stigma among staff to ensure that they are comfortable with taking the initiative to start the conversation about menstruation. Moreover, menstrual management is not only reliant on physical products, but also on access to safe and private sanitation facilities such as lockable toilets, showers, and laundry services.

In addition to the practical aspects of menstrual management, donation drives fail to make a critical assessment of menstruation, understanding that it is "a bodily process shaped by consumerism and controlled by corporations that disregard both human and environmental health" (Bobel 2010, 105). Disposable sanitary products may contain a host of dioxins, pesticides and chemical fragrances, which are not required to be labeled on the packaging (Spinks 2015). These can cause infections, rashes, and in some cases, death[5] (Bobel 2010, 108). Not only do disposable menstrual products pose a potential health threat to their users, the disposal of such products generates tons of landfill waste detrimental to the conservation of the planet. The Femcare industry is inherently anti-feminist: "shaming women through ad campaigns, polluting air and water supplies, and producing products that can cause microlacerations of the vaginal wall" (Bobel 2010, 109). The charitable provisioning solution, by encouraging the purchase of products from this industry for donations, offers a short-term stop-gap that benefits multinational corporations much more than the people these organizations try to help (Quint 2017). It is vital that we do not exclude marginalized menstruators from a health- and environmentally-aware menstrual consciousness, and that we actively move toward more emancipatory, transformative interventions in order to empower them to make critical choices about their menstrual management.

Two organizations that attempt to embrace feminist politics to galvanize change are UnTabooed, founded by Diandra Kalish in New York City in 2015 (Period, n.d.), and No More Taboo, founded by Chloe Tingle in Bristol, UK, in the same year. Both focus on providing educational workshops to socioeconomically marginalized women and introducing participants to reusable menstrual products such as cups and cloth pads (Period n.d., and author's own knowledge). This approach seeks to provide a comprehensive response to period poverty, incorporating the dissolution of the stigma around menstruation, taking steps to increase social and environmental well-being, and develop bodily self-confidence. Such participatory approaches can help menstruators feel more comfortable about approaching staff members for help, and provide a long-term solution to those who have a certain extent of stability—the use of a menstrual cup or cloth pads that can last years. However, this approach does not help those who are street homeless and constantly mobile. The participants considered their lack of spatial stability a hindrance to the use of reusable products. The paucity of safe, private spaces for hygiene management and other issues outlined in the text above can be, however, brought to the attention of homeless organizations, with the intent to create long-lasting change, and the amenities to manage menstruation in a safe and effective manner.

Toward a Vision of Structural Reform

Critics argue that the huge scale of resources invested by civil society simply alleviate the symptoms of homelessness without tackling the entrenched issues of destitution and housing, and serve solely as a distraction (Parsell and Watts 2017, 67). Parsell and Watts (2017) argue that the most effective way to ensure that people experiencing homelessness can access the products and services generously given by civil society, whether this be menstrual products, shower facilities or a warm space to rest, is through directly tackling the structural causes of homelessness. Governments must commit to reducing homelessness, transform housing policy and create better employment opportunities for those on the margins of society to be able to reintegrate and form an autonomous, stable part of society. Nevertheless, there are further actions that governments and third sector organizations can take while the problem of homelessness pervades. Menstrual health must be included in indicators of health for people experiencing homelessness, and proper training must be given to service providers about supporting their clients who menstruate. This includes not only the practical aspects, but also the emotional dimensions, taking into account the specific needs of transgender and non-binary people, menstruators with disabilities and the differences between cultures. Huge attitudinal shifts are required to ensure that the topic of menstruation is not erased under the shadow of stigma: one approach to ensuring that stigma is dissipated for further generations is a comprehensive, mandatory curriculum on the topic within schools.

Conclusions

This chapter has attempted to disrupt the clean incorporeality of academic discourse (Longhurst 2001, 2) and to engage with the messy worlds of menstruation as experienced by members of a socioeconomically marginalized group: women experiencing homelessness. It has been shaped by the subjectivities of the participants and their consciousness of their being-in-the-world, privileging their bodies as sites of material and emotional flux. A phenomenological lens has allowed a deeper understanding into the sensations, perceptions, and negotiations of the menstrual experiences of women without the means to access safe and stable accommodation.

The erasure of menstruation from public consciousness, due to its stigmatized condition, has uneven consequences on women experiencing homelessness. Menstruation is often overlooked and forgotten in reports addressing the health needs of women experiencing homelessness, and therefore fails to come to the attention of policy makers and service managers. This inattention means that women without adequate financial resources must use irregular, and at times, potentially pathologically unsafe methods of hiding their menstrual blood. The material and discursive narratives of menstruation, combined with the structural and financial exclusions of homelessness have real effects on the everyday realities of the participants' experiences.

The interviews revealed multiple narratives of participants' embodiment as menstruators who are homeless. The participants' sensual, emotional worlds of menstruation are fraught with negative emotion and pain. The stability and privacy needed for the emotional management and self-care that participants felt was necessary while menstruating was unattainable in many situations due to their time-space discontinuities (Rowe and Wolch 1990, 185) caused by homelessness. The dominant social, medical, and commercial scripts of cleanliness, pollution, and stigma that frame menstruation (Patterson 2013, 3) were reflected in their responses. However, the conversations also unveiled the multiplicity of strategic rationalities for the management of a fluid, emotional body in a homeless context, within which financial, material, and emotional resources are constrained. The threat of a stigmatic reaction to a bloodstain mobilizes women to undertake certain regimens to ensure that their menstrual blood is constantly hidden, such as prioritizing their spending, shoplifting, or finding alternative sources of menstrual management. However, the ontological devaluation that menstruation confers to women makes it more difficult for the participants to talk to employees of organizations, as it is a topic deemed embarrassing.

A number of initiatives to alleviate these effects on people experiencing homelessness have arisen since 2015, the majority focused on the provision of free disposable tampons and sanitary towels to shelters and other homelessness services. Despite the profound, positive short-term effect that this strategy could have on the socioeconomically marginalized, it creates a culture of dependence and does not tackle root causes of the stigma of menstruation, or seek solutions to the crisis of homelessness through more long-lasting solutions. Other interventions have sought to provide a longer-term outlook, attempting to educate both marginalized menstruators and service providers about the menstrual taboo and alternative methods of menstrual management. However, governments and civil society need to commit further to escalating menstruation in importance when taking into account the health of people experiencing homelessness, ensuring that menstruation is taught in a holistic manner to eliminate the menstrual taboo for following generations, and truly focus on tackling structural inequalities that create the conditions for poverty, homelessness, and destitution to exist.

NOTES

1. There is a growing base of literature on the experiences of menarche and menopause in non-Western societies, for example Aboriginal, Indian and Ethiopian groups, however, in Anglo-American texts, ethnic intersectionality is still limited.
2. A recent study (conducted in St. Louis, Missouri, USA) contributed greatly to diversifying this literature by conducting a study on the menstrual health needs of low-income women, including some participants that were experiencing homelessness (Sebert Kuhlmann et al. 2019).
3. "Being on" is a colloquial mannerism, meaning that she is currently menstruating.
4. This will be discussed in further detail below.
5. Toxic Shock Syndrome (TSS) can be fatal. According to NHS Choices (2014), TSS is "a rare but life-threatening bacterial infection caused by Staphylococcus aureus and Streptococcus pyogenes bacteria . . . a significant proportion of cases occur in women who are on their period and using a tampon".

REFERENCES

Atkin, Emily. 2018. *Why Do Americans Refuse to Give Up Tampons?* Accessed January 13, 2019. https://newrepublic.com/article/148432/americans-refuse-give-tampons.

Bobel, Chris. 2010. *New Blood: Third-Wave Feminism and the Politics of Menstruation.* New Brunswick: Rutgers University Press.

Bondi, Liz. 2005. "Making Connections and Thinking through Emotions: Between Geography and Psychotherapy." *Transactions of the Institute of British Geographers* 30 (4): 433–48.

Boosey, Robyn, Georgina Prestwich, and Toity Deave. 2014. "Menstrual Hygiene Management amongst Schoolgirls in the Rukungiri District of Uganda and the Impact on Their Education: A Cross-Sectional Study." *The Pan African Medical Journal* 19: 253.

Boyer, Kate. 2012. "Affect, Corporeality and the Limits of Belonging: Breastfeeding in Public in the Contemporary UK." *Health & Place* 18 (3): 552–60.

Butler, Judith. 1993. *Bodies That Matter: On the Discursive Limits of 'Sex'.* New York: Routledge.

Cloke, Paul, Jon May, and Sarah Johnsen. 2008. "Performativity and Affect in the Homeless City." *Environment and Planning D: Society and Space* 26 (2): 241–63.

D'Arcy, Conor. 2017. *A Western Union: Living Standards and Devolution in the West of England*. Accessed January 13, 2019. http://www.resolutionfoundation.org/publications/a-western-union-living-standards-and-devolution-in-the-west-of-england/.

DeVerteuil, Geoffrey, Jon May, and Jürgen Von Mahs. 2009. "Complexity Not Collapse: Recasting the Geographies of Homelessness in a 'Punitive' Age." *Progress in Human Geography* 33 (5): 646–66.

Dhingra, Rajni, Anil Kumar, and Manpreet Kour. 2009. "Knowledge and Practices Related to Menstruation among Tribal (Gujjar) Adolescent Girls." *Studies on Ethno-Medicine* 3 (1): 43–48.

Douglas, Mary. 1966. *Purity and Danger: An Analysis of the Concepts of Pollution and Taboo*. New York: Pantheon.

Ensign, Josephine. 2001. "Reproductive Health of Homeless Adolescent Women in Seattle, Washington, USA." *Women & Health* 31 (2–3): 133–51.

Farrugia, David. 2011. "The Symbolic Burden of Homelessness: Towards a Theory of Youth Homelessness as Embodied Subjectivity." *Journal of Sociology* 47 (1): 71–87.

Gerrard, Jessica, and David Farrugia. 2015. "The 'Lamentable Sight' of Homelessness and the Society of the Spectacle." *Urban Studies* 52 (12): 2219–33.

Gharib, Malaka. 2015. *Why 2015 Was the Year of the Period, and We Don't Mean Punctuation*. Accessed January 13, 2019. https://www.npr.org/sections/health-shots/2015/12/31/460726461/why-2015-was-the-year-of-the-period-and-we-dont-mean-punctuation.

Gouk, Annie. 2017. *In Numbers—The True Extent of Bristol's Homeless Crisis*. Accessed January 13, 2019. http://www.bristolpost.co.uk/news/bristol-news/numbers-true-extent-bristols-homeless-132603.

Grosz, Elizabeth A. 1994. *Volatile Bodies: Toward a Corporeal Feminism*. Bloomington: Indiana University Press.

Hall, Tom. 2017. "Citizenship on the Edge: Homeless Outreach and the City." In *Lived Citizenship on the Edge of Society*, 23–44. Cham: Palgrave Macmillan.

Homeless Link. 2017. "Annual Review 2017." Accessed January 13, 2019.

hooks, bell. 2000. *Where We Stand: Class Matters*. New York: Routledge.

Johnsen, Sarah, Paul Cloke, and Jon May. 2005. "Day Centres for Homeless People: Spaces of Care or Fear?" *Social & Cultural Geography* 6 (6): 787–811.

Johnston-Robledo, Ingrid, and Margaret L. Stubbs. 2013. "Positioning Periods: Menstruation in Social Context: An Introduction to a Special Issue." *Sex Roles* 68: 1–8.

Kissling, Elizabeth Arveda. 2006. *Capitalizing on the Curse: The Business of Menstruation*. Boulder: Lynne Rienner Publishers.

Kristeva, Julia. 1982. *Powers of Horror: An Essay on Abjection*. Translated by Leon Roudiez. New York: Columbia University Press.

Lahiri-Dutt, Kuntala. 2015. "Medicalising Menstruation: A Feminist Critique of the Political Economy of Menstrual Hygiene Management in South Asia." *Gender, Place & Culture* 22 (8): 1158–76.

Laws, Sophie. 1990. *Issues of Blood: The Politics of Menstruation*. London: Macmillan.

Lee, Janet, and Jennifer Sasser-Coen. 2015. *Blood Stories: Menarche and the Politics of the Female Body in Contemporary US Society*. New York: Routledge.

Longhurst, Robyn. 2000. "Corporeographies' of Pregnancy: 'Bikini Babes'." *Environment and Planning D: Society and Space* 18 (4): 453–72.

———. 2001. *Bodies: Exploring Fluid Boundaries*. London: Psychology Press.

MacDonald, Shauna M. 2007. "Leakey Performances: The Transformative Potential of the Menstrual Leak." *Women's Studies in Communication* 30 (3): 340–35.

Meijer, Irene Costera, and Baukje Prins. 1998. "How Bodies Come to Matter: An Interview with Judith Butler." *Signs: Journal of Women in Culture and Society* 23 (2): 275–86.

Merleau-Ponty, Maurice. 2002. *Phenomenology of Perception.* London and New York: Routledge & Kegan Paul.

Moran, Dermot. 2000. *Introduction to Phenomenology.* London: Routledge.

NHS Choices. 2014. *Toxic Shock Syndrome.* Accessed April 23, 2016. http://www.nhs.uk/conditions/Toxic-shock-syndrome/Pages/Introduction.aspx.

Padgett, Deborah K., Robert Leibson Hawkins, Courtney Abrams, and Andrew Davis. 2006. "In Their Own Words: Trauma and Substance Abuse in the Lives of Formerly Homeless Women with Serious Mental Illness." *American Journal of Orthopsychiatry* 76 (4): 461–46.

Parsell, Cameron, and Beth Watts. 2017. "Charity and Justice: A Reflection on New Forms of Homelessness Provision in Australia." *European Journal of Homelessness* 11 (2): 65–76.

Patterson, Ashly. 2013. "The Menstrual Body." MA diss., University of New Orleans.

Period. n.d. *Absorbing UnTabooed.* Accessed January 13, 2019. https://www.period.org/blog/untabooed.

Pile, Steve. 2010. "Emotions and Affect in Recent Human Geography." *Transactions of the Institute of British Geographers* 35 (1): 5–20.

Quint. 2017. "Never Mind Free Tampons—Schoolgirls Need Education about Their Periods." *The Guardian*, March 16. Accessed 27 April, 2019. https://www.theguardian.com/commentisfree/2017/mar/16/free-tampons-schoolgirls-menstruation-period-education.

Radley, Alan, Darrin Hodgetts, and Andrea Cullen. 2006. "Fear, Romance and Transience in the Lives of Homeless Women." *Social & Cultural Geography* 7 (3): 437–61.

Reeve, Kesia, Rionach Casey, and Rosalind Goudie. 2006. *Homeless Women: Still Being Failed Yet Striving to Survive.* Accessed 13 January, 2019. https://www4.shu.ac.uk/research/cresr/sites/shu.ac.uk/files/homeless-women-striving-survive.pdf.

Rowe, Stacy, and Jennifer Wolch. 1990. "Social Networks in Time and Space: Homeless Women in Skid Row, Los Angeles." *Annals of the Association of American Geographers* 80 (2): 184–204.

Scambler, Graham. 2009. "Health-Related Stigma." *Sociology of Health & Illness* 31 (3): 441–55.

Schutt, Russell K., and Gerald R. Garrett. 2013. *Responding to the Homeless: Policy and Practice.* New York: Springer Science & Business Media.

Sebert Kuhlmann, Anne, Eleanor Peters Bergquist, Djenie Danjoint, and L. Lewis Wall. 2019. "Unmet Menstrual Hygiene Needs among Low-Income Women." *Obstetrics & Gynecology* 133 (2): 238–44.

Shelter. 2018. "Homelessness in Great Britain: The Numbers Behind the Story." Accessed January 13, 2019. https://england.shelter.org.uk/__data/assets/pdf_file/0020/1620236/Homelessness_in_Great_Britain_-_the_numbers_behind_the_story_V2.pdf.

Smiles, Dana, Susan E. Short, and Marni Sommer. 2017. "'I Didn't Tell Anyone Because I Was Very Afraid': Girls' Experiences of Menstruation in Contemporary Ethiopia." *Women's Reproductive Health* 4 (3): 185–97.

Spinks, Rosie. 2015. *Disposable Tampons Aren't Sustainable, but Do Women Want to Talk about It?* Accessed January 13, 2019. http://www.theguardian.com/sustainable-business/2015/apr/27/disposable-tampons-arent-sustainable-but-do-women-want-to-talk-about-it.

Stöppler, Melissa Conrad. 2015. *Amenorrhea.* Accessed January 13, 2019. http://www.medicinenet.com/amenorrhea/article.htm.

US Department of Housing and Urban Development. 2018. "The 2018 Annual Homeless Assessment Report (AHAR) to Congress." Accessed January 13, 2019. https://www.hudexchange.info/resources/documents/2018-AHAR-Part-1.pdf.

Vijayaraghavan, Maya, Ana Tochterman, Eustace Hsu, Karen Johnson, Sue Marcus, and Carol L. M. Caton. 2012. "Health, Access to Health Care, and Health Care Use among Homeless Women with a History of Intimate Partner Violence." *Journal of Community Health* 37 (5): 1032–39.

Watson, Sophie, and Helen Austerberry. 1986. *Housing and Homelessness: A Feminist Perspective.* Vol. 86. London: Routledge & Kegan Paul.

Watts, Beth, Suzanne Fitzpatrick, and Sarah Johnsen. 2018. "Controlling Homeless People? Power, Interventionism and Legitimacy." *Journal of Social Policy* 47 (2): 235–52.

Williams, Jean Calterone. 1996. "Geography of the Homeless Shelter: Staff Surveillance and Resident Resistance." *Urban Anthropology and Studies of Cultural Systems and World Economic Development* 25: 75–113.

Yong, Michael. 2017. *Number of Rough Sleepers in Bristol Revealed but It's Only the 'Tip of the Iceberg'.* Accessed January 13, 2019. https://www.bristolpost.co.uk/news/bristol-news/number-rough-sleepers-bristol-revealed-916343.

Young, Iris Marion. 2005. *On Female Body Experience: "Throwing Like a Girl" and Other Essays.* New York: Oxford University Press.

CHAPTER 5

Opinion: Prisons that Withhold Menstrual Pads Humiliate Women and Violate Basic Rights

Chandra Bozelko

Everyone laughed when Piper Chapman emerged from the shower during the first season of Orange Is the New Black with bootleg shoes made of maxi pads – and inmates do sometimes waste precious resources like sanitary products with off-label uses.[1] At York Correctional Institution in Niantic, Connecticut, where I spent more than six years, I used the tampons as scouring pads – certainly not as sponges, because prison tampons are essentially waterproof– when I needed to clean a stubborn mess in my cell.

That should not lead anyone to think that sanitary products are easy to come by in jail. At York, each cell, which houses two female inmates, receives five pads per week to split. I'm not sure what they expect us to do with the fifth but this comes out to 10 total for each woman, allowing for only one change a day in an average five-day monthly cycle. The lack of sanitary supplies is so bad in women's prisons that I have seen pads fly right out of an inmate's pants: prison maxi pads don't have wings and they have only average adhesive so, when a woman wears the same pad for several days because she can't find a fresh one, that pad often fails to stick to her underwear and the pad falls out. It's disgusting but it's true.

The only reason I dodged having a maxi pad slither off my leg is that I layered and quilted together about six at a time so I could wear a homemade

C. Bobel et al. (eds.), *The Palgrave Handbook of Critical Menstruation Studies*, https://doi.org/10.1007/978-981-15-0614-7_5

diaper that was too big to slide down my pants. I had enough supplies to do so because I bought my pads from the commissary. However, approximately 80% of inmates are indigent and cannot afford to pay the $2.63 the maxi pads cost per package of 24, as most earn 75 cents a day and need to buy other necessities like toothpaste ($1.50, or two days' pay) and deodorant ($1.93, almost three days' pay). Sometimes I couldn't get the pads because the commissary ran out: they kept them in short supply as it appeared I was the only one buying them.

Connecticut is not alone in being cheap with its supplies for women. Inmates in Michigan filed suit last December alleging that pads and tampons are so scarce that their civil rights have been violated. One woman bled through her uniform and was required to dress herself in her soiled jumpsuit after stripping for a search.

The reasons for keeping supplies for women in prison limited are not purely financial. Even though keeping inmates clean would seem to be in the prison's self-interest, prisons control their wards by keeping sanitation just out of reach. Stains on clothes seep into self-esteem and serve as an indelible reminder of one's powerlessness in prison. Asking for something you need crystallizes the power differential between inmates and guards; the officer can either meet your need or he can refuse you, and there's little you can do to influence his choice.

When the York Correctional Institution became coed during my sentence – merging the old Gates Correctional Institution and the women's prison – a lieutenant who spent his career at York and was unaccustomed to working with male inmates told a group of inmates that the men would rather defecate in their pants than ask him for toilet paper and get jerked around for it.

To ask a macho guard for a tampon is humiliating. But it's more than that: it's an acknowledgement of the fact that, ultimately, the prison controls your cleanliness, your health and your feelings of self-esteem. The request is even more difficult to make when a guard complains that his tax dollars shouldn't have to pay for your supplies. You want to explain to him that he wouldn't have a paycheck to shed those taxes in the first place if prison staff weren't needed to do things like feeding inmates and handing out sanitary supplies – but you say nothing because you want that maxi pad.

The guards' reluctance to hand out the supplies is understandable because of inmates' off-label uses for the products. Women use the pads and tampons for a number of things besides their monthly needs: to clean their cells, to make earplugs by ripping out the stuffing, to create makeshift gel pads to insert under their blisters in uncomfortable work boots or to muffle the bang that sounds when a shaky double bed hits a cement wall whenever either of its sleepers move. The staff watches us waste a precious commodity. What they fail to acknowledge is that these alternative uses fill other unfulfilled needs for a woman to maintain her physical and mental health. If we had adequate cleaning supplies, proper noise control, band-aids for our blisters or stable beds, we would happily put the pads in our pants.

There are ways to restore dignity to America's inmates. For example, we could remove the entire sanitary supply problem if American prisons bought the newly-released Thinx for female inmates, which are super absorbent, stain-free underwear designed by a woman's start-up. Thinx are expensive – $200 for seven pair – but they still might be cost effective when you factor in the cost of buying disposable pads and the time and energy devoted to the pad power struggle in women's prisons. But I doubt that corrections systems in the United States will give up the forced scarcity of menstrual products in prison.

Though many argue that prisoners cannot be pampered in jail, having access to sanitary pads is not a luxury – it is a basic human right. Just like no-one should have to beg to use the toilet, or be given toilet paper, women too must be able to retain their dignity during their menstrual cycle. Using periods to punish women simply has no place in any American prison.

NOTE

1. "Prisons that Withhold Menstrual Pads Humiliate Women and Violate Basic Rights" by Chandra Bozelko was first published on June 12, 2015 in *The Guardian*. Reprinted with permission. No further reproduction or distribution of the material is allowed without permission from the publisher.

CHAPTER 6

Bleeding in Jail: Objectification, Self-Objectification, and Menstrual Injustice

Tomi-Ann Roberts

Between 20 and 60 female inmates are strip- and body-cavity searched in an outdoor bus garage with no privacy partitions, near the County Jail, upon booking and/or return from any medical appointments or court hearings. The strip- and body-cavity searches are conducted publicly (that is, en masse) and monitored by female deputies. The inmates stand together, sometimes in very close proximity to or even in physical contact with one another, and within view of one another during the procedure. While in this bus port, the women must remove their clothing and are told to lift their breasts and bellies, and then to bend over and spread their buttocks and labia to expose their rectums and vaginas for inspection with a flashlight. Women who are menstruating must remove their soiled tampons or menstrual pads in front of the entire group. In some cases, either because of heavy bleeding, or because they must stand long enough after removal of their menstrual products, women bleed down their legs and onto the floor of the bus port during the procedure. This occurs in the view of other inmates and of jail staff involved in the search process.

In 2015, I received an intriguing email from a young civil rights lawyer in Los Angeles, California, explaining that her firm was litigating a case challenging the way female detainees are strip searched. She provided me the above description of the procedure and asked whether I would be willing to serve as an expert for their class action. My first reaction to this request was frankly disbelief. When I called, I found I could neither fathom that this "procedure" was actually legal and that thousands of women had undergone it, nor that it should take "expertise" (mine, or anyone's) to convince a judge that such treatment violates the most basic of human rights to bodily integrity, dignity, and privacy. But she assured me that the procedure is indeed legal under California law and an expert was indeed needed to strengthen

© The Author(s) 2020
C. Bobel et al. (eds.), *The Palgrave Handbook of Critical Menstruation Studies*, https://doi.org/10.1007/978-981-15-0614-7_6

their case. The firm wished me to address how women are socialized to view menstruation as personal, private, and even disgusting, and how these socio-cultural attitudes contribute to feelings of shame and degradation when women are forced to remove soiled feminine hygiene products and potentially bleed openly in a large group setting.

I agreed to write a report based on the psychological science and philosophy of objectification, in which I argued that, among the many psychic consequences of self-objectification can be deeply negative attitudes toward the body's more corporeal features, including menstruation. These are revealed in women's own shame and self-disgust regarding their periods, which this degrading strip and body cavity search procedure only amplifies among the jailed inmates to the point of abject mortification.

I was not permitted in my expert report to speak much at all to the *treatment* of the inmates by the deputies for two reasons. First, the strip and body cavity search procedure as such is and will remain legal under California law. Our dispute was with the *conditions* under which it takes place—*en masse* in full view of others, and in a poorly maintained, potentially unsanitary setting (the bus port next door to the County Jail). Secondly, what we know of many deputies' abusive, shaming verbal commentary to the inmates during the procedure comes via inmate declarations, and hence is second-hand. But what I was not permitted to discuss in my report, I am eager to examine in this chapter.[1]

I will explore more in depth the issues raised around the problematic emotions of shame and disgust that ensue when we do as objectification and self-objectification ask of us, which is to repudiate our own or others' animal, biological bodies. First, I establish the theoretical background of objectification and self-objectification that informed my report. Next, I address research supporting the argument that deep shame and self-disgust is inevitably engendered in the inmates themselves who were subjected to the invasive procedure in full view of others. Third, I consider how the female deputies' mistreatment via shaming and degradation of the inmates during the procedure is an ironic reflection of their own wish to deny their animality, is therefore immoral, and ought to have no place anywhere, much less in a setting like a jail or prison where menstruators are already bereft of so-called "creature comforts." Fourth, I address another form of menstrual injustice widely meted out against incarcerated populations—the lack of access to and sometimes even withholding of menstrual management products. And finally, I close with some discomforts and conclusions.

OBJECTIFICATION, SELF-OBJECTIFICATION, AND THE CREATURELY MENSTRUATING BODY

Philosopher Martha Nussbaum (1995) described ways in which human beings can be treated as objects, clarifying seven properties of objectification, or the treating of persons as things. These properties include *instrumentality*, *denial of autonomy, inertness, fungibility, violability, ownership,* and *denial of subjectivity.* Rae Langton (2009) extended this analysis specifically to the objectification of female bodies, adding three more ways it is revealed: *reduction to a body or body parts, focus on appearance,* and *silencing.* The treatment of menstruating incarcerated women during the strip and body cavity search in the Los Angeles County Jail reflects many, if not all, of these properties of objectification.

My colleague Barbara Fredrickson and I offered objectification theory in 1997 to psychological science as a way of framing female embodied experience. We argued that Western cultures are saturated with heteronormative sexuality, and one feature of this is the pervasive evaluation of girls' and women's bodies, their worth primarily determined by their observable features, much in the same way that we might evaluate the worth of everyday, ordinary objects or tools. Such treatment, we posited, occurs along a continuum from the seemingly benign sexualized evaluation of their bodies to more extreme and undeniably brutal sex trafficking or rape (Fredrickson and Roberts 1997). We argued that this cultural milieu of sexual objectification accomplishes a colonization of the mind of many girls and women, who, as a consequence, become their own first surveyors—*self*-objectifying as a way of anticipating rewards and punishments likely to come from a culture that values their physical appearance above all else (Fredrickson and Roberts 1997; Roberts 2002).

So common as to be virtually normative, one consequence of the culture of sexual objectification and self-objectification is the widespread dissatisfaction women feel with their bodies; here we already see that this strip and body cavity search, involving the exposure of the most private parts of the inmates' bodies in public, would be particularly cruel and punishing for them. 80% of U.S. women are dissatisfied with how their bodies look, and most racial, ethnic and socioeconomic groups of American women appear to be increasingly similarly dissatisfied (Grabe and Hyde 2006). Women are socialized to value their physical appearance as the single most important element of their self-worth (Fredrickson and Roberts 1997). Because of the proliferation of mass media formats and the 24/7 delivery of these formats, all of us are socialized within this framework of idealized, sexually objectified female bodies (APA 2007). These images are often presented as the "normal" or average body, but they are in fact airbrushed, highly manipulated representations of an ideal. They are a fantasy. The use of these images sends the message that, in order for a woman (cis or trans) to be considered attractive, which is a prerequisite for female personhood, she must make tremendous efforts to look like the now-normalized ideal bodies she is inundated with by the media. This would not be harmful except for the fact that very few actual humans meet this photoshopped cultural ideal. Hence, like most women who aspire to this ideal, I argued that those in the Los Angeles County Jail already *felt* like failures, even before undergoing the procedure in question, surely setting them up for a more extreme emotional reaction during it.

Above all else, sexually objectified ideal female bodies are sanitized, hygienic, denuded, and deodorized, that is, devoid of any of the more "creaturely" or animal-like features such as body hair, genitals, or evidence of body products such as mucus or blood. Simone De Beauvoir wrote in *The Second Sex*, "In women are incarnated disturbing mysteries of nature . . . In woman dressed and adorned, nature is present but under restraint . . . A woman is rendered more desirable to the extent that her nature is more rigorously confined" (1952, 84). Building from this, my colleague Jamie Goldenberg and I (2004, 2011) provided an integration of objectification and terror management theories (for example, Greenberg, Pyszczynski, and Solomon 1986) to help explain the paradox that women's bodies are both idealized as objects of beauty and desire, but also derided for and typically required to regulate many of the creaturely functions of those very bodies. We argued that sexual objectification and self-objectification serve the purpose of distancing us from women's more creaturely functions, which are, ironically, existentially threatening. As Dinnerstein (1976) noted, the very functions that serve human existence—menstruation, birth, lactation—are, by extension, also reminders of human non-existence or mortality. In controlling women's bodies, patriarchal societies in effect control mortality itself. Objectifying and self-objectifying via sanitizing, deodorizing, denuding, dieting, surgery (the list goes on and on) then, serve as a psychic defense against the reminders of our animal (and hence mortal) natures that women's bodies, more than men's, engender.

Therefore, beyond normative body dissatisfaction, likely setting the inmates up for an experience of some discomfort at a minimum, forced to expose their own flawed bodies to deputies and one another, 20 years of research has established myriad far more extremely negative psychic consequences of sexual objectification and self-objectification (Roberts, Calogero, and Gervais 2018) predicted to arise, given this procedure's exposure of menstruating inmates' creaturely bodies to one another. It is to two particularly crushing emotions engendered by sexual and self-objectification—shame and disgust toward one's own corporeal, animal body—that I turn now.

The Inmates: Self-Objectification Yields Self-Disgust and Shame

The emotion of disgust is associated with the action tendency of putting distance between the self and the cause of disgust, which is considered contaminating. And indeed many religious traditions prohibit touching menstruating women, for example, and prescribe ritual cleaning after the period is over (Dunnavant and Roberts 2013). These prescriptions, proscriptions, and rituals reflect a belief that menstrual blood is contaminating. This is, of course, misguided thinking, because menstrual blood is not, in fact, a contaminant, but it is nevertheless deep-seated, because it is rooted in the ancient emotion of disgust. In a study of American mostly secular college students, male and female participants showed a disgust-like reaction to a woman who revealed her menstrual status ("accidentally" dropping a tampon out of her handbag) by sitting further away from her and derogating her competence. Participants also endorsed more sexually objectifying attitudes toward women's bodies in general after being reminded of menstruation by the dropped tampon (Roberts et al. 2002).

Self-objectification serves the function for women themselves of psychic distancing from their own creaturely nature, and several studies have found that indeed the more women self-objectify, the more they feel emotions such as disgust and shame toward their bodies' reproductive functioning, including menstruation (for example, Johnston-Robledo et al. 2007; Roberts 2004). Other studies show that women who score higher in self-objectification are more likely to endorse menstrual suppression, arguably a reflection of disgust's action tendency of removal of so-called contaminating entities (Johnston-Robledo et al. 2003). We see these attitudes and emotions reflected in inmates' testimonies about emotions the public strip and body cavity search engendered in them. One inmate stated that during the procedure, having her menstrual status publicly laid bare, made her feel "worthless as a woman," reflecting the ways the forced exposure of her "disgusting" creaturely body tore away the psychic defense of self-objectification that enabled her to be socially acceptable as a woman. Another inmate stated in her deposition: "I refuse to discontinue use of the Depo Provera shot for fear of menstruating at CRDF (the county jail)." This inmate found a way to ensure that her "worthlessness as a woman" would not be revealed during the strip searches because she would not ever have her period during them.

Most societies do not banish menstruating women from the community or otherwise require social quarantine, and most menstruators do not suppress their periods with pharmaceutical interventions, however menstrual management is associated with strong cultural taboos commanding that it not be seen, discussed or openly acknowledged, even among women themselves, reflecting a "concealment imperative" (Wood, this volume). Because of secrecy norms, having one's period is almost never openly acknowledged (Kissling 1996). This reflects the hiding action tendency that is commonly associated with shame. In many low- and middle-income countries, not knowing about menarche or understanding the process of menstruation leads to shame around menstruation, which in turn can lead girls to miss school, self-medicate and refrain from social interaction, effectively quarantining themselves (Chandra-Mouli and Patel 2017).

The menstruating inmates at the Los Angeles County Jail undoubtedly, therefore, felt profound shame, humiliation, and self-disgust during the strip and body cavity search. During the procedure, to determine which inmates would need clean menstrual products, deputies asked them to raise their hand in front of the entire group if they were menstruating, forcing women to "out" themselves to strangers as being on the bleeding days of their menstrual cycle. At this point, those having their periods were required to remove tampons or pads or other menstrual products in the presence of the group, and then wait some period of time—until completion of the visual body cavity inspection portion of the search—before replacing them. According to both inmates and deputies, some of the women were heavily bleeding. Given the norms of secrecy, concealment, and sanitation surrounding menstruation, revealing blood to others was surely mortifying. There is also no question that some heavily bleeding inmates would have had to handle heavily saturated tampons. Both deputies and inmates reported observing women drip blood or blood clots onto their legs and/or the ground while waiting to replace tampons or pads.

Statements from the inmates' depositions regarding their feelings during this experience include comments such as: "I felt like a big old hunk of meat," "I felt less than human," and "I felt like an animal," clear reflections of a profound blow to the psychic defenses constructed to keep one's creaturely self in check. These women likely wished they could hide or disappear, a common reaction to intense shame. They were unable to do so, and instead they turned their feelings of worthlessness and dehumanization inward, reporting that they were nothing but hunks of meat or—importantly—*animals*. This kind of language reflects a particular form of dehumanization as articulated by Haslam (2006), in which humans are denied uniquely human traits, and seen as animal-like. In *animalistic dehumanization*, people are considered unrefined, uncouth, incompetent, irrational, and undeserving of moral concern (for example, Loughnan, Haslam, and Kashima 2009).

I wrote in my report that for these women to be left standing with blood on their fingers, legs, and likely even dripping beneath them brings to mind images of the Stephen King novel and movie "Carrie." It would be a truly appalling humiliation, the stuff of the horror genre, inducing disgust in the other women witnessing, and self-disgust and profound shame in the menstruating women themselves, left to stand in their own stigma and "pollution." I further argued that the procedure would be nearly equally as mortifying for inmates who were not actively bleeding, for observing other inmates being required to endure the public exposure of their menstruation would likely have caused extreme anxiety and vicarious shame for them. Vicarious shame occurs when people identify with groups, experiencing emotions on their behalf (Welten, Zeelenberg, and Breugelmans 2012). In this case, even those who were not at that moment bleeding would have strongly identified with those who were, feeling vicarious shame at their public humiliation. Furthermore, the women who were not menstruating would have realized that they could be having their period the next time they were strip searched, and extreme anticipatory anxiety would very likely have ensued.

The Deputies: Shame and Disgust Are Used to Objectify and Humiliate

(An) analysis of disgust and shame shows us that human beings typically have a problematic relationship to their mortality and animality, and that this problematic relationship causes not just inner tension, but also aggression toward others. (Nussbaum 2004, 322)

When I read this passage in philosopher Martha Nussbaum's remarkable book *Hiding from Humanity: Disgust, Shame and The Law*, a framework for understanding the female deputies' abusive treatment of the inmates during the procedure emerged. Beyond the "inner tension" clearly reflected in the nearly 500 inmates' recurring references to feelings of shame, humiliation, and dehumanization, the female deputies' mistreatment via shaming and degradation of the inmates during the procedure is an ironic reflection of their own wish to deny their animality. Nussbaum argues in arenas such as sodomy, abortion, and pornography appeals to common understandings of the "revolting," "repugnant," or "shameful" are made to support law and punishment. However, disgust is irrational. It is based on "magical thinking" regarding contamination and is insensitive to information about actual risk.

Because we fear the disgusting and shameful within ourselves, Nussbaum argues that we cordon off, exclude, and often even punish groups of humans who exemplify animality as a way of elevating ourselves. Throughout history, then, certain disgusting and shameful properties such as smelliness, sliminess, and stickiness (all, incidentally, properties of menstrual blood) are repeatedly and monotonously projected onto certain groups: Jews, LGBTQ persons, Dalits, and women, providing a kind of emotional fuel for Anti-Semitism, homophobia, classism, caste prejudice, and misogyny (Nussbaum 2004).

In the case of the female deputies conducting the strip and body cavity search procedure, Nussbaum's analysis predicts a kind of Freudian reaction formation. That is, if I feel ashamed about my own animality (my knowledge that I, too, menstruate) then one solution is to transform that humiliation into rage-like idealization of myself and denigration of the inmates standing before me, to draw a line between us, and cordon off their polluting, contaminating stickiness away from me. Here is how shame and disgust provided the immoral fuel for the deputies to animalistically dehumanize their fellow female inmates. Indeed examples of deputies deploying disgust to objectify and dehumanize the inmates during the strip and body cavity search procedure abound. Heavier women were asked to "lift their stomachs" away from their genital region to expose it fully. One witness recounted a deputy remarking to another on her belly as she stood there, naked, holding it up and away, in the presence of the group, "Have you ever seen anything like that, how it hangs?"

For the visual body cavity inspection, deputies instructed inmates to turn back toward the wall, drop their underwear, bend over at the waist, reach behind their bodies and spread the lips of their labia to expose their vagina and anal aperture so that their body cavities could be inspected, one by one, by deputies using flashlights. In delivering the commands, deputies specifically ordered women to "open their vagina lips." Before doing so, they were told that "if you've got something in your pussy hole, take it out." Inmates reported that deputies yelled at them, criticizing them for not following directions, saying things like, "spread your lips, not your asshole," "spread your *pussy* lips." Several reported that as heavily bleeding women dripped blood onto their hands, clothing, legs, or on the ground while waiting to replace tampons, deputies derided them and refused to provide anything for cleaning up.

Deputies' testimonies included justifying their abusive language because inmates did not seem to understand them or had difficulty complying with instructions. This reflects animalistic dehumanization, which degrades perceptions of civility, rationality, competence (for example, Bongiorno, Bain, and Haslam 2013). Furthermore, my own and others' research has shown that indeed under conditions of body exposure or objectification, women (but not men) react with feelings of intense body shame and anxiety, which disrupts their cognitive concentration and diminishes their capacity to be effective (Fredrickson et al. 1998; Calogero 2004; Gervais, Vescio, and Allen 2011). It is not surprising that inmates had problems cooperating with the procedure; they were treated like animals. In their exposure of their own sticky, smelly, abject, corporeal bodies, they reminded *all* in the bus port that we are just animals. Such a reminder was surely flooding for the inmates themselves, and served as a trigger for objectification ("see how it hangs?") and animalistic dehumanization ("your *pussy* hole!") by the deputies to draw a distinct line between themselves and the inmates.

The elderly, white, male judge said something during the motion for class certification that I addressed in my report. He argued that, since female deputies were responsible for conducting the strip and body cavity search, the public fact of it could not have been meaningfully traumatizing. Indeed, he argued that the procedure was likely akin to how women might behave with one another in a locker room or spa. I will pass over here in silence this judge's seemingly willful naivety about the lengths to which most women and girls go to avoid public exposure of their bodies, even in locker rooms and spas, *and* the fact that the exposure in this case was forced, not voluntary. Instead I pointed to research that supports the opposite conclusion. Moral disengagement is the process of convincing the self that ethical standards do not apply to oneself in a particular context (Bandura 2016). Studies show that advantageous comparison within one's in-group can disable the mechanism of self-condemnation that would typically hold immoral action in check. In this case, the female deputies could compare themselves as morally superior to the female inmates, who are in jail after all for having broken the law, and this advantageous comparison likely fueled mistreatment during the procedure and even justified for them their own harsh, abusive language as necessary for compliance.

Secondly, other studies on collective threat show that intra-group hostility can be activated by "poor behavior" on the part of in-group members (Cohen and Garcia 2005). That is, when an in-group member reinforces a negative stereotype, then other in-group members may distance themselves from and even become hostile toward that person. In this case, bleeding inmates were reinforcing the negative stereotypes that abound about women's "monstrous" (that is, inhuman, creaturely, animal-like, out-of-control) bodies (Ussher 2006), likely generating disgust and hostility in the deputies toward them. Cohen and Garcia's (2005) collective threat framework helps explain the irony of the abusive commentary by the deputies who carried out the procedure which exposed the menstruators' monstrous "shame" in the first place. For they are also women who presumably also menstruate, but unlike the inmates, fully clothed ("dressed and adorned," as De Beauvoir put it), were not engaging in the "poor behavior" of reminding all of the negative stereotypes around menstrual status that are a fact of so many women's embodied lives.

OTHER MENSTRUAL INJUSTICE FOR INCARCERATED POPULATIONS

This case of the abusive strip and body cavity search at the Los Angeles County Jail for women is likely as shocking to most readers as it was to me when I first received the phone call from the civil rights lawyer. But there are other forms of menstrual injustice meted out on incarcerated women with which the public is likely more aware (see Bozelko [Chapter 5] in this volume). One of these is the commonly practiced restricted access to menstrual management products. A study by the Correctional Association of New York revealed that 54% of women in prison reported insufficient numbers of sanitary pads provided per month, and that these pads are flimsy and ineffective (Kraft-Stolar 2015). Purchasing menstrual products at the commissary typically comes at such a high cost as to be unaffordable by most inmates.

This restricted access was depicted in an episode in season 4 of the wildly popular television show *Orange is the New Black*. In it, we see inmates constructing makeshift pads and tampons using sleep masks and toilet paper and being dismissed by prison authorities when seeking help. In real life, in 2016, a detainee held on shoplifting charges was brought into a Kentucky courtroom for sentencing without pants, causing outrage in the female judge. The woman's attorney told the judge that she had been denied pants as well as menstrual hygiene products for days (Bever 2016). A video of the judge's phone call, presumably to the jail, asking "what the hell is going on?" while the pants-less woman stood behind a podium, went viral. In 2014, the ACLU of Michigan sued Muskegon County over hazardous and unconstitutional procedures at the county jail, including denial of access for female inmates to clean underwear and feminine hygiene products (ACLU of Michigan 2014). A former inmate testified that when she had her period, a guard—instead of giving her sanitary napkins—warned her that she'd "better not bleed on the floor."

Though perhaps not as outrageous as the degradation of bleeding inmates being strip searched *en masse*, I would argue that restricting access to menstrual products as a form of punishment is cut from the same cloth (pun intended). Denying incarcerated menstruators adequate products, facilities, and privacy is driven by the same animal-based dehumanization specifically engendered by objectifying female bodies. Putting the onus on inmates themselves to constrain and control their menstruating bodies as the objectifying culture requires, in a context in which access to the privacy and products typically used to do so is denied, plays a cruel trick on them. For who is to blame for their degradation and humiliation when they appear in court without pants or bleed on the floor of their cell but they themselves for being so "uncivil," for failing to uphold the requirements of sanitation, deodorization and hygiene of their creaturely bodies?

One plaintiff in the Michigan ACLU case regarding severely restricted access to menstrual products and private toilet facilities echoed testimony by our LA County plaintiffs, articulating this point poignantly: ". . . nobody deserves to be forced to live like an animal and be treated like one. We are women deserving of basic respect, sanitary conditions, bodily privacy, and simply to be treated like the women we are" (ACLU of Michigan 2014). To deny menstruating women adequate products and facilities is to treat them like "animals," publicly humiliating them, retaining psychic distance from them as though they are contaminating. Given that the United States is one of the top incarcerators of women in the world, with approximately **220,000** detained (Kajstura 2017), I am quite confident that countless incarcerated menstruators have indeed bled down their legs and onto jail and prison floors because of this denial. Furthermore, based on my own and others' work on objectification and self-objectification, I imagine many who have done so, further reinforce their own dehumanized status in the eyes of jail or prison officials and even the public, who react with disgust to their out-of-control, monstrous, animal bodies.

DISCOMFORTS: NAKED VS NUDE AND THE IS-OUGHT FALLACY

After I turned in my 30-page report to the law firm, I found myself wrestling with many discomforts as a feminist scholar. I knew my arguments about unique bodily privacy concerns for women essentialized gender differences in a way I would not do in my "real" scholarly work. Further, I did not *want* the world to be arranged in such a way that women's revelation of menstrual status ought to be shameful and degrading. I knew I was arguing it *was* the case, even though I wished it was not. These discomforts ended up getting thrown in my face, as lawyers for the defense found my testimony in a different case regarding the sexism of a public nudity ordinance requiring females, but not males, over the age of 10 to cover their chests. In that case, I testified that the sexes are not materially different in terms of breasts, that breasts are not genitals, and that therefore compelling one sex to cover their breasts but not the other amounts to sexism. The defendants took my testimony from the context of criminalizing women who voluntarily choose to expose their breasts (for example, while breastfeeding) and applied it to the context of compelling incarcerated women to publicly expose not only their breasts but their genitals, including while menstruating. I had to write a supplemental declaration in which I argued that my testimony in the two cases was completely consistent—to the extent women wish to voluntarily challenge gender stereotypes by exposing their breasts (or, I add here, their menstrual blood, as women such as Rupi Kaur and others in the "free bleed movement" have done) they should have that freedom, but they should never be forced to do so, especially given the shaming associated with such exposure.

Allow me to take a brief detour to Finland, where I spent much of my growing up, and where I imagine these distinctions between forced and voluntary exposure are perfectly obvious, because the distinction between the nude and the naked body is clear there. In Finland, where families and even strangers sit together naked in the sauna, the distinction art historian John Berger (1972) made between naked and nude is in high relief. He argued that to be naked is simply to be without clothes on, whereas to be nude is to have one's body put on display for scrutiny, commodification, or sexualization by others. In the sauna, people are not nude. They are naked, the better to sweat and experience the delights of a cold plunge all over the body. Sometimes extended families split up sauna time by gender. The girls and women go together, and the boys and men go together. My memories of gender-based sauna are some of the most cherished of my childhood—sitting naked alongside my younger sister and same aged cousins, older aunts, and my mother, and still older great aunts and my grandmother. Disgust and shame have no place in the sauna, because nakedness is not aware of itself. Here the naked truths of embodied living are revealed. And so here was where I saw what happens to bellies and breasts and vulvas with age and with illness. Here was where I was a witness to how much diversity there is among these precious parts, typically hidden from view, on different bodies. Here was where I learned about menstruating, seeing and asking about the blood on the towel beneath my mother, or the tampon string peeking out from between the labia of one of my aunts.

With this as my background, I found myself railing against the lawyers for the defense twisting my words, and also wishing that the Judge's presumption regarding women's open nakedness with one another was not ridiculous, but instead was true. If only it were the norm to be in non-self-conscious, naked togetherness, even changing our menstrual pads, tampons or cups in full view of one another! Indeed, this community-building aspect of menstruation was something that a colleague and I found to be more salient and even treasured among our more religious interviewees (Orthodox Jewish, Muslim, and Hindu) than our non-religiously affiliated ones in a study about attitudes toward menstruation (Dunnavant and Roberts 2013). Women who practice religious rituals of separation or cleansing (for example, mikveh bath) around their periods were ironically *less* shamed and secretive and more likely to feel that menstruation tied them to other women in a meaningful way than non-religious women, whose internalization of menstrual taboos, shame, and self-disgust so often isolate and alienate them from one another.

In writing this report, I was often put in mind of the "is-ought fallacy" (the assumption that because things are a certain way, they should always be that way, or that because something is not happening now it should never happen). Even though it *is* the case that this way of treating incarcerated women is a uniquely effective humiliation, given all that objectification and self-objectification have wrought in terms of our relationship to the female body, of course I do not think it *ought* to be so. Even though it *ought* to be possible for incarcerated menstruators to find it in themselves to stand tall and proudly naked, with their hands on their hips, staring deputies and guards in the eyes and bleeding magnificently down their legs, I see that a lifetime of introjecting shame and disgust toward their bodily, animal selves prevents them from this kind of defiance.

I asserted in my report that research supports the conclusion that strip and body cavity searches present unique privacy concerns for women that arise from socio-cultural representations of their bodies which engender emotions such as shame and disgust which are based not in fact, but in "magical thinking" about the creaturely process of menstruation. I argued that the experience of the forced violation of the fundamental right to privacy for these inmates was likely one of the most traumatizing of their lives and therefore was cruel and unusual. We won summary judgment regarding the conditions under which the search is conducted. The court ruled that the group strip searches were unconstitutional because privacy partitions always provided a feasible alternative. Although the decision did not fully capture the barbaric nature of the searches, the outcome was in some ways better than we anticipated in that the Court embraced the plaintiffs' most ambitious argument, which was that intrusive searches without privacy partitions violate the constitution (Amador v. Baca 2017, https://www.clearinghouse.net/detail.php?id=14457).

CONCLUSION

> Disgust and shame are dangerous social sentiments. We should be working to contain these, rather than building our legal world on the vision of human beings that these emotions contain . . . Society would do well to cast disgust and shame into the garbage heap where it would like to cast so many of us. (Nussbaum 2004, 171)

Today the strip and body cavity search procedure at the Los Angeles County Jail itself remains legal and continues to be conducted, only not in the bus port anymore, and for each woman, one by one behind a makeshift "dressing room" constructed of shower curtains. So though we technically won the case, and all bodies deserve privacy, the victory feels pyrrhic to me, because as I type this, I am quite certain that those privacy partitions are not doing much to address the internalized shame and self-disgust about their "monstrous," bleeding bodies the inmates are surely feeling during their procedure, and I imagine the deputies have already found creative ways to degrade the inmates from the other side of them.

In breaking news, as I was making my final edits to this chapter, the *Los Angeles Times* (Tchekmedyian 2019) reported that the parties have reached a settlement in which Los Angeles County will pay $53 million. Most of this will be paid as restitution to the thousands of women in the class who endured the procedure, and some will pay consultants to evaluate the whole strip and cavity search ordeal in light of the fact that such a high percentage of incarcerated women have histories of physical and sexual abuse. However, no sooner did I read this article than I received a phone call from another civil rights law firm in Chicago asking me to join in an amicus brief to help rectify a decision in the Seventh Circuit on prison/jail strip searches. The case in question was again a mass strip search of 200 women who were forced to undergo visual body cavity inspections in groups, naked, again with no accommodations for those who were menstruating. However, here they were rounded up by correctional officers in riot gear and wielding batons, and the ordeal was not done in the name of security, but as a gratuitous training exercise for incoming correctional officer cadets. As I typed my reply of "yes, count me in for this amicus brief" to the lawyer, I glanced at the poster in my office to renew my strength. It reads: "The work continues. Stay fierce."

My involvement in this case opened my eyes, and I hope this chapter will open others' eyes, to the ways in which the shame and disgust about our animal nature that menstruation engenders gets deployed inequitably in the service of a punishing debasement of disenfranchised women. This is a uniquely misogynist form of punishment, meted out against bodies and minds that have been colonized by objectification and self-objectification. We are far from there yet, but I yearn for a day when menstruation might no longer be the stigmatizing "mark" (Johnston-Robledo and Chrisler 2013) it is, both reflecting and contributing to women's lower social, political, and often even moral status, and providing the grotesquely ideal platform for this way to dehumanize those of us who landed on the wrong side of the law and who live in bodies that menstruate.

NOTE

1. This chapter uses statements made by women who had been detained in the L.A. County Jail, which I obtained through my role as expert witness in the Amador v. Baca case. The law firm which provided these inmate declarations, Kaye, McLane, Bednarski & Litt, confirmed that the statements could also be used for the purposes of this chapter. As the statements were completely anonymized and I never personally interacted with any of the women quoted, the Institutional Review Board at Colorado College deemed these materials as meeting the criteria for secondary research exemption.

REFERENCES

Amador v. Baca. 2017. *Civil Rights Litigation Clearing House.* University of Michigan Law School. https://www.clearinghouse.net/detail.php?id=14457.

American Civil Liberties Union of Michigan. 2014. "ACLU of Michigan Sues Muskegon County over Unconstitutional Policies, Hazardous Conditions at Jail," December 4. http://aclumich.org/article/aclu-michigan-sues-muskegon-county-over-unconstitutional-policies-hazardous-conditions-jail.

American Psychological Association [APA]. 2007. *Report of the APA Task Force on the Sexualization of Girls.* Washington, DC: American Psychological Association. http://www.apa.org/pi/wpo/sexualization_report_summary.pdf.

Bandura, Albert. 2016. *Moral Disengagement: How People Do Harm and Live with Themselves.* New York, NY: Worth Publisher.

De Beauvoir, Simone. 1952. *The Second Sex.* Translated by Howard Madison Parshley. New York, NY: Knopf.

Berger, John. 1972. *Ways of Seeing.* London: Penguin Books.

Bever, Lindsey. 2016. "She Has No Pants and She Is in Court: Judge Outraged over Inmate's Appearance." *Washington Post*, August 4, 2016.

Bongiorno, Renata, Paul G. Bain, and Nick Haslam. 2013. "When Sex Doesn't Sell: Using Sexualized Images of Women Reduces Support for Ethical Campaigns." *PLoS One* 8 (12): e83311.

Calogero, Rachel M. 2004. "A Test of Objectification Theory: The Effect of the Male Gaze on Appearance Concerns in Women." *Psychology of Women Quarterly* 28: 16–21.

Chandra-Mouli, Venkatraman, and Sheila Vipul Patel. 2017. "Mapping the Knowledge and Understanding of Menarche, Menstrual Hygiene and Menstrual Health among Adolescent Girls in Low- and Middle-Income Countries." *Reproductive Health* 14. https://doi.org/10.1186/s12978-017-0293-6.

Cohen, Geoffrey, and Julio Garcia. 2005. "'I am Us': Negative Stereotypes as Collective Threats." *Journal of Personality and Social Psychology* 89: 566–82.

Dinnerstein, Dorothy. 1976. *The Mermaid and the Minotaur: Sexual Arrangements and Human Malaise.* New York, NY: HarperCollins.

Dunnavant, Nicki, and Tomi-Ann Roberts. 2013. "Restriction and Renewal, Pollution and Power, Constraint and Community: The Paradoxes of Religious Women's Attitudes toward Menstruation." *Sex Roles* 68: 121–31.

Fredrickson, Barbara, and Tomi-Ann Roberts. 1997. "Objectification Theory: Toward Understanding Women's Lived Experience and Mental Health Risks." *Psychology of Women Quarterly* 21: 173–206.

Fredrickson, Barbara L., Tomi-Ann Roberts, Stephanie M. Noll, Diane M. Quinn, and Jean M. Twenge. 1998. "That Swimsuit Becomes You: Sex Differences in Self-Objectification, Restrained Eating, and Math Performance." *Journal of Personality and Social Psychology* 75: 269–85.

Gervais, Sarah J., Theresa K. Vescio, and Jill Allen. 2011. "What You See Is What You Get: The Consequences of the Objectifying Gaze for Women and Men." *Psychology of Women Quarterly* 35: 5–17.

Goldenberg, Jamie L., and Tomi-Ann Roberts. 2004. "The Beast Within the Beauty: An Existential Perspective on the Objectification and Condemnation of Women." In *Handbook of Experimental Existential Psychology*, edited by Jeff Greenberg, Sander L. Koole, and Tom Pyszczynski, 71–85. New York: Guilford.

———. 2011. "The Birthmark: An Existential Account of the Objectification of Women." In *Self-Objectification in Women: Causes, Consequences, and Counteractions*, edited by Rachel M. Calogero, Stacey Tantleff-Dunn, and J. Kevin Thompson, 77–99. Washington, DC: American Psychological Association Press.

Grabe, Shelly, and Janet Shibley Hyde. 2006. "Ethnicity and Body Dissatisfaction among Women in the United States: A Meta-Analysis." *Psychological Bulletin* 132: 622–40.

Greenberg, Jeff, Tom Pyszczynski, and Sheldon Solomon. 1986. "The Causes and Consequences of a Need for Self-Esteem: A Terror Management Theory." In *Public and Private Self*, edited by Roy F. Baumeister, 189–212. New York: Springer-Verlag.

Haslam, Nick. 2006. "Dehumanization: An Integrative Review." *Personality and Social Psychology Bulletin* 10: 252–64.

Johnston-Robledo, Ingrid, and Joan C. Chrisler. 2013. "The Menstrual Mark: Menstruation as Social Stigma." *Sex Roles* 68: 9–18.

Johnston-Robledo, Ingrid, Kristin Sheffield, Jacqueline Voigt, and Jennifer Wilcox-Constantine. 2007. "Reproductive Shame: Self-Objectification and Young Women's Attitudes toward Their Reproductive Functioning." *Women & Health* 46: 25–39.

Johnston-Robledo, Ingrid, Melissa Ball, Kimberley Lauta, and Ann Zekoll. 2003. "To Bleed or Not to Bleed: Young Women's Attitudes toward Menstrual Suppression." *Women and Health* 38: 59–75.

Kajstura, Aleks. 2017. "Women's Mass Incarceration: The Whole Pie 2017." *Prison Policy Initiative of the ACLU Campaign for Smart Justice*, October 19. https://www.aclu.org/sites/default/files/field_document/womenprisonreport_final.pdf.

Kissling, Elizabeth. 1996. "Bleeding Out Loud: Communication about Menstruation." *Feminism & Psychology* 6: 481–504.

Kraft-Stolar, Tamar. 2015. *Reproductive Injustice: The State of Reproductive Health Care for Women in New York State Prisons: A Report of the Women in Prison Project of the Correctional Association of New York*. New York, NY: Correctional Association of New York.

Langton, Rae. 2009. *Sexual Solipsism: Philosophical Essays on Pornography Objectification*. Oxford: Oxford University Press.

Loughnan, Steve, Nick Haslam, and Yoshihisa Kashima. 2009. "Understanding the Relationship between Attribute- and Metaphor-Based Dehumanization." *Group Processes and Intergroup Relations* 12: 747–62.

Nussbaum, Martha C. 1995. "Objectification." *Philosophy and Public Affairs* 24: 49–291.

———. 2004. *Hiding from Humanity: Disgust, Shame, and the Law*. Princeton, NJ: Princeton University Press.

Roberts, Tomi-Ann. 2002. "The Woman in the Body." *Feminism and Psychology* 12: 324–29.

———. 2004. "Female Trouble: The Menstrual Self-Evaluation Scale and Women's Self-Objectification." *Psychology of Women Quarterly* 28: 22–26.

Roberts, Tomi-Ann, Jamie L. Goldenberg, Cathleen Power, and Tom Pyszczynski. 2002. "'Feminine Protection:' The Effects of Menstruation on Attitudes toward Women." *Psychology of Women Quarterly* 26: 131–39.

Roberts, Tomi-Ann, Rachel M. Calogero, and Sarah J. Gervais. 2018. "Objectification Theory: Continuing Contributions to Feminist Psychology." In *APA Handbook of the Psychology of Women*, edited by Cheryl B. Travis and Jacquelyn W. White, 249–71. Washington, DC: American Psychological Association.

Tchekmedyian, Alene. 2019. "Women in Jail Endured Group Strip Searches. L.A. County to Pay $53 Million to Settle Suit." *Los Angeles Times*, July 16. https://www.latimes.com/local/lanow/la-me-ln-lasd-womens-jail-settlement-20190716-story.html?fbclid=IwAR0WeK7v474kral8Qk6Hxmnw0LCThBrwZLSyYldQ2Gm-RZak5-3UF1A_xKTc.

Ussher, Jane. 2006. *Managing the Monstrous Feminine*. Hove, East Sussex: Routledge.

Welten, Stephanie C. M., Marcel Zeelenberg, and Seger M. Breugelmans. 2012. "Vicarious Shame." *Cognition & Emotion* 26 (5): 836–46.

CHAPTER 7

Navigating the Binary: A Visual Narrative of Trans and Genderqueer Menstruation

S. E. Frank and Jac Dellaria

An understanding of trans and genderqueer perspectives on menstruation is essential to contribute to a non-pathologizing discourse about trans and genderqueer bodies and experiences.[1] Menstruation is embodied and intertwined with social expectations, norms and stereotypes of femininity.

The research on which these comics are based blends online ethnography and 19 interviews with trans and genderqueer emerging adults (ages 18-29) who recounted their embodied experiences with menstruation. Interviewees lived across the United States with a heavy concentration in Wisconsin and Chicago. Online articles and blogs (n=16) were analyzed along with public social media posts tagged #transmenstruation, #bleedingwhiletrans (n=530). The data yield that menstruation manipulates social interactions for trans and genderqueer people in four dominant realms of social life: (1) gender/sex identity (2) public bathroom attendance and navigation, (3) product marketing and messaging, and (4) healthcare. Each of these arenas is permeated by the biologically and socially constructed gender/sex binary, and as a result trans and genderqueer menstruators confront preexisting constraints ranging from social interactions to the built environment.

The first comic (Fig. 7.1) illustrates how menstruation can impact gender identity for trans and genderqueer people. Given the interactional development of the self and the body as a socio-historical politic, trans and genderqueer people who menstruate must contest their identity with the the persistance of the socially constructed female body in interactions with individuals and institutions. Trans and genderqueer people who menstruate face such discourse from family, friends, advertisers, product manufacturers and

C. Bobel et al. (eds.), *The Palgrave Handbook of Critical Menstruation Studies*, https://doi.org/10.1007/978-981-15-0614-7_7

Fig. 7.1 On Identity (Credit: Jac Dellaria. 2019)

Fig. 7.2 The Bathroom (Credit: Jac Dellaria. 2019)

Fig. 7.3 The Bathroom (Credit: Jac Dellaria. 2019)

Fig. 7.4 Product Problems (Credit: Jac Dellaria. 2019)

Fig. 7.5 At the Doctor's (Credit: Jac Dellaria. 2019)

signage throughout public spaces, which often serve as constant reminders of the contested self.

The second comic (Figs. 7.2 and 7.3) reveals bathrooms as sites of contested gender/sex identity in the context of menstruation. The rules and symbols that govern bathrooms as gender/sex policed spaces rouse fears and anxieties for many trans and genderqueer people. Menstrual products are gendered/sexed as products for women/females and so products are rarely made available in men's restrooms. Structurally, the stalls in men's rooms rarely have disposal receptacles, which signaled to participants that their bodies are unwelcome in men's spaces. In addition, the audible sound of opening a pad or tampon posits a risk that a trans/genderqueer person might reveal their status as someone who menstruates and thus does not adhere to the binary of gender and sex alignment. Trans and genderqueer persons employ coping strategies for avoiding this noise, such as opening a menstrual product before entering the bathroom, waiting to change a menstrual product until bathrooms are empty or seeking out a gender-neutral bathroom—strategies that burden menstruators with heightened awareness, vigilance, and anxiety.

The third comic (Fig. 7.4) shifts the focus to the feminization of menstrual products. Packaging, designs, commercials, and "feminine product" aisles in stores serve as regular reminders to trans and genderqueer people that a biological aspect of their bodies is deeply tied to social norms and expectations of femininity and womanhood.

The fourth comic (Fig. 7.5) illustrates generally negative interactions with healthcare providers. Even when doctors and healthcare staff attempt to include trans and genderqueer identities and bodies in their practices, efforts are often uncoordinated. Many trans and genderqueer persons experience misgendering by one or more of the medical staff or administration. Moreover, doctors and nurses ask about menstruation when it was irrelevant to the visit, which can be triggering for some people. Other people have been on testosterone for years making menstruation impossible.

Taken together, the research and comics illustrate the gendered symbolic interactions trans and genderqueer people must navigate in their daily lives. These micro social symbols of gender/sex distinction are symptoms of a larger gender regime in which gender/sex are interpreted, regulated, and policed.

Note

1. "Navigating the Binary: A Visual Narrative of Trans and Genderqueer Menstruation" S.E. Frank and Jac Dellaria was published in an extended form as Frank S. E. (2020) "Queering Menstruation: Trans and Non-Binary Identity and Body Politics." *Sociological Inquiry* 90 (2). Reprinted with permission. [OA CC-BY 4.0].

CHAPTER 8

The Human Rights of Women and Girls with Disabilities: Sterilization and Other Coercive Responses to Menstruation

Linda Steele and Beth Goldblatt

Introduction

Critical menstruation studies scholarship illuminates the politicized nature of menstruation and explores practices of feminist political activism surrounding menstruation (see, for example, Bobel 2010; Winkler and Roaf 2014). Building on the emerging discussions of disability and menstruation in that scholarship (see, for example, Przybylo and Fahs 2018; Winkler and Roaf 2014), in this chapter we introduce some domestic legal and international human rights dimensions of menstruation for women and girls with disabilities.[1] In particular, we focus on the use of non-consensual sterilization as a coercive response to menstruation. Menstruation is a key site for discrimination and violence against women and girls with disabilities, and on this basis has been the subject of longstanding activism by women with disabilities and their representative organizations (see, for example, Dowse et al. 2013; Ortoleva and Lewis 2012). We argue that it is time for critical menstruation studies scholarship to engage with the legal dimensions of menstruation in relation to women and girls with disabilities, and consider how mainstream menstruation activism can address the experiences and needs of women and girls with disabilities. This argument will be made on two bases. The first is that menstruation by women and girls with disabilities and menstrual behavior purportedly displayed by women and girls with disabilities (for example, poor hygiene management, erratic and emotional behavior, distress at blood) have been the basis for coercive interventions by parents, carers, medical professionals, and the courts, particularly through sterilization. The second is that while international human rights law provides a compelling

C. Bobel et al. (eds.), *The Palgrave Handbook of Critical Menstruation Studies*, https://doi.org/10.1007/978-981-15-0614-7_8

basis for preventing violence and discrimination related to menstruation, at the domestic level many governments have continued to support measures that enable sterilization and other coercive interventions targeting the menstruation of women and girls with disabilities. Such measures ignore human rights and at times even rationalize these interventions on human rights grounds.

We make our argument by reference to a case study of non-consensual sterilization of women and girls with disabilities in Australia as the most extreme form of coercive intervention, which is at the epicenter of contemporary political debates around reproductive justice for women and girls with disabilities. For example, Women with Disabilities Australia state:

> Forced and coerced sterilisation of women and girls with disability is a practice that violates multiple human rights treaties and instruments. It is an act of violence, a form of social control, and a clear and documented violation of the right to be free from torture and ill-treatment. Perpetrators are seldom held accountable and women and girls with disability who have experienced this egregious form of violence are rarely able to obtain any form of redress or justice. (2016, 10)

While this chapter uses a case study of sterilization, at the outset we note that it is important not to lose sight, in a critical and political sense, of 'less restrictive' alternatives to sterilization, particularly menstrual suppressant drugs. The 'temporary' and 'non-invasive' nature of menstrual suppressant drugs might render them subject to less legal and political scrutiny because they are not viewed as impacting as significantly on bodily integrity and autonomy because they do not involve the surgical cutting of the body, even though they are still non-consensual, involve entry of substances into the body and are similarly underpinned by problematic assumptions about disability, gender, and menstruation (see, for example, McCarthy quoted in Tilley et al. 2012, 422).

We have selected Australia for two reasons. In Australia, non-consensual sterilization of women and girls with disabilities is legal, which is common in many other jurisdictions including New Zealand, Germany, the United States of America, and France (Ortoleva and Lewis 2012, 43–44; Tilley et al. 2012, 415–18). Moreover, the practice and legal status of sterilization of women and girls in Australia have been the subject both of a long-standing campaign by women with disabilities and their representative organizations, and government scrutiny via a series of judicial decisions and parliamentary and law reform inquiries.

The chapter begins with an introduction to a critical framing of disability and menstruation. The chapter then shifts into an introduction to discrimination against women and girls with disabilities and its relationship to menstruation. The third section then discusses the use of sterilization of women and girls with disabilities to manage menstruation. Next, we explain the international human rights approach to disability, menstruation, and sterilization.

Lastly, we offer Australia as a case study of the challenges at a domestic level of engaging human rights to address issues surrounding menstruation and sterilization.

DISABILITY AND MENSTRUATION

Scholars engaged in critical menstruation studies have explored the place of menstruation in women's abjection and marginalization. This is exacerbated for women (including those with disabilities) who are positioned outside of normative constructs of the white, able, middle-class woman (see, for example, Bobel 2010, 28–41; Przybylo and Fahs 2018; Winkler and Roaf 2014, 3–7). An appreciation of these dynamics specifically in relation to women and girls with disabilities can be usefully developed by drawing on critical disability studies scholarship on normalcy.

Some critical disability studies scholars have argued that legal, social welfare, educational, and health service responses to disability can be understood through the lens of a medical model of disability. Pursuant to this medical model, disability is positioned as an individual, biomedical deficit which individualizes and pathologizes the disabled body as the natural target of medical and care interventions. Critical disability studies scholarship contests the medical model and some scholarship proposes that disability should instead be approached as constructed by reference to norms of ability. These norms reflect what is socially, culturally, politically, and economically valued within society, and intersect with norms pertaining to other social hierarchies such as gender, race, and class (Goodley 2014; Michalko and Titchkosky 2009). The medical model is premised on the assumed necessity of therapeutic intervention such that a disabled person's wishes are irrelevant and overridden if therapeutic intervention is considered medically beneficial. Yet, according to some critical disability studies scholarship, non-consensual medical and care interventions ostensibly for the benefit of people with disabilities can be reread as acts of violence directed toward enforcing normative orders and underlying hierarchies (Kafer 2013). For example, in the context of sterilization Tilley et al. state that "after the so-called discrediting of eugenic views associated with Nazism and the holocaust, [sterilization] was reframed on social or therapeutic grounds" (Tilley et al. 2012, 415).

In a context where women are constructed as deficient vis-à-vis men, feminist disability scholars have argued that women and girls with disabilities are positioned against norms of the *able* woman thus giving rise to greater degrees and different forms of discrimination, violence, and marginalization (Steele and Dowse 2016). Women and girls with disabilities are viewed as mentally and physically incapable of meeting gendered norms to conceal their menstruation (see Wood [Chapter 25] in this volume) and to control their sexuality and manage their fertility. Moreover, in being unable to meet gendered norms of motherhood and sexuality, women and girls with disabilities are viewed as burdens on those who provide care to them, with menstruation being seen as an additional and superfluous demand on labor and time for carers because women and girls with disabilities are viewed as not needing menstruation for reproductive reasons (Steele 2014, 23–30). As such, there are assumptions operating on at least three levels: (a) that women and girls with disabilities should conceal and control their menstruation, (b) that women and girls with disabilities are incapable of doing so, and (c) that it is inefficient and unnecessary for others to support women and girls with disabilities to menstruate. Having set out a critical framing for examining menstruation of women and girls with disabilities, we now overlay a legal framing by turning to situate menstruation and disability in the broader context of discrimination and violence against women and girls with disabilities.

MENSTRUATION AND DISCRIMINATION AGAINST WOMEN AND GIRLS WITH DISABILITIES

Discrimination may occur directly where, for example, a school refuses to admit a girl with disabilities who cannot manage menstruation unaided. Indirect discrimination could occur where a school does not refuse entry but fails to provide the facilities or supports for girls with disabilities. Here, although there is no policy overtly targeted at excluding menstruating girls with disabilities, the effect is to discriminate against this group. The concept of reasonable accommodation has been developed in discrimination law to require that adjustments are made to ensure the participation of the person who would otherwise be excluded. Disability often deepens individual and household poverty which means that lack of access to resources and facilities including water and sanitary protection can exacerbate discrimination against women and girls with disabilities (House, Mahon, and Cavill 2012, 158). In regions without household water supply, girls with disabilities may face more difficulties in accessing water sources or carrying heavy containers of water (House, Mahon, and Cavill 2012, 154). This means that they experience unequal access to water and sanitation such that they cannot access the personal care that is available to women without disabilities, and hence is a further dimension of the discrimination they experience in relation to menstruation.

Discrimination can be based on and perpetuate stigma and stereotyping. Stigma might result from the decision of a youth group to prevent girls with disabilities from attending excursions while menstruating. The exclusion of menstruating girls with disabilities might not be associated with any particular physical risk presented by the girls, but rather because the visibility of their menstrual blood or their perceived erratic behavior invokes in others feelings of disgust and fear (Steele 2016, 1031). As such, through their exclusion, menstruating girls with disabilities become abjected and dehumanized. Stereotyping would occur, for instance, when a health service provided menstrual pain management information to the carer of a woman with disabilities instead of to the woman herself. The service's assumption that the woman would be unable to understand or apply the information is based on stereotypes about people with disabilities lacking capacity or agency which leads to a devaluing of women's rights to dignity, bodily integrity, and reproductive and healthcare choice (Cusack and Cook 2009, 49–55).

Research on women with intellectual disabilities has found that this group experiences difficulties and discrimination in relation to menstruation in a variety of ways (Rodgers 2001; Tilley et al. 2012; Hamilton 2012, Stefánsdóttir 2014). Women may not be given adequate information about menstruation or menstrual management because it is assumed they are incapable of understanding this information. In one study (Rodgers 2001), women with disabilities avoided discussing menstruation with men due to embarrassment and fear that by providing evidence of female bodily functions they might expose themselves to abuse (529–30). They also avoided requesting pain medication from carers, particularly men, and were denied control to self-medicate for menstrual pain (526–27). Embarrassment and fear due to internalized stigma were encountered when women felt they had created a 'mess' or failed to meet perceived menstrual 'etiquette' (530). Their experiences of menstruation were generally negative and disempowering since their bodies were so often subject to control by carers and medical personnel. This assertion of control may result from the discomfort of society with seeing women with disabilities as sexual and as fertile (535).

These forms of discrimination against women and girls with disabilities that arise in relation to menstruation are usually 'intersectional' (Crenshaw 1989). This means that discrimination is not just based on sex/gender or disability alone but emerges from the coming together of both forms of discrimination to create a new type of unequal treatment (Sifris 2016, 55–56). In some cases, the discrimination described above arises from the intersection of one or both of these grounds and a third ground, a woman's menstrual status. The term 'menstrual-status discrimination' has been used by the UN Committee on the Rights of the Child (2016, para. 59e). A person may experience a specific form of discrimination as a menstruating woman with disabilities that a non-menstruating woman with disabilities might not and that a menstruating woman who does not have disabilities would not.

Sterilization, Disability, and Menstruation

Sterilization is one particularly extreme and severe manifestation of menstrual discrimination against women and girls with disabilities. The effects of sterilization on reproduction and menstruation are permanent. Sterilization procedures which have permanent effects on reproduction and menstruation include hysterectomy, tubal ligation, and endometrial ablation (Australian Senate Community Affairs References Committee 2013, 6). The relationship between disability, sterilization, and menstruation is twofold. First, steriliza-tion might be used specifically to prevent menstruation of women and girls with disabilities because of perceived issues with menstruation itself, includ-ing its impact on quality of life (for example, ability to participate in edu-cation and social events, receive good quality care), behavior (for example, distress and inability to cope with menstruation and 'poor hygiene practices') and existing health conditions (for example, hormonal impacts on epilepsy). Second, sterilization might be used to prevent reproduction and this inad-vertently requires preventing menstruation as a key factor in reproductive capacity. Typically, sterilization is rationalized as therapeutic and beneficial, in part because women and girls with disabilities are socially constructed as una-ble to manage their own menstruation and their menstruation is constructed as redundant and/or posing a threat to themselves or others (Steele 2008).

While there is no comprehensive international survey of laws on steriliza-tion of women and girls with disabilities, in their report on violence against women with disabilities Ortoleva and Lewis cite numerous examples of the legality of sterilization across a number of different countries (43–44; see also Rowlands and Amy 2017). In a variety of countries, non-consensual sterili-zation of women and girls with disabilities is legal where it is perceived to be in an individual's 'best interests' and occurs following decisions made by a substituted decision-maker (such as a guardian, parent, or court) pursuant to appropriate legal procedure (Ortoleva and Lewis 2012, 41–44). While some countries have more stringent legal tests and judicial safeguards that limit the bases on which sterilization can be authorized (for example, Canada and Germany), even in these jurisdictions laws fall short of complete prohibi-tion of sterilization (Dimopoulos 2016, 163–71; Rowlands and Amy 2017). Moreover, the consistent comments by various United Nations human rights bodies concerning the ongoing human rights violations through sterilization suggest these practices are systemic and widespread and there is yet to be a groundswell of countries absolutely prohibiting sterilization.

The legality of sterilization is concerning because sterilization contravenes fundamental rights related to autonomy, personal integrity, and bodily invio-lability. In law, each individual chooses what contact is made with their body. It is a criminal offense and a civil legal wrong if an individual touches anoth-er's body without their consent (commonly referred to as *assault* or *battery*). This is on the basis that individuals are assumed mentally capable to make decisions about their bodies. In contrast, the law denies this decision-making autonomy to individuals who are considered to lack mental capacity, for example, women and girls with disabilities. Instead, decisions about these women's and girls' bodies made by others such as judges, parents, and guard-ians are legally relevant in relation to the lawfulness of contact with their bod-ies. In the context of sterilization, women *without* disabilities might consent to undertake sterilization perhaps to prevent conceiving children or to mit-igate risks of some kinds of cancers. In making these decisions about their own bodies, their sterilization will not constitute unlawful contact. In contrast, it is irrelevant whether women and girls with disabilities want to be sterilized (or even know they are to be sterilized) because third parties such as judges, parents, and guardians possess the legal authority to consent to sterilization (Steele 2014). Sterilization in these circumstances is not unlawful and hence is not considered an 'injustice' deserving of redress. The lawfulness of sterilization is of particular concern since in addition to infringement of principles of personal integrity, bodily inviolability, and autonomy, there are multiple harms associated with sterilization including risks and potential side effects associated with surgical procedures, increased risk of some cancers, and ongoing grief and trauma (Australian Senate Community Affairs References Committee 2013, 8; Tilley et al. 2012, 420; Women with Disabilities Australia 2017, 8–9). In addition, sterilization is a totally disproportionate response to menstrual suppression, which in itself should be the choice of the woman herself.

Position Under International Human Rights Law

International human rights law provides some basis for asserting rights in relation to menstruation for women and girls with disabilities, including preventing sterilization. The rights of women and of people with disabilities are protected under international human rights law through the International Covenant on Civil and Political Rights (ICCPR) and the International Covenant on Economic, Social and Cultural Rights (ICESCR) and in specific conventions protecting the rights of women (the Convention on the Elimination of All Forms of Discrimination against Women, CEDAW) and the rights of people with disabilities (the Convention on the Rights of Persons with Disabilities, CRPD). This section will consider how international human rights law and its interpretation by treaty committees and special mandates holders apply to women and girls with disabilities in relation to menstruation.

Women with disabilities may experience violations that implicate their rights to bodily integrity and their rights to health. Non-consensual sterilization or administering menstrual suppressant drugs would result in a violation of ICCPR Article 7 that states "[n]o one shall be subjected to torture or to cruel, inhuman or degrading treatment or punishment." Forced sterilization of women with disabilities is considered to be cruel and degrading treatment and even torture (UN Special Rapporteur 2017, para. 30). The rights to non-discrimination, dignity, privacy, and the rights of children in the ICCPR would also be affected. At the same time, Article 12 of the ICESCR guarantees "the right of everyone to the enjoyment of the highest attainable standard of physical and mental health." Forced sterilization, as with any surgical procedure, involves dangers to a person's health and violates their right to consent to medical treatment.

The Special Rapporteur on violence against women, its causes, and consequences has noted that women with disabilities face an "intersecting confluence of violence which reflects both gender-based and disability-based violence" (2011, para. 28) and their marginalization has made them the "target of forced sterilization and other coercive birth control measures" (2011, para. 72). The UN CRPD Committee in its General Comment 3 (2016, para. 29) has noted the increased risk faced by women with disabilities to violence compared to other women. It categorizes such violence as interpersonal or institutional as well as structural that is: based on discrimination directed at a particular group. It gives an example of such violence as "the refusal by caregivers to assist with daily activities such as bathing, menstrual and/or sanitation management . . . which hinders enjoyment of the right to live independently and to freedom from degrading treatment" (para. 31). The Committee on the Elimination of All Forms of Discrimination Against Women understands gender-based violence against women as a form of discrimination (2017) and sees forced sterilization as a form of gender-based violence (para. 18).

The Special Rapporteur on torture and other cruel, inhuman or degrading treatment or punishment has noted that torture can be implied in situations where a person has been discriminated against on the basis of a disability, particularly in the context of a medical procedure (2008, para. 49), including sterilization of women and girls with disabilities (para. 60). The UN CRPD Committee considers forced sterilization to be "cruel, inhuman or degrading treatment or punishment" (2016, para. 32). Different treaty committees have raised concerns with the practice of forced sterilization that is unauthorized and non-therapeutic, for example the Committee on the Rights of Persons with Disabilities in relation to Jordan (CRPD Jordan 2017, paras. 35–36). Many countries have been the subject of recommendations to remove sterilization laws and end sterilization practices targeting women and girls with disabilities, particularly in recent years by the Committee responsible for CEDAW and the UN CRPD Committee including Japan, Kenya, the Czech Republic, Mauritius, Spain, Mexico, and many others (Special Rapporteur 2017, para. 30, footnote 36). The UN CRPD Committee has noted that even in countries where (non-therapeutic) forced sterilization is illegal such as Canada, "people with intellectual and/or psychosocial disabilities still experience involuntary sterilization through the manipulation of their consent" (CRPD Canada 2017, para. 35).

AUSTRALIA: A CASE STUDY IN MENSTRUATION, STERILIZATION, AND HUMAN RIGHTS

The recent political history of sterilization in Australia illuminates the challenges of engaging human rights in a domestic context in relation to menstruation and disability. Australian women and girls with disabilities and their representative organizations have for decades criticized sterilization as a state-sanctioned mode of discrimination, violence, and torture (Frohmader 2013). They have lobbied for its prohibition alongside redress for survivors and greater resources, information and services relating to menstruation, reproduction, and sexuality. In doing so, they have drawn on many of the dimensions of international human rights law discussed above. However, despite these efforts, sterilization of women and girls with disabilities remains lawful in Australia and recent government inquiries have demonstrated little political will to reform.

Sterilization of girls with disabilities is regulated by the Family Court of Australia (FCA). The FCA, operating in its 'welfare jurisdiction,' can authorize parental consent to sterilization of children with intellectual disabilities where the child lacks mental capacity to make decisions, the sterilization is in the 'best interests' of the child and there are no less invasive alternatives (for example, use of menstrual suppressant drugs) that are viable for that individual (for example, because of resistance to taking oral contraceptive) (*Secretary, Department of Health and Community Services v JWB* ('*Marion's Case*') (1992) 175 CLR 218). During the 1980s and 1990s, there were a number of FCA decisions and Australian High Court decisions concerning sterilization of girls with disabilities. These decisions illustrate the problematic associations between disability, gender, and menstruation discussed above. They portray girls as risky and dangerous by reason of their leaky bodies and irrational behavior attributed to their menstruation (Steele 2008, 2016). The girls are portrayed as being unable to comprehend menstruation as part of their bodies' processes. For example, in one decision the judge stated: "[during menstruation] L threw herself on the floor and scratched herself on the legs and face, . . . she would lash out if someone tried to assist her so they might be injured and she would claw her own body with her fingers until she drew blood" (*Re BW* (unreported, FamCA, Chisholm J, 10 April 1995) at 10). In this context, sterilization is viewed by judges as being in the girls' best interests because it will protect them from their erratic and risky behavior associated with menstruation and protect them from pregnancy and childbirth, including pregnancy arising from sexual abuse. Judges have also expressed need to use sterilization to avoid the "frightening and unnecessary experience" of being in public with visible bleeding (*Re Jane* (1988) 12 Fam LR 662 at 681). The judges also approach sterilization as being in the best interests of the child because it will protect parents and carers from the burden of care imposed by their superfluous menstruation and related behavior, and the burden of caring for any child born (Steele 2008). For example, Warnick J described sterilization of Katie as "lessen[ing] the physical burdens for the mother, in particular by decreasing the number of changes necessary in toileting" (*Re Katie* (unreported, FamCA, Warnick J, 30 November 1995) at 15). Moreover, in some decisions, the Family Court rejected alternatives to menstrual management on the basis that they would not be successful. For example, in one decision Cook J dismissed a menstrual management education program because he considered it "difficult to avoid the feeling, that here, perhaps too much reliance is being placed on the success of what are possibly imperfect programs, imperfectly administered and monitored upon, sadly, an imperfect subject" (*Re a Teenager* (1988) 13 Fam LR 85 at 94). These decisions reflect an institutionalizing of the stereotypes and stigmas associated with disability and menstruation that were introduced above, thus further embedding these as socially, and legally, acceptable and hence more difficult to contest.

In 2013, the Australian Government reported on a Senate Inquiry into sterilization of women and girls with disabilities (Australian Senate Community Affairs References Committee 2013). Menstruation figured prominently in the report, with the Committee noting at the outset that menstrual management is a common reason for sterilization (15). In its report the Committee observed that sterilization was being used to manage a broader array of care issues by reason of "lack of appropriate and adequate support for both people with disabilities and their carers" (31).

Reflecting observations made above about the stereotypes that inform a lack of information about menstruation, sexuality, and reproduction, the Committee also noted difficulty in discussing sexual and reproductive health with women and girls with disabilities to explain to them what was occurring during menstruation and how to manage menstrual hygiene (16). The Committee concluded that "there is a shocking lack of resources available for people with a disability" to assist them with "choices about relationships and sexuality" and "menstrual management" (48). It made a number of recommendations relating to access to education, training and information around sexuality and relationships for people with disabilities and families, medical and disability workers (ix), and about disability support planning addressing support for menstruation as well as 'support for relationships and sex education' (ix). Certainly, this is an improvement of the current position insofar as there is explicit recognition of educating women and girls with disabilities *and* their associates on menstruation. Yet, the report ultimately fell short of recommending that sterilization should not be permitted.

As part of its inquiry, the Committee was specifically tasked with considering Australia's compliance with its international obligations. Ultimately, the Committee was of the view that Australia's international human rights obligations did not require the prohibition of sterilization and sterilization could continue but recommended reforming the legal test from 'best interests' to 'best protection of rights.' This proposed test focuses on *particular* human rights, for example, to health and inclusion, and excludes the fundamental right of non-discrimination and equality. As such, it would be open for interpretations that are based on discriminatory ideas about disability (Steele 2016, 1004, 1036). It is troubling that sterilization has been affirmed as an appropriate intervention even in the face of the explicit attention to human rights of women and girls with disabilities in the CRPD.

In its review of Australia, in 2013, the UN CRPD Committee stated it was "deeply concerned that the Senate inquiry report . . . puts forward recommendations that would allow this practice to continue" (Committee on the Rights of Persons with Disabilities 2013, 5 [39]). The Committee urged Australia to adopt laws prohibiting sterilization "in the absence of their prior, fully informed and free consent" (6 [40]).

The resilience of discriminatory approaches to menstruating women and girls with disabilities, despite tireless efforts by advocates over many years, points to the persistence of stereotypes, power inequities, and limited recognition of the voices of women and girls with disabilities. Continued struggle is required, and critical menstruation scholars and activists could be useful allies.

CONCLUSIONS

This chapter has introduced some of the political and legal dynamics around menstruation facing women and girls with disabilities. We have argued that for women and girls with disabilities menstruation cannot be understood as an individual medical issue. Instead, menstruation needs to be understood in broader social and political contexts with consideration to the cultural and material dynamics that position women and girls with disabilities as not entitled to menstruate and, in turn, render menstruation a basis for discrimination and violence against women and girls with disabilities. The Australian case study has highlighted the endurance of cultural ideas about disability and menstruation in law, and the material impacts this has on women and girls with disabilities through violent, discriminatory and harmful practices of sterilization, as well as the resilience of these cultural ideas in the face of progressive human rights. Human rights violations relating to disability and menstruation track onto a broader longstanding problem of governments regularly violating human rights of people with disabilities and other marginalized groups. This arises from the limited enforceability of international human rights law in that governments, in exercising their state sovereignty, choose whether and how to meet their treaty obligations. Despite this problem, the capacity for international human rights law to frame sterilization as an issue of violence and discrimination and to foreground the equality and personal integrity of women and girls with disabilities provides a powerful ethical and moral force for challenging social assumptions about the inevitability of inequality of women and girls with disabilities (including in the context of menstruation) and the presumed therapeutic necessity of sterilization. In this way, international human rights law can be a vehicle for garnering widespread public support for domestic law reform on issues relating to menstruation and sterilization. Domestic law reform measures should be transformative in ensuring that structural change results for women and girls with disabilities (Fredman and Goldblatt 2014), addressing the systemic social, economic, and cultural issues that sustain the legitimacy and necessity of menstrual discrimination and violence.

Our chapter provides openings for critical menstruation scholars and activists to engage with ways in which domestic law and international human rights law might recognize menstrual injustice and menstrual harms and how these might be adequately redressed to ensure individual and transformative justice. Recalling that women and girls with disabilities are culturally and legally positioned as not meeting normative female gender roles, future engagement by critical menstruation studies scholars and menstrual activists should also consider how the situation of women and girls with disabilities relates to broader political issues around menstrual discrimination such as tampon taxes (see Weiss-Wolf [Chapter 41] in this volume) and menstrual leave (see Levitt and Barnack-Tavlaris [Chapter 43] in this volume) which are not necessarily of prime importance to women and girls with disabilities who are instead confronted with the possibility of removal of their very ability to menstruate per se. Therefore, such campaigns should not assume that all women are affected in the same way and should also address the diverse needs of women as well as trans men, intersex people, and others that experience discrimination in relation to menstruation (Goldblatt and Steele 2019).

Note

1. Przybylo and Fahs note that menstruation is 'complex: it is both highly gendered and not attached as a material reality to only one gender' (Przybylo and Fahs 2018, 209). We recognize that people who menstruate and are trans, intersex or gender non-conforming experience particular forms of discrimination and that addressing the full scope of menstruation discrimination goes beyond discrimination against women. In this article, however, we focus only on the legal dimensions relating to cisgender women and girls with disabilities.

References

Australian Senate Community Affairs References Committee. 2013. *Inquiry into Involuntary or Coerced Sterilisation of People with Disabilities in Australia*. Canberra: Commonwealth of Australia.

Bobel, Chris. 2010. *New Blood: Third-Wave Feminism and the Politics of Menstruation*. New Brunswick: Rutgers University Press.

Crenshaw, Kimberle. 1989. "Demarginalizing the Intersection of Race and Sex: A Black Feminist Critique of Antidiscrimination Doctrine, Feminist Theory and Antiracist Politics." *University of Chicago Legal Forum* 140: 139–67.

Cusack, Simone, and Rebecca J. Cook. 2009. "Stereotyping Women in the Health Sector: Lessons From CEDAW." *Washington & Lee Journal Civil Rights & Social Justice* 16: 47–78. http://scholarlycommons.law.wlu.edu/crsj/vol16/iss1/5/.

Dimopoulos, Andreas. 2016. *Issues in Human Rights Protection of Intellectually Disabled Persons*. London and New York: Routledge.

Disabled People's Organisation's Australia. 2017. *Submission to the Committee on the Rights of Persons with Disabilities List of Issues [Australia] to Be Adopted during the 18th Session of the Committee on the Rights of Persons with Disabilities*. Prepared by Women with Disabilities Australia (WWDA) on behalf of DPO Australia. Sydney: Disabled People's Organisations Australia (DPO Australia).

Dowse, Leanne, Karen Soldatic, Aminath Didi, Carolyn Frohmader, and Georgia van Toorn. 2013. *Stop the Violence: Addressing Violence Against Women and Girls with Disabilities in Australia: Background Paper*. Hobart: Women with Disabilities Australia. http://www.stvp.org.au/National-Symposium.htm.

Fredman, Sandra, and Beth Goldblatt. 2014. *Gender Equality and Human Rights*. Discussion Paper for Progress of the World's Women 2015. New York: UN Women, 1–60.

Frohmader, Carolyn. 2013. Women with Disabilities Australia, Submission No. 49 to Senate Community Affairs References Committee, Parliament of Australia. *Inquiry into Involuntary or Coerced Sterilisation of People with Disabilities in Australia*, March 2013. https://www.aph.gov.au/DocumentStore. ashx?id=0406ad6a-c555-47f2-8e3d-60a7cf8a041d&subId=16161.

Goldblatt, Beth, and Linda Steele. 2019. "Bloody Unfair: Inequality Related to Menstruation – Considering the Role of Discrimination Law." *Sydney Law Review* 41 (3): 293–325. http://classic.austlii.edu.au/au/journals/SydLawRw/2019/13. html.

Goodley, Dan. 2014. *Dis/Ability Studies: Theorising Disablism and Ableism*. London and New York: Routledge.

Hamilton, C. 2012. "Sterilisation and Intellectually Disabled People in New Zealand—Still on the Agenda?" *Kōtuitui: New Zealand Journal of Social Sciences Online* 7, no. 2 (November): 61–71. https://doi.org/10.1080/11770 83X.2012.724693.

House, Sarah, Thérèse Mahon, and Sue Cavill. 2012. *Menstrual Hygiene Matters: A Resource for Improving Menstrual Hygiene around the World*. Water Aid. https:// washmatters.wateraid.org/publications/menstrual-hygiene-matters.

Kafer, Alison. 2013. *Feminist, Queer, Crip*. Bloomington: Indiana University Press.

Michalko, Rod, and Tanya Titchkosky. 2009. "What Is and What Is Not Disability Studies." In *Rethinking Normalcy: A Disability Studies Reader*, edited by Rod Michalko and Tanya Titchkosky, 1–14. Toronto: Canadian Scholars Press.

Ortoleva, Stephanie, and Hope Lewis. 2012. *Forgotten Sisters: A Report on Violence Against Women with Disabilities, An Overview of Its Nature, Scope, Causes and Consequences*. Women Enabled and Northeastern University School of Law. https://womenenabled.org/pdfs/Ortoleva%20Stephanie%20%20Lewis%20 Hope%20et%20al%20Forgotten%20Sisters%20-%20A%20Report%20on%20 ViolenceAgainst%20Women%20%20Girls%20with%20Disabilities%20August%20 20%202012.pdf.

Przybylo, Ela, and Breanne Fahs. 2018. "Feels and Flows: On the Realness of Menstrual Pain and Cripping Menstrual Chronicity." *Feminist Formations* 30 (1): 206–29.

Rodgers, Jackie. 2001. "Pain, Shame, Blood and Doctors: How Women with Learning Difficulties Experience Menstruation." *Women's Studies International Forum* 24 (5): 523–39. https://www.sciencedirect.com/science/article/abs/pii/ S0277539501001959.

Rowlands, Sam, and Jean-Jacques Amy. 2017. "Sterilization of Those with Intellectual Disability: Evolution from Non-Consensual Interventions to Strict Safeguards." *Journal of Intellectual Disability*, forthcoming, online first. https://doi.org/10.1177/1744629517747162.

Sifris, Ronli. 2016. "The Involuntary Sterilisation of Marginalised Women: Power, Discrimination, and Intersectionality." *Griffith Law Review* 25 (1): 45–70.

Steele, Linda. 2008. "Making Sense of the Family Court's Decisions on the Non-Therapeutic Sterilisation of Girls with Intellectual Disability." *Australian Journal of Family Law* 22 (1): 1–23.

———. 2014. "Disability, Abnormality and Criminal Law: Sterilisation as Lawful and Good Violence." *Griffith Law Review* 23 (3): 467–97.

———. 2016. "Court-Authorised Sterilisation and Human Rights: Inequality, Discrimination and Violence Against Women and Girls with Disability?" *UNSW Law Journal* 39 (3): 1002–37. http://www.unswlawjournal.unsw.edu.au/sites/default/files/393-4.pdf.

Steele, Linda, and Leanne Dowse. 2016. "Gender, Disability Rights and Violence Against Medical Bodies." *Australian Feminist Studies* 31 (88): 117–24.

Stefánsdóttir, Guðrún V. 2014. "Sterilisation and Women with Intellectual Disability in Iceland." *Journal of Intellectual & Developmental Disability* 39 (2): 188–97.

Tilley, Elizabeth, Jan Walmsley, Sarah Earle, and Dorothy Atkinson. 2012. "'The Silence is Roaring': Sterilization, Reproductive Rights and Women with Intellectual Disabilities." *Disability & Society* 27 (3): 413–26.

Winkler, Inga, and Virginia Roaf. 2014. "Taking the Bloody Linen Out of the Closet: Menstrual Hygiene as a Priority for Achieving Gender Equality." *Cardozo Journal of Law and Gender* 21 (1): 1–37.

Women with Disabilities Australia (WWDA) *Position Statement 4: Sexual and Reproductive Rights*. WWDA, September 2016, Hobart, Tasmania.

Human Rights Documents

The Committee on the Elimination of All Forms of Discrimination Against Women. 2017. *General Recommendation No. 35 on Gender-Based Violence Against Women, Updating General Recommendation No. 19*, CEDAW/C/GC/35.

Committee on the Rights of Persons with Disabilities (CRPD). 2017. *Concluding Observations on the Initial Report of Jordan*, CRPD/C/JOR/CO/1.

Committee on the Rights of Persons with Disabilities (CRPD). 2017. *Concluding Observations on the Initial Report of Canada*, CRPD/C/CAN/CO/1.

Committee on the Rights of Persons with Disabilities (CRPD). 2013. *Concluding Observations on the Initial Report of Australia, Adopted by the Committee at Its Tenth Session (2–13 September 2013)*, 21 October, UN Doc CRPD/C/AUS/CO/1.

The Committee on the Rights of Persons with Disabilities. 2016. *General Comment No. 3 (2016) on Women and Girls with Disabilities*, CRPD/C/GC/3.

Committee on the Rights of the Child. 2016. *Concluding Observations: Nepal*, CRC/C/NPL/CO/3-5.

United Nations Special Rapporteur on Torture and Other Cruel, Inhuman or Degrading Treatment or Punishment. 2008. *Report to the General Assembly*, A/63/175.

United Nations Special Rapporteur on Violence Against Women, Its Causes and Consequences. 2011. *Report to the Human Rights Council*, A/HRC/17/26.

United Nations Special Rapporteur on the Rights of Persons with Disabilities. 2017. *Sexual and Reproductive Health and Rights of Girls and Young Women with Disabilities, Report to the Human Rights Council*, A/72/133.

CHAPTER 9

Personal Narrative: Let Girls Be Girls—My Journey into Forced Womanhood

Musu Bakoto Sawo

For most women, a wedding day is something to look forward to, a day to plan in excitement. It is a day to make great memories to reminisce about for many years. It is an event that many girls think about and start planning from a very early age.

However, for millions of young girls around the world, including in many African countries such as my own, The Gambia, marriage is far from being a choice. It abruptly interrupts or prematurely ends their childhood, forcing them into vicious cycles that are all too often impossible to break. Child brides are forced to abandon school and focus on building families with their husbands, who are usually complete strangers to them.

There are a few exceptions to this rule, and I was one of them. My story is an exception to a norm that would have changed my destiny. Unlike so many other child brides, I never abandoned my education after getting married at 14. As a survivor of child marriage, I vowed to fight against that practice, as well as other forms of violence against women—all driven by absurd prejudices that circumscribe women to limited fields of action in our patriarchal society.

Prior to reaching menarche, I remember having conversations about it with older girls. I was so curious to know what it entailed and how it felt. Although the explanations were blurry, with faint description of what menstruation actually was, I wasn't surprised the night mine finally happened. That night, I had a conversation with two older girls in my neighborhood. The idea that one could get pregnant once we reached menarche scared me.

Although I understood that menstruation was normal, I hid it from my mother for two years! Had my cousin not seen a blood stain on my shorts one evening, my mother might still not have known. Perhaps I hid it because

C. Bobel et al. (eds.), *The Palgrave Handbook of Critical Menstruation Studies*, https://doi.org/10.1007/978-981-15-0614-7_9

93

sex and sexuality aren't common topics Gambian mothers or parents openly discussed with their children, especially girls. But I was also terrified to let my mother know because I didn't want the responsibilities that come with it. Once a girl reaches menarche, in many communities, especially mine, she is considered fit for marriage. In my case, a suitor, a potential husband, would be all it would take to make me bury my dreams of getting an education and being a successful leader.

In contemporary Gambian society, mothers and aunts begin preparing their daughters and nieces for marriage as soon as they reach menarche. They hold regular sessions during which girls are taught their roles and responsibilities as wives. Since our society is patriarchal, women must become obedient partners who live in the shadows of their husbands. As most Gambian girls, I was given numerous sermons on how to be a good wife by pleasing my husband and his family. At some point, I couldn't help but wonder if all a wife was supposed to do is give and never receive anything in return.

In some communities, it is unacceptable for parents to consider marrying their daughters off before they reach puberty. Getting married before reaching menarche would translate to putting the cart before the horse. Menstruation in Gambian society symbolizes maturity, womanhood, and the capacity to conceive; it basically translates to the readiness of a girl to engage in sexual activities. It is also widely believed that the female body grows to a child-bearing stage at this time. Hence, it sends a signal that once a girl begins to have her period, she should be married off for fear of bringing shame to the family by getting pregnant out of wedlock. This perception, I believe, is the driving force behind child marriages. Nonetheless, there are other factors that fuel child marriage, such as poverty and the high dependency rate of women on men.

In my case, as I feared, the much-anticipated suitor came along, met my parents, and asked for my hand in marriage without consulting me, his bride-to-be. My parents, too, without consulting me (which was considered quite normal in the Gambian society at the time), decided to marry me off, even though I was just 14. I would have been another dropout, except I was so desperate that I threatened to take my own life if I was pulled out of school for marriage. Fortunately, I was allowed to continue my formal education, although under very difficult circumstances. I was required to perform my matrimonial duties while going to school. This meant that I had to cook, clean, do laundry and other domestic chores for my husband's family before going to school. This would normally have had a negative impact on my academic performance but I was determined to beat the odds. No child should be forced to bear such burden.

My story of becoming a child bride is not unique. There are hundreds of women in The Gambia who have been subjected to marriage immediately after they reach menarche. Some parents accept bride prices from suitors as early as when the brides-to-be are toddlers. In some cases, the suitors take care of all the financial and material needs of the girls from their childhood until they reach menarche, when they eventually get married off. Usually, some form of guarantee is provided to the suitor's family to ensure that their 'investments' do not go to waste. In July 2016, a presidential pronouncement, followed by the enactment of an Act of parliament, child marriage was banned in The Gambia. Despite the enactment of this law, child marriage remains prevalent in The Gambia. There are no proper mechanisms in place to monitor progress. And many parents feel that abject poverty, coupled with the high dependence of women on men in Gambian households, leave them no other choice.

Children's and women's rights activists, myself included, started the journey of campaigning to end child marriage many years ago. We have been creating awareness through the work we do with our various organizations on the health risks associated with getting married at such young ages. As a survivor, I am able to share my experiences with parents and potential child brides. I have traveled across The Gambia and have used my voice to speak against child marriage.

I now run Think Young Women, an organization that offers a girls' mentorship program where we train girls between the ages of 12 and 15 on sexual and reproductive health and rights. We also teach them about their bodies, especially about issues such as menstruation and the use of sanitary pads. This program enables these girls to become empowered and protect themselves from being at risk of child marriage. It teaches them life, communication, and advocacy skills with the objective of training them to be assertive so that they can protect themselves and other girls in their communities and schools from harmful traditional practices such as female genital mutilation (FGM) and child marriage. We also build their knowledge on where to report should they or girls in their communities be exposed to such risks.

Although some communities still hold the belief that once a girl reaches menarche she must be married off to avoid attracting unnecessary shame to the family, my organization and I are raising awareness and building capacities of these communities to understand the severe consequences of child marriage and other harmful traditional practices for girls. I continue to hold government accountable and push for policies that serve the interest of girls. This includes lobbying with government to put in place mechanisms that will ensure the effective implementation and enforcement of the law to ban child marriage. Most importantly, we advocate for establishing monitoring bodies that will investigate and prosecute perpetrators of child marriages. To achieve this, I also conduct trainings and capacity-building workshops for relevant government officials as well as raise awareness about the legal rights of girls, particularly their right to be protected from forced marriages, under domestic international and regional human rights treaties.

Understanding that poverty is one of the triggers of child marriage, my organization and I, as well as other relevant stakeholders, have been lobbying the Gambian government to create immediate and long-term economic opportunities for poor families. This, I believe, will prevent them from giving away their girls for marriage in exchange for economic gains. It would also increase employment opportunities for girls through microfinance schemes and livelihood skills, which will serve as alternatives to child marriage, especially for girls who are unable to continue their education. This also reduces girls' and women's dependence on men.

Bearing in mind that education plays a key role in curbing child marriages, my work also focuses on promoting the education of girls, building their leadership skills, and lobbying for increased investments in education. When education is accessible to and affordable for girls, it increases their chances of staying in school. Educating and engaging traditional and religious leaders through dialogue and lobbying them to discourage the practice of child marriage has also been an important part of my work. For a country, whose people draw inspiration and guidance from their religious and traditional leaders, this is essential.

In spite of the ongoing challenges registered by activists like me, we are making a lot of progress. With more concerted efforts, an increase in youth voices, commitment, and action, there is hope for a future where all girls are protected from the harms of child marriage and are able to fulfill their potentials and realize their dreams. I have a vision: to see a Gambia where women and girls have equal opportunities to dream and realize those dreams. I am on a mission to create safe spaces for women and girls and to empower, uplift, and protect them from child marriages. With support from my team and other stakeholders, I am going to change the social norm on child marriage and other harmful traditional practices in The Gambia and beyond. Nothing—not resource constraints, not backlash from communities or even my family—will deter me from achieving this.

CHAPTER 10

"I Treat My Daughters Not Like My Mother Treated Me": Migrant and Refugee Women's Constructions and Experiences of Menarche and Menstruation

Alexandra J. Hawkey, Jane M. Ussher, and Janette Perz

Menstruation is a material reality at some point in most women's lives. Yet, the discursive meaning assigned to menstruating bodies and the way in which they are experienced is dependent on the sociocultural and historical spaces which they occupy (Lee and Sasser-Coen 1996, 13). Across cultural contexts, menarche is constructed as a symbolic transition from childhood to womanhood, a period of growth and change, often linked with sexual maturation (Lee 2009, 622). While menstrual activists, artists, poets, and women's rights organizations are challenging negative representations and practices surrounding menstruation (Bobel 2010, 42), dominant discourses often still portray menstruation as something dirty and disgusting, and a bodily function to be silenced and concealed (Brantelid, Nilvér, and Alehagen 2014, 606; Mason et al. 2013, 4; see also Wood [Chapter 25] in this volume).

While globally there are similarities in the way menarche and menstruation are experienced, there are also cultural differences, including specific beliefs, practices, and restrictions placed on women during menses (Uskul 2004, 676). For instance, at menarche girls in Nepal may be required to undergo a period of seclusion (Crawford, Menger, and Kaufman 2014, 431; see also Rothschild and Shrestha [Chapter 66] in this volume), while women from rural India have reported that during menses, they are not allowed to attend to certain household chores, such as cooking or preparing food (Behera, Sivakami, and Behera 2015, 514). Many cultures also have positive or celebratory rituals and practices toward girls at menarche. For example, women from African countries such as the Republic of Benin, Cameroon,

© The Author(s) 2020
C. Bobel et al. (eds.), *The Palgrave Handbook of Critical Menstruation Studies*, https://doi.org/10.1007/978-981-15-0614-7_10

and Zambia have reported receiving special treatment at menarche, including gifts of perfume, underwear and jewelry (Uskul 2004, 676; see also Cohen [Chapter 11] as well as Gottlieb [Chapter 14] in this volume). Understanding these cultural differences is important as they shape the way in which menarche and menstruation are experienced by women (Hawkey et al. 2017, 1481).

Religious practices associated with menarche and menstruation also differ across sociocultural contexts (Guterman, Mehta, and Gibbs 2008, 5). For example, Orthodox Jewish women report being required to observe *niddah*, a practice which prohibits physical contact between men and women during menstruation, and for seven days thereafter (Hartman and Marmon 2004, 393; see also Mirvis [Chapter 12] in this volume). Islamic law also states that menstrual blood is impure and thus, women are restricted from attending mosques, touching religious texts, and abstain from pray or fasting during menstruation (Guterman, Mehta, and Gibbs 2008, 3). Such regulations may be experienced ambivalently by women, both seen as an inconvenience or burden, but also as a source of power (Hartman and Marmon 2004, 401), or a means for women to identify with each other, forming a sense of community (Dunnavant and Roberts 2013, 129; see also Cohen [Chapter 11] in this volume).

While there have been studies that explore heterogeneity of menstrual discourse and practice both within (see Mason et al. 2013), and across cultural contexts (see Uskul 2004), few studies explore how women negotiate menstrual discourses and practices when transitioning from one cultural context to another. Understanding migrant and refugee women's[1] experiences and constructions of menarche and menstruation is important, given women may have transitioned between two differing cultures and might need to negotiate conflicting cultural ideals associated with reproductive health (Salad et al. 2015, 8). Further, understanding women's embodied experiences of menarche and menstruation is important as they have significant implications for women's sexual and reproductive health more broadly. For example, women who have negative attitudes toward menstruation are more likely to support suppression of menstruation through long-term oral contraception use (Johnston-Robledo et al. 2003, 72). Menstrual shame has been linked to increased sexual risk-taking (Schooler et al. 2005, 329) and embarrassment toward other reproductive functions, such as childbirth (Moloney 2010, 156) and breastfeeding (Johnston-Robledo et al. 2007, 33). Mothers' attitudes toward menstruation shape the menstrual education girls receive, which if inadequate, might negatively affect their daughters experiences of menarche and ongoing perspectives toward menstruation and sexuality (Costos, Ackerman, and Paradis 2002, 56). However, to date, migrant and refugee women's attitudes and experiences of menarche and menstruation have largely been ignored, despite being intimately linked to fertility, sexual health, and a women's identity (Brantelid et al. 2014, 606; Teitelman 2004, 1300).

This chapter will explore this issue, drawing on the findings of a recent research study examining women migrating to Australia or Canada from a range of cultural backgrounds, the detailed methodology of which is published elsewhere (Ussher et al. 2017, 1904; Hawkey et al. 2017, 1475). In summary, we conducted eighty-four individual interviews and 16 focus groups with 85 participants (total $n = 169$) with women aged 18 years and over (average age 35), who had settled in Australia or Canada in the last ten years, having migrated from Afghanistan, Iraq, Somalia, South Sudan, Sudan, Sri Lanka (Tamil), India (Punjab) and varying South American countries (Latina). Women practiced a range of religions, including Islam, Christianity, and Hinduism. All participants, except for one Latina woman, identified as being heterosexual. Participants were recruited from Western Sydney, Australia, and Eastern or Greater Vancouver, Canada, regions that are typically of lower socio-economic status, with high concentrations of migrant communities. In this chapter, we draw on a material-discursive-intrapsychic theoretical framework, an approach that considers the biological, psychosocial, and discursive aspects of a phenomenon or experience, without one being privileged over the other (Ussher 2000, 207). Theorizing women's constructions and experiences within this approach allows for a detailed and integrated examination of the multiple factors that shape adult women's experiences of menarche and menstruation. For instance, this includes acknowledgment of the 'materiality' of menstrual blood and its 'discursive' meaning within specific cultural contexts, as well as 'intrapsychic' aspects of the embodied experience of menarche and menstruation, such as shame, fear, or anxiety. In this chapter we draw on retrospective accounts given by adult migrant and refugee women in relation to their own experiences of menarche and menstruation; we also explore how mothers address menstruation with their daughters today. Women will be referred to by their nationality/culture of origin to allow for examination of accounts within and across cultural groups.

Becoming a Woman: Menarche as a Marker of Womanhood

Menarche is a time of significant psychological and sociocultural adjustment, potentially leading girls to reconceptualize their identity as women within the patriarchal societies they live (Jackson and Falmagne 2013, 382). Menarche is discursively positioned as a marker of adulthood and reproductive maturity across many sociocultural contexts (Chang, Hayter, and Wu 2010, 457), as found in our study. A majority of women across all cultural groups described the material onset of bleeding as a discursively symbolic point in which they transitioned into womanhood. For example, participants told us, "you start bleeding and you become a woman" (Somali), "the day when the period comes, like she becomes a woman" (Iraqi). A number of participants positioned this as a positive transition they had been waiting for, with woman saying, "finally I am a woman" (Afghani) and "it was kind of a relief" (Somali).

When further questioned about what being a woman meant, a number of participants disclosed constructions centered on marriageability and child-bearing: "In South Sudan, when the girl has the first period . . . it's asso-ciated with marriage . . . you're going to get married and you are going to have babies" (South Sudanese). For a number of women, a direct outcome of menarche was immediate marriage and childbearing. As one woman told us, "I remember my uncle's wife told my dad [that I had reached menarche] and that is how I got engaged and married by 14. Before knowing anything I was already a mother" (Afghani). Early marriage was said to occur to protect women from the unwanted sexual advances of men and prevent women from engaging in premarital sex, or falling pregnant outside of wedlock, both of which were described by women as culturally and religiously forbidden (see Hawkey, Ussher, and Perz 2017). However, the majority of women did not position cultural norms of early marriage and childbirth positively, with one participant reporting it had caused her great anxiety: "I was scared because I knew that they are going to be forcing me to get married, and I wasn't prepared for it, I was scared to be a mum" (South Sudanese). All women rejected early marriage for their daughters, supported by legal restrictions on age of consent in Australia or Canada.

Menarche was positioned as a time in which young women's emerging sexuality was discursively positioned as problematic, both prior to and post migration. "Becoming a woman" was aligned with a woman's nascent sexu-ality, with participants repeatedly disclosing that they were warned to "avoid boys," "be more careful," and "watch your steps" after menarche. Such cau-tionary advice was predominantly delivered by mothers and was frequently at the forefront of girls' menstrual education, as one participant said, "My mum always told me . . . when you get the period, don't come closer to the men, don't sit with the men . . . you're going to fall pregnant" (Sudanese). However, warning messages received by menarcheal girls were often difficult to understand, given absent or incomplete information about the associa-tion between menstruation, sex, and pregnancy, as one woman commented, "they don't give any information . . . like any sexual relationship or any-thing . . . they won't tell" (Tamil). Focusing on warning messages and the avoidance of men following menarche, with no concomitant explanation as to how menstruation is linked to pregnancy, has been found to be confus-ing for young women (Costos, Ackerman, and Paradis 2002, 54): it may also lead to fears that any expression of sexuality would lead to pregnancy, which could result in young women associating their developing bodies and sexu-ality with shame, danger, or victimization (Mason et al. 2013, 4; Teitelman 2004, 1301).

Celebrating Womanhood: Ritualising the Menstrual Flow

Around the globe, menarche is acknowledged through cultural celebrations, ceremonies or rituals (Uskul 2004, 676); however, in many contexts these practices are undergoing change due to modernization and deviations to traditional belief systems (Crawford, Menger, and Kaufman 2014, 435). Among participants in our study, menstrual celebrations were most commonly discussed among Tamil and South Sudanese participants and included prayer, ceremonies, parties, and animal sacrifice. Participants told us, "the 30th day we celebrate and invite our cultural people, relatives and friends" (Tamil); "they celebrate it . . . young girls my age will come and you will be treated like you are getting married" (South Sudanese). Most participants positioned menstrual celebrations as a public recognition of their entry into womanhood: "just marking that she has become a woman" (South Sudanese). In other cases, participants disclosed that they "don't know" the reason for menstrual celebrations or believed their purpose has changed over time:

> Now it's like they do these things for fun . . . before, I think my parents' time . . . they do this sort of celebration to show the other people, I've got a girl . . . whenever you are ready, you can marry that girl. (Tamil)

Participants revealed obvious discomfort in the role that menstrual celebration played in announcing to the wider community that they were now menstruators. For example, participants described intrapsychic consequences of feeling "shy" and "embarrassed"; "I was really shy you know it's not good when people come over and say oh, she got [her] period, now she's a big girl" (South Sudanese); "I felt embarrassed . . . [they] look at you in like a different way" (Tamil). Menarche is generally considered a personal event, and many girls feel anxious about people knowing they are menstruating, thus go to great lengths to conceal it (Jackson and Falmagne 2013, 388). The discomfort with menarche celebrations reported in our study might therefore be associated with the public sharing of an intimate bodily process and being "viewed differently" among their communities. As argued by Johnston-Robledo and Chrisler (2013, 12), even though menarche celebrations attempt to promote positive messages, such as 'welcoming' into womanhood, it may be confusing to process as girls across cultural contexts are often simultaneously receiving stigmatizing messages about the taboo nature of menstruation, a bodily function to be contained and hidden.

Following migration, most Tamil and South Sudanese women positioned menarche celebrations as being redundant, outdated or inappropriate, with no sense of loss. For example, one participants said, "They [daughters] grow up in Australia and they see it's not [an] appropriate thing to announce" (South Sudanese). Other women told us,

> My daughter . . . she said, why do you want to have a function for getting a period . . . I also realized that it's true. It's the normal process in the body, so why should we have [a] party . . . I'm not going to follow it. (Tamil)

In another account, a woman living in Australia said, "they call it a Saree Party here" (Tamil), referring to the 'coming of age' party in a Western context where a girl wears a saree for the first time. These findings suggest that for a minority of women, where menstrual celebrations do continue to occur, they might do so in an adapted form.

Other traditional rituals that took place at menarche included ceremonies with leaves, dirt and water, slapping, dietary changes, flour hand dipping and wearing of new clothing. For example, one participant told us, "they put some tree leaves around her hand . . . to wish the girl luck to get married and have children" (Sudanese). Tamil women also reported specific dietary changes associated with menarche:

> They don't give any spicy food, no chilli . . . first they give us the raw egg . . . and the sesame oil . . . [in] our culture they believe there's a wound inside because of the new eggs produced and it [has to] come out, the blood and all that, to heal. (Tamil)

This account illustrates the cultural construction of menarche as a time when women have an internal wound that needs healing. Dietary changes were also positioned as a means to strengthen the reproductive system and avoid menstrual pain: "they think that it directly works with the womb you know the reproductive system will get the strength" (Tamil). One participant described continuing to give her daughter raw eggs at menarche: "here [Australia] I give my children only one egg a day, over there three eggs a day" (Tamil). However, another stated, "just a fresh egg . . . they think it's healthy, but after I came here I know it's bad, because it's not even boiled, it's not good for the health" (Tamil).

In this study, for a number of Tamil women, the menarche celebrations described above occurred following a period of seclusion, reflecting a complex cultural construction of menstruation: "I was made to stay in the room for one month until they had the ceremony" (Tamil). Although some participants who had experienced menstrual seclusion positioned it as being "natural," others found seclusion challenging: "you can't go outside that was tough" (Tamil). Women provided few explanations for menstrual seclusion; however, one participant described it in terms of the need for recuperation, drawing again on the concept of an internal wound: "there is some wound inside and the wound has to be healed, that's why they keep the girls in the room" (Tamil). It is possible however, that in the absence of a coherent explanation of menstruation prior to menarche or reasoning for menstrual seclusion, such practice might be confusing, and lead girls to associate their menarche with isolation (Crawford, Menger, and Kaufman 2014, 432). Participant accounts demonstrate how migrant and refugee women variably adopted, adapted, and questioned cultural practices surrounding menarche celebrations and rituals, demonstrating women's negotiation or navigation of differing cultural contexts following migration.

Silence and Secrecy: Education and Communication Around Menstruation

Preparedness for menarche plays an important role in how it is experienced, with girls who receive menstrual education prior to menarche reporting more positive experiences of this transition (Marván, Morales, and Cortés-Iniestra 2006, 327). However, many women in the present study described receiving little or no pre-menarcheal education in their countries of origin. Participants told us, "nobody tells us, nobody talk about it" (Somali); "we don't really talk about stuff like that" (Afghani). Participants stated that the reason menstruation was not discussed was because of shame and wider disapproval from family and friends: "they think it is shameful, it's disrespectful, you don't have respect for yourself" (Afghani).

Across cultural groups, many participants described receiving little or no preparatory menstrual education or support from their mothers, a finding that is not unique to migrant and refugee women alone (Cooper and Koch 2007, 65). When asked the reason for such reluctance on the part of mothers, it was described as a "cultural thing" which could not be challenged, as one woman said, "They are very secret about this stuff . . . my mum doesn't talk about periods or childbirth . . . maybe it's a cultural thing" (Tamil). In other instances, the information that participants described receiving was incorrect, or women were unsure of its meaning. For example, one participant's mother told her "not to sit anywhere dirty during our period because . . . everything is sort of open and you can get all kinds of infections" (Afghani). Despite being unsure of the meanings of the warnings women received, some participants passed this information on to their daughters who began menstruating following migration:

> I said ok if you want my advice when you got your period . . . you can't let the boy touch you . . . I told my daughter, she do it now . . . I don't know whether it is good or bad, I don't know. (South Sudanese)

This finding suggests that it may be useful to work alongside migrant and refugee women to understand where such cultural practices have evolved, and to highlight to women the possible consequences to their daughters if such beliefs are internalized, such as feelings of shame toward their bodies in relation to a normal healthy bodily function.

"When I Got My Period, My Heart Kind of Broke": Experiences of First Blood

In the absence of any framework to make sense of menarche, women associated their first menses with excrement, injury, and guilt. For example, one participant said, "I was kind of horrified that something was wrong with me or I might have hurt myself" (Afghani). Other women positioned their first menses as a form of punishment: "I thought I had done a sin or something really bad" (Afghani). Participants used strong emotive language such as "shocked," "scared," and "shame" when recalling these experiences. Many women also reported self-isolation; "there was a little dark room, and I would go there and I would lay out a mattress . . . and I would just sit there and cry" (Afghani), or described not disclosing to mothers and family that they had begun menstruation, as they felt "ashamed" or "shy," positioning their menarche experience in a negative light, "I kept praying, oh god never ever let this happen to me again" (Afghani). These findings reiterate the importance of adequate menstrual education prior to menarche (Teitelman 2004, 1298), particularly given negative constructions of menstrual blood lead women to feel humiliated and unclean, and might result in women developing ongoing associations between menstruation and contamination (Lee 2009, 621).

An additional consequence of inadequate education meant that at the time of menarche many women reported they had poor knowledge about the function of menstruation in relation to reproduction. For some, it was not until well into their menstruating years, or once they had migrated, that they became aware of the role menstruation had in childbearing: "Not until we got married, we came to know the whole thing, what happens" (Punjabi); "I really I didn't know until I became pregnant with my first son" (Iraqi).

Resisting Secrecy and Silence for Daughters Post Migration

As reported previously (Cooper and Koch 2007, 71), as a consequence of receiving little or no menstrual education themselves, many women disclosed being more open or wanting to be more open with their own daughters during menarche. For example, one participant said, "I want to avoid what happened to me when no one told me, so I told my daughter she already knows" (Sudanese). However, some of the mothers interviewed disclosed being "shy" to talk in depth with their daughters or "unsure" when the right time was to address menstruation:

> We're shy from those matters and even I can't talk to my daughter frankly and tell her what happened . . . I tell her about the period, and I tell her about the baby, but not the long procedure . . . I think I felt embarrassed. (Iraqi)

In another account, a woman described wanting to discuss menstruation with her 11-year-old daughter, but not knowing when or how: "until now I haven't said anything to her about it . . . I don't want her to experience the same as what I had . . . but I don't know when, and where and how" (Iraqi). These accounts highlight that although many mothers would like to educate and support their daughters through menarche, given their own poor experiences of menstrual education, they might lack the knowledge and confidence to do so at an appropriate time, prior to their daughter's menarche. At the same time, other participants actively sort information to better support their daughters. For example, one participant described attending a women's health course where she was provided with the appropriate information to support her daughter:

> I was scared and shy to talk about this topic . . . I went to a migrant resource centre and there was a lady . . . she talked about the periods and how to tell their daughters. I learnt from that session and it encouraged me to tell my daughter. (Sudanese)

In another example, a participant disclosed wanting to talk to their daughters in the future but only when they are "old enough" by cultural standards: "I will explain to my daughters . . . at an age of you know, nine, ten. No younger than that because I think I'm still following the culture" (Afghani). This account demonstrates how migrant women may navigate two differing cultural contexts, both resisting a taboo of silence and secrecy by wanting to provide their daughters with information, but also still adhering to cultural mores that require it to be done at an 'appropriate' age. These findings emphasize the need to provide newly arrived migrant and refugee women with access to menstrual education sessions to ensure they have a sound understanding of menarche as a biological function, but also an emotional transition. Women may benefit from specific guidance on how to broach the subject of menstruation with their daughters, including ways to relay information about menarche and menstruation that position it as a positive developmental stage.

Containment and Regulation of the Abject Menstrual Body

Although there are many cultural representations of blood, ranging from family and kinship, to violence and war, menstrual blood is almost always positioned negatively (Bramwell, 2001), as was reflected in women's accounts in this study. Participants repeatedly positioned blood as "disgusting," "dirty," "awful," and "not clean." One participant said, "You can smell there is something different . . . because the blood has come from the vagina, so I think it's dirty" (Iraqi). Menstrual blood was also positioned negatively by men in women's lives, including husbands, brothers, and fathers, requiring women to be discrete about their menstruation. As one participant disclosed, "he [father] and my brother, they preferred that I didn't throw out my feminine paper [pads] in the same bin that they were using. I don't understand, it was maybe some stupid thing that they had against blood" (Latina). One of the consequences of menstruation being constructed as disgusting and contaminating was women's desire to conceal their menstruating bodies from the wider world both prior to and following migration: "I started to wear dark colours when I get my period. I do not wear whites at all . . . it will look disgusting when it stains" (Iraqi). Many women described feelings of self-consciousness and greatly feared leakages, resulting in frequent visits to the bathroom, as one Sudanese participant told us, "I would go more frequently to the wash-room. I was afraid of my dress getting stained. It is a big problem . . . it was like a shame"; a Latina woman similarly disclosed, "I started to feel ashamed of my body . . . everybody is always telling you be careful, be extra cautious . . . you are going to be terrorised if you don't hide it." While self-surveillance is energy-consuming (Johnston-Robledo and Chrisler 2013, 14), these practices, in conjunction with negative cultural constructions of menstruation, are likely to result in women having negative attitudes toward their menstruating bodies (Roberts and Waters 2004, 18).

In addition, historically and cross-culturally, menstrual blood has been discursively constructed as being poisonous, magical, and polluting—a sign of the 'monstrous feminine' (Ussher 2006, 6). It is these negative representations of women and their menstruating bodies that have contributed to restrictions placed on women during menses (Buckley and Gottlieb 1988, 25). For instance, Sudanese women described the inability to enter the kitchen or carry out normal household duties while menstruating; a practice that a small number of women continued even following migration, "you can't cook, you can't wash dishes, you can't clean the house for one week until you are clean" (Sudanese). Although such restrictions might reinforce the notion of menstruation being dirty, one Sudanese woman, who continued to avoid cooking while menstruating, viewed such restraint in a positive light given it meant she had a break from her usually demanding household activities: "seriously for me, it's good, because I can relax" (Sudanese). Such examples were unique to Sudanese women and demonstrated how some women can position menstrual restrictions positively.

Women across cultures and religions are continually receiving paradoxical messages by which they are both demonized for their reproductive bleeding bodies, but praised for their ability to procreate (Goldenberg and Roberts 2011, 82). For example, while motherhood was highly valued across all cultural groups in this study, many women described prohibitions from religious activities, such as visiting the mosque, temple, or church, praying, touching the Koran or other religious texts, participating in religious ceremonies, and observing Ramadan, when menstruating. While the majority of women followed these restrictions due to their own interpretations of religious texts, other participants described that their mothers or older women in their family would relay these restrictions to their daughters. One Punjabi woman told us her mother had said, "not to touch the book [holy scriptures] during periods . . . since elders said [this], we followed through without questioning." Many women continued these avoidance practices during menstruation following migration, with women saying that, "these days we can't go to the temple" (Tamil) and that, "praying is for when you're pure and clean and you're respecting yourself in front of God" (Afghani).

A small number of Muslim women reported that they were required to undertake a cleansing bath before resuming religious activities, because menstrual blood was polluting; "when you have your period, before you've cleansed yourself, you're not allowed to pray or read the Koran" (Afghani). Religious prohibitions and the requirement of ritualized bathing may reinforce the construction of a woman's reproductive body as unclean and polluting, and thus herself as lacking purity: this might lead women to internalize feelings of shame and inferiority toward their own bodies (Crawford, Menger, and Kaufman 2014, 436). In contrast to these accounts however, some participants positioned religious restrictions as "traditional" practices and not something they themselves carried out today. As one Punjabi participant told us, "traditionally we are not meant to go to the temples . . . you don't do *puja* (prayer), but actually I don't worry about that . . . I think my mum follows it, but I don't." This account may suggest a change in religious practice following migration, however, it may also be reflective of modernisation or generational differences in relation to the continuation of restrictive religious practices associated with women and menstruation.

Nearly all women also described that sex during menstruation was strictly prohibited. Reasons for such restrictions on sexual activities included sex being "unhealthy," "harmful," and "dirty" when a woman is bleeding, and sexual abstinence being religiously sanctioned. For example, one woman said, "when it comes to religion, in Islam a man and a woman should not have sex when a woman is having their period, it is dirty, you are dirtying yourself" (Afghani). The impact of religious discourse on women's sexual practices was particularly evident among Muslim women's accounts, the majority of whom continued to avoid menstrual sex even following migration. In addition, menstrual sex was avoided as women considered it inappropriate for men to witness their menstruation "I never be near to my husband, this is a type of respect to him as a man. I don't like him to see something not good in me" (Iraqi). However, there were also exceptions to this, with a small minority of women stating that "sex is better during periods" (Punjabi) or that "it is not even an issue, having sex while I have my period" (Latina). These accounts suggest resistance to negative cultural discourses that position menstruation as unclean or disgusting, by both husbands, and participants themselves.

IMPLICATIONS AND CONCLUSIONS

Across cultural contexts, constructions of menarche and menstruation were strongly tied to notions of 'womanhood,' interlinked with reproduction and emergent sexuality. Nearly all women who took part in the study discursively positioned menarche and menstruation as shameful and abject, requiring associated regulatory practices of silencing and concealment. Silencing menarche and menstruation acts as a reinforcer of the discursive positioning of a woman's bleeding as a source of stigma (Johnston-Robledo and Chrisler 2013, 12), with material and intrapsychic consequences for women. Shame and silencing denied women the right to learn about the functioning of their reproductive bodies. As a result, women had no framework to make sense of their experiences at menarche resulting in negative attitudes toward their menstruation and poor knowledge of its link to fertility.

While there were a number of commonalities across cultural groups interviewed, there was also variation in cultural and religious discourse and practices. This was most evident in relation to culturally prescribed menarche celebrations, menstrual practices, restrictions and rituals, such as changes in diet reported by Tamil women and the avoidance of cooking described by Sudanese women. Furthermore, Muslim and Hindu women were more likely to describe religious rituals or restrictions associated with menses, compared to participants who followed other religions.

The findings of this study suggest however, that migrant and refugee women are not simply positioned within existing cultural discourses associated with menarche and menstruation, but can re-position themselves, variably adopting, resisting, negotiating, and tailoring discourses and practices associated with menstruation (Day et al. 2010, 238). While for some women the migration process facilitated such resistance or re-positioning, other women were still influenced by cultural discourse, particularly around disclosure of menstrual information to daughters. This suggests that a discourse of secrecy and silence, may be difficult to resist (Ussher et al. 2017, 1909) and new migrant and refugee girls and women need access to comprehensive menstrual support and education. Such information could be included alongside other sexual and reproductive health education, providing details about what menstruation is, its link to fertility and guidance on how menstruation can be celebrated and navigated in a healthy manner.

NOTE

1. In this chapter the term "migrant and refugee" is used to describe voluntary migrants and people of refugee or humanitarian background, who have a cultural heritage different from the dominant Anglo Australian/Canadian culture.

REFERENCES

Behera, Deepanjal, Muthusamy Sivakami, and Manas Ranjan Behera. 2015. "Menarche and Menstruation in Rural Adolescent Girls in Maharashtra, India." *Journal of Health Management* 17 (4): 510–19. https://doi.org/10.1177/0972063415612581.

Bobel, Chris. 2010. *New Blood: Third-Wave Feminism and the Politics of Menstruation*. New Brunswick, Canada: Rutgers University Press.

Bramwell, Ros. 2001. "Blood and Milk: Constructions of Female Bodily Fluids in Western Society." *Women & Health* 34 (4): 85–96. https://doi.org/10.1300/j013v34n04_06.

Brantelid, Ida Emilie, Helena Nilvér, and Siw Alehagen. 2014. "Menstruation during a Lifespan: A Qualitative Study of Women's Experiences." *Health Care for Women International* 35 (6): 600–16. https://doi.org/10.1080/07399332.2013.868465.

Buckley, Thomas, and Alma Gottlieb. 1988. *Blood Magic: The Anthropology of Menstruation*. London, UK: University of California Press.

Chang, Yu-Ting, Mark Hayter, and Shu-Chen Wu. 2010. "A Systematic Review and Meta-Ethnography of the Qualitative Literature: Experiences of the Menarche." *Journal of Clinical Nursing* 19: 447–60. https://doi.org/10.1111/j.1365-2702.2009.03019.x.

Cooper, Spring Chenoa, and Patricia Barthalow Koch. 2007. "'Nobody Told Me Nothin'": Communication about Menstruation among Low-Income African American Women." *Women & Health* 46 (1): 57–78. https://doi.org/10.1300/j013v46n01_05.

Costos, Daryl, Ruthie Ackerman, and Lisa Paradis. 2002. "Recollections of Menarche: Communication between Mothers and Daughters Regarding Menstruation." *Sex Roles* 46 (1): 49–59. https://doi.org/10.1023/a:1016037618567.

Crawford, Mary, Lauren M. Menger, and Michelle R. Kaufman. 2014. "'This Is a Natural Process': Managing Menstrual Stigma in Nepal." *Culture, Health & Sexuality* 16 (4): 426–39. https://doi.org/10.1080/13691058.2014.887147.

Day, Katy, Sally Johnson, Kate Milnes, and Bridgette Rickett. 2010. "Exploring Women's Agency and Resistance in Health-Related Contexts: Contributors' Introduction." *Feminism & Psychology* 20 (2): 238–41. https://doi.org/10.1177/0959353509359761.

Dunnavant, Nicki, and Tomi-Ann Roberts. 2013. "Restriction and Renewal, Pollution and Power, Constraint and Community: The Paradoxes of Religious Women's Experiences of Menstruation." *Sex Roles* 68 (1): 121–31. https://doi.org/10.1007/s11199-012-0132-8.

Goldenberg, Jamie L., and Tomi-Ann Roberts. 2011. "The Birthmark: An Existential Account of the Objectification of Women." In *Self-Objectification in Women: Causes, Consequences, and Counteractions*, edited by R. M. Calogero, S. Tantleff-Dunn, and J. K. Thompson, 77–99. Washington, DC: American Psychological Association.

Guterman, Mark A., Payal Mehta, and Margaret S. Gibbs. 2008. "Menstrual Taboos among Major Religions." *The Internet Journal of World Health and Societal Politics* 5 (2). https://doi.org/10.5580/1443.

Hartman, Tova, and Naomi Marmon. 2004. "Lived Regulations, Systemic Attributions: Menstrual Separation and Ritual Immersion in the Experience of Orthodox Jewish Women." *Gender & Society* 18 (3): 389–408. https://doi.org/10.1177/0891243204264810.

Hawkey, Alexandra J., Jane M. Ussher, and Janette Perz. 2017. "Regulation and Resistance: Negotiation of Premarital Sexuality in the Context of Migrant and Refugee Women." *The Journal of Sex Research*: 1–18. https://doi.org/10.1080/0 0224499.2017.1336745.

Hawkey, Alexandra J., Jane M. Ussher, Janette Perz, and Christine Metusela. 2017. "Experiences and Constructions of Menarche and Menstruation among Migrant and Refugee Women." *Qualitative Health Research* 27 (10): 1473–90. https://doi.org/10.1177/1049732316672639.

Jackson, Theresa E., and Rachel J. Falmagne. 2013. "Women Wearing White: Discourses of Menstruation and the Experience of Menarche." *Feminism & Psychology* 23 (3): 379–98. https://doi.org/10.1177/0959353512473812.

Johnston-Robledo, Ingrid, and Joan C. Chrisler. 2013. "The Menstrual Mark: Menstruation as Social Stigma." *Sex Roles* 68 (1): 9–18. https://doi.org/10.1007/s11199-011-0052-z.

Johnston-Robledo, Ingrid, Kristin Sheffield, Jacqueline Voigt, and Jennifer Wilcox-Constantine. 2007. "Reproductive Shame: Self-Objectification and Young Women's Attitudes toward Their Reproductive Functioning." *Women & Health* 46 (1): 25–39. https://doi.org/10.1300/j013v46n01_03.

Johnston-Robledo, Ingrid, Melissa Ball, Kimberly Lauta, and Ann Zekoll. 2003. "To Bleed or Not to Bleed: Young Women's Attitudes toward Menstrual Suppression." *Women & Health* 38 (3): 59–75. https://doi.org/10.1300/j013v38n03_05.

Lee, Janet. 2009. "Bodies at Menarche: Stories of Shame, Concealment, and Sexual Maturation." *Sex Roles* 60 (9–10): 615–27. https://doi.org/10.1007/s11199-008-9569-1.

Lee, Janet, and Jennifer Sasser-Coen. 1996. *Blood Stories: Menarche and the Politics of the Female Body in Contemporary US Society*. London, UK: Routledge.

Marván, Maria, Claudia Morales, and Sandra Cortés-Iniestra. 2006. "Emotional Reactions to Menarche among Mexican Women of Different Generations." *A Journal of Research* 54 (5): 323–30. https://doi.org/10.1007/s11199-006-9002-6.

Mason, Linda, Elizabeth Nyothach, Kelly Alexander, Frank O. Odhiambo, Alie Eleveld, John Vulule, Richard Rheingans, Kayla F. Laserson, Aisha Mohammed, and Penelope A. Phillips-Howard. 2013. "'We Keep It Secret so No One Should Know'—A Qualitative Study to Explore Young Schoolgirls Attitudes and Experiences with Menstruation in Rural Western Kenya." *PLoS One* 8 (11): e79132. https://doi.org/10.1371/journal.pone.0079132.

Moloney, Sharon. 2010. "How Menstrual Shame Affects Birth." *Women and Birth* 23 (4): 153–59. https://doi.org/10.1016/j.wombi.2010.03.001.

Roberts, Tomi-Ann, and Patricia L. Waters. 2004. "Self-Objectification and That 'Not so Fresh Feeling'." *Women & Therapy* 27 (3–4): 5–21. https://doi.org/10.1300/j015v27n03_02.

Salad, Jihan, Petra Verdonk, Fijgje de Boer, and Tineke A. Abma. 2015. "'A Somali Girl Is Muslim and Does Not Have Premarital Sex: Is Vaccination Really Necessary?' A Qualitative Study into the Perceptions of Somali Women in the Netherlands about the Prevention of Cervical Cancer." *International Journal for Equity in Health* 14 (1): 1–13. https://doi.org/10.1186/s12939-015-0198-3.

Schooler, Deborah, L. Monique Ward, Ann Merriwether, and Allison S. Caruthers. 2005. "Cycles of Shame: Menstrual Shame, Body Shame, and Sexual Decision-Making." *The Journal of Sex Research* 42 (4): 324–34. https://doi.org/10.1080/00224490509552288.

Teitelman, Anne M. 2004. "Adolescent Girls' Perspectives of Family Interactions Related to Menarche and Sexual Health." *Qualitative Health Research* 14 (9): 1292–308. https://doi.org/10.1177/1049732304268794.

Uskul, Ayse. K. 2004. "Women's Menarche Stories from a Multicultural Sample." *Social Science & Medicine* 59 (4): 667–79. https://doi.org/10.1016/j.socscimed.2003.11.031.

Ussher, J. M. 2000. "Women's Madness: A Material-Discursive-Intrapsychic Approach." In *Pathology and the Postmodern: Mental Illness as Discourse and Experience*, edited by F. Dwight. London, UK: Sage.

———. 2006. *Managing the Monstrous Feminine: Regulating the Reproductive Body.* London, UK: Routledge.

Ussher, J. M., Janette Perz, Christine Metusela, Alexandra J. Hawkey, Marina Morrow, Renu Narchal, and Jane Estoesta. 2017. "Negotiating Discourses of Shame, Secrecy, and Silence: Migrant and Refugee Women's Experiences of Sexual Embodiment." *Archives of Sexual Behavior* 46 (7): 1901–21. https://doi.org/10.1007/s10508-016-0898-9.

CHAPTER 11

Menstruation and Religion: Developing a Critical Menstrual Studies Approach

Ilana Cohen

INTRODUCTION

In September 2018, the Supreme Court of India overruled a ban preventing women and girls of menstruating age (between ten and 50) from entering the Sabarimala Temple in Kerala, India (Indian Young Lawyers Association & Ors. v. The State of Kerala & Ors, S.C.C. 1 [2018]). The ruling came three years after #happytobleed, a viral social media campaign, drew international outcry to a 2015 Sabarimala Temple board statement. The board stated that one day women between ten and 50, traditionally prohibited from entering the temple lest they distract the deity (perceived to be a celibate bachelor), may be able to enter the temple if they are not actually menstruating (BBC 2015). Nikita Azad, founder of the campaign, spoke out against the implied impurity of menstruation, and the campaign quickly became "an initiative against sexism" and perceived misogyny (Azad 2015). While the Supreme Court's ruling that it is illegal to prevent any woman from entering Sabarimala was welcomed by countless individuals and Kerala's state government, the ruling also set off "an uproar that has become a months-long battle over differences of caste, gender, party politics, and history" (Nair 2019). Indeed, the controversy exemplifies how the intersection of menstruation and religion engages not only contested classifications of the impure and the pure but urgent issues related to gender equality, hierarchies, delineations of communities, and boundaries of power as well.

This complex intersection is well-documented in an extensive body of literature exploring menstruation and religion from many disciplinary perspectives. While there is not space here to chronicle every way menstruation and religion have become topics of scholarship, activism, and reflection, this chapter discusses the possibilities and limitations of how the relationship

© The Author(s) 2020
C. Bobel et al. (eds.), *The Palgrave Handbook of Critical Menstruation Studies*, https://doi.org/10.1007/978-981-15-0614-7_11

between menstruation and religion is imagined and studied. Studies of menstruation and religion often concentrate on ritual impurity and associated prescriptive prohibitions to the extent that 'menstruation-and-religion' comes to connote a religion's classification of menstruation and instructions for what a menstruant should and should not do (see Bhartiya 2013; Dunnavant and Roberts 2012). Such references remove menstruation and menstrual practices from larger contextual frameworks which more targeted religion-specific scholarship may engage. They also tend to follow an established pattern of analysis—"the paradox approach" (Avishai 2008, 410)–which seeks to resolve the 'mystery' of compliance with what are perceived to be negative and restrictive traditions.

While existing scholarship also critiques and adapts "the paradox approach" to explore how women find meaning in and exert agency through observance of menstrual practices (Avishai 2008; Sharma 2014), this chapter offers an alternative approach. The chapter addresses the abstraction of religiously motivated menstrual practices from their wider contexts in order to more effectively engage with them and expand the discourse of menstruation and religion beyond a framework of negative restrictions. Recognizing these contextual larger systems demystifies menstrual practices and allows the discourse to include under-discussed themes such as the relationship between menstrual practices and the establishment and maintenance of personal and communal identities and hierarchies, gendered roles and expectations within religions, and how religiously motivated menstrual practices serve to channel women's sexuality.

In discussing these themes, I advance the menstruation and religion conversation beyond menstruation and a specific religion such as Hinduism, Christianity, or Judaism to religion as a cultural institution in which menstruation plays a practical and theoretical role. The relationships between menstruation and religion, and religion and culture suggest that menstruation is a cultural phenomenon just as much as it is a physiological one. Religiously and culturally motivated practices compel certain behaviors of menstruants, and those practices, and what menstruation symbolizes and communicates within a particular system, ultimately (re)produce religion and culture themselves (see Maharaj and Winkler [Chapter 15] in this volume).

To explore the ways in which menstruation is 'religion-producing,' I review literature on menstruation in Judaism and Hinduism, putting key ideas in conversation with each other. Both Judaism and Hinduism have defined menstrual traditions and prescriptions, often discussed within the framework of impurity and pollution. This framework, though worthy of attention, must be appropriately contextualized to avoid incorrectly implying that in each religious context menstruation is only a source of pollution on the one hand, and that menstruation is *the only* source of impurity and pollution on the other. The scholarship discussed here cannot provide comprehensive accounts of all perspectives on menstruation in Judaism and Hinduism. However, engaging examples from two capacious religious traditions offers both detailed accounts of specific practices and discussion of larger themes at play.

In the first section, I address the impurity framework and menstrual restriction discourse that dominates the discussion of menstruation and religion. In this first section, I show how in both Judaism and Hinduism menstruation is embedded within larger purity systems and codes of behavior which offer deeper understandings of the restriction discourse overall. With this understanding established, I show that the restriction discourse does not fully represent menstruation and religion through a discussion of how menstruation and menstrual practices can be associated with auspiciousness and positive, unique power. In the second section, I widen the scope of engagement and discuss how menstruation and religiously motivated menstrual practices can be read as defining factors and safeguards of the boundaries of specific religious communities in both traditions. Relevant scholarship shows the role menstruation plays in identity formation and explores the extent to which the onus of maintaining religious boundaries falls on women, channeled through their reproductive capabilities and sexualities. Finally, I discuss how menstruation and menstrual practices have become vehicles for wider religious transformation and activism and suggest arenas for further exploration.

THE IMPURITY FRAMEWORK AND RESTRICTIONS DISCOURSE: POSSIBILITIES AND LIMITATIONS

In both Judaism and Hinduism, menstruation is classified as a time of impurity according to systems that govern bodily fluids and interactions between individuals. It is important to study the context these systems create in addition to the top-level impurity classification. When removed from a specific religious context, the 'impurity' concept and its associated prescriptions often become framed as 'traditional practices,' 'myths, and 'religious superstitions' that seem to apply only to menstruation. Further, contextualizing menstrual practices offers more opportunity for discussion than the classification of impurity does alone, as demonstrated in the second part of this chapter.

Contexts for Menstruation in Judaism

In Judaism, menstruation and a menstruating woman fall under the Hebrew term *Niddah*, which is derived from the word meaning "separation" (Meacham 2009). Chapter 15 of Leviticus, a book in the Hebrew Bible, addresses menstruation within an instructional system concerned with impurity which, in the biblical context, "renders a person unfit to approach the altar" and engage in Temple-related rituals (Collins 2004, 145). There are three sources of impurity within this system: encountering a corpse, specific bodily emissions, and skin diseases (Collins 2004, 147). In the biblical context, bodily emissions include: menstruation, non-menstrual vaginal discharges, lochial bleeding, and seminal or non-seminal discharges

(Meacham 2009, "Leviticus 15"). While the biblical purity system treats men's and women's bodily emissions and resulting impurity similarly, there are differences regarding communicability and length of impurity between them (Hauptman 1998, 169). Menstrual bleeding, non-menstrual uterine bleeding, and non-seminal discharges all cause communicable impurity (understood to be transferred through physical contact or contact with something someone experiencing impurity has touched) while seminal discharge does not. The length of impurity for menstruation lasts for seven days but only one day for seminal discharge. As part of a larger list of prohibited sexual behaviors and partners, men are instructed to avoid a menstruating woman (Leviticus 18:19) in order to prevent sexual relations, which would transmit seven days of impurity, an undesired state that would limit ritual engagement, to the sexual partner (Meacham 2009, "Leviticus 15").

Extensive scholarship explores why specific bodily emissions cause impurity in the biblical context (see Whitekettle 1996; Boyarin 1993). For example, Howard Eilberg-Schwartz (1990) suggests that the system can be read as demarcating life and death. Certain fluids, like menstruation and semen, read as missed opportunities for conception, become symbolic of death and render impurity through connoting the impurity of death. Alternatively, impurity can be read as related to uncontrollability; the less controllable a bodily fluid, the more polluting. Menstrual blood and non-seminal discharge, released passively, render impurity for seven days while tears, sweat, and mucus, understood as more controllable, do not (Eilberg-Schwartz 1990, 182). In the biblical context, childbirth also results in impurity leading some to suggest that the key commonality among the three categories of impurity is liminality, or "the edges of life" (Collins 2004, 147). While many interpretations exist, some feminist readings suggest that the biblical purity system itself can be read as one that "privileges normal males and disadvantages all females and males with abnormal discharge" (Meacham 2009, "Leviticus 15").

The end of the biblical period and the destruction of the Second Temple in 70 C.E. rendered the purity system largely inapplicable, and the rabbinic period which followed saw changes in the practical significance of the purity system (Meacham 1999, 29).[1] During this period, sages expounded upon the contents of the Hebrew Bible through recorded discussions that comprise the Talmud, the codified body of Jewish law. While prescriptions for non-seminal emissions and various skin conditions lost their practical importance, prescriptions concerning menstruation remained. In fact, Talmudic discourse resulted in the addition of stringencies to the foundational guidelines regarding menstruation laid out in Leviticus 15 (Hauptman 1998, 169). These changes are recorded in discussions in Tractate Niddah, a book of the Talmud. These discussions produced the practices, referred to as the Laws of Niddah, observed by many Jewish individuals today. The Laws of Niddah prohibit contact between a married couple during the days of a woman's period and for seven 'clean' days following the cessation of bleeding until she immerses in a *miqvah* (ritual bath) for purification (Avishai 2008, 414; see also Mirvis [Chapter 12] as well as Maharaj and Winkler [Chapter 15] in this volume). Physical contact during menstruation is prohibited to avoid sexual arousal and intercourse which would communicate impurity to the man (Hauptman 1998, 161). Proper adherence is motivated in part because a child conceived during menstruation or before proper purification is categorized as "severely defiled" (Avishai 2008, 414). In fact, some scholars argue that the rabbinic concern in the Talmud is not so much with restrictions for menstruating individuals, but rather the conditions under which they become permissible to have sex with. Although the text is 'about' menstruation and women's bodies, it can be interpreted as a text focused more on the concerns of men (Hauptman 1998, 169), a point this chapter addresses in more detail later.

In summary, in Judaism, menstruation was once part of a larger system that governed many types of bodily emissions; over time and through rabbinic discussion, prescriptions regarding menstruation were adapted to remain relevant and observable, in part because the status of future generations depended upon their proper observance. The Laws of Niddah are observed in Orthodox Jewish communities where they are considered part of *Halacha* (binding divinely inspired Jewish law), and are "identified as the backbone of Jewish family life and as a mainstay of Jewish community" (Steinberg 1997, 11). Since most liberal Jewish denominations do not consider *Halacha* to be binding in the same way Orthodox denominations do, and since some may take issue with a function of the body being rendered impure (Meacham 1999, 33), they have rejected compulsory observance of the Laws of Niddah turning them into a personal choice.

Contexts for Menstruation in Hinduism: Brahmanical and Non-Brahmanical Approaches

Although recognized as a singular world religion, Hinduism "encompasses a broad array of traditions, sects, and religious-philosophical schools" (Apffel-Marglin 2008, 54). Each subset offers a distinct conceptualization of menstruation, but adequately discussing all traditions, sects, and religious-philosophical schools associated with Hinduism is well beyond the scope of this chapter. Because Brahmanical Hinduism (the dominant and subsuming form of Hinduism propelled by the priestly [Brahmin] caste) largely shapes prevailing attitudes toward menstruation (Hembroff 2010, 30), it is useful to understand the purity system inherent to it. Non-Brahmanical philosophies and framings, a selection of which this section also addresses, provide alternative explanations and expand the Brahmanical purity framework to explore how menstrual restrictions can be linked to auspiciousness and positivity.[2]

Dharma literature (religious law or sacred norms) provides Brahmanical codes of conduct for Hindu society. Manu Smriti (c. 100 CE) is considered the most definitive, authoritative work of Dharma literature (Olivelle 2004, 3) and offers "the most systematic of Brahmana ideologies" (Chakravarti 2003, 71). The text provides detailed guidance on everything from food preparation and hygiene routines to ritual observance and religious study. Primarily addressed toward Brahmin men, the text outlines proper behaviors for them vis-a-vis women and members of other castes and illustrates an 'ideal' society strictly ordered on personal and communal levels. Manu Smriti's treatment of menstruation is considered emblematic of the normative discourse around menstruation in Hinduism (Leslie 1994, 72) which must be understood within the larger systems of the text.

Interwoven within Manu Smriti are direct and indirect instructions for women and men from different castes of the four-tiered system and for those who fall outside the caste system (considered 'untouchable' according to the text, often referred to as Dalit). Some scholars even suggest that Manu Smriti's overarching intention is to maintain the boundaries of the text's idealized society and to preserve the purity of Brahmin men, reflecting the concerns of the Brahmin authors (Doniger 1991, xxiii). As this section demonstrates, the system that guides Brahmin men's behaviors regarding menstruating women overlaps with the system that governs interactions with 'low-caste' and Dalit individuals (see Sukumar [Chapter 13] in this volume).

Manu Smriti outlines 12 impurities of the body, including "oily exudations, semen, blood, (the fatty substance of the) brain, urine, feces, mucus, ear-wax, phlegm, tears, the rheum of the eyes, and sweat" (Sacred Texts, n.d., 5.135). Men and women are instructed to purify themselves of these substances according to their caste, and coming into contact with any impure entity, whether a substance or a person who is impure, can communicate temporary impurity (Ibid., 5.143). Indeed, the communicability of pollution and impurity through touch, sex, and the sharing of food is a key theme of the text.

In Manu Smriti, menstruating women are included within lists of individuals to be avoided and are likened to 'low-caste' individuals. Brahmin men are instructed to avoid sharing a bed with and to avoid having sex with a woman who is menstruating, no matter how much he might desire to (Ibid., 4.40); to avoid eating food that has been touched by a menstruating woman (Ibid., 4.208); and to avoid holding a conversation with a menstruating woman (Ibid., 4.57). According to Manu Smriti, shunning a woman when she menstruates increases a man's wisdom and long life (Ibid., 4.42). It is also noted that a 'high-caste' man should not touch food that belongs to 'lower caste' people, that has been touched by animals, or that has hair or bugs in it. The two strands of instructions are unified when Manu Smriti states that if a man touches an "untouchable," "a menstruating woman," "anyone who has fallen (from his caste)," "a woman who has just given birth," "a corpse," or anyone who has touched a corpse, he requires cleansing through a bath (Ibid., 5.85). The text suggests that menstruation is one of many substances and socially constructed hierarchies perceived to communicate impurity and to threaten the status of Brahmin men.

In addition to placing menstruation within the system governing bodily purity and caste distinction, Manu Smriti draws on an idea that menstruation is a sign of women's inherent impurity (Leslie 1994, 66), an idea established in a Vedic story that predates the Dharma text which links together menstruation, guilt, sexuality, and Brahmanicide. In this story, the god Indra commits Brahmanicide by killing Vicvarupa, a demon-like Brahmin. In order to absolve himself of this "sin," Indra appeals to the earth, trees, and women to each take a third of his guilt; they each do and receive a boon (blessing) in exchange. After the women take on Indra's guilt, "it [the guilt] became (a woman) with stained garments; therefore, one should not converse with (a woman) with stained garments, one should not sit with her, nor eat her food, for she keeps emitting the colour of guilt" (Keith 1914 [2-5-1]). In this story, menstruation is established as a sign of the worst sin in Brahmanical Hindu ideology.[3] Practices such as not entering the kitchen, not cooking for or touching others, avoiding sex, and refraining from entering temples during menstruation (Bhartiya 2013, 524), which fit within the instructions governing control of bodily substances and caste distinction recounted in Manu Smriti, are intensified by the Vedic story associating moral negativity with menstruation. These two approaches overlap in the Vedic story itself when the text explains that the progeny of a woman who has intercourse during menstruation will be "accursed" and that a man should therefore not have intercourse with a menstruating woman. A later Dharma text addressing the specific duties of a Brahmin wife emphasizes this idea, asserting that a child resulting from intercourse during menstruation will be an "untouchable" or "cursed" (Leslie 1989, 285). In summary, menstruation is associated with impurity in Brahmanical Hinduism due to its position within codified systems governing bodily secretions and caste distinction, as well as its association with sin and guilt.

While these Brahmanical approaches to menstruation may be considered normative, other schools point to menstrual restrictions as indicative of menstruation's auspicious and powerful nature, demonstrating the diversity of approaches to menstruation beyond of framework of restrictions even within what is perceived as a singular religion. Sangam literature (100–500 CE) offers one such example. Sangam literature represents a Dravidian worldview (as opposed to a Sanskrit-Aryan worldview) which was a precursor to the Brahmanical system that eventually became the religion associated with power throughout the subcontinent (Jenett 2005, 176). Originating from the geographic region now known as Tamil Nadu and Kerala in south India, Sangam poetry and literature reference *ananku*, a concept translated as a "sacred power" that is expressed through and fills women's bodies at menarche, during menstruation, and following childbirth (Jenett 2005, 177). *Ananku* is considered a precursor to the idea of Shakti ("divine vivifying female power" [Jenett, n.d.]) and the Shakta-Hindu tradition, in which the female body gives "meaning, form, and coherence to religious beliefs, acts, and relations" (Patel 1994, 69). In her analysis of Sangam poetry, Dianne Jenett notes parallels between menarchal girls, menstruants, and goddesses. She argues that menstrual taboos and practices "recognize the sacred power of the female and were instituted for reciprocal protection;" they allow a woman to access and use the "sacred power," especially potent during menstruation and after childbirth, to her community's benefit through appropriate self-restraint and separation (2005, 186), thus offering an alternate framework for understanding for menstrual restrictions.

The sacred power associated with menstruation is exemplified in the menstruating goddess at the Mahadevar Temple in Chengannur, Kerala. The goddess' menstrual cloth is considered so auspicious and powerful it is auctioned off whenever her menstruation occurs (Jenett 2005, 181; Joseph 2015). Jenett further notes parallels between south Indian menarche rituals and the ritual treatment of the menstruating goddess; these parallels suggest that the menarchal girl is considered to *be* a goddess (Jenett, n.d.). Contemporary enactment of the *Mañcaḷ Nīr Āṭṭu* (Turmeric Bathing Ceremony) in Tamil Nadu exemplifies such parallels. While removing ritual impurity, *tīṭṭu*, is one element of the ceremony, it is not the only dynamic at play. In this ceremony, the menarchal girl is showered in a traditional bath of turmeric and neem-infused water following a period of relative seclusion then dressed in a sari and jewels and presented to kin and neighbors who offer the girl blessings and gifts (Cohen 2017). Once a method for announcing marriageability, the ceremony includes rituals that purify, protect, and strengthen the menarcheal girl at a pivotal moment of transformation and celebrates the "happiness" of menarche (Cohen 2017).[4] Non-Brahmanical approaches associate an auspiciousness and sacred power with menstruation related to liminality and its generative qualities, which manifests in menarche ceremonies and practices that appear similar to those found in Brahmanical Hinduism but have different motivations.

This section has shown that in both Hinduism and Judaism menstruation is part of larger purity systems, which underlie menstrual restrictions, and that these systems are concerned with the boundaries of identity and community. Both Manu Smriti and the Talmudic discussion of *niddah* demonstrate a vested interest on the part of their male authors in maintaining their purity, especially regarding the highly communicable modes of sex and touch. While Tractate Niddah is 'about' women and menstruation, it can be read as a text about men's concern with purity, just as Manu Smriti can be read. Though the instructions in Manu Smriti are directed toward (Brahmin) men, ultimately they become institutionalized as behavior modifications women must observe in order to prevent communicating impurity to others. In both instances, menstruation is categorically associated with sex, sexuality, and reproduction and has direct bearing on the status of offspring conceived. Because of the significance such associations carry, the impurity and restrictions framework can end up dominating both practice and discourse; however, as this section has also shown alternative framings and interpretations, indigenous to various religious traditions, also exist and deserve the attention of a critical menstrual studies approach to religion and menstruation.

COMMUNICATING MORE THAN IMPURITY

Communicating Identities and Maintaining Boundaries

With the menstrual restrictions in Judaism and Hinduism contextualized in their respective systems, this section explores how menstrual practices and traditions, often engaged at the personal or individual level, can become vehicles for defining both individual and communal identities. This discussion deepens inquiry into the relationship between menstruation and religion beyond the personal, subjective, and experiential levels, to which "the paradox approach" may be limited, by engaging larger socio-political contexts and institutional religious dynamics.

In her foundational analysis of Tractate Niddah, Charlotte Elisheva Fonrobert argues that Talmudic discussions of menstruation do more than establish the Laws of Niddah; they contribute to discursive determinations of who belongs to the Jewish community. The rabbinic discourse on body fluids, menstruation, and resulting impurity—which, Fonrobert notes, no longer had bearing on ritual observance at the time of discussion—can be read as demarcating and defining the (Jewish) body politic on a discursive level by defining who should follow the laws and to whom the laws apply (2008, 264). In the period of Late Antiquity in which the Talmudic discussions are set, menstrual blood could signal both gender and Jewish group identity because it was only considered impure if it came from a Jewish woman (Fonrobert 2008, 261). Rabbinic imagination considered non-Jewish menstrual blood and sperm pure (Fonrobert 2008, 263) and thus unable to "communicate" anything. Fonrobert's reading of the Talmudic text shows that, historically, menstruation has been a vehicle for the establishment of group identity and suggests that observance of the Laws of Niddah can be a practical way of signaling group belonging. Contemporary ethnographic studies of women who observe the Laws of Niddah in Israel also demonstrate that adherence to them becomes a site for expressing religious Jewish identity and serves to differentiate the self from a secular other (Avishai 2008, 410). Taking issue with the suggestion that they are impure, some of these women prefer thinking of themselves as 'permissible' or 'forbidden' regarding sexual relations (Avishai 2008, 417), further highlighting the underlying purpose of the laws, in their religiously observant perspective, which is to govern sexual relations and ensure the ritual status of children conceived.

In Hinduism, the contextual background of Brahmanical male purity, understood to be threatened by menstruating women (and women in general) as well as 'low' caste individuals, suggests that menstrual restrictions are connected to the larger social inequalities of perpetuated caste-discrimination (see Sukumar [Chapter 13] in this volume). As explained earlier, menstruation is not the only classification that can pollute, and the caste-based classificatory system itself contributes to menstruation's association with impurity. Indeed, gender stratification and caste stratification are co-constituted through the system, and this awareness suggests that perhaps menstrual restrictions serve to preserve caste boundaries. Hindu menstrual restrictions, as presented in a subsection of Dharma literature for women, can be read as intended to make a menstruating woman less attractive and thus less likely to have sex (Hembroff 2010, 59) so as to avoid potentially conceiving a child classified as 'untouchable.' It has been well established that in Brahmanical patriarchy, "women are crucial in maintaining the boundaries between castes" (Chakravarti 2003, 34). Controlling women's sexuality preserves closed boundaries of caste and ensures 'generational purity' in terms of who they copulate with and, I argue, in terms of *when* they copulate, that is whether they are menstruating or not. For caste creation and control are dependent upon the control of some men and all women for specific (re)productive purposes (Chakravarti 2003, 45).

Analysis of religious menstrual practices at the communal, structural level show the role they play in determining, communicating, and maintaining identities, hierarchies, and culture itself. In both Jewish and Hindu contexts, there is an underlying motive of communicating sexual unavailability during menstruation. Given that in both Hinduism and Judaism intercourse with a menstruant results in a cursed or defiled state of being for a child should one be conceived, this motivation can be read as a patriarchal desire to ensure progeny of a specific identity and thus the assurance of a continued, bounded community. At the same time, since intercourse with a menstruant communicates impurity to a male partner in both contexts, it can be read as a tactic to prevent transmitting impurity through threatening the status of a child should one be conceived. Avishai's work shows how, regardless of how this underlying motivation is interpreted, observance of menstrual laws on an individual level can also communicate something about one's idea to a larger community (religious vs. secular identity). Similarly, the menarche ritual from Tamil Nadu is chiefly concerned with communicating the changed social status of the menarchal girl, who has become a young woman, to the larger community (Cohen 2017). While such rituals follow religious and cultural rules, in this example "Tamil culture is [also] being generated through the performance of a ritual in the form of the girl's coming-of-age [*Mañcaḷ Nīr Āṭṭu*] ceremony" as well (Sekine 2011, 184).

Menstruation as Catalyst for Religious Change

The avenues explored thus far help explain the Sabarimala Temple controversy, described at the beginning of the chapter. Initially, outcry centered on the sexism and misogyny perceived to be symbolized within the ban rather than a desire to actually enter the Temple (Azad 2015). Following the Supreme Court's 2018 ruling, however, women have attempted to actually enter Sabarimala as devotees (BBC 2019) demonstrating both a symbolic and literal alteration of a centuries-old tradition. But as this chapter's inquiry into Manu Smriti has shown, it is no surprise that "the Sabarimala conflict is also closely bound with India's reckoning with caste" (Nair 2019). In fact, the Supreme Court ruling was passed in the "egalitarian spirit" of the early twentieth-century Temple Entry movement, which sought to secure entrance for individuals from all castes into Hindu temples (Jeffery 1976). The ongoing debate between those who support the traditional Sabarimala ban and those in support of the Supreme Court's ruling, which has been described as pitting "dominant-caste Keralites" against competing interests (Nair 2019), underscores the deeply entwined relationship between menstruation-linked discrimination, gender-based discrimination, and caste-based discrimination. The Sabarimala controversy is a prime example of how pulling on the strings of menstruation and religion results in pulling on the strings of much more than 'impurity' and shows how activism responding specifically to menstruation has the potential to become activism responding to larger issues of systemic discrimination.

Menstruation has been a site for religious institutional transformation in a Jewish context as well. While the Talmudic text primarily locates authority on issues of Niddah with male rabbis, during the late twentieth and early twenty-first centuries, women have claimed Niddah as their own authoritative realm, producing a "niddah cultural industry" (Avishai 2008, 415). This industry includes resources for women with legal questions regarding the Laws of Niddah, marriage, sexuality, and women's health and includes a new leadership role for women within Orthodox communities, the *Yoetzet Halacha* (Women Halachic Advisor) who guides observance of the Laws of Niddah (Nishmat's Women's Health and Halacha, n.d.). In some liberal Jewish communities, the practice of monthly immersion in the *miqvah* for married and sexually active unmarried women has been claimed as a valued practice open to personal interpretation with an emphasis on renewal and reflection rather than purification. *Miqvaot* open to Jews of all denominational backgrounds and all sexual orientations, such as Mayim Hayim and ImmerseNYC, highlight immersion, whether monthly after menstruation or corresponding to any life transition, as a ritual of transformation rather than impurity. Menstruation and its associated practices, rituals, and restrictions have compelled the emergence of new ritual leadership roles for women (*Yoetzet Halacha*), the reclaiming, reinterpretation, and continuance of rituals (*miqvah* and menarche rituals), and Supreme Court rulings intended to promote egalitarianism and a spirit of equality. Indeed, these menstruation-focused examples affirm that women's agency can manifest in diverse ways and can motivate the protest against and rejection of menstrual restrictions as well as the reclaiming and adjusting of certain rituals—both of which ultimately have bearing on religious traditions and cultures, themselves, overall.

CONCLUSION

There is a rich relationship between menstruation and religion offering many avenues for inquiry. Religiously motivated menstrual observances have become a primary site through which this relationship is explored. The ways in which those practices are claimed as an inherent part of one's religion, rejected as too restricting and/or misogynistic, or re-claimed with alterations make them worthy of further study. Research into menstruation and religion should certainly include the following considerations: what codified and/or inherited explanations are provided for religious menstrual practices? Do women develop their own explanations? What are the larger contexts for understanding menstruation and menstrual practices in this religion? How do individuals engage with and feel about menstrual laws and practices? How have menstrual practices and observances changed over time?

But as this chapter demonstrates, a critical menstruation studies approach to menstruation and religion is situated in other axes of engagement as well. Through engaging menstruation as a theoretical concept and physiological reality in the broad contexts of Judaism and Hinduism, this chapter has contextualized menstrual practices within larger religious purity systems, exploring the complex intersection of menstruation and religion beyond a framework of restrictions. This chapter has approached menstrual practices through an inquiry into what and how they contribute to better understanding the ways a religious community defines and (re)produces itself. As such, it shifts away from the question of merely wondering how religious women may meaningfully navigate compliance with menstrual practices. It has demonstrated that menstruation can be read as a site through which women's sexuality—and by extension the boundaries of the religious community and maintenance of social hierarchies—are controlled according to particular ideologies, producing the idea that women are bearers of tradition and responsible for the wellbeing of the family, society, and religion itself. At the same time, menstruation and menstrual practices can become vehicles for change and innovation within religious contexts. The communal and institutional elements of menstruation within religion are important to include in menstruation and religion discourse, especially those concerned with policy issues and behavior modifications, for they highlight the larger referential systems that are so essential to any type of change involving menstruation within the context of religion.

NOTES

1. For a detailed account see "An Abbreviated History of the Development of the Jewish Menstrual Laws," by Tirzah Meacham in *Women and Water: Menstruation in Jewish Life and Law*, edited by Rachel Wasserfall (1999). For analysis of how theories of menstruation have changed over time within Judaism see Steinberg (1997).

2. Ayurveda offers an additional framework through which menstruation and menstrual restrictions can be understood, although there is not space here for discussion. Traditional Ayruvedic texts do not associate menstruation with pollution or negativity and instead discuss menstruation in the context of conception, noting that the days of menstruation are not conducive to conception (Leslie 1994, 69). For an Ayruvedic contextualization of menstruation see Hembroff (2010), Joseph (2015), Leslie (1994).
3. There are diverse interpretations of and ways of relating to this Vedic story; see Flueckiger (2013, 103) for an account of how women in Andhra Pradesh understand it positively, explaining it suggests women's unparalleled generosity, strength, and power.
4. For more on the relationship between menarche rituals and divinity see Judy Grahn's metaformic theory (1993); for further analysis of this menarche ceremony and its variations see Kapadia (1995), Bhattacharyya (1975), Ram (1996).

REFERENCES

Apffel-Marglin, Frédérique. 2008. *Rhythms of Life: Enacting the World with the Goddesses of Orissa*. Oxford Collected Essays. Oxford, New York: Oxford University Press.

Avishai, Orit. 2008. "'Doing Religion in a Secular World': Women in Conservative Religions and the Question of Agency." *Gender and Society* 22 (4): 409–33. https://doi.org/10.1177/0891243208321019.

Azad, Nikita. 2015. "#HappyToBleed: An Initiative Against Sexism." Countercurrents. org., November 23, 2015. http://www.countercurrents.org/azad231115.htm.

Bhartiya, Aru. 2013. "Menstruation, Religion and Society." *International Journal of Social Science and Humanity* 3 (6): 523–27. https://doi.org/10.7763/ijssh.2013.v3.296.

Bhattacharyya, N. N. 1975. *Ancient Indian Rituals and Their Social Contents*. London: Curzon Press.

Boyarin, Daniel. 1993. *Carnal Israel: Reading Sex in Talmudic Culture*. Berkley, CA: University of California Press.

Chakravarti, Uma. 2003. *Gendering Caste through a Feminist Lens*. Calcutta, India: STREE.

Cohen, Ilana. 2017. *More Than Blood: Menarche Ceremonies and Menstrual Hygiene in Tamil Nadu, India*. Unpublished Manuscript.

Collins, John. 2004. *Introduction to the Hebrew Bible*, 145–47. Canada: Fortress Press.

Dunnavant, Nicki, and Tomi-Ann Roberts. 2012. "Restriction and Renewal, Pollution and Power, Constraint and Community: The Paradoxes of Religions Women's Experiences of Menstruation." *Sex Roles*. https://doi.org/10.1007/s11199-012-0132-8.

Eilberg-Schwartz, Howard. 1990. *The Savage in Judaism: An Anthropology of Israelite Religion and Ancient Judaism*. Bloomington: Indiana University Press.

Fonrobert, Charlotte Elisheva. 2008. "Blood and Law: Uterine Fluids and Rabbinic Maps of Identity. *Henoch* 30 (2): 243–66.

Flueckiger, Joyce. 2013. "Female-Narrated Possibilities of Relationship." In *When the World Becomes Female: Guises of a South Indian Goddess*, 97–112. Indiana: Indiana University Press.

Grahn, Judy. 1993. *Blood, Bread and Roses: How Menstruation Created the World.* Boston: Beacon Press.

Hauptman, Judith. 1998. "Rereading the Rabbis: A Woman's Voice." Boulder, CO: Westview Press.

Hembroff, Nicole. 2010. "Orthodox Hindu Attitudes to Menstruation." MA Thesis, University of Lethbridge. https://www.uleth.ca/dspace/bitstream/handle/10133/2600/hembroff%2C%20nicole.pdf?sequence=1&isAllowed=y.

Jeffery, Robin. 1976. "Temple-Entry Movement in Travancore, 1860–1940." *Social Scientist* 4 (8): 3–27.

Jenett, Dianne E. 2005. "Menstruating Women/Menstruating Goddess: Sites of Sacred Power in South India." In *Menstruation: A Cultural History*, edited by Andrew Shail and Gillian Howie. New York, NY: Palgrave Macmillan.

———. n.d. "Menstruating Women/Menstruating Goddesses: Sites of Sacred Power in Kerala, South India, Sangam Era (100–500 CE) to the Present." Metaformia. http://www.metaformia.org/articles/menstruating-women-menstruating-goddesses/.

Joseph, Sinu. 2015. "Unearthing Menstrual Wisdom-Why We Don't Go to the Temple, and Other Practices." *Mythri: Imparting Awareness on Menstrual Hygiene to Adolescent Girls.* https://mythrispeaks.wordpress.com/2015/05/28/unearthing-menstrual-wisdom-why-we-dont-go-to-the-temple/.

Kapadia, Karin. 1995. *Siva and Her Sisters: Gender, Caste, and Class in Rural South India.* Boulder, CO: Westview Press.

Leslie, Julia. 1989. *The Perfect Wife: The Orthodox Hindu Woman According to the Stridharmapaddhati of Tryambakayajvan.* Delhi, India: Oxford University Press.

———. 1994. "Some Traditional Indian Views on Menstruation and Female Sexuality." In *Sexual Knowledge, Sexual Science*, edited by Roy Porter and Mikulas Teich, 63–81. Cambridge, Great Britain: Cambridge University Press.

Meacham (leBeit Yoreh), Tirzah. 1999. "An Abbreviated History of the Development of the Jewish Menstrual Laws." In *Women and Water: Menstruation in Jewish Life and Law*, edited by Rachel Wasserfall, 23–39. Hanover: Brandeis University Press.

———. 2009. "Female Purity (Niddah)." *Jewish Women: A Comprehensive Historical Encyclopedia*, March 1, 2009. Jewish Women's Archive. https://jwa.org/encyclopedia/article/female-purity-niddah.

Nair, Supriya. 2019. "Two Women Enter a Temple. A Country Erupts." *The New York Times*, January 8, 2019. https://www.nytimes.com/2019/01/08/opinion/india-women-sabarimala-temple.html.

Nishmat's Women's Health and Halacha. n.d. "Yoatzot Halacha." Accessed October 16, 2018. http://www.yoatzot.org/about-us/default.asp?id=593.

Olivelle, Patrick. 2004. *Manu's Code of Law: A Critical Edition and Translation of the Manava-Dharmasastra.* New York, NY: Oxford University Press.

Patel, Kartikeya C. 1994. "Women, Earth, and the Goddess: A Shākta-Hindu Interpretation of Embodied Religion." *Hypatia* 9 (4): 69–87.

Ram, Kalpana. 1996. "Uneven Modernities and Ambivalent Sexualities: Women's Constructions of Puberty in Coastal Kanyakumari, Tamil Nadu." In *Handbook of Gender*, edited by Raka Ray. New Delhi: Oxford University press.

"Sabarimala Temple: Indian Women Form '620 km Human Chain' for Equality." BBC News online. Last modified January 1, 2019. https://www.bbc.com/news/world-asia-india-46728521.

Sacred Texts. n.d. "Chapter V. *The Laws of Manu*." Accessed October 16, 2018. http://www.sacred-texts.com/hin/manu/manu05.htm.

Sekine, Yasumasa. 2011. *Pollution, Untouchability, and Harijans: A South Indian Ethnography*. New Delhi: Rawat Publications.

Sharma, Nitika. 2014. "From Fixity to Fluidity: Menstrual Ritual Change among Hindu Women of Nepalese Origin." PhD diss., University of Colorado Boulder. https://scholar.colorado.edu/cgi/viewcontent.cgi?article=1002&context=socy_gradetds.

Steinberg, Jonah. 1997. "From a 'Pot of Filth' to a 'Hedge of Roses' (And Back): Changing Theorization of Menstruation in Judaism." *Journal of Feminist Studies in Religion* 13 (2): 5–26.

The Laws of Manu. 1991. Translated by Wendy Doniger. London, England: Penguin Books.

The Veda of the Black Yajus School entitled Taittiriya Sanhita. 1914. Translated by Arthur Berriedale Keith. Cambridge, MA. http://www.sanskritweb.net/yajurveda/keith.pdf.

Whitekettle, Richard. 1996. "Levitical Thought and the Female Reproductive Cycle: Wombs, Wellsprings, and the Primeval World." *Vetus Testamentum* 46 (3): 376–91.

"Why Are Indian Women 'Happy to Bleed'?" BBC News online. Last modified November 23, 2015. http://www.bbc.com/news/world-asia-india-34900825.

CHAPTER 12

Personal Narrative: Out of the Mikvah, into the World

Tova Mirvis

The ritual bath was housed in the back of my parents' Orthodox synagogue, with a separate entrance to ensure privacy. Inside, there was a bathroom with a shower and tub, and in an adjacent room, the small pool—a *mikvah*—with enough space for one person to stand with her arms outstretched.[1] Above was a large round opening in the wall through which the attendant could watch and ensure that every part of the woman was fully under the water.

"Are you excited? Are you nervous?" my mother asked me as we walked in, a few nights before my wedding.

"Both," I admitted.

As a bride, I was required to immerse in order to be sexually permissible to my husband. As a wife, I would be required to do this every month.

In preparation, I'd soaked in a tub, cut my nails, scrubbed my calluses. I forced a comb through my thick hair. The comb ripped out strands of my hair but I wanted to follow the laws precisely.

In the months prior, I'd been studying the religious laws that would now apply to me, sitting around the dining room table of a rabbi's wife.

"This is beautiful," she told me and the dozen other engaged young women, about the rules of Jewish family purity. When we had our periods and for the seven days following, we were in a state of impurity: we couldn't touch our husbands—no sex, not a hug, not a handshake. Once our periods had ceased, we were to check ourselves for any remaining smudges or stains. When we believed ourselves to be clean, we were to leave the cloths inside us for thirty minutes, just to be sure, and then start counting seven clean days. Only at the end of these could we immerse in the mikvah and once again be permissible.

© The Author(s) 2020
C. Bobel et al. (eds.), *The Palgrave Handbook of Critical Menstruation Studies*, https://doi.org/10.1007/978-981-15-0614-7_12

In high school, equally strict rules of modesty had touched down on my body: high school, safety pins were kept in the office to fasten shut a low-cut blouse or a skirt with an offending slit. Mothers had to be called if a new skirt needed to be procured; a spare skirt was kept in the office for those times when a mother wasn't reachable. Our knees, elbows, and hair were discussed in black-scripted rabbinic texts, featured prominently in the school rules, in notes sent home reporting infractions. We were always subject to inspection, our bodies divided and measured and mapped. The rules were written across my body, mapped onto my skin, my hair, my thighs. Now that I was getting married, they were poised to enter my body as well.

You don't have to feel that way, I chided myself whenever I felt a slow burn of resistance. Contrary to how it might appear, this was not an invasion of the most private sphere of my body. This was not an issue of a woman being deemed impure. Shape it and twist it, change it and smooth it—some sort of machine inside my head, skilled at reprocessing and reconfiguring any torn bits into a smooth whole in whose billowing folds I could still seek comfort. Quibble, if necessary, with some of the details, parse the interpretations, summon various rabbinic figures to bolster or support—anything to prevent my body from whispering a small silent *no*.

I called the mikvah attendant so she could check me for any dangling cuticles or stray hairs that would constitute a separation between my body and the waters.

"I'm ready," I told her.

I descended the steps. Here was the portal to adult life—once a girl, now a woman.

**

I went to the mikvah every month of my marriage. I tallied the days of my period. I checked my underwear for any signs of blood. With the small white cloths, I inspected myself for staining.

"This is beautiful," I still told myself, but when I got to the mikvah, all I wanted to do was get in and out as quickly as possible.

It didn't matter how I felt about the rules, just as long as I followed them. I wanted to remain Orthodox, at all cost. Sometimes, in synagogue, I noticed that I stood with my arms folded across my chest, my fingers tightly digging into my arms as though I needed to hold myself intact. Sometimes, I felt like the hats I wore to synagogue were compressing my head, my thoughts, all of me. But even so, I was Orthodox, even though I sometimes doubted. It seemed less a statement of what I believed than a truth of who I was—its language, its rhythms, its customs, all part of me. Its weaknesses, its battlegrounds, its shortcomings, part of me as well.

Once I completed the required preparations, the mikvah attendant checked my back for any stray hairs that would constitute a prohibited separation between my body and the water. She examined my nails for any remnants of polish. She checked that my toenails had been clipped and scrubbed.

Everywhere you were supposed to be covered, yet as an Orthodox woman, you were always subject to inspection.

"Can you comb your hair a little better?" she asked me one time.

I was surprised—she'd never before said much to me, only picked a few hairs off my back or motioned to a hangnail I needed to snip.

I went back into the bathroom, held the comb to my hair, and looked in the mirror, feeling as though I'd been tasked with subduing the most resistant parts of myself.

Do you believe in it? I asked myself, a question I tried to avoid.

I looked at my hair. I wasn't going to comb it again.

"I can't," I told the attendant when I emerged from the room again.

She raised her eyebrows in confusion.

"I can't," I said again. Nothing in my life felt as certain as this one sentence.

With a small, perturbed shake of her head, she quickly inspected the rest of my body. Maybe she saw the resoluteness in my eyes. Maybe she was calculating that the sin would be on my ledger, not hers. Maybe I would be inspected more thoroughly in the future, the mikvah equivalent of a no-fly list.

With resigned approval, she watched as I went under the water, my fists loosely clenched, my eyes lightly closed. I felt pinned in place like the bugs in the collection I'd had to amass for my sixth-grade science class. I'd caught spiders and beetles and moths in a glass jar and placed a cotton ball soaked with nail-polish remover inside. I'd watched, horrified and fascinated, as they flittered and scurried then slowed, their legs no longer moving, their wings no longer flapping. When they were dead, I carefully emptied them onto a Styrofoam board and stuck a pin through each hard body.

"Kosher," she pronounced. "Kosher."

**

I couldn't go back. At the thought of it, my chest tightened, as though my ribs were curling, each into a small silent *no.*

But I couldn't not go either—the wheels of my marriage would have ground to a halt. Without the mikvah there could be no sex. And without shared observance, I couldn't imagine how we would exist together. My husband and I had signed a marriage contract, but another contract existed between us, equally binding and unchangeable, in which we agreed that we would always be Orthodox.

As a compromise, I started going to a non-Orthodox mikvah whose mission was to reinvent this ritual. Instead of inspecting me, the mikvah guide dimmed the lights and asked me how she could help make my experience more meaningful.

But I felt closed to the experience. I wasn't here in search of a meaningful ritual—I was here because I had to be, here to submit my body to rules, even if I didn't necessarily believe in them.

A few years later, a friend from our synagogue called me. "Do you want to do something a little crazy?" she asked.

To my surprise, she wanted me to go with her to a nearby lake and be the equivalent of the mikvah attendant as she immersed. Like me, she could no longer bring herself to go to the Orthodox mikvah where she usually went.

Going to the lake did sound a little crazy, but in a good way. It was night-time, and I could barely see my friend as she walked out into the lake, wriggled out of her bathing suit and went under. She had complied with the rules but also found a way around them. When I got home that night, I wondered how much longer I could continue to do that as well. I lay awake, thinking about street performers I'd seen a few months before, who folded themselves, arms over legs over necks, into smaller and smaller glass boxes: seemingly impossible feats but I knew all too well the feeling of having to contort yourself to fit inside.

The next month, when it was time for me to go to the mikvah, I also went to the lake. With my friend standing by the edge of the water, I waded out, slipped off my bathing suit and went under.

Alone in the water, my body made ripples that floated across the still surface. I lay on my back, took in the moon, which was low and full, and the sky lit with stars. I didn't think of myself as someone who would be moved by a lake or a night sky, yet I felt some softness and easing in my body, some relaxing of my always compressed state. If there was any sliver of meaning for me, it lay in the feeling of being away from the rules, away from the official eyes.

In the end, when I left, both the rules of Orthodox Judaism and my marriage, I remembered this feeling. The urge to leave had started to feel like a physical rising from inside. *No*, every part of me knew. *No*, I wasn't willing to live in accordance with the rules, and *no*, I didn't believe, really believe, their rules contained the ultimate truth, and *no*, I couldn't keep trying to tuck away this feeling, and *no*, I was no longer willing to follow without believing.

The next time I was in a lake, after I'd left, I swam out far out into the water, where I floated on my back, staring up at the sky domed above me and the trees circling all around. In the absence of the rules, my life felt as unmappable as the water I was in. But inside my chest, there was now a widened, no-longer-knotted feeling, as though more space has been created between my ribs. I was in this lake not to cleanse myself or purify myself but to open myself as wide as I could be.

NOTE

1. "Out of the Mikvah, into the World" by Tova Mirvas was first published on September 19, 2017 in *The Lenny Letter*. It is excerpted from the 2018 memoir *The Book of Separation*. Boston: Houghton Mifflin Harcourt. Reprinted with permission. No further reproduction or distribution of the material is allowed without permission from the publisher.

CHAPTER 13

Personal Narrative: Caste Is My Period

Deepthi Sukumar

The Sabarimala temple in Kerala is considered by Hindus to be one of the holiest of temples in India. According to Hindu mythology, Lord Ayyappan of the Sabarimala temple is believed to be a bachelor who has taken an oath of celibacy. The temple, which sees a large annual pilgrimage of more than 20 million people, prohibits the entry of women of menstruating age (between 10 and 50 years). A Supreme Court judgment in September 2018 granted women of all age groups entry into the temple, breaking the age-old tradition. The efforts of the state government to implement the Supreme Court order was met with widespread and violent protests supported by many mainstream political parties, including a large number of women. Even women who self-identified as feminists joined in the protest against allowing menstruating women into the Sabarimala temple. A prominent female member of parliament went so far as to say, "Would you take sanitary napkins seeped in menstrual blood and walk into a friend's home? You could not. And would you think it is respectable to do the same thing when you are walking into the house of God?" (NDTV 2018). The common belief in India is that menstruating women are impure and polluting. Religion and culture have placed taboos on menstruating women in every space they occupy—at home, at work, and at places of worship. At home, a menstruating woman is prohibited from preparing food, touching pickle, curd and spice containers, cooking utensils, and even entering the kitchen and puja room (a sacred room where idols are kept and worshipped). In agricultural work, she is told not to touch the crops as they will die. These taboos are so widespread that Procter and Gamble, a U.S. multi-national company, used the campaign 'Touch the Pickle' to draw attention to menstrual taboos, targeting urban middle-class women as a marketing strategy for their 'Whisper' menstrual pads. The campaign went viral and won the Grand Prix award in Glass Lion Category at Cannes in 2015.

© The Author(s) 2020
C. Bobel et al. (eds.), *The Palgrave Handbook of Critical Menstruation Studies*, https://doi.org/10.1007/978-981-15-0614-7_13

Though an urban middle-class Indian woman, I did not experience these menstrual taboos or restrictions. As a Dalit, my culture is different. Dalits do not practice menstrual taboos. I was never considered or treated as polluting and impure within my own household at any time even during my periods. But 'upper caste' people including menstruating 'upper caste' women considered me to be impure and polluting even when I was not menstruating. The caste system in India is the Hindu social structure that places people by birth in a hierarchy based on ritual purity. Brahmins (priestly class) are at the top of the Hindu social order, followed by Kshatriyas (soldiers), Vysyas (traders and merchants), and Shudras (slaves and servants for the three higher castes). The Dalits are outside the caste system and considered to be polluting, impure, and untouchable by all other caste groups. To maintain the social hierarchy and inequality, the caste system has historically discriminated against the Dalits using economic and social prohibitions based on the notions of purity and pollution. Dalits were not allowed to pursue an education or own articles or economic assets relating to wealth and property. The exclusion was complete by allowing only the lowest and unclean occupations such as agricultural work, removing carcasses of dead animals, cleaning human excreta and dead bodies. In the process, Dalits became culturally different. For example, pickle-making is an 'upper caste' culture. Pickle-making allows families to preserve food and avoid waste when there is abundance. It is not seen in Dalit households, where there is only scarcity and deprivation, so my mother, grandmothers, and members of my extended family did not make pickles.

Dalit housing is spatially segregated and is considered untouchable and polluting by the dominant caste groups. When Dalits themselves and their entire environs are considered to be polluting and untouchable, becoming impure only during our menstruation just did not arise. The women in my family did agricultural labor for daily wages and food. The menstrual taboo of not touching crops did not apply to us. In general, all menstrual taboos are based on the separation of purity and pollution and therefore could not be applied to Dalit women, as we are considered impure and polluting from birth until death.

But even with this different culture, I still felt guilt and shame when I got my first period. My vagina was making an uncontrollable mess. My menstrual blood was a nuisance, and I felt I had to hide it. The first bleeding was mild. I managed to wipe and dab with pieces of cloth, which I then threw away. When my white petticoat and school uniform skirt were stained, I hid them. My mother found these clothes and asked me about them. I told her I had cut my finger while sharpening a pencil and wiped the blood on my clothes. She accepted my explanation, but she kept a lookout. Her suspicion was confirmed the first day of my next period. She told my elder sister to explain to me what to do. My sister tore up an old sari into menstrual pad-sized bits and showed me how to wrap and fold them into a pad and wear it. I had to tie a cord, a strip of cloth torn from the same old sari, around my waist then place the cloth pad over my groin between my legs. Next, I had to tie the two ends of the pad to the string. This was all necessary to keep the pad in place. From then on it was a 'stressful period' every month. Feelings of shame and guilt came and went periodically unexplained. Bloodstains on my dress due to ill-fitting pads made me feel ashamed and the vaginal blood gave me feelings of guilt. Nobody explained menstruation to me, so I did not understand why blood came out of my vagina. Though my family members did not explicitly tell me anything, I just felt that I must hide it.

My own menstruation made me hypersensitive to other women's periods, including my many 'upper caste' classmates in school. In my first period, I hid it and in my next, I was told very quietly how to deal with it. In contrast, when my 'upper caste' peers got their first period, their families celebrated it with rituals, new clothes, and feasting. The puberty celebrations also known as 'coming of age' or 'becoming a big girl' are to inform the community about the availability of a girl for marriage. I would wonder at this because the same family that celebrates menstruation also deems it polluting. When I went to their houses for schoolwork or other peer-related business, I would always stand or sit outside their house on the verandah, never entering their homes. Though caste and untouchability were never explained to me, I somehow knew I was different from my 'upper caste' classmates. I knew there was some 'problem' with my identity and it meant that I was not allowed to go inside 'their' houses. But as I stood on the veranda, I would not be alone. Often a woman from the house would also be there on the veranda. My classmate explained that her mother was 'out-of-doors' because she was menstruating and she should not touch any article in the house except those kept aside for her 'out-of-doors' time—a water container, a plate, and a stick. I grew to feel sorry for these women when I saw them sitting outside with their plate and stick, signaling to everyone that they were menstruating. But then I would remember that these same women I pitied insisted I stand outside their home at *all* times, and after I left, they sprinkled 'holy' water on the place where I stood. Whether menstruating or not, Dalit women are considered impure and polluting.

In college, I went to an all-women hostel, where a new world of periods opened for me. Here we said, "I have the chums" when we had our periods. I began to wear store-bought sanitary pads—the cheaper brand without the plastic lining. Even with these pads, my days of blood-stained clothes continued. But here in the all-women hostel I stopped feeling ashamed or guilty. The college hostel was open to all. Girls from all castes and religions lived together in an equal space. We did not know or ask about each other's caste and class identities. We were aware that our different levels of self-esteem and confidence came from our different social backgrounds, but it did not go further than that. Everyone here had periods. Blood stains on dresses and disposing pads in the toilet bin were treated as matter of fact. There were no menstrual taboos. It was as if menstruation was an equalizer.

In the college hostel, we had a system to dispose of menstrual pads. I wrapped them in newspaper and put them in the trash bins. At that time,

I did not give a thought to what happened to them after that. But now I remember that there were women who moved silently, trying to be as invisible as possible. They cleaned the toilets and took away the soiled pads. I know now that these women belonged to the same Dalit community as mine.

But when I went back from the hostel, to my home or anywhere else, things remained the same. I still felt I had to be secretive about my periods. Hiding pads and newspapers while taking them into the toilet became an art. If there were no bins inside the toilet, I did not know what to do with the pad. I felt ashamed to carry the pad around looking for a bin. Once, at home, I flushed the pad down the toilet and almost committed a murder. The drain got clogged. The manual scavenger (a Dalit who is forced into unclean occupations of cleaning latrines and sewers) was engaged. He stripped down to his loin cloth, entered through the manhole, into the blocked drain with a rope tied to his waist and held by another man standing outside. Soon there was a tug on the rope and the scavenger was pulled up, holding my pad in his hand. He came out and looked straight at me, as I was the only person there who could have used that pad. Being from my community he used my mother tongue to tell me, "I could have died today in that manhole, thank God I came out alive." Though both of us were Dalits, we were divided further and set in a social hierarchy of our occupations. I also had the added advantage of my fore-parents having converted to Christianity, which helped my father to get a college education and escape from the caste-based occupations of cleaning human excreta and menstrual waste.

By the patriarchal and caste notions of purity and pollution, all women are considered impure and untouchable during menstruation. But for Dalits, pollution and untouchability starts from birth and never ends. The 'upper caste' women become polluting and untouchable only when they are menstruating. But, even during her periods, when she herself becomes an untouchable, the 'upper caste' woman holds the Dalits, both male and female, as untouchable and polluting, periods or no periods.

At their core, the menstrual taboos are designed to maintain the systems of caste and patriarchy for the dominance of the touchable caste men. To preserve and maintain caste purity, marriages are arranged within the same caste, and women are prohibited from having marital or sexual relations with men of 'lower castes.' Menstrual taboos that deem women impure and polluting in their periods contribute to the belief system that women are inferior. This menstrual shaming of women's bodies into impure and inferior objects has allowed the male to dominate and control women and their sexuality. Women are made to carry the burden of protecting the supremacy and purity of the male and his caste with deeply ingrained cultural practices such as menstrual segregation, ritual fasting by women to protect the men, and covering the head and face in a male presence.

Most 'upper caste' feminist campaigns do not address the role of caste in menstrual taboos. As a Dalit woman, I have never been able to engage or relate to them. They do not challenge that caste and patriarchy have designed and planted traditions and behavior for 'upper caste' women to diligently follow and be responsible to uphold the 'purity' of their caste. Menstrual behavior and taboos are part and parcel of the caste and patriarchal design to maintain the hierarchy of caste structure by propagating and using the belief system of purity and pollution. Feminists protesting for the entry of menstruating women to the Sabarimala temple are reiterating the untouchability reference made in the Supreme Court judgment with the hashtag '#women are not untouchable.' But they do not see untouchability and gender inequality through the lens of intersectionality. The focus is more on the inclusion of menstruating women than on the patriarchal features of the temple in the glorification and protection of male celibacy and the purification rituals therein. The discourse is silent on the problem of ritual purity being the patriarchal mechanism to maintain social inequality and the caste system. It ignores that the Dalit woman and her particular and complex problems—untouchability, caste discrimination, and oppression—are different from that of a menstruating 'upper caste' woman, yet linked to the same patriarchal and casteist scheme. The recent protests, which have raised menstruation to an unprecedented level on the gender debate, should become the entry point for addressing the role of caste and patriarchy in the complex oppression faced by Dalit women. The subject of menstrual taboos is the right context to understand gender inequality and oppression within the framework of intersectionality.

REFERENCE

NDTV. 2018. "Amid Sabarimala Row, Smriti Irani's Sanitary Pad Comment, and a Question." Last modified October 24, 2018. https://www.ndtv.com/india-news/smriti-irani-will-you-take-soaked-sanitary-pads-to-friends-smriti-irani-on-sabarimala-1936170.

CHAPTER 14

Menstrual Taboos: Moving Beyond the Curse

Alma Gottlieb

[Megyn Kelly] starts asking me all sorts of ridiculous questions . . . you could see there was blood coming out of her eyes, blood coming out of her wherever.
—U.S. presidential candidate, Donald Trump, commenting on hard-hitting questioning by journalist Megyn Kelly. *(Beckwith* 2015*)*

The day after a challenging U.S. presidential debate, then-candidate Donald Trump complained about a female journalist's tough questions by appealing to biological reductionism. His seemingly ambiguous reference to 'her wherever' clearly intended to signify 'vagina'—thereby evoking menstrual blood, and its presumed adverse effects.

In appealing to menstrual blood as the go-to explanation for a female journalist's emphatic interviewing style, Trump revealed that he (like many others) views women as different from men in two crucial ways: ruled by their biology, and naturally meek. If women behave assertively—in ways widely admired for men but disparaged for women—their supposedly out-of-character behavior must be dictated by something beyond their control. That 'something' is often assumed to be hormones organized around the menstrual cycle, prompting menstruating women to express out-of-control emotions, especially anger. In implicitly yet legibly evoking such long-standing gender stereotypes, Donald Trump signaled that menstrual taboos remain alive and well in the contemporary world.

In this chapter, I explore both historical and contemporary structures that undergird menstrual stereotypes and taboos. In the first section, I chart some religious foundations underlying widespread notions that menstruating

---

In memory of Franz Steiner—brilliant scholar of taboo who died before his time from the after-effects of stigma.

© The Author(s) 2020
C. Bobel et al. (eds.), *The Palgrave Handbook of Critical Menstruation Studies*, https://doi.org/10.1007/978-981-15-0614-7_14

143

women cause suffering both to themselves and others. I also explore some communities whose residents offer less negative interpretations. Here, I build on the work of *Blood Magic: The Anthropology of Menstruation* (Buckley and Gottlieb 1988b), a collection of feminist essays that helped inaugurate the modern anthropological study of menstruation practices and beliefs around the world. That collection introduced readers to a striking diversity of menstrual experiences, especially in the Global South. The essays argued that, despite shared biological roots, individuals and communities perceive and experience menstruation in enormously different ways, for reasons encompassing religious, political, demographic, and economic factors.

Today, lines between the Global South and the Global North are blurring. Examples of sometimes parallel, sometimes divergent menstrual experiences that appear below should remind readers that focusing on essentializing distinctions between 'us' and 'them' makes sense neither ethnographically nor ethically. In the second section, I explore briefly some important ways that diverse individuals and organizations are challenging classic stereotypes and taboos surrounding menstruation, forging an emerging global movement of menstrual activism (see Bozelko [Chapter 5]; Bobel and Fahs [Chapter 71]; Nyanzi [Chapter 42]; Weiss-Wolf [Chapter 41]; and Lewis [Chapter 58] in this volume). As with the meanings of taboos themselves, these acts of challenging taboos take various forms and involve diverse individuals across religion, ethnicity, class, caste, gender identity, and other factors.

Understanding and Respecting Taboos

In the United States, the catch-all term 'PMS'—short for 'premenstrual syndrome'—describes any otherwise-unexplained physical or emotional affliction that women of menstruating age experience before their menstrual period. (Though technically inaccurate, in many popular contexts, PMS is even evoked during and after menstrual periods.) In some accounts, a version of this condition reportedly affects up to 90% of all women (U.S. Department of Health and Human Services 2018). By definition, any 'syndrome' is something of a medical mystery (Mukherjee 2009). Anecdotally, patients report a set of co-occurring medical symptoms. However, science has not identified a shared cause. If and when such a cause is identified, the moniker 'syndrome' drops off, and a new name for an actual disease is assigned. With 'PMS,' more than 150 symptoms have found shelter under this very wide umbrella (Studd et al., n.d.; cf. Stolberg 2000). Since no scientific research has isolated a common etiology of these symptoms, the scientific justification for 'PMS' remains dubious (DeLuca 2015).

Even so, PMS has been creatively adapted into a popular neologism by being turned into a verb—as in, "she's PMS-ing." Despite the long list of symptoms, this neologism typically indexes just one symptom: the supposed tendency for a menstruating woman to lose control of her emotions in general, and to express annoyance, critique, or anger in particular (Gottlieb

1988a). Both women and men who highlight this focus may evoke hormones to explain why they oppose women's holding top political positions, which they claim would imperil citizens, due to unpredictable decisions (for example, Bradner 2015). To minimize the supposed risks of undisciplined behavior, modern menstruating women remain subject to new versions of ancient taboos, whose exploration will be at the center of this chapter.

Taboo

For many modern readers, the word 'taboo' unconsciously evokes 'primitive' peoples from long ago or far away—people who unthinkingly obey(ed) arbitrary rules that restrict their lives and thoughts. Yet, images of menstruation as symbolically polluting retain strong staying power in a world ostensibly ruled by science (cf. Coronil 1997; Goffman 1963; Little 2012; Miner 1956; Moeran 2014). Even highly educated women perpetuate menstrual taboos in intimate and public ways alike. Robust research across several decades on the many euphemisms women use for their periods signals how sociologically taboo it remains for women in many settings to discuss this basic biological function (for example, Cauterucci 2016; Chrisler 2011; Ernster 1975; Thornton 2013). No matter their metaphoric content, these euphemisms share one goal: to avoid clear biological descriptors such as 'menstruation,' 'menstrual period,' or 'period' (Kissling 1996; Newton 2016; for examples beyond English, see Escaja 2018; Ren, Simon, and Wu 2018).

Words tell a story. So do efforts to avoid words. The discursive act of substituting euphemisms for certain words brings us straight to the territory of taboo. Words to be avoided carry what the philosopher John Austin (1962) dubbed an "anti-illocutionary force"—the opposite of the illocutionary force conveyed by words that are spoken (rather than avoided). Words silenced by euphemisms put us square in the realm of the "magical power of words," as anthropologist S. J. Tambiah termed it (1968). Uttering phrases normally avoided violates the taboo—incurring emotional, sociological, spiritual, and/or political risks (cf. Allan and Burridge 2006; Pedraza 2018).

Why do words describing a biological process experienced by half our species have this symbolic power? I suggest that, in many parts of the world, the effort to circumvent speaking about menstruation in simple, neutral, or scientific terms and to rely, instead, on euphemisms that often involve shame and/or censure has deep roots in patriarchal ideology inherent in the Jewish and Christian traditions (and later adopted in Islam) (cf. Buckley and Gottlieb 1988a, 32). Of the many English-language euphemisms documented by scholars, one phrase recurs: "the curse." In a study conducted in Oregon, 50% of English-speaking women (aged 18–80) referred to menstrual periods as "the curse" (Lee 1994).

Menstruation as Curse

Why a curse? At one level, one might cite certain biological facts that could produce this dramatically negative view. Menstrual blood differs in many ways from venous blood. It cannot clot (Yang et al. 2012). It flows only from a single place in the body. It stops flowing of its own accord and should not be staunched; an obstructed flow (or amenorrhea) generally signals a problem such as polyps, fibroids, an eating disorder, a birth defect, or a genetic disorder (Pinkerton 2017). Although it may cause suffering, it rarely causes unstoppable hemorrhaging, hence it is rarely lethal. Menstrual blood is typically associated with only one gender—although, as discussed below, not universally. Perhaps a combination of these characteristics has endowed menstrual blood with mystical properties; hence, menstrual blood might readily become subject to taboos—mystically based rules that govern who may (and may not) touch, see, or speak about it (Steiner 1956).

However, none of these characteristics should inevitably produce a notion of a spiritual curse causing women's menstrual suffering. Rather, in Western/ized nations, the widespread concept of menstruation-as-curse likely derives from one specific religious tradition: the Jewish and Christian traditions' sacred text, the Bible. In the Hebrew Bible (a.k.a. the Pentateuch or Old Testament in Christianity), the first book, Genesis, explains that the mythical first woman, Eve, disobeyed her god by eating a forbidden apple. While Genesis names the pain of childbirth, not menstruation, as the curse for Eve's transgression, the Bible's third book, Leviticus, mentions the pain of menstruation and lists required and forbidden activities for menstruating women. Perhaps building on early Mesopotamian ideas about purity (Morrow 2017), generations of Bible readers—both Jewish and, later, Christian—have associated the two stories, such that menstruation is widely considered, and named, a curse like childbirth. As O'Grady has written (2003, 5):

> many biblical commentators throughout history have viewed the Levitical menstrual prohibitions as divine punishment for the sinful nature of woman, which, through the actions of Eve, effected the fall of humankind. Menstruation becomes the divine "curse" of women.

As early as the sixth century, Pope Gregory made this association, and across the ensuing 15 centuries, the popular interpretation has had authoritative theological foundations (Wood 1981, 713–14). One scholar argued that Hebrew Bible-based menstrual taboos account for the continuing exclusion of women from high office in many Jewish and Christian congregations, organizations, and communities (Phipps 1980; see also Cohen [Chapter 11] in this volume). As the third Abrahamic religion, Islam's sacred book, the Qu'ran, retained a version of these views, with menstruation deemed a "painful condition." In most modern Muslim communities, fasting during Ramadan, entering a mosque, praying, having sex, and making the full pilgrimage to Mecca are all forbidden for menstruating women (Ahmed 2015; Haleem 2011), although the textual basis of these injunctions is sometimes debated (Lizzio 2013; cf. Mazuz 2012; see also Maharaj and Winkler [Chapter 15] in this volume).

In short, the menstrual lessons of Genesis and Leviticus have cast a wide shadow across both time and space. Although Jews have always constituted a tiny proportion of humans (in 2015, a mere 0.2%—see Hackett and McClendon 2017), their religious heirs—first Christians, then Muslims—have expanded to constitute, collectively, some 54% of the world's contemporary population (Pew Research Center 2015). From many centuries of missionary and military activity, the impact of these two proselytizing religions has meant that virtually no community anywhere remains untouched by one or another of their teachings (for example, Comaroff and Comaroff [1991] 1997; Pawliková-Vilhanová 2007), and menstrual lessons stemming from the three religions are widely known and accepted (although Hinduism and other religions may have their own origins for menstrual taboos). My discussion of the Jewish-Christian-Muslim tradition of 'the curse' is intentionally broad here, precisely because it is a dominant (though not hegemonic) perspective across much of the world, despite local variations, interpretations, and contestations.

If the dominant legacy of the Jewish-Christian-Muslim heritage propagates an image of menstruation-as-curse, what, exactly, is meant by the term? The *Oxford English Dictionary* (OED) defines a curse as: "1a. An utterance consigning, or supposed or intended to consign, (a person or thing) to spiritual and temporal evil, the vengeance of the deity, the blasting of malignant fate, etc. It may be uttered by the deity, or by persons supposed to speak in his [*sic*] name, or to be listened to by him [*sic*]." This definition signals that certain unfortunate events presumably emanate from spiritual entities. A later definition further highlights the notion of divine punishment: "4a. The evil inflicted by divine (or supernatural) power in response to an imprecation, or in the way of retributive punishment." Implied in both definitions is an assumption of inevitability. The very notion of a god relies on the assertion of a power greater than that of humans. If a god curses a human, the suffering is, by definition, inexorable. A further OED definition specifically links women's menstrual experience with such inevitable suffering—"4d. *the curse*: menstruation. *colloquial*."

With the notion of a curse come specific behaviors and practices that, typically, communities require and women internalize. In communities influenced by the three Abrahamic religions and not (yet) experiencing challenges to their orthodoxies, notions of symbolic 'pollution' and 'stigma' typically join that of 'the curse' to ostracize menstruating girls and women on discursive, conceptual, and physical registers. 'Pollution' normatively becomes the community idiom through which 'stigma' assigns menstruators blame for the symbolic danger they represent. Even when religious discourses are absent, a powerful 'yuck factor' frequently teaches menstruators to maintain their distance from non-menstruators (and vice versa), both verbally and spatially. In these ways, conventional expectations concerning taboo behavior spread well beyond communities oriented around the Bible or the Qur'an. But are all taboos created equal?

Taboo as Morally Neutral

In its original context, the Polynesian word, *tapu*—from which we derive the English 'taboo'—refers to a state of being too powerful to discuss or act on (Steiner 1956). In Polynesian societies, neither positive nor negative associations inevitably apply; the word simply evokes the notion of morally neutral power. When applied to earlier menstrual practices in Polynesian societies, *tapu* often lacked negative associations of stigma.

For example, in some Austronesian language-speaking societies in Polynesia, such as the Micronesian atoll of Ulithi, menstruators were traditionally categorized as *tapu* and, as such, were required to distance themselves from the community, in offsite buildings (Lessa 1966). But such 'menstrual huts' normally drew several menstruating women at once. Taking advantage of what the philosopher Alfred Schutz (1967) might have classified as "consociality," co-menstruating women in the same structure often pursued crafts and other relaxing activities together. In this way, what an outsider might have perceived as a stigmatizing exile could feel like a welcome space of female sociability and rest. On the nearby island of Yap, women maintained a similar tradition. In the 1990s, they renamed the building for menstruating women the Faliyon Women Association's Cultural Center (Beardsley 1999). Among the Kwaio, yet another Austronesian-speaking community on the Solomon Islands, stringent menstrual taboos used to work to women's advantage by giving them access to spiritual power of ancestors (Keesing 1982); however, more recently these taboos have worked against women, for complicated reasons relating to anti-colonial protests and other components of modernity (Akin 2003). Their situation demonstrates the dynamic nature of taboos, which may change meanings across eras.

The notion of power as morally neutral, including as it relates to menstruation, exists far beyond Polynesia. In West Africa, the Beng people of Côte d'Ivoire traditionally partake of this view, using power in both supportive and destructive ways (Gottlieb 1989, 1992). This nuanced orientation contrasts markedly with a dominant Western ethos, which, rooted in the binary thinking of Plato and other ancient scholars, typically insists on moral decisiveness and excludes moral ambiguity: Things are either good or bad—not both or neither. By contrast, with a conceptual orientation rooted in non-binary thought, the Beng view taboos as morally neutral.

Menstruating Beng women are subject to several taboos, including prohibitions on entering the forest, where their fields are located (otherwise they may experience difficulty in their next childbirth), and on touching a corpse (otherwise they may suffer a state of perpetual menstruation).

However, these two prohibitions do not derive from a view of menstrual blood as inherently evil or polluting. Rather, menstrual blood is considered a symbol of human fertility, hence it should be separated from both vegetal fertility (agricultural fields) and death (corpses) (Gottlieb 1992). An indigenous male priest (or "Master of the Earth") of the Beng religion, Kouassi Kokla, explained:

> Menstrual blood is special because it carries in it a living being. It works like a tree. Before bearing fruit, a tree must first bear flowers. Menstrual blood is like the flower: it must emerge before the fruit—the baby—is born. Childbirth is like a tree finally bearing fruit, which the woman then gathers. (personal communication with author)

In evoking this metaphor, Kokla implied that menstrual taboos do not derive from pollution ideologies. Rather, they separate two kinds of fertility (human vs. vegetable), and they separate life (human fertility symbolized by menstrual blood) from death (corpses). This view of menstrual taboos speaks to broader ontological axioms, rather than patriarchal notions of female pollution. The fact that red palm nut sauce cooked for many hours by a menstruating woman is considered by most Beng people to be the most delicious of the many sauces locally available further signals a positive view of menstruation held by men and women alike (Gottlieb 1988b).

Taboo as Morally Ambivalent

Taboos may link to visions that are fully positive, fully negative, or neutral. They may also, in some communities, signal ambiguity, even ambivalence. As documented by a British colonial officer's report, in the early twentieth century the Asante people of Ghana held a morally ambivalent valuation of menstruation. According to Rattray's 1927 study, menstruating Asante girls and women traditionally maintained numerous taboos, including avoiding cooking for any adult men (including a husband), swearing an oath, crossing certain rivers deemed sacred, and touching certain drums and amulets. In earlier days, if a menstruating woman entered the shrine where ancestral stools were kept, she would have been immediately killed (74–75). These practices imply a view of menstrual blood and menstruating women as polluting and evil in the extreme.

However, these taboos, along with the draconian punishment for their violators, only told part of the story. Traditional Asante priests incorporated menstrual blood into mystically powerful brooms (*kunkuma*) that purportedly protected them from mystical harm (Rattray 1927, 14). This practice suggests a view of menstrual blood as both powerful and able to be harnessed for good. Indeed, the above-mentioned taboos were kept from fear that a menstruating woman's spiritual powers were so strong that they could annul men's powers (Rattray 1927, 75). More recently, an indigenous scholar reported that Asante women may still use menstrual blood to make "love charms and potions" (Agyekum 2002, 377).

Adding further complexity to menstruation's moral biography, Asante communities traditionally celebrated menarche with an empowering ritual. Publicly seated beneath an enormous, beautiful, hand-made umbrella of the sort normally reserved for kings, queen mothers, and chiefs, girls menstruating for the first time traditionally received gifts and congratulations, while community members sang and danced in their honor (Rattray 1927, 69–74). Some reports suggest that the ritual, called *bragoro*, remains vibrant today (Agyekum 2002, 380; Akwasi 2018). In the 1990s, this ritual proved important enough for some migrating families to bring to New York (Daniels 1991).

The ambivalence outlined above finds expression in language. Euphemisms for menstruation in the Akan languages spoken by the Asante and other linguistically related groups include phrases that emphasize "indisposition" and "seclusion" (Agyekum 2002, 372) and view menstruation as "toxic," "polluting," "revolting," and "dreadful" (ibid., 374). However, these negative terms are counterbalanced by phrases that emphasize "transition" (ibid., 379–81), "menstruation as a protective visitor" (ibid., 382), and "the *importance* (power, purification, and fertility) of menstruation" (ibid., 374; original emphasis), and that suggest pride in "the fertility and societal recognition of the female" (ibid., 367). The official euphemism to tell an Asante queen mother that an Asante girl under her jurisdiction has her first period is the phrase, ɔ-a-yɛbra—literally, "she has been made perfect" (ibid., 380). In short, the complex set of Asante practices surrounding menstrual blood includes both extremely negative and extremely positive associations. A multilayered view of menstrual symbolism among the Asante leads us far beyond a simple model of menstruation-as-pollution.

Such multilayered perspectives on menstruation exist in many other communities. For example, among the native Yurok of northern California, aristocratic women celebrated their periods with ten days of rituals that accrued prestige to them by heightening spiritual powers, while commoners lacked this privilege (Buckley 1988). In southern India, social class (inflected by the traditional Hindu caste system) shapes menstrual experiences differently. There, menarche rituals still practiced today teach young women ambiguous lessons about their sexuality and social position: Girls learn that their periods are sources of power that may either enhance or disrupt their families' sense of honor and caste standing and, for that reason, they must be monitored carefully (Cardiff 2016; see also Sukumar [Chapter 13] in this volume). On the Balinese island of Indonesia (historically influenced strongly by India), menstruating women refrain from entering a Hindu temple, cooking, having sex, and touching certain objects of men. In some circumstances, they must even sit atop a trash heap. At the same time, all menstruating women enjoy the symbolic status of a *raja*, or prince (Pedersen 2002). Even so, class further distinguishes menstruating women's experiences. High-caste Balinese women may enjoy prestige from their periods, yet they must also adhere to additional menstrual (and other) restrictions that significantly limit their autonomy. Low-caste women may not receive prestige from their periods nor face extensive menstrual restrictions, but their overall lower status may leave them feeling oppressed for different reasons (Pedersen 2002, 309–11). The complicated Yurok and Balinese cases argue for a nuanced approach that prioritizes local experiences and acknowledges multilayered value systems. Rather than starting from an assumption of menstrual experience as a biological given that identically shapes all menstruators in a community, the diverse anthropological record urges us to start inductively. How do individual menstruators experience their periods? How do others perceive them? And how do local value systems, power structures, and menstrual technologies shape these perceptions? These questions provide the most productive starting points for any menstrual inquiry.

Taboo and Shame

In contrast to the cases just discussed, Western interpretations of *tapu* have emphasized an exclusively negative moral valuation of taboo, for reasons, as noted earlier, related to the dominant philosophical orientation of dualism (Buckley and Gottlieb 1988a). With this emphasis on menstruation as taboo, a girl or woman in such communities who speaks openly of her period, especially if boys or men are present, is considered scandalous (Brumberg 1993; Houppert 1999).

In communities pervaded by such expectations, speaking of menstrual experiences even with medical staff may produce shame. In one study in Sweden, only 38% of women who suffered from excessive menstrual bleeding reported their condition to their doctors (Kadir, Edlund, and von Mackensen 2010). One medical researcher suggested, "Social taboos related to menstruation . . . may explain why women have a reluctance to discuss issues relating to menstruation with clinicians, especially male clinicians" (McLintock 2018, 24). Researchers point out that when the taboo on discussing menstruation applies to medical staff, health risks may result. Moreover, reluctance to seek help and information can lead to misunderstanding that can incline some women to seek unnecessary hysterectomies (O'Flynn 2006) or neglect detection of endometriosis (Seear 2009). A recent editorial in *The Lancet* (2018), notes: "In the UK, nearly 80% of adolescent girls have experienced concerning menstrual symptoms (such as unusually heavy or irregular bleeding) but hadn't consulted a health professional; 27% of those said they were too embarrassed to discuss the topic. The rooted silence surrounding periods is putting lives at risk."

It is not just words that are problematic. Equally powerful social expectations restrain behavior surrounding menstruating and menstruators. As I have suggested, the biblical story of menstruation-as-curse is widespread, especially in Western(ized) nations—but not universal. I have mentioned several communities in Africa and Asia in which the biblical/qur'anic interpretation of menstruation as a curse has no hold; many other such communities exist elsewhere (for example Baldy 2018; Hoskins 2002a; Maggi 2002). Even so, not all women accept dominant norms, even when influenced by the Bible or the Qur'an. These norms may appear universal because of their broad discursive power, but they are not hegemonic. Moreover, the ways that menstruators challenge taboos vary from community to community and from person to person. In the next section, I survey some contemporary challenges to normative, taboo-based menstrual regimes; many of these find fuller discussion in other chapters in this collection.

Menstrual Politics: Defying Taboos

Increasingly, women around the globe confront and contest inequalities that some menstrual taboos create. In this section, I briefly survey some of these social protests. The heterogeneity of this menstrual activism echoes the diversity of the taboos' meanings and social contexts. As with the act of following taboos, decisions to contest them originate in diverse motivations and have diverse repercussions.

An Emerging Menstrual Movement

Culturally rooted taboos have proven difficult to contest. However, once challenged systematically, taboos may erode surprisingly quickly. The groundswell of political action now protesting menstrual product taxes, sick leave inequity, unaffordability and environmental unsustainability of menstrual supplies, and toxicity in menstrual products becomes an instantly effective rejection of taboo—merely by publicly acknowledging menstruation in the first place. Put differently, challenging menstrual taboos revalues the experience of menstruation by normalizing it; menstrual blood becomes another ordinary bodily substance. Popular new texts challenge menstrual taboos (Stein and Kim 2009; Weiss-Wolf 2017), whether through memoirs (Farrell 2018), manifestos (Okamoto 2018), educational comic books (Gupta et al., n.d.), fiction (Walter 2016), or coloring books (Clemmer 2016). Menstruation-themed websites, podcasts, and blogs abound (for example, Bell 2014; Clancy, n.d.; Nilson, n.d.; Williams, n.d.). Smartphone apps help women track their cycles (Wortham 2014). "Menstrual Hygiene Day" promotes global conversations (About Menstrual Hygiene [MH] Day, n.d.). Menstruating athletes run marathons while 'free bleeding'—trailing 26 miles of blood to protest menstrual stigma (Gandhi 2015). Politicians position themselves publicly as 'period rights activists' (TedMed, n.d.). These diverse projects announce new approaches that, however distinct their specific origins and goals, together challenge longstanding menstrual taboos.

Challenging Menstrual Taboos

What happens when activists weaponize menstrual blood itself? In contesting inequalities, some activists use blood as a symbolic signifier, rendering public what normally remains private. Women of Northern Ireland harnessed this potent symbolism in a jailhouse protest. When 30 women who had been imprisoned for their fight for independence from Great Britain suffered egregious treatment in jail, they organized in 1980 to smear their menstrual blood on the prison walls (O'Keefe 2006). This protest took its force simply by violating standard menstrual taboos. The dramatic demonstration not only caught the attention of prison guards; one scholar argues that the 'Dirty Protest,' as it came to be called, later empowered a full-blown 'republican feminism' advocating a wide variety of women's rights beyond full independence from the UK. For example, it resulted in a landmark policy document approved by Sinn Féin (Northern Ireland's main nationalist party) arguing for "increased access to divorce, public childcare, childcare to be shared by both parents, free and accessible contraception [...], [and] non-directive pregnancy counseling and sex education" (ibid., 550–51). This case suggests that menstrual politics offer powerful options for women simultaneously exploiting and exploding menstrual taboos in support of broad social justice movements.

With the menstrual movement gaining force, some scholars warn of Eurocentric bias inadvertently introduced by efforts that ignore local menstrual culture and import culturally specific ideas about hygiene (Baldy 2016; Bobel 2019; Khoja-Moolji and Ohito 2018; Lahiri-Dutt 2014). As the menstrual movement spreads, such warnings will no doubt gain force. For example, not all menstruators endeavor to catch their flow of blood. Perhaps the most well-documented community of 'free bleeders' is that of the Rungus people of Borneo, whose longhouses elevated on stilts feature a space accommodating free-flowing menstrual blood:

> During the time of heaviest flow a woman chooses less strenuous tasks which can be performed while sitting on the longhouse verandah. She sits with her skirt discreetly pulled up and her legs covered with a cloth . . . If she gets up to move about she simply flushes the floor of [widely spaced] bamboo slats with water from a bamboo tube which is kept handy to clean up after all messes, including puddling babies [who urinate freely], and [other] spills. (Appell 1988, 110)

This publicly visible practice contributes to men and boys accepting menstruation as a normal and natural occurrence in women's lives (ibid.).

Such a space for free bleeding requires another luxury—the time to sit and bleed, without having to do strenuous work. Elsewhere, such moments characterize 'menstrual huts' as spaces for contemplation, relaxation, or spiritual renewal (Baldy 2018; Buckley 1988). However, some of these structures notoriously carry dangers because of their location. In some communities in the Far-Western region of Nepal, these include exposure to extreme weather, wildlife, and sexual assault (for example, Kadariya and Aro 2015; see also Rothchild and Piya [Chapter 66] in this volume). Although Nepal has criminalized these huts, and some Nepalese activists seek to abolish the institution—sometimes after observing menstrual-hut-based tragedy (White, Sharma, and Das 2013)—some communities continue to use them (Alayyan and Agence France-Presse 2017; Thorpe 2016). The Nepalese case reminds us to listen to the multiplicity of voices and perspectives seeking effective ways to address the strictures, and dangers, of some menstrual taboos.

Israel presents another scenario. There, some Ethiopian Jewish immigrants have brought their tradition of menstrual shelters to their new country—adapting immigration center caravans and, later, building small structures in their own back yards. For these women, retaining a modern version of menstrual structures becomes a means of asserting ethnic identity in a new land, and a counter-protest against a frequently unwelcoming environment (Cicurel and Sharaby 2007). In that sense, these women evoke the logic of Muslim feminists in Egypt, Iran, and elsewhere who have insisted on 're-veiling' as a political statement of independence, after colonial European powers compelled them to abandon their veils (Scott 2007). These and other cases attest to how defying taboos can take many forms—from opposing traditional menstrual practices to reclaiming them.

Getting Creative with Activism

It is now over 40 years since Gloria Steinem (1978) published her pathbreaking, counterfactual, feminist fantasy, "If Men Could Menstruate." Hundreds of thousands of women have likely read that essay; many of those readers are now raising their own daughters and granddaughters. These grand/mothers have rethought their menstrual experiences and are socializing girls to speak openly of all things period-related. Some host parties to celebrate first periods (for example, Bobel 2010), sometimes borrowing from empowering menarche rituals of communities they have read about, such as those of the Asante. These grand/mothers may simply aim to break the taboo against discussing menstruation. Others aim to implant in their girls a positive view of menstrual blood, even evoking women's sacred powers of fertility, perhaps taking inspiration from new menstrual rituals (Amberston 1994). Still others ally themselves loosely with a 'neo-pagan' identity, reimagining their relationship to the Christian icon Mary Magdalene, sometimes even leaving menstrual offerings while on pilgrimages in France and Spain (Fedele 2014). Elsewhere, artists produce paintings, mixed-media, video, and performance art pieces depicting menstrual experiences, sometimes using menstrual blood as their medium (Fahs 2016; Kutis 2019; Manica and Rios 2017). This new creative energy further signals the multiple connotations of menstrual blood among both menstruators and non-menstruators.

This variety of approaches also encompasses gender identity. Long ago, anthropologist Ian Hogbin ([1970] 1996) documented an "island of menstruating men" in New Guinea, where Wogeo men engaged in monthly "sub-incision" rituals. Cutting their penis, Wogeo men produced blood to purify themselves of what Wogeo religion claimed is pollution caused by the act of heterosexual sex. In this ritual, Wogeo men imitated the menstrual cycle of women—which, the Wogeo said, accomplished naturally the same aim of purifying women's bodies from the pollution of sex.

Although their motivations and strategies differ, some young menstrual activists beyond New Guinea likewise claim that women do not hold a monopoly on menstruation (nor do all women menstruate regularly). In a conference panel addressing transgender identities in menstrual experiences, Clemmer (2017) urged the audience: "Why have a 'feminine hygiene' aisle? . . . Don't assume women = menstruators (and vice versa). . . . We need to give up [using] 'feminine products' and other outdated phrases. . . . We need to avoid the conversation being just about cis-women!" In the same panel, another activist commented, "There are some companies that specifically make products for trans[-gendered] people. That's great, but NO companies should include gendered language that erases trans/non-binary identities. . . . ALL products should be gender-inclusive, including of trans people" (Pierce 2017).

With such comments, activist youth challenge the classic binary opposition distinguishing male from female. Echoing work on the cultural construction of gender by the farsighted Margaret Mead (1935, 1949), these menstrual activists are reimagining the menstrual experience from the ground up—and insisting (as anthropologists are wont to do) on its plural nature. The impressive variety of forms that individuals, communities, and organizations are taking to challenge debilitating menstrual taboos further supports the key point noted thirty years ago by anthropologists (Buckley and Gottlieb 1988b): menstrual blood, as with other bodily fluids and processes, holds different meanings for different individuals and across diverse communities. It follows that challenging, defying, and reimagining these meanings also takes a variety of forms.

Concluding Thoughts

In 2009, Nigerian writer Chimamanda Ngozi Adichie powerfully reminded a "TED Talk" audience of "the danger of a single story." As edited collections of anthropological work have demonstrated (Buckley and Gottlieb 1988a; Hoskins 2002b), research in communities around the world documents the striking variety of individuals' experiences of menstruation. Along with the variety of communities' normative attitudes and claims regarding menstrual blood, the global portrait of menstruation supports Adichie's general point.

It is true that biblical and, later, qur'anic views of menstrual blood as dirty, pain-inducing, and/or polluting—and of menstruating women as cursed—have traveled globally. However, as this chapter suggests, members of some indigenous communities continue to hold more positive, nuanced, or complicated views of menstruation. Moreover, in recent years, the quasi-hegemonic, Jewish and Christian view of menstrual blood as caused by a deity's curse—therefore polluting and stigmatizing— increasingly encounters resistance in the Western world. Furthermore, with ever-expanding global communications, both in person and online, the lines dividing experiences in the Global South and the Global North are blurring. One product of this impactful border-crossing is that negative attitudes regarding menstruation, along with hygienic and social practices, are beginning to change—in some places, both rapidly and for the better.

REFERENCES

About Menstrual Hygiene (MH) Day. n.d. http://menstrualhygieneday.org/about/about-mhday/.

Adichie, Chimamanda Ngozi. 2009. "The Danger of a Single Story." *TED Global Talk*. Filmed July. http://www.ted.com/talks/chimamanda_adichie_the_danger_of_a_single_story?language=en.

Agyekum, Kofi. 2002. "Menstruation as a Verbal Taboo among the Akan of Ghana." *Journal of Contemporary Ethnography* 58 (2): 367–87.

Ahmed, Beenish. 2015. "Bloody Hell: Does Religion Punish Women for Menstruating?" *Vice*, June 20. https://www.vice.com/en_us/article/7bdbw9/bloody-hell-menstruating-while-religious-235.

Akin, David. 2003. "Concealment, Confession, and Innovation in Kwaio Women's Taboos." *American Ethnologist* 30 (3): 381–400.

Akwasi, Tiffany. 2018. "Puberty Rites in Ghana–Types and Significance." Accessed June 4, 2019. https://yen.com.gh/108440-this-yen-gh-believes-our-manifesto.html#108440.

Alayyan, Sarra, and Agence France-Presse. 2017. "Nepal Bans Use of 'Menstrual Huts' to Banish Women from Homes." *The Telegraph*, August 9. https://www.telegraph.co.uk/news/2017/08/09/nepalbans-use-menstrual-huts-banish-women-homes/.

Allan, Keith, and Kate Burridge. 2006. *Forbidden Words: Taboo and the Censoring of Language*. Cambridge: Cambridge University Press.

Amberston, Celu. 1994. *Blessings of the Blood: A Book of Menstrual Rituals for Women*. Vancouver: Beach Holme Publisher.

Appell, Laura. 1988. "Menstruation among the Rungus of Borneo: An Unmarked Category." In *Blood Magic: The Anthropology of Menstruation*, edited by Thomas Buckley and Alma Gottlieb, 94–112. Berkeley: University of California Press.

Austin, John. 1962. *How to Do Things with Words*. Oxford: Clarendon Press.

Baldy, C. Risling. 2016. *"mini-k'iwh'e:n* (For that Purpose—I Consider Things) (Re)writing and (Re)righting Indigenous Menstrual Practices to Intervene on Contemporary Menstrual Discourse and the Politics of Taboo." *Cultural Studies ↔ Critical Methodologies* 17 (1): 21–29.

———. 2018. *We Are Dancing for You: Native Feminisms and the Revitalization of Women's Coming-of-Age Ceremonies*. Seattle: University of Washington Press.

Beardsley, Felicia R. 1999. "Yap Menstrual Hut–Yap State, Federated State of Micronesia." *Micronesian Diary: A Continuing Report from the Field*, April 10. http://www.intangible.org/Features/micronesia/text/Yap12.html.

Beckwith, Ryan Teague. 2015. "Trump: Debate Moderator Had 'Blood Coming Out of Her Wherever.'" *Time*, August 8.

Bell, Adey. 2014. "3 Ways to Have a More Potent Period . . . for Good." *YouTube*, April 29. https://www.youtube.com/watch?v=K_vwPtoTzMY.

Bobel, Chris. 2010. *New Blood: Third-Wave Feminism and the Politics of Menstruation*. New Brunswick: Rutgers University Press.

———. 2019. *The Managed Body: Developing Girls and Menstrual Health in the Global South*. New York: Palgrave Macmillan.

Bradner, Eric. 2015. "Businesswoman Stands by Her Man-Only View of Presidency." *CNN Politics*, April 18. https://www.cnn.com/2015/04/18/politics/ceo-women-shouldnt-be-president/index.html.

Brumberg, Joan Jacobs. 1993. "'Something Happens to Girls:' Menarche and the Emergence of the Modern American Hygienic Imperative." *Journal of the History of Sexuality* 4 (1): 99–127.

Buckley, Thomas. 1988. "Menstruation and the Power of Yurok Women." In *Blood Magic: The Anthropology of Menstruation*, edited by Thomas Buckley and Alma Gottlieb, 187–209. Berkeley: University of California Press.

Buckley, Thomas, and Alma Gottlieb. 1988a. "A Critical Appraisal of Theories of Menstrual Symbolism." In *Blood Magic: The Anthropology of Menstruation*, edited by Thomas Buckley and Alma Gottlieb, 1–53. Berkeley: University of California Press.

———, eds. 1988b. *Blood Magic: The Anthropology of Menstruation*. Berkeley: University of California Press.

Cardiff, Cassandra. 2016. "The Gendering Period: Menarche and Womanhood in Low-income Communities of Bengaluru, India." M.Phil. thesis, Development Studies, Department of International Development/Queen Elizabeth House/St. Cross/College, University of Oxford.

Cauterucci, Christina. 2016. "Embracing 'the Blob' and Other Period Euphemisms." *XX Factor: What Women Really Think*, March 1. http://www.slate.com/blogs/xx_factor/2016/03/01/euphemisms_for_periods_are_the_best.html.

Chrisler, Joan C. 2011. "Leaks, Lumps, and Lines: Stigma and Women's Bodies." *Psychology of Women Quarterly* 35 (2): 202–14.

Cicurel, Inbal, and Rachel Sharaby. 2007. "Women in the Menstruation Huts: Variations in Preserving Purification Customs among Ethiopian Immigrants." *Journal of Feminist Studies in Religion* 23 (2): 69–84.

Clancy, Kate. n.d. "Period. With Kate Clancy." http://kateclancy.com.

Clemmer, Cass. 2016. *The Adventures of Toni the Tampon: A Period Coloring Book*. N.p.: Bloody Queer Publishing.

———. 2017. "Comments in Panel on 'The Menstrual Movement: Gender Inclusivity'." PeriodCon (Conference Organized by Period: The Menstrual Movement), New York, NY, November 18.

Comaroff, John, and Jean Comaroff. (1991) 1997. *Of Revelation and Revolution*. 2 vols. Chicago: University of Chicago Press.

Coronil, Fernando. 1997. *The Magical State: Nature, Money, and Modernity in Venezuela*. Chicago: University of Chicago Press.

Daniels, Lee A. 1991. "Child Becomes Woman in Traditional Akan Rite." *New York Times*, July 15. Accessed June 3, 2019. https://www.nytimes.com/1991/07/15/nyregion/child-becomes-woman-in-traditional-akan-rite.html?searchResultPosition=4.

DeLuca, Robyn Stein. (2014) 2015. "The Good News about PMS." Ted Talk, March 17, 2015. https://www.youtube.com/watch?v=aI8KiPiVLyY.

Ernster, V. L. 1975. "American Menstrual Expressions." *Sex Roles* 1 (1): 3–13.

Escaja, Tina. 2018. "13 Lunas 13/13 Moons 13: A Video-Project about Sexuality and Menstruation." *Journal of International Women's Studies* 19 (3): 215–24.

Fahs, Breanne. 2016. "Smear It on Your Face: Menstrual Art, Performance, and Zines as Menstrual Activism." In *Out for Blood: Essays on Menstruation and Resistance,* 105–16. Albany: SUNY Press.

Farrell, Kate, ed. 2018. *Period: Twelve Voices Tell the Bloody Truth.* New York: Macmillan.

Fedele, Anna. 2014. "Reversing Eve's Curse: Mary Magdalene, Mother Earth and the Creative Ritualization of Menstruation." *Journal of Ritual Studies* 28 (2): 23–36.

Gandhi, Kiran. 2015. "Here's Why I Ran the London Marathon on the First Day of My Period—And Chose Not to Wear a Tampon." *The Independent,* August 14.   https://www.independent.co.uk/voices/comment/heres-why-i-ran-the-london-marathon-on-the-first-day-of-my-period-and-chose-not-to-wear-a-tampon-10455176.html.

Goffman, Erving. 1963. *Stigma: Notes on the Management of Spoiled Identity.* New York: Simon and Schuster.

Gottlieb, Alma. 1988a. "American Premenstrual Syndrome: A Mute Voice." *Anthropology Today* 4 (6): 10–13.

———. 1988b. "Menstrual Cosmology among the Beng of Ivory Coast." In *Blood Magic: The Anthropology of Menstruation,* edited by Thomas Buckley and Alma Gottlieb, 55–74. Berkeley: University of California Press.

———. 1989. "Witches, Kings, and the Sacrifice of Identity; or, The Power of Paradox and the Paradox of Power among the Beng of Ivory Coast." In *Creativity of Power: Cosmology and Action in African Societies,* edited by W. Arens and Ivan Karp, 245–71. Washington, DC: Smithsonian Institution Press.

———. 1992. *Under the Kapok Tree: Identity and Difference in Beng Thought.* Bloomington: Indiana University Press.

Gupta, Aditi, et al. n.d. *Menstrupedia Comic: The Friendly Guide to Periods for Girls.* Accessed June 5, 2019. https://www.menstrupedia.com.

Hackett, Conrad, and David McClendon. 2017. "Christians Remain World's Largest Religious Group, but They Are Declining in Europe." Pew Research Center, FacTank, News in the Numbers, April 5. https://www.pewresearch.org/fact-tank/2017/04/05/christians-remain-worlds-largest-religious-group-but-they-are-declining-in-europe/ft_17–04-05_projectionsupdate_globalpop640px/.

Haleem, M. A. S. Abdel. 2011. "Euphemism in the Qur'an: A Case Study of Marital Relations as Depicted in Q. 2:222–3." *Journal of Qur'anic Studies* 13 (1): 125–31.

Hogbin, Ian. (1970) 1996. *The Island of Menstruation Men: Religion in Wogeo, New Guinea.* N.p.: Waveland Press.

Hoskins, Janet. 2002a. "The Menstrual Hut and The Witch's Lair in Two Eastern Indonesian Societies." *Ethnology* 41 (4): 317–33.

———, ed. 2002b. "Blood Mysteries: Beyond Menstruation as Pollution." *Ethnology* 41 (4): 299–390 (special issue).

Houppert, Karen. 1999. *The Curse: Confronting the Last Unmentionable Taboo.* New York: Farrar, Strauss & Giroux.

Kadariya, Shanti, and Arja R. Aro. 2015. "*Chaupadi* Practice in Nepal: Analysis of Ethical Aspects." *Medicolegal and Bioethics* 5: 53–58. https://www.dovepress.

com/chhaupadi-practice-in-nepal-ndash-analysis-of-ethical-aspects-peer-reviewed-fulltext-article-MB.

Kadir, R. A., M. Edlund, and S. von Mackensen. 2010. "The Impact of Menstrual Disorders on Quality of Life in Women with Inherited Bleeding Disorders." *Haemophilia* 16 (5): 832–39.

Keesing, Roger. 1982. *Kwaio Religion: The Living and the Dead in a Solomon Islands Society.* New York: Columbia University Press.

Khoja-Moolji, Shenila, and Esther O. Ohito. 2018. "Containing the Leakiness of Impure Inhumans: Bleeding Third-World Bodies and the Confining Cultural Politics of Menstrual Hygiene Campaigns." In *Youth Sexualities: Public Feelings and Contemporary Cultural Politics*, edited by Susan Talburt, 107–28. Santa Barbara, CA: ABL-CIO Press.

Kissling, Elizabeth Arveda. 1996. "'That's Just a Basic Teen-age Rule:' Girls' Linguistic Strategies for Managing the Menstrual Communication Taboo." *Journal of Applied Communication Research* 24 (4): 292–309.

Kutis, Barbara. 2019. "The Contemporary Art of Menstruation: Embracing Taboos, Breaking Boundaries, and Making Art." In *Menstruation Now: What Does Blood Perform?* edited by Berkeley Kaite, 109–34. Ontario: Demeter Press.

Lahiri-Dutt, Kuntala. 2014. "Medicalising Menstruation: A Feminist Critique of the Political Economy of Menstrual Hygiene Management in South Asia." *Gender, Place and Culture: A Journal of Feminist Geography* 22 (8): 1–19.

Lancet, The. 2018. Editorial: "Normalising Menstruation, Empowering Girls." *The Lancet/Child & Adolescent Health* 2, no. 6 (June 1): 379. https://doi.org/10.1016/S2352-4642(18)30143-3.

Lee, Janet. 1994. "Menarche and the (Hetero)Sexualization of the Female Body." *Gender & Society* 8 (3): 43–62.

Lessa, William Armand. 1966. *Ulithi: A Micronesian Design for Living.* New York: Holt, Rinehart & Winston.

Little, Peter C. 2012. "Another Angle on Pollution Experience: Toward an Anthropology of the Emotional Ecology of Risk Mitigation." *Ethos: Journal of the Society for Psychological Anthropology* 40 (4): 431–52.

Lizzio, Celene Ayat. 2013. "Gendering Ritual: A Muslima's Reading of the Laws of Purity and Ritual Preclusion." In *Muslima Theology: The Voices of Muslim Women Theologians*, edited by Ednan Aslan, Marcia Hermansen, and Elif Medeni, 167–80. Bern, Switzerland: Peter Lang.

Maggi, Wynne. 2002. *Our Women Are Free: Gender and Ethnicity in the Hindukush.* Ann Arbor: University of Michigan Press.

Manica, Daniela Tonelli, and Clarice Rios. 2017. "(In)visible Blood: Menstrual Performances and Body Art" (Dossier: On Rituals and Performances). *Vibrant: Virtual Brazilian Anthropology* 14 (1). http://www.vibrant.org.br/lastest-issue-v-14-n-1-01-042017/.

Mazuz, Haggai. 2012. "Menstruation and Differentiation: How Muslims Differentiated Themselves from Jews Regarding the Laws of Menstruation." *Der Islam* (Berlin) 87 (1/2): 204–23.

McLintock, Claire. 2018. "Women with Bleeding Disorders: Clinical and Psychological Issues." *Haemophilia* 24 (S6): 22–28. https://onlinelibrary.wiley.com/doi/full/10.1111/hae.13501.

Mead, Margaret. 1935. *Sex and Temperament in Three Primitive Societies*. New York: William Morrow.

———. 1949. *Male and Female: A Study of the Sexes in a Changing World*. New York: William Morrow.

Miner, Horace. 1956. "Body Ritual among the Nacirema." *American Anthropologist* 58 (3): 503–7.

Moeran, Brian. 2014. "Business, Anthropology, and Magical Systems: The Case of Advertising." *2014 Ethnographic Praxis in Industry Conference Proceedings*, 119–32.

Morrow, Laura. 2017. "Israelite Ritual Law Concerning the Menstruant in Context: Embodiment and Meaning in Ancient Mesopotamia and Ancient Israel." Unpublished paper, Andrews University. https://digitalcommons.andrews.edu/cgi/viewcontent.cgi?article=1007&context=papers.

Mukherjee, Purpa. 2009. "Difference between Syndrome and Disease." *Difference between Similar Terms and Objects*, October 29. http://www.differencebetween.net/science/health/difference-between-syndrome-and-disease/.

Newton, Victoria Louise. 2016. "'Auntie's Come to Tea': Menstrual Euphemism." In *Everyday Discourses of Menstruation: Cultural and Social Perspectives*, 133–45. New York: Palgrave Macmillan/Springer.

Nilson, Ingrid. n.d. "Period Vids!" *IngridNilsen*. https://www.youtube.com/watch?v=bjHSxdR60b8&list=PLMli3KIg6MX9HXhBJA94GXwS_W_Du8tiE.

O'Flynn, Norma. 2006. "Menstrual Symptoms: The Importance of Social Factors in Women's Experiences." *British Journal of General Practice* 56 (533): 950–57.

O'Grady, Kathleen. 2003. "The Semantics of Taboo: Menstrual Prohibitions in the Hebrew Bible." In *Wholly Woman, Holy Blood: A Feminist Critique of Purity and Impurity*, edited by Kristin De Troyer, et al., 1–28. Harrisburg, PA: Trinity Press International.

Okamoto, Nadya. 2018. *Period Power: A Manifesto for the Menstrual Movement*. New York: Simon and Schuster.

O'Keefe, Theresa. 2006. "Menstrual Blood as a Weapon of Resistance." *International of Feminist Journal of Politics* 8 (4): 535–56.

Pawliková-Vilhanová, Viera. 2007. "Christion Missions in Africa & Their Role in the Transformation of African Societies." *Asian and African Studies* 16 (2): 249–60.

Pedersen, L. 2002. "Ambiguous Bleeding: Purity and Sacrifice in Bali." *Ethnology* 41 (4): 303–15.

Pedraza, Andrea Pizarro. 2018. "Introduction." In *Linguistic Taboo Revisited: Novel Insights from Cognitive Perspectives*, edited by Andrea Pizarro Pedraza, 1–9. Berlin/Boston: Mouton De Gruyter.

Pew Research Center. 2015. "The Future of World Religions: Population Growth Projections, 2010–2050." *Pew Research Center—Religion and Public Life*, April 2. http://www.pewforum.org/2015/04/02/religious-projections-2010-2050/.

Phipps, William E. 1980. "The Menstrual Taboo in the Judeo-Christian Tradition." *Journal of Religion and Health* 19 (4): 298–303.

Pierce, Mason. 2017. "Comments in Panel on 'The Menstrual Movement: Gender Inclusivity'." PeriodCon, New York, NY, November 18.

Pinkerton, Joanne V. 2017. "Absence of Menstrual Periods (Amenorrhea)." *Merck Manual—Professional Version*. Accessed May 29, 2019. https://www.merckmanuals.com/professional/gynecology-and-obstetrics/menstrual-abnormalities/amenorrhea.

Rattray, R. S. 1927. *Religion and Art in Ashanti.* Oxford: Clarendon Press.

Ren, Liqi, Denis Simon, and Jianfeng Wu. 2018. "Meaning in Absence: The Case of Tampon Use among Chinese Women." *Asian Journal of Women's Studies* 24 (1): 28–46.

Schutz, Alfred. 1967. *The Phenomenology of the Social World.* Evanston, IL: Northwestern University Press.

Scott, Joan Wallach. 2007. *The Politics of the Veil.* Princeton: Princeton University Press.

Seear, Kate. 2009. "The Etiquette of Endometriosis: Stigmatisation, Menstrual Concealment and the Diagnostic Delay." *Social Science and Medicine* 69 (8): 1220–27.

Stein, Elissa, and Susan Kim. 2009. *Flow: The Cultural Story of Menstruation.* New York: St. Martin's Press.

Steinem, Gloria. 1978. "If Men Could Menstruate." *Ms. Magazine,* October.

Steiner, Franz B. (1950–52) 1956. *Taboo.* London: Cohen & West.

Stolberg, Michael. 2000. "The Monthly Malady: A History of Premenstrual Suffering." *Medical History* 44: 301–22.

Studd, John, et al. n.d. "About PMS." *National Association for Premenstrual Syndrome.* http://www.pms.org.uk/About.

Tambiah, Stanley J. 1968. "The Magical Power of Words." *Man* 3, no. 2 (n.s.): 175–208.

TedMed. n.d. "Linda B. Rosenthal in a Nutshell." https://blog.tedmed.com/making-menstruation-matter/.

Thornton, Leslie-Jean. 2013. "'Time of the Month' on Twitter: Taboo, Stereotype and Bonding in a No-Holds-Barred Public Arena." *Sex Roles* 68: 41–54.

Thorpe, J. R. 2016. "Menstrual Huts Still Exist—And Here's Why That's A Problem." *Bustle,* December 9. https://www.bustle.com/articles/198540-menstrual-huts-still-exist-and-heres-why-thats-a-problem.

U.S. Department of Health and Human Services. 2018. "Premenstrual Syndrome (PMS)," March 16. Office on Women's Health, Office of the Assistant Secretary for Health, U.S. Department of Health and Human Services. https://www.womenshealth.gov/menstrual-cycle/premenstrual-syndrome.

Walter, Laura Maylene. 2016. "Museum of Menarche." *Ninth Letter* (Fall/Winter 2016). http://www.ninthletter.com/journal/147-13-2-fall-winter-2016.

Weiss-Wolf, Jennifer. 2017. *Periods Gone Public: Taking a Stand for Menstrual Equity.* New York: Arcade Publishing.

White, Pamela, Sunita Sharma, and Sunil Kumar Das. 2013. "Culture Clash: Menstruation Taboos and ODF in Nepal." Community-Led Total Sanitation, December 13. CommunityLedSanitation.org.

Williams, Laura. n.d. "The Best 10 Yoga Poses for a Painless Period." *Zenward.* http://blog.zenward.com/10-best-yoga-poses-painless-period/.

Wood, Charles T. 1981. "The Doctor's Dilemma: Sin, Salvation, and the Menstrual Cycle in Medieval Thought." *Speculum* 56 (4): 710–27.

Wortham, Jenna. 2014. "Our Bodies, Our Apps: For the Love of Period-Trackers." *New York Times,* January 23. https://bits.blogs.nytimes.com/2014/01/23/our-bodies-our-apps-for-the-love-of-period-trackers/.

Yang, Heyi, et al. 2012. "Proteomic Analysis of Menstrual Blood." *Molecular & Cellular Proteomics* 11 (10): 1024–35. www.mcponline.org.

CHAPTER 15

Transnational Engagements: Cultural and Religious Practices Related to Menstruation

Edited by Trisha Maharaj and Inga T. Winkler

Many religions and cultures have traditions and practices that influence the activities, experiences, and interactions of menstruators (see Cohen [Chapter 11] in this volume). This dialogue engages participants from various cultural and religious backgrounds to reflect on their personal experiences with menstrual practices. Very often, menstrual practices are presented as restrictive and coercive. However, the participants in this dialogue demonstrate understandings of menstrual practices that reveal a range of engagements with such expectations, including the exercise of agency in various ways.

Can you briefly introduce yourself and your background? Are there practices or traditions related to menstruation in the culture or religion you consider to be your own or with which you are otherwise familiar? Do you follow them?

Rosa Freedman: I live in the United Kingdom as a practicing Jew. In Judaism, the practices and rules about menstruation begin if or when a woman gets married, as those laws relate to what we call *taharat hamishpacha* (loosely translated as 'family purity'). Those laws involve menstruation, sexual relations, and other aspects of married life (physical, emotional, and spiritual). The simple way of explaining the rules is that a husband and wife do not have intimate relations from the time that menstruation begins until seven days after it ends, when the woman immerses in a ritual bath (*mikveh*) and is considered spiritually pure. I began following these rules when I got married. As part of the preparations for our wedding, I learned about the spirit and letter of the rules. The day before our wedding, I immersed in the

© The Author(s) 2020
C. Bobel et al. (eds.), *The Palgrave Handbook of Critical Menstruation Studies*, https://doi.org/10.1007/978-981-15-0614-7_15

mikveh and have done so every month since then, except when pregnant or breastfeeding (see also Mirvis [Chapter 12] in this volume).

Amina Darwish: I am the Muslim Chaplain at Columbia University in New York. Muslim acts of worship are divided into the worship of the heart, tongue and body. The physical acts of worship include *salah* (the physical prayer), fasting, and vaginal penetration. When women menstruate, we should focus on the worship of the heart and tongue and refrain from the physical acts of worship. My understanding is that our bodies are engaging in another physical form of worship during menstruation. After we finish our menses, we engage in a ritual bath called *ghusl* before resuming regular *salah*, fasting, and sexual activity.

Lubabah Helwani: I also live in the United States, and I grew up in a Muslim family. Growing up my family and myself took part in the religious traditions. The day I received my period was a defining moment for myself and my family. I was given a hijab and told that my actions are now my own. We talked about prayers, and a few days later my mother woke me up to pray the early morning prayer. I personally do not take part in any of the religious exclusions when I menstruate. I do not believe that I cannot fast or pray to God while I am on my period. I do believe that my bleeding is a form of prayer and that menstruation is a connection to God in this form.

Lina Mathew: I live in Kerala in India. I am a Syrian Orthodox Christian, while the majority population in Kerala is Hindu. In the Syrian Christian Orthodox and Jacobite churches in Kerala, menstruating women are traditionally not supposed to enter the church or partake of the Holy Communion, and are only permitted to sit outside. While no canon law explicitly bars women from fully participating in church services during their menstrual cycle, such prohibitions do exist within families to varying degrees, and include refraining from touching the Bible, lighting the lamp at home, or participating in religious ceremonies at home. In contrast, there are no religious restrictions for menstruating women in Syrian Catholic and Latin Catholic communities and the many protestant Christian communities in Kerala. As a young girl, based on my mother's protestant beliefs, I used to enter the Orthodox church of my father, even while menstruating. Yet, the two-hour service left me very tired and faint, especially when I got older.

Among many in the Hindu communities of Kerala, some believe that if a woman does not wash her menstrual cloth properly and leaves it lying around, a snake will come across the blood and die by beating its head on a stone. While killing itself, the snake will curse the woman. Many women believe themselves to have been cursed by this mythical snake because of their inability to bear children. Expensive prayers are performed in specific snake temples in Kerala in order to remove such snake curses and beget children. Further, many menstruating Hindu women in Kerala, from across the spectrum of castes, do not enter temples until seven days after the onset of menstruation. While women were formerly segregated from the household, nowadays such practices are not so rigid. In other parts of India, I understand that such strict menstrual practices are confined mostly to the upper castes.

Radha Paudel: I was born in a poor, rural family in the central part of Nepal. I have witnessed a variety of practices related to menstruation since I was seven years old, while my mother and three sisters were on their periods. They followed more than 40 types of restrictions related to what menstruating women can eat, touch, or participate in, as well as some related to menstrual blood itself. I was traumatized from seeing all of these restrictions. I also got the opportunity to witness an annual cleansing ritual called *Rishipanchami* when I was eight years old. *Rishipanchami* is an annual Hindu festival for purifying the contamination caused by menstruation throughout the year. Menstruating women and girls have to perform the purification ritual so that they can be rid of God's curse of contamination. It was painful seeing my three sisters put in a shed during their first and second period. So, I became determined to deny these restrictions at any cost. I have not followed any of these restrictions since I got my period at the age of 14.

Kalvi Karunanithy: I grew up in Tamil Nadu, India. When I got my first period I was asked to sit, sleep, and stay outside on the doorstep for three days. On the third day my aunty gave me a shower using yellow water. I felt weird, but was comforted by my parents who slept outside the house to take care of me. However, still I feel that it is a blind belief that people follow, as I didn't feel comfortable asking why we followed those practices at that time. Then my parents did a little *puja*, inviting all my relatives, and I felt that I got too much attention. This felt strange but I was comforted by many gifts like new dresses, bangles, and gold jewelry. One of my aunties introduced me to cloth as a menstrual absorbent and taught me how to use it during my period. But I was not really educated about menstruation itself. Everything was a bit of a blur for me at that time, and I was too shy to ask any kind of questions, so I kept quiet and blindly followed whatever were the instructions given to me:

I was asked to not to go to temple during my period
I was asked to shower for the first three days of my period and on the third day.
I was asked to take oil bath using ginger oil.
I was asked to not to talk with boys.
I was asked to sleep separately during my period and wash the mat the next day.
I was asked to not to enter the *puja* (prayer) room at my house.
I was asked not to touch or water the plants during my period.
I was asked to not to feed my own dog.

As I grew older, I no longer followed any of the orders imposed by my mother, as I didn't like them. For the first couple of months my mother was upset and used to scold me, but after some time, she was all right.

Krystal Ghisyawan: I grew up in a pretty religious Sanatanist Hindu family in Trinidad. It was therefore not uncommon to hear comments about someone being unclean, and having it explained to me that: "The blood is dirty, and *puja* things have to be clean, so you can't touch anything for the *puja*." I used to help my mom set up everything for *puja*. When she "wasn't clean," I did almost all the preparations myself, following her instruction. Sometimes, I too would not be clean, but out of habit, I would still touch things. I might wash a *thali* (platter) or pass the sugar from the aunties making sweet rice in the kitchen to my mom setting up in the living room. I would exclaim, "Oh I touched it! I'm not clean," but my mom would dismiss it, "It's alright. Didn't you wash your hands?" or "Don't worry about it." Perhaps it was this dismissal that planted the seed of doubt in my mind about how necessary it was to adhere to these practices of purity. This doubt has matured; I am no longer religious.

Jieun Choi: Growing up in South Korea (where I still live), menstrual blood was often considered filthy. In some communities, there was a local custom saying that women shouldn't sit on a broomstick because if it's tainted with their menstrual blood, it'll turn into a monster. Such notions made it clear to women that their menstrual blood needed to be taken care of or else, it'd be of harm to themselves and others. But on the other hand, as menstruation was inevitably linked to women's reproductivity, menstrual blood was considered sacred in some communities. Some say paper dabbed in menstrual blood was used as a charm for a tuberculosis patient in hopes it would help cure the disease. To an extent, such practices still exist in South Korea today. When I was in secondary school, my friends and I would whisper into each other's ears to ask for 'that' and without explicitly mentioning what 'that' is. Then one would discreetly hand over a menstrual pad to another as if she was an illegal drug dealer. Although changing slowly, there is a tremendous taboo surrounding using tampons and menstrual cups in Korea, supposedly due to their 'penetrating' properties. This has to do with the culture that emphasizes women's chastity.

Alfred Muli: I live and work in Kenya, and my impression is that most myths are actually quite similar across Kenya. The menstruating body is viewed as dirty, unclean and one that can contaminate others, hence the taboos that forbid certain activities such as interacting with men, cooking for the family, milking cows, and even going to the garden. One very common belief is that if you go to the garden when menstruating, the crops will dry up. Most of the restrictions draw from either traditional religions or Christianity, banning menstruators from going to places of worship or taking the lead in any religious activities such as standing in the pew or going to the shrines. Most people I know follow these practices. I was recently talking to some university students who believe that when your hair is plaited by someone who is menstruating, you get dandruff.

What do you think are the origins of these practices or traditions?

Rosa Freedman: The Jewish practices are grounded in rules found in religious scriptures. The rules of what a married woman and married man can and cannot do during menstruation and the seven days afterwards are set out in the *Gemarrah* (Talmud) and other Jewish texts that expand upon the rules that govern our everyday practices. The rules encourage a marital relationship to be based on more than physical or sexual connections, they also provide a method for encouraging fertility (as sexual relations usually begin the day before the most fertile days in the cycle), and they protect a woman at the time of the month when her body is going through significant hormonal and physical changes.

Amina Darwish: In Islam, the practices were narrated by Prophet Mohamed and transmitted to Muslim women through his wife Aisha and his daughter Fatima (peace be upon them). Aisha also narrates instances when Prophet Mohamed (peace be upon him) would show her extra affection while she was menstruating. This was also instructing her to not to try to control matters, but rather let her body do its thing.

Radha Paudel: Nepalese Hindu culture is influenced by Hinduism from India. There are a few Indian Hindu philosophies and epics such as *Manusmirti, Chanakyaniti, Garunpuran* and *Rishipanchami* that revealed that menstruation is bad luck, sin from god, dirty, impure and contaminated. In Nepal, almost all people believe that they have to continue the practice for the sake of God and culture. Otherwise some kind of mishap would happen to family members, especially the family head (referring to men members of the family).

Krystal Ghisyawan: South Asians were brought to the Caribbean from the mid-1800s, leaving behind their families, homes, and traditions. Even if they, as individuals, lacked the theoretical or scriptural religious knowledge, South Asians held on to traditional religious practices and passed them on intergenerationally. The traditional *purdah* practices (meaning partition or separation) kept menstruating women away from the rest of the family, and restricted her chores, like cooking. These restrictions were impractical for women in the Caribbean, who worked for a living and didn't have the support network needed for their chores to be taken on by someone else. While discarded socially, these restrictions still have religious significance; menstrual taboos are only relevant in relation to performing rituals, but are not seen as necessary in everyday life.

What are the religious bases for beliefs that influence menstrual practices? I have never witnessed or heard of a discussion of menstruation at a religious event, yet there are religious myths on the topic. One myth from Vedic time stated that women have accepted the sin of the God Indra having killed a Brahmin (highest caste of Hindu). The blood is the manifestation of that sin. This assertion precipitated many questions for me on the nature of sin and karma. Another reason I have been told by a pundit, was that when performing *puja*, the body is being symbolically given to God. When menstruating, it is as though you are offering the blood to God, too, and blood offerings are considered 'low' forms of worship. Santanists strive for 'high,' 'clean,' or 'pure' worship, the *sattvic*.

How are these practices generally perceived? And how do you perceive them?

Rosa Freedman: People who follow these rules tend to see them as liberating in terms of women's bodies, fulfilling in terms of marriage and relationships, and just part and parcel of our everyday practices and the rhythm and cycle of our month. I see it from a feminist perspective—during and after menstruation my body is going through hormonal and physical changes, and the rules enable me to focus solely on my body's needs and not to feel that by doing so I am disregarding my husband's needs or desires. There are women who find these rules restrictive, or others who find it intrusive to be naked in front of the *mikveh* attendant. But most people I know find it liberating and/or fulfilling.

Lubabah Helwani: Some women view the breaks in religious practices, such as the daily prayers, as an *intifal* or celebration. However, many others do not view menstrual practices as positively. Personally, I perceive the religious practices as controlling, fear-based, and a misinterpretation of the religious texts. Many of the religious practices were interpreted, defined, and preached by men. Even if perceived by some as a break, I feel that these breaks have been forced upon us and mean that we cannot take part in our connection with God. However, there are some more current interpretations that differ from those I was taught as a child. These interpretations are new; they are still being explored and discussed.

Amina Darwish: I am very familiar with these new interpretations. Hearing these narrations and gaining new perspective from female scholars was a very empowering experience. I now feel more spiritual during my menses and will focus more on prayer and having deeper conversations with God than I did before. Also, being mindful of my body and my regular menses makes me a better Muslim.

Kalvi Karunanithy: I do not follow all the practices that my elders asked me to follow, I only follow the practices that do not affect me in any way. I respect my mother's feelings and so I don't go to temple during my periods. However, if it were necessary for me to go to temple, I would go for sure. Generally, I find that some practices are restrictive, like asking the girls not to go outside or controlling their freedom by forcing them to say at home without providing them with proper reasoning. Other practices are celebratory like the first menarche function, as the girls are given special attention. I would say the practices can have positive aspects as well as negative impacts.

Radha Paudel: In my case, these practices were deeply traumatic. I was shocked when I saw fresh blood on my mother's leg when I was seven years old. My mother awkwardly told me that menstrual blood is sin from God and girls are born with this sin or curse.

There are a few communities that celebrate menarche, but then there are other forms of discrimination associated with menstruation in the same communities. In these instances, these rituals may not truly be celebrations and are instead a way of informally letting it be known that girls are ready for marriage or child bearing. Generally, I find that there is no association with liberty and dignity. They keep saying that now you (as a girl) have to do this and that due to maturity. I do not see any form of liberation or spirituality in this view of menstruation.

Krystal Ghisyawan: When I was religious and interested in participating in rituals, I sometimes used menstruation as an excuse when I wanted a break from all the rituals and prayers we were made to do. It's not like anyone was going to check my panties to determine the heaviness of the flow. There also isn't a chart correlating the heaviness of the flow to the level of impurity, yet I remember being allowed (sometimes forced) to participate in rituals (such as a *puja* at our home) when the flow was lighter. I also found it paradoxical that I was instructed not to talk about my menstruation, yet I was in a context where this 'private' bodily act was restricting my participation in the public act of worship, making my menstrual status apparent. Ideally, menstruation was supposed to be a secret, not something to advertise or brandish about, because having my period marked my body as a maturing one, preparing for sex and reproduction, other 'secret' and 'private' acts. I use quotations, as there is a false dichotomy between public and private, and none of these processes occur in a singular space or context, neither the *pujas* nor the menstruation. For instance, daily worship at home (*sandhya*) is not regulated in the same way as participating in communal rituals, such as *pujas* or *yajnas*. *Ramleela* (a dramatic reenactment of the Hindu epic *Ramayana*, in Trinidad, performed in improvised open-air theatres in parks and sporting fields) is considered comparable to performing a *yajna*. When I 'played' *Ramleela*, I would typically be on the field every night acting various roles, yet almost every year I would get my period during those days. I would dress in regular clothes and stand on the sidelines, finding someone else to fill my role. I was asked rhetorically: "You're not taking part today?" I remember feeling like a spectacle–very self-conscious, but not ashamed. I did not believe there was anything wrong or tainted about my menstruation. Yet hearing the sentence repeated in response to my lack of participation, over and over throughout my life, I too would respond in the coded way, "I'm not clean." Under my clothing, no one could see what was taking place, yet the ways in which I dressed and used my body in the space of the temple and the *Ramleela*, were socially read and understood.

Jieun Choi: I can very much relate to the mixed messages about 'keeping private' and being 'publicly' celebrated or exposed. Generally, I consider the cultural attitudes around menstruation quite restrictive. When I had my first period, my parents congratulated me with a bouquet of flowers for the beginning of my womanhood. But I was extremely ashamed of it because by then, I learned—both culturally and socially—to be ashamed of it.

Alfred Muli: The perceptions of menstrual practices are quite diverse but they are generally seen as restrictions. Menstruators are viewed as unclean and needing to be separated from the rest of society. They are faced with restrictions of what they can do or cannot do. Menstruation is also seen as a sign of physical and emotional maturity and sometimes it is an indicator of readiness for marriage. This is in itself damaging.

What do you see as the benefits of these practices and traditions? What do you see as the risks?

Krystal Ghisyawan: For women who want to worship, who love *puja*, or who enjoy practicing Hindu rituals, the rules of ritual purity can be annoying and even alienating. How do I continue to love a tradition that tells me my body is impure and unworthy to be brought before God?

Radha Paudel: Personally, I do not see any benefit at all (though there are some who follow these restrictions who do think there are benefits). I see all of the risks and dangers posed throughout the lifecycle for girls and women. Because of more than 40 types of restrictions related to touch, food, and participation/mobility, women have suffered from nutritional deficiencies, reproductive, and mental health issues, been deprived or absent from educational and economic opportunities, lost their dignity, lost their peace, and lived with chronic humiliation, inferiority complexes and even suicidal thoughts. In west Nepal, some girls and women die due to snake bites, animal bites, accidental fires, suffocation from carbon monoxide, or extreme cold in secluded menstrual sheds. They may encounter rape, sexual abuse, or even murder. More importantly, the development of power dynamics between boys and girls in between the ages of six and eight years old are shaped by menstrual restrictions, which are learned from senior female members in the family and neighborhood. Boys grow up feeling a deep sense of being of super 'powerful' and girls grow up feeling humiliated, inferior, and powerless because they have to go through menstrual restrictions. As was the case for me, they start to see themselves as lowly, accept anything and ask no questions. Meanwhile, boys start to exercise their power including abuse towards those they see as less powerful. Thus, gender-based violence including child marriage and sexual assault and rape are directly related to menstrual restrictions within many communities in Nepal. I strongly believe that the menstrual restrictions are one of the underlying causes of conflict and human rights violations.

Lubabah Helwani: I can see the benefits of clearly stating to society that a woman's place in society is a special one, in contrast to what we hear in our everyday life. When young women receive their periods for the first time, they view it as being 'let in on the secret' of being a woman.

Yet, at the same time this 'secret' can be perceived as embarrassing because when they start wearing the hijab, it generally signals to the entire community that they have received their period.

Amina Darwish: For me, the main benefit of our practices is viewing menstruation as Divinely ordained self-care. It also asks women to be mindful of our bodies. This can help women be healthier overall.

Lina Mathew: Men in traditional Hindu families in Kerala are equipped to carry on household chores, and not to complain when women cannot perform their usual household activities. The benefit of these traditions is that women get to rest during these days, especially if they are bleeding profusely and feel faint and tired. The custom of making women sleep separately is perceived to prevent men from having sexual intercourse with their wives, as menstrual blood is considered impure and sexual relations are considered taboo during this period. Yet, at the same time, women may consider separation practices demeaning which may make them hate womanhood.

Rosa Freedman: For me, the biggest benefits are the fact that my husband and I have 12 days per month where we communicate through words rather than touch, which means we talk about things (particularly problems) in detail. We also get excited each month about kissing or touching, let alone sexual relations, because we cannot have those things all of the time. I think this has kept our marriage strong and means we do not take one another for granted.

Who maintains these practices? What, if any, role do men play in these practices?

Rosa Freedman: It is the responsibility of the woman to check for uterine blood, but of both the woman and man obey the rules. Those include not being naked in front of one another when the woman is a *niddah* (menstruating or in the period before the *mikveh*) and not touching each other. My husband and I tend to talk to one another about it, but some couples are more reserved. As to societal pressure—no-one really knows who sticks to the rules, but there is a general expectation that most people do the bare minimum of not having sex during menstruation and the woman immersing in the ritual baths afterwards.

Lubabah Helwani: Menstruation and its practices and traditions are one of the few aspects that many Middle Eastern women are in control of in their daily lives. Men have no role in the practices and traditions of menstruation. In fact, they try to avoid all conversations that might allude to menstruation.

Radha Paudel: In Nepal, girls learn from friends, mothers, sisters, religious people and activities, both in informal ways and in grounded, deeper ways. The society, especially men as faith healers, teachers, health workers, and political leaders, create fear and anxiety around menstruation by saying that it is dirty and negatively linked with God. They blame menstruating girls and women if something happens to the family or community. No one likes to challenge these restrictions and they simply keep following for the sake of men, God, formality, or power. Men rarely ask women to follow the restrictions publicly, but it is passed down from wives or mothers for the sake of men. When we want to transform these practices, we have to acknowledge that men possess most of the power. Thus, men also have a key role in abolishing restrictions because men are part of the problem. In my experience, the transformative process is easier and faster where boys and men engage with it.

Alfred Muli: Yes, men have a role in maintaining menstrual practices. They tend to 'discipline' their wives or daughters in case they go against any of the traditions. This is mainly in protecting the family name as there would be repercussions in the event that one of the taboos were broken. For instance, it's taboo that anyone sees your menstrual blood. If it happens, you are thought to become barren. In most societies, being barren was seen as a curse, something with which no one would want to be associated. There is pressure to maintain such traditions given their link to other aspects of life that the society deems important such as food scarcity, marriage, or even avoiding upsetting the gods.

Krystal Ghisyawan: The majority of Hindu religious leaders are male, and so have never had to actually confront what it is like to menstruate or to have menstruation interrupt their religious practice. Yet, they play an important role in influencing how women experience their religious practice. Some younger male pundits who I have spoken with thought that practices related to menstruation are not relevant, as women and girls had no control over this part of their anatomy. They emphasized intentionality, that being pure was a matter of cleansing the mind and body.

In your communities, is there social pressure to maintain these practices? Is that changing?

Jieun Choi: The society at large maintains these practices. But amid the active feminist movement in South Korea in the recent years, more and more young women are claiming their rights. More women are informed about alternative menstrual products other than menstrual pads, which used to be a fixture among South Korean women. But this is not without dissenters, especially some men who think that using alternative products like tampons or menstrual cups would permanently affect the shape of women's genitalia.

Krystal Ghisyawan: Although within Hindu religious practice in Trinidad it is commonly accepted that menstruation is 'defiling,' 'impure,' and 'unsuited' for worship, observing these rules ultimately falls to women. Menstruation can be a very private act, which gives women and girls the opportunity to choose whether or not they wish to continue the tradition, since they can participate without anyone even knowing they are menstruating. This can give one the power to practice as they see fit, without leaving room for anyone to intervene or interfere. But as religious leaders and elder women reinforce these rules and rituals for younger girls, the practices persist, and girls often learn to regard their own bodies as tarnished and unclean. They may sacrifice their own desires (to participate in ritual) out of respect for this collective belief. What would happen if a girl or woman knowingly participated in ritual while they were having their period? What would happen if someone found out? The potential shaming could also be a deterrent. Ultimately, women police themselves, and in Trinidad and Tobago, most are equipped to resist the practices if they wish to do so, but the societal pressure to conform to traditional practices may be too strong a deterrent to outright rebellion.

Lina Mathew: The concept of menstruation has currently become a very sensitive political issue in Kerala. Recently, a case came up before the Supreme Court of India urging for a ban on the practice of prohibiting women between the ages of 10 and 50 from entering the Sabarimala Ayyappa temple in Kerala. The petitioners argued that this 'menstrual discrimination' leads to social stigma and shame based on gender. It was contended that a woman of menstrual age cannot be treated as 'polluted and untouchable.' In its judgment in *Indian Young Lawyers Association and Others* v. *State of Kerala and Others* delivered on September 28, 2018, the Supreme Court of India stated that the practice of banning women between the ages of 10 and 50 from entering the Sabarimala Ayyappa temple is discriminatory and the practice violates the rights of Hindu women (Supreme Court of India 2018). The Court stated that religion cannot be used as a cover to deny rights of worship to women. However, in the months ensuing the Supreme Court judgment, protests were organized by various groups to prevent women of menstrual age from entering the temple. On January 1, 2019 a women's wall, stretching from one end of the state to the other end, was organized by the ruling party of the Kerala Government pledging solidarity to women's rights and standing up against the perception of menstrual impurity. On January 2, 2019 two Hindu women entered the temple and offered prayers. Riots broke out the next day. However, these women have received support for braving the odds and standing up for women's rights. This case demonstrates that menstruation is private and public at the same time. It also demonstrates that it has deeply religious meaning that has a political dimension deemed important enough to spark large-scale protests. Our understanding of menstruation and the practices associated with it go to the core of our understanding of societal norms about gender roles. Active debates regarding the question of purity or impurity of menstruation can alone further dialogue on women's entry into places of worship and change of societal attitudes regarding practices of menstruation.

REFERENCE

Supreme Court of India. 2018. "Indian Young Lawyers Association v. The State of Kerala on 28 September, 2018." https://indiankanoon.org/doc/163639357/.

PART II

Menstruation as Embodied

CHAPTER 16

Introduction: Menstruation as Embodied

Tomi-Ann Roberts

This section begins where the subfield itself began—at the site of the human body. Of course we experience menstruation in the body, which is always already embedded in particular interactional and sociocultural discourses. The chapters in this section reflect on the power of institutions to subjugate and discipline bodies to probe the many ways the menstrual cycle becomes a site of sexualization, self-objectification, and abjection, of shame and shaming, of medicalization, disability and dysfunction, and even a source of moral panic. Thus the embodied experience of menstruation, from menarche to menopause, is rarely cause for celebration or even contentment, and is instead typically a project to manage properly as an essential component of "doing (feminine) gender." The most important of these, argues Jill Wood here, is concealment, and the menstrual hygiene industry steps into provide the tools necessary for this self-disciplining body project.

Research on perceptions of menstruating women reveal the widespread belief that the reproductive body has the power to cause disgust-like reactions in others, to alter women's personalities, making them "crazy," and that menstrual blood itself is a stigmatizing "mark" (see Johnston-Robledo and Chrisler). As well, the premenstrual phase of the cycle is widely associated with emotional lability, impulsivity, and irrationality. Sally King writes of how the "myth of the irrational female" has led to a prioritizing of psychological symptoms in diagnosis, treatment, and research on PMS, despite the fact that these are neither the most commonly experienced nor most disruptive of menstrual-related changes experienced by women.

Indeed Jane Ussher and Janette Perz describe how many women take up the position of the monstrous feminine (emotionally out of control, animal-like, and fat) in their own descriptions of their premenstrual embodiment. And yet, these self-conceptions need not only position the body as abject; they can also be a form of agentic subjectivity. Ussher and Perz find

© The Author(s) 2020
C. Bobel et al. (eds.), *The Palgrave Handbook of Critical Menstruation Studies*, https://doi.org/10.1007/978-981-15-0614-7_16

clues in women's PMS narratives that allow them to acknowledge the complexities in adopting the subject position "PMS sufferer," which *both* evokes connotations of the monstrous feminine *and* makes meaning of women's physical and emotional distress, through legitimizing their experiences as real and requiring support.

Centuries-old beliefs such as the wandering womb that causes hysteria and distress continue to be reflected in popular discourse, leading to real consequences not only for how we understand healthy menstruation, but also for identifying and treating menstruation-related disorders. Heather Guidone discusses how endometriosis affects an estimated 176 million people worldwide, causing pain and reduced quality of life, and yet this condition continues to be dismissed by both patients and practitioners alike. When symptoms are poorly understood, lengthy delays in diagnosis result, and the negative impact of endometriosis is exacerbated further.

Three chapters here discuss what all this mystification means for girls' embodiment as they mature into menstruating beings. Bobier, Piran, and Stubbs and Sterling each reveal the ways strongly enforced discourses of femininity, as well as girls' own self-objectification and self-sexualization "corset" (as Piran puts it) the way they can inhabit their bodies. Girls link menarche and sexuality and yet lack language to make this connection a source of embodied self-understanding, power, or enjoyment. Bobier's teen interviewees spoke of wishing to be "good girls" (that is, asexual), however, their fears of rape and pregnancy revealed their awareness that they are not fully in control of the way their developing bodies are seen or treated. Piran argues that the possibilities for positive embodiment at menarche hinge on girls' relational connections and Stubbs and Sterling offer concrete suggestions for menstrual education as an opportunity to provide girls with more accurate and positive views of their genitals and sexual selves.

The transnational engagement that concludes this section amplifies the voices of those whose menstruating bodies exist in fraught conditions. We hear from a formerly incarcerated woman, a woman with disabilities, and a woman who lives and works in disaster-prone regions. In each of these cases popular simple solutions to menstrual management ("let's give everyone free tampons and pads!") will not fly, either because they are unusable by some bodies or under certain conditions, but even more so because these interventions leave stigma—which these women face not only as menstruators but also as marginal—unaddressed. Their stories are a powerful reminder of the need for equitable and inclusive menstrual policies, diverse product designs, and above all, a framing of the menstruating body not as a site of humiliation and degradation, but rather of care.

Of course in matters of embodiment related to menstruation in a misogynist culture, one is damned if she does *and* damned if she doesn't. So even menopause is typically constructed not as a normal, healthy developmental transition, but instead as an illness. As I experience menopause myself, and fumble my way through such challenging body betrayals as heart palpitations and dizziness, wanting answers and not finding any, I take tremendous heart in Dillaway's closing words in her chapter: "Thinking about ourselves as living *in* this uncertain time, rather than just getting *through* it, may be the first step to understanding and owning the impact of menopause and reproductive aging."

The menstrual body is a complex reality that refuses simplification. It is obfuscated and degraded by misogynist assumptions that worm their way into our very own embodied experiences of our reproductive bodies. But I find hope in each of these chapters, all of which are saying, one way or another, that we deserve to *take care of* our menstruating and menopausal bodies, that we have a right to unburden ourselves from the project of concealing, medicating, or enduring pain or debasement of our bodies, and instead to take them up ourselves, whoever and wherever we are, gently, as our own.

CHAPTER 17

The Menstrual Mark: Menstruation as Social Stigma

Ingrid Johnston-Robledo and Joan C. Chrisler

Introduction[1]

The American artist Vanessa Tiegs (http://menstrala.blogspot.com) and the German artist Petra Paul (http://mum.org/armenpau.htm) are known for collecting their menstrual flow. When they have collected enough, they sprinkle, splash, and brush their blood across their canvases to create beautiful, and intriguing, works of art. Reactions to their work include shock at their audacity, amazement at their creativity, and disgust at their willingness to exhibit one of nature's most stigmatized fluids (www.truenuff.com/forums/show-thread.php?135-Menstrual-Art-by-Vanessa-Tiegs&p=1371&viewfull=1). One journalist (Heath 2007) wondered whether Tiegs' work should more properly be called art or a biohazard. Contemporary artists often aim to shock viewers (Stallabrass 2006), but these artists have a greater goal in mind (Chesler 2006; Cochrane 2009). They seem to want us to ask ourselves why a mundane product of nature is so shocking, given that most women experience the menses and manage their own menstrual flow for decades of their lives. They want us to consider why menstruation, a benign process essential to the production of human life, evokes fear, disgust, and comparison to toxic waste. We believe that viewers of Tiegs' and Paul's art react the way they do because menstrual blood is a stigmatized substance. In this theoretical paper, we review feminist scholarship regarding the attitudes and experiences of predominantly American girls and women to build the argument that menstruation is a source of social stigma for women. All studies cited in this article were conducted with American samples unless otherwise stated.

© The Author(s) 2020
C. Bobel et al. (eds.), *The Palgrave Handbook of Critical Menstruation Studies*, https://doi.org/10.1007/978-981-15-0614-7_17

WHAT IS STIGMA?

According to Goffman (1963), the word stigma refers to any stain or mark that sets some people apart from others; it conveys the information that those people have a defect of body or of character that spoils their appearance or identity. The word derives from a practice of the ancient Greeks, who branded criminals and slaves to mark their status. People reacted with disgust when they saw the brands associated with thieves or traitors, and citizens avoided interacting socially with criminals and slaves (Goffman 1963). Goffman (1963, 4) categorized stigmas into three types: "abominations of the body" (for example, burns, scars, deformities), "blemishes of individual character" (for example, criminality, addictions), and "tribal" identities or social markers associated with marginalized groups (for example, gender, race, sexual orientation, nationality). Social psychologists have conducted empirical studies of stigmatized conditions to determine which aspects of those conditions are most abhorrent to other people. The key dimensions are: peril (that is, the perceived danger to others; for example, HIV+ individuals), visibility (that is, the obviousness of the mark; for example, facial disfigurement), and controllability (that is, how responsible the individual is for the condition, such as whether the mark is congenital, accidental, or intentional; for example, obesity due to a medical condition or treatment vs. obesity due to "letting oneself go") (Crocker, Major, and Steele 1998; Deaux et al. 1995; Frable 1993). People's beliefs about the controllability of a stigmatized condition (for example, homosexuality) are important because they affect how much stigmatized people are disliked and rejected (Dovidio, Major, and Crocker 2000). For example, lesbians and gay men are better liked and more accepted by people who believe that sexual orientation is biologically based rather than freely chosen (Herek 2009).

MENSTRUATION AS A STIGMATIZED CONDITION

We argue that menstrual blood is a stigmatizing mark that fits all three of Goffman's categories. Menstrual rituals and hygiene practices imply that, like other bodily fluids (Rozin and Fallon 1987), menstrual blood is considered an abomination. Some have argued that menstrual blood is viewed as more disgusting or aversive than other bodily fluids such as breastmilk (Bramwell 2001) and semen (Goldenberg and Roberts 2004). In some cultures women are believed to be unclean during their menstrual periods, and they must take a ritual bath (for example, the Jewish Mikvah) to purify themselves before they can be intimate with a man (Cicurel 2000; Goldenberg and Roberts 2004). Given aversions to menstrual blood, a stain may be viewed as a blemish on one's character. From a content analysis of advertisements in Australian magazines, Raftos, Jackson and Mannix (1998) concluded that a powerful message these ads send to readers is that leaks of menstrual blood taint women's femininity because, through the proper choice of products, she should have kept the evidence of her menses out of sight. Lee (1994) found that 75% of the young women she interviewed had experienced or were afraid of experiencing leaks during menstruation. She concluded that visible signs of menstruation represent emblems of girls' contamination (Lee 1994). Roberts et al. (2002) were able to demonstrate empirically that even reminders of menstrual blood (for example, tampons) can lead to avoidance and social distancing, which suggests that menstrual blood may serve as a blemish on women's character. Because only girls and women menstruate, menstrual blood also marks a tribal identity of femaleness. When girls reach menarche (that is, experience their first menstruation), parents and others treat them differently than they did before (Lee and Sasser-Coen 1996). Post-menarcheal girls are cautioned about sexuality, told that they are now "grown-up," and urged to act "ladylike" in ways that restrict the freedom of behavior they had enjoyed in the past (Lee and Sasser-Coen 1996). Thus menstruation marks girls and women as different from the normative and privileged male body (Young 2005). Furthermore, if people hold cultural beliefs that the menstrual cycle causes women to be physically (menstrual phase) or mentally (premenstrual phase) disordered, then the stigma of menstruation also marks women as ill, disabled, out-of-control, unfeminine, or even crazy (Chrisler 2008; Chrisler and Caplan 2002).

Menstrual blood also reflects several of the key dimensions of a stigmatized condition. For example, it has been considered perilous—both magical and poisonous (Golub 1992). Many anthropologists have theorized about the origins and purposes of this symbolism, but, according to Buckley and Gottlieb (1988), there are few firmly established anthropological theories about why menstrual blood may have been viewed that way. Perhaps menstruation seemed magical because, before the physiology of the menstrual cycle was understood, individuals did not understand how women who were not wounded could bleed for five days without being seriously weakened or killed. Because men did not experience menses themselves, they must have been afraid of it, perhaps worried that close contact with menstrual blood might do them some physical damage or pollute them by its association with the mysterious female body. Thus, menstruation may have seemed poisonous.

These ideas are not to be dismissed as naïve or primitive; remnants of them persisted into modern times and remain present today. Cultural feminists (for example, Owen 1993; Stepanich 1992; Wind 1995) who advocate the celebration of menstruation with praise to the Moon Goddess continue the idea that menses and magic are connected. As late as the 1920s and 1930s, scientists (see Delaney, Lupton, and Toth 1987) were attempting to demonstrate that menstruating women exuded what were called menotoxins (that is, poisonous elements) in their menstrual blood, perspiration, saliva, urine, and tears. Images in popular culture of premenstrual women as out-of-control and likely at any moment to be verbally abusive or violent reinforce the ancient notion that menstruation constitutes a peril. In the 1990s, in his infamous "giraffe hunting" speech, Congressman Newt Gingrich commented that female soldiers do not belong in the trenches during times when they are highly susceptible to infections. His remarks imply that menstruating women somehow poison themselves and weaken their immune systems, but perhaps what really worried him is the idea that premenstrual American female soldiers might be even more dangerous than the enemy to their male comrades (Chrisler and Caplan 2002).

We assert that menstruation is more like a hidden than a visible stigma, but that is because women go to a great deal of effort to conceal it (Oxley 1998). Menstrual hygiene products (for example, tampons, pads) are designed to absorb fluid and odors, not to be visible through one's clothes, to be small enough to carry unobtrusively in one's purse, and to be discretely discarded in a bathroom container (Kissling 2006). It is usually not possible to know for certain that a woman is menstruating unless she says so . . . or unless menstrual blood leaks through her clothes and exposes her then stigmatized condition.

Until recently, menstruation was not controllable. The menstrual cycle is a force of nature; hormone levels ebb and flow in a regular (or irregular) rhythm. Unless women had an illness (for example, anorexia nervosa, polycystic ovary syndrome) or a temporary condition (for example, pregnancy, lactation, the low level of body fat frequently seen in long-distance runners) that halted the menstrual cycle, they could expect to menstruate at a time determined by their particular cycle. With the advent of oral contraceptives in the 1960s, however, scientists proved that menstruation could be controlled. Women have traditionally taken oral contraceptives daily for three weeks, then not taken pills for one week in order to allow for a form of "break-through bleeding" that resembles normal menstruation. However, in recent years continuous oral contraceptives have been marketed to women as a way to avoid menstruation altogether (Johnston-Robledo, Barnack, and Wares 2006). The ads suggest that women have the "freedom" to make a "choice" about whether to menstruate (Johnston-Robledo et al. 2003). However, against a back-drop of cultural messages that women should always be available (for example, to the men and children in their lives) and should avoid, if at all possible, anything that might discomfit others (Chrisler 2008), we might soon reach the point where most people believe that women should eliminate their menstrual cycles unless they are actively trying to become pregnant. This may increase the stigma attached to those who continue to menstruate regularly.

Transmission of Stigma of Menstruation

Most of the people who react with shock to Tiegs' and Paul's art have probably never been told that menstruation is a stigmatized condition, but their reactions suggest that they "know" it. The stigma of menstruation is conveyed to us everyday through a variety of sociocultural routes. For example, negative attitudes toward menstruation and cultural beliefs about menstruating and premenstrual women are transmitted through products and media (for example, advertisements, magazine articles, books, television) we see everyday (Chrisler 2008; Erchull 2010).

Advertisements are cultural artifacts that play an important role in the social construction of meaning (Merskin 1999). Ads for menstrual products have contributed to the communication taboo by emphasizing secrecy, avoidance of embarrassment, and freshness (Coutts and Berg 1993; Delaney, Lupton, and Toth 1987; Houppert 1999; Merskin 1999). Allegorical images, such as flowers and hearts, and blue rather than reddish liquid, have been used euphemistically to promote secrecy and delicacy (Merskin 1999). Ads play on women's fear of being discovered as menstruating because discovery means stigma (Coutts and Berg 1993). With the invention of panty-liners, advertisers began to tell women to use their products every day so that they can feel "confident" that they will always be "fresh" and untainted (Berg and Coutts 1994). When Oxley (1998) questioned 55 British women about their experiences with menstruation, she found that they echoed many of the themes in the ads. They felt self-conscious during the menses, preferred tampons because they are "less noticeable" than pads, believed that menstrual blood is distasteful to self and others, and supported the sex taboo.

Advertisements are not the only form of public discourse about menstruation. Attitudes are also conveyed through books, magazines and newspaper articles, jokes, and other cultural artifacts, such as "humorous" products like greeting cards and refrigerator magnets (Chrisler 2007, 2008). Most of the attitudes these media convey are negative, and together they have constructed a stereotype of menstruating women, especially premenstrual women, as violent, irrational, emotionally labile, out-of-control, and physically or mentally ill. We have seen bumper stickers (for example, "A woman with PMS and ESP is a bitch who knows everything"), buttons (for example, "It's not PMS, I'm always psychotic"), magnets (for example, "Be very careful: I have PMS and a gun"), cartoons, greeting cards, and books (for example, Raging Hormones: The Unofficial PMS Survival Guide, the cover of which pictures actress Joan Crawford as an axe murderer; Chrisler 2002). If this is what people think about women who are menstruating (or about to menstruate), it's no surprise that women try to conceal this stigmatized condition.

The stigmatized status of menstruation may also be transmitted through the educational booklets produced by sanitary napkin and tampon manufacturers; these booklets typically are written by nurses or health educators employed by the companies. We (Erchull et al. 2002) conducted a content analysis of 28 of these booklets, which were published between 1932 and 1997, and we found that the booklets placed much more emphasis on negative than on positive aspects of menstruation. Cramps, moodiness, and leaks were all mentioned frequently, but growing up was the only positive aspect mentioned. Descriptions of the menstrual cycle were kept vague for the most part. Estrogen and progesterone were mentioned in fewer than one-half of the booklets. Even the terms menstruation and ovulation were not used in every booklet, and only one booklet (produced by Planned Parenthood, not by a manufacturer) actually included the word menarche (the term for a girl's first menstrual period). The illustrations were also problematic. A few of the booklets did not show any external genitalia, and the diagrams of the female reproductive organs often were presented separately from any bodily reference or body outline, which makes it difficult for a girl to imagine the scale of the system if she does not know where it is located. These booklets are used to educate, but girls who read them might learn more about stigma than about their physiology. One booklet stated out-right that "your main concern will probably be avoiding accidents with an appropriately absorbent pad, avoiding a wet feeling, and using a pad that doesn't show." The emphasis on secrecy and the potential for embarrassment is present in all of the booklets, and this emphasis may contribute to negative attitudes toward menstruation (Hoerster, Chrisler, and Gorman 2003).

Finally, menstrual stigma is perpetuated indirectly through silence. Menstruation is typically avoided in conversation (Kissling 1996), except under certain circumstances (for example, in private with female friends and relatives, in a health education or biology class, in a doctor's office). The majority of American adults surveyed for The Tampax Report (1981) agreed that menstruation should not be discussed in "mixed company," and many thought that it should not be discussed with the family at home. Williams (1983) found that 33% of the adolescent girls she surveyed would not talk about menstruation with their fathers, and nearly all of her participants agreed that girls should not discuss menstruation around boys. Even psychotherapists have reported experiencing discomfort when their clients want to discuss some aspect of menstruation (Rhinehart 1989). When teachers separate girls and boys to view films about puberty, and when mothers arrange one-to-one, private, "facts of life" talks with their daughters, they are conveying not only facts but guidelines for communication; they are marking menstruation "as a special topic, not one for ordinary conversation" (Kissling 1996, 495). Exclusive talks held in private convey the notion that menstruation is an embarrassing event that must be concealed from others and never discussed openly.

The communication taboo is supported by the existence of dozens of euphemisms for menstruation (Ernster 1975; Golub 1992), and these euphemisms can be found in cultures around the world. Ernster (1975) examined a collection of American expressions in the Folklore Archives at UC-Berkeley, and she grouped them into categories. For example, some refer to female visitors (for example, "My friend is here," 6), others to cyclicity (for example, "It's that time again," "my time of the month/moon," "my period," 6), illness or distress (for example, "the curse," "the misery," "I'm under the weather," 6), nature (for example, "flowers," "Mother Nature's gift," 7), redness or blood (for example, "I'm wearing red shoes today," "red plague," "red moon," "bloody scourge," 6–7), or menstrual products (for example, "on the rag," "riding the cotton pony," "using mouse mattresses," 6). Some of these euphemisms are still in common use today (Chrisler 2011), and new ones have no doubt been invented. If menstrual blood were not stigmatized, there would be no reason to call it anything other than its formal name: menstruation or the menses.

Although feminist scholars and activists (for example, Owen 1993; Stepanich 1992; Taylor 2003; Wind 1995; see also Bobel 2006, 2010) have tried to promote the celebration of menarche and menstruation, their positive messages may be overshadowed by the stigmatizing messages. Even those women and girls who do internalize the positive messages may find themselves confused about how to celebrate something that is supposed to be hidden. Their concerns about the consequences of doing so may be well-founded.

Consequences of Stigma of Menstruation

The stigma of menstruation has negative consequences for women's health, sexuality, well-being, and social status. One of the consequences most frequently noted in the literature is self-consciousness and hypervigilance associated with concerns about the revelation of one's menstrual status. Oxley (1998) found that both undergraduate women and women employed in the medical professions reported high levels of self-consciousness during menstruation. The behaviors they engaged in, and activities they avoided, reflected their determination to hide their menstrual status from others. For example, they wore baggy clothes and preferred tampons over pads. They avoided swimming and sexual activities during menstruation, often because of their concern about how others would respond to their menstrual blood. The researcher concluded that women might feel unattractive during their menses because menstrual cycle effects (for example, bloating, acne) indicate that they have been betrayed by their bodies. She argued that, in order for women to accept themselves every day of the month, cultures must change the way menstruation is viewed, and women themselves must take more control over the way they experience and feel about menstruation. In other words, women must resist, and cultures must reduce, the stigma.

The self-monitoring that women do to be sure that they look their best and that their menstrual status is hidden is related to the Foucauldian concept of self-policing (Foucault 1979). In a study of women who met criteria for severe PMS, Ussher (2004) found that women understood, experienced, and interpreted PMS symptoms as violations of the norms for "appropriate" femininity (for example, resisting the need to nurture others at one's own expense, displaying anger or annoyance one would usually conceal, experiencing one's body as unruly or out-of-control). Ussher argued that women's tendency to pathologize premenstrual experiences and to apply the PMS label to themselves represents a form of behavioral self-policing that allows them to distance themselves from their embodied selves in an effort to retain their femininity. Lapses in self-policing such as choosing to say "no" to others can then be blamed on the body rather than on the woman's own desires.

Objectification theory (Fredrickson and Roberts 1997) may help to explain why certain women are self-conscious about menstruation and go to unusual lengths to hide or eliminate their periods. Sexual objectification occurs when a woman feels that she is separate from, or represented by, parts of her body that are deemed sexual, such as her breasts and buttocks (Bartky 1990). In a culture where women's bodies routinely are sexually objectified, women themselves can internalize the sexual objectification of their bodies and view themselves through the lens of a critical male gaze. This self-objectification may lead women to monitor themselves constantly and to alter their self-presentation accordingly. Looking at the self this way has negative implications for psychological and sexual well-being (Muehlenkamp and Saris-Baglama 2002; Szymanski and Henning 2007; Tylka and Hill 2004). Goldenberg and Roberts (2004) have applied principles of terror management theory (Greenberg, Pyszczynski, and Solomon 1986) to explain pervasive negative attitudes toward menstruation. They argued that menstruation and other reproductive functions serves as reminders of the creaturely and therefore mortal nature of humans and women's proximity to nature. In an effort to allay existential angst about mortality, women may distance themselves from menstruation by adhering to cultural beauty standards. Both of these theories shed light on explanations for women's self-consciousness during menstruation and the social stigma attached to menstruation.

Feminist researchers have begun to consider the impact of self-objectification on attitudes toward menstruation, a bodily function that is incompatible with the view of the body as a sex object or as sexually available to others. Women who tend to self-objectify have been found to have particularly negative attitudes toward menstruation (Johnston-Robledo et al. 2007; Roberts 2004). Undergraduate women with higher self-objectification tendencies also have said that they would prefer not to have menstrual cycles (Johnston-Robledo et al. 2003) and reported positive attitudes toward the elimination of menstruation through the use of continuous oral contraception (that is, menstrual suppression; Johnston-Robledo et al. 2007). Thus, self-objectification may lead women to maintain a sense of global shame about multiple reproductive events, including menstruation, birthing, and breastfeeding (Johnston-Robledo et al. 2007). The shame and lowered self-esteem is psychologically damaging and may lead women to make reproductive decisions (for example, menstrual suppression, elective cesarean section, high-risk sexual behavior) that could have negative ramifications for their physical health (Andrist 2008; Johnston-Robledo et al. 2007; Kowalski and Chapple 2000; Schooler et al. 2005).

Another consequence of menstrual stigma is observance of the sex taboo, that is, avoidance of intimate sexual relations during the menses. In a study of Latinas/os' sexual behavior during menstruation, the vast majority of women sampled reported that they avoided genital touching, oral sex, and sexual intercourse during menstrual bleeding; the men also reported that they avoided such activities with menstruating sexual partners (Davis et al. 2002). Why should women be bound by ancient fears about the uncleanliness of menstrual blood? Menstruation is a good time to have sex if the partners want to avoid pregnancy, and orgasm is said to relieve menstrual cramps (Boston Women's Health Book Collective 2005). Tanfer and Aral (1996) reported that women who had more lifetime sexual partners and more frequent sex were more likely to have sex during their menses than were women with fewer partners or less frequent sexual encounters. European American women were more likely than African American and Latin American women to say that they had had sex during their menses. Rempel and Baumgartner (2003) found that women who viewed menstruation as a normal and publicly acceptable event scored higher on a measure of personal comfort with sexuality and were more likely to have sexual relations during their periods than women who did not have such positive attitudes toward menstruation. On the contrary, Schooler et al. (2005) found that female undergraduate students who had feelings of shame regarding menstruation reported less sexual experience and more sexual risk-taking than did those who scored low on a measure of menstrual shame.

Finally, we believe that the stigma and taboo of menstruation both reflects and contributes to women's lower social status. In her classic, playful essay Gloria Steinem (1978) imagined that, if men could menstruate, menstruation would become an enviable, boastworthy, masculine event. She suggested, for example, that "sanitary supplies would be federally funded and free" (110). Her essay helps readers to understand that menstruation, as a biological, cultural, and political phenomenon, is only a "problem" because women do it.

Forbes et al. (2003) found that both male and female college students rated a menstruating woman as less sexy, more impure, and more irritable than women in general. Marván et al. (2008) asked college students in the U.S. and Mexico to list words that came to mind when they read the statements "A menstruating woman is . . . " and "A premenstrual woman is . . ." Only words that were mentioned by at least 50% of the 349 students were included in the analysis. Participants listed 92 negative words, which were grouped into the following categories: negative affect (for example, sad, frustrated), inactivity (for example, tired, weak), annoyance (for example, desperate, whining), instability (for example, unpredictable, moody), limitation/rejection (for example, incapable, unlovable), and physical symptoms (for example, crampy, bloated). In contrast, they could think of only 55 neutral words (for example, cyclical, using pads) and 33 positive words (for example, active, beautiful). Despite the stigma, 50% of the participants thought that women are active and beautiful even at "that time of the month."

Kowalski and Chapple (2000) investigated the consequences of the social stigma of menstruation on women's impression management behavior. They assigned young undergraduate women to be "interviewed" by a male confederate. Fifty percent of these women were menstruating at the time; the others were not. The male "interviewer" was aware of the menstrual status of 50% of the women in each group, and unaware of the menstrual status of the others. The menstruating participants interviewed by the man who was aware of their menstrual status believed that they had made a more negative impression on him than the women in the other three groups thought they had. They were also less concerned about making a positive impression on him than were the women in the other groups.

Roberts et al. (2002) primed menstrual status by manipulating whether their research assistant dropped a tampon or a hairclip where the participants in the study could see her do it. Both male and female undergraduate participants in the tampon condition later rated the research assistant as less competent and likeable than did the participants in the hairclip condition. Those who saw her drop the tampon also exhibited a tendency to sit farther away from her during the data collection than did those who saw her drop the hairclip. Results of this research show that the old ideas about stigma, taboo, and pollution are still operative. This work suggests that ruptures in women's concealment of their menstrual status lead to both social distancing and negative perceptions.

Clearly, more research is needed on how women's menstrual status may impact the way other people interact with and perceive them. However, it appears that women's desire and efforts to conceal their menstrual status may be well-founded. It would be interesting to study how people respond to women who actively subvert the cultural norm that menstruation should be hidden (for example, by discussing it openly or by washing out an alternative menstrual product, such as "the keeper," in a public restroom).

The self-monitoring for leaks and odors and the self-policing of behavioral or emotional clues to menstrual status is a waste of women's time and psychic energy that could be spent on more important or interesting pursuits. Young (2005) argued that menstruation is a source of oppression for women because of the shame attached to monthly bleeding and the challenges women face as menstruators in public spaces (such as work and school), and she argued that menstruation renders women "queer" in a society that identifies the male non-menstruator as the "normal" human. She suggested that menstruating women are, in effect, "in the closet" about their stigmatized menstrual status. "Social relations of somatophobia and misogyny continue to hold over women, in some circumstances, the threat of being 'outed' as menstruators, sometimes with serious consequences to their self-esteem or opportunities for benefits" (Young 2005, 113). Menstrual etiquette requires women to conceal their menstrual flow and to remain in the menstrual closet if they want to occupy public spaces along with men and nonmenstruating women (Laws 1990; Young 2005). But etiquette, like stigmatized conditions, depends on social, cultural, and historical context, and contexts can change.

CHALLENGING/RESISTING THE STIGMA OF MENSTRUATION

What would happen if more women like Vanessa Tiegs and Petra Paul were willing to violate cultural norms? We are not suggesting that the menstrual cycle should be romanticized, that all women should celebrate every menses they experience, or that menstruation is central to womanhood or femininity. However, we do believe that the stigma of menstruation limits women's behavior and compromises their well-being. There are many different ways to eliminate the stigma, an important step toward "menstrual justice" (Kissling 2006, 126).

Culpepper (1992) suggested that simply talking about menstruation can create more positive attitudes, and she designed workshops aimed at raising women's "menstrual consciousness" to facilitate these conversations. Issues girls and women discussed in her workshops included names for menstruation, attitudes toward and stories about menstruation, and customs surrounding menstruation. If menstruation were discussed more openly, it might be easier for girls and women to acknowledge the positive aspects of menstruation and to challenge others when they make assumptions that all women hate and want to eliminate their periods. When researchers bother to ask, women are forthcoming about positive aspects of menstruation (Chrisler et al. 1994; Johnston-Robledo et al. 2003) and express concerns about eliminating monthly bleeding through continuous oral contraceptive use (Johnston-Robledo et al. 2003; Rose, Chrisler, and Couture 2008).

There is some evidence to suggest that adolescent girls are attempting to resist and challenge traditional norms about menstruation through social interactions that take place online among peers. Polak (2006) explored chat rooms, message boards, websites, and individual girls' homepages to learn more about the ways adolescent girls, or "gURLs," are constructing and experiencing menstruation. Her observations indicate that girls are "rewriting" the dominant negative menstruation narrative that was transmitted by both the feminine hygiene product industry and adults in their lives, such as their mothers and grandmothers.

Polak found that American adolescent girls use online spaces to talk frankly and openly about menstruation. For example, they answered each other's questions, validated each other's experiences, and encouraged one another to talk to their boyfriends about menstruation. Polak noted an absence of euphemisms and even some open dialogue about extremely stigmatized aspects of menstruation, such as the various colors and consistency of menstrual blood. She argued that these new online conversations about menstruation could facilitate girls' identity development and healthy embodiment. Although Polak acknowledged that this forum may leave out girls who do not have immediate access to computers, she did not consider how ethnicity, social class, or sexual orientation might have influenced girls' ideas, dialogue, or posts. It is interesting to consider that the relative anonymity of chat rooms can make it easier to discuss topics that are taboo in face-to-face conversations.

However, face-to-face conversations about menstruation are also more common than they once were, especially among adolescent peers. Fingerson (2006) conducted a series of single-gender group interviews with predominantly European American adolescent boys and girls to explore their "menstrual talk." She concluded that some girls derived agency and empowerment from their menses. Themes that reflected this conclusion include girls' tendencies to embrace the challenge of managing their menstrual flow, to use and enjoy the privilege of having knowledge about their bodies that boys did not have, and to challenge the dominant and often negative social norms about menstruation. Although open talk about taboo topics is an important way to resist stigma, some of the girls attributed the empowerment derived from menstruation to their ability to embarrass boys with the mere mention of tampons or menstrual blood. Like the artists Tiegs and Paul, these girls are using shock to subvert the rule that menstruation must be hidden from the public square, but it is the stigma that allows them the power to embarrass boys at will. In a new print advertisement campaign for UKotex, consumers are encouraged to break the cycle of discomfort with tampons by being more open about them (Newman 2010). This goal is an admirable one, but the slogans for some ads (for example, "I tied a tampon to my keyring so my brother wouldn't take my car. It worked.") both challenge and reinforce the stigma of menstruation.

Girls living in the U.S. learn simultaneously that menstruation is important and natural and that they should hide and ignore it (Charlesworth 2001). How would this change if we celebrated menarche? Unlike Americans, individuals from countries around the world acknowledge this important rite of passage through various rituals such as a special gathering or party (Chrisler and Zittel 1998). The guest of honor may feel embarrassed initially, but a party could help her to realize that she, like other girls and women in her life, can overcome her embarrassment and have positive, even playful, attitudes toward menstruation. There are many organizations on the Internet, such as the Red Web Foundation (www.redwebfoundation.org) and First Moon (www.celebrategirls.org) that provide special kits to celebrate menarche and many other resources for raising menstrual consciousness.

The social stigma of menstruation can be challenged through the analysis of menstruation in popular culture. For example, social scientists have found that the popular press is rife with articles about menstruation that reinforce and perpetuate stigmatizing messages and provide inaccurate information about menstruation and premenstrual changes (Chrisler and Levy 1990; Johnston-Robledo, Barnack, and Wares 2006). Clearly, readers of popular magazines and newspapers should be encouraged to question and discuss what they read about menstruation in this material.

Others have resisted and challenged the stigma of menstruation through the creation of a menstrual counterculture. In his virtual museum, Harry Finley has collected women's stories about their experiences with menstruation as well as many images of advertisements, hygiene products, and other artifacts, which he displays on his website (www.mum.org). In her work on menstrual counterculture, Kissling (2006) noted that Finley's collection has a lot of educational potential because it challenges widely shared ideas about what is considered public and private. Artist/poet Geneva Kachman and several of her friends designated the Monday before Mother's Day as Menstrual Monday, a holiday to celebrate menstruation. She designs and distributes kits for this celebration including party blowouts made out of tampon applicators (www.moltx.org). Ani DiFranco's song, Blood in the Boardroom, is a rare example of popular music about menstruation. In her book Cunt, third-wave feminist Inga Muscio (2002) wrote about many different aspects of menstruation in a candid, humorous, and revolutionary way. Her writing on alternative menstrual products is especially compelling. Maybe it will inspire women to try an alternative to pads or tampons.

An important way to reduce stigma is social activism. Bobel (2006, 2008, 2010) has written extensively about the history of menstrual activism as well as the myriad ways contemporary menstrual activists are drawing attention to the health and environmental hazards of menstrual hygiene products through organizations, political action, zines, and other publications. This kind of work could help people to appreciate the extent to which the social stigma of menstruation fuels and is perpetuated by consumerism. Finally, health care providers are beginning to recognize and promote menstruation as an important indicator, even a vital sign, of girls' and women's overall health (Diaz, Laufer, and Breech 2006; Stubbs 2008). The mission of the Project Vital Sign (www.projectvitalsign.org) campaign is to raise awareness about the role of menstruation in women's psychological and physical health with the ultimate goal of encouraging an open dialogue about menstruation between health care providers and their female patients. Efforts to politicize and/or normalize menstruation could go a long way toward reducing its stigmatized status.

CONCLUSION

The consideration of menstruation as a source of social stigma has promising implications for theory, research, and practice. We have demonstrated that menstruation fits all three of Goffman's (1963) categories and reviewed a significant body of literature that supports the stigmatized status of menstruation, the means through which the stigma is transmitted, and the consequences of the stigma. All of these areas are worthy of continued theoretical development and empirical investigation. Clearly, the stigmatized status of menstruation has detrimental consequences for girls' and women's self-esteem, body image, self-presentation, and sexual health. Feminist therapists, educators, and healthcare providers can consider ways to alleviate these negative consequences and to assist girls and women in their efforts to resist the stigma of menstruation. Equally important is the evidence that suggests that menstrual status, both actual and symbolic, primes and elicits negative attitudes toward women. Challenging the stigma of menstruation and learning to appreciate, or at least not loathe, menstruation may have a positive impact on girls' and women's well-being as well as their social status.

NOTE

1. "The Menstrual Mark: Menstruation as Social Stigma" by Ingrid Johnston-Robledo and Joan C. Chrisler was first published in 2013 in *Sex Roles.* 68 (1–2): 9–18. Reprinted with permission. No further reproduction or distribution of the material is allowed without permission from the publisher.

REFERENCES

Andrist, L. C. (2008). The Implications of objectification theory for women's health: Menstrual suppression and "maternal request" cesarean delivery. Health Care for Women International, 29, 551–565. https://doi.org/10.1080/07399330801949616.

Bartky, S. L. (1990). Femininity and domination: Studies in the phenomenology of oppression. New York: Routledge.

Berg, D. H., and Coutts, L. B. (1994). The extended curse: Being a woman every day. Health Care for Women International, 15, 11–22. https://doi.org/10.1080/07399339409516090.

Bobel, C. (2006). Our revolution has style: Contemporary menstrual product activists "Doing Feminism" in the third wave. Sex Roles, 56, 331–345. https://doi.org/10.1007/s11199-006-9001-7.

Bobel, C. (2008). From convenience to hazard: A short history of the emergence of the menstrual activism movement, 1971–1992. Health Care for Women International, 29, 738–754. https://doi.org/10.1080/07399330802188909.

Bobel, C. (2010). New Blood: Third-Wave Feminism and the Politics of Menstruation. New Brunswick: Rutgers University Press. Boston Women's Health Book Collective. (2005). Our bodies, ourselves: A new edition for a new era. New York: Simon & Schuster.

Boston Women's Health Book Collective. (2005). Our Bodies, Ourselves: A New Edition for a New Era. New York: Simon & Schuster.

Bramwell, R. (2001). Blood and milk: Constructions of female bodily fluids in Western society. Women & Health, 34(4), 85–96. https://doi.org/10.1300/j013v34n04_06.

Buckley, T., & Gottlieb, A. (1988). Blood magic: The anthropology of menstruation. Berkeley, CA: University of California Press.

Charlesworth, D. (2001). Paradoxical constructions of self: Educating young women about menstruation. Women and Language, 24, 13–20.

Chesler, G. (Producer & director). (2006). Period: The end of menstruation [Motion picture]. New York: Cinema Guild.

Chrisler, J. C. (2002). Hormone hostages: The cultural legacy of PMS as a legal defense. In L. H. Collins, M. R. Dunlap, & J. C. Chrisler (Eds.), Charting a new course for feminist psychology (pp. 238–252). Westport: Praeger.

Chrisler, J. C. (2007). PMS as a culture-bound syndrome. In J. C. Chrisler, C. Golden, & P. D. Rozee (Eds.), Lectures on the psychology of women (pp. 154–171). Boston: McGraw Hill.

Chrisler, J. C. (2008). 2007 Presidential address: Fear of losing control: Power, perfectionism, and the psychology of women. Psychology of Women Quarterly, 32, 1–12. https://doi.org/10.1111/j.1471-6402.2007.00402.x.

Chrisler, J. C. (2011). Leaks, lumps, and lines: Stigma and women's bodies. Psychology of Women Quarterly, 35, 202–214. https://doi.org/10.1177/0361684310397698.

Chrisler, J. C., & Caplan, P. J. (2002). The strange case of Dr. Jekyll and Ms. Hyde: How PMS became a cultural phenomenon and a psychiatric disorder. Annual Review of Sex Research, 13, 274–306.

Chrisler, J. C., & Levy, K. B. (1990). The media construct a menstrual monster: A content analysis of PMS articles in the popular press. Women & Health, 16, 89–104. https://doi.org/10.1300/j013v16n0207.

Chrisler, J. C., & Zittel, C. B. (1998). Menarche stories: Reminiscences of college students from Lithuania, Malaysia, Sudan, and The United States. Health Care for Women International, 19, 303–312. https://doi.org/10.1080/073993398246287.

Chrisler, J. C., Johnston, I. K., Champagne, N. M., & Preston, K. E. (1994). Menstrual joy: The construct and its consequences. Psychology of Women Quarterly, 18, 375–387. https://doi.org/10.1111/j.1471-6402.1994.tb00461.x.

Cicurel, I. E. (2000). The Rabbinate versus Israeli (Jewish) women: The Mikvah as a contested domain. Nashim: A Journal of Jewish Women's Studies, 3, 164–190.

Cochrane, K. (2009, October). It's in the blood. The Guardian. Retrieved from http://www.guardian.co.uk/lifeandstyle/2009/oct/02/menstruation-feminist-activists.

Coutts, L. B., & Berg, D. H. (1993). The portrayal of the menstruating woman in menstrual product advertisements. Health Care for Women International, 14, 179–191. https://doi.org/10.1080/07399339309516039.

Crocker, J., Major, B., & Steele, C. (1998). Social stigma. In D. T. Gilbert, S. T. Fiske, & G. Lindzey (Eds.), Handbook of social psychology (4th ed., Vol. 2, pp. 504–553). Boston: McGraw-Hill.

Culpepper, E. E. (1992).Menstrual consciousness raising: A personal and pedagogical process. In A. J. Dan & L. L. Lewis (Eds.), Menstrual health in women's lives (pp. 274–284). Chicago: University of Illinois Press.

Davis, A. R., Nowygrod, S., Shabsigh, R., & Westhoff, C. (2002). The influence of vaginal bleeding on the sexual behavior of urban, Hispanic women and men. Contraception, 65, 351–355. https://doi.org/10.1016/s0010-7824(02)00279-2.

Deaux, K., Reid, A., Mizrahi, K., & Ethier, K. A. (1995). Parameters of social identity. Journal of Personality and Social Psychology, 68, 280–291. https://doi.org/10.1037/0022-3514.68.2.280.

Delaney, J., Lupton, M. J., & Toth, E. (1987). The curse: A cultural history of menstruation (rev. ed.). Urbana: University of Illinois Press.

Diaz, A., Laufer, M., & Breech, L. (2006). Menstruation in girls and adolescents: Using the menstrual cycle as a vital sign. Pediatrics, 118(5), 2245–2250.

Dovidio, J. F., Major, B., & Crocker, J. (2000). Stigma: Introduction and overview. In T. F. Heatherton, R. E. Kleck, M. R. Hebl, & J. G. Hull (Eds.), The social psychology of stigma (pp. 1–28). New York: Guilford.

Erchull, M. J. (2010). Distancing through objectification? Depictions of women's bodies in menstrual product advertisements. Sex Roles. Advance online publication. https://doi.org/10.1007/s11199-011-0004-7.

Erchull, M. J., Chrisler, J. C., Gorman, J. A., & Johnston-Robledo, I. (2002). Education and advertising: A content analysis of commercially produced booklets about menstruation. Journal of Early Adolescence, 22, 455–474. https://doi.org/10.1111/1471-6402.t01-2-00007.

Ernster, V. L. (1975). American menstrual expressions. Sex Roles, 1, 3–13. https://doi.org/10.1007/bf00287209.

Fingerson, L. (2006). Girls in power: Gender, body, and menstruation in adolescence. Albany: State University of New York Press.

Forbes, G. B., Adams-Curtis, L. E., White, K. B., & Holmgren, K. M. (2003). The role of hostile and benevolent sexism in women's and men's perceptions of the menstruating woman. Psychology of Women Quarterly, 27, 58–63. https://doi.org/10.1111/1471-6402.t01-2-00007.

Foucault, M. (1979). Discipline and punish: The birth of the prison. London: Penguin.

Frable, D. E. (1993). Dimensions of marginality: Distinctions among those who are different. Personality and Social Psychology Bulletin, 19, 370–380. https://doi.org/10.1177/0146167293194002.

Fredrickson, B. L., & Roberts, T.-A. (1997). Objectification theory: Toward understanding women's lived experiences and mental health risks. Psychology of Women Quarterly, 21, 173–206. https://doi.org/10.1111/j.1471-6402.1997.tb00108.x.

Goffman, E. (1963). Stigma: Notes on the management of spoiled identity. New York: Simon & Schuster.

Goldenberg, J. L., & Roberts, T.-A. (2004). The beast within the beauty: An existential perspective on the objectification and condemnation of women. In J. Greenberg, S. L. Koole, & T. Pyszczynski (Eds.), Handbook of experimental existential psychology (pp. 71–85). New York: Guildford.

Golub, S. (1992). Periods: From menarche to menopause. Newbury Park: Sage.

Greenberg, J., Pyszczynski, T., & Solomon, S. (1986). The causes and consequences of a need for self-esteem: A terror management theory. In R. F. Baumeister (Ed.), Public and private self (pp. 189–212). New York: Springer.

Heath, T.M. (2007). Vanessa Tiegs' menstrual blood painting journal: Art or biohazard? Retrieved from http://www.associatedcontent.com/article/280931/vanessa_tiegs_menstrual_blood_painting.html.

Herek, G. M. (2009). Sexual stigma and sexual prejudice in the U.S.: A conceptual framework. In D. A. Hope (Ed.), Contemporary perspectives on lesbian, gay, and bisexual identities (pp. 65–111). New York: Springer.

Hoerster, K. D., Chrisler, J. C., & Gorman, J. A. (2003). Attitudes toward and experiences with menstruation in the U.S. and India. Women & Health, 38(3), 77–95. https://doi.org/10.1300/J013v38n03_06.

Houppert, K. (1999). The curse: Confronting the last unmentionable taboo. New York: Farrar, Straus, & Giroux.

Johnston-Robledo, I., Ball, M., Lauta, K., & Zekoll, A. (2003). To bleed or not to bleed: Young women's attitudes toward menstrual suppression. Women & Health, 38(3), 59–75. doi:10.1300/J013v38n03_05. https://doi.org/10.1300/j013v38n03_05.

Johnston-Robledo, I., Barnack, J., & Wares, S. (2006). "Kiss your period good-bye": Menstrual suppression in the popular press. Sex Roles, 54, 353–360. https://doi.org/10.1007/s11199-006-9007-1.

Johnston-Robledo, I., Sheffield, K., Voigt, J., & Wilcox-Constantine, J. (2007). Reproductive shame: Self-objectification and young women's attitudes toward their bodies. Women & Health, 46(1), 25–39. https://doi.org/10.1300/j013v46n01_03.

Kissling, E. A. (1996). Bleeding out loud: Communication about menstruation. Feminism & Psychology, 6, 481–504. https://doi.org/10.1177/0959353596064002.

Kissling, E. A. (2006). Capitalizing on the curse: The business of menstruation. Boulder: Rienner.

Kowalski, R. M., & Chapple, T. (2000). The social stigma of menstruation: Fact or fiction? Psychology of Women Quarterly, 24, 74–80. https://doi.org/10.1111/j.1471-6402.2000.tb01023.x.

Laws, S. (1990). Issues of blood: The politics of menstruation. London: Macmillan.

Lee, J. (1994). Menarche and the (hetero)sexualization of the female body. Gender and Society, 8, 343–362. https://doi.org/10.1177/089124394008003004.

Lee, J., & Sasser-Coen, J. (1996). Blood stories: Menarche and the politics of the female body in contemporary U.S. society. New York: Routledge.

Marván, M. L., Islas, M., Vela, L., Chrisler, J. C., & Warren, E. A. (2008). Stereotypes of women in different stages of reproductive life: Data from Mexico and the U.S. Health Care for Women International, 29, 673–687. https://doi.org/10.1080/07399330802188982.

Merskin, D. (1999). Adolescence, advertising, and the idea of menstruation. Sex Roles, 40, 941–957. https://doi.org/10.1023/a:1018881206965.

Muehlenkamp, J. L., & Saris-Baglama, R. N. (2002). Selfobjectification and its psychological outcomes for college women. Psychology of Women Quarterly, 26, 371–379. https://doi.org/10.1111/1471-6402.t01-1-00076.

Muscio, I. (2002). Cunt: A declaration of independence. Emeryville: Seal Press.

Newman, A. A. (2010, March 16). Rebelling against the commonly evasive feminine care ad. New York Times, p. B3.

Owen, L. (1993). Her blood is gold: Celebrating the power and mystery of menstruation. San Francisco: Harper.

Oxley, T. (1998). Menstrual management: An exploratory study. Feminism & Psychology, 8, 185–191.

Polak, M. (2006). From the curse to the rag: Online gURLs rewrite the menstruation narrative. In Y. Jiwani, C. Steenbergen, & C. Mitchell (Eds.), Girlhood: Redefining the limits (pp. 191–207). New York: Black Rose Books.

Raftos, M., Jackson, D., & Mannix, J. (1998). Idealised versus tainted femininity: Discourses of the menstrual experience in Australian magazines that target young women. Nursing Inquiry, 5, 174–186. https://doi.org/10.1046/j.1440-1800.1998.530174.x.

Rempel, J. K., & Baumgartner, B. (2003). The relationship between attitudes toward menstruation and sexual attitudes, desires, and behavior in women. Archives of Sexual Behavior, 32, 155–163. https://doi.org/10.1023/a:1022404609700.

Rhinehart, E. D. (1989, June). Psychotherapists' responses to the topic of menstruation in psychotherapy. Paper presented at the meeting of the Society for Menstrual Cycle Research, Salt Lake City, UT.

Roberts, T.-A. (2004). Female trouble: The Menstrual Self-evaluation Scale and women's self-objectification. Psychology of Women Quarterly, 28, 22–26. https://doi.org/10.1111/j.1471-6402.2004.00119.x.

Roberts, T.-A., Goldenberg, J. L., Power, C., & Pyszczynski. (2002). "Feminine protection:" The effects of menstruation on attitudes toward women. Psychology of Women Quarterly, 26, 131–139. https://doi.org/10.1111/1471-6402.00051.

Rose, J. G., Chrisler, J. C., & Couture, S. (2008). Young women's attitudes toward continuous use of oral contraceptives: The effects of priming positive attitudes toward menstruation on women's willingness to suppress menstruation. Health Care for Women International, 29, 688–701. https://doi.org/10.1080/07399330802188925.

Rozin, P.,&Fallon,A. E. (1987).Aperspective on disgust. Psychological Review, 94, 23–41. https://doi.org/10.1037/0033-295x.94.1.23.

Schooler, D., Ward, M. L., Merriwether, A., & Caruthers, A. S. (2005). Cycles of shame: Menstrual shame, body shame, and sexual decision-making. Journal of Sex Research, 42, 324–334. https://doi.org/10.1080/00224490509552288.

Stallabrass, J. (2006). Contemporary art: A very short introduction. Oxford: Oxford University Press.

Steinem, G. (1978, October). If men could menstruate: A political fantasy. Ms., p. 110.

Stepanich, K. K. (1992). Sister moon lodge: The power and mystery of menstruation. Woodbury: Llewellyn.

Stubbs, M. L. (2008). Cultural perceptions and practices around menarche and adolescent menstruation in the United States. Annals of the New York Academy of Science, 1135, 58–66. https://doi.org/10.1196/annals.1429.008.

Szymanski, D. M., & Henning, S. L. (2007). The role of selfobjectification in women's depression: A test of objectification theory. Sex Roles, 56, 45–53. https://doi.org/10.1007/s11199-006-9147-3.

Tampax report, The. (1981). New York: Ruder, Finn, & Rotman. Tanfer, K., & Aral, S. O. (1996). Sexual intercourse during menstruation and self-reported sexually transmitted disease history among women. Sexually Transmitted Diseases, 23, 395–401.

Tanfer, K., and S. O. Aral. 1996. Sexual Intercourse during Menstruation and Self-Reported Sexually Transmitted Disease History among Women. Sexually Transmitted Diseases 23, 395–401.

Taylor, D. (2003). Red flower: Rethinking menstruation. Caldwell: Blackburn Press.

Tylka, T. L., & Hill, M. S. (2004). Objectification theory as it relates to disordered eating among college women. Sex Roles, 51, 719–730. https://doi.org/10.1007/s11199-004-0721-2.

Ussher, J. M. (2004). Premenstrual syndrome and self-policing: Ruptures in self-silencing leading to increased self-surveillance and blaming of the body. Social Theory & Health, 2, 254–272. https://doi.org/10.1057/palgrave.sth.8700032.

Williams, L. R. (1983). Beliefs and attitudes of young girls regarding menstruation. In S. Golub (Ed.), Menarche: The transition from girl to woman (pp. 139–148). Lexington: Lexington Books.

Wind, L. H. (1995). New moon rising: Reclaiming the sacred rites of menstruation. Chicago: Delphi Press.

Young, I. M. (2005). On female body experience: "Throwing like a girl" and other essays. New York: Oxford University Press.

CHAPTER 18

The Menarche Journey: Embodied Connections and Disconnections

Niva Piran

After I got my first period, they [three adult women] welcomed me into womanhood. They took me out to this beautiful waterfall, a beautiful place that I had never even heard of. And they each talked to me about being a woman and they each gave me something that had been given to them by another woman, and talked about the strengths and powers that she had . . . It made me feel really special . . . It's something I would really like to do for others . . . In different cultures people do more for that kind of getting your period, kind of growing up whatever for girls, but our culture doesn't really recognize that. And in some ways I think they kind of 'Oh it's one more headache,' but I think it's really important to recognize that, because it is a big moment . . . The ceremony made me feel more loved, like people will always be there for me.

Brenda, a 12-year-old White girl, living in a rural site in North America, in a home that vacillated between being financially strained and of middle class standing, participated in a study about the embodied life of adolescent girls (Piran 2017). In this narrative, Brenda describes a unique and meaningful welcoming ceremony conducted by three women during the month following her menarche. She highlights the positive embodying elements in this ceremony: the honoring of a big moment in her life marked by a biological transition, and feeling "special," "loved," and that "people will always be there" for her. Having gone through this ceremony, Brenda recognizes that she is situated differently from other girls and women that view menstruation as a "headache," making her motivated to conduct the same ceremony for others. Through this unique relational women-centered ceremony, taken place in a physical territory found and chosen by these women, patriarchy (and other sources of disruption to menstruating women) was marginalized. While, as we follow Brenda prospectively for the next five years, we find that she faces common challenges to owning menstruation positively in a patriarchal world, we also hear that she carries the welcoming ceremony with her, never feeling about menstruation as negatively as other girls. As this chapter

© The Author(s) 2020   201
C. Bobel et al. (eds.), *The Palgrave Handbook of Critical Menstruation Studies*, https://doi.org/10.1007/978-981-15-0614-7_18

unfolds, we return to examining particular elements in Brenda's narrative in relation to other girls at menarche.

Contrasting women's and men's experiences of inhabiting their bodies in the world in her book, *The Second Sex*, Simone de Beauvoir described the experience of (White, privileged) men as having ". . . direct and normal connection with the world" (de Beauvoir 1989, xxi). This chapter focuses on the experience of menarche and "connection to the world" through the lens of embodiment and the Developmental Theory of Embodiment (Piran 2017). This theory can be a productive frame for examining menarche. First, the theory centers on the social factors that shape the *quality of embodied lives*, all the way from 'positive body connection and comfort, embodied agency and passion, and attuned self-care' to 'disrupted body connection and discomfort, restricted agency and passion, and self-neglect or harm' (Piran 2016, 47). Second, the theory addresses both positive and disruptive social factors, thus resonating with concerns about the negative bias in the construction of menarche (and menstruation) in research (Burrows and Johnson 2005). Third, the theory, anchored in girls' and women's lived experiences, integrates a range of social factors that tend to be studied and conceptualized as separate phenomena, providing a wide lens through which to look at the intersection of body and culture at menarche.

In examining menarche in relation to "embodied connection with the world," I draw mainly from a five-year prospective interview study involving 87 interviews with 27 girls who were 9–14 in the first phase, that focused on girls' description of their embodied experiences and the social factors that shaped them (Piran 2017). The prospective lens helps clarify changes girls identify as particularly poignant for them at menarche. All girls resided in Canada during the time of the study and were from diverse social locations in terms of ethnicity and heritage, socioeconomic status, and family composition (Piran 2017). The chapter is also informed by a participatory action research with girls and boys, ages 10–18 (Piran 2001) and two retrospective life-history studies on embodied journeys with younger (ages 20–28) and older (ages 50–70) women, for a total of 84 interviews with 42 women.

The Developmental Theory of Embodiment (Piran 2017) suggests that the social experiences that shape embodiment, both facilitative and adverse factors, fall into three main categories. These include experiences in the physical domain (physical freedom vs. corseting), the mental domain of social discourses and expectations (mental freedom vs. corseting), and the social power and relational connections domain (social empowerment and relational connections vs. social disempowerment and relational disconnections). In particular, physical freedom, mental freedom, and social power and relational connections contribute to positive embodiment. In this chapter, we examine the experience of menarche narrated by girls going through puberty in these three categories, emphasizing the theme of "embodied connection to the world."

Embodied Connection and Agency
in the Physical Territory

They took me out to this beautiful waterfall, a beautiful place that I had never even heard of. And they each talked to me about being a woman.

One component Brenda highlights in her menarche welcoming ceremony relates to the beautiful place in nature, new to her, and shared with her by the three women. She emphasizes again the value she places on connection to the physical environment in a later interview where she singles out a challenging nature exploration with her mother as the time she has felt best ever in her body, feeling "excited, relaxed, and really good."

Freedom of physical engagement and movement (Piran 2017) is a category of physical experiences that includes engagement in joyful physical activities and the freedom to move freely in the private and public spheres— experiences that are centrally important to connection to the physical environment and to embodied agency. However, we find that girls at menarche commonly describe a highly disruptive *crisis of physical restriction* to continued engagement in physical activities and to free movement across a range of physical sites. The colonization of the public sphere by men and boys (as it also intersects with other dimensions of privilege related to ethnicity/ race, ability, social class, et cetera) is accentuated once girls menstruate and gendered boundaries become less permeable. For example 13-year-old Melissa describes the experience in the following way,

[Up to age 11] we stayed outside all the time. We'd go to ponds and try to catch mice and frogs and turtles and bugs. This boy was in my class and we hung out with his older brothers . . . We also played soccer and stuff . . . I would be out in the field *having fun*! . . . In grade 5 we stopped [nature explorations] . . . I spend more time with girls . . . Cause *when we started getting our periods*, then we'd talk more about that and about who do you like. [White, middle class, rural]

The schoolyard is another arena, we find, where the gender segregation becomes clear at menarche, around age 11, whereby boys remain physically active in the central space of the schoolyard while girls gather in the margins to talk (Edwards, Knoche, and Kumru 2001).

Concurrent with girls' expressions of protest and loss about their physical corseting, we find the glaring disappearance of narratives of fun and passion in leisure (often physical) activities, of physical agency, and of connection to the physical environment and to the co-ed non-sexualized social environment. Thirteen-year-old Lauren, for example, describes it in the following way, "The monthly thing! Hard to go camping, swimming. Guys don't have to worry about it . . . we used to play outside, fun!" The loss of the normative physical engagement in the public sphere, we find, is more profound among girls whose families struggle financially and who therefore cannot "buy" continued engagement in physical activities through paid organized sport activities.

The physical management of menstruation is one factor that shapes physical restriction, and girls often find that the social environment is not geared toward, attuned to, or supportive of menstruating bodies—a form of 'introduction' to living in women's bodies. Rather, different barriers are erected to their continued physical engagement with agency in the world. In our prospective study (Piran 2017) we find that, while some girls continue their engagement in physical activities with support from significant adults, most girls change to a more sedentary life style (Kimm et al. 2002).

Girls in our study relied on commercial menstruation supplies. Commercial supplies are costly, and trying to find cheaper alternatives may lead to greater discomfort, as Kyra describes,

> I was afraid to use tampons, but my friend had this pool party and I had my period and I was like, 'Ok, I am doing it' . . . But I use 'no name' and I thought it got stuck inside, and I was crying. 'No name' are not as good as Kotex, but I love Kotex cause they are easy.

Research around the world highlights challenges in accessing menstrual supplies within the context of poverty (for example, Crichton et al. 2013; Goldberg 2017). A meta-analysis of 138 studies conducted in India described multiple restrictions to girls' participation in the public sphere during menstruation, with a quarter of the girls missing school related to "menstrual hygiene management" (Van Eijk et al. 2016).

However, physical corseting during menstruation relates not only to barriers in accessing menstrual supplies, but also to the pervasive stigma of menstruation; as long as taboos and stigma surround menstruation, menstruating girls and women will not feel supported or safe in participating in the physical and social world (Bobel 2018). Even girls at menarche who have access to a range of commercial menstrual products commonly avoid physical activities they have previously cherished, such as swimming or camping, in order to avoid being found out as menstruating Lauren, an avid camper, for example explained that, "Camping would be fun. But I don't want to get my period during camping. Boys could not shower for days and they'd be fine. They don't have to worry about their periods so it is easier for them" [White, middle class, urban]. Hence, propagated messages by commercial manufacturer of menstrual supplies regarding the "successful" management and hiding of menstruation, serve to reinforce an oppressive status quo.

Access to physical territories reflects processes of social privilege, colonization, disenfranchisements, and oppression, bolstered by associated prejudicial belief systems, policies, and laws. Such processes impose physical restrictions and regulate the spaces where girls can engage post-menarche and during menstruation, with formal or informal policing. In our study, we found consistent peer, and at times parental, policing of girls' compliance with restricted physical activities post-menarche in line with 'femininity' discourses. These discourses are summarized aptly by 11-year-old Jackie, "Because girls, they are like, 'I'm a porcelain doll, I'm just gonna sit here and be perfect" [Aboriginal, working class, urban]. Girls who try to continue to play with boys physical games in the schoolyard, are at the risk of being ostracized.

Clothing norms post-menarche can reinforce processes of space colonization. In different Western countries, peer-enforced clothing norms for girls around menarche involve a shift to wearing 'feminine' tight and exposing clothing, which restrict physical action, as 12-year-old Madison described "I had a skirt on that was short and I jumped over a fence and the boys started laughing, like: what? They saw my skirt come up, I walked away. There are things girls are not supposed to do" [Aboriginal, working class, urban]. As girls cease their engagement in a range of physical activities, their physical skills and agency become compromised.

The expectation to clad the body of girls post-menarche in heavy layers of clothing comprises another challenge to physical engagement with the world. In Canada, we found a struggle around adaptation of public spaces to the needs of girls at menarche whose norms of attire differ from prevalent norms. For example, the lack of availability of women-only swim time at a local pool, prevented 13-year-old Hazel, who is a practicing Muslim and has started to wear a hijab upon her menarche, the opportunity to continue and pursue her favorite physical activity, swimming, "I love swimming so much. I mean, I haven't gone to swim in a while because I can't take the hijab off in a public pool" [Middle East heritage, working class, urban]. The physical restrictions on occupying particular spaces while menstruating or of engaging in varied activities while menstruating exist around the world (Van Eijk et al. 2016).

Still within the physical domain, menarche accentuates not only restrictions to engagement with the physical environment, but also the physical vulnerability of girls, as 12-year-old Madison describes, "I just got my period and it made me feel cautious. Just in case I ever turn stupid and do something. I don't want to get pregnant and be stuck with the baby . . . So it made me feel cautious" [Aboriginal, working class, urban].

Girls at puberty are exposed to a dramatic increase in sexual harassment and sexual violations (Piran 2017; "Sexual Violence: Facts at a Glance" 2012), and, in some cases, to early and forced marriages (Field and Ambrus 2008). Living in a rape culture (Buchwald, Fletcher, and Roth 1993), girls' awareness that sexual activity can, once they start to menstruate, lead to pregnancy, and that it is them who will endure the consequences, enhances the experience of vulnerability associated with the body, as Alice describes, "You need a lot more *precautions* in Grade 7 because of alcohol and drugs and sex with boys." Girls are entrusted with the responsibility of preventing pregnancy without the social conditions of safety, and, moreover, the lack of sanctioning and guidance in practicing their desires with self-attunement and boundaries (Piran 2017; Tolman 2002). Being socially abandoned to manage menstruation and sexuality with little support, guidance, or protection, it is not surprising the girls sometimes make potentially harmful decisions, such as having intercourse prior to menarche, 12-year-old Kyra describes,

> Girls start having sex in grade 5 because they don't have their period and they can't get pregnant. I think some of them do it because their boyfriends wanted to have sex and they think, 'what the hell, why not do it.'

Overall, then, overlapping experiences at menarche contribute to disrupted embodied connection and agency in the physical territory. As many girls' temporary freedom of engagement in the physical environment alongside boys ends at menarche and gender boundaries becomes less permeable, embodied joy and agency, as well as positive connection to the physical environment and to one's body—described by Brenda at the beginning of this section—are compromised.

Mental Corseting of the Embodied Self

> And they each talked to me about being a woman and they each gave me some-thing that had been given to them by another woman, and talked about the strengths and powers that she had . . . It made me feel really special.

Here Brenda highlights the value for her, as a girl who has just gone through menarche, of hearing the women speak of engaging in the world with strength and power while inhabiting women's bodies. Engaging with the world powerfully, meaningfully, and passionately by diverse women requires the maintenance of a critical stance toward constraining social discourses that regulate embodied lives of individuals of different social locations and that aim to maintain the social status quo (Piran 2017). Regarding femininity, our research suggests two sets of discourses that constrain the way girls and women inhabit their bodies and engage with the world and that exert accentuated pressures at menarche. We further find that menstruation-related discourses can be contextualized within other gender-related discourses narrated by girls at menarche.

The first set of constraining gender-related discourses, is *appearance-related*, entitled "*body as a deficient object*" (Piran 2017), and marks a woman's body both as an object of gaze, an adverse experience highlighted by Objectification Theory (Fredrickson and Roberts 1997), and as an inherently physically deficient site (Bordo 1993). While the discourse of objectification is narrated by girls prior to the onset of puberty, our prospective qualitative interviews reveal the emergence of new narratives, around the time of menarche, that mark multiple aspect of one's body as deficient, such as: "my hair needs to be tamed in the mornings, like a forest, a rat's nest," "I hate armpit hair, it's disgusting," "I hate the hair on my legs," "my stomach has like a bump," "my eyebrows are screwed up," "I'm not pretty without make up," "I am fat, I want to be small," and "it's weird to think about girls sweating, pretty girls do not sweat, I try not to." The internalized deficiency discourse compels girls at puberty to control and "repair" multiple aspects of their natural body, such as: weight, shape, hair (body and scalp), skin color, and facial features.

At menarche, menstruation compounds the experience of the body as a deficient object in its natural state (Roberts et al. 2002), as 13-year-old Kyra and 12-year-old Jackie described: "I hate my period, all these hormones, I am all bloated, I've got blood coming out of me everywhere, it's *disgusting*" [Kyra, White, working class, urban; "periods are *dirty* blood" [Hazel, Middle East heritage, working class, urban]. Within the context of engagement in varied body alterations, menstruation poses challenges to the wish to control and repair the body, leaving girls with the option of vigilance: being ever ready for their periods and finding strategies to hide their presence in their lives from their social world. They hold with terror the possible loss of such control and of facing the socially sanctioned response of disgust,

> It happened to my friend. She was running around and all of a sudden she got this red stuff on the back of her pants, and we're like "Eeuuww". It was gross and embarrassiong. She still denies it was her period. [Jackie, Aboriginal, working class, urban]

Though this response of disgust is shared among peers, girls specifically highlight boys' negative response to menstruation, "Boys probably think 'Ewww'" [Lauren; White, middle class, urban]. Thus, upon menarche, in particular, girls begin to engage with the world around them while viewing their bodies as deficient sites, experiences that inherently weaken their embodied agency in the world and direct them to channel energy into body repair.

The second set of constraining 'femininity' discourses is *comportment-related* and entitled *woman as docile* (Piran 2017). It directs girls and women to act in the world in demure and submissive ways; subvert their own needs, wishes and meaningful pursuits to that of boys and men; and avoid being too assertive, loud, dominant, needy, or simply 'too much.' One comportment-related discourse girls describe is the pressure to engage in '*feminine*' (or 'girlie') *activities*, once they go through menarche. Menarche problematizes a 'tomboy' way of inhabiting the world: a temporary option that allowed girls to have the freedom to act in the world similar to boys without social penalties, as 12-year-old Madison describes,

> Like I got my period 4 months ago, which is a big change of the body. A couple of months ago I started wearing pink and stuff. I used to wear real loose clothes, but now I wear clothes that are a bit closer to my body, tighter. People call [friend] a tomboy and they call her a man and stuff because she dresses in a

guy clothes. But they call me and [another friend] girlie girls because we dress now a lot in the color pink. I love sports but they usually say that girly girls don't like sports. [Aboriginal, working class, urban]

A second discourse from the *comportment-related 'woman as docile'* cluster that girls at puberty allude to is the *submissive/demure* discourse that reigns-in girls' voices, assertiveness, and anger (Piran 2017). Brenda, for example, describes that, "there is, like, a kind of thing for girls to be really nice" (Brown 2003), and contrasts, at age 12, being "nice" with being "a bitch" or "a butch," suggesting, for example, that "A butch speaks her mind. They are more comfortable with themselves." At menarche, the "PMS" discourse joins the "bitch" discourse and is contrasted with being compliant, nice, controlled, and subservient. At age 13, Kyra, for example, describes many sources of anger in her life. Having experienced lack of safety and fear from a male family member, his labeling of her anger and boundary-setting as 'PMS' and her 'bitch' self-labeling serve to problematize *her* body rather than challenge the sources of her anger in her home environment. Though owning being a "bitch" and a "psycho" while "on PMS" liberates Kyra to stand up to her abuser, this labeling ultimately maintains the abusive status quo,

> Some days you piss him off and he will flip. He's just way bigger than me and stronger, and when he yells, it scares the crap out of me. But, like period day is the day that I get pissed off. I would be in a grumpy mood and he's like, What's your problem? I'm just be like, 'Leave me alone!' and he' just be like, 'Oh, you're PMS-ing' . . . Whenever I'm getting near my period. I eat so much. Like I'm psycho. Like I'm actually insane. I am being such a bitch, but I can't stop myself. When you are on PMS then you are just a *bitch*.

We do not commonly hear from girls throughout puberty, including menarche, of relational connections at home or at school that support them in counteracting these discourses, a situation that likely emanates from adult women's own struggles with these discourses and the consistent impact of these discourses on peer norms.

Overall we find in our prospective study (Piran 2017), the tightening of 'femininity'-related discourses at menarche, both '*woman's body as deficient*' and '*woman as docile*,' clarify the way girls at menarche are expected to shift considerably the way they inhabit their bodies and comply with inequitable gendered relations (de Beauvoir 1989). At menarche, menstruation-related discourses, such as: *disgusting*, *dirty*, and *PMS*, weave into 'femininity' discourses and intensify their impact. Without relational connections that counteract 'femininity' stereotypes and their intersection with 'menstruation' discourses, as molds of femininity tighten, girls' connection to their bodies, their self-attunement, subjective immersion in passionate pursuits, and embodied agency, are affected adversely, disrupting embodiment.

Embodied (In)equity in Relational Communities

It's something I would really like to do for others . . . In different cultures people do more for that kind of getting your period, kind of growing up whatever for girls, but our culture doesn't really recognize that. And in some ways I think they kind of 'Oh it's one more headache,' but I think it's really important to recognize that, because it is a big moment . . . The ceremony made me feel more loved, like people will always be there for me.

Here Brenda alludes to the value of a community embracing menarche as an important landmark of growing up, and the tightening of relational connections that ensues. While most girls' experiences were starkly different from Brenda's, her observation of a close link between the response of the community and individual girls' experiences of menarche is apt; extending this thought, a more constructive experience of menarche relates to community transformations. Experiences of embodied equity in relational connections enhance experiences of embodied pride and self-worth, and, hence, positive embodiment; in contrast, exposure to prejudicial treatment and relational disconnections from desired communities disrupt embodiment (Piran 2017).

Reading the cultural landscape, in particular the stigma of menstruation and the lack of community embracement, or at least support, girls at menarche experience first-hand the demotion in social power associated with inhabiting a 'young woman's' body at the cusp of puberty. The intensity of the negative stigma attached to menstruation (Fahs 2011; Ginsburg 1996; Roberts et al. 2002) is matched by the intensity of *terror* as vividly described by premenstrual girls in the year preceding menarche from 'being found' menstruating, as ten-year-old Emma describes, "It's so scary getting your period, starting it and you won't be prepared, and you don't want anyone to know, you don't know what they'll think about you. Guys probably think 'Ew'" [White, middle class, rural]. Girls often make elaborate plans about how to be prepared for menarche, as ten-year-old Alice describes, "I am worried that if I have it [period] at school my teacher wouldn't have a tampon, but they have them upstairs in the girls' bathroom. I am always going to carry 25 cents with me. We are all worried" [Aboriginal, middle class, urban]. The stigma is as strong in all-girls schools, as 13-year-old Hazel describes, "I actually noticed that a lot of my friends in Islamic school were ashamed that they had it. It was this stigma thing." Hazel also raised the stigma associated with being barred from participation in prayer time during menstruation. The physical segregation of menstruating women from community, religious, and household activities has been reported around the world (Dunnavant and Roberts 2013; Van Eijk et al. 2016).

As girls internalize the stigma and comply with the social etiquette of not making the period a material presence in the social environment (Ginsburg 1996), a disempowering process, they experience further *embodied demotion* by the lack of social support, accommodation, or attunement to their needs in managing menstruation. Girls describe the school environment as not respectful to the needs of their maturing bodies: they face blocked access to washrooms during class time, lack of privacy associated with requests to access menstrual supplies, and empty tampon machines in washrooms, as Alice describes at the age of 12,

> Sometimes we need to go to the bathrooms, and the teacher is like, 'Why do you have to go to the bathroom'? . . . and if we go in the hallway to the locker people can see that, but we are not allowed to carry a purse to class. We opened up the tampon machine in the washroom. There are no tampons in it. They won't refill it.

School educational programs about puberty, while replete with diagrams and scientific explanations and conducted often in co-ed forums, do not address actual challenges in managing menstruation, as described earlier in the chapter.

Poverty challenges further the management of menstruation in all regions of the world (Van Eijk et al. 2016). In western countries that rely on manufactured sanitary supplies, the actual cost of menstrual supplies for people who live in poverty can be higher due to the difficulty of putting forward the money needed for a bulk purchase, or due to challenges to accessing transportation to larger stores and needing to rely on small stores in their own communities (Weiss-Wolf 2017). In most states in the US, there is sales tax on menstruation supplies, as they are not classified as "basic necessity" by these states (Weiss-Wolf 2017). This societal disavowal of a biological experience affecting half of the adolescent and adult populations relates centrally to social justice, and reflects the devaluation and disenfranchisement of adolescent girls, women, and their bodies.

As girls try to break their isolation around menstruation, they risk silencing at all levels of the social environment as, for example, Kyra and Hazel describe: "I was sitting in the car and I was asking my mom questions about tampons because I just started. And She's like, If you want to talk about tampons, you come to me privately. So I just asked my friend"; "Hazel, don't talk about it at the dinner table." Shame and isolation pervade even the closest girls' friendship networks during the first year post-menarche, as Lauren describes the situation at age 13, "Girls who have their periods at school, hide and change their shorts in the bathroom because it is embarrassing, and there are worries about embarrassing incidents, like blood soaking through" [White, middle class, urban]. The stigma and associated silencing therefor disrupt connections between girls, their mothers, and their friends (Stubbs and Costos 2004).

Girls use different strategies to reduce isolation and shame and restore some social power. They sometimes take a risk by reaching out to friends, as Lauren describes, "The first time you talk with other girls about that, you're putting yourself out there and you could have someone go, ew." A second approach to restoring social power and connections, utilized by girls who have been supported in learning how to manage menstruation, is to become a resource for other girls. Such experiences of functionality in relation to menstruation, such as confronting as a group of boys who have teased them about menstruation.

Girls' own initiatives to deal with stigma, isolation, and disempowerment, cannot replace the need for systemic transformations involving different levels of the social environment, from immediate relational connections to larger social organizations (Stubbs 2008). Having the experience of being part of at least one social system where equity is practiced (as Brenda's experience reflects) transmits embodied worth and therefore contributes to positive embodiment. It can further protect against the adverse impact of broader societal prejudices (Piran 2017).

Overall, within patriarchy and related ideological systems, the stigma, silencing, and lack of social and systemic support, marks menarche as a biological event associated with embodied demotion in social power and ruptures in relational networks and communities. Exposure to communities of equity, at any level of the social environment, can have an important protective impact and preserve important relational connections and embodied worth. Further, even within the context of social inequity, community connections among women around the experience of menstruation can have a positive, protective impact. For example, Dunnavant and Roberts (2013) found that women who followed restrictive religious rituals involving menstruation identified the heightened sense of community with other women as a positive aspect of their menstrual cycle; this element was lacking among non-religious women. The present study with girls at menarche and research with adult women therefore suggest positive possibilities inherent in providing girls and women with all-women spaces where connections can be forged around embodied experiences, comprising also fertile ground for the development of subversive knowledge and related activism. In contrast, the shamed isolation commonly reinforced by the dictates of silence and private management immobilizes girls and women from inhabiting their menstruating bodies with passion, agency, pride, and comfort.

In conclusion, the chapter suggests that menarche is associated with connections and disconnections in the life of girls in three domains of experiences that should be considered concurrently in order to understand girls' experiences of embodiment. While connection to the physical territory, established through engaging freely, actively, and safely in the public sphere, enhances embodied agency and positive connection to the body, girls commonly describe, protest, and grieve restrictions to such freedom at menarche; this change serves to reinforce boys' and men's colonization of the physical sphere once gender boundaries become less permeable. This disrupted connection to physical territories may not emerge as central in adult women's retrospective accounts of menarche, since restrictions to joyful physical engagement with the world becomes 'normative' post-puberty in western and non-western countries. Girls' experienced abandonment by adults and varied social systems (for example, schools, sports organizations) regarding their wish to stay engaged physically while menstruating—expressed, for example, through the lack of supportive guidance, forced secrecy, and infringed sexual safety—clarifies the colonization of their bodies and geographical territories by boys and men.

Examining the domain of social discourses at menarche suggests that the physical restrictions in the life of girls are reinforced by pressures to embody constraining discourses of 'femininity,' in particular, the woman's body as a deficient object and women as docile (Piran 2017). Menstruation-related discourses, such as menstrual bleeding as gross and menstruating women as bitchy or "on PMS," compound the deprecation of the natural body of girls and women and the problematizing of angry, assertive, or other 'unruly' behaviors. Girls at menarche are therefore under accentuated pressure to comply with living in altered and docile bodies, leading to girls' disrupted connections with their pre-menarche embodied selves.

Further disrupting embodiment is the demotion of menstruation and its association with stigma, silencing, and societal disavowal, to which girls are introduced at menarche. These social processes disrupt connections to social networks and communities, including among girls and women. Positive embodiment at menarche relates to the availability of relational connections and norms that can counteract adverse social experiences, and, ultimately, requires social transformations in all three domains.

References

de Beauvoir, Simone. 1989. *The Second Sex*. Translated by H. M. Parshley. New York: Random House.

Bobel, Chris. 2018. "Menstruators Need More Than Something to Bleed on, They also Need Information and Support." WASHfunders: Blog. http://washfunders.org/blog/menstruators-need-more-than-something-to-bleed-on-they-also-need-information-and-support/.

Bordo, Susan. 1993. *Unbearable Weight: Feminism, Western Culture, and the Body*. Berkeley: University of California Press.

Brown, Lyn Mikel. 2003. *Girlfighting: Betrayal and Rejection among Girls*. New York, NY: New York University Press.

Buchwald, Emilie, Pamela Fletcher, and Martha Roth. 1993. *Transforming a Rape Culture*. Minneapolis, MN: Milkweed Editions.

Burrows, Anne, and Sally Johnson. 2005. "Girls' Experiences of Menarche and Menstruation." *Journal of Reproductive and Infant Psychology* 23: 235–49. https://doi.org/10.1080/02646830500165846.

Crichton, Joanna, Jerry Okal, Caroline W. Kabiru, and Eliya M. Zulu. 2013. "Emotional and Psychosocial Aspects of Menstrual Poverty in Resource-Poor Settings: A Qualitative Study of the Experiences of Adolescent Girls in an Informal Settlement in Nairobi." *Health Care for Women International* 34: 891–916. https://doi.org/10.1080/07399332.2012.740112.

Dunnavant, Nicki C., and Tomi-Ann Roberts. 2013. "Restriction and Renewal, Pollution and Power, Constraint and Community: The Paradoxes of Religious Women's Experiences of Menstruation." *Sex Roles* 68: 121–31. https://doi.org/10.1007/s11199-012-0132-8.

Edwards, Carolyn P., Lisa Knoche, and Asiye Kumru. 2001. "Play Patterns and Gender." In *Encyclopedia of Women and Gender*, edited by Judith Worrell, 809–16. San Diego: Academic Press.

Van Eijk, Anna Maria., M. Sivakami, Mamita Bora Thakhar, Ashley Bauman, Kayla F. Laserson, Susanne Coates, and Penelope A. Phillips-Howard. 2016. "Menstrual Hygiene Management among Adolescent Girls in India: A Systematic Review and Meta-Analysis." *British Medical Journal Open*. https://www.ncbi.nlm.nih.gov/pmc/articles/PMC4785312/.

Fahs, Breanne. 2011. "Sex during Menstruation: Race, Sexual Identity, and Women's Accounts of Pleasure and Disgust." *Feminism & Psychology* 21: 155–78. https://doi.org/10.1177/0959353510396674.

Field, Erica, and Attila Ambrus. 2008. "Early Marriage, Age of Menarche, and Female Schooling Attainment in Bangladesh." *Journal of Political Economy* 116: 881–930. https://doi.org/10.1086/593333.

Fredrickson, Barbara L., and Tomi-Ann, Roberts. 1997. "Objectification Theory: Toward Understanding Women's Lived Experiences and Mental Health Risks." *Psychology of Women Quarterly* 21 (2): 173–206.

Ginsburg, Ruth. 1996. "'Don't Tell, Dear': The Material Culture of Tampons and Napkins." *Journal of Material Culture* 1: 365–75. https://doi.org/10.1177/135918359600100305.

Goldberg, Eleanor. 2017. "Why Many Native American Girls Skip School When They Have Their Periods." *Huffington Post*, August 25. http://www.huffingtonpost.ca/entry/what-its-like-to-be-a-teen-in-the-us-who-cant-affordtampons_us_597216d-de4b09e5f6cceddfb.

Kimm, Sue Y. S., Nancy W. Glynn, Andrea M. Kriska, Bruce A. Barton, Shari S. Kronsberg, Stephen R. Daniels, Patricia B. Crawford, Zak I. Sabry, and Kiang Liu. 2002. "Decline in Physical Activity in Black Girls and White Girls during Adolescence." *New England Journal of Medicine* 347 (10): 709–15. https://doi.org/10.1056/nejmoa003277.

Piran, Niva. 2001. "Re-Inhabiting the Body from the Inside Out: Girls Transform Their School Environment." In *From Subjects to Subjectivities: A Handbook of Interpretive and Participatory Methods*, edited by Deborah L. Tolman and Mary Brydon-Miller, 218–38. New York: New York University Press.

———. 2016. "Embodied Possibilities and Disruptions: The Emergence of the Experience of Embodiment Construct from Qualitative Studies with Girls and Women." *Body Image* 18: 43–60. https://doi.org/10.1016/j.bodyim.2016.04.007.

———. 2017. *Journeys of Embodiment at the Intersection of Body and Culture: The Developmental Theory of Embodiment.* San Diego, CA: Elsevier.

Roberts, Tomi-Ann., Jamie L. Goldenberg, Cathleen Power, and Tom Pyszczynski. 2002. "'Feminine Protection': The Effects of Menstruation on Attitudes towards Women." *Psychology of Women Quarterly* 26: 131–39. https://doi.org/10.1111/1471-6402.00051.

"Sexual Violence: Facts at a Glance." 2012. Division of Violence Prevention, National Center for Injury Prevention and Control, Center for Disease Control, US. https://www.cdc.gov/ViolencePrevention/pdf/SV-DataSheet-a.pdf.

Stubbs, Margaret L. 2008. "Cultural Perceptions and Practices Around Menarche and Adolescent Menstruation in the United States." *Annals of the New York Academy of Sciences* 1135: 58–66. https://doi.org/10.1196/annals.1429.008.

Stubbs, Margaret L., and Daryl Costos. 2004. "Negative Attitudes towards Menstruation: Implications for Disconnection within Girls and between Women." *Women & Therapy* 27: 37–54. https://doi.org/10.1300/j015v27n03_04.

Tolman, Deborah L. 2002. *Dilemmas of Desire: Teenage Girls Talk about Sexuality.* Cambridge, MA: Harvard University Press.

Weiss-Wolf, Jennifer. 2017. *Periods Gone Public: Taking a Stand for Menstrual Equity.* New York: Arcade Publishing.

CHAPTER 19

Resisting the Mantle of the Monstrous Feminine: Women's Construction and Experience of Premenstrual Embodiment

Jane M. Ussher and Janette Perz

The female reproductive body is positioned as abject, as other, as site of deficiency and disease, the epitome of the 'monstrous feminine.' Premenstrual change in emotion, behavior or embodied sensation is positioned as a sign of madness within, necessitating restraint and control on the part of the women experiencing it (Ussher 2006). Breakdown in this control through manifestation of 'symptoms' is diagnosed as PMS (Premenstrual Syndrome) or PMDD (Premenstrual Dysphoric Disorder), a pathology deserving of 'treatment.' In this chapter, we adopt a feminist material-discursive theoretical framework to examine the role of premenstrual embodiment in relation to women's adoption of the subject position of monstrous feminine, drawing on interviews we have conducted with women who self-diagnose as 'PMS sufferers.' We theorize women's self-positioning as subjectification, wherein women take up cultural discourse associated with idealized femininity and the reproductive body, resulting in self-objectification, distress, and self-condemnation. However, women can resist negative cultural constructions of premenstrual embodiment and the subsequent self-policing. We describe the impact of women-centered psychological therapy which increases awareness of embodied change, and leads to greater acceptance of the premenstrual body and greater self-care, which serves to reduce premenstrual distress.

© The Author(s) 2020
C. Bobel et al. (eds.), *The Palgrave Handbook of Critical Menstruation Studies*, https://doi.org/10.1007/978-981-15-0614-7_19

UNRAVELING PMS: PATHOLOGIZING FEMININITY AND THE FECUND BODY

It is estimated that around 8–13% of women meet a PMDD diagnosis each month, with around 75% meeting the lesser diagnosis of PMS—the same conglomeration of symptoms, just experienced to a lesser degree (Hartlage et al. 2012). However, the very notion of premenstrual change as deserving of diagnosis, and the inclusion of PMDD in DSM-IV, has met with widespread feminist opposition (Cosgrove and Caplan 2004). Feminist critics have dismissed this process of pathologization, arguing that premenstrual change is a normal part of women's experience, which is only positioned as "PMDD" or "PMS" because of Western cultural constructions of the premenstrual phase of the cycle as a time of psychological disturbance and debilitation (Chrisler 2004; Ussher 1989). This view draws on broader post-modern debates in critical psychology and psychiatry where all forms of mental illness or madness are positioned as social constructions that regulate subjectivity, disciplinary practices that police the population through pathologization (Fee 2000; Ussher 2011).

The process by which women take up the position of abjection personified, where premenstrual change is pathologized, and the fecund body is positioned as the *cause* of distress, can be described as a process of subjectification (Ussher 2003, 2006). The regimes of knowledge circulating within medicine, science, and the law, which are reproduced in self-help texts and the media (Fahs 2016; Bobel 2010; Chrisler and Caplan 2002), provide the discursive framework within which women come to recognize themselves as a 'PMS sufferer.'

In this vein, attention has been paid to women's internalization of the idealized 'good wife and mother,' combined with over-responsibility within the home, which can result in a pattern of self-silencing and self-sacrifice, leading to psychological distress. For some women, this self-silencing can lead to a build-up of emotion that erupts premenstrually (Perz and Ussher 2006; Ussher 2004). However, expression of anger or discontent is pathologized because women are deemed 'out of control,' resulting in legitimate emotion being dismissed as 'PMS' (Ussher and Perz 2013a, 2010). The self-positioning as PMS sufferer acts to maintain and reproduce the boundaries of femininity, with women judging themselves as bad, mad, or insane in relation to the ideal (Ussher 2006, 2011; Chrisler 2011).

DISCIPLINING THE UNRULY BODY: CONCEPTUALIZING PREMENSTRUAL EMBODIMENT

The positioning of woman as closer to nature, with subjectivity tied to a body deemed to be unruly or inferior, necessitating discipline and containment (Bordo 1990), is central to women's subjectification as PMS sufferers.

The bio-medical model which dominates research and treatment on PMS and PMDD, implicitly positions women's difference and deficiency as inevitable, and open to bio-medical 'intervention' (Ussher 2006). Thus in 1931, when 'Premenstrual Tension' first appeared in the medical literature, it was attributed to the 'female sex hormone' estrogen and regulated through correction of hormonal 'imbalance' (Frank 1931). In the intervening years, many different bio-medical theories of premenstrual symptomatology have been put forward, which have led to the adoption of a range of pharmacological interventions, with serotonin-reuptake inhibitors (SSRIs) currently recommended as "first line treatment" for PMDD (Ismaili et al. 2016).

At the same time, a gamut of psychological theories have been proffered to explain premenstrual distress, leading to the endorsement of Cognitive Behavior Therapy (CBT) as an effective solution (Kleinstäuber, Witthöft, and Hiller 2012). The body has a somewhat peripheral presence within this model of PMS, with distress assumed to arise from 'cognitive distortions,' and interventions focusing on women's psychological reappraisal of emotional and behavioral change premenstrually, alongside the development of behavioral coping mechanisms (Blake 1995). Common to both bio-medical and psychological models is that embodied change is positioned as a 'symptom' of PMS or PMDD, a material manifestation of disorder within. This is expressed as premenstrual bloating, swelling, breast tenderness, joint or muscle pain, headaches, and for some women, diarrhea and hot flushes (Endicott and Harrison 1990). However, such change has to be accompanied by a psychological 'symptom,' such as depression, anxiety or anger, to warrant diagnosis of a premenstrual 'disorder' (American Psychiatric Association 2013), implicitly positioning the body at the periphery of diagnosis of premenstrual disorders (PMDs).

RESEARCHING PREMENSTRUAL EMBODIMENT: POSITIVISM VERSUS CRITICAL REALISM

Psychologists have made some attempt to examine the nature and function of embodied premenstrual change, reporting that body image 'distortion' and body dissatisfaction is higher in the premenstrual phase of the cycle in the 'normal' female population who don't self-position as PMS sufferers (Kaczmarek and Trambacz-Oleszak 2016; Teixeira et al. 2013; Jappe and Gardner 2009; Carr-Nangle et al. 1994; Racine et al. 2012). In those women who do present with PMDs, levels of premenstrual symptom severity have been reported to be associated with body image disturbance (Muljat, Lustyk, and Miller 2007) and with body dissatisfaction (Kleinstäuber et al. 2016). Conducted through survey methods, which correlate menstrual cycle phase or premenstrual distress with perception of body size or body satisfaction, this body of research has proven inconclusive in determining whether "body dissatisfaction or a disturbed body image are vulnerability factors for, or consequences of premenstrual complaints" (Kleinstäuber et al. 2016, 761). This particular question, and the body of correlational research that informs it, is framed within a positivist epistemological standpoint (Keat 1979), which understands causality in terms of antecedent conditions and general laws governing phenomena, and utilizes the scientific method—in this case standardized survey instruments and statistical analysis—to 'objectively' examine variables of interest (Ussher 2005). What is absent from this analysis is the meaning and experience of embodied change from the perspective of women who inhabit the unruly premenstrual body, in the context of broader constructions of femininity and embodiment.

Feminist social constructionists have provided insight into the role of cultural discourse in the pathologization of the premenstrual woman (Chrisler 2004), as outlined above. However, social constructionism has been criticized for ignoring the "real" (Speer 2000), and marginalizing experience outside of the realm of language, in particular embodiment (Sims-Schouten, Riley, and Willig 2007). This is problematic, as a substantial proportion of women *do* perceive or experience emotional changes during the premenstrual phase of the cycle (Nevatte et al. 2013; Ussher and Perz 2013a, 2013b), as well as corporeral changes, including water retention and bloating (White et al. 2011), of that there is no doubt. It can also be seen to negate embodied or psychological change across the menstrual cycle, or other material aspects of women's existence that may be associated with their distress (Ussher 2005).

A critical realist epistemology (Bhaskar 1989) allows us to acknowledge the materiality of change across the menstrual cycle, including changes in corporeality, mood, or women's perception of embodied change, but also conceptualize this materiality as mediated by culture, language and politics. Described as a material-discursive standpoint (Ussher 2008b), critical realism has been positioned as a way forward for research examining embodiment in a sociocultural context (Williams 2003).

In the remainder of this chapter, we adopt a critical realist epistemology and a material-discursive framework to explore the implications of changes in premenstrual embodiment, and constructions of the idealized feminine body, on women's acceptance and resistance of the position of the monstrous feminine. We do this through drawing on interviews with women who self-identified as PMS sufferers, collected as part of a study examining the efficacy of a women-centered psychological therapy for moderate to severe premenstrual distress (which we henceforth define as PMS), the methodology and results of which are presented elsewhere (Ussher and Perz 2017). In summary, we interviewed 83 women, average age 35, who reported moderate-severe PMS, confirmed by three months of daily diary completion, about their subjective experience of premenstrual change. In the accounts below, we examine women's experience of premenstrual embodiment, prior to and after taking part in the psychological therapy.

INHABITING THE ABJECT PREMENSTRUAL BODY

"I Feel Fat and Ugly and Hate Myself": Self-Objectification and Dehumanization

The majority of women we interviewed reported negative feelings toward their bodies, and by implication their very selves, when they were premenstrual, describing themselves as "fat," "ugly," "a blimp," "gross," "frumpy," "sluggish," "disgusting," "lumpy," "sludgy," and "unattractive." In these accounts, negative feelings were attributed to perception of embodied change premenstrually, such as "bloating," "tenderness in the breasts," and "breasts that feel bigger," illustrated in the example below.

> I'm more bloated my boobs are already big so they're heaps bigger, my stomach's swollen and generally I feel quite puffy and fluid filled so I wouldn't say I feel particularly attractive at that time.

This bloating and self-positioning as "fat" was associated with perception of premenstrual weight gain. Women told us: "Two kilos goes on and it just makes me feel like crap, puffy in the face and round the guts, like right around my abdomen just puffs up"; and "physically I just feel about five times heavier than normal and bloated." Women explicitly described these changes as acting to annihilate their "self-confidence," "sense of being attractive," and "self-esteem"—their very sense of self as a woman. As one participant told us: "Yes I hate myself, I don't have any self-confidence and don't even want to look in any mirror." In contrast, women said that they felt "less concerned" about their bodies, or positioned them as "OK," when they were not premenstrual.

These accounts suggest a form of self-objectification (Fredrickson and Roberts 1997), wherein women have internalized a critical gaze that finds them wanting, because the "bloated," "fat" premenstrual body does not conform to the slim, contained, and feminine ideal. Similar accounts of surveillance and internalized judgment, have been found in interviews with women who position themselves as "overweight" or "obese" (Tischner 2013). Women's body fat is discursively positioned as ugly and stigmatizing within western culture, associated with loathing, disgust, and revulsion (Lupton 2013), with women expected to discipline and regulate the body, and thus the self, to maintain a slim, contained form (Chrisler 2011; Bordo 1993). Body fat is positioned as both a threat to health and morality (Lupton 2013), with "excess" fat a sign of women "letting themselves go" at both levels (Chrisler 2011, 205). Many women reported a disruption in their normal patterns of dietary retraint or "healthy eating" premenstrually, feeling "desperately in need of chocolate," or "down and depressed so I'd eat blocks of chocolate and chips." It is thus not surprising to find that women experience distress and self-loathing in relation to perceptions of a "fat" premenstrual body that "takes up more space."

Hatred of the fat body, and by implication the self, was evident in many women's accounts, with animalistic metaphors often being used. For example, "I feel like an elephant, very unattractive"; "I look at myself and I go 'You big fat pig,' I hate it"; "you're feeling revolting in yourself . . . you don't feel as feminine. I look like a dragon"; "I feel like a whale and hate my body during this time"; and "I feel like a frog . . . heavy, bloated, slow and lethargic." Animal metaphors are associated with dehumanization (Haslam, Loughnan, and Sun 2011) and social exclusion (Andrighetto et al. 2016), signifying a base and immoral nature, that lacks agency and rationality (Haslam 2006). Women who are animalized are positioned as creatures of emotion, nature and desire, and inferior to men (Tipler and Ruscher 2017), with pig and whale metaphors, in particular, signifying depravity (Haslam, Loughnan, and Sun 2011). Such dehumanization is also associated with the objectification of the female body (Morris, Goldenberg, and Boyd 2018), and thus self-positioning as animalistic serves to both denigrate the reproductive body and reinforce women's self-objectification during the premenstrual phase of the cycle. As the specific animal metaphors used by women signify fatness, self-hatred of premenstrual embodiment cannot be separated from the all-powerful cultural hatred of fatness.

"I Feel Really Exposed": Concealment and Separation of Self from the Unruly Premenstrual Body

People go to great lengths to distance themselves from or conceal their own 'beastly' animality (Haslam, Loughnan, and Sun 2011) or 'creatureliness' (Goldenberg et al. 2001). In this vein, visibility, and invisibility was central to the disciplining of the uncontained premenstrual body, associated with fear of surveillance from others, as well as constant self-surveillance. Many women attempted to conceal the premenstrual body from the critical gaze of others, reporting wearing "baggy clothes," "different clothes," "never leaving the house," or "staying away from the beach." This wasn't positioned as a form of coping or self-care in the face of discomfort—strategies that can reduce premenstrual distress (Ussher and Perz 2013b), but rather as a concealment of premenstrual abjection, and a resignation to making "less effort," all of which appeared to serve to add to women's distress.

> I do definitely feel bloated probably around that time, and so I won't wear the same clothes, *I'll wear more frumpy sort of clothes* and then I don't feel as good about myself as well. And *that probably adds a little bit to the negative moods.*

Women's attempts to conceal the fecund body reflects internalization of the discourse of the reproductive body as unclean and a source of pollution (Ussher 2006), which contributes to menstrual stigma and shame (Johnston-Robledo et al. 2007; Chrisler 2011). As one woman told us:

I feel that others are able to see my bloated stomach and recognise in me that I'm walking around premenstrual. It's like I'm carrying an extra burden of woman-ness around and I feel really exposed by that.

Another woman said "I change how I dress because I don't want to draw attention to the fact I'm about to bleed." Concealment of biological functioning is part of women's bodywork (Roberts 2004), and thus self-objectification serves as a "flight from corporeality" (Goldenberg et al. 2001) that 'thingifies' the premenstrual body and separates it from the self.

For many women, concealment was also focused on hiding "large," "swollen," "problematic breasts," that "go up a bra size." For example, "my breasts arrive a long time before I do if I've got PMS . . . I can be really self-conscious and embarrassed about it, so I try to cover it up"; "I feel my tits are so big that I can't put them in a particular shirt. So I'll want to hide them." These accounts reflect the positioning of a woman's breasts as signifiers of feminine sexuality (Young 1992), with large breasts associated with greater sexual objectification of women on the part of men (Gervais, Holland, and Dodd 2013). This can result in women feeling that they are constantly under surveillance and that their large breasts make them more noticeable and visible than other women (Millsted and Frith 2003). Whilst some women feel 'more attractive' as a result of premenstrual breast changes (King and Ussher 2013), or having large breasts (Millsted and Frith 2003), the accounts of women we interviewed reflect the greater body shame and social physique anxiety associated with an anticipated male gaze and objectification (Calogero 2004).

"I Feel Betrayed by My Body": Condemning Premenstrual Corporeality

Implicit in accounts of premenstrual embodiment is a body outside of the woman's control, undermining idealized femininity, wherein self-control is expected of 'good' women (Chrisler 2008). This was evident in accounts where women described the premenstrual body as a separate entity that was "doing" something to them, as evidenced in the following account: 'I hated my body very much for what it did to me . . . By 'hating my body for what it did' I mean everything, not just the physical effects." In this vein, many women condemned and further separated themselves from the premenstrual body, reporting feeling "betrayed," "disappointed," or "let down" by embodied changes. The 'out of control' premenstrual body is both positioned as cause of the woman feeling "fat" and "flabby," but also cause of her unruly emotions, illustrated in the example below.

I tend to put on a little bit of weight and stuff during that time too. So that makes me angry, because I am upset about that, and then *I tend to take it out on other people.*

The body is also implicated more broadly as a cause of premenstrual distress, described variously by women as caused by "crazy hormones," a "biological process" and "illness":

> I don't have control over how many hormones are flying about in my body, or anything like that.

> I feel like my hormones are not balanced, like they're completely out of whack. And – and then the brain whatever function. I definitely feel that it's, for it's biological, you know, affecting the way I think and feel.

The unruly premenstrual body therefore stands as a double assault on femininity—abhorrent, animalistic, fat, and "taking up more space," as well as out of control—the embodiment of the monstrous feminine (Ussher 2006). If women see themselves as uncontained and at the mercy of raging hormones or fatness, they position themselves as being attacked from within. The body becomes further objectified, alien to the woman, something that is acting against her (Ussher 2006). This blaming of the body may appear to function to exonerate the woman from judgments that attack her sense of self, as her abject corporeality and emotional transgressions are split off and projected onto a pathological condition, over which she has no control. Yet, as the focus of this projection is the reproductive body, which is implicitly positioned as disordered, unruly, and deviant, the outcome of this self-policing is a direct assault on the woman's corporeality (Ussher 2011). As Joan Chrisler argues, the fear of loss of control, and worry that others think we are out of control, serves as a form of "internalized oppression" that acts to "enforce gender roles and keep women from developing authentic selves" (Chrisler 2008, 8). However, this is not an inevitable process. Women can experience and acknowledge changes in premenstrual embodiment, without fear of loss of control or denigration of the self. The pull of the monstrous feminine can be resisted or reframed.

Reframing Premenstrual Embodiment: Resisting the Position of Monstrous Feminine

The women we interviewed all reported embodied change during the premenstrual phase of the cycle. However, these changes are not 'pure,' somehow beyond culture, beyond discourse. They are not simply *caused* by the reproductive body, by a syndrome called 'PMS.' And they are not inevitably experienced as distressing or problematic. It is important to acknowledge women's agency in negotiation of premenstrual change, and their ability to cope and make sense of premenstrual corporeality (Ussher and Perz 2013b).

For example, let us examine the debate about the reality of increases in body weight or "dimensions" during the premenstrual phase of the cycle. One study reported that whilst women reported premenstrual bloating, "objective" measurement could find no change, and the "discrepancy between the perceived body size and the actual body size (perception error) was significant" (Faratian et al. 1984). This may suggest that women are experiencing a "distortion of body image" premenstrually, a change in how they construct and position the body, rather than a material change in the body. Many women we interviewed appeared to construct embodied change in such a manner, aware that such change was more perceptual than material, and even describing corporeal self-condemnation as "irrational," as evidenced by the following extracts:

> Yes. I feel unattractive. *I know I still look the same* – it is all in my mind but that doesn't make me feel any better. I feel fat. I also will dress differently at that time of the month.

> I see all faults and *feel that they are larger than they are* (that is, my stomach, thighs) to the point that I can't stand to look at myself.

Some women told us that their partner reassured them that they "look as good today as you do any other time," but this had no impact, in the face of their "inner critic," which led one woman to say: "I feel fat. I feel ugly, I feel unattractive, unwanted. *I feel really paranoid*." In these accounts, women are both undermining the legitimacy of the embodied change that is the focus of their self-condemnation, and at the same time reinforcing self-criticism, by positioning hatred of the body as "irrational" or "paranoia," a manifestation of the pathology that is "PMS." However, women are also demonstrating awareness that they are *perceiving* the premenstrual body as "fat and ugly" and taking up the subject position of monstrous feminine as a result of this perception, which opens the door to the possibility of a reframing of both embodiment and the premenstrual self. This awareness is the first step in developing strategies of self-acceptance and self-care, and as a result, resisting self-objectification and self-positioning as the monstrous feminine (Ussher and Perz 2013b). It is a process that can be facilitated through women-centred psychological therapy.

Social constructionist and feminist critics have sometimes been critical of psychological 'intervention,' positioning it as a disciplinary practice that engenders self-policing through therapy, following a process of pathologization (Fee 2000; Foucault 1979; Ussher 2011). Women are told by experts within the 'psy-professions' others that they have a problem, and are then effectively positioned within the realm of psychiatric diagnosis and treatment, with all the regulation and subjugation that this entails (Ussher 2013).

However, we believe it is possible to simultaneously acknowledge the regulatory power of discourse and the role of the medical and psy-professions in women's subjectification, at the same time as recognizing the very real existence of distress, and the embodied or psychological changes women themselves experience associated with the fecund body. In order to do this, we have been involved in the development and evaluation of a non-pathologizing means of therapeutic support for women which acknowledges individual agency and the complex negotiations women engage in as they make sense of premenstrual change, with the aim of facilitating the adoption of strategies of self-care and coping (Ussher 2002). Drawing on both a narrative re-authoring framework (Guilfoyle 2014), and cognitive-behavioral models of PMS (Blake 1995), the specific aims of the therapy are to critically examine cultural constructions of femininity and PMS and how they impact women's premenstrual symptoms; to valorize women's expertise regarding their subjectivity and their bodies; to provide a non-pathologizing space for women to tell their story of PMS; to examine individual narrative constructions of PMS in the context of women's lives; to help women reframe their narrative to reduce distress; to identify and challenge negative cognitions associated with the body and with PMS; to examine perceptions of stress and of premenstrual symptoms to develop coping strategies for dealing with distress; and to encourage assertiveness and self-care throughout the cycle. This therapy has been found to be effective in significantly reducing premenstrual distress in a face to face one-to-one (Hunter et al. 2002) and couples format (Ussher and Perz 2017), as well as through self-help (Ussher and Perz 2006). In the face to face format, women discuss these issues with a therapist, over six to eight sessions, and engage in homework, such as doing things they enjoy, taking time-out, making note of the thoughts associated with premenstrual change and how these thoughts influence behavior, and practicing assertiveness. In the self-help format women are given information and exercises to practice at home.

Reevaluation of premenstrual embodiment is core to reduction in distress following this therapy (Ussher and Perz 2017; Ussher 2008a). In post-therapy interviews with women, we found marked reduction in reports of feeling "fat and ugly," or the use of animalistic metaphors. For example, women said "I don't feel bad about my body now. I couldn't care less about it now" and "I don't really have any negative feelings about my body anymore." This was associated with greater acceptance and understanding of embodied change, with less attention being paid to aspects of the body that had previously caused distress:

> What used to bother me before – bloating and not liking what I saw in the mirror, now doesn't seem to bother me as much, I do not dwell on it as much as I did before.

There were also accounts of awareness that such changes are normal and transitory, rather than a sign of pathology: "I know it's temporary and I know it's hormones and I know I'm bloated, so I'm not having as many issues with that."

Awareness of cyclical changes facilitated self-care: "I'm very aware of it when it is in the calendar and I can actually work my way around that with the knowledge that I might need a couple of days of rest, that I didn't used to do, and now I do." The development of active coping skills to deal with premenstrual changes included self-talk to reduce premenstrual negative moods, avoidance of conflict, changing perceptions of premenstrual emotion, and recognition of premenstrual needs: "taking the time-out to recognise my own needs has been very useful." Active engagement in coping strategies which focused on "looking after my body" or "feeling better about my body" included taking time to rest, engage in activities women enjoyed, exercise, meditation, improved diet, and reduction in alcohol and caffeine, illustrated in the example below.

> Physically I need a bit more rest is the main thing, um, so that's – that's a positive thing that it gives me that time to just slow down a bit and, um, have some time for myself. Like trying to do nice things for myself and do things that will make me feel good.

These self-care and coping strategies were reported to have a beneficial effect on women's moods, and their ability to control the experience and expression of negative emotion, resulting in significant reductions in premenstrual distress (Ussher and Perz 2017).

This form of psychological intervention does not remove premenstrual changes, but it can reduce and de-pathologize them, empower a woman to ask for appropriate support, and give her a greater sense of agency in relation to her body. This is no longer a passive docile body which requires medical management, but a body (and mind) which is positioned as being understood and accepted, potentially resulting in self-perceived "growth" through self-care:

> I am more sensitive around that period of time and I'm more susceptible to having old emotions and feelings that need to come up to leave me, but if I process it in the right way, it's a positive (and I've had) some growth out of it . . . I just think the self-care thing is a really big one for me, yeah.

Because of this, the majority of women reported that they felt confident that they could understand, and live with, their premenstrual changes, describing themselves as more "empowered," "energetic," and "creative" as a result. This is a movement away from the model of self-sacrificing femininity found to be associated with premenstrual distress (Ussher and Perz 2013a) to what has been described as a "mature model of care" (Pettersen 2012, 378), which acknowledges the importance of reciprocity and equality, and where self-care is be incorporated with care for others.

Conclusion

This form of women-centered psychological support can be effective in supporting women in the process of moving from an abject to an agentic subject position, without positioning her as needing to be managed, or her body as an unruly vessel that needs to be contained by experts (Ussher 2008a). 'PMS' is no longer positioned as an out of control illness, rather, as a label that makes sense of women's experience of psychological or embodied change in the premenstrual phase of the cycle (Ussher and Perz 2014). Women can resist the discursive positioning of the premenstrual woman as the epitomy of the monstrous feminine through positioning premenstrual emotions as "natural" or a reflection of "true feelings" about domestic, relationship or work issues, and embodied changes as something that can be understood and tolerated. This adoption of a PMS as normal/natural discourse served to facilitate women adopting an agentic position in relation to coping, through avoidance of stress and conflict, care of the self, and escaping relational demands and responsibilities (Ussher and Perz 2014). It can also function to engage partners in support, or facilitation of a woman's self-care (Ussher and Perz 2017). The body is central to this resistance of the monstrous feminine, as women can accept and acknowledge embodied change and psychological vulnerability, and even maintain the self-positioning as 'PMS sufferer,' without denigrating or pathologizing the body or the person.

This is analogous to the "tight-rope talk" identified by Sue McKenzie-Mohr and Michelle Lafrance, wherein women construct themselves as both "agents and patients: both active and acted upon" (McKenzie-Mohr and Lafrance 2011, 64), enabling women to take credit for agency in coping and deflect blame for "having" PMS. McKenzie-Mohr and Lafrance (2011) describe this adoption of a "both/and" position as enabling the re-authoring of emancipatory counterstories, which serve to challenge the oversimplification of "either/or" binaries, where women are "agent or patient," "powerful or powerless"; or in the case of PMS, premenstrual sufferer or non-sufferer/coper. As Catrina Brown (2007, 275) has argued, this "both/and" position "honors women's agency and power while not minimizing the impact of oppressive social discourses and social relations." This allows us to *both* acknowledge the materiality and discursive construction of premenstrual distress, *and* women's agency and power in understanding and coping with premenstrual change. It also allows us to acknowledge the complexities in women adopting the subject position "PMS sufferer," which *both* evokes connotations of the monstrous feminine *and* makes meaning of women's distress, through legitimizing their experiences as 'real' and as something that may require support. The reproduction and resistance of discourses associated with premenstrual embodiment are thus overlapping, rather than being discrete and separate processes (Day et al. 2010), and offering women-centered therapy for 'PMS' is not a form of regulation, but a feminist endeavor acknowledging women's need for understanding and support.

References

American Psychiatric Association. 2013. *Diagnostic and Statistical Manual of Mental Disorders, Edition V*. Washington DC: American Psychiatric Association.

Andrighetto, L., P. Riva, A. Gabbiadini, and C. Volpato. 2016. "Excluded from All Humanity: Animal Metaphors Exacerbate the Consequences of Social Exclusion." *Journal of Language and Social Psychology* 35 (6): 628–44.

Bhaskar, R. 1989. *Reclaiming Reality: A Critical Introduction to Contemporary Philosophy*. London: Verso.

Blake, Fiona. 1995. "Cognitive Therapy for Premenstrual Syndrome." *Cognitive and Behavioral Practice* 2 (1): 167–85.

Bobel, Chris. 2010. *New Blood: Third-Wave Feminism and the Politics of Menstruation*. Edited by Inc ebrary. New Brunswick, NJ: Rutgers University Press.

Bordo, S. 1990. "Feminism, Postmodernism and Gender-Scepticism." In *Feminism/Postmodernism*, edited by L. J. Nicholson, 133–56. New York: Routledge.

———. 1993. *Unbearable Weight: Feminism, Culture and the Body*. Berkeley: University of California Press.

Brown, C. 2007. "Feminist Therapy, Violence, Problem Drinking and Re-Storying Women's Lives: Reconceptualizing Anti-Oppressive Feminist Therapy." In *Doing Anti-Oppressive Practice*, edited by D. Baines, 128–44. Halifax: Fernwood Publishing.

Calogero, Rachel M. 2004. "A Test of Objectification Theory: The Effect of the Male Gaze on Appearance Concerns in College Women." *Psychology of Women Quarterly* 28 (1): 16–21.

Carr-Nangle, Rebecca E., William G. Johnson, Kimberly C. Bergeron, and Douglas W. Nangle. 1994. "Body Image Changes over the Menstrual Cycle in Normal Women." *International Journal of Eating Disorders* 16 (3): 267–73.

Chrisler, J. C. 2004. "PMS as a Culture-Bound Syndrome." In *Lectures on the Psychology of Women*, edited by J. C. Chrisler, C. Golden, and P. D. Rozee, 3rd ed., 110–27. Boston: McGraw Hill.

———. 2008. "2007 Presidential Address: Fear of Losing Control: Power, Perfectionalism and the Psychology of Women." *Psychology of Women Quarterly* 32 (1): 1–12.

———. 2011. "Leaks, Lumps, and Lines: Stigma and Women's Bodies." *Psychology of Women Quarterly* 35 (2): 202–14.

Chrisler, J. C., and P. Caplan. 2002. "The Strange Case of Dr. Jekyll and Ms. Hyde: How PMS Became a Cultural Phenomenon and a Psychiatric Disorder." *Annual Review of Sex Research* 13: 274–306.

Cosgrove, Lisa, and Paula J. Caplan. 2004. "Medicalizing Menstrual Distress." In *Bias in Psychiatric Diagnosis*, edited by Paula J. Caplan and Lisa Cosgrove, 221–32. Northvale, NJ, US: Jason Aronson.

Day, Katy, Sally Johnson, Kate Milnes, and Bridgette Rickett. 2010. "Exploring Women's Agency and Resistance in Health-Related Contexts: Contributors' Introduction." *Feminism & Psychology* 20 (2): 238–41.

Endicott, J., and W. Harrison. 1990. *Daily Rating of Severity of Premenstrual Problems Form*. New York: Department of Research Assessment and Training, New York Psychiatric Institute.

Fahs, B. 2016. *Out for Blood: Essays on Menstruation and Resistance*. New York: SUNY Press.

Faratian, B., A. Gaspar, P. M. S. O'Brien, I. R. Johnson, G. M. Filshie, and P. Prescott. 1984. "Premenstrual Syndrome: Weight, Abdominal Swellin, and Perceived Body Image." *American Journal of Obstetrics and Gynecology* 150 (2): 200–4.

Fee, Dwight. 2000. *Pathology and the Postmodern: Mental Illness as Discourse and Experience*. London: Sage.

Foucault, M. 1979. *Discipline and Punish: The Birth of the Prison*. London: Penguin.

Frank, Robert. 1931. "The Hormonal Causes of Premenstrual Tension." *Archives of Neurological Psychiatry* 26: 1053.

Fredrickson, Barbara L., and Tomi-Ann Roberts. 1997. "Objectification Theory: Toward Understanding Women's Lived Experiences and Mental Health Risks." *Psychology of Women Quarterly* 21 (2): 173–206.

Gervais, S. J., A. M. Holland, and M. D. Dodd. 2013. "My Eyes Are Up Here: The Nature of the Objectifying Gaze toward Women." *Sex Roles* 69 (11–12): 557–70.

Goldenberg, J. L., T. Pyszczynski, J. Greenberg, S. Solomon, B. Kluck, and R. Cornwell. 2001. "I Am Not an Animal: Mortality Salience, Disgust, and the Denial of Human Creatureliness." *Journal of Experimental Psychology: General* 130 (3): 427.

Guilfoyle, Michael. 2014. *The Person in Narrative Therapy: A Post-Structural, Foucauldian Account*. Edited by Connect Palgrave: Houndmills, Basingstoke: Palgrave Macmillan.

Hartlage, S., Sally Freels, Nathan Gotman, and Kimberly Yonkers. 2012. "Criteria for Premenstrual Dysphoric Disorder: Secondary Analyses of Relevant Data Sets." *Archives of General Psychiatry* 69 (3): 300.

Haslam, Nick. 2006. "Dehumanization: An Integrative Review." *Personality and Social Psychology Review* 10 (3): 252–64.

Haslam, N., S. Loughnan, and P. Sun. 2011. "Beastly: What Makes Animal Metaphors Offensive?" *Journal of Language and Social Psychology* 30 (3): 311–25.

Hunter, M. S., J. M. Ussher, S. Browne, M. Cariss, R. Jelley, and M. Katz. 2002. "A Randomised Comparison of Psychological (Cognitive Behaviour Therapy), Medical (Fluoxetine) and Combined Treatment for Women with Premenstrual Dysphoric Disorder." *Journal of Psychosomatic Obstetrics and Gynaecology* 23: 193–99.

Ismaili, E., S. Walsh, P. M. S. O'Brien, T. Bäckström, C. Brown, L. Dennerstein, E. Eriksson, E. W. Freeman, K. M. K. Ismail, N. Panay, T. Pearlstein, A. Rapkin, M. Steiner, J. Studd, I. Sundström-Paromma, J. Endicott, C. N. Epperson, U. Halbreich, R. Reid, D. Rubinow, P. Schmidt, K. Yonkers, and Disorders Consensus Group of the International Society for Premenstrual. 2016. "Fourth Consensus of the International Society for Premenstrual Disorders (ISPMD): Auditable Standards for Diagnosis and Management of Premenstrual Disorder." *Archives of Women's Mental Health* 19 (6): 953–58.

Jappe, L. M., and R. M. Gardner. 2009. "Body-Image Perception and Dissatisfaction Throughout Phases of the Female Menstrual Cycle." *Perceptual and Motor Skills* 108 (1): 74–80.

Johnston-Robledo, I., K. Sheffield, J. Voigt, and J. Wilcox-Constantine. 2007. "Reproductive Shame: Self-Objectification and Young Women's Attitudes toward Their Reproductive Functioning." *Women & Health* 46 (1): 25–39.

Kaczmarek, M., and S. Trambacz-Oleszak. 2016. "The Association between Menstrual Cycle Characteristics and Perceived Body Image: A Cross-Sectional Survey of Polish Female Adolescents." *Journal of Biosocial Science* 48 (3): 374–90.

Keat, R. 1979. "Positivism and Statistics in Social Science." In *Demystifying Social Statistics*, edited by J. Irvine, I. Miles, and J. Evans, 75–86. London: Routledge.

King, Marlee, and Jane M. Ussher. 2013. "It's Not All Bad: Women's Construction and Lived Experience of Positive Premenstrual Change." *Feminism & Psychology* 23 (3): 399–417.

Kleinstäuber, Maria, Katarina Schmelzer, Beate Ditzen, Gerhard Andersson, Wolfgang Hiller, and Cornelia Weise. 2016. "Psychosocial Profile of Women with Premenstrual Syndrome and Healthy Controls: A Comparative Study." *International Journal of Behavioral Medicine* 23 (6): 752–63.

Kleinstäuber, Maria, Michael Witthöft, and Wolfgang Hiller. 2012. "Cognitive-Behavioral and Pharmacological Interventions for Premenstrual Syndrome or Premenstrual Dysphoric Disorder: A Meta-Analysis." *Journal of Clinical Psychology in Medical Settings* 19 (3): 308–19.

Lupton, Deborah. 2013. *Fat.* Milton Park, Abingdon, Oxon; New York, NY: Routledge.

McKenzie-Mohr, S., and M. N. Lafrance. 2011. "Telling Stories Without the Words: 'Tightrope Talk' in Women's Accounts of Coming to Live Well After Rape or Depression." *Feminism & Psychology* 21: 49–73.

Millsted, Rachel, and Hannah Frith. 2003. "Being Large-Breasted: Women Negotiating Embodiment." *Women's Studies International Forum* 26 (5): 455–65.

Morris, K. L., J. Goldenberg, and P. Boyd. 2018. "Women as Animals, Women as Objects: Evidence for Two Forms of Objectification." *Personality and Social Psychology Bulletin* 44 (9): 1302–14.

Muljat, Alesha M., M. Kathleen B. Lustyk, and Andrea Miller. 2007. "Stress Moderates the Effects of Premenstrual Symptomatology on Body Image Reports in Women." *Annals of Behavior al Medicine* 33: S157–S157.

Nevatte, Tracy, Patrick O'Brien, Torbjorn Bäckström, Candace Brown, Lorraine Dennerstein, Jean Endicott, C. Epperson, Elias Eriksson, Ellen Freeman, Uriel Halbreich, Khalid Ismail, Nicholas Panay, Teri Pearlstein, Andrea Rapkin, Robert Reid, David Rubinow, Peter Schmidt, Meir Steiner, John Studd, and Inger Sundström-Poromaa. 2013. "ISPMD Consensus on the Management of Premenstrual Disorders." *Archives of Women's Mental Health* 16 (4): 279–91.

Perz, J., and J. M. Ussher. 2006. "Women's Experience of Premenstrual Syndrome: A Case of Silencing the Self." *Journal of Reproductive and Infant Psychology* 24 (4): 289–303.

Pettersen, Tove. 2012. "Conceptions of Care: Altruism, Feminism, and Mature Care." *Hypatia* 27 (2): 366–89.

Racine, Sarah E., Kristen M. Culbert, Pamela K. Keel, Cheryl L. Sisk, S. Alexandra Burt, and Kelly L. Klump. 2012. "Differential Associations between Ovarian Hormones and Disordered Eating Symptoms across the Menstrual Cycle in Women." *International Journal of Eating Disorders* 45 (3): 333–44.

Roberts, Tomi-Ann. 2004. "Female Trouble: The Menstrual Self-Evaluation Scale and Women's Self-Objectification." *Psychology of Women Quarterly* 28: 22–26.

Sims-Schouten, W., S. C. E. Riley, and C. Willig. 2007. "Critical Realism in Discourse Analysis: A Presentation of Systematic Method of Analysis Using Women's Talk of Motherhood, Childcare and Female Employment as an Example." *Theory Psychology* 17 (1): 101–24.

Speer, S. 2000. "Let's Get Real? Feminism, Constructivism and the Realism/Relativism Debate." *Feminism and Psychology* 10 (4): 519–30.

Teixeira, A. L. S., M. R. C. Dias, V. O. Damasceno, J. A. Lamounier, and R. M. Gardner. 2013. "Association between Different Phases of Menstrual Cycle and Body Image Measures of Perceived Size, Ideal Size, and Body Dissatisfaction." *Perceptual and Motor Skills* 117 (3): 892–902.

Tipler, C. N., and J. B. Ruscher. 2017. "Dehumanizing Representations of Women: The Shaping of Hostile Sexist Attitudes through Animalistic Metaphors*." *Journal of Gender Studies* 28: 1–10.

Tischner, Irmgard. 2013. *Fat Lives: A Feminist Psychological Exploration.* Abingdon, Oxon; New York, NY: Routledge.

Ussher, J. M. 1989. *The Psychology of the Female Body.* New York: Routledge.

———. 2002. "Processes of Appraisal and Coping in the Development and Maintenance of Premenstrual Dysphoric Disorder." *Journal of Community and Applied Social Psychology* 12: 1–14.

———. 2003. "The Role of Premenstrual Dysphoric Disorder in the Subjectification of Women." *Journal of Medical Humanities* 24 (1/2): 131–46.

———. 2004. "Premenstrual Syndrome and Self-Policing: Ruptures in Self-Silencing Leading to Increased Self-Surveillance and Blaming of the Body." *Social Theory & Health* 2 (3): 254–72.

———. 2005. "Unravelling Women's Madness: Beyond Positivism and Constructivism and towards a Material-Discursive-Intrapsychic Approach." In *Women, Madness and the Law: A Feminist Reader,* edited by R. Menzies, D. E. Chunn, and W. Chan, 19–40. London: Glasshouse Press.

———. 2006. *Managing the Monstrous Feminine: Regulating the Reproductive Body, Women and Psychology.* London: Routledge.

———. 2008a. "Challenging the Positioning of Premenstrual Change as PMS: The Impact of a Psychological Intervention on Women's Self-Policing." *Qualitative Research in Psychology* 5 (1): 33–44.

———. 2008b. "Reclaiming Embodiment within Critical Psychology: A Material-Discursive Analysis of the Menopausal Body." *Social and Personality Psychology Compass* 2 (5): 1781–98.

———. 2011. *The Madness of Women: Myth and Experience.* London: Routledge.

———. 2013. "Diagnosing Difficult Women and Pathologising Femininity: Gender Bias in Psychiatric Nosology." *Feminism & Psychology* 23 (1): 63–69.

Ussher, Jane M., and Janette Perz. 2006. "Evaluating the Relative Efficacy of a Self-Help and Minimal Psycho-Educational Intervention for Moderate Premenstrual Distress Conducted from a Critical Realist Standpoint." *Journal of Reproductive and Infant Psychology* 24 (2): 347–62.

———. 2010. "Disruption of the Silenced-Self: The Case of Pre-Menstrual Syndrome." In *The Depression Epidemic: International Perspectives on Women's Self-Silencing and Psychological Distress,* edited by D. C. Jack and A. Ali, 435–58. Oxford: Oxford University Press.

———. 2013a. "PMS as a Gendered Illness Linked to the Construction and Relational Experience of Hetero-Femininity." *Sex Roles* 68 (1–2): 132–50.

———. 2013b. "PMS as a Process of Negotiation: Women's Experience and Management of Premenstrual Distress." *Psychology & Health* 28 (8): 909–27.

———. 2014. "'I Used to Think I Was Going a Little Crazy': Women's Resistance of the Pathologization of Premenstrual Change." In *Creating Counterstories,* edited by S. McKenzie-Mohr and M. Lafrance, 84–101. London: Routledge.

———. 2017. "Evaluation of the Relative Efficacy of a Couple Cognitive-Behaviour Therapy (CBT) for Premenstrual Disorders (PMDs), in Comparison to One-to-One CBT and a Wait List Control: A Randomized Controlled Trial." *PLoS One* 12 (4): e0175068.

White, C. P., C. Hitchcock, Y. M. Vigna, and J. C. Prior. 2011. "Fluid Retention over the Menstrual Cycle: 1-Year Data from the Prospective Ovulation Cohort." *Obstetrics and Gynecology International* 2011: 1–7.

Williams, S. 2003. "Beyond Meaning, Discourse and the Empirical World: Critical Realist Reflections on Health." *Social Theory & Health* 1: 42–47.

Young, I. M. 1992. "Breasted Experience: The Look and the Feeling." In *The Body in Medical Thought and Practice*, edited by D. Leder, 215–32. Dordrecht, Netherlands: Kluwer Academic Publishers.

CHAPTER 20

Learning About What's "Down There": Body Image Below the Belt and Menstrual Education

Margaret L. Stubbs and Evelina W. Sterling

In Peggy Orenstein's (2016) book *Girls and Sex: Navigating the Complicated New Landscape* girls' reflections on their emerging sexuality are fumbling; they seem to lack a coherent language for describing their genitals. For example, one high school senior said, "I've never been comfortable with my parts down there" (Orenstein 2016, 65). Another reflected, "A guy is totally aware of what he looks like down there . . . but I don't know what they're seeing on me. I can't see it" (Orenstein 2016, 63). Preparation for menarche and its management compels a specific focus for girls on their genital area which has likely not occurred since learning about toileting and personal cleanliness in early childhood. As such, learning about periods *could* be an opportunity to help girls build a more accurate and positive foundation for understanding and thinking themselves as sexual beings. However, given girls' current reluctance or inability to name or see their genitals, developing agency in self-care or explorations of sexuality is unlikely. While the relationship between sexual behavior and menstruation is more apparent to older females, describing this relationship is challenging in educating young girls about menstruation. Nevertheless, providing an accurate and developmentally appropriate description of this relationship *is* possible and important in providing support for the development of girls' healthy sexuality.

C. Bobel et al. (eds.), *The Palgrave Handbook of Critical Menstruation Studies*, https://doi.org/10.1007/978-981-15-0614-7_20

GIRLS' DEVELOPING SEXUALITY

Feminist researchers have long provided insight into how best to study and characterize the development of girls' healthy sexuality (Fine 1988; Tolman 2002). This work acknowledges adolescent sexuality as a normative developmental process including recognition of sexual feelings, and exploration of sexual behavior over time (Tolman and McClelland 2011; Fortenberry 2014). Using both quantitative and qualitative methods, some focus on listening to girls as narrators, not just reporters, of their experiences (Tolman 2012). Distinctions between sexual agency and sexual subjectivity (Lamb and Peterson 2012) and the meaning of "empowerment" in the context of neoliberal messages that emphasize individual choice (Bay-Cheng 2015; Tolman, Anderson, and Belmont 2015) are also investigated.

Sexualization and objectification are features of the contemporary US society in which girls develop as sexual beings (American Psychological Association Task Force on the Sexualization of Girls 2007). These cultural features have also been noted as problematic outside of the United States as well (Crawford et al. 2008; Dakanalis et al. 2017). The influence of these features in the United States has been linked to negative attitudes toward menstruation, for example, the wish to eliminate periods (Johnston-Robledo et al. 2003), and more favorable attitudes to contraception that reduces or stops menstruation (Johnston-Robledo et al. 2007). Continuing research must now also address how self-sexualization and self-objectification are endorsed by girls themselves.

Shifting standards around genital appearance also impact girls' sexual body image. Research from developed countries has documented women's dissatisfaction with labia appearance, and the preference among older and younger females for what might be considered a prepubescent, non-descript "Barbie" vaginal area (Schick, Rima, and Calbrese 2011; Sharp, Tiggeman, and Mattiske 2014). This trend has also been widely noted in related public commentary (Bowerman 2017; Rabin 2016). The demand for surgical alteration of the labia, or labiaplasty, has increased dramatically in the United States (American Society for Aesthetic Plastic Surgeons [ASAPS] 2016; American Society of Plastic Surgeons [ASPS] 2016a, 2016b). In 2016, females aged 19–34 accounted for nearly half of all labia surgeries. While about 5% were performed on females under the age of 18, an increase in labiaplasty for these girls has grown since 2014 (ASAPS 2014, 2015, 2016).

Controversy surrounds the necessity for labia surgery. The wide variability of clitoral and labial presentation has been documented as normative (Lloyd et al. 2005). Further, no standardized criteria to define the medical need for labia surgery currently exist (Motakef et al. 2015). Aesthetic and plastic surgeons espouse women's (and girls') right to choose genital surgeries, including labiaplasty, and report client satisfaction after these procedures (Goodman et al. 2016; Hardwick-Smith 2011). The potential of therapeutic labiaplasty has also been acknowledged as an approach to curbing non-therapeutic female genital alteration where this is commonly practiced (Arora and Jacobs 2016). Nevertheless, the American College of Obstetrics and Gynecology [ACOG] (2007, 2017) argue against labiaplasty unless medically necessary for adolescent females under 18.

Some researchers identify easy access to pornography featuring hairless female genitalia and tucked-in labia minora as contributing to the demand for labiaplasty (Mowet et al. 2015). However, females' lack of awareness of the wide variation in genital appearance has also been named as contributing (Lloyd et al. 2005; Renganathan, Cartwright, and Cardozo 2009). It seems girls are aware of popular genital appearance *norms*, but unaware of the wide variation in typical, healthy female genital anatomical structure. Because learning and menstruation does not address this gap in body knowledge, it misses an early chance to provide evidence-based information about female genital structure and function.

Girls Talk About "Down There"

Of many topics related to the sexual behaviors and attitudes of those under 18, we focus here on girls' thoughts about engaging in oral sex. These yield important and specific information about their attitudes toward their own genitalia. We draw from two sources, heavily cited. Neither claim generalizability, but instead give voice to girls' own views. One is a peer-reviewed study of the oral sex experiences of 98 girls aged 12–17 (Burns, Futch, and Tolman 2011). A second, intended for the lay public, is Orenstein's (2016) report of her own journalistic investigation of girls' sex lives more broadly, from interviews with more than 70 girls and young women aged 15–20. We note, regrettably, that this material reflects the thoughts and experiences of heterosexual girls with male partners. More research is needed about the sexual (and menstrual) experiences of lesbian, bisexual, and transgender youth under 18.

Giving and Receiving (or Not) Oral Sex

In both sources, girls discussed how and when to perform oral sex on males, and what they felt about doing it. Girls acknowledged an expectation to do it. One explained: "It's always the same unspoken sequence . . . You make out, then he feels you up, then you give him head, and that's it" (Orenstein 2016, 58). Another said: ". . . there's pressure for girls to do it. So it's about how comfortable you are with resisting that pressure or not. It gets awkward to keep resisting" (Orenstein 2016, 58). Further, Orenstein's (2016) participants thought oral sex wasn't a big deal, wasn't sex (sex was sexual intercourse), and was safer than sex because girls weren't giving up their virginity, couldn't get pregnant, or get sexually transmitted infections.

In contrast, girls mentioned avoiding receiving oral sex, relating their reluctance to discomfort with and dislike of their own genital area. Given girls' ready access to current genital appearance norms, and the lack of information about the wide variation in female genital appearance, their genital dissatisfaction is not surprising. In a discussion with two girls, one said she'd "rather have sex before I'd do that" (Orenstein 2016, 63). Another who gave her boyfriend oral sex said she

> . . . never felt comfortable for him to return the favor. Because . . . okay this is weird to say, but a guy going down on you is more like a sacred thing. Like once you've done *that*, you must be really comfortable with the person, because it is not something that I'm just going to let you do . . . A guy is totally aware of what he looks like "down there" . . . but I don't know what they're seeing on me. I can't see it. (Orenstein 2016, 63)

When Orenstein (2016, 63) suggested "Well . . . there are these things called mirrors . . .," this interviewee replied: "Yeah . . . I'm not going to do that." Both girls were averse to seeing their own genitals.

Another, who prided herself on giving oral sex, was more direct. She offered that her boyfriend didn't want to do it and she'd never asked because "I don't like my vagina . . . I know that sucks. And I don't know why it should be so different, but I've internalized the idea" (Orenstein 2016, 63). Another, who self-identified as a feminist, said that when she received oral sex from her boyfriend, it was his idea:

> It was not fun for me. I was not comfortable with it at all. I guess I've never been comfortable with my parts "down there." It's not something I find attractive. So I don't like the idea of someone else "down there." (Orenstein 2016, 65)

Girls also mentioned potentially being embarrassed and teased by boys about

> . . . a fart with your vagina [featured on South Park] . . . now teenage boys have that as something they can say about girls and girls know they have that so you feel awkward . . . It's just there's this whole culture around making fun of female sexuality, you know? And it's super strong. (Orenstein 2016, 63–64)

Additionally, most of Orenstein's interviewees shaved or waxed their public hair, many since age 14. They never questioned it because they were already shaving their underarms and legs, and they'd seen older girls who were hairless. Girls also said it made them feel cleaner. Some mentioned that they shaved their public hair to avoid embarrassment, because they believed boys were disgusted by seeing it. Interestingly, neither Orenstein (2016) nor Burns, Futch, and Tolman (2011) reported comments from girls about encountering boys' pubic hair when giving oral sex.

Other thoughts about performing oral sex included describing it as a way to gain social status, or ". . . make friends with popular guys . . . It's how you rack up points for hooking up with someone without having to actually have sex" (Orenstein 2016, 54). Some mentioned using oral sex to avoid sexual intercourse in a hook up: ". . . if he expects and wants to be satisfied [but]

I want him to leave and I don't want anything to happen . . ." (Orenstein 2016, 54–55). Some girls recognized a lack of reciprocity related to oral sex:

> It's just expected that guys will get off . . . and then maybe he'll be like, 'Oh, do you want me to . . .?' It's never like he'll do something for me and *maybe* I'll do something for him. It's like *naturally* I do something and then he'll ask if I 'want' him too. (Orenstein 2016, 61)

Others linked the lack of reciprocity in oral sex to not expressing their own desires. One said: "It goes back to girls feeling guilty. If you go to a guy's room and are hooking up with him, you feel bad leaving him without pleasing him in some ways" (Orenstein 2016, 55). Another added: "I don't think girls are taught to express their wants. We're these docile creatures that just learn to please" (Orenstein 2016, 58). One noted a disconnect between being a strong assertive female in other arenas, but not in terms of sexual behavior: "I think I didn't realize . . . I guess no one ever told me that the strong female image also applies to sex" (Orenstein 2016, 58).

Did girls enjoy performing oral sex? One of Orenstein's (2016, 56) participants mentioned that it was "sort of fun getting in the rhythm of it . . ." but she also said that the physical part of it was "gross" and really hurt her throat: ". . . it's never *fun* fun." Girls interviewed by Burns, Futch, and Tolman (2011) also mentioned physical discomfort when performing oral sex. When asked about whether they themselves experienced excitement below the neck, girls offered no comments at all.

Girls interviewed by Burns, Futch, and Tolman (2011) mentioned being scared about not knowing how to do it and worrying about whether they would do it right. To get started, one girl described just doing it, like doing homework or taking a test, taking that first step, keeping on going, and then, finally, being finished. Eventually, with practice, and thinking about it while doing, girls did learn to do it. One said ". . . it wasn't intimate at all . . . It was more of, um, like an instructional video (laughs) I swear" (Burns, Futch, and Tolman 2011, 246). Some girls commented that they felt a sense of accomplishment in having learned how to give oral sex but even so, Burns, Futch, and Tolman (2011, 247) concluded that giving oral sex is not an "embodied experience for girls, but a cerebral one."

Girls willingness to learn how to give oral sex, even though finding it uncomfortable, even painful, and not particularly enjoyable, begs further analysis about agency, choice, and giving and receiving pleasure. That girls are much less interested in or even willing to receive oral sex seems, at least in part, due to their disdain for their own, but not boys' "down there."

Revisioning Menstrual Education

We take girls' comments as an opening to consider how preparation for menstruation can present young girls with accurate information about female genitalia as a foundation to buffer genital appearance and sexual expectation norms. Preparation for menstruation occurs in the context of approaching adolescence when changes in physical growth will be accompanied by sexual feelings and explorations. What better time to provide girls with a new vision of "down there," one that goes beyond a cursory explanation of menstruation and how to manage it, to one presenting a broader view of the cycle related to many aspects of health, including the healthy sexuality?

Such a revisioning of menstrual education first requires an acknowledgment of current cultural views of menstruation. Despite activist efforts to offset it (Bobel 2018; Fahs 2016; Weiss-Wolf 2017), stigma associated with menstruation, including menstrual sex as taboo (Fahs 2011, 2014), is still widespread throughout the world (Chrisler and Johnston-Robledo 2018; Dutta et al. 2016; Sommer et al. 2016). Negative views and representations of menstruation are not lost on adolescents or younger girls, who are avid readers of materials for adults (Stubbs 2008) and accomplished users of the internet (for example, Guttmacher Institute 2017). Especially during the transition from childhood to adolescence, girls become increasingly aware of cultural messages about both gender role expectations and discriminatory treatment of girls and women. They are vulnerable to espousing them, sometimes in conflict with their own views (Brown and Gilligan 1992; Piran 2017). With respect to menstruation, girls face a developmental dilemma—how to accept it as normative in the context of persistent menstrual stigma.

Given this situation, we now critique specific aspects of menstruation education gleaned from an exploratory review of 40 available books providing information on menstruation first undertaken by one of us (Stubbs 2013). Books were found in the children's section of a library located in an urban mid-Atlantic city, recommended by colleagues, or listed on several commercial websites such as Amazon.com. Since, we have perused updated versions of some reviewed in the original set and some more recent books for girls on the market.

How Menstrual Education Shortchanges Girls

Puberty as Problematic

Because information about menstruation is almost exclusively presented within a broader discussion of puberty, descriptions of puberty set the stage for information to follow. Unfortunately, puberty is often described as problematic—a time of upheaval and being out of control. Further, in descriptions of pubertal changes hormones are emphasized as in control, not girls themselves. While hormones *are* integral to pubertal processes, their role is often overdone and sensationalized. Hormones are described as racing around the body which girls can do nothing about (Thomas 2011). Rue (1995, 13) states: "What can I do about hormones? Well, nothing." Movsessian (2004, 25) fuels girls' powerlessness in the face of hormones and discourages the possibility of thinking about any other causes for their feelings or behavior:

Girls, when in doubt blame it on the hormones. I'm in a bad mood? It's the hormones. I want to eat that big slab of chocolate cake? It's the hormones. I'm getting oilier skin, hair and pimples? But of course, all together now, 'It's the hormones!' . . . You may start off being a mild-mannered little girl, minding your own business, getting on with things and then what happens? You fill out, get hairs growing all over the shop, get boobs and become . . . a puberty girl.

These introductory messages foreshadow uncertainty and incapacity as girls move forward to learn about menstruation and other pubertal events.

PMS: A Ubiquitous Cluster of Negative Symptoms

Negative physical and emotional symptoms (bloating, breast tenderness, acne, cramps, fatigue, irritability, and sadness) are mentioned as common menstrual experiences and typically categorized together as premenstrual syndrome (PMS). In the American Girl's *Care and Keeping of You* for younger girls aged eight to ten, Schaefer (2012, 76) explains:

> What is PMS? Once you begin to menstruate regularly you may notice some patterns in how you feel right before your period. Sometimes these physical and emotional symptoms are referred to as *premenstrual syndrome* or PMS for short. It's not a disease or an illness, just a natural part of your menstrual cycle.

Using the word "symptom" and naming the clustering of "symptoms" as a "syndrome" is problematic. These terms prime girls to associate these experiences with illness (Chrisler 1996). The assertion that these "symptoms" do *not* indicate illness or disease contradicts common usage of these terms and is confusing. Presenting negative "symptoms" in a defined category is especially problematic for younger girls since as concrete thinkers (Berger 2017) they are not yet cognitively able to critically evaluate this information.

The American Medical Association's *Girl's Guide to Becoming a Teen* also asserts that PMS symptoms are normal, but advises that ". . . it's important not to let them ruin your day (or your week!)" (Middleman and Pfeifer 2006, 71). This comment encourages girls to distance themselves from paying attention to their own experience, that is, by *not* thinking about it. Thomson (1995) goes a bit further suggesting that to deal with PMS, girls put things off until after their periods, thus undermining girls' agency and underscoring "loss-of-control" messages.

Finally, while many sources mention that hormones cause PMS, its etiology is still not clear (Hofmeister and Bodden 2016; Fisher et al. 2016). Also not mentioned are psychosocial factors which have been explored as contributors to PMS (King and Ussher 2012; Read, Perz, and Ussher 2014; Ussher and Perz 2013).

WHAT'S NORMAL AND WHAT'S NOT?

Some sources suggest that there is no normal way to have a period. Marzollo (1989, 3) writes: "There is no perfectly normal way to have your period. Instead there are lots of normal ways, and your way is one of them." Bloom (2017, 19) offers: "You'd be surprised how much is considered 'in the range of normal' when it comes to puberty." Yet juxtaposed is often at least a brief description of the "average" period which happens about every 28 days, includes 3–5 days of bleeding, and about six tablespoons of total blood loss. Thus, girls receive a double message: there *is* an average, "normal" period which includes PMS as *likely*, but one's unique menstrual experiences are *also* normal. How are young girls, especially concrete thinkers, to judge their own experience, in the context of these contradictory messages? Further, in the absence of detail about severity, how is a girl to judge when her experiences are important enough to be concerned about, when the primary message is that these changes are normal?

The notion that "there is no normal" may serve to reassure girls that there is a lot of variation in menstrual experience, which is certainly true. However, clinicians themselves acknowledge that it is difficult to diagnose serious underlying conditions in adolescents because menstrual cycles can take 2–3 years to become established in an individual (Hofmeister and Bodden 2016), and we must add, do *not* necessarily standardize to the parameters of the "average" period. Moreover, clinicians who assume that anything goes in early menstrual life may themselves not consider these experiences as problematic (Hillard 2014).

Two aspects of early menstrual experience often presented as normal, cramps, and irregularity, should be further considered. Gravelle and Gravelle (2006, 36) write: "Although cramps are uncomfortable, they don't mean that anything is wrong." Regarding irregularity, Dickerson (2005, 118) suggests that "For some girls being irregular is just part of who they are." Clear clinical guidelines *do* exist describing what's normal and not related to menstruation (Hillard 2014), including information about cramps and irregularity. These data warn that severe cramping or dysmenorrhea can indicate endometriosis (Hickey and Balen 2003; Hillard 2014). Similarly, cycle lengths lasting for more than 90 days, should be further investigated since irregularity can indicate polycystic ovarian syndrome, primary ovarian insufficiency, eating disorders and female athletic triad (Hillard 2014).

Also problematic is the alarmist language often used. For example, concerning toxic shock, Plaisted (2011, 24) warns (original emphasis):

> Most Important—tampons MUST be changed every few hours and should NEVER be left in the vagina for longer than eight hours. If you do, in rare cases, girls can develop something called Toxic Shock Syndrome, which will need urgent medical treatment.

Because they are salient, alarmist messages will stand out amid other information and can elicit girls' thoughts about personal vulnerability. The younger the reader, the less likely she will be able to think critically about such warnings.

Problematic Graphics and Stylistic Devices

The imagery in many menstrual educational materials can also be problematic. Many sources represent girls as cartoon figures, but some do not present varied body types. For example, while both slimmer and heavier girls were easily identified in figures presented in the older edition of the American Girls' *Body Book for Girls* (Schaefer 1998), in the newer version the drawings barely differentiate these body types (Schaefer 2012). At least varying ethnicities are now more often represented than not (including Schaefer 1998, 2012), but given the wide variety of body shapes and sizes among pre-menarcheal girls, those who can't identify their body type in these books may not see this material as relevant to them. When these representations include diagrams of internal or external genitalia, or how menstrual management products are used, they generally do not offer an accurate view of the size or location of these organs relative to a real girl's body. While the stand-alone line drawings or diagrams of the internal and external genitalia in many sources present a "standardized" view with accurate labels of the various parts, nowhere were representations or discussion of the wide variety of female external genital appearance.

Cartoon-bubble inserts or sidebars are also often used to explain a concept. The latter, as well as catchy chapter titles, are likely to appeal to young audiences but also often include stereotypically negative messages about menstruation. For example, Bailey (2016, 40–41) includes sidebar titles such as "Period Panic – No 1: Painful Periods" and "Period Panic – No 2: PMT [more commonly referred to as PMS]." Bloom (2017, 77) titles her chapter on menstruation "Shark Week."

No Mention of Menstruation as a "Vital Sign"

The most important failure of menstrual education is the omission from all materials reviewed that menstruation is a "vital sign" of general health to be considered along with body temperature, blood pressure, heart, and respiration rates. This perspective goes beyond stating that "menstruation is normal." First advanced over ten years ago by medical professionals, it has recently been reaffirmed by both the American Academy of Pediatrics (AAP) (2016) and ACOG's Committee on Adolescent Health Care (2015). Despite of the efforts of a committed few (for example, Hillard 2014), this perspective has not found its way into public discourse, menstrual education or even public health outreach.

Sterling (2018; Sterling, Karczmarczyl, and Ivabze 2017) has investigated the extent to which the menstrual cycle has been considered in core areas and outreach within public health initiatives, and finds it missing. She notes that menstrual management product manufacturers provide much of the menstrual education in schools but no program to evaluate their effectiveness. Critiques of these materials have long noted their presentation of mostly negative representations of menstruation (Johnston-Robledo and Chrisler 2013) and their continued emphasis on concealment and secrecy (Chrisler and Johnston-Robledo 2018). The omission of menstruation as a vital sign denies girls information about menstrual health and how that relates to other aspects of health and well-being, leaving the depiction of menstruation as a miserable menace unchallenged.

TOWARD A MORE POSITIVE VIEW OF MENSTRUATION AND "DOWN THERE"

We judge no one source as wholly negative. Rather, many sources include both problematic and admirable content and approaches to presentations of material. The problem is that no coordinated, evidence-based standards seem to guide what might be most helpful to girls about menstruation. Imagine what it might be like to create menstrual education content using evidence-based information about menstrual health as opposed to menstrual illness, acknowledging potential problems in a neutral way, and adjusting for cognitive ability and what we think of as a developmentally considered "need to know." Where is an evidenced-based scope and sequence chart of facts and concepts related to menstrual education? It has yet to be developed. Imagine, too, using menstruation as a vital sign as a starting point for menstrual education. Such a reframing could help to redress the burden of negativity in current materials. Toward that end, some specific suggestions for improving menstrual education follow.

Concerning Pubertal Change

Descriptions of puberty should provide positive aspects of change and growth and avoid overemphasizing the scariness of change or how immediately it occurs. Some books for girls do present this perspective. Marzollo (1989, 38) writes:

> One of the most amazing things your body does month after month, over and over again, is prepare to reproduce another human being. Although it is hoped that you won't use your reproductive system until you're truly ready to be a mother, knowing how amazing it is can give you new respect and even awe for yourself and your abilities. Such self-respect can help you in turn put menstruation into better perspective.

Feinmann (2004, 7) offers:

> Any big change can be scary. And let's face it, having your first period is proba-
> bly the biggest change in your life so far . . . Knowing what to expect will help
> you feel more in control. When you're in control, you're less anxious and better
> able to enjoy life.

We suggest drawing attention to emerging agency and self-reflection as
opposed to asserting an inevitable loss of personal control relegated instead
to an exaggerated role for hormones. We also advocate the use of neutral as
opposed to alarmist language. Indeed, some have found a way to describing
toxic shock using neutral as opposed to alarmist language (Loulan and
Worthen 2001; Middleman and Pfeifer 2006).

Concerning PMS

To replace the notion that PMS with its cluster of negative symptomatology
is inevitable, we advocate following Chrisler's (1996) suggestion that potential menstrual-related experiences be articulated as *changes* instead of *symp-*
toms to avoid embedding menstruation within the context of illness. Positive
premenstrual changes, which have been noted (King and Ussher 2012), can
also be mentioned. Given that biomedical and psychological guidelines now
suggest that premenstrual symptoms can persist even several days into one's
menstruation (Hofmeister and Bodden 2016), the term 'premenstrual' is confusing to us, but may explain why anecdotally, we have noticed that PMS now
seems to be a descriptor of both before and during at least part of one's menstrual cycle. We understand a need to mention PMS in some way in menstrual
education materials, since the term is used so frequently both in research and
by the public. Doing so, however, poses a challenge for educators who want
to provide a neutral perspective that supports girls in reflecting on their own
experiences as they enter menstrual life.

Concerning Genital Appearance

Many sources offer commentary on pubic hair, often featuring information
about how to groom or eliminate it. Many imply that pubic hair grooming is a matter of personal preference in the absence of scant information about potential health consequences related to various methods of
removal (DeMaria et al. 2014; Williamson 2015). These deserve more than
a mere mention, again, couched in neutral, non-alarmist language. Including
information about the variety of historical and cultural views on pubic hair
(and body hair in general) can further help girls contextualize current and
often changing beauty standards related to this aspect of genital appearance.

To help girls realize the size and shape of their internal genitalia, using
pictures of actual girls instead of cartoon figures should be considered.

Movsessian (2004) offers a clever way of presenting the internal organs. She superimposes a diagram of them on the abdomen of a fully clothed girl which shows the relative position and size of them more realistically than a stand-alone graphic. Pictures of real girls may make it easier for readers to engage with other content as well.

Finally, to encourage girls' acceptance for how their own genitals look, compared to how current beauty standards say they *should* look, representations of the wide variation in the appearance of the inner and outer labia and the clitoris should be presented as normative, along with information about historical, cultural, and biomedical preferences for how the shape and look of these parts of female anatomy are judged to be "acceptable" or not.

More discussion of alternative menstrual management products, such as menstrual cups, reusable pads, and menstrual underwear along with the standard pad and tampon examples is warranted, since very few books described these. This discussion could help girls think more about their genital area as they consider product choices. Inclusion here of information about environmental issues related to product manufacture and disposal, as well as their accessibility, or not, in the lives of girls worldwide would also be welcome.

Effective Approaches *Do* Exist

We found a variety of effective approaches in the sources reviewed that we hope to see again in new menstrual education materials. For example, inclusions of how to address practical problems, such as starting a period without a pad or tampon handy, will appeal especially to concrete thinkers, but also to older adolescents who have faced this challenge. Opportunities for interaction—quizzes about content, suggestions for reflective journaling, ways to document one's own development, and ways to track one's period—encourage readers' self-reflection and provide them with an active role in learning. Some sources provide a glossary, which is a quick resource for girls who may later come across terms already explained but can't quite remember the details. Insights about menstrual attitudes and practices from history, mythology, and psychosocial research can help girls contextualize their own experience and serve as a platform for using critical thinking to ponder their own practices related to menstruating. Finally, vignettes and comments from a variety of girls about their experiences will include their voices among those of adult authors for consideration.

Addressing Boys

Some books include both girls and boys as the audience and offer the chance to learn about their own and others' pubertal changes. However, if these books present negative representations of menstruation, boys as well as girls will also be subject to them. Accurate information about menstruation targeted just to boys is woefully absent (Allen, Kaestle, and Goldberg 2010; Lee 2009; Peranovic and Bentley 2017) and should be increased. This suggestion prompts addressing the menstrual experiences of trans individuals. Although an explanation of transgender identity is mentioned in some sources, none reviewed address the menstrual needs of these youth. For example, these youth may desire concealment for safety reasons, and not see it as a reflection of stigma as it might be for cis girls.

A Preface for Parents

Many sources also include a preface addressing both girls and mothers. Others are to be read by both mother and daughter. These are helpful approaches that could facilitate communication between both. The newer version of *Period!* (Loulan and Worthen 2001) comes with a detachable guide for parents and includes information about an approach a man might take as a menstrual educator. This is most welcome since men have not often been thought of in this role (Erchull and Richmond 2015). There are also books intended exclusively for mothers and other adult educators, yet another genre entirely in which information about menstruation should also be carefully reviewed.

Do mothers, (or fathers) actually buy or use the library to provide books about menstruation and puberty for their daughters? Which ones? Do daughters read these and/or appreciate having them? Today's girls and boys may be most comfortable learning about puberty and menstruation via websites. While future research should explore the content and use of these websites, we believe that our recommendations for improving menstrual education should inform any content regardless of the genre or venue in which it is presented.

Conclusion

Sexuality is inextricably linked to menstruation in terms of biological function, which is also understood psychosocially later in development. Yet we shy away from articulating this link in part because talking about sex and sexuality with children is challenging. Parents have reported that menstruation's relationship to reproduction, thus sexuality, is a stumbling block (Stubbs and Johnston-Robledo 2013). Similarly, in more formal menstrual education only scant attention to the connection between reproduction, sexuality, and menstruation is offered (Stubbs 2008). While girls may be told that menstruation occurs when an egg is not fertilized, what fertilization is or how it happens, begs detail.

Articulating how menstruation, reproduction, and sexuality intersect within menstrual education is long overdue and doable. Sheffield's (1988) book is an example of how even young children can be introduced to this linkage. Hers is an artistic portrayal intentionally designed as a starting point to later discussions rather than an explanation of details about the menstrual cycle. As girls get older, more information on the intersection of menstruation, reproduction, and sexuality can be added. Existing developmentally sequenced sex education curricula (Breuner et al. 2016; Flores and Barroso 2017) could help in creating a similar developmental approach to menstrual education curricula, which, if implemented, might actually provide parents (and others) with a more gradual way to address some aspects of sex.

In the sequence of development, the more intense focus on the genital area necessitated by preparation for menstruation provides a most opportune context for adding more fact-based information about the process of menstruation along with frank and accurate pictures or depictions and descriptions of the structures and functions of the reproductive organs involved. Inclusion at this time can serve as a knowledge base for girls to access later when they begin to explore becoming sexual. Accurate words (and pictures) of girls' vulvas, labia, clitorises, and vaginas earlier on in their development can help to normalize naming these body parts, as opposed to referring to them vaguely as "down there." The power to name and know their own bodies can give girls the power to name their sexual desires when they occur. To the extent that improved menstrual education can provide girls with a more accurate view of their genitals and a fuller understanding of these parts of the body, they will be positioned as more fully informed to evaluate and respond to the demands of fashion and make decisions about genital presentation, as well as those related to expectations about their sexual behavior.

REFERENCES

AAP (American Academy of Pediatrics). 2016. "Menstruation in Girls and Adolescents: Using Menstruation as a Vital Sign." *Pediatrics* 137 (3): e20154480. https://doi.org/10.1542/peds.2015-4480.

ACOG (American College of Obstetrics and Gynecology), Committee on Adolescent Health Care. 2015. "ACOG Committee Opinion No. 651: Menstruation in Girls and Adolescents: Using the Menstrual Cycle as a Vital Sign." *Obstetrics & Gynecology* 126 (6): e143–146. https://doi.org/10.1097/aog.0000000000001215.

———. 2017. "ACOG Committee Opinion No. 686: Breast and Labial Surgery in Adolescents." *Obstetrics and Gynecology* 129 (1): e17–e19. https://www.acog.org/-/media/CommitteeOpinions/Committee-on-Adolescent-HealthCare/co686.pdf?dmc=1&ts=20170217T0416002970.

ACOG (American College of Obstetrics and Gynecology), Committee on Gynecologic Practice. 2007. "ACOG Committee Opinion No. 378: Vaginal 'Rejuvenation' and Cosmetic Vaginal Procedures." *Obstetrics and Gynecology* 110 (3): 737–38. https://doi.org/10.1097/01.AOG.0000263927.82639.9b.

Allen, Kathrine R., Christine E. Kaestle, and Abbie E. Goldberg. 2010. "More Than Just a Punctuation Mark: How Boys and Young Men Learn about Menstruation." *Journal of Family Issues* 32 (2): 129–56. https://doi.org/10.1177/0192513x10371609.

APA (American Psychological Association), Task Force on the Sexualization of Girls. 2007. *Report of the APA Task Force on the Sexualization of Girls.* Washington, DC: American Psychological Association. http://www.apa.org/pi/women/programs/girls/report-full.pdf.

Arora, Kavita S., and Allan J. Jacobs. 2016. "Female Genital Alteration: A Compromise Solution." *Journal of Medical Ethics* 42 (3): 148–54. https://doi.org/10.1136/medethics-2016-103376.

ASAPS (American Society for Aesthetic Plastic Surgeons). 2014. "Cosmetic Surgery National Data Bank Statistics." *Aesthetic Surgery Journal* 34, no. 1 (1 July): 1–20. https://doi.org/10.1093/asj/34.Supplement_1.1.

———. 2015. "Cosmetic Surgery National Data Bank Statistics." *Aesthetic Surgery Journal* 35, no. suppl_2 (1 July): 1–24. https://doi.org/10.1093/asj/35.Supplement_2.1.

———. 2016. "Cosmetic Surgery National Data Bank Statistics." *Aesthetic Surgery Journal* 36, no. suppl_1 (1 May): 1–29. https://doi.org/10.1093/asj/36.Supplement_1.1.

ASPS (American Society of Plastic Surgeons). 2016a. *2016 National Plastic Surgery Statistics.* https://www.plasticsurgery.org/documents/News/Statistics/2016/2016-plastic-surgery-statistics-report.pdf.

———. 2016b. *Plastic Surgery Statistics Report.* https://www.plasticsurgery.org/documents/News/Statistics/2016/plastic-surgery-statistics-full-report-2016.pdf.

Bailey, Jacqui. 2016. *Sex, Puberty and All That Stuff.* New York, NY: Barron's Educational Series.

Bay-Cheng, Laina Y. 2015. "Living in Metaphors, Trapped in a Matrix: The Ramifications of Neoliberal Ideology for Young Women's Sexuality." *Sex Roles* 73 (7–8): 332–39. https://doi.org/10.1007/s11199-015-0541-6.

Berger, Kathleen S. 2017. *The Developing Person through the Lifespan.* 10th ed. New York, NY: Worth.

Bloom, Naama. 2017. *Hello Flo: The Guide, Period.* New York, NY: Dutton Children's Books.

Bobel, Chris. 2018. *The Managed Body: Developing Girls & Menstrual Health in the Global South.* London: Palgrave Macmillan.

Bowerman, Mary. 2017. "Increase in Labiaplasty as Women Strive for Normal Look." *USA Today*, March 3. https://www.usatoday.com/story/news/nation-now/2017/03/02/increase-labiaplasty-women-strive-normal-vaginas-what-does-normal-vagina-look-like-surgery/98629662/.

Breuner, Cora C., Gerri Mattson, AAP (American Academy of Pediatrics) Committee on Adolescence, and AAP Committee on Psychosocial Aspects of Child and Family Health. 2016. "Sexuality Education for Children and Adolescents." *Pediatrics* 138 (2): e20161348. https://doi.org/10.1542/peds.2016-1348.

Brown, Lyn M., and Carol Gilligan. 1992. *Meeting at the Crossroads: Women's Psychology and Girls' Development.* Cambridge, MA: Harvard University Press.

Burns, April, Valerie A. Futch, and Deborah L. Tolman. 2011. "'It's Like Doing Homework': Academic Achievement Discourse in Adolescent Girls' Fellatio Narratives." *Sexuality Research and Social Policy* 8 (3): 239–51. https://doi.org/10.1007/s13178-011-0062-1.

Chrisler, Joan. C. 1996. "PMS as a Culture-Bound Syndrome." In *Lectures on the Psychology of Women*, edited by Joan C. Chrisler, Carla Golden, and Patricia D. Rozee, 107–21. New York, NY: McGraw Hill.

Chrisler, Joan C., and Ingrid Johnston-Robledo. 2018. *Women's Embodied Self: Feminist Perspectives on Identity and Image*. Washington, DC: American Psychological Association.

Crawford, Mary, I-Ching Lee, Galina Portnoy, Alka Gurung, Deepti Khati, Pinky Jha, and Anajana C. Regmi. 2008. "Objectified Body Consciousness in a Developing Country: A Comparison of Mothers and Daughters in the US and Nepal." *Sex Roles* 60 (3–4): 174–85. https://doi.org/10.1007/s11199-008-9521-4.

Dakanalis, Antonio, Alix C. Timko, Masimo Clerici, Giuseppe Riva, and Giusepei Carrà. 2017. "Objectified Body Consciousness (OBC) in Eating Psychopathology: Construct Validity, Reliability, and Measurement Invariance of the 24-Item OBC Scale in Clinical and Non-Clinical Adolescent Samples." *Assessment* 24 (2): 252–74. https://doi.org/10.1177/1073191115602553.

DeMaria, Andrea, L., Marrisa Flores, Jacqueline M. Hirth, and Abbey B. Berenson. 2014. "Complications Related to Pubic Hair Removal." *Journal of Obstetrics and Gynecology* 210 (6): 528.e1–528.e5. https://doi.org/10.1016/j.ajog.2014.01.036.

Dickerson, Karle. 2005. *On the Spot: Real Girls on Periods, Growing Up and Finding Your Groove*. Avon, MA: Adams Media.

Dutta, Devashish, Chandler Badloe, Hyunjeong Lee, and Sarah House. 2016. *Supporting the Rights of Girls and Women through Menstrual Hygiene Management (MHM) in the East Asia and Pacific Region: Realities, Progress and Opportunities*. Bangkok, Thailand: UNICEF EAPRO (East Asia and Pacific Regional Office). http://www.cswashfund.org/sites/default/files/MHM_Realities_Progress_and_OpportunitiesSupporting_opti.pdf.

Erchull, Mindy J., and Katherine Richmond. 2015. "'It's Normal…Mom Will Be Home in an Hour': The Role of Fathers in Menstrual Education." *Women's Reproductive Health* 2 (2): 93–110. https://doi.org/10.1080/23293691.2015.1089149.

Fahs, Breanne. 2011. "Sex during Menstruation: Race, Sexual Identity, and Women's Accounts of Pleasure and Disgust." *Feminism & Psychology* 21 (2): 155–78. https://doi.org/10.1177/0959353510396674.

———. 2014. "Genital Panics: Constructing the Vagina in Women's Qualitative Narratives about Pubic Hair, Menstrual Sex, and Vaginal Self-Image." *Body Image* 11 (3): 210–18. https://doi.org/10.1016/j.bodyim.2014.03.002.

———. 2016. *Out for Blood: Essays on Menstruation and Resistance*. Albany, NY: State University of New York Press.

Feinmann, Jane. 2004. *Everything a Girl Needs to Know about Her Periods*. Portland, ME: Ronnie Sellers Productions.

Fine, Michelle. 1988. "Sexuality, Schooling, and Adolescent Females: The Missing Discourse of Desire." *Harvard Educational Review* 58 (21): 29–54.

Fisher, Carole, David Sibbrit, Louise Hickman, and Jon Adams. 2016. "A Critical Review of Complementary and Alternative Medicine Used by Women with Cyclic Perimenstrual Pain and Discomfort: A Focus Upon Prevalence, Patterns and Applications of Use and Users' Motivations, Information Seeking and Self-Perceived Efficacy." *Acta Obstetricia et Gynecologica Scandinavica* 95 (8): 861–71. https://doi.org/10.1111/aogs.12921.

Flores, Dalmacio, and Julie Barroso. 2017. "21st Century Parent-Child Sex Communication in the United States: A Process Review." *Journal of Sex Research* 54 (4–5): 532–48. https://doi.org/10.1080/00224499.2016.1267693.

Fortenberry, J. Dennis. 2014. "Sexual Learning, Sexual Experience, and Healthy Adolescent Sex Behavior." *Positive and Negative Outcomes of Sexual Behavior: New Directions for Child and Adolescent Development* 2014, no. 144 (Summer): 71–86. https://doi.org/10.1002/cad.20061.

Goodman, Michael P., Otto J. Placik, David L. Matlock, Alex F. Simopoulos, Teresa A. Dalton, David Veale, and Susan Hardwick-Smith. 2016. "Evaluation of Body Image and Sexual Satisfaction in Women Undergoing Genital Plastic/Cosmetic Surgery." *Aesthetic Surgery Journal* 36 (9): 1048–57. https://doi.org/10.1093/asj/sjw061.

Gravelle, Karen, and Jennifer Gravelle. 2006. *The Period Book: Everything You Don't Want to Ask (But Need to Know)*. New York: Walker & Company.

Guttmacher Institute. 2017. "American Adolescents' Sources of Sexual Health Information." New York, NY: Guttmacher Institute. https://www.guttmacher.org/fact-sheet/facts-american-teens-sources-information-about-sex.

Hardwick-Smith, Susan. 2011. "Examining the Controversy in Aesthetic Vaginal Surgery." *American Journal of Cosmetic Surgery* 28 (3): 106–13. https://doi.org/10.1177/074880681102800302.

Hickey, Martha, and Adam Balen. 2003. "Menstrual Disorders in Adolescence Investigation and Management." *Human Reproduction Update* 9 (5): 493–504. https://doi.org/10.1093/humupd/dmg038.

Hillard, Paula J. A. 2014. "Menstruation in Adolescents: What Do We Know? And What Do We Do with the Information?" *Journal of Pediatric & Adolescent Gynecology* 27 (6): 309–19. https://doi.org/10.1016/j.jpag.2013.12.001.

Hofmeister, Sabrina, and Seth Bodden. 2016. "Premenstrual Syndrome and Premenstrual Dysphoric Disorder." *American Family Physician* 94 (3): 236–40. http://www.gamgemai.co/Premenstrual%20Syndrome%20and%20Premenstrual%20Dysphoric%20Disorder.pdf.

Johnston-Robledo, Ingrid, and Joan C. Chrisler. 2013. "The Menstrual Mark: Menstruation as Social Stigma." *Sex Roles* 68 (1–2): 9–18. https://doi.org/10.1007/s11199-011-0052-z.

Johnston-Robledo, Ingrid, Kristin Sheffield, Jacqueline Voight, and Jennifer Wilcox-Constantine. 2007. "Reproductive Shame: Self-Objectification and Young Women's Attitudes toward Their Reproductive Functioning." *Women & Health* 46 (1): 25–39. https://doi.org/10.1300/j013v46n01_03.

Johnston-Robledo, Ingrid, Melissa Ball, Kimberly Lauta, and Ann Zekoll. 2003. "To Bleed or Not to Bleed: Young Women's Attitudes toward Menstrual Suppression." *Women & Health* 38 (3): 59–75. https://doi.org/10.1300/j013v38n3_05.

King, Marlee, and Jane M. Ussher. 2012. "It's Not All Bad: Women's Construction and Lived Experience of Positive Premenstrual Change." *Feminism and Psychology* 23 (3): 399–417. https://doi.org/10.1177/0959353512440351.

Lamb, Sharon, and Zoë D. Peterson. 2012. "Adolescent Girls' Sexual Empowerment: Two Feminists Explore the Concept." *Sex Roles* 66 (11–12): 703–12. https://doi.org/10.1007/s11199-011-9995-3.

Lee, Janet. 2009. "Bodies at Menarche: Stories of Shame, Concealment, and Sexual Maturation." *Sex Roles* 60 (9–10): 615–57. https://doi.org/10.1007/s11199-008-9569-1.

Lloyd, Jillian, Naomi S. Crouch, Catherine L. Minto, Lih-Mei Liao, and Sarah M. Creighton. 2005. "Female Genital Appearance: 'Normality' Unfolds." *BLOG: An*

International Journal of Obstetrics and Gynaecology 112 (5): 634–46. https://doi.org/10.1111/j.1471-0528.2004.00517.x.

Loulan, JoAnn, and Bonnie Worthen. 2001. *Period: A Girl's Guide.* Deephaven, MN: Book Peddlers.

Marzollo, Jean. 1989. *Getting Your Period: A Book about Menstruation.* New York: Dial Books for Young Readers.

Middleman, Amy B., and Kate G. Pfeifer. 2006. *American Medical Association's Girl's Guide to Becoming a Teen.* San Francisco, CA: Josey Bass.

Motakef, Saba, Jose Rodriguez-Feliz, Michael T. Chung, Michael J. Ingargiola, Victor W. Wong, and Ashit Patel. 2015. *Systematic Review of Labiaplasty* 135 (3): 774–88. https://doi.org/10.1097/prs.0000000000001000.

Movsessian, Shushann. 2004. *Puberty Girl.* Sydney, Australia: Allen & Unwin.

Mowet, Hayley, Karalyn McDonald, Amy S. Dobson, Jane Fisher, and Maggie Kirkman. 2015. "The Contribution of Online Content to the Promotion and Normalization of Female Genital Cosmetic Surgery: A Systematic Review of the Literature." *BMC Women's Health* 15 (110). https://doi.org/10.1186/s12905-015-0271-5.

Orenstein, Peggy. 2016. *Girls and Sex: Navigating the Complicated New Landscape.* New York, NY: Harper.

Peranovic, Tamara, and Brenda Bentley. 2017. "Men and Menstruation: A Qualitative Exploration of Beliefs, Attitudes and Experiences." *Sex Roles* 77 (1–2): 113–24. https://doi.org/10.1007/s11199-016-0701-3.

Piran, Niva. 2017. *Journeys of Embodiment at the Intersection of Body and Culture: The Developmental Theory of Embodiment.* Cambridge, MA: Academic Press.

Plaisted, Caroline A. 2011. *Girl Talk.* Irvine, CA: QEB Publishing.

Rabin, Roni C. 2016. "More Teenage Girls Seeking Genital Cosmetic Surgery." *The New York Times,* April 25. https://well.blogs.nytimes.com/2016/04/25/increase-in-teenage-genital-surgery-prompts-guidelines-for-doctors/.

Read, Jennifer R., Janette Perz, and Jane M. Ussher. 2014. "Ways of Coping with Premenstrual Change: Development and Validation of a Premenstrual Coping Measure." *BMC Women's Health* 14 (9000). https://doi.org/10.1186/1472-6874-14-1.

Renganathan, Arasee, Rufus Cartwright, and Linda Cardozo. 2009. "Genital Cosmetic Surgery." *Expert Review of Obstetrics & Gynecology* 4 (2): 101–4. https://doi.org/10.1586/17474108.4.2.101.

Rue, Nancy. 1995. *Everything You Need to Know about: Getting Your Period.* New York, NY: The Rosen Publishing Group.

Schaefer, Valorie L. 1998. *The Care and Keeping of You: The Body Book for Girls.* Middleton, WI: American Girl Publishing.

———. 2012. *The Care and Keeping of You (Revised): The Body Book for Younger Girls.* Middleton, WI: American Girl Publishing.

Schick, Vanessa R., Brandi Rima, and Sarah K. Calbrese. 2011. "Evulvalution: The Portrayal of Women's External Genitalia and General Physique across Time and the Current Barbie Doll Ideals." *The Journal of Sex Research* 48 (1): 74–81. https://doi.org/10.1080/00224490903308404.

Sharp, Gemma, Marika Tiggeman, and Julie Mattiske. 2014. "Predictors of Consideration of Labiaplasty: An Extension of the Tripartite Influence Model of Beauty Ideas." *Psychology of Women Quarterly* 39 (2): 182–93. https://doi.org/10.1177/0361684314549949.

Sheffield, Margaret. 1988. *Life Blood: A New Image for Menstruation*. New York, NY: Alfred A. Knopf.

Sommer, Marni, Bethany A. Caruso, Murat Sahin, Teresa Calderon, Sue Cavill, Therese Mahon, and Penelope A. Phillips-Howard. 2016. "A Time for Global Action: Addressing Girls' Menstrual Hygiene Management Needs in Schools." *PLOS Medicine* 13 (2): e1001962. https://doi.org/10.1371/journal. pmed.1001962.

Sterling, Evelina W. 2018. "Menstrual Education in the United States: Opportunities for Improvement and Barriers to Change." Paper. The American Public Health Association. San Diego, CA, November 2018.

Sterling, Evelina W., Diana Karczmarczyl, and Irenose Ivabze. 2017. "Menstrual Health Is Public Health: The Importance of Menstrual Cycle Education." Paper. The American Public Health Association. Atlanta, GA, November 2017.

Stubbs, Margaret L. 2008. "Cultural Perceptions and Practices Around Menarche and Adolescent Menstruation in the United States." *Annals of the New York Academy of Sciences* 1135 (1): 55–66. https://doi.org/10.1196/annals.1429.008.

———. 2013. "Current Menstrual Education Resources: Still Room for Improvement." Paper. 20th Biennial Conference of the Society for Menstrual Cycle Research, June 2013.

Stubbs, Margaret L., and Ingrid Johnston-Robledo. 2013. "Kiddy Thongs and Menstrual Pads: The Sexualization of Girls and Early Menstrual Life." In *The Sexualization of Girls and Girlhood: Causes, Consequences and Resistance*, edited by Eileen L. Zurbriggen, and Tomi-Ann Roberts, 213–34. New York, NY: Oxford University Press.

Thomas, Isabel. 2011. *Why Do I Have Periods?* Chicago, IL: Raintree.

Thomson, Ruth. 1995. *Have You Started Yet? Getting the Facts Straight*. New York, NY: Pan Macmillan.

Tolman, Deborah, L. 2002. *Dilemmas of Desire: Teenage Girls Talk about Sexuality*. Cambridge, MA: Harvard University Press.

———. 2012. "Female Adolescents, Sexual Empowerment and Desire: A Missing Discourse of Gender Equity." *Sex Roles* 66 (11–12): 746–57. https://doi. org/10.1007/s11199-012-0122-x.

Tolman, Deborah L., and Sara I. McClelland. 2011. "Normative Sexuality Development in Adolescence: A Decade in Review, 2000–2009." *Journal of Research on Adolescence* 21 (1): 242–55. https://doi.org/10.1111/j.15327795.2010.00726.x.

Tolman, Deborah L., Stephanie M. Anderson, and Kimberly Belmont. 2015. "Mobilizing the Metaphor: Considering the Complexities, Contradictions, and Contexts in Adolescent Girls' and Young Women's Sexual Agency." *Sex Roles* 73 (7–8): 298–310. https://doi.org/10.1007/s11199-015-0510-0.

Ussher, Jane M., and Janette Perz. 2013. "PMS as a Gendered Illness Related to the Construction and Relational Experience of Hetereo-Femininity. *Sex Roles* 68 (1–2): 132–50. https://doi.org/10.1007/s11199-011-9977-5.

Weiss-Wolf, Jennifer. 2017. *Periods Gone Public: Taking a Stand for Menstrual Equity*. New York, NY: Arcade Publishing.

Williamson, Heidi. 2015. "Social Pressures and Consequences Associated with Body Hair Removal." *Journal of Aesthetic Nursing* 4 (3): 131–33. https://doi. org/10.12968/joan.2015.4.3.131.

CHAPTER 21

Living in Uncertain Times: Experiences of Menopause and Reproductive Aging

Heather Dillaway

While there is existing clinical and attitudinal research on reproductive aging, we often forget to focus on the phenomenological experience of perimenopause and menopause—that is, *the lived, embodied, day-to-day experience of this reproductive and life course transition*. This chapter will explore the everyday experience of women living and passing through perimenopause and menopause, a transition that involves many day-to-day uncertainties and changes, including definitional uncertainties, uncertainties about signs and symptoms, uncertainties about aging, and uncertainties about motherhood and changing relationship statuses. Because the hallmarks of perimenopause and menopause are uncertainty and change, I will argue that experiencing menopause and reproductive aging is akin to living in uncertain times, and learning to live in and with this uncertainty is part of the everyday experience of this reproductive and life course transition. In fact, acknowledging and owning this uncertainty could be a new and different way of approaching and thriving during this reproductive transition.

Definitional Uncertainties

Women often say they are "in menopause," but what does that really mean? What are perimenopause, menopause, and postmenopause, and how do women recognize that they are in these stages or phases of reproductive aging? What marks the start and end of this transition, clinically and experientially? The term "menopause" refers to the permanent cessation of menstruation and is typically defined by the final menstrual period (FMP) (Utz 2011). Clinical studies suggest that the average age of cessation is 48–52 years of age, although women in Western countries end menstruation between their

© The Author(s) 2020

C. Bobel et al. (eds.), *The Palgrave Handbook of Critical Menstruation Studies*, https://doi.org/10.1007/978-981-15-0614-7_21

early 40s and late 50s (Mansfield et al. 2004). Menopause is caused by the fluctuation of hormone levels within the ovaries (Utz 2011). Despite popular belief, however, the decrease of estrogen hormones is not the sole cause of menstrual cessation; rather, many hormones fluctuate to cause this reproductive transition or "climacteric" (Fausto-Sterling 1992). These normal hormone fluctuations can occur anywhere from 8 to 10 years prior to cessation to several years afterwards (Fausto-Sterling 1992). What women experience is not just "menopause," then, but a complex reproductive aging transition that can sometimes span multiple decades.

Doctors and researchers commonly use a three-part clinical definition to make sense of women's experience of reproductive aging (Mansfield et al. 2004). According to clinical research and medical diagnosis guidelines, "menopause" is really defined by the *lack* of menstruation for 12 consecutive months and can only be defined in retrospect (Dillaway 2006). "Perimenopause" refers to the period leading up to menopause, and this is usually the time when signs or symptoms such as irregular bleeding, hot flashes, insomnia, and others may begin (Dillaway 2006). This stage is sometimes referred to as a "menopausal" stage simply because it precedes the time when a woman becomes menopausal (McElmurry and Huddleston 1991). This is likely why women may say they are "in menopause" when they are really in perimenopause. "Postmenopause" is the point after which a woman has not had a period for 12 months; often this stage is not diagnosed until one to two years after the cessation of periods (Dillaway 2006).

Women in their 40s and 50s in particular may find themselves wondering whether they have truly transitioned to perimenopause or reached menopause. Even after some signs and symptoms subside, women may ask, "Am I done or not?" (Dillaway and Burton 2011; Hyde et al. 2010). Moreover, all middle-aged women who menstruate may ask themselves and their doctors, "when will I be done?" Furthermore, when thinking about the stage or phase of reproductive aging that they might be "in," or "done with," women might make sense of their experiences differently than their doctors (even though they might rely on doctors for medical verification of their life stage [Hyde et al. 2010]). It is uncommon to hear individual women saying that they are "in postmenopause"; rather, they tend to say they are "done with menopause" during this stage. Or, in perimenopause, women might colloquially say they are "in the middle of menopause" (Dillaway and Burton 2011). Thus, women's language about this transition might vary from doctors' or researchers' language, and this sometimes makes it complicated for both women and doctors to understand each other. It is typically perimenopause that women are most concerned with, and women are often wondering when perimenopause will end. Nonetheless, one of the other definitional uncertainties for women is when perimenopause starts.

Some researchers have suggested that perimenopause begins when there is at least seven days in variability between one menstrual cycle and the next, then progresses to a separate stage characterized by skipped periods (at least a 60-day gap), and finally ends at 12 months past the FMP (for example, Harlow et al. 2008). However, Prior (2005) proposed instead that perimenopause can begin with symptomatic experiences like hot flashes and increased breast tenderness, and argued that because subjective changes precede menstrual irregularity women may initially find hot flashes and breast tenderness (as well as other signs and symptoms) just as problematic as changes in menstrual flow. Individual women also might not recognize clear transitions from one menopausal stage to another (Dillaway and Burton 2011). In a longitudinal study of 100 women, for instance, some women reported experiencing one stage for numerous years and then moving on quickly, but others reported flip-flopping between stages over many years (Mansfield et al. 2004). Flip-flopping between these stages can be quite confusing for individual women (Dillaway and Burton 2011). In Mansfield et al.'s study, though, one group of women progressed quickly through all three stages (Mansfield et al. 2004, 225).

Clinical definitions do not capture the variability in how women define reproductive aging for themselves, or how they experience the signs and symptoms of perimenopause, menopause, and postmenopause. Because of the focus on physiology only, existing clinical definitions of menopause are therefore partial at best. Social science and feminist researchers have suggested that women's personal definitions of reproductive aging—that is, how a woman herself recognizes, defines, and copes with her own experience of perimenopause, menopause, and postmenopause—are just as, if not more, important (for example, Dare 2011; Kafanelis et al. 2009). In fact, when individual women suggest that they are "in" menopause, this language validates women's experience of reproductive aging or perimenopause as a life stage, not just a retrospective moment or clinical diagnosis. Women's recognition that they live "in" an uncertain midlife reproductive stage (that is ill-defined for them at the current time) can be as helpful to their management of day-to-day signs and symptoms as the three-part clinical definition that remains static and often only retrospective. That is, women desire information that will help them understand their own day-to-day navigation of reproductive aging and its associated signs and symptoms. Broadening the working understanding of perimenopause, menopause, and postmenopause to include information on uncertain beginnings and ends, and varying signs and symptoms, means taking women's subjective experience of their reproductive bodies and midlife stages more seriously (Dillaway and Burton 2011). If we listen to women's discussions of being "in" this life stage, and living *in* uncertainty, then both clinical interventions and women's understanding of their day-to-day signs and symptoms could align better with women's experiences and needs.

UNCERTAINTY AROUND SIGNS AND SYMPTOMS

Defining the signs and symptoms of menopausal transitions has also proven difficult for clinicians and researchers, which makes for even more ambiguity for individual women. Medical websites and other authoritative sources, as well as women themselves, often attribute symptoms to perimenopause, including but not limited to mood changes, memory problems, fatigue, hot flashes (or flushes, depending on what country a woman lives in), insomnia, vaginal dryness, changes in libido, weight gain, hair growth, heart palpitations, headaches, and joint pain (for example, Avis et al. 2001). Indeed, more than 100 symptoms have been attributed to menopause (Cobb 1993; Fausto-Sterling 1992). Epidemiological studies have not found evidence of universal perimenopausal symptoms, however (Avis et al. 2001). Only bleeding changes, hot flashes, vaginal dryness, and insomnia are more frequently reported by perimenopausal and/or menopausal women, as compared with premenopausal women (Derry and Dillaway 2013). In Europe and the United States, approximately 70% of women will experience a hot flash (or flush) during perimenopause (Ayers, Forshaw, and Hunter 2010).

Epidemiological and social science research suggests that perhaps we blame more on perimenopause and menopause than we should (for example, Fausto-Sterling 1992) and that there may be very few bodily symptoms that can be tied specifically to menopause (and not, instead, to other life situations at midlife, lifestyle behaviors, or chronic conditions across women's life course). For instance, in a study of insomnia at perimenopause, Vigeta et al. (2012) found that many women who reported an increase in sleep disturbance in perimenopause also reported insomnia in earlier life stages. It was only after the effects of insomnia were more noticeable in their daily lives (at midlife) that they sought medical treatment for this symptom (Vigeta et al. 2012). Muhlbauer and Chrisler (2007) additionally suggested that vaginal dryness may simply indicate that women's bodies are not ready for sexual intercourse and that women take longer to become moist as they grow older; thus, vaginal dryness is neither a permanent condition for most middle-aged women, nor is it always directly attributable to menopause. Decreased sex drive may also be attributed to stress, fatigue, and over-burden at midlife, a time when women face multiple and overlapping responsibilities. Uncertainty around the attribution of symptoms to reproductive aging can be difficult for women as they attempt to evaluate and make sense of midlife experiences.

Women also have varied reactions to the potential signs and symptoms of certain stages of reproductive aging, and lack complete information about what signs or symptoms are like on a daily basis. For example, while biomedical studies describe the physiology of hot flashes, little is known about what individual women actually feel when hot flashes occur or how they affect their everyday lives. Because hot flashes can challenge norms about beauty and bodily control and sometimes alert others to women's menopausal status before women themselves are ready to disclose it (Dillaway 2011), individual women surely negotiate and think about hot flashes differently (especially when they happen in public settings). In particular, women may find hot flashes problematic in paid work settings and, depending on the type of paid work they do, they may define hot flashes negatively if they must negotiate them in front of coworkers (Dillaway 2011). Depending on the severity of a hot flash, women might need to attend to sweating, flushed skin, and the removal of layers of clothing, for instance, in the middle of a meeting or while cashing out a customer at a retail store. Depending on who is present during that hot flash an individual woman may feel stigmatized for the public nature of that symptom, or worry about whether others are thinking negatively about her while she is dealing with it (Dillaway 2011). Also at home, within family settings, male partners especially, but also children at times, become exasperated with women's "uncontrollable" bodies. Individual women must therefore handle others' perceptions of some signs and symptoms, which may make the individual experience of perimenopausal symptoms even more complicated (Dillaway 2011).

Biomedical definitions of "symptoms" are often purely negative, even pathological (Martin 1992). For instance, medical doctors often think of menopause as a cluster of symptoms caused by a "deficiency" in reproductive hormones (Lyons and Griffin 2003; Meyer 2003; Niland and Lyons 2011; Utz 2011). Ovarian failure and a deficiency of reproductive hormones are seen as the biological causes of any physical, physiological, or psychological changes during this time of life, and these changes are typically defined as negative (Lyons and Griffin 2003; Meyer 2003; Niland and Lyons 2011). This focus on biology eventually leads medical professionals to feel responsible for fixing women's "deteriorating" bodies and simultaneously encourages women to seek medical treatment to "replace" or "restore" their hormones (Martin 1992, 26; see also Meyer 2003; Niland and Lyons 2011).

However, research shows that women's subjective experience of these signs and symptoms are not always purely negative. Women may have many positive or neutral responses to symptoms. For instance, women in their 40s and 50s may have teenaged daughters who are just reaching menarche, and having irregular periods alongside daughters may be an opportunity to bond (Dillaway 2005a). Depending on the exact severity of hot flashes, individual women may also take the opportunity to define this symptom as simply their own "private summer" (Dillaway 2011). Not all bodily signs or symptoms represent physical decline, even if those signs or symptoms do cause hassle or uncertainty at times. Furthermore, women's previous reproductive experiences have prepared them well for how to handle both expected and unexpected bodily signs and symptoms. Upon reaching perimenopause, for instance, women who have menstruated are already accustomed to experiencing bodily signs and symptoms associated with menstrual cycles. Therefore, because menstruating women have learned to notice, wait for, track, monitor, and manage reproductive signs and symptoms for the entirety of their adolescent and adult lifetimes, the arrival of new and/or adjusted signs and symptoms at midlife may not startle women as much as clinicians might expect or suggest. Bodily changes also may happen gradually in many cases, giving women time to contemplate and adjust to their new life stage. On the other hand, adjusting to slightly new signs and symptoms can be difficult on a daily basis and, in some cases, the severity of midlife bodily changes may lead women to visits their doctors for relief. Women may have a range of reactions, attitudes, and experiences with the signs and symptoms of reproductive aging: some negative, some positive, some neutral, and some confusing, and one sign or symptom may bring forth multiple reactions, attitudes, and experiences. On the one hand, skipping a menstrual period may feel freeing but, on the other hand, it may induce anxiety. Likewise, having a hot flash may be embarrassing in public the first time, but one may learn quickly how to navigate public reactions and adjust diet and clothing in order to moderate this physical sign or symptom. In these cases, an initial reaction to a bodily sign or symptom might be positive or negative, but reactions, attitudes, and experiences may adjust over time (Dillaway 2005a, 2011).

Social science researchers have even proposed that menopausal symptoms are more the result of sociocultural attitudes and ideologies than of biology. Lock (1993) made this clear in her comparative study of menopausal women in North America and Japan. She found that "menopause is neither fact nor universal event but an experience that we must interpret in context" (370). For example, whereas North American ideology idealizes youth, this is not the case in Japan, where "middle-aged" individuals actually form the "backbone" of society, "support[ing] both old and young, keep[ing] the economy growing and build[ing] the new Japan – men and women alike" (370). Constricting gender ideology does exist in Japan, of course, but Japanese women conceptualize the life cycle as "part of a larger cyclical continuum rather than as a path of no return that fragments youth from age" (Lock 1993, 378). In support of the social construction of reproductive aging, Lock found fewer reports of symptoms, such as hot flashes, among Japanese women. Other anthropological studies suggest that increased medicalization of menopause in North America leads to more negative views among individual women in that setting as well (Lock 1993). Yet, even within North American countries, there is more variation in women's symptom experiences than we typically acknowledge.

Lock (1993) and Utz (2011) describe "local biologies," in which women think about and experience physical or physiological changes at midlife differently, depending on their expectations for reproductive aging, their knowledge about this transition, and other psychosocial life contexts. How women think about signs and symptoms must contextualized within their other life circumstances and social locations, and no two women may think about or experience signs or symptoms in the same way. Some feminist scholars have gone further and recognized that menopause is more than just a collection of uncomfortable signs and symptoms and is instead, at minimum, a neutral, and often even a positive life transition (Dillaway 2005a; Lyons and

Griffin 2003). For instance, women may welcome menopause as relief from the threat of pregnancy and the burdens of menstruation and contraception (Dillaway 2005a). The Baby Boomers are the first generation to have had widespread access to the birth control pill and other advances in contraceptive technologies that enable women to avoid biological motherhood (Dillaway 2005a). Contemporary women, then, may elect to end their reproductive years long before menopause. Within this context, menopause may be quite positive for many individuals. Regardless of the bodily changes they may face during menopause, then, individuals may view menopause as ushering in a "good" life stage, one that is better and more carefree than the one before it.

If we acknowledge women's vast experience with bodily signs and symptoms, we can recognize the limitless and varied reflections and responses to bodily moments and life course stages like reproductive aging. Signs and symptoms of perimenopause, menopause and postmenopause, as clinically defined, do not always represent the same thing for every woman, nor do they automatically signify deficit or decline. Living in and through the signs and symptoms of reproductive aging may indeed bring uncertainty for all, especially at the outset, and induce physical discomfort for some as well. Nonetheless, living in and through signs and symptoms may also introduce healthy and reflective adjustments to a new life stage.

Uncertainties About Aging

Whether or not women equate menopause with aging is also important for how individual women will view and experience menopause as a transition. Hall (1999) described an interview she conducted with one menopausal woman who explained,

> Those night sweats—and other symptoms I began to notice—suddenly made me feel old. One day I'm a young woman in her prime, and the next day I'm worrying about whether or not I'm prepared for retirement and thinking about getting my affairs in order. (17–18)

Because of its conflation with chronological aging, the onset of menopause may have resembled a "death sentence" for many women in the past (Dillaway 2006). A widely held misconception continues to exist that all women used to die before menopause, and thus postmenopausal women are currently living beyond what nature prescribes (Weed 1999). Even if menopause is not a death sentence in our imaginations anymore, we are still confronted with a reiteration of the idea that "old guys can be gorgeous, but old women pollute the landscape, so mask them, keep their 'decay' out of sight" (Gray 1996, 186). The message is that women lose their "feminine," "healthy," and "sexy" attributes in midlife, and begin a continual decline after that. "After 50," women supposedly take on "the grandma look [becoming] women in buns, girdles, and orthopedic shoes" (Jarrell 1999, 2; see also Utz 2011). This situation arises directly from the fusion of a sexist and ageist culture, one that insists on viewing "older" people as "ill" or "sick" (Zones 2005) and women primarily as "mothers" (and, if not mothers, then sex objects) (Dillaway 2005b).

The image of femininity *is* primarily associated with youth, physical beauty, fertility, and reproductive capabilities in Western industrialized countries (Brooks 2010; Wilson 1966; Winterich 2007). Women may therefore believe that bodily changes must be slowed down or halted; gray hair, wrinkles, dentures, age spots, drowsiness, thickening of the waistline, sagging abdomens and breasts, sagging muscles, or crinkling skin, and bifocal glasses may seriously challenge one's gendered self-identity (Brooks 2010; Berkun 1986; Dillaway 2005b; Grambs 1989; Winterich 2007; Zones 2005). Women have been taught since girlhood to control their bodies (Dillaway 2005b; Zones 2005), especially around others who might gaze upon their bodies, which could make menopause and other midlife bodily changes anxiety producing. The fear of "being let down by their bodies if they must maneuver in a world that favors youth" may be strong (Berkun 1986, 383; see also Dillaway 2005b).

On the other hand, age-based norms are changing as Baby Boomers age. For example, Baby Boomers may identify and behave in the same ways they did when they were chronologically younger. For example, they participate in similar leisure activities, wear similar clothing, and are as physically and socially active as they were in younger years (Featherstone and Hepworth 1991). As a result, some scholars now suggest that there is no current cultural consensus about what it means to be a midlife, mature, or menopausal woman and women no longer routinely report feeling old at midlife (Dillaway 2005a). Changing norms may leave room for middle-aged women to feel and act the same as they did in previous life stages.

In addition, women are increasingly in the paid labor force and have other meaningful social roles during midlife, so that menopause has less potency as a cultural marker of entry into a negative life stage. In fact, menopause may parallel women's greatest accomplishments in paid work in contemporary times. "Aging" stars (for example, Meryl Streep, Helen Mirren, or Diane Keaton) or "older" women in politics (for example, Hillary Clinton or Maxine Waters) call into question the idea that "older" women are merely stereotypical grandmothers who are past their prime (Dinnerstein and Weitz 1994; Featherstone and Hepworth 1991; Muhlbauer and Chrisler 2007). Thus, they may experience chronological aging or reproductive aging (or both) as less stressful and more representative of new possibilities than previous generations did. Self-esteem for contemporary middle-aged and older women may not only depend on physical appearance and biology, but also on non-reproductive and non-familial accomplishments—as is the case for men. Grambs (1989) argued that these developments make the current cohort of menopausal women more positive about experiences of reproductive aging than previous generations of women who had fewer opportunities outside the home.

The disruption of the equation of aging and gendered decline may lessen women's negative attitudes toward menopause and decrease some of women's attention to negative bodily experiences. Women can also be thankful to move past menstruation, the use of contraception or birth control, the threat of pregnancy, and other gendered hassles associated with managing their earlier reproductive lives. The sexual and reproductive freedoms gained via menopause (such as discontinuing birth control use) may be particularly poignant for Baby Boomer women, who lived through the fights for access to birth control and abortion, as well as paid parental leave, daycare, and other sexual freedoms (Dillaway 2005a). Furthermore, not having to buy pads, tampons, and other menstrual supplies anymore, or wearing white underwear whenever one desires can be liberating. Additionally, women sometimes find that other midlife health events, such as getting bifocals or having back pain, make them feel "older" than the onset of perimenopause or menopause. In this way, reaching menopause may be reaching a "good old" rather than the "bad old" that is signified by other health transitions (Dillaway 2005a). While women who have experienced reproductive difficulties (such as infertility or miscarriage) may feel extremely negative about losing their reproductive capacity, others may feel empowered to move beyond daily reproductive burdens.

Menopause and midlife is still a time when women's identities and experiences shift. Women may still "feel young" (or, at the very least, "not old yet") but they may still recognize that their bodily and other social transitions are marking a new life stage for them. Aging is something that is both revered and feared, and it is a process that seemingly speeds up in midlife. Midlife also brings other life transitions, such as adult children leaving the home, the death of parents, greater likelihood of chronic bodily conditions, and possibly retirement. Ageism may continue to result in middle-aged women or men watching younger coworkers receive promotions in paid work, instead of them. Thus aging is still a context that women might define as negative in part, even if they are positive or indifferent about reproductive aging. Reproductive aging can be seen as unique from other kinds of aging in its possibilities, however. Reproductive aging represents as much of an opportunity for freedom as for burden. Moving beyond reproductive capacity while still feeling young can mean making one's debut into uncertain and uncharted territory, yet territory that is less confined by gendered cultural norms that women have the power to make their own. Many other cultures have shown us that women can grow into their own power as they age out of reproductive capacity (Lock 1993), and the lessening of age-based norms in North America has also strengthened this possibility for women in the United States.

Uncertainties About Motherhood and Relationships

Family contexts can make the menopausal transition seem neutral, positive, or negative. For instance, women at midlife are often caring for elderly parents, partners, and children (often simultaneously); menopause can pale in importance to these caregiving responsibilities (Dare 2011). Women told Winterich and Umberson (1999, 61) that menopause was "last on [their] list" of things to worry about and "no big deal" in the context of other family problems (for example, divorce, children "acting out," or family death) (see also Dillaway 2005a, 2012). Dare (2011) also found that experiences of divorce, and the aging and death of parents weighed much more heavily on women than menopause. Thus, family contexts can make women feel neutral or positive about menopause. Bodily change may be noticed but might not be something for which women take time to seek medical help, given their busy and complicated family lives.

Menopause may seem more negative, however, in certain family contexts. Mansfield, Koch, and Gierach (2003) reported that husbands sometimes have negative views of menopause and/or know little about the experience overall. Even husbands who want to be supportive lack the information to be so or feel unsure about how to help (Mansfield, Koch, and Gierach 2003). Further, women sometimes want to share menopause experiences with their partners, but they avoid doing so because it is "a private thing" (Walter 2000, 117). Similar to what Fahs (2011) sounds about women's and partners' discomforts around having sex during menstruation, perimenopausal women have reported discomfort in talking with partners about perimenopausal signs and symptoms (Dillaway 2008). Due to a lack of information about menopause and women's reticence about sharing experiences, husbands' interactions with wives potentially cement negative meanings about menopause and encourage women to think about menopause as a collection of treatable symptoms.

We do know from previous research on social support that the older women are, the less support they receive from their spouses (Schwarzer and Gutierrez-Dona 2005). Partners can be supportive during other reproductive events (for example, pregnancy, birth, breastfeeding) but, because more secrecy surrounds menopause, Mansfield, Koch, and Gierach (2003) suggested that male partners may be less supportive at midlife than at previous stages (if only because they lack information about how to help). Further, because menopause sometimes remains invisible (or appears intangible) to others, partners and other family members may also not "see" reproductive aging as something with which to engage (unlike pregnancy, for instance, which typically results in publicly changing body and baby). Finally, lingering cultural views associating women with reproduction may make talking about menopause awkward, when it seems to represent an end to one of women's gendered roles (Dillaway 2008, 2012; Dinnerstein and Weitz 1994).

Negative images of menopause (and/or a lack of support for women at menopause) within families may affect women in adverse ways, lead them to internalize negative views, and incline them toward a medicalized view of menopause (Mansfield, Koch, and Gierach 2003). In this case, individual women may turn to medical relief for symptoms in order to appease intimate partners or important others. On the other hand, we know that sharing health experiences with intimate partners leads to better physical and emotional well-being (Mansfield, Koch, and Gierach 2003; Reid 2004; Walter 2000). "Women who are moving through the menopausal transition, even those who experience few adverse physical or emotional changes, still need to negotiate a change in status or a 'redefinition of self,' and married women may benefit from social support provided by their husbands during this transition" (Mansfield, Koch, and Gierach 2003, 103). Sharing health experiences in a positive way can increase individuals' reports of relationship satisfaction as well (Reid 2004), as family relationships are molded alongside health experiences.

In addition, however, perimenopausal women often rethink how they feel about intimacy and relationships. They may be on their own for the first time at midlife (because of divorce, widowhood, and empty-nesting), or they may finally be willing to end old or begin new relationships at this life stage. Those in new relationships may worry that menopausal status may deter others' interest in them, while others may feel much more "free" to engage in intimacy upon menopause (because of the lack of the threat of pregnancy). In sum, it is during this reproductive transition that many women may reconsider and question their motherhood and relationship statuses, and rethink their identities related to these statuses as well. This makes midlife and reproductive aging uncertain times for some women; in other cases, women might actually feel more certain about their identities than in any earlier period of their lives.

Reproductive Histories

Menopause can be neutral, positive, or negative for women as a result of earlier reproductive difficulties and reproductive histories. Women's feelings about menopause also cannot be separated from decisions that they and their partners make about whether to become parents. In some cases, menopause may take on more negative meanings, as seen in trends toward delayed childbearing and rising concerns about infertility (Dillaway 2012). Loss of fertility may be an important issue for a woman who has delayed childbearing until her 30s or 40s and then discovers that she has trouble conceiving. These women do not want to be finished with childbearing, either emotionally or in terms of life goals, and may feel that menopause has robbed them of their final chance to become biological parents. Conversely, coping with menopause may not seem as difficult as coping with infertility or miscarriage, and menopause may be a relief for a woman with a long history of difficulty becoming pregnant, who now has closure and can put decisions about infertility treatment and hopes for a family to rest (Dillaway 2012).

Other women may have never actively or consciously made the decision to stop having children (even if, for example, it has been more than 10 or 15 years since their youngest child was born). For these latter women, the onset of menopause may make it feel like they never had the choice to finalize their decision themselves—the onset of menopause took that final decision out of their hands (Dillaway 2012). This may happen in situations where women may have prioritized their careers and birthed a first child "late," in their late 30s or early 40s, for instance. Women with one biological child (or even two) may find themselves wishing they had had an additional child. In cases of divorce and remarriage, or delayed first marriage (especially if a new intimate partner desires children), menopause may also induce women's regret around the size of their families.

Regardless of women's exact feelings about reproductive aging, menopause does symbolize an end of their physical capacity to reproduce. While most women do not equate the end of fertility with menopause any longer (with the availability of contraception, fertility typically "ends" much earlier in contemporary times), women still often use this life course stage to recontemplate their motherhood status. Some women wish they had had more children, or that they could have had children at all. Women for whom motherhood was primary in importance may mourn the passing of stages of active caregiving. Others contemplate the relative lack of importance of motherhood in their lives, and see menopause as insignificant as a result. This means that women have wide-ranging responses to reproductive aging depending on their motherhood and family statuses. Unless we understand women's perceptions and experiences of menopause within the context of their choices and experiences around childbearing and childrearing, we cannot understand women's perceptions and experiences of reproductive aging. Because of women's mixed experiences of motherhood and family, experiences of reproductive aging will also vary. Individual women must find out for themselves how their reproductive histories might affect their perceptions and experiences of reproductive aging.

Uncertainty and Change Is Normal at This Life Stage

Regardless of how women feel about reproductive aging as a life course transition, there is no doubt that perimenopause and/or menopause involves biopsychosocial changes that women must navigate. Generally, then, this life stage brings both physical change and identity change. Because change can bring uncertainty, coping with uncertainty also becomes part of the everyday experience of reproductive aging (Kafanelis et al. 2009). Living through change can be good or bad or both, depending on women's perspectives and other life contexts. As outlined above, many of the uncertainties that women face can be attributed to confusion around the definitions of perimenopause and menopause, the uncertain signs and symptoms of this transition, uncertain feelings about aging, and reflections about motherhood and family experiences.

Bodily change, alongside other life course and identity changes, can be defined in varying ways by women, which means it is difficult to define the impact of reproductive aging on women in any uniform or standardized way. Women may feel very much in flux during this life stage. Perhaps this means that we need to talk more about the fact that this feeling of flux and uncertainty is actually normal and natural during reproductive aging. Learning that uncertainty and change are the hallmarks of reproductive aging experiences may be uniquely satisfying and validating to individual women living in this transition. In fact, acknowledging and welcoming this uncertainty could be a bold and different way of approaching and thriving during this inevitable transition. All women who reach midlife have already lived through considerable change and uncertainty—a lifetime of reproductive experiences alone have accustomed women to change and uncertainty. Understanding that this life stage simply represents an additional time for reflection and a new bodily challenge could be healthy and freeing. As our bodies find their "new normal" through reproductive aging, we are free to think about this time in the way that suits us best. Instead of confining ourselves to thinking about menopause and reproductive aging through a biomedical lens, we can think of this life stage more broadly, as a chance to redefine ourselves and our bodies one more time. Thinking about ourselves as living *in* this uncertain time necessarily, rather than just getting *through* it, may be the first step to understanding and owning the impact of menopause and reproductive aging. Individual women, as well as women as a whole, can make what they want of this uncertain time.

References

Avis, Nancy E., Rebecca Stellato, Sybil Crawford, Joyce Bromberger, Patricia Ganz, Virginia Cain, and Marjoria Kagawa-Singer. 2001. "Is There a Menopausal Syndrome? Menopausal Status and Symptoms across Racial/Ethnic Groups." *Social Science and Medicine* 52: 345–56.

Ayers, Beverley, Mark Forshaw, and Myra S. Hunter. 2010. "The Impact of Attitudes towards the Menopause on Women's Symptom Experience: A Systematic Review." *Maturitas* 65: 28–36.

Berkun, Cleo S. 1986. "On Behalf of Women Over 40: Understanding the Importance of the Menopause." *Social Work* 31 (5): 378–84.

Brooks, Abigail T. 2010. "Aesthetic Anti-Ageing Surgery and Technology: Women's Friend or Foe?" *Sociology of Health & Illness* 32: 238–57.

Cobb, Janine O. 1993. *Understanding Menopause.* New York: Penguin.

Dare, Julia S. 2011. "Transitions in Midlife Women's Lives: Contemporary Experiences." *Healthcare for Women International* 32 (2): 111–33.

Derry, Paula, and Heather Dillaway. 2013. "Rethinking Menopause." In *Women's Health Psychology*, edited by Mary Spiers, Pamela Geller, and Jacqueline Kloss, 440–66. Hoboken, NJ: Wiley.

Dillaway, Heather. 2005a. "Menopause Is the 'Good Old': Women's Thoughts about Reproductive Aging." *Gender & Society* 19 (3): 398–417.

———. 2005b. "(Un)Changing Menopausal Bodies: How Women Think and Act in the Face of a Reproductive Transition and Gendered Beauty Ideals." *Sex Roles* 53 (1–2): 1–17.

————. 2006. "When Does Menopause Occur, and How Long Does It Last? Wrestling with Age- and Time-Based Conceptualizations of Reproductive Aging." *Feminist Formations* 18 (1): 31–60.

————. 2008. "'Why Can't You Control This?' Women's Characterizations of Intimate Partner Interactions about Menopause." *Journal of Women & Aging* 20 (1/2): 47–64.

————. 2011. "Menopausal and Misbehaving: When Women 'Flash' in Front of Others." In *Embodied Resistance: Breaking the Rules in Public Spaces*, edited by Chris Bobel and Samantha Kwan, 197–208. Nashville, TN: Vanderbilt University Press.

————. 2012. "Reproductive History as Social Context: Exploring How Women Converse about Menopause and Sexuality at Midlife." In *Sex for Life: From Virginity to Viagra, How Sexuality Changes Throughout Our Lives*, edited by John DeLamater and Laura Carpenter, 217–35. New York: New York University Press.

Dillaway, Heather, and Jean Burton. 2011. "Not Done Yet?! How Women Discuss the 'End' of Menopause." *Women's Studies: An Interdisciplinary Journal* 40 (2): 149–76.

Dinnerstein, Myra, and Rose Weitz. 1994. "Jane Fonda, Barbara Bush, and Other Aging Bodies: Femininity and the Limits of Resistance." *Feminist Issues* 14: 3–24.

Fahs, Breanne. 2011. "Sex during Menstruation: Race, Sexual Identity, and Women's Accounts of Pleasure and Disgust." *Feminism & Psychology* 21, no. 2 (May): 155–78.

Fausto-Sterling, Anne. 1992. *Myths of Gender*. Revised edition. New York: Basic Books.

Featherstone, Michael, and Michael Hepworth. 1991. "The Mask of Ageing and the Postmodern Life Course." In *The Body: Social Process and Cultural Theory*, edited by Michael Featherstone, Michael Hepworth, and Bryan S. Turner, 371–89. London: Sage.

Grambs, Jean Dresden. 1989. *Women Over Forty: Visions and Realities*. Vol. 4. New York, NY: Springer.

Gray, F. du Plessix. 1996. "The Third Age: At a Certain Point, a Woman May Find That Men Stop Looking at Her and Start Listening to Her." *The New Yorker*, March 26, 186–91.

Hall, Lynne L. 1999. "Taking Charge of Menopause." *FDA Consumer* 33 (6): 17.

Harlow, Sioban D., Ellen S. Mitchell, Sybil Crawford, Bin Nan, Roderick Little, John Taffe, and ReSTAGE Collaboration. 2008. "The ReSTAGE Collaboration: Defining Optimal Bleeding Criteria for Onset of Early Menopausal Transition." *Fertility and Sterility* 89 (1): 129–40.

Hyde, Abbey, Jean Nee, Etaoine Howlett, Jonathan Drennan, and Michelle Butler. 2010. "Menopause Narratives: The Interplay of Women's Embodied Experiences with Biomedical Discourses." *Qualitative Health Research* 20 (6): 805–15.

Jarrell, Anne. 1999. "Noticed; Models, Defiantly Gray, Give Aging a Sexy New Look." *New York Times*, November 28. Accessed at http://www.nytimes.com/1999/11/28/style/noticed-models-defiantly-gray-give-aging-a-sexy-new-look.html.

Kafanelis, Betty V., Marion Kostanski, Paul A. Komesaroff, and Lily Stojanovska. 2009. "Being in the Script of Menopause: Mapping the Complexities of Coping Strategies." *Qualitative Health Research* 19, no. 1 (January): 30–41.

Lock, Margaret. 1993. *Encounters with Aging: Mythologies of Menopause in Japan and North America*. Berkeley, CA: University of California Press.

Lyons, Antonia C., and Christine Griffin. 2003. "Managing Menopause: A Qualitative Analysis of Self-Help Literature for Women at Midlife." *Social Science & Medicine* 56 (8): 1629–42.

Mansfield, Phyllis Kernoff, Molly Carey, Amy Anderson, Susannah Heyer Barsom, and Patricia Bartholow Koch. 2004. "Staging the Menopausal Transition: Data from the TREMIN Research Program on Women's Health." *Women's Health Issues* 14 (6): 220–26.

Mansfield, Phyllis Kernoff, Patricia Barthalow Koch, and Gretchen Gierach. 2003. "Husbands' Support of Their Perimenopausal Wives." *Women & Health* 38 (3): 97–112.

Martin, Emily. 1992. *The Woman in the Body: A Cultural Analysis of Reproduction.* 2nd ed. Boston, MA: Beacon Press.

McElmurry, Beverly J., and Donna S. Huddleston. 1991. "Self-Care and Menopause: Critical Review of Research." *Health Care for Women International* 12 (1): 15–26.

Meyer, Vicki. 2003. "Medicalized Menopause, US Style." *Health Care for Women International* 24 (9): 822–30.

Muhlbauer, Varda, and Joan C. Chrisler, eds. 2007. *Women Over 50: Psychological Perspectives.* New York: Springer Science & Business Media.

Niland, Patricia, and Antonia C. Lyons. 2011. "Uncertainty in Medicine: Meanings of Menopause and Hormone Replacement Therapy in Medical Textbooks." *Social Science & Medicine* 73 (8): 1238–45.

Prior, Jerilynn C. 2005. "Ovarian Aging and the Perimenopausal Transition." *Endocrine* 26 (3): 297–300.

Reid, Anne. 2004. "Gender and Sources of Subjective Well-Being." *Sex Roles* 51 (11–12): 617–29.

Schwarzer, Ralf, and Benicio Gutierrez-Dona. 2005. "More Spousal Support for Men Than for Women: A Comparison of Sources and Types of Support." *Sex Roles* 52 (7–8): 523–32.

Utz, Rebecca L. 2011. "Like Mother, (Not) Like Daughter: The Social Construction of Menopause and Aging." *Journal of Aging Studies* 25 (2): 143–54.

Vigeta, Sônia Maria Garcia, Helena Hachul, Sergio Tufik, and Eleonora Menicucci de Oliveira. 2012. "Sleep in Postmenopausal Women." *Qualitative Health Research* 22 (4): 466–75.

Walter, Carolyn Ambler. 2000. "The Psychosocial Meaning of Menopause: Women's Experiences." *Journal of Women & Aging* 12 (3–4): 117–31.

Weed, Susun S. 1999. "Menopause and Beyond: The Wise Woman Way." *Journal of Nurse-Midwifery* 44 (3): 267–279.

Wilson, Robert A. 1966. *Feminine Forever.* New York: Evans and Co. (distributed by J.B. Lippincott).

Winterich, Julie A. 2007. "Aging, Femininity, and the Body: What Appearance Changes Mean to Women with Age." *Gender Issues* 24 (3): 51–69.

Winterich, Julie A., and Debra Umberson. 1999. "How Women Experience Menopause: The Importance of Social Context." *Journal of Women & Aging* 11 (4): 57–73.

Zones, Jane Sprague. 2005. "Beauty Myths and Realities and Their Impact on Women's Health." In *Understanding Society: An Introductory Reader* edited by Margaret L. Andersen, Kim Logio, and Howard Taylor, 353–60. Belmont, CA: Thomson Learning, Inc.

CHAPTER 22

The Womb Wanders Not: Enhancing Endometriosis Education in a Culture of Menstrual Misinformation

Heather C. Guidone

THE ENDOMETRIOSIS ENIGMA

Described as "a riddle wrapped in a mystery inside an enigma" (Ballweg 1995, 275; Wilson 1987, 1), endometriosis is defined by the presence of endometrial-*like* tissue found in the extra-uterine environment (Johnson and Hummelshoj for the World Endometriosis Society Montpellier Consortium 2013). The disease elicits a sustained inflammatory response accompanied by angiogenesis, adhesions, fibrosis, scarring, and neuronal infiltration (Giudice 2010). The gold standard for confirmation of diagnosis is laparoscopy (D'Hooghe et al. 2019).

Characterized by marked distortion of pelvic anatomy (Kennedy et al. 2005), development of endometriomas and high association with

---

The vernacular of endometriosis is rooted in classic scholarship and the topic of menstruation itself is often cited as an example of biological reductionism: the medicalization of women and standardization of bodies (Rodríguez and Gallardo 2017). Hence, the author acknowledges that the terms "women" and "women's health" are enforcers of hetero-cisnormativity, gender binarism and gender essentialism. For the purposes of this chapter, incorporation of such terms is intended only as a theoretical framework, inclusive of all bodies who struggle with endometriosis and have suffered from the bias, negligence, misdiagnosis and medical misogyny which so often characterize the disease; such use is not intended to trivialize, equate or otherwise limit the scope of the condition to only lived experiences of those essentialized categories of "females." Furthermore, although often associated with the disease, "menstruation" is not synonymous with "endometriosis."

© The Author(s) 2020
C. Bobel et al. (eds.), *The Palgrave Handbook of Critical Menstruation Studies*, https://doi.org/10.1007/978-981-15-0614-7_22

269

comorbidities (Parazzini et al. 2017), endometriosis can result in significantly reduced quality of life. Although considered 'benign,' the disease may also be associated with higher risks of certain malignancies and shared characteristics with the neoplastic process (Matalliotakis et al. 2018; He et al. 2018).

Endometriosis is estimated to affect nearly 176 million individuals globally (Adamson, Kennedy, and Hummelshoj 2010), and ranks high among the most frequent causes of chronic pelvic pain (van Aken et al. 2017). A leading contributor to infertility, gynecologic hospitalization, and hysterectomy (Yeung et al. 2011; McLeod and Retzloff 2010; Ozkan et al. 2008), systemic influences of the disease can significantly impair physical, mental, emotional, and social health (Marinho et al. 2018). Definitive cause remains elusive, as does universal cure or prevention, and much of the discourse surrounding etiology and treatments remains ardently debated. Endometriosis imposes a staggering healthcare burden on society, with associated costs soaring into the billions (Soliman, Coyne, et al. 2017).

The complexities of this multidimensional condition remain poorly elucidated in current scientific works and little progress has been made toward deciphering endometriosis. Although research seems omnipresent, much of it is redundant in nature and the few qualitative studies conducted on the realities of living with the disease lack rigor (Moradi et al. 2014).

Though classically viewed as a 'disease of menstruation,' a uterus and routine menses are not de rigueur to diagnosis. The condition has been documented in post-hysterectomy/postmenopausal individuals (Ozyurek, Yoldemir, and Kalkan 2018; Soliman, Du, et al. 2017), rare cis males (Makiyan 2017, et al.), gender diverse people (Cook and Hopton 2017; Yergens 2016) and the human fetus (Schuster and Mackeen 2015; Signorile et al. 2010, 2012). Nevertheless, many continue to link the condition to simply 'painful periods' despite its profound impact far and apart from menses.

Comprehensive review of treatments for endometriosis, and the ensuing debates encompassing each, is outside the scope of this writing. However, timely diagnosis and multidisciplinary, integrative treatment are necessary to

---

Much of what is communicated about endometriosis, particularly in the scientific literature and media, reflects a stagnant belief system that perpetually confounds the diagnostic and treatment processes. Whilst medical knowledge, clinical experience and therapies are ever-evolving, the condition remains fundamentally mired in outdated assumptions that invariably lead to poor health outcomes. If we are to achieve real progress, we must strive towards an ideology which is truly reflective of modern concepts in order to elevate the condition to the priority public health platform it well deserves. To that end, though not intended as exhaustive or all-encompassing, the author has endeavored to incorporate the most current, authoritative facts about endometriosis herein—some of which run contrary to public doctrine.

effectively manage the condition—yet universal access to quality care remains limited in many settings, due in large part to dismissal of symptomology. *In brief:*

Laparoscopic excision is one of the most effective therapeutic options (Donnellan, Fulcher, and Rindos 2018; Franck et al. 2018; Pundir et al. 2017), affording biopsy-proven diagnosis and subsequent removal of lesions at the time of the surgical encounter. However, accuracy of diagnosis and treatment depends on ability of the surgeon to adequately identify the tissue in all affected areas.

Secondary to surgery are medical therapies. No drugs for endometriosis are curative; all have potential side effects (Rafique and Decherney 2017) and similar clinical efficacy in temporary reduction of pain. Menstrual suppression—which does not treat endometriosis, only symptoms—further supports the perception that menstruation is 'unhealthy' and requires pharmaceutical intervention.

Despite over 100,000 hysterectomies being performed annually as of this writing for a primary diagnosis of endometriosis and approximately 12% of individuals with the disease eventually undergoing hysterectomy as 'treatment,' there is an approximate 15% probability of persistent pain after hysterectomy, which may be due to incomplete disease removal, and a 3–5% risk of worsening pain or new symptom development (Rizk et al. 2014). Nor is menopause protective, with an estimated 2–4% of the endometriosis population being postmenopausal (Suchońska et al. 2018). In fact, postmenopausal endometriosis has demonstrated a predisposition to malignant change, greater tendency for extrapelvic spread, and development into constrictive and/or obstructive lesions (Tan and Almaria 2018).

Derived from the misogynist, antediluvian belief that painful menstruation was 'ordained by nature as punishment for failing to conceive' (Strange 2000, 616), pregnancy has long been suggested as a treatment or even cure for endometriosis. Nonetheless, pregnancy is not a 'treatment' option in any current clinical guidelines (Young, Fisher, and Kirkman 2016), nor does it prevent or defer progression of endometriosis (Setúbal et al. 2014). Moreover, the disease is linked to infertility, miscarriage, and potential complications in obstetrical outcomes (Shmueli et al. 2017) and ectopic pregnancy (Jacob, Kalder, and Kostev 2017).

Finally, adjuncts like pain management and pelvic physical therapy are also often recommended post-surgically to address secondary pain generators common with endometriosis that is, pelvic floor dysfunction. Other alternative and complementary measures may also be considered.

Defying Dogma: 'Killer Cramps' Are Not Normal

Classic presentations of endometriosis include but are not limited to abdominopelvic pain, infertility, dyschezia, dyspareunia, dysuria, physiologic dysfunction, and significantly reduced quality of life. Extrapelvic disease, while less common

(Chamié et al. 2018), may manifest in a variety of ways for example, catamenial pneumothorax. Among the most widely recognized of endometriosis symptoms is incapacitating menstrual cramping ('dysmenorrhea').

Indeed, menstrual pain without pelvic abnormality ("primary dysmenorrhea") is among the most common of gynecological disorders. Though accurate prevalence of dysmenorrhea is difficult to establish, it is estimated to impact up to 93% of adolescents (De Sanctis et al. 2015) and between 45 and 95% of all people with periods. When interviewed by Writer Olivia Goldhill (2016) for her heralded Quartz article on the lack of research into dysmenorrhea, Professor John Guillebaud went on record stating "period cramping can be almost as bad as having a heart attack." Though some have questioned the notion that *any* degree of menstrual pain is "normal" (Dusenbery 2018, 221), primary dysmenorrhea generally maintains a good prognosis.

Conversely, severe pain failing to respond to intervention ("secondary dysmenorrhea") is typically associated with conditions like endometriosis and warrants timely intervention (Bernardi et al. 2017). Moreover, a link between dysmenorrhea and the future development of chronic pelvic pain has been suggested (Hardi, Evans, and Craigie 2014), though symptoms are routinely misdiagnosed or otherwise dismissed (Bullo 2018). As a result, those suffering may be disparaged as 'menstrual moaners' or portrayed as simply unable to 'cope with normal pain' (Ballard, Lowton, and Wright 2006)—yet nearly 70% of adolescents with intractable dysmenorrhea or pelvic pain that fails to respond to initial therapy will later be diagnosed with endometriosis (Highfield et al. 2006).

Delays in the diagnosis of causative gynepathologies persist at the individual and medical level. To that end, healthcare professionals must engage patients in conversations which remain sensitive to cultural context, perceptions, and attitudes, yet draw out possible menstrual issues early so individuals are treated in timely and effective ways that harmonize with their specific needs.

Embodied Experience

A widespread lack of public education about pelvic pain and menstrual-related disorders persists. As a result, endometriosis remains under-diagnosed, inadequately treated and frequently marginalized. Inappropriate diagnostic tests, poor history taking, provision of temporary analgesics or hormonal suppression to merely treat symptoms—but not the disease itself–creates confusion in diagnosis, postponement in diagnostic confirmation and mismanagement (Riazi et al. 2014). Only a minority of studies adds to the contextual information required to understand what it means to actually struggle with endometriosis.

Misinformation about the disease remains ubiquitous, saturating the healthcare and public sectors. Affected individuals may delay seeking care for their symptoms, believing them to be a part of 'normal' menstruation, and healthcare workers may in turn dismiss their pain as "imaginary" (Bloski and Pierson 2008). To that end, healthcare encounters have been expressed as double-edged, both destructive and constructive; affecting not only the perception of the individual's physical condition, but her self-esteem, body, and sexuality (Grundström et al. 2018) as well. As a result, those with the condition must often become 'expert' or 'lead' patients; that is, those who are proactive with respect to their health and possess knowledge of their disease and symptoms in order to effectively direct and manage their own care.

Individuals with the endometriosis from all backgrounds have long described journeys characterized by ignorance, disbelief, and lack of knowledge on the part of their doctors and the public. Encountering attitudes that they 'exaggerated or imagined their symptoms or [have] low pain thresholds' and further insinuation that "psychological factors or former abuse enhanced the symptoms" (Grundström et al. 2018, 8) may compound feelings of vulnerability and anxiety. Many "feel angry and frustrated when they [have] experiences with doctors who misdiagnosed, did not diagnose, delayed diagnosis of endometriosis, or just generally did not listen to their concerns, symptoms, and experiences" (Moradi et al. 2014). Not surprisingly, some people with endometriosis may resort to maladaptive coping strategies as a result (Zarbo et al. 2018).

Clark (2012, 83) has described the impact endometriosis may have on a woman's sense of identity: "self-doubt plagued many . . . where they questioned their perception of the severity of the symptoms and ultimately their own sanity; mainly due to not being believed by medical practitioners and other lay people." Yet as Culley et al. demonstrated (2013), the distress so commonly experienced by those with endometriosis is in fact related in large part to dissatisfaction with care for the disease. The authors correctly suggest the negative social and psychological impacts of the condition could be improved by a number of strategies; not least of which include practitioner education efforts and raising awareness via education through schools and support groups.

Pandora's Jar: The Impact of the Wandering Womb and Hysteria on Endometriosis

From Greek *hysterikos* ('of the womb'), assumptions on the 'wandering' uterus have long influenced attitudes about women's health. Since the genesis of gynecology arose from the mythical first woman, Pandora, the womb was believed to have 'no natural home.' Identification of Pandora's jar (*pithos*) as a uterus has been widely represented in Hippocratic gynecology and Western art; its subsequent opening brought forth 'a range of evils including disease' (King 1998, 2, 47–48, 58).

Anxiety, sense of suffocation, tremors, convulsions, or paralysis and more have been attributed to the 'migratory uterus' (Tasca et al. 2012). Hippocrates remains largely credited for grouping such issues under the single designation of "*hysteria*," though King (1998) challenges such ascription (227, 237). Nonetheless, early physicians suggested that 'hysteria' could be counted among the '... symptoms of menstruation.' Some advised that women who frequently displayed nervous or hysterical symptoms in relation to menses 'ought to be incarcerated for their own safety and the good of society' (Strange 2000, 616); a tenuous correlation might even be drawn to today's menstrual huts.

Nezhat, Nezhat, and Nezhat (2012) further suggest there is irrefutable evidence that "hysteria, the now discredited mystery disorder presumed for centuries to be psychological in origin, was most likely endometriosis in the majority of cases ..." and as Jones (2015) proposes, discourse about the disease is "at least related to if not influenced by the social forces that shaped a diagnosis of hysteria" (1084).

Though 'hysteria' has been largely abandoned in modern nomenclature, the legacy of its impact persists. Today, symptoms of endometriosis may dismissed not as hysteria but 'somatization' (Pope et al. 2015). Women's pain is routinely under-treated, labeled inappropriately as having a sexually transmitted infection, told their symptoms are 'in their head' (Whelan 2007) or too often, simply not heard (Moradi et al. 2014).

Endometriosis also remains tethered to psychological profiling, with those suffering routinely described as high risk for anxiety, depressive symptoms, and other psychiatric disorders. In fact, however, it has been demonstrated that the presence of pain—versus endometriosis per se—is associated with such psychological and emotional distress (Vitale et al. 2017). Whelan (2007) further asserts what those with the endometriosis well know: "[c]ertainly, medical experts' ways of representing endometriosis often undermine the credibility of patient accounts . . . patients have often been represented in the medical literature as nervous, irrational women who exaggerate their symptoms" (958). Indeed, endometriosis is very much a corporeal condition with no regard for race, religious, sexual, socioeconomic, or mental health status.

SAMPSON AND THE ITINERANT UTERINE TISSUE

Reminiscent of the migrating womb, much of the dogma guiding endometriosis treatment and research today is rooted in the archaic supposition that the disease is caused by normal endometrium that has 'roamed' to distant sites. Just as the uterus does not wander, however, nor do fragments of entirely normal uterine tissue simply meander idly hither and yon resulting in endometriosis.

The premise of the condition arising from wholly normal albeit peripatetic endometrium sustains a century-old concept based on the works of Dr. John Sampson (1927). Essentially, he considered endometriosis lesions to be comprised of ordinary endometrial cells; in fact, while somewhat *resembling* native endometrium, they are *not identical* (Ahn et al. 2016)—an important distinction. An abundance of differential invasive, adhesive, and proliferative behaviors have been demonstrated in the eutopic and ectopic counterparts of endometrial stromal cells in patients with the disease (Delbandi et al. 2013), and the tissue is functionally dissimilar (Zanatta et al. 2010).

Contrary to Sampson's Theory, there is also evidence of endometriosis in cis males (Rei, Williams, and Feloney 2018, et al.), the human fetus (Signorile et al. 2009, 2010, 2012), females who have never menstruated (Suginami 1991; Houston 1984), and premenarcheal girls (Gogac et al. 2012; Marsh and Laufer 2005). The premise of 'retrograde periods' also fails to account for extrapelvic endometriosis in most cases. Moreover, though reflux menses is very common among people with periods, not all develop endometriosis; the incidence of disease is small compared to the occurrence of backflow experienced by most menstruators (Ahn et al. 2015). Similarly, as Redwine (1988) confirmed decades ago, endometriosis lacks the characteristics of an autotransplant (Khazali 2018).

Undeniably, pathogenesis remains rife with contention. Differing theories on varied mechanisms abound; stem cells, genetic polymorphisms, dysfunctional immune response, and an aberrant peritoneal environment have all been suggested in the establishment of endometriosis (Sourial, Tempest, and Hapangama 2014). The evidence also favors embryologic origins, with additional cellular and molecular mechanisms involved (Signorile et al. 2009, 2010, 2012; Redwine 1988). Nevertheless, no unifying theory to date accounts for all of described manifestations of endometriosis (Burney and Giudice 2012).

Unremitting Misinformation, Menstrual Taboos, and Diagnostic Delay

Much of society's derogatory view of menstruating individuals, including within the political sphere (*'blood coming out of her wherever ...'*), remains virtually unchanged, and the very normal physiological process of menstruation remains linked to unfavorable attitudes in all cultures (Chrisler et al. 2015). Periods are still considered taboo in many parts of the world, with persistent knowledge gaps resulting in part from poor puberty guidance (Chandra-Mouli and Patel 2017). Research on menstrual cycle-related risk factors is lacking (Harlow and Ephross 1995), and the media continues to reinforce misconceptions around social captivity, restrictions, professional inefficiency, physical, and mental discomfort (Yagnik 2012) related to menses. Menstrual bleeding continues to be portrayed as "messy, inconvenient, and [an] unnecessary phenomenon to be controlled or possibly eliminated" (McMillan and

Jenkins 2016, 1). Yet, with a nod to Bobel and Kissling (2011): "menstruation matters:" menstrual history is a key component in a comprehensive women's health assessment and an increasingly important variable in disease research (McCartney 2016).

For many, persistent taboos and perpetuation of 'period shaming' come at a high price: menstrual pain specifically, such as that often accompanying endometriosis, is routinely dismissed. Hence, the path to diagnosis is largely dependant upon the individual's own "knowledge and experience of painful menstruation and other symptoms and whether they know other people who have been diagnosed" (Clark 2012, 85).

Delayed diagnosis serves as a high source of stress responsible for an important psychological impact on individuals with endometriosis. Average diagnostic delays worldwide hover around 7.5 years (Bullo 2019) or even longer, with continued resistance to timely intervention and referrals. Indeed, several clinicians consider themselves inadequately trained to understand and provide psychosocial care for patients with the disease (Zarbo et al. 2018). Conversely, earlier diagnosis and efficient intervention decreases productivity loss, quality of life impairment, and healthcare consumption, consequently reducing total costs to patients and society alike (Klein et al. 2014).

Studies reveal a relationship between ambivalent sexism and more negative attitudes toward menstruation, which may also lead to reticence to report menstrual cycle-related symptoms (Marván, Vázquez-Toboada, and Chrisler 2014). Others may deliberately conceal concerns for fear of stigmatization, further leading to diagnostic delay (Riazi et al. 2014). Still others may seek to reduce stigma associated with menstruation through 'menstrual etiquette' (Seear 2009), perpetuating social rules and normative expectations of menstruating persons and fearing that disclosure would result in embarrassment or perception that they are 'weak' (Culley et al. 2013). The literature further suggests some patients may simply fail to seek timely medical help due to their own inability to identify symptoms as 'abnormal'—a failing of our menstrual education system.

To navigate the experiences of menstruation, endometriosis, and other episodes related to pain or vaginal bleeding, individuals "require factual and supportive information that enables them to differentiate between healthy and abnormal bleeding, to understand and take care of their bodies or those of dependents who may require assisted care, and to seek health advice appropriately" (Sommer et al. 2017, 2). Yet, menstrual teachings remain hampered by deficient cycles of misinformation. Education and perception are primarily communicated by mothers, sisters, or friends who themselves may lack accurate understanding (Cooper and Barthalow 2007), with resulting poor body literacy regarding reproductive anatomy, female hormones and their functions, effect of hormones on the menstrual cycle, ovulation, and conception (Ayoola, Zandee, and Adams 2016).

Likewise, menstrual health education programs in school and community settings remain deficient, particularly in low income settings, with many girls viewing school education about menstruation as *"inaccurate, negative, and late"* (Herbert et al. 2017, 14).

Conquering the Prevailing Ethos of Menstrual Shaming to Effect Positive Change

The perpetuation of menstrual shaming (for example, 'The Curse') has led to a prevailing ethos of generational taboos and lack of body literacy. There are consequences for such persistent bias, poor information systems, and practices; the resulting lack of education leads to delayed diagnosis and quality treatment of endometriosis and other gynepathologies with subsequent impact on fertility, loss of libido and pleasurable sex, chronic pain, diminished quality of life, loss of sense of self, body-negative thoughts, and more.

While disease knowledge has evolved, the deeply entrenched cultural norms surrounding both endometriosis and menstruation must continue to be challenged. Existing gaps must be bridged in order to eliminate the enduring barriers that persist. How and when girls learn about menses and its associated changes can impact response to the menstrual event and is critical to their knowledge, autonomy, and empowerment. Hence, it is necessary to overcome persistent myths, increase authoritative awareness of endometriosis, and articulate effective strategies to develop more robust literacy on the condition than presently exists.

Cooper and Barthalow (2007) previously established the need for menstrual education in schools, with the topic being offered even before menarche in order to better prepare girls for the experience and continuing throughout their educational career so that students can build upon their basic knowledge of the many themes involved with menstrual health. A three-pronged approach has been suggested (Subasinghe et al. 2016) to better inform individuals about dysmenorrhea specifically: having the school nurse provide educational leaflets to increase familiarity with the condition; encouraging health professionals to be more proactive in asking patients about the topic so that young menstruators with dysmenorrhea may be more likely to disclose their pain and symptoms; and finally, joint promotion by health professionals and schools of reliable, authoritative websites, and resources for additional guidance.

Oni and Tshitangano (2015) previously proposed that school health teams may also consider screening students for menstrual disorders in order to help diagnose underlying pathological causes and attend such issues accordingly. Similar findings on the need for adolescent education on the effective management of dysmenorrhea suggest that extending the educational program to parents and school leaders is beneficial as well (Wong 2011).

Evidence demonstrates that consistent delivery of a menstrual health education program in schools specifically increases awareness of endometriosis (Bush et al. 2017). Two successful examples of such programs are already underway:

The Endo What? Documentary team School Nurse Initiative (https://www.endowhat.com/school-nurse-initiative), founded by Shannon Cohn, is a collaborative effort to provide endometriosis education and awareness among school nurses and their students and

The New Zealand model and the first of its kind in the world, developed over two decades ago by Deborah Bush, MNZM, QSM, Dip Tchg. LSB, Chief Executive of Endometriosis New Zealand (http://www.nzendo.org.nz/how-we-help/all-about-me). Both efforts have served to educate countless individuals.

Building on the examples above, clinicians and the public alike will benefit from better understanding of endometriosis, thereby improving patient experiences and leading to improved outcomes. We must incorporate *correct* disease information along with ethical, social, cultural, economic, and diversity perspectives in emerging menstrual education curriculum.

In order to ensure appropriate intervention and reduce costly, unproven protocols, like-minded collaborators from practitioner, allied and mental health and others need to engage in associated efforts. There must be an emphasis placed on optimal pathways, evaluation of modern concepts, and cross-collaborative strategies. It is imperative that all individuals know when, where and how to obtain help when symptoms of menstrual-related disorders first arise, and it is vital that the public, including but not limited to, legislators, hospital administrators, gynecologists, and subspecialists become involved in these efforts.

Moreover, in that mothers often traditionally teach their daughters, we must rectify misperceptions and offer instruction on menstrual practices and disorders like endometriosis by providing compulsory education at school, in clinics, and kinship settings in order to encourage story-telling narratives and break the legacy of silence, misinformation, and fear. We must better elucidate the parameters of normal versus abnormal bleeding, pain, and related symptomology in order to recognize disorder and pain signaling throughout the cycle.

To address difficulties faced by low resource and medically underserved communities, use of participatory/community-based efforts, integrated messaging during clinic visits, and use of Information Technology (IT) and digital health tools where applicable can improve access to healthcare services and information in ways that enhance patient knowledge and self-management, thereby positively impacting health outcomes.

Through stakeholder partnerships, we can foster new menstrual educational programs to produce high-quality educational materials and afford better outcomes for all. A strong public health agenda for menstrual/endometriosis education must include a collaborative interface among public health, community and non-healthcare sectors.

SUMMARY

Endometriosis has the propensity to take away so many of an affected individual's choices: when and whether to engage in sex, when or if to pursue fertility, whether or not to undergo invasive procedures or to choose oft-ineffective menstrual suppressives that alter her cycle and more. We must strive toward early recognition and diagnosis, better understanding of pathophysiology and pain mechanisms, increased translational research and dissemination of authoritative facts on a widespread basis, starting with menstrual education among youth.

The current deficiency in quality menstrual education leads to confusion, inaccurate beliefs about and negative views on menstruation and related conditions. Though steps forward have been made, many individuals lack understanding of what constitutes menstrual dysfunction and *when, where and how* to seek care. It is imperative that patients and health professionals alike become better educated on the clinical characteristics of endometriosis, not least general practitioners and school nurses, who play crucial roles in early diagnosis. This is achievable through menstrual education programs that incorporate the disease as a leading cause of pain. Outlining optimal care pathways, encouraging timely recognition, improving research priorities, accepting modern concepts and emphasizing appropriate, cross-collaborative strategies to optimize outcomes can transform endometriosis care and reduce the role of 'menstrual silence' in its diagnosis and treatment.

Embarking on robust educational programs which begin in the primary setting and are shared across varied resources will enhance literacy on painful menstruation and gynepathologies, thereby affording access to better, earlier care and improving the lives of the millions suffering. By revitalizing menstrual communication and key conversations, we can put an end to the secrecy, silence, shame, and pain.

REFERENCES

Adamson, D., S. Kennedy, and L. Hummelshoj. 2010. "Creating Solutions in Endometriosis: Global Collaboration Through the World Endometriosis Research Foundation." *Journal of Endometriosis and Pelvic Pain Disorders* 2 (1): 3–6.

Ahn, S. H., K. Khalaj, S. L. Young, B. A. Lessey, M. Koti, and C. Tayade. 2016. "Immune-Inflammation Gene Signatures in Endometriosis Patients." *Fertility and Sterility* 106, no. 6 (November): 1420–1431.e7.de.

Ahn, S. H., S. P. Monsanto, C. Miller, S. S. Singh, R. Thomas, and C. Tayade. 2015. "Pathophysiology and Immune Dysfunction in Endometriosis." *BioMed Research International* 2015: 1–12. Article ID 795976. https://doi.org/10.1155/2015/795976.

van Aken, M. A. W, J. M. Oosterman, C. M. van Rijn, M. A. Ferdek, G. S. F. Ruigt, B. W. M. M. Peeters, D. D. M. Braat, and A. W. Nap. 2017. "Pain Cognition Versus Pain Intensity in Patients with Endometriosis: Toward Personalized Treatment." *Fertility and Sterility* 108, no. 4 (October): 679–86.

Ayoola, A. B., G. L. Zandee, and Y. J. Adams. 2016. "Women's Knowledge of Ovulation, the Menstrual Cycle, and Its Associated Reproductive Changes." *Birth* 43 (3): 255–62.

Ballard, K., K. Lowton, and J. Wright. 2006. "What's the Delay? A Qualitative Study of Women's Experiences of Reaching a Diagnosis of Endometriosis." *Fertility and Sterility* 86, no. 5 (November): 1296–301.

Ballweg, M. L. 1995. "The Puzzle of Endometriosis." In *Endometriosis*, edited by C. R. Nezhat, G. S. Berger, F. R. Nezhat, V. C. Buttram, and C. H. Nezhat. New York, NY: Springer.

Bernardi, M., L. Lazzeri, F. Perelli, F. M. Reis, and F. Petraglia. 2017. "Dysmenorrhea and Related Disorders." *F1000Rsearch* 7 (6): 1645.

Bloski, T, and R. Pierson. 2008. "Endometriosis and Chronic Pelvic Pain: Unraveling the Mystery Behind This Complex Condition." *Nursing for Women's Health* 12 (5): 382–95.

Bobel, C., and E. Arveda Kissling. 2011. "Menstruation Matters: Introduction to Representations of the Menstrual Cycle." *Women's Studies* 40 (2): 121–26.

Bullo, S. 2018. "Exploring Disempowerment in Women's Accounts of Endometriosis Experiences." *Discourse & Communication* 12 (6): 569–86.

———. 2019. "'I Feel Like I'm Being Stabbed by a Thousand Tiny Men': The Challenges of Communicating Endometriosis Pain." *Health* (London). February 19: 1363459318817943. https://doi.org/10.1177/1363459318817943.

Burney, R. O., and L. C. Giudice. 2012. "Pathogenesis and Pathophysiology of Endometriosis." *Fertility and Sterility* 98 (3): 511–19.

Bush, D., E. Brick, M. C. East, and N. Johnson. 2017. "Endometriosis Education in Schools: A New Zealand Model Examining the Impact of an Education Program in Schools on Early Recognition of Symptoms Suggesting Endometriosis." *The Australian and New Zealand Journal of Obstetrics and Gynaecology* 57, no. 4 (August): 452–57.

Chamié, L. P., D. M. F. R. Ribeiro, D. A. Tiferes, A. C. Macedo Neto, and P. C. Serafini. 2018. "Atypical Sites of Deeply Infiltrative Endometriosis: Clinical Characteristics and Imaging Findings." *Radiographics* 38, no. 1(January–February): 309–28.

Chandra-Mouli, V, and S. V. Patel. 2017. "Mapping the Knowledge and Understanding of Menarche, Menstrual Hygiene and Menstrual Health among Adolescent Girls in Low-and Middle-Income Countries." *Reproductive Health* 14: 30.

Chrisler, J. C., M. L. Marván, J. A. Gorman, and M. Rossini 2015. "Body Appreciation and Attitudes toward Menstruation." *Body Image* 12 (January): 78–81.

Clark, M. 2012. "Experiences of Women with Endometriosis: An Interpretative Phenomenological Analysis." Doctoral Thesis, Queen Margaret University, Edinburgh. Accessed January 2, 2018. http://etheses.qmu.ac.uk/1812.

Cook, A., and E. Hopton. 2017. "Endometriosis Presenting in a Transgender Male." *Journal of Minimally Invasive Gynecology* 24 (7): S126.

Cooper, S. C., and P. Barthalow Koch. 2007. "'Nobody Told Me Nothin': Communication About Menstruation among Low-Income African-American Women." *Women & Health* 46 (1): 57–78.

Culley, L., C. Law, N. Hudson, E. Denny, H. Mitchell, M. Baumgarten, and N. Raine-Fenning. 2013. "The Social and Psychological Impact of Endometriosis on Women's Lives: A Critical Narrative Review." *Human Reproduction Update* 19, no. 6 (November–December): 625–39.

Delbandi, A. A., M. Mahmoudi, A. Shervin, E. Akbari, M. Jeddi-Tehrani, M. Sankian, S. Kazemnejad, and A. H. Zarnani. 2013. "Eutopic and Ectopic Stromal Cells from Patients with Endometriosis Exhibit Differential Invasive, Adhesive, and Proliferative Behavior." *Fertility and Sterility* 100, no. 3 (September): 761–69.

D'Hooghe, T. M., A. Fassbender, F. O. Dorien, and A. Vanhie. 2019. "Endometriosis Biomarkers: Will Co-Development in Academia-Industry Partnerships Result in New and Robust Noninvasive Diagnostic Tests?" *Biology of Reproduction*. February 1. https://doi.org/10.1093/biolre/ioz016.

Donnellan, N. M., I. R. Fulcher, and N. B. Rindos. 2018. "Self-Reported Pain and Quality of Life Following Laparoscopic Excision of Endometriosis as Measured Using the Endometriosis Health Profile-30: A 5 Year Follow-Up Study." *Journal of Minimally Invasive Gynecology* 25 (7): S54.do.

Dusenbery, M. 2018. *Doing Harm: The Truth About How Bad Medicine and Lazy Science Leave Women Dismissed, Misdiagnosed, and Sick*. New York: HarperOne, pp. 221.

Franck, C., M. H., Poulsen, G. Karampas, A. Giraldi, and M. Rudnicki. 2018. "Questionnaire-Based Evaluation of Sexual Life After Laparoscopic Surgery for Endometriosis: A Systematic Review of Prospective Studies." *Acta Obstetricia et Gynecologica Scandinavica* 97: 1091–104

Giudice, L. C. 2010. "Clinical Practice: Endometriosis." *The New England Journal of Medicine* 362 (25): 2389–98.

Gogacz, M., M. Sarzyński, R. Napierała, J. Sierocińska-Sawa, and A. Semczuk. 2012. "Ovarian Endometrioma in an 11-Year-Old Girl Before Menarche: A Case Study with Literature Review." *Journal of Pediatric and Adolescent Gynecology* 25, no. 1 (February): e5–e7. https://doi.org/10.1016/j.jpag.2011.09.009.

Goldhill, O. 2016. "Period Pain Can Be 'Almost as Bad as a Heart Attack.' Why Aren't We Researching How to Treat It?" *Quartz Media*. Accessed October 15, 2017. https://qz.com/611774/period-pain-can-be-as-bad-as-a-heart-attack-so-why-arent-we-researching-how-to-treat-it.

Grundström, H., S. Alehagen, P. Kjølhede, and C. Berterö. 2018. "The Double-Edged Experience of Healthcare Encounters among Women with Endometriosis: A Qualitative Study." *Journal of Clinical Nursing* 27, nos. 1–2 (January): 205–11.

Hardi, G., S. Evans, and M. Craigie. 2014. "A Possible Link between Dysmenorrhoea and the Development of Chronic Pelvic Pain." *Australian and New Zealand Journal of Obstetrics and Gynaecology* 54, no. 6 (December): 593–96.

Harlow, S. D., and S. A. Ephross. 1995."Epidemiology of Menstruation and Its Relevance to Women's Health." *Epidemiologic Reviews* 17 (2): 265–86.

He, J., W. Chang, C. Feng, M. Cui, and T. Xu. 2018. "Endometriosis Malignant Transformation: Epigenetics as a Probable Mechanism in Ovarian Tumorigenesis." *International Journal of Genomics* 27 (March): 1465348.

Herbert, A. C., A. M. Ramirez, G. Lee, S. J. North, M. S. Askari, R. L. West, and M. Sommer. 2017. "Puberty Experiences of Low-Income Girls in the United States: A Systematic Review of Qualitative Literature from 2000 to 2014." *Journal of Adolescent Health* 60, no. 4 (April): 363–79.

Highfield, E. S., M. R. Laufer, R. N. Schnyer, C. E. Kerr, P. Thomas, P. M. Wayne. 2006. "Adolescent Endometriosis-Related Pelvic Pain Treated with Acupuncture: Two Case Reports." *Journal of Alternative and Complementary Medicine* 12, no. 3 (April): 317–22.

Houston, D. 1984. "Evidence for the Risk of Pelvic Endometriosis by Age, Race and Socioeconomic Status." *Epidemiologic Reviews* 6 (1): 167–91.

Jacob, L., M. Kalder, and K. Kostev. 2017. "Risk Factors for Ectopic Pregnancy in Germany: A Retrospective Study of 100,197 Patients." *German Medical Science* 19, no. 15 (December): Doc. 19.

Johnson, N.P., and L. Hummelshoj for the World Endometriosis Society Montpellier Consortium. 2013. "Consensus on Current Management of Endometriosis." *Human Reproduction* 28 (6): 1552–68.

Jones, C. 2015. "Wandering Wombs and 'Female Troubles': The Hysterical Origins, Symptoms, and Treatments of Endometriosis." *Women's Studies* 44 (8): 1083–113.

Kennedy, S., A. Bergqvist, and C. Chapron et al. 2005. "ESHRE Guideline for the Diagnosis and Treatment of Endometriosis." *Human Reproduction* 20 (10): 2698–704.

Khazali, S. 2018. "The BSGE Meets…David Redwine." *BSGE Newsletter/The Scope.* Issue 10, Autumn, p. 25.

King, H. 1998. *'Hippocrates Woman': Reading the Female Body in Ancient Greece.* London: Routledge, pp. 2, 47–48, 58, 227, 237.

Klein, S., T. D'Hooghe, C. Meuleman, C. Dirksen, G. Dunselman, and S. Simoens. 2014. "What Is the Societal Burden of Endometriosis-Associated Symptoms? A Prospective Belgian Study." *Reproductive BioMedicine Online* 28, no. 1 (January): 116–24.

Makiyan, Z. 2017. "Endometriosis Origin from Primordial Germ Cells." *Organogenesis* 3; 13, no. 3 (July): 95–102.

Marinho, M. C. P., T. F. Magalhaes, L. F. C. Fernandes, K. L. Augusto, A. V. M. Brilhante, and L. R. P. S. Bezerra. 2018. "Quality of Life in Women with Endometriosis: An Integrative Review." *Journal of Women's Health* 27, no. 3 (March): 399–408.

Marsh, E. E., and M. R. Laufer. 2005. "Endometriosis in Premenarcheal Girls Who Do Not Have an Associated Obstructive Anomaly." *Fertility and Sterility* 83, no. 3 (March): 758–60.

Marván, M. L., R. Vázquez-Toboada, and J. C. Chrisler. 2014. "Ambivalent Sexism, Attitudes towards Menstruation and Menstrual Cycle-Related Symptoms." *International Journal of Psychology* 49, no. 4 (August): 280–87.

Matalliotakis, M., C. Matalliotaki, G. N. Goulielmos, E. Patelarou, M. Tzardi, D. A. Spandidos, A. Arici, and I. Matalliotakis. 2018. "Association between Ovarian Cancer and Advanced Endometriosis." *Oncology Letters* 15 (5): 7689–92.

McCartney, P. 2016. "Nursing Practice with Menstrual and Fertility Mobile Apps." *The American Journal of Maternal/Child Nursing* 41 (1): 61.

McLeod, B. S., and M. G. Retzloff. 2010. "Epidemiology of Endometriosis: An Assessment of Risk Factors." *Clinical Obstetrics and Gynecology* 53, no. 2 (June): 389–96.

McMillan, C., and A. Jenkins. 2016. "'A Magical Little Pill That Will Relieve You of Your Womanly Issues': What Young Women Say About Menstrual Suppression." *International Journal of Qualitative Studies on Health and Well-Being* 23, no. 11 (November): 32932.

Moradi, M., M. Parker, A. Sneddon, V. Lopez, and D. Ellwood. 2014. "Impact of Endometriosis on Women's Lives: A Qualitative Study." *BMC Women's Health* 14: 123.

Nezhat, C., F. Nezhat, and C. Nezhat. 2012. "Endometriosis: Ancient Disease, Ancient Treatments." *Fertility and Sterility* 98 (6): S1.

Oni, T. H., and T. G. Tshitangano. 2015. "Prevalence of Menstrual Disorders and Its Academic Impact amongst Tshivenda Speaking Teenagers in Rural South Africa." *Journal of Human Ecology* 51 (12): 214–19.

Ozkan, S., W. Murk, and A. Arici. 2008. "Endometriosis & Infertility: Epidemiology and Evidence-based Treatments." *Annals of the New York Academy of Sciences* 1127: 92–100. Assessment of Human Reproductive Function.

Ozyurek, E. S., T. Yoldemir, and U. Kalkan. 2018. "Surgical Challenges in the Treatment of Perimenopausal and Postmenopausal Endometriosis." *Climacteric* 21, no. 4 (August): 385–90.

Parazzini, F., G. Esposito, L. Tozzi, S. Noli, and S. Bianchi. 2017. "Epidemiology of Endometriosis and Its Comorbidities." *European Journal of Obstetrics & Gynecology and Reproductive Biology* 209 (February): 3–7.

Pope, C., V. Sharma, S. Sharma, and D. Mazmanian. 2015. "A Systematic Review of the Association between Psychiatric Disturbances and Endometriosis." *Journal of Obstetrics & Gynaecology Canada* 37 (11): 1006–15.

Pundir, J., K. Omanwa, E. Kovoor, V. Pundir, G. Lancaster, and P. Barton-Smith. 2017. "Laparoscopic Excision Versus Ablation for Endometriosis-Associated Pain: An Updated Systematic Review and Meta-Analysis." *Journal of Minimally Invasive Gynecology* 24 (5) (July–August): 747–56.

Rafique, S, and A. H. Decherney. 2017. "Medical Management of Endometriosis." *Clinical Obstetrics and Gynecology* 60, no. 3 (September): 485–96.

Redwine, D. B. Mulleriosis. 1988. "The Single Best Fit Model of the Origin of Endometriosis." *Reproductive Medicine* 33 (11): 915–20.

Rei, C., T. Williams, and M. Feloney. 2018. "Endometriosis in a Man as a Rare Source of Abdominal Pain: A Case Report and Review of the Literature." *Case Reports in Obstetrics and Gynecology* 2008 (January 31): 2083121.

Riazi, H., N. Tehranian, S. Ziaei, E. Mohammadi, E. Hajizadeh, and A. Montazeri. 2014. "Patients' and Physicians' Descriptions of Occurrence and Diagnosis of Endometriosis: A Qualitative Study from Iran." *BMC Women's Health* 14: 103.

Rizk, B., A. S. Fischer, H. A. Lotfy, R. Turki, H. A. Zahed, R. Malik, C. P. Holliday, A. Glass, H. Fishel, M. Y. Soliman, and D. Herrera. 2014. "Recurrence of Endometriosis After Hysterectomy." *Facts, Views and Vision in* Obstetrics *and* Gynaecology 6 (4): 219–27.

Rodríguez, M. B., and E. B. Gallardo. 2017. "Contributions to a Feminist Anthropology of Health: The Study of the Menstrual Cycle." *Salud Colect* 13, no. 2 (April–June): 253–65.

Sampson, J. A. 1927. "Metastatic or Embolic Endometriosis, Due to the Menstrual Dissemination of Endometrial Tissue into the Venous Circulation." *The American Journal of Pathology* 3 (2): 93–110.43.a

De Sanctis, V., A. Soliman, S. Bernasconi, L. Bianchin, G. Bona, M. Bozzola, F. Buzi, C. De Sanctis, F. Tonini G, Rigon, and E. Perissinotto. 2015. "Primary Dysmenorrhea in Adolescents: Prevalence, Impact and Recent Knowledge." *Pediatric Endocrinology Reviews* 13, no. 2 (December): 512–20.

Schuster, M., and D. A. Mackeen. 2015. "Fetal Endometriosis: A Case Report." *Fertility and Sterility* 103, no. 1 (January): 160–62.

Seear, K. 2009. "The Etiquette of Endometriosis: Stigmatisation, Menstrual Concealment and the Diagnostic Delay." *Social Science & Medicine* 69, no. 8 (October): 1220–27.

Setúbal, A., Z. Sidiropoulou, M. Torgal, E. Casal, C. Lourenço, and P. Koninckx. 2014. "Bowel Complications of Deep Endometriosis During Pregnancy or In Vitro Fertilization." *Fertility and Sterility* 101, no. 2 (February): 442–46.

Shmueli, L., L. Salman, L. Hiersch, E. Ashwal, E. Hadar, A. Wiznitzer, Y. Yogev, and A. Aviram. 2017. "Obstetrical and Neonatal Outcomes of Pregnancies Complicated by Endometriosis." *Journal of Maternal-Fetal and Neonatal Medicine* 29 (October): 1–6.

Signorile, P. G., F. Baldi, R. Bussani, et al. 2010. "New Evidence of the Presence of Endometriosis in the Human Fetus." *Reproductive BioMedicine Online* 21 (1): 142–47.

Signorile, P. G., F. Baldi, R. Bussani, M. D'Armiento, M. De Falco, and A. Baldi 2009. "Ectopic Endometrium in Human Foetuses is a Common Event and Sustains the Theory of Müllerianosis In the Pathogenesis of Endometriosis, a Disease That Predisposes to Cancer." *Journal of Experimental and Clinical Cancer Research* 28 (1), article 49.

Signorile, P. G., F. Baldi, R. Bussani, R. Viceconte, P. Bulzomi, M. D'Armiento, A. D'Avino, and A. Baldi. 2012. "Embryologic Origin of Endometriosis: Analysis of 101 Human Female Fetuses." *Journal of Cellular Physiology* 227 (4): 1653–56.

Soliman, A. M., E. X. Du, H. Yang, E. Q. Wu, and J. C. Haley. 2017. "Retreatment Rates among Endometriosis Patients Undergoing Hysterectomy or Laparoscopy." *Journal of Women's Health* 26, no. 6 (June): 644–54.

Soliman, A. M., K. S. Coyne, K. S. Gries, J. Castelli-Haley, M. Snabes, and E. Surrey. 2017. "The Effect of Endometriosis Symptoms on Absenteeism and Presenteeism in the Workplace and at Home." *Journal of Managed Care & Specialty Pharmacy* 7: 745–54.

Sommer, M., P. A. Phillips-Howard, T. Mahon, S. Zients, M. Jones, and B. A. Caruso. 2017. "Beyond Menstrual Hygiene: Addressing Vaginal Bleeding Throughout the Life Course in Low and Middle-Income Countries." *BMJ Global Health* 2: e000405.

Sourial, S., N. Tempest, and D. K. Hapangama. 2014. "Theories on the Pathogenesis of Endometriosis." *International Journal of Reproductive Medicine* 2014: 179515.

Strange, J. M. 2000. "Menstrual Fictions: Languages of Medicine and Menstruation, c. 1850–1930." *Women's History Review* 9 (3): 607–28.

Subasinghe, A. K., L. Happo, Y. L. Jayasinghe, S. M. Garland, A. Gorelik, and J. D. Wark. 2016. "Prevalence and Severity of Dysmenorrhoea, and Management Options Reported by Young Australian Women." *Australian Family Physician* 45, no. 11 (November): 829–34.

Suchońska, B., M. Gajewska, A. Zyguła, and M. Wielgoś. 2018. "Endometriosis Resembling Endometrial Cancer in a Postmenopausal Patient." *Climacteric* 21, no. 1 (February): 88–91.

Suginami, H. 1991. "A Reappraisal of the Coelomic Metaplasia Theory by Reviewing Endometriosis Occurring in Unusual Sites and Instances." *American Journal of Obstetrics and Gynecolog* 165, no. 1 (July): 214–18.

Tan, D. A., and M. J. G. Almaria. 2018. "Postmenopausal Endometriosis: Drawing a Clearer Clinical Picture." *Climacteric* 21, no. 3 (June): 249–55.

Tasca, C., M. Rapetti, M.G. Carta, and B. Fadda. 2012. "Women and Hysteria in the History of Mental Health." *Clinical Practice & Epidemiology in Mental Health* 8: 110–19.

Vitale, S. G., V. L. La Rosa, A. M. Rapisarda, and A. S. Laganà. 2017. Comment on: "Impact of Endometriosis on Quality of Life and Mental Health: Pelvic Pain Makes the Difference." *Journal of Psychosomatic Obstetrics & Gynecology* 38, no. 1 (March): 81–82.

Whelan, E. 2007. "'No One Agrees Except for Those of Us Who Have It': Endometriosis Patients as an Epistemological Community." *Sociology of Health and Illness* 29, no. 7 (November): 957–82.

Wilson, E. A. 1987. *Endometriosis*. New York, NY: Alan R. Liss.

Wong, L. P. 2011. "Attitudes towards Dysmenorrhoea, Impact and Treatment Seeking among Adolescent Girls: A Rural School-Based Survey." *Australian Journal of Rural Health* 19, no. 4 (August): 218–23.

Yagnik, A. S. 2012. "Construction of Negative Images of Menstruation in Indian TV Commercials." *Health Care for Women International* 33 (8): 756–71.

Yergens, A. R. T. 2016. "Endometriosis and Gender Nonconformity." *Huffington Post,* June 9. Accessed September 1, 2018. https://www.huffingtonpost.com/ashley-rt-yergens/pumpkin-spice-lattes-endo_b_10265178.html.

Yeung, P. Jr., K. Sinervo, W. Winer, and R. B. Jr. Albee. 2011. "Complete Laparoscopic Excision of Endometriosis in Teenagers: Is Postoperative Hormonal Suppression Necessary?" *Fertility and Sterility* 95, no. 6 (May): 1909–12, 1912.e1.

Young, K., J. Fisher, and M. Kirkman. 2016. "Endometriosis and Fertility: Women's Accounts of Healthcare." *Human Reproduction* 31, no. 3 (March): 554–62.

Zanatta, A., A. M. Rocha, F. Carvalho, R. Pereira, H. S. Taylor, E. Motta, E. Baracat, and P. Serafini. 2010. "The Role of the Hoxa10/HOXA10 Gene in the Etiology of Endometriosis and Its Related Infertility: A Review." *Journal of Assisted Reproduction and Genetics* 27 (12): 701–10.

Zarbo, C., A. Brugnera, L. Frigerio, C. Malandrino, M. Rabboni, E. Bondi, and A. Compare. 2018. "Behavioral, Cognitive, and Emotional Coping Strategies of Women with Endometriosis: A Critical Narrative Review." *Archives of Women's Mental Health* 21, no. 1 (February): 1–13.

CHAPTER 23

Premenstrual Syndrome (PMS) and the Myth of the Irrational Female

Sally King

Ever since 'premenstrual' symptoms were first formally described in 1931, those most prominently cited in diagnostic criteria have been mood-based (Laws, Hey, and Eagan 1985; Ussher 2011). However, the reason for prioritizing psychological symptoms over physical ones remains unclear. For example, population studies have shown that they are not necessarily the most commonly experienced, uniquely determining, nor most disruptive of menstrual changes (for example, Dennerstein, Lehert, and Heinemann 2011; Choi et al. 2010). This chapter argues that this trend is influenced by a sexist historical assumption, referred to here as 'the myth of the irrational female'; the idea that women are *pathologically* emotional, and thus have a reduced capacity for reason, due to their reproductive biology.

I begin by outlining some historical shifts in medical descriptions of female-prevalent symptoms, to help explain why cyclical mood changes came to be prioritized over physical experiences, such as pain. I then use the latest Royal College of Obstetricians and Gynaecologists guidelines on PMS (RCOG 2016) to show that although clinical experts are trying to overcome this issue, they have not yet managed to completely disentangle PMS from the myth of the irrational female. I conclude by calling for a more integrated and rigorous approach to PMS definitions and research, in order to support people who *do* experience cyclical symptoms, without unintentionally pathologizing the menstrual cycle, or stigmatizing an entire gender.

© The Author(s) 2020
C. Bobel et al. (eds.), *The Palgrave Handbook of Critical Menstruation Studies*, https://doi.org/10.1007/978-981-15-0614-7_23

THE PSYCHOLOGIZING OF MENSTRUAL CYCLE-RELATED SYMPTOMS

Women, Wombs, Health, and Reason

Menstrual cycle-related symptoms have been recognized and treated by physicians for at least the past 3800 years. We know this because one of the oldest surviving medical texts, the *Kahun Gynaecological Papyrus* (c. 1800 BCE), describes some highly recognizable menstrual experiences (for example, period pain, heavy bleeding, lower back pain, and migraine) and ascribes them to '*discharges . . . clenches . . . wrappings . . . or wandering . . . of the womb*' (Quirke 2002, 1:1–29). Interestingly, this Ancient Egyptian scroll also lists other (less directly associated) female symptoms and similarly associates them with the womb for example, leg pain and toothache (Quirke 2002, 1:15–20).

In this way, the *Kahun Gynaecological Papyrus* perhaps exemplifies a recurring theme that has helped shape constructions of female health throughout the history of Western medicine, namely, the attribution of any and all (otherwise unexplained) female-prevalent symptoms to the womb (or menstruation by association), as *the* defining feature of embodied femininity (Ussher 1989; King 1998). While a causal association between the female reproductive system and ill health is not *necessarily* 'wrong' or sexist, a widespread (yet incorrect) assumption that a generalized converse logical position also holds true, is much more problematic. That is, 'if ill health is caused by the womb (in some women), then all women are ill (by virtue of having wombs).' Indeed, it is this logical fallacy that underpins much of the myth of the irrational female.

Right up until the sixteenth century, descriptions of female-prevalent symptoms were always physical in nature (King 1998). From the seventeenth century onwards, however, they came to be seen as more *emotional* in origin, and experience. In particular, the work of Thomas Sydenham (1624–1689) was influential in the reclassification of 'hysteria' (a term adapted from Classical references to *hysterikē pnix* that is, 'suffocation of the womb') from a gynecological condition, to a female-prevalent nervous condition, caused by a *weaker nervous constitution* and subsequent *emotional instability* (Gilman et al. 1993, 143–46).

Sydenham's new definition of 'hysteria' was in keeping with preexisting philosophical assumptions that women were inherently physically, spiritually, and intellectually inferior to men. Yet it meant that gender discrimination was thus now justified by a 'scientific' (that is, supposedly objective) medical claim that due to the pathological nature of the female (reproductive) body, women were also *pathologically* emotional (and, therefore, less capable of reason) and so inevitably subject to the control of men (Ussher 2005, 16).

By the late nineteenth century, the work of Sigmund Freud had firmly repositioned 'hysteria' as a type of *psychological neurosis* (Freud 1966). His work significantly influenced early twentieth century public discourse.

Indeed, one of the main arguments used against women's suffrage (the right to vote in parliamentary elections) was that (all) women were mentally unfit to make an informed and rational decision. Just such a belief was unambiguously conveyed by Sir Almroth Wright (1912), a famous physiologist and anti-suffragist; "*[On] Militant Hysteria- No doctor . . . can ever lose sight of the fact that the mind of woman is always threatened with danger from the reverberations of her physiological emergencies.*"

The Emergence of Premenstrual *Conditions*

The 'myth of the irrational female' is also a likely influence on the first formal medical description of 'Premenstrual Tension' ('PMT'—the precursor to 'PMS') by Frank (1931), a US gynecologist. For although Frank's paper initially documented cases of severe cyclical asthma, epilepsy, water retention, and cardiac irregularity, its main focus was on the experiences of a specific subset of patients, characterized by various signs of '*nervous tension*' (Frank 1931, 1054). Astonishingly, in several cases, Frank's symptom descriptions were simply value judgements on improper, or undesired, female behavior; and strikingly similar to contemporary descriptions of the 'hysterical woman' for example, "husband to be pitied," "unbearable, shrew," or "impossible to live with" (Frank 1931, 1055).

Frank was certainly not the first in the modern era to attribute changes in mood or emotions to the menstrual cycle. Various earlier sources describe apparently *well-known* expressions of cyclical emotional variability in a minority of menstruating women, although they were not previously considered to require medical intervention (for example, Hollick 1860, 91–92; Giles 1901, 27). Emily Martin (1987, 120) suggests that it is highly significant that Frank specifically discussed PMT in relation to women's (supposed lesser) ability to work, since he was writing during the Great Depression, a time when women workers were being pressured to leave paid employment in favor of men.

Also, the first estrogen hormone was discovered just three years before Frank published his paper. The identification of the so-called 'sex hormones' provided a new scientific explanation to support existing ideas about the female reproductive system as the origin of female-prevalent ill health, and implied 'natural' susceptibility to *irrational* behaviors (Ussher 2011, 21). Correspondingly, Frank associated such overtly *feminine* symptoms (that is, those affecting behavior, mood, or emotions) with an *excess* of the 'female' hormone (now disproved) (Frank 1931, 1056):

> It would thus appear that the continued circulation of an excessive amount of female sex hormone in the blood may in labile persons produce serious symptoms, some cardiovascular, but the most striking definitely psychic and nervous (autonomic). These periodic attacks . . . can be directly ascribed to the excessive hormonal stimulus.

During the 1950s, theories about Premenstrual Tension (PMT) continued to focus on the role of 'female sex hormones,' although symptoms were now attributed to a *deficiency* in progesterone (now disproved). PMT was renamed PMS (Premenstrual Syndrome), after Greene and Dalton (1953) argued that premenstrual symptoms were far more extensive than just 'nervous tension.' The most prominent 'PMS expert' at this time, Dr. Katharina Dalton, attempted to counter the undue emphasis on mood-based menstrual symptoms, and openly criticized what she called "*the hijacking of PMS by psychologists*" (Dalton and Holton 2000, 98). However, she also helped to perpetuate the myth of the irrational female by directly contributing to a somewhat dangerous (and unscientific) legal precedent for the mitigation of murder charges, if committed '*under the influence of PMS*' (Laws, Hey, and Eagan 1985, 65–79; Chrisler 2002).

Over the past thirty years, clinical descriptions of PMS have remained predominantly psychological in focus, especially since the inclusion of premenstrual disorders in the American Psychiatric Association's DSM (Diagnostic and Statistical Manual of mental disorders). First in the form of Late Luteal Phase Dysphoric Disorder (LLPDD) in 1987, and later its replacement, Premenstrual Dysphoric Disorder (PMDD), since 1994 (American Psychiatric Association 2000). Even reputable clinical sources sometimes refer to PMDD as '*severe* PMS'; implying that PMS is simply a less severe form of a mental health disorder (for example, Lopez, Kaptein, and Helmerhorst 2012; Maharaj and Trevino 2015; Naheed et al. 2017). In comparison, thyroid conditions, which are also 'hormonal' in origin and commonly cause severe mood changes, are not listed in the DSM.

So What? The Impact of Psychologising PMS

PMS research has, so far, been typified by contradictory, irreplicable, and usually highly contested, findings (Walker 1997; Knaapen and Weisz 2008; Halbreich 2007). This may partly be explained by the fact that for most of its history, it has been subject to a form of confirmation bias, or circular logic. By focussing on mood-based menstrual symptoms and neglecting those that are physical, PMS research *unavoidably* overlooks critical elements in the etiology (cause), prevalence, patient experiences, and treatments of menstrual cycle-related symptoms as a whole. For example, period pain is by far the most common menstrual cycle-related symptom, but it is not usually included in diagnostic criteria for PMS. This is despite the fact that period pain is known to have an effect on premenstrual mood, fatigue, and other symptoms (Balik et al. 2014; Smorgick et al. 2013).

Clinicians may struggle to identify menstrual cycle-related symptoms if PMS is understood to be essentially mood-based. For example, if certain physical symptoms are severe, a General Practitioner (GP) might simply diagnose a more chronic health condition, without first assessing if symptoms are cyclical in nature. Especially since even when PMS *is* suspected, many clinicians do not ask patients to record their symptoms over two cycles, as is required for formal diagnosis (Craner, Sigmon, and McGillicuddy 2014). (Menstrual migraine is an exception, since its high prevalence rate ensures that most clinicians are aware of a possible link to the menstrual cycle.)

Some clinicians may even question the validity of a PMS diagnosis, partly because symptoms are subjective and, therefore, difficult to qualify (for example, through clinical tests), but also because female-prevalent health issues in general, especially those that also affect mood (for example, fibromyalgia, chronic fatigue syndrome, irritable bowel syndrome [IBS]), are more likely to be dismissed, or assumed to be *psychosomatic* in origin (see Hoffmann and Tarzian 2001; Asbring and Närvänen 2002; Letson and Dancey 1996). This troubling situation is also likely to be influenced by the myth of the irrational female, since *hypochondria* (imagined or pretend illness) was emphasized in nineteenth century descriptions of 'female hysteria' (Veith 1965, 144–45).

Systematic reviews of published clinical trials are felt to be the '*gold standard for clarity, power, and precision*' in regard to the evidence-based management of PMS (O'Brien and Ismail 2007, 6). However, all five of the existing Cochrane systematic reviews of PMS treatments are based on clinical trials that almost universally selected participants using predominantly mood-based criteria (Jing et al. 2009; Lopez, Kaptein, and Helmerhorst 2012; Ford et al. 2012; Marjoribanks et al. 2013; Naheed et al. 2017). This is because the most widely used clinical tools for recording daily symptoms are either directly based on the diagnostic criteria for PMDD for example, the Daily Record of Severity of Problems (DRSP) (Endicott, Nee, and Harrison 2006), or otherwise over-emphasize emotional, behavioral, or psychological symptoms for example, variations of the Moos Menstrual Distress Questionnaire (MDQ) (Moos 1968). So, in effect, only a small subset of PMS patients/symptoms have ever been evaluated in the vast majority of clinical trials.

Some PMS researchers have focused on identifying cultural influences on clinical and scientific definitions of PMS. Their work is sometimes also limited by the assumption that menstrual changes are predominantly mood-based. For example, arguments put forward to oppose the unnecessary 'medicalization' of the menstrual cycle (while acknowledging the reality of the experiences of those who do experience distressing cyclical mood changes), may downplay, or neglect to mention physical menstrual cycle-related symptoms or conditions such as period pain, catamenial epilepsy, menstrual migraine, or cyclical asthma (for example, Chrisler and Gorman 2015). This omission might be seen to undermine their arguments (Kulkarni 2013), even though the point is still valid: The fact that a minority of people experience symptoms that are triggered by the menstrual cycle, does not justify the *pathologisation* of the cycle itself.

Research has found that people are more likely to report PMS if they have restricted access to social support (for example, Ussher, Perz, and Mooney-Somers 2007); an unequal share of household or childrearing responsibilities (for example, Coughlin 1990; Ussher 2003); or are experiencing relationship strain (for example, Kuczmierczyk, Labrum, and Johnson 1992). Any mention of 'psychosomatic,' or 'psycho-social' factors in relation to PMS, however, can trigger defensiveness on the part of the patient, their clinicians, or the wider public. Such terms are erroneously equated with *hypochondria*, even though this is not what is being described. Sadly, the resulting defensiveness can lead to the misinterpretation of clinical research findings. For example, a high-quality review that found inconclusive proof to support the existence of a '*specific premenstrual negative mood syndrome*' in the *general* menstruating population (Romans et al. 2012) was misunderstood by some as denying the existence of severe cyclical mood symptoms, entirely (for example, Kulkarni 2013).

REDEFINING PMS: WHERE ARE WE NOW?

The latest RCOG practitioner guidelines are a good representation of the current clinical definition of PMS (RCOG 2016). While it is clear that the guidelines are trying to move on from *psychologised* models of PMS, they also reveal the difficulty of doing so, especially within the parameters set by Evidence-Based Medicine (EBM) that is, ensuring that recommendations are supported by high-quality empirical evidence (Knaapen and Weisz 2008). Therefore, a close analysis of the guidelines is a useful exercise in identifying anything that could be hindering their application in clinical practice, or unintentionally reinforcing the myth of the irrational female.

To start with, the wording of the RCOG definition inadvertently evokes earlier descriptions of PMS (for example, 'a *vast array* of psychological symptoms ...') but otherwise takes extreme care to highlight the wide range of potential PMS symptoms, and the fact that it is their timing, and level of severity that indicate a legitimate diagnosis (RCOG 2016, 80):

> PMS encompasses a vast array of psychological symptoms such as depression, anxiety, irritability, loss of confidence and mood swings. There are also physical symptoms, typically bloatedness [sic] and mastalgia [breast pain]. It is the timing, rather than the types of symptoms, and the degree of impact on daily activity that supports a diagnosis of PMS. The character of symptoms in an individual patient does not influence the diagnosis ...

This seemingly simple definition is still not straightforwardly applied in the clinical context, however, or entirely free from evoking the myth of the irrational female.

Timing Is Everything

The term 'premenstrual' actually refers to the entire *luteal* phase of the menstrual cycle. This phase is typically 14 days long; beginning at ovulation and finishing when menstruation begins (Sonntag and Ludwig 2012). According to the RCOG guidelines, in order to qualify as PMS, symptoms *"...must be present during the luteal phase and abate as menstruation begins . . . then [be] followed by a symptom-free week"* (RCOG 2016, 80). However, cyclical symptoms often continue beyond the luteal phase, 2–4 days into menstruation (Hartlage et al. 2012). Also, as is typical of PMS research in general, this definition is based on an assumed 28-day menstrual cycle, which is merely an *average* cycle length. The 'normal' range of menstrual cycle length is usually cited as being anywhere between 21 and 35 days (for example, Simon et al. 2014, 702). This means that it may be impossible for someone with a 21-day cycle to ever experience a 'symptom-free' week, since they have only just finished menstruating when they are once again in the luteal phase of their cycle.

Also, by limiting the timing of symptoms to the luteal phase, those most closely associated with menstruation (for example, period pain, fatigue, lower back pain, other muscle/joint pain, or upset digestion) are potentially excluded from the otherwise apparently limitless list of possible menstrual cycle-related symptoms. As discussed in greater detail below, while this may seem clinically 'useful' on a superficial level (in terms of differentiating PMS from other more chronic health conditions), it also contributes to the somewhat arbitrary separation of undeniably *physiological* cyclical changes from those that are positioned as somehow more *psychological* in nature, despite their shared origin (Halbreich 2007, 13).

Spot the Difference—Normal Menstrual Change or PMS Symptom?

Differentiating 'normal menstrual changes,' or 'premenstrually exacerbated conditions,' from 'PMS symptoms' is perhaps the greatest problem affecting research on this topic (O'Brien 2007, xi). There is no clinical test, or known biomarker, that can be used to diagnose PMS, since the exact biological triggers of premenstrual symptoms remain uncertain, and they are nearly all subjective in nature, relying on patient description rather than any objective measurement of severity (Halbreich 2007, 17). Typically, patients are obliged to track daily menstrual changes over at least two cycles, usually involving an indication of symptom severity, before a doctor can formally diagnose PMS (RCOG 2016, 82).

Since the late 1990s, various attempts have been made to try to make symptom tracking more 'objective'; for example, by applying scores to individual symptoms and then having a *total score* diagnostic cut off point, or measuring the *degree of change* in severity over different menstrual phases (Steiner et al. 2011). However, the 'total score' approach can be misleading, especially if a patient experiences relatively few, but severe symptoms; and the 'degree of change' approach assumes linear changes in severity (that is, that changes get progressively worse toward menstruation), which is not reflective of many patient experiences (Halbreich 2007, 17).

This is why the RCOG guidelines provide activity-based criteria regarding symptom severity instead; *"... symptoms must be severe enough to affect daily functioning*" (RCOG 2016, 80). While this is certainly a useful and practical response to the problem of symptom subjectivity, issues remain. For example, even 'normal menstrual changes,' or relatively mild experiences of symptoms such as nausea, period pain, backache, or fatigue, can still affect daily functioning. Secondly, this approach still relies entirely on the quality of the daily rating tool, especially in regard to which symptoms define PMS, and how to quantify subjective experiences. Additionally, by not specifying any common symptoms, some that are severe in nature may be diagnosed as separate conditions, such as migraine, anxiety, or IBS, rather than possible PMS.

This approach also implies that *all* mild menstrual changes are 'subclinical symptoms,' that is, necessarily negative and pathological phenomena. This is an example of the medicalization of the menstrual cycle, a position that ignores any positive menstrual changes for example, higher energy levels, joyful moods, or increased creativity, and obscures the fact that it is a normal and healthy experience for the vast majority of menstruating people (Chrisler et al. 1994). It also makes value judgments on supposedly *feminine* attributes such as emotional sensitivity (Caplan, McCurdy-Myers, and Gans 1992; Ussher 2005, 2011; Chrisler and Gorman 2015), which could equally be positioned as a sign of *healthy* human behavior, given that failure to recognize or express emotions can result in serious health consequences such as suicidal or violent behavior (for example, Berke et al. 2016).

What Counts as a PMS Symptom?

Nearly all medical conditions are associated with a list of their most common or deterministic symptoms in order to aid diagnosis. However, such a list has always been difficult to establish for PMS (Halbreich 2007, 9). Without a shortlist of the most common experiences, any and every change in physical, psychological, or emotional state, becomes a potential 'symptom' of PMS. As a result, the number of supposedly distinct PMS symptoms has been highly exaggerated. This may be through the inclusion of non-pathological changes such as "loss of confidence" (for example, RCOG 2016, 80), value judgments on stereotypically 'feminine' behaviors such as "nagging" or "act spiteful" (for example, Halbreich et al. 1982, 48–49), or by vaguely referring to over 150 (for example, Moos 1968) or even 200 (for example, Dickerson et al. 2003, 1743) non-listed, but apparently 'previously associated' symptoms.

This situation also makes identifying the population prevalence, causal mechanism(s), or best approaches for the treatment of PMS, extremely technically challenging. For instance, it is impossible to meaningfully compare, or synthesize, population studies that have used different symptom criteria; or to isolate the most likely causes of, or treatments for, cyclical symptoms if they are so numerous, or varied, to lack common analytical properties (Halbreich 2007, 13). The highly individualized, subjective, and *biopsychosocial* (biological, psychological, and social) experience of menstrual cycle-related symptoms also makes them hard to measure, and difficult to control for, in the clinical research environment (Halbreich 2007, 17).

The Priming Effect of PMS Tracking Tools

Despite all of this confusion, some PMS symptoms really do still count more than others. In practice, it comes down to whichever PMS symptom tracking tool is used by the patient. In the case of the RCOG guidelines (RCOG 2016, 82), there is a clear steer toward promoting the most widely used tracking tools; the DRSP (Endicott, Nee, and Harrison 2006) or the Premenstrual Symptoms Screening Tool (PSST) (Steiner, Macdougall, and Brown 2003). Unfortunately, both of these tools are grounded in the specific diagnostic criteria for PMDD (APA 2000). As a result, emotional and psychological symptoms are over-represented across several supposedly distinct categories, and the majority of (the far more commonly experienced) physical menstrual changes are conflated into just one box (O'Brien and Ismail 2007, 5) (Table 23.1).

Research shows that experiences of PMS are highly susceptible to the 'priming effect' in which prior knowledge of what PMS is, and how it is viewed by wider society, plays a significant role in symptom perception (for example, Marván and Escobedo 1999). This creates an obvious problem for PMS tracking tools, which provide primes for such symptoms instead of open-ended symptom reporting. Indeed, Joan Chrisler has persuasively argued that PMS is a highly 'culture-bound' phenomenon, meaning that it is subject to a particular set of cultural beliefs and practices, rather than being a tangible medical 'truth' that is universally recognized, across all cultures (Chrisler 2012, 165):

> Data collected from women in Hong Kong . . . Taiwan . . . and mainland China . . . indicate that the most commonly reported premenstrual changes are fatigue, water retention, muscular tension, pain, and increased sensitivity to cold. Women in the United States do not report cold sensitivity, and women in China rarely report negative affect [low mood]. The results of these studies support the idea that culture shapes which variations in mood and physical sensations are noticed and which cause concern.

As if to prove the point, a few years after Chrisler's piece, an article assessing the reliability and validity of a directly translated version of the DRSP tool among Chinese women was published (Wu et al. 2013). It found that 126 Chinese women reported exactly the same symptoms as those listed on this Western (and PMDD-based) tracking tool. Since it is highly unlikely that Chinese women suddenly started experiencing completely new premenstrual symptoms, it seems reasonable to suggest that the DRSP tool itself predicted which type of symptoms were noticed, and now reported. This priming effect

Table 23.1   The most widely used PMS symptom tracking tools the 'DRSP' and 'PSST'

DRSP	PSST
1. Felt depressed, sad, "down" or "blue," or felt hopeless; or felt worthless or guilty	1. Anger/irritability
2. Felt anxious, tense, "keyed up," or "on edge."	2. Anxiety/tension
3. Had mood swings (i.e., suddenly feeling sad or tearful) or was sensitive to rejection or feelings were easily hurt	3. Tearful/increased sensitivity to rejection
4. Felt angry or irritable	4. Depressed mood/hopelessness
5. Had less interest in usual activities (work, school, friends, and hobbies)	5. Decreased interest in work activities
6. Had difficulty concentrating	6. Decreased interest in home activities
7. Felt lethargic, tired or fatigued; or had lack of energy	7. Decreased interest in social activities
8. Had increased appetite or overate; or had cravings for specific foods	8. Difficulty concentrating
9. Slept more, took naps, found it hard to get up when intended; or had trouble getting to sleep or staying asleep	9. Fatigue/lack of energy
10. Felt overwhelmed or unable to cope; or felt out of control	10. Overeating/food cravings
11. Had breast tenderness, breast swelling, bloated sensation, weight gain, headache, joint or muscle pain, or other physical symptoms	11. Insomnia
	12. Hypersomnia (needing more sleep)
	13. Feeling overwhelmed or out of control
	14. Physical symptoms: breast tenderness, headaches, joint/muscle pain, bloating, and weight gain

Credit: Sally King

of symptom tracking tools is therefore a major source of confirmation bias within PMS research, and one that sadly contributes to contradictory research findings, while reinforcing the myth of the irrational female.

WHERE DO WE GO FROM HERE?

I would argue that if we are really serious about implementing EBM, then we need to significantly improve the rigor of PMS research methods and tools. Only then might we avoid the confirmation bias created by historical assumptions insisting that cyclical symptoms, as experienced by a minority of menstruating people, are somehow more *psychological* than *physical* in origin and experience, or evidence of an *inherently* pathological female reproductive system. Based on the analysis of the RCOG guidelines above, there are a few issues that could be easily resolved with greater linguistic and logical precision, and the integration of knowledge from 'biopsychosocial' PMS research.

Firstly, in regard to the *timing* and *naming* of cyclical symptoms, the term 'premenstrual syndrome' could be changed to 'peri-menstrual symptoms.' This would make explicit the fact that cyclical symptoms are not technically a 'syndrome' and do not only occur 'just before' menstruation. This would also formally recognize, the role of period pain, by far the most common cyclical symptom, which is known to interact with, if not cause, several others.

Secondly, to help differentiate 'normal menstrual changes' from 'cyclical symptoms,' we could develop an indicative list of the *most common* cyclical symptoms. This list would exclude any that are: value judgments on femininity, normal human behaviors, duplications, unlikely to ever require medical intervention, or not supported by (unbiased) epidemiological data. This list could then inform symptom tracking tools, and the selection criteria used in clinical trials. As things stand, by stating that "*there is no limit on the type or number of symptoms experienced*" (RCOG 2016, 80), the RCOG guidelines imply that PMS is somehow 'different' to other illnesses, and unintentionally reinforces the notion that the female (reproductive) body is inherently pathological.

Thirdly, we could integrate known (that is, evidence-based) contributing factors to PMS into symptom tracking tools and clinical guidelines. For example, by including space to note any social, dietary, or lifestyle 'triggers' of symptoms on tracking tools, and by ensuring that clinical guidelines endorse the identification of any such factors and describe a range of treatment options through which they might best be managed. Of course, this integration would need to be communicated in a way that does not accidentally imply that this is a gender-specific phenomenon, or one akin to *hypochondria*. We could do so by sharing examples of other, well-known, psycho-social factors in ill health, such as nausea or vomiting when feeling nervous, or irritability when feeling hungry or tired.

Finally, we could all consistently remind our clinical and public audiences that only a minority of people who menstruate experience severe cyclical symptoms requiring medical support, and that it does not follow that the menstrual cycle is itself a form of illness, or proof of any 'natural' inferiority of women. This includes: never overstating the implications of research findings, never saying 'women' or 'people who menstruate' when actually meaning 'people who experience PMS,' never presenting PMS as a mood disorder, and no longer selecting 'PMS' research participants or diagnosing 'PMS' using the diagnostic criteria for 'PMDD.'

Explicitly recognizing and countering the myth of the irrational female and its influence on the way in which PMS is clinically described and managed, is an important step toward better supporting those who do experience cyclical symptoms, without unintentionally implying that the menstrual cycle is itself a form of illness, or any sort of 'biological' justification for gender inequality.

References

American Psychiatric Association. 2000. *DSM-IV TR. Diagnostic and Statistical Manual of Mental Disorders.* 4th edition TR.

Asbring, Pia, and Anna-liisa Närvänen. 2002. "Women's Experiences of Stigma in Relation to Chronic Fatigue Syndrome and Fibromyalgia." *Qualitative Health Research* 12 (2): 148–60. https://doi.org/10.1177/104973230201200202.

Balik, Gülşah, Işik Üstüner, Mehmet Kağitci, and Figen Kir Şahin. 2014. "Is There a Relationship between Mood Disorders and Dysmenorrhea?" *Journal of Pediatric and Adolescent Gynecology* 27 (6): 371–74. https://doi.org/10.1016/j.jpag.2014.01.108.

Berke, Danielle S., Dennis E. Reidy, Brittany Gentile, and Amos Zeichner. 2016. "Masculine Discrepancy Stress, Emotion-Regulation Difficulties, and Intimate Partner Violence." *Journal of Interpersonal Violence*, May. https://doi.org/10.1177/0886260516650967.

Caplan, Paula J., Joan McCurdy-Myers, and Maureen Gans. 1992. "Should 'Premenstrual Syndrome' Be Called a Psychiatric Abnormality?" *Feminism & Psychology* 2 (1): 27–44. https://doi.org/10.1177/0959353592021003.

Choi, Dooseok, Dong Yun Lee, Philippe Lehert, Im Soon Lee, Seok Hyun Kim, and Lorraine Dennerstein. 2010. "The Impact of Premenstrual Symptoms on Activities of Daily Life in Korean Women." *Journal of Psychosomatic Obstetrics and Gynecology.* https://doi.org/10.3109/01674820903573920.

Chrisler, Joan C. 2002. "Hormone Hostages: The Cultural Legacy of PMS as a Legal Defense." In *Charting a New Course for Feminist Psychology*, edited by Lynn H. Collins, Michelle R. Dunlap, and Joan C. Chrisler, 238–52. London: Praeger.

———. 2012. "PMS as a Culture-Bound Syndrome." In *Lectures on the Psychology of Women*, edited by J. C. Chrisler, C. Golden, and P. D. Rozee, 4th ed., 155–71. Long Grove: Waveland Press.

Chrisler, Joan C., Ingrid K. Johnston, Nicole M. Champagne, and Kathleen E. Preston. 1994. "Menstrual Joy: The Construct and Its Consequences." *Psychology of Women Quarterly* 18 (3): 375–87. https://doi.org/10.1111/j.1471-6402.1994.tb00461.x.

Chrisler, Joan C., and Jennifer A. Gorman. 2015. "The Medicalization of Women's Moods: Premenstrual Syndrome and Premenstrual Dysphoric Disorder." In *The Wrong Prescription for Women: How Medicine and Media Create a 'need' for Treatments, Drugs, and Surgery*, 77–98. Santa Barbara, CA: Praeger.

Coughlin, Patricia C. 1990. "Premenstrual Syndrome: How Marital Satisfaction and Role Choice Affect Symptom Severity." *Social Work (United States)* 35 (4): 351–55. https://doi.org/10.1093/sw/35.4.351.

Craner, Julia R., Sandra T. Sigmon, and Morgan L. McGillicuddy. 2014. "Does a Disconnect Occur between Research and Practice for Premenstrual Dysphoric Disorder (PMDD) Diagnostic Procedures?" *Women and Health* 54 (3): 232–44. https://doi.org/10.1080/03630242.2014.883658.

Dalton, Katharina, and Wendy M. Holton. 2000. *The PMS Bible: The Guide to Understanding and Treating PMS.* 6th ed. London: Vermilion.

Dennerstein, Lorraine, Philippe Lehert, and Klaas Heinemann. 2011. "Global Study of Women's Experiences of Premenstrual Symptoms and Their Effects on Daily Life." *Menopause International* 17 (3): 88–95. https://doi.org/10.1258/mi.2011.011027.

Dickerson, Lori M., Pamela J. Mazyk, and Melissa H. Hunter. 2003. "Premenstrual Syndrome." *American Family Physician* 68 (6): 1743–52.

Endicott, Jean, John Nee, and W. Harrison. 2006. "Daily Record of Severity of Problems (DRSP): Reliability and Validity." *Archives of Women's Mental Health* 9 (1): 41–49. https://doi.org/10.1007/s00737-005-0103-y.

Ford, Olive, Anne Lethaby, Helen Roberts, and Ben W. J. Mol. 2012. "Progesterone for Premenstrual Syndrome." *Cochrane Database of Systematic Reviews*. March. https://doi.org/10.1002/14651858.CD003415.pub4.

Frank, Robert T. 1931. "The Hormonal Causes of Premenstrual Tension." *Archives of Neurology Psychiatry* 26: 1053–57.

Freud, Sigmund. 1966. "Hysteria." In *The Standard Edition of the Complete Psychological Works of Sigmund Freud, Volume I (1886–1899): Pre-Psycho-Analytic Publications and Unpublished Drafts*, 37–59. New York: Vintage

Giles, A.E. 1901. *Menstruation and Its Disorders*. London: Baillière, Tindall and Cox.

Gilman, Sander L., Helen King, Roy Porter, G. S. Rousseau, and Elaine Showalter. 1993. *Hysteria Beyond Freud*. Berkeley, CA: University of California Press.

Greene, Raymond, and Katharina Dalton. 1953. "The Premenstrual Syndrome." *British Medical Journal* 1 (4818): 1007–13. https://doi.org/10.1136/bmj.1.4818.1007.

Halbreich, Uriel. 2007. "The Diagnosis of PMS/PMDD—The Current Debate." In *The Premenstrual Syndromes: PMS and PMDD.*, edited by Patrick Michael Shaughan O'Brien, Andrea J. Rapkin, and Peter J. Schmidt, 9–19. Boca Raton, FL: CRC Press.

Halbreich, Uriel, Jean Endicott, S. Schacht, and J. Nee. 1982. "The Diversity of Premenstrual Changes as Reflected in the Premenstrual Assessment Form." *Acta Psychiatrica Scandinavica*. https://doi.org/10.1111/j.1600-0447.1982.tb00820.x.

Hartlage, S. Ann, Sally Freels, Nathan Gotman, and Kimberly Yonkers. 2012. "Criteria for Premenstrual Dysphoric Disorder." *Archives of General Psychiatry* 69 (3): 300. https://doi.org/10.1001/archgenpsychiatry.2011.1368.

Hoffmann, Diane E., and Anita J. Tarzian. 2001. "The Girl Who Cried Pain: A Bias Against Women in the Treatment of Pain." *Journal of Law, Medicine and Ethics*, no. 29: 13–27. https://doi.org/10.1111/j.1748-720X.2001.tb00037.x.

Hollick, Frederick. 1860. *The Marriage Guide or Natural History of Generation; A Private Instructor for Married Persons and Those about to Marry, Both Male and Female*. 200th ed. New York: T. W. Strong.

Jing, Zheng, Xunzhe Yang, Khaled MK Ismail, Xiao Y. Chen, and Taixiang Wu. 2009. "Chinese Herbal Medicine for Premenstrual Syndrome." *Cochrane Database of Systematic Reviews*, January. https://doi.org/10.1002/14651858.CD006414.pub2.

King, Helen. 1998. *Hippocrates' Woman: Reading the Female Body in Ancient Greece*. London: Routledge.

Knaapen, Loes, and George Weisz. 2008. "The Biomedical Standardization of Premenstrual Syndrome." *Studies in History and Philosophy of Science* 39 (1): 120–34. https://doi.org/10.1016/j.shpsc.2007.12.009.

Kuczmierczyk, Andrzej R., Anthony H. Labrum, and Carolyn C. Johnson. 1992. "Perception of Family and Work Environments in Women with Premenstrual Syndrome." *Journal of Psychosomatic Research* 36 (8): 787–95. https://doi.org/10.1016/0022-3999(92)90137-Q.

Kulkarni, Jayashri. 2013. "PMS Is Real and Denying Its Existence Harms Women." *The Conversation*. https://theconversation.com/pms-is-real-and-denying-its-existence-harms-women-11714.

Laws, Sophie, Valerie Hey, and Andrea Eagan. 1985. *Seeing Red; the Politics of Pre-Menstrual Tension*. London: Hutchinson.

Letson, Sue, and Christine P. Dancey. 1996. "Nurses' Perceptions of Irritable Bowel Syndrome (IBS) and Sufferers of IBS." *Journal of Advanced Nursing* 23 (5): 969–74. https://doi.org/10.1046/j.1365-2648.1996.10416.x.

Lopez, Laureen M., Adrian A. Kaptein, and Frans M. Helmerhorst. 2012. "Oral Contraceptives Containing Drospirenone for Premenstrual Syndrome." *Cochrane Database of Systematic Reviews*, February. https://doi.org/10.1002/14651858.CD006586.pub4.

Maharaj, Shalini, and Kenneth Trevino. 2015. "A Comprehensive Review of Treatment Options for Premenstrual Syndrome and Premenstrual Dysphoric Disorder." *Journal of Psychiatric Practice* 21 (5): 334–50. https://doi.org/10.1097/PRA.0000000000000099.

Marjoribanks, Jane, Julie Brown, Patrick Michael Shaughn O'Brien, and Katrina Wyatt. 2013. "Selective Serotonin Reuptake Inhibitors for Premenstrual Syndrome." In *Cochrane Database of Systematic Reviews*, edited by Julie Brown. Chichester, UK: Wiley. https://doi.org/10.1002/14651858.CD001396.pub3.

Martin, Emily. 1987. *The Woman in the Body: A Cultural Analysis of Reproduction*. Milton Keynes: Open University Press. https://doi.org/10.1080/14616700220145650.

Marván, Ma Luisa, and Claudia Escobedo. 1999. "Premenstrual Symptomatology: Role of Prior Knowledge about Premenstrual Syndrome." *Psychosomatic Medicine* 61 (2): 163–67. https://doi.org/10.1097/00006842-199903000-00007.

Moos, Rudolf H. 1968. "The Development of a Menstrual Distress Questionnaire." *Psychosomatic Medicine* 30 (6): 853–67. https://doi.org/10.1097/00006842-196811000-00006.

Naheed, Bushra, Jan Herman Kuiper, Olalekan A. Uthman, Fidelma O'Mahony, and Patrick Michael Shaughn O'Brien. 2017. "Non-Contraceptive Oestrogen-Containing Preparations for Controlling Symptoms of Premenstrual Syndrome." *The Cochrane Database of Systematic Reviews* 3 (March). https://doi.org/10.1002/14651858.CD010503.pub2.

O'Brien, Patrick Michael Shaughn. 2007. "Preface." In *The Premenstrual Syndromes: PMS and PMDD*, edited by Patrick Michael Shaughan O'Brien, Andrea J Rapkin, and Peter J Schmidt, xi–xii. Boca Raton, FL: CRC Press.

O'Brien, Patrick Michael Shaughn, and Khaled M. K. Ismail. 2007. "History of the Premenstrual Disorders." In *The Premenstrual Syndromes: PMS and PMDD*, edited by Patrick Michael Shaughan O'Brien, Andrea J. Rapkin, and Peter J. Schmidt, 1–8. Boca Raton, FL, USA: CRC Press.

Quirke, Stephen. 2002. "Manuscript for the Health of Mother and Child." Trans. The Kahun Gynaecological Papyrus. 2002. http://www.ucl.ac.uk/museums-static/digitalegypt/med/birthpapyrus.html.

RCOG. 2016. "Management of Premenstrual Syndrome: Green-Top Guideline No. 48." *BJOG: An International Journal of Obstetrics & Gynaecology* 124 (3): 73–105. https://doi.org/10.1111/1471-0528.14260.

Romans, Sarah, Rose Clarkson, Gillian Einstein, Michele Petrovic, and Donna Stewart. 2012. "Mood and the Menstrual Cycle: A Review of Prospective Data Studies." *GENM* 9: 361–84. https://doi.org/10.1016/j.genm.2012.07.003.

Simon, Chantal, Hazel Everitt, Francoise Van Dorp, and Matthew Burke. 2014. *Oxford Handbook of General Practice*. Oxford: Oxford University Press.

Smorgick, Noam, Courtney A. Marsh, Sawsan As-Sanie, Yolanda R. Smith, and Elisabeth H. Quint. 2013. "Prevalence of Pain Syndromes, Mood Conditions, and Asthma in Adolescents and Young Women with Endometriosis." *Journal of Pediatric and Adolescent Gynecology* 26 (3): 171–75. https://doi.org/10.1016/j.jpag.2012.12.006.

Sonntag, Barbara, and Michael Ludwig. 2012. "An Integrated View on the Luteal Phase: Diagnosis and Treatment in Subfertility." *Clinical Endocrinology* 77 (4): 500–507. https://doi.org/10.1111/j.1365-2265.2012.04464.x.

Steiner, Meir, M. Macdougall, and E. Brown. 2003. "The Premenstrual Symptoms Screening Tool (PSST) for Clinicians." *Archives of Women's Mental Health* 6 (3): 203–9. https://doi.org/10.1007/s00737-003-0018-4.

Steiner, Meir, Miki Peer, Mary MacDougall, and Roger Haskett. 2011. "The Premenstrual Tension Syndrome Rating Scales: An Updated Version." *Journal of Affective Disorders* 135 (1–3): 82–88. https://doi.org/10.1016/j.jad.2011.06.058.

Ussher, Jane M. 1989. *The Psychology of the Female Body*. London and New York: Routledge.

———. 2003. "The Ongoing Silencing of Women in Families: An Analysis and Rethinking of Premenstrual Syndrome and Therapy." *Journal of Family Therapy* 25: 388–405. https://doi.org/10.1111/1467-6427.00257.

———. 2005. *Managing the Monstrous Feminine: Regulating the Reproductive Body*. London and New York: Routledge. https://doi.org/10.4324/9780203328422.

———. 2011. *The Madness of Women: Myth and Experience*. London and New York: Routledge. https://doi.org/10.4324/9780203806579.

Ussher, Jane M, Janette Perz, and Julie Mooney-Somers. 2007. "The Experience and Positioning of Affect in the Context of Intersubjectivity: The Case of Premenstrual Syndrome." *International Journal of Critical Psychology* 21: 144–65.

Veith, Ilza. 1965. *Hysteria; The History of a Disease*. Chicago: University of Chicago Press.

Walker, Anne E. 1997. "Premenstrual Syndrome." In *The Menstrual Cycle*, 144–90. London and New York: Routledge.

Wright, Almroth. 1912. "Militant Hysteria." *Spectator Magazine*, March 30, 1912.

Wu, Liping, Zhong He, Hong Zhao, Di Ma, Sijia Zhang, Hanyu Deng, and Tao Liang. 2013. "Chinese Version of Daily Record of Severity of Problems: Reliability and Validity." *Journal of Advanced Nursing* 69 (2): 449–56. https://doi.org/10.1111/j.1365-2648.2012.06070.x.

CHAPTER 24

The Sexualization of Menstruation: On Rape, Tampons, and 'Prostitutes'

Lacey Bobier

Menarche is a pivotal event in the development of one's sexuality. For girls, it marks a transition from childhood (culturally constructed as asexual), to womanhood (a sexualized, objectified other). Bodies become sexually marked through reproductive potential and observable pubertal developments. Research has shown that girls experiencing menarche celebrate this transition (Fingerson 2006; Lee 2009). These same studies have shown that girls rarely make the connection between menstruation and reproduction or sexuality (Fingerson 2006) and, when they do, react negatively to this association (Lee 2009).

Previous studies have relied on surveys, short-answer questionnaires, or written narratives (see Brooks-Gunn and Ruble 1982; Morse and McKinnon Doan 1987; Koff and Rierdan 1995; Moore 1995; Schooler et al. 2005), and are thus limited by the researchers' interpretations. Work using interviews has often accessed post-adolescent populations (see Lee 1994; Beausang and Razor 2000), thereby relying on reconstructed narratives that may not accurately depict thought processes as they occur during the early years of menstruation; initial understandings are likely reinterpreted in light of later accumulated experiences. Janet Lee (1994) found some evidence linking menarche and sexuality that relied on narratives from interviewees ages eighteen to eighty. Finally, many important texts in this literature pre-date the turn of the century. Margaret Stubbs and Ingrid Johnston-Robledo (2013) explored how the phenomenon of the sexualization of girls might affect girls' experiences of menstruation. Given that studies on the sexualization of girls emerged in the early nineties, corresponding with the increasing cultural presence of the process itself, the fact that few salient studies postdate this era may offer some insight into why themes of sexuality have not been prominent.

© The Author(s) 2020
C. Bobel et al. (eds.), *The Palgrave Handbook of Critical Menstruation Studies*, https://doi.org/10.1007/978-981-15-0614-7_24

Adolescent girls raised in this cultural context may be more likely to highlight the sexual dimension of menses.

Though gender scholars debate the difference between sex and gender, theorists have argued that sex, rather than being a "naturally" determined biological fact, is socially constructed (see Kessler 1990; Anne Fausto-Sterling 2000). A substantial portion of such work draws on cases of intersex individuals to elucidate how one's supposedly biological sex can, in fact, be based on cultural definitions of the gender categories man and woman (see, for example, Kessler 1990; Anne Fausto-Sterling 2000). This literature shows that central to the definition of "female" is reproductive potential (Anne Fausto-Sterling 2000) and, therefore, menarche and menstruation. For this reason, sex and gender blur when discussing menstruation. Consequently, I chose to interchange the terms female and woman, referring to one sex/gender category for the sake of this paper. Despite my terminology here, it is important to recognize that many people born with "female" physiology do not identify as women. Moreover, many people born without this physiology do identify as women. However, the sample of this study is comprised of women-identified "females" and maintains a focus on this population, though the findings have implications for the aforementioned groups to be explored by future research.

The extant research concerning young adolescents and menarche have primarily concerned attitudes, knowledge, preparation, and outcomes. Though pre- and post-menarcheal girls have reported that they feel/felt prepared for menarche (Morse and McKinnon Doan 1987; Koff and Rierdan 1995b), studies have shown that both groups lack cognitive knowledge about the physiological process and subjective, experiential knowledge (Brooks-Gunn and Ruble 1982; Koff and Rierdan 1995a; Moore 1995; Fingerson 2006). Preparation for menstruation positively corresponds to more positive attitudes and experiences of menarche (see Rierden 1983; McPherson and Korfine 2004). However, even these girls associated menarche and menstruation with ambivalent and negative sentiments (Koff and Rierdan 1995b). They saw menstruation as something embarrassing, dirty, and gross, and experienced resulting shame and anxiety (Ruble and Brooks-Gunn 1982; Morse and McKinnon Doan 1987; George and Murcott 2011; Lee 1994; Moore 1995; Kissling 1996). These feelings largely stem from societal taboos concerning female bodily processes and sexuality, the corresponding medicalization, and the consequent pressure to sanitize and conceal these functions, including the supposed hygienic crisis that is menstruation (Brumberg 1997). The most frequently mentioned positive reaction to menstruation was a sense of maturity (Ruble and Brooks-Gunn 1982; Morse and McKinnon Doan 1987; Moore 1995). The framework of menarche as instrumental to the transition to womanhood provides context for the development of associations between menarche and sexuality.

Other studies exploring girls' thoughts at this developmental stage provide further reason to suspect they may connect menarche and sexuality. Research relates physical evidence of pubertal developments, specifically in the form of breast growth, to girls' awareness of increased attention to their bodies, alterations in responses to their bodies and behaviors, and discomfort with these reactions (Martin 1996; Summers-Effler 2004). This increased bodily awareness and discomfort could easily translate into and parallel feelings that accompany the transition in social and sexual standing that menarche signifies. Accordingly, Lee (2009) explained that, at the time of menarche, the several participants in her study who did connect menarche to sexual maturation "felt disempowered by the way people were starting to look and relate to them" (622). This sentiment likely reflects the process of sexual objectification.

Sexual objectification occurs when women are reduced to and treated as bodies, specifically, bodies intended for use and consumption by others (Fredrickson and Roberts 1997). Self-objectification involves the internalization of an outsider's view of one's body, or judging bodily attractiveness and value through an external lens, and the resulting treatment of oneself as an object to be viewed and evaluated (Martin 1996; Fredrickson and Roberts 1997). Self-objectification is deeply tied to menarche as the onset of menstruation is linked with a transition to the sexual and reproductive realm (Lee 1994). Girls also begin the process of self-objectification at puberty because their bodies become more overtly sexual and, thus, sexualized and objectified by others (Thorne 1994; Martin 1996). Accordingly, women are especially at risk of adopting an observer's perspective of their body during times of physiological change, such as the onset of menstruation (Fredrickson and Roberts 1997).

Moreover, as Karin Martin (1996) indicated, while girls may not consistently make explicit links between intercourse, sexuality, and menstruation, puberty and sexuality are intertwined, and parents and teachers often present information that contain a sexual subtext, intentionally or not. The fact that menstruation is normally incorporated into *sex education* classes underscores this point.

While Martin (1996) found that her female interviewees connected menarche to pregnancy and sex, few have replicated this discovery. As recently as 2006, Fingerson found that her interviewees only connected menstruation to reproduction (and sexuality) when asked to explain the purpose of menstruation. This finding is consistent with previous research that revealed girls did not spontaneously reference reproduction in connection to menstruation (Ruble and Brooks-Gunn 1982; Koff and Rierdan 1995b). Havens and Swenson's (1988) review of educational advertisements resulted in the discovery that only three of thirty-one videos on menstruation mentioned the relationship between menstruation and reproduction in the form of the relationship between tampons and virginity. Stubbs and Johnston-Robledo further discovered that mothers, a primary source of information on menses (Whisnant and Zegans 1975; Havens and Swenson 1988; Koff and Rierdan 1995a, 1995b; Moore 1995; Kissling 1996; Beausang and Razor 2000), failed to contemplate the relationship between the sexualization of girls and their daughters' pubertal development. These same authors hypothesized a possible connection between sexuality and girls' attitudes toward menstruation, suggesting that girls might embrace menstrual suppression as a way to self-sexualize and reject the menstruating body that is incompatible with the sexualized ideals to which girls are encouraged to aspire.

Despite sparse evidence that adolescent girls link menarche, menstruation, and sexuality, themes concerning maturational development, indicated that girls may yet make associations between these topics and experiences. Following Stubbs and Johnston-Robledo, I expected millennials to have incorporated sexuality into their understandings and experiences with menstruation more so than previous generations. Since the late 1980s, research has found that girls are sexualized in a myriad of ways including sexualized portrayals in media and representation in merchandise, parental messages concerning the primary importance of appearance, peer policing, self-sexualization through adherence to these standards and, at the extreme, sexual assault, abuse, traffic, and prostitution (Zurbriggen et al. 2007). This corresponds to a moral panic in which parents and news media depict certain media outlets as inappropriately positioning girls as "sexy," while consumer culture targets increasingly younger audiences, supposedly robbing them of their innocence and leading to future promiscuity (Egan 2013). In opposing the sexualization of girlhood there is an increasing emphasis on innocence and purity via virginity (Valenti 2009). Such concerns may heighten the awareness of the sexual implications surrounding girls' change in menstrual status.

I chose to interview young adolescent girls as they offered a unique lens to understand the process of developing sexual subjectivity as it occurs. By looking at this moment, we can see how girls navigate various discourses and cultivate the language to represent their thoughts. Because these girls lack the vocabulary to describe their relationship with menstruation, they are less able to articulate their understanding of menses. We can appreciate how they unpack ideas as girls interpret menstruation through frameworks they have developed separate from formal discourses. This led me to ask: What meanings do girls attach to menstruation and menarche? Do they see menstruation as related to sexuality, explicitly or implicitly? How do they navigate the various meanings?

The Study

Methods

I interviewed ten middle school girls ages eleven through thirteen, recruited from three private middle schools in Michigan that were comprised primarily of white, middle to upper middle-class pupils. My sample reflected these statuses, though one student identified as African American. All but one girl was post-menarcheal. The median age of onset was eleven (SD: 1.5 years), with the youngest being nine, and the oldest thirteen.

Two schools had small student bodies (<30 girls each), were coed and had liberal atmospheres. One was single-sex, conservative, larger, and affiliated with the Catholic Church. Originally, I thought the contrasting liberal and conservative bents might manifest in my findings, however, there was little variation by school. This consistency is something to be further explored in future research, along with the roles of class and race variation (my sample, lacking in diversity, cannot speak to these factors of potential import).

Recruitment involved a mix of introducing myself and my research to the student body, distributing informational letters for the girls and their parents in person and via email (these letters were accompanied by consent forms for the parents, assent forms for the girls, and a short questionnaire on the students' backgrounds). For the conservative school, recruitment ultimately turned into snowball sampling as a girl from one social group participated, and her friends followed. This may have affected my findings as friends may share opinions. The topic of my research likely limited my sample size as some girls relayed that others were too embarrassed to participate and/or their parents had refused to consent.

Before interviews, girls were asked to complete questionnaires with their parents. This form included questions concerning: age; grade; socioeconomic status; religion; the number and gender of siblings; the number of years they had attended their current school; and race/ethnicity. Interviews occurred one-on-one in available classrooms or offices. The familiar setting, combined with the privacy of the locations, allowed the girls to feel more relaxed. Interviews ranged from half an hour to an hour and fifteen minutes; most lasted around an hour. Interviews began with conversation to build rapport (about girls' athletic activities, summer plans, et cetera.), followed by a quick preface concerning topics the discussion might cover and students' rights as participants. With the approval of the Institutional Review Board, all interviews were audiotaped, and interviews and questionnaires were transcribed. Identifiers were removed from transcriptions, participants were assigned pseudonyms, and schools given aliases. The interviews were coded inductively.

I performed semi-structured interviews for which I came prepared with a list of questions, though the girls often answered my questions without my prompting. Interview questions covered a range of topics including: introductions to the subject of menstruation; the roles of friends, siblings, boys, schools, parents, the medical establishment, the internet, books, and the media; the biological process of menstruation; feminine hygiene products; symptomatology; age of menarche; menstrual suppression; managing menstruation; and (dis)likes concerning the menstrual experience. I also offered a scripted anecdote concerning menstruation hoping girls would respond with their own stories.

Aware that my questions broached awkward topics, I carefully monitored my self-presentation, attempting to appear in a way that would warrant the respect, and, thus, cooperation of school administrators, without alienating my interviewees. I hoped my youthful look and demeanor would lead the girls to view me as a near-peer in whom they could confide. Given the intimate details the girls disclosed, I believe I succeeded in earning their trust. When asking a participant at the end of our session to please send her friends my way, she responded, "definitely, I'll tell my friends about the cool college lady researching PMS."

Because of the demographics of my sample, as well as the fact that participation was voluntary, the sampling involved in this project is not representative. These factors, as well as the size of my sample, limit the generalizability of my data. However, the consistency of my findings across all three schools would indicate that the trend of the sexualization of menstruation extends beyond my sample.

Findings

Tampons, the Fear of "Down There," and "Waiting Until You Are Ready"

As in previous studies, my interviewees referenced reproduction and sexuality when asked about the purpose of menstruation. However, this sample of girls stood out in the way they interwove these topics throughout the entire interviews. A quintessential example occurred when I asked Sophia (11, white) how she felt about commercials for feminine hygiene products and she went on to describe how commercials for Trojan condoms made her feel uncomfortable. This incident highlights a salient dimension of this association between menarche and sexual development during these interviews: girls experienced this terrain with anxiety and trepidation. This is evident in the way participants explained their reasons for avoiding or delaying the use of tampons. All but one of my interviewees regularly employed pads instead of tampons, but would occasionally make exceptions when they wanted to swim. For instance, when asked whether she had ever used a tampon, Isabella (13, white) explained that:

> The problem is, I'm a full-time swimmer, so I didn't want to use a tampon, but eventually I had to . . . like my first time, at the end of the week, it [the menstrual flow] was really light. I barely needed one but I did [use a tampon] just in case. It took a few tries [to insert it] and it wasn't really right . . . I was like freaking out . . . I freaked out because sticking something in [the vagina] was creepy.

Isabella's terminology was characteristic of these stories. Other standard words and phrases included: "awkward;" "uncomfortable;" and "scary." For all of the participants, this experience made a lasting impression. They were able to recount, in detail, where they were, who was with them, what was said, even what brands were chosen. This moment was powerful because it was the first time they had explored, as Emma (12, white) and Olivia (13, white) were prone to say, "down there." This discomfort was amplified by the fact that many of them did not actually know where to put the tampon or the proper anatomical terminology. Olivia elaborated by stating that "you never really see it [the vagina] until you actually experience it." While these young girls considered these stories representative of a traumatic developmental moment, these memories will likely fade relative to other life experiences. This may be one reason such accounts have not appeared in previous studies that rely on retrospective accounts.

While previous research has framed fears surrounding tampons as resulting from lack of bodily knowledge (Martin 1996; Fingerson 2006) (which surely remains a contributing factor in this sample), my interviewees also consistently employed sexual language when narrating their thoughts of and experiences with tampons. Both indirectly and directly, my interviewees revealed that this experience was partly distressing because they were forced to confront (and, consequently, reject) their sexuality. The language used to frame their relationships with tampons reflected this confrontation. Girls explained how pads operated as place holders, something to help you adapt to menstruation without having to go "down there." My interviewees referred to pads as "starters" and unanimously agreed that one should wait to try tampons until "she is ready." Interestingly, the language used here is similar to phrases that surround the topic of first sex. Girls are relentlessly told to wait "until they're ready." Apparently, adolescent girls associate the penetration of the penis into the vagina with the insertion of a tampon. Girls define sex in phallocentric terms and, in this characterization, neglect numerous, non-penetrative forms of sexuality. For example, when questioned as to why one would avoid tampons, Sophia launched into a monologue about her sister's refusal to use tampons and how "that's kind of a good sign because [she] doesn't want anything to go in there." Here, Sophia associates tampons with phallocentric sexual activity.

Although the definition of "sex" has been and remains contested, one consistent definition concerns the penetration of a vagina by a penis (Maines 1999). Historically, many consider virginity to cease when the hymen is "broken"[1] (The Swedish Association for Sexuality Education). However, exercise itself can tear, thin or stretch the hymen (The Boston Women's Health Collective 2011) resulting in physical alterations before puberty. Tampons can also affect the structure (The Boston Women's Health Collective), which may further cement the association between tampons and sex for girls. Penetration is consistently associated with a particularly condemned form of sex for girls in this age group. In avoiding penetration girls are valuing virginity above all other sexualities.

While defending her choice to abstain from tampon usage, Madison (13, white) explained that:

> It's scary, like, what if you put it in the wrong hole? . . . It's also like the idea of sex, which freaks me out because I don't like thinking of another person inside of me. It's just like, nope, that's a future Madison problem. I'm not going to think about that. It's just like this is so weird. So weird.

Madison describes how she associates tampons with sexual penetration. In order to navigate her discomfort with the sexuality she sees as associated with menstruation, she employs a strategy which I refer to as the splintering of the self. Like the other girls in my study, Madison is adjusting to the idea of being a menstruating woman. For her, this is enough of a challenge. Trying to integrate a sexual aspect into her identity is, for the moment, too overwhelming. Thus, she splits herself into two different people: a girl/woman who has just reached menarche, and a future sexual woman. By assigning the latter to the future, she can focus on the present.

The Splintering of the Current and Future Self

This strategy of splintering was employed by many of the participants. Each girl assured me that she was not sexually active, yet most addressed the topic of a future sexually active self. They did not think it was necessary to "deal with it" presently and assigned that frightening duty to a future self. Splintering, and the intensity with which girls rejected the need to address their sexuality in the present, shows the extent to which adolescent girls are intimidated by and even afraid of their sexuality. Not yet ready to grapple with such issues, girls avoid things that might remind them of their sexuality (such as tampons).

The Present Rejection of All Things Sexual: The Pill and Intercourse

All but one girl vehemently asserted that they were not sexually active without any prompting from me. This relates back to Sophia applauding her sister for not "want[ing] anything in there." They were invested in the idea of the sexually innocent "good girl." Along with rejecting tampons, only one girl said she would consider using the pill to control cramps or extend time between cycles. Many feared that birth control pills could cause infertility, but even with the caveat of no harmful side effects, interviewees were still disconcerted by this idea. There were, however, other factors at play. Though all but one of the participants knew that birth control pills were used to prevent pregnancy, none of them knew how the pill worked, nor did they know that the pill could also be used to mitigate the severity of cramps or extend time between periods (although, after I explained this, a few commented that they had heard that mentioned somewhere). However, they primarily associated birth control pills with intercourse.

When asked who should use the pill, Mia responded that "sexually active teenagers" should. Similarly, Sophia argued that the pill is for people who are "sexually unrestrained," "provocative girls" (that is, sexual appearance correlates with sexual behavior), and, finally, "prostitutes." Again, we see girls holding onto their innocence and good-girl standing by separating themselves from sexuality as represented by the pill and its users.

While forcefully denying that they would ever take the pill in the near future, again, the girls realized that a future self may. Sophia stated that "I plan on not taking birth control at this point. I might need it for my future self, but I'm not going to have sex until I'm married or engaged." She divides her identity and assigns the sexual part to her future self (that is, splintering). Sophia is so frightened by her own sexuality that she cannot even imagine its existence without the framework of a traditional, potentially procreative, and certainly far-off relationship.

My findings contradicted Stubbs' and Johnston-Robledo's hypothesis; my interviewees rejected menstrual suppression through the use of birth control pills. They tried to distance themselves from the position of a sexualized object and the accompanying social and psychological complexities that position forces them to confront. However, this reaction may be complicated by the age of my interviewees and their confusion about the process of menstrual suppression.

Of note is the heteronormativity underlying the imagining of the future self. Supposedly, Sophia would be using the pill to avoid pregnancy, implying that her partner would be male. The same applies to Madison's discomfort with the idea of having "another person inside of me" and categorizing that as a "future Madison problem." She too seems to be invoking a future heterosexual self.

Fear of Rape and Resulting Pregnancy

One girl, who was not sexually active, expressed concern that she was pregnant because she was a few weeks late in her cycle. She was unaware that irregularity is typical for girls in their first years of menstruation and unclear as to how girls become pregnant. Her anxiety exposes a sense of helplessness and confusion that pervaded the interviews.

While it is evident that adolescent girls resist the sexualization of their bodies, they appear to recognize the choice is not theirs alone. They can abstain from using tampons, the pill, or engaging in sexual activities, but, nevertheless, they perpetually exist under the scrutiny of the male gaze. Their bodies come to signify sexuality for others, if not for themselves. When questioned as to whether she found anything about the onset of her period to be unsettling, Olivia launched into a story in which she recounted: "I know it's kind of gruesome but like people get raped and there was this one girl who was like ten and she had her period and she got raped around the same time that her eggs were releasing and she got pregnant. And that's scary." Apparently, Olivia thought something similar could happen to her. Sophia, in responding to the same prompt, explained: "afterward [after her first period] I had this one concern that I would get raped when I was on my period and then get pregnant ... I think everyone will be scared once they get their first period." Again, we see confusion about the physiology of menstruation as it would be nearly impossible for her to get pregnant when she is in the midst of her period. It is worth exploring whether demystifying the bodily process of menstruation (and sex) would mitigate some of girls' anxieties. However, even if girls did understand menstruation and sex better, it is unlikely this would fully address underlying concerns about lack of control over their new sexual positioning.

When I first formulated the question about the burdensome aspects of menstrual onset, given prior findings, I had expected answers about cramps or worries about leaking. Rape is a subject that girls in previous studies on menstruation had never broached. At first, I thought this was an isolated event, but a few other girls hinted at the same fear. Additionally, Olivia and Sophia did not go to the same school or know each other. I probed to see whether either could identify the source of this concern, but neither could remember a specific incident. Both also claimed that they had not discussed the topic in health class, nor with parents or teachers. Therefore, I concluded that these statements hint at a subconscious concern over what it means to be a menstruating woman in our society. Reproductive capacity and bodily transformations indicate to others that these girls have developed into sexual beings. As women, their sexuality is influenced and controlled by others, through ideology (in the media, in health class, et cetera) (Garcia 2009; Ward et al. 2019), interpersonal relations (slut-shaming, et cetera) (Armstrong et al. 2014) and, at an extreme, through rape (Valenti 2009). Notably, my interviewees did not worry about rape before menarche, nor did they believe girls who have not begun menstruating needed to worry about it. This is strong evidence of my interpretation of these concerns as representing awareness of insertion into, what Lee (1994) calls, "the dominant patterns of sexuality" (346).

The girls' articulation of this emerging fear was interesting in that they were not simply concerned about being raped, but about being raped *and getting pregnant* as a result. Pregnancy functions as perhaps the most obvious signal of one's sexuality. This indicates an added layer of apprehension regarding judgment. As the girls consistently noted, they feel they are not the appropriate age for intercourse. Even if they were, female sexuality is still taboo in our culture, and, therefore, a source of shame. Insecure in their sexuality, the girls feared the judgment of others. They wanted to be seen as "good" girls. Thus, while they believed they were more vulnerable to rape after their transition in sexual status, the potential consequences post-menarche seemed to trump the concern of rape itself.

Summary

My participants made clear the implicit connection between menarche, menstruation, and sexuality as they fluidly transitioned between subjects of menstruation and sex. Girls talked about tampons and Trojans in the same breath, used rhetoric surrounding sex such as "waiting until you are ready" to describe tampon usage, pointed to "provocative girls" as their opposites, and, in one case, feared pregnancy due to irregular cycles. These concerns demonstrated the girls' desire to be "good girls" (that is, asexual), however, their fears of rape and pregnancy highlighted their realization that they exist within a set of gender and power relations; they do not fully control the way their bodies are seen or treated.

Discussion

The sexualization of girlhood brings to the fore what Deborah Tolman terms the slut/prude/virgin continuum, whereby behavior, appearance, and morals (the former two supposedly being indicative of the latter) are characterized as existing somewhere between that of a "good girl" and a "slut." Given the pressure to maintain innocence during childhood and fend off the forces of (hyper)sexualization, it is easy to understand why girls may experience trepidation about exiting this space into one of developing sexuality. The particular life stage of these interviewees may help to explain why they were so attached to their self-image as good girls, intimidated by reminders of sexuality, and employed the strategy of splintering.

Sophia most clearly constructed this good girl/bad girl dichotomy. She put a positive value judgment on her sister for wanting to avoid things that reminded her of sex (such as tampons), said that "sexually reckless" girls are the kind of girls who should take the pill, and used her virginity in opposition to the sexuality of the "whore" by saying that she would not take the pill at this point in her life, but "prostitutes" should. In rejecting tampons, the pill, and "provocative girls," my interviewees are investing in the idea of the good girl/virgin. Virginity serves as their form of sexuality and, therefore, strategy to delay dealing with the integration of the sexual self. This is further facilitated by the splintering of the self, which allows them to acknowledge a future sexual self, without needing to resolve the implications of their new social standing.

Labeling itself is an interactional process involving stigma and status (Armstrong et al. 2014). To end up on either extreme of this continuum can result in various formal and informal modes of punishment, though some of the material penalties that may accompany this label, such as rape, are more severe. This process of labeling and related consequences constitutes a form of social control that affects girls' thoughts and behaviors. It is likely that menarche makes real the entrance into these sexual and social relations, and thus discourse on the virgin/whore dichotomy impacts the meaning girls make of this physiological event. Evidencing the impact of this paradigm on their thoughts and behaviors, interviewees themselves reinforced this construct in the way they created a sense of duality between "provocative girls" and themselves. Jessica Valenti (2009) explains that, for girls and women, morality is dependent on sexuality. Thus, in reproducing such dyads, girls claimed their moral superiority and defended their character to the extent possible in the realm of female sexuality.

Such efforts to present themselves as good girls demonstrated interviewees demonstrated internalization of an external gaze. They did not want to appear like "promiscuous" girls and avoided this through choices that served as immediate symbols of their separation from their sexuality. These choices served a deeper hope that these strategies would indicate a societally acceptable form of sexual female character, that is, virginity. However, they realized that this self-presentation could only provide so much protection. Roberts (2002), suggested that we can look at the sexualization of girls and women across a continuum and that "...the most extreme act on the continuum of self-objectification is (. . .) rape" (326–27). Accordingly, my participants revealed their self-objectification and insight into said phenomenon when they cited rape as a potential outcome from their menarcheal transition. Regarding menarche as a transition in sexual status allowed them to make this connection.

Interestingly, my participants were not just concerned by rape itself, but by the potential result: pregnancy. The fear of pregnancy seemed to supersede the fear of rape. Pregnancy is perhaps the most obvious marker of deviant sexuality at this age. For these girls, being labeled a bad girl ranked above the act of rape, demonstrating the extent to which they have internalized a system of value that hierarchizes and reduces them to their gendered sexuality.

CONCLUSION

It is important to understand how menarche and menstruation are perceived by adolescent girls because of the immediate impact, long-term consequences, and girls' roles in defining what it means to be a woman/female. Lee (1994) summarized the broad importance of research on girls' understandings of menarche when she wrote:

> Menarche is a physiological happening, framed by the biomedical metaphors of current scientific knowledge, yet also a gendered sexualized happening, a transition to womanhood as objectified other. What is crucial here is that this juncture, menarche, is a site where girls become women and gender relations are produced. Such relations are about power and its absence; power to define the body and live in it with dignity and safety; power to move through the world with credibility and respect. (360)

My interviewees felt the acute shift in their position within gender relations and their lack of power in this moment.

Though my sample was small, the consistency of the packaging of menarche, menstruation, and sexuality across and throughout interviews indicates that this phenomenon is not limited to this group of girls. Contrary to previous work, this study shows that menarche and menstruation are bound up with sexuality in the minds of young adolescent girls. It is likely that the age of the interviewees contributed to these findings. In exiting a space emphasizing childhood innocence, girls are preoccupied with dissociating themselves from sexuality. We see this in the rejection of tampons, birth control, promiscuity, as well as through the use of splintering. Through these methods, girls resist entrance into broader sexual relations where they lose further control of their new bodies. They cling to the image of the good girl, an attempt to hold on to some sense of childhood innocence (that is, asexuality) and to approach teenage sexuality through the safest avenue. Nevertheless, these strategies play into the reproduction of the same cultural ideals and power relations that obstruct the development of girls' sexual subjectivity.

Girls' awareness of their reproductive bodies' status as public bodies, and the importance of their appearance as good girls, may further reflect understandings of rape discourse. Specifically, they may internalize the message that rape occurs when girls "ask for it," a message reinforced by rape legislation that often considers the question of who is "rape-able" using idealized notions of female purity (Valenti 2009).

The sexualization of menstruation performed by these girls reveals that this process is problematic not because of the physiological process of menstruation itself but because of the reproductive and sexual implications of menarche and the construction of female sexuality. Future research must explore how the intersection of other identities, such as race and class, relate to the sexualization of menstruation. As intersecting identities alter the construction and experience of female sexuality, so too may they shape the meanings of menarche and the strategies employed to navigate this transition. For this age group in particular, more research into girls' experiences with and understandings of menstruation is required. Further research in this area can inform endeavors to guide girls through this transition in a way that allows them to engage their bodies and sexuality with positivity and ownership. Intervening in this moment may be crucial, not only to regain individual sexual subjectivity, but to redefine womanhood and gendered power relations.

Note

1. The hymen, or vaginal corona, is actually a membrane that cannot be "broken" and is a permanent, though changing, structure located one to two centimeters inside the vaginal opening (The Swedish Association for Sexuality Education).

References

Armstrong et al. 2014. "'Good Girls': Gender, Social Class, and Slut Discourse on Campus." *Social Psychology Quarterly* 77 (2): 100–22.

Beausang, Carol C., and Anita G. Razor. 2000. "Young Western Women's Experiences of Menarche and Menstruation." *Health Care for Women International* 21 (6): 517–528.

The Boston Women's Health Collective. 2011. *Our Bodies, Ourselves*. New York: Touchstone.

Brooks-Gunn, Jeanne and Diane N. Ruble. 1982. "The Development of Menstrual-Related Beliefs and Behaviors during Early Adolescence." *Child Development* 53 (6): 1567–77.

Brumberg, Joan Jacobs. 1997. *The Body Project: An Intimate History of American Girls*. New York: Vintage Books.

Egan, R. Danielle. 2013. *Becoming Sexual: A Critical Appraisal of the Sexualization of Girls*. Malden, MA: Polity Press.

Fausto-Sterling, Anne. 2000. *Sexing the Body: Gender Politics and the Construction of Sexuality*. New York, NY: Basic Books.

Fingerson, Laura. 2006. *Girls in Power: Gender, Body, and Menstruation in Adolescence*. Albany, NY: State University of New York Press.

Fredrickson, Barbara L. and Tomi-Ann Roberts. 1997. "Objectification Theory: Toward Understanding Women's Lived Experiences and Mental Health Risks." *Psychology of Women Quarterly* 21: 173–206.

Garcia, Lorena. 2009. "'Now Why Do You Want to Know About That?': Heteronormativity, Sexism, and Racism in the Sexual (Mis)education of Latina Youth." *Gender and Society* 23 (4): 520–41.

George, Alison, Anne Murcott. 2011. Research Note: Monthly Strategies for Discretion: Shopping for Sanitary Towels and Tampons. *The Sociological Review* 40 (1): 146–162.

Havens, Beverly and Ingrid Swenson. 1988. "Imagery Associated with Menstruation in Advertising Targeted to Adolescent Women." *Adolescence* 23 (89): 87–97.

Kessler, Suzanne J. 1990. "The Medical Construction of Gender: Case Management of Intersexed Infants." *Signs: Journal of Women in Culture and Society* 16 (1): 3–26.

Kissling, Elizabeth Arveda. 1996. "'That's Just a Basic Teen-age Rule': Girls' Linguistic Strategies for Managing the Menstrual Communication Taboo." *Journal of Applied Communication Research* 24: 292–309.

Koff, Elissa and Jill Rierdan. 1995a. "Early Adolescent Girls' Understanding of Menstruation." *Women & Health* 22 (4): 1–19.

———. 1995b. "Preparing Girls for Menstruation: Recommendations from Adolescent Girls." *Adolescence* 30 (120): 795–811.

Lee, Janet. 1994. "Menarche and the (Hetero)Sexualization of the Female Body." *Gender & Society* 8: 343–62.

———. 2009. "Bodies at Menarche: Stories of Shame, Concealment, and Sexual Maturation." *Sex Roles* 60: 615–27.

Maines, Rachel P. 1999. *The Technology of Orgasm: "Hysteria," the Vibrator, and Women's Sexual Satisfaction*. Baltimore, MD: The Johns Hopkins University Press.

Martin, Karin A. 1996. *Puberty, Sexuality, and the Self: Girls and Boys at Adolescence*. New York: Routledge.

McPherson, M. E., and L. Korfine. 2004. "Menstruation Across Time: Menarche, Menstrual Attitudes, Experiences, and Behaviors." *Women's Health Issues* 14: 193–2004.

Moore, Susan M. 1995. "Girls' Understanding and Social Constructions of Menarche." *Journal of Adolescence* 18(1): 87–104.

Morse, Janice M. and Helen McKinnon Doan. 1987. "Adolescents' Response to Menarche." *Journal of School Health* 57 (9): 385–89.

Rierdan, J. 1983. "Variations in the Experience of Menarche as a Function of Preparedness." In *Menarche: The Transition from Girl to Woman*, edited by Sharon Golub, 119–25. Lexington, MA: Lexington Press.

Roberts, Tomi-Ann. 2002. "The Woman in the Body." *Feminism & Psychology* 12 (3): 324–29.

Ruble, Diane N. and Jeanne Brooks-Gunn. 1982. "The Experience of Menarche." *Child Development* 53 (6): 1557–66.

Schooler et al. 2005. "Cycles of Shame: Menstrual Shame, Body Shame, and Sexual Decision-Making." *The Journal of Sex Research* 42 (4): 324–34.

Stubbs, Margaret L. and Ingrid Johnston-Robledo. 2013. "Kiddy Thongs and Menstrual Pads: The Sexualization of Girls and Early Menstrual Life." In *The Sexualization of Girls and Girlhood: Causes, Consequences, and Resistance*, edited by Eileen L. Zurbriggen and Tomi-Ann Roberts, 213–34. New York: Oxford University Press.

Summers-Effler, Erika. 2004. "Little Girls in Women's Bodies: Social Interaction and the Strategizing of Early Beast Development." *Sex Roles* 51 (1/2): 29–44.

The Swedish Association for Sexuality Education. *The Vaginal Corona*. Translated by Jonas Hartelius. Stockholm: RFSU. https://www.rfsu.se/globalassets/pdf/vaginal-corona-english.pdf

Thorne, Barrie. 1994. *Gender Play: Girls and Boys in School*. New Brunswick, NJ: Rutgers University Press.

Tolman, Deborah et al. 2015. "Mobilizing Metaphors: Considering Complexities, Contradictions, and Contexts in Adolescent Girls' and Young Women's Agency." *Sex Roles* 73: 298–310.

Valenti, Jessica. 2009. *The Purity Myth: How America's Obsession with Virginity Is Hurting Young Women*. Berkeley, CA: Seal Press.

Ward et al. 2019. "Following Their Lead? Connecting Mainstream Media Use to Black Women's Gender Beliefs and Sexual Agency." *The Journal of Sex Research*: 1–13.

Whisnant, Lynn, and Leonard Zegans. 1975. "A Study of Attitudes toward Menarche in White Middle-Class American Adolescent Girls." *American Journal of Psychiatry* 132 (8): 809–814.

Zurbriggen et al. 2007. *Executive Summary: APA Task Force on the Sexualization of Girls*. Washington, DC: American Psychological Association.

CHAPTER 25

(In)Visible Bleeding: The Menstrual Concealment Imperative

Jill M. Wood

Researchers and scholars from various and diverse backgrounds have long discussed the degree to which society views women's menstruation[1] as taboo, shameful, disgusting, and consequently is shrouded in secrecy (for example, Bobel 2006; Brooks-Gunn and Ruble 1982; Golub 1992; Johnston-Robledo et al. 2003; Stubbs and Costos 2004; Ussher 1989). Menstrual taboos that stigmatize women still abound in the US, serving to limit women's ability to fully participate in their lives and in society (for example, Johnston-Robledo and Chrisler 2013; Patterson 2014; Ussher 1989, 2006). Houppert (1999) coined the phrase "the culture of concealment" to explain how menstrual taboos and stigma shape women's experience of menstruation and manipulate women into menstrual shame and secrecy, often via menstrual hygiene products. Similarly, the commodification of menstruation through "feminine hygiene" care continues to promise new and improved ways for women to keep themselves "clean" and their periods a secret—either by opting out of menses altogether via menstrual suppression or through products that render menses invisible.

Simultaneously, the medicalization of menstruation conceptualizes women as deficient, ill, and diseased, and menses in need of medical treatment and management (for example, Lippman 2004; Wood et al. 2007). Through medicalization, menstruation is constructed as a "health" issue from a biomedical perspective and jurisdiction for how to treat and monitor it is assigned to the medical system. This conflation of biology as ideology serves to limit women's participation both in their private and public lives because menses marks women not only as physically ill but also as emotionally impaired. In this way, medical social control results through ideological frameworks that define and discipline menstruation as a health issue.

© The Author(s) 2020
C. Bobel et al. (eds.), *The Palgrave Handbook of Critical Menstruation Studies*, https://doi.org/10.1007/978-981-15-0614-7_25

The conceptualization of menstruation as disease and the association of menstruation with femaleness has established menstruation as a political issue (for example, Bobel 2010; Ussher 2006).

While Houppert has described the significance of the culture of concealment surrounding menstruation, this chapter explains menstrual concealment an as imperative that women adopt through their internalization of and adherence to menstrual discourse. "The menstrual concealment imperative" is a conceptual framework to explain how women's internalized perceptions of menstruation as diseased, taboo, and stigmatized contribute to their disembodiment and self-objectification (Roberts 2004). It suggests that women's vigilance about menstrual concealment is a form of self-surveillance and self-objectification that is fostered by the medicalization of women's bodies and neoliberal approaches to women's health. The potential for menstruation to be oppressive is rooted in a complex, multifaceted, and all-encompassing imperative for women that functions as gendered body politics to (re)produce the very conceptualization of women's bodies as othered (de Beauvoir 1952). Offering a critique of the medical system and menstrual hygiene industry, this paper analyzes menstrual discourse that establishes women as diseased and as unable to know their bodies. Using a neoliberal rhetoric of "choice" the menstrual hygiene industry cleverly posits menstrual concealment as "freedom" and thereby facilitates women's complicity in their own subjugation. I offer "the menstrual concealment imperative" as a theory to explain how women's internalization of the culture of concealment is form of social control and a body project (Brumberg 1998) that keeps women disembodied and oppressed.

Menstrual Discourse

In this section, I use Foucault's work on discourse to explain the significance of the production of menstrual knowledge as oppressive to women. First, I will discuss Foucault's conceptualization of discourse and biopower as it relates to menstrual discourse. Next, I will discuss the significance of menstrual taboos and menstrual stigma. Finally, I will discuss medicalization in terms of how menstrual discourse constructs women as diseased by virtue of their menses, and therefore how knowledge production by the medical field and the associated female hygiene industry establishes women as perpetually disempowered.

Foucault's (1984) concept of biopower explains how social norms and expectations embedded in microlevels of everyday life coalesce into powerful discourses that shape what is considered normal on both individual and societal levels. Arguing that biopower is produced through discourse (and individuals' desire to adhere to it), he explains how pressure to conform to cultural norms produces individuals' voluntary self-subjugation. In this way, biopower is a form of social control enacted through individuals' internalization of dominant discourses that result in self-disciplinary practices. Foucault argues that biopower is especially salient in terms of regulating bodies through medicine and technology because of discourse structures social and individual attitudes, thoughts, and behaviors as if the knowledge is objective. However, this knowledge is characterized by a specific epistemological stance in order to maintain ideological control. Ussher (2006) argues that biopower constructs the female body as deviant, polluted, and "monstrous" based on reproductive processes like menstruation, regulating women's bodies through discourse that positions it as in need of surveillance and treatment. Menstrual discourse constructs women's bodies as diseased, shameful, and polluted. Foucault's concept of biopower elucidates how women's bodies are a site of self-discipline and how these practices produce 'docile bodies' (for example, Bartky 2014; Bordo 1989; Pylpa 1998; Patterson 2014). Pylpa explains: "… medicine creates the discourse that defines which bodies, activities, and behaviors are normal; at the level of practice, medical procedures are a principal source of the institutional regulation and disciplining of bodies" (30). In this way, menstrual discourse creates biopower at both micro and macro levels. Individuals voluntarily conform to disciplinary strategies of their bodies through their own desire (Foucault 1977), and in doing so biopower is produced and (re)produced. Foucault's work prompts an analysis of menstrual discourse that asks questions about how menstrual knowledge is produced and by whom, what constitutes knowing about menses, and who has authority and power to produce menstrual knowledge. Later in this chapter, I will use feminist analyses of Foucault's work on biopower to argue that women's internalization of menstrual discourse results in their self-surveillance of their menstruation and ultimately explains women's internalized need for menstrual concealment.

Menstrual Taboos, Stigma, and Silence

The prominence and significance of menstrual taboos and stigma in women's lives has been well documented by scholars in various disciplines for decades (for example, Bobel 2006; Brooks-Gunn and Ruble 1982; Buckley and Gottlieb 1988; Golub 1992; Roberts 2004; Stubbs and Costos 2004; Ussher 1989). Evidence of how menstrual taboos stigmatize women is evidenced in the description of how menstruation and menstrual blood have been variously described, as: simultaneously magical and poisonous (Golub 1992); an abomination (Rozin and Fallon 1987), disgusting and aversive (Bramwell 2001), contaminating (Laws 1990) unclean and unpure (Cicurel 2000), a threat to femininity (Lee 1994), and a blemish to a one's character (Johnston-Robledo and Chrisler 2013). Menstruating women are also perceived as a danger to men and a threat to male power (Delaney et al. 1988; Guterman et al. 2007).

The consequence of menstrual taboos for women's lives is significant and varied as prohibited behavior for menstruating women is contextualized culturally, geographically, and according to religious and other social practices. Many women report their need to maintain cleanliness during menses, and most women report altering their usual activities during menses (Jurgens and Powers 1991). Indeed, women's motivation to change their behavior during their menses may be a reasonable precaution as menstruating women are harshly judged and characterized as irritable and unsexy (Forbes et al. 2003). Roberts et al. (2002) found that by virtue of simply having a tampon visible in her bag, a woman is perceived as less likeable, less competent and that observers made an effort to physically distance themselves from her. Menstrual taboos function to separate, exclude, and even banish menstruating women from public and private spheres preventing their full participation in public life as well as in their own full subjectivity (for example, Johnston-Robledo and Chrisler 2013; Roberts 2004; Thomas 2007; Thornton 2013; Ussher 2006).

Silence perpetuates menstrual stigma and is a key indicator of the culture of concealment (Delaney et al. 1988; Houppert 1999; Kissling 2006). Menstruation is considered inappropriate public conversation to the extent that girls and women are often too uncomfortable to discuss the topic even with each other, healthcare providers, or family members (for example, Golub 1992; Houppert 1999; Johnston-Robledo and Chrisler 2013). Ussher (2006) describes women's "unspeakable bodies" referring to how the silence surrounding menstrual shame results in women's self-isolation. Similarly, Houppert's (1999) "culture of concealment" explains how sociocultural influences construct menstruation not just as taboo, shameful, and debilitating but also as invisible in US culture. She details the powerful influence that the menstrual product industry has to conceptualize menstruation as an illness, and how the development and advertisement of menstrual products create and reinforce women's insecurities around their periods as a hygiene crisis. In this way, menstrual product companies convince women to conceal their periods, and then provide their own products to enable that concealment, reinforcing cultural attitudes that menstruation is embarrassing and should be kept secret. Menstrual discourse disempowers women by not only constituting menstruation as a negative, taboo, and stigmatized event that women must conceal, but also by enabling others to produce knowledge about women's bodies that is not based on their own situated knowledge and experiences.

The Medicalization of Women's Bodies

In addition to the conceptualization of menstruation as taboo and stigmatized, menstrual discourse is also characterized by the medicalization of menstruation that further constructs menstruation as a disease process in need concealment via medical management. Conrad (1992) explains medicalization as a sociocultural process that functions as a form of social control. Women are especially vulnerable to medicalization and an overwhelming majority of women's natural life processes have been medicalized

(Conrad 1992; Ehrenreich and English 2005). As such, feminist researchers and women's health activists critique the over-medicalization of normal, natural, healthy body processes that characterize women's everyday lives (for example, Kaufert and Gilbert 1986; Lippman 2004; Tiefer 1995). These scholars have demonstrated how women's bodies, especially reproductive processes, are medicalized as a form of political and social control to the detriment of women's health and lives (for example, Ussher 2006; Ruzek 1978; Wood et al. 2007).

The medicalization of women's bodies has significant consequences for women's lives. First, women are ideologically constructed as deficient, ill, and diseased to legitimate the need for medical treatment and constant medical surveillance. Secondly, medicalization functions as a form of social control by establishing medical practitioners as the experts on women's bodies based on women's illness as defined by the medical model. Kaufert and Gilbert (1986) argue that when women are not considered capable of knowing their bodies then their subjective experience of themselves shifts to that of a patient. As patients, women are morally obligated to be treated and medical practitioners are ethically required to diagnose and treat their assumed diseased state. In this way, the medicalization of menstruation contributes to menstrual discourse that positions women in a constant state of disease. As Ehrenreich and English (2005) explain, "Not only [are] women seen as sickly – sickness [is] seen as feminine" (22).

While scholarship on medicalization as a theoretical framework has existed for decades, biomedicalization has been more recently described as distinct from medicalization both in terms of historical context and an increased focus on techno-scientific processes (Clarke et al. 2010). Like medicalization, the biomedicalization of menstruation shapes menstrual discourse through the production of knowledge that establishes women as unable to know their own bodies. While medicalization controls bodies through defining disease, biomedicalization encourages the transformation of bodies based on the construct of health, so that biomedicalization is broader, more invasive, and reaches into lifestyle decisions around health, risk, illness, and wellness as a moral imperative. The biomedicalization of health is characterized by a more complex and insidious structure of knowledge production and dissemination, including the corporatization of medicine. No longer is medical discourse produced solely by medical experts, but also by pharmaceutical companies, media outlets, alternative medicine practitioners, patient self-help groups, and research conglomerates that often advertise products, drugs, technologies, and information targeted directly to consumers. In biomedicalization, it is not an illness, disease, or, dysfunction that is treated as in medicalization, but the risk of these (Armstrong 1995). Because "health" is so broadly conceptualized, so poorly defined by so many "health experts," and health recommendations constantly change, individuals no longer need to be sick to be treated; simply the risk of eventual poor health is sufficient for intervention. When health is no longer conceptualized as the absence of disease, the possibilities for diagnosis, treatment, and intervention to address the risk of illness is literally endless (Clarke et al. 2010).

Despite theoretical distinctions between medicalization and biomedicalization, I use the term (bio)medicalization in this chapter to refer to the simultaneous ways that women's menstrual bodies are transformed, regulated, and controlled through menstrual discourse. Because menstruation is conceptualized from a biomedical perspective as a form of illness, women are encouraged to transform their bodies to prevent potential hazards of menstruation. Technological and pharmaceutical interventions promise menstrual concealment to women as an individual "choice" (for example, menstrual suppression) by transforming women's menstrual bodies into non-bleeding ones. Similarly, menstrual products are marketed to women as hygiene products so that women can manage (conceal) their menses as part of their individual responsibility for their own health. In this way, (bio)medicalization contributes to menstrual discourse by establishing the amorphous healthcare industry as the experts on menstruation while assigning women to the perpetual role of patient. Together, the (bio)medicalization of menstruation and menstrual stigmas and taboos function to create a menstrual discourse which controls women's bodies and lives based on epistemologically flawed biomedical ideologies. Menstrual concealment is constituted in menstrual discourse as an individual women's "choice" as part of her own pursuit of her health. In order to understand how the menstrual concealment becomes an imperative for individual women, next I will consider how women internalize menstrual discourse through self-surveillance.

Self-Surveillance and Self-Objectification

In this section, I explain how menstrual discourse and biopower facilitate women's internalization of menstrual self-surveillance as imperative menstrual concealment practices. According to Foucault (1977), individuals come to desire conformity to discourse through biopower and engage in a resultant process of self-surveillance through panoptical power. The panopticon is a paragon of how to socialize masses into a state of constant self-policing so that individuals conform to "normal" behaviors and attitudes without the need for external enforcement. Through discourse, these norms become so desirable for individuals to obey that they then voluntarily self-monitor their adherence to them. Knowledge produces discourse that establishes cultural norms that individuals desire to conform to; because the knowledge is presented as objective and "true," individuals organize their behavior around the discourse, thereby enacting their own practice of self-surveillance and self-discipline. This constant self-surveillance and self-regulation is posited as individualism, despite the fact that it is culturally created, and in this way biopower is difficult to see as external to the individual (Bartky 2014; Foucault 1977).

Bartky (2014) builds on Foucault's work, noting how gendered notions of power in self-surveillance practices are especially problematic for women and their bodies. She argues that the panoptical gaze is male, and therefore that women's self-surveillance based on this patriarchal view results in their disembodiment. Barkty discusses her work in the context of women's self-monitoring of their own appearance, explaining that women internalize and embody patriarchal notions of beauty and subsequently adopt body projects (Brumberg 1998) to change their bodies to adhere to cultural beauty norms. Women readily enact these body projects as part of their own self-surveillance and self-disciplinary practices resulting in the production of their own docile bodies. Women's discipline of their own bodies via an internalized patriarchal panoptical view is insidious because the authority to maintain control of women's bodies is both nowhere and everywhere, and it is neither natural nor completely voluntary. That is, when women practice self-disciplining body projects they do so without coercion but not by their own free will either.

Just as Bartky uses Foucault's work on biopower, discourse, and self-surveillance to analyze women's self-subjugation around appearance, this chapter applies those concepts to menstruation. Menstrual discourse is characterized by negative views of menstruation that encourage girls and women to self-surveil and manage their bodies to maintain menstrual secrecy (for example, Chrisler 2004; Erchull et al. 2002; Martin 1992; Stubbs and Costos 2004). Whether learned through mothers, educators, product advertisements, or other girls, menstrual discourse encourages girls and women to perceive their bodies as polluted and shameful, and as such, out of their control (Chrisler 2004; Jackson and Falmagne 2013). Ussher (2006) argues that when menstruation is positioned as a form of embodied pathology menstrual discourse encourages women's self-surveillance, self-policing, self-silencing, self-blame, self-sacrifice, and contributes to women's guilt, shame, and blaming of the body.

Other feminist scholars have also established that women's bodies are sites of discipline and that subjectivity is tied to the body, which is constantly in need of management, containment, and discipline (for example, Bartky 2014; Bordo 1990; Lee 1994; Martin 1992; Young 1997). Moreover, because women are primarily valued based on their appearance, self-disciplinary body projects (Brumberg 1998) are strongly associated with femininity. Such body projects reify patriarchal constructs of femininity as women judge themselves as "good" or "bad" women based on how well they conform to standards of femininity that require them to distance their bodies from their selves (Roberts and Waters 2004). In order to be "good," women are necessarily disembodied, objectified, and self-silenced from their menstrual bodies (Roberts and Waters 2004; Ussher 2006). Ussher argues that menstrual discourse posits women as closer to nature due to their bodily subjectivity, and in this way women's reproductive bodies are a marker of "hegemonic constructions of femininity" (2). As such, menstrual discourse constructs

women's bodies as pathological and defines women based on their reproductive capacity, associating femininity with women's ability to maintain
the secrecy of their polluted bodies. Ussher explains that the implications of
defining women as a held hostage by their menstrual bodies are significant in
how women can inhabit and know their own bodies as well as the development of women's subjectivity.

Similarly, Persdotter (2020, this volume) introduces the concept of menstronormativity to expand menstrual discourse as a more all-encompassing
process and system that describes the ordering of menstruation on both sociocultural levels and in individual's lives. Menstronormativity, the aggregate of
menstrual norms, stigmas, etiquette, and discourse, describes the regulation of
some menstrual subjectivities as "good" and others as "bad." Women's acceptance of menstronormativity fuels self-surveillance and self-disciplinary body
projects. The process through which women adopt this internalized male gaze
of their bodies and selves can be understood through objectification theory
and ultimately, women's self-objectification.

Objectification theory explains how the sexual objectification of girls and
women functions to separate their bodies from their personhood so that
female bodies are viewed in terms of how they serve others, often the sexual interests of men (Bartky 1990; Fredrickson and Roberts 1997). When
women internalize this objectification, they adopt an outsider's view of themselves, evaluating their bodies and appearance from a sexually objectified gaze;
this is the process of self-objectification (Fredrickson and Roberts 1997).
Fredrickson and Roberts argue that self-objectification explains a woman's
sense of self-detachment from her own body. As such, self-objectification is
one-way women unwittingly participate in their own oppression.

Self-objectification is a common practice for women, especially around
reproductive functions like menstruation, and it is associated with a host
of negative health effects including increased body self-surveillance, body
shame, and negative attitudes about body functions like menstruation
(Johnston-Robledo et al. 2003; Roberts and Waters 2004; Roberts 2004).
Roberts (2004) applies objectification theory to menstruation, arguing that as
women internalize US culture's sexual objectification of themselves that menstruation must be concealed in order to appear as adequately feminine, attractive, and sexually desirable. Women who engage in self-objectifying menstrual
practices also report more self-surveillance and associated feelings of shame,
self-loathing, and self-disgust. In this way, self-objectification prevents women
from inhabiting their bodies in an emotionally and physically authentic way.
This contributes to women's alienation from their subjective experiences and
is a form of de-selfing as women replace their own sense of self with an outsider's (male) gaze (Roberts and Waters 2004, 13).

Because menstruation is viewed as the antithesis of a sexually desirable
feminine body, women learn that to be sexually desirable, attractive, and
feminine menstruation must be concealed (Grose and Grabe 2014). For
instance, Erchull's research found that women's bodies are portrayed as

highly sexualized even in ads for menstrual products, illustrating the need for women to use such products for menstrual concealment in order to ensure their constant sexually availability. Other researchers have also found that women's attitudes toward their menstrual bodies are incompatible with their internalized valuing of their bodies as sexually desirable (Johnston-Robledo et al. 2007). In this way, women's self-objectification contributes to their desire to distance themselves from their bodies via menstrual concealment (Erchull 2013; Grose and Grabe 2014; Roberts 2004; Roberts et al. 2002) or menstrual suppression (Johnston-Robledo et al. 2007).

As problematic as self-objectification and its associated risks are for women, adherence to idealized female body standards may be logistically beneficial to women in a patriarchal culture. For example, Fredrickson and Roberts (1997) propose that self-objectification might appear to women as a strategy to claim power in a patriarchal system in which attractiveness is currency. Just as women benefit economically from being attractive, women who distance themselves from their bodies, especially menstruating bodies that are feared and abhorred, have more opportunities in the public sphere (Roberts and Waters 2004). In this way, menstrual concealment serves as a tool to distance oneself from the feminine, and it does benefit women in terms of their acceptance in a patriarchal society. In this way self-objectification can be considered a survival strategy to present their bodies in an idealized form, and in this way self-objectification is about self-surveillance (Roberts and Waters 2004).

The Menstrual Concealment Imperative

As women self-objectify through a patriarchal body-hating view of themselves, menstrual concealment offers women a way to "free" themselves from their monstrous body. Women not only feel obligated to render their periods invisible, but when framed as an empowering choice, menstrual concealment falsely offers women a sense of control over their out of control bodies. As a theory, the "menstrual concealment imperative" explains how women internalize menstrual discourse and willingly practice self-surveillance and self-disciplining body projects, even though such practices are self-subjugating and disempowering. This section of the chapter will explain menstrual concealment as a required form of self-surveillance in which women become disembodied through this self-disciplinary practice in their search for "freedom" from and control over their bodies.

The menstrual concealment imperative is about freedom and control to women; "freedom" from their bodies that mark them as othered and are a significant source of their oppression. The concealment imperative offers a solution to women's objectified, pathologized, and then self-objectified bodies, yet women become disembodied through these self-disciplinary concealment practices. In this way, the concealment imperative is a panopticon-like form of social control that women willingly participate in, and as they do so women become complicit in a menstrual discourse that requires them to be disembodied and objectified. When women internalize these negative perspectives via self-objectification, menstrual concealment is no longer a cultural norm but a moral imperative to exist in a patriarchal society. Without agency to contextualize their own menstrual experiences, women desire distance from their menstrual bodies and menstrual concealment offers this disembodiment. The menstrual concealment imperative is a self-perpetuating cycle of self-surveillance, self-discipline, and self-subjugation.

Menstrual concealment is imperative for menstruating women for several reasons. First, menstrual concealment is required for women to be considered as competent (Roberts et al. 2002), attractive and sexually appealing (Erchull 2013). In order to succeed in public life women must transform their bodies to meet patriarchal expectations for how their bodies appear to others and how their bodies impact others' feelings about them in terms of comfort, sexual attractiveness, and hygiene. Menstrual concealment benefits women socially, politically, and personally because menstruation marks bodies as feminine and therefore as weak. Practically speaking, women are more successful in their lives if they appear unencumbered by their menses. The menstrual concealment imperative explains practical benefits in women's public and private lives that may result from their concealment practices.

Secondly, women perceive menstrual concealment as imperative because menstrual discourse dictates how women experience their menstruation as polluted, unclean, disgusting, and as an illness to be managed. Menstrual discourse conceptualizes menstruation as pathological and posits the transformation of the diseased body as the "right" way to avoid possible risks associated with menstruation. In (bio)medicalization terms, menstrual concealment is both control over and transformation of the female body into one that is less stigmatized. Surveillance medicine requires the management of the menstrual body through menstrual concealment as a moral obligation for women as patients and health care consumers to avoid ambiguous risks associated with the illness of menstruation. Menstrual concealment is imperative for women to avoid illness and consider themselves "healthy." Moreover, women may feel out of control in their bodies when their bodies are positioned as monstrous, disgusting, and diseased; menstrual products are offered as a way for women to "control" their bodies. As such, menstrual products are a technology used to transform the dysfunction of the menstruating female body into a non-menstruating one (Vostrel 2008). Thus, the menstrual concealment imperative is constructed and (re)produced through menstrual discourse and menstronormativity to allow women to dissociate from their bodies that mark them in oppressive ways. Because stigma surrounding the menstrual body threatens women's full access to the public sphere (Thomas 2007), it is understandable that women willingly become disembodied as a potentially liberating tactic in patriarchal culture.

Finally, menstrual concealment is imperative because in women's private lives it marks them as "good women." Patriarchal standards of femininity are rooted in how women's bodies serve others; women's bodies must be clean, sexually attractive, and not inconvenient or uncomfortable for others. Girls adopt the concealment imperative very early; at menarche, they learn how to manage their menstrual shame by the concealment of their menstruation in order to prevent others' discomfort (Kissling 1996; Jackson and Falmagne 2013). Through self-objectification of themselves as monstrous, girls and women adopt self-surveillance and self-disciplinary practices to conceal their menses. Thus, menstrual concealment is imperative for women to consider themselves "good" based on patriarchal standards of femininity that require women's docile bodies; there is little possibility for women to avoid menstrual concealment and still claim an identity as "a good woman," "healthy," "attractive," or even "smart." When women self-objectify and internalize hegemonic requirements for their bodies based on patriarchal standards of femininity, women must necessarily become disembodied or risk self-hatred (Roberts 2004). That is, the inability for women to avoid self-hatred without menstrual concealment illustrates the imperative nature of menstrual concealment.

Women may interpret menstrual management and concealment as a form of empowerment and control over their bodies, especially when menstrual concealment is marketed to women as convenience that is characterized as "freedom." The menstrual product industry has created a market for their own products based on the culture of concealment (Houppert 1999), referring to these products, as "feminine hygiene" and "sanitary protection" to reinforce the notion that menstruation is an unsanitary condition that girls and women need to protect themselves and others from (Vostrel 2008). Menstrual product advertising and direct to consumer (product) education reinforces (bio)medicalization and reifies women's need to conceal their menstruation. Products are marketed to girls and women as convenience and "freedom" from their bodies because of how effective they are at enabling women to conceal their menstruation. For instance, Proctor and Gamble advertise "Always My Fit" to women as their allies in "better period protection" through a custom fit sizing chart, now available on the top of all pad packages. The brand claims that, "60% of women wear the wrong size pad and 100% can change that!" (always.com 2017, "Tips and Advice Choosing a Pad"). Using a neoliberal approach to target women's self-loathing of their menstrual bodies, Always My Fit offers pseudo control, choices, and power to women, "… when many women experience a leak they often blame themselves . . . the truth is that a lot of women do not know that leak free periods are possible [if you find] the right pad coverage."

For these reasons, menstrual concealment may feel empowering to women, especially as the commodification of menstruation offers women the ability to purchase freedom from their bodies through menstrual products that claim to be specially designed for them. The pressure for body transformation, like menstrual concealment, as a form of individualism and "choice" is characteristic of (bio)medicalization, panoptical forms of social control, and neoliberalism. As such, it is hard to identify as imperative. Bartky (2014) argues that the lack of an enforcer in the disciplining of female docile bodies makes women's subordination seem isolated, normal, and appears as more of an individual choice than an institutional mandate. In this way, the menstrual concealment imperative is both invisible and self-sustaining.

Yet, because menstrual concealment is imperative for women's acceptance and success in both their public and private lives, this practice of self-discipline is not a true choice. I argue that women "choose" to become disembodied and self-subjugating as a form of false consciousness due in part to the conceptualization of concealment as a cultural norm instead of as an imperative. Reframing menstrual concealment practices as imperative self-disciplining behaviors offer a framework to understand women's "choice" to conceal menstruation as a false one for several reasons. First, the risk for women not to conceal is tremendous including being judged as incompetent, emotional, unattractive, unclean, and diseased. Women may prudently judge that given other forms of gender oppression, menstrual concealment benefits them in important logistical ways like obtaining or maintaining employment and/or long-term partnerships or marriage. For example, Bartky (2014) discusses that women risk the refusal of male patronage and related intimacy as well as success in their economic and social livelihood when they avoid forms of bodily self-discipline. Moreover, a woman's sense of herself will likely be compromised by avoiding self-disciplinary practices because they are so critical to social constructions of herself as a woman and individual (Bartky 2014). Secondly, women are often not aware that menstrual concealment is a self-disciplinary practice as a result of their own self-objectification of their bodies. When menstrual concealment is marketed to women as convenience or empowerment, the imperative nature of concealment is rendered invisible. Third, women cannot make a true choice about their menstruation when they are distanced from their bodies. Without agency and subjectivity, women's ability to make decisions, as is characteristic of true choice, is impossible. Fahs (2014) distinguishes between the 'freedom to' and the 'freedom from' in regard to women's subjectivity and agency, arguing that a feminist understanding of freedom must involve both aspects of freedom. In this way, women's freedom to choose menstrual concealment is dependent on women's freedom from menstrual stigmas that mandate menstrual concealment. Finally, menstrual concealment cannot be a true choice for women when alternatives to it are not presented. For menstrual concealment to be a viable choice, women must be able to choose to claim their menstrual realities just as freely as they opt to conceal menses.

THE FUTURE

Feminist menstrual researchers have remarked on "unspeakable womanhood" (Ussher 2017) and a missing discourse around women's reproductive bodies (Roberts 2017). I offer the menstrual concealment imperative as a conceptual tool for menstrual scholars and researchers to refer to the totality of the various interrelated processes and layered structural barriers that contribute to women's oppression via menstrual discourse. The invisibility of the menstrual concealment imperative contributes to how insidious it is in women's lives; when women internalize menstrual discourse they become disembodied, self-objectify and willingly engage in their own self-surveillance and self-discipline. One possible implication of the menstrual concealment imperative as a theoretical tool is that by describing and naming it, then the imperative for women to conceal their menstruation is visible and less insidious. This visibility lends legitimacy to women's experiences and therefore creates the possibility for resistance to menstrual concealment as imperative for women's freedom and success in the private and public spheres. Resistance to the menstrual concealment imperative must begin with making it visible, as Ussher explains: "Identifying self-policing practices allows women to develop more empowering strategies for reducing or preventing . . . distress, developing an ethic of care for the self, and no longer blaming the body ..." (2).

Notably, women's voices and experiences are largely missing from menstrual discourse because of their disembodiment, and therefore women's own voices and positive experiences of menstruation can be seen as a form of resistance. Patterson (2014) argues that resistance to normative menstrual discourse can range from being "period positive" to more radical forms of menstrual activism, as is characteristic of menarchists: "Menarchists argue that women need to take back the power of their bodies by publicly undermining patriarchal attempts at control that lead to women's bodily self-loathing. They call on women to reclaim their bleeding bodies, and the entitlement to bleed without secrecy and shame" (105–6). Bobel (2010) explains how menstrual activists, acting in their individual lives, can create change at level of menstrual discourse: "The activists subvert the precepts of the dominant narrative of menstruation and strive for an authentic autonomous embodiment. Their aim is to seize agentic menstrual consciousness from the docile, disciplined body and stimulate new ways of knowing and being that neither shame nor silence" (41).

Yet, to resist the menstrual concealment imperative on an individual level, a woman has to resist the internalization of her objectified menstrual body and resulting self-discipline in the form of "menstrual management." Thus, I argue that menstrual management of any kind, even with environmentally conscious do-it-yourself, reusable products, is a defining characteristic of the menstrual concealment imperative because "management" is a form of concealment. As Persdotter (2020, this volume) argues with her concept of menstronormativity, we exist in and simultaneously produce menstrual norms so that it is hard to operate outside the boundaries of this power. Foucault also struggled with the possibility of how to transgress the power of discourse while inside discourse; one possibility for imagining resistance to the menstrual concealment imperative is via his work on resistance as counter-power (Pickett 1996). Free bleeding, or the refusal to use products to collect menstrual blood, is one possible form of women's resistance to the menstrual concealment imperative. In fact, free bleeding as a movement is a form of collective unity and activism among menstruators against menstrual stigma, shame, and the culture of concealment that fuels the need for menstrual management and menstrual "hygiene" products that are increasingly commodified in capitalist cultures (for example, Bobel 2006; Fahs 2016; Lapekas 2013).

In conclusion, in order to resist the concealment imperative at the level of discourse, we must be able to locate it as just one possibility of relating to our menstrual bodies; in order to contest menstrual concealment as imperative, we must locate the imperative as a false truth that appears as all-encompassing because it serves to keep women simultaneously tied to and alienated from their bodies as part of what it means to be "good." The menstrual concealment imperative is a body project (Brumberg 1998) that keeps women in a psychological state of self-hatred and constantly preoccupied with their physical bodies as a way to keep women busy and "in their place." After all, the menstrual concealment imperative is rooted in menstrual taboos and stigmas based on men's fear of women's menstruation (Delaney et al. 1988; Guterman et al. 2007) and women's own self-internalized fear of their menstrual bodies. The menstrual concealment imperative has implications to understand the various ways in which women's bodies are regulated at sociocultural and individual levels. As women's ability to control their own bodies is increasingly under political attack, it is critical to illuminate the ways in which women's disembodiment and willingness to distance themselves from their authentic experiences feeds patriarchal control of women's bodies and therefore their lives. If menstrual concealment can be disentangled from menstrual discourse that dictates self-surveillance and self-objectification of women's self-shamed bleeding bodies, the possibility exists for women to navigate their menstrual experiences with embodied subjectivity.

NOTE

1. I acknowledge the inherent risks associated with essentializing 'women' as menstruators, and yet the feminization of women's reproductive bodies as polluted and diseased contributes to menstrual concealment as imperative for female bodies. See Bobel (2010, 11–13) for a discussion of the gendered language around menstruators.

REFERENCES

Armstrong, David. 1995. "The Rise of Surveillance Medicine." *Sociology of Health and Illness* 17 (3): 393–404.

Bartky, Sandra Lee. 1990. *Femininity and Domination: Studies in the Phenomenology of Oppression*. New York: Routledge.

———. 2014. "Foucault, Femininity, and the Modernization of Patriarchal Power." In *The Politics of Women's Bodies: Sexuality, Appearance, and Behavior*, 4th ed., edited by Ruth Weitz and Kwan, Samantha, 64–85. Oxford: Oxford University Press.

de Beauvoir, Simone. 1952. *The Second Sex.* Translated by H. M. Parshley. New York: Vintage Books.

Bobel, Chris. 2006. "Our Revolution Has Style: Contemporary Menstrual Product Activists 'Doing Feminism' in the Third Wave." *Sex Roles* 54: 331–45.

———. 2010. *New Blood: Third Wave Feminism and the Politics of Menstruation.* New Brunswick, NJ: Rutgers University Press.

Bordo, Susan R. 1989. "The Body and the Reproduction of Femininity: A Feminist Appropriation of Foucault." In *Gender/Body/Knowledge: Feminist Reconstructions of Being and Knowing,* edited by Susan Bordo and Jaggar, Allison, 13–33. New Brunswick: Rutgers University Press.

———. 1990. "Feminism, Postmodernism and Gender Skepticism." In *Feminism/ Postmodernism,* edited by L. Nicholson, 133–56. New York, NY: Routledge.

Bramwell, R. 2001. "Blood and Milk: Constructions of Female Bodily Fluids in Western Society." *Women & Health* 34 (4): 85–96.

Brooks-Gunn, J., and D. N. Ruble. 1982. "The Development of Menstrual-Related Beliefs and Behaviors during Early Adolescence." *Child Development* 53 (6): 1567–77.

Brumberg, Joan. *1998. The Body Project: An Intimate History of American Girls.* New York: Random House Inc.

Buckley, Thomas, and Alma Gottlieb. 1988. *Blood Magic: The Anthropology of Menstruation.* Berkeley: University of California Press.

Chrisler, Joan (editor). 2004. *From Menarche to Menopause: The Female Body in Feminist Therapy.* New York: Routledge.

Cicurel, I. E. 2000. "The Rabbinate Versus Israeli (Jewish) Women: The Mikvah as a Contested Domain." *Nashim: A Journal of Jewish Women's Studies* 3: 164–90.

Clarke, Adele E., Laura Mamo, Jennifer Fosket, Jennifer Fishman, and Janet K. Shim. 2010. *Biomedicalization: Technoscience, Health, and Illness in the U.S.* Durham: Duke University Press.

Conrad, Peter. 1992. "Medicalization and Social Control." *Annual Review of Sociology* 18: 209–32.

Delaney, J., M. J. Lupton, and E. Toth. 1988. *The Curse: A Cultural History of Menstruation.* Urbana: University of Illinois Press.

Ehrenreich, Barbara, and Deidre English. 2005. *For Her Own Good: Two Centuries of the Experts Advice to Women.* New York: Anchor Books.

Erchull, Mindy J. 2013. "Distancing Through Objectification? Depictions of Women's Bodies in Menstrual Product Advertisements." *Sex Roles* 68 (1–2): 32–40.

Erchull, Mindy, Joan Chrisler, J. A. Gorman, and Ingrid Johnston-Robledo. 2002. "Education and Advertising: A Content Analysis of Commercially Produced Booklets about Menstruation." *Journal of Early Adolescence* 22: 455–74.

Fahs, Breanne. 2014. "'Freedom to' and 'Freedom from': A New Vision for Sex-Positive Politics." *Sexualities* 17 (3): 267–90.

———. 2016. *Out for Blood: Essays on Menstruation and Resistance.* Albany: SUNY Press.

Forbes, Gordon, Leah E. Adams-Curtis, Kay B. White, and Katie M. Holmgren. 2003. "The Role of Hostile and Benevolent Sexism in Women's and Men's Perceptions of the Menstruating Woman." *Psychology of Women Quarterly* 27 (1): 58–63.

Foucault, Michel. 1977. *Discipline and Punish: The Birth of the Prison.* Translated by Alan Sheridan. New York: Pantheon Books.

————. 1984. *The Foucault Reader*. Edited by Paul Rainbow. New York: Pantheon.

Fredrickson, B. L., and Tomi-Ann Roberts. 1997. "Objectification Theory: Toward Understanding Women's Lived Experiences and Mental Health Risks." *Psychology of Women Quarterly* 21: 173–206.

Golub, Sharon. 1992. *Periods: From Menarche to Menopause*. Riverside, CA: Sage.

Grose, Rose G., and Shelly Grabe. 2014. "Sociocultural Attitudes Surrounding Menstruation and Alternative Menstrual Products: The Explanatory Role of Self-Objectification." *Health Care for Women International* 35: 677–94.

Guterman, Mark, Payal Mehta, and Margaret Gibbs. 2007. "Menstrual Taboos among Major Religions." *The Internet Journal of World Health and Societal Politics* 5 (2): 1–7.

Houppert, Karen. 1999. *The Curse: Confronting the Last Unmentionable Taboo: Menstruation*. New York: Farrar, Strauss, and Giroux.

Jackson, Theresa E., and Rachel J. Falmagne. 2013. "Women Wearing White: Discourses of Menstruation and the Experience of Menarche." *Feminism & Psychology* 23 (3): 379–98.

Johnston-Robledo, Ingrid, M. Ball, K. Lauta, and A. Zekoll, 2003. "To Bleed or Not to Bleed: Young Women's Attitudes toward Menstrual Suppression." *Women & Health* 38 (3): 59–75.

Johnston-Robledo, Ingrid, J. Voigt, K. Sheffield, and J. Wilcox-Constantine. 2007. "Reproductive Shame: Self-Objectification and Women's Attitudes toward Their Reproductive Functioning." *Women & Health* 46 (1): 25–39.

Johnston-Robledo, Ingrid, and Joan Chrisler. 2013. "The Menstrual Mark: Menstruation as Social Stigma." *Sex Roles* 68: 9–18.

Jurgens, Janice J., and Bethel A. Powers. 1991. "An Exploratory Study of the Menstrual Euphemisms, Beliefs, and Taboos of Head Start Mothers." In *Menstruation, Health, and Illness*, edited by Diana L. Taylor and Nancy F. Woods. New York, NY: Hemisphere Publishing Corp.

Kaufert, Patricia, and Penny Gilbert. 1986. "Women, Menopause, and Medicalization." *Culture, Medicine, and Psychiatry* 10: 7–21.

Kissling, Elizabeth. 1996. "'That Just a Basic Teen-Age Rule': Girls' Linguistic Strategies for Managing the Menstrual Communication Taboo." *Journal of Applied Communication Research* 24: 292–309.

————. 2006. *Capitalizing on the Curse: The Business of Menstruation*. London: Lynne Rienner Publishers.

Lapekas, J. M. 2013. "Red Moon Rising: Breaking the Cycle of Menstrual Shaming through Countercultural Rhetoric." Retrieved from Proquest Dissertations Publishing

Laws, S. 1990. *Issues of Blood: The Politics of Menstruation*. London, UK: Macmillan.

Lee, Janet. 1994. "Menarche and the (Hetero)Sexualization of the Female Body." *Gender and Society* 8 (3): 343–62.

Lippman, Abby. 2004. "Women's Cycles for Sale: Neomedicalization and Women's Reproductive Health." *Canadian Women's Health Network Magazine* 6/7: 41.

Martin, Emily. 1992. "Medical Metaphors of Women's Bodies: Menstruation and Menopause." In *The Woman in the Body* by Emily Martin, 16–40. Boston: Beacon Press.

Patterson, Ashly. 2014. "The Social Construction and Resistance of Menstruation as a Public Spectacle." In *Illuminating How Identities, Stereotypes and Inequalities Matter through Gender Studies*, edited by Farris, D. Nicole, D'Lane R. Compton, and Mary Ann Davis, 91–108. New York: Springer.

Persdotter, Josefin. 2020. "Towards a Definition of Menstronormativity." In *The Palgrave Handbook of Critical Menstrual Studies* (this volume).

Pickett, Brent L. 1996. "Foucault and the Politics of Resistance." *Polity* 28 (4): 445–66.

Pylpa, Jen. 1998. "Power and Body Practice: Applying the Work of Foucault to an Anthropology of the Body." *Arizona Anthropologist* 13: 21–36.

Roberts, Tomi-Ann. 2004. "Female Trouble: The Menstrual Self-Evaluation Scale and Women's Self-Objectification." *Psychology of Women Quarterly* 28: 22–26.

———. 2017. Personal communication, The Society for Menstrual Cycle Research, Kennesaw State University Atlanta GA, June 23, 2017.

Roberts, Tomi-Ann, Jamie Goldenberg, Cathleen Power, and Tom Pyszczynski. 2002. "'Feminine Protection': The Effects of Menstruation on Attitudes towards Women." *Psychology of Women Quarterly* 26: 131–39.

Roberts, Tomi-Ann, and Patricia L. Waters. 2004. "Self-Objectification and That 'Not So Fresh' Feeling': Feminist Therapeutic Interventions for Healthy Female Embodiment." *Women & Therapy* 27 (3/4): 5–21.

Rozin, Paul, and April Fallon. 1987. "A Perspective on Disgust." *Psychological Review* 94 (1): 23–41.

Ruzek, C. B. 1978. *The Women's Health Movement: Feminist Alternative to Medical Control.* New York: Praeger.

Stubbs, Margaret, and Daryl Costos. 2004. "Negative Attitudes toward Menstruation: Implications for Disconnection Within Girls and Between Women." *Women & Therapy* 27 (3/4): 37–54.

Thomas. 2007. "Tips and Advice on Choosing a Pad." Accessed June 22, 2017. https://always.com/en-us/tips-and-advice/choosing-a-pad.

Thornton, Leslie-Jean. 2013. "Time of the Month" on Twitter: Taboo, Stereotype and Bonding in a No-Holds-Barred Public Arena." *Sex Roles* 68: 41–54.

Tiefer, Leonnore. 1995. *Sex Is Not a Natural Act and Other Essays.* New York: Routledge.

Ussher, Jane M. 1989. *The Psychology of the Female Body.* London: Routledge.

———. 2006. *Managing the Monstrous Feminine: Regulating the Reproductive Body.* New York: Routledge.

———. 2017. "Unspeakable Womanhood: Experiences and Constructions of Menstruation in Migrant and Refugee Women." Paper presented at the Society for Menstrual Cycle Research, Kennesaw State University Atlanta GA, June 24, 2017.

Vostrel, Sharra. 2008. *Under Wraps: A History of Menstrual Hygiene Technology.* Maryland: Lexington Books.

Wood, Jill M., Patricia Barthalow Koch, and Phyllis Kernoff Mansfield. 2007. "Is My Period Normal? How College-Aged Women Determine the Normality or Abnormality of Their Menstrual Cycles." *Women & Health* 46 (1): 41–56.

Young, Iris Marion. 1997. "Menstrual Meditations." In *On Female Body Experience: Throwing Like a Girl and Other Essays by Iris Marion Young*, 97–122. New York: Oxford University Press.

CHAPTER 26

Transnational Engagements: From Debasement, Disability, and Disaster to Dignity—Stories of Menstruation Under Challenging Conditions

Edited by Milena Bacalja Perianes and Tomi-Ann Roberts

INTRODUCTION

While there has been a significant rise in attention paid to the menstrual experience, there are a variety of voices and bodies that remain inaudible and invisible. Those neglected are typically the most vulnerable and therefore less able to comfortably and confidently manage their menstrual health. The double stigma they face—as both menstruator and marginal—exacerbates their precarity and creates further boundaries to their health and wellbeing.

This chapter sheds light on the experiences of menstruators under a few such challenging conditions. Through the lens of three women, one living with disabilities, a second incarcerated and a third living in a disaster zone, we can begin to understand how they manage their menstrual needs. Their stories are a powerful (and at times disquieting) reminder of the need for equitable menstrual policies, diverse product design and more thoughtful and inclusive responses to women and girls at all stages, and in all situations.. Because situations constantly shift and evolve, innovative, inclusive and iterative responses must also be dynamic.

Ultimately, these first-person narratives challenge us to think about others muted in our current menstrual discourse and how we might bring them from margin to center. While we should be proud of the increased visibility of menstruation as an issue worthy of research and investment, we must redouble our efforts to be truly and profoundly inclusive.

© The Author(s) 2020
C. Bobel et al. (eds.), *The Palgrave Handbook of Critical Menstruation Studies*, https://doi.org/10.1007/978-981-15-0614-7_26

From Behind the Wall: Menstruating While Incarcerated

Janelle Chambers

Let me tell you a bit about myself. I am a 33-year-old mother of three children aged 2–13 and am happily married to a very loving wife. I had a very strong Christian upbringing and I love the Lord. I was raised in the church and had to deal with a lot in life including being constantly judged for my lifestyle. There were a lot of things that were not spoken about when I was young because it was not considered the "right speech." This has had a large impact on growing up, and I think, my experience of menstruating while being incarcerated.

No matter what a woman's upbringing, religious beliefs, sexual orientation, values or morals are, in life the one thing we have in common is menstruating every month regardless if we like it or not. So what is it like to have your period while incarcerated? The one thing we as women all have in common is that periods suck! It sucks even more when you are in a place you never dreamed of being. So not only do you have a million and one thoughts racing through your head, being fearful of the unknown and unexpected, but then BAM here comes Aunt Flo to visit.

When your period arrives you suddenly find yourself in a room with a toilet that is shared with another woman watching you. Are there tampons or pads under the bathroom sink . . . umm I think not, this is jail. The door is locked so what do you do? Do you bleed everywhere or wrap toilet paper up and lay it on your panties? Sometimes incarcerated women choose to lay toilet paper in their underwear until the dayroom opens. That's *if* it does.

Managing your menstruation is a very public experience when incarcerated. In order to get some no-name brand sanitary napkins women have to be good. We can only access products in the dayroom, which is a shared space were inmates congregate and where necessary or luxury items are purchased or distributed. Everyone knows when you are on your period because your only option to get a product is to go to the dayroom to request them, then take them back to your cell. How embarrassing! Imagine, being with hundreds of women you don't know. You don't know how many of them will have an off day that may lead to a fight. You don't know who will be making comments among themselves about the "bleeding inmate." There are deputies that like to belittle you and treat you like a caged animal. We have to learn to be okay with it, to be desensitized to this kind of treatment—this is the price we pay for the crimes we committed. There is a lot of shame here. "Oh the convict needs a pad for her bloody period," the guards broadcast to the whole dayroom. The way we are treated makes us feel more shame than for what we have already done.

To make matters worse the dayroom visit usually takes up your time so you can't shower, which means everyone remembers you as the stinky bloody chick that didn't get to take her (5 minute) shower when the dayroom was open. When you are back in your cell you have to hope your roommate is cool and gives you space to deal with it.

I was fortunate when I was first incarcerated to have good roommates. They each taught me valuable things that helped me on my journey while in jail. Learning how to manage your period was a part of that. I learned how to wash up in the small sink in our room that was connected to the toilet. This same sink was used to brush our teeth, wash our face and hands. We had a routine when one of us got our period which helped. We were lucky to have a good rapport with each other. An inmate was even luckier when they developed a good relationship with one or more deputies to be able to shower when other inmates were sleeping.

Hygiene was a big concern for me, for us all. Of course problems arise when we have low-quality products, and limited access to water to manage our menstruation. I believe my fibroids were caused by my incarceration and having to use things no woman should ever have to use. Who thinks of the health risks at a time and place like that? All I was concerned about was having my hygiene up to par and not having a bloody mess. As we only get to change out of our county blues once a week, it's important to keep them as clean as possible. Can you imagine if you had an accident from Aunt Flo and you now have to wait a week for a clean pair of clothes? With only one pair of panties to hold your napkin in place, you have to be careful.

Menstruation is an ongoing issue you regularly have to handle. Think about it. It's your first court hearing, now you're thinking, "yes! I finally get a break. I can breathe some fresh air and hopefully see my family members, or use the phone in the holding tank." Of course that sounds fantastic until reality hits you when you get off the bus from court back at the jailhouse. You are going to be strip-searched, but you have your period.

The strip search process when you reenter jail after a court hearing is particularly humiliating. They always want to make sure you are not bringing anything back into the facility after having contact with the outside world. Sometimes they line up 100 women and make all of us remove our clothes and stand next to each other, in the nude, being degraded and humiliated. Now you are told to run your dirty hands through your hair. Lift up your breasts while the deputy runs their hands underneath. They tell you to run in place to see if anything will fall from anywhere. They make you bend over, squat and cough. They tell you to take out anything in your Va-jay-jay, like pads or tampons, and throw it in front of you in the center of the room where all can see it.

So the center of the room becomes one big open trash bin of bloody products that some of the inmates have to later clean up. "Nasty, dirty, filthy pigs" was a famous line the deputies would say to us. "You B***** stink, no wonder you're here because nobody out there wanted you." It never matters who you were, us women inmates were treated the same. Not only was our freedom taken from us, but now our dignity ripped right from underneath us. Some women there had never even been intimate with another person and then were forced to expose their most sacred place. We didn't mean anything so why should anyone care. I stood next to a nun once and all she could do was cry and repeat *The Lord's Prayer*. If you were menstruating it didn't matter if blood just ran down your legs while you stood there. You just wanted the humiliation and nightmare to end and hope nobody remembered what you looked like.

I can probably sit here and tell you of the 1546 days of torment and torture I endured, but this is a very hard state of mind to go back to that I have spent many years dealing with, on top of my many other medical issues. I really hope and pray that I have shed some light on the many different emotional effects of dealing with menstruating while being incarcerated. I am grateful to be able to be the voice of so many of my sisters who have, and who are still, dealing with these challenges. We now have a voice for us that may be heard a little better with the help of this book.

DESIGNING MENSTRUAL PRODUCTS FOR DISABLED BODIES

Jane Hartman Adamé

It somehow slips society's mind that disabled people can menstruate, be considered sexually attractive, have sex, and even reproduce. Of course, some of these are not true of all disabled people, but neither are they true of all able-bodied individuals. All of these same complexities that an able-bodied person can experience are also possible for disabled individuals like myself.

Adding to the complexity of menstruating with a disability is that products that are made to manage and support our menstrual health and wellness are often made without any consideration of our specific experiences. As a result, it can be a struggle to find products to meet your individual needs. Luckily, changes are happening in this arena, with more innovative, unique, problem-solving products arriving to market at an increasing pace. However, we have a lot of ground to make up for.

When we consider the concept of disability and menstruation, one often-neglected concept is that menstruation itself looks a lot like disability. I'm not speaking just about dysmenorrhea, or endometriosis, although those certainly can qualify as well—but even the "simple, normal" menstrual cycle. To wrap our heads around this, let's look at two common definitions of disability, the medical model and the social model.

As a medical condition, menstruation requires intervention with (usually) approved medical devices to manage the collection of the menses itself. It can also require care and treatment for the symptoms that come along, such as pain and inflammation, by over-the-counter pain meds, warm compresses or packs, or pain creams applied to the lower abdomen or even the low back.

In a more social context, a menstruating person may need to alter the clothes they wear for the day to accommodate bloating and provide comfort, which could alter how they are perceived (or feel they are perceived) outside of the home. They may need to miss a day or two of work with severe cramps, or heavy bleeding that requires frequent product changes. These frequent bathroom trips or distracting bouts of pain can inhibit job performance, eat up sick days, and have a myriad of other impacts on a person's life. Due to the cultural shame and taboo surrounding the period, attempts are made to hide these changes, such as tucking one's tampon into a sleeve to slip unnoticed into the bathroom. Additionally, there is social pressure from advertisements that a person should be able to go forth and be productive-period or not. There is also pressure, due to the normalization of this process, to not seek medical care for menstrual issues. It's seen as something common, but even common conditions can have significant impacts on us.

The menstrual cycle, when viewed through the lens of disability, is a relapsing, recurring condition (in that menstruation itself and its uncomfortable symptoms occur approximately monthly for a number of days, and then recede for the remainder of the month) that just so happens to occur with such prevalence and regularity that its very essence is dismissed, and complaints about changing symptoms can be overlooked or ignored. Even just the fear of the possibility of being dismissed (in addition to the shame and taboo aspects of this area of our health) keeps many of our menstruating community from setting foot in a doctor's office to talk about these issues.

Evidence of delayed diagnoses of endometriosis and other serious life-affecting menstrual conditions shows us that menstruation is not taken seriously. When you add menstruation to disability, then you have a double whammy. Women's pain is expected, from all processes from typical menstruation, to atypical menstruation, to childbirth and beyond. If we allowed ourselves to look at the taboo concept of menstruation with the lens of the even more taboo concept of disability, we could start to treat both of these states of being with more care, both medically and socially.

Although I can't speak for all disabilities, living with hypermobile Ehlers-Danlos Syndrome certainly poses some challenges when it comes to managing my menstrual cycle. Chronic pain, alone, can be greatly exacerbated by menstrual pain. Add to that the risk factor of changing out menstrual products with dislocation risk, and it's clear that periods are extra challenging to manage. Another challenge that my disability poses, and which is certainly not unique to me, is pelvic floor dysfunction. Whether you have a weakened pelvic floor, prolapse, vaginismus, or other eccentric pelvic muscle activation, internally worn products can be a pain to use. Anything requiring bearing down (applying downward force to the pelvic floor by way of muscle activation) can be a risky thing to do if the pelvic floor is weakened or any organs are prolapsed. If muscles are overly tight, internally worn products can be uncomfortable to wear or pose a serious challenge when it comes to removal.

Outside of my personal experience, I've also come to learn of the challenges faced by folks who utilize carers to manage their menstrual hygiene. This can be a challenging and sensitive subject, as the carer has to directly interact with the user's menses. Of course, menstrual products are not made to be managed by anyone other than the user, which is where we find ourselves having to get creative.

Fortunately, many of us don't have to be relegated to pads—although there is nothing inherently bad about a pad. It's my belief that all menstrual care options are good options, and nobody other than you can decide what the right choice is for your body.

Personally, I do prefer internally worn products, especially those that require less frequent changes than tampons. Menstrual discs and cups can both be worn for up to 12 hours—even if you can wear your tampons for the maximum amount of time, which is 8 hours, these newer options cut down the number of product changes per day from 3 to 2. This may seem like an insignificant difference, but if you have mobility challenges, fatigue, or chronic pain, one less change per day can mean the ability to use that energy for something else.

When I lost the ability to use my menstrual cup, I was devastated. For myself and many others, the complicated maneuver required to remove a cup makes it nearly impossible (or very risky) to use. Many times, I subluxed or dislocated joints and/or set muscles into spasm, all from trying to position my body and arm in such a way that I could grasp the cup to break the seal in order to safely remove it. Instead of admitting defeat, I decided to create a cup that I and others like me could use.

With my design, now known as the FLEX Cup, I added a ReleaseRing™ which is a piece that threads from the top of the cup through the middle, ending in a soft silicone ring that moves with your body when worn, so the cup can be worn low with the ring accessible in the labial area, or more internally. Either way, since this piece is attached to the top of the cup, it indents the side when pulled and makes cup removal more akin to removing a tampon.

When we launched the cup on Kickstarter, it made huge waves—many people never considered the challenges that those of us with disabilities faced during menstruation. Others who were nervous to try cups found ease in this design. With many people, both disabled and able-bodied, the risk of potentially ending in the ER to have the cup removed was a deal breaker. I knew that eliminating this risk would be of great benefit to so many people. During the Kickstarter, The Flex Company saw value in our design and eventually bought the patent. They also added my cofounder Andy Miller and myself to their team, and together we are making great strides to make menstrual products comfortable and accessible to all, having landed on retail shelves just about a year after inception.

The other product we make, the menstrual disc, can be a great option and is a top choice for those with Ehlers-Danlos Syndrome and other mobility-limiting conditions. Since insertion doesn't require any complex folding or grip strength, insertion of a disc can be easier than a cup for some. Since this product can also be worn for up to 12 hours but through its single use, it eliminates the cleaning process, which can be taxing for many.

Additionally, the FLEX menstrual disc is the only disc available on a subscription cycle—we don't usually consider the process of going to the store to restock menstrual products as an access issue, but it is. Here, reusable options or subscription services can be great access solutions.

So, how do you determine what is the right solution for your body? I know many people flock to YouTube to watch reviews of menstrual products, but in truth, there is so much variance in our anatomies, regardless of additional conditions and factors, that there is only one true way to determine what will work best for you—and that is to give new things a try.

Menstruation in Emergencies: Developing a Period-Friendly Emergency Response

Mayuri Bhattacharjee

I am Mayuri Bhattacharjee, a menstrual health educator with experience of teaching more than 8000 women and girls about menstrual health in India, especially in the flood-affected regions in Assam, a state in India's northeast. I run a nonprofit which works on the intersections of public health and climate action and I am a Climate Reality Leader at the Climate Reality Project.

I grew up in Assam, India. A state which faces floods every year and when I attained menarche, I, like many girls in India, was subjected to menstrual taboos which made me feel impure. So, now to change this narrative, I work at the crossroads of natural disasters and menstrual health because of my own background and lived experience in Assam.

Women and adolescent girls in disaster-prone and fragile contexts face many challenges and Assam's floods are no exception to this. As you might have seen in news covering disasters, toilets, and other sanitation and hygiene facilities are hard hit during floods across the world. Menstrual hygiene practices are often compromised in the tough conditions surrounding a natural disaster. This is in part due to a lack of proper shelter, water supply, disposal and washing facilities for women and girls to safely and securely manage their periods.

In Assam, a lack of knowledge around the biological process of menstruation has exacerbated negative experiences around menstruation. Many myths exist around the topic including the common notion that menstruation is the passing of bad blood from a woman's body. In some communities, the practice of segregation during menstruation is observed even in a flood shelter. This is a risky practice because being separated during floods can expose those menstruating to health and security risks. In a worst-case scenario, kidnapping or human trafficking can occur as floods are a fertile ground for traffickers who are on the lookout for easy prey. Apart from physical risks, the notion that menstrual blood is impure and so is a menstruating woman, puts unwanted mental stress on women who are already carrying a significant burden in emergency situations.

Fundamentally, more attention needs to be paid to the intersection between climate change and women's changing health needs. While a wide variety of information and materials are being made available to women regarding menstruation, the experiences of women in emergencies remain an underserved topic.

With climate change increasingly impacting weather patterns, like many other places, annual flooding in Assam will continue to be a challenge. As researchers and practitioners, we need to think about how to address women's and girl's menstrual health needs in this context and prepare them to be able to manage their menstrual cycles during emergency situations. This is a public health issue often overlooked by local authorities. This should not be an afterthought during a crisis, rather governments and public bodies need to work on mitigation strategies which target women's specific health needs. In Assam, this can include pre-disaster planning and interventions in education and infrastructure including the construction of season embankments, and disaster specific plans for the disposal of menstrual waste. Menstrual waste disposal is a challenge even in normal situations and the problem is only exacerbated during floods.

As a result of this lack of planning or foresight from state governments, there is a tendency to apply a Band-Aid or a one size fits all approach to menstrual health gaps. Most often that solution is the blanket distribution of sanitary pads. This process often leads to a glamorization of pads and the demonization of local cloth alternatives. Rarely, are tampons or menstrual cups a part of hygiene kits because of the taboos around virginity and limited knowledge among those procuring and distributing relief materials. Instead of promoting only one product—usually non-compostable, disposable pads—government and humanitarian agencies should look at other eco-friendly options such as low-cost cloth pads and menstrual cups. Thoughtful and considered emergency responses remain a barrier to adequate menstrual health in emergency settings because menstrual health remains a women's issue, a minor issue, and not a public health concern.

I believe an important first step toward developing a period-friendly emergency response is to normalize conversations around menstruation especially among those who plan mitigation and emergency responses. We must move the conversation from just women and adolescent girls, but ensure men are also empowered to talk about menstruation without shame. The most effective step toward this is menstrual health education at a community-level and special orientation sessions for frontline humanitarian response workers.

Addressing menstrual health needs in contexts in which natural disasters occur regularly, requires a sensitive response to women's diverse experiences. At the Sikun Relief Foundation, we don't believe in promoting one kind of menstrual hygiene product, but we believe in informed choice. Women in emergency settings deserve the same choices as every other woman therefore our interventions focus on introducing a basket of menstrual hygiene products and giving women the knowledge they need to make informed decisions for the context they are in. Unlike traditional menstrual health education, our interventions start well before the flood season with a major focus on how women and girls can understand and manage their cycles before and during an emergency. Rather than dictating one solution, we believe in women and girls coming up with their own personal strategies to deal with their periods.

Co-creating WASH and MHM solutions and frameworks together with women and adolescent girls would be more effective in developing a gender—sensitive response mechanism during emergencies because women's experiences are central to the effectiveness of disaster management planning and response.

PART III

Menstruation as Rationale

CHAPTER 27

Introduction: Menstruation as Rationale

Breanne Fahs

Menstruation has been used as an excuse in many contexts. There is much legal and historical precedent linking menstruation to the category *woman*, and the limited political rights offered to persons therein. This section looks, in part, at the ways that menstruation has been used as a rationale to purposely curtail women's political rights, access to legal processes, and/ or benefits of citizenship. The way that stories of exclusion are built, often through norms of constructing menstruation as disabling and as a liability, constitute the core of this section.

Politicians have invoked the menstrual rationale to make clear that women do not belong: G. Gordon Liddy on Sonia Sotomayor and her confirmation to the US Supreme Court; Newt Gingrich on women having "biological problems" being out on the battle field for thirty days; Donald Trump on debate moderator Megyn Kelly who had "blood coming out of her whatever." They all used menstruation to signal *woman*, and in particular, woman "out of place." This rationale also appears in court cases, such as the use of PMS as a legal defense, to rules about when women can and cannot participate in sports or participate or enter religious sites. Further, the ways that these stories getting stacked upon one another—the military feeding stories of exclusion to the media, schools feeding stories of exclusion to the institutional practices of the family—mean that pushing women into ever-more-narrow boxes has resulted in a diminishing of the menstrual experience.

Conversely, the rationale can also be used as a source of bodily assertion and inspiration to challenge the menstrual *status quo*. No one accomplished this with more wit and incisiveness than Gloria Steinem with her short satirical essay, "If Men Could Menstruate," which succinctly underscored meanings of menstruation by turning the tables on men, revealing the ironic cultural constructions argued as innate. Steinem's essay was both

© The Author(s) 2020
C. Bobel et al. (eds.), *The Palgrave Handbook of Critical Menstruation Studies*, https://doi.org/10.1007/978-981-15-0614-7_27

funny and furious, drawing upon both humor and a serious consideration of the risks of leaving the status quo intact. This approach includes the rapidly growing constellation of activist efforts that seek to wrest menstruation from, for example, the Femcare industry. These alternative discourses empower menstruators themselves to redefine the meaning of menstruation and its embodiment at political, spiritual, and social levels.

Chapters in this section, then, explore the menstrual rationale, as well as critiques of the uses and abuses of the persistent and global menstrual taboo to market an ever expanding array of menstrual care products, from tampons to pharmaceuticals such as cycle-stopping contraceptives, a.k.a., menstrual suppressants and hormone therapies for menopausal women. These chapters include theoretical essays, cultural studies analyses, substantive literature reviews, historical reviews, and personal reflections, all of which reveal the breadth of ways in which scholars and writers have conceptualized the menstrual rationale.

We begin with Josefin Persdotter when she evokes the figure of the "menstrual monster" to challenge and reorient ideas about menstrunormativity, particularly as she welcomes and celebrates the idea of the "monstrous" menstruator as a figure that smashes rationales for excluding women from social and political rights. To imagine menstrual "normativity" is, in a sense, to also conceptualize menstrual *non*-normativity, a concept with many hazardous implications for the diverse group of people who menstruate.

In this light, Ela Przybylo and Breanne Fahs show the insidious workings of menstrual stories through menstrual product advertisements, revealing the ways that corporations and capitalism draw upon "empowerment" rhetoric only to further stigmatize menstruators. By foregrounding happy, able-bodied, celebratory periods, menstrual product advertisements ultimately shape narratives that excludes menstrual crankiness and any body/shape/size outside of the narrow "ideal." In essence, only certain forms of emotional and affective experience are welcomed and allowed within the corporate framing of menstruation. Connected to this embrace of menstrual crankiness, Maureen McHugh argues for the importance of critically examining "menstrual moaning," or the ways that women talk negatively about their periods. Linking menstrual moaning to fat talk, McHugh asks whether we should embrace menstrual moaning or move toward more positive visions of menstrual experiences. In both chapters, notions of how women are excluded from *disliking* or *liking* their periods, and how emotional experiences of menstruation are regulated closely and intensely, constitute a core question in these chapters.

We next consider from a psychological perspective the much-understudied framework of men and boy's attitudes about menstruation, particularly as they internalize ideas of menstruation as a rationale for sexist notions of women's "out of control" moods, dismissal of women's pain or discomfort, and silence around the menstrual experience. The limited ways that boys learn about menstruation—or are expected to know about it—shows both how boys and men are not framed as invested in women's bodies and health, but also how menstruation becomes a key criteria for how men learn to discount and insult women.

We then move to a more global focus on the menstrual rationale, starting with Milena Bacalja Perianes and Dalitso Ndaferankhande reflecting on menarche rituals in Malawi and how starting to menstruate situates girls in a double bind: they both *become* women and thereby gain power and status, but they also face narrow social scripts about how they can behave and what menstruation means. In the following chapter, Sheryl Mendlinger offers a personal reflection essay in which she considers the span of her 30-year career studying menstrual health alongside ideas about cultural differences between Israel and the United States. Drawing on her work interviewing 48 mother–daughter dyads from Israel, including many immigrant women, she encourages a better understanding of the ways that knowledge of menstruation is acquired and how menarche stories contain a wide range of emotional experiences that speak not only to women and their bodies but also to the movement of immigration itself. This powerful back and forth—between empowerment and constraint, newfound freedoms and entrenched traditions—represents broader stories about menstruation as rationale.

This section concludes with a transnational engagements chapter that looks at menstrual health education throughout the world, detailing the wide range of approaches to studying, managing, and disseminating menstrual knowledge that is situated in diverse cultural contexts. This chapter features people speaking together from different cultural and geographical backgrounds to better understand how to teach people about menstruation. Ultimately, this transnational engagements chapter shows the power of an "all hands on deck" mentality to menstrual inclusion, as we learn about the fusion in menstrual health education between public and private sector, formalized non-profits and scrappy activism, medical doctors and local teachers, and fancy technology and folk wisdom. It is a reminder about how to best topple the barriers that women (and all menstruators) face when menstruation is used as a rationale for exclusion: go to the root of the problem and work collaboratively and in the spirit of social justice for all.

CHAPTER 28

If Men Could Menstruate

Gloria Steinem

Living in India made me understand that a white minority of the world has spent centuries conning us into thinking a white skin makes people superior, even though the only thing it really does is make them more affected by ultraviolet rays and wrinkles.[1]

Reading Freud made me just as skeptical about penis envy. The power of giving birth makes "womb envy" more logical, and an organ as external and unprotected as the penis makes men very vulnerable indeed.

But listening recently to a woman describe the unexpected arrival of her menstrual period (a red stain had spread on her dress as she argued heatedly on the public stage) still made me cringe with embarrassment. That is, until she explained that, when finally informed in whispers of the obvious event, she said to the all-male audience, "and you should be *proud* to have a menstruating woman on your stage. It's probably the first real thing that's happened to this group in years!"

Laughter. Relief. She had turned a negative into a positive. Somehow her story merged with India and Freud to make me finally understand the power of positive thinking. Whatever a "superior" group has will be used to justify its superiority, and whatever and "inferior" group has will be used to justify its plight. Black men were given poorly paid jobs because they were said to be "stronger" than white men, while all women were relegated to poorly paid jobs because they were said to be "weaker." As the little boy said when asked if he wanted to be a lawyer like his mother, "Oh no, that's women's work." Logic has nothing to do with oppression.

So what would happen if suddenly, magically, men could menstruate and women could not?

Clearly, menstruation would become an enviable, boast-worthy, masculine event:

C. Bobel et al. (eds.), *The Palgrave Handbook of Critical Menstruation Studies*, https://doi.org/10.1007/978-981-15-0614-7_28

Men would brag about how long and how much.

Young boys would talk about it as the envied beginning of manhood. Gifts, religious ceremonies, family dinners, and stag parties would mark the day.

To prevent monthly work loss among the powerful, Congress would fund a National Institute of Dysmenorrhea. Doctors would research little about heart attacks, from which men were hormonally protected, but everything about cramps.

Sanitary supplies would be federally funded and free. Of course, some men would still pay for the prestige of such commercial brands as Paul Newman Tampons, Muhammad Ali's Rope-a-Dope Pads, John Wayne Maxi Pads, and Joe Namath Jock Shields—"For Those Light Bachelor Days."

Statistical surveys would show that men did better in sports and won more Olympic medals during their periods.

Generals, right-wing politicians, and religious fundamentalists would cite menstruation ("*men*-struation") as proof that only men could serve God and country in combat ("You have to give blood to take blood"), occupy high political office ("Can women be properly fierce without a monthly cycle governed by the planet Mars?"), be priests, ministers, God Himself ("He gave this blood for our sins"), or rabbis ("Without a monthly purge of impurities, women are unclean").

Male liberals and radicals would insist that women are equal, just different; and that any woman could join their ranks if only she were willing to recognize the primacy of menstrual rights ("Everything else is a single issue") or self-inflict a major wound every month ("You *must* give blood for the revolution").

Street guys would invent slang ("He's a three-pad man") and "give fives" on the corner with some exchange like, "Man you lookin' *good!*"

"Yeah, man, I'm on the rag!"

TV shows would treat the subject openly. (*Happy Days:* Richie and Potsie try to convince Fonzie that he is still "The Fonz," though he has missed two periods in a row. *Hill Street Blues:* The whole precinct hits the same cycle.) So would newspapers. (SUMMER SHARK SCARE THREATENS MENSTRUATING MEN. JUDGE CITES MONTHLIES IN PARDONING RAPIST.) And so would movies. (Newman and Redford in *Blood Brothers!*)

Men would convince women that sex was *more* pleasurable at "that time of the month." Lesbians would be said to fear blood and therefore life itself, though all they needed was a good menstruating man.

Medical schools would limit women's entry ("they might faint at the sight of blood").

Of course, intellectuals would offer the most moral and logical arguments. Without the biological gift for measuring the cycles of the moon and planets, how could a woman master any discipline that demanded a sense of time, space, mathematics—or the ability to measure anything at all? In philosophy and religion, how could women compensate for being disconnected from the rhythm of the universe? Or for their lack of symbolic death and resurrection every month?

Menopause would be celebrated as a positive event, the symbol that men had accumulated enough years of cyclical wisdom to need no more.

Liberal males in every field would try to be kind. The fact that "these people" have no gift for measuring life, the liberals would explain, should be punishment enough.

And how would women be trained to react? One can imagine right-wing women agreeing to all these arguments with a staunch and smiling masochism. ("The ERA would force housewives to wound themselves every month": Phyllis Schlafly. "Your husband's blood is as sacred as that of Jesus—and so sexy, too!": Marabel Morgan.) Reformers and Queen Bees would adjust their lives to the cycles of the men around them. Feminists would explain endlessly that men, too, needed to be liberated from the false idea of Martian aggressiveness, just as women needed to escape the bonds of "menses-envy." Radical feminists would add that the oppression of the non-menstrual was the pattern for all other oppressions. ("Vampires were our first freedom fighters!") Cultural feminists would exalt a bloodless female imagery in art and literature. Socialist feminists would insist that, once capitalism and imperialism were overthrown, women would menstruate, too. ("If women aren't yet menstruating in Russia," they would explain, "it's only because true socialism can't exist within capitalist encirclement.")

In short, we would discover, as we should already guessed, that logic is in the eye of the logician. (For instance, here's an idea for theorists and logicians: If women are supposed to be less rational and more emotional at the beginning of our menstrual cycle when the female hormone is at its lowest level, then why isn't it logical to say that, in those few days, women behave the most like the way men behave all month long? I leave further improvisation up to you.)

The truth is that, if men could menstruate, the power justifications would go on and on.

If we let them.

Note

1. "If Men Could Menstruate" by Gloria Steinem was first published in October 1978 in *Ms.* Magazine. Reprinted with permission. No further reproduction or distribution of the material is allowed without permission from the publisher.

CHAPTER 29

Introducing Menstrunormativity: Toward a Complex Understanding of 'Menstrual Monsterings'

Josefin Persdotter

MEET THE MENSTRUAL MONSTER

You've met it before. "The menstrual monster" pops up in popular culture as the stereotypical (pre)menstrual woman, "a frenzied, raging beast . . . prone to rapid mood swings and crying spells, bloated and swollen from water retention, out of control, craving chocolate, and likely at any moment to turn violent" (Chrisler et al. 2006, 371). Many have also used the terms "monster" and "monstrous" to refer to otherized, marginalized, liminal beings, who in one way or another stand outside the realms of "normality" (for example, Hughes 2009; Shildrick 2002). Many critical menstrual scholars implicitly or explicitly argue that menstruators and menstruating have in that more metaphorical sense been positioned as monstrous through communication taboos (Kissling 1999), pathologization and medicalization (Mamo and Fosket 2009; Martin 2001; Gunson 2010; Oinas 1998; Ussher 2006), stigmatization (Kowalski and Chapple 2000; Johnston-Robledo and Chrisler 2013), and concealment imperatives (Wood 2019; Young 2005; Laws 1990). Such menstrual *monstrosity* is often burdensome to the monstrous subject as they are positioned as abnormals, outsiders, deemed less human (more monstrous) than that which adheres to normative ideals. Marion Yong has argued that since menstruating is a stigmatized and monstrous position, menstruators are effectively positioned "in the closet" trying to "pass" as non-menstruators (Young 2005). Correspondingly, Jane Ussher argues that women and menstruators are tasked with controlling the menstrual "unruly" monstrous body, and if they fail they are "at risk of being positioned as mad or bad, and subjected to discipline or punishment" (Ussher 2006, 4). Ussher also argues

© The Author(s) 2020
C. Bobel et al. (eds.), *The Palgrave Handbook of Critical Menstruation Studies*, https://doi.org/10.1007/978-981-15-0614-7_29

that managing this monstrosity is a cumbersome, time-consuming, self-oppressive task (Ussher 2006). Menstrual monstrosity has concrete negative effects, for example impacting body-shame (Johnston-Robledo et al. 2007), increasing sexual risk behavior (Schooler et al. 2005; Rembeck 2008), and even negatively effecting the experience of birth (Moloney 2010). By discussing and deconstructing such consequences, the menstrual countermovement (Persdotter 2013) has shed light on the possibility of alternatives, that it can be otherwise. In this text I seek to assemble, assist, and advance this tradition through trying to make visible the complex ways in which menstrual monsterings work through introducing a concept that I call *menstrunormativity*. Because while menstruation is surely "having its moment" (Bobel 2017) all around the world, it is crucial that we maintain a critical eye on dominations and marginalizations of menstruation and menstruators and explore more ways of understanding how they work.

Through introducing menstrunormativity I make four interlinked arguments: (1) that normativities work in clustered, complex, and contradictory ways, (2) the cluster of normativities that surround menstruation produce an impossible ideal subjectivity (the imagined *menstrunormate*) which follows that we are all actually *menstrual monsters*, that (3) normativities are *continuously co-produced* by every*one* and every*thing*, which means we are all, always, culpable in creating normativities (and monsters) and lastly (4) embracing ourselves (we scholars, activists, retailers, menstruators, feminists, parents, children) as both Doctor Frankenstein and as monsters carries significant potential: producing more possibilities for livable lives for both the menstrual countermovement (who can see itself and others in more nuance and imagine ever stranger bedfellows) and menstruators (who can imagine more ways of being menstruator, and feel less bad about their inevitable menstrual monstrosity).

INTRODUCING MENSTRUNORMATIVITY

I propose we use the neologism *menstrunormativity* to refer to the hegemonic social system of multiple and contradictory normativities that order and stratify menstruation and menstruating. It is the multitude of entwined social/medical/statistical norms, discourses and imperatives that construct certain ways of understanding and experiencing menstruation as ideal/correct/healthy/normal and morally superior, and others as wrong/unhealthy/abnormal, or *monstrous*. The term menstrunormativity draws from conceptual predecessors such as *heteronormativity*, *cisnormativity*, *homonormativity*, and *bodynormativity* that all highlight ways in which dominant social systems stratify certain aspects of life: positioning some sexualities, subjectivities, and bodies as Other, unnatural, abject and some (heterosexuals, cisgendered, et cetera.) as natural, correct, and privileged, that is the norm (Robinson 2016; Nord, Bremer, and Alm 2016). I argue that norms, ideals, and imperatives around menstruation do very similar things: they position some menstrual subjectivities, some menstrual bodies, some menstrual behavior as ideal, correct and good, and some as abnormal, unhealthy, disgusting. And—I think this is an important accentuation—I argue that these (as all) normativities work in clustered ways: coming at menstruators from all sides, pushing, compressing, and limiting menstrual existence in contradictory and even paradoxical ways. Menstruators are told to simultaneously: Don't tell anyone you're menstruating! But be proud of your functional body! It's perfectly natural to bleed! That's gross, conceal!

Conceptual Building Blocks

The Latin term *norma*, the root of words such as norm, normal(ity) and normativity, means approximately rule, pattern, precept, or standard (Folkmarson Käll 2009). In everyday life "norm" and "normal" are often understood as "standard" and "usual" in a statistical sense, also called *statistic normality* (Tideman 2000). In the case of menstruation, statistic normality could be the most common way to menstruate in terms of cycle length; an average level of discomfort, or that most women adhere to the local *menstrual etiquette* (Laws 1990). For example:

"most women menstruate"

"the normal menstrual cycle is 28 days"

"most menstruators do not speak of their menstruation"

The terms "normality," "norms," and "normativity" have different meanings in different disciplines. In sociology (which is my discipline) norms are understood as socio-cultural rules that regulate appearance and behavior in social systems, sanctioning behavior that goes against the norm (Johnson 2000). That is: norms—in the sociological sense—are always *normative* as they stipulate a right and a wrong way, how something *ought to be, should* be done; an endorsed and authoritative moral ideal (Folkmarson Käll 2009). Moreover, the *statistically* normal often becomes *normatively* normal in that the "normal" becomes the "good" as "the normal" turns into "an attractive normative position, which other positions are viewed against" (Niklasson 2014, 13, my translation). Thereby, the most common way to menstruate becomes the "right" way to menstruate whereas other ways are sanctioned:

"most women menstruate" → *"women who do not menstruate are not real women"*

"the normal menstrual cycle is 28 days" → *"the 32-day cycle could be a sign of illness"*

"most menstruators do not speak of their menstruation" → *"menstruators who do speak of their menstruation are weird"*

Importantly, normativity is also intimately intertwined with constructions of the *natural* in producing "bodily imperatives that are deemed socially and ethically acceptable" (Weiss 2014, 106); that it is natural for women to menstruate, that a natural cycle is 28 days, et cetera. Further, constructions of normality within medicine categorize that which is positioned as abnormal as *unhealthy* (Zeiler and Folkmarson Käll 2014, 7) this has been called *medical normality* and is often coupled with an idea of treatment as a technology to achieve normality (Tideman 2000).

Please note that I do not mean to dismiss that deviations from menstrual medical normality might be signs of poor health and a source of much physical and psychological suffering, but instead highlight their social constructions and social and emotional consequences an try to explore how they could be otherwise.

Conceptual Siblings

Heteronormativity, Cisnormativity, Homonormativity, and *Bodynormativity* are all concepts similar to how I seek to develop menstrunormativity highlight the ways in which normative powers works in a certain sphere of life. Heteronormativity has been defined as "a hegemonic social system of norms, discourses, and practices that constructs heterosexuality as natural and superior to all other expressions of sexuality" (Robinson 2016). *Cisnormativity,* a related concept (and interlinked system) is in the same way defined as a hegemonic social system but instead of sexuality focuses on the binary gender system that fundamentally assumes two static, stable, and un-mixable gender-categories: man and woman. In doing so, cisnormativity positions all other gendered/sexed identities unintelligible (Nord, Bremer, and Alm 2016). These theoretical concepts share a (trans and queer) feminist interest in understanding, conceptualizing, and critically engaging with social systems "that constitute some subject-positions as recognizable and others as abject" (Nord, Bremer, and Alm 2016, 5) and enable us to see the characteristics of large, complex, multidimensional structures, or networks that positions phenomena people and their experiences as either right or wrong, norm-conforming or norm-countering. Sociologist Susan Leigh Star (1990) has described similar systems as sets of conventions or standards that seek to order and stratify life; creating "members" and "non-members" of certain networks or social worlds. They produce standardized modes of existence that create insiders and outsiders, "normals" and "abnormals," which "often involve the private suffering of those who are not standard" (Star 1990, 94). Queer theorist Judith Butler has written of the "constraints" that produce "intelligible bodies" as well as "unthinkable, abject, unlivable bodies" (Butler 1993, x). Building on Star's phrasing, menstrunormativity circles something like *sets of menstrual standards and conventions that seek to order and stratify menstrual life*; creating hierarchies of existence; menstrual insiders and outsiders, menstrual "normals" and "abnormals," regulating what menstrual worlds come into being, what menstrualities become possible or impossible; which menstrualities become uncomplicated, effortless and easy and which become difficult, burdensome, and painful. Many menstrual scholars have shown how normative pressures work on menstruation. For example menstruating or behaving outside of normative assumptions of menstruality can result in stigmatization and marginalization (Crawford, Menger, and Kaufman 2014; Kowalski and Chapple 2000; Johnston-Robledo and Chrisler 2013) and feelings of disgust and shame (Lee 2009; Schooler et al. 2005; Ussher 2006; Johnston-Robledo et al. 2007; Fahs 2011; Moloney 2010; Young 2005), where for example those who bleed too much, too seldom, or those who fail to conceal their menstruation are socially sanctioned externally ("That's not normal!" "You're repulsive!"), and internally ("There's something wrong with me," "I feel unfeminine," "I'm disgusting").

Menstrunormativity differs from the concepts heteronormativity and cisnormativity in that it does not entail the norm-positon within its semantic morphology. Whereas *hetero*sexuals and *cis*gendered are the ideal subject positions of the respective systems, *menstru*ating is not the sought position in menstrunormativity. The ideal normative subject-position has been called *the normate* by crip-scholar Rosmarie Garland-Thomson (1997). That which I suggest we call the *menstrunormate* is thus the menstrual equivalent. But unlike the cases of hetero- and cisnormativity the menstrunormate is not the menstruator, but *the "right" kind* of menstruator, say: a ciswoman with a regular 28-day cycle, who obscures all evidence of menstruation.

Instead of the norm-position, the initial *menstru-* in menstrunormativity encircles the area in question: menstruation and menstruating. In that sense it is similar to the crip concept *bodynormativity* which refers to ideas about hegemonic notions of the body, that is, social constructions of what is considered normal or deviant embodiment (Malmberg 2008, 2009). Malmberg argues that bodynormativity stipulates that correct/natural/beautiful bodies should not leak, shake uncontrollably, or be asymmetrical. Correspondingly, menstrunormativity stipulate normative (normal/natural/healthy) menstruation, as, for example, controllable and concealed, a menstrunormate that does not leak, show, smell, or tell of their menstruation.

The linguistic logics of *homonormativity*, another terminological sibling to menstrunormativity, differs somewhat. Homonormativity builds directly from heteronormativity and conceptualizes the ways in which homosexuals conform specifically to *hetero*normative orders and institutions, such as marriage (Robinson 2016). Such interlinkages are an important feat of all the systems of normativity mentioned here. For example, cisnormativity and heteronormativity are part of each other; they interlink and overlap through that which Butler has called the *heterosexual matrix* (Butler 1990): cisgendered women desire cisgendered men. One hegemonic social system intersects with another. Likewise, menstrunormativity is entangled with hetero-, cis, and bodynormativity: the menstrunormate lives up to certain standards of embodiment and "hetero-femininity" (see for example, Ussher and Perz 2013).

The Paradoxical Creatures of Menstrunormativity

As illustrated above, the menstrunormate menstruates according to menstrual norms. But what menstrual norms are those? This is where it starts to get complex and contradictory. Viewing menstrunormativity as a cluster of different sets of normativities entails understanding how sometimes wholly contradictory ideas impact a subject in synchronicity. The menstrunormate can thus for example in a paradoxical way both be bleeding regularly and never bleed at all. Star argues (1990) that as we are situated in many different networks we always have *multiple memberships* in many social worlds at once, and as such we are always marginal and monstrous in some regard. Normative ideals come from several directions, and impact the subject in multiple ways, creating a *multiple marginality* (Star 1990). These multiple marginalities can relate to each other in thoroughly contradictory ways. For example, we can take the menstrual suppression debate where (I simplify for clarity) one side argues that modern amounts of menstrual cycles within a lifespan is "unnatural" and potentially "unhealthy," as menarche comes earlier and menstruators are more seldom pregnant compared to other—"more natural"—historical eras, following that it is right and moral to reduce the number of cycles medically (Coutinho and Segal 1999). The other side argues that menstruation is a natural part of life and therefore healthy, following that it is (potentially) unhealthy (and immoral) to interfere with it medically (see Hasson 2012b), for example constructing "pill-periods" as "not real" (Hasson 2016). The menstrunormate is in these two different discourses two fundamentally conflicting things:

"the menstrunormate is that which does not menstruate"

"the menstrunormate is that which does menstruate"

Many menstruators are part of both discourses exemplified above, as well as many others, simultaneously. Menstruating means being affected by a clustered multitude of different and sometimes conflicting normativities; biomedical ideas of physical normalcy, "Femcare-industry" ideas of "normal flows," patriarchal ideas of man as norm, feminist ideas of "menstrutalk" as emancipatory, et cetera. When we see all these different normativities working together I argue the menstrunormate is unveiled as a mirage. There is no menstrunormate, only an abundance of menstrual monsters.

The Impossible Menstrunormate

When we see menstrunormativity as a clustered multitude of normativities we can see the menstrunormate as simultaneously non-bleeding (living up to the masculine, non-leaky body as ideal) and bleeding (living up to ideas of what is natural and healthy, as well as of certain femininity); at the same time "menstrusilent" (according to contemporary local menstrual etiquette) and "menstrutalking" (mirroring contemporary feminist role models). These multiple and contradictory normativities and marginalities produce an impossible menstrunormative space, creates an illusion of a menstrunormate, a non-achievable model-menstruator. Rosemarie Garland-Thomson argues the normate is an illusion and a "phantom figure" outlined by the "array of deviant others whose marked bodies shore up the normate's boundaries" (2009, 45). Her illustrative phrasing places bodies of abnormals, of monsters, around a void, marking up the boundaries of the ideal. I emphasize that when you view normativities as clustered and potentially contradictory—at least when it comes to menstruation—there is not even a void. There is nothing at all. When you draw it out and put all the "monstrous menstruators" beside each other, there is no empty ideal space in the middle. Menstrunormative ideals are everywhere around us, but the subject-position it propagates is a nothingness as dominant ideals overlap and contradict each other. It is impossible to attain it, as it is nowhere to be found. Instead it morphs to monster at every turn.

The Boundless Menstrual Monster

The monster as a figure has been used with great excellence to discuss female, reproductive, menstrual, and non-normconforming embodiment (Braidotti 1994; Shildrick 2002; Ussher 2006; Young 2005). In a society where man is considered norm "[w]oman, as a sign of difference, is monstrous" (Braidotti 1994, 81); "her seeping, leaking, bleeding womb [stand] as a site of pollution and source of dread" (Ussher 2006, 1), as pathology (Lie 2012; Strange 2000; Johannisson 2005; Martin 2001), and menstruation is—as in the part of the menstrual suppression debate—positioned as unnatural and unnecessary (Hasson 2016; Mamo and Fosket 2009). If menstruation itself is positioned as disease, unnatural, and pollution, then all menstruators are menstrual monsters. Within these discourses the normative ideal is to *not bleed*, to bleed is to be Other. As such menstruation is a foil to femininity, to modernity, to the docile, controllable body. But simultaneously, other hegemonic ideas stipulate that women *should bleed*. Strong "menstrual imperatives" positions menstruating as part of womanhood and natural, real, correct, healthy, natural, and normal aspect of life. The assumption that all healthy women should menstruate renders those who do not outside the bonds of correct femininity/womanhood, thus all women who do not menstruate are menstrual monsters. In other words, women cannot escape the status of being "menstrual monsters" regardless of whether they menstruate or not. So, from the get-go and fundamentally, the menstrunormate is non-achievable and the menstrual monster is boundless. But the cluster of menstrunormativity encompasses many more paradoxes and contradictions.

I have already touched upon medical normativities which sets up specific standards for menstrual non-monstrosity. Defining what constitutes normal menstruation is imperative and ubiquitous for contemporary medical scientists and clinicians as well as their patients (see for example, AAoP and ACoOaG 2006; Hasson 2012a). In their striving to identify, demarcate, and categorize (and ultimately cure) illness/pathology the medical sciences set up crucial (narrow) standards of menstrual existence. For example the idea of the normalcy of the 28-day cycle can be argued to be a medico-social construct (Oudshoorn 1994) as can the idea of the healthiness/naturalness of reliable regular menstrual cycle (Martin 2001), and the definition of normal amounts of menstrual fluid as more than 20 but less than 80 milliliters per period (see Janson and Landgren 2015) is often described as the normal scope but is by some argued to be too narrowly defined (Clancy 2016). These medical normativities however effectively draws out boundaries of menstrual existence, stipulating that the menstrual monster bleeds incorrectly, simultaneously bleeding too little and too much, too often and too seldom, and irregularly, as the medical construct of the normal becomes the ideal. The menstrual monster also feels incorrectly: is in too much pain (diagnosed with dysmenorrhea) or it is too angry, too sad, too depressed (diagnosed with PMDD). Yet, menstruating with too much ease or even pleasure is also monstrous, because while pain and suffering are often categorized as menstrual disorders (and as such pathology, anomaly, monstrous, treatable) they are simultaneously understood as part of what constitutes a normal menstruation. Menstrual pain, severe PMS, and other menstrual problems are normalized by health care practitioners (Oinas 1998)—as well as pop culture (Rosewarne 2012)—as normal aspects of healthy menstrual bodies, implicitly arguing that "menstruation should be a little painful" (Malm 2014). So, the menstrual monster has painful gut-wrenching period cramps and simultaneously take little notice of their periods. Continuing along the line of emotions, the paradoxical creature that is the menstrual monster is neither appropriately ashamed nor proud enough of their periods. On the one hand, it violates menstrual etiquette and does not perform menstruation as social stigma; talks openly and proudly about their menstruation, carries their menstrual products openly, and does not adhere to the concealment imperative (Wood 2019; Young 2005). On the other hand, and at the same time, the menstrual monster violates (menstru)normative ideas of the menstrual countermovement: it is not proud enough of their periods, does not want to talk publicly about periods.

The menstrual monster also has the wrong body. Its vagina is too small, too large, it smells too much, it is too moist and too dry, it is too deep and too shallow; it is allergic to cotton, and silicone, and rubber; and it doesn't have a vulva, vagina, or a uterus. The transwoman (Berg 2017) and women with ambiguous female genitals (Guntram 2014) become menstrual monsters herein. Relatedly, the menstrual monster identifies incorrectly, or rather, though it menstruates, it identifies as other than ciswoman (for example, transmen, non-binary). Recent scholarship highlights how transgendered menstruators become monstrous as their bleeding bodies conflict with their non-woman identities (Chrisler et al. 2016). The menstrunormate is cisgendered, but the cisgendered (woman) menstruator is also monstrous as that subject-position is positioned as wrongfully privileged in many feminist spaces. Therein, you are monstrous if you menstruate and do not identify as woman, *and* if you menstruate and identify as woman.

The menstrual monster is thus simultaneously menstruating and non-menstruating, feminine and un-feminine, bleeding too much and too little, too early and too late, in too much and too little pain, is too proud and too embarrassed, transgendered and cisgendered. All these are features of this paradoxical creature. When we see it laid out, in a cluster of conflicting normativities, it is clear that the menstrunormate is not only unattainable, but non-existent. It is simply not possible to be both non-bleeding and bleeding, and so on. Crip-scholar Mary Shildrick has argued the importance of undoing the singular category of the monster (Shildrick 2002) contending that the monster is not a singular but a multitude. Others have underlined that the dichotomies of absolute standards—no standards, or monstrous—normate, are false (Law 1991, 5). No one is standard, no one is normate, we are all multiply marginal (Star 1990). Viewing normativities as clustered positions the menstrunormate as chimera and makes monsters of all, but it also, importantly, makes visible that we are all culpable of creating monsters.

We Are All, Always, Doctor Frankenstein

Let me share a personal experience. When the menstrual countermovement began to gain momentum in Sweden, we Swedish menstrual activists were eagerly opposing ideas of menstruation as abject ("menstruation can be fun and beautiful!"), pathologized ("PMS doesn't have to be negative!"), medicalized ("menstrual suppressants shuts of the signal-system that is your natural cycle!"), and silenced ("break the communication taboo!"). But in the process of dismantling these normative powers we also created new menstrunormativities where for example talking about menstruation was positioned as better than not talking about it; loving one's period was more feminist than hating one's period; cups were cooler than pads; and not using hormonal birth control was healthier than "pill-popping." In that, the movement created many a menstrual monster: those who think periods are gross, who have serious problems with cyclical depression, who use menstrual suppressants, et cetara. Monstering was never our intent but it was a consequence all the same. Some were, and are, left out, rendered less "real," less "feminist," less "possible." With this example I want to say that we all, even grass-root menstrual activists with the best of intentions, create monsters.

Menstrunormativity is not a so called uniform "homogeneous, global external entity that exists outside of us" (Brown 2012, 1066) that represses menstruators from above/outside but a multiply paradoxical and diverse cluster that is continuously co-produced by us all: the medical sciences; the hygiene product industry, the kids in the hallway; the public health informants, the movie industry, the scholars by their computers; and the activists in their red tents: we all create menstrunormativity, all the time. It is also important to consider that it is not only human actors that take part in the co-production. Several critical menstrual scholars have shown how technologies matter in the configuring of menstrual correctness and normalcy (Hasson 2012b; Malmberg 1991; Vostral 2008). Menstrual technologies co-produce "boundaries of health and illness;" "subjectivities and gendered forms of embodiment;" "cultural ideas about bodies and identities" (Mamo and Fosket 2009, 925). Menstrunormativity is built into menstrual technologies such as pads, tampons, cups, toilets, pharmaceuticals, and menstrual charting apps. The ways these technologies function and malfunction, fit and chafe, produce (illusions of) normates and (actual) monsters. The menstrunormate use "normal sized" tampons, they don't make menstrual messes in the toilet, their menstrual cup does not leak. The menstrual monsters' tampons do not stay in place, the cramps become worse when they use the cup, the sponge leaks, the pad slips, the menstrual charting apps give faulty suggestions on when their next period is due. All these technologies carry menstrunormativity within them and position all who cannot use them with ease as menstrual monsters. Menstrunormativity is co-produced by every*one* and every*thing*.

Additionally, menstrunormativity is a perennial and inescapable. Menstrual monsters will always be created. Now, when the menstrual countermovement celebrates so many successes, this proposition of the continuance of normativities is perhaps particularly indispensable as it offers an important reconfiguration of how we perceive the goal and role of the movement.

Since the early days of menstrual scholarship, the idea of "the menstrual taboo" has been very pervasive in social studies of menstruation (see for example, Frazer 1922; Douglas 1966; Delaney, Lupton, and Toth 1976; Buckley and Gottlieb 1988) as well as in menstrual activist accounts (Bobel 2010; Persdotter 2013). The prevalence of the idea of the menstrual taboo is problematic in several ways. Some for example argue it is too often accepted as fact rather than challenged as theory (Newton 2016, 42), others argue it is often wrongly thought of as a homogenous universal entity (Buckley and Gottlieb 1988). I would like to stress that the idea of the menstrual taboo is problematic because it creates a false narrative of there being a post-taboo society that is completely "free and inclusive." As the "fight to end period shaming is going mainstream," (Jones 2016) and some menstrual communication taboos rapidly (finally!) deteriorate in front of us we risk concluding prematurely and without caution that we are done. Because what happens when those taboos are gone? When we can indeed talk openly about menstruation? What remains? There are still forces at play that have negative effects for menstruators, to which we need to attend to critically: new or persistent ideas with the power to create livable and unlivable lives.

Furthermore, the idea of the menstrual taboo also carries with it a risk of positioning other cultures, contexts, and actors (the Global South; the product industry; the conservatives) as still left in that objectionable, pitiful state of tabooing menstruation. But in fact, we find repressive ideas of correct menstrual behavior in all camps, both before and after the fight to end period-shaming went mainstream. I say it again: we will all, always, create menstrual monsters. "We" are just as Frankensteinian as "they" are. Thus, I argue the idea of the menstrual taboo builds in false dichotomies of us–them, good–evil, repressed–emancipated, and risks masking the continuance of menstrunormative constructions that go on around us. In this duality of being both Frankenstein and monster I argue there are transformative potentials through which we can view—and craft—the world a little differently.

Monstrous Potentials—The Transformative Futures of Menstrunormativity

> We need the power of modern critical theories of how meanings and bodies get made, not in order to deny meanings and bodies, but in order to build meanings and bodies that have a chance for life. (Haraway 1988, 580)

If we see menstrunormativity as a cluster of multiple normativities that surround menstruation we can see how it comes at us from all sides. Wherever we try to fit we get excluded; pushed away, expelled, turned monstrous. Menstrunormativity produces an ideal that is thoroughly impossible, a non-space of menstrual existence. No one can menstruate (or have a uterus and not menstruate) in accordance with menstrunormativity. Menstruators, women, the menstrual countermovement, as well as the menstrual product industry, work hard to pass as non-monstrous, as "right," as "good," as "normal," "real," and "emancipated." The multiply paradoxical nature of the menstrunormative ideals reveals the menstrual experience as always already monstrous (we all are menstrual monsters), and the creation of monsters as inevitable (all are Frankenstein, none of us are innocent bystanders). Within that there is substantial transformative potential. Star argues that "refusing to discard any of our selves . . . refusing to 'pass' or become pure" and acknowledging our multiple marginality is a source of power (1990, 82). Thus, recognizing our shared inescapable monstrosity and monstering could transform the way menstruators and the menstrual countermovement view themselves and each other. Through making visible the multitude of conflicting ideas of the menstrunormate it ceases to be idol/standard/model and sought-after position. Through highlighting the impossibility of doing menstruation "correctly" and instead making visible the multiplicity in the menstrual experience we can lessen the pressure on what it is to be menstruator, allowing more of a polyphony and dissonance in our menstrual and embodied experiences (cf. Nord, Bremer, and Alm 2016, 6). It makes it possible to build meanings and bodies that have greater chances for livable, recognizable life (Haraway 1988). Because while we are all inevitably monsters, some monsters suffer more from their monstrosity than others: "some monsters find it so easy that they scarcely look like monsters at all; . . . some monsters are truly wretched, subjected to pain, deprived of all hope and dignity" (Law 1991, 18). All those actors that take part in the co-production of menstrunormativity have the potential to reduce the effects of monstrosity; lessen the load and ease the pain, expand the space for menstrual existence, making more lives more livable.

Hopeful Frankensteins

In late 2017, the pad-brand Libresse (known also as Bodyform, Saba and Nosotras) launched a campaign called "#bloodnormal," including a YouTube-video in which they showed a red liquid (at last!) and portrayed a wide array of ways to do and be menstruator: menstruators doing hard-core sports; talking publicly about periods; swimming; partying; laughing, crying; having painful periods; reading period poetry, et cetera. "I wanted to posit a view unfettered by judgement" tweeted the add creator David Wolfe. Long gone seems the blue liquid, the white pants, and their immeasurable mon-sterings of menstruators. Similarly, in the moment of writing there is on my desk, beside my computer, a pair of pink underwear from H&M-owned clothing company with a text that says "Periods are cool. And painful. And messy. And great." The text joyfully allows for a heterogeneous complexity, ambiguity and paradoxicallity of thought and experience. These two examples show periods have indeed gone public, that the fight to end period shame has gone mainstream. And as is clear here: not only civil society and hard-core menstrual activists but even large corporate actors, with a wide reach and consequent power, seem to grasp the importance (and of course potential goodwill, legitimization, and profit—but I leave that to others to disentangle) of depicting inclusive and diverse experiences. In this they assist in loosen-ing menstrunormativity's grip on the menstrual experience, making menstrual monstrosity less miserable. That these destabilizations of normativities come from corporate actors might be unexpected by menstrual activists and critical scholars as we have long positioned "the industry" as the "bad guy" behind the scenes of the construction of the menstrual taboo (see for example, Kissling 2006). But "they" need not be worse than "us." While seemingly strange bedfellows to the menstrual countermovement these corporate actors play a crucial role in increasing space for menstrual life, If we see all as both Frankenstein and monster—as the dichotomy of "us-them" and "good-evil" wither—companionships between "them" and "us" (as if we could ever make such distinctions) may have greater chances of making greater things.

As the menstrual scholarship, activism, and politics grow stronger, it is more important than ever that we keep a keen eye on processes of exclu-sion and restriction that go on within and alongside our work. We need to try and understand the complex and multiple ways in which menstrunor-mativities are *still* and *always* developed and enacted—even when they seem to be "unfettered with judgement" and look like universal emancipation. Menstrunormativity is changing continuously, but there will always be sets of standards, rules, and ideas that order and stratify menstrual life. I hope that thinking with the concept menstrunormativity, along with the attached principles discussed in this paper, strengthens critical menstrual scholarship to employ a continuous critical gaze on the normativities and how they are (re)created in all menstrual practice and discourse; in medicine, industry, activism, and critical menstrual studies alike. I hope that the concept helps us ask continuously: What is constructed as menstrunormative and ideal? Who are excluded, marginalized, or invisibilized in the commercials; in the menstrual countermovement; in Kathmandu; Gothenburg, and New York; in our research practices, and for that matter: in this book? And—most importantly— *what can we do to change that?*

References

American Academy of Pediatrics (AAoP), and American College of Obstetrician and Gynecologists (ACoOaG). 2006. "Menstruation in Girls and Adolescents: Using the Menstrual Cycle as a Vital Sign." *Pediatrics* 118: 2245–50.

Berg, Beatrice. 2017. "Transgendering Menstruation: En kvalitativ studie av uppfattningar kring menstruation i relation till transidentitet i Sverige." Bachelor thesis, Universtiy of Gothenburg.

Bobel, Chris. 2010. *New Blood: Third-Wave Feminism and the Politics of Menstruation.* New Brunswick: Rutgers University Press.

———. 2017. "Introduction speech at the 22nd Biennial Conference of the Society for Menstrual Cycle Research." Conference themed "*Menstrual Health: Research, Representation and Re-Education.*" Kennesaw, Atlanta, USA.

Braidotti, Rosi. 1994. *Nomadic Subjects: Embodiment and Sexual Difference in Contemporary Feminist Theory.* New York: Columbia University Press.

Brown, Gavin. 2012. "Homonormativity: A Metropolitan Concept That Denigrates 'Ordinary' Gay Lives." *Journal of Homosexuality* 59: 1065–72.

Buckley, Thomas, and Alma Gottlieb (eds.). 1988. *Blood Magic: The Anthropology of Menstruation.* Berkeley: University of California Press

Butler, Judith. 1990. *Gender Trouble: Feminism and the Subversion of Identity.* New York: Routledge.

———. 1993. *Bodies That Matter: On the Discursive Limits of "sex".* New York: Routledge.

Chrisler, Joan. C., J. A. Gorman, J. Manion, M. Murgo, A. Barney, A. Adams-Clark, J. R. Newton, and M. McGrath. 2016. "Queer Periods: Attitudes toward and Experiences with Menstruation in the Masculine of Centre and Transgender Community." *Cult Health Sex* 18: 1238–50.

Chrisler, Joan C., Jennifer Gorman Rose, Susan E. Dutch, Katherine G. Sklarsky, and Marie C. Grant. 2006. "The PMS Illusion: Social Cognition Maintains Social Construction." *Sex Roles* 54: 371–76.

Clancy, Kate. 2016. "Period Podcast." *Episode 2 Monkey Trainers and Tampon Squeezers.* Accessed October 16, 2016. https://periodpodcast2.libsyn.com/episode-2-monkey-trainers-and-tampon-squeezers.

Coutinho, Elsimar M., and Sheldon J. Segal. 1999. *Is Menstruation Obsolete?* New York: Oxford University Press.

Crawford, M., L. M. Menger, and M. R. Kaufman. 2014. "'This Is a Natural Process': Managing Menstrual Stigma in Nepal." *Culture Health & Sexuality* 16: 426–39.

Delaney, Janice, Mary Jane Lupton, and Emily Toth. 1976. *The Curse: A Cultural History of Menstruation.* New York: Dutton.

Douglas, Mary. 1966. *Purity and Danger: An Analysis of Concepts of Pollution and Taboo.* London: Psychology Press.

Fahs, B. 2011. "Sex during Menstruation: Race, Sexual Identity, and Women's Accounts of Pleasure and Disgust." *Feminism & Psychology* 21: 155–78.

Folkmarson Käll, Lisa. 2009. *Normality/Normativity.* Uppsala: Centre for Gender Research, Uppsala University.

Frazer, James George. 1922. *The Golden Bough: A Study in Magic and Religion.* London: Macmillan.

Garland-Thomson, Rosemarie. 1997. *Extraordinary Bodies: Figuring Physical Disability in American Culture and Literature.* New York: Columbia University Press.

———. 2009. *Staring: How We Look.* Oxford: Oxford University Press.

Gunson, Jessica Shipman. 2010. "'More Natural but Less Normal': Reconsidering Medicalisation and Agency through Women's Accounts of Menstrual Suppression." *Social Science & Medicine* 71: 1324–31.

Guntram, Lisa. 2014. "Ambivalent Ambiguity? A Study of How Women with "Atypical" Sex Development Make Sense of Female Embodiment." PhD diss., University of Linköping.

Haraway, Donna. 1988. "Situated Knowledges: The Science Question in Feminism and the Privilege of Partial Perspective." *Feminist Studies* 14: 575–99.

Hasson, Katie Ann. 2012a. "From Bodies to Lives, Complainers to Consumers: Measuring Menstrual Excess." *Social Science & Medicine* 75: 1729–36.

———. 2012b. *No Need to Bleed: Technologies and Practices of Menstrual Suppression.* Berkeley: University of California.

———. 2016. "Not a 'Real' Period? Social and Material Constructions of Menstruation." *Gender & Society.*

Hughes, Bill. 2009. "Wounded/Monstrous/Abject: A Critique of the Disabled Body in the Sociological Imaginary." *Disability & Society* 24: 399–410.

Janson, Per Olof, and Britt-Marie Landgren. 2015. *Gynekologi.* Lund: Studentlitteratur.

Johannisson, Karin. 2005. *Den mörka kontinenten: kvinnan, medicinen och fin-de-siècle.* Stockholm: Norstedt.

Johnson, Allan G. 2000. *The Blackwell Dictionary of Sociology: A User's Guide to Sociological Language.* Oxford: Wiley.

Johnston-Robledo, I., K. Sheffield, J. Voigt, and J. Wilcox-Constantine. 2007. "Reproductive Shame: Self-Objectification and Young Women's Attitudes toward Their Reproductive Functioning." *Women & Health* 46: 25–39.

Johnston-Robledo, Ingrid, and Joan C. Chrisler. 2013. "The Menstrual Mark: Menstruation as Social Stigma." *Sex Roles* 68: 9–18.

Jones, Abigail. 2016. "The Fight to End Period Shaming Is Going Mainstream." *Newsweek.*

Kissling, Elizabeth Arveda. 1999. "When Being Female Isn't Feminine: Uta Pippig and the Menstrual Communication Taboo in Sports Journalism." *Sociology of Sport Journal* 16: 79–91.

———. 2006. *Capitalising on the Curse: The Business of Menstruation*. London: Lynne Rienner Publishers.

Kowalski, Robin M., and Tracy Chapple. 2000. "The Social Stigma of Menstruation: Fact or Fiction?" *Psychology of Women Quarterly* 24: 74–80.

Law, John. 1991. *A Sociology of Monsters: Essays on Power, Technology, and Domination*. London: Routledge.

Laws, Sophie. 1990. *Issues of Blood: The Politics of Menstruation*. Basingstoke: Macmillan.

Lee, J. 2009. "Bodies at Menarche: Stories of Shame, Concealment, and Sexual Maturation." *Sex Roles* 60: 615–27.

Lie, Anne Kveim. 2012. "Syke blødinger og 1800-tallets sykelige kvinner." In *Rødt og hvitt - Om blod og melk i fortid og samtid*, edited by Hilde Bondevik and Anne Kveim Lie. Oslo: Unipub.

Malm, Christoffer. 2014. "Mensvärken kan vara dold sjukdom." *Aftonbladet*, 6th July, section General, Del 1.

Malmberg, Denise. 1991. "Skammens röda blomma? menstruationen och den menstruerande kvinnan i svensk tradition." PhD diss., University of Uppsala.

———. 2008. "The Dysfunctional Female Body as Seen from the Perspective of Body-Normativity." In *Gender and the Interests of Love: Essays in Honour of Anna G. Jónasdóttir*, edited by G. Jónasdóttir Anna, Kathleen B. Jones, and Gunnel Karlsson. Örebro: Örebro University.

———. 2009. "Bodynormativity – Reading Representations of Disabled Female Bodies." In *Body Claims*, edited by Janne C. H. Bromseth, Lisa Folkmarson Käll, and Katarina Mattson. Uppsala: Centre for Gender Research, Uppsala University.

Mamo, Laura, and Jennifer Ruth Fosket. 2009. "Scripting the Body: Pharmaceuticals and the (Re)Making of Menstruation." *Signs: Journal of Women in Culture and Society* 34: 925–49.

Martin, Emily. 2001. *The Woman in the Body: A Cultural Analysis of Reproduction*. Boston: Beacon Press.

Moloney, Sharon. 2010. "How Menstrual Shame Affects Birth." *Women and Birth* 23: 153–59.

Newton, Victoria Louise. 2016. *Everyday Discourses of Menstruation*. London: Palgrave Macmillan.

Niklasson, Grit. 2014. "Når det normale bliver det gode: normalitet, normativitet og graviditet." *Dansk Sociologi* 25.

Nord, Iwo, Signe Bremer, and Erika Alm. 2016. "Redaktionsord: Cisnormativitet och feminism." *Tidskrift för genusvetenskap TGV* 37: 2–13.

Oinas, E. 1998. "Medicalisation by Whom? Accounts of Menstruation Conveyed by Young Women and Medical Experts in Medical Advisory Columns." *Sociology of Health & Illness* 20: 52–70.

Oudshoorn, Nelly. 1994. *Beyond the Natural Body: An Archaeology of Sex Hormones*. New York: Routledge.

Persdotter, Josefin. 2013. "Countering the Menstrual Mainstream—A Study of the European Menstrual Countermovement." Master's thesis, University of Gothenburg.

Rembeck, Gun. 2008. "The Winding Road to Womanhood: Adolescents' Attitudes towards Menstruation, Womanhood and Sexual Health: Observational and Interventional Studies." PhD diss., University of Gothenburg.

Robinson, Brandon Andrew. 2016. "Heteronormativity and Homonormativity." In *The Wiley Blackwell Encyclopedia of Gender and Sexuality Studies*. Hoboken, NJ: Wiley.

Rosewarne, Lauren. 2012. *Periods in Pop Culture: Menstruation in Film and Television*. Lanham, MD: Lexington Books.

Schooler, Deborah, L. Monique Ward, Ann Merriwether, and Allison S. Caruthers. 2005. "Cycles of Shame: Menstrual Shame, Body Shame, and Sexual Decision-Making." *The Journal of Sex Research* 42: 324–34.

Shildrick, Margrit. 2002. *Embodying the Monster: Encounters with the Vulnerable Self*. London: Sage.

Star, Susan Leigh. 1990. "Power, Technology and Phenomenology of Conventions— On Being Allergic to Onions." In *Technoscience: The Politics of Interventions (2007)*, edited by Kristin Asdal, Brita Brenna, and Ingunn Moser. Oslo: Unipub.

Strange, J. M. 2000. "Menstrual Fictions: Languages of Medicine and Menstruation, c.1850–1930." *Womens History Review* 9: 606–28.

Tideman, Magnus. 2000. *Normalisering och kategorisering: om handikappideologi och välfärdspolitik i teori och praktik för personer med utvecklingsstörning*. Lund: Studentlitteratur.

Ussher, Jane M. 2006. *Managing the Monstrous Feminine: Regulating the Reproductive Body*. London and New York: Routledge.

Ussher, Jane. M., and Janette Perz. 2013. "PMS as a Gendered Illness Linked to the Construction and Relational Experience of Hetero-Femininity." *Sex Roles* 68: 132–50.

Vostral, Sharra Louise. 2008. *Under Wraps: A History of Menstrual Hygiene Technology*. Lanham: Lexington Books.

Weiss, Gail. 2014. "Uncosmetic Surgeries in an Age of Normativity." In *Feminist Phenomenology and Medicine*, edited by Kristin Zeiler and Lisa Folkmarson Käll. Albany: SUNY Press.

Wood, Jill M. 2019. "Menstrual Politics: The Menstrual Concealment Imperative." In *The Palgrave Handbook of Critical Menstrual Studies*, edited by Chris Bobel, B. Fahs, I. Winkler, T. Roberts, E. Kissling, and K. A. Hasson. London: Palgrave.

Young, Iris Marion. 2005. *On Female Body Experience: "Throwing Like a Girl" and Other Essays*. New York: Oxford University Press.

Zeiler, Kristin, and Lisa Folkmarson Käll. 2014. *Feminist Phenomenology and Medicine*. Albany: SUNY Press.

CHAPTER 30

Empowered Bleeders and Cranky Menstruators: Menstrual Positivity and the "Liberated" Era of New Menstrual Product Advertisements

Ela Przybylo and Breanne Fahs

INTRODUCTION

Menstrual product advertisements (Mpads), since their early years, have been reliant on misogynist discourses around menstruation, portraying menstruation as shameful and unmentionable, vaginas themselves are the source of squalor and filth, and menstrual blood as requiring containment, concealment, and sanitization through euphemism and blue-liquid advertising (Saz-Rubio and Pennock-Speck 2009; Kissling 2002; Thornton 2013). Grounded in representations of menstruating bodies on white, cisgenderist, ableist, fatphobic, and heteronormative terms, menstrual product advertising has a long history that spans from the early commercialization of menstrual products in the 1920s and across the mediums of print (since 1921), television (since 1970), and the internet. In recent years, in the context of increasing menstrual activism, fights for menstrual justice and access to menstrual products, and feminist efforts at shifting discourses around what it means to bleed, it seems that a "new" era of Mpads is surfacing—one more willing to both poke fun at histories of menstrual shame and to envision bleeding on empowering and glittery terms.

Yet as we explore in this chapter, while new menstrual product advertising certainly does envision menstrual bleeding on celebratory and liberated terms, major brands such as Always, U by Kotex, and Libresse/Bodyform do so by co-opting feminist discourses and the energies of menstrual activism. Continuing to rely on narrow understandings of menstruating bodies as

© The Author(s) 2020   375
C. Bobel et al. (eds.), *The Palgrave Handbook of Critical Menstruation Studies*, https://doi.org/10.1007/978-981-15-0614-7_30

feminine and female bodies, these ads tend to push able-bodied fitness as well as obligatory positivity onto those who menstruate, while foregoing questions of menstrual product access and the more painful aspects of bleeding. Swept up by the flows of menstrual positivity, and distancing themselves from previous commitments to shaming menstruators, companies such as Always, U by Kotex, and Libresse/Bodyform seem newly unable to come to terms with the fact that "periods don't have glitter in them" (as a HelloFlo ad we later examine jokingly proclaims) or that they are not by default empowering and fun.

In this chapter, we examine a series of new Mpads, including "Reality Check" by Kotex (2013), "First Moon Party" by HelloFlo (sponsored by U by Kotex 2014), "#LikeAGirl" by Always (2014), and "Blood" by Bodyform UK (2016). We have selected these ads in particular because they are produced by large Western megacorporations that have historically relied on menstrual shaming even while also co-opting feminist efforts of menstrual and bodily positivity for profit. We thus argue that these Mpads co-opt feminist discourses and energies of body positivity and menstrual positivity toward pushing consumer capitalist goals of selling menstrual gear with an "empowered" message. Drawing on feminist menstrual scholarship, we argue that parallel to other co-optations of feminist self-love and body positivity rhetoric, menstrual positivity is thinned, whitened, and transformed when utilized toward the sale of menstrual gear. We argue that "positivity"—while an important starting point from which to undertake feminist menstrual activism, praxis, and theorizing—is easily co-optable within neoliberal marketing cultures. Acknowledging the importance of affirmative menstrual messaging, we nevertheless develop a "menstrual crankiness" that draws on positivity but also holds it critically at bay. Aligned with queer theoretical work on the political import of negative affects such as that of Sara Ahmed (2010), we agitate for the importance of menstrual crankiness in pushing at sexist and transphobic discourses around bodies and embodiment—arguing that menstrual crankiness is vital to thinking the material pains and pleasures of menstrual bleeding.

In thinking about the value of moving toward a crankier approach to the menstruating body as one that pushes beyond the framework of empowerment, we centralize recognition of pain, access, and justice. In short, we advocate a cranky menstrual experience as a way to embrace a more complete and justice-oriented perspective of menstruation attuned to the difficult aspects of bleeding. Menstrual justice, a riff on reproductive justice (Ross et al. 2016), in this sense includes thinking about menstruation, menstrual pain, and access to menstrual products intersectionally, and in dialogue with one's position in relation to power structures. Menstrual justice also includes seeing access to menstrual products, including pain control, as a human right as well as understanding menstrual pain as a form of chronic pain that necessitates proper access to care (Winkler and Roaf 2014; Kissling 2006, 126; Jones 2016; Przybylo and Fahs 2018).

Notably, this orientation also extends to the language we use in this piece. As in our earlier piece on menstrual pain (Przybylo and Fahs 2018), we interchangeably move between using the language of women and the language of menstruators. This recognizes the value of expanding the circle of "who menstruates" beyond women and troubling binary generalizations that all women menstruate and all men do not while also recognizing the deeply misogynistic impulses of menstrual negativity and stigma. Further, while much of the language used by advertisers references the "feminine"—with, for example, "feminine hygiene," "feminine products," "feminine odor," and so on—we shift toward a language of "menstrual products" and "menstrual product ads" to move away from a feminized language of the abject feminine and toward a language of menstrual care (Bobel 2019).

BLUE BLOODSTAINED PANTIES: A BRIEF OVERVIEW OF MENSTRUAL PRODUCT ADVERTISING

Menstruators have faced a plethora of images and ideas that treat menstruation as disgusting, contaminating, and debilitating. Menstrual product advertising has drawn upon and created this culture of negativity around menstrual experiences for decades. The phrase *feminine hygiene*—a relic from 1930s advertisements for birth control—itself suggests the inherent dirtiness of the female body and the relative cleanliness of the "well-managed" menstrual body (hidden, clean, out of sight), while also failing to imagine how bodies who are not feminine or female may menstruate (Fahs 2016; Tone 1996). A quick look at today's Mpads shows a range of bizarre and at times absurd paradoxes and contradictions around what menstruators should do and feel in regards to their periods.

Mpads rely centrally on shame and embarrassment to sell products, teaching the viewer that one can only properly manage menstruation by making individual decisions to reduce the possibility of disaster. That is, by buying certain brands of tampons and pads, menstruators can prevent embarrassing and shame-producing leaking, staining, and limits to their daily activities which supposedly are a result of menstruation. In short, menstrual products help to alleviate the "stain" of menstruation on women's femininity (Coutts and Berg 1993). Menstrual education products also rely on these narratives; one analysis of menstrual product booklets from 1932 to 1997 found that narratives of promoting secrecy did not change for this 50+year span (Erchull et al. 2002).

When selling single-use menstrual products, advertisers depict the menstruating body as unclean and dirty; these tactics are used to market panty liners, pads, and tampons to mainstream audiences (Berg and Coutts 1994; Kissling 2006). This is especially true in U.S. contexts, but it also has appeared in the Global South; for example, India's leading brand of single-use pads is tellingly called *Whisper* (Bobel 2019). Common tropes in advertising include menstrual blood depicted as "blue liquid," hyper-hygienic bleeding (where menstrual blood is never actually depicted), out of control bleeding where the specter of leaks and stains lurks everywhere and at all times, and narratives of intense shame and secrecy (Luke 1997). Little acknowledgment of the environmental destructiveness of single-use products is shown, thus promoting hyper-consumerism and teaching menstruators that they can solve the menstrual taboo by using disposable, single-use products and hiding their menstrual cycles altogether (Davidson 2012). Further, the ads are designed to heighten insecurities in the viewer, particularly in adolescent menstruators (Simes and Berg 2001). Mpads depict menstrual containment as only achieved through buying single-use (typically Procter and Gamble) products, through fear of menstrual contagion, and by placing the responsibility for hiding periods on individuals (Przybylo and Fahs 2018).

Further, alongside disgust narratives, women are portrayed as sporty, active, happy, and "liberated" by good/effective menstrual products (Fingerson 2012). This leaves menstruators with a variety of contradicting messages, such as: that menstruation is normal but it must be kept hidden and secret; that menstruation is a state integral to "being a woman" but that it is also a tainted femininity; that menstruation must be overcome through menstrual products which enable women to imagine themselves as active and full of "empowered" femininity; and that protection failure and incorrect product choice are the ultimate problems but freedom is achieved through the correct product choice (Raftos, Jackson, and Mannix 1998). Menstruation, then, is depicted in advertisements as the ultimate mark of shame and as the avenue to liberation via product consumption.

Ironically, at the same time, Mpads have consistently benefited from feminist advances and co-opted feminist vocabularies toward the selling of products. Sharra Vostral (2008) discusses the ways in which menstrual product companies have absorbed feminist advances toward selling their wares, such as in the 1920s as well as the 1970s and 1980s, when the feminist work of asserting bodily autonomy and not being constrained by bleeding was capitalized on, so that products were framed as having "freed the body" (Saz-Rubio and Pennock-Speck 2009, 2547). For example, in the context of 1960s civil rights and feminist struggles, menstrual products began to co-opt the language of liberation with the branding of new products such as "Newfreedom" and "Stayfree" (Vostral 2008).

This absorption of feminist discourses by menstrual product manufacturers is part of a larger ongoing phenomenon of "commodity feminism," "marketplace feminism," or "femvertising" (Goldman, Heath, and Smith 1991; Gill 2008; Heath and Potter 2005; Zeisler 2016). Scholars have argued that feminist gains are taken up, distilled, and diminished when the mainstream marketplace takes hold of them, such that larger social inequalities remain unchallenged while a veneer of progress and advancement on individualistic and consumerist terms is provided (Banet-Weiser and Portwood-Stacer 2017; Zeisler 2016). In terms of advertising, as Andi Zeisler argues, feminism tends to be framed less as "a set of values, ethics, and politics, [and more as] an assessment of whether or not a product is worthy of consumption" (2016, 32). Similarly, the Mpads we consider in this chapter absorb feminist discourses as a means to shake off the tawdry, retrograde, sexist ideas of history and to step into a new "liberated" era of empowered bleeding that can best sell their products.

EMPOWERED BLEEDERS: THE "LIBERATED" ERA OF MENSTRUAL PRODUCT ADVERTISING

While both large scale and small-scale feminist-oriented companies (such as Lunapads) seek to shed the sexism of the past toward reenvisioning a newly liberated menstrual subject, we will focus on two types of new Mpads emerging from the big players of menstrual products, such as Kotex, Always, and Bodyform/Libresse. As mentioned, we focus on ads put forward by large Western corporations that have profited and continue to profit from menstrual shaming, even while also profiting from co-opting feminist discourses and political advances. We group our analysis into two trends we see as arising in contemporary menstrual advertising, and draw on several examples to discuss how these trends, while at first glance conveying an "empowered" approach to menstruating, continue to sneak in negative and limiting ideas around who menstruators are and what they can do.

The first trend in recent Mpads we consider involves a model that seeks to create distance from past tendencies of menstrual advertising by a direct and narrated positioning of a liberated, implicitly feminist, and explicitly empowered bleeder. We call this form of menstrual product advertising the "peppy performance" model since it works hard to put forward a peppy performance of happy bleeding. The second form of menstrual product advertising we elaborate on involves the demonstration of empowerment through the envisioning of femininity as a site of sportiness, toughness, and ability; this model we identify as the "fit bleeding" model. In what follows we examine each of these approaches to envisioning the newly liberated menstruator and menstrual product brand and the extent to which these approaches revitalize tropes around bleeding and menstrual product advertising.

In the "peppy performance" approach to Mpads, sarcastic and humorous approaches to various aspects of the commercialized experience of menstruation are honed through a knowing feminist discourse that is informed by activists' "radical menstrual humor" (Bobel 2010, 127–32). One example of this approach is "Reality Check," which advertises a line of "new" menstrual products named "U" by Kotex. Owned by Kimberly-Clark (the producer of Kleenex), Kotex launched in 1920 drawing on its stock of excess cellucotton bandages after the first world war, with a successful advertising campaign launched in 1921 in *Ladies' Home Journal* (Vostral 2008, 65). U by Kotex was in turn launched in 2009 and is directed at young adult menstruators aged 14–21 (Newman 2010). In a series of ads spanning several years, a knowing tone of ridicule is developed by the narrators, one that seeks to distance U by Kotex from other menstrual products through establishing a knowingly liberated and skeptical menstruating subject.

For instance, in "Reality Check" (2013), a twenty-something white, apparently middle class cisgender woman pokes fun at the euphemistic tradition of Mpads. She asks herself, "How do I feel about my period?" and answers this question by saying "I love it. I want to hold really soft things like my cat; it makes me feel really pure. Sometimes I just want to run on the beach. I like to twirl, maybe in slow motion, and I do it in my white spandex. And usually by the third day I really just want to dance." Throughout this narration clips are shown from other Kotex commercials that feature pink butterflies, women running on beaches or doing yoga in white clothing while holding a red ball. The ad ends with the narrator poking fun at the biggest euphemism in Mpads—the blue liquid—by saying "the ads on TV are really helpful cause they use that blue liquid and I'm like oh *that's* what's supposed to happen." As the commercial concludes, Kotex asks us "Why are tampon ads so ridiculous?" proceeding to propel us "to break the cycle" and use U by Kotex. Through referencing knowledge about the ridiculousness of menstrual product advertising, Kotex strives to situate U by Kotex as an elevated, "with it" line of menstrual products that speak for and to a newly empowered feminine subject. The message resonates as knowingly enlightened and "empowered" while Kotex washes its hands of its legacy as a producer of harmful discourses around menstrual blood and women's bodies throughout its nearly 100-year long history of Mpad production. In this sense, "Reality Check" subtly places the blame for retrograde ideas around bleeding on menstruators themselves while undertaking the peppy performative work necessary to distance itself from its own role in menstrual shame.

Another form of peppy performance is visible in the Mpad advertising around the HelloFlo Period Starter Kit. HelloFlo is an e-commerce company "focused on providing products, content and services for women and girls through every life stage" (helloflo, "FAQs" 2018). The HelloFlo website strives to include a greater variety of individuals in menstrual conversations and features posts in relation to such topics as bleeding while in a wheelchair, menstrual pain and bloating, and menstrual activism. Menstrual studies scholar Chris Bobel (2006) has argued that "menstrual product activism" of smaller companies such as Lunapads effectively directs public attention to the environmental and health hazards of menstrual product disposability, and to "the promotion of healthier, less expensive, and less resource-intensive alternatives" (331). While HelloFlo in many ways is characteristic of what Bobel outlines as "menstrual product activism," HelloFlo is also partnered with Kotex, suggesting a more insipid co-optation of menstrual product activism.

HelloFlo, in partnership with U by Kotex, has created a series of three Mpads to promote the Period Starter Kit—"The Camp Gyno," "First Moon Party," and "The Period Fairy." In imaginative ways, these ads draw on feminist humor to sell products to young girls and their parents through reimagining the period on empowering and positive terms. The first period appears in these ads less as a "curse" than as a "red badge of courage," "cherry slush club," "blood sisterhood" and "vagical" experience. Through the Mpads, in other words, U by Kotex and HelloFlo imaginatively seek to remake and rename the first period as a happy, celebratory, and above all humorously feminist experience for young feminized and feminist-attuned menstruators. For example, "The Period Fairy" features a young black girl in her quest for the "shero" that is the "period fairy." Obliterated out of history, the period fairy nonetheless has a "herstory" that is related throughout the ad. While our narrator never manages to find her, she learns that she can be her own "shero" while wearing glittery tampon earrings (a feminist staple of menstrual embrace). Along similarly inventive lines, "Camp Gyno" features a young white girl who attains power and popularity at camp after she is the first to get her period and "First Moon Party" features a young white girl with such a deep desire to get her period that she fakes it by painting rubylicious nail polish on a pad in hopes of fooling her mom and friends (see Fig. 30.1). The mother, who is not convinced since "what she thinks I wouldn't know? Periods don't have glitter in them," decides to organize a menstrual party for her daughter replete with pin the pad on the period, a uterus piñata, marshmallows dipped in a red fondue fountain, and a vagician (see Fig. 30.2).

Wildly imaginative and hilarious, these Mpads play with feminist rhetoric to reimagine the first period as fun, humorous, and ultimately positive and empowering. Yet even as they aspire to multiculturalism and diversity as well as to reimagining the period experience, the Mpads function to uphold the value of the commercialization of bleeding through selling the Period Starter Kit as (in their words) "Santa for your vagina." Mobilizing feminist language and gynocentric imagery, the ads perform a peppy commercialization of the period experience, suggesting that it is ultimately the period starter kit that can change experiences around bleeding while continuing to remain

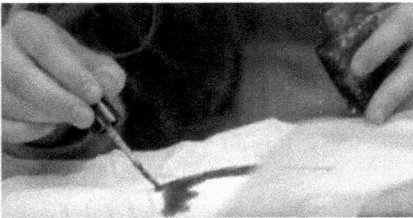

Fig. 30.1 a and b. "First Moon Party" (2014) by HelloFlo Period Starter Kit (partnered with Kotex). The film stills feature the young white girl painting rubylicious nail polish on a pad in hopes of fooling her mom and friends that she has gotten her period

Fig. 30.2   a and b. "First Moon Party" (2014) by HelloFlo Period Starter Kit (partnered with Kotex). The film stills feature the blood-themed menstrual party the mother throws when her daughter tries to fool her that she's experiencing menarche. Notice the playful approach of the "First Moon Party" exhibited by the bleeding red fondue fountain and the menstruation-man exploding from the cake

inattentive to issues of menstrual injustice (such as uneven access to menstrual products) as well as menstrual pain, discomfort, and gender dysphoria. Also, these Mpads continue to feminize menstruation as an experience that is only pertinent to young girls who present in feminine ways, failing to imagine how variously gendered children might also experience menstruation as well as be incapable of experiencing it on liberating, empowering, or positive terms. Further, in subtle ways these Mpads continue to focus on bringing ease and positivity to the white menstrual experience as they continue to focus on white girlhood, even while including girls of color.

The second approach to "liberated" menstrual product advertising moves away from a feminized representation of period bleeding. It involves a discourse of "fit bleeding" that strives to distance itself from the feminization of menstruation and move toward a masculinizing discourse around bleeding rooted in toughness and sportiness on able-bodied terms. The bodies represented are fit, capable, thin, and non-disabled, grounded in ideals of the perfectly optimized feminine body that exists in itself as a symbol of social equality and liberation. These ads tap into decade-old discourses around situating sportiness as an aspirational mode of femininity, a desirable way to embody "being a girl." Discourses of sportiness and activeness have been rehearsed in menstrual product advertising since the early days of Mpads such as with the use of various sports including golf in the later 1920s, as well as swimming, cycling, and more recently tennis (featuring Serena Williams) (Jutel 2005; Rice 2014, 220). Promoting activity as a form of freedom from female biology and/or menstruating, advertisers have for decades relied on selling products through the trope of bodily freedom, implying that "products would free them from the bondage of female bodies and preserve femininity" (Vostral 2008, 74; Rice 2014, 220; Kissling 2006). Drawing on sports to prove the maximum protection and dynamism of a particular menstrual product brand, this tradition of ads stipulates women's and

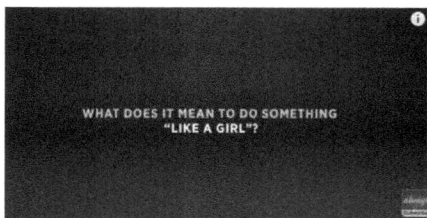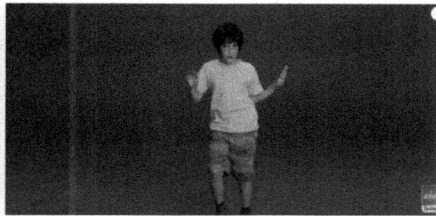

Fig. 30.3 a and **b**. "#LikeAGirl" first launched by Always in 2014. The film stills feature a white young boy performing running "like a girl"

girl's sportiness as an achievement of liberation, as a demonstration of gender equality, and as an accomplishment facilitated by menstrual products themselves.

For example, "#LikeAGirl" first launched by Always in 2014, actively distances itself from sexist notions around gender while remaining caught up in envisioning girlhood in terms of bodily capacity, ability, and sport performance. Parallel to "Reality Check" by U by Kotex, Always develops a strategy that involves distancing itself and absolving itself of responsibility for menstrual shaming, sexism, and body negativity that it has capitalized on since its introduction by Procter and Gamble in 1983. In the original "#LikeAGirl" Mpad from 2014, a series of white and of color cisgender girls, young women, one boy, and one man are asked to perform actions such as "run like a girl," "fight like a girl," and "throw like a girl" (see Fig. 30.3). Always presents this pseudo-feminist lesson as a moment to assert that "confidence plummets during puberty" and that "Always wants to change that." Again, without providing a contextualization of the sexist, racist, and fatphobic landscape in which girls come to hate their bodies in the first place, and without taking responsibility for its role in perpetuating sexist stereotypes in the past through its advertising, Always appears here as an enlightened, pseudo-feminist company. Benefiting directly from menstrual activism and feminist involvement in shifting ideas around bodies and menstruation, Always instead presents itself as a vanguard brand at the head of the feminist struggle. In doing so, it directly co-opts feminist discourses to rebrand its previously sexist advertising.

Further, Always uses sportiness to gauge social equality, progress, and bodily empowerment. Drawing on discourses that situate sportiness as necessarily an achievement, Always denigrates so-called imagined "feminine" behaviors such as lack of strength, unsportiness, and performance failure on ableist, sexist, and implicitly cisgenderist and fatphobic terms. Failing to see how top sport performance and a "fit" body might not be a desired goal for all, Always minimizes the pain, discomfort, and dysphoria that might be associated with menstruation for many and especially for those who experience menstrual pain, menstruators with disabilities, and gender nonconforming persons. Eliding menstruation altogether, "#LikeAGirl" by Always hones menstrual invisibility, making confusing connections between socialization, sexism, and periods that leave viewers unclear as to the role periods and Always are supposed to play in empowered subjecthood. If anything, "fitness" is conceived of as the ability to somehow overcome one's period *and* femininity with the enlightened Always brand leading the way.

The brand Libresse, going by Bodyform in the UK and other names elsewhere, is owned by the Swedish Essity. In recent years, Libresse has moved to advertising that taps more directly into the empowered and liberated rhetoric of the other ads we have been examining. "Blood" by Bodyform UK (2016) draws on a yet different visual language to portray fit bleeding and menstrual empowerment. The commercial features athletic and sporty bodies outperforming expectations of femininity through boxing, cycling, rugby, running, surfing, rock climbing, ballet, horse sword fighting, and skateboarding. The bodies are all fit, thin, and able-bodied though significantly not all white and with some gender variance, as masculine-presenting menstruators are featured in the ad (see Fig. 30.4). Here, all the bodies are shown to be bleeding due to their high sport performance—we see bleeding hands, bleeding toes, blood trickling down the face, scraped knees, et cetera. The slogan in turn reads "no blood should hold us back" and "live fearless bodyform" while an appropriated Native American "war song" plays in the background. Recasting menstruating on fit and tough terms, blood is effectively reconceived as a muscular and masculinized experience akin to other forms of high-performance bleeding. Empowerment operates here on tougher and more astutely marketable terms as menstruating is rebranded as both an experience that does not stand in the way of optimal performance as well as part and parcel of the dirty, gritty life that characterizes excelling at any sport. Drawing on a masculinizing rhetoric of celebrating sport injuries and a dark, gritty, serious color scheme (see Fig. 30.4), "Blood" remakes bleeding into a heroic pursuit in a way that no other ad has managed to do. At the same time, "Blood" invites us to imagine menstrual blood as part of a life of

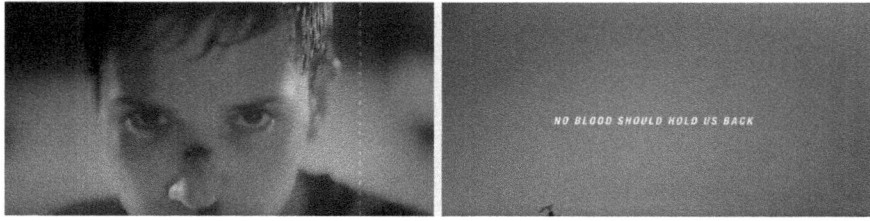

Fig. 30.4 **a** and **b.** "Blood" by Bodyform UK//Libresse (2016). The film stills feature a masculine-presenting boxer bleeding from the nose after a boxing match as an example of how "no blood should hold us back" whether it be menstrual or nonmenstrual blood

fitness on thin and largely able-bodied terms. Also, even as multiple forms of bleeding are shown on heroic terms toward selling menstrual products, menstrual blood or bleeding is nowhere visually indexed in this ad. Thus, even while "Blood" argues for the similarity between "all blood" in its message, it refrains from doing so in its visuals.

In various ways, the Mpads we have explored foreground a new era of menstrual product advertising and aim to speak to an empowered menstruator who, for the most part, is enticed to envision menstruation on positive, celebratory, and liberated terms. Always, U by Kotex, and Libresse/Bodyform, position themselves as leaders of this new vanguard of empowered menstrual positivity, tending to critique and distance themselves from their role in upholding and creating sexist menstrual shaming discourses. At the same time, the "new" empowered menstrual subject continues to be envisioned on cisgenderist terms often directly rooted in an ableist approach to bodily performance and sportiness. Not only are disabled bodies not portrayed, ableism is imbued in the fit performance of an achievement-oriented sporty femininity that is masculine, rough, and above all imagined as socially equal to men through the disavowal of continuing socially and capitalistically sanctioned misogyny and inequality. Drawing on Sarah Banet-Weiser and Laura Portwood-Stacer's (2017) discussion of commodity feminism, these advertisements "[seem] to explicitly recognize that inequality exists while stopping short of recognizing, naming, or disrupting the political economic conditions that allow that inequality to be profitable" (886). In conceiving of menstruation on predominately celebratory lines, an "empowered" menstruating subject is developed, one that is girly and peppy, feminine yet sporty, and that is distanced from menstrual pain, discomfort, and gender dysphoria. While reinvisioning menstruation is a vital aspect of challenging sexism, these ads co-opt feminist discourse and are strategically ignorant of the pernicious roles that capitalism itself plays in the menstrual experience. These Mpads tout a new "empowered," "liberated," and "reflexive" approach to menstrual products that claims to be critical of itself and co-opts the language of menstrual and bodily positivity in a thinned way and ultimately toward pushing consumer capitalist goals of selling menstrual gear. Even more troublingly, the "empowered" message sells a happy and mood-controlled menstrual experience that foregoes questions of pain and menstrual justice.

CRANKY MENSTRUATORS: BEYOND BODY AND MENSTRUAL POSITIVITY

In many ways, some of the Mpads we reviewed push the envelope of advertising, dismantle harmful ideas around menstruation, and challenge sexist discourses around body shaming. Yet they do so not because Kotex, Always, or Bodyform are themselves liberating products to use, but rather because feminist and menstrual activisms have paved the way for a rhetoric of body positivity and menstrual positivity to take hold. Body positivity draws on decade long traditions that originate in Black Power, disability justice, transgender organizing, and fat activism. Maxine Leeds Craig (2002) discusses how the politics of Black Power in the sixties and seventies were firmly grounded in aesthetic styles that established that "black is beautiful." Remaking the very concept of beauty, black activists and thinkers of the era laid claim to self-love in a black-loathing U.S. context. This in turn expanded the possibilities of representation in advertising "to include some black women" (108), with, for example, the first black woman appearing in an Mpad in 1962 in *Ebony* magazine. Further, "[t]he message embedded in the styles and practices [of the era] was throw off the old and love yourself. Self-love became understood as a political act" (Craig 96; also see Hobson 2016). In other words, black activism conceptualized body positivity and self-love as a deeply political terrain central to revolutionary struggle.

Yet self-love and body positivity are necessarily transformed in the hands of advertisers. Relying on activist body work, ad campaigns draw on the cultural and political resources of feminism and other social movements both indirectly and directly, such as for example with the hiring of feminist experts as consultants for ad campaigns as with Dove's Real Beauty Campaigns. Rosalind Gill and Ana Sofia Elias (2014) argue that "love your body" discourses in advertising contain elements of psychic regulation since a new disciplinary norm has emerged that requires self body-love for feminine-presenting subjects in particular. It is ironic to think that just as a greater variety of bodies are represented in advertising, the spectrum of acceptable moods and dispositions toward one's body is narrowing.

"Love your body" themes in advertising, Gill and Elias outline, are forms of "emotional capitalism" (Illouz 2007), and while they portray bodies on more positive terms, this is often simultaneously undermined by an imperative to love one's body without a full consideration of social contexts of misogynism, transphobia, fatphobia, ableism, and ageism. Rather, an individualized responsibility emerges to overcome social hatred within oneself, to change one's relationship with oneself and one's body because one's "relationship to the self has gone bad or been broken" (Gill and Elias 2014, 181). Redefined beauty and abstract love are held open as possibilities to consumers even as political struggle and collective organizing that might challenge capitalistic self-realization are not. As such, a necessarily shallow version of body positivity emerges, one that reroutes its original politicized energies toward consumer culture and individualized "overcoming."

In this section, we argue that body and menstrual positivity need to be complimented by bodily and menstrual crankiness. We hold that while bodily celebration, including menstrual positivity, is an important component of the feminist landscape today, it is equally important to retain a cranky approach to the body and embodiment—one attuned to pain, leaking, trauma, discomfort, and dysphoria. A cranky approach to menstruation, in particular, asks that we take stock of the uncomfortable and dissident feelings that are part and parcel of the menstrual experience as much as the ads we reviewed suggest otherwise. Rather than celebrating menstruation and advances in menstrual ads, we call for a cranky approach to bleeding that is invested in social justice, demands access to menstrual products for all, and that is alerted to the pain and discomfort involved in bleeding.

Reconceiving of menstruation on positive terms has been central to shifting women and menstruators' embodied experiences of bleeding. Menstrual bleeding has a long history of being stipulated on profoundly negative terms and has been historically conceived of as a pathology, debility, and biological proof of gendered inferiority (Steinem 1978; Aristotle in Dean-Jones 1994; Martin 2001; Vostral 2008). Menstrual activists and feminists have led the way in challenging such negative understandings of menstruation through art, zines, protest signage, performance, and online discourse (Bobel 2006, 2010; Clemmer 2017; Fahs 2016; Truax 2017). At the same time, menstruation can be a difficult and taxing embodied experience entwined with pain (such as dysmenorrhea and endometriosis), discomfort, dips in mental health, and it can shore up gender dysphoria for transmasculine, genderqueer, and gender variant bleeders (Chrisler et al. 2016; Fahs 2016). While menstrual pain is routinely denied the status of "real" pain, studies suggest that up to 85% of menstruators experience pain, sometimes as serious as the pain experienced during a heart attack (Przybylo and Fahs 2018; Fenton 2016; Jones 2016; Balbi et al. 2000; Banikarim, Chcacko, and Kelder 2001; Grandi et al. 2012; Ju, Jones, and Mishra 2014). Though some report positive experiences with menstruation, overwhelmingly, menstruating women report menstrual negativity (likely because they exist in a culture of shaming menstruation), including emotions like shame, disgust, embarrassment, and horror (McPherson and Korfine 2004; Roberts 2004). These feelings are at once somatic, psychic, and rooted in social stigma. The first period is also often experienced as traumatic in itself, and is characterized by distress, shock, and dread (Rice 2014, 227–28).

Periods can also trigger gender dysphoria, or the distress associated with one's gender not aligning with one's body appearance (Bell 2017; Reading 2014). Notably, advertising for menstrual products can itself have a disorienting and triggering effect on transmasculine and genderqueer bleeders, since it "enforces the otherness of [one's] period" and one can "feel mocked by the celebratory ads and the excited-looking packaging" (Zulch 2017). Women who do not bleed, including postmenopausal women, post-hysterectomy women, and transgender women, are in turn publicly left out of conceptualizations of womanhood through Mpads that serve to draw up the contours of "womanhood" in terms of whether one menstruates or not (Bobel 2010). Further, most MPads visually exclude transgender women from the menstrual celebrations of womanhood. Alongside blatantly transexclusionary advertising, the effusively celebratory tone common among the Mpads we examined has a deleterious effect on many bleeders who experience mental unwellness and physical pain during their periods. In this regard, even while an overly celebratory and peppy tone of menstrual product advertising seems to advocate for a more positive approach to the menstruating body, it fails many types of bodies including trans, genderqueer, gender variant folks, menstruators with disabilities, and those who experience pain and dips in mental wellness during their periods. We thus see the need to develop an approach to periods that creates space for the difficult aspects of bleeding and fosters a "cranky" approach to menstruation.

We see menstrual crankiness as an approach to bleeding that hones and acknowledges the negative, troubling, painful, and traumatic aspects of the period experience. First, this approach challenges the overwhelming social call to be happy and positive, including in regards to bleeding. Sara Ahmed (2010) outlines how happiness has been used as a form of social regulation, especially in regards to oppressed people, so that happiness and positivity emerge as "not so much a right as a responsibility. We have a responsibility for our own happiness insofar as promoting our own happiness is what enables us to increase other people's happiness" (9; see also Frye 1983, 2–3). Positivity has emerged as an American paradigm rooted in U.S. exceptionalism and a pull yourself up by the bootstraps orientation based in the false idea that "success happens to good people and failure is just a consequence of a bad attitude" (Halberstam 2011; see also Ehrenreich 2009). In relation to menstruation this is rooted in the idea that positive thinking can positively affect the menstrual experience and wish away bodily trauma and pain. It thus deduces that one's menstrual experience—whether it is happy or difficult—is based on one's psychology and "outlook" rather than in the materialities and structural inequalities of pain and trauma.

A cranky approach to menstruation asks that we question broader projects around body positivity and menstrual positivity and the ways in which they have been co-opted. As Jack Halberstam (2011) writes, crankiness, like other "negative affects [. . .] poke[s] holes in the toxic positionality of contemporary life" (3). In honing a cranky approach to menstruation, we are interested in challenging the overwhelming call to "love our periods" and "love our bodies" even while the "our" of these statements is always more available to some more so than others. The cranky menstruator is positioned in opposition to these celebratory scripts, understanding them as just another way to milk menstruators' pockets for money. In this sense, the cranky menstruator is, drawing again on Ahmed (2010), a "troublemaker [who] might be trouble because [they get] in the way of the happiness of others" by refusing to make bleeding palatable and celebratory (60). Like Ahmed's killjoy who "might bring others down [. . .] by talking about unhappy topics such as sexism" the cranky menstruator is cranky about menstrual pain, menstruatory exclusion, menstrual capitalism, and uneven access to menstrual products (66). Ahmed argues that "happiness provides as it were a cover, a way of covering over what resists or is resistant to a [particular] view of the world" (83–84); we see menstrual positivity in advertising functioning in a similar way, covering up particular truths about bleeding in pursuit of a peppy, fit, and utilizable bleeding body. In this sense, crankiness is both metaphoric and embodied, grounded in a cranky approach to injustice fueled by feminist anger as well as by the material aspects of menstrual crankiness.

Yet even as the Mpads we reviewed hold out happiness in the form of menstrual positivity to menstruating bodies, they implicitly draw a line as to who can share in this menstrual happiness by excluding bodies of size, masculine of centre and genderqueer bodies, transfeminine bodies, and disabled bodies from its cornucopias of menstrual celebration. Drawing on Ahmed's words, these differently menstruating bodies are "banished" from happiness in the menstrual advertising landscape, with menstrual happiness and celebration not evenly extended to all (17). Or in Halberstam's language, menstrual experiences and bodies not represented in advertisement celebrations come to form the "failed" subjects of menstruation—those who cannot, will not, or are not invited to celebrate in the blood party.

A body and menstrual positivity that wants to guarantee that we love our periods and bodies flattens and disregards some of the more difficult and troubling aspects of bleeding. Using crankiness to critique the co-optation and wholesale application of menstrual positivity creates space to recognize the realness of menstrual pain and menstruation-related trauma as well as the ways in which menstruation can actually stimulate bodily hatred and loathing rather than bodily love. Allowing for a recognition of the self-loathing that bleeding can ignite suggests that working toward body-love and menstrual-love are not always practical, feasible, or even preferred models for coping with periods. Rather, body-love and menstrual-love, are conveniently utilizable by capitalist schemas of round-the-month productivity that encourage menstruators to "overcome" pain and trauma at all costs toward better serving the labor force and ideals of a contained, non-leaky, fit, and non-disabled body. A cranky approach to menstruation also challenges the idea of the celebrated "fit" bleeder proffered in many Mpads. A menstrual crankiness thus is cranky about a model of bleeding that adheres to a fitness paradigm and represents particular bodies toward selling the period as something that can and must be overcome. Instead, menstrual crankiness recognizes that period pain *can* prevent sport performance, that periods as they relate to dysphoria, trauma, and depression, can and do incite anger, discontent, disavowal, and dissonance rather than happy positive feelings.

Finally, a cranky approach to menstruation also suggests that we need to be very, very cranky about how menstrual products advertise themselves as "liberated" and their menstrual gear as "liberating" while menstruators in North America and around the world are not provided access to menstrual supplies such as tampons, pads, cups, and pain killers. With prices of menstrual products increasing annually, and with taxes placed on menstrual products that mark them as an "inessential" commodities (though this is starting to change), access to menstrual products operates as a capitalist privilege aligned with class privilege, white privilege, and first world privilege. Groups in vulnerable situations such as incarcerated people, homeless people, and even hospitalized people are routinely denied and restricted access to this basic human necessity, just as menstruators in the Global South are portrayed as "uncivilized" for using rags, bark, or other non-commercial materials for managing their menstrual cycles (Bobel 2019). If menstrual commodities, as Mpads have suggested for decades, are emblems of "empowerment" that make civic participation more possible for menstruators through containing bleeding, access to menstrual products and recognition that many ways of managing menstruation are possible, needs to be extended far and wide. Activists have been getting cranky about this throughout the last decade, as is demonstrated by efforts ranging from eliminating the "tampon tax" (Crawford and Spivack 2017), to providing menstrual products to incarcerated women (Marusic 2016; White 2018), to questioning the (largely white) efforts to "give" menstrual products to women and girls in the Global South (Bobel 2019).

A cranky approach to menstruation thus needs to be committed to forgoing menstrual celebration in favor of greater access to menstrual supplies and products and a broader challenging of consumer capitalism. Menstrual activism itself must continue its cranky quest to move beyond products-as-the-end-goal and instead look deeper into the root structures of why Mpads exist and why they "sell" menstruators their own empowerment through highly problematic frameworks of narrow femininity. And finally, a radical analysis of *why women lack such access* must underlie all efforts to disseminate menstrual products on a wider scale; to ignore the bigger stories surrounding the traveling, marketing, and usage of menstrual products reduces the potential justice implications of radical menstrual activist work. Such activist work seeks to not only transform the way people "manage" their periods, but also to link up with other social justice movements that think and work together on ways to reduce stigma, improve the material conditions of people's lives, and transform thinking about bodies and embodiment.

REFERENCES

Ahmed, Sara. 2010. *The Promise of Happiness*. Durham, NC: Duke University Press.

Balbi, C., R. Musone, A. Menditto, L. Di Prisco, E. Cassese, M. D'Ajello, D. Ambriosio, and A. Cardone. 2000. "Influence of Menstrual Factors and Dietary Habits on Menstrual Pain in Adolescence Age." *European Journal of Obstetrics & Gynecology and Reproductive Biology* 91 (2): 143–48.

Banet-Weiser, Sarah, and Laura Portwood-Stacer. 2017. "The Traffic in Feminism: An Introduction to the Commentary and Criticism on Popular Feminism." *Feminist Media Studies* 17 (5): 884–88.

Banikarim, Chantay, Mariam R. Chcacko, and Steve H. Kelder. 2001. "Prevalence and Impact of Dysmernorrhea on Hispanic Female Adolescents." *Archives of Pediatrics and Adolescent Medicine* 154 (12): 1226–29.

Bell, Jen. 2017. "Gender Dysphoria and Your Cycle." *Clued In*. https://medium.com/cluedin/gender-dysphoria-and-your-cycle-5c9922bad7ef.

Berg, D. H., and L. Block Coutts. 1994. "The Extended Curse: Being a Woman Every Day." *Health Care for Women International* 15 (1): 11–22.

Bobel, Chris. 2006. "'Our Revolution Has Style': Contemporary Menstrual Product Activists 'Doing Feminism' in the Third Wave." *Sex Roles* 54 (5-6): 331–45.

———. 2010. *New Blood: Third Wave Feminism and the Politics of Menstruation.* Camden, NJ: Rutgers University Press.

———. 2019. *The Managed Body: Developing Girls and Menstrual Health in the Global South.* London: Palgrave.

Chrisler, Joan C., Jennifer A. Gorman, Jen Manion, Michael Murgo, Angela Barney, Alexis Adams-Clark, Jessica R. Newton, and Meaghan McGrath. 2016. "Queer Periods: Attitudes toward and Experiences with Menstruation in the Masculine of Centre and Transgender Community." *Culture, Health & Sexuality* 18 (11): 1238–50.

Clemmer, Cass. 2017. *Toni the Tampon: A Period Coloring Book.* New York: Bloody Queer.

Coutts, L. Block, and D. H. Berg. 1993. "The Portrayal of the Menstruating Woman in Menstrual Product Advertisements." *Health Care for Women International* 14 (2): 179–91.

Craig, Maxine Leeds. 2002. *Ain't I a Beauty Queen? Black Women, Beauty, and the Politics of Race.* New York: Oxford University Press.

Crawford, Bridget J., and Carla Spivack. 2017. "Tampon Taxes, Discrimination, and Human Rights." *Wisconsin Law Review* 491. http://digitalcommons.pace.edu/lawfaculty/1070/.

Davidson, Anna. 2012. "Narratives of Menstrual Product Consumption: Convenience, Culture, or Commoditization?" *Bulletin of Science, Technology, & Society* 32 (1): 56–70.

Dean-Jones, Lesley. 1994. "Medicine: The 'Proof' of Anatomy." In *Women in the Classical World,* edited by Elaine Fantham, Helene Peet Foley, Natalie Boymel Kampen, Sarah B. Pomeroy, and H. A. Shapiro. 183–205. New York: Oxford University Press.

Ehrenreich, Barbara. 2009. *Bright-Sided: How the Relentless Pursuit of Positive Thinking Has Undermined America.* New York: Metropolitan Books.

Erchull, Mindy J., Joan C. Chrisler, Jennifer A. Gorman, and Ingrid Johnston-Robledo. 2002. "Education and Advertising: A Content Analysis of Commercially Produced Booklets about Menstruation." *The Journal of Early Adolescence* 22 (4): 455–74.

Fahs, Breanne. 2016. *Out for Blood: Essays on Menstruation and Resistance.* Albany, NY: SUNY Press.

Fenton, Siobhan. 2016. "Period Pain Is Officially as Bad as a Heart Attack—So Why Have Doctors Ignored It? The Answer Is Simple." *The Independent,* February 19, 2016. http://www.independent.co.uk/voices/period-pain-is-officially-as-bad-as-a-heart-attack-so-why-have-doctors-ignored-it-the-answer-is-a6883831.html.

Fingerson, Laura. 2012. *Girls in Power: Gender, Body, and Menstruation in Adolescence.* Albany, NY: SUNY Press.

Frye, Marilyn. 1983. *The Politics of Reality: Essays in Feminist Theory.* Trumansburg, NY: Crossing Press.

Gill, Rosalind. 2008. "Empowerment/Sexism: Figuring Female Sexual Agency in Contemporary Advertising." *Feminism and Psychology* 18 (1): 35–60.

Gill, Rosalind, and Ana Sofia Elias. 2014. "'Awaken Your Incredible': Love Your Body Discourses and Postfeminist Contradictions." *International Journal of Media and Cultural Politics* 10 (2): 179–88.

Goldman, Robert, Deborah Heath, and Sharon L. Smith. 1991. "Commodity Feminism." *Critical Studies in Mass Communication* 8: 333–51.

Grandi, Giovanni, Serena Ferrari, Anjeza Xholli, Marianna Cannoletta, Federica Palma, Cecilia Romani, Annibale Volpe, and Angelo Cagnacci. 2012. "Prevalence of Menstrual Pain in Young Women: What is Dysmenorrhea?" *Journal of Pain Research* 5 (2012): 169–174.

Halberstam, Jack. 2011. *The Queer Art of Failure*. Durham: Duke University Press.

Heath, Joseph, and Andrew Potter. 2005. *The Rebel Sell: How the Counterculture Became Consumer Culture*. Chichester and Toronto: Wiley.

Helloflo. 2018. "FAQs." http://helloflo.com/faq/.

Hobson, Janell. 2016. "Black Beauty and Digital Spaces: The New Visibility Politics." *Ada: A Journal of Gender, New Media, and Technology* 10. http://adanewmedia.org/2016/10/issue10-hobson/.

Illouz, Eva. 2007. *Cold Intimacies: The Making of Emotional Capitalism*. Cambridge: Polity.

Jones, Cara. 2016. "The Pain of Endo Existence: Toward a Feminist Disability Studies Reading of Endometreosis." *Hypatia* 31 (3): 554–71.

Ju, Hong, Mark Jones, and Gita Mishra. 2014. "The Prevalence and Risk Factors of Dysmenorrhea." *Epidemiologic reviews* 36 (1): 104–13.

Jutel, Annemarie. 2005. "Cursed or Carefree? Menstrual Product Advertising and the Sportswoman." In *Sport, Culture, and Advertising: Identities, Commodities and the Politics of Representation*, edited by Steven J. Jackson and David L. Andrews, 213–26. London: Routledge.

Kissling, E. 2002. "On the Rag on Screen: Menarche in Film and Television." *Sex Roles* 46 (1-2): 5–12.

———. 2006. *Capitalizing the Curse: The Business of Menstruation*. Boulder: Lynne Rienner.

Luke, Haida. 1997. "The Gendered Discourses of Menstruation." *Social Alternatives* 16 (1): 28–30.

Martin, Emily. 2001. *The Woman in the Body: A Cultural Analysis of Reproduction*. Boston: Beacon Press.

Marusic, Kristina. 2016. "The Sickening Truth about What It's Like to Get Your Period in Prison." *Women's Health*. July 7, 2016. http://www.womenshealthmag.com/life/women-jail-periods.

McPherson, Marianne E., and Lauren Korfine. 2004. "Menstruation across Time: Menarche, Menstrual Attitudes, Experiences, and Behaviors." *Women's Health Issues* 14 (6): 193–200.

Newman, Adam Andrew. 2010. "Rebelling against the Commonly Evasive Feminine Care Ad." *The New York Times*, March 15, 2010. http://www.nytimes.com/2010/03/16/business/media/16adco.html.

Przybylo, Ela, and Breanne Fahs. 2018. "Feels and Flows: On the Realness of Menstrual Pain and Cripping Menstrual Chronicity." *Feminist Formations* 30 (1): 206–29.

Raftos, Maree, Debra Jackson, and Judy Mannix. 1998. "Idealised Versus Tainted Femininity: Discourses of the Menstrual Experience in Australian Magazines That Target Young Women." *Nursing Inquiry* 5 (3): 174–86.

Reading, Wiley. 2014. "My Period and Me: A Trans Guy's Guide to Menstruation." *Everyday Feminism*. https://everydayfeminism.com/2014/11/trans-guys-guide-menstruation/.

Rice, Carla. 2014. *Becoming Women: The Embodied Self in Image Culture*. Toronto: University of Toronto Press.

Roberts, Tomi-Ann. 2004. "Female Trouble: The Menstrual Self-Evaluation Scale and Women's Self-Objectification." *Psychology of Women Quarterly* 28 (1): 22–26.

Rosewarne, Lauren. 2012. *Periods in Pop Culture: Menstruation in Film and Television.* New York: Lexington Books.

Ross, Loretta, Elena GutiŽrrez, Marlene Gerber, and Jael Silliman. 2016. *Undivided Rights: Women of Color Organizing for Reproductive Justice.* Chicago: Haymarket Books.

Saz-Rubio, Del, Ma Milagros, and Barry Pennock-Speck. 2009. "Constructing Female Identities Through Feminine Hygiene TV Commercials." *Journal of Pragmatics* 41 (12): 2535–2556.

Simes, Mika R., and D. H. Berg. 2001. "Surreptitious Learning: Menarche and Menstrual Product Advertisements." *Health Care for Women International* 22 (5): 455–69.

Steinem, Gloria. 1978. "If Men Could Menstruate: A Political Fantasy." *Ms. Magazine*: 110.

Thornton, Leslie-Jean. 2013. "'Time of the Month' on Twitter: Taboo, Stereotype and Bonding in a No-Holds-Barred Public Arena." *Sex Roles* 68 (1–2): 41–54.

Tone, Andrea. 1996. "Contraceptive Consumers: Gender and the Political Economy of Birth Control in the 1930s." *Journal of Social History* 29 (3): 485–506.

Truax, Raegan. 2017. "Sloughing." https://raegantruax.com/durational-performances/sloughing/.

Vostral, Sharra L. 2008. *Under Wraps: A History of Menstrual Hygiene Technology.* Lanham, MD: Lexington Books.

White, Kaila. 2018. "Arizona Prisons Will Now Give Female Inmates Free Tampons." *Arizona Central.* https://www.azcentral.com/story/news/politics/legislature/2018/02/21/arizona-prisons-give-free-tampons-female-inmates/361075002/.

Winkler, Inga T., and Virginia Roaf. 2014. "Taking the Bloody Linen out of the Closet: Menstrual Hygiene as a Priority for Achieving Gender Equality." *Cardozo Journal of Law and Gender* 21: 1–37.

Zeisler, Andi. 2016. *We Were Feminists Once: From Riot Grrrl to CoverGirl®, the Buying and Selling of a Political Movement.* New York: PublicAffairs.

Zulch, Sebastian. 2017. "How Having a Period Negatively Impacts My Mental Health." *Helloflo.* http://helloflo.com/period-negatively-impacts-mental-health/.

CHAPTER 31

"You Will Find Out When the Time Is Right": Boys, Men, and Menstruation

Mindy J. Erchull

The majority of menstruators are women—although not all women menstruate and not all menstruators are women. Given this, it is not surprising that there has been little focus on the attitudes about and experiences with menstruation among men. This chapter reviews the research about menstruation and boys and men who are not themselves menstruators in order to shed light on how they learn about menstruation, what they actually know about menstruation, and what beliefs and attitudes boys and men hold about both menstruation and menstruators. It is important to note that nearly all of the research in this area reflects an underlying assumption of a gender binary. Because of this, binary gendered language is often used in this chapter to reflect the nature of the research that has been done. Moreover, this chapter focuses cisgender boys and men. While a critically important topic to study, the experiences of men and non-binary people who menstruate is not covered in this chapter—largely because, to date, there is little academic work on this topic (for notable exceptions, see Chrisler et al. 2016; Fahs 2016; Rydström, this volume; Frank and Dellaria, this volume).

Education About Menstruation

Menstruation is typically framed as a normal, natural biological process, but at the same time, it is also framed as a process that should be kept private or secret—particularly from boys and men (Stubbs 2008; White 2013). This ambivalent cultural context is the backdrop in which girls and boys learn about menstruation, and it continues to influence the beliefs and attitudes of women and men throughout their lives. While girls are seen as those who need to know about menstruation because they will, typically, experience it,

© The Author(s) 2020
C. Bobel et al. (eds.), *The Palgrave Handbook of Critical Menstruation Studies*, https://doi.org/10.1007/978-981-15-0614-7_31

nonetheless, many adolescent girls included in one study in the United States reported feeling unprepared for menarche (White 2013). Research performed in the United States indicates that mothers are a primary source of information about menstruation for girls, but they also learn about menstruation from their peers, the media, and school-based health education programs—the latter of which are often gender segregated (Allen, Kaestle, and Goldberg 2011; Fingerson 2005; Koff and Rierdan 1995; Lee 2008). Also, other research from the United States has shown that some girls report not feeling well-educated about or prepared for menarche (Cooper and Koch 2007; Costos, Ackerman, and Paradis 2002). Even when girls report receiving information about the biology of menstruation and how to manage it hygienically, they want additional information about more subjective aspects of menstruation such as how menstruation feels physically (Koff and Rierdan 1995).

Like girls, boys in the United States typically report incomplete and/or inaccurate knowledge about menstruation, and boys generally receive less education about menstruation than do girls (Allen, Kaestle, and Goldberg 2011). Research from Taiwan (Chang, Hayter, and Lin 2012), the U.K. (Lovering 1995), and Australia (Peranovic and Bentley 2017) indicates that gender-segregated health and sexual education curricula rarely include menstruation in the content covered with boys. Even when these programs are not gender segregated as was the experience in a sample of participants from New Zealand, the information provided about female pubertal development and menstruation is often framed less positively than is true for the information about male pubertal development (Diorio and Munro 2000). Of course, formal education programs are not the only place boys learn about menstruation. Research conducted in the United Sates (Allen, Kaestle, and Goldberg 2011) and the U.K. (Lovering 1995) shows that, like girls, boys sometimes learn about menstruation from their mothers. Research more broadly about sexual education in the United States indicates that peers and the media are often key sources of information for boys (Epstein and Ward 2008).

When boys do learn about menstruation, they often report gaining knowledge in relatively informal ways. David Linton (2019) described a theme that emerged from his discussions about menstruation with men: They often first encountered menstruation through a "mysterious incident" or "encounter that left them asking, 'What's going on here? What am I not supposed to know?'" (Linton 2019, 8). For example, in one study, some young men from the U.S. reported having learned about menstruation for the first time when a sister experienced menarche (Allen, Kaestle, and Goldberg 2011). This does not mean that they were actually learning a great deal, however. For example, one participant recalled being frightened when his sister had her first period and had "a huge blood stain in the back of her night clothes" (Allen, Kaestle, and Goldberg 2011, 139). When he subsequently asked his mother for more information, "she only stated that 'you will find out when the time is right'" (139). In another study conducted in the United States, boys reported learning about menstruation through overhearing discussions among female friends and classmates (Fingerson 2005). Men in both the United States (Allen and Goldberg 2009; Allen, Kaestle, and Goldberg 2011; Erchull and Richmond 2015) and Australia (Peranovic and Bentley 2017) often report that they learned information about menstruation from sexual and romantic partners. One Australian man said that "if a man lives with a female partner he needs to understand as much as it is possible about all aspects of her life including menstruation" (Peranovic and Bentley 2017, 120). A participant in another study from the United States discussed how he learned about menstruation in order to help identify when his partner would be most likely to conceive when they decided to have a child (Allen, Kaestle, and Goldberg 2011).

Given the lack of education boys receive about menstruation, the fact that research consistently shows that men report feeling less knowledgeable about menstruation than women is to be expected. Studies done in the United States (Brooks-Gunn and Ruble 1986) and Italy (Amann-Gainotti 1986) in the 1980s indicated that adolescent boys and young men generally reported learning less about menstruation than female peers reported, and they often did not hold an accurate understanding of what menstruation was and how it worked. For example, they might understand that it involves blood coming out of the vagina, but they might attribute to something other than the shedding of the uterine lining, such as "the breaking of the ovaries" or to expel a harmful microbe (Amann-Gainotti 1986, 706). In other cases, they may misunderstand even more of what is involved as was the case with a 14-year-old boy who thought that menstruating involved urinating "blood because the bladder breaks" (Amann-Gainotti 1986, 706).

Similar findings about cursory and inaccurate knowledge about menstruation have since been found in studies with samples from around the globe. High school boys in the United States were described in one study as "half-knowing" about menstruation where they were familiar with it but were unsure of the details such as the differences between pads and tampons or the PMS stood for *pre*menstrual syndrome rather than *post*-menstrual syndrome (Fingerson 2005, 101). In part because of receiving information from myriad informal sources, a number of participants in one study of male undergraduate students in the U.S. reported that they "pieced together" information about menstruation as children and adolescents (Allen, Kaestle, and Goldberg 2011, 141). In a study of junior high school students in Taiwan, despite most of the male participants being familiar with menstruation, they had less accurate knowledge of menstruation than did their female peers, and they were more likely to endorse cultural myths about menstruation and restrictions on menstruating girls and women (Cheng, Yang, and Liou 2007). These included not eating or drinking iced foods or going to temple while menstruating. Similarly, in another study of Taiwanese early adolescents, boys reported a lack of knowledge and held misinformation about menstruation including that girls have periods every one to two weeks, that menstrual fluid is comprised of "blood, ovum and . . . sperm," and that when girls "have their periods, they can give both to a baby by themselves" (Chang, Hayter, and Lin 2012, 517).

Research conducted in Australia, however, does indicate that men do often encounter, and even seek out, additional information about menstruation as they get older (Peranovic and Bentley 2017). Much of this added knowledge seems to come from discussions with female partners as was the case with a sample of participants from the United States (Allen, Kaestle, and Goldberg 2011). Given this, it is not clear that men actually know more about menstruation through these types of discussions. It may well be that they are just more familiar with menstruation. Other research conducted in Brazil (Caçapava Rodolpho et al. 2016) and the U.K. (Liao, Lunn, and Baker 2015) focusing specifically on the experiences of men with female partners in perimenopause indicates that there is still a lack of knowledge. Men reported that they were only familiar with menopause in general terms and lacked key information. Some men did, however, feel that they had a responsibility to seek out information about menopause to be better able to support their partners. However, this is another case where men are haphazardly piecing together information from a number of, largely informal, sources.

BELIEFS AND ATTITUDES ABOUT MENSTRUATION

Among girls and women, lack of knowledge about and inadequate preparation for menstruation has been shown to be related to more ambivalent and negative attitudes about menstruation in both Italy and the United States (for example, Amann-Gainotti 1986; Scott et al. 1989; White 2013). Given that boys and men typically receive less education about menstruation and have less accurate knowledge, it should be unsurprising that negative beliefs and attitudes are commonly found in research done with diverse groups of participants. For example, older research from the United States showed that college men were more likely to perceive menstruation as debilitating than was true of their female peers (Brooks-Gunn and Ruble 1980, 1986). Other older research conducted in the United States showed that negative attitudes were even held by young adolescent boys and that these were more negative than those of their female peers who had begun to menstruate (Clarke and Ruble 1978). However, another third older study conducted in the United States showed that, while younger men were more likely to view menstruation as debilitating than young women, there was no such gender difference found for an older sample (Chrisler 1988). So, while these negative attitudes may develop early, they may also soften as men age and gain more knowledge of and experience with menstruation.

More recent research with samples from throughout the world have shown a similar pattern of more negative attitudes being held by men than by women. Specifically, in a sample of young adults from the United States, men more strongly endorsed the idea that menstruation requires secrecy than did women (Marván et al. 2006). They also more strongly endorsed the idea that there are things women should and should not during menstruation. Research with Mexican samples has also shown that men are more likely than women to view menstruation negatively, but like the earlier research by Chrisler (1988) in the United States, these negative views were more pronounced for younger rather than older men (Marván, Cortés-Iniestra, and González 2005; Marván et al. 2006). Research conducted in East Asia also shows similar patterns. Male adolescents in Taiwan reported more negative attitudes about menstruation than did their female peers (Cheng, Yang, and Liou 2007). In a study of Chinese undergraduates in Hong Kong, men more strongly endorsed the idea that menstruation should be kept secret than women, but they were actually less likely than the women in the sample to view menstruation as disabling (Wong et al. 2013).

Beliefs and Attitudes About Menstruators

It has been suggested that men's attitudes are particularly important factors to consider in relation to the perpetuation and dissemination of stereotypes about menstruation and menstruators (Fromme and Emihovich 1998). Menstrual ideologies do both reflect and reinforce patriarchal social structures privileging men and boys, and research shows that boys start using menstruation as a means of asserting dominance when they are young. Research conducted in both New Zealand (Diorio and Munro 2000) and the U.K. (Lovering 1995) indicate that this is largely done through teasing and ridiculing girls about their periods. This is not a practice limited to young boys, however. More research conducted in the U.K. suggests that explicit teasing transitions to jokes about menstruation or menstruating women in general (Newton 2016). These jokes, however, often center on menstruating women being unclean or disgusting and can serve to reinforce a patriarchal gender hierarchy.

Research conducted in both Mexico and the United States has also shown that negative attitudes about menstruators are related to endorsement of sexist ideologies—particularly hostile sexism which involves explicitly negative attitudes about women and beliefs about female inferiority—among both female and male participants (Chrisler et al. 2014; Forbes et al. 2003). Both women and men also seem to evaluate women perceived as menstruating negatively. In one study conducted in the United States, a female confederate dropped either a wrapped, unused tampon, or a hairclip in the presence of a participant (Roberts et al. 2002). Confederates who dropped a tampon were subsequently rated as both less competent and less likable than those who dropped a hairclip. Women seem to anticipate this type of negative evaluation. In another study using a United States sample, female participants who were menstruating and thought a male confederate was aware of their menstrual status believed that he liked them less than was true for both participants who thought he did not know that they were menstruating or participants who were not menstruating (Kowalski and Chapple 2000).

These negative evaluations of women are made in a cultural context where negative attitudes and stereotypes about menstruation and menstruators are common. As demonstrated in a study conducted in Australia, many cultural stereotypes surrounding menstruation involve beliefs about excessive emotionality and pain being part of menstruation—experiences that are typically viewed negatively and as undesirable (Christensen and Oei 1990). Generally, pain is perceived as part of the experience of menstruation, and men's beliefs about this are reflected in the findings reviewed above where men are generally more likely than women to perceive menstruation as debilitating. Emotionality, however, is generally more strongly connected to beliefs about the premenstrual phase of the menstrual cycle and is generally thought of as a core component of premenstrual syndrome (PMS; Chrisler and Gorman 2015).

Western cultures, in particular, hold a medicalized view of premenstrual changes women may experience (Chrisler and Gorman 2015). Both women and men endorse this perspective, but stereotypical views about premenstrual women seem to be a key part of men's understanding of menstruation. Men often view PMS as an excuse women make use of to avoid unwanted tasks or to excuse emotional reactivity (King, Ussher, and Perz 2014; Newton 2016; Reberte et al. 2014; Ussher and Perz 2013). In one study conducted in the U.K., men reported that they often heard *other* men attribute women's anger, frustration, or general emotionality to "that time of the month," implying menstruation or being in the premenstrual phase (Laws 1992, 125). They also reported being aware of these types of menstrual cycle-related changes in their female partners. It is also not uncommon for women to be explicitly asked, often by men, if they are *on the rag* or if it's *that time of the month* when they react emotionally, as was found in reports from a different sample of participants from the U.K. (Newton 2016). While both women and men use these phrases, there is a negative and dismissive quality to the way they are often used by men. However, some men did indicate that this type of thinking was reductionist and served to dismiss women's actual concerns and experiences (Laws 1992). Moreover, research conducted in Brazil and Australia indicates that some men do see the premenstrual and menstrual phases as well as perimenopause as times when it is important to be particularly supportive of their partners (Caçapava Rodolpho et al. 2016; King, Ussher, and Perz 2014; Reberte et al. 2014).

DIS/COMFORT WITH MENSTRUATION

An underlying theme in much of the research about boys, men, and menstruation is the idea that not only is menstruation not a domain for men, it is not a topic with which they are comfortable. Given this, menstruation should be discussed and handled discreetly among women. Both Sophie Laws (1992) an David Linton (2019) have described this pattern of keeping menstruation outside of men's attention as part of the etiquette of menstruation. Boys, such as those included in a study conducted in Taiwan, have described learning about menstruation as uncomfortable and embarrassing (Chang, Hayter, and Lin 2012). This discomfort does not necessarily go away as boys mature, however. For examples, while some Australian men in one study reported that they had developed a comfort with discussing menstruation and viewed it as "no big deal," others reported continued discomfort with the topic (Peranovic and Bentley 2017, 119). Other men did not see it as something they need to know about or be involved with.

For men who report comfort with menstruation as part of research conducted in the United States and Australia, intimate relationships with women are often discussed as part of developing this attitude (Allen, Kaestle, and Goldberg 2011; Peranovic and Bentley 2017). They learn more about menstruation from their partners, but they also just come to see menstruation as more normal and routine. While comfort with menstruation may be part of some men's intimate relationships, that is not the case for all men, however. Some men from United States and U.K. samples still expect their partners to conceal evidence of menstruation from them (Allen, Kaestle, and Goldberg 2011; Laws 1992).

A specific context where this dichotomy shows up is sexual contact during menstruation. For men who are uncomfortable talking about menstruation or seeing evidence that menstruation happens (for example, pads and tampons), disgust is often a common reaction to the idea of coming into contact with menstrual fluid. As one man from a United States sample put it, "the idea of getting menstrual blood on me horrifies me" (Allen, Kaestle, and Goldberg 2011, 147). In a study conducted in the United States, men reported being concerned with the messiness inherent in menstrual sex, but they also reported that engaging in this behavior was shameful—particularly for the menstruating women (Allen and Goldberg 2009). However in multiple studies conducted in the United States, women do report desiring and having sex during menstruation even though actual rates of sexual activity go down during this phase compared to others across the menstrual cycle (Fahs 2011; Hensel, Fortenberry, and Orr 2007). However, some of the reason why women do not have sex with men while menstruating may be due to their concerns about disgusting or discomforting male partners (Fahs 2011). For men in a United States sample who did report having menstrual sex regularly, they talked about it as a sign of maturation and viewed it as part of being in a trusting and caring intimate relationship (Allen and Goldberg 2009).

THE SPECIAL CASE OF FATHERS

While intimate relationships are one context in which men are often confronted with menstruation, parenting is another such context. Research conducted in both Australia and the United States indicated that fathers generally struggle to talk with their children about sexuality and sexual health (Kirkman, Rosenthal, and Feldman 2002; Wright 2009). There is often a gendered division of labor in how these conversations are handled where mothers are expected to talk with daughters and fathers are expected to talk with sons. In a United States sample, fathers often reported that menstruation was a topic they are not in a good position to discuss because it was outside of their own experience (Wilson, Dalberth, and Koo 2010). As was found with an Australian sample, however, some men do believe that they need to understand menstruation so that they can take an active role in parenting their children and educating both daughters and sons about menstruation (Peranovic and Bentley 2017). Other research conducted in the United States has found that menstruation was generally part of discussions between mothers and daughters but not fathers and daughters, and it was not part of discussions between either mothers or fathers with sons (Wyckoff et al. 2008).

Fathers can find themselves facing a paradox: their daughters may want their support and want them to be open if daughters want to talk, but they are also uncomfortable actually talking about menstruation with them (Koff and Rierdan 1995). Research conducted in North America does show that girls generally do not talk with their fathers about menstruation and that they may even avoid their fathers starting at menarche (Koff and Rierdan 1995; Uskul 2004). Research conducted in Canada showed that when mothers told fathers about their daughters beginning to menstruate, some girls became upset (Uskul 2004), and daughters in a United States sample reported embarrassment when their fathers tried to make celebratory gestures related to the experience of menarche (Chrisler and Zittel 1998). Not all girls have or live with mothers, however, so single, custodial, and gay fathers cannot necessarily foist menstruation-related interactions onto a female co-parent. Given this, they can find themselves being the parent who is present to provide education about and support around menstruation despite the discomfort both they and their daughters might experience. Girls still may not have these interactions with their fathers, however. In a study conducted in the United States, girls not living with their mothers reported having to wait to get information until they could talk with their mother or another woman with whom they were close, such as a grandmother or older sister (Kalman 2003). They generally reported embarrassment discussing menstruation with their fathers, but for girls who did not have ready access to a woman to talk with, they would talk with their fathers despite reporting discomfort with the process.

While daughters may be uncomfortable talking about menstruation with their fathers, sons just generally do not receive much information about menstruation from either mothers or fathers (Brooks-Gunn and Ruble 1986; Wyckoff et al. 2008). One study conducted in the United States has specifically explored fathers' attitudes about and experiences with educating both their sons and daughters about menstruation (Erchull and Richmond 2015). The fathers in this study did believe that fathers were important sources of information for both sons and daughters (ranked 6th and 10th out of 20 sources, respectively), but they saw themselves as a significantly more important source of this information for their sons. They did not report different levels of comfort with talking to sons and daughters about menstruation, however. That said, there were some differences in what fathers reported as part of these conversations. With daughters, there was a greater focus on positive framing of menstruation, and it was only in regard to talking to their daughters that fathers mentioned avoidance. With sons, fathers were more likely to mention including a discussion of mood swings, and it was only with sons that they reported focusing on having empathy for menstruators as well as providing advice for how to cope with menstruators.

WHERE DO WE GO FROM HERE?

Overall, while there is some consistency in the research findings, there is actually very little research about boys, men, and menstruation to draw on. More research is needed to gain a better understanding of what boys and men actually know about menstruation, how they have gained that knowledge, and how this relates to their attitudes and behaviors. At the heart of this is a problematic cycle where menstruation is viewed as the purview of women, so men are perceived as being uncomfortable with the topic. It is also seen as unimportant to educate men about menstruation in much detail because it is not something they are expected to deal with. This lack of knowledge can contribute to stereotypical beliefs, negative attitudes, and general discomfort. This cycle needs to be interrupted.

Even looking beyond the fact that some men menstruate and not all women do, men do interact with menstruators. They have mothers and sisters who do or did menstruate. They may have partners who menstruate. They parent daughters who do or will menstruate. They share communities with menstruators. When people, regardless of gender identity, understand what menstruation is, what is involved in the menstrual cycle, and how it can (and does not) impact women, menstruation can truly be made normative rather than just being framed in education contexts as a normal biological process that then needs to not be discussed publically because it is taboo. Most men may not menstruate, but knowledge can facilitate comfort, and comfort can facilitate open discussion, and openness can facilitate cultural positivity.

REFERENCES

Allen, Katherine R., and Abbie E. Goldberg. 2009. "Sexual Activity during Menstruation: A Qualitative Study." *Journal of Sex Research* 46 (6): 535–45. https://doi.org/10.1080/00224490902878977.

Allen, Katherine R., Christine E. Kaestle, and Abbie E. Goldberg. 2011. "More Than Just a Punctuation Mark: How Boys and Young Men Learn about Menstruation." *Journal of Family Issues* 32 (2): 129–56. https://doi.org/10.1177/0192513X10371609.

Amann-Gainotti, Merete. 1986. "Sexual Socialization during Early Adolescence: The Menarche." *Adolescence* 21 (83): 703–10.

Brooks-Gunn, Jeanne, and Diane N. Ruble. 1980. "The Menstrual Attitude Questionnaire." *Psychosomatic Medicine* 42 (5): 503–12. https://doi.org/10.1097/00006842-198009000-00005.

———. 1986. "Men's and Women's Attitudes and Beliefs about the Menstrual Cycle." *Sex Roles* 14 (5-6): 287–99. https://doi.org/10.1007/BF00287580.

Caçapava Rodolpho, Juliana Reale, Bruna Cid Quirino, Luiza Akiko Komura Hoga, and Patrícia Lima Ferreira Santa Rosa. 2016. "Men's Perceptions and Attitudes toward Their Wives Experiencing Menopause." *Journal of Women & Aging* 28 (4): 322–33. https://doi.org/10.1080/08952841.2015.1017430.

Chang, Yu-Ting, Mark Hayter, and Mei-Ling Lin. 2012. "Pubescent Male Students' Attitudes towards Menstruation in Taiwan: Implications for Reproductive Health Education and School Nursing Practice." *Journal of Clinical Nursing* 21 (3-4): 513–21. https://doi.org/10.1111/j.1365-2702.2011.03700.x.

Cheng, Ching-Yu, Kyeongra Yang, and Shwu-Ru Liou. 2007. "Taiwanese Adolescents' Gender Differences in Knowledge and Attitudes towards Menstruation." *Nursing & Health Sciences* 9 (2): 127–34. https://doi.org/10.1111/j.1442-2018.2007.00312.x.

Chrisler, Joan C. 1988. "Age, Gender-Role Orientation, and Attitudes toward Menstruation." *Psychological Reports* 63 (3): 827–34. https://doi.org/10.2466/pr0.1988.63.3.827.

Chrisler, Joan, C., and Carolyn B. Zittel. 1998. "Menarche Stories: Reminiscences of College Students from Lithuania, Malaysia, Sudan, and the United States." *Health Care for Women International* 19 (4): 303–12. https://doi.org/10.1080/073993398246287.

Chrisler, Joan C., and Jennifer A. Gorman. 2015. "The Medicalization of Women's Moods: Premenstrual Syndrome and Premenstrual Dysphoric Disorder." In *The Wrong Prescription for Women: How Medicine and Media Create a "Need" for Treatments, Drugs, and Surgery*, edited by Maureen C. McHugh and Joan C. Chrisler, 77–98. Santa Barbara, CA: Praeger.

Chrisler, Joan C., Jennifer A. Gorman, Jen Manion, Michael Murgo, Angela Barney, Alexis Adams-Clark, Jessica R. Newton, and Meaghan McGrath. 2016. "Queer Periods: Attitudes toward and Experiences with Menstruation in the Masculine of Centre and Transgender Community." *Culture, Health & Sexuality* 18 (11): 1238–50. https://doi.org/10.1080/13691058.2016.1182645.

Chrisler, Joan C., Jennifer A. Gorman, Maria Luisa Marván, and Ingrid Johnston-Robledo. 2014. "Ambivalent Sexism and Attitudes toward Women in Different Stages of Reproductive Life: A Semantic, Cross-Cultural Approach." *Health Care for Women International* 35 (6): 634–57. https://doi.org/10.1080/07399332.2012.740113.

Christensen, A. P., and T. P. S. Oei. 1990. "Men's Perception of Premenstrual Changes on the Premenstrual Assessment Form." *Psychological Reports* 66 (2): 615–19. https://doi.org/10.2466/pr0.1990.66.2.615

Clarke, Anne E., and Diane N. Ruble. 1978. "Young Adolescents' Beliefs Concerning Menstruation." *Child Development* 49 (1): 231–34. https://doi.org/10.2307/1128615.

Cooper, Spring Chenoa, and Patricia Barthalow Koch. 2007. "'Nobody Told Me Nothin': Communication about Menstruation among Low-income African American Women." *Women & Health* 46 (1): 57–78. https://doi.org/10.1300/J013v46n01_05.

Costos, Daryl, Ruthie Ackerman, and Lisa Paradis. 2002. "Recollections of Menarche: Communication between Mothers and Daughters Regarding Menstruation." *Sex Roles* 46 (1–2): 49–59. https://doi.org/10.1023/A:1016037618567.

Diorio, Joseph A., and Jennifer A. Munro. 2000. "Doing Harm in the Name of Protection: Menstruation as a Topic for Sex Education." *Gender and Education* 12 (3): 347–365.

Epstein, Marina, and L. Monique Ward. 2008. "'Always Use Protection': Communication Boys Receive about Sex from Parents, Peers, and the Media." *Journal of Youth and Adolescence* 37(2): 113–26. https://doi.org/10.1007/s10964-007-9187-1.

Erchull, Mindy J., and Katherine Richmond. 2015. "'It's Normal… Mom Will Be Home in an Hour': The Role of Fathers in Menstrual Education." *Women's Reproductive Health* 2 (2): 93–110. https://doi.org/10.1080/23293691.2015.1089149.

Fahs, Breanne. 2011. "Sex during Menstruation: Race, Sexual Identity, and Women's Accounts of Pleasure and Disgust." *Feminism & Psychology* 21 (2): 155–78. https://doi.org/10.1177/0959353510396674.

———. 2016. *Out for Blood: Essays on Menstruation and Resistance*. Albany, NY: SUNY Press.

Fingerson, Laura. 2005. "Agency and the Body in Adolescent Menstrual Talk." *Childhood* 12 (1): 91–110. https://doi.org/10.1177/0907568205049894.

Forbes, Gordon B., Leah E. Adams-Curtis, Kay B. White, and Katie M. Holmgren. 2003. "The Role of Hostile and Benevolent Sexism in Women's and Men's Perceptions of the Menstruating Woman." *Psychology of Women Quarterly* 27 (1): 58–63. https://doi.org/10.1111/1471-6402.t01-2-00007.

Fromme, Rebecca E., and Catherine Emihovich. 1998. "Boys Will be Boys: Young Males' Perceptions of Women, Sexuality, and Prevention." *Education and Urban Society* 30 (2): 172–88. https://doi.org/10.1177/0013124598030002003.

Hensel, Devon J., J. Dennis Fortenberry, and Donald P. Orr. 2007. "Situational and Relational Factors Associated with Coitus during Vaginal Bleeding among Adolescent Women." *Journal of Sex Research* 44 (3): 269–77. https://doi.org/10.1080/00224490701443940.

Kalman, Melanie. 2003. "Taking a Different Path: Menstrual Preparation for Adolescent Girls Living Apart from Their Mothers." *Health Care for Women International* 24 (10): 868–79. https://doi.org/10.1080/07399330390244275.

King, Marlee, Jane M. Ussher, and Janette Perz. 2014. "Representations of PMS and Premenstrual Women in Men's Accounts: An Analysis of Online Posts from PMSBuddy.com." *Women's Reproductive Health* 1 (1) 3–20. https://doi.org/10.1080/23293691.2014.901796.

Kirkman, Maggie, Doreen A. Rosenthal, and S. Shirley Feldman. 2002. "Talking to a Tiger: Fathers Reveal Their Difficulties in Communicating about Sexuality with Adolescents." *New Directions for Child and Adolescent Development* 2002 (97): 57–74. https://doi.org/10.1002/cd.50.

Koff, Elissa, and Jill Rierdan. 1995. "Preparing Girls for Menstruation: Recommendations from Adolescent Girls." *Adolescence* 30 (120): 795–812.

Kowalski, Robin M., and Tracy Chapple. 2000. "The Social Stigma of Menstruation: Fact or Fiction?" *Psychology of Women Quarterly* 24 (1): 74–80. https://doi.org/10.1111/j.1471-6402.2000.tb01023.x.

Laws, Sophie. 1992. "'It's Just the Monthlies, She'll Get over It': Menstrual Problems and Men's Attitudes." *Journal of Reproductive and Infant Psychology* 10 (2): 117–28. https://doi.org/10.1080/02646839208403944.

Lee, Janet. 2008. "'A Kotex and a Smile' Mothers and Daughters at Menarche." *Journal of Family Issues* 29 (10): 1325–47. https://doi.org/10.1177/0192513X08316117.

Liao, Lih-Mei, Sarah Lunn, and Martyn Baker. 2015. "Midlife Menopause: Male Partners Talking." *Sexual and Relationship Therapy* 30 (1): 167–80. https://doi.org/10.1080/14681994.2014.893290.

Linton, David. 2019. *Men and Menstruation: A Social Transaction*. New York, NY: Peter Lang.

Lovering, Kathryn Matthews. 1995. "The Bleeding Body: Adolescents Talk about Menstruation." In *Feminism and Discourse: Psychological Perspectives*, edited by Sue Wilkinson and Celia Kitzinger, 10–31. Thousand Oaks, CA: Sage.

Marván, Ma Luisa, Dyana Ramírez-Esparza, Sandra Cortés-Iniestra, and Joan C. Chrisler. 2006. "Development of a New Scale to Measure Beliefs about and Attitudes toward Menstruation (BATM): Data from Mexico and the United States." *Health Care for Women International* 27 (5): 453–73. https://doi.org/10.1080/07399330600629658.

Marván, Maria Luisa, Sandra Cortés-Iniestra, and Regina González. 2005. "Beliefs about and Attitudes toward Menstruation among Young and Middle-Aged Mexicans." *Sex Roles* 53 (3-4): 273–79. https://doi.org/10.1007/s11199-005-5685-3.

Newton, Victoria Louise. 2016. *Everyday Discourses of Menstruation: Cultural and Social Perspectives*. London: Palgrave Macmillan.

Peranovic, Tamara, and Brenda Bentley. 2017. "Men and Menstruation: A Qualitative Exploration of Beliefs, Attitudes and Experiences." *Sex Roles* 77 (1-2): 113–24. https://doi.org/10.1007/s11199-016-0701-3.

Reberte, Luciana Magnoni, José Henrique Cogo de Andrade, Luiza Akiko Komura Hoga, Trudy Rudge, and Juliana Reale Caçapava Rodolpho. 2014. "Men's Perceptions and Attitudes toward the Partner with Premenstrual Syndrome." *American Journal of Men's Health* 8 (2): 137–47. https://doi.org/10.1177/1557988313497050.

Roberts, T. A., J. L. Goldenberg, C. Power, and T. Pyszczynski, T. 2002. "'Feminine Protection': The Effects of Menstruation on Attitudes towards Women." *Psychology of Women Quarterly* 26 (2): 131–39. https://doi.org/10.1111/1471-6402.00051.

Scott, Clarissa S., Danette Arthur, Maria Isabel Panizo, and Roger Owen. 1989. "Menarche: The Black American Experience." *Journal of Adolescent Health Care* 10 (5): 363–68. https://doi.org/10.1016/0197-0070(89)90212-X.

Stubbs, Margaret L. 2008. "Cultural Perceptions and Practices around Menarche and Adolescent Menstruation in the United States." *Annals of the New York Academy of Sciences* 1135 (1): 58–66. https://doi.org/10.1196/annals.1429.008.

Uskul, Ayse K. 2004. "Women's Menarche Stories from a Multicultural Sample." *Social Science & Medicine* 59 (4): 667–79. https://doi.org/10.1016/j.socscimed.2003.11.031.

Ussher, Jane M., and Janette Perz. 2013. "PMS as a Gendered Illness Linked to the Construction and Relational Experience of Hetero-femininity." *Sex Roles* 68 (1-2): 132–50. https://doi.org/10.1007/s11199-011-9977-5.

White, Lisandra Rodriguez. 2013. "The Function of Ethnicity, Income Level, and Menstrual Taboos in Postmenarcheal Adolescents' Understanding of Menarche and Menstruation." *Sex Roles* 68 (1-2): 65–76. https://doi.org/10.1007/s11199-012-0166-y.

Wilson, Ellen K., Barbara T. Dalberth, and Helen P. Koo. 2010. "'We're the Heroes!': Fathers' Perspectives on Their Role in Protecting their Preteenage Children from Sexual Risk." *Perspectives on Sexual and Reproductive Health* 42 (2): 117–24. https://doi.org/10.1363/4211710.

Wong, Wing Chi, Mei Kuen Li, Wai Ying Veronica Chan, Yuen Yu Choi, Chi Hung Sandra Fong, Ka Wah Kara Lam, Wun Chi Sham et al. 2013. "A Cross-Sectional Study of the Beliefs and Attitudes towards Menstruation of Chinese Undergraduate Males and Females in Hong Kong." *Journal of Clinical Nursing* 22 (23-24): 3320–27. https://doi.org/10.1111/jocn.12462.

Wright, Paul J. 2009. "Father-Child Sexual Communication in the United States: A Review and Synthesis." *Journal of Family Communication* 9 (4): 233–50. https://doi.org/10.1080/15267430903221880.

Wyckoff, Sarah C., Kim S. Miller, Rex Forehand, J. J. Bau, Amy Fasula, Nicholas Long, and Lisa Armistead. 2008. "Patterns of Sexuality Communication between Preadolescents and Their Mothers and Fathers." *Journal of Child and Family Studies* 17 (5): 649–62. https://doi.org/10.1007/s10826-007-9179-5.

CHAPTER 32

Menstrual Shame: Exploring the Role of 'Menstrual Moaning'

Maureen C. McHugh

Communication about menstruation, like other taboo topics, is culturally constrained. Girls are taught to maintain a silence about menstruation; there is a taboo that makes it socially unacceptable to talk about menstruation (Lee and Sasser-Coen 1996b, 10). Research confirms that people do not feel comfortable talking about menstruation (Houppert 1999). When women break the silence taboo, the talk is typically negative in tone (Lee and Sasser-Coen 1996b, 42). In her research on communication about menstruation, Kissling (1996, 308) reports that as a result of menstrual taboos, the only way for girls and young women to talk about it is to complain about menstrual pain and symptoms. Questioning the silence, secrecy, and negativity toward menstruation found repeatedly in the literature, Burrows and Johnson (2005, 239–43) conducted focus groups with teenage women in the United Kingdom; the respondents similarly expressed largely negative views and in their accounts menstruation was constructed as embarrassing, shameful and something not to discuss in public, or even in private.

Research has shown that women identify PMS symptoms as a common way to communicate about their menstrual cycles (Chrisler and Gorman 2015, 87). Similarly, all the respondents in the Burrows and Johnson study (2005, 243) used an illness narrative, and talked about the negative physical and emotional symptoms associated with menstrual cycles. Even before their first experiences with menstruation, girls already believe menstruation is accompanied by physical discomfort, increased emotionality and mood changes (Burrows and Johnson 2005, 244) These results are consistent with those reported in an exploratory study of menstrual management by Oxley (1998) who found women emphasized pain and discomfort. Classroom discussions and exercises conducted in 2018 in Psychology of Women

© The Author(s) 2020
C. Bobel et al. (eds.), *The Palgrave Handbook of Critical Menstruation Studies*, https://doi.org/10.1007/978-981-15-0614-7_32

classes confirmed that young women's conversations regarding menstruation addressed solely negative aspects of menstruation, and focused primarily on PMS symptoms (Sopko et al. 2018). Chrisler and her colleagues suggested that such complaining and negative talk around PMS can be a bonding experience for women (Chrisler et al. 2006, 371). As such, sisterhood may be experienced as a shared expression of bloating, cramping, and irritability (Arden, Dye, and Walker 1999, 264); women's accounts of shared misery may create a sense of connection (Fahs 2016, 27).

In this chapter, I examine women's negative communications regarding menstruation which I have termed "menstrual moaning." Menstrual moaning is employed here to refer to women sharing negative experiences of menstruation. I use the word *moaning* as a descriptive term for negative talk about menstruation; it is not meant to have a pejorative connotation. Examining the cultural taboos regarding menstruation that privilege negative, debilitating, and uncontrollable aspects of menstruation, I argue that menstrual moaning reflects shame about our menstruating bodies. Applying insights from shame theory, I consider the relation of secrecy/silence and menstrual moaning to the perpetuation of menstrual shame.

MENSTRUAL STIGMA AND SHAME

Society, through negative messages about women's cycles and negative responses to any mention of the menses socially constructs menstruation as negative. Lack of knowledge about the body and the menstrual cycle may put girls and young women in a position where they experience negative feelings such as shame about their reproductive body functions and lower self-esteem (White 2013, 67). Johnston-Robledo and Chrisler (2013, 2–6) demonstrated that menstruation is a source of social stigma for women, and documented many ways that menstruation is viewed negatively in our culture. Roberts and her colleagues (Roberts et al. 2002) demonstrated empirically that even reminders of menstrual blood (for example, seeing a wrapped, unused tampon) can lead to avoidance and social distancing, which suggests that menstrual blood may serve as a blemish on women's character. Negativity about menstruation subjects women to ridicule, dismissal, and trivialization (Fahs 2016, 4)

Most media messages in the United States represent menstruation as undesirable, and together they constitute a stereotype of menstruating women, especially premenstrual women, as violent, irrational, emotionally labile, out-of-control, and physically or mentally ill (Chrisler and Gorman 2015, 87). McKeever (1984, 39) argued that girls arrive at puberty with negative attitudes toward menstruation, which are perpetrated through education, media, and folk wisdom.

Stigmatization and negativity toward menstruation is also reflected in communication via contemporary social media. Thornton (2011, 51) reported a content analysis of references to menstruation in over 2000 "tweets," (messages limited to 140 characters) posted in a two-week period through Twitter in 2010; the tweets posted reflected the social stigma attached to menstruation, viewing menstruation as debilitating. Similarly, recent research on public discourse about menstruation provided evidence for continued negativity and censorship of (online) communication about menstruation. Using critical discourse analysis, Lese (2016, 59) examined how menstruation is constructed (negatively) by public media discourse which reinforced norms of menstruation stigma. The use of language such as "leaks" isolated menstrual blood as outside of the acceptable norm, and reaffirmed that menstruation should not be seen. The stigma that women feel about hiding menstruation from others leads to overwhelming feelings of embarrassment (Cafolla 2015; Driscoll 2015; Sanchez 2015). Stigmatized language creates a culture of shame that teaches women that getting a period stain in public is one of the most embarrassing things that can happen to them (Sanchez 2015).

In 2015 Rupi Kaur posted on social media a photo of herself lying in a bed with her back to the camera with a visible menstrual leak/stain on her pants and another on the sheet. The photo was removed from the social media platform twice (Kaur 2015) because it allegedly violated the social media app's norms. In taking down Kaur's photos, Instagram actively continued to "promote a long tradition of shaming people who menstruating . . . as though their bodies are naturally dirty" (Brodsky 2015), consistent with the view that women's bodies are seen as "gross" and "offensive" because they are outside of the normative, patriarchal frame that shapes dominant discourse (Cafolla 2015). Silence plays an important role in the stigmatization of menstruation; shame is engendered when being told not to speak. Kaur's photos themselves were silenced by dominant discourse, as controlled by Instagram, when they were deleted. Many of the public discourse authors analyzed by Lese (2016, 59–60) echoed the role silence played in their own experiences with menstruation. Menstruation is an experience that women are taught to hide and never speak of (Cafolla 2015; Driscoll 2015).

Period shaming is a consequence of the social construction of menstruation as an undesirable bodily event (Bobel 2008). Periods are perceived as a strictly negative process that is dirty, disgusting, and icky (Driscoll 2015). In a recent national poll of 1500 women conducted for Thinx (makers of period underwear), 42% of women reported having experienced period-shaming (Siebert 2018). Women polled reported feeling uncomfortable with references to the vagina, and 62% were irked by the term period. The poll responses are consistent with the position of Fahs (2016) who connects menstrual shame with body shame including shame of nakedness, smells, and body hair. Shame is extended to the vagina which is considered to be unpleasant. Use of euphemisms to refer to genitals (Braun and Kitzinger 2001) and to menstruation (Lee and Sasser-Coen 1996b, 78) reflects shameful attitudes. When women communicate about their periods, the conversations employ euphemisms to skirt around having to say "blood" or "menstruation" (Kissling 1996). Using euphemisms to discuss menstruation hides the shame of periods and the "offensive feelings associated with it" (Lee 2007, 12). The communication taboo is supported by the existence of dozens of euphemisms for menstruation (Johnston-Robledo and Chrisler 2011). In a crowd sourced study, Clue found 5000 euphemisms across many languages (https://helloclue.com/articles/culture/top-euphemisms-for-period-by-language). If there was no menstrual stigma, there would be no need for euphemisms, no "reason to call it anything other than its formal name: menstruation or the menses" (Johnston-Robledo and Chrisler 2013, 29).

Menstrual shame is a component of reproductive shame according to Johnston-Robledo and her colleagues (Johnston-Robledo et al. 2007). In their research, participants who expressed negative attitudes toward menstruation voiced similar negativity toward other aspects of women's reproductive bodies. For example, women who viewed menstruation as shameful also reported shameful attitudes toward breastfeeding. Women who reported higher levels of body shame and self-surveillance had significantly more shameful attitudes toward reproductive functions than women with lower levels of body shame and body surveillance.

Schooler et al. (2005) examined menstrual shame as a factor in sexual decision making, concluding that menstrual shame may inhibit sexual agency, and the ability of women to acknowledge and advocate for pleasure; they found that menstrual shame was linked with sexual risk-taking and decreased sexual experience. Qualitative studies of women's attitudes toward menstrual sex indicated widespread negative attitudes (Fahs 2011); women reported disgust and shame associated with the menstrual messiness. Chrisler and Johnston-Robledo (2018, 76) conclude that even when women are not menstruating, sexual subjectivity is impacted by menstrual negativity, and they suggest that reducing menstrual shame may contribute positively to women's sexual agency.

The Menstrual Closet

The stigma associated with menstruation is reinforced through our cultural practices of secrecy; keeping menstruation a secret supports views of menstruation as dirty (Martin 1992). Girls report being embarrassed if they are seen with a menstrual product (Lee and Sasser-Coen 1996b, 60). Thus, menstruation is a condition that requires women to engage in surveillance and control practices to keep it hidden from others. Women and girls go to a great deal of effort to conceal their menstrual status (Oxley 1998).

Young (2005, 106) referred to this cultural directive for secrecy as the "menstrual closet": "From our earliest awareness of menstruation until the day we stop, we are mindful of the imperative to conceal our menstrual processes.... Do not discuss your menstruation . . . leave no bloodstains on the floor, towels, sheets, or chairs. Make sure that your bloody flow does not visibly leak through your clothes, and do not let the outline of a sanitary product show" (106–7).

Menstrual products (for example, tampons, pads) are designed to absorb fluid and odors, not to be visible through one's clothes, to be small enough to carry unobtrusively in one's purse, and to be discretely discarded in a lined bin in a public restroom (Kissling 2006). Because menstruation continues to be presented as a taboo topic, product manufacturers characterize menstruation as a "hygiene crisis" that requires the manufacturers' "expert" advice as to how to properly care for the menstruating body (Costos, Ackeman, and Pardis 2002). An analysis of media use of taboo-related words showed that advertisements of menstrual products reinforce the secrecy and taboo regarding menstruation (Thomas 2007). Moreover, they highlight the qualities of the product that mostly help consumers avoid shame and embarrassment (Raftos, Jackson, and Mannex 1998). Roberts and Waters (2004) described these cultural practices as relaying the message that women's bodies are unacceptable and need to be sanitized. In other words, menstruation must be kept under wraps. That is, it is a hidden rather than a visible stigma, but that is because women go to great lengths to conceal it (Oxley 1998). Lee and Sasser Cohen contend that "menstrual etiquette requires women to uphold taboos against themselves through their own behaviors of silence and concealment" (1996a, 78). Relatedly, Field and Woods (2017) discussed these practices in relation to what they term the "concealment imperative," a form of social control in which women constantly engage in their own medical surveillance so that women's experiences of menstruation become disembodied. Menstruators' subjugation to the menstrual closet is essentially an oppressive process of concealment and abjection of menstruation that constitutes a normalizing experience. In Young's (2005, 111) analysis, the menstrual closet is maintained by and contributes to women's experience of menstrual shame.

The social oppression of women as menstruators and as owners of a body regarded as unclean, deficient, and disembodied includes the mandate of silence and concealment (Perz and Ussher 2006) which contributes to menstrual shame, and feelings of shame maintain compliance. An analysis of the practices involved in the operation of concealment and secrecy helps to comprehend the functions of menstrual shame, and how menstrual shame might be resisted. To resist menstrual shaming, we must come out of the menstrual closet.

Menstrual Shaming in Relation to Shame Theory

The stigma of menstruation has negative consequences for women's physical and mental health, sexuality, well-being, and social status. The menstruation taboo, like body shaming, is deeply ingrained. Sanchez (2015) argued that we will not eliminate hateful behavior toward women until we reject the notion that menstruation is shameful.

The elimination of menstrual shame may involve an understanding of the dynamics of shame. Here, the approach to women's shame articulated by

Brene Brown based on her interviews with 200 women (2004) is employed as a framework for understanding and resisting menstrual shame. In her analysis, women experience shame when they fail to meet extensive, layered, complex and competing demands and expectations. Brown (2004) describes shame as an intensely painful experience of believing we are flawed and thus unworthy of acceptance and belonging. She observed that shame is experienced as a profound sense of inadequacy, and often results in women avoiding social contact, leaving women feeling trapped, isolated, and powerless. In other words, when feeling shame, women often hide from others to avoid further shame. When we avoid talking about shame, shame unravels our connection to others. Therefore, sharing our story is a strategy to establish connection. According to Brown (2007), the greatest challenge to developing shame resilience is the way shame actually makes us less open to giving or receiving empathy. Shame is maintained because it is very difficult for women to access its antidote, connection, and empathy. Empathy helps to heal shame, and empathy comes from community, a community of people who are facing the same struggle.

In Brown's later analysis (2006), women's shame is often tied to isolation and secrecy, and interrupts women's sense of community and connection. Brown's perspective on shame resilience is that women should share their stories, including their pain, with other women (Brown 2006). Woman-to-woman connections can resolve shame. But menstrual shame undermines women's capacity to speak about menstruation and build those necessary connections. And yet, finding the courage to speak about our experiences, and anticipating and receiving an empathetic response are key to women's resilience to shame (Brown 2006). Feminist activists encourage women to resist the expectation of silence through, first, examining shame and shame-making experiences in a social context. Recognizing our shared experiences of shame can help us to recognize our collective vulnerability and women can come to realize that they are not alone. Once the shame is exposed as a sexist social construction, feminist activists urge women to break the taboo regarding talking about menstruation. Recognizing our shared experiences of shame can help us to recognize our collective vulnerability and women can come to realize that they are not alone. Thus, sharing our menstrual stories with empathetic listeners might help to reduce menstrual shame. This perspective also has been suggested by Schooler (2001), who found that communication about menstruation among a community of women may lessen girls' shame (Schooler 2001). In fact, talking openly about menstruation is one point of menstrual activism (Fahs 2016).

Menstrual Moaning

Cultural attitudes that require girls and women to maintain secrecy and silence regarding menstruation contribute to the production and maintenance of menstrual shame. Breaking the taboos against talking about menstruation may be viewed as a form of resistance. Talking about the symptoms, discomfort, and moodiness associated with the menstrual cycle may represent (young) women's attempts to use connection and community created through communication to resist the culturally imposed directives for secrecy and silence regarding menstruation. But negative talk may not be an effective way to resist menstrual shame.

As previously reviewed, young women's talk about menstruation is typically negative. Burrows and Johnson (2005, 236) reported their respondents in a focus group study revealed a negative perspective. When pressured to present a positive perspective, the respondents relayed negative symptoms associated with the menstrual cycle (Burrows and Johnson 2005, 243). Stubbs and Costos (2004, 42) describe the negativity of their students toward menstruation observed over twenty years of teaching. Students are only able to list negative symptoms associated with the menstrual cycle. Shame toward menstruation is so deeply rooted that women could not even name one positive aspect of menstruation, with some noting they thought, "the answer could not exist" (Stubbs and Costos 2004, 41). My students, over three decades, also easily report an array of negative symptoms associated with menstruation, such as cramps, acne, or lack of energy and/or concentration. However, the students do not recognize any time in the menstrual cycle when they have a good appetite or digestion, when skin is clear or when they have high energy and focus. They do not perceive menstruation as a cycle, but a series of symptoms experienced premenstrually or menstrually. This point was made effectively in Chrisler's et al. (1994) development of the Menstrual Joy Questionnaire. Participants who completed the MJQ often reported surprise, as most had not considered any positive aspects of the menstrual cycle before (Chrisler et al. 1994).

When students in my class were asked to report the kinds of statements about menstruation that they might make, all of the statements were negative (Sopko et al. 2018) referring to embarrassment, disgust, leakages, cramps, pain, and inconvenience. Like fat talk, menstrual moaning may actually reinscribe shame rather than resist it. The term fat talk was coined by Nichter and Vuckovic (1994) to refer to the tendency of girls and young women to disparage their own bodies as fat, a tendency observed in their ethnographic study of high school girls. Young women engage in fat talk frequently and openly. Research has tied fat talk to social norms, and young women's acceptance of social values about thinness and weight management (Clarke, Murnen, and Smolak 2010, 2). Fat talk may make participants feel better when they share concerns about their bodies with others, and communicate to others that they perceive being thin as ideal; fat talk may open the door for an empathic response and connection to another woman. Fat talk and menstrual moaning may be viewed as parallel phenomena as both involve young women (in conversations with other young women) making negative comments about their bodies, as reported through ethnographic research. Bonding with other women is one possible motivation for both fat talk and menstrual moaning.

But there are differences between the two discourses. Menstrual moaning separates from fat talk because, due to menstrual stigma, many young women do not talk openly and frequently about their menstrual experiences or cycles. But when they do, the talk is inevitably negative much like fat talk. In moaning or complaining about menstruation, women communicate a shared experience, and more specifically they express shared cultural perspectives on the negativity of menstruation, bleeding, premenstrual experiences, and the female body. Sharing accounts of discomfort, pain, and embarrassment may create connections with other women, as discussed earlier.

Fat talk is typically a clearly self-shaming discourse (for example, my butt is too big to wear yoga pants) whereas menstrual moaning typically involves negativity toward "it" (the period) not the self (Lee and Sasser-Coen 1996b, 43); menstruation is viewed as an external entity rather than as aspect of oneself. Kissling (1996, 300) similarly observed that girls avoided descriptive language such as "menstruate" "blood," "menarche," or even "period"; by using terms such as "it" when discussing menstruation, girls may maintain a level of distance. Therefore, it may be less clear to the women who engage in menstrual moaning that such talk is self-shaming.

Importantly, research on fat talk has demonstrated that participating in fat talk, in spite of Brown's research on the power of connection to reduce shame, does *not* make women feel better about their body concerns shared with others. In fact, fat talk has been shown to have deleterious effects. For example, Salk and Engeln-Maddox (2011, 27) found that students overhearing fat talk were likely to engage in fat talk themselves, and the level of fat talk engaged in by the women then related to their reported levels of negative self-esteem and body dissatisfaction. The analysis of fat talk as a social practice contributing to women's negative body attitudes, and as a form of self-shaming is easily appreciated. Possibly the analysis of young women's motivation to voice negative menstrual attitudes and experiences, and the impact of those negative statements is less apparent.

Menstrual moaning may be viewed as women's resistance to the secrecy, taboos, and concealment imperatives. In this sense, talking about one's menstrual experiences may be viewed as a form of activism, a breaking of the menstrual taboo, and/or a form of resistance to the patriarchal norm. But, like fat talk, engaging in menstrual moaning may have a deleterious impact on women's menstrual attitudes. Engaging in menstrual moaning reiterates negative cultural construction of women's bodies as flawed, deficient, and diseased. Research has indicated that communication about menstruation must be negative. Typically, women's negative talk about menstruation views women's experiences as both biological and personal, (Lee and Sasser-Coen 1996b, 47) and fails to recognize the socio-cultural factors that contribute to menstrual stigma, especially the social construction of menstruation. Thus, even as women share and connect with other women about the menstrual experience, menstrual moaning may not be constructive and may actually reinforce shame rather than encourage shame resilience and women's empowerment.

Further, talking about menstruation as PMS, in particular, reinforces medicalization and medical experts' control of women's bodies and experiences (Chrisler 2018, 23). The medicalization of menstruation frames periods as a problem needing management rather than a natural process of the female body (Chrisler and Gorman 2015, 91). The biological and medicalized approach to discussing menstruation frames the body as a problem, promotes women's silence, and reinforces period shaming.

Menstrual moaning may reify the patriarchal and medicalized construction of menstruation as problematic, pathological, and painful, and contribute to women's negative experience of menstruation. Women's participation in menstrual moaning, that is negative menstrual conversations, reflects and reinforces negative attitudes toward menstruation and women's levels of body shame and body surveillance. Frequently engaging in negative menstrual conversations may be associated with higher levels of body shame and more negative attitudes toward menstruation. In research on partner communications regarding menstrual experiences and distress, Ussher and Perz (2008, 147) reported that empathy, support, and understanding offered by partners, common in lesbian relationships, reduced guilt and self-pathologizing of women in relation to menstruation. The shame framework suggests discussions of menstruation, including sharing our own menstrual experiences can build shame resilience when the discussions are positive or at least balanced. Stubbs and Costos (2004, 41) describe their students as both ignorant of and resistant to any positive connotations and experiences connected to menstruation. Before proposing positive attitudes and talk about menstruation as a form of activism and shame resilience, it may be important, even necessary to understand why girls' and young women's reactions and expressions regarding menstruation are negative.

Menstrual Activism

Menstruation is an everyday experience for women[1] and should be acknowledged and understood as a natural bodily function for which our (negative) attitudes toward menstruation are neither natural, nor inherent. In fact, the negativity and stigma surrounding menstruation are cultural products created in a social context in which women and women's bodies are viewed as inferior to men's bodies (Young 2005, 110), and reproduction as a system is viewed as a deficient, dysfunctional, and disgusting (McHugh and Chrisler 2015, 3). Ignorance and invisibility of menstruation create fertile grounds for the development of disparagement and abhorrence of menstrual blood and the menstruating woman. Menstrual scholars and activists argue that we need to challenge, upset, and reverse the silence and shame surrounding women's menstruation (for example, Fahs 2016, 94–96). Recognizing, respecting, and even celebrating menstruation "disrupts the boundaries of patriarchy, and underscores the realities of misogyny" (Fahs 2016, 115). On the other hand, by keeping our menstrual cycles closeted, we are participating in the patriarchal system (Sangre Menstrual 2009, cited by Fahs 2016, 110). In this essay, I have examined talking about menstruation as an attempt by women to resist the silence and taboo surrounding menstruation. However, when our talk regarding menstruation remains negative, that is menstrual moaning, we may not resist negativity regarding menstruation. Rather, we may affirm it. This hypothesis, of course, requires empirical verification. If menstrual moaning is confirmed as contributing to, or even increasing, menstrual negativity and shame, we might encourage women to recognize and talk more positively about menstruation, recognizing that they may be resistant to such suggestions. In research cited here, young women have been surprised at the idea of positive aspects of the menstrual cycle, and resistant to adopting a positive stance toward menstruation. Young women may not be likely to adopt a positive attitude toward menstruation as long as our culture views women's menstruating bodies as deficient, disgusting, and/or diseased. And yet, the most productive path toward a culture free of menstrual stigma may depend on a shift in the way we talk about periods.

Note

1. It is important to note that not only cisgendered women (and girls) menstruate, but also some transmen, gender non binary and intersex individuals do as well. And of course, not all women menstruate.

References

Arden, Madelynne A., Louise Dye, and A. Walker. 1999. "Menstrual synchrony: Awareness and Subjective Experiences." *Journal of Reproductive and Infant Psychology* 17 (3), 255–65.

Bobel, Chris. 2008. "From Convenience to Hazard: A Short History of the Emergence of the Menstrual Activism Movement, 1971–1992." *Health Care for Women International* 29 (7): 738–54. https://doi.org/10.1080/07399330802188909

Braun, Virginia, and Celia Kitzinger. 2001. "'Snatch,' 'Hole,' or 'Honey-Pot'? Semantic Categories and the Problem of Nonspecificity in Female Genital Slang." *Journal of Sex Research* 38, no. 2 (May): 146–58.

Brodsky, A. 2015. "Instagram Bans Photo for Showing Menstruation." *Feministing,* March 27. Retrieved from http://feministing.com/2015/03/27/instagram-bansphotos-for-showing-menstruation/#comment-1933779612.

Brown, Brene. 2004. *Women and Shame: Reaching Out, Speaking Truths, and Building Connection.* Austin: 3CPress.

———. 2006. "'Shame Resilience Theory': A Grounded Theory Study on Women and Shame." *Families in Society* 87, no. 1 (January): 43–45.

———. 2007. *I Thought it Was Just Me: Women Reclaiming Power and Courage in a Culture of Shame.* West Hollywood, CA: Gotham Books.

Burrows, Anne, and Sally Johnson. 2005. "Girls' Experiences of Menarche and Menstruation." *Journal of Reproductive and Infant Psychology* 23, no. 3 (August): 235–49. https://doi.org/10.1080/02646830500165846.

Cafolla, Anna. 2015. "I'm a Girl and I Have Periods, Whether Instagram Likes It or Not." *The Independent*, April 1. Retrieved from http://www.independent.co.uk/voices/comment/i-m-a-girl-and-i-have-periodswhether-instagram-likes-it-or-not-10149494.html.

Chrisler, Joan, C. 2018. "PMS as a Culture Bound Syndrome." In *Lectures on the Psychology of Women*. 5th ed. edited by Joan. C. Chrisler and Carla Golden, 227–47. New York, NY: McGraw Hill.

Chrisler, J., and Ingrid Johnston-Robledo. 2018. *Women's Embodied Self: Feminist Perspectives on Identity and Image*. Washington, DC: American Psychological Association.

Chrisler, Joan C., and Jennifer A. Gorman. 2015. "The Medicalization of Women's Moods: Premenstrual Syndrome and Premenstrual Dysphoric Disorder." In *The Wrong Prescription for Women: How Medicine and Media Create a "Need" for Treatments, Drugs, and Surgery*, edited by Maureen C. McHugh and Joan C. Chrisler, 77–98. Santa Barbara, CA: ABC-CLIO.

Chrisler, Joan C., Jennifer Gorman Rose, Susan E. Dutch, Katherine G. Sklarsky, and Marie C. Grant. 2006. "The PMS Illusion: Social Cognition Maintains Social Construction." *Sex Roles* 54 (5–6): 371–376.

Chrisler, Joan C., Ingrid K. Johnston, Nicole M. Champagne, and Kathleen E. Preston. 1994. "Menstrual Joy: The Construct and its Consequences." *Psychology of Women Quarterly* 18 (3): 375–387.

Clarke, Paige MacDonald, Sarah K. Murnen, and Linda Smolak. 2010. "Development and Psychometric Evaluation of a Quantitative Measure of 'Fat Talk'." *Body Image* 7, no. 1 (January): 1–7.

Costos, Daryl, Ruthie Ackerman, and Lisa Paradis. 2002. "Recollections of Menarche: Communication between Mothers and Daughters Regarding Menstruation." *Sex Roles* 46, no. 1-2 (January): 49–59.

Driscoll, B. 2015. "I've Got a Confession to Make About My Body And You May Not Like It." *Huffington Post*, March 26. Retrieved from 70 http://www.huffingtonpost.co.uk/brogan-driscoll/periods-ive-got-a-confession-tomake_b_6946630.html.

Fahs, Breanne. 2011. "Sex during Menstruation: Race, Sexual Identity, and Women's Qualitative Accounts of Pleasure and Disgust." *Feminisms & Psychology* 21 (2): 155–78

———. 2016. *Out for Blood: Essays on Menstruation and Resistance*. Albany NY: SUNY Press.

Field, M., and Jill Wood. 2017. "Menstrual Politics: Resisting the Concealment Imperative through RUMPS." Paper presented to the Biennial Conference of the Society for Menstrual Cycle Research, Kennesaw State, Georgia, June.

Houppert, Karen. 1999. *The Curse: Confronting the Last Unmentionable Taboo-Menstruation*. New York, NY: Farrarm Straus & Giroux.

Johnston-Robledo, Ingrid, and Joan C. Chrisler. 2011. "The Menstrual Mark: Menstruation as Social Stigma." *Sex Roles* 68, no. 1 (January): 1–10.

Johnston-Robledo, Ingrid, and Joan C. Chrisler. 2013. "The Menstrual Mark: Menstruation as Social Stigma." *Sex Roles* 68 (1–2): 9–18.

Johnston-Robledo, Ingrid, Kristin Sheffield, Jacqueline Voigt, and Jennifer Wilcox-Constantine. 2007. "Reproductive Shame: Self Objectification and Young

Women's Attitudes toward Their Reproductive Functioning." *Women and Health* 46, no. 1 (February): 25–39.

Kaur, Rupi. 2015. "If the Photo of My Period Made You Uncomfortable, Ask Yourself Why." *Huffington Post*, March 30. Retrieved from http://www.huffing-tonpost.ca/rupikaur/period-picture-instagram_b_6971228.html.

Kissling, Elizabeth A. 1996. "That's Just a Basic Teen-age Rule: Girls' Linguistic Strategies for Managing the Menstrual Communication Taboo." *Journal of Applied Communication Research* 24, no. 4 (November): 292–309.

———. 2006. *Capitalizing on the Curse: The Business of Menstruation*. Boulder, CO: Lynne Rienner.

Lee, J. 2007. "Exposing Longstanding Taboos around Menstruation: A Cross-Cultural Study of the Women's Hygiene Product Advertising." In *Conference Papers–International Communication Association* 1.

Lee, Janet, and Jennifer Sasser-Coen. 1996a. *Blood Stories: Menarche and the Politics of the Female Body in Contemporary U.S. Society*. New York, NY: Routledge.

———. 1996b. "Memories of Menarche: Older Women Remember their First Period." *Journal of Aging Studies* 10 (2): 83–101.

Lese, Kathryn M. 2016. "Padded Assumptions: A Critical Discourse Analysis of Patriarchal Menstruation Discourse." Unpublished master's theses. https://commons.lib.jmu.edu/master201019/103.

Martin, Emily. 1992. *The Woman in the Body: A Cultural Analysis of Reproduction*. Boston, MA: Beacon.

McHugh, Maureen, and Joan Chrisler. 2015. "The Medicalization of Women's Bodies and Everyday Experiences." In *The Wrong Prescription for Women: How Medicine and Media Create a Need for Treatments, Drugs, and Surgery*, edited by Maureen McHugh and Joan Chrisler, 1–15. Santa Barbara, CA: Praeger.

McKeever, Patricia. 1984. "The Perpetuation of Menstrual Shame: Implications and Directions." *Women and Health* 9, no. 14 (October): 33–47.

Nichter, Mimi, and Nancy Vuckovic. 1994. "Fat Talk." In *Many Mirrors: Body Image and Social Relations*, edited by Nicole Sault, 109–31. New Brunswick, NJ: Rutgers University Press.

Oxley, Tanith. 1998. "Menstrual Management: An Exploratory Study." *Feminism & Psychology* 8, no. 2 (May): 185–91.

Perz, Janette, and Jane M. Ussher. 2006. "Women's Experience of Premenstrual Syndrome: A Case of Silencing the Self." *Journal of Reproductive and Infant Psychology* 24, no. 4 (November): 289–303.

Raftos, Maree, Debra Jackson, and Judy Mannix. 1998. "Idealised Versus Tainted Femininity: Discourses of the Menstrual Experience in Australian Magazines That Target Young Women." *Nursing Inquiry* 5, no. 3 (September): 174–86.

Roberts, Tomi-Ann, Jamie L. Goldenberg, Cathleen Power, and Tom Pyszczynski. 2002. "'Feminine Protection': The Effects of Menstruation on Attitudes towards Women." *Psychology of Women Quarterly* 26, no. 2 (June): 131–39.

Roberts, Tomi-Ann, and Patricia L. Waters. 2004. "Self- Objectification and That "Not so Fresh Feeling." *Women & Therapy* 27, nos. 3–4 (September): 5–21.

Salk, Rachel, H., and Renee Engeln-Maddox. 2011. "'If You're Fat, Then I'm Humongous!' Frequency, Content, and Impact of Fat Talk among College Women." *Psychology of Women Quarterly* 35, no. 1 (March): 18–28.

Sanchez, Erika. 2015. "OPINION: Menstruation Stigma Is a Form of Misogyny." May. Retrieved October 25, 2017. http://america.aljazeera.com/opin-ions/2015/5/menstruation-stigma-is-a-form-of-misogyny.html.

Sangre Menstrual. 2009. "Manifesto por la Visibilidad de la Regla." http://sangre-mensraul.wordpress.com/2009/05/17/manifesto-por-la0visibilidad-de-la-regla/.

Schooler, Deborah. 2001. "Messages about Menstruation: The Role Menstrual Education in Shaping Young Women's Attitudes about Menstruation and Their Sexual Decision Making." Unpublished master's thesis, University of Michigan.

Schooler, Deborah, L. Monique Ward, Ann Merriwether, and Allison S. Caruthers. 2005. "Cycles of Shame: Menstrual Shame, Body Shame, and Sexual Decision-Making." *The Journal of Sex Research* 42, no. 4 (November): 324–34.

Siebert, Valerie. 2018. "Nearly Half of Women Have Experienced Period Shaming." *New York Post*, January 3. Retrieved from https://nypost.com/2018/01/03/nearly-half-of-women-have-experienced-period-shaming/.

Sopko, Cynthia, Maureen C. McHugh, Nava Sedaghat, and Kathleen M. DiMattia. 2018. "Studying Negative Menstrual Conversations." Paper presented to the Association for Women in Psychology, Philadelphia, PA.

Stubbs, Margaret L., and Daryl Costos. 2004. "Negative Attitudes toward Menstruation: Implications for Disconnection Within Girls and between Women." *Women & Therapy* 27, nos. 3–4 (January): 37–54.

Thomas, Erika M. 2007. "Menstruation Discrimination: The Menstrual Taboo as a Rhetorical Function of Discourse in the National and International Advances of Women's Rights." *Contemporary Argumentation and Debate* 28 (September): 65–90.

Thornton, Leslie Jean. 2011. "'Time of the Month' on Twitter: Taboo, Stereotype, and Bonding in a No-Holds-Barred Public Arena." *Sex Roles* 68, no. 1 (January): 41–54. https://doi.org/10.1007/s11199-011-0041-2.

Ussher, Jane M. and Janette Perz. 2008. "Empathy, Egalitarianism and Emotion Work in the Relational Negotiation of PMS: The Experience of Lesbian Couples." *Feminism & Psychology* 18, no. 1 (February): 87–111.

White, Lisandra. 2013. "The Function of Ethnicity, Income Level, and Menstrual Taboos in Post-Menarchal Adolescents' Understanding of Menarche and Menstruation." *Sex Roles* 68, nos. 1–2 (January): 65–76.

Young, Iris M. 2005. *On Female Body Experience: Throwing Like a Girl and Other Essays.* New York, NY: Oxford.

CHAPTER 33

Becoming Female: The Role of Menarche Rituals in "Making Women" in Malawi

Milena Bacalja Perianes and Dalitso Ndaferankhande

African women are not problems to be solved. (Nnaemeka 2005)

Menstruation signifies the beginning of adulthood, and the female body's biological changes which allow for reproduction. Around the world it is accompanied by a variety of socio-cultural practices which both celebrate the transition from child to adult, but also fear the onset of sexual activity. At menarche, girls come to understand their reproductive capacities and future responsibilities constituting a 'female consciousness' in direct response to biological functionality (Johnston-Robledo and Stubbs 2013; Allen and Goldberg 2009; Lee 2002). Menarche rituals, and the resulting social and hygienic customs around menstruation (known as menstrual practices), form the basis of a variety of disciplinary practices which come into play during puberty to legitimize certain narratives around female identity and constitute and validate specific modes of being female (Bobel 2010). Menarche rituals and menstrual practices therefore are central to codifying and regulating female behaviour, and consequently manifesting a collective female consciousness.

Significant efforts have been made by Western feminists to understand and critique the process of making women, such as Simone de Beauvoir who famously wrote: "A woman is not born, she is made" (1949, 283). Historically, the biological ability of women to give birth has become one of the 'most powerful social and cultural constructs of feminine identity' (Harcourt 2009, 33). As a result, the materiality of the female body is viewed as a driver for why patriarchal societies relegate women to an inferior role within society, that of mother and wife. Such a perspective understands gender as inherently relational, oppositional, and ultimately hierarchal (Dietz 2003; Irigaray 1991). In such a context the normalization of

© The Author(s) 2020
C. Bobel et al. (eds.), *The Palgrave Handbook of Critical Menstruation Studies*, https://doi.org/10.1007/978-981-15-0614-7_33

motherhood is often posited as synonymous and instrumental in women's oppression.

Based on qualitative research methods, this chapter seeks to understand the role that menarche rituals play in 'making women' in Malawi. To be a woman is collectively ascribed, and individually understood, through the active and intelligible performance of menarche rituals, and consequent menstrual practices. Through such traditions girls become women with their newly ascribed gender imbuing them with a locus of power within their communities. This chapter will examine the role of menstruation in constructing and validating female identity within the relevant cultural and symbolic meanings which are attached to it (Young 2004). Cognizant of the politics of western feminist thought, it does not seek to critique the validity of biological determinism or the ontological irreducibility implicit in local gender narratives in Malawi. Rather, it seeks to determine the extent to which menarche rituals in Malawi create a collective female consciousness that governs individual bodies and behaviour, which ultimately cement women's role in society.

This chapter will explore the specific roles that menstrual rituals and practices play in facilitating the integration of girls into social structures and normalizing what it means to be collectively female in this context. It reads these rituals through an African ontological position in order to move beyond understanding African women's subjectivity through the lens of Western gender-based hierarchies and oppression.

An African ontology recognizes the history of distortion and dispossession of African subjectivity and seeks to understand social phenomena and formulate theory which is situated in, and through African centred knowledge (Oyewùmí 2005; Nnaemeka 2005; Cooper and Morrell 2014). Mazama's (2003, 6) model of Afrocentricity emphasizes the significance of moving beyond a historical legacy of colonial oppression on the continent which ultimately creates one-dimensional female figures (Lazreg 2005). Therefore central to an African ontological position is the use of African lexicon and matrices to speak to local phenomena (Cooper and Morrell 2014). Such an exercise is critical to understanding gendered social dynamics in Malawi, and in contextualizing results within this research project. Most pertinent is the importance of the specificity of African discourse and understandings of gender. Lazreg (2005) suggest that in seeking to understand the foreign subject of the African female, we must not fetishize, or exoticize her, but seek to create an intelligible understanding of alternate modes of being female. In recognizing such modes we are able to discern how individuals mediate self-narratives and elucidate a more nuanced understanding of knowledge production around what it means to be female in an African context.

Western feminism has a tendency to read African women's subjectivity in terms of gender-based hierarchies and gender oppression. Oyewùmí (2005) suggests that such a standpoint uses a western framework of gender construction which implies male privilege and female subordination, while neglecting other forms of social stratification such as race, ethnicity, class, or age (Oyewùmí 2005; Steady 2005). African scholars have sought to challenge this imposed hierarchy on African subjects, and suggest that both men and women in Africa have been inferiorized by the colonial process (Oyewùmí 1997). Hudson-Weems (2003) asserts that men are not the enemy of women in African discourse, as is the case for white western feminism. While Nnaemeka (2005) suggests that abuse of the female body in the developing world can be studied within the context of oppressive conditions under patriarchy, there remains a distinct need to realize that such distinctions in social hierarchy emerged out of the colonial experience and is not an indigenous construct. As a result, not all phenomena regarding female subjects can be understood as gendered experiences, but can be explained by examining a variety of other social practices and stratification (Oyewùmí 1997, 2005, 2011).

By situating menstruation in African narratives and epistemologies (with particular reference to Malawi), one can develop an understanding of how gender comes to be understood at a personal level, through the collective and relational experience of menstruation in Malawi which is guided by social and cultural practices. In this context, the onset of menarche equips girls with the socially gendered expectations for womanhood in her community. While some of the more traditional practices and narratives remain harmful, the process of developing a female consciousness through these rituals serves as a means of gaining community-based power and status, and ultimately enables women to individually and collectively navigate social spaces. In understanding the lived experiences of menstruating women in Malawi from an African ontological position, this chapter pushes us to move beyond viewing women as passive recipients of gendered practices at menarche, rather identifying the ways women actively and passively [re]produce menstrual practices to develop both personal collective notions of gender identity.

Research Methodology

A core tenet of this research is its use of feminist methods as a framework to inform research design, data collection, and analysis, placing women and their experiences at the center of the research. As a methodological approach, constructivist grounded theory enabled experiential knowledge to uncover basic social processes with participants validating findings on an ongoing and iterative basis to become co-creators of the research.

This chapter is grounded in findings from qualitative research conducted in 2016 which used a phenomenological approach to understand how women's menstrual experiences constitute knowledge. Using constructivist grounded theory, and a mixed methods design, 35 participants took part (28 women, and 7 men) through interviews and an open-ended survey using random snowballing sample. Several Non-Government Organization (NGOs) were used as gatekeepers to participants to enable the researchers to establish a relationship with participants and more easily overcome stigma and taboo associated with speaking about menstruation.

DATA ANALYSIS

All data was coded using NViVo (a qualitative data coding software package) employing multiple layers of coding and categorization to consolidate, analyze, and conceptualize the data. Incomplete data, mostly due to recording and transcription challenges was removed from the data set. The remaining data from interviews and surveys was analyzed using five levels of coding (Hernandez 2009): memos from interviews described first impressions of interviews; open coding allowed for phenomena to be categorized thematically; axial coding related codes to categories using inductive and deductive thinking; selective coding facilitated the process of identifying core concepts and themes to narrate around; lastly theoretical memos were used to link together data, literature, and ideas. A variety of semantic[1] and latent[2] themes were identified and validated with participants through a 'cheat sheet.' This iterative process enabled the researcher to recalibrate the analysis of data and informed the findings sections.

As result three overarching motifs where noted: menarche rituals, menstrual narratives, and menstrual etiquette which each constitute menstrual practices in Malawi. In tandem or individually, they govern how women understand their bodies, and ultimately how they behave. This chapter will focus specifically on the first motif—menarche rituals—and will explore the varying Malawian rites of passage to determine how they create a collective female consciousness that governs individual bodies and behavior. The findings are explored through several key themes: Marking femaleness through menarche; Ascribing gender to girls through ritual; Becoming female through consciousness; Making women through sexuality; and Imbuing power through motherhood. All direct quotations from participants can be found in italics within the chapter.

MARKING FEMALENESS THROUGH MENARCHE

The onset of menarche, and rituals and practices that accompany it, play a key role in producing and governing gender identity in Malawi. They signify and celebrate adulthood while validating certain ways of being female, marking girls as women, and enabling culturally specific, gendered social dynamics. As one informant explained: *"To be a woman, it means that you should have a husband who will take care of you and [your] children. You should also know how to take care of your husband, like in bed."* Similar to the Gikuyu of Kenya[3] (Arnfred 2004) rather than being determined at birth, in Malawi, gender is ascribed and made intelligible to the individual and communicable to the community at menarche and through a complex process of formal and informal rituals and practices that insert girls in social patterns in society.

Arnfred (2004, 29) suggests that "initiation rituals [are] the ritual celebration of a shared female gender identity" itself. In Malawi, menstruation plays an important role in creating a sense of belonging, and collective identity as a female. It is both a biological and social marker of femaleness, with menarche rituals as a means of "turning" girls into women. It provides the opportunity for girls to develop a sense of their individual, relational, and collective responsibilities as women.

Ascribing Gender to Girls Through Ritual

'Ritual is a culturally constructed system of symbolic communication' with societies cosmologies[4] embedded in them (Jeyaraja 2014, 1). Rituals have been perceived as a mechanism by which societies can commune with the supernatural or immaterial world, Jeyaraja (2014) suggests that rituals are a means of codifying and regulating the myths and customs of a community. Malawi is a country governed by a set of cosmologies expressed through specific rituals which play an important and influential role in community life (MHRC 2005). They are often used to celebrate occasions, such as in the case of menarche, but also as a tool to formulate and govern social behaviors. In many African societies, rites of passage mark the three main stages of one's life: rites of separation, the rites of transition, and the rites of incorporation (Munthali and Zulu 2007). Initiation ceremonies are not specific to Malawi; they have been studied in various African countries, most notably in Mozambique, Kenya, and Zambia.

Arnfred (2004, 29) claims that 'the rituals instruct young women in the rules of decent female behavior such as exhibiting self-control, downcasting the eyes, and displaying respect to men and elders.' Njambi and O'Brien (2005) proposes that puberty-related rituals are critical in social practice. Young people are prepared and instructed in their new responsibilities as adults, with a particular emphasis on entering the sexual world (Munthali and Zulu 2007). Within this context, menarche rites and rituals are an integral element in constituting women's place in society (Delaney, Lupton, and Toth 1988). They facilitate the integration of children into social structures, marking them as part of the community.

Menarche rituals are a means of both recognizing and celebrating the sexual maturation of girls and boys in Malawi. They take two main forms: the formal practice of an initiation ceremony,[5] or more informal rites of passage. Both play a core role in ascribing gender and imbuing women with status. The specific practices depend on ethnic group and religious affiliation and can be performed pre-puberty, or at the onset of puberty for boys and girls, with distinct messaging and purposes.

Munthali and Zulu (2007) propose that traditional initiation ceremonies come in various forms ranging from structured rituals including large numbers of children organized at specific times a year, and others who are loosely organized and adapted to meet the ages and needs of the people or communities involved. For girls specifically, both types seek to mark the beginning of adulthood (marked by the onset of menstruation) and are infused with socio-cultural meanings centred around the roles and responsibilities of their new position in society.

Formal initiation ceremonies play a central role in determining social status, formulating identity, and creating a sense of group unity in Malawi, particularly in rural areas. For girls, they center around defining adulthood in terms of marriage, sexual activity, and reproduction. In different parts of the country formal initiation ceremonies or camps are held to mark girls and boys transition to adulthood by taking children aside from the community and instructing them. There are a multitude of culturally specific ceremonies which are referred to by different names depending on locality: *Chindakula* (in Mangochi), *Maseseto* (in Nsanje), *Masosoto* (in Mulanje), *Chinamwali cha atsikana* (Northern Malawi), *Unamwali* (Chiradzulu) (Johnson 2013).

It is important to note that unlike most Western societies where menarche is seen as the beginning of adulthood, in communities where initiation ceremonies are commonly practiced in Malawi, it denotes adulthood. That is, the ceremony serves to inform the community of the girl's new status as woman. As one informant explained: "*It was the symbol of showing that now you have grown up. Everyone should know.*" Upon initiation, girls are seen as matured and introduced into the community as ready for marriage. At this point, they are often referred to as a *namwali* (bride).

Informal rites of passage also occur across Malawi and are usually centred on community and religious gatherings reflecting a decentralization of rites, rituals, and ceremonies around menarche. While traditional initiation ceremonies are seen by the educated and urban elite as a traditional practice reflecting a "*backward culture . . . they do it in villages,*" informal community or religious-based customs have come to replace formal initiation in many parts of Malawi. These informal gatherings vary across Malawi but are considered more of a celebratory or educational event among kinship groups or religious communities. "*It's not that people do initiation ceremonies like the way they do there in the villages. For us you do it in your own family. So it's not an initiation itself. It's family or religious groups.*"

BECOMING FEMALE THROUGH CONSCIOUSNESS

At the onset of menarche, women play a key role in disseminating information with female relatives, community members, and village elders coming to induct a girl. Where mothers are often the primary source of knowledge regarding menstruation and puberty in Western societies, in Malawi such a practice is rare. As one informant shared: *"most mothers do not talk about sensitive issues [with their children]."* Whether within the confines of formal or informal menarche rituals, mothers rarely take part in their own daughters' rite of passage. Across the sample, urban and rural participants spoke of the inability to speak to mothers about menstruation. The two most common sources of knowledge were grandmothers and aunts with older sisters and other females from the community also routinely stepping into assist girls at menarche. *"Some aunties came by, my mom's older close friends, and they came to talk to me about menstruating."*

In analyzing the aforementioned experiences through an African lexicon, we are able to discern broader kinship patterns in Malawi. Motherhood or mothering is not confined to biology. How information is distributed regarding menstruating reflects the decentered role of motherhood in the menstrual process. *"I told her about it and she referred me to my aunt who would assist me."* While it would be easy to understand such experiences as mirroring the deeply ingrained nature of menstrual taboo in women's psyche, it also speaks to broader local patterns of family and social kinship. The ritual avoidance of mother and daughter in regards to menstrual conversations may be a vestige of a matrifocal society whereby women are not only the figurehead of a family, but also imbued with moral power (Amadiume 1997). Therefore maintaining that status and power requires physical and social separation from those within the family at times. Coupled with the intentional commitment to hand over instruction to other female elders, the work of mothering in Malawi is distributed across the community with biological mothers responsible for moral nurturing, and social nurturing by those within the community at large. Such a practice inducts girls into the community of womanhood as a whole, while reinforcing the notion of her femaleness as relational to those around her.

As previously mentioned, through both formal initiation ceremonies and informal rites of passage, girls become accepted into distinct places within the social structure complete with corresponding responsibilities. It is important to note that this place is not inherently inferior to the place occupied by their male counterparts. Oyewùmí (2005) suggests that the western framework of gender construction implies male privilege and female subordination which does not necessarily apply in an African context. Through this lens, one could argue that the socially gendered dynamics codified upon menarche are not necessarily perceived as restrictive or oppressive. Rather, they are integral to the perpetuation of a cohesive society in which a sense of community, and consequent responsibilities, is of greatest value.

This can be best understood through the core principle of an Afrocentric ontology of commonality or relationality between beings, (Harris 2003). Commonality in this context refers to *ubuntu* or *umunthu*,[6] which is a philosophy in Southern Africa emphasizing an attitude of community and solidarity (Tamale 2011). It highlights a communal notion of existence which Harris (2003, 114) refers to as 'we are, therefore I exist.' This philosophy juxtaposes a western, liberal, tradition of individualism, with value and worth of individuals in *umunthu* cultures, being relative and relational. *Umunthu* is a way of being, which is deeply ingrained in local ideologies and manifests itself in a variety of social dynamics. Initiation into the female collective is essential to *umunthu*.

Through an indigenous lens, we can come to understand these rituals as forming a gendered practice through which girls come to understand what it means to be female, and symbolic of their inclusion into a community as a whole. As a result of the notion of *umunthu*, a girl's consciousness is only possible through relative and relational interactions with those of the same, and opposite sex.

MAKING WOMEN THROUGH SEXUALITY

Themes of motherhood and sexuality are central tenants to understanding femaleness for Malawian women. These manifest in different ways between formal and informal menarche rituals with religious and ethnic considerations manifesting into complex messaging around sexual conduct and gendered behavior. Though menstruation has marked girls as female, sexuality is central to obtaining social status as a woman. Sexual activity is not only for the purpose of reproduction but also the pleasure of the participants involved. Interestingly, within Malawi, the ability to give pleasure is important to the notion of femaleness, which is rather antithetical to a Judeo-Christian position where by chasteness is one of the highest virtues in a woman.

In formal initiation ceremonies, young women at once become inserted and insert themselves into the dominant patterns of sexuality in Malawi. A particular emphasis is placed upon instructing girls and boys in a variety of social conduct which seeks to enhance sexual pleasure and inform young people how to engage in acceptable social patterns of behavior. Girl initiates are often taught about their roles and responsibilities regarding adulthood, including wifely duties including how to sexually pleasure men, perform traditional dances to attract men, and labia pulling.[7] Similarly, boys are also taught how to behave in a "*masculine way*" including how to respect elders, and "*take care of their wives*" (including sexual and financial responsibilities). Several participants spoke to the expectations placed on women to both pleasure and "*hold a man*," with labia pulling and "*ritual defloration*" as techniques to prepare for sexual activity, and "*keep a man interested.*" Furthermore, traditional practices regarding sexual activity are not viewed as antithetical with Christian teachings with virginity not always viewed as a necessity for marriage. Though participants suggested that sexual skills were meant to be used within marriage, pre-marital sex and subsequent child bearing were viewed as common, and not explicitly frowned upon. In some rural areas girls who have born children were considered more desirable. "*No. It's not important, it's not even necessary . . . For us, they will go for the girls that are sexually active. Sometimes these men, they could laugh for maybe if you are a girl, you're maybe 18 years old and you had never had sex and this man is having sex with you and he finds that you are a virgin. He's going to go in the community, say, 'that woman, just look at her that one, she's just grown up but she was a babe.*"

These aforementioned examples begin to paint a picture of how young people in Malawi are instructed in appropriate gendered behavior, with particular respect to relational and sexual conduct. For boy and girl initiates, there is an emphasis on practicing what has been learnt in initiation. "*Once you have been circumcised . . . you are a grown up man and you should actually practice.*" Evidence such as this highlights the cultural aspects of initiation ceremonies which encourage young people to demonstrate their adulthood through sexual activity.

Within communities practicing formal initiation ceremonies, ritual defloration, and social exclusion can be a significant part of initiation ceremonies at the onset of menarche

Girls are sexually initiated by a fisi (hyena) after initiation ceremonies in some cultures, in some districts, they talk of that fisi, that hyena man, to be like it's his job, he's an expert, so he's the one who actually has to sleep with several.

The practice of a "*fisi,*" a male adult who has sexual intercourse with newly initiated girls, is practiced most commonly among the Chewa and Yao within formal initiation ceremonies and has been documented in studies (MHRC 2005; Munthali and Zulu 2007). Initiation, and particularly ritual defloration, remain enshrouded in secrecy with initiated girls often instructed, "*not to tell anyone what they have learned or their mothers will die.*" Delaney, Lupton, and Toth (1988) suggest that the actual or symbolic act of ritual defloration can be seen in a multitude of cultures around the world and is an integral aspect of puberty-related rituals. Ritual defloration in this context become a means of controlling and facilitating how a girl enters the sexual world. The secrecy surrounding initiation ceremonies is used to reinforce the significance of the ritualistic practice. In doing so, it emphasizes the collective identity and status of those have been initiated and 'become women,' to the uninitiated other. "*They come out of it feeling more woman and there are some girls from my tribe who now feel less woman because they didn't.*"

Within the cohort of participants informing this research, almost all referenced sexual activity and sexual behaviour in tandem with menstrual rituals and practices. This suggests that the onset of menarche in Malawi is closed tied to sexual activity with further research needed to understand the positive and negative impact of this on sexual and reproductive health outcomes. Of note, is that while the act of ritual defloration was more commonly practiced among communities who practiced formal initiation ceremonies, sexual conduct remained a recurring theme throughout menarche rituals, menstrual narratives, and menstrual practices in different parts of Malawi.

With informal rites of passage practiced through community-based gatherings or celebrations, information regarding appropriate social and sexual conduct, and hygienic practices were also dispersed to girls. Interestingly, depending on location, some traditional beliefs and customs around sexuality continue to permeate these informal celebrations.

> *In my setting, there's nothing like initiation camps, but the elders come and they assign you some girls, big girls to now teach you how to start pulling the labia. Pulling of the labia. They start telling you, "Now you have to pull these labia's in preparation for a man. If you don't have this, then you have problems in your marriage. The man is going to do such a thing if he doesn't find this on your private parts," so you have to pull them so you start that kind of process.*

There remains a distinct tone of conflict within discourses around menarche rituals, and broader menstrual practices between traditional beliefs and modern narratives. With the introduction of monotheistic religions, there have been significant shifts in socio-cultural practices across the country (MHRC 2005). However, indigenous ways of doing and knowing still remain, co-existing and at times clashing, with more puritanical Christian and post-colonial narratives. This requires Malawians to find ways to navigate between traditional and modern practices, reconciling discord or simply accepting it.

This is demonstrated through the way that religious communities (particularly Christian and Muslim) have taken on the responsibility of menarche rituals in parts of Malawi. Similar beliefs to initiation ceremonies are expressed in these celebrations in terms of social conduct and respect for elders where there is a distinct focus on celibacy and avoidance of sexual activity within Christian-centered rituals. Girls are explicitly told "*they are no longer allowed to play with male friends*," with an ethos of separation of the genders emphasized except in religious contexts. While sexual activity is technically forbidden outside of marriage, it (and consequent child-bearing) remain a common practice and tolerated among the participants in the research. "*In some rural areas girls who have born children are more desirable, men go for them because they are women who can please a man.*"

This shifting dynamic in menarche rituals illustrates the conflicting messages around the acceptability of sexual activity, as Christian values have begun to erode more traditional beliefs. Munthali and Zulu (2007) argue that Christian groups have developed their own form of initiation ceremonies to instruct young people in appropriate behaviour and expected responsibilities, partially as a response to the explicit sexual content in traditional ceremonies. While religion has played a core role in changing social dynamics, it is not necessarily an individual's religious background which determines the type of initiation or messaging around sex received. Without exploring the complex intersection between ethnicity[8] and religious identity in Malawi, it is problematic to assume that one plays a more dominant role than another. For example, Munthali and

Zulu (2007) suggest that initiation ceremonies are strongly promoted and adhered to among Muslims, as the largest number of initiation ceremonies are conducted in Mangochi (which has the highest percentage of Muslims in the country). However, Mangochi is predominantly a Yao ethnic region, therefore Muslim and Yao traditions mutually influence local practice of the ritual. Regardless of ethnic or religious background, sexual and social conduct remain core to both formal and informal menarche rituals.

With the rise of Christian tradition across the country, tension has risen in how women experience and express their sexuality. Conflicting messages regarding the acceptability of sexual activity appear to be a by-product of Christian values which restrict sexual freedom outside the confines of marriage, and a reticence around speaking about female pleasure. Yet, traditionally a more flexible mindset enabled greater sexual exploration, as still seen in some initiation ceremonies. Malawian women navigate this tension through different means.

Consequently, for initiated women initiation is a source of pride reflecting their ability to fulfil this requirement of femaleness and validate their female identity. *"They come out of it feeling more woman and there are some girls from my tribe who now feel less woman because they didn't, they're not as cultural and not as trained in those things."*

IMBUING POWER THROUGH MOTHERHOOD

Sexuality and sexual activity is more important than simply pleasuring and holding a male; it can also be understood within the context of the motherhood paradigm. Marriage and motherhood are core aspects of women's identity in Malawi. As previously mentioned, in some Chewa tribes, from the onset of menarche, girls are referred to as *namwali* meaning that they are a bride and therefore mature enough for marriage. Women who do not conform to these standards usually have lesser positions in Malawian society.

> *They're not counted. If you don't have children, it's like you are, I don't know a better way to say it, but they look at you as somebody stupid. Somebody who is not normal. Similar with the women that are not married.*

Within Malawi, Motherhood has great importance to individual female identity, and it is also viewed as a communal responsibility. Achieving motherhood is such a critical element of Malawian culture that within legal tradition the inability to bear children is a legitimate ground for divorce. There is a level of status placed upon women from motherhood and excluded from those who do not have children. This is connected to women's relational responsibility to bear children for her husband and the community, *"if you have no children in the community, a man is seen [as less than]. .... they want that kind of respect. If you are barren, you are a disappointment."* Childbearing therefore raises the status of both women and men.

The pressures of the motherhood paradigm lingers urgently within Malawi, as a necessary reality for girls to achieve. As Kopytoff (2005, 31) claims 'childrearing is a great weight of realizing female identity.'

In terms of people deciding that, 'I don't want to have kids, I don't want to get married,' I think I have not met a lot of women here who say those things, who give those views. I don't know if it's a Malawian thing or whatever but I think that a lot of African women don't usually, I don't think their mind ever really goes there. It is very rare to find a woman who will voluntarily say that, 'I have made that choice. I don't want to have kids. I don't wanna get married ever and that's just my choice.'

It is important to recognize the ways in which women's power in Africa has been predicated on the logic and centrality of motherhood in African discourse (Amadiume 1997; Dove 2003). Amadiume (1997) argues that while being a wife and mother have been viewed as means of disempowerment for western feminists, in African tradition motherhood is a culturally recognized autonomous unit, and central social category. In realizing the importance of the 'motherhood paradigm,' scholars can begin to recognize that the feminine is not afforded lower status in African society, rather women's power and status emerges from such a position (Amadiume 1997; Chodorow 1978; O'Reilly 2007). The echoes of matrifocality[9] remain within the motherhood paradigm, that women are invested in power by right of the status that motherhood affords them (Nnaemeka 2005). Though western feminists (Rosaldo 1974; Ortner 1974) may critique that women's power still derives from their reproductive capacity, and limits them to it, African scholars speak to this as a practical conclusion, rooted in the irreducible reality that reproduction remains a biological process that women's reproductive systems enable (Amadiume 1997). As a matrifocal perspective puts women as the central (and most powerful figure) at the heart of the family, motherhood is a source of power. Nnaemeka (2005, 319) highlights the importance of this alternate source of social hierarchy and explores 'the cyclical ordering of social life' in Africa. She claims that status and power ebbs and flows in African societies depending on a number of other factors including age, marriage, and class. Similarly, Oyewùmí (2005) suggests that prior to colonization, gender was not the organizing principle in African society, rather seniority was critical to social hierarchy. Therefore, within the context of menstrual rituals, it could be argued that collectively 'making women' is critical to enabling girls to gain wider social status and decision-making power in Malawi.

Recognizing the historical legacy of the motherhood paradigm and matrifocality in Malawian consciousness, assists in speaking to some of the inherent tensions in menarchal rituals today, and how social patterns of behavior are experienced and legitimized at an individual level. Lee (1994) suggests that menarche becomes a signifier of reproductive potential and embodied womanhood, and as a result it becomes irrevocably intertwined with sexuality. The findings from this research affirm that women in Malawi obtain power through motherhood. Rather than women being reduced to mothers, through the identity of motherhood establishes their place in the social hierarchy over non-mothers. It affirms what African womanist theorist Dove (2003) argues, that woman are revered within the role as mother and bearer of life and as the moral compass of family and society. She is not just a mother in terms of bearing and rearing children, but also a figurehead and central point for social organization and understanding one's subjectivity. Nnaemeka (2005) proposes that full elderhood for women is only achieved once a woman has had a successful pregnancy. The motherhood paradigm illustrates the complex cultural beliefs and social practices which influence how Malawian women orient their individual self-narratives around a collective understanding of the female body and identity. Furthermore, an understanding of the personal self is deeply ingrained within the aforementioned *umunthu* logic, and the social responsibility of the individual to a whole.

It is important to note that using matrifocality as a specific matrix by which to understand everyday practices of gender identify and behaviour in Malawi, differs significantly from reading Malawi through a matriarchal lens. A matriarchal society is that in which power is invested in, and exercised by women. This is not the case in post-colonial Malawi. There is, however, extensive debate over the extent to which matrilineal principles continue to exist in Malawi, and if women exert real power under such social systems (Phiri 1983; Arnfred 2004). Phiri (1983) speaks to the tensions between Malawian traditions, in particular matrilineal principles, and Christian teachings. He suggests that Christian missionaries played a central role in undermining matrilineal customs and practices, particularly around marriage and kinship dynamics. He claims that by the 1960s matrilineage ceased to exist as a social and economic unit in Malawi. In contrast, Johnson (2013) argues that matrilineage remains an important feature of social life, particularly in rural life in Southern Malawi. Rather than assert in favour of either position, this research seeks to draw attention to the vestiges of such traditions and how they remain embedded in cultural institutions and practices, specifically menarchal rituals. While there has been a decline in women's authoritative position, the idea of motherhood remains powerful in cultural imagery and in socializing behavior, particularly around menstruation.

Conclusion: Listening to Women's Voices

Menstruation, as both biological and social marker, plays an important role in creating a sense of belonging and collective female identity in Malawi. Menarche rituals (both formal and informal) function to welcome and induct women into the social order ultimately creating a sense of group unity among women. Furthermore, the process of becoming female at menarche, and 'making women' through menarche rituals and, are a means of codifying and regulating female behaviour and bestowing corresponding responsibilities. These enable young people to understand socially gendered dynamics and facilitate their integration into a cohesive and communal society. Therefore within a *umunthu* culture, collective consciousness precipitates individual identity—'we are women, therefore I am a woman.'

Seen through an afrocentric lens, this process cannot be assumed to be inherently negative, hierarchal, or oppressive. It has different meaning across different social, religious, and ethnic strata with women actively and passively (re)producing these practices in order to navigate social spaces and integrate into social structures over a lifetime. While at menarche girls are passive recipients of this gendered knowledge, they become more active role in [re]producing female identity as they come to induct other girls as they age. Gender in such a context is not static, it is ascribed and enacted by navigating through collective practices and personal meaning including a locus of power which corresponds to their newly gendered body.

This chapter seeks to move beyond theorizing or problematizing African women's subjective experiences within oppressive gender discourses. It elevates women's own voices recognizing women as 'knowers or agents of knowledge about themselves and their menstrual experiences' (Wood et al. 2007, 44). Its emphasis is on resisting applying western models of gender to the Malawian context that reduce the female experience to men exerting power over women. Rather it asks us to think critically about how the gendered experiences of menstruation in Malawi is historically grounded and culturally bound within an African ontology, In other words, as Oyewùmí asserts: 'If gender is socially constructed, then gender cannot behave in the same way across time and space' (2005, 11).

This shift in thinking results in several key findings from the research to highlight a nuanced understanding of the gendered experience of menarche ritual in Malawi. First and foremost menarche rituals enable girls to achieve a state of womanhood with menstruation playing a central role in creating a sense of belonging and collective identity as female. These practices are essential to validating specific modes of being female, with the primary modality of femaleness in Malawi one which is closely linked to sexual activity and motherhood. Learning gendered appropriate behaviors at menarche enables young people to insert themselves into social and sexual patterns in society. Consequently, being ascribed a gender is not necessarily an oppressive practice but a mode of enabling young girls to understand their gender as relational to both other women and men as well as society as whole.

There is much to be gained by situating menstrual experiences within their relevant cultural, symbolic, and ontological contexts, particularly when researching women who have been historically and discursively dispossessed. In feminist analysis of menstruation we must be cautious of our tendency to reproduce the social categorization's and prejudices that are inherent in society and knowledge production at large (Lazreg 2005). This chapter has sought to disrupt the ways in which *thinking* and *knowing* African women's bodies and experiences have become a neo-colonial practice in itself. This is not to suggest that the onset of menarche in Malawi is in some way a wholly empowering experience, rather it reflects a complex social process where women (as individuals and as a collective) come to navigate narrative, meaning, and practice within their own lexicon and matrices. As a result, the process of developing a female consciousness through menarche rituals serves as a means of becoming seen as a gendered being and gaining community-based power and status.

Notes

1. Semantic themes emerged naturally in conversations with participants using this specific language. For example: taboos, cleanliness, female identity, motherhood, shame, sexuality, and exclusion.
2. Inductive logic has been used to connect semantic themes to the underlying ideas, patterns, and assumptions (latent themes) which was situated within the literature review and theoretical framework.
3. The Gikuyu tribe is a Bantu tribe that neighbours the Embu, Kikuyu woman harvesting tea Mbeere and Meru tribes around Mount Kenya.
4. Cosmologies in this sense refers to beliefs, values, customs, and social practices which order and govern a society.
5. In different parts of Malawi formal initiation ceremonies or camps are held to mark girls and boys transition to adulthood. There are a multitude of culturally specific ceremonies and can be referred to by different names depending on locality: *Chindakula* (in Mangochi), *Maseseto* (in Nsanje), *Masosoto* (in Mulanje), *Chinamwali cha atsikana* (Northern Malawi), *Unamwali* (Chiradzulu).
6. *Ubuntu* is commonly used in South Africa, where as in Chewa it is referred to as *Umunthu*.
7. Labia pulling is the act of elongating the labia minora through manual manipulation (pulling) or physical equipment (such as weights). The results are that is belied to enhance sexual pleasure for men.
8. This refers to both the identification with, and enactment of customs and practices of a particular ethnic group, but also acknowledges the ways in which beliefs from an ethnic group can permeate village life as a result of direct association with that group.
9. Matrifocality refers to a family or social structure whereby the mother is the central figure.

References

Allen, Katherine, and Abbie Goldberg. 2009. "Sexual Activity during Menstruation: A Qualitative Study." *The Journal of Sex Research* 46 (6): 535–45.

Amadiume, Ifi. 1997. *Reinventing Africa: Matriarchy, Religion and Culture*. London: Zed Books Ltd.

Arnfred, Signe. 2004. *Re-Thinking Sexualities in Africa*. Uppsala: Almqvist & Wiksell Tryckeri.

de Beauvoir, Simone. 1949. *The Second Sex*. London. Random House.

Bobel, Chris. 2010. *New Blood: Third-Wave Feminism and the Politics of Menstruation.* New Brunswick: Rutgers University Press.

Chodorow, Nancy. 1978. *The Reproduction of Mothering: Psychoanalysis and the Sociology of Gender.* Berkeley: University of California Press.

Cooper, Brenda, and Robert Morrell. (Ed.). 2014. *Africa-Centred Knowledges: Crossing Fields and Worlds.* Suffolk: Boydell and Brewer.

Delaney, Janice, Mary Jane Lupton, and Emily Toth. 1988. *The Curse: A Cultural History of Menstruation.* Chicago: University of Illinois Press

Dietz, Mary G. 2003. "Current Controversies in Feminist Theory." *Annual Review of Political Science* 6 (1): 399–431.

Dove, Nah. 2003. "Defining African Womanist Theory." In *The Afrocentric Paradigm,* edited by Ama Mazama. Asmara: Africa World Press Inc.

Harcourt, Wendy. 2009. *Body Politics in Development: Critical Debates in Gender and Development.* London: Zed Books.

Harris, Norman. 2003. "A Philosophical Basis for an Afrocentric Orientation." In *The Afrocentric Paradigm,* edited by Ama Mazama. Asmara: Africa World Press Inc.

Hernandez, Cheri Ann. 2009. "Theoretical Coding in Grounded Theory Methodology." *Grounded Theory Review: An International Journal* 3 (November): 51–60.

Hudson-Weems, Clenora. 2003. "African Womanism." In *The Afrocentric Paradigm,* edited by Ama Mazama. Asmara: Africa World Press Inc.

Irigaray, Luce. 1991. *The Irigaray Reader.* Edited by Margaret Whitford. Oxford: Blackwell

Jeyaraja, Stanley. 2014. *Culture, Thought and Social Action: An Anthropological Perspective.* Boston: Harvard University Press.

Johnson, Jessica. 2013. *Chilungamo? In Search of Gender Justice in Matrilineal Malawi.* Cambridge: University of Cambridge.

Johnston-Robledo, Ingrid, and Margaret L. Stubbs. 2013. "Positioning Periods: Menstruation in Social Context: An Introduction to a Special Issue." *Sex Roles* 68: 1–8.

Kopytoff, Igor. 2005. "Women's Roles and Existential Identities." In *African Gender Studies Reader,* edited by Oyèrónké Oyěwùmí, 127–44. New York: Palgrave Macmillan.

Lazreg, Marnia. 2005. "Decolonizing Feminism." In *African Gender Studies Reader,* edited by Oyèrónké Oyěwùmí, 67–80. New York: Palgrave Macmillan.

Lee, Janet. 1994. "Menarche and the (Hetero) Sexualization of the Female Body." *Gender & Society* 8 (3): 343–62.

Lee, Shirley. 2002. "Health and Sickness: The Meaning of Menstruation and Premenstrual Syndrome in Women's Lives." *Sex Roles* 46 (1-2): 25–35.

Mazama, Ama. (Ed.). 2003. *The Afrocentric Paradigm.* Asmara: Africa World Press, Inc.

MHRC. 2005. *Cultural Practices and Their Impact on the Enjoyment of Human Rights, Particularly the Rights of Women and Children in Malawi.* Lilongwe: Malawi Human Rights Commission.

Munthali, Alister, and Eliyah Zulu. 2007. "The Timing and Role of Initiation Rites in Preparing Young People for Adolescence and Responsible Sexual and Reproductive Behaviour in Malawi." *African Journal of Reproductive Health* 11 (3): 150–67.

Njambi, Wairimu Ngaruiya, and William E. O'Brien. 2005. "Revisiting 'Woman–Woman Marriage': Notes on Gikuyu Women." In *African Gender Studies Reader*, edited by Oyèrónké Oyĕwùmí, 145–65. New York: Palgrave Macmillan.

Nnaemeka, Obioma. 2004. "Nego-Feminism: Theorizing, Practicing, and Pruning Africa's Way." *Signs: Journal of Women in Culture and Society* 29 (2): 357–85.

Nnaemeka, Obioma. 2005. "Bringing African Women into the Classroom: Rethinking Pedagogy and Epistemology." In *African Gender Studies Reader*, edited by Oyèrónké Oyĕwùmí, 51–65. New York: Palgrave Macmillan.

O' Reilly, Andrea (Ed.). 2007. *Maternal Theory: Essential Readings*. Bradford, Canada: Demeter Press.

Ortner, Sherry. 1974. "Is Female to Male as Nature Is to Culture?" In *Woman, Culture, and Society*, edited by M. Z. Rosaldo, and L. Lamphere. Stanford, CA: Stanford University Press.

Oyĕwùmí, Oyèrónké. 1997. *The Invention of Women: Making An African Sense of Western Gender Discourses*. Minneapolis: University of Minnesota Press.

———. 2005. *African Gender Studies: A Reader*. New York: Palgrave Macmillan.

———. 2011. *Gender Epistemologies in Africa: Gendering Traditions, Spaces, Social Institutions, and Identities*. New York: Palgrave Macmillan.

Phiri, Kings M. 1983. "Some Changes in the Matrilineal Family System Among the Chewa of Malawi Since the Nineteenth Century." *The Journal of African History* 24 (2): 257–74.

Rosaldo, Michelle. 1974. "Woman, Culture and Society: A Theoretical Overview." In *Woman, Culture, and Society*, edited by M. Z. Rosaldo, and L. Lamphere. Stanford, CA: Stanford University Press.

Steady, Filomina. 2005. *Women and Collective Action in Africa: Development, Democratization, and Empowerment, with Special Focus on Sierra Leone*. New York: Springer.

Tamale, Sylvia (Ed.). 2011. *African Sexualities: A Reader*. Cape Town: Pambazuka Press.

Wood, Jill M., Patricia Barthalow Koch, and Phyllis Kernoff Mansfield. 2007. "Is My Period Normal? How College-Aged Women Determine the Normality or Abnormality of their Menstrual Cycles." *Women & Health* 46 (1): 41–56.

Young, Iris M. 2004. *On Female Body Experience: Throwing Like a Girl and Other Essays*. New York: Oxford University Press.

CHAPTER 34

Researcher's Reflection: Learning About Menstruation Across Time and Culture

Sheryl E. Mendlinger

In 1969, at the age of 18, after graduating high school in Houston, Texas, I packed a steamer trunk with a year's supply of shampoo, toilet paper, and tampons and went to study abroad in Israel. All these products were outrageously expensive, and tampons and pads were not readily available at that time. I soon found that many women in Israel were still using wads of cotton wool to absorb their menstrual blood, something I had never seen in the USA. I had no idea that 30 years later the idea of tampons would lead to a topic of research for a Ph.D. dissertation. Over the course of that year I started to learn the language, adapt to the culture of the host society, and made my decision to remain in the country. Several years later, I married, had children, and was constantly navigating this duality through the challenges of bilingualism and biculturalism within the family as well as the general society. As a new immigrant I began adapting to the new culture yet holding onto many of the beliefs and traditions that I brought from my home country.

The impetus for this research began when my teenage daughter came home from school in the early 1990s (born in Israel and raised in Beer Sheva, in the Negev region) hearing from her friends "if you use tampons you could lose your virginity." I thought back to my mother who had introduced me to tampons in the 1960s, as a teenager growing up in the USA, she certainly would not have given me a product that would cause me to lose my 'virginity.' Because we lived in a diverse community in southern Israel (Negev) and many of the parents of my daughter's friends had immigrated from North Africa, Iraq, and Eastern Europe to Israel, I began thinking about what we learn from our mothers, even about menstrual products, is significant in understanding generational and cultural differences that can impact our lives in so many ways. Years later when I chose my topic for my Ph.D., I examined mother–daughter dyads from North Africa, Ethiopia,

C. Bobel et al. (eds.), *The Palgrave Handbook of Critical Menstruation Studies*, https://doi.org/10.1007/978-981-15-0614-7_34

Europe, North America, the Former Soviet Union, and Israel in order to better understand how mothers transmit knowledge about health behaviors, specifically about menstruation, to their daughters.

Israel is an immigrant ethnically diverse society. The mass immigration to Israel that began in the 1950s, and continues today, brought a wide variety of cultural traditions to the country. One of the long-term effects of immigration has been the ongoing differentiation of the Jewish population based on countries of origin within an evolving Israeli identity and norm. The data collection took place at a time in Israel's history when there was a mass immigration from two countries in particular. Between 1989 and 2000, over a million people immigrated to Israel from the Former Soviet Union and tens of thousands from Ethiopia. Over 10% settled in the Negev area where we lived, and that became a living sociological laboratory in which disparities of health could be examined.

Through my research[1] I was able to capture the acculturation process of some of the daughters in real time. At the time of my data collection, there was limited research conducted on intergenerational aspects of mothers and adult daughters who took part in the same study, and even fewer studies that looked specifically about intergenerational *menstrual* knowledge transmission, especially between mothers and daughters who immigrated from a diverse set of populations to the same country. This rapid increase in immigrant population, in such a short time span, affected various cultural aspects of the community and specifically in the area of women's health. Long-range effects of immigration are often seen not only among the first generation of immigrants, but in the second generation as well.

The qualitative ethnographic study included data collected from 48 in-depth interviews from four mother–daughter dyads from six groups that were defined by the origin of the mothers, who were born in (1) Israel; or immigrated to Israel from: (2) North Africa; (3) Europe; (4) the Former Soviet Union (FSU); (5) the United States or Canada; and (6) Ethiopia. These groups represent a cross section of the local Jewish population in the Negev, Israel. Although these women are Israeli citizens, I refer to them as Israeli, North African, European, FSU, USA, and Ethiopian for clarity sake only.[2]

Considering that I had emigrated years before from the USA, I occupied a unique researcher's vantage point that enabled me to better understand the challenges that immigrants have when moving between countries. When the women were asked to talk about their experience at puberty, they often constructed their story related to their physical body development and integrated it within the context of the menarche experience. For the most part, women remembered the actual day, even when it was decades ago. To capture the types of knowledge that mothers passed on to their daughters, I developed a model of knowledge acquisition for learning about menstruation that articulated four types: traditional, embodied, technical, and authoritative. A central thread that runs in this model is that knowledge acquisition is different in an ethnically pluralistic context than in one more homogenous.

The "Slap" Heard 'Round the World' and Other Menarcheal Traditions

One of the most interesting aspects of this research is the stories the women shared about their menarcheal (first period) experiences. In many traditional cultures there were ceremonies, rituals, and traditions that were passed from one generation to another that may give either a positive or negative valence to the way women view menstruation. Traditional knowledge and rituals often provide strong emotional support for daughters allowing a comfortable transition through this key developmental stage. Several women whose origins were from Europe spoke about "the slap." One mother told the story that when she got her first period her mother slapped her across the face, which she did to both of her daughters, "my family always passed down traditions," yet there was uncertainty what it meant. While the mother said the reason was "something about the blood coming back," the daughter said it was something about "the blood not going to your mind." The actual historical reasons for the slap are unknown. However different explanations include: the manner in which the act of slapping took place could determine the duration of menstruation; it was necessary for a girl when she becomes a woman as protection against disgrace; and the rush of blood will make the girl have a wonderful color in her cheeks throughout her life. The uncertainty surrounding this ritual exemplifies that even as traditions continue, their rationales are often lost.

Women who immigrated from North Africa and Ethiopia looked at the onset of menstruation as a milestone that symbolized fertility, therefore achieving status in the traditional society. Several of the women born in North Africa spoke with joy and excitement when they reminisced about getting their periods for the first time and the celebrations that surrounded this event with mothers playing an integral role in this transition. Some of the traditions included mothers giving their daughters pieces of jewelry and preparing special foods and performing the ritual of putting three of the daughters' fingers in flour so that they should only get their period for three days. Another tradition is the oil ceremony in which mothers told their daughters to look and smile at their reflection in a bowl of olive oil, and then their faces were rubbed with oil. These girls were told that the image they saw on that day should continue and they should enjoy a happy life. The women noted that the oil would smooth a woman's passage into womanhood. This oil ceremony was accompanied by a festive meal with traditional foods including honey-dipped, oil-fried cakes. When these women, now elders, were asked if they continued this tradition with their daughters, they all answered an emphatic no, of course not. It appears that the change and adaptation to the new culture and environment took precedence over retaining certain cultural practices. But there were exceptions. For instance, I interviewed one daughter of American immigrants who grew up on a Kibbutz, a collective community. She spoke with excitement as she recalled when each girl in her age cohort got her period, they celebrated with gifts and a special party.

CULTURAL AND RELIGIOUS PROHIBITIONS

There are many religious customs and restrictions during menstruation related to the Jewish laws of family purity that include *niddah*, the time of separation; *teuma*, being unclean and untouchable; and *tahara*, the cleansing process following menstruating that religious Jewish women continue to observe today. An Ethiopian daughter told me that when you menstruate "you can't touch food for the Sabbath . . . my mother will not let me prepare food that is for the Sabbath." A North African mother shared, "it is forbidden to go into the Synagogue or hold the Torah Scrolls, or go to the cemetery since you are 'ritually impure', these were the things they always told us."

The older Ethiopian women talked about their experiences at menarche which included special foods that their mother's prepared, and then the young girl would go to the 'menstruation hut' where she would remain until the completion of her period. An Ethiopian mother noted "when you go to the hut someone brings you food and if you touch them, they must stay there as well . . . ." They looked forward to this time to be taken care of and pampered by the other women of the community. As another mother noted "in Ethiopia there is no rest until you go to the hut, only during menstruation the women rests," and her daughter explained, "the week away from chores was fun and a brief vacation from everyday life and being with friends and not having to work so hard." However, the mothers did not continue traditions, such as preparing special foods or going to the menstruation hut after they immigrated to Israel.[3]

TECHNICAL KNOWLEDGE—"THERE WEREN'T PADS, THERE WERE RAGS..."

Several mothers from Israel, North Africa, and Europe reported using rags when they first got their periods. They told me "there weren't pads, there were rags, we just cut rags and we began with that, we would cut all sorts of strips of material and we would wash them and use them again . . ." and "we would buy small white towels . . . we would wash and boil them in some sort of box,. . . they would come out white like snow. . . ." Another daughter from Ethiopia said she used cotton until she read in a magazine about pads. Many of the mothers and their daughters from almost all the countries reported using cotton and then over the years, they started to use pads and or tampons when they became more accessible. Often the daughters taught the mothers about products.[4]

The mothers from the former Soviet Union (FSU) immigrated either in the 1970s or late 1980s–1990s began menstruating before immigration and spoke about using cotton wool when they first got their periods because "we didn't have other things." Mothers who immigrated in the 1970s also noted that "after immigration I used pads . . . I never heard about pads or tampons in Russia, only here [in Israel]," and another mother noted "back then there weren't any, when I came to Israel [in 1971] tampons were already here." It appears that pads and tampons were not available to the mass population in the FSU in the 1970s. On the other hand a mother from the FSU who only immigrated to Israel in 1994 spoke about using other products: "I knew about tampons when my daughter was about 14–15 [about 1990–1991], they were already in the stores [in the FSU], when there was something new, we started to experiment on ourselves." It appears that accessibility of menstrual products has a great influence on what women use, and the methods of menstrual care have changed over the last several decades.[5]

There were some interesting differences based on country of origin. The more traditional the women, especially those from North Africa and Ethiopia, the less they used tampons. The reasons they gave included: not wanting to touch oneself; not putting foreign objects inside their bodies; perceived loss of virginity; and hearing in the media that tampons caused infections and cancer. All the mothers I interviewed from the United States and Canada, however, used pads and soon moved to tampons during their teenage years. Their daughters followed similar patterns. None of these women reported using cotton wool which led me to assume that since the mothers never used cotton wool, they did not recommend it to their daughters. These daughters began using tampons at a relatively early age, often younger than their Israeli contemporaries.

The use of products can have a profound effect on how women deal with menstruation in their daily lives. Many of the mothers spoke about restrictions and the inability to take part in certain activities that appear to be a direct result of how they handled menstruation. As an example, if one used cotton or pads it was almost impossible to swim or take part in certain sports activities. This may explain why women told me that it was not possible to swim while menstruating. However, the introduction of tampons enabled women greater mobility and therefore, freedom. For example, one daughter who immigrated from the FSU was a gymnast who never practiced or competed while menstruating. The first time she even heard about tampons was in the late 1990s when an Israeli woman introduced her to tampons. She was so delighted to have access to tampons she exclaimed, "the person who invented the tampons should get the Nobel prize."

The model of knowledge acquisition of health has been applied to other aspects of women's health behaviors that include: family planning and birth control,[6] informed decision making for breast cancer patients,[7] and could be extended into further studies of learning about birth, breastfeeding, infant care, coping with marital conflict, and mid-life or older age health challenges. Furthermore, the results should allow health professionals to tailor the educational programs to the ways in which women form their approaches to critical health behavior. These educational programs can be developed to focus on various target groups within the population and to include health education that takes into account the broad spectrum of cultural aspects of the attitudes of health behavior in ethnically pluralistic cultures.

My research was the first undertaking of its kind that examined the intergenerational transfer of menstrual health knowledge and practices in Israel. Although the data was collected almost two decades ago, it remains relevant because it provides unique insight into the effects of mass immigration to a culturally diverse country. In addition, at the time of this study, the internet as a major source of information was not readily available in Israel to the general public, especially in the Hebrew language. Therefore, learning about menstruation was primarily based on verbal communication or written information. Today, as women are moving from one country to another, either by choice or through forced political or economic displacement, it becomes very important to understand the varied cultural beliefs and traditions of people on the move as they adjust to their new countries and home.

NOTES

1. Sheryl E. Mendlinger and Julie Cwikel. 2005/2006. "Learning about Menstruation: Cultural Diversity and Knowledge Acquisition." *International Journal of Diversity in Organizations, Communities and Nations*, Vol. 5.
2. The qualitative ethnographic study included data collected from 48 in-depth interviews from four mother–daughter dyads from six groups that were defined by the origin of the mothers, ages 55–65, who were born in (1) Israel; or immigrated to Israel from: (2) North Africa in the 1950s–1960s; (3) Europe in the 1950s–1960s; (4) the Former Soviet Union (FSU) in the 1970s–1980s; (5) the United States or Canada in the 1970s–1980s; and (6) Ethiopia in the 1980s. The daughters, ages 25–35, met the following criteria: (1) born in Israel or immigrated as children or teens; (2) educated in schools in the Negev; (3) completed undergraduate degrees or equivalent; and (4) given birth to their first child. These daughters were chosen since giving birth marks a critical period in the evolving mother–daughter relationship.
3. Sheryl E. Mendlinger and Julie Cwikel. 2006. "Health Behaviors over the Life Cycle among Mothers and Daughters from Ethiopia." *Nashim: A Journal of Jewish Woman's Studies and Gender Issues* 12 (Fall): 57–94.
4. Sheryl E. Mendlinger, Julie Cwikel, Rachel Gershaw, and Patricia Ann Farrell. 2011. "Blood Is Thicker than Water: Menarche as a Trigger for Memories—Mother-Daughter Dyads across Cultures." Unpublished manuscript, Society for Menstrual Cycle Research (SMCR), Pittsburgh, PA.
5. Ibid.
6. Sheryl E. Mendlinger and Julie Cwikel. 2005. "Menstruation and Reproductive Practices—Religion and Traditions and the Influence of Immigration on Mother-Daughter Dyads across Cultures." Unpublished manuscript, Society for Menstrual Cycle Research (SMCR), Boston, MA.
7. Laura E. Warren, Sheryl E. Mendlinger, Katherine A. Corso, and Caprice C. Greenberg. 2012. "A Model of Knowledge Acquisition in Early Stage Breast Cancer Patients." *The Breast Journal* 18 (1): 69-72.

CHAPTER 35

Transnational Engagement: Designing an Ideal Menstrual Health (MH) Curriculum—Stories from the Field

Breanne Fahs and Milena Bacalja Perianes

INTRODUCTION

How people learn about menstruation has a deep impact on the broader social justice implications of menstrual health, particularly as education becomes a force working against stigma, shame, and silence around menstruating bodies. The specific tools people use to implement menstrual education also play a central role in shaping the public face of menstrual health. If menstrual education is fun and light-hearted, people learn that menstruation is "no big deal" and something that does not differ much from other bodily processes. If menstrual health education can be accessed more readily, more people learn about periods and reproductive health, including (and importantly) men and boys. We think together here about the specific ways that menstrual health education works to undermine stigma, expand who learns about menstruation, and forge alliances between new organizations, partners, groups, and individuals.

The dialogues in this chapter on menstrual health education suggest that the world of teaching about menstruation is rapidly changing, expanding, and growing. Once confined to the narrow field of the Water Sanitation and Hygiene (WASH) sector, menstrual education has evolved to encompass the varied and complex social, cultural, political, economic, and environmental determinants of health which facilitate or inhibit women and girl's well-being. Through diverse and holistic approaches to menstrual health, we are increasingly moving away from discursive practices that ground menstruation within a hygiene and cleanliness mindset and open up new possibilities for applying a menstrual lens to, and through a variety of sectoral approaches.

C. Bobel et al. (eds.), *The Palgrave Handbook of Critical Menstruation Studies*, https://doi.org/10.1007/978-981-15-0614-7_35

First and foremost, the dialogues in this chapter suggest that menstrual education has moved far beyond the confines of schools, and as such, beyond the limitations of traditional models of a relatively limited, short-term intervention given only to girls in a traditional educational setting. Instead, menstruators are taught about menstrual health and menstrual decision-making outside of school, with emphases on health literacy and education through a human-rights or justice-based learning experience. These dialogues explore just some of the integrated approaches to menstrual health that currently exist which draw upon new technologies, new modes of education, and new kinds of curriculum and learning.

With renewed attention to body literacy, menstrual health education has expanded to include more nuanced understandings of menstrual experiences paying particular attention to the connection between how education can, and does, impact health decision-making over a women or girls lifetime. In such a context, menstrual education is no longer simply focused on the need to "manage menstruation" hygienically and periodically, rather menstruation as an opportunity to increase bodily autonomy and as central to closing the gender gap. As a result, doctors work in tandem with public health officials. Community organizers work in partnership with schools. WASH actors work together with menstrual activists. The public sector works together with the private sector. These fusions and overlaps have resulted in alliances and collaborations that reflect exciting new models for social change that draw from the strengths of each of these entities while minimizing their weaknesses. For example, businesses get to innovate, while activists get to ensure access to those typically ignored or excluded. Doctors relay important health information and people in local communities have increased ownership over translating these messages to meet their needs and contexts. Most importantly, these new models of menstrual education ensure that we continue to see menstrual health as central to the ability of women and girls to exercise their human rights. We enlist the support of *all* genders in *all* parts of the world in these efforts.

Shaping the Dialogue: Clue's Approach to Menstrual Health Education

– Anna Druet

Clue is a menstrual health app that uses science and data to help users discover the unique patterns in their individual cycle. It reminds users about their period, PMS and fertile window.

Explain your approach and intervention toward menstrual health? How do you implement your menstrual health curriculum?

The menstrual tracking app is, above all, a tool. It allows people to create a record of their menstrual cycles and periods, to track their health and identify cycle-related patterns. It facilitates connections with others and enables individualized healthcare. These are active elements of the tool itself, however it also has a variety of equally valuable passive elements, particularly education and learning. At Clue, we aim to enable learning as people use it, resulting in topics that are often taboo to become normalized, neutral, and reframed. Our approach to menstrual health is actually quite *zoomed-out* from the period. The period is viewed as a gateway into the rest of the cycle. Often, people start using our app as a period tracker alone however by interacting with the app and reading content, we aim to instill in our users an understanding of their cycle as much more than just the period. Promoting body literacy is an important value for us.

I sometimes think of the period as an (easily overlooked) black box. The contents of this box can go relatively unnoticed, or adversely perceived, until something happens that triggers a person to look deeper. Maybe they've had trouble becoming pregnant, or maybe they have symptoms of a reproductive health condition. By opening it, they learn what has been sitting there all along—a fifth vital sign, a signal of health, a process that can connect a person to their body, or to others. People suddenly learn that their cycle may influence their acne, pain tolerance, headaches, or social lives. It may give them a better idea of their risk level for pregnancy if a condom breaks and they are considering emergency contraception. It may play into their muscle-building capacity and influence their workout routine. It may influence how a certain sex position feels on a given day. Learning about the cycle can go far beyond fertility and bleeding. At Clue, we have found people are surprised and excited to unleash the potential of this new information!

What makes your approach unique?

Our communication style is very important to our approach. We follow a carefully selected, ever evolving set of goals and rules to create the Clue tone and feel. Normalizing taboo topics is an overall goal, but the approach can shift from topic-to-topic. We aim to make things approachable and easily understandable, while also communicating the importance of complexity in some instances. We know that oversimplified statements about health, a research finding, treatment, or a little-understood disease, can be harmful. This approach is shaped by our research, academic collaborators, and the people who use our app. We receive hundreds of emails each month from people asking questions about their cycle, health concerns, and basic anatomy. These questions help us understand what is most concerning for people, and where gaps might be in our own materials.

How did you develop your curriculum and who did you consult?

Our menstrual curricula is designed collaboratively. We work directly with some of the world's top clinical researchers on menstrual health via our medical board and research partners. We use services like Uptodate, which incorporate new peer-reviewed research (when sufficient) into current clinical gold standards of practice. We do internal literary reviews, of varying depths, and have extensive research and fact-checking protocol to ensure rigor and accuracy.

What kind of impact have you had?
Our approach has resonated with people. Today we have over ten million active users in 190+ countries. Our app has been translated into 15 languages and counting. We have seen shifts in the questions sent to our support team as we have incorporated certain educational goals into the app and content more broadly. Our dedicated users have led to the development of an unprecedented dataset on cycles and health, which we use in collaboration with academic partners for menstrual health research. We are conscious to choose research topics that can benefit the people who give access to their (amalgamated, anonymous) data for our carefully selected research collaborations. Our website, where we publish weekly content on menstrual and reproductive health topics, gets 2 million readers each month, most of whom are coming to the site after searching for health info in a search engine. We learned that people in areas with higher gender inequality are more likely to use their mobile devices to access sexual health information.

What languages do you use and are there any tensions or conflicts about the language you use for menstrual health education?
Challenges for us have been in getting language right. Our app is not currently localized, meaning the same version of the app is being used globally (translated into different languages). The base and context of knowledge varies dramatically across Clue users, as does the status of evolving language around gender and health.

In terms of language, there are also differences in opinion in where the line sits between approachability and euphemisms—fluid vs. mucus, for example. We are grateful to have so many engaged Clue-users who give us a lot of feedback and have an ongoing process where we work with consultants to help in the development of our language.

Communicating on under-researched topics and bridging the gap between diverse menstrual-health spaces can also be challenging. We support access to comprehensive, reputable healthcare services, and information for all. However emerging dialogues on women's health have also brought voice to considerable and important disciplinary gaps. These include: the impact of female physiology being left out of research; the ignoring or minimization of female pain; the dearth of satisfactory answers on how to treat common but complex disorders like Polycystic Ovary Syndrome (PCOS) or endometriosis; the arguably disproportionate under-testing of gynecological medical devices, and undervaluing of female health issues. These facts have led to an important and necessary movement and conversation around women's health. On the other hand, they have also led to the creation of some resources and services using misinformation, a lack of credible research, or entities looking to profit from desperation. For us balancing clinical understanding with new research and evolving trends can be tricky. When writing on evolving health topics, we look out for newer findings and examine the researchers conducting them, to see what recent contributions are being made to burgeoning bodies of knowledge. Sometimes this is all of what is out there. We make it clear when we share any findings which are controversial, unreplicated, or require further validation. It is surprising how much research fits into this category. All sides of these emerging conversations need to take women's health seriously enough to vet and ethically portray the information they are offering to people in vulnerable positions.

What does the future of MH education and curricula look like?
In the future, I would like to see educational initiatives that focus on the cycle overall, cycle phases, and the cycle as a health indicator tool akin to a fifth vital sign. I would like to see boys and men included in menstrual educational initiatives. Like many, I would also like to see a shift in handling of the topic of premenstrual syndrome (PMS)—the 200+ symptoms associated with an inconsistently defined syndrome cover a window of time in the cycle with several hormonal shifts, and diverse experiences. We also need more educational resources dedicated to perimenopause. Ultimately, I believe that more research into hormonal diversity, experiences around the ovulatory phase, and into positive (or perhaps not-only-negative) changes associated with the cycle would be incredibly valuable.

MENSTRUPEDIA: MAKING PERIODS FUN AND ACCESSIBLE

– Carla Giacummo
   Menstrupedia is a comic which acts as guide to educate people around the world on menstruation.

Explain your approach and intervention toward menstrual health? How did you implement your menstrual health curriculum?
I began working on menstrual health while running Eco-Ser in 2010, an NGO which increases women's knowledge of menstrual cups as a healthy product option for women. Within this project we worked with Mooncup Ltd., to educate and distribute cups to girls in rural areas or from disadvantaged backgrounds. After interacting with hundreds of women, I realized there was a clear lack of resources to talk about periods with girls aged 9 and up in the Latin America region, as well as tools for parents and educators to generate natural and easy interaction around the menstrual cycle. When I met Aditi Gupta, the Founder of Menstrupedia India, I realized that Menstrupedia was exactly the resource I was looking for to enable my own work in Latin America. Therefore I worked to translate Menstrupedia for a Spanish-speaking audience with sensitivity to the local context.

What makes your approach unique?
Within an Indian context, girls are often in a very vulnerable situation regarding the menstrual cycle and puberty related changes. The Menstrupedia comic contents captures different experiences to address all that a girl or a boy should understand regarding this time in a very easy, clear, and

funny way. I recognized that cartoons are helpful to address complex concepts which through storytelling help young people grow up with positive messages around puberty and changes in their life. Within a Latin America context, there was no material available like this. Most sexual education focuses on contraceptive methods, abortion, and biological differences. Most resources I encountered either skipped the menstrual cycle or failed to acknowledge the importance of menstruation in little girls lives. The work of Menstrupedia in India and now the Spanish translation, connects the biological impact of puberty and menstruation with the emotional, social, and psychological aspects in order to help girls on their journey to womanhood.

What kind of impact have you had?
While working on the Spanish translation, I tested the English version at the biggest ecological event in our country. At this event a variety of different projects, alternative health workshops, and ecological products are presented to the public with approximately 15,000 people in attendance. I was able to sell several of the resources immediately and realized that mothers felt an incredible sense of relief to find a tool to help them start conversations with their daughters. However, there was clearly a need, and demand, for them to be readily available in Spanish. Teachers and midwives also showed interest in using the resource in educational and clinical settings. The first edition in Spanish was available in October 2015.

How did you develop your curriculum and who did you consult?
Menstrupedia's curriculum was developed with input from health and educational experts. The content has been carefully designed and created with expert input, therefore the Spanish version duplicated the content, with minor adjustments for a Latin American context. After the first edition was produced with a positive response from Non-Government Organization's in Latin American countries (as well as Spanish speakers from across Spain, USA, and Europe) I collaborated with Daniel Goldman, CEO of Inmmentor, an expert in innovation. We both understood Menstrupedia as fundamentally an innovative educational tool, which supports an improved understanding of menstrual and reproductive health for girls in rural areas and indigenous communities in Latin America. With high rates of adolescent pregnancy in these populations we worked to increase distribution with the hope that in the long term we would see its impact on these pressing issues, among others. With strong interest from Mexico, Guatemala, Nicaragua, Colombia, Perú, Puerto Rico, Argentina, Chile, and Brazil we have worked to toward internationalization of Menstrupedia including adding several companion resources such as audiovisual, and trainers guide.

What languages do you use and are there any tensions or conflicts about the language you use for menstrual health education?
While there are nuances across different Spanish-speaking countries, Menstrupedia was translated into *neutral* or the most common form of Spanish

with minimal colloquialisms. We have found few tensions or conflicts, as the content of the book is already culturally sensitive. Regardless of a readers beliefs or context, those who use the resource are not made to feel uncomfortable.

To ensure the accuracy of the translation we worked with Dana Riensi, Spanish Teacher, Library and Educational Technology Specialists (Oregon) as well as NGO partners from Guatemala, México, and Perú. This helped to translate and implement not only the text, but the audiovisual guides.

What do you wish could be included in future curricula about menstrual health?

After two years of working with different international and national organizations, as well as school teachers, social educators, and health professionals, I believe that there is a need for the development of a menstrual health workshop for adults. Menstrual health knowledge remains a gap in a variety of adult populations therefore those implementing our resources, also require capacity building regarding the importance of integrated menstruation health education. We have received many requests for further guidance beyond the audiovisual and trainers guide. This resource would need to be tailored to the needs of those working in rural areas, where populations are being affected by a broader lack of resources, hygienic products, and access to water, sanitation and hygiene. I also hope one day to be able to create a Portuguese and sign language (LSU) version of Menstrupedia

What does the future of MH education and curricula look like?

I believe MH education and curricula should be simple and easily accessible by all. Governments seem unaware of the importance of MH education in spite of its social, economic, and cultural impact. There is a strong need for improved puberty-related education delivered at a younger age. It will enable girls to feel more confident and help boys to understand what happens not only in their own bodies, but also those of girls. I hope to see plenty of information available that looks at the great biological and emotional changes that occur at this time.

I believe that early MH education is an opportunity for girls' interaction with their bodies and cycle. Menstrual dignity has to do with their own confidence, self-esteem, and healthy growth. However, it is also an opportunity for greater awareness and respect for human beings and their right to make decisions over the bodies and lives. We cannot express with words what we do not know. So, MH education is a great opportunity for adults and children to learn about their emotional and physical health and how they can talk about it.

WoMena: Educating Around Innovative Products—
Developing a Curriculum to Support Menstrual
Cups Programming

– Marianne Tellier, Shamirah Nakalema, Siri Tellier, and Andisheh Jahangir

WoMena is a Danish founded NGO that promotes the use of evidence-based, effective reproductive health technologies and solutions such as menstrual cups through research, advocacy, and project implementation in Uganda.

Explain your approach (SRHR, WASH, gender), and intervention?
WoMena has a broad view on menstrual health including physical, mental, social, and structural aspects, which we translate into a set of intervention activities:

1. Menstrual health facility assessments to understand local logistical challenges.
2. Menstrual health training (and if we have time, SRHR training) with a comprehensive curriculum including: puberty, female anatomy, vaginal coronas, menstrual cycle, menstrual hygiene, nutrition, pain management, and methods (pros/cons, use, disinfection and storage).
3. Creating support structures for example involving parents, community members, teachers, training of trainers, health workers, and political/local leaders in communities.
4. Educating local rights bearers on advocacy tools aimed at encouraging local duty bearers to effectively manage water and sanitation facilities.
5. Emphasizing male involvement for example fathers and male community members, male teachers, and boys in schools.
6. Monitoring the use of the products and evaluating the acceptability and following up to address the challenges in relation to the use of the product.
7. Scaling up projects by designing pilots with the next steps in mind (for example, cost-efficiency).
8. Educating at the top through national advocacy.
9. Effective knowledge management including active review of scientific literature to enable internal learning.

To do this, we employ both paid local and international staff and volunteers, which contributes to a richness of perspectives and energy to implement our programs, and achieve social impact.

What makes your approach unique?
WoMena's educational approach is unique in several ways. Our approach is context based, and our standard curriculum is usually tailored to the needs of the local population. We also use locally available materials, for easier accessibility and affordability of the menstrual management materials we distribute. We believe that when something is so new, like the menstrual cup, and knowledge is so low, you need to go into the fundamentals, personal stories, as well as a lot of honesty, to build people's trust, such as: how many holes women have, pros and cons of cups, and how difficult it is when you take out your cup the first time. Furthermore, we believe in a life-skills based educational approach. Such an approach recognizes that different people learn in

different ways—therefore we like to make our trainings interactive, with lots of questions, physical props that people can touch, illustrations, energizers, learning through movement, friendly competitions and time for participant demonstrations.

We train trainers and educate many groups in the community, so that everyone feels included. We are conscious that men and communities have a role to play so that the new user has a supportive environment. We do pre- and post-training tests to see whether our training has had an impact and monitor how well we are doing. Last, but not least, we have also designed a whole training module with different training methods, as this is an essential part of quality training at WoMena.

WoMena's five training principles include:

1. Create trust and good environment for discussion
2. Ensure people <u>understand</u> what you are teaching them (and why)
3. Ensure people <u>remember</u> what you teach them
4. Interactivity and participatory
5. Plan and manage your training well.

What kind of impact have you had?
Our approach has had many layers of impact. The projects we support measure 75–94% uptake and 82% long-term continued use of menstrual cups among girls/women receiving our education and programming. Through our community involvement model we have seen considerable break down of taboos about menstrual health, opening up new avenues for important discussions in communities, while catalyzing change to male participants' perceptions of menstrual health (WoMena 2018a; Gade and Hytti 2017). Through our impact evaluations we know that thousands of women and girls have expressed increased feelings of freedom and comfort during their periods after starting to use the products, which research demonstrates can have a positive, trickle-down effect on girls' school performance (WoMena 2018b; Gade and Hytti 2017).

We also work at a systemic level to address menstrual equity issue including effectively advocating for the inclusion of menstrual cups in a Uganda's national menstrual hygiene management curriculum in 2018 training manual drafted by the Ministry of Education, as well as the waiver of value-added tax for menstrual cup importation. We have built the capacity of ten NGOs to deliver MHM trainings around cup-related projects. We have also partnered to develop training materials for AFRIpads, which led to us has a potential reach of 1 million girls and women. We are currently completing similar work and developing educational materials for Ruby Life. Lastly, we have built capacity and provided meaningful learning experiences for +100 volunteers from Denmark, Sweden, Iran, Finland, Uganda, US, UK, Lithuania, Germany, and Spain.

How did you develop your curriculum and who did you consult?
WoMena's training curriculum was developed in 2012 with the Red Cross and further improved upon through research and project testing in 2013 and 2014. However, our first official comprehensive curriculum was produced and implemented in 2015 based on examples from organizations working with similar environments and target groups. This included resources produced or used by the Red Cross, school readers from Uganda, Kenya, and Tanzania, menstrual cup supplier user materials, group participation materials, as well as expert materials on specific topics such as hymens from the Swedish Association for Sexuality Education. The initial draft was developed for an internal training of trainers by a group of six volunteers in consultation with actors in the field of reproductive health, such as Ruby Life, Menstrual Cup Uganda, and Reproductive Health Uganda. As a relatively new area of research and programming, we consulted with our internal volunteer Medical Advisory Team on particular topics to ensure it met the highest quality standards. We ensured our educational materials were useful in a variety of settings, so we designed the material to be user friendly and flexible, with suggestions on how the content could be tailored for different contexts.

Later updates to our main curriculum included lessons learned and questions from the field, brainstorms and curriculum development projects with partners such as AFRIpads, and updates based on new publications of relevant curriculum from the Uganda Ministry of Education and Sports. We also further developed our material on hymen and virginity with inputs from experts dedicated to this topic.

Now, when we implement new projects, our training curriculum is tailored to the specific context by involving community health professionals and local leaders as part of the design process. Also, as education is not limited to training, our knowledge management team prepares FAQ sheets to answer in depth questions asked by implementers, women/girls, and community leaders on topics such as Toxic Shock Syndrome, hymens, menstrual irregularities, and infections.

What languages do you use and are there any tensions or conflicts about the language you use for menstrual health education?
Our main curriculum is in English, although we have almost finished developing a Danish curriculum for Danish contexts. We are also finalizing the translation of our training tool designed for northern Uganda humanitarian contexts into Juba Arabic. In Uganda, we have different trainers who speak various local languages. In most cases they are able to deliver the training in the local language, though we sometimes use translators to deliver our trainings. Local trainers (such as teachers or other community members) are always trained to deliver the training and are also provided with menstrual products for their personal use so they are personally familiar with how the products work. WoMena's trainers then support the local trainers in teaching beneficiaries with teachers sometimes translating the content of the training where necessary.

We rarely experience tensions related to the language we use, although people are often surprised to hear others speak so openly about menstruation and vaginas. Usually when we experience hesitancy around product use, we engage in open dialogue regarding the products (and any issues around its use) and ensure that all beneficiaries understand that uptake is totally optional.

What do you wish could be included in future curricula about menstrual health?

There is a need to increase the focus of MH education on other parts of women's bodies especially myths around hymens, vaginal coronas, and virginity. This must first focus on increased knowledge around basic female anatomy by health professionals as well as the social connotations surrounding it. All training curricula should ensure versatile activities and active participation of beneficiaries, and move toward explanations through illustrative means, rather than simply text and words. Lastly, it would greatly improve implementation to have the best training curriculums translated into different local languages.

What does the future of MH education and curricula look like?

In Uganda and many other countries, the future for MHM education is bright. For example, MHM and menstrual cups are now included in the national curriculum in Uganda. There is more advocacy work being done by implementers of MHM with government and local leaders receiving more information regarding menstrual health issues. Male involvement in MHM education is also increasing, with the participation of men and boys in educational and advocacy initiatives at the grass root levels. Last but not least, WoMena is concentrating more on low-literacy populations in its curricula development which helps us to meet the needs of the most vulnerable populations. We can't wait!

SRHR Africa Trust: Integration and Mainstreaming Menstruation into Health Programming

– Vicci Tallis

SRHR Africa Trust (previously known as Southern African AIDS Trust) is an advocacy organization working around Sexual and Reproductive Health Rights (SRHR) and Gender Equality across Eastern and Southern Africa (ESA).

Explain your approach (SRHR, WASH, gender), and intervention?

We work with communities, regional Non-Governmental Organizations (NGOs) and activists working for universal sexual and reproductive health and rights (SRHR) focused on girls, adolescents and young women to promote inclusive, equitable systems for health. We work in communities, districts, countries, and the region, while also having a global footprint in the policy and advocacy space.

Why menstrual health? We believe MH is a major concern for adolescent girls, and young women in the ESA region who lack access to adequate materials which restricts their choices and impacts negatively on their school attendance. We believe that contributing to the improved MH of girls and young women can reduce the risk of HIV infections. Evidence suggests that a lack of access to proper water and sanitation during menstruation and using non-appropriate materials for absorbing menstrual blood, can lead to vaginal imbalances and infections which have been associated with an increased risk of acquiring HIV, human papillomavirus as well as adverse pregnancy outcomes. At the SRHR Africa trust, we see MH as core to women and girl's health more broadly.

What makes your approach unique? How did you develop your curriculum and who did you consult?
Our programming around menstrual health happens in different ways: (1) #TheGirlPlan; (2) Supporting local community based organizations; and (3) Youth hubs.

#TheGirlPlan
We developed #TheGirlPlan to address the unmet needs of adolescent girls and young women and to improve the trajectory of their lives. #TheGirlPlan has six key, interlinked elements: keeping girls in safe schools, ending child marriage, addressing gender based violence, universal access to SRH services including modern contraception, comprehensive sexuality education and menstrual health.

We aim to promote adequate menstrual health through advocacy, capacity building and improvement of menstrual health facilities as critical parts of #TheGirlPlan. We have developed tools to equip learners and teachers to implement #TheGirlPlan in schools. These tools include booklets, posters and a YouTube video. The Menstrual health curriculum covers the following areas: (1) Education about girls' bodies and puberty; (2) Facilities and commodities needed for menstrual health—clean water, safe spaces et cetera; (3) Menstrual Products; and (4) Stigma and Discrimination.

As a result of this innovation, an increased number of schools now have emergency sanitary pads available and school attendance registers also report an increase of female school attendance. Girls also report that promoting menstrual hygiene has boosted their confidence and ability to participate in sports, despite having their period.

Support for organizations
Production of reusable sanitary pads is supported in three countries. SRHR Africa Trust (SAT) Zimbabwe was the pioneer, and designed and developed the reusable pad together with adolescent girls and young women through a participatory process. The reusable pad has passed scientific tests of the Standards Association of Zimbabwe which checked for the safety of materials used and microbiological activity after use, washing and drying. Since then, Chiedza and

SAT Zimbabwe have become part of the technical committee of the Standard Association of Zimbabwe.

It is important to support community partners with the provision of materials for making reusable sanitary pads in rural areas where levels of poverty contribute to a lack of access to menstrual products including pads. In Malawi, organizations have created mothers groups and identified male champions to create an enabling environment for menstrual health. These groups are usually comprised of mothers, fathers, sisters and brothers of the girls who access the pads. Community based organizations are also involved in raising awareness regarding menstrual health through creative means. For example, Chiedza, our partner in Zimbabwe who is based in Manicaland, works with young people to express their menstrual concerns through drama, dance and music which provides a platform for young people to discuss different issues they encounter and create community-based solutions in response.

Youth Hubs
We develop programs for and with youth, paying particular attention to their voices and experiences. We are focused on growing, nurturing and supporting advocates across the continent in SRHR among the variety of projects and programs that SRHR Africa Trust runs. The primary purpose of our Youth Hubs is to bring together like-minded youth activists to convene, network and collaborate around SRHR advocacy issues in the region. The Hubs operate in Botswana, Malawi, Zambia and Zimbabwe. However, we also understand the need to include all youth and have virtual Hubs which attract young leaders from across the continent including Madagascar, South Sudan and Tanzania. Our Hubs carry out large-scale, out-of-the box thinking initiatives around SRHR and menstrual health issues. As they have grown and evolved their continued impact is seen at the grassroots level as well as in public policy as they advocate for change with their local governments.

What kind of impact have you had?
Young activists from the hubs have lobbied for increased and improved efforts of government and stakeholders to ensure that young girls have clean and safe toilets and washrooms with access to a range of menstrual products. Youth hubs have also engaged in media advocacy to address taboos associated with menstruation and starting conversations that de-mystify menstruation.

These conversations are driven by young people—whether they focus on advocacy around reducing sanitary pad tax, the implementation of hygienic washrooms in primary and secondary schools, inclusion of personal menstrual hygiene discussions in schools, or opening up discussions on the menstrual cycle to include management of menstrual cramps and the myths that the society creates pertaining to the management of menstrual cramps.

What languages do you use and are there any tensions or conflicts about the language you use for menstrual health education?

In most countries, the target audience determines our choice of language. Often discussions are held in multiple languages. Our materials are initially developed in English and then translated into local languages if needed with all materials tested in each country we work in. We also try and encourage discussions across the Southern African region—building regional movements that reflect the diverse voices of young people who demand improved information and services.

What do you wish could be included in future curricula about menstrual health?

The Youth Hubs play an active role in Menstrual Health day where we have adopted an inclusive, leave no-one behind approach. We actively work to reach out to, and raise issues affecting those often left behind—such as prisoners, women living with HIV, trans communities and refugees. We believe they are important to conversations around menstrual health and hope to see them included more in the future of menstrual health programming.

THE CUP: EDUCATION WITH A SPECIAL FOCUS ON THE BOY METHODOLOGY

– Steve Nganga Wambui, Camilla Wirseen

The Cup Foundation is a California based non-profit organization founded in 2015 with a mission to empower underprivileged girls worldwide by providing them with sustainable menstrual cups and comprehensive education on sexuality and reproductive rights.

The Cup runs a community empowerment program where underprivileged girls and their needs are at the center. The program is called The Cup Program and targets girls, boys, parents, teachers, elders and informal leaders. Girls are provided with three educational sessions and a menstrual cup while the boys are trained over the course of two sessions. The trainings usually take place in school settings with a focus on the trainers building a relationship with young people through role modelling. Teachers at the school are not allowed to participate in the trainings as they often have a complicated relationship with youths. It is instead young men and women who come from similar backgrounds who are facilitators. Our facilitators are often individuals who will have an easier time to connect with youths and who are able to provide an intimate and safe space.

Explain your approach (SRHR, WASH, gender), and intervention?

The Cup believes there are three core issues facing adolescent girls growing up in challenging environments; poverty, cultural taboos around sex and menstruation, and gender inequality that often are inbedded in their culture. These challenges leave girls alone and with many gaps in basic knowledge putting them at risk for a variety of issues such as: teenage pregnancy, HIV, school dropout, sexual abuse, health infections, unsafe abortions, and low self-esteem. The Cup has developed a holistic manual which has a strong human rights perspective that targets many of the aforementioned challenges facing girls.

We believe that the only way to break down menstrual and sex related taboos taboo is by talking and including boys and men in the conversations around menstruation in order to break the stigma surrounding the issue. Closing girls off into one room and only educating them about menstruation will not challenge cultural taboos which are ingrained in society. During our boy focused sessions, the trainers engage in conversations about puberty and menstruation. We examine how reproduction is a natural part of human development with menstruation as natural, healthy and a normal part of girls and women's lives. We have found that boys are usually very curious, not shy and ask many questions.

We work toward building empathy and greater understanding among boys regarding period pain, discomfort, lack of solutions, leakages, shame, transactional sex, relationships, and parenting. We touch upon other important issues such as human rights, sex education, gender roles, hygiene, abuse, and HIV. We focus on empowering boys to be important allies to the women in their lives as brothers, boyfriends, husbands, and fathers. The boy curriculum also targets challenges that are specific to boys own development and other issues such as drugs and crime.

What makes your approach unique?
The Cup Life Education Program is unique because it provides comprehensive education to adolescent boys and girls that goes beyond menstruation in order to tackle challenges that arise from cultural taboos and gender inequality more broadly. The program does not only have a comprehensive curriculum but is also holistic in its nature by including; informational meetings with the parents, teachers, and elders; a youth helpline to provide long-term support; and the links to access medical services and counseling sessions if needed. Our trainers play an important role within our educational program and are trained to communicate sensitively regarding these complex subjects and perform peer counseling. They are often selected based on their similar backgrounds and experiences as the youths they educate which improves the efficacy of the trainings. The manual and methodology is developed so it can be adapted to a variety of cultural contexts.

What kind of impact have you had?
The Cup has trained 20,000 girls and over 10,000 boys in Nairobi slums as well as other parts of Kenya since 2015. The decision to include boys in the training was seen as a necessary step to reach a healthier social environment among girls and boys and a way to challenge gender inequality. The cup has a general user acceptance of 80% and girls who are users express that they feel more comfortable, happy, relaxed and are able to concentrate in school. The female trainers are almost in every session able to identify girls who are

in need of counseling or medical attention and the team supports them to get help whenever they can. We have also found that starting a conversation with the menstrual cup makes it is easier to speak about taboo topics that are usually avoided as the cup helps with entering into difficult conversations with parents, teachers, and elders. Lastly, the number of girls missing classes due to lack of proper menstrual solution has gone down with the introduction of menstrual cup program in our intervention areas.

What languages do you use and are there any tensions or conflicts about the language you use for menstrual health education?
Our trainers usually use Swahili and English depending on the context. However, they often make a point of talking the local slang called 'sheng' which makes the boys and girls feel relaxed and at home. The trainers are often dressed in a casual and relatable manner with jeans and sneakers. This casual dress code and use of local language helps to encourage the youths and create a safe space.

How did you develop curriculum and whom did you consult?
The Founder of the Cup, Camilla Wirseen, developed the curriculum after having read several health manuals and books on different topics aimed at youths from different origins and after having worked in the field in Kibera for over 8 years with Peepoople, a sanitation solution of which she is the co-founder. Most ideas for the methodology came from interviewing the trainers and women who grew up in similar environments to identify the key relevant problems facing different groups as well as reading and analyzing the many anonymous questions that have been gathered since start of program from the girls and boys that have participated in the program. The manual was developed over a one-year period and is regularly revised and updated to reflect any new learning from the field. The curriculum is leaning on the International technical guidance on sexuality education by UNESCO.

What do you think could be included in future curriculum about menstrual health?
We believe that in the future most menstrual health curriculums and programs will become more holistic as menstruation cannot be separated from sexuality education. We see menstrual knowledge as central to health more broadly. For example, a lack of period is one of the signs of pregnancy. By only giving girls a basic menstrual education and solution, girls are left vulnerable to HIV and teenage pregnancy. Comprehensive and holistic menstrual health education can help to combat these related issues. At The Cup, we hope to men and boys as a central part of the menstrual conversation, and their involvement as key tackling taboos.

What does the future of menstrual education and curriculum look like?
As the founder of The Cup my hope for the future is that all adolescents, whether rich or poor, male or female will be reached with similar programs that provide comprehensive sexuality education with a human rights approach

where the MHM is only one of the topics. I would also like to see programs that could be connected to health clinics so children and parents also have access to health care services in their communities. I don't believe that teachers are best equipped to talk to youths about such sensitive subjects wherever they are. Communication and trust is key in this regard. Much like people who are in need of counseling, require support from someone outside their circle and environment where there is no dependency, power relationship, or risk of abuse. I hope all young people have access to networks who can support them through puberty.

References

Gade, Anna, and Laura Hytti. 2017. "Menstrual Health in Rhino Camp Refugee Settlement, West Nile, Uganda: Pilot Project Intervention Report." Kampala, Uganda: WoMena and ZOA.

WoMena. 2018a. "Imvepi Menstrual Cup Implementation Pilot Report, West Nile, Uganda." Kampala, Uganda: WoMena (unpublished manuscript).

———. 2018b. "Programmatic Approach." Accessed September 19, 2018. http://womena.dk/goals-strategies-policies/.

PART IV

Menstruation as Structural

CHAPTER 36

Introduction: Menstruation as Structural

Inga T. Winkler

To recognize menstruation as structural, we must tune into the very political dimensions that undergird our institutions, laws, policies, budgets, guidelines, taxation, programs, and data collection. Historically, decision-makers have paid limited attention to menstruation—either due to oversight and neglect or due to deliberate exclusion. Yet, the last decade has seen enormous developments; at various levels, menstruation is rising to the level of global awareness. This might be what most distinguishes the current state of the menstrual movement from its past. Menstruation is gaining traction.

Examples of this shift abound. We see an emerging field of 'menstrual hygiene management' as development practice. Policy advocacy is burgeoning, for example addressing 'period poverty' in the United Kingdom and 'menstrual equity' in the United States. Efforts to improve menstrual hygiene and health range from innovating and distributing products, building girl-friendly facilities in schools, raising awareness through broad-based campaigns such as establishing May 28 as Menstrual Hygiene Day, campaigning to lift the 'tampon tax,' developing national menstrual health policies in India, Kenya, Senegal, and many other countries, to embedding menstrual health in the Sustainable Development Goals.

Against this background, this section offers an overview and early assessment of these developments at various levels including those driven by practitioners, policy-makers, activists, and civil society actors. It seeks to capture these trends and initiatives through a combination of practice-based and research-based chapters that bring together different perspectives, voices, and experiences. This diversity is essential to engage different types of emerging knowledge in this field and to combine practical experience with critical reflection.

The section starts with a reflection by Catarina de Albuquerque and Virginia Roaf on the origins of raising awareness on menstrual hygiene in the

© The Author(s) 2020
C. Bobel et al. (eds.), *The Palgrave Handbook of Critical Menstruation Studies*, https://doi.org/10.1007/978-981-15-0614-7_36

context of human rights. The account demonstrates that the water, sanitation and hygiene (WASH) sector has played a critical role in the early development of these efforts as it provided an entry point. However, they also acknowledge that there are obvious, but underutilized linkages with other fields including health, education, work, sexual and reproductive health and rights, culture, religion, among many others.

While these global developments have played a vital role in raising the profile of menstruation, governments and other actors at the *national* level have been instrumental in developing policies, programs, tax reforms, and advocacy campaigns. Archana Patkar traces the development of such efforts in multiple countries, exploring the pathways, challenges, and successes of embedding menstrual health in formalized national agendas. Swatija Manorama and Radhika Desai complement this policy analysis by taking a critical look at the gaps in Indian health policies, while Rockaya Aidara and Mbarou Gassama Mbaye offer insights into a menstrual hygiene management program in West Africa. Jennifer Weiss-Wolf offers an account of her involvement with the 'menstrual equity' movement in the United States, including her reflections on the elimination of the 'tampon tax' as a gateway for activism. The strongest testament to the deeply political nature of menstruation is revealed by Stella Nyanzi's essay in which she details her bold efforts to hold the Ugandan government accountable for its promise to provide menstrual products to girls leading to her imprisonment for her critique of the Ugandan President, Yoweri Museveni.

Another key policy debate is about menstrual leave, which Rachel Levitt and Jessica Barnack-Tavlaris address in their chapter. This debate, possibly more than any other, shows that policies and other structural changes must not be developed in isolation, but accompanied by sociocultural resistance to menstrual stigma and gendered stereotypes. Introducing a menstrual leave policy in a particular context that is not yet equipped for it might have unintended negative repercussions for menstruators. On the other hand, the introduction of menstrual leave policies and other legislative, policy, and judicial developments serve to normalize menstruation and, in that way, can be very powerful.

In the subsequent two chapters, Libbet Loughnan et al. turn to exploring menstruation specifically in the context of the Sustainable Development Goals and the monitoring frameworks they provide, and Siri Tellier et al. address humanitarian responses that point to the challenges that arise in the immediate aftermath of disasters as well as long-term protracted crises.

While menstrual hygiene and health programming is a rapidly developing and dynamic field, this section seeks to provide an overview of the current state of knowledge, which also prompts us to ask 'where do we go from here?'. Venkrataman Chandra-Mouli and Sheila Vipul Patel map the knowledge of menarche and menstrual health among adolescent girls and generally find low levels of knowledge in many low- and middle-income countries, and the same holds true for many girls living in the Global

North. Just as importantly, the section also acknowledges and draws attention to the fact that there is a lot we still do *not* know, while program and policy responses are moving at a rapid pace. Julie Hennegan provides a critical analysis of the evidence for menstrual health interventions in low- and middle-income countries, discussing the approaches which have been trialed to date and highlighting key gaps and new potential pathways. The "Transnational Engagements" edited by Victoria Miller and Inga T. Winkler on the emergence of the field of menstrual health and hygiene conclude this section and similarly raise important questions about the status, development, and future directions of the field, which warrant further exploration. The rapid momentum around menstruation as focus of intervention brings significant opportunities, but it also poses risks.

First, several of the chapters highlight the centrality of WASH in spiraling attention to menstrual hygiene. WASH is a largely technical sector focused on facilities and infrastructure. Moreover, many of the initiatives described in this section are focused on products: their distribution, use, regulation, safety, and taxation. While these initiatives can make valuable contributions (and surely most menstruators assert the need for something to bleed on or into), we observe a narrow focus on the provision of menstrual products as 'quick fixes,' losing sight of structural injustices, which can be addressed through the lens of menstruation. This narrowed focus risks obscuring broader questions of gender justice and body politics which underlie sociocultural perceptions of menstruation. Too few programs tackle the embarrassment and shame associated with menstruation at a deeper level and instead often reiterate (sometimes inadvertently) that menstruation is best kept concealed. We often hear that, for instance, initiatives in the WASH sector or the elimination of the 'tampon tax' can provide *entry points* for tackling the broader, structural challenges. It is time to honor that promise and to move from the entrance into the deeper structure. Only then will we be able to normalize menstruation.

Second, there is a risk of rushing toward poorly conceived policies that are developed on the basis of sensationalized accounts instead of reliable evidence. The widely cited claim that one in ten 'African' girls allegedly miss school because of menstruation, is one recent example of a claim without backing. The contributions in this section caution us to develop a solid evidence base for policy and program development. While many (in particular qualitative) studies detail the challenges women and girls face with regard to inadequate facilities, infrastructure, products, education, and other means of support, we must acknowledge that experiences of menstruation are very context-specific and be careful not to overgeneralize from existing (small-scale) studies.

Third, there is a need to challenge the underlying assumptions of some interventions. As is often the case in global development, we must avoid the imposition of Western ideas, perceptions, or products without adequate regard for local contexts, such as the wholesale dismissal of menstrual cloth as a viable option. Such assumptions lose sight of the aim to enable empowered and informed decision-making, a move that can patronize menstruators. Because participation of the people concerned is key for something as personal as menstruation, many programs would benefit from devoting more time to understanding menstruators' priorities and preferences. Acknowledging the agency of women and girls and any menstruators requires that *they* must be able to decide how they want to cope on the days they menstruate, what activities and cultural practices they want to engage or avoid, and what materials they want to use to catch the flow (if at all).

Finally, there is a risk of rushing decisions and entrenching existing patterns of exclusion and marginalization. In our efforts to move menstruation from the margins to the center, do we leave some individuals behind? For instance, by framing menstrual health as an 'adolescent girls' issue,' do we ignore other women's health issues across the life course? Do we acknowledge women who do not menstruate as well as the needs and experiences of menstruators who do not identify as female? By focusing interventions on girls in schools, do we reinforce the marginalization of girls who are out of school? To respond to these challenges, future initiatives should consider how to most productively support, inform, and normalize the menstrual health of marginalized populations including people who are trans or gender nonconforming, live in poverty, experience homelessness, and/or are deprived of their liberty. We must tune into the experiences of people who are on the move or live with a disability, as well as others living in various conditions of precarity (see "Part I: Menstruation as Fundamental" of this Handbook). We need more emphasis on the lived realities and experiences of *all* menstruators to inform policy-making, programming, innovation, and data collection to ensure that no menstruators are excluded.

Policies and other instruments can be a powerful normalizer that recognize menstruation as requiring political attention, but such normalization and structural change can only be achieved when policy-making and programming include initiatives to challenge sociocultural norms and menstrual stigma. In order to bring about such structural change we need to think about menstruation in a comprehensive way and use it as a lens to identify and address injustices at the intersection of gender and a range of other identities.

CHAPTER 37

Practice Note: Why We Started Talking About Menstruation—Looking Back (and Looking Forward) with the UN Special Rapporteur on the Human Rights to Water and Sanitation

Virginia Roaf and Catarina de Albuquerque

Virginia Roaf: Catarina, we have been asked to reflect on your work as Special Rapporteur on the human rights to water and sanitation,[1] and how it was that we started to work on menstruation as a human rights issue over a decade ago. What do you remember of that time?

Catarina de Albuquerque: It is a great opportunity to look back and see how far we have come! It is already over a decade since I became Special Rapporteur—and so much has moved on since then in terms of discussions on menstruation and menstrual health and the importance of this for achieving gender equality. I started my mandate on human rights, water, and sanitation with a consultation on sanitation, as this was the aspect that needed the most attention. At that time, the right to sanitation was very much the poor cousin of the right to water. Further, 2008 was the first UN International Year of Sanitation, and as such, sanitation was being recognized as a 'problem child' of the Millennium Development Goals, always relegated to second place in comparison to water.

Virginia Roaf, specialist in the human rights to water and sanitation, was an advisor to Catarina de Albuquerque from 2010 to 2014.

Catarina de Albuquerque was the UN Special Rapporteur on the human rights to water and sanitation from 2008 to 2014.

C. Bobel et al. (eds.), *The Palgrave Handbook of Critical Menstruation Studies*, https://doi.org/10.1007/978-981-15-0614-7_37

Virginia Roaf: I remember that your focus on sanitation, both in your first report to the Human Rights Council on sanitation in 2009, as well as during your country visits to Costa Rica, Egypt, and Bangladesh of that year started you thinking about the need for better access to sanitation in schools, in particular for girls.

Catarina de Albuquerque: During those visits, I was particularly looking at girls' sanitation needs and, in visiting schools in Bangladesh, I learnt that girls' needs for privacy, adequate sanitation, clean water, and somewhere to change and wash themselves and menstrual products was preventing some adolescent girls from attending school. I also visited several community initiatives to raise awareness about menstrual hygiene, especially among adolescent girls. Menstruation was in Bangladesh, as elsewhere, taboo, and so I was impressed by the efforts of the communities I visited to ensure that girls are aware of adequate menstrual hygiene and that they have access to appropriate facilities. In my report to the Human Rights Council, I recommended that the government should ensure that women and girls have access to information about menstrual hygiene, and that sanitation facilities enable them to practice good menstrual hygiene (de Albuquerque 2009, Paras. 18, 69, 125).

Virginia Roaf: I remember that this was a pretty fertile time for those of us working with you on your mandate—we started working on menstruation at a broader level in response to what we learnt from women and girls following your country visits.

Catarina de Albuquerque: Absolutely, my visit to Bangladesh informed the rest of my mandate in terms of the needs of menstruating girls and women, not just in schools but across all aspects of their lives, at home, in the workplace and beyond. Menstrual hygiene and health became an additional consideration to be included within the requirements of the human right to sanitation, and all subsequent visits and most thematic reports included a consideration of the needs of women and girls due to their menstrual cycle as a matter of course.

Virginia Roaf: As I remember, one of the messages you wanted to convey over the course of your mandate was that human rights are universal and are to be applied equally all around the globe. People working within the development discourse often imagine that human rights are more likely to be violated in developing countries. What did you find with respect to menstruation in this regard?

Catarina de Albuquerque: While many may consider the taboos and stigma of menstruation to be something to be concerned about in developing countries, I found the same stories of discrimination when I visited Slovenia in 2010, when visiting a Roma community. One Roma woman told me about the shame she felt about the conditions in which she had to practice her menstrual hygiene. Even when the women were able to wash in the river in the spring and summer months, they risked being chased away by the local police, which they considered humiliating. This lack of adequate hygiene perpetuated discriminatory stereotypes about Roma among Slovenians, and I was keen to shine a light on that (de Albuquerque 2010, Para. 36).

The toll of poor menstrual hygiene and the stigma associated with menstruation on women and girls was also apparent in my mission to the Pacific Islands of Tuvalu adolescent girls told me that they sometimes skipped school during their periods because they could not afford to buy menstrual pads, and that families sometimes had to compromise on other daily necessities in order to provide their daughters with menstrual pads (de Albuquerque 2013b, Para. 32). Moreover, on the Pacific Islands (and I also visited Kiribati), disposal of all waste, including menstrual pads is a problem in atoll islands; Kiribati has no land on which to dump waste and does not recycle its waste (de Albuquerque 2013c, Paras. 35–36).

Virginia Roaf: What other experiences do you remember that women have shared about how they managed their menstruation?

Catarina de Albuquerque: On mission in Thailand, I remember asking one migrant worker how she bathed at the open bathing point especially during menstruation, and she told me: "I just shut my eyes and wash myself as quickly as possible, so that it is done" (de Albuquerque 2013a, Para. 25).

But the lack of attention to menstruation is something I recall everywhere I went. When I visited refugee camps in Jordan in 2014, one girl told me that the menstrual pads are such poor quality that she could not leave home for fear of leakage. I believe that this issue has since been addressed, and the menstrual pads provided by the humanitarian services have improved (de Albuquerque 2014a, Para. 53; see also Tellier et al. [Chapter 45] in this volume).

But the problem is the same whether among people in so-called developed or developing nations, refugees or nationals. On my visit to Kenya, women and girls shared their stories of how difficult it is to reconcile their menstruation days with attending school or working. A small pack of menstrual pads in Kenya costs KSh.80 on average; women in informal settlements and girls in communities living in poverty all said they could not afford them and therefore use a piece of cloth instead that can be difficult to wash out due to scarcity of water.

I was impressed by the courage of one girl in a remote area in Kanyadhiang who openly said, after being asked what she wanted most to have a better life "I want one more pair of panties because I only have one and I need to wash them daily and I want sanitary napkins" (de Albuquerque 2014b).

Virginia Roaf: Would you say that you had moments of hope, where women's and girls' needs were being considered?

Catarina de Albuquerque: Absolutely, yes. In Turkana, the poorest county in Kenya, I learnt that the county government was providing menstrual pads to secondary school girls. This has since been expanded to cover all 4.2 million school girls in Kenya. I would love to see all countries in the world adopt similar measures, to give menstrual hygiene greater political and budgetary visibility for menstrual hygiene (de Albuquerque 2014a, Paras. 61, 80–81).

Virginia Roaf: I visited Kenya in late 2018, and the people you traveled with then are still talking of your visit and the recommendations that you made. Impressively, they have been incorporated into the work of the Human Rights Commission in Kenya, which now has a unit dedicated to the human rights to water and sanitation, and on an even better note, a menstrual hygiene policy is going through the final stages of approval in Kenya, and as you mentioned, free menstrual pads are now available in schools (Global Citizen 2018). Things do indeed change for the better (see Patkar [Chapter 38] in this volume)!

Catarina de Albuquerque: Meeting with girls and women and talking about menstruation on country visits was always a highlight. The discussions we had elicited much laughter, but also embarrassment. The stories that girls were telling about their experience of getting their first period, the lack of information on how their bodies were developing, what this meant for their lives, their worries about attending school, their lack of medical or social support with their periods rang true with me, and I think with all of us in the team. Perhaps because we were mostly (sometimes exclusively) female, we could relate to what women and girls were experiencing.

Virginia Roaf: I think that in our sector, we often focus on the material solutions, better construction, the technologies, even better menstrual products. But within these stories, it became clear that the problem also contained an enormous need for education, understanding, and for acceptance—not just for girls, but for boys, for women and men, for the whole of society to be better informed on menstruation.

Catarina de Albuquerque: Absolutely, it was through these discussions with girls and women on country visits, that I identified the stigmatization of women due to their periods as a significant barrier to resolving a whole host of other issues related to menstruation, such as reproductive health, education, and gender equality. And since the unique mandate of a Special Rapporteur is listening, I was able to experience the lived experiences of women and girls and then take these to the level of the UN in an attempt to influence policy-making and programming.

Virginia Roaf: I remember that one of the most significant issues was a lack of data—and the information that existed was full of assumptions and anecdotal evidence. Considering women and girls' needs relating to menstruation, and their negative experiences relating to menstruation, all over the world, certainly not limited to developing countries, we identified the need for more research, as data on women and girls' menstrual needs was, and continues to be extremely limited. We also identified taboos suggesting that people consider menstrual blood to be 'dirtier' than other blood. Specifically, I remember working with you to prepare your keynote at a session at the World Water Forum in Marseille, in 2012, and that was one of the first times that menstruation had a dedicated session. What do you remember of that event?

Catarina de Albuquerque: I was invited by the UN Water Supply and Sanitation Collaborative Council to give the keynote at their session on menstruation, in recognition of the work that I had already done to popularize menstrual hygiene as a significant barrier to the realization of the human rights to water and sanitation. In my address, I identified the taboo of menstruation preventing progress on the achievement of gender equality. In fact, I remember starting that keynote stating that I, too, menstruate, and it was seen as a brave intervention!

Virginia Roaf: I believe it was partly in response to these discussions that we decided to dedicate your 2012 report to the Human Rights Council to the impact of stigma on the realization of the human rights to water and sanitation.

Catarina de Albuquerque: Yes, it was clear to me that we needed to look at discrimination in a more nuanced way, and we decided to focus on stigma as one of the key underlying factors that influences so many of our perceptions of women and girls. Of course, menstruation had to be part of that. At the time, I wrote:

> The silence and stigma surrounding menstruation makes finding solutions for menstrual hygiene management a low priority. Menstruating women and girls often lack a private place to change or wash the rags used. Menstruation has many negative cultural attitudes associated with it, including the idea that menstruating women and girls are contaminated, dirty, impure or polluted. These manifest in practices such as the seclusion of women and girls, reduced mobility, dietary restrictions, and/or women and girls being required to use different water sources or prohibited from preparing food for others during menstruation practices that are often deeply rooted in sociocultural and patriarchal interpretations of religious prescriptions. Even where such restrictions are not followed, women and girls may continue to harbour internalized stigma and are embarrassed to discuss menstruation. The lack of privacy for cleaning and washing, the fear of staining and smelling, and the lack of hygiene in school toilets are major reasons for being absent from school during menstruation, and have an education. (de Albuquerque 2012, Para. 25)

Virginia Roaf: As I remember, it was about this time that we started engaging with discussions on the post-2015 agenda (which became the Sustainable Development Goals or SDGs). Our goal was to integrate attention to menstrual hygiene at yet another level. Your team was represented in all the water, sanitation, and hygiene preparatory technical working groups, including the sanitation and hygiene working groups in particular. At the end of the process, in November 2012, we recommended that menstrual hygiene to be included in the future post-2015 goals, targets, and indicators (UNICEF/WHO Joint Monitoring Programme 2013, 2–3, 6). While this recommendation was not included in the Sustainable Developments Goals themselves, different UN agencies, including the UNICEF/WHO Joint Monitoring Programme, have embarked on significant efforts to improve monitoring of menstrual hygiene and to collect more comprehensive data on menstruation (see Loughnan et al. [Chapter 44] in this volume).

Virginia Roaf: It is now nearly ten years later and it is interesting to reflect how far the conversation has come. A film on menstruation has even won an Oscar this year—what should we take from this?

Catarina de Albuquerque: It's true, and gratifying that this issue has been taken up so widely and with such enthusiasm across the globe. Menstrual hygiene and health have become part of mainstream policy-making, programming, and advocacy efforts. Organizations and governments are working on a range of aspects related to menstrual health including sexual and reproductive rights. There is a marked difference to when we started talking about menstruation. The menstrual health agenda is starting to reach far beyond the Water, Sanitation and Hygiene (WASH) sector which is essential to address the full range of human rights issues related to menstruation.

Virginia Roaf: Yes. And yet, over a decade ago, the WASH sector was an ideal entry point for starting the conversation on menstrual health. What prompted us to consider menstruation as part of your mandate on water and sanitation were the stories women and girls told you. They mentioned their very real needs in terms of access to facilities and materials. The predominantly male engineers working in water and sanitation were surprised to hear that menstrual hygiene was something that also lay within their remit, but once the initial surprise was overcome and the silence around menstruation was broken, most WASH sector professionals were open and eager to address the issue and to start finding solutions for the immediate needs by women and girls. This focus on tangible and concrete measures is a huge part of what triggered the increasing efforts on menstrual hygiene management.

Frustrating as it remains that many women have the primary role as homemakers and as caregivers, water and sanitation are still particularly critical issues for women. Therefore, I do feel that considering menstrual health through the lens of the human rights to water and sanitation still makes sense, although I think it is not the only lens that we should be using. Through access to adequate water and sanitation services, which include consideration of privacy and dignity, and which comply with the standards of the human rights to water and sanitation, many issues relating to menstrual hygiene and health will be addressed. Adequate water for washing ourselves and our menstrual pads, a place to safely dispose of used menstrual materials, sanitation facilities available at school, at the workplace, and elsewhere, these services address requirements for menstrual hygiene. And precisely because the WASH sector is perceived as rather technical, it provided an 'innocuous cover' to start addressing issues that are ultimately about deeply entrenched gender norms.

Catarina de Albuquerque: I absolutely agree. But at the same time these benefits point to the risks associated with addressing menstruation exclusively through the lens of WASH due to the heavy focus on tangible outputs, facilities, and the provision of materials. While the WASH sector proved to be an excellent entry point, we must get past this narrow understanding of menstrual hygiene to fully embrace menstrual health. The World Health Organization defines health as "a state of complete physical, mental and social well-being and not merely the absence of disease or infirmity" (WHO 1946) and we must embrace all these dimensions in relation to menstruation. This entails education, reproductive and mental health, working conditions, as well as sociocultural taboos. To manage our periods and menstrual health more generally, it is essential that we understand how our bodies work and are able to fluently express what we need in terms of help or support when things go wrong. The taboos regarding menstruation, for example, extend to the lack of attention that the medical profession pays to poor menstrual health. For example, the symptoms of endometriosis are not well understood and too often dismissed by health care providers. Women and girls need better education about their bodies. Men and boys must also be educated on the menstrual cycle and women and girls' needs to manage their menstruation, so that they can respond with understanding and not fear. Menstruation is also deeply enmeshed with cultural and religious traditions, and we need to approach the discussion of such practices with more nuance. I am emboldened by the efforts of women and girls across the globe to treat menstruation as a fact of life and to resist attempts to restrict them, for instance from accessing cultural or religious sites.

Reflecting on where we began over a decade ago and how far we have come, I think we are truly beginning to realize that menstruation is a gender equality issue that affects the realization of all human rights. Menstruation is about so much more than blood and the materials and services (in the context of WASH) that we need to deal with it. I am glad that we were able to identify this entry point in the WASH sector and that I could contribute to normalizing menstruation all over the world, beginning with my discussions with the girls in Bangladesh to the Palais de Nations in Geneva.

NOTE

1. An independent expert mandated by the UN to promote compliance with and examine violations of the human rights to water and sanitation.

REFERENCES

de Albuquerque, Catarina. 2009. "Report from the UN Independent Expert on the Issue of Human Rights Obligations Related to Access to Safe Drinking Water and Sanitation on Her 2009 Mission to Bangladesh." Human Rights Council, A/HRC/15/55.

de Albuquerque, Catarina. 2010. "Report of the Special Rapporteur on the Human Right to Safe Drinking Water and Sanitation on Her 2010 Mission to Slovenia." Human Rights Council, A/HRC/18/33/Add.2.

de Albuquerque, Catarina. 2012. "Report of the UN Special Rapporteur on the Human Rights to Water and Sanitation on Stigma and the Realization of the Human Rights to Wand Sanitation." Human Rights Council, A/HRC/21/42.

de Albuquerque, Catarina. 2013a. "Report of the Special Rapporteur on the Human Right to Safe Drinking Water and Sanitation on Her Mission to Thailand." Human Rights Council, A/HRC/24/44/Add.3.

de Albuquerque, Catarina. 2013b. "Report of the Special Rapporteur on the Human Right to Safe Drinking Water and Sanitation on Her Mission to Tuvalu." Human Rights Council, A/HRC/24/44/Add.2.

de Albuquerque, Catarina. 2013c. "Report of the Special Rapporteur on the Human Right to Safe Drinking Water and Sanitation on Her Mission to Kiribati." Human Rights Council, A/HRC/24/44/Add.1.

de Albuquerque, Catarina. 2014a. "Report of the Special Rapporteur on the Human Right to Safe Drinking Water and Sanitation on Her Mission to Jordan." Human Rights Council, A/HRC/27/55/Add.2.

de Albuquerque, Catarina. 2014b. "Statement by the United Nations Special Rapporteur on the Human Right to Water and Sanitation on Her Visit to Kenya." July 22–28. Accessed April 20, 2019. https://www.ohchr.org/EN/NewsEvents/Pages/DisplayNews.aspx?NewsID=14912&LangID=E.

Global Citizen. 2018. "4 Million Kenyan Schoolgirls Are Going to Receive Free Sanitary Pads." Accessed at: https://www.globalcitizen.org/en/content/kenya-schoolgirls-free-pads/.

UNICEF/WHO Joint Monitoring Programme. 2013. "Proposal for Consolidated Drinking Water, Sanitation and Hygiene Targets, Indicators and Definitions."

World Health Organization. 1946. "Constitution."

CHAPTER 38

Policy and Practice Pathways to Addressing Menstrual Stigma and Discrimination

Archana Patkar

INTRODUCTION

I vividly remember being shushed at any mention of menstruation as a potential factor in absenteeism and poor learning achievements among adolescent girls in school in India. As recently as 1994, menstruation was the ultimate unmentionable, shrouded in multiple veils of forbidden sexuality and silent patriarchy. It did not help that there were no national policies that included menstruation, and none of the international human rights instruments explicitly mentioned it (Winkler and Roaf 2014). Working on education, adolescent health, and water and sanitation, I was struck by the systematic neglect of the human life course in the design of the built-up environment and seized every opportunity to understand and address menstrual stigma in policy and practice.

People at home, school, in the fields, offices, while traveling or playing a sport all ached with silent resentment at the ignominy of the monthly period because they were forced to suffer it in shame and silence. Adolescent girls were often uninformed and unprepared for menarche. Exclusion and shame led to misconceptions and unhygienic practices during menstruation (Chandra-Mouli and Patel 2017). The pattern repeated itself when conversations expanded to include persons with disabilities or HIV; those who had been genitally cut or who suffered from fistula; or those negotiating questions of gender and sexual identity. Discussing menstrual flows with the transgender community revealed a complex web of intersecting stigma around identity, occupation, class, age, caste, and ethnicity. With every conversation I engaged in, people of all ages poured out their thoughts, ideas, fears, hopes, and demands around menstruation.

© The Author(s) 2020
C. Bobel et al. (eds.), *The Palgrave Handbook of Critical Menstruation Studies*, https://doi.org/10.1007/978-981-15-0614-7_38

It was evident early on that, freed of stigma and mysticism, menstruation and its practical challenges might actually be nonthreatening and universal enough to stride across geographical context and sector silos as a powerful, positive disruptor for gender equality. Engineers, administrators, medical practitioners, nurse-midwives, hygiene promoters, educators, adolescent girls and boys, women and men from all walks of life and at different stages of life shared, listened, learned, and took a pledge to break the silence and walk without shame. Across national boundaries, from village, ward, county, province, and state, these voices rose to a crescendo that could not be ignored.

This chapter seeks to share a transformative policy pathway to gender equality using noncontroversial and pragmatic yet universal entry points, such as water, sanitation, and hygiene (WASH), to make explicit women's silent and neglected needs around menarche and menstruation. It draws on my professional experiences: first, from 1994 to 2010, as an independent adviser on education, health, WASH, and urban poverty projects to many governments and development partners; then, from 2010 to 2018, as head of policy at the Water Supply and Sanitation Collaborative Council (WSSCC), a small organization within the United Nations working on sanitation and hygiene, with national governments in Asia and Africa.

The approaches employed in this work included a judicious mix of advocacy related to policy, systems behavior change, and the media (see Naeve et al. 2017). They often operated simultaneously and at different levels to build critical mass and reach tipping points for change. The process relied on making the voices of thousands of women and girls in Asia and Africa heard. These voices rang clear and true; they changed the perceptions, understanding, attitudes, and openness of diverse populations and their policy makers, who cast away age-old taboos to ask questions, learn, and embrace the universal truth of menstruation.

As a result, between December 2013 and 2019, India and Senegal published and Kenya approved menstrual hygiene/health policies and guidelines that broke the silence on menstruation and recognized menstruators' needs for basic services. Afghanistan, Bangladesh, Bhutan, Cameroon, Maldives, Nepal, Niger, Sri Lanka, and Tanzania followed suit, initiating trainings and institutional changes to health, education, and WASH government guidelines, budgets, and projects. A corpus of facilitators[1] trained in gender equality and menstrual hygiene management (MHM), built a cadre of government officers who championed a carefully thought-through policy advocacy and practice agenda.

Sanitation and Hygiene Are Powerful Entry Points for Change

The language of WASH has long been one of taps, toilets, and soap, with people absent from the conversation. The recognition of the right to water and sanitation in 2010 (UNGA 2010) and the subsequent distinction and separation of the right to sanitation in 2015 were crucial milestones to move the sector forward (UNGA 2015; Winkler 2016). However, there remains a wide gulf between the rudimentary sanitation infrastructure available to those living in poverty and the obligations enshrined in these rights (Patkar 2016).

Far from being a pleasant experience, toilets, hand-washing facilities, and water points are often wet, damp, ill-smelling, dimly lit, unhygienic spaces where one can slip, fall, hurt oneself, or fall ill. Worse, sanitation services are heavily gendered, and women across the world report a fear of being followed, seen, touched, or violated (Kulkarni, Bhat, and O'Reilly 2017). Several studies have shown the intimate connection between the sanitation environment and the physical and mental well-being of women and girls (O'Reilly 2016; Sahoo et al. 2015). Furthermore, trans men and women feel unwelcome in single-sex toilets worldwide and are at risk of violence in these spaces (Boyce et al. 2018).

Urination, defecation, bathing, changing, and washing are basic, everyday needs that require privacy, comfort, and dignity. Women's and girls' sanitation behaviors are more expansive than urination and defecation (Sahoo et al. 2015), but these are often overlooked in policy, investments, and initiatives (WaterAid 2013). There has been surprisingly little analysis when it comes to the question of whether toilet access, design, or programs satisfy female needs around menstruation, menopause, and infant or elder care (Burt, Nelson, and Ray 2016). More generally, Phadke, Khan, and Ranade (2011) present a powerful reflection of women's experiences, encounters, and negotiations in urban public spaces in India. They eloquently illustrate that public spaces—streets, parks, playing fields, and courts—have been designed without women in mind and are therefore primarily frequented by men, excluding women and restricting their presence, visibility, and voices.

But it is just as possible to imagine a reversal of these gendered environments. Reimagine, if you will, public spaces redesigned to accommodate women and men of all ages, including menstruators, pregnant women, elderly persons, and caregivers with babies or young children. Imagine spaces where menstruators are able to change their menstrual materials, wash their cloths freely without being ashamed or ridiculed, dry them in the sun, and inhabit public spaces with confidence because they are sure of finding facilities to wash and change wherever they go. Imagine the breaking down of unconscious biases so that one day we can think of unisex washrooms that all genders can use without fear.

This simple yet inclusive reimagining of the world around us provided the underlying principles for the policy and practice change I sought. Because sanitation and hygiene require spaces that come with a certain conceptualization of use, toilets, washbasins, and their surroundings have the potential to help us think differently about human needs. There is something about the very basic human need to relieve oneself that is indisputably universal. Basic sanitation cuts across context, culture, and religion, and it remains a human need throughout life, with a significant daily impact on people's lives (Essity 2017a).

PROOF OF CONCEPT: FROM LISTENING TO POLICY DEVELOPMENT

Embarking on the Journey

The journey began in India in 1994, when I worked with the state government on the District Primary Education Programme of Madhya Pradesh. The program was launched to achieve the objective of universalization of primary education. In the 1990s, Madhya Pradesh was a state plagued with a deteriorating male-to-female ratio, high maternal mortality, large gender gaps in education, and poor female participation in the workforce (Prabhu, Rampal, and Shukla 1995). Endowed with visionary leadership in the Ministries of Education and Finance,[2] the powerful combination of human and financial resources and political will, positioned Madhya Pradesh to be a flagship state for achieving parity in primary education.

Madhya Pradesh was not alone in grappling with girls' frequent absenteeism, low motivation, and eventual dropping out of school. The links with menarche were unmistakable: restricted interaction, play, and mobility; reduced confidence; withdrawal and other behavioral changes; and low motivation to continue secondary school (Sommer et al. 2016). In my own state of Maharashtra, I witnessed first-hand the immense water and sanitation burden that women and girls experienced every day as water collectors, carriers, managers, and household and street cleaners. Sanitation duties were reserved for women and 'lower-caste' men, bringing with them distaste, discrimination, and exclusion. Menstrual blood—the mere mention of it—triggered the same reactions (see Sukumar [Chapter 13] in this volume). No one wanted to talk about it. Menstruation was socially, politically, and culturally unmentionable. Anyone articulating it was politely ignored. This complete denial meant that parents and teachers alike could not show interest in the topic. The 'menstrual lag' was systematic and seemed to persist even in schools with decent infrastructure. Teachers simply skipped over the reproductive health chapter or asked students to read it at home (Singh 2019). The human body and how it reproduced was a mystery to be magically unlocked once a couple tied the knot. Concepts of sexual attraction were entirely absent from the conversation, as was the link to reproductive anatomy and function (Thirunavukarasu and Simkiss 2013).

I came up against the same silence, reticence, and opposition when seeking answers in other sectors. Specialists in reproductive health, life skills, and WASH had never really thought about menstrual health (Bharadwaj and Patkar 2004). Privately, policy makers acknowledged that the issue 'may' be important but complained about the lack of data, social mores, and worries about 'corrupting' our girls if we talk about 'it'—'it' being closely related to sexuality. While everyone I spoke to agreed in principle,[3] many enlightened health and WASH professionals at the time simply felt that there were so many other important issues, this one could wait (Bharadwaj and Patkar 2004).

But women could wait no longer. They wanted to talk about menstruation. They wanted to learn about their bodies. They wanted to have information and make choices. In focus group after focus group over the next ten years across India, and later in Cambodia, Senegal, Kenya, Niger, Nigeria, Tanzania, Bangladesh, Bhutan, Nepal, and Sri Lanka, women and girls opened up after the initial silence and embarrassment. Young girls and older married women alike, bombarded us with questions that they had never been allowed to ask. So did men and boys.

Listening: The Menstrual Hygiene Management Lab

The Nirmal Bharat Yatra, also called the Great WASH Yatra, a traveling carnival organized by the Government of India in partnership with WASH United, was a golden opportunity to make this silent demand heard (Desai 2012; Chanam 2012; Fernandez 2012). Conceived originally to deliver two important messages—stop defecating in the open and wash your hands at critical times with soap and water—we added a third: Manage your menstruation hygienically and with safety and dignity (Patkar 2012). This was a never-before-heard message in the government's health and WASH arsenal and a critical first step in breaking the silence. Neither cautious nor careful, it was broadcast with fanfare and the endorsement of Bollywood celebrities who talked about menstruation and sanitation in the same breath, with the blessings of the then-minister for Drinking Water and Sanitation, Jairam Ramesh (Gottipati 2012; YouTube 2012).

In 2012, a small team[4] designed the Menstrual Hygiene Management Lab (MHM Lab) as part of the Great WASH Yatra that journeyed through five Indian states and 2000 km over 51 days (WaterAid, WSSCC, and Unilever 2013). Some 12,000 girls and women took part in focus group discussions, and the Yatra team collected preliminary data from 747 participants. Women, girls, men, and boys—everyone wanted to know about their bodies, learn to express and manage their secret fears and worries, feel clean, smell good, understand their sexual needs better, and be in control of their body and mind. The original MHM Lab design was conceived as a safe space for women, but we realized at our first stop, in Wardha, Maharashtra, that we had to include men and boys.

The questions did not stop at menstruation, as grandmothers, young men, and boys flocked to the MHM Lab to participate in focus groups and individual interviews conducted by our volunteers and partners.[5] The results showed that 70.9% of the girls had no idea what was happening when they began to bleed and boys could not understand why girls suddenly stopped talking to them or playing with them (WaterAid, WSSCC, and Unilever 2013). All Lab participants, male and female, understood the menstrual cycle, made and wore a 28-bead sparkling saffron-and-red bracelet that represented the menstrual cycle and took the pledge to break the silence (Chanam 2012; George 2012).

Thus, the MHM Lab served two interlinked objectives: It was a safe space to talk about menstruation at a time when this was still unthinkable and, it allowed us to build the evidence for policy change and refine the prototype training toolkit. The methodology of the MHM Lab was later refined and published as *The MHM Lab Manual*, endorsed by the Government of India as a training methodology for raising awareness and replacing stigma around menstruation with practical information and pride in one's own body (WSSCC 2013c). National, state, and district officials witnessed the evidence in action, which paved the way for policy reform.

Evidence for Policy and Capacity Development

The Union Minister, H.E. Jairam Ramesh, visited the MHM Lab and experienced first-hand the intensity of discussions and the massive queues before the tent (Gupta 2013). The news traveled quickly. Everywhere the Great Wash Yatra pitched its tents, the facilitators and managers of the MHM Lab met with state and district officials, principals, school teachers, and opinion leaders in addition to thousands of men, women, boys, and girls to talk about the issue and seek joint solutions to destroy the stigma and neglect that menstruators faced (Patkar and Deschaine 2014). This laid the groundwork for policy discussions at the state and national level.

Campaigns to build awareness and advocacy by powerful politicians and celebrities are important in initiating change. Certainly, they create a buzz, prod the media, and bring unmentionables to the agenda (Ma 2016). But as a long-term skeptic of advocacy's unfinished business and largely unmeasured effectiveness (see Puri 2016), I was determined to hitch the learning outcomes of the Yatra, impressive in terms of diversity and scale, to the readiness of the Ministry of Water and Sanitation to break the silence on menstruation in national policy.

On my next visit to the Ministry of Drinking Water and Sanitation in New Delhi in early 2013, Joint Secretary Jitendra Mathur, Directors Vijay Kumar and later Sujoy Mojumdar, and their colleagues were fully aware of the success of the MHM Lab. The conversation with the ministry moved rapidly: What would the integration of MHM into the national policy on water and sanitation look like? What were the budgetary implications? What capacity to implement change was needed, and how would this be built within the government system? How could the change on the ground be monitored for uptake, learning, and improvement?

Working with slim resources and the existing normative mandate of the Government of India, government partners and my team, we sought opportunities within existing government schemes to break the silence on menstruation. Putting together an amendment to the Nirmal Bharat Abhiyan (Total Sanitation Campaign) (Indian National Congress 2018) was a straightforward solution: Simple, easily understandable language that could immediately be translated into budgets and local plans across the country's 29 states, 725 districts, and 7 union territories. The three paragraphs deliberately contained no human rights language and almost erred on the side of brevity. They simply stated that existing funds may be used "to raise awareness, information and skills on menstrual hygiene management" and to "implement safe disposal solutions for menstrual waste" (Ministry of Drinking Water and Sanitation 2013). The language used matched that of the existing water and sanitation policy and was therefore practical, output-oriented, and designed for use by planners and engineers. The aim was to get this off the ground—and it worked.

In September 2013 Pankaj Jain, the secretary in the Ministry of Drinking Water and Sanitation, Government of India, endorsed and launched the MHM toolkit at the first national MHM training for government officials from health, WASH, social welfare, and public administration held in New Delhi (WSSCC 2013a, 2013b, 2013c). Sixty-two participants from the states of Uttar Pradesh, Jharkhand, Bihar, Madhya Pradesh, Chhattisgarh, Uttarakhand, and Delhi attended the four-day training. The response was tremendous, resulting in demand and follow-up trainings in southern and western India.

In 2014, under the Modi government, menstrual hygiene was maintained under the national Clean India Mission (Swachh Bharat Mission)[6] guidelines, which state, "Issues relating to women's personal hygiene namely menstrual hygiene are to be focused under the Swachh Bharat Mission (SBM). Girls and women have hygiene and sanitation needs linked to their menstrual cycle. Women suffer in the absence of knowledge about safe practices on Menstrual Hygiene Management (MHM)" (MDWS 2017). The guidelines provided funds to conduct information, education, and communication activities to raise awareness among all stakeholders about MHM, including why menstruation occurs and how to manage it safely (washing, cleaning, using, and changing clean absorbents regularly) and menstrual waste management (safe disposal solutions for used cloths and pads).

EXTRAPOLATING A PATHWAY AND A METHODOLOGY

With MHM now firmly in national policy, a mammoth task lay ahead. Developing a pathway for menstrual health involved (1) deepening the engagement in India with a strategy to percolate the change holistically to local levels; (2) leveraging the momentum to expand menstrual health policy development to other countries and regions; (3) expanding the agenda to address other taboo topics; and (4) addressing the diversity of menstruators' needs.

Deepening the Engagement in India

Breaking the silence in policy was seen as a real victory by the ministry staff, who became passionate advocates for MHM. The language of the policy amendment was deliberately short and sharp, but it prioritized the first prong of increased information, education, and communication through sequencing of activities and funding. However, the tangible quickly takes precedence over slower, deep-rooted changes. The proxies that proliferated were menstrual absorbents, subsidies, handouts, unsustainable projects, and the endless rounds of training that poor women in communities often undergo to learn about their own stark realities. Despite the publication of MHM Guidelines with a focus on girls (MDWS 2015), MHM remained on the periphery of the Swachh Bharat Mission, whose primary aim was to make India open-defecation-free by October 2019 (MDWS 2018). As it became easier to talk about menstrual hygiene, sanitary pad distribution schemes became the flavor of the moment. Without complementary information, counseling, and support mechanisms in place, adolescent girls were unable to or chose not to use them (Chandra Sharma 2018). These schemes did little to change the narrative or to recognize and value the diverse experiences of menstruators across the life course. As Chris Bobel argues, "the largely product-based solutions that often follow fail to challenge the social construction of the menstrual body as dirty and in need of concealment" (2019).

The persistence of menstrual stigma points to the challenges of using the WASH sector, with its focus on hygiene and tangible solutions, as an entry point. Working with policy makers, it was prudent to couch activist actions—for example, breaking the silence on menarche and menstruation—within the instrumentalist language linked to budgets and measurable outputs. The water and sanitation sector, ever pragmatic and action oriented, offered the ideal launch pad. However, without parallel mechanisms and advocacy stressing participation, voice, and choice, there was always a risk that change would be limited to physical design around taps and toilets, pads and cups. This would ignore the crucial need for reflection, spaces for dialogue, learning and sharing between men and women, and the myriad expressions of gender and sexual identity that are needed to replace body stigma and shame with knowledge and pride. MHM can too easily be co-opted by a training, pad-manufacturing, and distribution army, leaving notions of voice and agency far behind. Once the initial euphoria linked to a holistic approach dies down, champions leave and the dust settles, policies intended to be transformational are often narrowly interpreted. The menstrual hygiene movement seems to have settled on WASH facilities, tax-free pads, and incinerators as key indicators of action and success. Encompassing broader questions of menstrual health has thus far proved challenging in India, but the policy change can only be applauded as progressing a transformative agenda in a complex and diverse country of 1.4 billion people—a country in which the right to reproduction is traditionally seen as a family health and well-being issue (Kosgi et al. 2011) and menstruation, especially for young girls, means that marriage and childbirth are the next rites of passage into adulthood (CREA 2005). The developments to date are important early steps that have unleashed language, conversations, and demands that have been suppressed for centuries. And with them come a flurry of innovative strategies, resources, and programs to make up for lost time in India and beyond.

Broadening: Testing the Proof of Concept for Universality and Scale in Africa

If senior, male government policy makers could talk about menstruation in the patriarchal states of Northern India, could the same be achieved in other geographies using the same methodology? Building on the success in India and momentum created at the international level through a series of advocacy events (ANEW 2013; Naafs and Brewer 2013), it was time to test the waters in Africa. In September 2013, a small team,[7] in partnership with UNICEF WCARO in Dakar, Senegal, organized the first-ever multicountry training workshop on menstruation for government officers from 22 countries in West and Central Africa (ANEW 2013). Regional Sanitation Adviser for UNICEF WCARO, Jane Bevan, brought critical commitment and resources to break the silence in the region. We invited officers from health, education, WASH, and environment ministries, as well as teachers, trainers, practitioners, and activists. Participants shared personal stories of shame and stigma and forged a strong commitment, and the movement was born. Several meetings and discussions later, the Government of Senegal launched The Joint Programme on Gender, Sanitation and Hygiene, cofinanced by WSSCC and UN Women in Louga, Senegal, on March 8, 2014, to listen and respond to the needs of women and girls regarding sanitation and hygiene in policies, laws and regulations, and related budgets in Senegal, Niger, and Cameroon (see Aïdara and Gassama Mbaye [Chapter 40] in this volume).

The policy pathway was similar to India and yet different. Senegal, with a population of 14 million, of which 60% were under age 25, was rattled by the raw novelty of the term 'MHM.' From ward to village to ministry to agency, I was told that this would need to be handled extremely sensitively. However, a mix of essential ingredients—in this case, action research in urban and rural areas to fill the baseline and evidence gap, rapidly adapted tools following local interviews and focus groups, advocacy with parliamentarians and an interministerial committee—unleashed tremendous support for MHM (WSSCC and UN Women 2016).

Leading the charge were champions from the Health, Education, Environment, and WASH[8] ministries, with the able support of the UN Women team in Dakar.[9] For example, Amadou Diallo from the Ministry of Water and Sanitation in Senegal spoke eloquently about how the work on MHM had opened the eyes of the ministry to a new way of working—one that made complete sense within the ethos and culture of fairness, respect, and dignity innate in Senegalese tradition. MHM was able to weave a common thread across health, education, gender, environment, and WASH—within government as well as development agencies—reflecting realities that are not confined to one sector and therefore must be solved through convergence. The Joint Programme has successfully integrated MHM in the national strategy of the Ministry of Water and Sanitation, which now mandates that investments in WASH take into consideration women's needs, particularly menstrual management (Ministère de l'Hydraulique et de l'Assainissement, République du Sénégal 2016).

English and Swahili-speaking Kenya presented an interesting contrast. In 2016, Kenya was already a hotbed of MHM activity, with large UN agencies and a plethora of national NGOs working on raising awareness and producing pads. However, it had yet to articulate a coherent approach to menstrual health and hygiene, as the focus was largely on keeping girls in school. Girls who had dropped out of school, HIV-positive women and girls, and female refugees were not yet a focus of the discussion (Geertz et al. 2016).

The first national MHM training was organized with the Ministry of Health in 2016 for 70 officers working in health, environment, water, and sanitation. It served to galvanize latent champions into action (Okwaro et al. 2016). 12 County first ladies were crowned MHM champions. An initially skeptical Ministry of Health was convinced about the relevance of the issue and the need to train all 47 counties and move to the logical next step: a coherent national policy. Building a cadre of trainers to work across East Africa was an important investment and critical link in the policy pathway.[10]

Kenya's—and the world's—first dedicated MHM policy, approved in November 2019, is a bold statement of recognition of its approximately 25 million menstruators and their needs. A series of multi-stakeholder consultations were convened by an impressive cast of characters: Director of Public Health Kepha Ombacho; his deputy, Jackson Muriithi; WASH Alliance and WSSCC national coordinator, Tobias Omufwoko; Knowledge and M&E officer, Neville Okwaro from the Ministry of Health WASH Hub; and WSSCC.[11] The meetings involved a range of NGOs, researchers, development partners, manufacturers, and county governments, and they were marked by numerous power plays: between national control and newly decentralized counties fighting for fiscal control; institutional fragmentation and confusion between which government authority would manage which aspect of MHM; competing interests on generating funds for particular pet projects or research; and lots of excitement around disposable and reusable pad innovation. However, the meetings were largely positive, with a shared desire for coherence in policy and practice. Translating the policy into meaningful budgets, actions, and monitoring across 47 districts is the next step, along with urgently needed product regulation, an area of concern for the Ministry of Health.

The Kenyan Cabinet, chaired by President Uhuru Kenyatta, approved the National Menstrual Hygiene Management (MHM) Policy highlighting MHM as a rights issue and bringing it into the mainstream of the country's health and development agenda (Government of Kenya 2019). The policy intends to highlight MHM as a rights issue and bring it into the mainstream of health and development. It examines the social, economic, cultural, and demographic context of MHM, including its implications for menstruators' health. Menstruation is a critical indicator of female health and vitality, and a healthy menstrual cycle indicates overall well-being. This policy aims to break the silence around the biological phenomenon of menstruation and menstrual blood to enable Kenya's women and girls to access information, make informed choices, and participate fully in all walks of life, every day of the month.

Regional Declarations and Benchmarking

Creating regional MHM hubs of activity on policy and practice was intended to engage neighbors, catalyze innovation and action, and break the silence on the ground and within WASH and health policy. The first countries chosen—India, Senegal, and Kenya—were also regionally recognized influencers, innovators, and leaders well placed to drive transformation beyond their borders. They then found regional counterparts for sharing, benchmarking, learning, and dialogue: Niger and Cameroon in West and Central Africa; Tanzania and Uganda in East Africa; and Afghanistan, Bangladesh, Bhutan, the Maldives, Nepal, Pakistan, and Sri Lanka in South Asia.

Regional sanitation conferences and their declarations were an essential part of this policy advocacy, as all countries report on their achievements regionally every two years (Naafs and Brewer 2013; Narayan et al. 2011). Bringing MHM into regional declarations was therefore crucial for advocacy, action, benchmarking, regional learning, and sharing. The fifth South Asian Conference on Sanitation (SACOSAN V) in Kathmandu, Nepal, in October 2013 ended with a declaration urging compliance measures on MHM facilities and explicitly calling for an end to discrimination across its eight member states (Kathmandu Declaration 2013). At SACOSAN VI in Dhaka, ministers and government officials from the Maldives, Sri Lanka, and Bhutan—as well as Sri Lanka's Minister of City Planning and Water Supply, Rauff Hakeem, and Member of the Legislative Assembly in India, Prem Lata Singh—responded to the concerns and demands expressed by adolescent girls, transgender persons, and persons with visual and hearing impairments. These were then reflected as commitments in the Dhaka Declaration (2016). Similarly, at AfricaSan V, Ministers Mansour Faye from Senegal and Wassalke Boukhari from Niger co-chaired a powerful session on MHM and spoke about their countries' commitments to breaking the taboo in policy and providing MHM-friendly public facilities and awareness programs (WSSCC and UN Women 2015a).

In addition, building capacity and equipping policy makers and practitioners with information and tools were essential. To deliver the SACOSAN commitments on MHM, WSSCC pledged to train government staff of key national ministries in all eight countries in the South Asian Association for Regional Cooperation (Theseira 2016). National trainings were complemented by regional trainings that brought together officials from several countries to listen, share, and learn in neighboring host countries (Okwaro and Davis 2017; Theseira 2016). Expanding regionally and providing opportunities for mutual sharing, learning, and benchmarking was crucial to reach a critical mass of champions and believers.

Regional voices were important not just for sharing and learning but to validate global advocacy and policy. These same ministers, researchers, and practitioners from Africa and Asia brought commitments and demands to sessions at the Commission on the Status of Women in New York, the Water and Health Conference in North Carolina, International Women's Day, the World Health Assembly in Geneva, and other global platforms. They helped break the silence from the bottom up (George 2013; Patkar et al. 2017; WaterAid 2013; WSSCC and UN Women 2015b; WSSCC 2016a, 2016b, 2017a, 2017b; UNC 2017; Essity 2017b).

Expanding the Conversation to Other 'Unmentionables'

As ministers and senior bureaucrats, village chiefs, parliamentarians, and the media started openly discussing menstrual health and hygiene, would they also be more open to conversations on other hitherto taboo topics? In Senegal, Cameroon, and Niger, menstruators who talked about their first period went on to describe menstrual experiences across their lives (WSSCC and ONU Femmes 2015), including stories of genital cutting and injuries during childbirth (Mimche et al. 2017). These were taboo topics, shameful for perpetrators, accomplices, and victims alike, that commanded humble listening.

These conversations—across different countries and cultures and often initiated around health-seeking behaviors, child spacing, and contraception—provided a natural space to disrupt notions of voice, control, choice, and even gender identity and expression. Most important was the discussion around fertility, especially in countries in West Africa, where women themselves perceived that a woman's fecundity made them more attractive to men. Surrounded by myths, perceptions, and social stigma, women's information, support, and health-seeking behavior around menopause is not that different from menarche (Wambua 1997; see also Bacalja Perianes and Ndaferankhande [Chapter 33]; Dillaway [Chapter 21]; and Singh and Sivakami [Chapter 70] in this volume), as is the subject of female urinary incontinence (Aoki et al. 2017). The burden on individuals at these life stages is disproportionate to the limited attention in policy, research, financing, and care.

Catering to the Diversity of Menstruators: Work in Progress

Globally, menstruators suffer in silence, but many experience stigma and discrimination on multiple grounds: socioeconomic status, choice of occupation, gender identity, caste, class, ethnicity, and many other factors (see McCarthy and Lahiri-Dutt [Chapter 3]; Bozelko [Chapter 5]; Roberts [Chapter 6]; Vora [Chapter 4]; Sukumar [Chapter 13]; Steele and Goldblatt [Chapter 8]; Hawkey, Ussher, and Perz [Chapter 10]; Persdotter [Chapter 29]; Steward et al. [Chapter 55]; Frank and Dellaria [Chapter 7]; and Rydström [Chapter 68] in this volume). Disability, gender, and sexual identity were almost entirely absent from the emerging MHM conversation, as was the intersection of multiple stigmas throughout life, with particular vulnerabilities at puberty, pregnancy, childbirth, and menopause.

The Sanitation Action Summit organized by WSSCC in Mumbai in November 2016 created a platform for listening, learning, and joint action (Sanitation Action Summit 2016). The summit brought together citizens from across India who are often silenced and/or made invisible—sanitation workers, adolescent girls and boys, transgender persons, older persons, and persons with hearing and/or visual impairments. It concluded: "The challenge is not just to build infrastructure like toilets but to fight stigma and prejudice that discriminate against women, girls, transgender persons, sanitation workers and many more" (WSSCC 2016b). The summit was followed by the publication of transformative national *Guidelines on gender issues in sanitation* (MDWS 2017), marking an important departure from narrow WASH messaging to explicit acknowledgment of the dangers of instrumentalist, stereotypical messages for behavior change. The guidelines state:

> It is noted that behaviour-change messaging [in Swachh Bharat Mission (Gramin)] often includes subjects like 'shame and dignity of women.' While these may be useful for entry-point messaging they carry risks of lack of ownership by men and the reinforcing of gender stereotypes (like women should not step out of the house, men as custodians of women's dignity, etc.) The ... messaging should therefore be gender sensitive and targeted at both men and women, particularly focusing on men, who are often the primary decision makers in rural households where household expenditure is involved.

These guidelines are fairly radical in a culture that places women at two ends of a complex spectrum: the first, as the stereotypical goddess to be revered—a wife, daughter, or sister to be protected from the male gaze; and the second, a workhorse without agency. Significant in these guidelines is the emphasis on dignity versus patriarchal messaging that reinforces widely accepted perceptions of a woman's modesty, dignity, and therefore place in society (MDWS 2017). The guidelines go on to advocate for inclusion and adaptation to ensure that the third gender,[12] elderly persons, children, and persons with disabilities have unfettered access to facilities. They end with detailed guidance to address MHM-related taboos and superstitions and specifically acknowledge women's sanitation stress and insecurity and call for better siting, ventilation, and consideration of age and disability in facilities design and management.

Persons with Disabilities

MHM poses particular challenges for persons with disabilities (Majumdar 2019). At the Sanitation Action Summit 2016, visually impaired girls talked about these struggles. One girl explained, "Since there are no dustbins or other disposal facilities, we are dependent on our relatives to throw away used pads" (WSSCC 2016c). Another girl added: "When there are no toilets we go out to change our sanitary materials. This is unsafe as we cannot see stray animals or other hazards" (WSSCC 2016c). Finally, yet another girl stated: "We are often pressured by our families to remove our uterus in order to stop menstruating" (WSSCC 2016c; see Steele and Goldblatt [Chapter 8] in this volume).

Through learning and innovation partnerships with the Centre of Excellence in Tactile Graphics of the Indian Institute of Technology and the Saksham Trust, our team[13] published an MHM toolkit for persons with visual impairments (WSSCC 2018). This toolkit comprised a tactile book, facilitators manual, tactile apron, audiobook, EPUB, and a video in Indian Sign Language. Before publication, it was tested in Sri Lanka and India through five regional workshops for 250 trainers who work with persons with visual impairments. Of those trained, 30% subsequently used the materials with 4748 visually impaired girls (*Newz Hook* 2019). In February 2018, Minister Uma Bharati and Secretary Parameswaran Iyer, of India's Ministry of Drinking Water and Sanitation, launched the toolkit on MHM, which was endorsed by the Ministries of Social Welfare and Drinking Water Supply and Sanitation.

Transgender Persons
Nuanced policies and practice that address menstruators in all their diversity is a rare opportunity to embrace nonbinary definitions of identity and sexuality. The silence and taboo around periods is further exacerbated for trans men, who report stigma, discomfort, annoyance, disgust, and pain linked to menstrual periods (Bell 2019). Expanding the MHM work to the trans community met with opposition from WSSCC's Board in Geneva despite the huge demand in the country and the support of WSSCC's Executive Director.[14] Early conversations and workshops with the trans community in South Asia, have uncovered persistent menstrual stigma among trans men (Sarfaraz 2016) underlining the value of a human, holistic approach to sexual reproductive health and rights, with menstruation as a powerful entry point to redefine ancient feminine and masculine boundaries.

Essential Ingredients to Reach the Tipping Point

Malcolm Gladwell described what he termed the 'tipping point' (2000) and identified three key variables for social epidemics: (1) The power of the few (a few individuals who are convinced and clear); (2) the stickiness factor (the change must make sense to people); and (3) the power of context (the change must be relevant to the context and culture). Gladwell's framework can be extrapolated to explain some of what worked in our efforts to transform the deafening silence around menstruation into language and action.

While the countries were different, the methodology was consistent: a deliberate recipe designed to produce an 'aha moment' of realization that menstruation is not 'dirty' or 'polluting,' but entirely natural. Women's and girls' voices and participation had to be matched with empathy, understanding, and action from duty bearers. Many MHM trainings and dialogues created a shared reality.

The interaction, role plays, honest voices, and tough talk by many who had suffered ultimately produced changemakers, champions, movers, and shakers.

This journey revealed five key lessons:

1. Changes in social norms, institutions, and behaviors are best facilitated by simple, evidence-based policies anchored in voice and participation.
2. Men can be powerful, supportive agents of change when presented with a practical, action oriented, compelling arguments in the voices of those affected most.
3. Basic infrastructure, such as taps and toilets, by their sheer essential and ubiquitous, noncontroversial presence serve as powerful entry points to exploring and addressing gender issues and wider human rights at scale.
4. Human needs such as sanitation and hygiene—including those related to menstruation—are universal. But the services to address these basic needs are traditionally gendered, catering to a young, virile male. They ignore the life course and associated physical and psychosocial factors through which users perceive comfort, safety, convenience, and dignity. Acknowledging the diversity of changing needs in service delivery is transformative and quite doable.
5. Failing to do so will leave those who are currently unheard and often forgotten, even further behind.

I believe the changes in India, as well as in other countries, were more than just a response to an issue whose time had come. Rather, the evolution from stigma to openness about MHM brings together all the elements of Gladwell's framework. It was the result of sustained advocacy by a multitude of stakeholders with a belief in and commitment to change, an appetite for risk, and a long-term view; coupled with careful strategy, local champions, and systematic, multipronged interventions so that the seed takes root and grows into a dense forest. On reaching this tipping point, menstruation went from the silent unmentionable to being 'en vogue.'

The Power of the Few

The spokespersons in India in the 1990s who were instrumental in sowing the seeds of the dynamic field of menstrual health and hygiene were little-known movers and shakers, not Bollywood celebrities or UN officials. They understood very early the importance of matching advocacy with action to achieve change at scale. They were individuals and organizations who worked in this area, with no resources and no Internet to garner virtual support or share experiences. I must share a few as important sources of inspiration, courage, creativity, and leadership.

Santa Sheela Nair as the Principal Secretary of Municipal Administration and Water Supply in the Indian state of Tamil Nadu (2001) was one of the first to address the menstrual hygiene needs of women and girls in practice

(Alok 2010). Known widely as India's water woman for her successful rainwater harvesting strategy that rescued the state from drought, she instituted a practical strategy to raise awareness and provide basic safe management and disposal options for schools in the state. A powerful and consistent advocate, she reminded global audiences at Rio+20 that women menstruate and menstrual hygiene is an inseparable aspect of WASH services.

Lakshmi Murthy founded Vikalp Design in 1988, focusing on sexual and reproductive health communication for rural adolescents. An early advocate for menstrual health, Lakshmi designed the *Lace wallah kapda*, a reusable cloth pad with straps, which evolved in 2000 into improved *Uger* pads (Murthy 2017). She developed simple tools—including the *mahavari chakra*, or menstrual wheel, later adapted by the WSSCC Yatra team—to teach young people about menstrual health (VikalpDesign 2005).

Anshu Gupta of Goonj developed the three pillars of awareness, access, and affordability to enable dignity during menstruation for millions of rural women (Murthy 2017). A pioneer in the quest for sustainable, planet-friendly menstrual products, Goonj's two-decade-old campaign, "It's not just a piece of cloth," has helped women bleed with dignity and safety while quietly providing an alternative to expensive and polluting menstrual products.

The Bangladesh Rural Advancement Committee (BRAC), one of the world's largest NGOs, boasted a sophisticated production and distribution system of affordable menstrual pads, produced locally by women as early as 2000 (Bharadwaj and Patkar 2004). In 2004, working with A. Mushtaque Chowdhury, Babar Kabir, and Milon Kanti Barua, on BRAC's gender strategy, I received encouragement to improve the integration and institutionalization of menstrual health into BRAC's program. Presented with women's very practical demands for materials, convenience, and privacy, BRAC management moved quickly to improve production and distribution of absorbents at an affordable price along with facilities redesigned to make menstruation a safe, comfortable, and dignified experience for women and girls.

The Stickiness Factor

Doing something about menstruation made sense to men and women of all ages and nationalities. WASH policies were universally focussed on infrastructure, so thinking about how menstruators would use facilities was a novel exercise. It made sense to policy makers and practitioners seeking to understand why women did not use facilities even when they existed (Winter et al. 2019), or why girls chose to stay away from school some days (Oster and Thornton 2011), and what menstruators actually did to cope. Most importantly it made sense to menstruators. It made sense to grandmothers and mothers who had been denied this simple information while growing up to see their charges glowing with new confidence. And it made sense to men and boys, who dispassionately considered this new information, reacting more often than not with empathy and support.

The Power of Context

Yes, context is key and MHM advocates and practitioners adjusted programs, materials, and language to local realities. Yet, concerns of 'culture' and 'context' were submerged in the universality of biology, patriarchy, stigma, trauma, and coping strategies. Life is a lived experience with common threads and encounters across geographical, linguistic, sociocultural, and economic boundaries. While context shapes the unique experience of each menstruator, it is against a universal backdrop of stigma and restrictions, of ignorance of the human body and the reproductive cycle (see Chrisler 2013); of recurrent patterns restricting mobility, voice, and choice; and a ubiquitous external environment that denies a woman the right to participate fully alongside men in public life. The neglect of menstruators' needs was indeed universal. Local context nuanced the idioms and expressions, the restrictions placed at menarche, and the traditional practices in menstrual management, but the patterns were unmistakable and knew no geopolitical boundaries. Every time, in every context, the topic was relevant and long overdue. Each time, silence was replaced by stories of trauma and repression, quickly giving way to hope and ideas for change.

Conclusion: What Lies Beyond the Tipping Point?

These are early, heady days. Activists, entrepreneurs, and researchers are busy generating creative approaches, products, and evidence to bolster the case for menstrual health (USAID, Kiawah Trust, and Dasra 2014). This may be a good time to pause and take stock. There is no doubt that menstruation provides a powerful entry point to open up forbidden themes, questions, and perceptions and to renegotiate agreements on public goods and services. However, even when negotiated, owned, and published, normative frameworks must continue to be embedded in the voices that formed them to remain alive, relevant, and accountable. Otherwise they leave those who were central to the transformation far behind. Pragmatic solutions that translate policy into practice can move the dialogue on rights and entitlements at speed, yet, in the rush to reach scale, they risk losing the focus on voice and agency of the most marginalized.

Development policies, guidelines, and schemes tend to fit into two categories: (1) broad, sweeping statements of goodwill and change that are hard to implement and where no one is directly accountable, or (2) very specific guidelines that assign a budget and responsibility, often with tangible, measurable outcomes. The policy changes in India, Senegal, and Kenya fall into the second category. The subtext and wider frame around women's empowerment, however, did not fit neatly into subnational budget proposals and were not easily understood by finance ministries seeking visible accountability and tangible outcomes. Practitioners found it easier to mention the unmentionable and to distribute free menstrual pads than to ensure a holistic, transformative approach that included pride in one's body, knowledge of one's sexuality, and the information and freedom to make choices. The silence and taboo around menstruation was too easily replaced by gendered messaging on how women could manage their periods in line with a western image of the ultraconfident, modern woman who juggles a billion things, including her periods, effortlessly.

At the same time, human rights language is notoriously unpopular with governments in many countries, including in countries with active media and civil society networks. Opposition is based on a range of predictable reactions ranging from fear of public interest litigation to simple aversion to overtly feminist or transformative propositions that challenge 'how we do things here.' Even when human rights language is incorporated into policies, translating these far-reaching policies into practice is often elusive. Normative frameworks cannot be seen outside the sociocultural and economic contexts within which they are expected to bring about change. Those who engage in that context must tread carefully to be allowed to continue their work without shaking the status quo too much. For transformative change to take hold, only the voices of those most concerned can deepen the discourse and direct the winds of change.

Menstrual hygiene and health integrated into WASH and health policies have bordered heavily on practical needs, using binary definitions and largely 'safe' parameters. But they provide a stronger hook than well-meaning text on women's empowerment, impossible for finance and sector ministries to translate into action. Published policies are an important milestone and serve to direct the next steps in the journey. Without them, practice will continue to be eulogized as model villages or pilot projects, but defy scale. Generic language in policy around women's needs, suffering, and marginalization is unhelpful; it risks instant dilution in strategy and budgets, followed by business as usual at the implementation stage. It is important to spell out clearly stigmatized concepts, lay bare the reasons for stigma, and replace them with practical solutions that are universally acceptable. But that is just the beginning. Redesigned public spaces, open conversations at home across generations, and a new vocabulary spoken without fear bring confidence and new vistas. Menarche and menstruation symbolize much more than a monthly period to be managed: sexual and gender identity, desire, sense of self, voice, choices, womanhood, and rites of passage. They offer a powerful hook for the realization of multiple human rights.

NOTES

1. Urmila Chanam, Krishna Ramawat, and Komal Ramdey in India.
2. Amita Sharma and Sumit Bose.
3. Tamsyn Barton, Peregrine Swann, Richard Montgomery, Alison Barret, Brian Baxendale, Nigel Kirby, and many others.

4. Zelda Yanovich, David Trouba, Archana Patkar, WSSCC, Geneva with national coordinator Vinod Mishra and local consultants Vijay Gawade, Lakshmi Murthy, and Maria Fernandes.
5. Urmila Chanam, Vaishali Chandra, and Aneesha Menon (Volunteers from Arghyam), Krishna Ramawat (Goonj), and Maria Fernandes (Consultant, WSSCC).
6. The Swachh Bharat Mission or Clean India Mission, was launched by Prime Minister Modi for the period 2014–2019 with the aim of cleaning up the streets, roads, and infrastructure of India's cities, towns, and rural areas.
7. Rockaya Aidara and Archana Patkar, WSSCC.
8. Minister Mansour Faye, Amadou Diallo, Arouna Traore, Mingue Ndiatte Ndiaye, and many others.
9. Mbarou Gassama Mbaye, Dienaba Wane Ndiaye, Josephine Odera, Corinne Delphine N'daw.
10. Carol Koi, Abdiwahit Jama, Eva Muhia, Irene Gai, Christine Mwaka, Patricia Mulongo Neville Okwaro, Daniel Kurao, and Jane Kiminta in Kenya; Khady Sonko, Seynabou Sarr, Rokhaya Ngom, Khoudia Mbengue Fall in Senegal; Director, Labo Madougou in Niger; Beatrice Eyong, Jocelyne Alice Ngo Njiki, and Olivia Mvondo Boum in Cameroon.
11. Chaitali Chattopadhyay, Virginia Kamowa, and Archana Patkar.
12. In April 2014, the Supreme Court of India recognized hijras, transgender people, eunuchs, and intersex people as a 'third gender' in law. See National Legal Services Authority v Union of India, Writ Petition No. 400 of 2012 with Writ Petition No. 604 of 2013, Supreme Court of India (15 April 2014), available at: https://indiankanoon.org/doc/193543132/.
13. Kamini Prakash, Komal Ramdey, and Archana Patkar with Saksham and Noida Deaf Society. Publication by the Communications team, WSSCC Geneva.
14. Christopher Williams.

REFERENCES

Alok, Kumar. 2010. *Squatting with Dignity: Lessons from India*. New Delhi: Sage.
ANEW. 2013. "Opening of the WSSCC Training on Equity and Inclusion in Sanitation and Hygiene in Senegal." *ANEW Africa*, September 13. http://www.anewafrica.org/news/article_article_115.html.
Aoki, Yoshitaka, Heidi W. Brown, Linda Brubaker, Jean Nicolas Cornu, J. Oliver Daly, and Rufus Cartwright. 2017. "Urinary Incontinence in Women." *Nature Review Disease Primers* 3 (17097). https://doi.org/10.1038/nrdp.2017.42.
Bell, Jen. 2019. "What It's Like to Get Your Period When You're Trans." *Clue*, February 18. https://helloclue.com/articles/cycle-a-z/what-it's-like-to-get-your-period-when-you're-trans.
Bharadwaj, Sowmyaa, and Archana Patkar. 2004. "Menstrual Hygiene and Management in Developing Countries: Taking Stock." *Junction Social, Social Development Consultants* 3: 1–20.
Bobel, Chris. 2019. *The Managed Body: Developing Girls and Menstrual Health in the Global South*. Cham, Switzerland: Palgrave Macmillan.
Boyce, Paul, Sarah Brown, Sue Cavill, Sonalee Chaukekar, Beatrice Chisenga, Mamata Dash, Rohit K. Dasgupta, et al. 2018. "Transgender-Inclusive

Sanitation: Insights from South Asia." *Waterlines* 37 (2): 102–17. https://doi.org/10.3362/1756-3488.18-00004.

Burt, Zachary, Kara Nelson, and Isha Ray. 2016. "Towards Gender Equality through Sanitation Access." UN Women. https://www.unwomen.org/en/digital-library/publications/2016/3/towards-gender-equality-through-sanitation-access.

Chanam, Urmila. 2012. "The Nirmal Bharat Yatra—Goodwill on Wheels!" *India Water Portal*, November 11, 2018. https://www.indiawaterportal.org/articles/nirmal-bharat-yatra-goodwill-wheels.

Chandra-Mouli, Venkatraman, and Sheila Vipul Patel. 2017. "Mapping the Knowledge and Understanding of Menarche, Menstrual Hygiene and Menstrual Health among Adolescent Girls in Low- and Middle-Income Countries." *Reproductive Health* 14 (30). https://doi.org/10.1186/s12978-017-0293-6.

Chandra Sharma, Neetu. 2018. "Use of Sanitary Pads Sparse Despite Govt Schemes: Studies." *Live Mint*, February 10. https://www.livemint.com/Politics/zbk4JLAnsoHjmbvp1rgeNJ/Use-of-sanitary-pads-sparse-despite-govt-schemes-studies.html.

Chrisler, Joan. 2013. "Teaching Taboo Topics: Menstruation, Menopause, and the Psychology of Women." *Psychology of Women Quarterly* 37 (1): 128–32. https://doi.org/10.1177/0361684312471326.

CREA. 2005. "Adolescent Sexual and Reproductive Health and Rights in India." New Delhi: Creating Resources for Empowerment in Action (CREA). http://www.nipccd-earchive.wcd.nic.in/sites/default/files/PDF/18%20adolescent%20working%20paper.pdf.

Desai, Priya. 2012. "The Nirmal Bharat Yatra (The Great WASH Yatra)—A Travelling Carnival Over 6 Weeks Spreading Awareness on Sanitation and Hygiene in India." *India Water Portal*, September 24. https://www.indiawaterportal.org/articles/nirmal-bharat-yatra-great-wash-yatra-travelling-carnival-over-6-weeks-spreading-awareness.

Essity. 2017a. "Joining Forces for Progress: Hygiene Matters Report 2016/17." http://reports.essity.com/2016-17/hygiene-matters-report/hygiene-matters-survey-2016-17.html.

Essity. 2017b. "Hygiene—A Circle of Life." https://www.essity.com/company/essentials-initiative/stories-and-videos/hygiene-a-circle-of-life/quotes-about-the-initiative/.

Fernandez, Maria. 2012. "The Nirmal Bharat Great WASH Yatra: What a Long, Strange and Wonderful Journey It Has Been." Water Supply and Sanitation Collaborative Council. https://www.wsscc.org/2012/11/16/the-nirmal-bharat-great-wash-yatra-what-a-long-strange-and-wonderful-journey-it-has-been/.

Geertz, Alexandra, Lakshmi Iyer, Perri Kasen, Francesca Mazzola, and Kyle Peterson. 2016. "Menstrual Health in Kenya: Country Landscape Analysis." FSG. https://menstrualhygieneday.org/wp-content/uploads/2016/04/FSG-Menstrual-Health-Landscape_Kenya.pdf.

George, Rose. 2012. "When Neelam Started Her Period, She Thought She Was Dying Like Her Mother." *New Statesman America*, November 9. https://www.newstatesman.com/world-affairs/2012/11/when-neelam-started-her-period-she-thought-she-was-dying-her-mother.

George, Rose. 2013. "Celebrating Womanhood: How Better Menstrual Hygiene Management Is the Path to Better Health, Dignity and Business." Geneva: Water Supply and Sanitation Collaborative Council. https://www.issuelab.org/resources/19523/19523.pdf.

Gottipati, Sruthi. 2012. "Vidya Balan and Jairam Ramesh Ream Up for Toilets." *India Blogs-New York Times.* September 29, 2012. https://india.blogs.nytimes. com/2012/09/29/vidya-balan-and-jairam-ramesh-team-up-for-toilets/.

Government of Kenya. 2019. "Cabinet Approves Policies to Stimulate Economic Growth, Empower the Youth." November 21, 2019. https://www.president. go.ke/2019/11/21/cabinet-approves-policies-to-stimulate-economic-growth-empower-the-youth/.

Gupta, Aditi. 2013. "Practising Hygiene Is as Important as Performing Pujas." *Menstrupedia*, November 4. http://menstrupedia.com/blog/practising-hygiene-is-as-important-as-performing-pujas-2/.

Indian National Congress. 2018. "Nirmal Bharat Abhiyan: Everything You Need to Know about Nirmal Bharat Abhiyan." October 29. https://www.inc.in/en/in-focus/nirmal-bharat-abhiyan-everything-you-need-to-know-about-nirmal-bharat-abhiyan.

Kathmandu Declaration. 2013. "Sanitation for All: All for Sanitation." Fifth South Asian Conference on Sanitation, October 22–24. Kathmandu, Nepal. https:// sacosan.com/wp-content/uploads/2017/10/Fifth-SACOSAN-Khatmandu-Declaration-2013.pdf.

Kosgi, Srinivas, Vaishali N. Hegde, Satheesh Rao, Shrinivasa Bhat Undaru, and Nagesh B. Pai. 2011. "Women Reproductive Rights in India: Prospective Future." *Online Journal of Health and Allies Sciences* 10 (1): 1–5. http://ro.uow.edu.au/medpapers/130.

Kulkarni, Seema, Sneha Bhat, Kathleen O'Reilly. 2017. "No Relief: Lived Experiences of Inadequate Sanitation Access of Poor Urban Women in India." *Gender and Development* 2: 167–83. https://doi.org/10.1080/13552074.2017.1331531.

Ma, Julie. 2016. "25 Famous Women on Periods." *The Cut*, October 21. https:// www.thecut.com/2016/10/25-famous-women-on-periods.html.

Majumdar, Swapna. 2019. "How a Tactile Book Helps Visually Impaired Women Deal with Their Period." *The Hindu*, March 16. https://www.thehindu.com/society/how-a-tactile-book-helps-visually-impaired-women-deal-with-their-period/article26544937.ece.

Mimche, Pr Honoré, Seke Desyg, Maxime Tiembou Noumeni, Moïse Tamekem, Rockaya Aidara, and Mbarou Gassam Mbaye. 2017. "Menstrual Hygiene Management and Female Genital Mutilation: Case Studies in Senegal." Water Supply and Sanitation Collaborative Council. https://www.wsscc.org/wp-content/uploads/2018/01/20171227_Policy-GHM-Senegal-EN-WEB-1.pdf.

Ministère de l'Hydraulique et de l'Assainissement, République du Sénégal. 2016. "Lettre de Politique Sectorielle de Développement. 2016–2025." https://www. pseau.org/outils/ouvrages/mha_lettre_de_politique_sectorielle_de_developpement_2016_2025_2017.pdf.

Ministry of Drinking Water and Sanitation (MDWS). 2013. *Office Memorandum.* Government of India, No. SMG-1/2013-RDD (NBA), December 10. http:// hptsc.nic.in/M3.pdf.

———. 2015. "Menstrual Hygiene Management: National Guidelines." December. Government of India. https://jalshakti-ddws.gov.in/sites/default/files/Menstrual %20Hygiene%20Management%20-%20Guidelines.pdf.

———. 2017. "Guidelines on Gender Issues in Sanitation." Government of India, No. S-11018/2/2017-SBM. March 4. https://jalshakti-mdws.gov.in/sites/default/files/Guidelines%20on%20Gender%20issues%20in%20Sanitation.pdf.

———. 2018. "Guidelines for Swachh Bharat Mission (Gramin)." December 31. Government of India. https://jalshakti-ddws.gov.in/sites/default/files/SBM%28G%29_Guidelines.pdf.

Murthy, Lakshmi. 2017. "The Journey of Menstrual Hygiene Management in India. #ThePadEffect." *Feminism India*, May 19. https://feminisminindia.com/2017/05/19/journey-menstrual-hygiene-management-india/.

Naafs, Arjen, and Tim Brewer. 2013. "Progress on SACOSAN v. Commitments." *Traffic Lights Paper*. Freshwater Action Network South Asia, WaterAid. http://www.freshwateraction.net/sites/freshwateraction.net/files/Final%20Traffic%20Lights%20Paper%202016.pdf.

Naeve, Katie, Julia Fischer-Mackey, Jyotsna Puri, Raag Bhatia, and Rosaine N. Yegbemey. 2017. "Evaluating Advocacy: An Exploration of Evidence and Tools to Understand What Works and Why." International Initiative for Impact Evaluation, Working Paper 29. https://doi.org/10.23846/wp0029.

Narayan, Ravi, Henk Van Norden, Louisa Gosling, and Archana Patkar. 2011. "Equity and Inclusion in Sanitation and Hygiene in South Asia." A Regional Synthesis Paper. UNICEF/WSSCC/WaterAid.

Newz Hook. 2019. "World Menstrual Hygiene Day—Tactile Kit Explains Periods to Visually Impaired Girls." May 29. https://newzhook.com/story/21724.

Okwaro, Neville and Isobel Davis. 2017. "Master Training of Trainers on Menstrual Hygiene Management Nepal & Pakistan." WSSCC, Workshop Report, February 20–26. https://www.wsscc.org/wp-content/uploads/2017/04/Nepal-MHM-ToT-Report-Final-2017.pdf.

Okwaro, Neville, Sailas Nyareza, Jenny Karlsen, and Inga Winkler. 2016. "First National Training of Trainers on Menstrual Hygiene Management—Kenya." *Workshop Report*. Water Supply and Sanitation Collaborative Council. https://www.wsscc.org/wp-content/uploads/2016/08/Kenya-MHM-ToT-Report.pdf.

O'Reilly, Kathleen. 2016. "Surviving as an Unequal Community: WASH for Those on the Margins." In *Eating, Drinking: Surviving the International Year of Global Understanding (IYGU)*, edited by Peter Jackson, Walter E. L. Spiess, and Farhana Sultana, 51–56. Cham: Springer.

Oster, Emily, and Rebecca Thornton. 2011. "Menstruation, Sanitary Products and School Attendance: Evidence from a Randomized Evaluation." *American Economic Journal: Applied Economics* 3 (1): 91–100. https://doi.org/10.1257/app.3.1.91.

Patkar, Archana. 2012. "Speech by Archana Patkar at the Nirmal Bharat Yatra Launch." Water Supply and Sanitation Collaborative Council. https://www.wsscc.org/2012/10/01/speech-by-archana-patkar-at-the-nirmal-bharat-yatra-launch/.

———. 2016. "Leave No One behind: Equality and Non-Discrimination in Sanitation and Hygiene." In *Sustainable Sanitation for All*, 267–80. Rugby, UK: Practical Action Publishing. https://doi.org/10.3362/9781780449272.016.

Patkar, Archana, and Emily Deschaine. 2014. "Translating Silence into Action—Progress on Policy and Practice in Menstrual Hygiene Management." In *WASH in Schools Empowers Girls' Education: Proceedings of the Menstrual Hygiene Management in Schools Virtual Conference 2013*. New York: United Nations Children's Fund and Columbia University.

Patkar, Archana, Rockaya Aidara, and Anthony Dedouche. 2017. "Water Sanitation and Hygiene in the Informal Sector: Case Studies from Africa and Asia." CSW61

NY, Water Supply and Sanitation Collaborative Council. https://www.wsscc.org/wp-content/uploads/2017/05/CSW61-Session-Summary-Feb17-1.pdf.

Phadke, Shilpa, Sameera Khan, and Shilpa Ranade. 2011. *Why Loiter? Women and Risk on Mumbai Streets*. New Delhi: Penguin Books India.

Prabhu, K. Seeta, Anita Rampal, and Alok Shukla. 1995. "The Madhya Pradesh Human Development Report." UNDP. https://www.undp.org/content/dam/india/docs/human_development_report_madhya_pradesh_1995_full_report.pdf.

Puri, Jo. 2016. "Evaluating Advocacy—What Are We Learning?" Geneva: International Initiative for Impact Evaluation. https://genevaevaluationnetwork.files.wordpress.com/2015/05/advocacy_geneva_april_5_2016_3ie.pdf.

Sahoo, Krushna Chandra, Kristyna R. S. Hulland, Bethany A. Caruso, Rojalin Swain, Matthew C. Freeman, Pinaki Panigrahi, and Robert Dreibelbis. 2015. "Sanitation-Related Psychosocial Stress: A Grounded Theory Study of Women Across the Life-Course in Odisha, India." *Social Science & Medicine* 139: 80–89. https://doi.org/10.1016/j.socscimed.2015.06.031.

Sarfaraz, Kainat. 2016. "Trans-Men: A Minority Within the Marginalised." *Indian Express*, December 10. https://indianexpress.com/article/india/transgender-transmen-third-gender-minority-rights-empowerment-4415111/.

Sikri, A. Supreme Court of India. *National Legal Services Authority v Union of India*, Writ Petition No. 400 of 2012 with Writ Petition No. 604 of 2013, Supreme Court of India. April 15, 2014. https://indiankanoon.org/doc/193543132/.

Singh, Manvi. 2019. "It's Time Teachers Stopped Skipping the Reproduction Chapter." *Youth Ki Awaaz*, May. https://www.youthkiawaaz.com/2019/05/as-the-children-are-growing-shouldnt-sex-education-grow-with-them/.

Sixth South Asian Conference on Sanitation. 2016. "Better Sanitation Better Life." *The Dhaka Declaration*. https://www.endwaterpoverty.org/sites/default/files/oldfiles/SACOSAN_VI_declaration.pdf.

Sommer, M., B. Caruso, M. Sahin, T. Calderon, S. Cavill, T. Mahon et al. 2016. "A Time for Global Action: Addressing Girls' Menstrual Hygiene Management Needs in Schools." *PLoS Med* 13 (2): e1001962. https://doi.org/10.1371/journal.pmed.1001962.

Theseira, Julian. 2016. "WSSCC Fulfils Its SACOSAN VI Commitments to Train Governments and Influencers in South and South-East Asia with 'MHM & Gender in WASH' Training in Sri Lanka." Water Supply and Sanitation Collaborative Council. https://www.wsscc.org/2017/11/16/wsscc-fulfils-sacosan-vi-commitments-train-governments-influencers-south-south-east-asia-mhm-gender-wash-training-sri-lanka/.

Thirunavukarasu, A., and Doug Simkiss. 2013. "Developments in Reproductive Health Education in India." *Journal of Tropical Pediatrics* 59 (4): 255–57. https://doi.org/10.1093/tropej/fmt066.

United Nations General Assembly Resolution 64/292. The Human Right to Water and Sanitation, A/Res/64/292, July 28, 2010. https://undocs.org/A/RES/64/292.

United Nations General Assembly Resolution 70/165. The Human Rights to Safe Drinking Water and Sanitation, A/Res/70/169, December 17, 2015. https://undocs.org/A/RES/70/169.

University of North Carolina. 2017. "Water and Health Conference: Where Science Meets Policy." https://waterinstitute.unc.edu/conferences/waterandhealth2017/.

USAID, Kiawah Trust, and Dasra. 2014. "Spot On! Improving Menstrual Health and Hygiene in India." https://www.dasra.org/resource/improving-menstrual-health-and-hygiene.

VikalpDesign. 2005. "Mahwari Chakka." https://www.vikalpdesign.com/mahwari_chakka.html.

Wambua, L. T. 1997. "African Perceptions and Myths about Menopause." *East African Medical Journal* 74 (10): 645–46. https://www.ncbi.nlm.nih.gov/pubmed/9529747.

WaterAid. 2013. "Making Connections: Women, Sanitation and Health." April 29, 2013. https://sswm.info/sites/default/files/reference_attachments/Making_Connections__Women_Sanitation_and_Health_290413.pdf.

WaterAid, WSSCC, and Unilever. 2013. "We Can't Wait: A Report on Sanitation and Hygiene for Women and Girls." https://www.unilever.com/Images/we-can-t-wait---a-report-on-sanitation-and-hygiene-for-women-and-girls--november-2013_tcm244-425178_1_en.pdf.

Winkler, Inga T. 2016. "The Human Right to Sanitation." *University of Pennsylvania Journal of International Law* 37: 1331–1406. https://scholarship.law.upenn.edu/jil/vol37/iss4/5.

Winkler, Inga T., and Virginia Roaf. 2014. "Taking the Bloody Linen Out of the Closet—Menstrual Hygiene as a Priority for Achieving Gender Equality." *Cardozo Journal of Law and Gender* 21 (1): 1–37.

Winter, Samantha Cristine, Robert Dreibelbis, Millicent Ningoma Dzombo, and Francis Barchi. 2019. "A Mixed-Methods Study of Women's Sanitation Utilization in Informal Settlements in Kenya." *Plos One.* https://doi.org/10.1371/journal.pone.0214114.

WSSCC. 2013a. "Menstrual Hygiene Management: Training of Master Trainers." Water Supply and Sanitation Collaborative Council. https://www.wsscc.org/wp-content/uploads/2016/04/WSSCC-MHM-India-Training-of-Trainers-Report-New-Delhi-India-24-27-September-2013.pdf.

———. 2013b. "The Menstrual Wheel." Water Supply and Sanitation Collaborative Council. https://www.wsscc.org/resources-feed/menstrual-wheel/.

———. 2013c. "MHM Lab Convenor's Manual." *WSCC Learning Series: Menstrual Health Management.* Water Supply and Sanitation Collaborative Council. https://www.wsscc.org/wp-content/uploads/2015/10/MHM-lab-manual-EN-LowRes.pdf.

———. 2016a. "CSW60—Achieving Gender Equality through WASH—Event Programme." Water Supply and Sanitation Collaborative Council. https://www.wsscc.org/resources-feed/achieving-gender-equality-wash-event-programme/.

———. 2016b. "Changing Hearts and Minds to Leave No One behind. Sanitation Action Summit 2016: Reflections." https://www.wsscc.org/wp-content/uploads/2017/02/Sanitation-Action-Summit-2016-India.pdf.

———. 2017a. "CSW61-Water Sanitation and Hygiene in the Informal Sector: Session Summary." Water Supply and Sanitation Collaborative Council. https://www.wsscc.org/wp-content/uploads/2017/05/CSW61-Session-Summary-Feb17-1.pdf.

———. 2017b. "Health and Hygiene Across the Life Course." World Health Assembly, Side Session Report. The Water Supply and Sanitation Collaborative Council, SCA, Government of Kenya and WaterAid. https://www.wsscc.org/wp-content/uploads/2017/06/Health-and-Hygiene-Across-the-Life-Course-WHA-2017-Report-final.pdf.

————. 2018. "As We Grow Up: A Tactile Book on Menstrual Hygiene Management Facilitator's Manuel (EN and HIN)." Water Supply & Sanitation Collaborative Council. https://www.wsscc.org/resources-feed/grow-tactile-book-menstrual-hygiene-management-facilitators-manual/.

WSSCC and ONU Femmes. 2015. "Gestion de L'Hygiene Menstruelle: Comportements et Pratiques Dans la Region de Kedougou, Senegal." https://menstrualhygieneday.org/wp-content/uploads/2016/12/UN-Women-GHM-Comportements-et-Pratiques-K%C3%A9dougou-S%C3%A9n%C3%A9gal.pdf.

WSSCC and UN Women. 2015a. "Information Letter no. 5, April–June 2015." The Joint Programme on Gender, Hygiene and Sanitation, WSSCC/UN Women. https://www.wsscc.org/wp-content/uploads/2016/10/Info-letter-5-Joint-Programme-on-Gender-Hygiene-and-Sanitation.pdf.

————. 2015b. "Information Letter No. 7, October–December 2015." The Joint Programme on Gender, Hygiene and Sanitation, WSSCC/UN Women. https://www.wsscc.org/wp-content/uploads/2016/03/GHS_Info_Letter_7_EN.pdf.

————. 2016. "Best Practices of the Joint Programme on Gender, Hygiene and Sanitation." WSSCC/UN Women. https://www.wsscc.org/wp-content/uploads/2016/11/Best-Practices-of-the-Joint-Programme-on-Gender-Hygiene-and-Sanitation-UN-Women-WSSCC.pdf.

YouTube. 2012. "Jairam Ramesh and Vidya Balan Launch 'Nirmal Bharat Yatra.'" https://www.youtube.com/watch?v=mUwKqj7gUFI&feature=youtu.be.

CHAPTER 39

Menstrual Justice: A Missing Element in India's Health Policies

Swatija Manorama and Radhika Desai

INTRODUCTION: A MENSTRUAL JUSTICE FRAMEWORK

Menstruation is a key process in a woman's life integral to her well-being and an indicator of her fertility. One expects that menstruation's contribution to the propagation of the species would lead to the valorization of both deed and doer—namely, menstruation and women.[1] Instead, globally, and in particular in India (our site of research), menstruation is largely shamed and silenced. In India, the demarcation of female bodies as menstruating bodies is embedded deeply within religious, social, cultural, and political milieu and is customarily stigmatizing (Outlook 2018; Bhartiya 2013; Johnston-Robledo and Chrisler 2011). Against this background, our chapter explores how Indian health and related policies address menstruation and menstrual health.

Within Hinduism there are two contradictory streams of belief regarding menstruation. While menstrual blood and menstruating women are seen as 'polluting' because all bodily excretions are "ritually impure" (Bhartiya 2013, 524–25; Eichinger Ferro-Luzzi 1974; Garg and Anand 2015, 184–86), the alternative 'tantric' view is that, through their menses, menstruating women embody infinite creative power and immense energy (Chawla 2002; Zsigmond 2012). The onset of menarche signifies fertility and is celebrated within many Hindu homes (Bhartiya 2013, 525; Eichinger Ferro-Luzzi 1974). For girls themselves, menarche signals the transition from childhood into the gendered construction of 'womanhood' (Manorama and Hora 2002). Girls recognize that the imposition of cultural and social controls over their pubescent bodies, their sexual and nonsexual deployment, and the manifold

---

Both authors have contributed equally to this work.

© The Author(s) 2020
C. Bobel et al. (eds.), *The Palgrave Handbook of Critical Menstruation Studies*, https://doi.org/10.1007/978-981-15-0614-7_39

increased restrictions on their physical mobility in time and space ensure that they grow to be 'women of a specific deportment' (Kågesten et al. 2016, 2; Manorama and Hora 2002).

We argue that the state, rather than confronting this complex web of beliefs and practices centered on menstruation, chooses to endorse it. Most recently, the state support for continued exclusion of women of menstruating age from the Sabarimala temple (Outlook 2018) is the most egregious instance of the state's role in the perpetuation of menstrual stigma. To signal the critical role of menstruation in the denial of gender justice, we propose a 'menstrual justice'[2] approach. Emerging from India's women's health and people's health movements (Manorama and Shah 1996; Saheli Women's Resource Centre 2001), this approach is based on an alternative understanding of women's biology and health. 'Menstrual justice' is a holistic approach that entails listening with sensitivity and respect to girls' and women's menstrual health needs that emerge from their sociocultural location and gendered everyday experiences (SAMA Team 2005; Rishyasyringa 2000).

The central tenet of the menstrual justice approach is that menstruation is a physiological process directly linked to psychosocial and cultural–religious aspects. Its objective is twofold: First, it seeks to make explicit all aspects of women's lives that are linked to menstruation beyond fertility and reproduction. Second, it helps to delineate the links between this complex web of thought and practices and women's experiences of indignity, discrimination, inequality, and injustice. It considers the specific ways in which the sociocultural–religious discourse of menstruation and its associated practices are reflected in national policies that effect violations of women's human rights, discrimination, and inequality. In doing so, the approach directs attention to the role of the political institutions and state policies in this process. Following an overview of the interconnections between menstruation and women's health, we examine state policies with reference to women's basic, psychosocial, gynecological, reproductive, and menopausal health in the context of the menstrual justice framework.

An Overview of Interconnections Between the Menstrual Cycle and Health

Anchoring of the menstrual justice framework in meanings associated with a bodily process makes it particularly suited to address gaps in women's health. We begin with a brief overview of the impacts of women's health and social factors on the menstrual cycle, demonstrating the need for policy and programming that recognizes and values the full spectrum of menstruality among women throughout India's diverse communities.

Basic Health

Basic physical and psychosocial health are crucially tied to healthy childhood development, puberty, and the onset of menarche, as well as a predisposition to menstrual health-related morbidities and mortality throughout adulthood (UNICEF 2012; Biro and Deardorff 2013; Boutot 2018). Nutritional deficiencies affect the onset of menarche and menopause, intensity of PMS symptoms, and the duration, heaviness, and frequency of one's flow (Jeejebhoy 2000; Gokhale 1996; Patle et al. 2015; Jahangir 2018; Jungari and Chauhan 2017).

Girl-children encounter unequal access to the nutritional, medical, and emotional support systems essential to basic health (Guilmoto et al. 2018; Inamdar, Inamdar, and Sachdeva 2011). Early childhood malnutrition is the leading risk factor for disease nationwide and follows girls into adolescence and adulthood (India State-Level Disease Burden Initiative Collaborators 2017, 2446). Approximately 38% girls under five are stunted, 21% are wasted, and 36% are underweight (Indian Institute of Population Studies (IIPS) and ICF 2017). In the age group 15–19, 47% are underweight (UNICEF 2012), compared to 22.9% in the age group 15–49. Anemia is a little over 50% in both groups (Indian Institute of Population Studies (IIPS) and ICF 2017; UNICEF, n.d.). Given their scale, these nutritional imbalances have a significant impact on menstrual health, as poor nutrition can delay menarche and disrupt the menstrual cycle.

Psychosocial Health

Discrimination against the girl-child leads to the 'normalization' of malnourishment, physical abuse and a sense of powerlessness and low self-esteem among preadolescent and early adolescent girls (Manorama and Hora 2002). Indian adolescent girls already tend to experience lower self-esteem and self-confidence, diffidence, and feelings of subordination compared to boys (Karki and Espinosa 2018, 113; Manorama and Hora 2002). Their branding as 'impure' after the onset of menarche undermines the healthy development of self-confidence and emotional well-being (Bharatwaj and Sindu 2014; Kumar and Srivastava 2011; Soumya and Sequira 2016, 12). Their lack of body and menstrual literacy and knowledge affirms and amplifies the negativity toward menstruation (FSG 2016; Raut et al. 2015, 63).

Additional social control at this critical developmental juncture is one more blow to the adolescent girl's fragile psychosocial state (Blum, Mmari, and Moreau 2017). Worldwide, depression is a common consequence of these restrictions (Chandra-Mouli et al. 2017, S6). In India, psychosocial problems are a more likely cause of suicides than diagnosable psychiatric problems (Mythri and Ebenezer 2016, 493–98), and 56% of suicides of women occurred in the age group of 15–29, compared to 40% for men (Rane and Nadkarni 2014, 77).

Gynecological and Reproductive Care

Under sociocultural silence and shame and misconceptions of what constitutes a 'normal' period, Indian women often refrain from reporting symptoms associated with gynecological morbidities; they consider symptoms such as severe pain to be part and parcel of being women (Inamdar, Sahu, and Doibale 2013, 9; Ramasubban and Jeejebhoy 2000a, 24–25). Reproductive Tract Infections (RTIs) and menstrual disorders are among the most commonly observed gynecological morbidities (Gosalia et al. 2012; Gulati, Chaurasia, and Singh 2009; Inamdar, Sahu, and Daibole 2013; Oomman 2000) and show increased incidence in older women, poor women, women with lower literacy, married women, and women with higher gravidity and parity (Tulasi and Babu 2018, 6; Oomman 2000, 251–52). Early age at menarche has been associated with gynecological morbidities as well (Das et al. 2015, 12; Oomman 2000, 250–51).

Menopause

India's mean age for the onset of menopause ranges between 41.9 and 49.4 years—significantly lower than the global mean of 45 to 55 (Pallikadavath et al. 2016, 367). Poor nutrition, lower socioeconomic status such as membership in a scheduled caste or tribe, adolescent pregnancy, lack of education, and exposure to strenuous work have been found to contribute to a higher prevalence of early menopause (Ahuja 2016; Jungari and Chauhan 2017; Syamala 2010, 254–55).

A high percentage of perimenopausal and postmenopausal women suffer from gynecological morbidities such as vaginal irritation, incontinence, reduced sexual desire, and mental health issues such as depression (Mishra 2011; Susila and Roy 2014, 55; Syamala 2010, 255). The symptoms are compounded by a burden of noncommunicable disease in the Indian population (Ahuja 2016; Susila and Roy 2014, 55). Despite these preliminary findings, studies of gynecological morbidity among menopausal women are rare.

The above discussion reveals high levels of ill-health among females from birth to old age that can be tied to menstrual health and body awareness. Across the life course, girls, women of reproductive age, perimenopausal, menopausal, and postmenopausal women exhibit a low level of basic and psychosocial health and a high incidence of gynecological morbidities, including menstrual health issues. In the next section, we examine India's national health policies to assess the manner in which the state has addressed these widespread basic and menstrual health problems.

ANALYSIS OF HEALTH POLICIES: A MENSTRUAL JUSTICE PERSPECTIVE

While our analysis focuses on health policies, we also include interrelated policies in adjacent sectors such as nutrition and sanitation. This section analyzes the policies introduced above: primary healthcare, psychosocial health, reproductive health and fertility control, and menopause. We also examine how menstrual health is addressed in sanitation policies.

Primary Healthcare

Until the formulation of India's National Health Policy (NHP) in 1983, healthcare was addressed through Five-Year Plans. The NHP 1983 endorsed the idea of "universal primary healthcare services" but rejected its concomitant goal of "free healthcare provision" on the grounds of unaffordability (Ministry of Health and Family Welfare, Government of India (MoHFWGOI) 1983). The NHP 1983 emphasized the curative approach at the expense of preventive, promotive public health, and rehabilitative aspects of healthcare (Duggal and Gangolli 2005). The NHP 1983 led to the expansion of health infrastructure (Duggal 2005, 33–34), but its expected benefits to women's health did not materialize. Health workers were not trained to cater to women's needs, and family planning and immunization took a disproportionate amount of their time (Duggal 2005, 34).

The NHP 2002 (MoHFWGOI 2002) moved away from the goal of universal comprehensive primary healthcare to a regime of public health characterized by selective care targeted at specific groups (Duggal 2005, 35). Recognizing women's poor health status (Sarojini et al. 2006, 23) and the withering of rural health capacity except for "family welfare activities" (MoHFWGOI 2002, 9), the NHP 2002 promised "top funding priority to programmes relating to women's health" (Sarojini et al. 2006, 23; MoHFWGOI 2002, 32). However, health benefits still did not accrue because the policy's approach to women's health lacked specificity (Sarojini et al. 2006, 24). While it mentioned the need to "attend to specific requirements of women in a more comprehensive manner" (MoHFWGOI 2002, 32), it did not identify a single women's health issue. The only women-specific goal it set was to reduce the Maternal Mortality Rate (MMR), revealing its reductive understanding of women's health *as* maternal health (MoHFWGOI 2002, 21).

The policy prescriptions of the NHP 2002—namely, improvement in availability and access to quality healthcare for rural populations, especially women and children—were reflected in National Rural Health Mission (NRHM) 2005 (MoHFWGOI 2015). In reality, programs of reproductive and child health and HIV/AIDS in NRHM 2005 got greater funding support over primary healthcare (Duggal and Gangolli 2005, 11). The Accredited Social Health Activists (ASHA), introduced as an interface between the village and the public health system, were given insufficient training for primary and menstrual healthcare functions, and their incentive structure was aligned to promote reproductive health and family planning (Sarin et al. 2016; Hussain 2011, 56).

Women's health was also purportedly targeted through national nutrition interventions, which can have a significant impact on improving menstrual health. These include the Integrated Child Development Services Scheme introduced in 1975 and the Nutrition Policy of 1993. The scheme initially provided nutrition supplements only to undernourished pregnant women, lactating mothers, and children under age six (Ministry of Women and Child Development, Government of India (MoWCDGOI) 1975). The Nutrition Policy of 1993 (MoWCDGOI 1993) extended nutritional support to malnourished adolescent girls to redress high levels of undernourishment (Kanani 2002). However, the policy did not propose a plan to address intrahousehold gender discrimination toward girl-children such as underfeeding, little or no provision of nutritional foods, denial of medical treatment for ill-health, et cetera, which is the primary cause of malnourishment. The policy lacked a perspective of "women's nutrition for its own sake" (Kanani 2002) and continued to adopt the myopic view of equating women's health with reproduction. It failed to suggest measures to compensate the specific nutritional shortfalls of menstruating or menopausal women and to institute a research program to examine the impact of malnourishment on women's menstrual health across the life cycle (Jeejebhoy 2000, 145; Oomman 2000, 260; Sandoiu 2017).

The most recent National Health Policy (2017) does not address the gaps in the provision of healthcare to women either (MoHFWGOI 2017), although it has a sprinkling of 'women's health issues' across its 30-odd pages. Women's health needs are specifically addressed in the sections on reproductive health, maternal health, child and adolescent health (RMNCH+A), malnutrition, population stabilization, gender-based violence (GBV), and women's health and gender mainstreaming. Under the subheading of "Women's Health & Gender mainstreaming" it mentions "enhanced provisioning for reproductive morbidities and health needs of women beyond the reproductive age group (40+)" (MoHFWGOI 2017, 14); however, there is no mention of how this is to be achieved through changes in the current health programs.

The trajectory of India's national health and nutrition policies shows encouraging growth in awareness of the need to rectify gender imbalances and to devote more resources toward improving women's health. However, we do not yet see sufficient programmatic changes to support the implementation of these policies.

Psychosocial Health

India's first National Mental Health Policy, adopted in 2014, mentions in its preamble the necessity to cater to vulnerable groups (MoHFWGOI 2014). Yet it fails to address the mental health consequence of high levels of stress women suffer because of gender discriminatory practices such as neglect, child marriage, infertility, failure to give birth to sons, witch-hunting, and sexual assault within marriage (Ramasubban and Jeejebhoy 2000a, 34; Sarojini et al. 2006, 41). The policy also seems blind to the emotional anguish women suffer because of chronic issues related to menstruation and the impact of other diseases on menstruation (Oomman 2000, 253–55; Alvergne, Wheeler, and Tabor 2018). For instance, it fails to recognize the negative psychosocial health impacts of the perception that girls and women are 'polluting' during menstruation (FSG 2016). Rather than mandate access to counselors and health workers equipped with gender-sensitive training on menstrual health as demanded by women's health activists (Oomman 2000, 257–58), the policy medicalizes mental health conditions in psychiatric terms and adopts a 'special case' approach that does not cater to the mental health of menstruating women (Varma 2014, 45).

Reproductive Healthcare and Fertility Control

Population growth led to the integration of Family Planning Program (FPP) and health services since the Third Five-Year Plan (1961–1966) (Ministry of Health and Family Welfare, Government of India (MoHFWGOI) 2000; Visaria 2000, 335). FPP is part of a primary healthcare system whose beneficiaries were illiterate, poor, and 'lower' caste women. Women were provided contraceptives with little regard for their needs, and the quality of services was poor (Visaria 2000). The result was an increase in menstrual problems and RTIs from intrauterine devices and gynecological morbidities post-sterilization (Oomman 2000, 247–48). It was only in 1997 that the government accepted a women's empowerment and comprehensive reproductive healthcare paradigm through its Reproductive and Child Health (RCH) program. The RCH included some services demanded by women activists: prevention and treatment of RTIs and sexually transmitted infections (STIs); reproductive health services for adolescents; and information, education, and counseling on health, sexuality, and gender (Sarojini et al. 2006, 28). But the government's own review of RCH showed that critical components of reproductive and sexual health were ignored in training and implementation, resulting in a lack of treatment for contraceptive side effects and post-delivery complications. There was also a total neglect of health needs of adolescents outside marriage (Santhya 2003, 28; Sarojini et al. 2006, 31). Failure to address critical factors of the RCH approach—such as the right to information on associated risks and the unavailability of services or skilled personnel to deal with side effects of contraception—led to exacerbating women's already poor health, including gynecological morbidities (Pachauri 2004, 18; Santhya 2003, 25–26).

Even in programs to prevent STIs, menstrual and gynecological care was not given precedence (WHO 2007, 2). There were no processes to empower women and enable them to choose their contraceptive methods or treatments for infection (SAMA Team 2005, 157; Santhya 2003, 28). The subsequent phases of RCH—RCH II (2005) and Reproductive, Maternal and Child Health, Plus Adolescents (RMNCHA+) (2013)—have made no midcourse corrections; instead, they've shifted the focus once again to maternal, infant, and child care (Ministry of Health and Family Welfare, Government of India (MoHFWGOI), n.d.). The implementation of RCH shows an absence of promotion of adolescent and adult women's knowledge of their own bodies, issues of sexuality, menstrual health and hygiene, and the right to contraceptive choice without coercion. The above overview of reproductive health policies demonstrates that population control continues to be bundled within reproductive health and is least concerned with menstrual health. The RCH program does not recognize the continuities between reproductive health, general health, and women's social location.

Menopause

The NHP 2017 recognizes the health needs of women beyond reproductive age (40+) (MoHFWGOI 2017). As such, it has taken the first steps toward heeding the demands of women's health activists for policies that address women's health across the life cycle rather than merely in their reproductive years (SAMA Team 2005). However, the policy homogenizes older women; it includes women in late reproductive age, perimenopausal, recently postmenopausal, and beyond age 60, and it does not give any indication of the strategy to address health issues of this 40+ age group. The health needs of this group are also covered in the 1999 National Policy on Older Persons, which was drastically revised in 2011. Like the NHP 2017, it shows awareness of the need for increased attention on older women, but its approach to issues of aging is gender-neutral (Ministry of Social Justice, Government of India 2011). Neither policy recognizes the consequences of menstrual conditions experienced before menopause and the ramifications of menopause on physical and mental health; neither articulates which treatments or procedures within existing infrastructure can be used to treat postmenopausal morbidities. Also missing is any recognition of the need for gender-specific research on issues such as detriments to women's postmenopausal longevity and long-term consequences of gynecological morbidities (ibid.; Syamala 2010; Jani and Manorama 2007).

An analysis of India's past policies on physical and mental health reveals historical approaches to women's health that are based on a myopic understanding of women as 'female reproducing bodies' who need to be managed for population stabilization and reproductive health. This limited focus has perpetuated neglect of the health needs of girl-children, prepubescent girls, and perimenopausal and postmenopausal women. It does nothing to interrupt and dismantle the cycles of menstrual stigma that compromise women's and girls' mental and physical health.

Swachh Bharat Abhiyaan (SBA): The 'Clean India Mission'

The SBA is the first large-scale government program that includes a strategy to bring out the taboo subject of menstruation (Swachh Bharat Mission 2019). This section briefly analyzes the component of menstrual hygiene management (MHM) in SBA through the lens of menstrual justice (see also Patkar

[Chapter 38] in this volume). The goals of SBA are (1) achievement of dignity for adolescent girls and women and (2) retention of adolescent girls in school (FSG 2016) through a strategy of providing sanitation infrastructure, access to menstrual products, and information, education, and communications (IEC) for MHM awareness among adolescent school-going girls, boys, and community. The aim of IEC, per its technical guidelines, is "to create awareness in order to overcome the silence around MHM and break the taboos within the broader society, communities, and also among family members" (Ministry of Drinking Water and Sanitation, Government of India, n.d., 1). The IEC material is mandated to include:

> Facts about menstruation, biology and process; Frequently-asked questions and answers; Myths about menstruation and address them with facts; Case studies/ experience from girls—How to stay healthy during menstruation—what protection to wear, what to eat, what exercise to take, how to keep clean, how to deal with cramps, how to clean, dry or dispose of sanitary materials, etc. (Ministry of Drinking Water and Sanitation, Government of India 2015, 13)

A close reading of the SBA guidelines for MHM shows two critical gaps, namely (1) absence of culturally embedded gender-specific understandings of menstruation and (2) linkages with public health. Both of these have consequences for girls' and women's menstrual health. Embodied shame, guilt, and negativity among Indian adolescent girls is not linked merely to the biology of menstruation but to the complex web of beliefs and practices of menstruation and its sociocultural and religious meanings that result in indignity and injustice. The onset of menarche is accompanied with increased disciplining on 'how to be a woman,' and adolescent girls' mental health and sense of self-esteem are negatively affected (FSG 2016; Garg and Anand 2015; Karki and Espinosa 2018). The SBA does not include any programmatic interventions to address these core gender-specific aspects of menstruating adolescent girls. As argued in a 2016 landscape analysis of menstrual health in India, "the lack of psycho-social support and limited facilitator capacities miss the opportunity to build the girl's confidence and shift inherent discriminatory social norms that define a girl's role in Indian society" (FSG 2016, 18). SBA's 'theory of change'—that awareness of menstruation as a natural physiological process will remove silence, stigma, and shame—is simplistic and grossly inadequate.

India's public health system has a vast infrastructure and network of workers at the village level—the site at which SBA is being implemented. One of the central features of any water, sanitation and hygiene (WASH) program is the connection to health. Yet the MHM Guidelines in SBA do not have any component to make visible the relationship of the menstrual cycle to basic health, psychosocial well-being, and menstrual health. The SBA envisions MHM in a restrictive manner, as an issue to be addressed *only* in the context of the period of bleeding, not across the menstrual cycle. Therefore, although its IEC material includes information on the biology of the menstrual cycle, it does not have any programmatic component to link adolescent girls with public health.

The SBA's shortsightedness missed an excellent opportunity to link the MHM and WASH agenda with the public health agenda in spite of the RCH focus on the public health system for over two decades. The focus on 'hygiene' rather than 'health' means a lost chance to address health issues associated with the disorders of the menstrual cycle, such as dysmenorrhea, amenorrhea, menorrhagia, pelvic infections, and endometriosis, among others. The SBA, currently housed in two separate ministries (the Ministry of Drinking Water and Sanitation and the Ministry of Housing and Urban Affairs), should have invited the Ministry of Health to colead the SBA or to build the missing components required for a comprehensive menstrual health approach through revisions of the NHP 2017. Instead the NHP 2017, formulated a full three years after the launch of the SBA, only includes a stated 'intention of introducing school health programs to address issues of health and hygiene' (MoHFWGOI 2017, 11), possibly a veiled reference to menstrual health and hygiene. The NHP 2017 too has failed to grasp the opportunity offered by SBA to incorporate discourse and action to address barriers to women's menstrual health.

CONCLUSION

India's health policies (1983, 2002, and 2017) and adjacent policies on nutrition, mental health, older persons, population, and rural health have failed to address the inextricable linkage between menstrual stigma and women's basic health. The recent SBA is no exception, evident in the fact that it restricts its intervention to MHM and does not extend it to menstrual health.

We propose a new construct, 'menstrual justice,' to make explicit the links between the marking of women's bodies as inferior and the discrimination, inequality, and injustice they suffer. In particular, we oppose the use of menstruation to control women and their bodies, including the use of categories that compartmentalize women and girls: preadolescents and adolescents, women of reproductive age, and postmenopausal women. These artificial divisions ignore continuities in the underlying causes of women's experiences of ill-health across the life cycle. We contend that, in practice, women physically and psychosocially experience the health effects of having their bodies marked as 'impure' well before the bleeding begins and well after it ceases. Thus, we propose a nonreductionist approach to women's health that goes beyond the narrow confines of fertility and menstrual health: menstrual justice in health.

The menstrual justice framework provides a lens through which to comprehend the discrimination and human rights violations that are borne by women and that result from marking women primarily and exhaustively as 'menstruating bodies' in the specific sociocultural and religious contexts of India. Rooted in women's rights and gender equality, this lens can serve as the basis for compelling the state to dismantle edifices built on the designation of menstruating Indian women's bodies as 'impure.'

Notes

1. We recognize that the term menstruators is more gender inclusive, and nonessentializing. However, our chapter focuses on discrimination against women embedded in India's partriachal society, so we opted to use the terms woman and women here. The complex processes of affirmation, challenge, and discrimination against individuals who identify as Third Gender in India are beyond the scope of this chapter.
2. Margaret Johnson (2019) captures the multiplicity of axes of domination that impact women's experiences of menstruation and suggests that Menstrual Justice is a 'structural intersectionality.' Our concept of Menstrual Justice has emerged from the bottom up, from Indian women's everyday experiences of menstruation in the sociocultural, religious, and political conditions of their living.

References

Ahuja, Maninder. 2016. "Age of Menopause and Determinants of Menopause Age: A PAN India Survey by IMS." *Journal of Midlife Health* 7 (3): 126–31. https://doi.org/10.4103/0976-7800.191012.

Alvergne, Alexandra, Marija Vlajic Wheeler, and Vedrana Hogqvist Tabor. 2018. "Do Sexually Transmitted Infections Exacerbate Negative Premenstrual Symptoms? Insights from Health." *Evolution, Medicine, and Public Health* 2018: 138–50. https://doi.org/10.1093/emph/eoy018.

Bharatwaj, R. S., K. Vijaya, and T. Sindu. 2014. "Psychosocial Impact Related to Physiological Changes Preceding, at and Following Menarche among Adolescent Girls." *International Journal of Clinical Surgical Advances* 2 (1): 42–53.

Bhartiya, Aru. 2013. "Menstruation, Religion and Society." *International Journal of Social Science and Humanity* 3 (6): 523–27.

Biro, Frank, and Julianna Deardorff. 2013. "Identifying Opportunities for Cancer Prevention during Preadolescence and Adolescence: Puberty as a Window of Susceptibility." *Journal of Adolescent Health* 52: S15–S20. http://dx.doi.org/10.1016/j.jadohealth.2012.09.019.

Blum, Robert, Kristin Mmari, and Caroline Moreau. 2017. "It Begins at 10: How Gender Expectations Shape Early Adolescence Around the World." *Journal of Adolescent Health* 61: S3–S4. https://doi.org/10.1016/j.jadohealth.2017.07.009.

Boutot, Meagan. 2018. "The Immune System and the Menstrual Cycle." *Clue*, March 7. https://helloclue.com/articles/cycle-a-z/the-immune-system-and-the-menstrual-cycle.

Chandra-Mouli, Venkatraman, Marina Plesons, Emmanuel Adebayo, Avni Amin, Michal Avni, Joan Marie Kraft, Catherine Lane, et al. 2017. "Implications of the Global Early Adolescent Study's Formative Research Findings for Action and for Research." *Journal of Adolescent Health* 61: S5–S9.

Chawla, Janet. 2002. "Celebrating the Divine Female Principle." *Boloji*, September 16. http://www.boloji.com/articles/6151/celebrating-the-divine-female-principle.

Das, Padma, Kelly Baker, Ambarish Dutta, Tapoja Swain, Sunita Sahoo, Bhabani Sankar Das, Bijay Panda, et al. 2015. "Menstrual Hygiene Practices, WASH Access and the Risk of Urogenital Infection in Women from Odisha, India." *PLoS One* 10 (6): e0130777. https://doi.org/10.1371/journal.pone.0130777.

Duggal, Ravi. 2005. "Historical Review of Health Policy Making." In *Review of Health Care in India*, edited by Leena V. Gangolli, Ravi Duggal, and Abhay Shukla, 21–40. Mumbai: Center for Enquiry into Health and Allied Themes (CEHAT).

Duggal, Ravi, and Leena V. Gangolli. 2005. "Introduction to Review of Healthcare in India." In *Review of Health Care in India*, edited by Leena V. Gangolli, Ravi Duggal, and Abhay Shukla, 3–18. Mumbai: Center for Enquiry into Health and Allied Themes (CEHAT).

Eichinger Ferro-Luzzi, Gabriella. 1974. "Women's Pollution Periods in Tamil Nadu (India)." *Anthropos* Bd. 69 H. 1/2: 113–161. http://www.jstor.org/stable/40458513.

FSG. 2016. "Menstrual Health in India—Country Landscape Analysis." FSG. http://menstrualhygieneday.org/wp-content/uploads/2016/04/FSG-Menstrual-Health-Landscape_India.pdf.

Garg, Suneela, and Tanu Anand. 2015. "Menstruation Related Myths in India: Strategies for Combating It." *Journal of Family Medicine Primary Care* 4, no. 2 (April–June): 184–86. https://doi.org/10.4103/2249-4863.154627.

Gokhale, Leela. 1996. "Curative Treatment of Primary (Spasmodic) Dysmenorrhoea.' *The Indian Journal of Medical Research* 103: 227–31. https://www.ncbi.nlm.nih.gov/pubmed/8935744.

Gosalia, Vibha, Pramodkumar Verma, Vikas Doshi, Manindrapratap Singh, Sanat Rathod, and Mehul Parmar. 2012. "Gynecological Morbidities in Women of Reproductive Age Group in Urban Slums of Bhavnagar City." *National Journal of Community Medicine* 3 (4): 657–60. http://www.njcmindia.org/home/download/337.

Guilmoto, Christophe Z., Nandita Saikia, Vandana Tamrakar, and Jayanta Kumar Bora. 2018. "Excess Under-5 Female Mortality Across India: A Spatial Analysis Using 2011 Census Data." *The Lancet Global Health* 6 (6): e650–58. https://www.thelancet.com/pdfs/journals/langlo/PIIS2214-109X(18)30184-0.pdf.

Gulati, S. C., Alok R. Chaurasia, and Raghubansh M. Singh. 2009. "Women's Reproductive Morbidity and Treatment-Seeking Behaviour in India." *Asian Population Studies* 5 (1): 61–84. https://doi.org/10.1080/17441730902790131.

Hussain, Zakir. 2011. "Health of the National Rural Health Mission." *Economic and Political Weekly of India* 46 (4): 53–60.

Inamdar, I. F., C. Sahu Priyanka, and Doibale, M. K. 2013. "Gynaecological Morbidities among Ever Married Women: A Community Based Study in Nanded City, India." *IOSR Journal of Dental and Medical Sciences (IOSR-JDMS)* 7 (6): 5–11. www.iosrjournals.org.

Inamdar, Madhuri, Sameer Inamdar, and N. L. Sachdeva. 2011. "Health Status of Rural Girls." *National Journal of Community Medicine* 2 (3): 388–93. http://www.njcmindia.org/home/issue_download/2/3.

India State-Level Disease Burden Initiative Collaborators. 2017. "Nations Within a Nation: Variations in Epidemiological Transition Across the States of India, 1990–2016 in the Global Burden of Disease Study." *Lancet* 390: 2437–60. http://dx.doi.org/10.1016/.

Indian Institute of Population Studies (IIPS) and ICF. 2017. "National Family Health Survey (NFHS-4), 2015–16: India." Mumbai: IIPS. http://rchiips.org/Nfhs/NFHS-4Reports/India.pdf.

Jahangir, Andisheh. 2018. "Do Nutritional Deficiencies Lead to Menstrual Irregularities?" *International Journal of Nutritional Science and Food Technology* 4, no. 5 (July): 27–31.

Jani, Anju, and Swatija Manorama. 2007. "Ageing Women and Health Care Technologies, Public Health and Policies." *Second National Bioethics Conference, Indian Journal of Medical Ethics*, NIMHANS Convention Centre, Bangalore, India, December 6–8.

Jeejebhoy, Shireen. 2000. "Safe Motherhood in India: Priorities for Social Science Research." In *Women's Reproductive Health in India*, edited by Radhika Ramasubban and Shireen Jeejebhoy, 236–79. Jaipur: Rawat Publications.

Johnson, Margaret. 2019. "Menstrual Justice." *UC Davis Law Review* 53 (1): 1–79.

Johnston-Robledo, Ingrid, and Joan C. Chrisler. 2011. "The Menstrual Mark: Menstruation as Social Stigma." *Sex Roles*, July 31. https://doi.org/10.1007/s11199-011-0052-z.

Jungari, Suresh, and Bal Govind Chauhan. 2017. "Prevalence and Determinants of Premature Menopause among Indian Women: Issues and Challenges Ahead." *Health & Social Work* 42 (2): 79–86. https://doi.org/10.1093/hsw/hlx010.

Kågesten, A., S. Gibbs, R. W. Blum, C. Moreau, V. Chandra-Mouli, A. Herbert, and Avni Amin. 2016. "Understanding Factors That Shape Gender Attitudes in Early Adolescence Globally: A Mixed-Methods Systematic Review." *PLoS One* 11 (6): e0157805. https://doi.org/10.1371/journal.pone.0157805.

Kanani, Shubhada. 2002. "How Gender Sensitive Is the National Nutritional Policy of India?—A View of the Policy through the Gender Lens." *Medico Friend Circle Bulletin* (January–February): 292–93.

Karki, Richa, and Cristina Espinosa. 2018. "Breaking Taboos: Menstruation, Female Subordination and Reproductive Health, the Case of India." *Insights Anthropology* 2 (1): 111–20. https://scholarlypages.org/Articles/anthropology/iap-2-011.php?jid=anthropology.

Kumar, Anant, and Kamiya Srivastava. 2011. "Cultural and Social Practices Regarding Menstruation among Adolescent Girls." *Social Work in Public Health* 26: 594–604. https://doi.org/10.1080/19371918.2010.525144.

Manorama, Swatija, and Chayanika Shah. 1996. "Towards a New Perspective on Women's Bodies: Learning and Unlearning Together." *Economic and Political Weekly* 31 (16–17): WS-35-38. https://www.epw.in/journal/1996/16-17/review-womens-studies-review-issues-specials/towards-new-perspective-womens.

Manorama, Swatija, and Nischint Hora. 2002. "Subterranean Violence in the City: Working with Pre-Adolescent Schoolgirls in Mumbai." Abstract of Paper presented at the Conference, "Gender and Culture: Leisure, Consumption and Women's Everyday Lives." Cheltenham and Gloucester, UK, 12–14 July 2001. Published at *Women's Global Network for Reproductive Rights Newsletter 75-Atria.* https://www.atria.nl/ezines/email/WomensGlobalNetwork/2002/No75.pdf.

Ministry of Drinking Water and Sanitation, Government of India (MoDWSGOI). 2015. "Menstrual Hygiene Management—National Guidelines." Accessed July 16, 2018. https://jalshakti-ddws.gov.in/sites/default/files/Menstrual%20Hygiene%20Management%20-%20Guidelines_0.pdf.

————. n.d. "Technical Guide 1. IEC for MHM." Accessed July 16, 2018. http://unicef.in/CkEditor/ck_Uploaded_Images/img_1512.pdf.

Ministry of Health and Family Welfare, Government of India (MoHFWGOI). 1983. "National Health Policy 1983." Accessed July 16, 2018. https://www.nhp.gov.in/sites/default/files/pdf/nhp_1983.pdf.

————. 2000. "National Population Policy 2000." Department of Family Welfare, Ministry of Health and Family Welfare, Government of India. Reprint 2002. https://mohfw.gov.in/sites/default/files/26953755641410949469%20%281%29.pdf.

————. 2002. "National Health Policy 2002." Ministry of Health and Family Welfare, Government of India. Accessed July 16, 2018. https://mohfw.gov.in/sites/default/files/18048892912105179110National.pdf.

————. 2014. "New Pathways New Hope." National Mental Health Policy of India. October 2014. https://www.nhp.gov.in/sites/default/files/pdf/national%20mental%20health%20policy%20of%20india%202014.pdf.

————. 2015. "National Rural Health Mission 2005." Accessed July 16, 2018. https://www.nhp.gov.in/the-national-rural-health-mission_pg.

————. 2017. "National Health Policy 2017." Ministry of Health and Family Welfare, Government of India. Posted April 28, 2017. https://mohfw.gov.in/sites/default/files/9147562941489753121.pdf.

————. n.d. "National Health Mission." https://nhm.gov.in/index1.php?lang=1&level=1&sublinkid=794&lid=168.

Ministry of Social Justice. Government of India (MoSJGOI). 2011. "National Policy for Senior Citizens." March 2011. http://socialjustice.nic.in/writereaddata/UploadFile/dnpsc.pdf.

Ministry of Women and Child Development, Government of India. 1975. "Integrated Child Development Services (ICDS) Scheme." Accessed July 16, 2018. https://icds-wcd.nic.in/icds.aspx.

Ministry of Women and Child Development, Government of India (MoWCDGOI). 1993. "Nutrition Policy 1993." Accessed July 16, 2018. https://wcd.nic.in/sites/default/files/nnp_0.pdf.

Mishra, Shailendra Kumar. 2011. "Menopausal Transition and Postmenopausal Health Problems: A Review on Its Bio-cultural Perspectives." *Health* 3 (4): 233–37. https://doi.org/10.4236/health.2011.34041.

Mythri, Starlin, and Johann Ebenezer. 2016. "Suicide in India: Distinct Epidemiological Patterns and Implications." National Guidelines MHM. Accessed July 16, 2018. https://jalshakti-ddws.gov.in/sites/default/files/Menstrual%20Hygiene%20Management%20-%20Guidelines_0.pdf.

Oomman, Nandini. 2000. "A Decade of Research on Reproductive Tract Infections and Other Gynaecological Morbidity in India: What We Know and What We Don't Know." In *Women's Reproductive Health in India*, edited by Radhika Ramasubban and Shireen Jejeebhoy, 236–79. Jaipur: Rawat Publications.

Outlook. 2018. "Editor's Essay: Why Menstruation Is Outlook's Issue of the Year." December 27. https://www.outlookindia.com/magazine/story/issue-of-the-year-menstruation-why-it-was-the-natural-choice/301037.

Pachauri, Saroj. 2004. "Expanding Contraceptive Choice in India: Issues and Evidence." *The Journal of Family Welfare* 50 (Special Issue): 13–25.

Pallikadavath, S., R. Ogollah, A. Singh, T. Dean, A. Dewey, and W. Stones. 2016. "Natural Menopause among Women Below 50 Years in India: A Population-Based Study." *The Indian Journal of Medical Research* 144 (3): 366–77. https://doi.org/10.4103/0971-5916.198676.

Patle, A. Rupali, and Sanjay S. Kubde. 2015. "Anemia: Does It Have Effect on Menstruation?" *Scholars Journal of Applied Medical Sciences* 3 (1G): 514–17. https://pdfs.semanticscholar.org/5cbc/6c989933a25419bbb41a63f332715f866abf.pdf.

Ramasubban, Radhika, and Shireen J. Jeejebhoy. 2000a. "Introduction." In *Women's Reproductive Health in India*, edited by Radhika Ramasubban and Shireen Jeejebhoy, 13–39. Jaipur: Rawat Publications.

———, eds. 2000b. *Women's Reproductive Health in India*. Jaipur: Rawat Publications.

Rane, Anil, and Abhijit Nadkarni. 2014. "Suicide in India: A Systematic Review." *Shanghai Archives of Psychiatry* 26 (2): 69–80. http://dx.doi.org/10.3969/j.issn.1002-0829.2014.02.003.

Raut, Prerana, V. V. Wagh, S. G. Choudhari, and A. B. Mudey. 2015. "Need for Counseling Services to Get Rid of Negativity Linked with Menstruation—A Study among Late Adolescent Rural Girls in Central India." *International Journal of Health Sciences & Research* 60 (5): 60–65.

Rishyasyringa, Bhanwar. 2000. "Social Policy and Reproductive Health." In *Women's Reproductive Health in India*, edited by Radhika Ramasubban and Shireen J. Jeejebhoy, 418–45. Jaipur: Rawat Publications.

Saheli Women's Resource Centre. 2001. "Reproductive Rights in the Indian Context: An Introduction." English Version of chapter, In *Nariwadi Rajaneeti—Sangharshaevam Mudde* (Feminist Strategies—Struggles and Issues), edited by Sadhana Arya, Nivedita Menon, and Jinee Lokaneeta. Delhi: Delhi University.

SAMA Team. 2005. "Reproductive Health Services—The Transition from Policy Discourse to Implementation." In *Review of Health Care in India*, edited by Leena V. Gangolli, Ravi Duggal, and Abhay Shukla, 153–69. Mumbai: Center for Enquiry into Health and Allied Themes (CEHAT).

Sandoiu, Ana. 2017. Being Underweight May Trigger Early Menopause. *Medical News Today*, October 26. Accessed July 25, 2019. https://www.medicalnewstoday.com/articles/319865.php.

Santhya, K. G. 2003. *Changing Family Planning Scenario in India: An Overview of Recent Evidence*. New Delhi, India: Population Council.

Sarin, Enisha, Sarah Smith Lunsford, Ankur Sooden, Sanjay Rai, and Nigel Livesley. 2016. "The Mixed Nature of Incentives for Community Health Workers: Lessons from a Qualitative Study in Two Districts in India." *Frontiers of Public Health* 4: 38. https://doi.org/10.3389/fpubh.2016.00038.

Sarojini, N. B., Suchita Chakraborty, Deepa Venkatachalam, Saswati Bhattacharya, Anuj Kapilashrami, and Ranjan De. 2006. *Women's Right to Health*. New Delhi, India: National Human Rights Commission.

Soumya, Leena, and Leena Sequira. 2016. "A Descriptive Study about Cultural Practices about Menarche and Menstruation." *Nitte University Journal of Health Science* 6 (2): 10–13.

Susila, T., and G. Roy. 2014. "Gynecological Morbidities in a Population of Rural Postmenopausal Women in Pondicherry: Uncovering the Hidden Base of the Iceberg." *The Journal of Obstetrics and Gynecology of India* 64 (1): 53–58. https://doi.org/10.1007/s13224-013-0475-2.

Swachh Bharat Mission—Gramin, Dept. of Drinking Water and Sanitation. Ministry of Jal Shakti 2019. "About SBM." Last Modified March 12, 2019. http://swachhbharatmission.gov.in/SBMCMS/about-us.htm.

Syamala, T. S. 2010. "Reaching the Unreached: Older Women and the RCH Programme in India, the Challenges Ahead." *Journal of Health Management* 12 (3): 249–60. https://doi.org/10.1177/097206341001200303.

Tulasi, Naga P., and G. Krishna Babu. 2018. "Self Reported Gynaecological Morbidity among Currently Married Women in Urban Slums of Kakinada City, Andhra Pradesh." *IOSR Journal of Dental and Medical Sciences (IOSR-JDMS)* 17 (7): 4–7. https://doi.org/10.9790/0853-1707170407.

United Nations Children's Fund (UNICEF). 2012. "Progress for Children: A Report Card on Adolescents." Number 10, April 2012. http://www.unicef.org/media/files/PFC2012_A_report_card_on_adolescents.pdf.

———. n.d. "Press Releases: Progress for Children." Accessed July 7, 2018. http://unicef.in/PressReleases/61/Progress-for-Children-A-Report-Card-on-Adolescents.

Varma, Ruchika. 2014. "Absence of Women's Needs in the National Mental Health Programme of India." *International Journal of Health Sciences* 2 (1): 37–59. http://ijhsnet.com/journals/ijhs/Vol_2_No_1_March_2014/4.pdf, https://doi.org/10.15640/ijhs.

Visaria, Leela. 2000. "From Contraceptive Targets to Informed Choice: The Indian Experience." In *Women's Reproductive Health in India*, edited by Radhika Ramasubban and Shireen Jejeebhoy, 331–82. Jaipur: Rawat Publications.

World Health Organization (WHO), Regional Office for South-East Asia. (2007). "Adolescent Sexual and Reproductive Health & HIV/AIDS among Young People: Compendium of Institutions in India". WHO Regional Office for South-East Asia. https://apps.who.int/iris/handle/10665/204767.

Zsigmond, André. 2012. "Ancient Tantric Goddess Worship—Past and Present." *Goddess Pages Blog*, April 17, 2012. Accessed May 20, 2018. https://www.goddess-pages.co.uk/blog/2012/04/17/ancient-tantric-goddess-worship-past-and-present-2/.

CHAPTER 40

Practice Note: Menstrual Hygiene Management—Breaking Taboos and Supporting Policy Change in West and Central Africa

Rockaya Aidara and Mbarou Gassama Mbaye

When the Water Supply and Sanitation Collaborative Council (WSSCC) and UN Women started working on menstrual hygiene management (MHM) in West and Central Africa, many people were unfamiliar with the acronym. During our first meetings in the region, officials from ministries, parliamentarians, and social entrepreneurs were surprised to hear MHM being spoken about, even if they suspected or knew that many women and girls faced challenges related to menstruation. One social entrepreneur told us after a training of trainers "we've been working on this for so long but we were not able to conceptualize it this well." We proceeded to hold a first 'MHM Lab,' a simple but effective training session held in tents aimed at breaking the silence and learning about the menstrual cycle (Fig. 40.1). We also carried out a study in Louga, Senegal, which provided vital quantitative and qualitative information and a solid basis from which to start conversations on menstrual hygiene. The interest, passion, and enthusiasm of the people we met were overwhelming. From the little girl in a village to the minister in a capital city, we have encountered people who wanted to change the status quo, and we have enjoyed going on this journey with them.

C. Bobel et al. (eds.), *The Palgrave Handbook of Critical Menstruation Studies*, https://doi.org/10.1007/978-981-15-0614-7_40

Fig. 40.1   MHM lab (Credit: © WSSCC/Javier Acebal 2016)

Our Approach to Menstrual Hygiene Programming: Training, Research, and Policy Development

In many West and Central African societies, menstruation is a taboo subject rooted in longstanding beliefs and myths that consider menstrual blood to be impure. Menstruation is a sensitive topic, and practices vary from one context to another. In order to respond effectively to the needs of each population, our work is country- and context-specific. However, there are agreed principles and standards as well as a modus operandi that we follow. These include (1) to leave no one behind and to contribute to the reduction of inequalities rather than reinforcing them, (2) to be collaborative and work across sectors, (3) to build evidence and fill research gaps, (4) to design inclusive policies based on the collected evidence, and (5) to train local professionals and actors who will ensure continuity and strengthen this work through their actions and initiatives at the local and national levels.

Throughout the program, we have provided technical assistance to governments. We did not impose rules or guidelines but helped strengthen systems and design policies with the respective governments that would be of benefit to all, including women and girls. Indeed, WSSCC put women's and girls' sanitation and hygiene needs and rights at the core of its strategic plan. We provided expertise on policy design and training, and helped articulate research gaps and technical needs, and provided solutions to address them.

CRITICAL ELEMENTS FOR THE FULFILMENT OF WOMEN AND GIRLS HUMAN RIGHTS' DURING MENSTRUATION

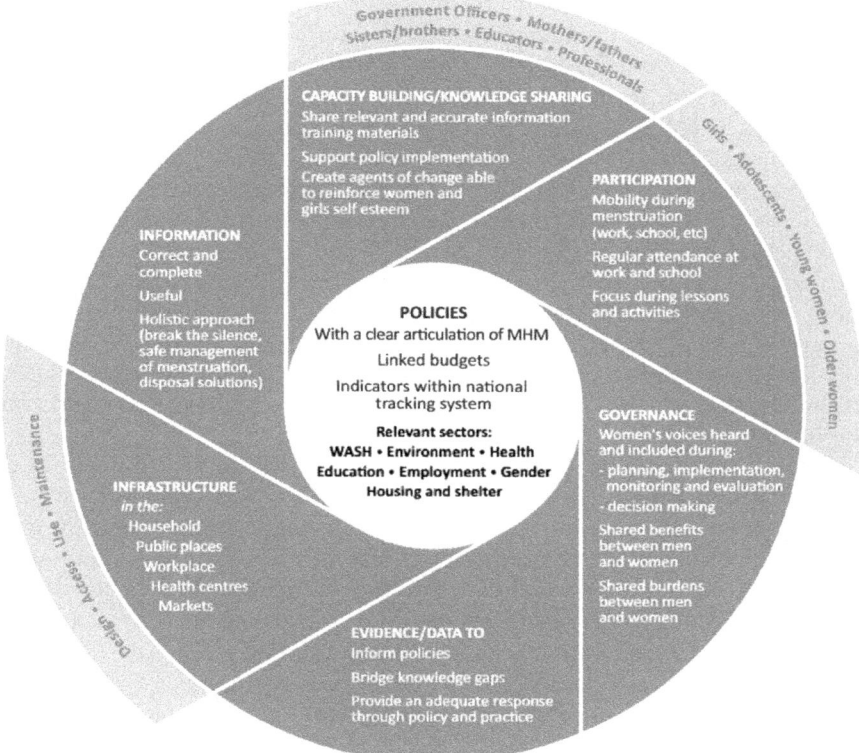

Fig. 40.2 Graph of training, research and policy circle (Credit: © WSSCC 2016)

In particular, we worked with local, national, and international actors to propose a plan of action for each country. These planning activities brought together ministries (education, health, labor, planning, and environment), other UN agencies, civil society organizations, nongovernmental organizations, and the media. For example, in Niger, the media played a key role in breaking the silence on menstruation. The training delivered to journalists by the 'WASH[1] Journalist Network' provided them with the tools to appropriately report on MHM.

Hence, the building blocks of our programs are policy-relevant research, capacity strengthening, systems strengthening, monitoring, and inter-sectoral collaboration. The training is an important part of the policy work as it fosters knowledge and a common understanding of overlooked issues. Combined with further research, it is also the first phase of the analytical work required when designing the policies. During the training and research, we gain a better understanding of the MHM practices in a country, the sociocultural context, and the political, physical, and institutional barriers (Fig. 40.2).

Trainings

Addressing menstruation as a taboo subject was facilitated by lively dialogue and exchanges with girls, women, boys, and men, including in ministries, NGOs, schools, and communities-at-large (Joint Programme 2016). The first major activity organized by the program was an 'MHM Lab' in the city of Louga (300 km from Dakar, the capital of Senegal). The purpose of the Lab was multifold: It was an opportunity (1) to learn more about women's and girls' practices, (2) to train the first round of Lab facilitators, and (3) to collect crucial data which would consequently be presented to ministers and members of parliament. In the five-day training of trainers and MHM Lab, tools such as the booklet "As We Grow Up" and the menstrual wheel, which visualize the changes in the body that take place from childhood to adulthood, were effective in providing participants with enough confidence to be able to offer these trainings themselves (Fig. 40.3). Participants of the trainings felt empowered to speak out during ministries' planning activities. As a result, MHM was made explicit in education, health, gender, water and sanitation, and environment public policy.

Today, participants in the trainings form a community of trainers in an online platform, where they regularly share resources on menstrual hygiene management. Though trainers are very active online in some countries, joining the platform has proven to be more difficult in Niger due to connectivity issues. However, the trainers continue to meet face-to-face, to run MHM Labs and advocacy sessions, and are involved in key menstrual hygiene management related processes at the national and local level.

The menstrual hygiene management trainers in the region are a vital force for the program. They are leading labs and facilitating advocacy

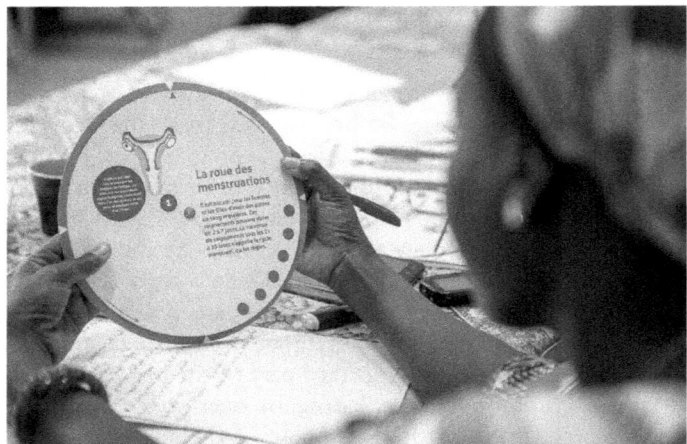

Fig. 40.3   Menstrual wheel (Credit: © WSSCC 2016)

sessions—with support from the Joint Programme's team—targeting government officials, members of parliament, the population at large, including women's groups, children in schools, persons with disabilities, teachers, social entrepreneurs, and NGOs, among others. In total, more than 26,000 people have been reached by the 620 trainers trained by the Joint Programme's team (Joint Programme 2018). People have told us how they have changed as individuals and as fathers, mothers, teachers, and supervisors (YouTube 2018). Even after the end of the program, trainings and MHM Lab sessions continue, demonstrating the sustainability of the approach. Social entrepreneurs have also picked up the methodology. Different actors like Kmerpad, Santé Mobile, and Speak Up Africa partner regularly to run MHM Labs with support from the program's MHM trainers (Joint Programme 2017a).

Data Collection and Applied Research

Data is important to build evidence-based strategies and to inform future policies throughout the program. The Minister of Water and Sanitation of Senegal realized that menstrual hygiene was a significant challenge when he received our first research report on Louga in 2014 (Joint Programme 2014). It encouraged him to take action within the ministry by reviewing the sector policy and related guidelines. In turn, the position of the Government of Senegal toward MHM encouraged the former Minister of Water and Sanitation of Niger to take action.

Studies provided key information and statistics on attitudes toward menstruation and menstrual hygiene behaviors and practices (Joint Programme 2014, 2015a, 2015b). They also examined public policies and found that none addressed menstrual hygiene management at the time. This laid the groundwork for subsequent policy developments. The studies addressed several areas, including the silence around menstruation, women's and girls' beliefs and attitudes, and the fear and stress experienced by women and girls when menstruating.

For instance, the study in Senegal found that a significant share of women work in the informal sector face challenges when menstruating because sanitation facilities and private spaces are largely unavailable to them (Joint Programme 2015a). In refugee camps in Cameroon, 99% of women indicated that they do not feel safe enough to use sanitation facilities at night. With regard to education, the study in Cameroon found that the advice girls receive—whether from mothers or at school—was often limited to practical information about the use of menstrual products and did not extend to understanding physiological factors and other aspects of menstrual literacy (Joint Programme 2015b). Based on a review of policies, the study in Niger found that menstruation is largely absent from sectoral policy documents and that women lack a forum to express concerns and indicate their needs in relation to menstruation (Joint Programme 2017a). The findings from these studies provided the rationale and the necessary data and statistics for menstrual hygiene management interventions.

Policy Development and Budgeting

The Joint Programme responded to a genuine demand of young girls and women who wanted to know more about their bodies and the menstrual cycle, a subject that is rarely discussed in families or in public. Through the trainings, research, and advocacy, the Joint Programme successfully mobilized governments in West and Central Africa around menstrual hygiene management (Joint Programme 2018). Menstrual hygiene management provided an entry point for discussing other women's rights issues such as family planning, early and forced marriage, gender-based violence, female genital cutting. The MHM Labs organized by local actors in villages, communes, schools, health centers, public places, and in refugee camps provided the space for intergenerational dialogue around these issues.

The involvement of decision-makers at the highest level during the MHM Labs enabled them to pave the way in speaking out and ensuring that their respective governments accorded priority to menstrual hygiene through developing policies and allocating budgets. The program contributed significantly to gender-sensitive budgeting in water, sanitation and hygiene (WASH) in Senegal, Niger, and Cameroon. For example, in Senegal, the WASH ministry undertook a review of sanitation facilities' design, which led to increased public spending for gender-sensitive facilities. The new design of toilets includes a systematic separation of toilets (women/men, boys/girls). It also includes more spacious sanitation facilities for women (Joint Programme 2017b). The supplementary cost for the new toilets was about 200 USD per facility. Similarly, in Cameroon, the budget of five municipalities' pilots of the program increased to address the need for gender-sensitive toilets. Gender-sensitive toilets were also built in two refugee camps in Minawao and Ngam in Cameroon (Joint Programme 2017a).

Looking beyond toilets, education, health, and environment ministries from the region (especially in Senegal, Niger, and Cameroon) have reviewed their policies to include menstrual hygiene management and menstrual health more broadly. Over the years, the acronym MHM (which provided our entry point) has been replaced in some contexts by menstrual health, which we believe is a more holistic term that goes beyond the simple management of menses. Teachers in Cameroon have developed courses based on WSSCC training materials to address MHM with students in training centers. All stakeholders agreed that working on menstrual hygiene and health required collaboration across sectors. Interministerial task forces were set up in Cameroon, Niger, and Senegal, bringing together the ministries in charge of WASH, health, education, environment, and labor to address menstrual health in a holistic manner. The ministries that led the policy analysis were very effective, especially in Niger. As a result, several policies and implementation plans have been reviewed or newly designed to include menstrual hygiene. The Government of Senegal made menstrual hygiene management a mandatory feature of future sanitation projects to be funded by the Government through its coordinating mechanism. At the local level, municipalities in Cameroon and Senegal increased their budgets to review the design of the facilities in public spaces (Joint Programme 2016). Overall, the Joint Programme has affected change in policies, institutions, and budgets at various levels.

The Way Forward: Strong Institutions, Policies, and Budgets for Long-Term Results

The adoption of new policies is not an end in itself, though it is a significant step in advancing women's rights, equality, and empowerment. Policies have implications for budgets, programs, and service provision, and we have seen their implementation across many countries. They also provide a basis for accountability: governments make public commitments in their policies to which they can be held accountable by civil society, the media, and communities. On a different but equally important level, policy development also plays a significant symbolic role in creating spaces for dialogue about menstruation. Such talk normalizes discussion and makes menstrual health a component of government activity.

Overall, our program was met with an overwhelmingly positive response. Initial skepticism quickly yielded to the open sharing of experiences about menstruation. The MHM Labs provided ideal spaces for these conversations and served multiple purposes: to learn about experiences of menstruation, to provide training on the menstrual cycle, and to engage policy-makers who quickly grasped the fundamental importance of menstruation. In addition to creating these spaces for interaction, the choice of the WASH sector as an entry point proved to be ideal in our target contexts. The seemingly technical nature of WASH and the use of the abbreviation 'MHM' provided an opening for beginning to address broader concerns about gender equality and other stigmatized conditions (see Patkar [Chapter 38] in this volume).

However, a lot still needs to be done. The policies and budgets now need to be streamlined with effective action plans and strengthened systems that recognize menstrual hygiene management, and more broadly, as we recommend, menstrual *health*. And they must provide a space to voice women's needs at all stages, from the design of programs to their monitoring and evaluation. Smart action plans, monitoring, systems strengthening, budget increases, and the review of the design of some facilities are essential to increase access to gender-sensitive and MHM-friendly facilities in public spaces, schools, health centers, and any other places where people live, work, play—all spaces where people menstruate. Infrastructure is important, but equally important is one's capacity to talk, walk, and move freely during menstruation, without fear or shame, which must be our top priority.

Note

1. WASH is the acronym for Water, Sanitation and Hygiene, a sector in global development.

References

Joint Programme on Gender, Hygiene and Sanitation. 2014. "Menstrual Hygiene Management: Behavior and Practices in the Louga Region, Senegal." Accessed April 28, 2019. https://www.wsscc.org/wp-content/uploads/2015/09/Louga-Study-EN-LoRes.pdf.

———. 2015a. "Menstrual Hygiene Management: Behaviour and Practices in the Kedougou Region, Senegal." Accessed April 28, 2019. https://www.wsscc.org/wp-content/uploads/2015/05/kedougou_study_en_lores.pdf.

———. 2015b. "Menstrual Hygiene Management: Behaviour and Practices in Kyé-Ossi and Bamoungoum, Cameroon." Accessed April 28, 2019. https://www.wsscc.org/wp-content/uploads/2016/04/Study-%E2%80%93-Menstrual-Hygiene-Management-Behaviour-and-Practices-in-Kye-Ossi-and-Bamoungoum-Cameroon-WSSCC-UN-Women.pdf.

———. 2016. "Field Mission in Senegal Highlights Strengths and Weaknesses of Sanitation Programmes." Accessed April 28, 2019. https://www.wsscc.org/wp-content/uploads/2016/12/GHS-newsletter-Issue-9-EN-1.pdf.

———. 2017a. "The Inception of a Political Movement for MHM in Niger." Accessed April 28, 2019. https://www.wsscc.org/wp-content/uploads/2017/08/Joint_Programme_InfoLetter_10_ENG.pdf.

———. 2017b. "Menstrual Hygiene Management Included in Senegal's New Sectoral Policy Paper." Accessed April 28, 2019. https://www.wsscc.org/wp-content/uploads/2018/01/20180122_MHM-newsletter_Issue_11_EN.pdf.

———. 2018. "End-Term Evaluation of the Joint Programme on Gender, Menstrual Hygiene and Sanitation." Accessed April 28, 2019. http://gate.unwomen.org/Evaluation/Details?EvaluationId=11311.

YouTube. 2018. "Presentation du programme conjoing Genre, Hygiene, Assainissement." Accessed April 28, 2019. https://www.youtube.com/watch?v=8zsgvpHODOE.

CHAPTER 41

U.S. Policymaking to Address Menstruation: Advancing an Equity Agenda

Jennifer Weiss-Wolf

In the United States, mainstream acknowledgment of menstruation as a source of struggle—in the media and among policymakers—is a relatively recent phenomenon (Zraick 2018). As with most hardship, marginalized populations, including those who are poor, incarcerated, and homeless, are at greatest risk. For instance, a 2019 study found that nearly two-thirds of women surveyed in St. Louis, Missouri could not afford menstrual products during the previous year and nearly one in five could not afford them each month (Sebert Kuhlmann et al. 2019). Those experiencing homelessness report using tampons and pads for longer than recommended or improvising with items such as paper bags or newspapers (Upadhye 2016). And incarcerated individuals and those caught in the criminal justice system must often beg or bargain with corrections staff for basic hygiene needs, part of a degrading and dehumanizing power imbalance—and a failure of the state to provide for those in its custody (Bozelko 2015).

Decades of menstrual activism have laid the groundwork for the recent popular engagement (see Bobel 2008). This newest wave includes a new policy frame that colleagues and I introduced in 2015 intended to put menstruation at the heart of U.S. politics, with arguments rooted in core concepts of equity and civic participation. The deliberate choice of the phrase 'menstrual equity' as a frame harkens to the democratic creed and aspiration of civic, economic, and political engagement and opportunity in the United States. It is broader than a sanitation, hygiene or public health frame, and

Jennifer Weiss-Wolf gratefully acknowledges Victoria Miller for her editorial contributions and invaluable support.

more applicable to U.S. policymaking than a human rights frame and hence provides a tangible basis for advocacy. It has proven to be a frame that has massive potential as evidenced by its recent and rapid success.

U.S. Law and Policy Agenda 2015–2018

Forging meaningful change through law- and policy-making has distinct advantages. To start, implementation of laws is an efficient way to benefit the widest swath of the population. The laws we pass make an affirmative and deliberate statement about the values for which we stand—an especially critical ingredient in dismantling the stigma of menstruation.

Prior to initiation of the menstrual equity frame and advocacy campaign, menstruation had only been at the center of policymaking very briefly. For a short period of time in the 1990s, activists pressed Congress to consider menstruation—largely in response to a spate of deaths from Toxic Shock Syndrome caused by high-absorbency tampons (see Vostral 2018) followed by a later (unrelated) report by the Food and Drug Administration (FDA) of toxicity in the synthetic fibers and chemicals used in regular tampons (Fetters 2015). In 1997, U.S. Congresswoman Carolyn Maloney introduced federal legislation that would require funds for independent research into potentially hazardous ingredients in menstrual products, as well as mandate transparency in sharing and publicizing testing results (Tampon Safety and Research Act of 1997). Her bill, the Tampon Safety and Research Act of 1997 (H.R. 2900), was unsuccessful. In 1999, she renamed it the Robin Danielson Act, in honor of a woman who died from Toxic Shock Syndrome. She continued to reintroduce the bill, most recently in 2017, but it has never advanced or succeeded in garnering broad support.

There are myriad ways in which the material dimensions of menstruation continue to be disregarded as a matter of public policy: Tampons and pads have not been designated as allowable budgetary expenses for publicly funded shelters or crisis and emergency centers; they are not provided in a consistent or fully accessible way in correction and detention facilities; menstrual products are not covered by public health and nutritional benefits programs; they are not made uniformly available in schools or workplaces; in 30 states (as of April 2020), menstrual products are not sales tax exempt; inconsistent tax classification status of menstrual products prevented them from being included in Health Savings or Flexible Spending Account allowances, a change that was finally made in the Coronavirus Aid, Relief, and Economic Securities (CARES) Act, signed by President Trump on March 27, 2020 (Weiss-Wolf 2017).

Consider how different governmental agencies acknowledge and address menstrual products—not only with a disparate, inconsistent vocabulary and recognition of purpose but always to the detriment of those who menstruate:

- The federal government bears responsibility for approving and disclosing the ingredients in menstrual products via the FDA. The FDA designates tampons a Class II "medical device" (U.S. Food and Drug Administration 2014). The only affirmative obligations required of manufacturers are that they provide basic instructional labeling on packaging and keep records of adverse events caused by their products. Transparency in testing results is not required, nor is labeling or detailed disclosure about product ingredients on packaging.
- At the U.S. Department of Labor, the Occupational Safety and Health Administration (OSHA) mandates that employers provide all workers with safe, sanitary toilet facilities as well as products required for washing and drying (United States Department of Labor, n.d.a). Menstrual products are conspicuously excluded from the list of hygiene necessities, though addressed in meticulous detail in disposal protocols for facilities (United States Department of Labor, n.d.b).

It was not until 2015, and the ascent of arguments to eliminate the "tampon tax"—shorthand for the fight to exempt menstrual product purchases from sales tax—that the issue drew meaningful attention and action. That campaign was soon joined by targeted municipal, state, and federal advocacy to ensure that products are freely accessible via public agencies that serve those who are young, have low income, homeless, or incarcerated. I describe these initiatives below, with a focus on the variety of movement-building strategies and legal arguments that have been employed.

THE 'TAMPON TAX' AS A STRATEGIC ENTRY POINT FOR U.S. ADVOCACY

The 'tampon tax' is regular sales tax or value-added tax applied to menstrual products—or, conversely, the failure of the state to exempt these products from sales tax by placing them in the category of necessity. This is not a special or additive levy, or "luxury tax" as some assume, but is regular sales tax applied to menstrual products—ranging from roughly 4–10% depending on the state tax code. Generally, and as a matter of historical practice since sales taxes were first implemented in the United States in the 1930s, states exempt food and other items deemed necessities of life, such as medicine and prescription drugs, from sales tax.

Prior to 2015, there were five states that already exempted menstrual products, though not as a result of a concerted campaign. Another five states collect no sales tax at all and therefore do not tax these items. By 2018, a further five states had successfully eliminated the 'tampon tax.'

A year's supply of tampons and pads for a person costs in the range of $70 to $120, depending on where one lives, the heaviness of one's flow, and the ability to take advantage of cost-cutting measures. (Reusable alternatives like menstrual cups, absorbent underwear, and cloth pads often have high upfront costs but are more cost-effective over time.) All told, the expense can add up to more than $5000 over a lifetime per person; sales tax adds several hundred dollars to this total. However, this is not the only reason why campaigns have focused on lifting this tax burden.

Advocating for removal of the 'tampon tax' is a smart strategic entry point because the arguments are straightforward for the general public, salient for the media, and manage to transcend partisan politicking. This approach challenges laws that are archaic, unfair, and discriminatory and helps inch toward a model of economic parity and gender equality. From my perspective as a menstrual activist, it accomplishes all of this—and even more: It offers a gateway for enabling people to talk and think about the wider implications of menstruation—social, economic, and otherwise—in our policymaking. As a result, numerous campaigns to address the taxation of menstrual products have long been underway around the world. Notable examples are Kenya, Canada, England, Malaysia, India, Australia, and South Africa, all countries where the equivalent of sales tax has been successfully challenged (see Patkar [Chapter 38] in this volume).

In the United States, the 'tampon tax' has offered a ripe forum for domestic policy advocacy and a chance to develop and test equity-based language for addressing the economics of menstruation. Among its most distinct characteristics in the United States, the tampon tax argument has unusually strong trans-partisan appeal. There are persuasive perspectives from all sides—left, right, libertarian—variously focused on social justice, gender equity, tax relief, and/or limiting the scope of government reach. One unique challenge we face as activists is the federal structure of the United States. Because there is no national sales tax in the United States, but instead a multitude of municipal- and state-specific tax codes, it is difficult to organize one overarching nationwide campaign.

Despite this, in October 2015, our inaugural campaign launched with the first national petition, co-sponsored by *Cosmopolitan* magazine, "Stop Taxing Our Periods. Period" (Change.Org, n.d.). Our primary goal for making the petition national in focus, rather than state-by-state, was to elevate the issue with a singular call for action, and then pressure state lawmakers across the country to act simultaneously when the 2016 legislative sessions commenced. The petition sparked immediate and high-level attention by the media, policymakers, and other influential voices. *The New York Times* editorialized in favor of not only repealing sales tax on menstrual products but also ensuring their affordability and accessibility (Editorial Board 2016). Policy and fiscal analyses were published by *The Economist* (2016) and the *Wall Street Journal* (Larimer 2016). The American Medical Association, the largest association of physicians in the United States, issued a position statement urging states to exempt menstrual products from sales tax as sound health and medical practice (American Medical Association 2016). An op-ed by constitutional scholar, now dean at the University of California, Berkeley, School of Law, Erwin Chemerinsky, put forth legal criteria for abolishing the tax, rooted in the argument that it would alleviate discriminatory impact (Chemerinsky 2016). Even President Obama weighed in, reflecting that the issue's historic neglect was indicative of our nation's less-than-fully representative leadership, stating, "I suspect it's because men were making the laws when those taxes were passed" (Rhodan 2016).

Legislators responded. Between 2016 and 2018, proposals to exempt menstrual products from sales tax have been introduced or debated in 24 state legislatures (Weiss-Wolf 2017, 137). The city of Chicago was the first jurisdiction in 2016 to eliminate the tampon tax. The states of New York and Illinois unanimously passed laws in 2016 that were signed by those states' respective governors—in New York, a Democrat, in Illinois, a Republican— demonstrating that there was bipartisan commitment to the issue. In the 2017 legislative session, Florida successfully passed a tax exemption for menstrual products, also signed by a Republican governor. And in 2018, Connecticut became the next state to eliminate the tax by legislation. In the nation's capital, the Washington, DC Council exempted menstrual products from sales tax in 2016. On Election Day 2018, Nevada voters approved the first-ever ballot measure on the tampon tax. The lawmakers who proposed the measure made clear their intention that the state should not be funding its needs "on the backs of women" (Kasperkevic 2018). Given the unnecessary hurdles that have emerged in several state legislatures, some detailed below, ballot measures are an important lever for making change, and a strategy to replicate going forward.

In states where bills failed, legislative rationale has varied. In California, a bill passed in 2016, only to be vetoed by the governor on grounds of fiscal restraint; it was reintroduced in 2017 and 2018, but has been unable to advance. Utah's Hygiene Tax Act did not leave the committee (an all-male panel) in 2016 and 2017. Among the reasons for rejecting the bill were fears of a too-subjective tax code and concerns about recouping $1 million in lost revenue (Close 2016).

And yet, legislators did not raise the prospect of making up the difference in lost revenue by taxing alternative items that could be used or purchased equally by both men and women. Nor did they demonstrate concern about non-necessity items already chosen for sales tax exemption. The variety of allowable exemptions from state to state are extraordinarily subjective—ranging from kettle corn in Iowa, to gun club memberships in Wisconsin, to Mardi Gras beads in Louisiana, to arcade game tokens and potato chips purchased from a vending machine in Utah (Weiss-Wolf 2016).

Among the handful of denouncers, *Washington Post* opinion writer Catherine Rampell published "The Tampon Tax Fraud" (Rampell 2016). In *The Daily Beast*, Samantha Allen belittled "the mainstream feminist issue *du jour*" in her essay, "The 'Tampon Tax' Outrage is Overblown" (Allen 2015). Both framed their opposition on the argument that menstrual products are not subject to any specific, additive tax (a common retort). Wrote Rampell: "Politicians didn't decide one day that periods were gross and therefore ought to be made more expensive." And from Allen: "It is not a state-sanctioned war on periods." These statements miss the point. Whether or not it was a question of nefarious intent (doubtful) or clueless oversight (likely), the reality is that legislators had not ever before been called upon to consider menstruation at all either in the evolution of tax code exemptions or any other legislation.

In fact, the menstrual equity agenda must be taken beyond simple sales tax reform. That the sales tax question has shone a light on the issue of the economics of menstruation—and has gotten media outlets to report and editorialize on it at all—is alone a tremendous advance and fuels our ability to develop more nuanced policy arguments.

Beyond the Tampon Tax: The Case for Universal Access

Policies need to go deeper than sales tax reform, beyond ensuring equitable access to products to delve further into the many ways stigma and silence around menstruation has a harmful impact. Making the case that menstrual products be freely available to certain target populations defines the simultaneous policy agenda that we have been advancing since 2015. Three populations—students and low-income teenagers, those dependent on public shelters (see Sebert Kuhlmann et al. 2019; see also Vora [Chapter 4] in this volume), and people who are incarcerated or in the custody of the government (see Bozelko [Chapter 5] as well as Roberts [Chapter 6] in this volume)—stand out as strategic targets for advocacy—both because the need is great and there are clear policy options for addressing them.

In 2016, the New York City Council became the first jurisdiction to unanimously pass the most comprehensive legislation of its kind in the world in the form of three laws. The first requires the city to make tampons and pads freely accessible via dispensers in bathrooms at its eight hundred public schools reaching approximately three hundred thousand students. The second law provides a budget for all shelters overseen by the Department of Homeless Services and the Department of Health and Mental Hygiene to provide free menstrual products—an estimated 2 million tampons and 3.5 million pads yearly. And finally, a third law eliminates the cap on the number of pads given to women in custody of the Department of Correction, which had previously amounted to a mere eleven thin, poor quality pads per period (NYC.gov 2016).

New York City's campaign to pass these laws was swift, highly public, and extraordinarily effective. Legislators and activists worked collaboratively and focused on obtaining mainstream media coverage—high in quality and quantity—that featured an array of compelling personal stories. This led to resounding public support and pressure to advance legislation. The campaign unfolded quickly, over the course of a single year, and the bills passed seamlessly with the unanimous support of City leadership. The widespread support for these policies helped to spur other cities and local jurisdictions to take similar action.

A state-by-state and federal agenda has seen similar popular support and more bipartisan buy-in throughout the following two years, covering availability of menstrual products in public schools, homeless shelters, and in detention centers. In the latter case, numerous state legislators have taken up the cause for those who are incarcerated because it is an extraordinarily popular reform. 2018 polling research by the Justice Action Network shows that 90% of voters are in favor of providing free menstrual products in prisons, with robust bipartisan support (Meng, Ferreras-Copeland, and Weiss-Wolf 2018). In one state, Arizona, social media became a powerful tool for calling out the deeply ignorant commentary made by committee members, which only fueled more public support for the reform (Held 2018).

Even a very polarized Capitol Hill has taken a stand. First, after the Dignity for Incarcerated Women Act (S. 1524) was introduced in July 2017 by four U.S. Senators, which included a prominent menstrual access provision, the Department of Justice issued a rule that tampons and pads be freely provided in all federal correctional facilities (Dignity for Incarcerated Women Act 2017). By December 2018, bipartisan prison reform legislation, the FIRST STEP Act, passed and was signed into law by President Trump. It specifically includes the provision of menstrual products at federal facilities: the first-ever federal menstrual access mandate.

Menstrual equity has one of its fiercest federal champions in U.S. Congresswoman Grace Meng (D-NY), who has developed creative interventions to ease the financial burdens of menstruation, including at shelters and crisis centers, allowing them to purchase menstrual products with FEMA grant funds (Meng 2016). The Congresswoman also introduced the first-ever federal menstrual equity legislation, the Menstrual Equity for All Act of 2017 (H.R. 972) (Menstrual Equity for All Act of 2017). She reintroduced the bill in March 2019. Both versions would require state prisons to provide menstrual products as a condition of receiving federal funding and require employers with more than 100 employees to provide menstrual products in workplace bathrooms, among other provisions.

Collectively, these legislative provisions will help make menstruation more affordable for millions. And federal activity stands to vastly elevate the politics of periods on the national and global stage.

What's Next

As this recent menstrual equity agenda has won public support and continues to win in legislatures, our next step must be a bold one. That is, we must elevate menstruation as an actual catalyst for how we consider *all* of the laws and policies by which we live. Menstruation provides an extraordinary lens for discerning whether these measures enable full, fair societal participation. This means going further than product accessibility and safety, which has proven a valuable entry point for getting menstruation the attention it needs and deserves, but is not the end game for a holistic menstrual agenda, or indeed agenda for gender equality. Rather, it is a first step in addressing menstrual stigma and has sparked a nationwide discussion because the 'tampon tax' and lack of access to products are visible and tangible examples of inequalities that many people can easily relate to.

However, a comprehensive menstrual equity agenda entails more than that. This agenda envisions a world where patients do not hesitate to speak with their health care providers about menstrual pain and other disorders and providers do not dismiss women's symptoms; a world where no woman is perceived as unfit or too 'emotional' to do her job while menstruating; a world where all individuals who menstruate have the proper information and agency to make decisions regarding care of their bodies; and ultimately a world where menstruation is not stigmatized.

At this stage, the challenge is to move from the tangible, visible, and easily presentable to the underlying menstrual equity challenges that are much more deeply engrained in society and accentuate gender inequalities. At the top of the possibilities to address these structural challenges are educational opportunities that expand curricula on menstruation—beyond health and hygiene to integrate relevant content in lesson in history, literature, and science. This would go a long way helping to reduce the stigma and create an equitable learning environment. The opportunities also reach into labor policies that include menstruation when tackling issues ranging from workplace safety, to paid leave (see Levitt and Barnack-Tavlaris [Chapter 43] in this volume), to addressing more sweeping gender discrimination and stereotyping. They include working with health care providers to include attention to menstruation and menstrual irregularities in standard practice. The menstrual equity lens provides us with an opportunity to spot and address inequities in all spheres of life.

Embracing a menstrual equity policy frame is a new way to at once recognize the political power and absolute normalcy of periods.

References

Allen, Samantha. 2015. "The 'Tampon Tax' Outrage Is Overblown." *The Daily Beast*, October 22. http://www.thedailybeast.com/articles/2015/10/22/the-tampon-tax-outrage-is-overblown.html.

American Medical Association. 2016. "AMA Adopts New Policies on Final Days of Annual Meeting." AMA-ASSN.org, June 15. https://www.ama-assn.org/ama-adopts-new-policies-final-day-annual-meeting.

Bobel, Chris. 2008. "From Convenience to Hazard: A Short History of the Emergence of the Menstrual Activism Movement, 1971–1992." *Health Care for Women International* 29 (7): 738–54.

Bozelko, Chandra. 2015. "Prisons That Withhold Menstrual Pads Humiliate Women and Violate Basic Human Rights." *The Guardian*, June 12. https://www.theguardian.com/commentisfree/2015/jun/12/prisons-menstrual-pads-humiliate-women-violate-rights.

Change.Org. n.d. "Stop Taxing Our Periods!" https://www.change.org/m/end-the-tax-on-feminine-hygiene-products-notaxontampons.

Chemerinsky, Erwin. 2016. "In Tampon Tax, a Discriminatory California Policy Lives On." *Los Angeles Daily News*, September 21. https://www.dailynews.com/2016/09/21/in-tampon-tax-a-discriminatory-california-policy-lives-on-erwin-chemerinsky/.

Close, Kerry. 2016. "An All-Male Committee in This State Just Voted to Keep Taxing Tampons." *TIME*, February 11. http://time.com/money/4217234/tampon-tax-utah/.

"Dignity for Incarcerated Women Act (S. 1524)." https://www.congress.gov/bill/115th-congress/senate-bill/1524/text.

Editorial Board. 2016. "End the Tampon Tax." *The New York Times*, February 8. https://www.nytimes.com/2016/02/08/opinion/end-the-tampon-tax.html.

Fetters, Ashley. 2015. "The Tampon: A History." *The Atlantic*, June 1. https://www.theatlantic.com/health/archive/2015/06/history-of-the-tampon/394334/.

"FIRST STEP Act (H.R. 5682)." 2017. https://www.congress.gov/bill/115th-congress/house-bill/5682.

Held, Amy. 2018. "Arizona Department of Corrections Changes Sanitary Pad Policy Following Backlash." National Public Radio, February 15. https://www.npr.org/sections/thetwo-way/2018/02/15/586134335/arizona-department-of-corrections-changes-sanitary-pad-policy-following-backlash.

Kasperkevic, Jana. 2018. "Nevada's 'Tampon Tax' Ballot Initiative Brings Up Questions about Fairness and Gender Equality." *Marketplace*, November 1. https://www.marketplace.org/2018/11/01/elections/nevada-s-tampon-tax-ballot-initiative-brings-questions-about-fairness-and.

Larimer, Sarah. 2016. "The 'Tampon Tax' Explained." *Washington Post*, January 8. https://www.washingtonpost.com/news/wonk/wp/2016/01/08/the-tampon-tax-explained/.

Meng, Grace. 2016. "Meng: FEMA to Permit Homeless Assistance Providers to Purchase Feminine Hygiene Products—Such as Tampons and Pads-with Federal Grant Funds." Press release, March 1. https://meng.house.gov/media-center/press-releases/meng-fema-to-permit-homelessassistance-providers-to-purchase-feminine.

Meng, Grace, Julissa Ferreras-Copeland, and Jennifer Weiss-Wolf. 2018. "Women Are Finally Winning the Period Rights Fight." *Newsweek*, January 25. https://www.newsweek.com/women-finally-winning-period-rights-fight-790990.

"Menstrual Equity for All Act of 2017 (H.R. 972)." https://www.congress.gov/bill/115th-congress/house-bill/972.

NYC.gov. 2016. "Mayor de Blasio Signs Legislation Increasing Access to Feminine Hygiene Products for Students, Shelter Residents and Inmates." NYC.gov, July 13. http://www1.nyc.gov/office-of-the-mayor/news/611-16/mayor-de-blasio-signs-legislation-increasing-access-feminine-hygiene-products-students-.

Rampell, Catherine. 2016. "The 'Tampon Tax' Fraud." *Washington Post*, January 25. https://www.washingtonpost.com/opinions/the-tampon-tax-fraud/2016/01/25/fb9c7e68-c3a8-11e5-8965-0607e0e265ce_story.html?utm_term=.ad6a992c1126.

Rhodan, Maya. 2016. "President Obama Doesn't Understand the 'Tampon Tax' Either." *TIME*, January 15. http://time.com/4183108/obama-tampon-tax-sanitary/.

Sebert Kuhlmann, Anna, Eleanor Peters Bergquist, Djenie Danjoint, and L. Lewis Wall. 2019. "Unmet Menstrual Hygiene Needs among Low-Income Women." *Obstetrics & Gynecology*, February. https://journals.lww.com/green-journal/Fulltext/2019/02000/Unmet_Menstrual_Hygiene_Needs_Among_Low_Income.2.aspx.

"Tampon Safety and Research Act of 1997 (H.R. 2900)." https://www.congress.gov/bill/105th-congress/house-bill/2900/text.

The Economist. 2016. "Removing the Tampon Tax Is Good Politics, but What about the Economics?" *The Economist*, March 18. https://www.economist.com/free-exchange/2016/03/18/removing-the-tampon-tax-is-good-politics-but-what-about-the-economics.

U.S. Food and Drug Administration. 2014. "Classify Your Medical Device." Updated July 29. https://www.fda.gov/MedicalDevices/DeviceRegulationandGuidance/Overview/ClassifyYourDevice/default.htm.

United States Department of Labor. n.d.a. "Most Frequently Asked Questions Concerning the Bloodborne Pathogens Section." https://www.osha.gov/laws-regs/standardinterpretations/1993-02-01-0.

———. n.d.b. "Restroom and Sanitation Requirements." https://www.osha.gov/SLTC/restrooms_sanitation/.

Upadhye, Janet. 2016. "This Is How Homeless Women Cope with Their Periods." *Bustle*, October 18. https://www.bustle.com/articles/190092-this-is-how-homeless-women-cope-with-their-periods.

Weiss-Wolf, Jennifer. 2016. "Menstrual Products Are Taxed in 40 States: Here's What You Can Buy Tax-Free." *Ms*, February 19. http://msmagazine.com/blog/2016/02/19/menstrual-products-are-taxed-in-40-states-heres-what-you-can-buy-tax-free/.

———. 2017. *Periods Gone Public: Taking a Stand for Menstrual Equity*. New York: Arcade Publishing.

Vostral, Sharra. 2018. *Toxic Shock: A Social History*. New York: New York University Press.

Zraick, Karen. 2018. "It's Not Just the Tampon Tax: Why Periods Are Political." *The New York Times*, July 22. https://www.nytimes.com/2018/07/22/health/tampon-tax-periods-menstruation-nyt.html.

CHAPTER 42

Personal Narrative: Bloody Precarious Activism in Uganda

Stella Nyanzi

Editor's Note: I am honored and humbled to include Stella Nyanzi's essay in our Handbook. When we refer to menstruation as structural, we often mean that menstruation is *political*. Menstruation exemplifies how women's and menstruators' bodies and sexualities are politicized. No one can speak better to that than Stella Nyanzi whose research and activism bridges gender, sexuality, marginalized communities, LGBTIQ rights, and freedom of expression. In her essay, she describes and analyzes her dissident activism against the president's unfulfilled promise of providing sanitary pads to school-girls in resource-poor communities in Uganda through the #Pads4GirlsUg campaign.

Stella Nyanzi began drafting this essay after being released from prison on bail in May 2017, which she describes below. She had been arrested for her criticism of Janet Museveni in her role as Minister of Education and the President, Yoweri Museveni, related to his unkept promise to provide menstrual pads to girls. In November 2018, she was arrested again and was accused of "cyber harassment and offensive communication," again related to a critical poem she wrote about the Ugandan President, in which she employs a graphic description of his mother's vagina. Amnesty International and PEN International have called for the charges against her to be dropped—thus far to no avail. Stella Nyanzi, a mother of two twin boys and a girl, turned down the option of posting bail explaining that it would be just an illusion of freedom. We jointly finalized her essay with the support of Adrian Jjuuko who discussed the revised version with her when visiting her in prison as she is denied any reading materials. At the time this Handbook goes to print, Stella Nyanzi is still imprisoned. Yet, her powerful words permeate the prison walls—Inga Winkler.

C. Bobel et al. (eds.), *The Palgrave Handbook of Critical Menstruation Studies*, https://doi.org/10.1007/978-981-15-0614-7_42

Introduction: Severely Punished for Menstrual Activism

On April 7, 2017, I presented a pitch about the urgent need for sustainable menstrual health and hygiene to a widely publicized Rotary Club Fellowship in Kampala City. On my way home, ten unidentified armed individuals abducted me from a vehicle. They drove around the dark city for three hours, and then dumped me in a cell at Kira Division Police Station. My captors neither had an arrest warrant, nor informed me of the charges. However, on the long circuitous drive around the city, they repeatedly condemned my acerbic criticisms of the ruling party in Uganda for various failed promises. Specifically, they ridiculed me for publicly challenging the president and his wife for reneging on a promise to provide free sanitary pads to schoolgirls. On the third day of my detention with neither charge nor trial, I finally learnt that I was charged with using social media for public solicitation of money and fundraising for sanitary pads which I distributed across the country.

I was not surprised by these attempts to criminalize my social media activity. The month before my arrest, I had already been subjected to grueling interrogation at the headquarters of the Criminal Intelligence and Investigation Department (CIID). I had also been indefinitely suspended from my permanent employment at Makerere University—allegedly because I insulted the First Lady who was also the Minister of Education. My home was under intense surveillance. My family's movements were monitored and our family vehicle was trailed. My phone communication was tapped. I was barred from boarding a plane to attend an academic conference at the University of Amsterdam and indefinitely banned from traveling abroad. And I received several anonymous death threats via social media, phone, and radio.

After spending three nights in detention, I was arraigned before the Chief Magistrate's Court amidst heightened security and charged with cyber harassment and offensive communication against the president of Uganda. Before I was able to enter a plea, the state prosecutor submitted an application to the court to subject me to involuntary mental examination and treatment before proceeding with the hearing. After I plead not guilty of the charges, the Chief Magistrate refused my application for bail. Instead, he remanded me to Luzira Women's Prison, a maximum-security prison, where I spent 33 days and nights.

The President's Broken Promise

At the height of his presidential campaign in November 2015, incumbent President Yoweri Museveni promised that if he was re-elected into power, he would provide free text books, pens, computers, and sanitary pads to students attending Universal Primary Education in Uganda. He pointed out that access to pads would be crucial for girls to stay in school. At a campaign rally in November 2015, he explained

I want all our daughters to attend school and remain there until they complete their studies. One of the reasons that force our daughters out of school is that when their periods start, they do not have sanitary pads. When they are in class, they soil their dresses. So they run away from school.[1]

However, after being sworn in as president of Uganda, Museveni reneged on his promise of free pads. His wife, Janet Museveni, who was appointed Minister of Education, announced that the Ministry of Education had no money in the 2016–2017 financial year to fulfill the president's promise of free sanitary pads for schoolgirls.

As a Ugandan citizen, a woman who is still menstruating, and a member of the opposition *voce populi*, I intensified my criticism of the regime's failed promises—specifically focusing on the cruel exploitation of schoolgirls' lack of menstrual products. I was horrified by the blatant manipulation of poor girls' menstruation in order to enhance election votes, only to discard the promise after being sworn in office as president. I posted many critical pieces denouncing Janet Museveni's position as 'Mama of the nation,' questioning her intellectual ability to formulate relevant education policies, and challenging Yoweri Museveni's unending list of failed promises. Characteristically deploying sexual metaphors and allegorical comparisons between genitalia and other body parts, my posts were aimed at provoking thought, discussion, debate, and action. The posts went viral.

#Pads4GirlsUg: Popularizing Menstrual Hygiene Education and Materials

After being summoned to the headquarters of the CIID for interrogation, I decided subvert this interrogation into a protest by inviting interested individuals to bring sanitary pads to the CIID offices. I posted a photo on Facebook of the summons letter alongside an appeal for contributions of sanitary pads that I would deliver to schoolgirls. The Facebook post was widely shared and attracted the interest of local journalists, activists, and politicians. Notably, Hon. Betty Bakireke Nambooze, a prominent female Member of Parliament from the opposition party came to the CIID headquarters and gave several interviews condemning President Museveni's failure to fulfill his campaign promise to provide free sanitary pads to schoolgirls and spoke in favor of my criticism of the regime's failures and misgovernance. To those who opposed my language of protest because they claimed it was vulgar, obscene, profane, disrespectful, and thus culturally unacceptable, Hon. Betty Bakireke Nambooze explained why I was particularly authorized to speak in this way. First, my studies focused on sexualities, and second, among the Baganda there is customary license for a Nalongo like myself—the mother of twins—to express herself graphically. She stated,

I have come to see Stella. Where are the Women Members of Parliament who represent women in the districts? They should be here in solidarity with Stella because she is doing their work . . . . People should understand Stella. She is not vulgar just for the sake of being vulgar. This is what she studied. And in the struggle to liberate oneself, one uses what one has. Those are the tools she has. People who think that she speaks vulgar words should also speak their own words because this is an opportunity that they have to raise their decent words besides Stella's words that they say are indecent so that the world chooses their own as the correct ones . . . . Also people should know that Stella is a Nalongo as a mother of twins. I learnt that in Buganda the mothers of twins have special permission to use whatever language to express their grievances . . .[2]

After my five-hour long interrogation, I exited from the offices of the CIID Director and made a televised appeal to individuals to continue donating menstrual hygiene materials. This, in effect, was the public launch of the #Pads4GirlsUg—a citizen-led campaign to collect and distribute pads to schoolgirls in Uganda and to provide menstrual education as a direct challenge to President Museveni's failed promise.

When I first requested citizens to donate pads and funding for #Pads4GirlsUg, I was overwhelmed by the donations that began coming in. People giving money ranged from notable politicians to ordinary Ugandans struggling to make ends meet. For example, Hon. Muhammad Nsereko, a Member of Parliament for Kampala District, enthusiastically encouraged us and gave us 1 million Ugandan Shillings. An elderly farmer sold her chicken and contributed 15,000 Shilling. Random *bodaboda men* (motorcycle taxi drivers) donated as little as 3000 Uganda Shillings—which was sufficient to procure a packet of twelve disposable sanitary pads. The unplanned but extensive publicity through both public media and social media spread the word about the #Pads4GirlsUg campaign, not only in Uganda but around the world. Through these donations, we were able to provide menstrual hygiene materials to almost 3000 schoolgirls, young mothers, and teachers in five districts of Uganda.

Although menstruation is a natural biological process that most adolescent girls undergo, it is widely shrouded in secrecy and shamed in Uganda. When students shared their personal experiences with the #Pads4GirlsUg team, many revealed that they began their menstrual period without having received any relevant education or talks. Given this widespread lack of relevant information, it was important for us to include menstrual health education in the package that we delivered to schools and communities. We gave talks to both male and female students and available staff about menstruation, menstrual health, and hygiene. We distributed age-appropriate booklets to students. Alongside the teaching, we demonstrated how to use, clean, store, and dispose of both the re-useable and disposable sanitary pads. We foregrounded questions and queries of clarification from the students. In each of the school and community meetings, we collected questions and individually addressed these drawing from our expertise, experience, and research. To shatter the stigma and dismantle the secrecy around menstruation, I also introduced the students to a cheerful easy chant, to which we all danced and publicly enacted the motions of wearing menstrual hygiene materials.

BACKLASH AGAINST THE #PADS4GIRLSUG CAMPAIGN: UNDERMINING AND CRIMINALIZING

Our opponents greatly criticized the #Pads4GirlsUg campaign, specifically condemning my insistence on offering menstrual education to both girls and boys because leaving out the boys entrenched their ignorance about the normality of menstruation. In a rare television interview, First Lady and Minister of Education, Janet Museveni, criticized me for publicly teaching girls and boys about menstruation.

> And I thought that we cannot have a teacher really do that. And as an example to children she is teaching. And even go in front of a classroom and try to demonstrate wearing a sanitary pad . . . in front of a classroom. I didn't understand what character that is. So I wanted to tell people of good will who really were hearing that and reading it from I think social media that I forgive her. And I let the people of Uganda judge.[3]

As such, Janet Museveni, shamed me for being 'an educationist' who was publicly using offensive language to criticize the government's failure to fulfill the president's campaign promise. Instead of apologizing to Ugandans because her husband reneged on his promise and appropriated rural poverty to garner more votes by manipulating voters, and instead of addressing the critical issues at hand, she audaciously declared that she was forgiving me for insulting her.

Similarly, Rosemary Nansubuga Seninde, a Member of Parliament and Minister of State for Primary Education in Uganda also critiqued the public discussion of menstruation because it broke taboo and secrecy. In a public address on April 7, 2017 she said,

> For a girl-child to tell people that 'I'm menstruating,' it was a taboo. You would not tell anyone that your period is now. That you are in the period, because it was regarded as something secret. It was only supposed to be known by the mother only. Not anyone else. Not even your sisters. Not even your brothers. Only your mother. You would whisper to the mother. Then she would help you, teach you how to do it, how to clean yourself, how to wash your pants, how to prepare your pad and so on and so forth. And you would go to school. Nobody would even know you are in your period. Are our girls here? Can you stand up our dear girls? Come and learn the culture of Buganda . . . . Nobody is interested in knowing that you are in your period. Who is interested? But what is so shocking and what is so surprising is that Ugandans are waiting for President Museveni to buy pads for Uganda children . . .[4]

These attempts to shame and denigrate our provision of menstrual education to students generated debates within Uganda. Rather than align themselves with sensitization and public education about menstrual issues, it was disheartening that many powerful public officials adopted moralizing tones steeped deep in cultural shaming, taboo, and secrecy. Many public officials criticized our education methods and processes—claiming (correctly) that we were breaking social norms and abusing cultural taboos which they claim forbid the discussion of menstruation outside the intimate circles of paternal aunts and their nieces.

The #Pads4GirlsUg was heavily politicized in Uganda to the point of criminalization. As the person at the forefront of the campaign, I was maligned as an obscene, vulgar, profane, rude, radical, and angry woman who was dangerous to the status quo and a source of pollution for respectable people. My language of protest on Facebook, Twitter, and public media was criticized for being overly graphic and appropriating sexual imagery. My previous political participation in demonstrations and LGBTIQ pride parades were alluded to as proof of my lewdness, immorality, and vulgarity. Therefore, different public officials, religious clerics, traditional leaders, and even some women leaders publicly condemned the fact that I was at the forefront of the #Pads4GirlsUg campaign. Conservative individuals from my ethnic group, clan, religious denomination, and professional affiliations categorically disassociated themselves from me because they argued that a foul-mouthed woman was a-cultural, un-African, ungodly, immoral, unprofessional, and irreverent.

Given that most of the members of the #Pads4GirlsUg Working Group were publicly known as local LGBTIQ rights activists, those who wanted to discourage support for our activities strategically capitalized on associating our project with a 'homosexualizing mission.' I was portrayed as a fervent supporter of homosexuality who was collaborating with local LGBTIQ rights activists to recruit schoolgirls into homosexual behavior. These claims were so preposterous that some homophobic opponents went as far as claiming that our #Pads4GirlsUg campaign was providing 'homosexual pads' that would convert the recipients into lesbians. These accusations of recruitment of minors into homosexuality were further enhanced with assertions that I was taking lesbians into secondary schools under the pretext of distributing free sanitary pads.

Opposition to our campaign increased by seeking to influence radio and television stations, the schools meant to benefit from the campaign, as well as the pad suppliers. Radio and television stations that were previously receptive and supportive of our activities changed their open policies toward us. As the campaign grew in popularity among disgruntled citizens, public officials' narrative about #Pads4GirlsUg became more negative, and public media houses were less willing to engage with us. With time, it became clear that these public media houses were under instructions not to host any members of the #Pads4GirlsUg Working Group.

While some schools were initially scheduled to receive #Pads4GirlsUg products, they later withdrew their willingness to host us. Even in private schools, there was real pressure to comply with the orders from the Ministry of Education which ultimately had the power to withdraw licenses of operation and exclude the school from direct benefits of the government. After negotiations in some districts and schools, we agreed that it would be foolhardy to deny the students the menstrual materials but it would also be reckless to force the school to disobey the instructions. Instead we drove our loaded vehicle outside the school grounds and set up our #Pads4GirlsUg paraphernalia just outside the school gate. All female students and staff members who were interested in participating in the campaign, met us outside the school premises and received a comprehensive package with educational information and materials.

As part of the backlash, the Minister of Education also raised doubts against the distribution processes, distributors' credibility and intentions, and the quality of distributed sanitary pads. Given that the campaign necessarily confronted and challenged the most powerful individuals in Uganda, our sanitary pad suppliers who were initially willing business partners subsequently became fearful to be associated with us. Although inevitable, it was shocking when a key supplier of the cheapest disposable sanitary pads used in the campaign turned down our cash payments for a bulk purchase. It became apparent that suppliers who provided us with menstrual products faced a real or imagined fear of surveillance by state agents, targeting for political witch-hunting, cancelation of trading licenses, imposition of unfair exorbitant taxes, and unnecessary product assessments.

Initially unbeknownst to us, police officers began trailing members of the #Pads4GirlsUg Working Group who mostly appeared in media awareness programs. These police officers went as far as physically tracking us through our mobile phone numbers. As the surveillance intensified, police officers started following, surveilling, and hounding Working Group members. In my case, they went to my home in my absence and interrogated the house help, went to my children's school, trailed the personal vehicles of family members, and were listening into my conversations on my mobile phone. This culminated in several overt and covert threats of death, and the threat of torture of my children if I did not stop the #Pads4GirlsUg campaign. Thus, as described above, it was not surprising when I was inundated with summons for interrogation and was detained without trial.

Refusing to Be Silenced: Radical Rudeness

Public officials sought to undermine our #Pads4GirlsUg campaign by influencing schools meant to benefit from the campaign, radio and TV stations, and pad suppliers. The efforts they undertook to discredit our campaign demonstrate how deeply political—or rather politicized—decisions about pad distribution and menstrual health education are in Uganda. They painted a picture of me and other #Pads4GirlsUg Working Group members as 'spreading' homosexuality, as un-African, and as vulgar and profane. Yet, I see radical rudeness as the only viable means to challenge those in power. As I have said elsewhere, challenging power isn't polite or beautiful. On June 16, 2019, my 45th birthday, I released 45 poems I have written in the Luzira Prison in Kampala where I am imprisoned once again since November 2018, including this one:

FREE SANITARY PADS

In the twilight of the dictatorship
The dictator went really berserk.
He dipped the tight fingers of his iron fist
Into the menstrual blood of poor Ugandan girls.
In a frenzy of cheap popularity
He made another empty promise.
"Vote for me and I will give your daughters free sanitary pads to keep them in school!"
The dictator is a delinquent lying swine.

Gullible peasants with bleeding daughters
Switched their votes to the promise giver.
As they voted they remembered how
His tight fist was covered in fresh menses.
They recalled the belligerent sneer he wore
As he exposed the plight of their bleeding daughters.
"Vote for me and I will give your daughters free sanitary pads to keep them in school!"
The dictator is an exploitative thief of votes.

After swearing back into power
The dictator proved himself a liar.
He publicly washed the menstrual blood
Off his blood-stained thieving hands.
He delegated his drunk wife to lie about
The lack of funds for free sanitary pads.
Does he think we forgot his fake promise?
"Vote for me and I will give your daughters free sanitary pads to keep them in school!"
The dictator's biography is stained with menstrual blood.

Notes

1. Excerpt of verbatim speech reprinted in Wesonga, N., and B. Oketch. 2015. "Museveni Promises Pupils Free Books, Pads." *Daily Monitor*, November 13, 2015.
2. Excerpt of translated interview. Interview telecast on television is available at: https://www.youtube.com/watch?v=W0P8_VH3wME.
3. Transcription of the television interview by the author.
4. Transcription of the public address by the author. For a newspaper report of her public address in Karamoja district, see Kolyangha, M. 2017. "Minister Blames Parents Over High School Dropout Rate in Karamoga." *Daily Monitor*, April 7, 2017.

CHAPTER 43

Addressing Menstruation in the Workplace: The Menstrual Leave Debate

Rachel B. Levitt and Jessica L. Barnack-Tavlaris

INTRODUCTION

Menstrual stigma socially conditions menstruators[1] to conceal menstruation and avoid discussion about it (Johnston-Robledo and Chrisler 2013, 11). Both concealment and secrecy can have detrimental consequences for a menstruator's psychological well-being, such as holding negative attitudes toward menstruation and partaking in self-objectification (Johnston-Robledo and Chrisler 2013; Roberts, Calogero, and Gervais 2018). The oppressive beliefs about and attitudes toward menstruation that permeate heteropatriarchal culture can extend beyond menstruators' psychological health; they may negatively affect their overall well-being including their personal and professional achievements and success, physical health, and the right to feel empowered and experience equality.

Menstrual leave, which allows a menstruator to take time off if they are unable to attend work due to menstruation, is a policy that could affect menstruators in many ways, including their status in the workplace. The question is whether such a policy would benefit or disadvantage a menstruator's well-being. In spite of progressive aims, could menstrual leave policies actually increase discrimination and negative attitudes toward menstruators?

Proponents of menstrual leave argue that such policies have the potential to de-stigmatize discussion of menstruation in the workplace (*CBC Radio* 2017), and may be helpful to those who experience menstrual cycle-related illnesses such as endometriosis and dysmenorrhea. However, there is a paucity of research on the effects of these policies on menstruators and menstrual stigma, as well as on the places of employment that are offering them. Furthermore, there is no research to date on the ways in which menstrual leave may be *counterproductive* to the overall welfare of menstruators through

© The Author(s) 2020

C. Bobel et al. (eds.), *The Palgrave Handbook of Critical Menstruation Studies*, https://doi.org/10.1007/978-981-15-0614-7_43

unintended consequences, such as discrimination in hiring practices, salaries or wages, and promotion. Thus, the purpose of this chapter is to (1) present the controversy surrounding menstrual leave through discussion of the potential implications of such policies, (2) argue that without thoughtful implementation, menstrual leave could have negative effects on menstruators, and (3) highlight the gaps and call for more research in this area. Throughout the chapter, we integrate findings from a study we conducted with a national U.S. sample, which examined attitudes toward and perceptions of menstrual leave (Barnack-Tavlaris et al. 2019).

Existing Menstrual Leave Policies

To date, menstrual leave policies exist across the globe in places such as Japan, Taiwan, China, South Korea, Indonesia, Zambia, and Mexico (Chang et al. 2011; Dan 1986; Forster 2016; Matchar 2014; Worley 2017). Additionally, some professional organizations/companies have chosen to implement menstrual leave, including Coexist (U.K.), Culture Machine (India), Gozoop (India), and the Victorian Women's Trust (Australia). It is difficult to ascertain how these policies are implemented and the extent to which women use the leave because there is often little public data and limited access to human resources policies and procedures which companies typically do not make publically available.

Japan's national policy was introduced in 1947 after American Occupation Forces advised the country to provide women with days off during menstruation (Dan 1986, 8). Japan's menstrual leave policy allows any female worker to use leave if she experiences physical distress from menstruation so severe it makes it difficult to attend work (Dan 1986, 1). The policy does not specify the number of days or whether the leave is paid or unpaid (Dan 1986, 1; Japanese Labor Standards Act 1947). Dan found that women's divisions of labor unions pushed for paid menstrual leave because "they argued that menstruation is a 'barometer' for reproductive ability, and that even women without symptoms ought to take leave to protect their future motherhood" (Dan 1986, 8). However, women who used leave for this purpose were faced with negative consequences such as discrimination and harassment by employers. This resulted in the government proposing to remove menstrual leave from the law to safeguard female workers from discrimination, and led to tension between women's labor unions and the government (Dan 1986, 9–11). While many working women supported the policy in principle, use of the policy varied greatly. Many women rarely used leave due to a variety of factors, including a lack of need, improvements in working conditions (that is, less physical labor demands and more paid holidays), the fact that menstrual leave was unpaid for non-union workers, and making the choice between creating a work life and potential motherhood (Dan 1986, 9). Due to a lack of publicly available data, it is unclear to which extent menstrual leave is presently used among Japan's menstruating population.

Publicly available information about menstrual leave policies in other countries (mostly in Asia) is limited. Three provinces in China (Hubei, Shanxi, and Ningxia) currently offer menstrual leave (Worley 2017), and in Indonesia, women are allotted two days per month of menstrual leave (Matchar 2014). In 2001, South Korean female workers were granted one day of menstrual leave per month (Matchar 2014). Additionally, a menstrual leave amendment was added to Taiwan's Act of Gender Equality in Employment in 2002, which allotted female employees the right to apply for menstrual leave (Chang et al. 2011).

In Zambia, women are given one menstrual leave day per month. The policy is referred to as a "Mother's Day," which stresses women's potential of becoming mothers (Worley 2017). In 2017, Italy proposed a bill for a menstrual leave policy to parliament which would ensure that companies grant three paid days per month to female employees who experience painful periods, as long as they provide a medical certificate from a doctor (Momigliano 2017). In Mexico, women working in the federal court are eligible for one day of menstrual leave if they experience physical complications, however, the policy does not state if this day is offered per year or per month (MH Hub, n.d.a). In Chile, a bill is being proposed to allot women paid leave if they have been diagnosed with endometriosis and/or dysmenorrhea (MH Hub, n.d.b).

At the level of companies and organizations, Coexist, a social enterprise organization in the U.K., with apparent progressive values, offers their female workers the option of one paid day of menstrual leave per month (Quarshie 2017). Bex Baxter, Coexist's former People Development Manager, dealt with dysmenorrhea for years before developing a flexible menstrual leave policy that would support Coexist's female staff (Quarshie 2017). The policy was developed together with Lara Owen, who explains that "Bex and her Board felt it was important to formalise the policy to legitimise and think through women's needs at work, and perhaps also to act as a model and inspiration for other organisations to do the same" (Owen 2018, 26). The following statement appears in an online FAQ about Coexist's policy

> Menstruating staff who opt into the policy are entrusted to respect their cycle and take responsibility for their own well-being. . . they need to check in with their line manager regarding their individual well-being requirements, and in any instance, any time off or alteration to their working hours must be communicated and signed off with their manager. Some roles allow menstruating staff the option to work from home, or alternatively to use a quiet space away from the main office. . . Coexist recognises the importance of difference and debate, therefore should an employee not wish to take part in the policy, they can request to opt out with no judgement or discussion. (Quarshie 2017)

Coexist also offers all of their employees "well-being rooms" where menstruators and non-menstruators alike can take a moment to focus on their health and wellness during work hours (Quarshie 2017). As of 2017, 7 out of 13 menstruating employees have used the menstrual leave policy (Quarshie 2017). Additionally, the company asserts that there was no backlash from men. They report that "men expressed no resentment, and instead said they liked the fact that menstruation was addressed openly, and felt it gave them permission to also adjust their working day to their bodies when needed" (Owen 2018, 28).

In India, two companies offer their employees menstrual leave. Gozoop, a digital communications agency, first introduced their policy in 2017. This policy states that women may work from home one day per month during menstruation.

> The thought behind this policy is that a woman needs to rest her body while on her period, so she has the facility to work from her comfort zone. We take away the stress of travel, crowded environments, uncomfortable chairs, fear of stains, etc. which will enable women to complete work in an efficient and effective manner from home. While drafting this policy we particularly ensured for it to be a work from home so that this does not make the gender less hire-able. . . About 76% of the (female) workforce have used their menstrual leave since implemented in March 2017. (Bansi Raja, Email to author, June 7, 2018)

This policy differs from other menstrual leave policies in that it offers flexibility in the workplace instead of actual time off. For example, employees are able to work from home instead of in the office. Another company, Culture Machine, a digital media firm, offers their female employees one paid day off per month, referred to as 'First Day of Period Leave' (Blush Originals 2017). The company created a YouTube video that showed female employees finding out about their menstrual leave policy for the first time. Women in the video were excited about the policy and felt that it would contribute positively to female employees' well-being (Blush Originals 2017).

In Australia, the Victorian Women's Trust (VWT), a women's advocacy agency with 15 employees, offers their staff a menstrual policy that extends to employees in menopause (the first of its kind); furthermore, they have created a policy template to encourage other companies to offer menstrual leave (Melican and Mountford 2017). Their policy began with the *Waratah Project*, which explored issues surrounding menstruation and menopause in order to end stigma (Melican and Mountford 2017). In an online survey, they found that 58% of respondents across Australia and globally supported the idea of menstrual leave, believing that it would contribute to a better menstrual experience (Melican and Mountford 2017, 1). VWT's menstrual leave policy provides employees with multiple options, including working from home, working in the office in a more quiet and comfortable area, or using one paid menstrual leave day per month (Melican and Mountford 2017). The VWT hopes that their policy will send the message that menstruation and menopause are natural and normal biological processes (Melican and Mountford 2017).

Given the existence of diverse menstrual leave policies across the globe and the current discussion of whether or not to extend such policies, it is important to think critically about both the potential benefits and drawbacks that could result from implementing menstrual leave. In these next sections, we will explore this tension, incorporate some findings from our qualitative study that speak to these potential outcomes, and highlight that more research is needed in order to determine the full extent of the implications of menstrual leave on menstruators' overall well-being and status in the workplace. Further, we argue that the potential impact of these policies will be determined by how and where they are implemented.

Theorizing Menstrual Leave: Using the Framework of Objectification and Ambivalent Sexism

The discussion on menstrual leave must be understood in the context of societies in which menstruation has been used to marginalize and oppress women and other menstruators through the process of objectification and the ideology of sexism. Objectification dehumanizes women by reducing their worth to the surface of their bodies (Bordo 1993, 309; Fredrickson and Roberts 1997). Objectification is achieved under the male gaze, an invisible yet unavoidable presence that encompasses heteropatriarchal, and often Eurocentric, views of how women are expected to behave and appear that disassociates women from their own well-being, success, needs, and desires (Bartky 1988; Roberts, Calogero, and Gervais 2018; Roberts et al. 2002). In order to preserve status in a heteropatriarchal society, women must internalize the male gaze and practice self-monitoring body disciplines such as dieting, makeup, the use of menstrual care products (often advertised as "feminine hygiene products"), and by concealing all matters of menstruation (Bartky 1988; Johnston-Robledo and Chrisler 2013; Roberts et al. 2002). Researchers have found that in order for women to meet the unrealistic standards of femininity, they must choose to "adopt a menstrual etiquette that reinforces fear about the female body and its functions" (Grose and Grabe 2014, 679). As a result, unconcealed menstruation becomes inherently unfeminine and socially unacceptable.

Much like objectification, sexism also uses menstruation to marginalize women. Attitudes toward and beliefs about women can be simultaneously hostile and benevolent; this interlocking ideological belief system is referred to as ambivalent sexism (Forbes et al. 2003; Glick and Fiske 1996, 2001). Hostile sexism is the explicit and aggressive or violent prejudice and discrimination against women (for example, rape culture and the belief that women are inferior to men; Forbes et al. 2003; Glick and Fiske 1996). Hostile sexist beliefs and attitudes elicit negative responses toward attempts to dismantle patriarchal systems (Good and Rudman 2010, 482). Benevolent sexism is a way of viewing sex and gender in stereotypical, traditional ways (Forbes et al. 2003; Glick and Fiske 1996). Benevolent sexism implies that menstruation is a sign of womanhood and feminine fragility (Forbes et al. 2003; Glick and Fiske 2001), and reinforces the stereotype that women's purpose in life is to bear children. Benevolent sexism can be hard to detect and may not activate negative or hostile feelings, however, it does perpetuate a patriarchal system of gendered power dynamics (Forbes et al. 2003; Glick and Fiske 1996; Good and Rudman 2010). Menstrual leave may then masquerade as a way of protecting the well-being of menstruators, and like benevolent sexism, may still work to undermine menstruators' well-being. An example of benevolent sexism occurring in the context of menstrual leave is evident in Japan's menstrual leave policy, which was created to protect women's capacity to bear children (Dan 1986), and in Zambia's policy, which highlights women's presumed role as mothers/future mothers. In the following sections, we use the framework of objectification and ambivalent sexism to discuss the potential benefits and drawbacks of menstrual leave.

POTENTIAL BENEFITS OF MENSTRUAL LEAVE

In this section, we consider whether menstrual leave has the potential to positively contribute to the well-being of menstruators in two key interrelated ways: by promoting menstrual health and by de-stigmatizing menstruation.

Menstrual leave may benefit the health of individuals who menstruate, including those who experience mild, moderate, or severe discomfort from menstruation, and those who experience menstrual cycle-related illnesses such as dysmenorrhea, endometriosis, ovarian cysts, and mood disorders. For some women, menstrual symptoms may interrupt their daily lives, making it more difficult to participate in normal activities. For example, in one cross-sectional study with 762 participants, 71.5% believed that dysmenorrhea was a normal part of women's life, and on average, reported that their menstrual symptoms moderately affected their daily lives (Chen, Kwekkeboom, and Ward 2016, 268–69). However, the social unacceptability of the discussion of menstrual symptoms can result in societal pressure to keep menstrual distress a secret from coworkers and health professionals alike (Johnston-Robledo and Chrisler 2013). Therefore, menstrual leave may offer women the opportunity to speak up about their menstrual cycle-related health issues, and to take time to recover or seek treatment.

Some of the existing menstrual leave policies were clearly written with these positive intentions in mind (for example, Coexist, Gozoop, Culture Machine, and VWT). For example, some policies provide women with the opportunity to take leave *or* provide workplace flexibility so that women can reserve their medical leave for non-menstrual related illness. Additionally, these policies promote well-being by providing additional opportunities to care for oneself (for example, "well-being rooms," workplace flexibility, normalized discussion of menstruation, et cetera). In our study conducted with a national U.S. sample ($N=600$), participants were asked what they thought about a menstrual leave policy and the effect the policy would have in the

U.S.: 23% of participants believed there would be only positive effects for women in the workplace including time needed to cope with symptoms, and overall improved well-being (Barnack-Tavlaris et al. 2019).

Some proponents of menstrual leave argue that such policies have the potential to benefit women by reducing stigma and encouraging more open discussion of menstruation (*CBC Radio* 2017). For example, as discussed earlier, at Coexist, female employees are encouraged to discuss menstruation with their line managers (Quarshie 2017) and menstrual leave is discussed in employee meetings among all staff members (Owen 2018; Quarshie 2017). Discussion of menstruation has the potential to be framed in a light that normalizes or neutralizes it, and therefore de-stigmatizes the conversation surrounding it and thus has the potential to reduce (self)objectification and the objectifying gaze of others.

Another positive implication for menstrual leave may aid individuals who identify as gender queer[2]/non-binary or as transmen. If offered this type of leave, gender queer/non-binary individuals and transmen may feel an increased amount of safety because menstruation may be a time in which these individuals face increased amounts of transphobia and other types of gender discrimination. Chrisler and colleagues (2016) examined transmen's attitudes toward menstruation, and found that some participants reported that they do not feel safe and/or comfortable using a public restroom when they are menstruating because they fear that menstruation will no longer be concealed. If their menstrual status is revealed, this can result in both discrimination and violence. By having the option to use menstrual leave, transmen and individuals who identify as gender queer/non-binary may be able to avoid having their menstrual status revealed (that is, 'outed,' discussed in more depth below).

POTENTIAL DRAWBACKS OF MENSTRUAL LEAVE

Though menstrual leave may benefit menstruators in some ways, it may also produce a number of unintended negative implications for menstruators, such as perpetuating sexist beliefs and attitudes, contributing to menstrual stigma and perpetuating gender stereotypes, negatively impacting the gendered wage gap, and reinforcing the medicalization of menstruation.

Perpetuating Sexist Beliefs and Attitudes

Both hostile and benevolent sexism have been found to result in evaluating a menstruating woman more negatively than a non-menstruating woman. For example, Roberts and colleagues (2002, 136) found that participants rated a woman more negatively (for example, less competent, less likeable, and more irrational) when they believed that she was menstruating. Furthermore, Forbes and colleagues (2003, 60–61) found that both female and male college participants rated a menstruating woman as more impure, less sexually desirable, more irritable, and rated her higher on adjectives associated with neuroticism than non-menstruating women. By extension, workers who take advantage of menstrual leave policies could be perceived more negatively than those who do not use the policy. This can be explained by benevolent and hostile sexism which assume that women do not belong in the workplace. Additionally, benevolent sexism may undermine activist intentions by pacifying intentions for progressive change (Becker and Wright 2011, 62, 74). As a result, benevolent sexist attitudes and beliefs may undermine women's and other menstruators' desire to change existing gendered power dynamics.

Contributing to Menstrual Stigma and 'Outing' in the Workplace

Menstrual leave brings menstruation to light in the workplace. As such, it violates the cultural mandate to keep menstruation concealed. Thus, those who opt to use menstrual leave may put themselves at risk because they may violate cultural norms about menstrual nondisclosure. This unmasking of menstruation in public spaces might contribute to rather than dismantle menstrual stigma. Menstrual leave may also perpetuate the belief that menstruation is something about which to be ashamed and to be kept private, and therefore avoided in the workplace.

The term 'outing' is commonly used to refer to an occurrence in which a member of the LGBTQ+ community's gender identity or sexual orientation is revealed by another person or entity without their permission (Hunter 2007, 154), which may have a number of negative consequences including discrimination, shaming, and violence. It can also be used to refer to the disclosure of one's menstrual status without the menstruator's consent. Enforcing a work place practice concerning menstruation may require "explicit and public definitions for matters that usually remain implicit and private" (Dan 1986, 2). When an employee chooses to use their leave, at the very least, their supervisor will know that they are menstruating, a disclosure that could activate objectification, sexism, and discrimination, even if at an implicit level.

It is also important to acknowledge the diverse and intersectional way in which prejudice and discrimination play out, which may prevent all menstruators from having equal access to menstrual leave. Not all menstruators are women, therefore, fair policies would need to include queer individuals. At the same time, non-binary, gender queer, and trans menstruators may be uncomfortable disclosing their menstrual status; doing so may even result in endangerment (for example, violence toward queer individuals). As such, menstrual leave policies could 'out' queer individuals who must disclose their menstrual status to use the leave provided, or it may, conversely, prevent them from feeling entitled to using menstrual leave. Additionally, there is no current information on whether menstrual leave has been or will be offered to menstruators who do not identify as women. Therefore, fair policies must be inclusive and provide options for confidential disclosure.

Gender and Menstruation in the Workplace

Due to societal expectations of women related to childcare and domestic duties, their absence from work is often ascribed to their gender role (Patton and Johns 2007). Thus, menstrual leave may have the potential to perpetuate the assumptions that women are unfit for the workplace and thus, justify gender discrimination. Women may actively choose not to use menstrual leave (if provided) in an attempt to gain workplace advancement (that is, promotions, job training, and social networks) and avoid negative social judgments (Grose and Grabe 2014). For example, it has been reported that women in China are hesitant to use menstrual leave days because they fear that the disclosure of their menstrual status could reinforce stereotypes of female fragility and unproductivity (Forster 2016).

In our study on the anticipated effects of menstrual leave, 49.3% of participants thought the policy would have *only* negative effects in the U.S. including discrimination and creating division between genders. Furthermore, 20.7% of participants believed such policies were unfair to men, or that people would perceive them as unfair to men (Barnack-Tavlaris et al. 2019).

As a result of the same dynamics, menstrual leave may also contribute to the gendered wage gap and the lack of women's advancement in the workplace, a.k.a, 'the glass ceiling.' As previously discussed, researchers have found that women's absence from work perpetuates stereotypes of women as less worthy and reliable employees (Patton and Johns 2007, 1587–601), thus undermining their career development progress.

Men could even use menstruators' absence to their own advantage, undermining menstruators' success in the workplace. In a study on the gendered impact of long working hours on managers, Simpson (1998, 37) found that men purposefully use women's absence from work to progress their own position; this creates a competitive masculine culture that purposefully excludes, silences, and appropriates women's work and gives women a tokenized status in the workplace. Simpson (1998, 43) also found that men used women's absences as a way to make women appear less committed to their jobs, which helped men increase their own workplace resources while simultaneously hindering the achievements of their female coworkers.

The Medicalization of Menstruation

The medicalization of menstruation, which portrays menstruation as a disease, contributes to convincing women that their bodies are in need of 'fixing' with the use of various pharmaceutical products (Stein Deluca 2017, 37–38). The medicalization of menstruation also objectifies the female body by commodifying it through capitalist maintenance and care. The patriarchy profits off the menstruating body by making menstruators believe that they need to buy specific menstrual cycle-related products (for example, diuretics for bloating and high tech menstrual product innovations such as super absorbent underwear or scented menstrual pads and tampons) for their bodies to remain and/or become socially acceptable, as such furthering the objectification of women.

Through the medicalization of menstruation, menstrual leave may perpetuate the idea that menstruation is 'debilitating' for all or most women, and thus women are not capable of working (efficiently or at all) while menstruating. Menstrual leave may reinforce beliefs that menstruation is an illness that impedes all women, making it appear as a health issue that is associated with both symptoms and risks (Barnack-Tavlaris 2015). In Barnack-Tavlaris et al.'s (2019) study, some participants resisted the idea that menstruation is debilitating, and used that argument to explain why they would not support such a policy. For example, approximately 11% of participants' qualitative responses pertained to menstrual leave being unnecessary (for example, menstruation is not a disease and should not be treated as such, and that such policies were "ridiculous"). On the other hand, one could argue that creating a separate category of menstrual leave that distinguishes it from medical or sick leave could combat the medicalization of menstruation.

Implementing Menstrual Health Policies and Alternatives

The potential implications of menstrual leave may be determined by how a policy is both worded and implemented; that is, the intentions that underlie the creation of such a policy can determine the outcome. For example, discussion about menstruation among all employees may foster support for women's reproductive health and reproductive rights, and may help educate people about menstruation since so many misconceptions circulate throughout heteropatriarchal culture (such as the "hormone myth," the idea that women became hormonal monsters when menstruating, and demonstrate characteristics such as unreliable decision-making; Stein Deluca 2017, 1–7). Additionally, discussion of menstruation may produce more positive, or at least neutral, attitudes toward menstruation that do not elicit sexist attitudes, and this may in turn reduce self-objectification (Johnston-Robledo et al. 2007). Therefore, the language used in menstrual leave policies must be created with the intention of ensuring that the policy fosters normalization and open discussion of menstruation, and does not perpetuate objectification, sexism, and patriarchal misconceptions about menstruation. Among the examples presented above, it is apparent that Coexist, Gozoop, Culture Machine, and VWT used progressive feminist values to create their policies in order combat menstrual stigma, sexism, and objectification.

While the effects of menstrual leave policies on menstruator's overall well-being are unclear and in large part depend on *how* policies are developed and implemented, what seems evident is that some menstruators would benefit from workplace flexibility more generally (for example, more time off, the ability to work from home, customized work schedules). The proclaimed need for menstrual leave may be an opportunity for some societies and employers to reevaluate attitudes surrounding absenteeism and work ethic, and consider the creation of sick, parental, and vacation leave policies that are responsive to worker needs, including the provision of additional paid leave. Indeed, in order for any type of leave (that is, menstrual, sick, parental, et cetera.) to effectively aid employee's well-being, some countries may need to reshape their attitudes toward workplace productivity and presenteeism (working while sick).

In one study of Dutch workers, Hansen and Anderson (2008, 956) found that over 70% of employees worked while sick at least once over the course of a year. The researchers note that possible predictors of presenteeism include employment conditions (such as job insecurity) and financial need (Hansen and Anderson 2008, 958), which implies that employees must choose financial stability and job security over their health and well-being. Individuals should not feel pressured to attend work when their health and wellness are at risk. Many cultures stress presenteeism and hyper-productivity in the workplace over health and well-being, which can endanger employees' health (Callen, Linley, and Niederhauser 2013, 1316). In workplaces that stress these qualities, the ideal employee becomes defined as someone who is always available to their employer and does not "bring family concerns or other outside matters to work" (Stein Deluca 2017, 74).

Alternatives to menstrual leave to ensure menstruators' well-being may include equipping workplaces with rest/break rooms for anyone who is feeling under the weather—physically, mentally, or emotionally, such as "well-being rooms," so that employees can "take some time away from work, whether this is for a cup of tea on the sofa or an hour away from their desk to rest . . . or to work in a quiet space away from the busy office environment, if needed" (Quarshie 2017). Relatedly, workplaces should consider stocking bathrooms or breakrooms with menstrual products, hot pads, and pain relievers.

Conclusion

There is a paucity of research that explores the ways in which menstrual leave may potentially benefit or disadvantage menstruators in the workplace, but plenty is known about how gender oppression shapes perceptions of menstruators. As researchers explore the dimensions of menstrual leave policies, we recommend a number of considerations. First, future research must be intersectional and attuned to diverse menstruators and varied menstrual experiences. Second, it should examine the extent to which menstrual leave reinforces the medicalization of menstruation. For example, does the presence of menstrual leave policies inadvertently pressure menstruators to either medically suppress their cycles or avoid taking leave? Further, do coworkers construe a woman's behaviors differently and more negatively (for example, by attributing emotions or behaviors to menstruation/PMS) when she takes menstrual leave? In order to create a policy that fully contributes to menstruators' well-being rather than undermining it, menstrual stigma and myths must be challenged and dismantled actively both inside and outside of the workplace.

Assessing the pros and cons of menstrual leave can serve as an entry point to discussions about workplace culture and accommodations more generally. Are sick leave and parental leave provisions ample? How else can workplaces be designed with diverse worker needs in mind? It is important to consider if flexibility can be extended to other stigmatized health conditions (for example, irritable bowel syndrome, mental health disorders, HIV), so that the menstrual policies do not perpetuate discrimination. In advance of a body of research, companies and organizations who choose to implement menstrual leave policies should consider including menstrual discrimination in their anti-discrimination policies, along with training.

In order for menstrual leave policies to be fully beneficial to menstruators' well-being, cultural beliefs about and attitudes toward menstruation must change. Menstrual stigma must be continuously challenged and heteropatriarchal beliefs dismantled. Menstrual leave policies will only advance gender equality if they are adopted in spaces committed to challenging menstrual stigma and dismantling gender-based oppression.

NOTES

1. The term 'menstruator' is used with purpose; this term acknowledges that people who menstruate are not all women, and that not all women menstruate.
2. Queer is an umbrella term used to recognize and include the many identities within the gender spectrum.

REFERENCES

Barnack-Tavlaris, Jessica L. 2015. "The Medicalization of the Menstrual Cycle: Menstruation as a Disorder." In *The Wrong Prescription for Women: How Medicine and Media Create a 'Need' for Treatments, Drugs, and Surgery*, edited by Maureen C. McHugh and Joan C. Chrisler, 61–75. Santa Barbara: Praeger.

Barnack-Tavlaris, Jessica L., Kristina Hansen, Rachel B. Levitt, and Michelle Reno. 2019. "Taking Leave to Bleed: Perceptions and Attitudes toward Menstrual Leave Policy." *Health Care for Women International* 40 (12): 1355–1373. https://doi.org/10.1080/07399332.2019.1639709.

Bartky, Sandra L. 1988. "Foucault, Femininity, and the Modernization of Patriarchal Power." In *Femininity and Foucault: Reflection of Resistance*, edited by I. Diamond and L Quinby, 93–111. Boston: Northeastern University Press.

Becker, Julia C., and Stephen C. Wright. 2011. "Yet Another Dark Side of Chivalry: Benevolent Sexism Undermines and Hostile Sexism Motivates Collective Action for Social Change." *Journal of Personality and Social Psychology* 101 (1): 62–77.

Blush Originals. 2017. "First Day of Period Leave [Video File]." Retrieved from https://www.youtube.com/watch?time_continue=11&v=avPgUxGC1Sg.

Bordo, Susan. 1993. "The Body and the Reproduction of Femininity." In *Unbearable Weight: Feminism, Western Culture, and the Body*, edited by Susan Bordo, 309–26. Berkeley: University of California Press.

Callen, Bonnie L., Lisa C. Lindley, and Victoria P. Niederhauser. 2013. "Health Risk Factors Associated with Presenteeism in the Workplace." *Journal of Occupational and Environmental Medicine* 55 (11): 1312–17.

CBC Radio. 2017. "Should Women Get Paid Menstrual Leave?" *CBC Radio.* Retrieved from https://www.cbc.ca/radio/thecurrent/the-current-for-june-14-20171.4158414/should-women-get-paid-menstrual-leave-1.4158421.

Chang, Chueh, Fen-Ling Chen, Chu-Hui Chang, and Ching-Hui Hsu. 2011. "A Preliminary Study on Menstrual Health and Menstrual Leave in the Workplace in Taiwan." *Taiwan Gong Gong Wei Sheng Za Zhi* 30 (5): 436–50.

Chen, Chen X., Kristine L. Kwekkeboom, and Sandra E. Ward. 2016. "Beliefs about Dysmenorrhea and Their Relationship to Self-Management." *Research in Nursing & Health* 39: 263–76. https://doi.org/10.1002/nur.21726.

Chrisler, Joan C., Jennifer A. Gorman, Jen Manion, Michael Murgo, Angela Barney, Alexis Adams-Clark, Jessica R. Newton, and Meaghan Mcgrath. 2016. "Queer Periods: Attitudes toward and Experiences with Menstruation in the Masculine of Centre and Transgender Community." *Culture, Health, & Sexuality* 18 (11): 1238–50.

Dan, Alice. 1986. "The Law and Women's Bodies: The Case of Menstruation Leave in Japan." *Healthcare for Women International* 7 (1–2): 1–14.

Forbes, Gordon B., Leah E. Adams-Curtis, Kay B. White, and Katie M. Holmgren. 2003. "The Role of Hostile and Benevolent Sexism in Women's and Men's Perceptions of the Menstruating Woman." *Psychology of Women Quarterly* 27 (1): 58–63.

Forster, Katie. 2016. "Chinese Province Grants Women Two Days 'Period Leave' a Month." *Independent.* Retrieved from http://www.independent.co.uk/news/world/asia/chinaperiod-leave-ningxia-womentwodays-a-month-menstruation-a7197921.html.

Fredrickson, Barbara L., and Tomi-Ann Roberts. 1997. "Objectification Theory: Toward Understanding Women's Lived Experiences and Mental Health Risks." *Psychology of Women Quarterly* 21 (2): 173–206.

Good, Jessica J., and Laurie A. Rudman. 2010. "When Female Applicants Meet Sexist Interviewers: The Costs of Being a Target of Benevolent Sexism." *Sex Roles* 62(7–8): 481–93.

Glick, Peter, and Susan T. Fiske. 1996. "The Ambivalent Sexism Inventory: Differentiating Hostile and Benevolent Sexism." *Journal of Personality and Social Psychology* 70 (3): 491–512.

———. 2001. "An Ambivalent Alliance: Hostile and Benevolent Sexism as Complementary Justifications for Gender Inequality." *American Psychologist* 56: 109–18.

Grose, Rose G., and Shelly Grabe. 2014. "Sociocultural Attitudes Surrounding Menstruation and Alternative Menstrual Products: The Explanatory Role of Self-Objectification." *Healthcare for Women International* 35 (6): 677–94.

Hansen, Claus D., and Johan H. Andersen. 2008. "Going Ill to Work—What Personal Circumstances Attitudes and Work-Related Factors Are Associated with Sickness Presenteeism?" *Social Science & Medicine* 67 (6): 956–964.

Hunter, S. 2007. *Coming Out and Disclosures: LGBT Persons across the Life Span.* Binghamton, NY: The Haworth Press, Inc.

Johnston-Robledo, Ingrid, and Joan C. Chrisler. 2013. "The Menstrual Mark: Menstruation as Social Stigma." *Sex Roles* 68 (1–2): 9–18.

Johnston-Robledo, Ingrid, Kristin Sheffield, Jacqueline Voight, and Jennifer Wilcox-Constantine. 2007. "Reproductive Shame: Self-Objectification and Young Women's Attitudes toward Their Reproductive Functioning." *Women & Health* 46 (1): 25–39.

Matchar, Emily. 2014. "Should Paid 'Menstrual Leave' Be a Thing?" *The Atlantic.* Retrieved from https://www.theatlantic.com/health/archive/2014/05/should-women-get-paid-menstrual-leave-days/370789/.

"Labor Standards Act". 1947. Accessed February 24, 2019. Japan. https://www.ilo.org/dyn/travail/docs/2021/Labor%20Standards%20Act%20-%20www.cas.go.jp%20version.pdf.

Melican, Casimira, and Grace Mountford. 2017. "Why We've Introduced a Menstrual Policy and You Should Too [Web Log Post]." *Victorian Women's Trust.* Retrieved from https://www.vwt.org.au/blog-menstrual-policy/.

MH Hub. n.d.a. "Mexico: Agreement by Which Public Servants Are Granted a One-Day Leave of Absence Due to Complications of a Physiological Nature". Accessed February 24, 2019. https://mhhub.org/hive/policy/menstrual-leave/recMpMmKwVUJsgWWd/.

———. n.d.b. "Chile: Menstrual Law." Accessed February 24, 2019. https://mhhub.org/hive/policy/menstrual-leave/recT8pla1LMHmAV6K/.

Momigliano, Anna. 2017. "Italy Set to Offer 'Menstrual Leave' for Female Workers." *The Independent.* Retrieved from https://www.independent.co.uk/news/world/europe/italy-menstrual-leave-reproductive-health-women-employment-a7649636.html.

Owen, Lara. 2018. "Menstruation and Humanistic Management at Work: The Development and Implementation of a Menstrual Workplace Policy." *e-Organizations & People* 25 (4): 23–31.

Patton, Eric, and Gary Johns. 2007. "Women's Absenteeism in the Popular Press: Evidence for a Gender-Specific Absence Culture." *Human Relations* 60 (11): 1579–612.

Quarshie, Adam. 2017. "Coexist Pioneering Period Policy [Web Log Post]." *Hamilton House.* Retrieved from https://www.hamiltonhouse.org/coexist-pioneering-period-policy/.

Roberts, Tomi-Ann, Rachel M. Calogero, and Sarah J. Gervais. 2018. "Objectification Theory: Continuing Contributions to Feminist Psychology." In *APA Handbook of the Psychology of Women: Vol. 1: History, Theory, and Battlegrounds.* Washington, DC: American Psychological Association.

Roberts, Tomi-Ann, Jamie L. Goldenberg, Cathleen Power, and Tom Pyszcynski. 2002. "'Feminine Protection': The Effects of Menstruation on Attitudes towards Women." *Psychology of Women Quarterly* 6: 131–39.

Simpson, Ruth. 1998. "Presenteeism, Power, and Organizational Change: Long Hours as a Career Barrier and the Impact on Working Lives of Women Managers." *British Journal of Management* 9: 37–50.

Stein Deluca, Robyn. 2017. *The Hormone Myth: How Junk Science, Gender Politics & Lies about PMS Keep Women Down.* Oakland: New Harbringer Publications, Inc.

Worley, Will. 2017. "The Country Where All Women Get a Day Off Because of Their Period." *Independent*. Retrieved from https://www.independent.co.uk/news/world/africa/zambia-period-day-off-women-menstruation-law-gender-womens-rights-a7509061.html.

CHAPTER 44

Monitoring Menstrual Health in the Sustainable Development Goals

Libbet Loughnan, Thérèse Mahon, Sarah Goddard, Robert Bain, and Marni Sommer

INTRODUCTION: THE SUSTAINABLE DEVELOPMENT GOALS 2015–2030

This chapter examines the relationship between menstruation and the Sustainable Development Goals (SDGs), which are the main global development framework for 2015–2030. We examine this relationship by focusing on monitoring progress toward these goals.

The global development community in partnership with national governments and global institutions proposed "a universal call to action to end poverty, protect the planet, and ensure that all people enjoy peace and prosperity" (UNDP 2018, para. 2) as represented by the SDGs, which were adopted by United Nations Member States in 2015. The SDGs "seek to realize the human rights of all and to achieve gender equality and the empowerment of all women and girls" (Sustainable Development Knowledge Platform 2015, 1). They envisage a transformative approach to global development and advancement and have embedded within them the human rights principles of universality, indivisibility, and interdependence (United Nations 2015b, 1). The broad development agenda is codified into 17 SDGs (Fig. 44.1).

© The Author(s) 2020
C. Bobel et al. (eds.), *The Palgrave Handbook of Critical Menstruation Studies*, https://doi.org/10.1007/978-981-15-0614-7_44

Goal 1. No Poverty
Goal 2. Zero Hunger
Goal 3. Good Health and Well-Being
Goal 4. Quality Education
Goal 5. Gender Equality
Goal 6. Clean Water and Sanitation
Goal 7. Affordable and Clean Energy
Goal 8. Decent Work and Economic Growth
Goal 9. Industry, Innovation and Infrastructure
Goal 10. Reduced Inequalities
Goal 11. Sustainable Cities and Communities
Goal 12. Responsible Consumption and Production
Goal 13. Climate Action
Goal 14. Life Below Water
Goal 15. Life on Land
Goal 16. Peace, Justice and Strong Institutions
Goal 17. Partnerships for the Goals

Fig. 44.1 List of Sustainable Development Goals (*Source* United Nations Department of Public Information 2019)

WHY THE SDGS MATTER FOR MENSTRUAL HEALTH

In evaluating progress on the SDGs, analyzing whether the menstruation needs of populations are being met can yield valuable information. First, the menstrual cycle's regular occurrence between puberty and menopause is an important indicator of good reproductive health. Second, overcoming menstrual-related stigma and ensuring that women and girls can manage their menstruation is key to achieving SDGs that touch on women's and girls' comfort, agency, participation, safety, well-being, and dignity.

The ability of people who are menstruating (for example, adolescent girls, women, and also others who do not identify as women or girls and menstruate[1]) to manage monthly menstruation requires three distinct components: tailored *assets* (for example, knowledge, confidence, and awareness about how to manage menstruation; an adequate supply of hygienic absorbent materials to collect menstrual blood; water and soap for washing the body), *services* (for example, information and education), and *spaces* (for example, safe and convenient sanitation facilities to change and dispose of materials with privacy and dignity as often as necessary for the duration of a menstrual period) (WHO/UNICEF JMP 2015, 45; Patkar 2012). These collective components have been termed "menstrual hygiene management" (MHM) by the global community of experts working to improve access to drinking water, sanitation, and hygiene (the WASH sector). The major domains of life—including home, schools, health centers, and public sanitation facilities—should be supportive of MHM by addressing the need for these assets, services, and spaces. But often they do not. This negatively affects women and girls, and all those who menstruate, throughout their lives. Ongoing stigma around menstruation hinders action on adequately addressing the three components in many contexts, including the perpetuation of many menstrual-related restrictions in certain cultural contexts.

The SDGs are translated into policies and programs at global, national, and local levels; thus, whether the relevant SDGs are supportive of menstruation-related needs or not will be reflected in myriad ways in adolescent girls' and women's lives. As Dr. Jyoti Sanghera, Head of Economic, Social and Cultural Rights at the Office of the High Commissioner for Human Rights, articulates: "Stigma around menstruation and menstrual hygiene is a violation of several human rights," including "the right to human dignity, but also the right to non-discrimination, equality, bodily integrity, health, privacy and the right to freedom from inhumane and degrading treatment from abuse and violence" (George 2013, 5). Menstruation-related tailored assets, services, and spaces are essential to meeting the rights articulated in the SDGs and addressing the persistent stigma women and girls face.

The Scope of This Chapter: Indicators for Monitoring

The 17 SDGs are represented by 169 targets and 232 indicators. The targets, designed to guide development efforts until 2030, were negotiated by UN member states taking into consideration inputs from a wide range of stakeholders, with progress monitored by one or two globally applicable and measurable indicators per target. The indicators are the main monitoring apparatus of the SDGs, so we focus on them in this chapter. They represent an agreed-upon evidence base by which progress on the targets and goals will be monitored. This monitoring is essential to ensure that real progress is being made at the local, national, and global levels. It also will assist governments and development partners with guiding investments and targeted action.

Our analysis provides perspective on the extent to which the indicators adequately capture whether adolescent girls' and women's menstruation-related needs are being met. We also examine the potential of other large-scale menstruation-related monitoring initiatives to help understand whether the SDGs are being achieved and how they might address weaknesses of the existing indicator framework. Through this assessment, we offer a goal-by-goal overview of the national-scale monitoring that substantiates the connections between menstrual health and the SDGs. We do not attempt to assess each of these connections, but rather to identify the strongest links and stimulate discussion so that we may collectively strengthen linkages that will lead to progress on menstrual health and the SDGs.

The Links Between Monitoring Progress Toward the SDGs and Menstruation

As MHM is not explicitly mentioned in any of the SDGs, targets, or indicators, global development actors need to more explicitly articulate why monitoring on MHM contributes to the assessment of progress on multiple goals. Here we articulate links between menstrual health and SDG indicators under goals that address the following issues: poverty, health, education, gender equality, water and sanitation, and others.

SDG 1, and particularly Target 1.4, recognize the gendered nature of poverty. The phrasing of the corresponding Indicator 1.4.1 is gender-neutral (see Fig. 44.2). However, its definition and monitoring should not be. First, disaggregation by sex is an explicit aim for the monitoring of all

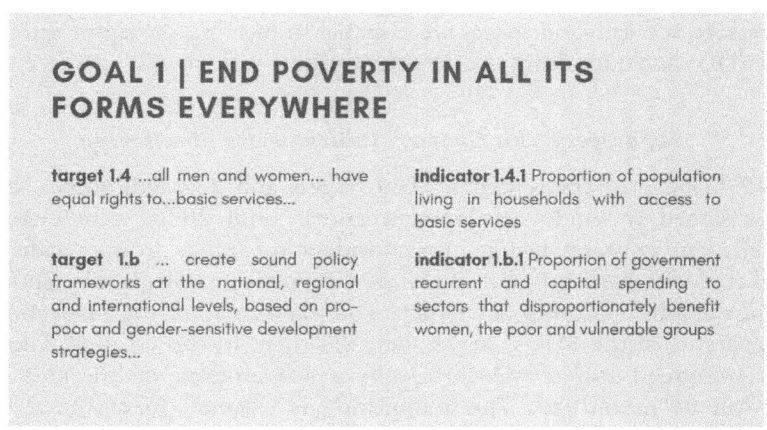

Fig. 44.2 SDG 1 indicators with the highest relevance to menstruation (*Source* UN Statistics Division [2018]. Design Credit: Sydney Amoakoh 2019)

SDGs (United Nations 2015b, 37). Second, the metadata explaining the rationale for the indicator recalls that member states commit to "ensuring that these basic services are responsive to the rights and needs of women" (United Nations 2016, 8).

Despite this, the definition of "basic services" does not appear to specify services that directly respond to women's and girl's menstrual hygiene needs. Basic services include: "drinking water, sanitation and hygiene, energy, mobility, waste collection, health care, education and information technologies" (UN Statistics Division 2018). The sanitation ("the use of improved facilities that are not shared with other households") and hygiene ("handwashing facility on premises with soap and water") aspects are linked to SDG 6 definitions of "access to basic services," suggesting that unless these explicitly incorporate services (and assets and spaces) needed to manage menstruation then Goal 1 will fall short of reflecting the basic needs of menstruating girls and women.

Indicator 1.b.1 measures government spending that disproportionately benefits women, which could be interpreted with relevance to menstruation. Increasing government spending on any of the critical components of MHM could be claimed as progress on MHM—for example, dissemination of information about how to manage menstruation. Reducing taxes on sanitary products might also arguably be considered as progress (Fig. 44.3).

The focus of the denominator in Indicator 3.7.1 is the ability to reproduce, using an age bracket typifying the ages women menstruate. The indicator will miss important information about women whose menarche comes earlier, or menopause later. Indicator 3.8.1 brings WASH and health together, with the custodian agency (World Health Organization, or WHO) making an index from 14 indicators of universal health coverage. The

GOAL 3 | ENSURE HEALTHY LIVES & PROMOTE WELL-BEING FOR ALL AT ALL AGES

target 3.7...ensure universal access to sexual and reproductive health-care services...including for family planning, information and education, and the integration of reproductive health into national strategies and programmes

indicator 3.7.1 Proportion of women of reproductive age (aged 15-49 years) who have their need for planning satisfied with modern methods

target 3.8 Achieve universal health coverage, including...access to quality essential health-care services...for all

indicator 3.8.1 Coverage of essential health services

Fig. 44.3 SDG 3 indicators with the highest relevance to menstruation (*Source* UN Statistics Division [2018]. Design Credit: Sydney Amoakoh 2019)

NOTES | A toilet can be considered to have menstrual hygiene facilities if it

- has a bin with a lid on it for disposal of used menstrual hygiene products, and
- water and soap available in a private space for washing

Fig. 44.4 Measurements already operationalized for SDG monitoring in healthcare facilities (*Source* WHO/UNICEF JMP [2016], 5. Design Credit: Sydney Amoakoh 2019)

metadata notes that the WASH aspect of this broad concept is basic WASH as defined under SDG 6 (UN Statistics Division 2018).

MHM is clearly of utmost importance to women's health, and despite the links above, it is not directly mentioned in SDG 3. It is under other SDGs where the most explicit links between health and menstruation currently sit. Healthcare centers and schools have been identified as places other than home where people spend critical or lengthy periods of time, so they have priority in the monitoring of indicators related to SDG 6, which we discuss further below. In recognition of this, the WHO/UNICEF Joint Monitoring Programme (JMP) is expanding its global databases to include WASH in institutional settings. To be considered to meet the SDG 'basic' sanitation service criteria, healthcare centers are required to have at least one usable improved toilet designated for women and girls, which provides facilities to manage menstrual hygiene (see Figs. 44.4 and 44.5).

SDG 4 focuses on education. United Nations Educational, Scientific and Cultural Organization (UNESCO), the custodian agency for monitoring Indicator 4.a.1 and Indicator 4.3.1, makes it clear that the indicators fall short of representing the relevance menstruation has to the goal. Of the many changes during puberty, UNESCO considers menstruation to have the most pronounced effect on the quality and enjoyment of education (UNESCO 2014, 10) and that "primary school is [thus] the right place and the right time to reach young people with puberty education" (UNESCO 2014, 10). Adequate assets, services, and spaces for menstrual hygiene are likely to be essential to enable full engagement and concentration (Sommer et al. 2016, 3). However, similar to SDG 1, the metadata sheet notes the "basic" WASH component of Indicator 4.a.1 defers to the JMP "(as per the WASH indicator

GOAL 4 | ENSURE INCLUSIVE & EQUITABLE QUALITY EDUCATION & PROMOTE LIFELONG LEARNING OPPORTUNITIES FOR ALL

target 4.3 ...ensure equal access for all women and men to affordable and quality technical, vocational and tertiary education, including university

target 4.a ... build and upgrade education facilities that are... gender sensitive and provide safe, non-violent, inclusive and effective learning environments for all

indicator 4.3.1 Participation rate of youth and adults in formal and non-formal education and training in the previous 12 months, by sex

indicator 4.a.1 Proportion of schools with access to: ... (e) basic drinking water; (f) single-sex basic sanitation facilities; and (g) basic handwashing facilities (as per the WASH indicator definitions)

Fig. 44.5 SDG 4 indicators with the highest relevance to menstruation (*Source* UN Statistics Division [2018]. Design Credit: Sydney Amoakoh 2019)

definitions)"—that is, the proportion of schools with single-sex basic sanitation and handwashing facilities with water and soap. These indicators go some way to understanding whether schools are providing gender-sensitive environments but do not include many relevant aspects of MHM—for example, menstrual hygiene facilities or materials. It should be noted that, in addition to monitoring the "basic" WASH in schools indicators, for contexts where there is a greater capacity for monitoring, the JMP outlines an "advanced service" of sanitation (including menstrual hygiene facilities) and hygiene (including menstrual hygiene materials) in school settings (WHO/UNICEF JMP 2018, 11), accompanied by an expanded set of recommended questions. However, unless the services are considered "basic," they remain outside the cross-cutting SDG monitoring. The metadata sheet for Indicator 4.3.1 does not mention menstruation at all, but the participation rates will be disaggregated by sex, so where inequities are found, stakeholders might do well to check whether inadequate MHM helps explain it (Fig. 44.6).

Indicators and targets for SDG 5 do not explicitly mention menstruation, but that does not mean it is not relevant. As the SDGs have embedded the human rights framework, an understanding of SDG 5 must incorporate the notion of *substantive* equality, examining both the explicit and direct discriminatory practices by states that treat women and girls differently and also those that may appear gender-neutral yet practically disadvantage women and girls (Winkler and Roaf 2014, 20). Societal taboos and embarrassment around menstruation result in neglect of women's menstrual hygiene needs. This is compounded by gender inequality, which often excludes women and girls from decision-making in development processes (Mahon and Fernandes 2010, 100).

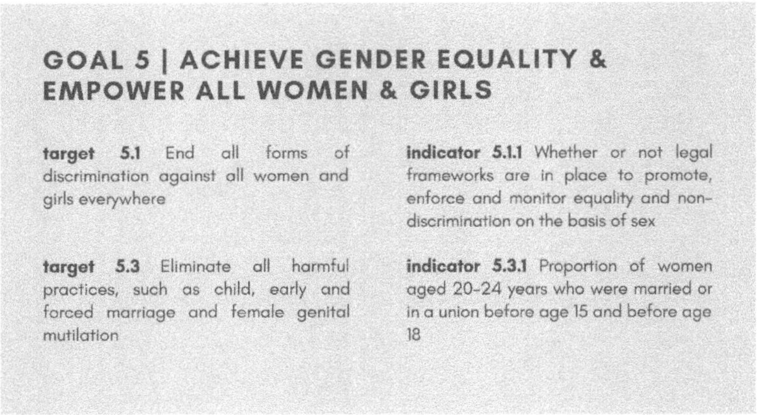

GOAL 5 | ACHIEVE GENDER EQUALITY & EMPOWER ALL WOMEN & GIRLS

target 5.1 End all forms of discrimination against all women and girls everywhere

indicator 5.1.1 Whether or not legal frameworks are in place to promote, enforce and monitor equality and non-discrimination on the basis of sex

target 5.3 Eliminate all harmful practices, such as child, early and forced marriage and female genital mutilation

indicator 5.3.1 Proportion of women aged 20–24 years who were married or in a union before age 15 and before age 18

Fig. 44.6 SDG 5 indicators with the highest relevance to menstruation (*Source* UN Statistics Division [2018]. Design Credit: Sydney Amoakoh 2019)

According to custodian agencies of Indicator 5.1.1, national counterparts assess the indicator by scoring on 45 binary yes/no questions. They include, for example, "Does the law mandate nondiscrimination on the basis of gender in employment?" and "Are there laws that explicitly require the production and/or dissemination of gender statistics?" (UN Statistics Division 2018). Being able to answer "yes" to these questions represents progress under this indicator. This bodes well for efforts to increase access to MHM services, assets, and spaces, as the indicator clearly incentivizes generating evidence on gender-based inequality and a legal framework that those perceiving discrimination can seek help from.

Circumstances are not always straightforward, as an example from China illustrates. China's State Council issued the "Special Rules on Labor Protection for Female Employees" in 2012 "to reduce and resolve the special difficulties which female employees may have in the course of their labor due to their physiological characteristics, and to protect their health." The rules define the "scope of prohibited labor for female employees during the menstrual period" (ACFTU 2014, 1). Although phrased about protection, the broad range of prohibited labor may also incentivize an employer to recruit males instead of females (see Levitt and Barnack-Tavlaris in the volume). We hope for healthy debate on what constitutes progress or regression under each of these questions used in monitoring.

Under Indicator 5.3.1, which addresses early marriage, we see a link to menstruation. In some cultures, due to a combination of social or economic reasons and anxieties, menarche can be a trigger for forced or early marriage. Because of the relationship between menstruation and gender inequality, any tracking of improvements in menstruation-related discrimination and provision of services to address adolescent girls' and women's needs could also

GOAL 6 | ENSURE AVAILABILITY & SUSTAINABLE MANAGEMENT OF WATER & SANITATION FOR ALL

target 6.2 By 2030, achieve access to adequate and equitable sanitation and hygiene for all and end open defecation, paying special attention to the needs of women and girls and those in in vulnerable situations

indicator 6.2.1a Proportion of population using safely managed sanitation services

indicator 6.2.1b Proportion of population with a hand-washing facility with soap and water at home

Fig. 44.7 SDG 6 indicators with the highest relevance to menstruation (*Source* UN Statistics Division [2018]. Design Credit: Sydney Amoakoh 2019)

provide proxy measures for progress toward gender equality. Therefore, the indicators discussed in relation to other SDGs may also relate to SDG 5 (Fig. 44.7).

Target 6.2 has strong relevance to menstrual hygiene: ". . . *paying special attention to the needs of women and girls. . . .*" We argue that this implies addressing the menstrual hygiene needs of women and girls, yet Indicators 6.2.1a and 6.2.1b, with a focus on safely managed sanitation services and handwashing facilities at home, fails to address menstrual hygiene directly.

More background may help readers understand more about the SDG 6 wording "*paying special attention to the needs of women and girls.*" During development of the SDG 6 monitoring framework, the JMP convened a technical working group on hygiene that focused on three areas: food hygiene, handwashing with soap, and MHM. The MHM sub-group formulated the following target: "By 2040 all women and adolescent girls are able to manage menstruation hygienically and with dignity." It recommended three MHM indicators: (1) % of schools and primary health facilities distributing/disseminating accurate, contextually appropriate, pragmatic menstrual management information; (2) % of teachers and healthcare workers who can answer a set of basic questions on MHM; and (3) % of public facilities, schools, institutions, transport hubs, and markets that provide gender-separated latrines with water and soap and disposal facilities for menstrual materials (WHO/UNICEF JMP 2016). While the official SDG indicator list did not incorporate these indicators, there has been good progress in monitoring some aspects of these indicators.

The JMP's monitoring already surpasses the limited scope of the official Indicators 6.2.1a and 6.2.1b. The JMP monitors drinking water, sanitation, and hygiene through service "ladders," differentiating more levels or "rungs" of service than called for in the SDG indicators. Most relevant for MHM are the estimates of the proportion of the population practicing open defecation (no sanitation service), the proportion using basic sanitation services, and the proportion with basic hygiene (Loughnan et al. 2016). Not practicing open defecation implies having access to a sanitation facility, which may go some way to offering privacy for changing menstrual materials, part of the critical components of MHM. The use of basic or safely managed sanitation services (by definition not shared with other households) may increase the likelihood women and girls have private facilities for MHM.

Indicator 6.2.1b focuses on the presence of basic hygiene—here, hand-washing facilities on the premises with soap and water available. The indicator can provide information on MHM, but only partially so. It suggests an adolescent girl or woman would have some access to water and soap, two of the critical components of MHM. However, having a place for general handwashing is not the same as having sufficient water, soap, and space for washing and bathing themselves and their menstrual materials.

As mentioned under the section on SDG 3, the JMP's recommendations for SDG monitoring in health care settings explicitly includes in "basic" sanitation tracking that toilets must be ". . . separated for women, [and] provide menstrual hygiene facilities. . ." (WHO/UNICEF JMP 2017b, 46). In school settings, the JMP considers menstrual hygiene facilities, but only in monitoring at an "advanced" level of access. These healthcare centers and school-specific applications are the only, and limited, instances in the SDG monitoring framework where menstruation is directly addressed.

This still falls short of monitoring the intended scope of the SDG target. Monitoring whether women and girls have the necessary information, confidence, and awareness about how to manage their menstruation—and an adequate supply of hygienic absorptive materials—would also need to be monitored to pay "special attention to the needs of women and girls" in ensuring "adequate and equitable sanitation and hygiene for all."

Other SDG Targets

At-scale monitoring of menstrual health may hold benefits for understanding changes in a range of indicators under other SDGs. Due to space constraints, we give just a few brief additional examples. Goal 10, the "inequality" goal, brings in a range of indicators examining whether opportunities and participation is leveling out between traditionally disadvantaged and advantaged populations. Some of the monitoring explained above would be relevant under the broad theme of SDG 10. The SDG 10 inequality indicator we examined was Indicator 10.3.1: "Proportion of population reporting having personally felt discriminated against or harassed in the previous 12 months on the basis of a ground of discrimination prohibited under international human rights law" (UN Statistics Division 2018). Discrimination based on an individual's sex is prohibited under international human rights law. Some nationally representative surveys like Nepal Multiple Indicator Cluster Surveys (MICS) 2014 survey have included questions to understand discriminatory practices against menstruating women (Central Bureau of Statistics 2015, 151). Such findings might be recognized under monitoring Indicator 10.3.1 (methodology for this indicator is under development).

Indicator 1.5.3, 11.B.1, and 13.1.2 are all "Number of countries that adopt and implement national disaster risk reduction strategies in line with the Sendai Framework for Disaster Risk Reduction 2015-2030" (UN Statistics Division 2018). The Sendai Framework was developed to reduce risks and losses stemming from disasters and to "Build Back Better." The framework explicitly recognizes that women are disproportionately affected in disaster settings and requires "all-of-society engagement and partnership" that pays "special attention to people disproportionately affected by disasters" (United Nations 2015a, 13). Omitting MHM needs in an emergency would be a failure to meet the specifications of the Sendai Framework.

As our findings across this chapter highlight, there are multiple linkages between menstruation and gender inequalities, and these occur across different domains, including poverty, health, education, and WASH. As menstruation is a biological characteristic of women and adolescent girls, SDG 5 may be the fulcrum around which other levers for change relating to social and biological experiences of menstruation coalesce.

Main Sources of Primary Data on Menstrual Health

The earlier sections of this chapter moved systematically through the global SDG framework to examine links with menstruation and noted the limited extent to which the SDG indicator framework addresses the menstrual hygiene needs of women and girls. This section describes some of the major data collection pioneers implementing at-scale monitoring of menstruation-related indicators, which can aid understanding of SDG progress, fill gaps in the indicator framework under target 6, and support monitoring of SDGs 1 and 4. However, it will be some time before a sufficient number of surveys are available to allow systematic reporting for a reasonable number of countries over time.

Data About the General MHM Circumstances of Women and Girls

UNICEF-supported MICS have been conducted in over 100 countries and cover a wide range of topics focusing on children's and women's health and well-being. MICS already includes a set of questions that can be used to collect data for SDG Indicators 6.2.1a and 6.2.1b. They have also long asked questions of mothers on whether their menstrual period has returned since giving birth and on the existence and date of their last period (WHO/UNICEF JMP 2017a, 16). In collaboration with the WHO/UNICEF JMP and in consultation with MHM experts, MICS have now included new questions on menstrual hygiene in surveys conducted since 2017. As of February 2019, there are over 60 MICS surveys at various stages of planning, fieldwork, or analysis that will include these new questions (UNICEF 2019). The questions, based on a pilot in Belize (Khan et al. 2017), focus on the ability to wash and change in privacy at home and access to appropriate reusable or single-use menstrual hygiene materials—two of the essential elements of MHM—as well as participation in social activities, school, or work during a women's menstrual period. The questions pertain to multiple SDGs and hence collect new evidence for monitoring these elements. Furthermore, the datasets contain a wealth of information that can readily be analyzed in conjunction with the menstrual well-being variables.

The Performance Monitoring and Accountability 2020 (PMA2020) surveys led by Johns Hopkins University have been conducted in 11 countries and cover two main topics: family planning and WASH (PMA2020 2018, 9). These surveys have a greater emphasis on MHM and have included a broad evolving range of questions on menstrual hygiene including the main environments for MHM, safety, cleanliness and privacy of MHM facility, disposal of materials, and washing and drying of reusable materials. The latest questionnaires also ask women to recall their age of menarche (Sommer 2013, 401). If PMA2020 consistently collects good-quality data, this may represent a great source for monitoring those essential aspects.

While these questions are standard in the MICS and PMA2020 survey networks, other nationally representative surveys have the option to include them. Household surveys may further offer an opportunity to examine the costs of accessing materials for managing menstruation. A recent review by the International Household Survey Network (IHSN) of income and expenditure surveys for 100 countries found almost all surveys include some information on personal care products (for instance, soap or sanitary pads) (WHO/UNICEF JMP 2017b; IHSN 2019). A lack of consistency across the existing datasets makes them unamenable to the production of comparable international estimates.

Data About the MHM Provisions at Schools and Healthcare Centers

As noted earlier, WASH in schools and healthcare centers have been identified as priorities for monitoring of SDGs 4 and 6. In August 2018, the JMP released a baseline assessment of WASH in Schools, finding that "few countries had data on the proportion of schools providing menstrual hygiene management education, sanitary towels and facilities for the disposal of used materials" (WHO/UNICEF JMP 2018, 8, 50). Nevertheless, a number of countries have already adopted the JMP-recommended questions for schools, either integrating these questions in Education Management Information Systems (EMIS) (for example, Papua New Guinea) or conducting dedicated surveys (for example, in Liberia, and Nigeria's WASH Norm 2018), and more are expected to in the near future.

The main international surveys of health centers include WHO's Service Availability and Readiness Assessment (SARA), DHS's Service Provision Assessment (SPA), and the World Bank's Service Delivery Indicators (SDI) (UNICEF and WHO 2015). They include questions about the availability of water and sanitation facilities in health centers and would be the logical existing mechanisms for collecting this data. Now that consensus and clear collection and analysis methods have been defined, it is expected that an increasing number of these surveys, along with national Health Management Information Systems, will begin to use the question noted in Fig. 44.4 and generate critical information on MHM in the coming years.

New Forms of Data to Aid Understanding of SDG Progress

In the run-up to Menstrual Hygiene Day 2017, UNICEF polled users of U-report about menstrual hygiene across 19 countries (Wijesekera 2017, para. 8). Such data are not representative, as a phone or internet access was needed for participation, but they may provide a useful "snapshot" of the situation that is relevant across the SDGs, particularly to SDGs 4, 5, and 6.

LOOKING FORWARD

Although the current global SDG monitoring framework does not comprehensively address the full range of essential MHM components, this chapter makes clear how related monitoring efforts and a nuanced interpretation of the existing targets and indicators can support progress in addressing women's and girls' unmet menstrual hygiene needs. Existing and emerging menstruation-related monitoring could be more effectively used to understand changes measured by a range of SDG indicators and to address their limitations—and a large number of countries are preparing to collect data on menstrual health in the coming years. There is an urgent need for cross-sectorial discussions and action based on the evidence consolidated here to conduct further inquiry into the intersection between the topics of menstruation and the SDGs with an explicit focus on what is monitorable. Furthermore, there is a need to develop monitoring approaches for the essential elements of menstrual hygiene that have yet to be included in monitoring mechanisms (the availability of information about how to manage menstruation, as well as confidence and awareness). With concerted efforts to roll out standardized data collection on all elements of menstrual hygiene, in a few years it may be possible to compare progress between and within countries with a view to tracking progress on menstrual hygiene needs.

We hope this chapter helps breathe new life into the discussion on menstrual hygiene and sustainable development targets. As the SDG indicators are adopted more widely and more data are amassed and interpreted, the generation of sex-disaggregated data under all indicators may itself provide additional unanticipated utility for understanding whether menstruation management needs are being met in the specific setting. The many connections between MHM and the range of SDGs will likely become even more evident.

Note

1. The scope of this chapter is limited, focusing on monitoring menstrual hygiene for women and adolescent girls. To the extent that the content is useful to others who menstruate, we would encourage a more inclusive interpretation of the text. The full spectrum of people menstruating needs to be addressed with dedicated research, policy, and programming beyond the scope of this chapter.

References

ACFTU. 2014. "Special Rules on the Labor Protection of Female Employees (2012)." All-China Federation of Trade Unions (Blog), August 26. http://en.acftu.org/28616/201408/26/140826131330762.shtml.

Central Bureau of Statistics. 2015. "Nepal Multiple Indicator Cluster Survey 2014." Kathmandu, Nepal: Central Bureau of Statistics and UNICEF Nepal. https://mics-surveys-prod.s3.amazonaws.com/MICS5/South%20Asia/Nepal/2014/Final/Nepal%202014%20MICS_English.pdf.

George, Rose. 2013. "Celebrating Womanhood: How Better Menstrual Hygiene Management Is the Path to Better Health, Dignity and Business." Geneva: Water Supply and Sanitation Collaborative Council. http://wsscc.org/wp-content/uploads/2016/05/Celebrating-Womanhood-How-better-menstrual-hygiene-management-is-the-path-to-better-health-dignity-and-business-%E2%80%93-WSSCC-2013.pdf.

International Household Survey Network (IHSN). 2019. "Measuring Non-Food Expenditures." IHSN. http://www.ihsn.org/projects/non-food-assessment.

Khan, Shane M., Robert E. S. Bain, Karsten Lunze, Turgay Unalan, Bo Beshanski-Pedersen, Tom Slaymaker, Richard Johnston, and Attila Hancioglu. 2017. "Optimizing Household Survey Methods to Monitor the Sustainable Development Goals Targets 6.1 and 6.2 on Drinking Water, Sanitation and Hygiene: A Mixed-Methods Field-Test in Belize." Edited by Asim Zia. *PLoS One* 12 (12): e0189089. https://doi.org/10.1371/journal.pone.0189089.

Loughnan, Libbet C., Rob Bain, Rosemary Rop, Marni Sommer, and Tom Slaymaker. 2016. "What Can Existing Data on Water and Sanitation Tell Us about Menstrual Hygiene Management?" *Waterlines* 35 (3): 228–44. https://doi.org/10.3362/1756-3488.2016.019.

Mahon, Thérèse, and Maria Fernandes. 2010. "Menstrual Hygiene in South Asia: A Neglected Issue for WASH (Water, Sanitation and Hygiene) Programmes." *Gender & Development* 18 (1): 99–113. https://doi.org/10.1080/13552071003600083.

Patkar, Archana. 2012. "Preparatory Input on MHM for JMP Equity and Non-Discrimination Working Group by Archana Patkar." Water Supply and Sanitation Collaborative Council (WSSCC).

PMA2020. 2018. "Performance Monitoring and Accountability 2020: Measuring Performance, Informing Policy, Empowering Communities." Report Summary. Baltimore, MD: Johns Hopkins University and Bill and Melinda Gates Foundation. https://www.pma2020.org/sites/default/files/PMA2020_Brochure_100818_SPREADS.pdf.

Sommer, Marni. 2013. "Menarche: A Missing Indicator in Population Health from Low-Income Countries." *Public Health Reports* 128 (5): 399–401. https://doi.org/10.1177/003335491312800511.

Sommer, Marni, Bethany A. Caruso, Murat Sahin, Teresa Calderon, Sue Cavill, Therese Mahon, and Penelope A. Phillips-Howard. 2016. "A Time for Global Action: Addressing Girls' Menstrual Hygiene Management Needs in Schools." *PLOS Medicine* 13 (2): e1001962. https://doi.org/10.1371/journal.pmed.1001962.

Sustainable Development Knowledge Platform. 2015. "Transforming Our World: The 2030 Agenda for Sustainable Development." United Nations. https://sustainabledevelopment.un.org/post2015/transformingourworld.

UN Statistics Division. 2018. "SDG Indicators Metadata Repository." SDG Indicators, October 2018. https://unstats.un.org/sdgs/metadata/.

UNDP. 2018. "What Are the Sustainable Development Goals?" Sustainable Development Goals, 2018. https://www.undp.org/content/undp/en/home/sustainable-development-goals.html.

UNESCO. 2014. *Puberty Education & Menstrual Hygiene Management.* Good Policy and Practice in Health 9. Paris: United Nations Educational, Scientific and Cultural Organization. http://unesdoc.unesco.org/images/0022/002267/226792e.pdf.

UNICEF. 2019. "MICS 6 Surveys." http://mics.unicef.org/surveys.

UNICEF and WHO. 2015. "Water, Sanitation and Hygiene in Health Care Facilities: Status in Low- and Middle-Income Countries and Way Forward." UNICEF and WHO.

United Nations. 2015a. "Sendai Framework for Disaster Risk Reduction 2015–2030." New York: United Nations. http://www.unisdr.org/files/43291_sendaiframeworkfordrren.pdf.

———. 2015b. "Transforming Our World: The 2030 Agenda for Sustainable Development." New York: United Nations. https://sustainabledevelopment.un.org/content/documents/21252030%20Agenda%20for%20Sustainable%20Development%20web.pdf.

———. 2016. *Resolution Adopted by the General Assembly on 23 December 2016.* http://www.un.org/en/development/desa/population/migration/generalassembly/docs/globalcompact/A_RES_71_256.pdf.

United Nations Department of Public Information. 2019. *Sustainable Development Goals.* https://sustainabledevelopment.un.org/sdgs.

WHO/UNICEF JMP. 2015. *Progress on Sanitation and Drinking Water: 2015 Update and MDG Assessment.* Geneva: World Health Organization and United Nations Children's Fund.

———. 2016. "Monitoring WASH in Health Care Facilities: Final Core Indicators and Questions." World Health Organization and UNICEF. https://washdata.org/report/jmp-2016-core-questions-and-indicators-monitoring-winhcf.

———. 2017a. "MICS6 Questionnaire for Individual Women." UNICEF Multiple Indicator Cluster Surveys. http://mics.unicef.org/files?job=W-1siZiIsIjIwMTcvMTEvMzAvMTkvMTIvNDIvNDg5L01JQ1M2X1F1ZX-

N0aW9ubmFpcmVfZm9yX0luZGl2aWR1YWxfV29tZW5fMjAxNzExMj-
kuZG9jCJdXQ&sha=6328aec5f4566776.

———. 2017b. *Progress on Drinking Water, Sanitation and Hygiene: 2017 Update and SDG Baselines.* Geneva: World Health Organization and United Nations Children's Fund.

———. 2018. "Drinking Water, Sanitation and Hygiene in Schools." *Global Baseline Report 2018.* New York: UNICEF and WHO. https://data.unicef.org/wp-content/uploads/2018/08/JMP-WASH-in-Schools-WEB.pdf.

Wijesekera, Sanjay. 2017. "#MenstruationMatters . . . to All of Us." *HuffPost* (Blog), May 27, 2017. https://www.huffingtonpost.com/entry/menstruationmatters-to-all-of-us_us_59298550e4b07d848fdc0415.

Winkler, Inga T., and Virginia Roaf. 2014. "Taking the Bloody Linen Out of the Closet: Menstrual Hygiene as a Priority for Achieving Gender Equality." *Cardozo Journal of Law and Gender* 21 (1): 1–37.

CHAPTER 45

Practice Note: Menstrual Health Management in Humanitarian Settings

Marianne Tellier, Alex Farley, Andisheh Jahangir,
Shamirah Nakalema, Diana Nalunga, and Siri Tellier

INTRODUCTION[1]

The 2015 earthquake in Nepal affected an estimated 1.4 million women of reproductive age (Chaudhary et al. 2017, 1). Women ranked the need for menstrual hygiene products as a top priority (after food, but ahead of soap and medicine). However, none reported receiving such products during the first month after the earthquake, while more than 50% reported receiving soap (Budhathoki et al. 2018, 3). Meanwhile, in Uganda, about 1 million refugees have lived in refugee settlements for years or even decades, and rely on food rations for daily sustenance. But a local trainer reports that women have difficulties walking to distant food distribution points while menstruating, as they chafe and stain. Some women sell their food rations to buy pads (CARE International and WoMena Uganda 2018, 9).

We are volunteers or staff with WoMena, an NGO which has worked since 2012 to improve menstrual health and management in over 100 settings in Ugandan communities, including in refugee settlements. This review reflects on the challenges associated with meeting menstrual health needs in various humanitarian settings through an analysis of the 2015 Nepal earthquake, a short-term emergency setting, and some of our findings to date from the long-term settings of Kitgum, Rhino, Bidibidi, and Imvepi.

© The Author(s) 2020
C. Bobel et al. (eds.), *The Palgrave Handbook of Critical Menstruation Studies*, https://doi.org/10.1007/978-981-15-0614-7_45

While the term 'menstrual hygiene management' has long been widely used (WHO and UNICEF 2012, 14), we refer to 'menstrual health and management' since menstruation is not only an issue of 'hygienic' management of something 'dirty,' but should be expanded to the concepts of wider physical, mental, and social well-being, as conceptualized in the WHO Constitution (WHO 1946, 1). We use the term 'menstruators,' in recognition that not all who menstruate are girls or women, and not all who are girls or women menstruate. Finally, we use the term 'humanitarian settings' to refer to situations where a major 'human-made' or 'natural' disaster has disrupted the livelihoods and/or support systems of a community to an extent that the community cannot cope without outside assistance (Tellier et al. 2017, 120). The affected population may be dependent on outside organizations for basic needs such as food and housing.

Disasters affect millions of people every year. They may result in forced displacement, and the UNHCR estimated there were over 25 million refugees and 40 million internally displaced persons in 2017 (UNHCR 2018b, 2). Humanitarian settings vary widely. One misconception is that disasters develop suddenly and revert to a more 'normal' state within months. In fact, many last for decades. UNHCR estimates as of 2015 showed that the average duration of 32 protracted refugee situations around the globe was 26 years (UNHCR 2016, 20). Evidence related to health in humanitarian settings is generally poor (Blanchet et al. 2017, 1). The topic of menstrual health is no exception, but the following summarizes some main findings from the limited literature that is available.

Menstrual Health Challenges in Humanitarian Settings

Menstruators living in humanitarian settings face many of the same challenges as those in more stable situations, but may find that their challenges are exacerbated. Barriers to menstrual health are numerous, complex, and interconnected, and one person's lived experiences with menstruation cannot be used to fully describe the experiences of all other menstruators. Therefore, we have attempted to structure the following section according to simple, overarching themes which tend to underpin menstrual health management.

Menstrual Shame and Stigma

Stigma is at the root of many challenges related to menstrual health, and often manifests in menstruators experiencing shame and embarrassment (Johnston-Robledo et al. 2013, 1). In refugee settings, common fears of chafing, staining, and odor are exacerbated, for instance, when menstruators walk long distances to food distribution points (CARE International and WoMena Uganda 2018, 11). Such discomforts may induce stress: 60% of Syrian refugees reported stress due to lack of underwear (Pujol-Mazzini 2017, 1), refugees in Iran reported that asking unfamiliar male health providers for help caused stress (Sohrabizadeh et al. 2018, 4) and, for refugees in Jordan, drying materials where unfamiliar males may see them is a major problem (ACF et al. 2017, 4–12).

In essence, the negative impacts of stigma bleed into menstruators' daily lives.

Health, Including Sexual and Reproductive Health and Rights (SRHR)

Circumstances typical to humanitarian settings and the general stigma and lack of information that surround menstruation often converge to act as barriers to the physical and mental health of menstruators. Menstrual stigma prevents many women from seeking medical attention from aid agencies (Pujol-Mazzini 2017), and the stress associated with menstrual disorders may be prompted or exacerbated by the volatility of emergency settings (VanLeeuwen and Torondel 2018, 356). Such restrictions on menstruators' comfort and ability to access health services prevents them from receiving diagnoses and treatment for health issues, and cuts them off from streams of information on health and hygiene, menstrual or otherwise.

Urogenital infections have a high prevalence (20–30% or more), in high-, middle-, and low-income countries. However, whereas it seems plausible that there is a connection between different materials or practices, there is as yet little evidence, either in humanitarian or more stable settings (Balls et al. 2017, 1; WoMena 2019a, 1). Some studies are beginning to shed some light on this, but further studies are needed to draw general conclusions (Torondel et al. 2018, 8–9; Phillips-Howard et al. 2016, 4–5).

Limitations on resources and economic opportunities in emergency settings may also compromise health. Menstruators may take recourse to the strategy of engaging in sex work to pay for basic needs, including menstrual materials (Phillips-Howard et al. 2016, 1). Such strategies place menstruators at higher risk for gender-based violence and contracting sexually transmitted infections (Women's Refugee Commission 2016, 5, 7), and are reported to increase in humanitarian settings where the accessibility of menstrual materials is poor (Samari 2017, 256).

Access to Menstrual Materials, Including Affordability and Acceptability

Affordability of menstrual materials is a widespread problem, sometimes referred to as 'period poverty.' That problem may be exacerbated in refugee contexts, where many people have lost their livelihood. Humanitarian organizations often hand out menstrual materials in kits, which are intended to last for three months. The cost to humanitarian agencies is high: UNFPA kits, for example, cost on average 22 USD (Abbott et al. 2011, 44). Some organizations continue with distributions for decades, but standards dictating frequency and overall duration of distribution programs vary greatly. Many organizations do so only for the first three months of an emergency

(Sommer 2012, 91; Kragelund Nielsen 2017, 11) with the expectation that, when they stop distributions, menstruators will revert to customary sources, whatever they may be.

Apart from affordability, determining which products are 'acceptable' presents a major challenge to aid workers. Some authors assume that acceptability can be inferred from 'usual practice' before the emergency, but usual practices may be difficult to identify and implement (Parker et al. 2014, 449–50). The Sphere minimum standards[2] recommend that assistance for a long list of activities, including the planning and provision of "menstrual hygiene supplies and facilities" and waste disposal mechanisms (Sphere Association 2018, 100, 102), should be culturally and contextually appropriate. However, it provides little guidance on how to achieve this, given that the text is a set of minimum standards, rather than guidelines (Sphere Association 2018, 104). Community consultations on the cultural and contextual nuances of local menstrual practices are likely to cause programming delays in the early phases of an emergency (Abbott et al. 2011, 59), and are therefore typically carried out in later phases, when menstrual materials may no longer be distributed.

Without contextual insight from consultations, humanitarian organizations may distribute products that are unfamiliar to menstruators or which may be unsuitable for the setting. For instance, recipients of disposable pads after the Pakistan earthquake used them for various household purposes (Sommer et al. 2016, 255). Older menstruators may see reusable pads as a sustainable option, whereas adolescent girls may find it difficult to manage washing and drying (Robinson and Obrecht 2016, 17; Kågesten et al. 2017, 15). Moreover, culture is a fluid concept, but there is little literature focusing on the trends and influencing factors behind changes in menstruators' views over time. One humanitarian staffer notes that refugee women are eager to try new products and that, "they are all the time on the phones and they watch Instagram, Facebook, and YouTube, and they do have advertisement about the cup, for example, and they ask me what it is" (VanLeeuwen and Torondel 2018, 355). In addition to these complexities in obtaining acceptable menstrual materials, menstruators in refugee settings also typically struggle to find the safe and adequate spaces needed for changing, washing, or disposing of said materials, as well as for carrying out other menstrual practices.

Water, Sanitation, and Hygiene, Including Environmental Concerns

Sanitation facilities in camps are often inadequate for menstruators: they are neither sex segregated, adequately private nor sufficiently lighted (Atuyambe et al. 2011, 1). Obtaining water for washing materials and soiled garments is a problem, as is finding inconspicuous spaces for drying. Menstruators may dispose of materials in toilets or latrines instead of bins because they fear being observed (Elledge et al. 2018, 9).

Populations with Particular Needs

The cycle of issues that feed and characterize gaps in the menstrual health of refugees is amplified for populations with particular needs. Menstruators with disabilities and their caretakers are likely to have special, unmet needs (Myers et al. 2018, 7; Rasanathan et al., forthcoming). Unaccompanied or orphaned girls may also face unique challenges, as they often lack a social support network. Moreover, existing discrimination based on indigenous or minority status may be exacerbated in humanitarian emergencies (Sommer, Schmitt, and Clatworthy 2017, 63). From the margins of their communities, these groups encounter unique and profound difficulties accessing the benefits of international and local relief efforts.

Response at the International Level

Following a crisis, a government's capacity to lead and coordinate relief efforts within its borders may be temporarily compromised. The chaos means not only that pre-existing problems of the affected populations are multiplied, but also that government capacities to meet their populations' needs are decimated: staff fatalities soar, and buildings, computer systems, and electric supplies are demolished. A wide range of international actors may rapidly begin operations to fill infrastructural gaps, therefore at such times, international coordination becomes key. In 1991, a UN resolution was adopted to step up and improve coordination in international humanitarian action (UN General Assembly 1992, 1). Subsequently, an extensive set of structures, policies, and guidelines has been developed by UN organizations, while the aforementioned *Sphere Standards* have been developed by the Red Cross/Red Crescent and over 600 humanitarian practitioners representing non-governmental organizations of various sizes from over 80 countries.

Water, sanitation, and hygiene received early attention in international coordination efforts, and the sector was one of the first to identify menstruation as an issue (Sommer 2012, 84). Menstruation would seem to be a good fit for the reproductive health field. However, it took time before reproductive health in itself was recognized as a component of humanitarian assistance, since it was not perceived as life-saving (Chynoweth 2015, 2). It took even longer before menstruation was perceived as part of overall reproductive health. One of the first acknowledgements was in 2000, when the UNFPA recommended that the response to reproductive health should include 'dignity kits' with menstrual materials, and successfully advocated that such services are life-saving and therefore eligible for humanitarian funding (Abbott et al. 2011, 14, 16).

Gradually, the scope of menstrual health response was widened beyond materials. In 2016 Sommer et al. proposed that such response needs to entail the, "provision of safe, private, and hygienic water and sanitation facilities for changing menstrual materials and bathing, easy access to water inside or near toilets, supplies (e.g. laundry soap, separate basin) for washing and drying menstrual materials discreetly, disposal systems through waste management, and access to practical information on MHM (menstrual hygiene management), for adolescent girls in particular" (Sommer et al. 2016, 247). The 2018 version of the *Sphere Standards* includes guidance on the provision of menstrual health 'hardware,' such as the suggestion of providing 15 disposable pads per month or of providing a bin and incinerator for waste; and it features guidance on providing software, such as information about menstruation for students and the consultation of women on their preferred menstrual health practices (Sphere Association 2018, 104). In 2017, Columbia University and the International Rescue Committee published a toolkit for integrating Menstrual Hygiene Management (MHM) into humanitarian response (Sommer et al. 2017). Their work also helped to build recognition that higher level buy-in was necessary for tangible changes in menstrual health programming to take place. In 2018, the UNFPA convened a symposium calling for more attention to menstruation in general within reproductive health, and specifically in humanitarian settings, with a focus on longer-term sustainability in regards to policy and financing (Department of Women, South Africa, and UNFPA 2018, 6).

Thus, much progress has been made at the international level. However, "[d]espite a growing dialogue around MHM (menstrual hygiene management) in emergencies, there remains a lack of clarity on the key components for a complete MHM response, the responsible sectoral actors to implement MHM activities, the most effective interventions to adapt in emergency contexts, and insufficient guidance on monitoring and evaluation" (VanLeeuwen and Torondel et al. 2018, 1). One prominent contributor to the lack of global consensus in these areas is the fact that beneficiary needs and the determinants of 'appropriate' response can vary greatly from one national context to another.

RESPONSE AT THE NATIONAL LEVEL

To provide an indication of the great variation in humanitarian settings, we list some lessons learned from two different contexts: the 2015 earthquake in Nepal and long-term refugee settlements in Uganda.

Nepal

In the aftermath of the earthquake, organizations found it challenging to provide all menstruators, estimated at around 1.4 million, with menstrual materials (seemingly mostly disposable pads) (Chaudhary et al. 2017, 37). One issue was affordability. In order to have provided all menstruators with disposable pads according to the Sphere minimum standards, 750,000 pads would have to have been financed, produced, distributed, and disposed of, per day. In reality, materials were unevenly provided, and there was little attention given to issues beyond provision, such as the logistics of disposal (Myers et al. 2018, 2–9). The Nepal Red Cross Society imported a machine from India for making affordable disposable pads. However, the machine was found to require high maintenance, and the cost (around 0.05 USD per pad) was not sufficiently low to be affordable (Bharati, email to Siri Tellier, 13 April 2018). Some authors suggest that reusable cloth was popular both pre- and post-earthquake, and should also be used in emergencies instead of introducing disposable pads (Budhathoki et al. 2018, 6). One response implemented in the 14 most affected districts included teaching adolescents to make reusable pads using old clothes. Beyond the focus on materials, the program trained school teachers and health workers on how to educate adolescents about menstruation, and on providing adolescent-friendly spaces for engaging in discussions (Chaudhary et al. 2017, 37). Such interventions are key to combatting menstrual stigma, which remains prevalent. One woman explained, '[m]y brother asked me if there was a problem as I was staying inside for a long time. I just couldn't say out loud that I started my periods. I stayed inside the tent . . .' (Budhathoki et al. 2018, 4).

Most organizations paid little attention to the needs of particular groups, such as persons living with disabilities (Myers et al. 2018, 2–9). Dalit communities were marginalized by authorities and had particularly low access to services (Dalit Civil Society Massive Earthquake Victim Support and Coordination Committee 2015, 3, 34), demonstrating how caste-based discrimination is perpetuated in times of disasters and puts already marginalized individuals at even greater disadvantages.

This large-scale, sudden onset disaster poignantly demonstrates some of the issues faced by aid workers in administering menstrual health programming in the initial stages of an emergency: the enormous magnitude of the problem, the need to identify priorities of the affected population, and assumptions about what is 'culturally acceptable,' within a short time frame.

Uganda

Refugees in Uganda number around 1 million, with many settlements having existed for at least a decade (UNHCR 2018a). Uganda has adopted an approach which is more sustainable than in many other countries, giving refugees more freedom to develop livelihoods. For example, income, and imposes fewer restrictions on travel outside the settlements. However, a large part of the cost is borne by international donors (UNHCR 2018a).

Women and girls in Uganda in general face significant stress in dealing with menstruation due to taboos, feelings of shame, and unavailability of products (Miiro et al. 2018, 1). Presumably such shame exists not only in the minds of menstruators, but also in the minds of those around them (Miiro et al. 2018, 5–7). Therefore, WoMena has adopted an approach based on the Ecological Model, which goes from the individual level to the societal

(Dahlberg and Krug 2002). The approach generally takes around six months to implement, and starts with the first point of contact with a community to assess whether there is interest in engaging with the program. If the community agrees to participate, next steps include training trainers and local supporters, including self-assigned 'role model men,' helping menstruators develop their skills and knowledge in using materials over a minimum of three months, then performing end-line surveys and consultations, and conducting follow-up after some months or years. Within this overall approach, WoMena frequently tries out small-scale variations as inspired by community suggestions (WoMena n.d.). For example, in the Bidibidi settlement pictorial training tools were used alongside relevant translations of the materials (WoMena 2019b; Elrha 2018). WoMena provides information on a wide range of menstrual materials, but solely distributes reusable materials, particularly menstrual cups and reusable pads. Cups are becoming increasingly popular in high income countries, but also in many countries in Africa, due to their low cost, high comfort, and sustainability. Reusable pads (particularly AfriPads) are provided as an alternative option to cups in many cases, as they have many of the same qualities (although they are less used in High Income Countries).

Menstrual Shame and Stigma

As mentioned above, in Uganda there is much shame, misinformation, and fear surrounding menstruation. Not surprisingly, this also seems to hold true for Uganda's refugee communities. Menstruators feel particular shame in washing and drying products. In Kitgum, for example, some women were ashamed that the menstrual blood stains on their cloth pads would be visible to others (Tellier, Hyttel, and Gad 2012, 8).

In many places around the world, parents, teachers, and girls see low understanding of menstruation as a problem, but are often too shy to address it. However, girls indicate that they are less shy to ask their parents about menstruation than about other issues related to SRHR (Ivanova et al. 2019, 1). WoMena's training approach and experience is that focusing on puberty education can be an entry point for sparking wider discussions on reproductive health (CARE International and WoMena Uganda 2018, 7), and for helping families and communities to become more open and less shameful. By the end of the Imvepi pilot, 94% of girls had discussed menstruation with family members and friends after the project, whereas before it was considered taboo.

Community, especially male, involvement is also key to addressing stigma. Men may have concerns about menstrual irregularities caused by contraception (Hyttel et al. 2012, 151) and are often curious to learn more about menstruation, particularly because they are often the ones who pay for materials (CARE International and WoMena Uganda 2018, 6). One role model man from Imvepi stated, "I have understood that this (menstruation) is not a disease. Before I used to see it as something very bad, but nowadays, no problem" (CARE International and WoMena Uganda 2018, 7).

Perhaps most importantly, WoMena's community-based model has helped to address the implications of stigma-induced stress, including heightened fear, as well as lack of agency and mobility (Tellier, Hyttel, and Gad 2012, 11). One of the most frequent reports by menstruators who have participated in WoMena supported activities is that they have gone from a state of fear to one of freedom. One beneficiary reported, "I feel free . . . Now I can drive my bike (to the food distribution site) and I don't have to look back" (Tellier, Hyttel, and Gad 2012, 6). These outcomes evidence the power that positive social attitudes toward menstruation have in improving menstruators' access to humanitarian programming.

Health, Including Sexual and Reproductive Health and Rights (SRHR)
One of the biggest concerns of menstruators relates to menstrual problems including cycle irregularities and pain management. In one study, only around a third of adolescents had ever in their lives consulted a medical doctor, with issues related to HIV and menstrual problems being the top reasons for seeking a consultation (Ivanova et al. 2019, 1). In response, WoMena has included information about such problems in educational material, including our FAQs explaining what is 'normal,' and when to see a health provider (WoMena 2018, 1–5). The training and kits have been adjusted to include more information and guidance on pain alleviation (for example, using a towel and bottle for warm compresses).

Menstrual health-related interpersonal violence is another significant concern and can take many forms. CARE reports that, in some cases, women were beaten by their husbands because they cut up distributed blankets to make pads (CARE International and WoMena Uganda 2018, 4). There are repeated reports by parents that puberty leads to early marriage, due not least to the fact that some cannot afford to buy pads for their daughters (CARE International and WoMena Uganda 2018, 4). Girls report engaging in transactional sex to buy pads, for example Rachel (a Kitgum resident) says, "it (the project) should be extended to younger girls in school . . . a young girl can even get HIV-positive by looking for money to buy pads, if she prostitutes herself" (Tellier, Hyttel, and Gad 2012, 7). This demonstrates just some of the ways in which threats to physical well-being and inadequate health and management often go hand-in-hand.

Access to Menstrual Materials, Including Affordability and Acceptability
Affordability of menstrual materials is an overwhelming obstacle in Ugandan settlements once kits are no longer distributed. According to one trainer, "[s]ome women will sell their (food) rations but you can only get about [4 USD] from that. You then spend [1.6 USD] of that on pads alone and now you only have [2.4 USD] left" (CARE International and WoMena Uganda 2018, 24). This is confirmed by a mother in Rhino settlement, who shared that, "when you arrive in the camp they give you three packets of pads which each

packet contains 10 pieces but if it gets over and we don't receive other pads we use piece of clothes [sic]" (Norelius 2017, 47).

To save money, menstruators report resorting to changing disposable pads less often than suggested by manufacturer guidelines (Norelius 2017, 47), sharing with others, or washing and reusing disposable pads (WoMena 2019a, 1). Disposable pads cost 18–35 USD per year, reusable pad kits such as AfriPads cost around 5 USD per kit and last 1–2 years, while cups sell at 5–15 USD and last up to 10 years (Tellier and Hyttel 2018, 20). Pilot projects supported by WoMena and its partners indicate that menstruators are willing to pay 4–7 USD for a cup (Tellier and Hyttel 2018, 29).

At the same time, acceptability of products is equally important. There are often questions and concerns (which are at times posed by humanitarian staff rather than menstruators) regarding whether menstrual cups or other reusable materials are 'culturally acceptable': Will it affect the hymen? Will it hurt? Will it be difficult to use? These legitimate concerns must be taken seriously, and both the negative and positive potential results of each product choice should be openly considered.

The end-line study from Kitgum indicates that the overall experience for the participants using cups was either very good (75%, $n = 18$) or good (25%, $n = 6$) (Tellier, Hyttel, and Gad 2012, 2). The pilot project in Rhino settlement indicated that initial uptake was higher for reusable pads than cups (100% versus 69%). Consistent use was slightly higher for cups than reusable pads (87% versus 74%), as users overcame initial difficulties. Beneficiary satisfaction rates were high for both materials, and increased over time as menstruators got accustomed to each. The highest satisfaction rates were found among beneficiaries who were provided both materials. They could alternate, for example, using cups when there was insufficient water or on heavy flow days (Gade and Hytti 2017, 8, 32). The presence of experienced users who could answer questions was key in continued use, especially cups. Kits and training were also modified to meet concerns. For example, in response to some experiencing challenges boiling the cup, a metal container for boiling was added to the kit (Tellier, Hyttel, and Gad 2012, 8).

Regarding the quality and acceptability of kits, it is also important to ask: Acceptable to whom? At what time? How (for example, what type of interaction with the community)? WoMena has found that refugee populations may be more open to changing their practices, and possess more of the time needed to engage in consultations, than populations living in more stable conditions. While this is only a formative impression, it has been confirmed by others (Sphere Association 2018, 75; VanLeeuwen and Torondel 2018, 355). Outside of menstruators' personal preferences, as well as the quality and contextual acceptability of menstrual products, practicalities like those concerning water, sanitation, and hygiene must also be taken into consideration.

Water, Sanitation, Hygiene, Including Environmental Concerns

One estimate from the Imvepi project is that it takes each menstruator around 15 liters per cycle to wash reusable products, including hands and soiled underwear. Here, cups have an advantage, since they can be used without underwear, need an estimated 1–2 liters per cycle, and do not need to be dried. The main complication is that they need to be boiled once a month (Hyttel et al. 2017). One Imvepi participant noted she was happy to use a cup, "because it uses less water and [her] grandmother supports [her] in using it because of that" (CARE International and WoMena Uganda 2018, 10).

Populations with Particular Needs

WoMena's consultative approach helps to identify menstruators with special needs. This includes a wide range of disabilities which must be taken into account, as menstruators with limited mobility (for example, if they have lost a leg) or cognitive disabilities, as well as their caregivers, have particular needs. This is an area where a systematic response might be further developed (WoMena n.d.).

In sum, long-term emergencies present very different needs than acute situations. WoMena's experiences in the Ugandan context suggest refugees may be relatively open to trying out innovative approaches. Working with a more developmental approach and engaging widely with communities not only seems to afford greater possibilities of overcoming the practical issues related to materials, water, and sanitation, but also presents opportunities to change social attitudes toward menstruation.

Conclusion

Over the last few decades there has been increased attention to menstruation in humanitarian settings. The reproductive health sector has been a relative newcomer to this discussion, but even at this early stage of engagement, we find that discussions on menstrual health are proving to be a potent entry point for reducing shame and ignorance about a host of sexual and reproductive health and rights.

Challenges remain to achieving standardized, coordinated, and sustainable programming, namely in: reaching policy agreements on what the basic contents of emergency response should be, and determining who would be responsible for different tasks at different stages of response. The intention should be that, over time, a transition can be made from a narrow humanitarian 'three-month kit' culture to one of the more sustainable solutions. What is more, determining what is 'culturally acceptable' plays a key role in developing such solutions. Culture is a fluid concept, and 'acceptability' needs to address questions such as: acceptable to whom? when? and how? WoMena's experience is that refugees may be quite open to trying out innovative approaches, and have more time to do so, than people in stable settings. After all, they have already demonstrated their adaptability and agency by fleeing and adjusting to new living conditions.

In a broader perspective, the transition from dealing with acute emergencies to a longer term, sustainable approach is a high profile and persistent challenge for the entire humanitarian field. At times, these challenges are partially presented by the competing approaches utilized by actors from different fields. In humanitarian assistance, 'neutrality' is traditionally a key concept (ICRC 1979, 7), meaning that one may need to work *around* local institutions, to ensure impartiality in administering aid. Meanwhile, in development assistance, one central principle is usually 'participation,' which includes working *through* local institutions to build long-term solutions, after the initial stages of an emergency have passed (UNDG 2003).

Finally, context matters. In humanitarian action, the approach is often to have a blueprint for the first few months, with the intention that the approach should be 'contextualized' in the long term. Uganda hosts one of the largest long-term refugee populations in sub-Saharan Africa, and employs more sustainable approaches to providing aid than many other countries in the world (UNHCR 2016). This can help create an environment for a developmental approach that is tailored to the needs of menstruators. In conclusion, comparing the situations in Nepal and Uganda may help in finding some small-scale, practical example of how to bridge the gap between short- and longer-term approaches in humanitarian assistance.

NOTES

1. We would like to acknowledge the excellent research assistance provided by Sydney Amoakoh.
2. An internationally recognized set of common principles and universal minimum standards in humanitarian response. Source: https://www.spherestandards.org/about.

REFERENCES

Abbott, L., B. Bailey, Y. Karasawa, D. Louis, S. McNab, D. Patel, C. Lopez, R. Rani, C. Saba, and L. Vaval. 2011. *Evaluation of UNFPA's Provision of Dignity Kits in Humanitarian and Post-Crisis Settings.* School of International and Public Affairs, Columbia University.

ACF, UNHCR, and UNICEF. 2017. *Results from a Pilot with the Disinfecting Menstrual Hygiene Pad 'Safepad™' Azraq Refugee Camp—Jordan.*

Atuyambe, Lynn M., Michael Ediau, Christopher G. Orach, Monica Musenero, and William Bazeyo. 2011. "Landslide Disaster in Eastern Uganda: Rapid Assessment of Water, Sanitation and Hygiene Situation in Bulucheke Camp, Bududa District." *Environmental Health* 10 (1): 38.

Balls, Emily, Robert Dreiselbis, and Belen Torondel. 2017. *Menstrual Hygiene Management: Policy Brief.* London: SHARE Consortium. http://www.susana.org/en/resources/library/details/2787.

Blanchet, K., A. Ramesh, et al. 2017. "Evidence on Public Health Interventions in Humanitarian Crises." *Lancet* 390 (10109): 2287–96.

Budhathoki, Shyam Sundar, Meika Bhattachan, Enrique Castro-Sánchez, Reshu Agrawal Sagtani, Rajan Bikram Rayamajhi, Pramila Rai, and Gaurav Sharma. 2018. "Menstrual Hygiene Management among Women and Adolescent Girls in the Aftermath of the Earthquake in Nepal." *BMC Women's Health* 18 (1): 33.

CARE International and WoMena Uganda. 2018. *Ruby Cups: Girls in Imvepi Refugee Settlement Taking Control.*

Chaudhary, Pushpa, Giulia Vallese, Meera Thapa, Valerie Broch Alvarez, Latika Maskey Pradhan, Kiran Bajracharya, Kazutaka Sekine, Shilu Adhikari, Reuben Samuel, and Sophie Goyet. 2017. "Humanitarian Response to Reproductive and Sexual Health Needs in a Disaster: The Nepal Earthquake 2015 Case Study." *Reproductive Health Matters* 25 (51): 25–39.

Chynoweth, S. K. 2015. "Reproductive Health in the Changing Humanitarian Context—Findings from the IAWG on Reproductive Health in Crises' 2012–14 Global Evaluation." *Conflict and Health* 9: 1–154.

Dahlberg, L. L., and E. G. Krug. 2002. "Violence—A Global Public Health Problem." In *World Report on Violence and Health*, edited by E. Krug, L. L. Dahlberg, J. A. Mercy, A. B. Zwi, and R. Lozano. Geneva, Switzerland: World Health Organization.

Dalit Civil Society Massive Earthquake Victim Support and Coordination Committee. 2015. *Waiting for "Justice in Response": Report of Initial Findings from Immediate Needs Assessment and Monitoring Responses towards Affected Dalit Communities in Nepal Earthquake 2015.*

Department of Women, South Africa, and UNFPA. 2018. *First East and Southern Africa Regional Symposium Menstrual Health Management for Adolescent Girls and Women.* Johannesburg. https://esaro.unfpa.org/sites/default/files/pub-pdf/MHM%20Report%20final%20digital.pdf.

Elrha. 2018. "Ndrelmba, Perotiyapa and Dora: Translating Menstruation." Accessed February 26, 2019. https://www.elrha.org/project-blog/ndrelmba-perotiyapa-and-dora-translating-menstruation/.

Elledge, Myles, Arundati Muralidharan, Alison Parker, Kristin Ravndal, Mariam Siddiqui, Anju Toolaram, and Katherine Woodward. 2018. "Menstrual Hygiene Management and Waste Disposal in Low and Middle Income Countries—A Review of the Literature." *International Journal of Environmental Research and Public Health* 15 (11): 2562.

Gade, Anna, and Laura Hytti. 2017. *Menstrual Health in Rhino Camp Refugee Settlement, West Nile, Uganda, Pilot Project Intervention Report.* Kampala: WoMena Uganda and ZOA.

Hyttel, Maria, Jennifer J. K. Rasanathan, Marianne Tellier, and Willington Taremwa. 2012. "Use of Injectable Hormonal Contraceptives: Diverging Perspectives of Women and Men, Service Providers and Policymakers in Uganda." *Reproductive Health Matters* 20 (40): 148–57.

Hyttel, Maria, Camilla Faldt Thomsen, Bianca Luff, Halvor Storrusten, Viola Nilah Nyakato, and Marianne Tellier. 2017. "Drivers and Challenges to Use of Menstrual Cups among Schoolgirls in Rural Uganda: A Qualitative Study." *Waterlines* 36 (2): 109–24.

International Committee of the Red Cross (ICRC). 1979. *The Fundamental Principles of the Red Cross: commentary.*

Ivanova, Olena, Masna Rai, Wendo Mlahagwa, Jackline Tumuhairwe, Abhishek Bakuli, Viola N. Nyakato, and Elizabeth Kemigisha. 2019. "A Cross-Sectional Mixed-Methods Study of Sexual and Reproductive Health Knowledge, Experiences and Access to Services among Refugee Adolescent Girls in the Nakivale Refugee Settlement, Uganda." *Reproductive Health* 16 (1): 35. https://doi.org/10.1186/s12978-019-0698-5.

Johnston-Robledo, Ingrid, and Joan C. Chrisler. 2013. "The Menstrual Mark: Menstruation as Social Stigma." *Sex Roles* 68 (1–2): 9–18. https://link.springer.com/article/10.1007/s11199-011-0052-z.

Kågesten, Anna E., Linnea Zimmerman, Courtland Robinson, Catherine Lee, Tenaw Bawoke, Shahd Osman, and Jennifer Schlecht. 2017. "Transitions Into Puberty and Access to Sexual and Reproductive Health Information in Two Humanitarian Settings: A Cross-Sectional Survey of Very Young Adolescents From Somalia and Myanmar." *Conflict and Health* 11 (1): 24.

Kragelund Nielsen, Johanne. 2017. "Women and Girls' Experiences with Safety, Hygiene and Sanitation in Connection with Menstrual Health Management in Nyarugusu Refugee Camp." Master's thesis, University of Copenhagen.

Miiro, George, Rwamahe Rutakumwa, Jessica Nakiyingi-Miiro, Kevin Nakuya, Saidat Musoke, Juliet Namakula, Suzanna Francis, et al. 2018. "Menstrual Health and School Absenteeism among Adolescent Girls in Uganda (MENISCUS): A Feasibility Study." *BMC Women's Health* 18 (1): 4.

Myers, Anna, Samira Sami, Monica Adhiambo Onyango, Hari Karki, Rosilawati Anggraini, and Sandra Krause. 2018. "Facilitators and Barriers in Implementing the Minimum Initial Services Package (MISP) for Reproductive Health in Nepal Post-Earthquake." *Conflict and Health* 12 (1): 1–9.

Norelius, Hanna Maria. 2017. "Managing Menstruation During Displacement: A Mixed Methods Study Investigating Menstrual Hygiene Management in Rhino Refugee Settlement. Uganda." Master's thesis, University of Copenhagen.

Parker, Alison H., Jen A. Smith, Tania Verdemato, Jeanette Cooke, James Webster, and Richard C. Carter. 2014. "Menstrual Management: A Neglected Aspect of Hygiene Interventions." *Disaster Prevention and Management* 23 (4): 437–54.

Phillips-Howard, Penelope A., Elizabeth Nyothach, Feiko O. ter Kuile, Jackton Omoto, Duolao Wang, Clement Zeh, Clayton Onyango, et al. 2016. "Menstrual Cups and Sanitary Pads to Reduce School Attrition, and Sexually Transmitted and Reproductive Tract Infections: A Cluster Randomised Controlled Feasibility Study in Rural Western Kenya." *BMJ Open* 6 (11): 1–11.

Pujol-Mazzini, Anna. 2017. "For Refugee Women, Periods a Dangerous, Shameful Time." *Reuters*, March 8. https://www.reuters.com/article/us-womens-day-refugees-periods-feature/for-refugee-women-periods-a-dangerous-shameful-time-idUSKBN16F1UU.

Rasanathan Jennifer, J. K., Andisheh Jahangir, Mary Grace Sandy, Marianne Tellier, and Siri Tellier (WoMena). 2019. *Doubly Left Behind at the Intersection of Disability and Menstruation*. Manuscript submitted for publication.

Robinson, Alice, and Alice Obrecht. 2016. "Improving Menstrual Hygiene Management in Emergencies: IFRC's MHM Kit." HIF/ALNAP Case Study. London: ODI/ALNAP.

Samari, Goleen. 2017. "Syrian Refugee Women's Health in Lebanon, Turkey, and Jordan and Recommendations for Improved Practice." *World Medical & Health Policy* 9 (2): 255–74.

Sohrabizadeh, Sanaz, Katayoun Jahangiri, and Reza Khani Jazani. 2018. "Reproductive Health in the Recent Disasters of Iran: a Management Perspective." *BMC Public Health* 18 (1): 389.

Sommer, Marni. 2012. "Menstrual Hygiene Management in Humanitarian Emergencies: Gaps and Recommendations." *Waterlines* 31 (1–2): 83–104.

Sommer, Marni, Margaret L. Schmitt, David Clatworthy, Gina, Bramucci, Erin Wheeler, and Ruwan Ratnayake. 2016. "What Is the Scope for Addressing Menstrual Hygiene Management in Complex Humanitarian Emergencies? A Global Review." *Waterlines, Practical Action Publishing* 35 (3): 245–64. https://doi.org/10.3362/1756-3488.2016.024.

Sommer, M., M. Schmitt, and D. Clatworthy. 2017. *A Toolkit for Integrating Menstrual Hygiene Management (MHM) into Humanitarian Response.* New York: Columbia University, Mailman School of Public Health and International Rescue Committee.

Sphere Association. 2018. *The Sphere Handbook: Humanitarian Charter and Minimum Standards in Humanitarian Response.* 4th ed. Geneva: Sphere Association. www.spherestandards.org/handbook.

Tellier, M., M. Hyttel, and M. Gad. 2012. *Assessing Acceptability and Hygienic Safety of Menstrual Cups as Menstrual Management Methods for Vulnerable Young Women in Uganda Red Cross Society's Life Planning Skills Project.* Kampala: WoMena Ltd.

Tellier, Siri, and Maria Hyttel (WoMena). 2018. "Menstrual Health Management in East and Southern Africa: A Review Paper." Johannesburg: UNFPA.

Tellier, Siri, Andreas Kiaby, Lars Peter Nissen, Jonas Torp Ohlsen, Wilma Doedens, Kevin Davies, Dan Brun Petersen, Vibeke Brix Christensen, and Niall Roche. 2017. "Basic Concepts and Current Challenges of Public Health in Humanitarian Action." In *International Humanitarian Action*, 229–317. Cham: Springer.

Torondel, Belen, Shalini Sinha, Jyoti Ranjan Mohanty, Tapoja Swain, Pranati Sahoo, Bijaya Panda, Arati Nayak, et al. 2018. "Association between Unhygienic Menstrual Management Practices and Prevalence of Lower Reproductive Tract Infections: A Hospital-Based Cross-Sectional Study in Odisha, India." *BMC Infectious Diseases* 18 (1): 1–12. https://doi.org/10.1186/s12879-018-3384-2.

UN General Assembly. 1992. Strengthening of the Coordination of Humanitarian Emergency Assistance of the United Nations: Resolution, Adopted by the General Assembly, April 14, 1992, A/RES/46/182. https://www.refworld.org/docid/3b00f18620.html.

UN High Commissioner for Refugees (UNHCR). 2016. "Global Trends: Forced Displacement in 2015." June 20. Accessed September 14, 2018. http://www.refworld.org/docid/57678f3d4.html.

UN High Commissioner for Refugees (UNHCR). 2018a. "Global Focus—Uganda." Accessed January 7, 2019. http://reporting.unhcr.org/node/5129.

———. 2018b. "Global Trends—Forced Displacement in 2017." Accessed September 14, 2018. http://www.unhcr.org/globaltrends2017/.

UN Sustainable Development Group (UNDG). 2003. "The Human Rights Based Approach to Development Cooperation towards a Common Understanding among UN Agencies." May 3–5.

VanLeeuwen, Crystal, and Belen Torondel. 2018. "Improving Menstrual Hygiene Management in Emergency Contexts: Literature Review of Current Perspectives." *International Journal of Women's Health* 10: 169–86. https://doi.org/10.2147/IJWH.S135587.

World Health Organization (WHO). 1946. Preamble to the Constitution of the World Health Organization as Adopted by the International Health Conference, New York, June 19–22 (Official Records of the World Health Organization, no. 2, p. 100).

World Health Organization (WHO), and the United Nations International Children's Emergency Fund (UNICEF). 2012. Report of the Second Consultation on Post-2015 Monitoring of Drinking-Water, Sanitation and Hygiene, WHO/UNICEF Joint Monitoring Programme for Water Supply and Sanitation (JMP), Hosted by the Royal Government of the Netherlands in the Hague, December 3–5, 2012: 89.

WoMena. 2018. "WoMena FAQs: Menstrual Irregularities." http://womena.dk/womena-faqs-what-are-menstrual-irregularities/.

WoMena. 2019a. "WoMena FAQs: Is There Any Connection between Menstrual Cups and Infections?" http://womena.dk/is-there-any-connection-between-menstrual-cups-and-infections/.

WoMena. 2019b. "Photo Story: Meeting Linguistic Challenges with Image." http://womena.dk/photo-story-meeting-linguistic-challenges-with-images/.

WoMena. n.d. *Programmatic Approach*. Accessed September 19, 2018. http://womena.dk/goals-strategies-policies/.

Women's Refugee Commission. 2016. *Mean Streets: Identifying and Responding to Urban Refugees' Risks of Gender-Based Violence*. https://www.womensrefugeecommission.org/images/zdocs/mean-streets-urban-gbv-sex-workers.pdf.

CHAPTER 46

Mapping the Knowledge and Understanding of Menarche, Menstrual Hygiene and Menstrual Health Among Adolescent Girls in Low- and Middle-Income Countries

Venkatraman Chandra-Mouli and Sheila Vipul Patel

PLAIN ENGLISH SUMMARY[1]

Our paper maps the knowledge, attitudes, beliefs and practices surrounding menarche, menstrual hygiene and menstrual health among adolescent girls in low and middle income countries in order to inform the future design of relevant policies and programming.

Our study of over 80 journal articles from a number of low and middle income countries confirmed that:

- Many adolescent girls start their periods uninformed and unprepared
- Mothers are the primary source of information, but they inform girls too-little and too-late and often communicate their own misconceptions
- Because menstruation is widely seen as polluting and shameful, girls are often excluded and shamed in their homes and in their communities
- Many do not have the means for self-care and do not get the support they need when they face problems, which hinders their ability to carry on with everyday activities and may also establish a foundation for life-long disempowerment.

© The Author(s) 2020

C. Bobel et al. (eds.), *The Palgrave Handbook of Critical Menstruation Studies*, https://doi.org/10.1007/978-981-15-0614-7_46

Efforts to respond to girls' needs are fragmented and piece-meal. There is growing acknowledgement that efforts are more likely to be successful if they come together in a whole-of-community approach that involves schools, health facilities, and homes and communities to:

- Educate girls about menstruation
- Create norms that see menstruation as healthy and positive, not shameful and dirty
- Improve access to sanitary products, running water, functional toilets and privacy for self-care
- Improve care for and support by girls' families when they have their periods
- Improve access to competent and caring health workers when they experience menstrual health problems

BACKGROUND

Girls in many low- and middle-income countries (LMIC) enter puberty with knowledge gaps and misconceptions about menstruation, unprepared to cope with it and unsure of when and where to seek help. This is because the adults around them, including parents and teachers, are themselves ill-informed and uncomfortable discussing sexuality, reproduction and menstruation (which frequently comes laden with dirty, polluting and shameful connotations).

To respond to the increased international attention on empowering girls through the United Nation's Sustainable Development Goals, this review aims to map the knowledge, attitudes, beliefs and practices surrounding menarche, menstrual hygiene and menstrual health among adolescent girls in LMIC in order to inform the future design of relevant policies and programming. To do this, our objectives are to answer the following questions: (1) how knowledgeable are girls in LMIC about menstruation and how prepared are they for reaching menarche, (2) who are their sources of information regarding menstruation, (3) how well do the adults around them respond to their information needs, (4) what negative health and social effects do adolescents experience as a result of menstruation, and (5) how do adolescents respond when they experience these negative effects and what practices do they develop as a result?

METHODS

Our literature search aimed to identify articles that evaluated the knowledge of girls regarding menstruation, their information sources, the health and social effects of menstruation, and how adolescents and adults responded to these effects. We searched Google Scholar, PubMed and EBSCO's Global Health database for articles in peer-reviewed journals published between 2000 and 2015. To identify relevant literature, we used the following search strategy: (menarche or menstruation or menstrual health or menstrual hygiene or menstrual management) and (adolescence or adolescent or youth or young) and (female or girl or women) and (knowledge or belief or practice or experience).

Through a title and abstract review, papers in English that addressed the experiences of adolescent girls (ages 10–19) in LMIC were retained. Full text articles were reviewed to determine whether studies addressed one or more of our five questions. Given the limited research available, descriptive overviews and interventions using quantitative, qualitative, or mixed methods of any sample size were all included.[2] While the focus of this paper is on menstrual experiences, studies that reported on the preparedness and attitudes of pre-menarcheal girls were included so long as data were stratified by those who had and had not reached menarche. To complement our search, we reviewed the reference lists of the included articles and identified a small number of additional studies that met these broad criteria. Finally, we searched and included publications by United Nations agencies and international non-governmental organizations that responded to how organizations and their LMIC partners are responding to the needs of girls.

RESULTS

A total of 81 articles were identified after discarding duplicate articles and those that did not meet inclusion criteria (Table 46.1).

Table 46.1 Study characteristics

Characteristic	Frequency
Design	
Descriptive	70
Intervention	11
Method	
Mixed methods	16
Qualitative	7
Quantitative	58
Region	
East Africa (Ethiopia, Kenya, Malawi, Tanzania, Uganda)	10
North Africa (Egypt)	4
West Africa (Ghana, Nigeria)	10
North/Central America (Mexico)	3
South America (Brazil)	1
East Asia (China)	1
Southeast Asia (Malaysia)	6
South Asia (Bangladesh, India, Nepal, Pakistan, Sri Lanka)	39
West Asia (Iran, Jordan, Lebanon, Turkey)	7
Setting[a]	
Mix	21
Rural	23
Urban	30
School status[a]	
Mix of school-going and out-of-school	12
School-going	63

[a]All included studies did not specify setting or girls' school status

HOW KNOWLEDGEABLE ARE GIRLS ABOUT MENSTRUATION AND HOW PREPARED ARE THEY FOR REACHING MENARCHE?

Girls across LMIC have limited knowledge and understanding about menstruation prior to reaching menarche. The proportion of girls that were aware ranged from 2.8% of rural girls questioned in Rajasthan, India [1] to all urban girls in Turkey [2] (Table 46.2). Village-based meetings for girls in a Maharashtra, India were tested as a platform for disseminating health messages, and significantly contributed to an increase from 35.1% of girls interviewed in 2003 to 55.4% of girls interviewed in 2007 being aware of menstruation before its onset (p-value < 0.05) [3].

Three quarters of 1,573 Chinese girls surveyed rated their menstrual knowledge as inadequate or very inadequate [4]. Even so, girls with any knowledge often hold misconceptions about menstruation. For example, a study conducted in rural Nepal reported that 6.0% of 150 girls surveyed recognized menstruation as a physiological process while 82.0% believed it was a

Table 46.2 Awareness of menstruation prior to menarche

First author, Year	Country	Setting	School status	N	Aware
North Africa					
Eswi 2012 [45]	Egypt	Urban	School-going	200	74.0%
South Asia					
Bosch 2008 [28]	Bangladesh	Rural	Unclear	156	35.0%
Khanna 2005 [1]	India	Mix	Out-of-school	358	5.6%
Dambhare 2012 [32]	India	Mix	School-going	561	75.6%
Juyal 2013 [65]	India	Mix	School-going	453	64.5%
Khanna 2005 [1]	India	Mix	School-going	372	9.8%
Thakre 2011 [12]	India	Mix	School-going	387	37.0%
Khanna 2005 [1]	India	Rural	Mix	NR[a]	2.8%
Dasgupta 2008 [8]	India	Rural	School-going	160	67.5%
Shanbhag 2012 [10]	India	Rural	School-going	329	57.9%
Sudeshna 2012 [15]	India	Rural	School-going	190	47.4%
Dhingra 2009 [29]	India	Rural	Unclear	200	64.0%
Tiwari 2006 [11]	India	Unclear	School-going	763	62.7%
Khanna 2005 [1]	India	Urban	Mix	NR[a]	12.1%
Omidvar 2010 [46]	India	Urban	School-going	336	64.5%
Yasmin 2013 [13]	India	Urban	School-going	147	42.2%
Bobhate 2011 [7]	India	Urban	Unclear	241	20.3%
Udgiri 2010 [23]	India	Urban	Unclear	342	18.4%
Ali 2010 [6]	Pakistan	Urban	Government school	425	47.8%
Ali 2010 [6]	Pakistan	Urban	Out-of-school	425	38.8%
Ali 2010 [6]	Pakistan	Urban	Private school	425	34.1%
West Asia					
Reis 2011 [27]	Turkey	Urban	Mix	310	67.4%
Ersoy 2004 [2]	Turkey	Urban	School-going	1017	100.0%

NR Not reported
[a]Of 730 girls, the number in rural versus urban settings was not specified

curse [5]. Understanding that menstruation is a natural bodily function was higher at 19.3% in Pakistan [6], 18.3–86.3% in five Indian states [1, 7–13], and 96.7% in Nigeria [14]. Menstruation was considered a curse, disease, or representation of sin by some girls in five Indian states [7–12, 15] and Uganda [16]. Prior to receiving health education at school, 72.4% girls in India considered menstrual blood impure [17].

An additional knowledge gap among girls is a lack of awareness regarding the origins of menstrual blood—no more than a third of girls correctly identified the uterus as the source of menstrual blood in four Indian states [7, 12, 15, 18, 19] and rural Nepal [5]. One study in a fourth Indian state reported almost no girls being aware of the source of their blood (2.5%) [8], while another nearly two-thirds being aware (63.3%) [13]. In cities in Pakistan [6] and Nigeria [14], 37.2 and 78.7% identified the uterus as the source, respectively, compared to 82.9% of school-going girls in rural Uganda [16]. Only a third of rural-living, high school girls surveyed in India associated the attainment of menarche with the capacity to conceive [10].

Age had a significant influence on slum dwellers' knowledge in India, with older girls more knowledgeable about menstruation than their younger counterparts (p-value <0.05) [7]. Similar findings were reported among Nigerian schoolgirls (p-value <0.05) [20]. Compared to those not attending school, awareness was greater among schoolgirls in India [1] and Pakistan [6]. Education level had a significant influence on menstrual knowledge in India [7] and Nigeria [14] (p-value <0.05).

Who Are Girls' Sources of Information?

Across LMIC studied, mothers were often the most frequently cited source of information and advice for girls regarding menstruation (Table 46.3). Compared to girls residing in urban parts of Ethiopia [21] and India [22], those in rural settings reported their mothers as an information source less often (possibly because there were other female relatives they could turn to). Following mothers, sisters were the next most common resource in four Indian states [1, 11, 12, 23], Mexico [24], Nepal [25], Nigeria [20, 26], Pakistan [6], and Turkey [27], though they were utilized by less than a quarter of girls. In some contexts, sisters and friends surpassed mothers as the primary source of information [6, 13, 28, 29].

A majority of studies which examined the roles of teachers and/or health professionals as providers of menstrual information ranked them as the least common sources compared to female relatives and friends (Egypt [30], Ghana [31], India [1, 7, 11, 12, 15, 32–35], Jordan [36], Malaysia [37–39], Nepal [25], Nigeria [14, 20, 26, 40–42], Sri Lanka [43], and Turkey [27].) Teachers were cited as a source by less than 5.0% of girls questioned in three Indian states [1, 32, 33], Nepal [25], and Sri Lanka [43]. At most, a third of subjects in urban Nigeria cited teachers as a source [41]. While students in

Table 46.3 Most commonly reported sources of menstrual information

First author, Year	Country	Setting	School status	N	Most common source (%)
East Africa					
Zegeye 2009 [21]	Ethiopia	Mix	School-going	564	Mother (39.7%)
North Africa					
Abd El-Hameed 2011 [44]	Egypt	Mix	School-going	160	Mother (59.4%)
El-Gilany 2005 [30]	Egypt	Mix	School-going	642	Mother (92.2%), mass media (92.2%)
Eswi 2012 [45]	Egypt	Urban	School-going	200	Mother (53.0%)
West Africa					
Gumanga 2012 [31]	Ghana	Urban	School-going	456	Parent (80.2%)
Adinma 2009 [26]	Nigeria	Urban	School-going	550	Mother (48.4%)
Ajah 2015 [42]	Nigeria	Urban	School-going	482	Mother (81.5%
Aniebue 2009 [40]	Nigeria	Urban	School-going	495	Mother (71.5%)
Lawan 2010 [20][f]	Nigeria	Urban	School-going	385	Mother (35.3%)
Oche 2012 [14]	Nigeria	Urban	School-going	122	Mother or grandmother (56.6%)
Central America					
Marván 2012 [24][f]	Mexico	Urban	School-going	405	Mother (78.0%)
South Asia					
Bosch 2008 [28]	Bangladesh	Rural	Unclear	86	Sister (29.0%)
Dambhare 2012 [32][a]	India	Mix	School-going	561	Mother (38.2%)
Thakre 2011 [12][a]	India	Mix	School-going	143	Mother (71.3%)
Khanna 2005 [1]	India	Rural	Mix	–	Mother (55.1%)
Dasgupta 2008 [8][a]	India	Rural	School-going	160	Mother (37.5%)
Kanotra 2013 [92]	India	Rural	School-going	323	Mother (94.4%)
Kotecha 2009 [33][a]	India	Rural	School-going	340	Mother (32.9%)
Mudey 2010 [34]	India	Rural	School-going	300	Mother (40.7%)
Shanbhag 2012 [10]	India	Rural	School-going	506	Mother (55.1%)
Sudeshna 2012 [15][a]	India	Rural	School-going	80	Mother or sister (45.0%)
Dhingra 2009 [29]	India	Rural	Unclear	200	Friend (83.0%)
Tiwari 2006 [11]	India	Unclear	School-going	486	Mother (60.7%)
Khanna 2005 [1]	India	Urban	Mix	–	Mother (66.8%)
Sharma 2008 [64][b]	India	Urban	Mix	156	Mother (73.7%)
Omidvar 2010 [46][a]	India	Urban	School-going	215	Mother (54.0%)
Yasmin 2013 [13]	India	Urban	School-going	147	Friend (20.4%)
Bobhate 2011 [7][c]	India	Urban	Unclear	241	Mother (75.9%)
Udgiri 2010 [23][a]	India	Urban	Unclear	63	Mother (63.5%)
Adhikari 2007 [5]	Nepal	Rural	School-going	150	Coursebook (14.7%)
Sharma 2003 [25]	Nepal	Urban	School-going	96	Mother (37.5%)
Ali 2010 [6][a]	Pakistan	Urban	Government school	203	Sister (35.5%)
Ali 2010 [6][a]	Pakistan	Urban	Out-of-school	165	Sister (49.7%)
Ali 2010 [6][a]	Pakistan	Urban	Private school	145	Mother (37.9%)
Chandraratne 2011 [43][b]	Sri Lanka	Urban	School-going	473	Mother (67.0%)

{continued}

Table 46.3 (continued)

First author, Year	Country	Setting	School status	N	Most common source (%)
Southeast Asia					
Lee 2006 [37]	Malaysia	Mix	School-going	2247	Mother (80.0%)
Wong 2011 [47][e]	Malaysia	Rural	School-going	984	Mother (62.3%)
Wong 2011 [38][d]	Malaysia	Rural	School-going	577	Mother (31.7%)
Wong 2011 [48][d]	Malaysia	Urban	School-going	407	Mother (62.7%)
West Asia					
Jarrah 2012 [36]	Jordan	Urban	School-going	408	Mother (57.1%)
Reis 2011 [27]	Turkey	Urban	Mix	310	Mother (18.7%)
Ersoy 2004 [2]	Turkey	Urban	School-going	1017	Mother (55.7%)

[a]Pre-menarcheal sources
[b]Sources among those with menstrual problems
[c]Post-menarcheal sources
[d]Sources among those who ever received information
[e]Sources among those with dysmenorrhea
[f]First source

urban Malaysia were more likely than those in rural settings to cite teachers as a source, a considerable number had never encountered menstrual-related topics in school [39]. Less than 1.0% of girls in a rural part of India [33] and urban parts of Jordan [36] reported having received information regarding menstruation from health professionals. At most, a quarter of study participants in urban Nigeria cited health professionals as a source [41].

Some studies reported mass media, such as radio, television, newspapers, magazines, books, and the Internet either as the only resource available to girls or as supplements to other sources of information (Egypt [30, 44, 45], Ghana [31], India [15, 29, 32, 34, 35, 46], Jordan [36], Malaysia [37–39, 47, 48], Nepal [25], Nigeria [40–42], Sri Lanka [43], and Turkey [2, 27]). In a few instances, such sources were reported by more than a quarter of girls: 72.4% in Nigeria [41], 92.2% in Egypt [30], and 29.2–43.6% in Malaysia [37, 38, 47, 48].

Girls reported not having received information from any source in some studies. As few as 6.8 and 7.0% of girls in urban Nigeria [42] and Egypt [45], respectively, and 7.8% of a mix of urban- and rural-living girls in Ethiopia [21] reported having no source. A study of urban- and rural-living girls in India reported a quarter without a source [32]. In rural Nepal, 76.0% of girls reported having no menstrual information source [5].

How Well Do Adults Respond to Girls' Information Needs?

Whether by a relative, friend, or other community member, the information on menstrual health and hygiene provided to adolescents is not always timely nor is it adequate.

Researchers found that mothers and other relatives in India [1, 7] and Tanzania [49] who did provide girls with information often did not do so until after menarche. In Mexico, however, 94.0% of girls reported that they had discussed menstruation with their mothers prior to menarche [24]. A study in Nigeria reported 55.2% of school-going girls were "trained" prior to reaching menarche, which included being made aware of what to expect at menarche and how to collect blood and dispose of materials [40]. Further, parents' education level was found to have a significant influence on pre-menarcheal knowledge in Nigeria (p-value <0.05) as girls whose parents had received tertiary education were the most likely to have been trained [40]. In India [15, 33] and Kenya [50, 51], girls reported that little information about menstruation was provided and nearly no explanation. Sources of information may have their own misconceptions about menstruation, which they may pass on. Mothers interviewed in Bangladesh attributed menstruation to God [28]. During initiation rites in Malawi, misconceptions, like men can get hurt if they come in contact with menstrual blood, are told to girls by female relatives [52].

Given the link between menstruation and the ability to conceive, mothers interviewed in Bangladesh did not consider it appropriate to discuss the matter with their pre-menarcheal daughters [28]. Both mothers and teachers, most of whom were male, in Kenya cited discomfort as an impediment to discussing menstruation with girls [53]. Teachers in rural Tanzania warned their students that their mothers would be very upset if told about their reaching menarche [49]. This may be a result of cultural taboos that prevent parents from discussing sex-related topics with their daughters. Taboos were also cited as a by the few teachers in Tanzania who wanted to provide support to their students [54]. Teachers in Kenya did not perceive menstrual education as part of their role nor did they feel properly prepared to share information with their students [53].

A majority of teachers (70–90%) at schools in Ghana who had been trained to use play-based approaches to promote menstrual knowledge and practices were confident in discussing menstruation with their students compared to their counterparts at schools not using similar approaches who had limited conversations [55]. Overall, 82.4% of study participants in Jordan felt they were not adequately prepared for reaching menarche [36]. Of girls in rural Nepal who received information from a parent, friend or a coursebook, an overwhelming majority felt menstrual-related topics were not properly taught [5].

Four-fifths of school-going girls questioned in Egypt wanted more information [30]. However, girls in Malaysia [39] and Tanzania [54] reported feeling ashamed, embarrassed and uncomfortable when inquiring about menstruation from adults.

WHAT ARE THE EMOTIONAL, PHYSICAL AND SOCIAL IMPACTS OF MENSTRUATION ON GIRLS?

Anticipated and experienced menstruation-related impacts were mostly negative (with some positive impacts when girls were better informed/prepared). While eight out of every ten study participants in Mexico expected at least one positive change to occur upon reaching menarche, all expected at least one negative change to occur; the nine most expected changes among urban and rural girls were negative, e.g., discomfort, worry, cramps [56]. Overall, 89.4% of anticipated changes reported by pre-menarcheal girls and 88.7% of experienced changes reported by post-menarcheal girls were negative [57].

EMOTIONAL IMPACTS

A quarter to eight out of every ten girls questioned in various LMIC reported not being mentally prepared for reaching menarche (Brazil [58], China [4], India [11], Jordan [36], Mexico [24, 56], Nepal [25], Nigeria [40],). Many girls had negative reactions to their first period (Table 46.4). For example, a majority of school-going girls in one study in India described menarche as a shocking or fearful event and many cried upon seeing their blood [18].

Some school-going girls perceived menstrual blood to be dirty or described feeling disgusted by their period: 30.5% in Lebanon [59], 48.9% versus 72.8% in rural versus urban Malaysia [38, 48], and 10.0–23.4% in two Indian states [11, 17, 34]. Additionally, girls in Kenya revealed that "the girl with her period is the one to hang her head" because she becomes the target of unwanted and sometimes unkind attention [53]. Mood swings and irritability connected to menstruation were both reported by more than two-thirds of schoolgirls in India [60], Lebanon [59], and Malaysia [38, 48].

Not all feelings about reaching menarche were negative; more than half of schoolgirls questioned in China [4], India [9], and Malaysia [38, 48] felt pride in maturing. Focus groups in rural and urban settings in Kenya [51] and Tanzania [61] with female students revealed a similar sentiment. The more school-going girls in Mexico knew about menstruation, the less negative their attitudes (p-value < 0.05); and the more prepared they felt, the more positive their attitudes (p-value < 0.0001) [24]. A later menarcheal age and higher socioeconomic status seemed to further reduce negative reactions among girls in Malaysia [39] and Turkey [2], respectively.

Table 46.4   Negative reaction upon reaching menarche

First author, Year	Country	Setting	School status	N	Negative reaction[a] (%)
West Africa					
Aniebue 2009 [40]	Nigeria	Urban	School-going	495	50.3%
Oche 2012 [14]	Nigeria	Urban	School-going	122	53.3%
Central America					
Marván 2001 [57]	Mexico	Urban	School-going	98	15.3%
East Asia					
Tang 2003 [4]	China	Unclear	School-going	1,573	72.0%
South Asia					
Bosch 2008 [28]	Bangladesh	Rural	Unclear	86	64.0%
Mudey 2010 [34]	India	Rural	School-going	300	43.7%
Shanbhag 2012 [10]	India	Rural	School-going	329	44.1%
Tiwari 2006 [11]	India	Unclear	School-going	763	20.6%
Bobhate 2011 [7]	India	Urban	Unclear	241	64.7%
Udgiri 2010 [23]	India	Urban	Unclear	342	31.0%
Adhikari 2007 [5]	Nepal	Rural	School-going	150	96.7%
Ali 2010 [6]	Pakistan	Urban	Government school	425	55.8%
Ali 2010 [6]	Pakistan	Urban	Out-of-school	425	62.6%
Ali 2010 [6]	Pakistan	Urban	Private school	425	55.1%
West Asia					
Reis 2011 [27]	Turkey	Urban	Mix	310	43.9%
Ersoy 2004 [2]	Turkey	Urban	School-going	1,017	49.8%

[a]Reported shock, panic, confusion, tension, fear, shame or embarrassment at menarche

Physical Impacts

Physical impacts of menstruation that were commonly reported across studies included premenstrual symptoms or syndrome and painful periods. These outcomes were almost always reported by at least half of the sample (Table 46.5). At most, 93.2% of rural-living girls in India reported experiencing a premenstrual symptom [62] and 94.4% of school-going girls in Egypt reported experiencing dysmenorrhea [44]. In Ethiopia, girls with premenstrual symptoms suffered more often from dysmenorrhea than those without (82.4% versus 40.3%, respectively) [21]. Of girls with dysmenorrhea in Ghana, nearly two-thirds experienced symptoms during most or all cycles [31]. A majority of rural-living girls surveyed in Malaysia considered dysmenorrhea a normal aspect of menstruation [47].

Table 46.5 Physical impacts of menstruation

First author, Year	Country	Setting	School status	N	PMS[a]	Severe pain[b]	Headache	Swelling[f]	Fatigue[d]
East Africa									
Zegeye 2009 [21]	Ethiopia	Mix	School-going	565	75.4%	72.0%	NR	NR	NR
North Africa									
Abd El-Hameed 2011 [44]	Egypt	Mix	School-going	160	NR	94.4%	NR	NR	NR
West Africa									
Gumanga 2012 [31]	Ghana	Urban	School-going	456	NR	74.4%	NR	NR	NR
Ajah 2015 [42]	Nigeria	Urban	School-going	482	75.1%	64.1%	NR	NR	NR
Aniebue 2009 [40]	Nigeria	Urban	School-going	495		26.7%	NR	NR	NR
Central America									
Marván 2001 [57]	Mexico	Urban	School-going	98	NR	NR	6.1%	NR	14.3%
South America									
Pitangui 2013 [58]	Brazil	Urban	School-going	174	NR	73.0%	14.4%	61.5%	27.6%
South Asia									
Dambhare 2012 [32]	India	Mix	School-going	561	56.3%	56.2%	26.7%	NR	NR
Thakre 2012 [22]	India	Mix	School-going	387	55.8%	61.0%	NR	NR	NR
Baidya 2014 [77]	India	Rural	Mix	200	8.5%	59.5%	16.0%	NR	4.0%
Rana 2015 [63]	India	Rural	Mix	400	NR	46.8%	NR	NR	NR
Bodat 2013 [67]	India	Rural	School-going	622	NR	58.1%	NR	NR	NR
Kanotra 2013 [92]	India	Rural	School-going	323	3.1%	18.3%	25.7%	NR	NR
Mudey 2010 [34]	India	Rural	School-going	300	NR	NR	NR	NR	NR
Shanbhag 2012 [10]	India	Rural	School-going	329	NR	61.3%	NR	NR	NR
Wasnik 2015 [72]	India	Rural	School-going	435	17.9%	62.3%	6.7%	NR	NR
Patil 2013 [62]	India	Rural	Unclear	440	93.2%	28.0%	1.8%	NR	NR
Chaudhuri 2012 [60][e]	India	Unclear	School-going	224	NR	59.8%	28.1% (of 128)	NR	70.3% (of 128)
Sharma 2008 [64]	India	Urban	Mix	198	63.1%	67.2%	16.7%	11.1%	48.0%
Chandraratne 2011 [43][f]	Sri Lanka	Urban	School-going	594	66.2%	61.3%	28.1% (of 393)	NR	29.1% (of 393)

(continued)

Table 46.5 (continued)

First author, Year	Country	Setting	School status	N	PMS[a]	Severe pain[b]	Headache	Swelling[f]	Fatigue[d]
Nair 2012 [73]	India	Urban	School-going	3443	NR	72.4%	13.9%	NR	36.1%
Sharma 2003 [25]	Nepal	Urban	School-going	96	NR	69.8%	NR	NR	NR
Southeast Asia									
Lee 2006 [37]	Malaysia	Mix	School-going	2247	74.7%	69.3%	NR	NR	NR
Wong 2011 [47]	Malaysia	Rural	School-going	1295	NR	76.0%	NR	NR	NR
Wong 2011 [38]	Malaysia	Rural	School-going	1295	63.1%	NR	47.3%	12.6%	81.1%
Wong 2010 [78]	Malaysia	Urban	School-going	1075	NR	74.5%	NR	NR	NR
Wong 2011 [48]	Malaysia	Urban	School-going	1076	56.5%	NR	38.4%	13.1%	75.4%
West Asia									
Poureslami 2002 [79]	Iran	Urban	School-going	250	NR	71.2%	NR	NR	NR
Jarrah 2012 [36]	Jordan	Urban	School-going	490	NR	NR	50.4%	NR	80.6%
Santina 2012 [59]	Lebanon	Urban	School-going	389	NR	74.3%	22.8%	34.6%	NR
Eryilmaz 2009 [80][g]	Turkey	Unclear	School-going	1951	NR	72.2%	26.1% (of 1408)	NR	11.9% (of 1408)
Reis 2011 [27]	Turkey	Urban	Mix	310	NR	23.9%	NR	NR	NR

PMS Premenstrual symptoms/syndrome, NR Not reported

[a] Premenstrual symptoms or premenstrual syndrome
[b] Including dysmenorrhea
[c] Swelling or bloating
[d] Fatigue or dizziness
[e] Headaches and fatigue reported among girls with dysmenorrhea who agreed to provide more information
[f] Headaches and fatigue reported among girls with PMS
[g] Headaches and fatigue reported among girls with dysmenorrhea

SOCIAL IMPACTS

Activities of daily living or daily routines were restricted by menstruation among a quarter of girls in rural India [63], a third of female students in Brazil [58] and Egypt [44], and among 60.0% of slum dwellers in India [64]. In urban Sri Lanka, schoolgirls with premenstrual syndrome had significantly more disruptions to their daily routines than those without [43]. Daily activities are further limited by taboos related to what and who menstruating girls are able to come in contact with. Menstruating girls in India and Nepal are sometimes limited from entering kitchens or bedrooms to ensure menstrual blood does not contaminate food or others [1, 5, 11, 12, 63, 65]. Household work such as cooking was often cited as 'not allowed' for menstruating girls in India [1, 12, 63], Kenya [53], and Nepal [5]. Female students from a mix of rural and urban settings in India reported limitations on who they could touch while menstruating [12, 18]. Other social limitations frequently reported include avoiding physical or social activities (e.g., sports and functions), abstaining from religious activities or missing school (Table 46.6).

Girls in Malaysia [38, 48] and Pakistan [6] reported restrictions on religious activities due to menstruation. Studies reporting the complete abstinence of religious activities come mostly from India; this practice was reported by 44.7–94.2% of girls interviewed [7–10, 12, 13, 15, 34, 60, 63, 64]. Education level had a significant influence on the practice of avoiding holy places in India (p-value < 0.05) [23]. Mothers interviewed in Nigeria revealed that they advised their daughters to refrain from praying during their periods [41]. This aligns with another study in Nigeria, in which 43.4% of girls reported abstaining from religious activities [14]. Health education interventions in Bangladesh [66] and India [19] did not result in significant declines in religious restrictions among girls during their periods.

When asked if girls can go to school while menstruating, 70.7% of girls in rural Nepal responded 'no' [5]. Actual absenteeism reported in various LMIC did not reach that level, instead ranging from 2.0% of urban-living girls in Nigeria [40] to 61.7% of rural-living girls in Uganda [16]. Focus groups in Malaysia revealed that dysmenorrhea may have a greater impact on school absenteeism for girls in urban settings than those in rural settings [38]. Dysmenorrhea was significantly associated with missing school among urban-living girls in Lebanon [59] as was pain severity among those in Brazil [58] (p-value < 0.05). Menstrual disorders in general was significantly associated with missing school among rural-living girls in India (p-value < 0.001) [67]. In Kenya, male teachers reportedly teased girls about menstruation when they returned to school after being absent for a few days [53]. Although the teachers denied this, they noted that they were concerned with girls being distracted in class [53]. Teachers interviewed in Ghana shared similar concerns about girls being distracted and missing school [68]. Girls, themselves, in India [60], Malaysia [47, 48], and Uganda [16] associated menstruation with poor academic performance and low grades.

Table 46.6 Social impacts of menstruation

First author, Year	Country	Setting	School status	N	Avoid physical or social activities[a] (%)	Abstain from religious activities (%)	Miss school or work (%)
East Africa							
Zegeye 2009 [21][b]	Ethiopia	Mix	School-going	407	NR	NR	48.2%
Boosey 2014 [16]	Uganda	Rural	School-going	140	NR	NR	61.7% (of 133)
North Africa							
Abd El-Hameed 2011 [44]	Egypt	Mix	School-going	160	54	NR	NR
West Africa							
Adinma 2009 [26]	Nigeria	Urban	School-going	550	NR	NR	4.5%
Ajah 2015 [42]	Nigeria	Urban	School-going	482	NR	NR	12.2%
Aniebue 2009 [40]	Nigeria	Urban	School-going	495	37.6%	NR	2.0%
Oche 2012 [14]	Nigeria	Urban	School-going	122	4.9%	43.4%	NR
South America							
Pitangui 2013 [58][b]	Brazil	Urban	School-going	127	NR	NR	30.7%
South Asia							
Dambhare 2012 [32][b]	India	Mix	School-going	561	NR	NR	13.9%
Juyal 2013 [65]	India	Mix	School-going	453	8.6%	87.4%	NR
Thakre 2011 [12]	India	Mix	School-going	387	NR	44.7%	5.2%
Rana 2015 [63]	India	Rural	Mix	400	28.0%	53.2%	26.4%[c]
Bodat 2013 [67]	India	Rural	School-going	622	NR	NR	43.2%
Dasgupta 2008 [8]	India	Rural	School-going	136	36.3%	60.0%	13.8%
Kanotra 2013 [92][b]	India	Rural	School-going	59	76.6%	87.0%	NR
Mudey 2010 [34]	India	Rural	School-going	300	NR	94.2%	NR
Shanbhag 2012 [10]	India	Rural	School-going	329	NR	75.8%	37.9%
Sudeshna 2012 [15]	India	Rural	School-going	190	NR	NR	25.8%
Chaudhuri 2012 [60][b]	India	Unclear	School-going	128	53.5%	NR	17.2% (of 116)[c]
Sharma 2008 [64]	India	Urban	Mix	156	25.6%	NR	14.0%
Goel 2011 [9]	India	Urban	School-going	478	42.7%	76.2%	NR
Sharma 2003 [25]	Nepal	Urban	School-going	67	20.0%	NR	NR

(continued)

Table 46.6 (continued)

First author, Year	Country	Setting	School status	N	Avoid physical or social activities[a] (%)	Abstain from religious activities (%)	Miss school or work (%)
Yasmin 2013 [13]	India	Urban	School-going	147	18.4%	90.5%	NR
Bobhate 2011 [7]	India	Urban	Unclear	241	24.1%	90.0%	NR
Ali 2010 [6]	Pakistan	Urban	Government school	425	67.3%	NR	NR
Ali 2010 [6]	Pakistan	Urban	Out-of-school	425	58.1%	NR	NR
Ali 2010 [6]	Pakistan	Urban	Private school	425	58.1%	NR	NR
Southeast Asia							
Lee 2006 [37]	Malaysia	Mix	School-going	2247	NR	NR	7.0%
Wong 2011 [47][b]	Malaysia	Rural	School-going	984	58.6%	NR	18.1%
Wong 2010 [78][b]	Malaysia	Urban	School-going	801	50.2%	NR	NR
Wong 2011 [48]	Malaysia	Urban	School-going	1076	61.5%	NR	NR
West Asia							
Poureslami 2002 [79]	Iran	Urban	School-going	250	33.0%	NR	15.2%
Santina 2012 [59]	Lebanon	Urban	School-going	389	NR	NR	41.4%

[a]Avoid or reduce physical activities (including playing and sports) or social activities (including functions and friendships with males)
[b]Among girls with dysmenorrhea
[c]Among school-going girls

A puberty education intervention with provision of sanitary pads in a non-randomized trial in Ghana significantly improved girls' school attendance (p-value < 0.001) [69]. Alternatively, a randomized trial in Nepal [70] demonstrated that providing menstrual cups may improve convenience and mobility, and one in Kenya [71] demonstrated that they can reduce distractions associated with leakage and improve school attendance.

HOW DO GIRLS RESPOND TO NEGATIVE EFFECTS AND WHAT PRACTICES DO THEY DEVELOP?

To address the physical impacts of menstruation described in Table 46.5, some girls reported using traditional medicine or remedies (Bangladesh [72], Brazil [58], India [63, 64, 73, 74], Malaysia [37, 47, 48, 75], Sri Lanka [43],) and others reported taking medication to relieve pain, often by self-medicating or consulting pharmacies (Bangladesh [72], Brazil [58], Egypt [44], India [60, 63, 64, 74], Iran [76], Malaysia [39, 47, 75], Nigeria [42], and Turkey [27, 77]). Consultation of health professionals for menstrual-related problems was minimal, generally reported by less than a fifth of girls (Bangladesh [72], Brazil [58], Ethiopia [21], India [60, 63, 64, 73, 74], Iran [76], and Malaysia [37, 47, 48],). However, one study did report that 69.8% of Indian girls with problems sought attention from a health professional [7]. Another study in India reported that 19.2% of girls with a problem never discussed it with anyone—a health professional, relative, or friend [64]. Girls in Bangladesh were significantly more likely to consult someone regarding their problems after participating in 12 health education sessions over the course of six months than at baseline (p-value < 0.01) [66].

General lack of adult guidance related to menstruation may contribute to the variation in basic hygiene management practices such as use of sanitary absorbents and bathing daily across LMIC (Table 46.7). Use of sanitary pads to absorb blood ranged from 2.0% of schoolgirls in rural Nepal [5] to 69.1–93.8% of urban-living girls in Nigeria [14, 20, 40, 41]. All but one study of girls in rural parts of seven Indian states reported greater proportions of girls using cloth compared to sanitary pads [1, 8, 10, 12, 15, 18, 29, 34, 63, 67, 78]. Sanitary pad use was significantly higher among urban-living girls in India [22] and Ethiopia [21] as was use of sanitary pads or new cloth among school-going girls in India [1] (p-value < 0.01). A quasi-experimental study testing village-based meetings for girls in India as a platform for disseminating health messages contributed to significant increases in the use of sanitary pads and a decrease in the reuse of cloth (p-value < 0.05) [3]. While fewer girls using sanitary pads in one study in rural India reported poor fit and rashes than did girls using cloth [63], the cost of sanitary pads were a concern for some girls questioned in other studies in India [12, 15, 34], Tanzania [61], and Uganda [16]. Almost all of the 102 urban-living girls questioned in Kenya preferred sanitary pads for their convenience and reliability, but nearly

Table 46.7 Menstrual hygiene management practices (%)

First author, Year	Country	Setting	School status	N	Use sanitary pads (%)	Use sanitary pads and cloth (%)	Use old or new cloth (%)	Use othermaterial[a] (%)	Bathe daily (%)
East Africa									
Zegeye 2009 [21]	Ethiopia	Mix	School-going	565	37.6%	NR	62.5%	NR	NR
North Africa									
Abd El-Hameed 2011 [44]	Egypt	Mix	School-going	160	NR	NR	NR	NR	100.0%[b]
El-Gilany 2005 [30]	Egypt	Mix	School-going	642	66.8%	NR	27.9%	5.3%	70.9%[b]
West Africa									
Iliyasu 2012 [41]	Nigeria	Urban	Mix	184	81.0%	NR	NR	NR	NR
Aniebue 2009 [40]	Nigeria	Urban	School-going	495	69.1%	NR	9.1%	21.8%	NR
Lawan 2010 [20]	Nigeria	Urban	School-going	371	93.8%	NR	6.2%	NR	NR
Oche 2012 [14]	Nigeria	Urban	School-going	122	86.9%	NR	9.0%	4.1%	NR
Boosey 2014 [16][c]	Uganda	Rural	School-going	140	47.1%	NR	87.1%	37.1%	NR
South Asia									
Khanna 2005 [1]	India	Mix	Out-of-school	304	2.0%	NR	90.9%	0.3%	NR
Juyal 2013 [65]	India	Mix	School-going	453	38.4%	26.7%	34.9%	NR	63.6%
Khanna 2005 [1]	India	Mix	School-going	307	6.2%	NR	68.4%	0.7%	NR
Khanna 2005 [1]	India	Rural	Mix	281	3.2%	NR	92.2%	0.7%	NR
Rana 2015 [63]	India	Rural	Mix	400	39.0%	NR	61.0%	NR	NR
Bodat 2013 [67]	India	Rural	School-going	622	48.1%	NR	51.9%	NR	NR
Dasgupta 2008 [8]	India	Rural	School-going	160	11.3%	40%	48.8%	NR	NR
Kanotra 2013 [92]	India	Rural	School-going	323	89.5%	NR	10.5%	NR	NR
Mudey 2010 [34]	India	Rural	School-going	300	15.7%	NR	46.7%	NR	NR
Narayan 2001 [18]	India	Rural	School-going	327	1.7%	48%	82.5%	NR	NR
Shanbhag 2012 [10]	India	Rural	School-going	329	44.1%	21.2%	34.7%	NR	88.8%
Sudeshna 2012 [15]	India	Rural	School-going	190	13.2%	24.2%	62.6%	NR	NR
Thakre 2011 [12]	India	Rural	School-going	146	30.8%	NR	69.2%	NR	NR
Wasnik 2015 [72]	India	Rural	School-going	435	33.6%	9.2%	57.2%	NR	NR

(continued)

Table 46.7 (continued)

First author, Year	Country	Setting	School status	N	Use sanitary pads (%)	Use sanitary pads and cloth (%)	Use old or new cloth (%)	Use other material[a] (%)	Bathe daily (%)
Dhingra 2009 [29]	India	Rural	Unclear	200	NR	NR	NR	NR	0.0%[2]
Khanna 2005 [1]	India	Urban	Mix	330	4.8%	NR	69.0%	0.3%	NR
Goel 2011 [9]	India	Urban	School-going	478	NR	NR	NR	NR	92.9%
Nair 2012 [73]	India	Urban	School-going	3443	45.5%	38.2%	16.3%	NR	97.6%
Narayan 2001 [18]	India	Urban	School-going	292	8.3%	17.1%	72.2%	NR	NR
Omidvar 2010 [46]	India	Urban	School-going	350	68.9%	NR	19.1%	11.1%	81.7%[b]
Thakre 2011 [12]	India	Urban	School-going	241	60.6%	NR	39.4%	NR	NR
Yasmin 2013 [13]	India	Urban	School-going	147	82.3%	1.4%	16.3%	NR	85.7%
Bobhate 2011 [7]	India	Urban	Unclear	241	43.2%	NR	41.5%	15.4%	NR
Adhikari 2007 [5]	Nepal	Rural	School-going	150	2.0%	NR	98.0%	NR	4.0%
Ali 2010 [6][c]	Pakistan	Urban	Government school	425	17.9%	NR	87.5%	3.9%	44.2%[b]
Ali 2010 [6][c]	Pakistan	Urban	Out-of-school	425	13.2%	NR	81.0%	6.6%	45.9%[b]
Ali 2010 [6][c]	Pakistan	Urban	Private school	425	33.8%	NR	62.6%	4.4%	45.2%[b]
West Asia									
Pouresmali 2002 [79]	Iran	Urban	School-going	250	NR	NR	NR	NR	66.0%[b]

NR Not reported

[a]Other materials include various types of tissue and cotton

[b]Proportion of girls that bathe at all during their period

[c]Multiple responses accepted for materials used

half used cloth or a combination of cloth and sanitary pads to save money [50]. Tissue paper and cotton were also cited as absorbents for girls in various LMIC, with tissue paper reported by as many as 37.1% of rural-living schoolgirls in Uganda [16].

A minority of girls in Egypt [30] and India [12, 46], no more than one in five, compared to 56.5% of girls in Nigeria [20] changed absorbents while at school. Most girls in Egypt [30] and Uganda [16] felt schools lacked privacy and most in India [12] and Nigeria [20] preferred to change materials at home. Insufficient latrines, water supplies and disposal infrastructure further presented a barrier for students in India [15, 67], Tanzania [61], and Uganda [16] to manage their periods at school.

Methods for disposing of materials beyond throwing them away with other trash included burning, burying, and flushing materials. Very few (2.5%) girls in Egypt dispose of absorbents by burning them compared to 17.0–76.0% in India [12, 63, 73] and Nigeria [14, 40]. Indian girls in rural settings were significantly more likely to report burning materials than those in urban settings (p-value < 0.05) [22]. Of those who reuse cloth, drying washed materials in sunlight rather than in hiding varied from 30.7% of girls in urban Pakistan [6] to 44.3–72.4% of schoolgirls in India [10, 12, 15, 18, 73]. School-based health education in India led to significant improvements in washing cloth with soap, in drying them in the sun and in disposing of them safely [17].

Reported bathing practices in India ranged from all 200 rural-living girls in one state abstaining during menstruation [29] to nearly all 3,443 girls in urban areas of another state bathing daily [73]. One study found that the practice of daily bathing was significantly higher among urban-living girls than rural-living girls (p < 0.05) [65], and another found that both a regular source of water and a private bathroom exclusive to a family had significant relationships with taking a daily bath (p < 0.001) [13]. In Turkey [27] and Nigeria [20], 11.9% and 72.5% of urban-living girls reported increasing the number of baths they take. Girls in rural Kenya revealed that they wanted to bathe more frequently during their period, but were concerned about using limited water supplies and feared revealing to their family that they were menstruating [51]. A quasi-experimental study involving 698 girls in Iran showed that those who participated in 10 two-hour teacher-led sessions on pubertal changes engaged in regular bathing more than those who did not, and the difference was significant (p-value < 0.01) [79]. Another quasi-experimental study in Egypt found a significant increase in the number of girls bathing daily during their periods after participating in four 30–45 min health education sessions focused on menstruation (p-value < 0.001) [80]. Four studies in India defined satisfactory cleaning of genitalia to mean washing two or more times a day while menstruating; one-third to three-quarters of girls met this criterion [12, 13, 29, 34]. Unsatisfactory cleaning was significantly higher among rural-living girls than those in urban parts [22]. In a study evaluating the impact of school-based health education, the proportion of girls using soap to clean their genitalia significantly increased from 30.0 to 94.3% (p-value < 0.01) [17].

Girls in Mexico who had previous knowledge of the physiology of menstruation were significantly more likely to know what was happening in their bodies and what to do, in terms of hygiene management, upon reaching menarche (p-value < 0.001) [24]. For schoolgirls in Jordan, being prepared prior to menarche resulted in more positive attitudes, and attitude had a significant positive correlation with practices (p-value < 0.05) [36]. Schooling status among girls in Iran also had a significant positive correlation with practices (p-value < 0.01) [6, 66].

CONCLUSION

An important limitation of this review is that vague measures are often used to describe the menstrual experiences of girls, which impede data aggregation and direct comparisons. For instance, studies used different yardsticks for adequate or inadequate knowledge, and used the terms premenstrual syndrome and dysmenorrhea loosely. Further, many studies had small sample sizes and relied heavily on self-report. Some studies had low response rates due to discomfort or limitations to discussing menstruation. Another limitation is that a majority of relevant data from included studies come from a limited number of countries and are not representative of all LMIC. Among the included countries and across all LMIC is great cultural variation, and the results presented here should be considered in light of these unique perspectives. Despite these limitations, the evidence presented allows for the following conclusions:

- Substantial numbers of girls in many countries have knowledge gaps and misconceptions about menstruation. This leaves them unprepared when they reach menarche and causes fear and anxiety.
- Mothers, other female relatives and female peers are their main sources of information and advice on menstruation. The information they receive, however, is not always timely nor adequate. Only some have access to additional information from sources such as mass media and the Internet.
- Girls experience a variety of symptoms during menstruation—pain, headaches and fatigue. These symptoms combined with taboos result in their not being able to participate in household, school, or social activities.
- Very few girls seek health care when they experience menstrual health problems. If anything, they may resort to household remedies.
- Girls in poor urban and rural communities of LMIC are less likely to obtain and use sanitary pads. Instead, they use materials made at home

with scraps of old cloth, cotton, paper, etc. Lack of privacy, access to clean water and functional toilets make it harder for them to manage their periods.

It is clear that far too many girls across LMIC are struggling with nearly complete ignorance of their normal biological maturation and its consequences, and when they do receive education, still struggle with inadequate sanitary materials and insufficient physical and emotional support. Although there is no convincing evidence that poor menstrual hygiene management leads to ascending reproductive tract infections [81] or causes lasting sequelae, this review underscores that coming to terms with menarche and navigating the shame and practical challenges associated with its management may cause girls great anxiety and sadness. There remains a need for further research into the physical, mental and social impacts of such distress. For example, being unprepared for menarche, being excluded and shamed during monthly periods, being hindered in self-care and uncared for when unwell, undermines a girl's sense of being in charge of her life, her sense of self-worth and her sense that the individuals and institutions around her are responsive to her needs. The huge and lasting impacts this can have on girls' lives remain to be studied.

In the short term, however, there are intervention studies that demonstrate the ability to improve girls' menstrual knowledge and hygiene management. Health education interventions like school-based sessions tested in India have resulted in improved understanding post-intervention [17], and similar programs in Egypt [80] and Iran [79] have improved the bathing practices of girls during their periods. Additionally, a quasi-experimental study in India that involved training of medical officers and providing reference tools led to statistically significant improvements in their case management of menstrual health problems for female patients between the ages of 15 and 24 [82].

Some exciting initiatives led by academics, international agencies and the private sector are also under way. Educating and encouraging parents to communicate with their daughters and sons about puberty and menstruation is being implemented by the Families Matter Program [83]. A five-year initiative by Columbia University has launched locally designed puberty booklets for girls and for boys in Tanzania, Ghana, Ethiopia, and Cambodia; these have been embraced by the Ministries of Education and of Health in all four countries [84]. Save the Children has also developed workbooks, modeled on those by Columbia University, for girls and boys in Nepal, Uganda and Malawi and is carrying out puberty education programs in multiple countries [85]. Similar efforts, such as CycleSmart™ [86] and GrowUp Smart [87], are being implemented by Georgetown University's Institute of Reproductive Health in Rwanda and Guatemala. In 2014, UNESCO published a policy booklet with guidance to improve school administrators and teachers' abilities to educate and support girls and boys in classrooms [88]. Procter & Gamble, a major producer of sanitary products, has launched communication programmes in several of the countries where it sells products with marketing approaches aimed at legitimizing family discussion of menstruation, and engaging and educating girls while building their self-esteem [89]. During its Celebrating Womanhood event in 2013, the Water Supply & Sanitation Collaborative Council defined menstrual hygiene as a priority and outlined a 3-pronged approach that includes breaking the silence around this topic, ensuring hygienic management, and identifying mechanisms for safe reuse and disposal of materials [90]. Linking menstruation education with efforts to improve water, sanitation and disposal facilities in schools has also been actively promoted and implemented by UNICEF at the country and global levels [91].

While these initiatives are important and promising first steps, greater uptake and commitment is needed to fulfill the rights of girls related to menstrual knowledge, health and hygiene. Concerted multi-level efforts are required to achieve this. At the individual level, girls and boys need to be educated about puberty. At the family level, girls need support during their menstrual cycles. At the community level, we must improve access to sanitary products, running water, functional toilets and privacy. We need competent and caring health care workers who can respond to girls' questions and concerns, and to provide care when they have menstrual health problems. Finally, we need leaders who can change the perception of menarche and menstruation to one of normalcy and promise rather than of shame.[3]

Author Contributions This review was conceived by VCM. He supported SVP in carrying out the review and preparing the first draft of the paper. Initial findings were presented virtually by VCM and SVP at UNESCO's International Technical Consultation on Menstrual Hygiene Management and the Education Sector in July 2013. SVP updated the search, incorporated the new data and prepared a revised draft of the paper. VCM and SVP reviewed and approved the final draft.

Funding Sheila Patel was engaged by WHO to research and coauthor the paper. Venkatraman Chandra-Mouli is a staff member of the World Health. Organization. This is part of his work.

NOTES

1. "Mapping the Knowledge and Understanding of Menarche, Menstrual Hygiene and Menstrual Health among Adolescent Girls in Low- and Middle-Income Countries" by Venkatraman Chandra-Mouli and Sheila Vipul Patel was first published in *Reproductive Health* 14 (30): 1–16. Reprinted with permission [OA CC-BY 4.0].
2. Only descriptive studies with quantitative results are included in Tables 46.2, 46.3, 46.4, 46.5, 46.6 and 46.7.
3. Marni Sommer, Mailman School of Public Health, Columbia University reviewed and provided inputs on an earlier version of the paper.
   Asanthi Balapitiya, Ministry of Health, Sri Lanka, who interned with WHO in 2015, updated the search in July 2015 and completed initial data abstraction.

ABBREVIATIONS

LMIC: Low- and middle-income countries; MHM: Menstrual hygiene management

REFERENCES

1. Khanna A, Goyal RS, Bhawsar R. Menstrual practices and reproductive problems: a study of adolescent girls in Rajasthan. J Health Manag. 2005;7(1):91–107.
2. Ersoy B, et al. Effects of different socioeconomic conditions on menarche in Turkish female students. Early Hum Dev. 2004;76(2):115–25.
3. Dongre AR, Deshmukh PR, Garg BS. The effect of community-based health education intervention on management of menstrual hygiene among rural Indian adolescent girls. World Health Popul. 2007;9(3):48–54.
4. Tang CS, Yeung DY, Lee AM. Psychosocial correlates of emotional responses to menarche among Chinese adolescent girls. J Adolesc Health. 2003;33(3):193–201.
5. Adhikari P, et al. Knowledge and practice regarding menstrual hygiene in rural adolescent girls of Nepal. Kathmandu Univ Med J (KUMJ). 2007;5(3):382–6.
6. Ali TS, Rizvi SN. Menstrual knowledge and practices of female adolescents in urban Karachi, Pakistan. J Adolesc. 2010;33(4):531–41.
7. Bobhate P, Shrivastava S. A cross sectional study of knowledge and practices about reproductive health among female adolescents in an urban slum of Mumbai. J Fam Reprod Health. 2011;5(4):117–24.
8. Dasgupta A, Sarkar M. Menstrual hygiene: how hygienic is the adolescent girl? Indian J Community Med. 2008;33(2):77–80.
9. Goel MK, Kundan M. Psycho-social behaviour of urban Indian adolescent girls during menstruation. Australas Med J. 2011;4(1):49–52.
10. Shanbhag D, Shilpa R, D'Souza N, Josphine P, Singh J, Goud BR. Perceptions regarding menstruation and Practices during menstrual cycles among high school going adolescent girls in resource limited settings around Bangalore City, Karnataka, India. Int J Collab Res Inter Med Public Health. 2012;4(7):1353–62.
11. Tiwari H, Oza UN, Tiwari R. Knowledge, attitudes and beliefs about menarche of adolescent girls in Anand district, Gujarat. East Mediterr Health J. 2006;12(3-4):428–33.
12. Thakre SB, Thakre SS, Reddy M, Rathi N, Pathak K, Ughade S. Menstrual hygiene: knowledge and practice among adolescent school girls of Saoner, Nagpur district. J Clin Diagn Res. 2011;5(5):1027–33.
13. Yasmin S, Mallik S, Manna N, Ahmed A, Paria B. Menstrual hygiene among adolescent school students: an indepth cross-sectional study in an urban community of West Bengal, India. Sudan J Public Health. 2013;8(2):60–4.
14. Oche MO, Umar AS, Gana GJ, Ango JT. Menstrual health: the unmet needs of adolescent girls' in Sokoto, Nigeria. Sci Res Essays. 2012;7(3):410–8.
15. Ray S, Dasgupta A. Determinants of menstrual hygiene among adolescent girls: a multivariate analysis. Natl J Commun Med. 2012;3(2):294–301.
16. Boosey R, Prestwich G, Deave T. Menstrual hygiene management amongst schoolgirls in the Rukungiri district of Uganda and the impact on their education: a cross-sectional study. Pan Afr Med J. 2014;19:253.

17. Nemade D, Anjenaya S, Gujar R. Impact of health education on knowledge and practices about menstruation among adolescent school girls of Kalamboli, Navi-Mumbai. Health Popul Perspect Issues. 2009;32(4):167–75.

18. Narayan K, Srinivasa D, Pelto P, Veerammal S. Puberty rituals, reproductive knowledge and health of adolescent schoolgirls in South India. Asia-Pac Popul J. 2001;16(2):225–38.

19. Arora A, Mittal A, Panthania D, Mehta C, Bunger R. Impact of health education on knowledge and practices about menstruation among adolescent school girls of rural part of district Ambala, Haryana. Ind J Comm Health. 2013;25(4):492–7.

20. Lawan UM, Yusuf NW, Musa AB. Menstruation and menstrual hygiene amongst adolescent school girls in Kano, Northwestern Nigeria. Afr J Reprod Health. 2010;14(3):201–7.

21. Zegeye DT, Megabiaw B, Mulu A. Age at menarche and the menstrual pattern of secondary school adolescents in northwest Ethiopia. BMC Womens Health. 2009;9:29.

22. Thakre SB, et al. Urban-rural differences in menstrual problems and practices of girl students in Nagpur, India. Indian Pediatr. 2012;49(9):733–6.

23. Udgiri R, Angadi MM, Patil S, Sorganvi V. Knowledge and practices regarding menstruation among adolescent girls in an urban slum, Bijapur. J Indian Med Assoc. 2010;108(8):514–6.

24. Marvan ML, Molina-Abolnik M. Mexican adolescents' experience of menarche and attitudes toward menstruation: role of communication between mothers and daughters. J Pediatr Adolesc Gynecol. 2012;25(6):358–63.

25. Sharma M, Gupta S. Menstrual pattern and abnormalities in the high school girls of Dharan: a cross sectional study in two boarding schools. Nepal Med Coll J. 2003;5(1):34–6.

26. Adinma ED, Adinma JI. Menstrual characteristics amongst south-eastern Nigerian adolescent school girls. West Afr J Med. 2009;28(2):110–3.

27. Reis N, Kilic D, Engin R, Karabulutlu O. Sexual and reproductive health needs of adolescent girls from conservative and low-income families in Erzurum, Turkey. Health Popul Perspect Issues. 2011;3(6):370–7.

28. Bosch AM, Hutter I, van Ginneken JK. Perceptions of adolescents and their mothers on reproductive and sexual development in Matlab, Bangladesh. Int J Adolesc Med Health. 2008;20(3):329–42.

29. Dhingra R, Kumar A, Kour M. Knowledge and practices related to menstruation among tribal (Gujjar) adolescent girls. Ethno-Medicine. 2009;3(1):43–8.

30. El-Gilany AH, Badawi K, El-Fedawy S. Menstrual hygiene among adolescent schoolgirls in Mansoura, Egypt. Reprod Health Matters. 2005;13(26):147–52.

31. Gumanga SK, Kwame-Aryee RA. Menstrual characteristics in some adolescent girls in Accra, Ghana. Ghana Med J. 2012;46(1):3–7.

32. Dambhare DG, Wagh SV, Dudhe JY. Age at menarche and menstrual cycle pattern among school adolescent girls in Central India. Glob J Health Sci. 2012;4(1):105–11.

33. Kotecha PV, et al. Reproductive health awareness among rural school going adolescents of Vadodara district. Indian J Sex Transm Dis. 2009;30(2):94–9.

34. Mudey A, Kesharwani N, Mudey GA, Goyal RC. A cross-sectional study on awareness regarding safe and hygienic practices amongst school going adolescent girls in rural area of Wardha District, India. Glob J Health Sci. 2010;2(2):225–31.

35. Ray S, et al. Knowledge and information on psychological, physiological and gynaecological problems among adolescent schoolgirls of eastern India. Ethiop J Health Sci. 2011;21(3):183–9.

36. Jarrah SS, Kamel AA. Attitudes and practices of school-aged girls towards menstruation. Int J Nurs Pract. 2012;18(3):308–15.

37. Lee LK, et al. Menstruation among adolescent girls in Malaysia: a crosssectional school survey. Singapore Med J. 2006;47(10):869–74.

38. Wong LP. Attitudes toward menstruation, menstrual-related symptoms, and premenstrual syndrome among adolescent girls: a rural school-based survey. Women Health. 2011;51(4):340–64.

39. Wong LP. Premenstrual syndrome and dysmenorrhea: urban-rural and multiethnic differences in perception, impacts, and treatment seeking. J Pediatr Adolesc Gynecol. 2011;24(5):272–7.

40. Aniebue UU, Aniebue PN, Nwankwo TO. The impact of pre-menarcheal training on menstrual practices and hygiene of Nigerian school girls. Pan Afr Med J. 2009;2:9.

41. Iliyasu Z, et al. Sexual and reproductive health communication between mothers and their adolescent daughters in northern Nigeria. Health Care Women Int. 2012;33(2):138–52.

42. Ajah LO, et al. Adolescent reproductive health challenges among schoolgirls in southeast Nigeria: role of knowledge of menstrual pattern and contraceptive adherence. Patient Prefer Adherence. 2015;9:1219–24.

43. Chandraratne NK, Gunawardena NS. Premenstrual syndrome: the experience from a sample of Sri Lankan adolescents. J Pediatr Adolesc Gynecol. 2011;24(5):304–10.

44. Abd El-Hameed NA, Mohamed MS, Ahmed NH, Ahmed ER. Assessment of dysmenorrhea and menstrual hygiene practices among adolescent girls in some nursing schools at lL-Minia governorate, Egypt. J Am Sci. 2011;7(9):216–23.

45. Eswi A, Helal H, Elarousy W. Menstrual attitude and knowledge among Egyptian female adolescents. J Am Sci. 2012;8(6):555–65.

46. Omidvar S, Begum K. Factors influencing hygienic practices during menses among girls from south India: a cross sectional study. Int J Collab Res Intern Med Public Health. 2010;2(12):411–23.

47. Wong LP. Attitudes towards dysmenorrhoea, impact and treatment seeking among adolescent girls: a rural school-based survey. Aust J Rural Health. 2011;19(4):218–23.

48. Wong LP, Khoo EM. Menstrual-related attitudes and symptoms among multi-racial Asian adolescent females. Int J Behav Med. 2011;18(3):246–53.

49. Sommer M. Ideologies of sexuality, menstruation and risk: girls' experiences of puberty and schooling in northern Tanzania. Cult Health Sex. 2009;11(4):383–98.

50. Crichton J, et al. Emotional and psychosocial aspects of menstrual poverty in resource-poor settings: a qualitative study of the experiences of adolescent girls in an informal settlement in Nairobi. Health Care Women Int. 2013;34(10):891–916.

51. Mason L, et al. 'We keep it secret so no one should know'–a qualitative study to explore young schoolgirls attitudes and experiences with menstruation in rural western Kenya. PLoS One. 2013;8(11):e79132.

52. Munthali AC, Zulu EM. The timing and role of initiation rites in preparing young people for adolescence and responsible sexual and reproductive behaviour in Malawi. Afr J Reprod Health. 2007;11(3):150–67.

53. McMahon SA, et al. 'The girl with her period is the one to hang her head' Reflections on menstrual management among schoolgirls in rural Kenya. BMC Int Health Hum Rights. 2011;11:7.

54. Sommer M. An early window of opportunity for promoting girls' health: Policy implications of the girl's puberty book project in tanzania. Int Electron J Health Educ. 2011;14:77–92.

55. Dorgbetor G. Mainstreaming MHM in schools through the play-based approach: lessons learned from Ghana. Waterlines. 2015;34(1):41–50.

56. Marvan ML, Vacio A, Espinosa-Hernandez G. Menstrual-related changes expected by premenarcheal girls living in rural and urban areas of Mexico. Soc Sci Med. 2003;56(4):863–8.

57. Marvan ML, Vacio A, Espinosa-Hernandez G. A comparison of menstrual changes expected by pre-menarcheal adolescents and changes actually experienced by post-menarcheal adolescents in Mexico. J Sch Health. 2001;71(9):458–61.

58. Pitangui AC, et al. Menstruation disturbances: prevalence, characteristics, and effects on the activities of daily living among adolescent girls from Brazil. J Pediatr Adolesc Gynecol. 2013;26(3):148–52.

59. Santina T, Wehbe N, Ziade F. Exploring dysmenorrhoea and menstrual experiences among Lebanese female adolescents. East Mediterr Health J. 2012;18(8):857–63.

60. Chaudhuri A, Singh A. How do school girls deal with dysmenorrhoea? J Indian Med Assoc. 2012;110(5):287–91.

61. Sommer M. Where the education system and women's bodies collide: The social and health impact of girls' experiences of menstruation and schooling in Tanzania. J Adolesc. 2010;33(4):521–9.

62. Patil MS, Angadi MM. Menstrual pattern among adolescent girls in rural area of Bijapur. Al Ameen J Med Sci. 2013;6(1):17–20.

63. Rana B, Prajapati A, Sonaliya KN, Shah V, Patel M, Solanki A. An assessment of menstrual hygiene practices amongst adolescent females at Kheda district of Gujarat state, India. Healthline J. 2015;6(1):23–9.

64. Sharma P, et al. Problems related to menstruation amongst adolescent girls. Indian J Pediatr. 2008;75(2):125–9.

65. Juyal R, Kandpal SD, Semwal J. Social aspects of menstruation related practices in adolescent girls of district Dehradun. Indian J Community Health. 2013;25(3):213–6.

66. Haque SE, et al. The effect of a school-based educational intervention on menstrual health: an intervention study among adolescent girls in Bangladesh. BMJ Open. 2014;4(7):e004607.

67. Bodat S, Ghate MM, Majumdar JR. School absenteeism during menstruation among rural adolescent girls in Pune. Natl J Community Med. 2013;4(2):212–6.

68. Joshi D, Buit G, González-Botero D. Menstrual hygiene management: education and empowerment for girls? Waterlines. 2015;34(1):51–67.

69. Montgomery P, et al. Sanitary pad interventions for girls' education in Ghana: a pilot study. PLoS One. 2012;7(10):e48274.

70. Oster E, Thornton R. Menstruation, sanitary products, and school attendance: evidence froma randomized evaluation. Am Econ J. 2011;3(1):91–100.
71. Mason L, Laserson K, Oruko K, et al. Adolescent schoolgirls' experiences of menstrual cups and pads in rural western Kenya: a qualitative study. Waterlines. 2015;34(1):15–30.
72. Kabir H, et al. Treatment-seeking for selected reproductive health problems: behaviours of unmarried female adolescents in two low-performing areas of Bangladesh. Reprod Health. 2014;11:54.
73. Nair MK, et al. Menstrual disorders and menstrual hygiene practices in higher secondary school girls. Indian J Pediatr. 2012;79 Suppl 1:S74–8.
74. Baidya S, Debnath M, Das R. A study of reproductive health problems among rural adolescent girls of Mohanpur block of West Tripura district. Al Ameen J Med Sci. 2014;7(1):78–82.
75. Wong LP, Khoo EM. Dysmenorrhea in a multiethnic population of adolescent Asian girls. Int J Gynaecol Obstet. 2010;108(2):139–42.
76. Poureslami M. Assessing knowledge, attitudes, and behavior of adolescent girls in sub-urban districts of Tehran about dysmenorrhea and menstrual hygiene. J Int Womens Stud. 2002;3(2):51–61.
77. Eryilmaz G, Ozdemir F. Evaluation of menstrual pain management approaches by Northeastern Anatolian adolescents. Pain Manag Nurs. 2009;10(1):40–7.
78. Wasnik VR, Dhumale D, Jawarkar AK. A study of the menstrual pattern and problems among rural school going adolescent girls of Amravati district of Maharashtra, India. Int J Res Med Sci. 2015;33(55):1252–6.
79. Fakhri M, et al. Promoting menstrual health among persian adolescent girls from low socioeconomic backgrounds: a quasi-experimental study. BMC Public Health. 2012;12:193.
80. Allah ESA, Elsabagh EEM. Impact of health education intervention on knowledge and practice about menstruation among female secondary school students in Zagazig City. J Am Sci. 2011;7(9):737–47.
81. Sumpter C, Torondel B. A systematic review of the health and social effects of menstrual hygiene management. PLoS One. 2013;8(4):e62004.
82. Nanda PMA, Mukherjee S, Barua A, Mehl GL, Venkatraman CM. A study to evaluate the effectiveness of WHO tools: orientation programme on adolescent health for health care providers and adolescent job aid in India. Geneva: International Center for Research on Women; 2012.
83. Vandenhoudt H, et al. Evaluation of a U.S. evidence-based parenting intervention in rural Western Kenya: from parents matter! To families matter! AIDS Educ Prev. 2010;22(4):328–43.
84. Sommer M, Ackatia-Armah N, Connolly S, Smiles D. A comparison of the menstruation and education experiences of girls in Tanzania, Ghana, Cambodia and Ethiopia. Compare. 2014;45(4):589–609.
85. Children, S.t. Adolescent Sexual and Reproductive Health. 2015. Available from: http://www.savethechildren.org/site/c.8rKLIXMGIpI4E/b.9080949/k.F576/Adolescent_Sexual_and_Reproductive_Health.htm.
86. Health, I.f.R. Meeting the needs of adolescents: Introducing CycleSmart™. 2013. Available from: http://irh.org/blog/meeting-the-needs-of-adolescentsintroducing-the-cyclesmart-kit/.

87. Health, I.f.R. A3 project. Available from: http://irh.org/projects/a3_project/. Accessed 15 oct 2014.
88. Kettaneh APS, Todesco M. Good policy and practice booklet no. 9: puberty education and menstrual hygiene management. Paris: United Nations Educational, Scientific and Cultural Organization; 2014.
89. Always. Tips & advice: "the talk". Available from: http://always.com/en-us/tips-and-advice/the-talk. Accessed 15 oct 2014.
90. George R. Celebrating womanhood: how better menstrual hygiene management is the path to better health, dignity and business. Geneva: Water Supply & Sanitation Collaboration Council; 2013.
91. Sommer M. V. E., Worthington, N., Sahin M. WASH in schools empowers girls' education: proceedings of the menstrual hygiene management in schools virtual conference 2012. in Menstrual Hygiene Management in Schools Virtual Conference. New York, NY: United Nations Children's Fund and Columbia University; 2012.
92. Kanotra SK, Bangal VB, Bhavthankar DP. Menstrual pattern and problems among rural adolescent girls. International Journal of Biomedical and Advance Research. 2013; 4(8):551–4.

CHAPTER 47

Interventions to Improve Menstrual Health in Low- and Middle-Income Countries: Do We Know What Works?

Julie Hennegan

MENSTRUATION MATTERS

Menstrual health interventions have been motivated by growing recognition of the difficulties women face during menstruation, and the consequences of unmet menstrual needs. Qualitative studies have highlighted negative psychosocial outcomes associated with menstruation including distress, fear, shame, and anxiety, and have linked difficulties in managing menstrual bleeding with school absenteeism and disengagement (Sommer et al. 2015; Mason et al. 2013). Past quantitative studies on the links between menstrual management and school attendance have provided more mixed results (Grant, Lloyd, and Mensch 2013; Tegegne and Sisay 2014). However, more recent trials and cross-sectional studies have provided more support for this association (Montgomery et al. 2016; Miiro et al. 2018). Consistent with absences from school, there is self-reported evidence that girls avoid other activities such as physical exercise, work, or household tasks while menstruating (Hennegan et al. 2016a; Chandra-Mouli and Patel 2017). Furthermore, poor menstrual hygiene has been posited to lead to negative physical health outcomes including reproductive and urinary tract infections, with some emerging evidence of this association (Das et al. 2015; Phillips-Howard et al. 2016b). A study in Kenya has suggested that girls may be at risk of further harm through engaging in transactional sex to obtain menstrual supplies (Phillips-Howard et al. 2015).

Multiple actors have responded to reports of women's challenges with menstruation. A growing list of governments, international organizations, and NGOs have invested in menstrual health. Most of this attention has

C. Bobel et al. (eds.), *The Palgrave Handbook of Critical Menstruation Studies*, https://doi.org/10.1007/978-981-15-0614-7_47

accelerated within the last decade, with many programs now advertising extensive reach. This sharp increase in the number of organizations seeking to intervene means increasing numbers of women and girls are involved in menstrual health interventions (Bobel 2019).

How Do We Know What Works?

While swift action to address unmet menstrual needs is laudable, good intentions do not guarantee positive outcomes. Seemingly intuitive community programs or policies often fail to have the anticipated results and can risk unintended harms (Hennegan 2017). Further, restricted budgets calling for a need to do more with less, place a premium on the identification and improvement of efficient interventions. For the objective evaluation of the effectiveness of an intervention, randomized controlled trials have long represented the gold standard (Gottfredson et al. 2015). While such studies are not sufficient to fully understand a program, and other complementary methods such as in-depth interviewing to capture rich qualitative data on experiences and contextual influences are needed (Gambrill 2006; Lewin, Glenton, and Oxman 2009), high-quality trials are able to minimize the many biases inherent in other designs. By collating multiple trials, systematic reviews can draw together this evidence to assess '*what works.*' Systematic reviews apply a structured methodology to appraise, critique, and synthesize extant findings. They can estimate the pooled effect of interventions and can identify gaps in current research, methodological weaknesses, and provide guidance for future improvements.

This chapter draws on the findings of a systematic review which aimed to collate and appraise trials of the effectiveness of menstrual health interventions for improving education and psychosocial outcomes in low- and middle-income countries (Hennegan and Montgomery 2016). The review did not include studies exclusively assessing impacts on reproductive tract infection outcomes as another systematic review had focused on this issue (Sumpter and Torondel 2013). Study designs eligible for inclusion in the review included randomized and cluster randomized controlled trials, as well as non-randomized controlled trials including more than one intervention and control site. The review undertook a systematic search of 15 databases, as well as handsearching, grey literature search,[1] checks of reference lists, and contact with experts. Searches were undertaken in January 2015. For this chapter, database searches, and citation tracking of papers included in the review were undertaken in December 2017 to identify additional eligible studies published between 2015 and 2017. The original review identified a total of eight eligible trials, with a further four trials published since the initial search. Thus, this chapter evaluates a total of 12 trials.

The subsequent sections of this chapter will summarize menstrual health interventions that have been trialed to date and provide insights into their effectiveness. It will reflect on what we do, and more often do not, know about what works in meeting the menstrual needs of women and girls in low- and middle-income settings.

INTERVENTIONS

Menstrual health is multifaceted. Hygienic menstrual practices and positive experiences of menstruation are dependent on the physical and social environment. Women and girls report needing knowledge and education about menstruation to understand their bodies, inform effective and hygienic practices, and to dispel unhelpful myths or taboos (Chothe et al. 2014). They also need clean materials to absorb or catch menses, as well as infrastructure to support safe locations for changing absorbents, and access to water, soap, and locations for washing reusable materials and their bodies (Sommer and Sahin 2013). Menstruation occurs in a context of social norms that may positively or negatively influence experience and dictate behavior. Further, the presence or lack of support from key individuals such as parents, teachers, and friends shapes experiences (Geertz et al. 2016; Hennegan et al. 2017). Menstrual needs are interconnected, and women and girls may face deprivations across several contributing factors. For the purposes of categorizing interventions, there has been a distinction between the types of interventions that provide a physical resource such as a menstrual absorbent or water, sanitation and hygiene (WASH) facilities, and programs providing psychosocial intervention which may include education, or efforts to address harmful taboos and stigma (Sumpter and Torondel 2013; Hennegan and Montgomery 2016; Geertz et al. 2016). The latter aimed at addressing psychosocial deprivations have been termed *software interventions,* while the former targeted at material deprivations have been termed *hardware interventions.* These interventions may include multiple components, and software and hardware interventions may be combined.

Software Interventions

A lack of menstrual knowledge has been identified across countries as a challenge to positive menstrual experiences (Chandra-Mouli and Patel 2017). Adolescent girls, particularly around the time of menarche, have been found to lack information on the function and origin of menstruation (Crichton et al. 2013). Many girls report being unaware that menstruation would occur, contributing to distress and fear at onset (Chandra-Mouli and Patel 2017). A lack of understanding of menstruation and strategies for hygienic menstrual practices are likely to perpetuate distress over subsequent menses and may contribute to negative effects on health and education. Deficits in knowledge are often combined with misinformation evoking further fear. Studies have reported taboos around the disposal of menstrual blood or consequences of others seeing used menstrual materials (Sommer et al. 2015), others have documented cultural expectations which explicitly restrict participation in social settings, bathing, cooking, or religious practices while menstruating (Chandra-Mouli and Patel 2017). At the same time, the stigmatized nature of menstruation across settings results in more implicit behavioral norms such as the expectation to keep menstrual status hidden and secret at all times (Jewitt and Ryley 2014). This prevents women and girls from discussing menstruation openly and receiving advice, support, or direction to health services where abnormalities are present (McMahon et al. 2011). Silence around menstrual issues may also mean that these needs are not prioritized in household budgets and girls are not equipped with the resources to assist the management of menstrual bleeding. Men and boys similarly lack knowledge and accurate information about menstruation and may perpetuate negative attitudes (Mason et al. 2017).

Software interventions include all interventions designed to address these highlighted psychosocial deprivations. Five trials of software interventions were included in the 2016 systematic review, with a subsequent three trials with an independent software condition published since this time. Greater detail regarding the design of these studies and risk of bias assessments are available in the original review (see Hennegan and Montgomery 2016). Software interventions to date have focused on the provision of menstrual education via a range of modalities, including; written materials such as leaflets and posters (Mbizvo et al. 1997), school-based education sessions provided by social workers ranging from one 120-minute session (Fetohy 2007) to ten 2 hour sessions (Fakhri et al. 2012), and peer, school, or health provider-led education approaches (Djalalinia et al. 2012; Abedian et al. 2011). An additional trial undertaken in Iran, published in 2017 (Afsari et al. 2017), compared the effectiveness of educating mothers to that of educating girls. Limited information was provided in these studies on the contents of the education offered. Most included statements that education included information on the definition and purpose of menstruation, the cyclic nature and timing, and other physical changes associated with puberty. Broad statements about education on 'self-care' and management practices were included with limited detail. There was a high risk of bias across these studies, with deficits in reporting, preregistration, and issues with participant attrition (dropout) and inappropriate treatment of outcome variables.

A more recent cluster-randomized trial provided a pilot evaluation of the *Growth and Changes* puberty education book in Ethiopia, based on qualitative research in multiple countries (Blake et al. 2017). This study identified improvements in knowledge about menstruation and decreases in fear and shame associated with menstruation at four-week follow-up. The trial used a clustered design including 20 schools. However, intervention and control conditions were assigned at the district, rather than school level. As such, the study only allocated one intervention and one control condition and results should be interpreted with caution. Qualitative components of the study triangulated quantitative findings with girls reporting the books reduced shame and confusion and helped to overcome secrecy around menstruation (Blake et al. 2017).

One additional pilot trial in Ghana included in the 2016 systematic review (Hennegan and Montgomery 2016) included both hardware and software conditions, with an education-alone condition allowing for a comparison of the effectiveness to a control. This study was the only one to assess impacts of education on school attendance rather than knowledge outcomes, however it was a small pilot with one education-alone and one control site (Montgomery et al. 2012). The study observed improvements in school attendance at 5-month follow-up. This Ghana pilot was extended to a larger study of eight schools in Uganda, two of which received an education-only intervention (Montgomery et al. 2016). The larger trial replicated the positive effects on school attendance. Qualitative follow-up of the trial identified avenues for improvement in the education provided (Hennegan et al. 2017). This study found that, beyond providing an awareness of menstruation prior to menarche, girls reported the practical information around managing menstrual bleeding to be particularly useful, and that more was desired. In interviews at the conclusion of the study, girls felt they still lacked information about vaginal discharge, menstrual irregularities, and reproductive tract symptoms. The study also found that social support, likely resulting from a shared language and education around menstruation, was particularly important and a possible driver of the positive effects identified in the quantitative results. Support from teachers and peers dictated girls' menstrual experience at school and contributed to their comfort and confidence to manage menstruation (Hennegan et al. 2017).

Taken together, extant trials of software interventions indicate that education interventions improve menstrual knowledge. It is unclear if this translates to impacts on psychosocial outcomes or wellbeing as identified in qualitative research as most studies failed to assess these outcomes. Of those that did, two identified improvements (Blake et al. 2017; Montgomery et al. 2012) and one reported no difference although suffered significant limitations due to participant dropout (Montgomery et al. 2016). Less evidence is available to support impacts on school attendance, although one program of work including a pilot and larger study indicates some promising results (Montgomery et al. 2012, 2016). More recent studies highlight the value of including complementary qualitative work alongside trials in the evaluation of menstrual health interventions (Montgomery et al. 2016; Blake et al. 2017). Hennegan et al. (2017) suggested that two levels of menstrual education may be needed. The first, and what appears to have received more attention in trials to date, is to provide girls with information about the basics of menstruation; what it is, why it happens, when it happens, and that this is a healthy and natural process linked to reproduction. This information helps to dispel fear of illness and shame by emphasizing the healthy nature of menstrual bleeding. The second level of education, which appears less prominently in trials, is more detailed information about menstrual experience and practical guidance. This may include suggestions on how to record or "chart" menstrual timing and bodily symptoms that might indicate the onset of menstruation so that girls may be aware, normal and abnormal vaginal discharge, irritation and pain, pain management strategies, and advice on absorptive materials and their hygienic management. Such education may be well placed to support effective practice, the identification of abnormalities, and equip girls to feel more informed, confident, and in control of their menstrual care and their bodies.

Hardware Interventions

To effectively and hygienically manage menstruation, women and girls need access to sufficient quantities of clean, safe, comfortable materials fit for the purpose of collecting menses (Sommer and Sahin 2013). Materials take many forms, from commercial disposable or reusable sanitary pads, tampons, and menstrual cups, to homemade pads or cloth. Women and girls may use different strategies depending on available resources, needs, and personal and cultural preferences. In low- and middle-income contexts, some women and girls may struggle to access or purchase sufficient materials supporting their preferences. This may lead to the use of alternatives which are not fit for purpose and are experienced as uncomfortable or ineffective (Ellis et al. 2016). Beyond materials, women and girls need access to private, safe locations to change menstrual materials, to wash and dry materials if they are reusable, and to wash their body with soap and water (Sommer and Sahin 2013). As such, supportive infrastructure, and WASH facilities are an essential enabler. These facilities are often unavailable or inadequate to meet women's and girls' physical and psychological needs (Fisher 2006; Hennegan et al. 2018). Due to the stigmatized nature of menstruation, women and girls have emphasized the need for a private space where they feel safe and comfortable to undertake menstrual practices such as changing or washing materials (Budhathoki et al. 2018). Locations for drying reusable absorbents are also needed. Hygiene guidelines have recommended drying absorbents in the sun to benefit from the antimicrobial properties of UV light, however, the perceived need to conceal menstruation means many girls report drying absorbents under mattresses or in other, hidden locations that are unlikely to facilitate hygiene (House et al. 2012; Hennegan et al. 2016b). Disposal of single-use menstrual absorbents has been largely overlooked in research and intervention to date. Few studies have investigated disposal practices, even when implementing single-use product interventions. In absence of waste management strategies, used products may be thrown into fields and waterways, pit latrines, or sanitation systems where they may cause blockages or health risks (Geertz et al. 2016). In addition to these environmental impacts, disposal presents a challenge for women and girls. Cultural taboos around others seeing menstrual blood, or the burning of menstrual blood add strain on finding acceptable disposal options and may contribute to distress and avoidance of social settings during menstruation (Sommer et al. 2015).

Hardware interventions have been conceptualized as programs providing material resources such as menstrual materials, infrastructure, or disposal facilities. However, to date, only the provision of menstrual materials has been trialed. Three eligible trials of hardware interventions were identified through systematic review (Hennegan and Montgomery 2016), with a further two studies published by the end of 2017 (Montgomery et al. 2016; Phillips-Howard et al. 2016b). Of the total five trials, two focused on the effectiveness of providing menstrual cups, two reusable sanitary pads, and one on disposable sanitary pads.

In the first trial of hardware provision for improving girls' menstrual management and school attendance, Oster and Thornton (2011) provided girls in Nepal, and their mothers, with a menstrual cup and instructions for use. They found no improvement for the girls provided with cups, compared to those who were not, however, school attendance was high for both groups at baseline and its results likely reflect a ceiling effect. A second study of menstrual cups was undertaken in Kenya and published in 2016 (Phillips-Howard et al. 2016b). The largest trial to date, this study included girls from 30 primary schools randomized to one of three conditions; a control, the provision of disposable menstrual pads, or the provision of a menstrual cup. Education was provided in all conditions. The study did not find any significant difference between conditions on the risk of school dropout over the one year follow-up period and did not report differences in school attendance as this data was collected using calendars given to participants and absences were very rarely reported, thus not analyzed. The study did not assess psychosocial outcomes but focused on sexually transmitted and reproductive tract infections as secondary outcomes, finding lower risks for those receiving either menstrual product (Phillips-Howard et al. 2016b).

The Ghana pilot study including both hardware and software components provided disposable sanitary pads in hardware conditions and found a moderate improvement in school attendance at three- and five-month follow-ups (Montgomery et al. 2012). This was followed by a larger study of reusable sanitary pads in Uganda published in 2016 (Montgomery et al. 2016). This study included four arms: two schools receiving education as noted above, two control schools, two receiving reusable pads alone, and two receiving combined education and product provision. The larger trial replicated results of the pilot with significant benefits for school attendance identified over the 24-month period, with no differences in the effects of the three different intervention conditions. The trial included only a small number of clusters and suffered from significant attrition with girls transferring or dropping out of school, with the longest follow-up period of any menstrual health intervention to date. Attrition levels suggest specific effect estimates of the trial should be interpreted with caution, with imputed analyses suggesting a beneficial effect of the pads or education conditions between 5.2 and 24.5% (Montgomery et al. 2016). Qualitative follow-up supported the positive quantitative results (Hennegan et al. 2017). As noted above, social support due to the increased visibility of menstruation in schools may have contributed to intervention effects across schools. Girls reported largely positive experiences of the reusable pads which they felt were more comfortable and reliable, reducing fears of garment soiling. Another pilot study of homemade reusable sanitary pads in Uganda revealed a moderate, although nonsignificant effect likely due to the small sample size and number of clusters (Wilson, Josephine, and Pitt 2014).

Taken together, these studies suggest that providing menstrual materials *may* improve girls' school attendance, but that more research is needed. Studies varied in the outcomes assessed, duration, products provided, and design weaknesses making it difficult to draw conclusions. In comparison to the software interventions, hardware interventions focused more heavily on school attendance outcomes, and often neglected to assess more proximal change such as menstrual experience and distress. This could be facilitated through complementary qualitative research, and mediation and moderation methods undertaken with quantitative data collected as part of trial baseline and follow-up surveys. Research mapping connections between menstrual management and schooling and refinement of attendance measure techniques may improve outcome assessment in future trials (Miiro et al. 2018).

Do We Know What Works?

Menstrual health programming is rapidly increasing in scale (Bobel 2019). At the same time, trials of the effectiveness of interventions are limited in number and beset with limitations. There is some emerging evidence of the effectiveness of education interventions for improving knowledge, and for product provision interventions to encourage greater school attendance. However, methodological weaknesses across the evidence base mean no strong conclusions can be drawn (Hennegan and Montgomery 2016). More rigorous research is needed to evaluate interventions to improve menstrual health. Future trials must improve upon the limitations of past work. This includes improved reporting of studies, the use of adequate sample sizes, and strategies to mitigate attrition. Greater breadth is needed in the outcomes included in trials, for example, the assessment of menstrual experience in addition to school attendance. Across the studies presented in this review, there were limitations to the types of outcomes assessed and the way they were measured. Complementary quantitative and qualitative studies alongside trials may unearth additional understanding of participant experiences and lessons for intervention improvement. Moreover, as complex interventions within complex settings, understanding how the different aspects of interventions act and interact with the study context is key. This process can be facilitated by explicit theory on the mechanisms at play in interventions, and targeted investigation of these pathways during evaluation (Moore et al. 2015). Further, understanding the interaction between mechanisms of impact and the study context is critical for examining whether we would expect intervention effects to replicate in another setting, or at scale. For example, one currently registered trial lists sexually transmitted infections as primary outcomes in a trial of a menstrual health intervention but has not yet started an explicit framework for testing the pathways through which improved menstrual experiences will impact sexually transmitted infection rates (Clinical Trials 2017). One mechanism may be through reduced engagement in transactional sex to obtain menstrual materials (Phillips-Howard et al. 2015). If this is the case, positive effects identified in the trial are unlikely to replicate in settings where transactional sex is less prevalent.

One series of studies included in the systematic review included both hardware and software conditions. In both the Ghana pilot (Montgomery et al. 2012) and the follow-up in Uganda reported in Montgomery et al. (2016), education interventions and product provision interventions showed equal improvements in school attendance by the end of the follow-up period. It is unclear if these interventions influenced attendance and menstrual experience through different pathways, and future work is needed to unpack the potential of these interventions in combination. Careful attention to evaluate different parts of interventions is needed in development, piloting, and trial evaluation (Craig et al. 2008). This might mean the combination of hardware (for example, menstrual product or infrastructure provision) and software (for example, education or social norm) interventions, and also attention to more nuanced details such as education delivery mechanisms (for example, group-based, health provider presence), and content must be paid. More thorough evaluations may be able to identify best practices and suggest minimal or standard intervention packages which could be disseminated more broadly in programming and policy. For example, the identification of core educational components as has been helpful for guidance on the conduct of comprehensive sexuality education (UNESCO 2015). There is some evidence from sexual education studies that greater effect sizes may be achieved when adapting established programs, compared to those that are 'homegrown' (Kirby, Laris, and Rolleri 2006; Fonner et al. 2014). Similarities to sexual education for menstrual education mean that identifying and core components of education, that can be adapted to local needs, could be valuable in improving effectiveness. This may also reduce the time needed to invest in development; an important consideration for resource-constrained organizations.

Beyond improving methodology in the evaluation of menstrual health interventions, more research and funding for intervention development is needed. Many factors contributing to women's and girls' menstrual experiences have not yet been addressed by tested interventions. Current education interventions may help dispel myths and stigma around menstruation, however, there have been no trials of interventions designed to target this important barrier to menstrual health. The perceived need to conceal menstrual status to minimize embarrassment contributes to women's and girls' distress during menstruation, avoidance of social settings (including school), and heightens the need for private locations for menstrual management and restricts washing, drying, and disposal choices. More research is needed to unpack the potential for interventions targeting menstrual stigma and their influence on hygiene practices, help-seeking, and menstrual experience more broadly. Importantly, education about menstruation alone is unlikely to be sufficient to change social norms (Bicchieri 2017). Mass media and other community programs to normalize menstruation may have positive influences at a population or community level and provide more supportive environments for women and girls. Studies of the influence of mass-media campaigns have been undertaken for other health and hygiene challenges, such as water and sanitation products, diarrheal disease, immunization, nutrition, mother-to-child transmission of HIV, and reproductive health, and could inform studies of these approaches to menstrual health (Naugle and Hornik 2014; Evans et al. 2014). Interventions that target parental or male attitudes to menstruation have received little attention. One upcoming cluster RCT with multiple intervention arms will include a condition providing both school level and community level education (Sol et al. 2019) and may provide new insights on the effectiveness of this approach. No studies have trialed improvements to WASH or other infrastructure to support menstrual management. These infrastructure challenges are relevant in home, school, and work environments and interventions have the potential to be integrated into other campaigns such as those aiming to improve sanitation coverage (Hennegan et al. 2018).

New intervention approaches to address menstrual health are varied. They range from more detailed and comprehensive education on menstrual health and body literacy, to reducing stigma and taboo at all levels, and infrastructure and WASH facilities to support menstrual activities, as well as integrated packages of these components. Moreover, menstrual pain and disorders have been largely neglected in trials and cross-sectional studies and populations such as those in humanitarian settings, those living with disabilities, girls outside of schools, and adult women at home and in the workplace have received inadequate attention (Sommer, Chandraratna, et al. 2016).

To support the development and evaluation of such interventions an improved evidence base is needed. This should include funding and conduct of non-trial methods such as cross-sectional and longitudinal designs to inform the breadth of menstrual needs, women's and girls' experiences and preferences, and to build theory around the links between menstruation and target outcomes such as health, education, and wellbeing (Hennegan 2017; Phillips-Howard et al. 2016a). This would inform intervention design as well as outcome assessment in trials. Greater integration of NGO, multinational programming, and academic research efforts would be of great value. NGO programs often only include case-study assessments of effectiveness but have often described more integrated approaches to intervention development (Sommer, Sahin, et al. 2016). Sufficient funding must be allocated to evaluation, to ensure that policies and programs are evidence-based and effective prior to large-scale dissemination. In the absence of evidence for the effectiveness of interventions, implementing organizations and advocates should proceed with caution. Integrated monitoring strategies should pay special attention to unintended harms. With considerable attention to methodological challenges, studies can improve understanding of menstrual health and develop and evaluate interventions which are well placed to improve women's and girls' menstrual health.

NOTE

1. Handsearching refers to manual searches of the table of contents of relevant journals or conference proceedings. Grey literature searching includes searches on websites and databases for 'grey literature,' that is, research that has been reported outside of peer-reviewed publications.

REFERENCES

Abedian, Zahra, Maryam Kabirian, Seyed Reza Mazlom, and Behroz Mahram. 2011. "The Effects of Peer Education on Health Behaviors in Girls with Dysmenorrhea." *Journal of American Science* 7 (1): 431–38.

Afsari, Atousa, Mojgan Mirghafourvand, Sousan Valizadeh, Massomeh Abbasnezhadeh, Mina Galshi, and Samira Fatahi. 2017. "The Effects of Educating Mothers and Girls on the Girls' Attitudes Toward Puberty Health: A Randomized Controlled Trial." *International Journal of Adolescent Medicine and Health* 29 (2): 984.

Bicchieri, Cristina. 2017. *Norms in the Wild: How to Diagnose, Measure, and Change Social Norms*. New York: Oxford University Press.

Blake, Sarah, Melissa Boone, Alene Yenew Kassa, and Marni Sommer. 2017. "Teaching Girls about Puberty and Menstrual Hygiene Management in Rural Ethiopia: Findings from a Pilot Evaluation." *Journal of Adolescent Research* 33 (5): 623–46. https://doi.org/10.1177/0743558417701246.

Bobel, Chris. 2019. *The Managed Body*. Cham: Palgrave Macmillan.

Budhathoki, Shyam Sundar, Meika Bhattachan, Enrique Castro-Sánchez, Reshu Agrawal Sagtani, Rajan Bikram Rayamajhi, Pramila Rai, and Gaurav Sharma. 2018. "Menstrual Hygiene Management among Women and Adolescent Girls in the Aftermath of the Earthquake in Nepal." *BMC Women's Health* 18 (1): 33.

Chandra-Mouli, Venkatraman, and Sheila Vipul Patel. 2017. "Mapping the Knowledge and Understanding of Menarche, Menstrual Hygiene and Menstrual Health among Adolescent Girls in Low-and Middle-Income Countries." *Reproductive Health* 14 (1): 30.

Chothe, Vikas, Jagdish Khubchandani, Denise Seabert, Mahesh Asalkar, Sarika Rakshe, Arti Firke, Inuka Midha, and Robert Simmons. 2014. "Students' Perceptions and Doubts about Menstruation in Developing Countries a Case Study from India." *Health Promotion Practice* 15 (3): 319–26. https://doi.org/10.1177/1524839914525175.

Clinical Trials. 2017. "Cups or Cash for Girls Trial to Reduce Sexual and Reproductive Harm and School Dropout (CCg)." *ClinicalTrials.gov Identifier: NCT03051789.* Retrieved from: https://clinicaltrials.gov/ct2/show/record/NCT03051789?term=NCT03051789&rank=1 [Accessed March 2017].

Craig, Peter, Paul Dieppe, Sally Macintyre, Susan Michie, Irwin Nazareth, and Mark Petticrew. 2008. "Developing and Evaluating Complex Interventions: The New Medical Research Council Guidance." *BMJ* 337: a1655.

Crichton, Joanna, Jerry Okal, Caroline W. Kabiru, and Eliya Msiyaphazi Zulu. 2013. "Emotional and Psychosocial Aspects of Menstrual Poverty in Resource-Poor Settings: A Qualitative Study of the Experiences of Adolescent Girls in an Informal Settlement in Nairobi." *Health Care for Women International* 34 (10): 891–916.

Das, Padma, Kelly K. Baker, Ambarish Dutta, Tapoja Swain, Sunita Sahoo, Bhabani Sankar Das, Bijay Panda, Arati Nayak, Mary Bara, and Bibiana Bilung. 2015. "Menstrual Hygiene Practices, WASH Access and the Risk of Urogenital Infection in Women from Odisha, India." *PLoS One* 10 (6): e0130777.

Djalalinia, Shirin, Fahimeh Ramezani Tehrani, Hossein Malek Afzali, Farzaneh Hejazi, and Niloofar Peykari. 2012. "Parents or School Health Trainers, Which of Them Is Appropriate for Menstrual Health Education?" *International Journal of Preventive Medicine* 3 (9): 622.

Ellis, A., J. Haver, J. Villasenor, A. Parawan, M. Venkatesh, M. C. Freeman, and B. A. Caruso. 2016. "WASH Challenges to Girls' Menstrual Hygiene Management in Metro Manila, Masbate, and South Central Mindanao, Philippines." *Waterlines* 35 (3): 306–23.

Evans, William Douglas, S. K. Pattanayak, S. Young, J. Buszin, S. Rai, and Jasmine Wallace Bihm. 2014. "Social Marketing of Water and Sanitation Products: A Systematic Review of Peer-Reviewed Literature." *Social Science & Medicine* 110: 18–25.

Fakhri, Moloud, Zeinab Hamzehgardeshi, Nayereh A. Hajikhani Golchin, and Abdulhay Komili. 2012. "Promoting Menstrual Health among Persian Adolescent Girls from Low Socioeconomic Backgrounds: A Quasi-Experimental Study." *BMC Public Health* 12 (1): 193.

Fetohy, Ebtisam M. 2007. "Impact of a Health Education Program for Secondary School Saudi Girls about Menstruation at Riyadh City." *Journal of the Egyptian Public Health Association* 82 (1–2): 105–26.

Fisher, Julie. 2006. "For Her It's the Big Issue: Putting Women at the Centre of Water Supply, Sanitation and Hygiene." Available from: https://www.wsscc.org/resources-feed/big-issue-putting-women-centre-water-supply-sanitation-hygiene/ [Accessed June 2016].

Fonner, Virginia A., Kevin S. Armstrong, Caitlin E. Kennedy, Kevin R. O'Reilly, and Michael D. Sweat. 2014. "School Based Sex Education and HIV Prevention in Low-and Middle-Income Countries: A Systematic Review and Meta-Analysis." *PLoS One* 9 (3): e89692.

Gambrill, Eileen. 2006. "Evidence-Based Practice and Policy: Choices Ahead." *Research on Social Work Practice* 16 (3): 338–57.

Geertz, Alexandra, Lakshmi Iyer, Perri Kasen, Francesca Mazzola, and Kyle Peterson. 2016. "An Opportunity to Address Menstrual Health and Gender Equity." FSG. Online. Available from: https://www.fsg.org/publications/opportunity-address-menstrual-health-and-gender-equity [Accessed January 2017].

Gottfredson, Denise C., Thomas D. Cook, Frances E. M. Gardner, Deborah Gorman-Smith, George W. Howe, Irwin N. Sandler, and Kathryn M. Zafft. 2015. "Standards of Evidence for Efficacy, Effectiveness, and Scale-Up Research in Prevention Science: Next Generation." *Prevention Science* 16 (7): 893–926.

Grant, Monica, Cynthia Lloyd, and Barbara Mensch. 2013. "Menstruation and School Absenteeism: Evidence from Rural Malawi." *Comparative Education Review* 57 (2): 260–84.

Hennegan, J. 2017. "Menstrual Hygiene Management and Human Rights: The Case for an Evidence-Based Approach." *Women's Reproductive Health* 4 (3): 212–31.

Hennegan, J., Catherine Dolan, Laurel Steinfield, and Paul Montgomery. 2017. "A Qualitative Understanding of the Effects of Reusable Sanitary Pads and Puberty Education: Implications for Future Research and Practice." *Reproductive Health* 14 (1): 78.

Hennegan, Julie, Catherine Dolan, Maryalice Wu, Linda Scott, and Paul Montgomery. 2016a. "Schoolgirls' Experience and Appraisal of Menstrual Absorbents in Rural Uganda: A Cross-Sectional Evaluation of Reusable Sanitary Pads." *Reproductive Health* 13 (1): 143.

———. 2016b. "Measuring the Prevalence and Impact of Poor Menstrual Hygiene Management: A Quantitative Survey of Schoolgirls in Rural Uganda." *BMJ Open* 6 (12): e012596.

Hennegan, Julie, Linnea Zimmerman, Alexandra K. Shannon, Natalie G. Exum, Funmilola OlaOlorun, Elizabeth Omoluabi, and Kellogg J. Schwab. 2018. "The Relationship between Household Sanitation and Women's Experience of Menstrual Hygiene: Findings from a Cross-Sectional Survey in Kaduna State, Nigeria." *International Journal of Environmental Research and Public Health* 15 (5): 905.

Hennegan, J., and Paul Montgomery. 2016. "Do Menstrual Hygiene Management Interventions Improve Education and Psychosocial Outcomes for Women and Girls in Low and Middle Income Countries? A Systematic Review." *PLoS One* 11 (2): e0146985.

House, Sarah, Sue Cavill, and Thérèse Mahon. 2012. *Menstrual Hygiene Matters: A Resource for Improving Menstrual Hygiene around the World.* WaterAid.

Jewitt, Sarah, and Harriet Ryley. 2014. "It's a Girl Thing: Menstruation, School Attendance, Spatial Mobility and Wider Gender Inequalities in Kenya." *Geoforum* 56: 137–47.

Kirby, Douglas, B. A. Laris, and Lori Rolleri. 2006. "The Impact of Sex and HIV Education Programs in Schools and Communities on Sexual Behaviors among Young Adults." Family Health International, Research Triangle Park, N.C., 17.

Lewin, Simon, Claire Glenton, and Andrew D. Oxman. 2009. "Use of Qualitative Methods alongside Randomised Controlled Trials of Complex Healthcare Interventions: Methodological Study." *BMJ* 339: b3496.

Mason, L., E. Nyothach, K. Alexander, F. O. Odhiambo, A. Eleveld, J. Vulule, R. Rheingans, K. F. Laserson, A. Mohammed, and P. A. Phillips-Howard. 2013. "'We Keep It Secret so No One Should Kno'—A Qualitative Study to Explore Young Schoolgirls Attitudes and Experiences with Menstruation in Rural Western Kenya." *PLoS One* 8 (11): e79132.

Mason, Linda, Muthusamy Sivakami, Harshad Thakur, Narendra Kakade, Ashley Beauman, Kelly T. Alexander, Anna Maria van Eijke, Kayla F. Laserson, Mamita B. Thakkar, and Penelope A. Phillips-Howard. 2017. "'We Do Not Know': A Qualitative Study Exploring Boys Perceptions of Menstruation in India." *Reproductive Health* 14 (1): 174.

Mbizvo, M. T., J. Kasule, V. Gupta, S. Rusakaniko, S. N. Kinoti, W. Mpanju-Shumbushu, A. J. Sebina-Zziwa, R. Mwateba, and J. Padayachy. 1997. "Effects of a Randomized Health Education Intervention on Aspects of Reproductive Health Knowledge and Reported Behaviour among Adolescents in Zimbabwe." *Social Science & Medicine* 44 (5): 573–77.

McMahon, Shannon A., Peter J. Winch, Bethany A. Caruso, Alfredo F. Obure, Emily A. Ogutu, Imelda A. Ochari, and Richard D. Rheingans. 2011. "'The Girl with Her Period Is the One to Hang Her Head' Reflections on Menstrual Management among Schoolgirls in Rural Kenya." *BMC International Health and Human Rights* 11 (1): 7.

Miiro, George, Rwamahe Rutakumwa, Jessica Nakiyingi-Miiro, Kevin Nakuya, Saidat Musoke, Juliet Namakula, Suzanna Francis, Belen Torondel, Lorna J. Gibson, and David A. Ross. 2018. "Menstrual Health and School Absenteeism among Adolescent Girls in Uganda (MENISCUS): A Feasibility Study." *BMC Women's Health* 18 (1): 4.

Montgomery, Paul, Caitlin R. Ryus, Catherine S. Dolan, Sue Dopson, and Linda M. Scott. 2012. "Sanitary Pad Interventions for Girls' Education in Ghana: A Pilot Study." *PLoS One* 7 (10): e48274.

Montgomery, Paul, Julie Hennegan, Catherine Dolan, Maryalice Wu, Laurel Steinfield, and Linda Scott. 2016. "Menstruation and the Cycle of Poverty: A Cluster Quasi-Randomised Control Trial of Sanitary Pad and Puberty Education Provision in Uganda." *PLoS One* 11 (12): e0166122–e0166122.

Moore, Graham F., Suzanne Audrey, Mary Barker, Lyndal Bond, Chris Bonell, Wendy Hardeman, Laurence Moore, et al. 2015. "Process Evaluation of Complex Interventions: Medical Research Council Guidance." *BMJ* 350 (1): h1258.

Naugle, Danielle A., and Robert C. Hornik. 2014. "Systematic Review of the Effectiveness of Mass Media Interventions for Child Survival in Low-and Middle-Income Countries." *Journal of Health Communication* 19 (supl): 190–215.

Oster, Emily, and Rebecca Thornton. 2011. "Menstruation, Sanitary Products, and School Attendance: Evidence from a Randomized Evaluation." *American Economic Journal: Applied Economics* 3 (1): 91–100.

Phillips-Howard, Penelope A., Bethany Caruso, Belen Torondel, Garazi Zulaika, Murat Sahin, and Marni Sommer. 2016a. "Menstrual Hygiene Management among Adolescent Schoolgirls in Low-and Middle-Income Countries: Research Priorities." *Global Health Action* 9 (1): 33032.

Phillips-Howard, Penelope A., Elizabeth Nyothach, Feiko O. Ter Kuile, Jackton Omoto, Duolao Wang, Clement Zeh, Clayton Onyango, et al. 2016b. "Menstrual Cups and Sanitary Pads to Reduce School Attrition, and Sexually Transmitted and Reproductive Tract Infections: A Cluster Randomised Controlled Feasibility Study in Rural Western Kenya." *BMJ Open* 6 (11): e013229–e013229.

Phillips-Howard, Penelope A., George Otieno, Barbara Burmen, Frederick Otieno, Frederick Odongo, Clifford Odour, Elizabeth Nyothach, Nyanguara Amek, Emily Zielinski-Gutierrez, and Frank Odhiambo. 2015. "Menstrual Needs and Associations with Sexual and Reproductive Risks in Rural Kenyan Females: A Cross-Sectional Behavioral Survey Linked with HIV Prevalence." *Journal of Women's Health* 24 (10): 801–11.

Sol, Lidwien, V. Schölmerich, K. Liket, H. Alberda. 2019. *The Ritu Study Protocol: A Cluster Randomized Controlled Trial of the Impact of Menstrual Health Programs on School Attendance and Wellbeing of Girls in Rural Bangladesh.* Simavi: Amsterdam. Available from: https://simavi.org/what-we-do/programmes/ritu/ [Accessed January 2019].

Sommer, M., and Murat Sahin. 2013. "Overcoming the Taboo: Advancing the Global Agenda for Menstrual Hygiene Management for Schoolgirls." *American Journal of Public Health* 103 (9): 1556–59.

Sommer, M., M. Sahin, L. Paloubis, M. Troung, and J. Sinden. 2016. *WASH in Schools Empowers Girls' Education: Proceedings of the Menstrual Hygiene Management in Schools Virtual Conference 2015.* New York: United Nations Children's Fund and Columbia University.

Sommer, M., Nana Ackatia-Armah, Susan Connolly, and Dana Smiles. 2015. "A Comparison of the Menstruation and Education Experiences of Girls in Tanzania, Ghana, Cambodia and Ethiopia." *Compare: A Journal of Comparative and International Education* 45 (4): 589–609.

Sommer, M., S. Chandraratna, S. Cavill, T. Mahon, and P. A. Phillips-Howard. 2016. "Managing Menstruation in the Workplace: An Overlooked Issue in Low- and Middle Income Countries." *International Journal for Equity in Health* 15: 86.

Sumpter, Colin, and Belen Torondel. 2013. "A Systematic Review of the Health and Social Effects of Menstrual Hygiene Management." *PLoS One* 8 (4): e62004.

Tegegne, Teketo Kassaw, and Mitike Molla Sisay. 2014. "Menstrual Hygiene Management and School Absenteeism among Female Adolescent Students in Northeast Ethiopia." *BMC Public Health* 14 (1): 1118.

UNESCO. 2015. *Emerging Evidence, Lessons and Practice in Comprehensive Sexuality Education: A Global Review.* Paris: United Nations Educational, Scientific and Cultural Organization.

Wilson, Emily, Josephine Reeve, and Alice Pitt. 2014. "Education. Period. Developing an Acceptable and Replicable Menstrual Hygiene Intervention." *Development in Practice* 24 (1): 63–80.

CHAPTER 48

Transnational Engagements: Menstrual Health and Hygiene—Emergence and Future Directions

Edited by Victoria Miller and Inga T. Winkler

Over the last decade, menstrual health and hygiene has emerged as a set of efforts to dissolve menstrual-related barriers. As the field continues to grow and receive increasing attention from organizations, donors, governments, and the media, it is important to assess both the risks and rewards of this work. In conversation with one another, representatives from key organizations from around the world reflect on the efforts to address menstrual health and hygiene and what it requires to productively move forward.

Menstrual hygiene management (MHM) has been closely associated with the water, sanitation and hygiene (WASH) sector. What do you think are the benefits of MHM's close relationship with WASH? Do you also see any potential risks?

Danielle Keiser: Without the initial push of the WASH sector, most of us working on MHM would be lightyears behind where we are now in terms of getting local communities, the media, and the international community to acknowledge the importance of menstrual health and hygiene.

The benefits of the association to WASH are clear: menstrual hygiene involves clean water and soap for cleansing the body and reusable menstrual materials as well as adequate sanitation facilities that are private, safe, and clean and include a way of disposing blood or used materials.

But there are also clear risks of the association with WASH. Menstruation is not just about maintaining cleanliness, it is about a much larger view of the menstrual cycle as a vital sign for health, a source of biological education and an entry point to female health across the lifecycle. Pigeonholing

C. Bobel et al. (eds.), *The Palgrave Handbook of Critical Menstruation Studies*, https://doi.org/10.1007/978-981-15-0614-7_48

any menstrual-related international development work as merely MHM work limits the opportunity to position the reality of the menstrual cycle as a 4-phase cycle and omnipresent process that occurs across many years of women's lives. Menstrual health is so much more than just hygiene management; it's a body of knowledge, a practice, a compass, a yardstick, and an indicator all at the same time.

Neville Okwaro: It is important to look at the strong linkages between menstruation and WASH. We need to consider the safety of various products used to manage menstruation as well as access to safe water to bathe during a girl's or woman's period. Coupled with these WASH-related challenges is the disposal of used sanitary pads and other menstrual products which is a public health concern.

However, reducing menstruation to a WASH issue is narrowing the scope of the human rights that girls and women enjoy. Before even thinking of access to water and sanitation facilities, we need to consider how girls and women can go through their periods each month unimpeded and without disrupting any aspects of their lives. Menstruation is also a multi-sectoral issue that permeates the very essence of life. Menstruation ought to be integrated in all sectors of life—education, gender, water, workforce, reproductive health, and other issues affecting women and girls. Therefore, it cannot be reduced to a WASH issue only; menstruation is a human rights issue.

Ina Jurga: The WASH sector was one of the first sectors to advocate for and integrate MHM, as many organizations work in schools and approached it from the perspective of assuring provision of appropriate facilities for girls coupled with hygiene education. WASH in Schools has many cross-sectorial aspects of providing facilities (usually through WASH sector), and/or health education (through health or education sector).

Jocelyne Alice Ngo Njiki: The association of menstrual hygiene and the WASH sector make it possible to directly confront the government with its responsibilities. In the areas of health or the advancement of women, for example, MHM is one issue among many. In WASH, MHM highlights gender inequalities and provides guidance on adequate infrastructure for MHM that the other sectors could hardly address. Therefore, the WASH sector provides a unique entry point.

Kamini Prakash and Vinod Mishra: We agree that menstruation is a human rights issue. Let's stop looking at women and girls in silos. The focus needs to be on addressing the taboos and transforming the shame and stigma into confidence and pride. This cannot be done if we continue to look at menstruation through sectoral lenses. Even if a woman has access to water and sanitation facilities, she may not be empowered to use them because of the stigma around menstruation. The Bageshwar district in the Himalayan state of Uttarakhand is a case in point. Although households have latrines, menstruators are not allowed to use them or the community latrines during their period due to social taboos. Instead, they have to go out in the open to defecate and manage their menstruation.

Gerda Larsson: During the last decade, we have seen increased attention toward menstruation, also at the global level. In September 2018, the U.N. Human Rights Council adopted a resolution adding important new language on water, sanitation, and gender. The resolution makes an unprecedented call to "address the widespread stigma and shame surrounding menstruation and menstrual hygiene by ensuring access to factual information thereon, addressing the negative social norms around the issue and ensuring universal access to hygienic products and gender-sensitive facilities, including disposal options for menstrual products." While this is a tremendous win and the first time a resolution of this magnitude included menstruation, other global strategic documents such as the Sustainable Development Goals (SDGs) and Global Strategy for Women's, Children's and Adolescent Health developed by the World Health Organization (WHO) still lack specific goals and indicators around menstruation. Without such guiding documents and policies there is a risk that governments, organizations, and companies forget to include and think about how menstruation affects their work.

What does it mean to move menstrual hygiene and health beyond the WASH sector?

Danielle Keiser: Our organization, the Menstrual Health Hub, links menstrual health with WASH, but also with Sexual and Reproductive Health and Rights (SRHR), as well as education, work, human rights, economics, et cetera. Because of the omnipresence of the cycle, menstrual health and hygiene are truly intersectional and multi-sectoral issues and should be taken as such and not forced to fit into any one sector. Therefore, the most appropriate context is *all contexts* in order to mainstream menstrual health. The overarching theme though is aiming for gender equality, since female biology has been a significant cause of women's poverty and oppression. We will only achieve broad-based menstrual health if it is mainstreamed across all areas of life, including business, policy, education, media, and the home.

Julitta Onabanjo: The advocacy and technical contribution from the WASH sector has been and continues to be a critical part of the effort that has ensured today's recognition of MHM as a development and rights issue. That said, at policy and program level most of the attention given to MHM in WASH programs has focused on facilities, hygiene, products, and education for girls in school. This has resulted in girls and women that are out-of-school and marginalized populations, as well as the essential components of community mobilization and engagement of parents, religious leaders, and men and boys being for the most part excluded. The prioritization of a lifecycle approach and situating menstruation as a sexual and reproductive health and rights matter allows for a broadening of partnerships beyond the WASH sector. This allows for efforts to ensure that no one is left behind, and addresses other aspects critical to menstrual health, including social norms, stigma, and discrimination. The prioritization of such approaches is key if we are to ensure gender transformation and normalize menstruation.

Gerda Larsson: It means taking on a wider and more holistic perspective. A move toward menstrual health also requires addressing issues such as menstrual pain and gaining a much better understanding of the impacts of menstrual stigma. Through research that The Case for Her funded, we learned that menstruation-related pain can be a primary reason to begin contraception use. By silencing periods or only addressing them in the context of WASH, we risk under-exploring different strategies to address sexual and reproductive health and rights as well as social inclusion.

Do you think there would be more pushback if menstrual health was more closely linked to sexual and reproductive health?

Jocelyne Alice Ngo Njiki: On the contrary, it would certainly depend on the context, but based on our experience with the sensitization of populations, establishing a link between MHM and sexual and reproductive health would make such programming more attractive. We find a lot of interest in these linkages during our trainings. In fact, using menstrual health as an entry point might help us tackle well-known challenges in the context of sexual and reproductive health. On the other hand, I do think there is a risk that we lose the specificities of menstrual health when it becomes part of the more general sexual and reproductive health discourse.

Kamini Prakash and Vinod Mishra: Even if we wanted to limit our conversations to just MHM, participants invariably ask questions about sexual and reproductive health. If our work is to remain relevant, we need to allow our training sessions to be learner-led. It is difficult to discuss menstruation in isolation of sexual and reproductive health. For example, one of the female participants in Muzaffarnagar in Uttar Pradesh asked us how she could prevent her husband from having sex with her during her period. Instead of limiting the conversation to menstruation in this case, we made it a point to highlight that a woman has a right over her own body and can refuse sex—even when she is not on her period. Trainers should be equipped to talk about these issues—regardless of a possible pushback—because they are fundamental to gender equality.

Julitta Onabanjo: We know that across the continent, African societies and their views around sexuality and menstruation are diverse, some remaining relatively static and others quickly evolving. While menstruation is embraced by some societies as a rite of passage to womanhood, there are others in which menstruation is veiled in secrecy and conservatism, or in notions of impurity—as is often the case with many other sexual and reproductive health matters. In the twenty-first century, we must recognize and appreciate changing gender roles and identities and what this means to policy and programming. We must also take into consideration traditional and cultural practices that have held us back or that could move us forward. Overall, we need to consider menstruation in the context of healthy sexuality, rights, and choices.

As we normalize menstruation, we must not trivialize the reality of menstrual irregularities and disorders; premenstrual syndrome and premenstrual dysphoric disorders are real and contraceptive choices are linked to menstrual experiences. In many countries, unmet need for family planning remains high, in part due to women's fears or uncertainties about the effect particular contraceptives may have on their menstrual health. This highlights the importance of rethinking product information to enable women to make an informed choice. As we strengthen health systems to be integrated and accessible to all, we must also focus on a competent health workforce that can promote positive menstrual health, identify early menstrual irregularities and prevent unnecessary discomfort, while recognizing more serious menstrual disorders and referral for specialized care.

Do you think an MHM approach that includes all challenges throughout the reproductive lifecycle would provide a more inclusive approach to menstrual health? Specifically, how can MHM begin to address the needs of perimenopausal and menopausal women for example?

Ina Jurga: Absolutely. But for our organization, WASH United, it makes sense to start by focusing on adolescent girls: menarche is not only a key signal for entering puberty but also a critical time to inform about menstrual hygiene and share that menstruation is normal. Furthermore, interventions in schools are generally easy to organize. But yes, it is important to be more inclusive: persons with disabilities, transgender/nonbinary people, and also older women have periods and their specific challenges. One entry point for reaching them could be health workers who have the authority and access to reach more diverse populations.

Julitta Onabanjo: Indeed, with the increased focus on MHM and recognition of menstruation as a sexual and reproductive health issue, we must also recognize that menstrual health is broader than the monthly vaginal bleeding episodes. Vaginal bleeding outside menstruation includes, for example, postpartum bleeding, bleeding after a miscarriage, endometriosis, fibroids, and different types of cancers that are associated with heavy bleeding. Likewise, we have an opportunity to address the needs of perimenopausal and menopausal women in our policy and programmatic responses, such as through increased investment in research and education and integration of these topics in pre- and in-service training for service providers to improve access to and quality of services.

Within UNFPA, the United Nations sexual and reproductive health and rights agency, work on menstrual health has evolved significantly over the past decade. UNFPA promotes a lifecycle approach to menstrual health, and to sexual and reproductive health in general, with adequate responses to address the different experiences and challenges encountered throughout the reproductive lifecycle, from menarche to menopause.

Neville Okwaro: It is important that girls and women understand how and why menstruation occurs. This can be best explained through the life cycle approach. Understanding sexual and reproductive health will promote an understanding of menstruation and menstrual health. In many communities, sexuality is riddled with myths, taboos, and mystery. It is not to be talked about in the open. These myths and taboos in turn have a direct effect on the community's view of menstruation. Menstruation can be a stepping stone to challenge the myths and taboos around sexuality while sexual and reproductive health education can be integral in demystifying menstruation and challenging the existing myths, taboos, and beliefs around menstruation.

Menstrual hygiene and health programming in the context of international development has experienced a fast rise over the last years. This is encouraging, but do you think this rapid growth might pose any risks?

Ina Jurga: My experience is that once the challenges have been unveiled, backed by a growing evidence base and champions for the cause, many organizations, governments, and individuals realized that menstrual hygiene was a very neglected aspect and that now it must be considered in programming related to women and girls. Among those now paying attention are donors.

Gerda Larsson: Since 2012, The Case For Her has provided catalytic funding to organizations, research institutions, companies, and individuals tackling menstruation as a means to develop as holistic experts across the space, create evidence of what works and advocate for effective solutions. We are excited to see the growth of the space and believe in sharing the evidence and lessons learned we have gathered. We encourage all organizations to contribute to sharing information and new initiatives to seek out this evidence.

Neville Okwaro: I somehow attribute the fast rise of Menstrual Hygiene and Health programming partly to organizations trying to remain relevant. Donor resources have been dwindling and programs that used to receive huge chunks of funding cannot access the same enormous funding. Therefore, organizations have learned to mutate and change strategy to attract funding. MHM programming is rather a new field and aligning an organization's strategy with MHM program translates to more funding. That is why most organizations are involved in sanitary pad distribution—to them, products are the core of MHM.

Ina Jurga: Indeed, with MHM being perceived as new and exciting, there is the risk that it will keep attention until another 'hot' topic emerges in a few years. And because it is 'hot' today, it tends to attract organizations to pitch projects, even if they lack a solid base of experience. From a programming perspective, I hope, MHM will be streamlined and considered as a minimum package for any kind of hardware and software intervention in schools, institutions, and workplace. I hope it will move from trendy to mainstreamed.

Neville Okwaro: At the moment, almost all organizations are coming up with their own understanding of MHM and various training tools and manuals have been developed. There are no standardized MHM Manuals, training tools, and guidelines. Therefore, everyone out there believes that the MHM programming they are running is the best. This is slowly eroding the quality of MHM because many are not so clear on what MHM programming should entail.

Julitta Onabanjo: I agree. A challenge caused by the rapid rise in attention given to MHM is that we now see many parallel programs and initiatives, which may not be well-coordinated, and may instead be driven by resources and donor requirements at a particular point in time, with limited geographical scale, and inadequate attention to the diverse and long-term needs of girls, women, and other people who menstruate, who they intend to serve. In addition, MHM is often positioned as a stand-alone issue, with a relatively narrow programmatic focus on increasing product availability and limited integration into other programs. This highlights the need for concerted efforts to ensure government commitment and ownership within a strong policy environment, enhanced partnership building and multi-sectoral coordination, with strong engagement of policy makers, the private sector, civil society organizations (CSOs), media, and people who menstruate. We need to connect those representing a diverse range of sectors such as education, water and sanitation, trade and industry, gender and social development, innovation, and technology. We must be careful not to silo menstrual health in our efforts to bring it to the fore.

Ina Jurga: Another risk is that without solid research, a lot of communication is based on unverified and not sufficiently robust enough quantitative data and simply repeating information on the alleged impact. Yet, studies based on qualitative research reveal that many women and girls report that inadequate resources and the sociocultural context including menstrual stigma have an impact on their menstrual experiences, which negatively impact their physical and mental health, education, employment, and social participation. Evaluations of programs seeking to improve women's and girls' lives suggest that interventions can make a difference, but we must be careful to draw on the right data for the right purpose in the right context.

Neville Okwaro: I agree, there is a dearth of data on MHM. For the data we have, we need to ask ourselves, is the data reliable and representative to the extent that it can be used to make inferences about the true situation? If yes, how many practitioners have tailored their programming based on the available data? How many have generated further evidence and used this evidence for decision-making to inform their MHM programs? We have seen zombie statistics (such as one out of ten girls not attending school when menstruating) circulating. To counter these tendencies, we need to make more responsible use of the existing data and acknowledge its limitations, and we need more research that builds a stronger evidence base.

Gerda Larsson: The question of school absenteeism is just part of a much more complex picture that has yet to be made clear. Funding is needed to solidify the research and evidence base past missed days of school and consider, for example, a student's ability to concentrate in class. To date, the scope of measures has been simplified, and we realize that we have been asking the wrong questions.

Jocelyne Alice Ngo Njiki: We need research addressing menstrual hygiene and health in a range of different disciplines including sociology, biology, physiology, medicine, education, and economics. For example, we would like to know what a society stands to lose financially if it disregards MHM in its public policies and its planning and budgeting. Such studies would help elevate the status of our work.

In other sectors, we have seen that efforts often reach the "low-hanging fruit." Addressing the needs of the most visible risks leaving behind marginalized groups. Have you observed this neglect of certain populations in MHM work?

Julitta Onabanjo: With the rapid rise of MHM programming, most of the existing WASH efforts have focused on girls in school, and left out girls and women not in school, as well as other populations such as transgender people and those with disabilities, as well as the critical components of community mobilization and engagement of parents, religious leaders, and men and boys to address social norms, stigma and discrimination, and normalizing menstruation.

Kamini Prakash and Vinod Mishra: We agree, there are many groups whose special needs have not been considered and addressed, such as girls and women with disabilities, menopausal women, women in institutions, homeless women, trans men, women living in disaster settings, to name a few. These groups too often remain unseen and unheard. We need to understand their challenges and see how they can be addressed.

It is also important to note that women do not stay within the home but also go out to work and use public spaces. Toilets in worksites, educational institutions communities, public institutions, and public spaces must be designed to take into consideration menstrual hygiene needs.

Danielle Keiser: With the rapid growth of MHM I also see the risk of leaving boys behind. If we continue to frame menstrual health and puberty changes as girls' and women's issues, there will be major consequences for men and boys, who were 'left out' or considered unimportant to involve. Education around menstruation changes everything—for *everyone*. After all, teaching about menstrual health is a gateway to talking about bodies more generally, and boys are certainly interested in how their bodies work, too.

Neville Okwaro: It is imperative that researchers capture the voices of vulnerable and marginalized individuals and groups. In that context, numbers cannot just be used as a measure of success. Other qualitative and inclusive data need to be generated. Girls and women will continue to menstruate irrespective of their social status and community setting. Inclusive data will promote equity in decision-making and reach out to marginalized and vulnerable individuals.

Finally, recent critics argue that access to menstrual products is too large a priority in menstrual hygiene and health programming and advocacy. Do you agree?

Ina Jurga: From the perspective of menstruating women and girls, who cannot afford or access suitable products, receiving products for free or accessing cheaper products does fulfill a need for hygiene and personal well-being every month. If this keeps girls in schools, or women at the workplace—great. But product distribution needs to come with education including how to use, change, clean, and dispose of products correctly, and even more, education for everyone to empower girls/women/boys/men and to remove negative attitudes and restrictions, and that cannot be done by only providing products.

Neville Okwaro: Concentrating on menstrual products is jumping the gun. Menstrual education is more important than distribution of sanitary pads and assuming menstruators know how to use them. It is vital to pass knowledge on all the available options available to manage menstruation and let users choose for themselves.

Danielle Keiser: People want quick fixes. They want to see that a problem has an easy answer. And pads and other products seemingly provide that answer. Some may think 'No more suffering for "those poor girls" now that they have pads.' But that is akin to putting a band-aid on a gaping wound—it is only going to bring short-term relief because it does not actually engage with the more complex issues of education, poverty, and inequity at the heart of menstrual health as a social issue. So when we address educational needs, we need to think far beyond menstrual hygiene education (often with a focus on product use) and extend to broader aspects of menstrual health literacy.

Gerda Larsson: It is easy to think that products are the silver bullet, but free pads, tampons, cups, and underwear cannot address menstrual stigma. While products definitely play a role, and absorbents can help a person manage their period, it is not a holistic approach. We need educational programs to tackle insufficient knowledge around menstruation, offset stigma surrounding vaginal bleeding, and provide menstruators with tools to make informed choices.

Kamini Prakash and Vinod Mishra: We agree. There is a risk that the rapid rise of menstrual hygiene and health programming in the recent past may lead to quick-fix solutions such as the provision of menstrual pads, without looking at the other interlinked dimensions of menstrual hygiene such as menstrual taboos, gender inequities, and menstrual waste disposal.

All too often, menstrual hygiene is equated only with the use of menstrual pads. Instead it is important to look at the entire MHM value chain—from breaking the silence and looking at inequitable gender norms, spreading awareness on MHM, providing clean and private WASH facilities and menstrual absorbents, and finally providing safe facilities for disposal of menstrual waste. These are interrelated and interdependent aspects in the MHM chain. For instance, we cannot promote menstrual hygiene without breaking the silence and creating awareness.

In that context, we need more data on the health impact of commercial, disposable sanitary pads which use super absorbent polymers, gels, and other chemicals, so that women and girls can make an informed choice about the menstrual absorbents they use. We also need more data on the environmental risks of disposable pads. With an increasing number of adolescent girls preferring to use disposable pads, the environmental concerns are just coming to light. Schools in India have started installing incinerators to dispose of menstrual pads, but they do not necessarily adhere to emission norms of the Central Pollution Control Board and thus, can be hazardous for school children and staff. By promoting disposable pads, we may be solving the problem of menstrual hygiene but we may also be creating new challenges due to unsafe menstrual waste disposal.

You all mention the importance of education, but mostly in relation to the use of products. What kind of education do we need to ensure menstrual health comprehensively, combat the stigma around menstruation, and enable menstruators to take informed decisions not just in relation to materials but any aspect of menstruation?

Danielle Keiser: The kind of education that is needed is the kind that does not feel like education, but urgency of biological reality and absolute common sense. Separating this from culture is the hard task. Good education starts with the first contact point anyone has with learning about human reproductive biology. Our bodies are not gross and menstruation, while sometimes unpleasant, is natural, normal, and necessary for human life to continue existing as we know it.

Ina Jurga: Our objective has always been to achieve scale and thus the design of our training sessions and educational materials has evolved from providing comprehensive MHM education to offering a very compact, easy-to-use, and low-cost guide (which can also serve as an introduction to more comprehensive or additional materials). Many teachers lack both capacity and comfort to teach about MHM, so we wanted to create something that is very simple, yet covers the basics. Furthermore, and maybe most important, we focus on changing attitudes and empowering girls so they don't feel lonely with their periods and feel empowered to ask questions. The next steps for us will be to explore how we can sustain support and communication beyond the initial training, as well as how to include boys as well as mothers who are often left out, even though they are key informants.

Vinod Mishra and Kamini Prakash: It's not enough to educate menstruators alone when most of the challenges come from the community elders, religious leaders, mothers, and male family members. Mothers, for example, are a primary source of information about menstruation for adolescent girls. Research has shown that 70% mothers in India consider menstruation as dirty and polluting. Little wonder then that girls grow up observing social taboos which are often harmful and prevent girls from realizing their true potential. We therefore need to extend educational initiatives beyond girls in schools.

At WSSCC, we have started focusing on making our trainings more inclusive. In keeping with the SDG principle of Leave No One Behind, we have developed MHM materials for menstruators with visual and hearing impairments. For example, *As We Grow Up: About Menstruation* is an instructional video on MHM in Indian Sign Language. We have engaged intensively with the disability sector in India, especially with organizations that represent the visually impaired. To date, we have trained over 70 organizations across India on MHM.

Julitta Onabanjo: Menstruation impacts educational outcomes and the education system can be an impactful channel to promote and ensure menstrual and other sexual and reproductive health well-being. The importance of access to comprehensive sexuality education and health services becomes vital conduits to ensuring menstrual literacy and normalization. It is also vital to guarantee self-esteem, self-worth, and empowerment that can influence other aspects of sexuality, reproduction as well as general decision-making in life. Menstrual literacy therefore should ideally happen before menarche and move beyond the schoolgirl to include all girls, women, and people who menstruate as well as parents, teachers, community leaders, and other influencers, such as traditional and social media to change the public discourse.

One of the challenges mentioned earlier is the fragmentation between different organizations and approaches. Do you see any initiatives to counter that?

Kamini Prakash and Vinod Mishra: Yes, the coming together of organizations and individuals working on MHM is a sign of progress. In India, the Menstrual Health Alliance India (MHAI) aims to engage with government, civil society, media, and corporations to use evidence-based research to raise awareness and promote inter-sectoral action on MHM. In the last year it has convened a media workshop, given technical recommendations to the government on the Goods and Services tax (GST) on menstrual pads, developed a policy brief and resource book on menstrual waste disposal for the Ministry of Drinking Water and Sanitation and also engaged with Bureau of Indian Standards to develop standards for reusable cloth pads.

Julitta Onabanjo: There are similar developments in the African context. To support efforts to strengthen coordination of existing MHM programs in Africa, UNFPA cohosted the first-ever African Symposium on Menstrual Health Management in Johannesburg, South Africa in May 2018. The event identified knowledge management and information sharing as critical elements of an effective response to menstrual health management in Africa, which led to the establishment of the African Coalition for Menstrual Health Management. The Coalition aims to strengthen advocacy for multi-sectoral policy making, support efforts to fill in existing research gaps, develop product standards and ensure constant and regular availability and diversification of menstrual programs, and products and supplies. The Coalition also supports efforts to ensure that there is proper policy and program guidance to integrate MHM across all relevant sectors.

Given the developments on MHM over the last 5 years and the risks you identified, what do you think needs to happen to strengthen comprehensive programming on menstrual health?

Danielle Keiser: Talking about menstruation openly is the entry point to strengthening comprehensive programming on menstrual health. This will be reflected in a thriving ecosystem of powerful players beyond the development sector who say and do something about it in their everyday work. Investors can fund menstrual health, as it is the foundation for future investments that address female health across the reproductive journey. Policy makers and public servants can advocate for gender-smart policies that look at equity, access, and safety of period products and WASH facilities. Educators can integrate menstrual health in their health, education, and life skills curriculums. Mothers and fathers can prepare themselves by accessing helpful suggestions for guiding their daughters and sons into adulthood.

Neville Okwaro: There is a need to define what constitutes a menstrual health program. It should be holistic and not just address one aspect. It should start from access and move to the provision of practical information which will demystify menstruation including information about menstrual management options and how to dispose of the various materials and products.

Julitta Onabanjo: The need for MHM standards, regulations, unified frameworks, and effective coordination mechanisms cannot be overemphasized. Our point of advocacy is for MHM to be integrated across sectors and areas to ensure that the focus on MHM is expanded from just being about WASH or product availability as a stand-alone initiative, to a more integrated, cross-sectoral policy, and programmatic issue involving adolescents and women's sexual and reproductive health, education and services, and water and sanitation and waste disposal in a broad range of contexts: schools, workplaces, humanitarian settings, and more.

PART V

Menstruation as Material

CHAPTER 49

Introduction: Menstruation as Material

Katie Ann Hasson

The varying meanings attached to menstruation, the stories we tell about it, the ways we manage it, and even how we experience menstruation differ depending on time, place, culture, and individual embodiment. But menstruation *itself* should seem self-evident. Menstruation is a biological reality, after all—a material fact, a fluid produced by the body that can be seen and felt, a reality that gives rise to a range of materials and products designed to absorb or contain it. And yet, as the chapters in this section show, despite its seemingly obvious materiality, menstruation must nonetheless be made to "matter."

The material production surrounding menstruation includes a variety of menstrual management technologies such as pads, tampons, cups, and newer product innovations, such as the more than 100 cycle-tracking phone applications on the current market, "period-friendly" underwear, and subscription services that send consumers boxes of menstruation supplies on a monthly basis. It also includes a growing number of artistic productions, from Judy Chicago's work of the 1970s, including the lithograph "Red Flag" and the installation "Menstruation Bathroom," through more contemporary menstrual art shows, films, documentaries, and aesthetic objects—each of which shed light upon the social, political, and medical meanings of menstruation.

This section starts and ends with one of the most obvious manifestations of menstruation as material: menstrual management technologies. As Elizabeth Kissling asserts, "Within the current cultural logic of late capitalism, a woman's relationship to her menstrual cycle is largely defined through consumer products" (2006, 123). It begins with examinations of the material of these products themselves—what should a tampon or pad be made of, and how do scientists determine what is safe?—and closes with an examination of the economic interests engaged with producing and distributing these products.

© The Author(s) 2020
C. Bobel et al. (eds.), *The Palgrave Handbook of Critical Menstruation Studies*, https://doi.org/10.1007/978-981-15-0614-7_49

A number of chapters explore different ways of translating the materiality of menstruation into the languages of science, technology, and data. Science is one arena that has attempted to reckon with menstrual material in numerous, often problematic ways. The development and testing of tampons is one of these (where the outcomes are sometimes quite humorous: tiny mouse tampons! that ubiquitous blue fluid!). How can the messy materiality of menstruation as embedded in the bodies, lives, and experiences of menstruators be reconciled with demands for precise, objective data in the lab? This is particularly tricky when health and safety are at issue and experiments attempt to analyze tampon ingredients and absorption apart from the messy menstrual environment—that is, menstruators' bodies, which must be "instrumentalized," or turned into tools for scientific research. Vostral and Reame variously explore these questions in the case of tampon development and safety, raising questions and offering cautionary advice for today's menstrual health advocates.

Measuring and monitoring menstruation turns out to be complicated in a number of ways as well. From Reame's problematizing the use of saline solution to test tampon absorption, the chapters shift to other difficulties with measurement. What aspects of the menstrual cycle ought to be measured and how? Measuring and tracking hormone levels, menstruation-related school absences, and the length and characteristics of individual menstrual cycles all turn out to be less straightforward than one might presume. Measurements of hormone levels must be timed and matched by day of the menstrual cycle (Houghton and Elhadad); school absences are difficult to reliably track (Benshaul-Tolonen et al.); and period-tracking apps in their current form rely on a range of assumptions about gender, sexuality, behavior, and menstrual cycles that reduce their usefulness for a wide range of users (Fox and Epstein).

Several chapters examine the experience of menstruation in relation to its most concrete materiality. Steward et al. provide insights into both bodily sensations and menstrual management tasks and how they affect people with sensory and cognitive disabilities. My own research (in Chapter 58) examines the mismatches between experiences of monthly bleeding and definitions of menstruation that are changing to accommodate menstrual management technologies that aim to reshape the biology of menstruation using ingested hormones.

Artistic production has also proven fertile ground for exploring menstrual matters. The very substance of menstruation can be translated into the symbolic language of artistic representation and performance. How is blood portrayed in painting? How does this differ when that blood is "gendered," as in menstruation and childbirth? Cole explores the changing meanings and portrayals of gendered blood in art. Relatedly, in a series curated by Jen Lewis that pairs several artists' statements with examples of their work, we engage the process and product of creating art using menstrual materials—whether as medium or subject or both.

The last few chapters return to a discussion of menstrual technologies, particularly the multi-billion dollar industry that produces, markets, and sells menstrual management products around the world. Tarzibachi examines how Latin American countries are targeted as markets for menstrual products originating in the US, as well as how the meanings and use of these products in the Latin American context have changed over time. Punzi and Werner explore how social enterprises combine a social mission with a for-profit model to address the social problems surrounding menstruation, while investigating some of the challenges and critiques these hybrid ventures face. The final chapter engages representatives of a range of these "caring corporations" through a discussion of the tension between making a profit and promoting social change. Together, these concluding chapters provide a glimpse of how a new generation of companies are attempting to foster new relationships between women and girls and their bodies, and between wealthy consumers in the Global North and beneficiaries of free or reduced-cost products in low- and middle-income countries.

From high technology to classical art, scientific research to menstrual health education, the chapters in "Menstruation as Material" use a critical lens to explore the many manifestations of, sometimes quite literally, menstrual *matters*. This wide range of explorations necessitates a variety of approaches, and therefore the chapters in this section take many forms. Alongside social science and humanities pieces (drawing on a range of qualitative and quantitative research and analysis), we find personal reflections woven into a historical account of research on Toxic Shock Syndrome (TSS); artists' notes and narratives; and firsthand accounts of building businesses that aim not only to address the menstrual needs of individual consumers but also to foster tangible social transformation. Considered together, the contents of this section represent some of the ways that the field of critical menstruation studies simultaneously grapples with the materiality of—and the process of *materializing*—menstruation.

In the end, what remains clear is how little is actually self-evident after all.

References

Kissling, Elizabeth. 2006. *Capitalizing on the Curse: The Business of Menstruation.* Boulder: Rienner.

CHAPTER 50

Of Mice and (Wo)Men: Tampons, Menstruation, and Testing

Sharra L. Vostral

The tampon, a uniquely gendered technology to absorb menstrual fluid, has a long history of use and modification during the twentieth century. The familiar cotton, but also rayon, silk, polyester, and even hydrophilic salts have been used at one point or another. By 1980, 70% of women in the United States used tampons during their fertile years to absorb their menses (Okie 1981). Women still rely on them to contain a bodily fluid in an American society where the norm is to keep it hidden (Vostral 2008). Tampons have come to belong to a liberation toolbox, providing a means for physical mobility and bodily freedom in a society where menstruation has neither been privileged nor celebrated. What is less recognized is that women's bodily experiences with tampons are directly linked to developments in material science, corporate research, and gynecological observations about menstrual cycles. In the face of multiple variables, how did researchers devise tests to improve designs while keeping an eye on women's health?

The practical work of tampon testing has been predicated upon broad cultural conditions: prevailing ideas about women's bodies, notions about physiological and cultural gender differences, and concepts of the role of science and medicine in optimizing progress and well-being in the United States. I argue that to capture these conditions, the relationship of tampons to testing and to women's bodies is best seen as a three-way recursive loop (Fig. 50.1). However, in this loop women's bodies must be instrumentalized and subjected to a scientific rationality, which is no easy task methodologically. Sometimes women are "the right tool for the job," to borrow from Adele Clarke and Joan Fujimura, and sometimes they are not (1992). Seen through the lens of socially informed cultural analysis, it is also

© The Author(s) 2020
C. Bobel et al. (eds.), *The Palgrave Handbook of Critical Menstruation Studies*, https://doi.org/10.1007/978-981-15-0614-7_50

Fig. 50.1 The recursive
loop of testing (Credit:
Sharra L. Vostral)

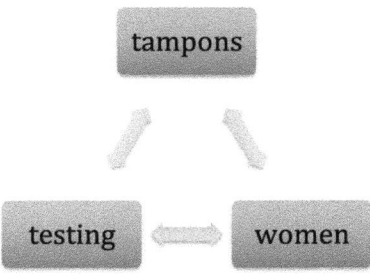

politically problematic to strip away agency and humanity in the name of collecting data for the purposes of science and evidence-based medicine.

The pressures to collect empirical data through experimentation furthered a scientization of menstruation, which began late in the nineteenth century (Vostral 2008; Hasson 2012). Patterns of social power and privilege configured this science, with evidence taking different forms and shapes during the twentieth century. For example, during the 1920s, ideas about scientific management informed data collection about sanitary napkin distribution and use, while the 1980s witnessed tightly controlled laboratory conditions to simulate vaginal anatomy and tampon absorption. Ideas of efficacy and safety also shifted depending on notions of corporate responsibility and market opportunity in a given setting or era. The conditions under which menstrual technologies emerged and perished on the American market varied on many levels.

In this essay, I divide approaches to the material testing of tampons into three major categories: (1) cultivating women test subjects; (2) medicalizing tampons; (3) measuring for safety. The first section demonstrates how women test subjects held a prominent role in the generation of empirical data from the 1920s to the 1950s. The inclusion of women in systematic "field" trials of industrial products marked a radical departure from the use of anecdotes and hearsay to improve sanitary napkins. The medicalization of tampons, discussed in section two, demonstrates how by mid-century, medical models gained traction and some physicians emerged to offer evidence for tampon safety. Nurses and gynecologists became the mouthpieces of assurance and helped to allay fears as well as endorse the efficacy of tampons. Part three focuses upon new measurements for safety, ushered in with formalized Food & Drug Administration (FDA) regulations set in place in the late twentieth century for measuring and assessing tampons as medical devices. However, the testing often failed to predict imminent health hazards, such as Toxic Shock Syndrome (TSS) associated with superabsorbent tampons. Design modifications to tampons, changes in material composition, and the cultivation of women test subjects exposed assumptions in scientific and medical practice, ideas about safety, and attitudes concerning gendered and menstruating bodies.

HUMANS AS TEST SUBJECTS

In order to appreciate the contours of how tampons have been tested, it is important to acknowledge the context of Western-style medicine that rose to acceptance during the late nineteenth and early twentieth centuries. Medicine began to incorporate scientific practices and methods into the assessment of treatments and surgeries. Some of these treatments were perceived as harmless despite their consequences, with efficacy figuring as the central concern for the analysts, while others raised serious concerns about the ethics of the very practices (Lederer 1997). There is a growing body of literature addressing this unsavory side of medicine in which imprisoned and institutionalized men and women, orphans, and disadvantaged "others" became the objects of questionable medical experimentation, conducted usually without their consent (Goodman 2008; Hornblum 1999). Changes in the mid-to-late twentieth century in human subjects testing guidelines, institutional research board (IRB) review, and informed consent are certainly improvements, though problems still remain (Epstein 2009; Wailoo 2008; Braun 2005).

Tampons, however, provide a case for examining not only the role of human test subjects, but also how researcher and subject gender identities influence the ways that testing is constructed and implemented. The gender identity of the person creating this technology begins to reveal itself as central to the details of scientific development, and starkly so. The tampon was seen by researchers as a female-specific technology that could not reasonably be tested upon men, meaning that any effort at reaching associative results was not seen to be scientifically valid. Here, women are seen as different from the ideal white male test subject. Unlike cosmetics that were tested on male prisoners (Hornblum 1999), tampons were not, and scientists were hard pressed to imagine a male body correlate to the cervix and vagina during menstruation. On the sliding scale of gender difference, testing tampons on men crossed an imaginary binary, and the decoupling of menstruation from the category of women had not yet been conceptualized with the term "menstruators" (Bobel 2010).

As such, male scientists could not "self-experiment" as had been the practice for so many in the past. The simple taste test in chemistry determining sour acids from bitter bases demonstrates one level of this first-person knowledge. Rebecca Herzig outlines the gruesome ways that scientists offered their own bodies and well-being for the pursuit of knowledge and information, particularly in the area of x-ray experimentation (Herzig 2001, 2005). In part, this ability to self-experiment relied on a level of personal agency and decision-making embodied by privileged white men of the time.

The relationship of male researchers and female test subjects was often particularly laced with gendered power structures of patriarchy and paternalism. While male inventors could not sacrifice themselves, though they may have wanted to, many could request or insist that their wives volunteer as test subjects. If the male scientists did not have their own first-hand experiences, sensations, and feelings with tampons to rely upon as trustworthy data, wives were their proxies. This opens up an interesting window into their marriages with wives seeming to be not only complicit but also active proponents of the designers' patents. Years invested in marriage and the familiarity offered by that relationship provided a trusted interpreter for some scientists' first-person observations. Though the patents do not mention these collaborations, other sources do, and the invisible wives did indeed have a say in how the menstrual management technologies emerged. Not only were wives drafted into the service of science, but other nearby women including secretaries, receptionists, and relatives also served as testers of new menstrual products.

Cultivating Female Test Subjects

The first person to give serious attention to women users of sanitary napkins was Lillian Gilbreth, who deployed the scientific method in creating a hypothesis and collecting statistically significant data along with qualitative responses to interviews and surveys. Known simply as "Report of Lillian Gilbreth, January 1, 1927," it was the first to examine the entire market and pose recommendations about sanitary napkins and their use, design, and advertising. Well known for her work with husband Frank on time-motion studies, Gilbreth set out to continue the couple's work after Frank's death in 1924 (Lillian passed away years later in 1972), only to find that as a woman she was tracked by her clientele to study "feminine" things. Yet, Gilbreth was also deeply concerned with improving the quality of women's everyday lives, serving on committees under nearly every president from Hoover to Kennedy, consulting about women's labor, women in defense services, and even the Girl Scouts. Establishing her own credentials within that constrained field, however, she landed one of her first big accounts with Johnson & Johnson, who sorely needed advice about women's feminine hygiene practices, as they were called. Kotex's 75% market share walloped Johnson & Johnson's Nupak, which held only 2%. With gusto, she dove into the study of sanitary napkins and provided a searing and revealing report to R. W. Johnson.

Gilbreth and her two female staff members conducted a multipronged study about the state of sanitary pads, contemporary products on the market, their effectiveness, and women's wants (Gilbreth 1927; Fouché and Vostral 2011). She interviewed educators, deans, and instructors teaching health and hygiene classes to ascertain their opinions about the design and materials for active young women; her staff conducted interviews and sent out surveys to more than 3000 women about their menstrual hygiene use; finally, her team studied the characteristics, strengths, and flaws of 53 different sanitary napkin brands available for purchase at the time. She was particularly blunt in her assessments of what she saw as bad branding. "L. B. Sanitary #24: If you have to say *sanitary*, you may as well add napkin; Eagle Brand #25: A buyer would be lucky if she got what she asked for under this name instead of cheese or baby's milk" (Gilbreth 1927, 37–44). The "S. S. Napkin" so obviously referenced the name of an imaginary boat that the staffers refrained from even commenting about the ludicrous branding.

The important thing about Gilbreth's methodology was that it was the first systematic study of women's preferences about menstrual management technologies that focused on the end users, not just the needs of the corporation. This scientific method was in stark contrast to the anecdotal evidence male scientists gleaned from letters of complaint received by customer service, and more provincially, asking their female secretaries, typists, and wives (Graham 1998). It was clear that women did not like the commercial menstrual products, and Gilbreth ascertained why. Everything from the marketing, to the sizes of the boxes, to the actual product and its materials mattered. She learned that Kotex, though still short on women's praise, held appeal because it could be modified, in the best sense of "appropriating technology" (Eglash et al. 2004). Vaseline rubbed on the corners softened the edges and improved comfort. Tabs could be cut and shortened, and even the cellucotton filler could be thinned to make it less bulky (Gilbreth 1927).

Based upon Gilbreth's detailed report, R. W. Johnson took the criticisms and made substantive changes to a new sanitary pad called Modess, a name no doubt sanctioned by Gilbreth as unobtrusive and feminine sounding, like *modest*. In his patent of 1929, there is no mention of Gilbreth by name, but her recommendations set forth in the report were embodied in a new sanitary pad, one based upon prototypes she tested. Importantly, her suggestion for "cut off corners" based upon women's responses emerged as "feathered edges" in the patent, reflecting a discourse of an industrial process rather than a seamstress at a home. The filling, though compared to the "soft feel of cotton," was not composed of it. Johnson instead incorporated a water repellent "cellulosic vest" to act as a barrier between the pad and an undergarment. Next, "filmated gauze" enclosed the vested pad, and Johnson also had a pending patent on that fabric: "surgical gauze the interstices and one face of which are clothed with a superficial film of free fibrous material, preferably cotton slivers" for absorption and soft texture (Johnson 1927, patent #1,705,366). This layering also contributed to a pad with less bulk.

Though Johnson incorporated many of Gilbreth's suggestions, materialized in the patent and new product, a more prescient recommendation centered upon predicting, testing, and analyzing materials. She warned Johnson & Johnson not to follow the trend of designing "laboratory devices for testing." In essence, Gilbreth declared limits to the utility of technology as a source of information. Machines could not so easily replicate women. Only women held the unique qualifications to test sanitary pads because they knew how they felt and how they fared. In so doing she granted the women voice, or subjective legitimacy. Real-life field tests were required, and moreover, should be managed by a woman researcher. She recommended that "a woman be added to the staff of Johnson and Johnson and that all products be submitted to women for inspection of design and tested for actual use." But she warned not to bother with hiring a woman with no smarts, for "an unintelligent tester would be worthless or worse because their findings might encourage the development of a poor product or discourage the development of a good one" (Gilbreth 1927, 133–34). She predicted that random testing practices and insufficient interpretation of the results from testing could yield devastating effects upon a brand and therefore a company.

As it stood, the survey generated interest in a new product. During her interviews, "questions were frequent as to when it would appear, what name it would bear, and who was manufacturing it" (Gilbreth 1927, 14). Gilbreth capitalized upon women's desires for a better sanitary napkin as an incentive to enroll them as test subjects; the only remuneration was satisfaction in the knowledge that their responses would contribute to the design of a new product. The success of the investigation was owed to Gilbreth's adept skills and survey methods in generating woman-to-woman answers and responses. Together, these female researchers contributed to a growing science of menstruation, in which their own position as scientists leveraged certain privileges but their status as women circumscribed their research. They shared an interest in the test subjects as a part of improving an essentialized understanding of "the women's experience," yet they still instrumentalized them in assessing the materiality of products and female bodies as measured by tools, thus bolstering their expertise. They provided an avenue to hear many women's voices, but it was their privilege that legitimated the stories of the test subjects.

Medicalizing Tampons

Asking women questions about their menstrual practices, and even requesting that they report back about sanitary napkin use may have been quite personal, but a new level of testing emerged when the "internal sanitary napkin"—the tampon—emerged as a viable menstrual management technology. In this new era, medical practitioners began to dominate the design and testing of the material components of tampons. No doubt because they were used vaginally, this fell to the realm of gynecologists. The expert stature brought by physicians and nurses helped to legitimate not only the concept of the tampon but its reputation for efficacy and safety. This was especially important for the younger clientele that began to use tampons, with many questioning whether or not they were threatening the virginity of pubescent girls by breaking the hymen (a task purportedly for a husband on a wedding night).

Physicians from the late nineteenth to early twentieth centuries sometimes inserted tampon-like absorbent materials into body orifices to stop bleeding, and any tampon patented between 1890 and 1921 referred solely to medical and surgical application (Farrell-Beck and Klosterman Kidd 1996). The notion of a specially designed, disposable tampon specifically for menstrual fluid emerged in the late 1920s and early 1930s. Earle Cleveland Haas, an osteopath, patented both a compressed cotton tampon and a telescopic applicator. His design was unique as much for the tampon composed of cotton stitched lengthwise and then compressed by old candy machines, as for the disposable cardboard telescopic applicator to deliver the tampon into the vaginal canal without women's fingers needing to touch any body parts. Rather pragmatic in his assessment, he reported that he had seen so many patients in his practice struggle with the mess of pads, he felt he could do better. He also enlisted his wife (who remained unnamed), a nurse, to wear many of the prototype tampons for his 1931 patent that later became Tampax (Bailey 1987). Through her connections, they distributed the tampons to other female colleagues, enrolling them for feedback. So convinced was he of his command of female anatomy, earned through years of medical practice, that he felt there was no need for data collection, such as measuring the size of the vaginal canal. He noted that he had "seen so damn many of them I had an idea" of the various shapes and sizes (*Kehm v. Procter & Gamble* 1982, 1179). Believing women had standardized anatomy, they were interchangeable.

Furthermore, he dismissed the idea of broader testing exactly because of the standardization of the vagina. He continued, "Don't make any difference where [the tampon] was in there, it would soak it up" and "I tried so many of them I knew about how much they would absorb" (*Kehm v. Procter & Gamble* 1982, 1179). Thus, he dispelled any need for broader testing beyond that of his wife and her opinion, so strong were his convictions and his hubris. Like many engineers of the time, even in unrelated fields such as the cement industry (Slaton 2003), he exuded confidence and arrogance about the certainty of his science, which was not even really "science" at all. This often worked to stifle naysayers and claim intellectual rights to knowledge, dismissing critique and bluntly asserting scientific superiority. Once Haas received his patents, he sold them to Gertrude Tenderich for $32,000 and it was she who marketed and developed the Tampax branding and founded Tambrands corporation (*Kehm v. Procter & Gamble* 1982). Tenderich, however, made sure to include Haas in her marketing by making claims that a physician designed Tampax and that the *Journal of the American Medical Association* had accepted the new product for advertising, a claim that had no substantive value. She also advertised the tampon at medical conventions with a female X-ray technician who dressed as a nurse to further legitimate the product (Bailey 1987).

The issue of testing women's bodies for tampon efficacy, however, gained momentum during the 1940s and 1950s when empirical medical data became preferred to anecdotal or even qualitative evidence. Corporations led the way by enrolling physicians and gynecologists at medical colleges to devise trials, locate women test subjects, and manage and publish the results of the experiments. In this way, the corporations could affix supposedly neutral results, affirmed by objective scientific methods, to their products and garner approval for a legitimate and functional medical device. The Personal Products Corporation, a division of Johnson & Johnson which manufactured Meds tampons, sponsored one such test. The name of Meds captured its medicinal overtones to make it sound sterile, safe, and salubrious. Irja Widenius, a physician in the Department of Obstetrics and Gynecology at Columbia University, led the study, using the cutting-edge technology of X-rays to determine absorption capability and position of tampons in situ. This was a remarkable, and very visual, method. Instead of measuring how much water a cotton tampon might hold, or even collecting used tampons to determine their absorption capacity, she instead X-rayed women's pelvic areas while they were menstruating and wearing different brands of tampons to "see" how well they worked, and how well they were positioned to avoid leakage. "A Study of Commercially Manufactured Catamenial Tampons" (1944) was published in the *American Journal of Obstetrics and Gynecology.*

The main challenge that Widenius encountered was that menstrual blood and cotton tampons were not particularly suited for this sort of visualization since they were transparent when viewed with X-rays. Widenius directed technicians to brush barium sulfate, a radiopaque powder, onto the tampons prior to the female test subjects inserting them into their vaginal canals. She took a series of before-and-after X-rays, including half an hour after insertion, to determine the ideal position of the tampons and how well they were absorbing fluids. Widenius was delighted with the results. Not only did she conclude that tampons were safe, but she addressed concerns that they might cause infertility by pointing out that of the twenty-five women who participated, five became pregnant during the study and three went on to deliver babies. Of course, with our current understanding of the dangers posed by radiation to DNA threads, and especially to sensitive reproductive organs and vulnerable fetuses, the test in retrospect is rather horrific and points to the need for safeguards delivered by institutional review board protocols. Yet, Widenius interpreted the pregnancies as data points in relation to the tampons and surmised that they did not cause infertility and were safe. By turning women's bodies into tools, she legitimated the cotton tampon, its safety, and the practice of testing itself as a means to quell cultural concerns.

Measuring and Its Limitations

From the very early days of the FDA, tampons received little attention and were classified as cosmetics, falling into a feminized category that had little to no oversight in terms of labeling, materials content, safety, or advertising claims. Because the Medical Device Amendments (MDA) of 1976 categorized tampons as Class II medical devices—products that had already been sold on the market and (in the eyes of those deciding such matters) proven safe by their track record—they did not need to undergo safety testing. Furthermore, companies could establish "substantial equivalence" for any new product, thus bypassing a premarket approval process at the FDA (Vostral 2018). For the most part, the design of newly patented tampons was still not substantially different from the old Tampax patent of 1931, even though new materials were incorporated. Semi-synthetics such as rayon, derived from cellulose wood pulp, found their way into the composition of tampons such as Tampax, Playtex, Kotex, and O.B. Such materials were cheaper than cotton, had a soft texture, and blended well to form a cotton/rayon plug. Tampon patents emerged during the 1960s calling for a great range of materials that could be used: hemp, silk, Dacron, polyester, nylon, acrylic fibers, viscose rayon, protein fibers, and glass fibers, to name a few (Graham et al. 1960, patent #2,934,068). Another material such as polyacrylate, a polymer structure that forms a gel when introduced to water, seemed like a perfect medium to incorporate into tampons. The era of the "superabsorbent" tampon ushered in synthetic materials, but these additions did not trigger safety concerns because the MDA "grandfathered" their approval.

However, the assumption that synthetic materials were inert came back to haunt scientists. By the 1970s, Procter & Gamble had begun to design, test, and market Rely, an entirely synthetic tampon devoid of cotton. Its components were synthetic: a polyester bag containing small polyester foam cubes and the thickening agent carboxymethylcellulose that formed a gel once exposed to fluid. The design was entirely innovative, and it truly absorbed more liquid than its cotton counterparts; uncomfortably so, according to many women who complained Rely dried them out, making additional tampon changes painful.

In 1978, during the time period that researchers at Procter & Gamble were honing the final design and marketing of Rely tampons, medical researchers in Colorado were naming a new disease called TSS caused by the bacterium *Staphylococcus aureus*. Because TSS was first associated with children, it took extensive work by state and federal epidemiologists to establish and clarify its links to superabsorbent tampons, and furthermore, that women who had used them were suffering and dying from TSS ("Toxic-Shock Syndrome—United States" 1980; "Follow-up on Toxic-Shock Syndrome—United States" 1980; "Follow-up on Toxic-Shock Syndrome" 1980). Findings from the Centers for Disease Control (CDC) that identified Rely as the tampon associated with the most cases of TSS precipitated its withdrawal from the market in September of 1980. Despite its very dramatic departure and the bad publicity for Procter & Gamble, the company insisted that it had surpassed federal expectations for testing at every level, as dictated by the MDA.

Testing components separately using deliberately circumscribed procedures meant that few problems with the tampon could be detected. Individually, polyester or carboxymethylcellulose could be sterile, non-carcinogenic, and hypoallergenic. Procter & Gamble scientists tested for toxicity, carcinogenic effects, and potential harm to fetuses by implanting polyurethane (one of the initial components) under animal skin. They also ran tests in animals to check for inflammation or an allergenic irritation to the vaginal tract. When the polyurethane composition became a safety concern for women in the test market of Rochester, New York, managers decided to make a switch to polyester. This polyester was tested again, fed to lab animals, and furthermore, shaped into tiny mouse tampons called pledgets that lab technicians inserted into the vaginal orifices of mice to have them wear for weeks on end (*Kehm v. Procter & Gamble* 1982).

The mouse substitutes, though problematic on multiple levels, were not necessarily a bad thing. The commonly held practice to use lab mice, and the metaphorical guinea pig test subject, protected humans from a wide array of toxic exposure and potential ethical abuses. Yet, one must question the usefulness of these mouse pledgets, since rodents do not menstruate or experience monthly periods. They are decidedly not similar to human women in this way. As a substitute, they seem to lack numerous commonalities, from menstrual fluid to a vaginal microbiome, which both turned out to interact with tampons. The mouse tests may have served as a functional test activity, but not a good replication of field conditions.

Both artifacts and bodies—of animals and women—were tested in different ways to gauge product efficacy during a menstrual period. Neither may have provided much sound evidence to make safety claims because of assumptions that the tampon remained inert in the vagina, and that (for a moment in suspended imagination) mouse and woman body parts were interchangeable. After the animal studies were complete, Procter & Gamble tested Rely in women. There were 1332 women in the study, and though the results were not made public, they likely enumerated women's preferences about the tampon. In addition, based on the sample size they would not find the problem of TSS that would ultimately undermine the product. The sample size was far too small to predict TSS, when only one to 17 women out of 100,000 contracts it (USFDA 1989). They did not know to look for it and did not have a sufficient sample size to predict it.

The tests conducted by Rely's developers, which sought to analyze and control specific conditions and predict outcomes, failed in the most immediate sense because they were not constructed to account for bacterial interactions with supposedly inert materials, which was indeed a new variable with TSS. But on a broader level we might see that scientists did not find danger because they were not testing for that kind of safety hazard, and did not even think to do so. It may be an unfair expectation of them, however it is equally troubling that the synthetic materials, necessary to produce the new line of superabsorbent tampons, were deemed equivalent to cotton. Problems with vaginal sensitivity emerged and were accounted for by researchers, but not the unimaginable notion that the components would coalesce as an in vivo petri dish, turning the vagina into an incubator.

Along with the superabsorbent materials introducing sufficient oxygen to change the vaginal canal from anaerobic to aerobic, the conditions were prime for the bacterial growth of *S. aureus* and its expression of toxins. A correlating shift in pH from acidic to neutral during menstruation, and a lack of the requisite antibodies to the toxin in some women, meant the conditions could turn deadly (Tierno 2001). The superabsorbent and synthetic material changes made to many tampons, and not just Rely, accelerated cases of TSS during the late 1970s and early 1980s, though Rely bore the brunt of blame. Women had unknowingly endangered their health during their periods by using synthetic and superabsorbent tampons; testing had not adequately predicted this damaging bodily harm.

In part, the trust placed into laboratory simulacrum by researchers and marketers overwhelmed findings detected in the broader environment by some state and federal epidemiologists, almost rendering them moot. Because the science could not definitively prove the chain reaction leading to tampon-associated TSS, for example, it was very difficult to regulate tampons or provide guidelines about who should use them and under what conditions. In fact, it is exactly because of the vaginal vagaries that another test came to be developed and subsequently standardized to reduce variables and instrumentalize the menstruous vagina: the syngyna (to see it in use, go to https:// youtu.be/KYXUQDzSg4o). A lab instrument composed of a glass vacuum tube, it flattened differences in women's physiology and utilized simulated menstrual fluid or saline, dyed blue no less, to measure tampon absorption.

In light of superabsorbent tampons being a cofactor in TSS, and lacking a system to label the actual gram-weight absorbency capability of tampons, the syngyna had an important place in tampon safety. Yet the syngyna, like many of these other tests, raises questions about the value of the data produced when no actual menstrual fluid was used to test the tampons (Vostral 2017, 2018). The tools for this test had imitated and sublimated women's bodies to the point of their very expunction. At the behest of Esther Rome of the Boston Women's Health Book Collective and her involvement on the Tampon Task Force for tampon safety and labeling, Nancy Reame, a professor of nursing at the University of Michigan, ran absorption tests with the syngyna using both saline and heparinized blood (see Reame, in this volume). She discovered that superabsorbent tampons, of any brand, held more blood than saline; this meant that the standardizing process itself was not accurate, and many brands of tampons were destined to be mislabeled as less absorbent than they actually were (Vostral 2017). Furthermore, they were more dangerous exactly because they were more absorbent. Her work raises unsettling questions about the validity of data derived from simulated vaginas and laboratory menses, where women have been erased from menstruation itself.

CONCLUSION

Assumptions about women as test subjects, the infallibility of testing, and the inert nature of new materials converged with tampon development during the twentieth century in the United States. Scientific intentions notwithstanding, narrowly defined tests begot narrowly defined results, which often were neither sufficiently applicable to human bodies nor particularly useful in any other knowledge-making project about the technology. In part, the testing provided a sort of ritualistic comfort that the new products were safe, further sanctified by limited approvals granted by the governing body of the FDA and the Bureau of Medical Devices. In this regulated system, the generation of data became an analgesic and a comforting routine by which trust was engendered. The data may have revealed the efficacy of a single component but by definition could not account for the complexity of the whole, and this sort of synthesis was not a strong suit of the scientific method practiced.

How both tampons and female bodies were tested to ascertain product effectiveness reveals changing ideas about menstruation, women as instrumentalized test subjects, and concerns about health and safety. Though researchers and corporate managers looked to results from lab tests to eliminate certain kinds of dangers, they were often unable to test or predict problems posed by multivariable systems and real-world conditions. However, contingency is everywhere. At some historical moments, substituting lab animals for a human vagina followed guidelines about not using human bodies for research, as per Procter & Gamble. But at other times, as with Gilbreth, testing devices were no substitute for protocol allowing women to speak about their own bodies. With the health risk of TSS associated with tampons, further testing was necessary; measuring absorbency and devising a standard to represent tampon sizes and ranges provided important information to women. The variety of methods used to gauge efficacy shines light upon the constructions of menstrual health, women's bodies, and technologies of menstrual management, as well as how despite best efforts, evidence-based facts do not always provide the answers to practitioners' questions.

If testing is to remain a component of menstrual management technology development, the conclusion drawn from this history is to acknowledge the significant contributions of thoughtful women researchers participating in the scientific process. This is not to essentialize their knowledge, but to show how women's menstrual experiences inform the science. From Lillian Gilbreth insisting upon a "woman's experience" to Nancy Reame using empirical evidence to argue for a feminist science methodology in tampon absorption tests, women's voices and feminist approaches matter in cultivating better and more rigorous evidence about the safety of menstrual management technologies.

REFERENCES

Bailey, Richard. 1987. "Small Wonder: How Tambrands Began, Prospered and Grew." Tambrands Inc.

Bobel, Chris. 2010. *New Blood: Third-Wave Feminism and the Politics of Menstruation.* New Brunswick: Rutgers University Press.

Braun, Lundy. 2005. "Spirometry, Measurement, and Race in the Nineteenth Century." *Journal of the History of Medicine Allied Science* 60 (2): 135–69.

Clarke, Adele, and Joan Fujimura, eds. 1992. *The Right Tools for the Job: At Work in Twentieth-Century Life Sciences.* Princeton: Princeton University Press.

Eglash, Ron, Jennifer Croissant, Giovann Di Chiro, and Rayvon Fouché. 2004. *Appropriating Technology: Vernacular Science and Social Power.* Minneapolis: University of Minnesota Press.

Epstein, Steven. 2009. *Inclusion: The Politics of Difference in Medical Research.* Chicago: University of Chicago Press.

Farrell-Beck, Jane, and Laura Klosterman Kidd. 1996. "The Roles of Health Professionals in the Development and Dissemination of Women's Sanitary Products, 1880–1940." *The Journal of the History of Medicine and Allied Sciences* 51 (July): 325–52.

"Follow-up on Toxic-Shock Syndrome." 1980. *MMWR* 29, no. 37 (September 19): 441–45.

"Follow-up on Toxic-Shock Syndrome—United States." 1980. *MMWR* 29, no. 25 (June 27): 297–99.

Fouché, Rayvon, and Sharra Vostral. 2011. "'Selling' Women: Lillian Gilbreth, Gender Translation, and Intellectual Property." *Journal of Gender, Social Policy and the Law* 19 (3): 825–50.

Gilbreth, Lillian. 1927. "Report of Gilbreth, Inc., January 1, 1927." Papers of Frank and Lillian Gilbreth, Special Collections Purdue University, N-File, Box 95.

Goodman, Jordan, ed. 2008. *Useful Bodies: Humans in Service of Medical Science in the 20th Century.* Baltimore: The Johns Hopkins University Press.

Graham, George C., et al. 1960. United States Patent Office, 2,934,068, April 26, 1960.

Graham, Laurel. 1998. *Managing on Her Own: Dr. Lillian Gilbreth and Women's Work in the Interwar Era.* Norcross, Georgia: Engineering & Management Press.

Hasson, Katie Ann. 2012. "From Bodies to Lives, Complainers to Consumers: Measuring Menstrual Excess." *Social Science & Medicine* 75 (10): 1729–36.

Herzig, Rebecca. 2001. "In the Name of Science: Suffering, Sacrifice, and the Formation of American Roentgenology." *American Quarterly* 53 (4): 563–89.

———. 2005. *Suffering for Science: Reasons and Sacrifice in Modern America.* Piscataway: Rutgers University Press, 2005.

Hornblum, Allen M. 1999. *Acres of Skin: Human Experiments at Holmesburg Prison.* New York: Routledge.

Johnson, R. W. 1927. United States Patent Office, 1,705,366, May 24, 1927.

Lederer, Susan. 1997. *Subjected to Science: Human Experimentation in America before the Second World War.* Baltimore: The Johns Hopkins University Press.

Michael L. Kehm v. Procter & Gamble. 1982. United States Courthouse, Cedar Rapids, Iowa.

Okie, Susan. 1981. "Toxic-Shock Syndrome Cases Show Decline." *Washington Post*, May 26, 1981.

Slaton, Amy. 2003. *Reinforced Concrete and the Modernization of American Building, 1900–1930*. Baltimore: Johns Hopkins University Press.

Tierno, Philip. 2001. *The Secret Life of Germs: Observations and Lessons from a Microbe Hunter*. New York: Atria Books.

"Toxic-Shock Syndrome—United States." 1980. *MMWR* 29, no. 20 (May 23): 229–30.

US Food and Drug Administration. 1989. *Labeling for Menstrual Tampons* [21 CFR Part 801, Section 801.430]. Federal Register 54: 43771.

Vostral, Sharra. 2008. *Under Wraps: A History of Menstrual Hygiene Technology*. Lanham: Lexington Books.

———. 2017. "Toxic Shock Syndrome, Tampon Absorbency, and Feminist Science." *Catalyst: Feminism, Theory, Technoscience* 3 (1): 1–30.

———. 2018. *Toxic Shock: A Social History*. New York: New York University Press.

Wailoo, Keith. 2008. *The Troubled Dream of Genetic Medicine: Ethnicity and Innovation in Tay-Sachs, Cystic Fibrosis, and Sickle Cell Disease*. Baltimore: The Johns Hopkins University Press.

CHAPTER 51

Toxic Shock Syndrome and Tampons: The Birth of a Movement and a Research 'Vagenda'

Nancy King Reame

Toxic Shock Syndrome and Tampons: The Birth of a Movement

Long before 2015 was lauded by the mainstream media as "the year of the Period," the menstrual health movement was launched, in part, by the discovery in 1980 that tampons were linked to some 800 cases of severe illness and 20 deaths caused by the made-in-America tragedy of Toxic Shock Syndrome (TSS) (Shands et al. 1980). Menstrual TSS (mTSS) is a severe, life-threatening bacterial infection that gives rise to flu-like symptoms early on, but rapidly escalates to falling blood pressure, organ failure, and death (5–10% mortality) if not recognized in time (Reingold et al. 1982). For those who recover, morbidity is high with a 30% recurrence rate (Hajjeh et al. 1999) (Fig. 51.1).

Once the connection between tampons and TSS was exposed, it became clear to the Food and Drug Administration (FDA) and tampon producers that a huge gap existed in their understanding of how menses interacted with the vaginal environment. In the public domain, prospective studies had yet to be done examining how menses degraded during its absorption by what was considered at the time to be an "inert" feminine hygiene product, and research on the basic elements of menses was relatively rare. The limited

This paper is dedicated to the memory of Esther Rome, co-founder of the Boston Women's Health Book Collective and one of the original authors of *Our Bodies, Ourselves*. Not only a leading women's health activist, she was a true pioneer in the menstrual health movement.

© The Author(s) 2020
C. Bobel et al. (eds.), *The Palgrave Handbook of Critical Menstruation Studies*, https://doi.org/10.1007/978-981-15-0614-7_51

number of studies available were mostly related to abnormal uterine bleeding and postpartum hemorrhage in obstetric patients. All we knew about the role of tampons in TSS was based on the Centers for Disease Control (CDC) reports and phone interviews with victims after the fact. No wonder the industry and the public health world were blindsided by TSS.

As a newly minted PhD in reproductive physiology and former OB nurse, I was becoming involved in menstruation research. I had received a $25,000 grant from Kimberly-Clark Corporation, makers of Kotex products, to conduct a study of menstrual fluid components and flow characteristics as a preliminary step in their development of a synthetic menstrual fluid that could be used for testing product absorbency. At the time, the industry standard was blue-dyed saline.

I wondered why such a straightforward study could not have been accomplished "in-house." When I visited Kimberly-Clark headquarters to deliver my findings, I learned that most corporate scientists (besides being all male) were trained as paper chemists; they started their research and development (R&D) careers in the "feminine hygiene" product line before progressing to more prestigious divisions such as Kleenex and paper towels. The social stigma of menses and its products appeared to be entrenched even in the tampon industry. Chris Bobel would later refer to such period shaming as

Fig. 51.1 Nancy Reame with the Syngina she used in her study of tampon absorbency (circa 1982) (Credit: Advance Magazine/ Peter Yates c.1982. Used with the permission of Michigan Medicine)

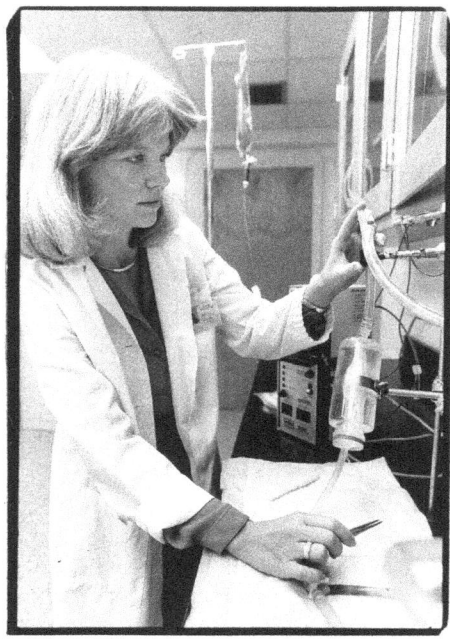

evidence of "disgust directed at the female body" (2018). Undoubtedly, this bias shaped both industrial and federal responses to the TSS epidemic.

As tampon makers rushed to mount an all-out defense of their brands against wrongful death lawsuits,[1] there was public outcry for the federal government to improve oversight of tampon production. Consumers wanted guidance from the CDC on choosing the lowest absorbency size of tampons in order to avoid mTSS. Without a comparative labeling guide on every box, women had no way of knowing how one company's "regular" compared to another company's "super."

Under mounting public pressure, in 1982 the FDA requested that the American Society for Testing of Materials (an independent medical device research society that develops consensus guidelines) convene a Task Force composed of representatives from each tampon company, consumer organizations, and women's health advocacy groups. The goal was to build a set of Federal standards for tampon absorbency ranges and nomenclature similar to those established for sunscreen products, as well as industry-wide testing procedures for tampon safety. As a start, each company agreed to undertake a series of laboratory tests anonymously comparing each other's products to determine, for the first time, how absorbency ranked across brands.[2]

As one of the few nonindustry researchers who had actually worked with menses, I was opportunely positioned to serve as the scientific consultant for the consumer groups on the Task Force and to conduct identical testing in our university labs. The FDA protocol called for the use of the "Syngina" (synthetic vagina) test to measure absorbency of all the products of US tampon makers, which were provided in a blinded fashion by the FDA to each lab.

The Syngina method (Fig. 51.2) specifies a test procedure and test fluid for the laboratory measurement of the absorbency of menstrual tampons. Considered the gold standard by the tampon industry worldwide even today, it is used in production sites for quality control and ensuring compliance with FDA regulations. Designed to simulate body temperature, vaginal pressure, and flow rates, the Syngina apparatus introduces defined amounts of test fluid (the blue saline solution) until the tampon leaks. The tampon weight is taken before and after the test to calculate the weight in grams of fluid absorbed.

To make ours a meaningful study, I assumed we would not only use the standard test fluid, but also conduct comparative studies with menstrual blood, given the stark differences in their chemical and physical properties (not to mention the color!). Here was an opportunity to gain valuable information that might shed light on basic tampon–menses interactions under controlled conditions. But I knew the industry folks would poo-poo this idea based on its presumed variability in composition from day-to-day and woman to woman,[3] not to mention the likelihood that the apparatus and flow rates would have to be recalibrated, given the differences in viscosity.

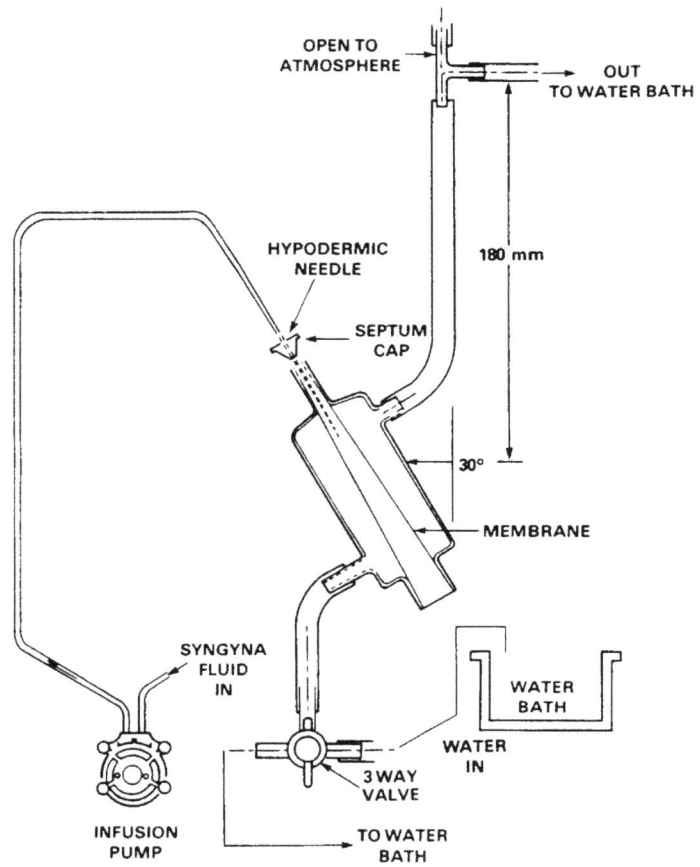

Fig. 51.2 The Syngina test instrument: The industry standard for measuring tampon absorbency (*Source* Code of Federal Regulations Title 21, Volume 8 (2018), figure 2, accessed March 1, 2018, https://accessdata.fda.gov/scripts/cdrh/cfdocs/cfCFR/CFRSearch.cfm?fr-801.430. Credit: The US Food and Drug Administration [FDA] 2018)

As a compromise, I proposed using the next best thing: outdated heparinized blood from the hospital blood bank. Here was a test solution easily obtainable by all the labs and much more uniform in its consistency, yet similar to menses in its de-clottable state. Industry members of the Task Force emphatically refused this suggestion based on the presumption that the blood would be too viscous and would clog up the machine.

With the full support of the consumer groups and women's health advocates on the Task Force, our all-woman lab[4] at the University of Michigan

Children's Hospital (thank you Dr. Robert Kelch, Chair of Pediatrics) went ahead as the sole study site to perform the Syngina tests using both salt water and venous blood. We discovered that our results were consistent, easily reproducible, and demonstrated marked differences across all brands and absorbency categories (junior, regular, super, and super plus). In both cases, the test clearly documented what the CDC had already shown: one brand's "regular" was more absorbent than another's "super." But, most importantly, in all cases the absorbency of the surrogate menstrual fluid exceeded that of the blood-free test fluid, especially for the superabsorbent brands (Reame, Delonis, and Lewis 1987).

Years later, this early "menstrual moment" would be described by Vostral (2017b) as an example of feminist science challenging the industry practice of deconstructing the flesh and blood of the human vagina to simulate the "biological materiality of menstruating bodies" with a Syngina device and salt water. In the 1980s and just starting my career in women's health, I wouldn't have recognized the term "feminist science." I just knew, as a nurse and a menstruating woman, that comparing salt water to blood simply made good sense.

I submitted my final report to the National Consumers' League in 1983 (Reame 1983). But it remained confidential for several years as the Task Force became polarized about how best to formulate a labeling standard that would satisfy the competing needs of tampon makers while addressing the concerns of consumer groups. After the Task Force disbanded in 1985, the lack of transparency about the absorbency guidelines, as well as the FDA's failure to require independent research or the listing of all tampon ingredients, served as stinging reminders of how much more work was left to be done. Sadly, our demands for the release of industry testing data for such properties as shredding, string strength, and contaminants remained unaddressed. Moreover, follow-up attempts by menstrual activists to sustain national interest in mTSS and the need for greater consumer protections were met with inconsistent attention by the media. As a way to make the broader medical world aware of the basic flaws in the proposed federal standard, a group of us wrote up a summary of my findings as a Letter to the Editor of the *Journal of the American Medical Association* in 1988 (Rome, Wolhandler, and Reame 1988).

In 1990, the demands of the Task Force's women's health activists and consumer members were ultimately achieved, in part, with the establishment of a comparative numbering system for absorbency across all brands (Table 51.1). However, there would never be any mention on package labeling that the laboratory test fluid was blue-dyed saline. Women would *falsely* assume that the chart depicting the ranges of absorbency for light, regular, super, and super plus referred to grams of menstrual fluid, not grams of salt water. Thus, they would underestimate the true physiologic conditions of the

Table 51.1   FDA-required absorbency ranges for labeling of tampon products sold in the US

Ranges of absorbency in grams[a]	Corresponding term
6 and under	Light absorbency
Greater than 6–9	Regular absorbency
Greater than 9–2	Super absorbency
Greater than 12–15	Super plus absorbency
Greater than 15–8	Ultra absorbency
Above 18	No term

[a]Refers to Syngina Fluid (10 grams sodium chloride, 0.5 gram Certified Reagent Acid Fuchsin, 1000 milliliters distilled water)
Source Data adapted from sec. 801.430 User labeling for menstrual tampons, *Code of Federal Regulation*, Title 21, Vol. 8 (2018), sec. 801.430, accessed March 1, 2019, https://www.accessdata.fda.gov/scripts/cdrh/cfdocs/cfCFR/CFRSearch.cfm?fr=801.430
Credit: Nancy Reame

tampon-vaginal environment with normal wear—one of the many omissions that persists due to gaps in assessment protocols, regulatory oversight, and common practices of the tampon industry today.

Contemporary Period Products, Regulations, and Manufacturing Practices

The Regulated Tampon: Or Is It?

Although there are still no federal laws requiring manufacturers' universal disclosure of tampon (or pad) ingredients or by-products,[5] most tampon producers, under pressure from menstrual activists, now voluntarily disclose the chemical makeup of their brands. But much more specific information and research about the manufacturing process is needed. As an example of the harm caused when industry testing is inadequate, in 1996 Patrick Schlievert's lab demonstrated that pluronic L92, the normally nontoxic substance used for the outer coating of Rely tampons, could dramatically magnify the amount of toxin produced by the TSS-strain of bacteria (Schlievert et al. 2010).

Other flaws in the oversight process may also be at play. Since 1980, new tampon products must be reviewed by the FDA and demonstrate equivalency in safety and effectiveness with those already on the market. The review process, known as premarket approval (PMA), requires results of testing on the safety of all materials in tampons and applicators; tampon absorbency, strength, and integrity; and whether tampons alter the growth of both normal and harmful bacteria in the vagina. Once approved, any changes affecting the safety or effectiveness of tampons must be reviewed via a PMA supplement. Such changes include those made to the indication for use; facilities to manufacture, process, or package the device; and sterilization procedures, labeling, and packaging. Perhaps most noteworthy in these regulations is the leeway manufacturers are afforded in the oversight of product modifications. It rests on PMA holders to determine whether changes in safety or effectiveness are substantial enough to trigger a supplemental government review (C.F.R. § 814.20). To what extent this practice has led to suboptimal oversight of manufacturing and marketing practices remains unclear. Not surprisingly, the FDA webpage which describes this process to consumers is silent on these omissions in regulatory oversight (http://www.fda/gov/ForConsumerUpdates/ucm612029.htm).

Industry Practices: Behind Closed Doors

While government standards for testing tampon safety and defining ingredients have stagnated, tampon producers seeking a competitive edge on their rivals have introduced innovations in tampon testing methods, ingredients, and manufacturing, as evidenced by a review of patents, industry websites, medical literature, and court cases. For example, in 1999 Kimberly-Clark was granted a patent for the invention of an "artificial menses fluid," which they hoped would provide a safer, more cost-effective menses substitute to better emulate real menstrual blood (Achter et al. 1999). The simplicity of their patent was stunning: a recipe of defibrinated blood (ranging from 10–60%) and egg white (20–60%). Although the conditions of the application stated that human blood could be used, the patent authors cited concerns that using it would expose researchers to viruses or diseases (column 1, line 21). Instead, the inventors proposed swine blood from the slaughterhouse (no diseases there!).

Producers frequently shift the composition of ingredients in their tampons, which makes gaps in standards on the disclosure of menstrual product ingredients particularly troubling. Most producers readily acknowledge that tampons in the US are usually made of nonorganic cotton, rayon, or a combination of these materials. However, they seldom disclose exact proportions or the cost motivations and implications of these choices. Viscose rayon (a semisynthetic fiber) is highly absorbent and very cheap, making it preferable to cotton for boosting profits (Kumeh 2010). Moreover, the proportion of cotton to synthetics within tampons is known to vary with the price of cotton (Rosenthal 2019). Thus, because manufacturers are not required to disclose the ingredients, such dynamic lot-to-lot variations are not likely to be viewed as substantial enough to be reported to the FDA. It is unknown whether these differences in the ratio of materials alter product performance or chemical exposures, but they demonstrate another reason why more transparency is needed.

Tampon producers are also forging into the "green" product market. Perhaps the most striking example of this is P&G's acquisition in early 2019 of the small US company *L.,* makers of organic period products (tampons and pads) advertised as, "free of pesticides, chlorine, fragrances or dyes" (https://thisisl.com). It is anyone's guess to what extent P&G's new-found commitment to all natural products is a strategy to counter concerns about by-products formed in the bleaching processes used to make their nonorganic cotton products, like *Tampax Pure and Clean.* Alternatively, it could represent recognition that consumer products with natural ingredients make up the fastest growing segment in the industry (Meyersohn 2019). In any case, by acquiring a leader in the Femtech industry, corporate America has now become fully immersed in the "go green" movement.

MTSS IN THE TWENTY-FIRST CENTURY: NEW PRODUCTS, NEW WORRIES, AND GAPS IN RESEARCH

Today, some 30 years after the mTSS epidemic, one might presume there is little cause to pursue tampon safety in regards to regulation, production, and research. Indeed, as shown in Fig. 51.3, there was a direct correlation between the dramatic decline in TSS cases and the FDA actions relating to tampon absorbency over the 1980s.

Yet, despite the very low incidence of contemporary mTSS (1–17 cases/yr; Code of Federal Regulations 2017), the disease remains of compelling interest for several reasons: the widespread use of tampons in the Global North;

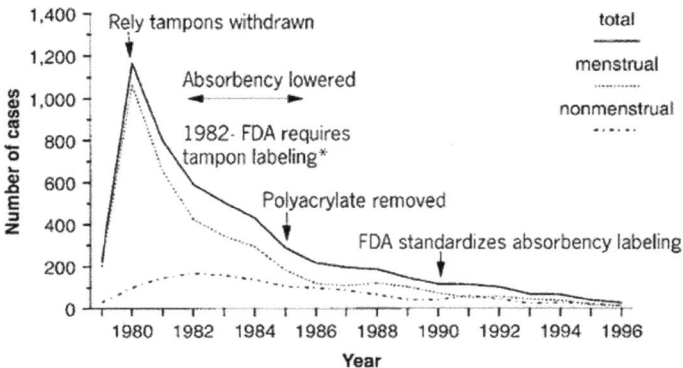

Fig. 51.3 Profile of the decline in both TSS cases and tampon absorbency, 1980–1996 from the CDC website. Accessed July 26, 2019 at https://wwwnc.cdc.gov/eid/article/5/6/99-0611-f1 (*Original source* Hajjeh RA et al. Toxic Shock Syndrome in the United States: Surveillance Update, 1979–1996. Emerging Infectious Diseases Vol. 5, No. 6, November–December 1999, page 807. Credit: The US Center for Disease Control 1999)

growing evidence that menstrual cups may also pose risks (Mitchell et al. 2015; Sharma et al. 2018; Jacquemond et al. 2018; van Eijk et al. 2019); and mounting concern that methicillin-resistant staphylococcal aureus (MRSA) organisms pose a new threat (Sada et al. 2017; Sharma et al. 2018).

What is especially troubling is that the underlying epidemiology of TSS remains equivocal. Despite the industry-wide reduction of tampon absorbency, cases of mTSS persist, with occasional flare-ups in state-wide reports (Cowart 2016). In early 2016, five cases of mTSS in Michigan were reported to the CDC in a span of four months, far outpacing the usual number of cases nationwide in any one year (24 cases in 2017; CDC 2017; Eisner 2016). In Europe, where mTSS is not a "notifiable illness" to the health authorities, health professionals have warned that the number of cases today may be much higher than 15 cases per year, due to misdiagnoses or "near-misses" in meeting the number of diagnostic criteria.[6] Even in the US, it is argued that the use of overly stringent diagnostic criteria may be substantially underestimating the total TSS burden (DeVries et al. 2011). Despite these concerns, researchers have yet to definitively clarify the risk factors associated with the disease.

Several labs over the years have found evidence which debunks the long-standing notion that lower absorbency brands and all-cotton products might be safer or provide greater protection from TSS (Spaulding et al. 2013). Most recently this idea was challenged by findings from a French laboratory study which demonstrated that both the TSS-linked bacteria as well as its deadly toxin could be easily grown on tampon fibers across all brands and absorbencies as well as in menstrual cup materials. The investigators concluded that menstrual cups may be even riskier due to their shape and volume, which may allow more oxygen to enter, transforming the anaerobic environment to an aerobic one in which the bacteria can thrive (Nonfoux et al. 2018). Moreover, because a bacterial biofilm adheres readily to the cup's surface, the investigators recommended that they be sterilized between uses, rather than just washed with soap and water. In contrast, epidemiologic evidence suggests menstrual cups may not impose any more risk than tampons, while having a high level of acceptance as an affordable, single-use product in both high- and low-income countries. A recent meta-analysis by van Eijk and colleagues (2019) of 43 studies involving more than 3000 participants in 99 countries concluded that menstrual cups "seem to be an effective and safe alternative to other menstrual products," although more research is needed.

The CDC has also recently expressed doubts about some of their original data. During the TSS epidemic, they conducted telephone interviews with verified TSS patients and with control cases to compare sensitive information about menses management, including duration of tampon wear (Shands et al. 2017). Unsurprisingly, when assessed this way, there were no differences by study group. One CDC panelist confirmed that, "we don't know if more frequent changing reduces risk (of mTSS) . . . there was never any science done to support this recommendation" (Shands et al. 2017). Despite this, the FDA considered it prudent to include in TSS label warnings the advice to avoid wearing tampons for longer than 8 hours.

Such uncertainties and gaps in research may have disproportionately higher impacts on younger women and adolescent girls (Shands et al. 1980). It is now believed mTSS is triggered by the overgrowth of bacteria (Staphylococcus aureus) capable of producing toxins in the presence of the menses usually retained in a tampon, particularly in young individuals who lack adequate antibodies to mount a sufficient immune response (Spaulding et al. 2013). At the same time, one study found that relevant antibody levels in adolescent girls as young as age 12 were already elevated to adult levels, implying that they should be well protected (Parsonnet et al. 2005). Nonetheless, the average age of mTSS in the UK is 19 (Sharma et al. 2018), some 40% of mTSS cases in the US occur in 13–19 year-olds (Hajjeh et al. 1999), and a hospital-based medical record review suggested that the rates of TSS in adolescents is rising (Brite and Cope 2004).

It is unclear to what extent other demographic factors, such as race and ethnicity, contribute to the risk of contracting TSS. This means women of color are another group unduly affected by gaps in research on period product safety (Parsonnet et al. 2005). There is some evidence which suggests Black women may be less likely to mount an adequate antibody response to the toxin (Parsonnet et al. 2005), however, most US studies involve inadequate numbers of women of color. The well-documented ethnic and cultural differences in menstrual management practices (Finkelstein and Von Eye 1990), coupled with emerging evidence for ethnic differences in the vaginal environment and disease rates (for example, vaginitis), further add to the need for more diverse study populations (Nicole 2014).

As mTSS persists, the evidence gathered from independent researchers linking the broader spectrum of period products to its etiology is steadily mounting (Fig. 51.4). Add to this an array of emerging vaginal products that expand the definition of "tampon," including disposable tampons for urinary bladder leakage, medicated tampons for menstrual cramp relief, and tampons containing capsules for incubation of sperm and egg for intravaginal fertilization (Reame 2018). Without careful scrutiny of these and other new vaginal products, one must ask the question: Is this the next biotechnology accident waiting to happen? It is hard to tell.

Because of the rare nature of mTSS, research on the illness is especially difficult to undertake, is largely supported by industry when conducted, and is limited to cross-sectional designs that give just snapshot impressions of the vaginal milieu. Moreover, there have been no additional case-controlled, epidemiologic follow-up studies in TSS victims, or laboratory-based, cross-brand comparisons undertaken by federal agencies to provide ongoing assessments of the absorbency performance of contemporary products (Shands et al. 2017).

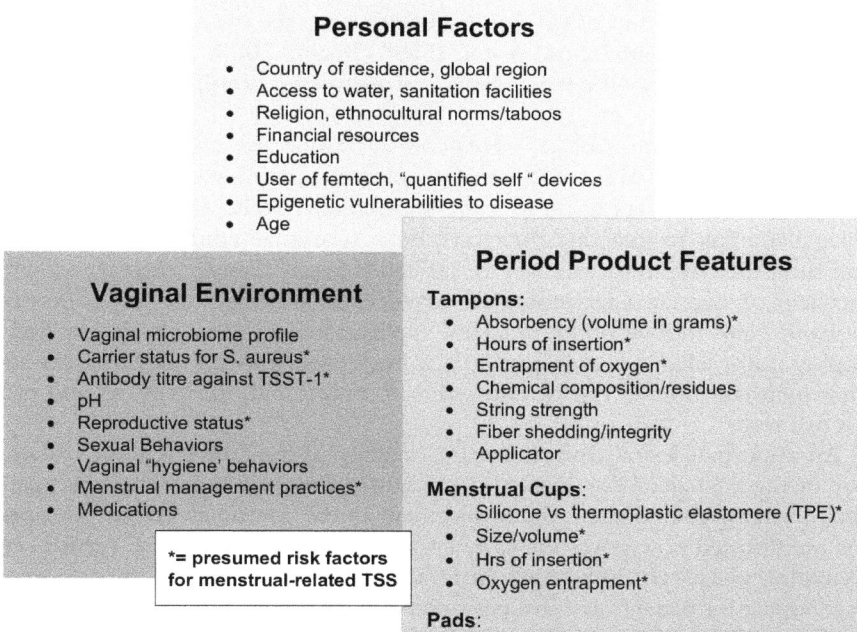

Fig. 51.4 Factors influencing the vaginal microbiome and menstrual health (*Source* Content summarized from Schlievert et al. (2010), Spaulding et al. (2013), Davis et al. (2014), Jacquemond et al. (2018), Nonfoux et al. (2018). Credit: Nancy Reame)

The present state of assessment protocols, regulatory oversight, and common practices in the tampon industry are unlikely to change this any time soon. Thus, should a future mTSS outbreak occur, federal agencies may not be equipped to help consumers understand the risk profiles of new products. Encouragingly, the contemporary atmosphere of menstrual health advocacy offers unprecedented opportunity to change that.

Using Menstrual Activism to Define the Research 'Vagenda'[7] in Menstrual Health

In a 2018 commentary, I reflected on the lessons learned through efforts to draw broader attention to the public health campaigns which sought improved tampon testing and safety standards in the 1980s. I noted that, "[e]mbedded in a culture of menstrual stigma, and without the benefit of podcasts, tweets, blogs, and other social media amplifiers, these efforts were unable to gain the kind of traction needed to keep the message in front of the public. We were clearly ahead of our time" (Reame 2018, 252). Fortunately, these are just some of the many tools that menstrual health advocates have in their arsenals today.

Just as the Boston Women's Health Book Collective and other consumer groups were during the mTSS crisis, menstrual activists today are at work as citizen scientists. In addition to savvy social media skills, they draw on medicolegal training to generate their own data, scrutinize consumer complaints, and monitor symptoms as a way to influence social change.[8] Moreover, with the help of Big Data technologies, advances in precision medicine, massive biobanks, and mobile phone tracking applications, we have clearly reached a tipping point when research gaps in the lived experience of menstruators and their menstrual health can be addressed in more rapid, meaningful, and productive ways.

An important lesson from the mTSS tragedy of the 1980s was the recognition by the US federal government of consumer groups as legitimate expert partners. In many disease areas, the involvement of the consumer/patient as expert has transformed how government-sponsored biomedical research is carried out. Examples, such as the web-based *Patients Like Me* project, reflect this important new respect for patients as equal partners in the research enterprise. Indeed, one private research institute with a major focus on women and children, the Patient-Centered Outcomes Research Institute (PCORI), even requires involvement by patients and other stakeholders across all stages of the research process and mandates the use of patient-driven outcomes (www.pcori.org). Such an approach is clearly overdue for menstrual health research. An encouraging sign of interest by the federal government came in 2018 with the NIH invitational conference on menstruation science, which covered an array of promising new discoveries in 'omic' analysis of the endometrium, new sampling or imaging modalities, smart technologies/apps and mHealth platforms, and health literacy and dissemination frameworks (https://www.ncbi.nlm.nih.gov/pmc/articles/PMC5782905/).

Despite these collective sociopolitical forces, I would argue that the need for continuing vigilance and improved tampon safety protections is as great as ever, perhaps even more so, given the current anti-woman, anti-environment political environment. With that in mind, I offer below some suggestions for ways that the activists in this "second menstrual moment" can partner with the federal research enterprise to improve the research 'vagenda' in menstrual health (updated from Reame 2018). US health policies and regulations governing the safety of menstrual hygiene products remain relatively unchanged, despite the fact that the precise mechanism for tampon-related TSS is still debated, superabsorbent products continue to be introduced, and occasional spikes in TSS cases occur. The 2018 NIH Conference held on Menstrual Science was an encouraging first step, taken partially in response to a worldwide call to action by grassroots organizations to focus on the long-ignored menstrual health agenda so in need of attention.

RECOMMENDATIONS TO IMPROVE THE RESEARCH 'VAGENDA' IN MENSTRUAL HEALTH CARE AND PERIOD PRODUCT SAFETY

1. Tell legislators it is time to pass H.R. 3865, the Robin Danielson Feminine Hygiene Product Safety Act (116th US Cong. 2019), with language that includes menstrual cups and bladder support products in all testing and research requirements (absorbency, chemical composition, string strength, fiber shedding, and tampon integrity).

2. Demand through the Freedom of Information Act that the FDA release its PMAs for new inventions that were subsequently approved, as well as any post-marketing data on health and safety problems reported to the CDC by product brand and absorbency.

3. Petition relevant NIH Institutes and Offices (for example, National Institute of Child Health and Human Services, Office of Research on Women's Health) to partner with the CDC and FDA to make tampon safety a priority for future strategic plans for women's health research.

4. Petition federal government research agencies (for example, National Institute of Environmental Health Sciences, National Cancer Institute) to undertake studies of the vaginal microbiome and characterize menstrual fluid as has been done with saliva for its potential as an exposome biomarker of health (Bessonneau, Pawliszyn, and Rappaport 2017).

5. Petition the NIH to fund studies that use Big Data approaches to study race/ethnicity variation in the interactions between vaginal health practices, the vaginal microbiome, and mTSS by taking advantage of the evolving million-patient database and biorepository of the NIH "*All of Us*" research initiative (https://allofus.nih.gov).

6. Capitalize on the menstrual health movement to lead a public health campaign to "Make the Menstrual Cycle a Vital Sign." Partner with relevant medical, nursing, and health care organizations (for example, the American College of Obstetricians-Gynecologists, the Society for Women's Health Research) to develop practice guidelines for making the menstrual cycle a vital sign across the reproductive life span. The current ACOG guidelines, for example, are silent on risk factors for mTSS in women and girls, and say nothing about menopause-related health impacts.[9]

NOTES

1. Proctor & Gamble's superabsorbent, all-synthetic product *Rely* proved to be the most lethal and was quickly taken off the market.
2. Sharra Vostral has written extensively about these events (Vostral 2017a).
3. In agreement with Epstein and colleagues (2017), I use the term "woman" here to refer to anyone who has ever experienced a menstrual cycle. At the same time, I acknowledge that not all individuals who have a menstrual cycle identify as a woman, and not all those who identify as a woman have a menstrual cycle.
4. Pediatric Endocrine Lab staff members: Maura Baun, Katherine Kersey; Nursing masters' students, Ruth Lewis, Susan DeLonis.

5. The Robin Danielson Feminine Hygiene Product Safety Act of 2019, H.R. 3865, remains in committee at the time of this writing.
6. The Toxic Shock Syndrome Information Service (TSSIS) is a UK organisation funded and supported by Lil-Lets UK Ltd., Johnson & Johnson Consumer Services EAME Ltd., Kimberly-Clark Ltd., Ontex Retail Ltd., Procter & Gamble UK and Toiletry Sales Ltd. www.toxicshock.com/tssfacts/.
7. I attribute the word "vagenda" to comedienne Amanda Seales, who used this in her HBO special "I Be Knowin'" (2019) in reference to her vagina's agenda.
8. For example, see the *ChemFatale* report by the Women's Voices for the Environment (www.womensvoices.org/feminine-care-products/whats-in-your-tampon/).
9. During 2018 and 2019, while writing this chapter, Dr. Reame served as a paid consultant to Diva International, Inc. (makers of the Diva menstrual cup), unpaid advisor to the menstrual tracking app *Clue*, the blog *WomenLivingBetter*, and the Femtech company DropletHealth. In the 1980s, Dr. Reame served as the independent laboratory for the consumer representatives on the FDA Task Force that tested tampon absorbency and set labeling requirements for the US tampon industry. She also serves as an advisor to the women's health advocacy organization, *Our Bodies, Ourselves*. Part of this commentary originally appeared in the keynote paper, "A Research Program on the Menstrual Cycle: Looking Back, Moving Forward" presented at the 2017 meeting of the Society for Menstrual Cycle Research. Other sections were adapted from "The Legacy of Tampon-Related Toxic Shock Syndrome: Feats, Failures, and Future Challenges for Women's Health Scholars" (Reame 2018).

References

Achter, et al. 1999. Artificial Menses Fluid. US Patent 5,883,231, filed August 21, 1991 and issued March 16, 1999.

Bessonneau, Vincent, Janusz Pawliszyn, Stephen M. Rappaport. 2017. "The Saliva Exposome for Monitoring of Individuals' Health Trajectories." *Environmental Health Perspectives* 125 (7): 077014.

Bobel, C. 2018. "Menstrual Pads Can't Fix Prejudice." *The New York Times*. https://www.nytimes.com/2018/03/31/opinion/sunday/menstrual-periods-prejudice.html.

Brite, R. A., and J. U. Cope. 2004. "Tampon-Related Toxic Shock Syndrome (TSS) Continues to Peak among Adolescent Girls: A Nationwide Hospital Study." *Journal of Pediatric and Adolescent Gynecology* 17 (3): 222.

Centers for Disease Control and Prevention. 2017. *Morbidity and Mortality Weekly Report (MMWR)*. https://www.cdc.gov/mmwr/volumes/66/wr/mm6652md.htm.

Code of Federal Regulations. 2017. Title 21, Volume 8, Sec. 801.430, April 1, 2017. "User Labeling for Menstrual Tampons."

Cowart, L. 2016. "Women Are Still Getting Toxic Shock Syndrome and No One Knows Why." *Washington Post*, March 21, 2016. https://www.washingtonpost.com/news/speaking-of-science/wp/2016/03/21/women-are-still-getting-toxic-shock-syndrome-and-no-one-quite-knows-why/?noredirect=on.

Davis, C. C., M. Baccam, M. J. Mantz, T. W. Osborn, et al. 2014. "Use of Porcine Vaginal Tissue Ex-Vivo to Model Environmental Effects on Vaginal Mucosa to

Toxic Shock Syndrome Toxin-1." *Toxicology and Applied Pharmacology* 274 (2): 240–48.

DeVries, A. S., L. Leshe, P. M. Schlievert, et al. 2011. "Staphylococcal Toxic Shock Syndrome 2000–2006: Epidemiology, Clinical Features, and Molecular Characteristics." *PLoS One* 6 (8): e22997.

Eisner, Jennifer. 2016. "Michigan Confirms Increase in Toxic Shock Syndrome Cases." *Michigan Department of Health and Human Services.* https://www.michigan.gov/mdhhs/0,5885,7-339--378024--,00.html.

Epstein, Daniel A., Nicole B. Lee, Jennifer H. Kang, Elena Agapie, Jessica Schroeder, Laura R. Pina, James Fogarty, Julie A. Kientz, and Sean Munson. 2017. "Examining Menstrual Tracking to Inform the Design of Personal Informatics Tools." *Proceedings of the 2017 CHI Conference on Human Factors in Computing Systems,* 6876–6888.

Finkelstein, J. W., and A. Von Eye. 1990. "Sanitary Product Use by White, Black and Mexican American Women." *Public Health Reports* 105: 491–96.

Hajjeh, A., A. Reingold, A. Weil, K. Shutt, A. Schuchat, and B. A. Perkins. 1999. "Toxic Shock Syndrome in the United States: Surveillance Update, 1979–1996." *Emerging Infectious Diseases* 5 (6): 807–10.

Jacquemond, J., A. Muggeo, G. Lamblin, A. Tristan, Y. Gillet, P. A. Bolze, et al. 2018. "Complex Ecological Interactions of Staphylococcus Aureus in Tampons during Menstruation." *Scientific Reports* 8: 9942.

Kumeh, T. 2010. "What's Really in That Tampon?" *Mother Jones.* https://www.motherjones.com/politics/2010/10/whats-really-tampon-and-pad/.

Meyersohn, Nathaniel. 2019. "Procter & Gamble Buys Organic Tampon Brand L." *CNN,* February 5, 2019. https://www.cnn.com/business/live-news/stock-market-news-today-020519/h_5036bff9491116ae94555c198bbfc2a0.

Mitchell, M., S. Bisch, S. Arntfield, and S. Hosseini-Moghaddam. 2015. "A Confirmed Case of Toxic Shock Syndrome Associated with Use of a Menstrual Cup." *Canadian Journal of Infectious Diseases and Medical Microbiology* 26: 218–20.

Nicole, W. 2014. "A Question for Women's Health: Chemicals in Feminine Hygiene Products and Personal Lubricants." *Environmental Health Perspectives* 122 (3): A70–A75.

Nonfoux, L., M. Chiauzzi, C. Badiou, et al. 2018. "Impact of Currently Marketed Tampons and Menstrual Cups on Staphylococcus Aureus Growth and TSST-1 Production In Vitro." *Applied and Environmental Microbiology* 84 (12): e00351–18.

Parsonnet, J., et al. 2005. "Prevalence of Toxic Shock Syndrome Toxin 1-Producing Staphylococcus aureus and the Presence of Antibodies to This Superantigen in Menstruating Women." *Journal of Clinical Microbiology* 43: 4628–34.

Reame, Nancy. 1983. Nancy Reame to David Swankin, with Attached Report. Boston Women's Health Book Collective, Folder 34, Box 1. Cambridge, MA: Radcliffe Institute for Advanced Study Harvard University—Schlesinger Library, May 23, 1983.

Reame, Nancy. 2018. "The Legacy of Tampon-Related Toxic Shock Syndrome: Feats, Failures, and Future Challenges for Women's Health Scholars." *Women's Reproductive Health* 5 (4): 250–61. https://doi.org/10.1080/23293691.2018.1523118.

Reame, N. E., S. Delonis, and R. Lewis. 1987. "Menstrual Tampons: Changes in Absorbency and Styles since 1982." In *Seventh Conference of the Society for Menstrual Cycle Research*, University of Michigan, Ann Arbor, June 1987.

Reingold, A. L., N. T. Hargrett, K. N. Shands, et al. 1982. "Toxic Shock Syndrome Surveillance in the United States, 1980 to 1981." *Annals of Internal Medicine* 96: 875–80.

Rome, E. R., J. Wolhandler, and N. Reame. 1988. "The Absorbency of Tampons." *JAMA* 259: 685–86.

Rosenthal, Linda. 2019. "Assemblymember Linda B. Rosenthal Announces Passage of First-in-the-Nation Menstrual Product Ingredient Labeling Bill in both Assembly and Senate." News Release, June 21, 2019, Office of NY State Assemblymember Linda B. Rosenthal. Accessed April 23, 2020. https://nyassembly.gov/mem/Linda-B-Rosenthal/story/87707.

Sada, R., S. Fukuda, and H. Ishimau. 2017. "Toxic Shock Syndrome Due to Community-Acquired Methicillin-Resistant Infection: Two Case Reports and a Literature Review in Japan." *ID Cases* 8: 77–80.

Schlievert, P. M., K. A. Nemeth, C. C. Davis, M. L. Peterson, and B. E. Jones. 2010. "Staphylococcus Aureus Exotoxins Are Present In Vivo in Tampons. *Clinical and Vaccine Immunology* 17: 722–27.

Seales, Amanda. 2019. "I Be Knowin." HBO, January 26, 2019.

Shands, K., A. Reingold, G. Matthews. 2017. Toxic Shock Syndrome: A Lasting Legacy, October 19, 2017, Q&A response (approx. 1hr 12 mins–1hr 20 mins). https://www.cdc.gov/od/science/wewerethere/toxicshock/.

Shands, K. N., G. P. Schmid, B. B. Dan, D. Blum, R. J. Guidotti, N. T. Hargrett, R. L. Anderson, D. L. Hill, C. V. Broome, J. D. Band, and D. W. Fraser. 1980. "Toxic-Shock Syndrome in Menstruating Women: Association with Tampon Use and Staphylococcus Aureus and Clinical Feature in 52 Cases." *The New England Journal of Medicine* 303: 1436–42.

Sharma, Hema, Debra Smith, Claire E. Turner, Laurence Game, Bruno Pichon, Russell Hope, Robert Hill, Angela Kearns, and Shiranee Sriskandan. 2018. "Clinical and Molecular Epidemiology of Staphylococcal Toxic Shock Syndrome in the United Kingdom." *Emerging Infectious Diseases* 24 (2): 258–266.

Spaulding, A. R., W. A. Salgado-Pabón, P. L. Kohler, A. R. Horswill, D. Y. M. Leung, and P. M. Schlievert. 2013. "Staphylococcal and Streptococcal Superantigen Exotoxins." *Clinical Microbiology Reviews* 26 (3): 422–47.

van Eijk, A. M., et al. 2019. "Menstrual Cup Use, Leakage, Acceptability, Safety, and Availability: A Systematic Review and Meta-Analysis." *Lancet Public Health* 4 (8): e376–93. http://dx.doi.org/10.1016/S2468-2667(19)30111-2.

Vostral, S. 2017a. "Toxic Shock Syndrome, Tampons and Laboratory Standard-Setting." *CMAJ* 189 (20): E726–E728.

———. 2017b. "Toxic Shock Syndrome, Tampon Absorbency, and Feminist Science." *Catalyst: Feminism, Theory, Technoscience* 3 (1). http://dx.doi.org/10.28968/cftt.v3i1.28788.

CHAPTER 52

Measuring Menstruation-Related Absenteeism Among Adolescents in Low-Income Countries

Anja Benshaul-Tolonen, Garazi Zulaika, Marni Sommer, and Penelope A. Phillips-Howard

INTRODUCTION

It is frequently articulated that menstruation may impede girls' educational attainment in low-income countries. A range of factors have been implicated, including the lack of suitable changing opportunities and latrines in schools, the lack of resources resulting in girls' use of unhealthy or ineffective products, and the fear of leaking leading to reduced concentration. Alongside these physical factors, shame and stigma are commonly reported as impediments to girls' fullest participation in their schooling. There is a vast body of qualitative work that illustrates these issues, such as Adinma and Adinma (2008), El-Gilany, Badawi, and El-Fedawy (2005), Johnson et al. (2016), Mason et al. (2013), McMahon et al. (2011), Sommer (2009, 2010a, 2013), Sommer and Ackatia-Armah (2012), Sommer et al. (2015), and an overview by Kirk and Sommer (2006).

In recent years, a few quantitative studies have been conducted in low- and middle-income countries that aim to understand the association between school attendance and menstruation (Grant, Lloyd, and Mensch 2013; Benshaul-Tolonen et al. 2019; Tegegne and Sisay 2014) or the impact of menstrual hygiene management (MHM) interventions on schoolgirls, using non-randomized (Montgomery et al. 2012, 2016) and randomized (Oster and Thornton 2011; Phillips-Howard et al. 2016a) study designs. Hennegan and Montgomery (2016) conducted a systematic review of eight studies and found moderate reductions in absenteeism from menstrual

© The Author(s) 2020
C. Bobel et al. (eds.), *The Palgrave Handbook of Critical Menstruation Studies*, https://doi.org/10.1007/978-981-15-0614-7_52

hygiene interventions but point out considerable heterogeneity in study design and risk of bias in the underlying studies. Another systematic review and meta-analysis focusing on MHM in India found that absenteeism during periods is common, but when the analysis was adjusted for region, the relationship was not significant (van Eijk et al. 2016). MHM researchers have advised that greater attention be placed on improving the scope and robustness of research to reduce the risk of absenteeism being considered the sole or predominant indicator of a successful MHM intervention (Phillips-Howard et al. 2016b).

Impact evaluation techniques such as randomized control trials (RCT), lauded as among the most reliable methods for understanding development policy impact, offer useful insights for allocating funding to interventions that provide the most positive impact per dollar spent. Determining cost-effectiveness (CE), when outcomes are correctly captured and measured, is financially and ethically prudent in the resource-constrained development policy world. However, correctly identifying and measuring CE is tricky because the researcher needs to determine a priori which outcomes to measure, how to measure them (variable definition), and, importantly, how to define the sample population and size.

This chapter discusses how school enrollment and absenteeism behavior while enrolled in school can be useful outcomes for MHM interventions, as well as how an overreliance on these outcome measures may limit MHM policy impact. While research priorities for MHM have previously been spelled out in Phillips-Howard et al. (2016b) and Sommer et al. (2016), with appropriate methodologies discussed, to date there have been few properly designed analytical studies that have focused on school absenteeism. This chapter does not intend to summarize the qualitative literature on menstruation and schooling. Instead it discusses how school enrollment and absenteeism behavior are consequences of lack of MHM. The chapter also discusses externalities, pre-analysis plans, and CE as they relate to impact evaluation that is relevant to MHM research.

OVERVIEW OF EXISTING STUDIES

We explored the literature on MHM and school absenteeism and included studies that investigated absenteeism using recall data, diary data, school records, or spot check data. Studies have generally defined absenteeism as any child who was not documented to be present at school at the time of study, which thus can include schoolchildren who have migrated, transferred, or dropped out as well as those temporarily absent at the time of study. Table 52.1 provides an overview of 11 quantitative studies and two systematic reviews that explore the link between menstruation and education, revealing strong heterogeneity in methods, samples, target groups, and findings, which has hindered our understanding of the effectiveness of menstrual-related policies aimed at increasing educational attainment.

Table 52.1 Overview of absenteeism studies

Authors and Year	Design	Sample Size, Grades	Country	Findings	Limitations
Adukia (2017)	Quasi-experimental study using the roll-out of a latrine building program	17,796 schools, grades 1–8	India	Construction of latrines increased enrollment of young students, both boys and girls. Pubescent girls benefit from single-sex latrines. Mixed-gender latrines had no effect on girls' enrollment	Focus on enrollment only. Does not specifically explore menstruation. Findings could be due to safety concerns, hygiene, comfort, et cetera
Alam et al. (2017)	Nationally representative cross-sectional study. Students self-report if they missed school during last three menses	700 schools, 2332 schoolgirls, ages 11–17	Bangladesh	41% report missing school, on average 2.8 days per menstrual cycle	Self-reported absenteeism, using recall method (last three menses)
Freeman et al. (2012)	Cluster-randomized trial with three treatment arms: (i) water treatment and hygiene promotion, (ii) sanitation improvement, and (iii) control	135 schools, 6036 students, grades 4–8	Kenya	WASH improvements reduced absenteeism among girls but had no effect on boys	Study was affected by postelection violence. No specific menstrual hygiene intervention. Pupil-reported absence during two weeks. Sample is repeated cross-section (not longitudinal) at the school level
Grant et al. (2013)	Cross-sectional survey	1675 students aged 14–16, and 845 adolescents out of school, ages 14–16	Malawi	High levels of absenteeism among boys and girls. No significant gender difference. Only 2.4% of absent days reported by girls due to periods in face-to-face interviews, but 1/3 of girls report being absent during last period when reporting to a computer	Cross-sectional data using recall. Unclear classification of cramps as illness or period-related. Inconsistent results depending on mode of data collection

(continued)

Table 52.1 (continued)

Authors and Year	Design	Sample Size, Grades	Country	Findings	Limitations
Montgomery et al. (2012)	Non-randomized trial with three treatments: (i) sanitary pad provision with MHM education, (ii) only MHM education, and (iii) control, at the village level	120 schoolgirls, from 4 villages, ages 12–18	Ghana	Pads and education increased attendance by 9%, and education alone also increased attendance	Non-randomized study. Small sample sizes per treatment group (60, 21, 35, respectively)
Montgomery et al. (2016)	Quasi-randomized cluster trial at the school level with four treatment arms: (i) reusable pads + education, (ii) reusable pads, (iii) education, and (iv) control. Attendance recorded at baseline and at 18 months follow up	8 primary schools, 1124 schoolgirls. Grades 3–5, ages 10–13 (approximately)	Uganda	No effect on psychosocial well-being outcomes. High mobility within the sample, with girls in the control group the most likely to transfer schools. Reduced attendance over time, with smaller reductions in intervention groups (5.2–24.5%, depending on sample correction)	Quasi-randomized, high mobility and drop out of the sample, infrequent measurement of attendance (twice). Did not specify menarche as eligibility criteria at baseline
Oster and Thornton (2011)	Pilot randomized control trial with two arms: menstrual cup and control. Cross-validates self-reported data with school records	4 schools, 199 students, grades 7–8	Nepal	No baseline differences in absenteeism across period days and non-period days, and miss 0.19% of schooldays, or 0.4 schooldays per year because of periods, most commonly because of cramps (43.8%). Menstruation associated with "extremely small" and statistically significant effect on attendance. Program did not reduced absenteeism	Small sample size, limited external validity, program evaluation in nonrelevant context

(continued)

Table 52.1 (continued)

Authors and Year	Design	Sample Size, Grades	Country	Findings	Limitations
Phillips-Howard et al. (2016a)	Randomized feasibility study with three treatment arms: (i) menstrual cup, (ii) menstrual pads, and (iii) control	30 primary schools, 751 girls enrolled (644 followed), grades 5–8	Kenya	Reductions in STI/RTIs, but no effect on absence rates	Risk of misclassification in student diary data to capture absenteeism. Potential Hawthorne effect
Benshaul-Tolonen et al. (2019)	Same context as Phillips-Howard et al. (2016a) with additional attendance data on 6000 female and male students	32,349 spot checks and school records for male and female students, of which 3417 were spot checks on study participants in Phillips-Howard et al. (2016a), grades 5–8	Kenya	High levels of absenteeism due to transfers across schools. School records, validated against spot checks, are inaccurate. Girls are no more likely to miss school than boys. The menstrual cup arm does not reduce absenteeism rates, but the sanitary pad arm reduced absenteeism by 5.4 percentage points	Study does not calculate absenteeism on days with periods. High level of transfers out of study schools. Reasons for absence reported by third party
Sivakami et al. (2019)	Cross-sectional surveys	43 schools, 2561 girls ages 12 and above, grades 8–10	India	Menstruation reported leading to concentration issues (40–45% of students), pain (31–38%), and absenteeism among subset (6–11% of students). Mobility and concentration issues less prevalent among students using disposable pads	Self-reported absenteeism

(continued)

Table 52.1 (continued)

Authors and Year	Design	Sample Size, Grades	Country	Findings	Limitations
Tegegne and Sisay (2014)	Mixed methods with cross-sectional survey	Quantitative survey of 595 schoolgirls ages 10–19, grades 7–8	Ethiopia	Over 50% report having been absent from school during period, less common among students using disposable sanitary pads. 58% of schoolgirls report a decline in school performance post menarche. Qualitative link between teasing and school dropout. 86.4% agree that a girl can go to school during her period. 3.52% report disposing menstrual pads in waste bins, with 33.4% in open field, and remainder in latrines	Self-reported absenteeism
Hennegan and Montgomery (2016)	Systematic review of 8 studies	Education interventions (5 studies); Sanitary products (2 studies); Education and sanitary products (1 study). Sample sizes ranging from 120–1823 per study, varying grades	Iran, Saudi Arabia, Zimbabwe, Ghana, Nepal, and Kenya	Menstrual hygiene interventions lead to moderate reductions in absenteeism	Study heterogeneity, risk of bias in underlying studies

(continued)

Table 52.1   (continued)

Authors and Year	Design	Sample Size, Grades	Country	Findings	Limitations
van Eijk et al. (2016)	Systematic review and meta-analysis. 77.5% of studies were cross-sectional surveys, 14% before–after analysis, 12% utilized mixed methods	138 studies with 97,070 girls	India	Inappropriate disposal of sanitary products common (23%), 24% report missing school during periods with variability across regions; absence associated with pad use but significance lost when data analyses controlled for region in India	Quality issues in included studies, and variability in reported MHM effects preventing aggregate analyses of differing risks

For purposes of brevity, only quantitative studies focusing on educational outcomes were included. This Table, then, is incomplete. We refer to the systematic reviews for a more complete overview

An overview of the cross-sectional literature exploring the link between MHM and absenteeism illustrates large differences across contexts and studies. Girls aged 11–17 in Bangladesh report high levels of absenteeism during periods (41%, Alam et al. 2017), and a systematic review of studies in India similarly find high levels of absenteeism during periods (24%, van Eijk et al. 2016). In the latter study, differences in absence relating to pad use were no longer significant when taking region of India into account. In a recent study from three states in India, absenteeism rates were reportedly 6–11% among girls in grades 8–10 (Sivakami et al. 2019), while in neighboring Nepal, period-related absenteeism was almost non-existent among girls in grades 7–8 according to one study (0.19%, Oster and Thornton 2011).

Three studies from Malawi, Kenya, and Uganda find high overall levels of absenteeism among girls, especially due to transfers and mobility (for grades 5–8 in Benshaul-Tolonen et al. 2019; for grades 3–5 in Montgomery et al. 2016), but low menstrual-related absenteeism (2.4% of absent days, Grant, Lloyd, and Mensch 2013, for ages 14–16). Because menstruation is limited to 0–5 days per month, absence within these few days may be hard to isolate in a high-absenteeism context, such as those found in Malawi (Grant, Lloyd, and Mensch 2013) and western Kenya (Benshaul-Tolonen et al. 2019). In addition, two studies found no or weak gender differences in school absenteeism behavior (Benshaul-Tolonen et al. 2019; Grant, Lloyd, and Mensch 2013). However, the method used to collect such data may influence reporting. This is illustrated by Grant, Lloyd, and Mensch (2013), where one-third of girls reported having missed at least one day of school during their last period when answering using an audio-computer assisted survey instrument (ACASI) instead of reporting face-to-face. (The same question was not collected using face-to-face interviewing.)

Building on the cross-sectional evidence of absenteeism, a subset of studies evaluates policies aiming to reduce school absenteeism among girls. A pilot non-randomized intervention in Ghana (Montgomery et al. 2012) and a quasi-randomized intervention in Uganda (Montgomery et al. 2016) that provided education and sanitary pads were associated with increased school attendance, but notably showed a similar rate of change in absence among girls provided education only. The study designs and program effects may thus not be interpreted as causal. A larger cluster-randomized study from Kenya that provided sanitary pads or menstrual cups and followed 644 girls over an average of ten months found no or weak evidence of reductions in drop-out rates or absenteeism (Phillips-Howard et al. 2016a), although sanitary pads appear to have marginally reduced absenteeism (Benshaul-Tolonen et al. 2019).

Two studies focusing on latrine availability and quality are included. Latrine building and latrine improvements positively impact school enrollment and school attendance, especially for pubescent girls (grades 6–8 in Adukia 2017; grades 4–8 in Freeman et al. 2012). The gender differential could stem from menstruation-related absenteeism among girls;

however, many factors differ between pubescent girls and boys, such as safety and privacy concerns while using latrines. The two studies (Adukia 2017; Freeman et al. 2012) did not specifically explore how latrine availability interacts with MHM needs, which limits our understanding of this potential channel. In fact, no study has been identified that evaluates the effect of latrine improvement programs specific to menstrual-related absenteeism.

Beyond school absenteeism, studies report several issues that girls face relating to MHM, such as lack of suitable disposal possibilities (Tegegne and Sisay 2014; van Eijk et al. 2016) and concentration issues (40–45%, Sivakami et al. 2019). In focus groups, some girls reported that being teased and humiliated after leaking led some girls to drop out of school (Tegegne and Sisay 2014). Cramps and pain stand out as a common issue, as 31–38% of girls interviewed in three government schools in India reported suffering from abdominal pain during their period (Sivakami et al. 2019). In fact, pain may be one of the main reasons for missed schooling; in one study in Nepal, almost half of missed days during periods were due to cramps (Oster and Thornton 2011). No program evaluation studies have been published to date on tackling cramps and pain as a means to increase school attendance and participation, although testing of pain relief has been conducted in pilot schools in Uganda.

Choosing Outcomes

One of the most complex aspects of study design, especially in impact evaluations, is choosing the right measurable and objective outcome variables, and defining them in a transparent and intelligible way that will accurately reflect program effects. This is preferably done *a priori* (Head et al. 2015) and publicly registered[1] (Miguel et al. 2014) to avoid choosing definitions that yield statistically significant results, so-called cherry-picking and/or p-hacking. Journals tend to favor publishing studies that intentionally or unintentionally overreport statistically significant results (Head et al. 2015; Miguel et al. 2014; Brodeur 2016), leading to publication bias. To caution against this kind of bias in the published body of research, Miguel et al. (2014) show how small changes to variable definition for educational outcomes can yield different program effects. Preregistration of studies with a predefined statistical analysis plan thus emerges as a best practice to be adopted by all quantitative, menstrual-related intervention studies, especially, but not limited to, impact evaluations.

Thinking About the Margin: Extensive and Intensive Margins

The field of economics is interested in the *extensive* and *intensive margins* when discussing policy impact. In the context of schooling, one easy way to measure impact is to focus on school enrollment. Enrollment is an *extensive margin* metric that answers how many students are enrolled in school, or the likelihood that a given student is enrolled in school. While enrollment is fairly simple to measure, making it a good contender for evaluating a menstrual health intervention, it tells us little about students' actual learning, making it a fairly crude measure. Alternatively, the *intensive margin* measure, one that helps us understand how often pupils attend school or what share of school hours are missed by a given student, may provide a better option. Intensive margins often involve more decisions regarding variable definition. Here, researchers must specify how to measure school attendance (for example, share of days in a school week missed or share of hours in the school day missed). In the context of MHM-related absence, it also raises other questions. Should absence due to period cramps *before* the onset of a period be classified as menstrual-related absence? More sophisticated temporal analysis than the extant "Period=Yes or No" is needed to detect behaviors across the menstrual cycle.

Hard vs Soft Metrics

In contrast to qualitative studies, quantitative studies rely on outcomes that are seemingly easy to measure and quantify. In the interest of precision, we propose making further distinctions between *hard* and *soft metrics*. *Hard metrics*— such as physical attendance in the classroom or exam scores—are more readily available and observable than *soft metrics*—such as concentration while in the classroom or absorption of knowledge. Test scores are often used in school-related impact evaluations, but they are an imperfect metric of learning, as even the most comprehensive and well-designed exam rarely reflects true knowledge. Impact evaluations in the MHM arena often limit their outcomes to "hard" metrics, a focus that may neglect positive impacts on equally important "soft" outcomes, such as concentration, participation, learning, self-esteem, enjoyment of learning, and comfort. While "soft" outcomes are not necessarily impossible to measure, doing so in a satisfactory way may require more complex, time-consuming, and expensive studies.

When Measuring "Hard" Outcomes Is Fraught

Even the most readily observable and measurable "hard" outcomes can be fraught. Consider the following example: The research team wants to understand if providing menstrual pads helps adolescent girls in rural areas in developing countries increase their educational attainment. They meet with principals of 40 schools in the study region, who agree to participate. The research team implements the program following the "ideal" setup, with randomizing treatment at the school level because of potential externalities. Analysis of the baseline data shows that girls are just as likely to be absent from school on days when they have their period in the treatment schools as in the control schools.

Scenario 1: The administrative school records used have a lot of mistakes, so the program evaluation does not show any significant effects. In the presence of random measurement error in the dependent variable, the program effect will be correctly estimated but the variance will be larger. This could lead to a type II error, where we fail to reject the null hypothesis, that there is no difference across groups. We may wrongly conclude that the program didn't work, and the policy will be under-utilized from a social desirability standpoint.

Scenario 2: The school record captures presence ("present"), but when the student is absent ("absent"), it either results in an entry in the record as "absent" or no record ("missing record"). To further complicate matters, sometimes the head teacher gets interrupted and not all attendance calls are completed, leading to missing records in the attendance book for some students who were present. This data suffers from sample selection bias, and regression analysis using the absence data as the dependent variable will lead to bias.

Scenario 3: The principal is worried about the school's attendance data and whether it could be accessed by other parties, such as the regional government, which might have implications for future decisions about funding or permits. The principal decides to "clean" the data before sharing it. Because the principal is nonrandom in her application of the "cleaning," the measurement error likely follows a pattern that biases any results based on this data. While the randomization was successful at baseline and independent of principal behavior, the treatment came to interact with the principal behavior *ex post*.

Scenario 4: Girls in the control group receive extra attention which encourages them to attend school in ways similar to the intervention group (that is, a Hawthorne effect).

Self-Reporting Bias

Asking respondents directly may be the easiest way to understand certain behaviors. But relying on self-reported measures of school attendance may lead to biased estimates and wrong conclusions, due to *recall bias* and *social desirability bias*. Recall bias is a real threat to surveys that ask participants to recall absenteeism behavior during previous menstrual periods. Indeed, recall bias has been shown to be common within epidemiological and medical studies relying on recall data (Althubaiti 2016; Coughlin 1990). Solutions to recall bias include asking participants to keep a diary and reducing the length of the recall period (Althubaiti 2016), although the data may still suffer from self-report bias. Social desirability bias is a threat in surveys when the topic is associated with shame or stigma and anonymity cannot be guaranteed at the time of the data collection. It is most easily overcome with validation of the instrument before the intervention (Althubaiti 2016). In the context of MHM, Grant, Lloyd, and Mensch (2013) complemented face-to-face survey questions with a more "private" option using a computer system (ACASI) and found higher levels of reported menstruation-related absenteeism in the more "private" option.

Some further attempts have been made to cross-validate absenteeism data. For example, Oster and Thornton cross-validated self-reported absenteeism from diaries with school records in Nepal and found high levels of agreement. Benshaul-Tolonen et al. (2019) cross-validated school records against spot check data in 30 schools in Kenya and found a fair amount of inconsistency across the two data sources. We argue that studies focusing on absenteeism during periods should consider how to elicit truthful reporting, be prospective or current rather than retrospective, and aim to cross-validate any metric of absenteeism.

Externalities

The term *externalities*, as used in economics, refers to effects on third parties that are not actively involved in consumption or production choices. Externalities are important to consider in any impact evaluation; failure to do so can lead to an overestimation or underestimation of the program impact. Menstrual hygiene evaluations must consider a range of potential positive and negative externalities. For example, a large-scale, free, single-use menstrual pads program could put non-participants' health at risk if these used pads are disposed of in latrines, causing the latrines to malfunction and preventing their use among the wider population (*negative externality*). However, the same program may be effective in reducing transactional sex for pads among impoverished girls, in this way reducing the prevalence of sexually transmitted infections (STIs) and lowering the risk of STI transmission in the community (*positive externality*). Similarly, a menstrual health information program rolled out to 6th graders may have a positive impact on 4th graders, if many 6th graders have younger siblings to whom they pass on the information (*positive externality*). Therefore, a well-designed program will, firstly, *randomize* treatment at a level where no contamination of the control group is expected. Often with menstrual hygiene programs, randomization should be done at the school level with a sufficient number of schools included (more than 45 schools [so-called clusters] according to Angrist and Pischke 2008), ensuring balance on student characteristics at baseline. A well-designed program will measure externalities where appropriate and be explicit about when it does not. A study that does not measure externalities should make that clear, especially when conducting any cost–benefit (CB) or CE calculations.

Cost–Benefit and Cost-Effectiveness Analysis

CB and CE analyses are two distinct tools that help measure policy success. CB analysis measures the benefit from a policy intervention in terms of monetary gains. In the health and education sphere, CB analysis can be complex because it necessitates making assumptions about the statistical value of life, expected longevity, and change in future and lifetime earnings in the study population. CE analysis, on the other hand, avoids many of these assumptions by simply comparing two or more policies against each other. For example, providing menstrual pads might reduce absenteeism by one day per ten USD spent on pads. Within the same or a different program, providing a menstrual cup may be found to reduce absenteeism by 1.2 days per ten USD spent. In this case, the latter policy is more cost-effective (note that these numbers are hypothetical). Additionally, CE analysis allows comparison of completely different policies as long as the studies have similarly defined outcome variables. This is well illustrated in the J-PAL policy bulletin (2012) which compares the CE of a deworming program to alternative policies with the same aim of raising attendance, for example providing free school uniforms or scholarships.

A main limitation with CB and CE analyses is that they are limited to measured outcomes and do not fully reflect the costs and benefits of aspects that were not monitored within the research program. For example, if a program reduces time spent on washing cloths and increases money spent on soap but these outcomes were not monitored within the program, the CE and CB analyses will not fully reflect these costs and benefits. Similarly, externalities (such as impacts on siblings and friends who are not study participants) may not be captured in the study design and, therefore, not reflected in CE or CB analysis. These issues are inherent to all CE and CB analyses, of course; they are not specific to MHM studies.

Pre-analysis Plans

Pre-analysis plans, where the researcher registers the research protocol and planned analysis in advance of collecting the data, are tools available for impact evaluations that can increase transparency in research.[2] The use of pre-analysis plans has been lauded within the social sciences for reducing cherry-picking of outcomes (Casey, Glennerster, and Miguel 2012; Miguel et al. 2014) and mitigating donor/funder pressure to publish only positive results. Research by Casey, Glennerster, and Miguel (2012) provided support for the use of pre-analysis plans by demonstrating how researchers could alter their choice of outcome and show that a program was either a failure or a success. The pre-analysis plan limited donor-pressure to show a successful program.

However, pre-analysis plans are sometimes said to stifle researcher creativity and the adaptability of research programs in the wake of unexpected findings and events. Medical and clinical sciences, where RCTs and pre-analysis plans caught a foothold much earlier than in the social sciences, adopted pre-analysis plans to ensure quality in drug trials. In the case of MHM studies focusing on school absenteeism, it is important to register the research questions, the study design, the definition of absenteeism, and the collection method for absenteeism data. The use of pre-analysis plans is recommended for all impact evaluation analyses, not only those related to MHM.

The Benchmark Study

Grounded in this review of the state of research examining the relationship between educational attainment and menstruation, we propose the features of an ideal study. This study would:

- Allow for mixed methods to capture attitudes, beliefs, practices, and ideal scenarios at baseline and how these attitudes, beliefs, practices, and ideal scenarios differ at the endline
- Differentiate between the extensive margin (drop-out rates) and the intensive margin (absenteeism)
- Take into account seasonal and cyclical temporal variations in outcomes, such as premenstrual absenteeism due to cramps and differential absenteeism rates per term during harvest seasons or near school exams. (For example: Before–after studies may need to evaluate outcomes at the same time in a calendar year.)
- Use a before–after study design with randomized allocation of treatment and predefined outcomes (see pre-analysis plans) that are objective and observable.
- Consider externalities (effects on third parties). For example, a student in the treatment group who received sanitary pads decides to share them with a friend who was randomized into the control group. If spillover effects are likely, randomization should be done at the cluster level, dividing schools into treatment and control schools, rather than at the individual level. Cluster randomized trials need to have sufficient number of clusters to ensure statistical inference.
- Include a post-study roll-out of benefits to all participants for ethical reasons. For example, the control group should receive reusable menstrual pads or a menstrual cup upon the termination of the study.
- Measure changes in the external environment. A menstrual health program may change school-level administration or student use of the latrines. Therefore, a careful survey of the external environment is recommended.

Discussion

There is mixed evidence on whether menstruation leads to higher absenteeism rates in low- and middle-income countries, with some studies finding weak or nonexistent links between periods and absenteeism (Grant, Lloyd, and Mensch 2013; Oster and Thornton 2011; Phillips-Howard et al. 2016a). Other studies confirm high levels of menstruation-related absenteeism (for example, Tegegne and Sisay 2014), especially among girls using traditional materials (Mason et al. 2015). While menstrual-related absence remains an important and unanswered topic, many factors hinder our understanding of how menstruation affects educational attainment and the psychosocial aspects of schoolgirls' lives. We call for the use of a broader set of outcomes in studies that explore the links between menstruation and education in low- and middle-income countries.

Furthermore, stigma and taboos can make the measurement of menstrual-related absenteeism hard. In one study in Kenya, girls reported that other girls, but not themselves, miss school because of their periods (Mason et al. 2013). In Malawi, girls were much more likely to report being absent from school during their period if they reported in private to a computer instead of face-to-face to an enumerator (Grant, Lloyd, and Mensch 2013). Data collected on sensitive topics such as menstrual experiences and menstrual-related school absenteeism should therefore be cross-validated before use.

Because of the potential role of stigmas, taboos, and varying levels of poverty, we must refrain from overinterpreting the external validity of studies (the extent to which the results of a study generalize to other contexts, populations, and times). For example, a pilot study conducted in Nepal showed that providing menstrual products to schoolgirls did not improve school attendance, possibly because the girls did not report missing school because of their periods in the first place (Oster and Thornton 2011). While the study was well designed and implemented, it does not show that menstrual hygiene interventions *cannot* reduce school absenteeism. It merely shows that in a sample of 199 schoolgirls in Nepal, who do not miss school because of their periods, such interventions are ineffective in improving attendance. The external validity of these results to contexts with high menstrual-related absenteeism is likely low.

We assert that the majority of studies to date are informative but suffer from three main methodological limitations. First, the limited focus on menstrual-related absence hampers our understanding of the wider threats that menstruation poses to participation in school and the psychosocial aspects of schoolgirls' experiences. Second, most studies use data sources that may suffer from self-reporting and recall bias, and few studies validate their data using alternative methods. Third, the external validity of even the best menstrual-related studies must be considered, given the influence of stigma and taboos in determining behavior and experiences. Therefore we caution policy makers against relying on school absence as the sole outcome variable. We also caution against the overinterpretation of results from existing studies that often lack both scope and precision.

Furthermore, we recommend that future studies explore the importance of pain management. School-age girls from several different contexts have reported abdominal pain as an issue. A majority of students in an Indian study reported that menstrual pains reduced participation (Sivakami et al. 2019), 66% of interviewed students in Nigeria reported abdominal pain and discomfort (Adinma and Adinma 2008), and pain was considered an issue by both students and teachers in Cambodia (Connolly and Sommer 2013). Other issues that future studies should explore include the impact of menstruation on concentration, test scores, and self-esteem. To date, qualitative studies have best demonstrated an understanding of the relevance of these margins, and future quantitative studies need to build on this body of work to support the design of evidence-based programs.

Lastly, studies find that absenteeism is common among both boys and girls in some contexts (Benshaul-Tolonen et al. 2019; Grant, Lloyd, and Mensch 2013) and that puberty education is lacking for both girls and boys (Sommer 2013). Further research is needed to understand the underlying causes of high rates of absenteeism among boys and girls and to identify cost-efficient policies to reduce absenteeism among all students.

In summary, questions of internal and external validity are important to consider when deciding whether to employ or reject a potential policy, especially since there may be differences in menstrual stigma and taboos across different populations. To avoid misinformed conclusions regarding program success or failure, we recommend employing mixed methods for MHM interventions, designing each program with utmost care, validating the survey instrument, and replicating the study design in multiple contexts.

Finally, more research is needed on the topic of menstrual-related educational attainment, especially research that is sensitive to the context and particular intervention, in line with recommendations made by Sommer (2010b). We encourage attempts to conduct streamlined studies that would allow for comparisons across contexts, as recommended by Hennegan and Montgomery (2016). Reproducible and homogenous research designs with standardized metrics and definitions that are meaningful to educationalists across several distinct contexts (Phillips-Howard et al. 2016b) could provide a fruitful way forward in overcoming the shortcomings of the limited and heterogeneous body of evidence on this question of MHM in educational settings.[3]

NOTES

1. Trial registration sites such as clinicaltrials.gov require that trial details, including defined outcomes, be uploaded prior to enrollment of participants. Trial site and date of registration are reported in the resulting publication.
2. For social sciences: https://www.socialscienceregistry.org/, for medical sciences: http://www.isrctn.com/. Three recent survey articles published in *Science* (Miguel et al. 2014), *Journal of Economic Perspectives* (Olken 2015) and *Journal of Economic Literature* (Christensen and Miguel 2018) further discuss the need for transparency within the social sciences.
3. Thanks to Masih A. Babagoli for research assistance.

REFERENCES

Adinma, Echendu Dolly, and J. I. B. Adinma. 2008. "Perceptions and Practices on Menstruation amongst Nigerian Secondary School Girls." *African Journal of Reproductive Health* 12 (1): 74–83.

Adukia, Anjali. 2017. "Sanitation and Education." *American Economic Journal: Applied Economics* 9 (2): 23–59.

Alam, Mahbub-Ul, Stephen P. Luby, Amal K. Halder, Khairul Islam, Aftab Opel, Abul K. Shoab, Probir K. Ghosh, Mahbubur Rahman, Therese Mahon, and Leanne Unicomb. 2017. "Menstrual Hygiene Management among Bangladeshi Adolescent Schoolgirls and Risk Factors Affecting School Absence: Results from a Cross-Sectional Survey." *BMJ Open* 7 (7): e015508.

Althubaiti, Alaa. 2016. "Information Bias in Health Research: Definition, Pitfalls, and Adjustment Methods." *Journal of Multidisciplinary Healthcare* 9: 211.

Angrist, Joshua D., and Jörn-Steffen Pischke. 2008. *Mostly Harmless Econometrics: An Empiricist's Companion*. Princeton, NJ: Princeton University Press.

Benshaul-Tolonen, Anja, Garazi Zulaika, Elizabeth Nyothach, Clifford Oduor, Linda Mason, David Obor, Kelly T. Alexander, Kayla F. Laserson, and Penelope A. Phillips-Howard. 2019. "Pupil Absenteeism, Measurement, and Menstruation: Evidence from Kenya." CDEP-CGEG 2019 Working Paper No. 74, March 2019.

Brodeur, Abel, Mathias Lé, Marc Sangnier, and Yanos Zylberberg. 2016. "Star Wars: The Empirics Strike Back." *American Economic Journal: Applied Economics* 8 (1): 1–32.

Casey, Katherine, Rachel Glennerster, and Edward Miguel. 2012. "Reshaping Institutions: Evidence on Aid Impacts Using a Preanalysis Plan." *The Quarterly Journal of Economics* 127 (4): 1755–1812.

Christensen, Garret, and Edward Miguel. 2018. "Transparency, Reproducibility, and the Credibility of Economics Research." *Journal of Economic Literature* 56 (3): 920–80.

Connolly, Susan, and Marni Sommer. 2013. "Cambodian Girls' Recommendations for Facilitating Menstrual Hygiene Management in School." *Journal of Water, Sanitation and Hygiene for Development* 3 (4): 612–22.

Coughlin, Steven S. 1990 "Recall Bias in Epidemiologic Studies." *Journal of Clinical Epidemiology* 43 (1): 87–91.

El-Gilany, Abdel-Hady, Karima Badawi, and Sanaa El-Fedawy. 2005. "Menstrual Hygiene among Adolescent Schoolgirls in Mansoura, Egypt."*Reproductive Health Matters* 13 (26): 147–52.

Freeman, Matthew C., Leslie E. Greene, Robert Dreibelbis, Shadi Saboori, Richard Muga, Babette Brumback, and Richard Rheingans. 2012. "Assessing the Impact of a School-Based Water Treatment, Hygiene and Sanitation Programme on Pupil Absence in Nyanza Province, Kenya: A Cluster-Randomized Trial." *Tropical Medicine & International Health* 17 (3): 380–91.

Grant, Monica, Cynthia Lloyd, and Barbara Mensch. 2013. "Menstruation and School Absenteeism: Evidence from Rural Malawi." *Comparative Education Review* 57 (2): 260–84.

Head, Megan L., Luke Holman, Rob Lanfear, Andrew T. Kahn, and Michael D. Jennions. 2015. "The Extent and Consequences of P-Hacking in Science." *PLoS Biology* 13 (3): e1002106.

Hennegan, Julie, and Paul Montgomery. 2016. "Do Menstrual Hygiene Management Interventions Improve Education and Psychosocial Outcomes for Women and Girls in Low and Middle Income Countries? A Systematic Review." *PLoS One* 11 (2): e0146985.

Johnson, Lynn, Teresa Calderón, Caroline Hilari, Jeanne Long, and Claudia Vivas. 2016. "Menstrual Hygiene Management Impacts Girls' School Experience in the Bolivian Amazon." United Nations Children's Fund, La Paz Bolivia, May 2016.

J-PAL Policy Bulletin. 2012. "Deworming: A Best Buy for Development." Abdul Latif Jameel Poverty Action Lab, Cambridge, MA.

Kirk, Jackie, and Marni Sommer. 2006. "Menstruation and Body Awareness: Linking Girls' Health with Girls' Education." *Royal Tropical Institute (KIT), Special on Gender and Health*, 1–22. http://www.susana.org/_resources/documents/default/2-1200-kirk-2006-menstruation-kit-paper.pdf.

Mason, Linda, Elizabeth Nyothach, Kelly Alexander, Frank O. Odhiambo, Alie Eleveld, John Vulule, Richard Rheingans, Kayla F. Laserson, Aisha Mohammed, and Penelope A. Phillips-Howard. 2013. "'We Keep It Secret so No One Should Know'—A Qualitative Study to Explore Young Schoolgirls Attitudes and Experiences with Menstruation in Rural WESTERN KENYA." *PLoS One* 8 (11): e79132.

Mason, Linda, Kayla Laserson, Kelvin Oruko, Elizabeth Nyothach, Kelly Alexander, Frank Odhiambo, Alie Eleveld, et al. 2015. "Adolescent Schoolgirls' Experiences of Menstrual Cups and Pads in Rural Western Kenya: A Qualitative Study." *Waterlines* 34 (1): 15–30.

McMahon, Shannon A., Peter J. Winch, Bethany A. Caruso, Alfredo F. Obure, Emily A. Ogutu, Imelda A. Ochari, and Richard D. Rheingans. 2011. "'The Girl With Her Period Is the One to Hang Her Head' Reflections on Menstrual Management among Schoolgirls in Rural Kenya." *BMC International Health and Human Rights* 11 (1): 7.

Miguel, Edward, Colin Camerer, Katherine Casey, Joshua Cohen, Kevin M. Esterling, Alan Gerber, Rachel Glennerster, et al. 2014. "Promoting Transparency in Social Science Research." *Science* 343 (6166): 30–31.

Montgomery, Paul, Caitlin R. Ryus, Catherine S. Dolan, Sue Dopson, and Linda M. Scott. 2012. "Sanitary Pad Interventions for Girls' Education in Ghana: A Pilot Study." *PLoS One* 7 (10): e48274.

Montgomery, Paul, Julie Hennegan, Catherine Dolan, Maryalice Wu, Laurel Steinfield, and Linda Scott. 2016. "Menstruation and the Cycle of Poverty: A Cluster Quasi-randomised Control Trial of Sanitary Pad and Puberty Education Provision in Uganda." *PLoS One* 11 (12): e0166122.

Olken, Benjamin A. 2015. "Promises and Perils of Pre-analysis Plans." *Journal of Economic Perspectives* 29 (3): 61–80.

Oster, Emily, and Rebecca Thornton. 2011. "Menstruation, Sanitary Products, and School Attendance: Evidence from a Randomized Evaluation." *American Economic Journal: Applied Economics* 3 (1): 91–100.

Phillips-Howard, Penelope A., Elizabeth Nyothach, Feiko O. ter Kuile, Jackton Omoto, Duolao Wang, Clement Zeh, Clayton Onyango, et al. 2016a. "Menstrual Cups and Sanitary Pads to Reduce School Attrition, and Sexually Transmitted and Reproductive Tract Infections: A Cluster Randomised Controlled Feasibility Study in Rural Western Kenya." *BMJ Open* 6 (11): e013229.

Phillips-Howard, P. A., B. Caruso, B. Torondel, G. Zulaika, M. Sahin, and M. Sommer. 2016b. "Menstrual Hygiene Management among Adolescent Schoolgirls in Low- and Middle-Income Countries: Research Priorities." *Glob Health Action* 9 (December 8): 33032. https://doi.org/10.3402/gha.v9.33032.

Sivakami, M., A. M. van Eijk, H. Thakur, N. Kakade, C. Patil, S. Shinde, N. Surani, A. Bauman, G. Zulaika, Y. Kabir, A. Dobhal, P. Singh, B. Tahiliani, L. Mason, K. T. Alexander, B. T. Mamita, K. F. Laserson, and P. A. Phillips-Howard. 2019. "Effect of Menstruation on Girls and Their Schooling, and Facilitators of Menstrual Hygiene Management in Schools: Surveys in Government Schools in Three States in India, 2015." *Journal of Global Health* 9 (1): 010408.

Sommer, Marni. 2009. "Ideologies of Sexuality, Menstruation and Risk: Girls' Experiences of Puberty and Schooling in Northern Tanzania." *Culture, Health & Sexuality* 11 (4): 383–98.

———. 2010a. "Where the Education System and Women's Bodies Collide: The Social and Health Impact of Girls' Experiences of Menstruation and Schooling in Tanzania." *Journal of Adolescence* 33 (4): 521–29.

———. 2010b. "Putting Menstrual Hygiene Management on to the School Water and Sanitation Agenda." *Waterlines* 29 (4): 268–78.

———. 2013. "Structural Factors Influencing Menstruating School Girls' Health and Well-Being in Tanzania." *Compare: A Journal of Comparative and International Education* 43 (3): 323–45.

Sommer, Marni, and Nana Mokoah Ackatia-Armah. 2012. "The Gendered Nature of Schooling in Ghana: Hurdles to Girls Menstrual Management in School." *JENdA: A Journal of Culture and African Women Studies* 20: 63–79.

Sommer, Marni, Nana Ackatia-Armah, Susan Connolly, and Dana Smiles. 2015. "A Comparison of the Menstruation and Education Experiences of Girls in Tanzania, Ghana, Cambodia and Ethiopia." *Compare: A Journal of Comparative and International Education* 45 (4): 589–609.

Sommer, Marni, Bethany A. Caruso, Murat Sahin, Teresa Calderon, Sue Cavill, Therese Mahon, and Penelope A. Phillips-Howard. 2016. "A Time for Global Action: Addressing Girls' Menstrual Hygiene Management Needs in Schools." *PLoS Medicine* 13 (2): e1001962.

Tegegne, Teketo Kassaw, and Mitike Molla Sisay. 2014. "Menstrual Hygiene Management and School Absenteeism among Female Adolescent Students in Northeast Ethiopia." *BMC Public Health* 14 (1): 1118.

van Eijk, A. M., M. Sivakami, M. B. Thakkar, et al. 2016. "Menstrual Hygiene Management among Adolescent Girls in India: A Systematic Review and Meta-Analysis." *BMJ Open* 6: e010290. https://doi.org/10.1136/bmjopen-2015-010290.

CHAPTER 53

Practice Note: 'If Only All Women Menstruated Exactly Two Weeks Ago': Interdisciplinary Challenges and Experiences of Capturing Hormonal Variation Across the Menstrual Cycle

Lauren C. Houghton and Noémie Elhadad

Menstrual health is important in its own right, but menstrual characteristics, including the signs and symptoms that accompany the cycle and their underlying hormonal patterns, are also risk factors for chronic diseases such as breast cancer and endometriosis, two diseases to which we have dedicated our own research careers. One such risk factor for both diseases is early age at menarche (Bodicoat et al. 2014; Parazzini et al. 2017). When one of us got her first period at age 12.11 years (LCH knows her exact age because it was on Christmas Eve), she was ashamed. Now 24 years and 288 menstrual cycles later, she has personally interviewed over 1000 girls and women about their menstrual cycles and the shame has subsided (Houghton et al. 2014, 2018, 2019). When the other got her first period at age 13.9 years, she soon realized that periods were going to play a painful part of her life. She was later diagnosed with endometriosis and has since designed technology to give a voice to women living with endometriosis (McKillop, Mamykina, and Elhadad 2018; Urteaga et al. 2018). Since their first periods, secular changes in societal views have also been dramatic: Teenage girls are coding their own video games "to rid the world of the menstrual taboo" ("Tampon Run," n.d.), governments are passing "Menstrual Equity" legislation (Zraick 2018), and women downloading period tracking apps have made them the second most used health app (PRIORI DATA). These societal and technological changes have also changed the way we and other anthropologists,

C. Bobel et al. (eds.), *The Palgrave Handbook of Critical Menstruation Studies*, https://doi.org/10.1007/978-981-15-0614-7_53

epidemiologists, and data scientists interested in characterizing hormonal variation in menstrual patterns conduct menstruation research.

Anthropologists have been instrumental in demonstrating that there is global variation in menstruation, its timing across the life course, the cultural practices surrounding it, and the underlying hormones that regulate it. Starting in the 1980s, biological anthropologists went to the field in various populations and compared menstrual cycles among women living in vastly different ecological contexts with starkly different norms surrounding reproductive timing. They demonstrated that the Dogon women of West Africa menstruate three times less than American and European women in their lifetime (Strassmann 1999). This dramatic difference in lifetime number of menstrual cycles is explained by the later age at menarche, earlier age at first pregnancy, higher parity, and earlier age at menopause in the Dogon women.

The variation in number of lifetime menstrual cycles is also reflected in underlying variation in reproductive hormones. For example, the overall pattern of progesterone across the menstrual cycle is similar among American, Polish, Lese, and Nepali women, with levels low and flat during the follicular phase (days 1–14), rising during ovulation (day 14), peaking during the mid-luteal phase (day 21), and slowly declining in the late luteal phase until day 28. However, the absolute levels of hormone at each of these phases is three-fold higher in American and Polish women compared to the Lese and Nepalese (Ellison 1994). The variation in hormone profiles demonstrates natural variation in ovarian function across populations, and this variation should be considered normal rather than pathological (Ellison et al. 1993).

Technological advances in hormone measurement made this important research possible. Peter Ellison and his colleagues leveraged the advancement in assays (Riad-Fahmy et al. 1982) to measure hormones in saliva for field settings (Ellison 1993). The advantage of using saliva was that anthropologists could collect daily samples from women and, using preservatives, store the samples at room temperature in isolated field locations where refrigeration was not possible (Lipson and Ellison 1989). It was important to collect daily samples to capture the hormonal profile of the entire cycle and to detect day of ovulation in order to align cycles.

Epidemiologists have also been interested in the variation of menstrual patterns (timing and frequency) and reproductive hormones because of their associations with breast cancer (Mumford et al. 2012; Whelan et al. 1994; Terry et al. 2005). As a result, hundreds of studies have compared hormones between women with and without breast cancer; between women with different risk factors, such as reproductive events, diets, and physical activity levels; and between women living in populations with different breast cancer incidence rates. Epidemiologists, who typically work with much larger study populations than anthropologists, conducted these comparisons using serum or urine samples collected on specific days of the menstrual cycle. In order to correctly time collection, epidemiologists asked women to recall the date of their last menstrual period (LMP) and then count forward to schedule a specimen collection on the targeted day. In most cases the target day would be menstrual cycle day 21, when both progesterone and estradiol are relatively high.

Restricting comparison of hormones to specific days is cost-effective for large-scale studies, but it comes at a scientific cost in studies of premenopausal women. Characterizing a woman's "dose" of estrogen based on only one or two timed samples may not capture the full variation in hormones over the menstrual cycle. For example, we know elevated estrogen levels are associated with increased risk for breast cancer among postmenopausal women (from whom a single sample can be collected since there is no menstrual cycle variation to account for), but we are less confident this association holds in premenopausal women (Key 2011). Additionally, follicular phase estradiol is associated with increased breast cancer risk, whereas luteal phase estradiol is not (Eliassen et al. 2006). These gaps in understanding regarding premenopausal hormones and their relationship with breast cancer are particularly problematic given that breast cancer incidence is on the rise, specifically in US women under 40 years old (Johnson, Chien, and Bleyer 2013). So while epidemiologists have also contributed to knowledge through large cohort studies, their narrow focus on measuring hormone levels on specific days has limited our understanding of the full variability of hormones in menstruation.

Further, calibrating the hormone samples to the cycle day relies on women's ability to recall the date of their LMP, but cycle lengths vary within and between women and women's recollections may not be accurate. For example, some epidemiologists have followed up with women to confirm that the sample collection in fact occurred on or near the targeted day. They provided participants, who in this case were highly motivated nurses, with a pre-addressed postcard to return with the date when their next period began. They found that specimens were collected as many as four days before or after the target day in the best-case scenarios (Eliassen et al. 2006).

In our experience, some women record their menstrual period in their own personal calendar, but many women do not. Over the course of interviewing women and asking them to recall their LMP, we have observed several estimation techniques. If her period just finished, she would count back five days to the day it began. Or, she would think of where she was the last time it occurred and then work out the date based on where she was that month. If her period had not been within the last week or if there was no significant event to help jog her memory, the most common response was to default to "oh, about two weeks ago." This suggests that self-reported LMP is a rather rough estimate.

Advancements in statistical methods can overcome some of these challenges when collecting daily specimens is not feasible. In one study, one of us took an alternative approach to targeting a specific cycle day. We collected specimens from all of the women in our study over the same two calendar days and later asked them to confirm when their next menstrual period

(NMP) started (Houghton et al. 2016; Troisi et al. 2014). Thus, by chance, women were at different days of their menstrual cycle and this yielded a collection of samples randomly distributed across a menstrual cycle. We then took this random distribution and modeled (using cubic splines) the hormone profile at the population level using individual data on specific days. Without giving the model a specified shape, the curves mimicked biological curves seen in individual women. We were then able to compare hormone concentrations between populations by calculating the area under each curve. In this specific case, we were able to demonstrate that estrogens were actually higher in Mongolian women than in British women, a finding that was counter to what most people would predict, given that Mongolia has some of the lowest breast cancer rates in the world (Troisi et al. 2014). Confirming the date of the NMP was of the utmost importance, to make sure we knew the exact menstrual day on which the sample had been collected. Data scientists use even more sophisticated modeling and prediction techniques to predict the hormone profiles underlying women's menstrual cycle characteristics. For example, the state-of-the-art model developed by Clark and colleagues accurately captures the same hormonal patterns as a dataset of daily hormonal measurements from healthy women (Harris Clark, Schlosser, and Selgrade 2003).

Our own experience in collecting and analyzing menstrual cycle and hormonal data mirrors the methodologies in the literature, ranging from small-scale studies with "in-depth" hormone measurements (daily readings throughout the menstrual cycle) to large-scale cohort analysis with "shallow" hormone measurements (under-sampling in time, or sampling only at known phases of the cycle). The "deepest" examples can be found in reproductive endocrinology, where researchers have established basic knowledge about the cycle through small-scale studies of 12 women and 15-minute interval blood serum measurements (Murdoch et al. 1985). But such a deep dive is not the route we have chosen to follow. Rather, we are both now excited to incorporate mobile apps into our research toolbox to mitigate the tension between the need for multiple measurements and the invasive nature of taking such measurements. Mobile health is a game changer for investigating the menstrual cycle, making data collection possible at an unprecedented scale. Women can now use menstrual tracking apps to track their own cycles, along with a range of signs and symptoms—data which previously did not get recorded. It is not only possible, women are doing it. Cycle tracking apps are some of the fastest-growing health apps and have a loyal base of users (Fox and Duggan 2012; Wartella et al. 2016). In 2018, 48% of US females ages 18- to 22-years-old and 25% of teen girls reported having used a period tracking app (Fox 2018). Some apps have as many as 10 million users worldwide ("Clue Wants Your Research Proposals on Reproductive Health," n.d.). There is a great opportunity to leverage self-tracking technology to enable characterizing the menstrual cycle at scale and for individuals.

One promising aspect of apps used for period tracking is their flexibility in capturing a range of experiences related to menstruation. Popular trackers enable users to self-track their cycle and period lengths, as well as specific symptoms (such as, headaches, sleep, bowel movements, pain levels); psychosocial elements (such as, moods or socializing patterns); and sexual activity. Many apps further enable users to customize their tracking options, thus getting closer to capturing the broader context and the narrative around menstruation. While cycle characteristics on their own might not be enough for algorithms to model cycle variations, the additional data elements can significantly boost these models' ability to infer underlying hormonal variations.

Beyond self-tracked data, emerging analytics can help us determine the ideal hormone-sampling rate for future studies or improved diagnostics. For example, Urteaga and colleagues (2017, 2019) showed that machine learning-based approaches that combine the aforementioned state-of-the art model for daily average hormone concentrations with cycle data collected using a mobile phone application can accurately identify the phases within the menstrual cycle, even without daily measurements. Furthermore, the model allows for generation of hormone patterns with different characteristics, such as varying cycle length. Once these methods are validated with in-depth hormonal datasets, they have the potential to revolutionize how we study the variation in menstrual cycle and hormonal patterns in healthy and unhealthy women.

As app developers try to perfect their algorithms to predict each user's next period more accurately, and researchers try to use app tracking data to predict underlying hormonal profiles, this momentum must be accompanied by some caution and critical reflections. First, there is the risk that only using data collected by menstruators who use apps may further marginalize those menstruators who don't use apps for biological (irregular periods) or socially patterned reasons. For example, while smart phone and app use is global, within emerging markets people with less education are still less likely to use apps than those with more education (Taylor and Silver 2019). Second, new mHealth data collection methods may lead to large menstrual cycles studies at unprecedented scales, but the cultural norms with which the technology was developed may obscure the cultural context of the populations under study (Fox and Epstein [Chapter 54] in this volume). Principles of citizen science might be helpful in engaging on-the-ground perspectives from diverse groups of menstruators. From the collective work of anthropologists, epidemiologists, and data scientists, we know that our culture and biology both influence our menstrual health. We should ensure that our methods are equally attuned to biological plausibility and cultural context.

REFERENCES

Bodicoat, Danielle H., Minouk J. Schoemaker, Michael E. Jones, Emily McFadden, James Griffin, Alan Ashworth, and Anthony J. Swerdlow. 2014. "Timing of Pubertal Stages and Breast Cancer Risk: The Breakthrough Generations Study." *Breast Cancer Research* 16 (1): R18. https://doi.org/10.1186/bcr3613.

"Clue Wants Your Research Proposals on Reproductive Health." n.d. Accessed March 19, 2019. https://helloclue.com/articles/about-clue/clue-research-innovation-program-call-for-proposals.

Eliassen, A. Heather, Stacey A. Missmer, Shelley S. Tworoger, Donna Spiegelman, Robert L. Barbieri, Mitch Dowsett, and Susan E. Hankinson. 2006. "Endogenous Steroid Hormone Concentrations and Risk of Breast Cancer among Premenopausal Women." *Journal of the National Cancer Institute* 98 (19): 1406–15. https://doi.org/10.1093/jnci/djj376.

Ellison, P. T. 1993. "Measurements of Salivary Progesterone." *Annals of the New York Academy of Sciences* 694 (September): 161–76. http://www.ncbi.nlm.nih.gov/pubmed/8215052.

———. 1994. "Salivary Steroids and Natural Variation in Human Ovarian Function." *Annals of the New York Academy of Sciences* 709 (February): 287–98. http://www.ncbi.nlm.nih.gov/pubmed/8154721.

Ellison, P. T., C. Panter-Brick, S. F. Lipson, and M. T. O'Rourke. 1993. "The Ecological Context of Human Ovarian Function." *Human Reproduction* 8 (12): 2248–58. http://www.ncbi.nlm.nih.gov/pubmed/8150934.

Fox, Susannah. 2018. "Digital Health Practices among Teens and Young Adults in the U.S." 2018. https://susannahfox.com/2018/07/31/digital-health-practices-among-teens-and-young-adults-in-the-u-s/.

Fox, S., and M. Duggan. 2012. "Mobile Health 2012 | Pew Research Center." n.d. Accessed March 19, 2019. http://www.pewinternet.org/2012/11/08/mobile-health-2012/.

Harris Clark, L., Paul M. Schlosser, and James F. Selgrade. 2003. "Multiple Stable Periodic Solutions in a Model for Hormonal Control of the Menstrual Cycle." *Bulletin of Mathematical Biology* 65 (1): 157–73. https://doi.org/10.1006/bulm.2002.0326.

Houghton, L. C., G. D. Cooper, G. R. Bentley, M. Booth, O. A. Chowdhury, R. Troisi, R. G. Ziegler, R. N. Hoover, and H. A. Katki. 2014. "A Migrant Study of Pubertal Timing and Tempo in British-Bangladeshi Girls at Varying Risk for Breast Cancer." *Breast Cancer Research* 16 (1). https://doi.org/10.1186/s13058-014-0469-8.

Houghton, Lauren, Davaasambuu Ganmaa, Philip Rosenberg, Dambadarjaa Davaalkham, Frank Stanczyk, Robert Hoover, and Rebecca Troisi. 2016. "Associations of Breast Cancer Risk Factors with Premenopausal Sex Hormones in Women with Very Low Breast Cancer Risk." *International Journal of Environmental Research and Public Health* 13 (12): 1066. https://doi.org/10.3390/ijerph13111066.

Houghton, L. C., J. A. Knight, M. J. De Souza, M. Goldberg, M. L. White, K. O'Toole, W. K. Chung, et al. 2018. "Comparison of Methods to Assess Onset of Breast Development in the LEGACY Girls Study: Methodological Considerations for Studies of Breast Cancer." *Breast Cancer Research* 20 (1). https://doi.org/10.1186/s13058-018-0943-9.

Houghton, Lauren C., Julia A. Knight, Ying Wei, Russell D. Romeo, Mandy Goldberg, Irene L. Andrulis, Angela R. Bradbury, et al. 2019. "Association of Prepubertal and Adolescent Androgen Concentrations with Timing of Breast Development and Family History of Breast Cancer." *JAMA Network Open* 2 (2): e190083. https://doi.org/10.1001/jamanetworkopen.2019.0083.Johnson.

Johnson, Rebecca H., Franklin L. Chien, and Archie Bleyer. 2013. "Incidence of Breast Cancer with Distant Involvement among Women in the United States, 1976 to 2009." *JAMA* 309 (8): 800–805. https://doi.org/10.1001/jama.2013.776.

Key, Timothy J. 2011. "Endogenous Oestrogens and Breast Cancer Risk in Premenopausal and Postmenopausal Women." *Steroids* 76 (8): 812–15. https://doi.org/10.1016/J.STEROIDS.2011.02.029.

Lipson, S. F., and P. T. Ellison. 1989. "Development of Protocols for the Application of Salivary Steroid Analysis to Field Conditions." *American Journal of Human Biology* 1 (3): 249–55. https://doi.org/10.1002/ajhb.1310010304.

McKillop, Mollie, Lena Mamykina, and Noémie Elhadad. 2018. "Designing in the Dark." In *Proceedings of the 2018 CHI Conference on Human Factors in Computing Systems—CHI '18*, 1–15. ACM Press, New York, New York, USA. https://doi.org/10.1145/3173574.3174139.

Mumford, Sunni L., Anne Z. Steiner, Anna Z. Pollack, Neil J. Perkins, Amanda C. Filiberto, Paul S. Albert, Donald R. Mattison, Jean Wactawski-Wende, and Enrique F. Schisterman. 2012. "The Utility of Menstrual Cycle Length as an Indicator of Cumulative Hormonal Exposure." *The Journal of Clinical Endocrinology & Metabolism* 97 (10): E1871–79. https://doi.org/10.1210/jc.2012-1350.

Murdoch, A. P., P. J. Diggle, W. Dunlop, and P. Kendall-Taylor. 1985. "Determination of the Frequency of Pulsatile Luteinizing Hormone Secretion by Time Series Analysis." *Clinical Endocrinology* 22 (3): 341–46. http://www.ncbi.nlm.nih.gov/pubmed/3978835.

Parazzini, F., G. Esposito, L. Tozzi, S. Noli, and S. Bianchi. 2017. "Epidemiology of Endometriosis and Its Comorbidities." *European Journal of Obstetrics & Gynecology and Reproductive Biology* 209 (February): 3–7. https://doi.org/10.1016/j.ejogrb.2016.04.021.

"PRIORI DATA—Mobile App Data Intelligence, ASO Tool & Market Research." n.d.

Riad-Fahmy, D., G. F. Read, R. F. Walker, and K. Griffiths. 1982. "Steroids in Saliva for Assessing Endocrine Function." *Endocrine Reviews* 3 (4): 367–95. https://doi.org/10.1210/edrv-3-4-367.

Strassmann, B. I. 1999. "Menstrual Cycling and Breast Cancer: An Evolutionary Perspective." *Journal of Women's Health* 8 (2): 193–202. http://www.ncbi.nlm.nih.gov/pubmed/10100133.

"Tampon Run." n.d. Accessed March 19, 2019. http://tamponrun.com/.

Taylor, Ken, and Laura Silver. 2019. "Smartphone Ownership Is Growing Rapidly Around the World, but Not Always Equally | Pew Research Center." Pew Research Center, 2019. http://www.pewglobal.org/2019/02/05/smartphone-ownership-is-growing-rapidly-around-the-world-but-not-always-equally/.

Terry, Kathryn L., Walter C. Willett, Janet W. Rich-Edwards, David J. Hunter, and Karin B. Michels. 2005. "Menstrual Cycle Characteristics and Incidence of Premenopausal Breast Cancer." *Cancer Epidemiology Biomarkers & Prevention* 14 (6): 1509–13. https://doi.org/10.1158/1055-9965.EPI-05-0051.

Troisi, Rebecca, Daavasambuu Ganmaa, Isabel dos Santos Silva, Dambadarjaa Davaalkham, Philip S. Rosenberg, Janet Rich-Edwards, Lindsay Frasier, et al. 2014. "The Role of Hormones in the Differences in the Incidence of Breast Cancer between Mongolia and the United Kingdom." *PLoS One* 9 (12): e114455. https://doi.org/10.1371/journal.pone.0114455.

Urteaga, Iñigo, David J. Albers, Marija Vlajic Wheeler, Anna Druet, Hans Raffauf, and Noémie Elhadad. 2017. "Towards Personalized Modeling of the Female Hormonal Cycle: Experiments with Mechanistic Models and Gaussian Processes," Machine Learning for Health Workshop at NIPS (ML4H'17).

Urteaga, Iñigo, Mollie McKillop, Sharon Lipsky-Gorman, and Noémie Elhadad. 2018. "Phenotyping Endometriosis through Mixed Membership Models of Self-Tracking Data," November. Machine Learning for Healthcare (MLHC'18).

Urteaga, Iñigo, Tristan Bertin, Theresa Hardy, David Albers, and Noémie Elhadad. 2019. "Multi-Task Gaussian Processes and Dilated Convolutional Networks for Reconstruction of Reproductive Hormonal Dynamics," Machine Learning for Healthcare (MLHC'19).

Wartella, Ellen, Vicky Rideout, Heather Montague, Leanne Beaudoin-Ryan, and Alexis Lauricella. 2016. "Teens, Health and Technology: A National Survey." *Media and Communication* 4 (3): 13. https://doi.org/10.17645/mac.v4i3.515.

Whelan, E. A., D. P. Sandler, J. L. Root, K. R. Smith, and C. R. Weinberg. 1994. "Menstrual Cycle Patterns and Risk of Breast Cancer." *American Journal of Epidemiology* 140 (12): 1081–90. http://www.ncbi.nlm.nih.gov/pubmed/7998590.

Zraick, Karen. 2018. "It's Not Just the Tampon Tax: Why Periods Are Political." *The New York Times*, July 22, 2018. https://www.nytimes.com/2018/07/22/health/tampon-tax-periods-menstruation-nyt.html.

CHAPTER 54

Monitoring Menses: Design-Based Investigations of Menstrual Tracking Applications

Sarah Fox and Daniel A. Epstein

INTRODUCTION

In this chapter, we[1] describe our efforts to examine and reimagine menstrual tracking technology—or, mobile applications (apps) designed to support the documentation and quantification of menstrual cycle data. In their current format, these technologies encourage those who menstruate to extract intimate information about the body. Users of these applications are commonly asked to record menstrual cycle start and end dates, consistency or color of menstrual flow, physical and emotional symptoms, and details of sexual behavior. In return, these apps promise to predict the beginning and duration of one's next cycle and "fertile days," and to offer insight into managing one's period (for example, tips on forms of self-care and material preparedness through the carry of pads and so on).

Technology design as a broad industry has often ignored practices and issues associated with women's health, including menstruation. For instance, Apple Health launched in 2014 with the promise of monitoring "all of your metrics that you're most interested in," yet it did not include the ability to track menstruation until a year later after online backlash (Perez 2015). This lack of attention appears to be changing, as start-ups have formed to fill in this gap: the number of menstrual applications has grown rapidly over the last decade, with an estimated 200 million downloads worldwide by 2016 (Dreaper 2016). More recently, companies such as Clue, Kindara, and Glow have developed standalone mobile applications for menstrual tracking and integrated hardware meant to connect the collection of sensor-based data with predictive models (for example, a connected thermometer encouraging

© The Author(s) 2020
C. Bobel et al. (eds.), *The Palgrave Handbook of Critical Menstruation Studies*, https://doi.org/10.1007/978-981-15-0614-7_54

daily collection of basal body measurements and automatic storage of such data [Magee 2015]). The start-ups tout precise data analytics to inform users about their cycles (Magee 2015; Lomas 2015), while they garner millions of dollars in venture capital. Yet, there remain questions about ways the data being collected constructs a narrow and sometimes instrumentalized view of menstrual experiences—questions we puzzle over the course of this chapter.

In examining menstrual tracking technology, we take up a designerly lens (Rosner 2018; Zimmerman, Forlizzi, and Evenson 2007), highlighting the ways in which apps inscribe particular visions of what the body is capable of and how to make sense of menstruation. In doing so, we identify openings to explore how these technologies could exist differently, to encourage self-knowledge and affirm and support the needs of different kinds of menstruating bodies (that is, beyond a fertility or contraceptive focus that occupies many apps on the market). Toward the aim of examining and reimagining, we identify expanded forms of *monitoring menses*—ones that emphasize *modules*, which work to align with people's goals and identities rather than the models that come with algorithmic ways of knowing, and *dimensionality*, rather than a user's relation to averages or norms.

In what follows we describe two case studies: one examines the design of existing applications and how they are interpreted by users, and the other uses participatory approaches that draw on the experiences of menstruators to introduce alternatives to dominant menstrual app protocols. With the first, we uncover core issues of usability and inclusion in existing processes of menstrual tracking—for instance, apps assume users have a regular cycle, are interested in tracking for fertility, identify as female, and have one male partner. Here, the focus is on how current apps support people's needs and what designers of personal tracking tools might learn from people who record and make use of their menstrual cycle data. The other study examines tracking as an under-interrogated form of recordkeeping about the body and traces how it has evolved to serve people differently. It further takes up collaborative techniques of design to invite members of the menstruating public to reimagine these technologies to better serve their own goals of menstrual sensemaking.

Background

Before describing these cases in detail, we first turn to a backdrop of design activity within and just outside the space of menstrual tracking that both animates and motivates our discussion.

The Personal (Informatics) is Political

Health is a big information problem waiting for data analytics and wearable sensors. I wanted to start somewhere to make a difference [. . .] I found it in procreation. (Max Levchin [Co-founder, Glow; former Chief Technology Officer, Paypal])

Within the realm of technology design, the past decade has seen a turn toward big data. Though not a wholly new phenomenon, innovation in terms of cloud storage and machine learning has introduced an era where "value comes from the patterns that can be derived by making connections between pieces of data, about an individual, about individuals in relation to others, about groups of people, or simply about the structure of information itself" (boyd and Crawford 2012). Practitioners and "tech evangelists" alike preach the almost limitless potential of data to tell us things about the world—with enough of it, we can cast away uncertainty and focus the fuzziness associated with forms of risk. From forecasts on market performance to understanding food inequity (De Choudhury, Sharma, and Kiciman 2016), data is seen as the answer to some of the world's most elusive concerns.

There is hardly a more apt embodiment of this utopian view of technology than the above statement from Max Levchin, Co-Founder of the menstrual tracking app Glow (Goode 2013). Prior to Glow, Levchin's interests largely focused on building finance companies, including a startup that enables people to send money to others (Paypal) and another to predict the risk associated with lending (Affirm). As he explained during Glow's debut at the technology conference D: All Things Digital, concerns for menstrual tracking and financial transaction are not so distinct—predictive models could guide both this modern form of fertility awareness and market decision-making (Goode 2013). In a world where infertility is largely left uncovered by health insurance, Levchin insists, menstrual tracking fills a gap, offering a form of technologically aided assurance in place of medical attention. If one carefully collects data, such as cervical mucus texture and menstrual cycle dates, and follows the guidance of the predictive model, then medical or health advice (through a doctor or doula) is no longer necessary to achieve pregnancy—all one needs is a smartphone, an internet connection, $47.99 (for a premium account), and a willingness to track.

In applying a design-oriented lens to menstrual tracking, we sought to understand the ways these contemporary technological infrastructures are not formed de novo (Star 1999; Ribes and Jackson 2013), but are instead inscribed with particular histories. For instance, many of the measurements and predictive models upon which popular menstrual apps rely are drawn from fertility awareness-based methods, or recording techniques designed to encourage "natural family planning." Promoted by Catholic physicians, endorsed by the Pope, and taken up by parishioners over most of the last century, these birth control methods aimed at allowing Catholic couples to have more control over the size of their families (Ashley 2006). Here, then, menstrual tracking apps not only inherit modes of counting but also moral orientations.

Beyond 'Fertile Windows'

A contemporary political climate particularly unfriendly to forms of reproductive healthcare and education has placed new weight and sense of importance on data-rich menstrual tracking technologies. A leaked White House memo, for instance, described a US Department of Health and Human Services proposal to remove all forms of family planning programming for teens aside from the promotion of fertility awareness-based methods (Beutler 2017). In Europe, Natural Cycles was the first menstrual tracking application to be certified by the European Union as a method of contraception (a designation later also conferred by the United States Food and Drug Administration). This move has since garnered renewed attention and scrutiny, as a Swedish clinic raised flags after 37 of their patients reported unwanted pregnancies despite using the app (Pardes 2018; England 2017). These cases begin to expose the limits to a singular vision on the role of technology in menstrual tracking and the ways in which these apps—of both political and personal consequence—are ripe for critical, empirical investigation.

Toward developing techniques for conducting such critical investigation, we draw on recent approaches within the field of Human–Computer Interaction that argue for a feminist practice of technology design. For example, civic media scholar Catherine D'Ignazio and colleagues use the format of the large-scale hackathon to focus engineering attention on the task of redesigning the breast pump, a technology the researchers identify as long overlooked by industry (D'Ignazio et al. 2016). Similarly, Almeida et al. (2016) use familiar smartphone technology to promote pelvic floor fitness and everyday engagement with reproductive health concepts. Together these projects have helped scaffold a growing literature concerned with offering direct ways for researchers and practitioners alike to critically and productively explore alternatives to current design situations. In the cases described below, we take up methods common in processes of technology development and evaluation (for example, surveys, heuristic evaluation, interviews, app review analysis, participatory design) to investigate and conceptualize alternatives to available apps that reinforce particular visions of the menstruating body as realizing a reproductive capacity or as associated with a particular brand of femininity.

CASE ONE: EXAMINING EXISTING MENSTRUAL TRACKING TECHNOLOGIES

In the first case study, I (Epstein) conducted research with Human–Computer Interaction scholars Lee, Kang, Agapie, Schroeder, Pina, Fogarty, Kientz, and Munson to understand how contemporary mobile apps support or neglect those tracking their menstrual cycles—accounts we, in turn, used to inform recommendations for improving the designs (Epstein et al. 2017). The approach in this work was to characterize people's in situ experiences and challenges tracking their menstrual cycles as a means to critique the design choices of contemporary apps. To do this, we drew from three key streams of data: 2000 app store reviews of the 12 most reviewed Android and iPhone apps, a survey of 687 people who menstruate (recruited via Facebook, Twitter, and Reddit), and 12 interviews with individuals from socioeconomic, race, and identity groups underrepresented in our survey sample. Elsewhere, we discuss in more detail our study's methods, as well as offer longer form descriptions of our findings (Epstein et al. 2017). In what follows, we summarize our study, describing why participants chose to use mobile applications to track their menstrual cycles and the challenges posed by commercial apps.

Motivations for Monitoring

Many of the participants in our study reported turning to digital methods for tracking their menstrual cycle as part of a broader evolution of digitizing personal data (for example, schedule planning and financial management moving from paper to digital systems). Some respondents described searching for an app immediately after obtaining a smartphone or beginning to use a digital calendar for the rest of their schedule planning. Participants who continued to use paper-based systems such as diaries and journals preferred the flexibility and privacy they associated with paper. Other participants tracked more implicitly by following pill-based hormonal birth control, noticing early symptoms of the arrival of their period, or simply remembering when their period last occurred and counting forward in their head.

Pitfalls of Prediction

To enable preparation, most apps include the ability to send push notifications a day before a person's period is expected to arrive or they are next expected to ovulate. Participants reported finding this feature the most compelling reason for using a digital app. They evaluated an app primarily on its ability to predict where they were in their cycle, such as whether they were about to have their period or were about to ovulate. 18% of app reviews, for instance, mentioned the accuracy or inaccuracy of prediction. This prediction is sometimes presented as a single-day estimate, as with the app Life (Fig. 54.1a). Other apps, such as Clue, provide a range of possible affected days and describe a sense of uncertainty in their prediction (Fig. 54.1b) represented by the bubbles at the tail ends of the fertile window and confidence intervals (for example, ±1) in the text description. Participants described switching apps in search of more accurate alternatives, similar to the way people abandon apps and devices they use to track other aspects of their wellbeing (Epstein et al. 2015).

To scaffold prediction, apps often ask people to enter their average cycle length and flow duration upon installation, updating these predictions once

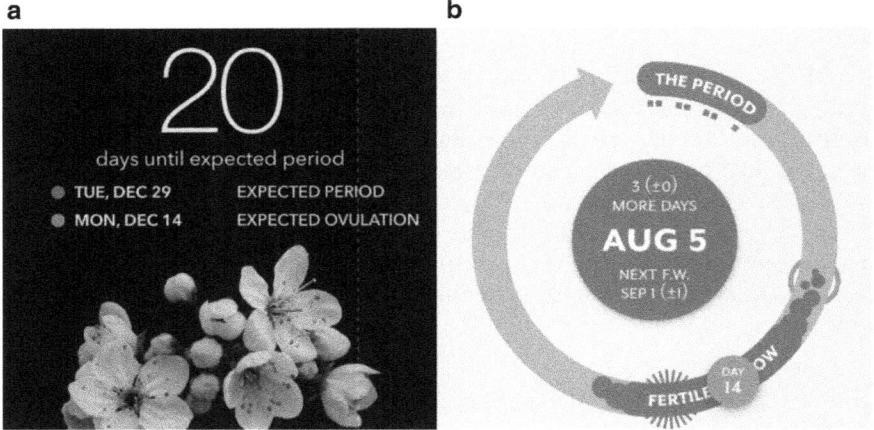

Fig. 54.1 Phone apps predict when someone is next expected to have their period or ovulate. The Life app (left) presents this production through single-day estimates, while the Clue app (right) provides a range of potential dates for the event (Credit: © Life Fertility Tracker IVS 2017 and © BioWink GmbH 2017. Photo Credit: Screenshots taken by Daniel Epstein in 2017)

Fig. 54.2 Apps such as My Cycles (left) and Period Tracker (right) typically ask for average cycle length and flow duration to aid prediction. Although this prediction may later be improved by journaled data, it is not resilient to variations due to factors such as irregular cycles, stress, birth control, or forgetting to journal (Credit: © StayWell Company LLC 2017 and © ABISHKKING LIMITED 2017)

the user has experienced their cycle and logged it in the app (Fig. 54.2). Unfortunately, these predictions encode assumptions about the regularity and frequency of a person's cycle. Apps assume minimal cycle variation, when in reality a person's cycle can vary by days or even weeks (Fraser et al. 2007). Moreover, apps fail to account for variations caused by life factors such as stress, sleep, and changing birth control methods. Some apps even fail to account for pregnancy: one survey participant noted that "*a pregnancy, a baby, and a year and a half of breastfeeding later, the app thinks my normal cycle length is about every 700 days!"* Apps also assume diligence in use and entry, which is known to be unrealistic in personal tracking (Epstein et al. 2015). Participants noted that logging their period's late arrival, or forgetting to record when their period ended, impacted the app's ability to make future predictions accurately. The realities of everyday use of cycle tracking apps often impede the success of the predictive models which underlie the app's core function and reinforce natural family planning.

To address these discrepancies, designers of menstrual tracking technology could examine additional techniques for modeling and communicating predictions. For example, evaluating interfaces which describe ovulation and period arrival as probabilities, rather than less-reliable binary predictions. At minimum, apps should allow people to correct a prediction when it falls out of line with reality, in order to be resilient to the myriad reasons why a prediction might be inaccurate (for example, changes in cycle, forgetting to track).

Menstrual Tracking Aesthetics

Similar to trends observed in the context of apps to support pregnancy (Peyton et al. 2014), most menstrual tracking apps we analyzed used stereotypically feminine attributes throughout their interfaces, such as predominantly pink color palettes or frequently used images of flowers and hearts (Fig. 54.3). Though some people reported appreciating the particular brand of femininity represented, most viewed it as a negative design trait. One participant described the design as trying to *"dumb it down,"* leading her to wonder *"why can't keeping track of my menstruation be a professional and organized task?"*

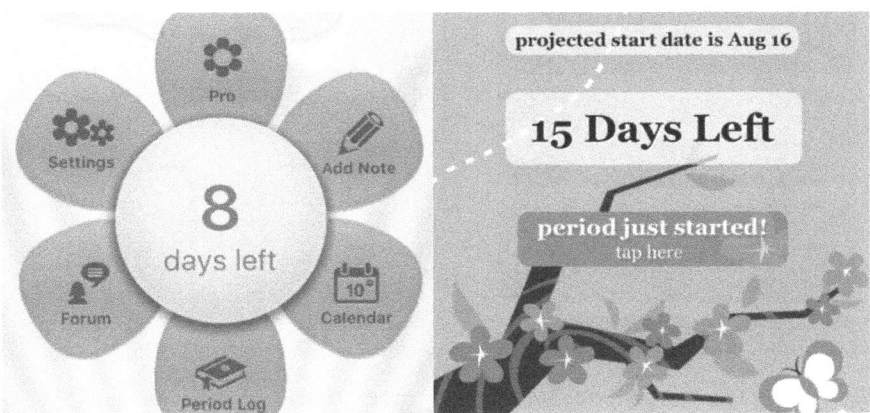

Fig. 54.3 Most period tracking apps we observed employ flowery and pink visual features, such as the main screens of Period Diary (left) and P. Tracker Lite (right) (Credit: © Bellabeat, Inc. 2017 and © GP Apps 2017)

Many participants sought out apps which had more neutral color schemes, with 38 out of 200 reviews praising Clue's relatively neutral visual design (Fig. 54.1, right).

Though some participants felt comfortable sharing their menstrual information, for others, the feminine aesthetics and often-obvious naming of apps (for example, Period Tracker, Period Diary) sometimes interfered with their desire to keep menstruation private. For example, one participant mentioned she "*used to be embarrassed when other people looked at my phone and saw a bright pink tracking app*," which prompted her to switch to an app with a more neutral aesthetic. Others felt the notifications interfered with their ability to keep their menstruation private, and disabled them as a result. Options for a more neutral design aesthetic and more subtle phrasing in notifications, as well as less conspicuous app names, could respond to this desire to keep menstruation and tracking information private.

Representations of Gender and Sexuality

Further, we observed that the apps we reviewed often make heteronormative assumptions about people tracking their menstruation and reinforce binary conceptions of gender. Participants who identified as non-binary or male struggled to find apps which "*didn't misgender me*," as one user described, or that featured gender-neutral language. For example, at the time of the study, Glow (Fig. 54.4, left) directed anyone who identified male upon setup to an alternate view of the app which focused on penile and testicular health.

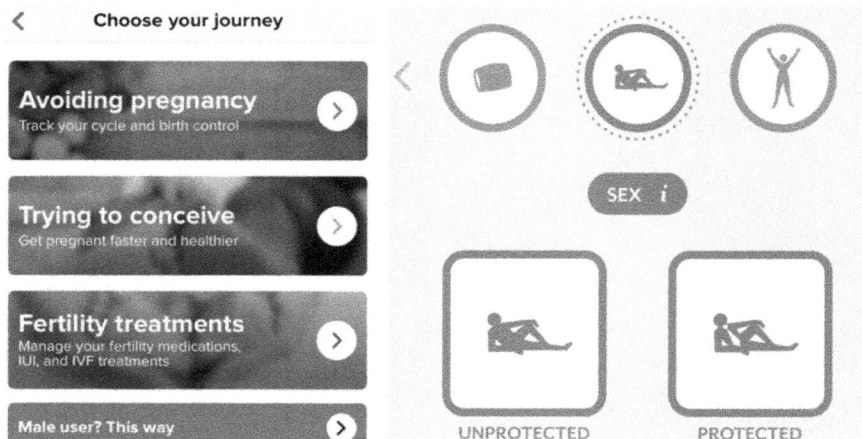

Fig. 54.4 The design and language in many menstrual tracking apps encode heteronormative assumptions. In Glow (left), people who identify as male are directed to an alternative view of the app. Clue's iconography (right) suggests a male sexual partner. We note that since conducting this research, Clue has updated their icons for logging sexual activity to be abstract rather than anthropomorphized (Credit: © Glow, Inc. 2017 and © BioWink GmbH 2017)

When apps support logging sexual activity or sharing data, they often assume that a person's sexual or relationship partner identifies as male. For example, at the time of the study, Clue provided two options for logging sex, both of which used icons suggesting a male partner (Fig. 54.4, right; we note that since conducting this research in 2016, the icons have been updated to be abstract representations). Other apps, such as My Period Tracker, only support asymmetric sharing of menstrual data (for example, sharing information about one's cycle with exactly one partner who is not also collecting their own data).

Fertility Focus

The emphasis on ovulation information (see Fig. 54.1) and inclusion of fertility tips led participants to feel that apps were primarily designed to support people in trying to conceive, rather than the range of goals they brought to tracking (for example, a general health check or avoiding pregnancy). Beyond feeling their goals were secondary, some participants felt uncomfortable with this focus. For example, some teenage participants stressed that they felt "*too young*" to care about fertility information. Participants who struggled with infertility, on the other hand, felt the ovulation information served as a reminder of their struggles. One participant said, "*I am no longer trying to get pregnant and I don't like the reminder of TTC [trying to conceive] or the tiny glimmer of hope that maybe by magic this will be the month when a miracle happens.*"

To design apps that avoid heteronormative and reductive assumptions about gender and sexuality, we suggest a more modular approach, one that allows people to align designs with their identity and goals. We believe more people would benefit from a gender-agnostic aesthetic that avoids gender-suggestive iconography or text. At minimum, apps should offer multiple themes or profiles from which to choose. Designs should also enable users to hide or remove ovulation and sexual activity markers, to allow those who do not wish to see this information to avoid it.

As they are currently designed, apps and tools for menstrual tracking fall short of addressing the needs and expectations people have expressed are important, including prediction, aesthetics, and tracking goals. The desire for alternatives is expressed succinctly by a respondent who lamented "*I've tried 4 apps. They all suck . . . I would think a creative woman would've created something better by now.*"

CASE TWO: REIMAGINING MENSTRUAL TRACKING THROUGH PARTICIPATORY DESIGN

To pursue avenues for challenging narrow conceptions of menstrual tracking, I (Fox) collaborated with information scholars Menking and Eschler and bioinformatics scholar Backonja to conduct a multipart research program, described here as the second case (Eschler et al. 2019; Fox et al. 2020). During the initial stages of research, we first reviewed perspectives on menstrual literacy from popular and publicly available health information-related websites, alongside the data collection techniques and interface characteristics of a sample of menstrual tracking apps (described in more detail in Eschler et al. 2019). One of the most striking disconnects between the menstrual literacy resources and the apps we examined concerned the focus on fertility (echoing the fertility focus highlighted in the previous case). To further explore design potentials in the space of menstrual tracking, we then took up collaborative methods of design (described further below) to imagine how tracking technologies might be developed differently to incorporate the concerns of a multiplicity of menstruators. In what follows, we focus on the results of these participatory encounters.

Period Packets: Reimagining the App Protocol

Inviting further reimagination of the menstrual tracking app protocol, we released a set of participatory design objects in the form of design packets (Gaver et al. 2001; Pierce and DiSalvo 2017)—here called Period Packets—asking participants to make their own sense of menstruation through a series of open activities. Drawing on the methodological tradition of research through design (Zimmerman, Forlizzi, and Evenson 2007), and more specifically the approach of cultural probes (Gaver et al. 2001), we took up ambiguity and provocation as a resource for conversation and collaborative design. In an early example of this method, design researchers Bill Gaver and Tony Dunne's design team used packages of materials with postcards and disposable cameras to engage with groups of older adults across three geographically distant retirement communities. In doing so, the design team aimed to reverse the promise of improving social life through technology, a promise "that tries so hard to be rational" (Gaver et al. 2001, 7). Instead, they sought a means of supporting the idiosyncrasies of everyday life, from chatting to creative expression. The outgrowths of these encounters are meant to be "generative of design potentials and possibilities, ones loosely directed toward more preferred states" (Pierce and DiSalvo 2017), rather than discrete solutions for all. In the Period Packets, we asked how respondents formed their own sense of the menstrual process, including prompts such as: "How do you know your period is coming? What does it mean to know your period? How does your period fit into your life? Record a week of your period, focusing on what's interesting to you." With the physical packets, we included markers, paint, and other craft materials, as well as a link to an online folder meant to receive digital text, image, video, and audio files, offering opportunity for expression through any means respondents saw fit—from sketches to sound files.

We distributed the packets online, via the project's own website, Instagram account,[2] and through email, sending the call for participation to our own personal and professional networks and snowballing from there. We also circulated the packets through the physical mail—initially, sending sets of packets to queer and feminist bookstores across the country and later fulfilling requests for physical packets received through the project's website. Ten people completed and returned the packet, either through an online form or physical mail, offering long-form responses to the open-ended questions and describing their own mechanisms for menstrual sensemaking both visually and textually (Fig. 54.5).

Through their reflections, respondents troubled notions of the body that configure it as wholly knowable and controllable, or presuppose that women's bodies, in particular, are primarily for reproduction. For instance, some described existing methods they used to collect information about their menstrual experience that did not match the data types and forms of capture that are featured in tracking apps. One respondent noted regularly, "*[recording] the amount of fluid captured in my diva cup or pad or underwear throughout the day, color, viscosity, presence of any clots or cervical fluid*" in order to compare relative volume across their cycle. Others suggested they had a general sense for where they were in their cycle based on corporeal experience and avoided formal data collection. One respondent reported being able to "*[. . .] tell based on discharge and other bodily changes. I don't take my temp each morning cuz [Sic] i am the worst at consistency but i can tell other ways, like body aches, getting emotional, etc.*" These

Fig. 54.5 Circulated online and through the mail, the Period Packet invited participants to reimagine the period tracking app by illustrating their own menstrual sensemaking practices, through both textual description and craft techniques

responses suggest existing practices of sensemaking that might evade current apps, but could complement and extend them by allowing for more flexibility and variability across individuals' experiences.

In what follows, we highlight two vignettes to illustrate expanded notions of monitoring menses introduced and illustrated across the pages of the packets we received.

Modes of Reflection

One respondent, Jenna,[3] reported charting and journaling on her menstrual cycle experience every day (see Fig. 54.6) and described, *"organizing my life around my cycle."* She further noted that this recognition of the body in daily practice fostered existential exploration and reflection:

> *I think knowing my period involves understanding why I get it, and situating it in the context of other female-bodied people/creatures and the elements. I used to think that ovulation happened during a period and that the whole deal was just a pain in the butt. Now I understand the process a lot better and feel glad when I shed my uterine lining because it means my body is working as it should. Like the cycles of birth, death, and rebirth that we see in nature (phases of the moon, tide, seasons) with each cycle I am letting go of old stuff my body doesn't need and beginning anew.*

Here, Jenna gestures toward forms of sensemaking practice that currently fall outside of what designers and technologists developing apps recognize. Rather than tracking to find optimal days for conception, for instance, she hoped to foster a relational perspective through recordkeeping—a practice as much about making sense of her place in the world as it was about predicting a next period.

Fig. 54.6 Respondent Jenna charts and journals about her menstrual experience every day. Within the pages of the Period Packet, she describes her motivations for pursuing this practice and offers an example of one such entry

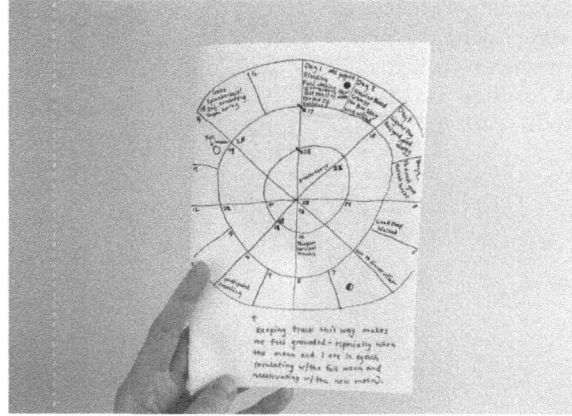

Affirmative Design

For others, prediction played a crucial role; not necessarily for the ways in which Levchin and others might imagine (such as ovulation, for those trying to conceive), but with no less consequence. One respondent, Robert, who identified as a transgender man, detailed through text and imagery his traumatic experiences with menstruation (see Fig. 54.7), which he described as causing "*the most most dysphoria*" for him. Tracking, to Robert, was a means to prepare, both materially and emotionally for what was to come with his period:

> *[My period] used to be a thing I had to emotionally prepare for (either by closing off entirely and pointedly not thinking about it much more than "okay time to change tampons, don't look at it, etc.") So knowing when it was coming via when I realized the aching correlation, using a tracker app, etc. made it a lot easier on me since I could reliably know when and not have to look at a calendar in dread.*

Yet, feelings of dread had not totally subsided, even with transition, Robert noted. Instead, he described having occasional periods since beginning testosterone shots, which could be more frequent if for some reason he was not able to take the prescription:

> *If I forget/can't get it refilled on time it becomes more and more of an anxiety trigger until I can. If I <u>do</u> just completely forget and then get a period I'll immediately (if I can) do my shots to end it ASAP [. . .] Oddly, starting T has made the occasional accidental period easier to deal with emotionally compared to before. Not sure why but I'm not complaining! Now, it's just a matter of "oh opps," go do a tampon, change underwear if necessary, then go back to what I was doing. I know that occasionally my depression would be worse the week/a few days before when I'd look back in retrospect, so sometimes [that] would explain it.*

Fig. 54.7 Through illustration, Robert recalls how tracking was a matter of materially and emotionally preparing for what was to come with menstruation—exacerbated feelings of dysmorphia

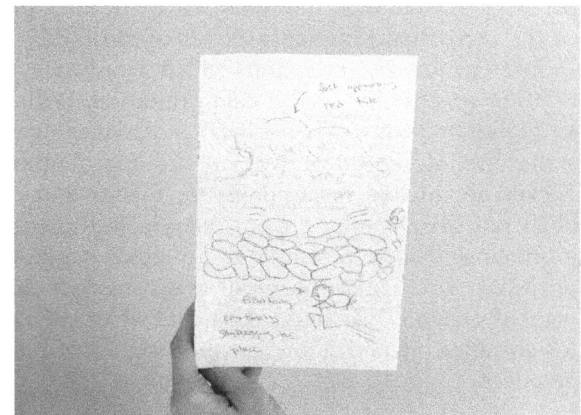

In the above statement, Robert moved through his changing relationship to menstruation, one that had triggered intense experiences throughout his life, but was becoming easier to handle as it became more sporadic. This might seem paradoxical, at first: how would an "accidental period" be easier to deal with? That it was occasional and accidental made all the difference to Robert. It was not a constant trauma anymore. If anything, it was a reminder to take care of himself and to take time to get the testosterone he needed.

Through the circulation of the Period Packets we aimed to acknowledge a sense that no two menstruators are the same, while also recognizing all are experts of their own experience. Queer, unpartnered, infertile women, those uninterested in procreation, and transgender folks are all too often left from consideration in the design of menstrual tracking apps. Yet, throughout their packets, respondents offered new types of data and forms of interaction such as recording endometriosis pain and charting testosterone levels alongside period frequency. In doing so, respondents introduced new layers of depth and understanding to the period app protocol. This depth, we argue, should not be limited by technological prescription, but rather supported through further design experimentation.

CONCLUSION

Across the two cases detailed in this chapter, we call for more and different sorts of designs in the space of menstrual tracking—ones that might be more reflective of the variety of interests and needs of menstruators (concerns only partially embodied in the accounts we have shared here). With the first case, we aimed to identify points where existing menstrual tracking apps do not match people's needs, preferences, and identities. Through analysis of app reviews, survey responses, and interviews, we uncovered a range of characteristics that cause people discomfort or feelings of exclusion, including assumptions about tracking goals (fertility), aesthetic preferences (pink and flowery imagery), gender identity (female), and sexual partner (one male partner). Rather than modeling an application on a particular understanding of the menstrual experience or set of attributes, we argue future technology should opt for design defaults which avoid such assumptions by, for instance, offering adaptability in the data presented. With the second case, we sought to highlight forms of menstrual recordkeeping and sensemaking already in active use, alongside or beyond the app. Through the form of the Period Packet, we invited respondents to engage and express these sometimes-implicit practices—ranging from recognizing growing aches and pains to daily journaling. The responses they returned suggest that though their existing practices of sensemaking might evade current apps, technologies may have a place if carefully managed. Using flexible and dynamic sorts of accounting, for instance, would allow technology to complement and extend existing practices.

Though it may seem that there is a long way to go before mainstream technological development supports varying approaches to monitoring, our two case studies suggest there may be productive overlaps, ways in which menstrual tracking technology can better align with people's needs and preferences. By critiquing existing technologies, our aim is not to point at a problem and walk away, but rather to offer generative openings on the sorts of designs that could exist alongside the existing terrain of apps.

NOTES

1. In this chapter, Fox and Epstein discuss a set of studies done in collaboration with fellow researchers at University of Washington and Northwestern University. The first case describes a collaboration between Daniel Epstein, Nicole Lee, Jennifer Kang, Elena Agapie, Jessica Schroeder, Laura Pina, James Fogarty, Julie Kientz, Sean Munson. The second case draws on a collaboration between Sarah Fox and Amanda Menking, Jordan Eschler, and Uba Backonja. More details about each individual study can be found in the articles referenced below.
   Daniel A. Epstein, Nicole B. Lee, Jennifer H. Kang, Elena Agapie, Jessica Schroeder, Laura R. Pina, James Fogarty, Julie A. Kientz, and Sean Munson. 2017. Examining Menstrual Tracking to Inform the Design of Personal Informatics Tools. In Proceedings of the 2017 CHI Conference on Human Factors in Computing Systems (CHI '17). ACM, New York, NY, USA, 6876–6888.
   Jordan Eschler, Amanda Menking, Sarah E. Fox, and Uba Backonja. 2019. "Defining Menstrual Literacy with the Aim of Evaluating Mobile Menstrual Tracking Applications." *CIN: Computers, Informatics, Nursing* 37 (12): 638–646.
   Sarah E. Fox, Amanda Menking, Jordan Eschler, and Uba Backonja. 2020. "Multiples over Models: Interrogating the Past and Collectively Reimagining the Future of Menstrual Sensemaking." *ACM Transactions on Computer-Human Interaction (TOCHI)* 27 (5).
2. The project website with the digital version of the Period Packet can be accessed at http://periodpacket.com/. The Instagram account can be found under the handle "PeriodPacket": https://www.instagram.com/periodpacket/.
3. To protect respondents' privacy, we use pseudonyms throughout the accounts in this chapter.

REFERENCES

ABISHKKING LIMITED. 2017. *Period Tracker – Period Calendar Ovulation Tracker* (V.1.4.84). [Mobile App].

Almeida, Teresa, Rob Comber, Gavin Wood, Dean Saraf, and Madeline Balaam. 2016. "On Looking at the Vagina through Labella." In *Proceedings of the 2016 CHI Conference on Human Factors in Computing Systems*, 1810–1821. CHI '16. New York, NY, USA: ACM. https://doi.org/10.1145/2858036.2858119.

Ashley, Benedict. 2006. *Health Care Ethics: A Catholic Theological Analysis*. 5th ed. Washington, DC: Georgetown University Press.

Bellabeat, Inc. 2017. *Bellabeat Period Diary* (V.3.4). [Mobile App].

Beutler, Brian. 2017. "Leaked Memo Reveals White House Wish List." *Crooked Media* (blog), October 19, 2017. https://crooked.com/article/leaked-memo-reveals-white-house-wish-list/.

BioWink GmbH. 2017. *Clue Period & Cycle Tracker* (V.3.4). [Mobile App].

boyd, danah, and Kate Crawford. 2012. "Critical Questions for Big Data." *Information, Communication & Society* 15 (5): 662–79. https://doi.org/10.1080/1369118X.2012.678878.

De Choudhury, Munmun, Sanket Sharma, and Emre Kiciman. 2016. "Characterizing Dietary Choices, Nutrition, and Language in Food Deserts via Social Media." In *Proceedings of the 19th ACM Conference on Computer-Supported Cooperative Work & Social Computing*, 1157–1170. CSCW '16. New York, NY, USA: ACM. https://doi.org/10.1145/2818048.2819956.

D'Ignazio, Catherine, Alexis Hope, Becky Michelson, Robyn Churchill, and Ethan Zuckerman. 2016. "A Feminist HCI Approach to Designing Postpartum Technologies: 'When I First Saw a Breast Pump I Was Wondering if It Was a Joke.'" In *Proceedings of the 2016 CHI Conference on Human Factors in Computing Systems*, 2612–2622. CHI '16. New York, NY, USA: ACM. https://doi.org/10.1145/2858036.2858460.

Dreaper, Jane. 2016. "Warning over Period Tracker Apps." *BBC News*, August 11, 2016, sec. Health. http://www.bbc.com/news/health-37013217.

England, Rachel. 2017. "Natural Cycles Says Contraceptive App Is More Effective Than the Pill." *Engadget* (blog), September 13, 2017. https://www.engadget.com/2017/09/13/natural-cycles-contraceptive-app-more-effective-than-pill/.

Epstein, Daniel A., Nicole B. Lee, Jennifer H. Kang, Elena Agapie, Jessica Schroeder, Laura R. Pina, James Fogarty, Julie A. Kientz, and Sean Munson. 2017. "Examining Menstrual Tracking to Inform the Design of Personal Informatics Tools." In *Proceedings of the 2017 CHI Conference on Human Factors in Computing Systems*, 6876–6888. CHI '17. New York, NY, USA: ACM. https://doi.org/10.1145/3025453.3025635.

Epstein, Daniel A., An Ping, James Fogarty, and Sean A. Munson. 2015. "A Lived Informatics Model of Personal Informatics." In *Proceedings of the 2015 ACM International Joint Conference on Pervasive and Ubiquitous Computing*, 731–742. UbiComp '15. New York, NY, USA: ACM. https://doi.org/10.1145/2750858.2804250.

Eschler, Jordan, Amanda Menking, Sarah E. Fox, and Uba Backonja. 2019. "Defining Menstrual Literacy with the Aim of Evaluating Mobile Menstrual Tracking Applications." *CIN: Computers, Informatics, Nursing* 37 (12): 638–646.

Fox, Sarah E., Amanda Menking, Jordan Eschler, and Uba Backonja. 2020. "Multiples over Models: Interrogating the Past and Collectively Reimagining the Future of Menstrual Sensemaking." *ACM Transactions on Computer-Human Interaction (TOCHI)* 27 (5).

Fraser, Ian S., Hilary O. D. Critchley, Malcolm G. Munro, Michael Broder, and Writing Group for this Menstrual Agreement Process. 2007. "A Process Designed to Lead to International Agreement on Terminologies and Definitions Used to Describe Abnormalities of Menstrual Bleeding." *Fertility and Sterility* 87 (3): 466–76. https://doi.org/10.1016/j.fertnstert.2007.01.023.

Gaver, William, Ben Hooker, Anthony Dunne, and Paul Farrington. 2001. *The Presence Project*. London: Art Books Intl. Ltd.

Glow. 2017. *Glow Period, Fertility Tracker* (V.6.3.0). [Mobile App].

Goode, Lauren. 2013. *Max Levchin's Glow Fertility App: The Full D11 Session*. All Things Digital.

GP Apps. 2017. *Period Tracker by GP Apps* (V.9.5.1). [Mobile App].

Life Fertility Tracker IVS. 2017. *Life – Period Tracker Calendar* (V.4.0.2). [Mobile App].

Lomas, Natasha. 2015. "Period Tracker App Clue Gets $7M to Build a Platform for Female Health." *TechCrunch* (blog), October 9, 2015. http://social.techcrunch.com/2015/10/09/clue-series-a/.

Magee, Christine. 2015. "More Than Just a Period Tracker, Kindara Raises $5.3M to Understand Women's Health." *TechCrunch* (blog), August 19, 2015. http://social.techcrunch.com/2015/08/19/more-than-just-a-period-tracker-kindara-raises-5-3m-to-understand-womens-health/.

Pardes, Arielle. 2018. "In Contraceptive Tech, the App's Guess Is as Good as Yours." *WIRED* (blog), January 19, 2018. https://www.wired.com/story/natural-cycles-contraceptive-apps/.

Perez, Sarah. 2015. "Apple Stops Ignoring Women's Health with IOS 9 HealthKit Update, Now Featuring Period Tracking." *TechCrunch* (blog), June 9, 2015. http://social.techcrunch.com/2015/06/09/apple-stops-ignoring-womens-health-with-ios-9-healthkit-update-now-featuring-period-tracking/.

Peyton, Tamara, Erika Poole, Madhu Reddy, Jennifer Kraschnewski, and Cynthia Chuang. 2014. "'Every Pregnancy Is Different': Designing MHealth for the Pregnancy Ecology." In *Proceedings of the 2014 Conference on Designing Interactive Systems*, 577–586. DIS '14. New York, NY, USA: ACM. https://doi.org/10.1145/2598510.2598572.

Pierce, James, and Carl DiSalvo. 2017. "Network Anxieties Design Packets." In *Proceedings Research through Design Conference*, 277–91. Edinburgh, UK.

Ribes, David, and Steven J. Jackson. 2013. "Data Bite Man: The Work of Sustaining a Long-Term Study." In *"Raw Data" Is an Oxymoron*, edited by Lisa Gitelman, 147–66. Cambridge, MA: MIT Press.

Rosner, Daniela K. 2018. *Critical Fabulations: Reworking the Methods and Margins of Design*. Cambridge, MA: The MIT Press.

Star, Susan Leigh. 1999. "The Ethnography of Infrastructure." *American Behavioral Scientist* 43 (3): 377–91. https://doi.org/10.1177/00027649921955326.

StayWell. 2017. *My Cycles Period and Ovulation* (V.5.9.2). [Mobile App].

Zimmerman, John, Jodi Forlizzi, and Shelley Evenson. 2007. "Research Through Design as a Method for Interaction Design Research in HCI." In *Proceedings of the SIGCHI Conference on Human Factors in Computing Systems*, 493–502. CHI '07. New York, NY, USA: ACM. https://doi.org/10.1145/1240624.1240704.

"Life is Much More Difficult to Manage During Periods": Autistic Experiences of Menstruation

Robyn Steward, Laura Crane, Eilish Mairi Roy, Anna Remington, and Elizabeth Pellicano

INTRODUCTION[1]

The onset of menses (menarche) is an important—and often challenging (Burrows and Johnson 2005)—transition in any woman's life. This is especially the case for developmentally-disabled women, who may experience menarche and menstruation differently—and more negatively—compared to non-disabled women (Ditchfield and Burns 2004; Rodgers and Lipscombe 2005). These include frequent reports of dysmenorrhoea (painful periods), menorrhagia (heavy periods), menstrual hygiene issues and cyclical mood and behavioral changes, akin to premenstrual syndrome (PMS) [and its more severe form, premenstrual dysphoric disorder; American Psychiatric Association (APA) 2013] (Burke et al. 2010; Jeffery et al. 2013; Mason and Cunningham 2008; Rodgers et al. 2006).

Remarkably little is known, however, about the menstrual experiences of women on the autism spectrum. This paucity of research may be unsurprising given the male predominance in autism (see Loomes et al. 2017) but the few existing studies in this area give cause for concern. Although there are apparently no significant differences in the age of menarche between autistic girls and girls with other developmental conditions (Burke et al. 2010), there are

C. Bobel et al. (eds.), *The Palgrave Handbook of Critical Menstruation Studies*, https://doi.org/10.1007/978-981-15-0614-7_55

several reports (including case studies) of marked changes linked to menarche and menstruation in autistic girls and women (the majority with additional intellectual disabilities), including cyclical self-injurious behaviors (Lee 2004), mood symptoms and emotional dysregulation (Burke et al. 2010; Hamilton et al. 2011; Lee 2004; Obaydi and Puri 2008), and an amplification of autistic symptoms (sensory issues and repetitive behaviors; Hamilton et al. 2011; Lee 2004).

One observational study with women with additional intellectual disabilities living in residential homes and care units in England reported an alarming result: of the 26 autistic women sampled, 92% (n = 24) fulfilled DSM-IV (APA 2000) criteria for late luteal phase dysphoric disorder (a severe form of PMS), compared with only four (11%) of the 36 non-autistic women (Obaydi and Puri 2008). The lack of research on, and awareness of, these purportedly higher rates of premenstrual symptoms in autistic women means both that their potential cause(s) remain unknown and that, worryingly, autistic women are unlikely to receive the gynaecological care they may need. Moreover, to our knowledge, no existing study has directly examined whether these apparent premenstrual symptoms are a problem for the person concerned, *from their own perspective*.

This preliminary study—co-produced by an autistic woman (RS) and non-autistic female researchers (LC, ER, AR, LP)—sought to redress the imbalance in research, by asking autistic and non-autistic people about their experiences of menstruation through an online survey. Specifically, we sought to understand the kind of information they would have liked to have known at menarche, and whether, for autistic participants specifically, they felt being autistic affected, or was affected by, menstruation and its manifestations.

METHOD

The survey began with a series of background items, including participants' age, gender and connection with autism. These items were followed by questions on their experiences of menstruation and of growing up more broadly. The three most relevant to the current report are analysed below,[2] including (1) "How did you first learn about periods?" (closed question), (2) "What information do you think would have been important to have before you started your period?" (open question), and (3) "Do you think that you have experienced difficulties with periods that are related to autism?" (open question).

Participants were recruited via a convenience sampling method (using website posts and social media). In all, 459 people completed the survey. People were allowed to identify more than one connection—as being autistic themselves, a parent, a professional or a sibling—to reflect potential multiple roles, but were also asked to identify the perspective from which they responded to the survey (e.g., as autistic). For this report, we focus on those participants who identified themselves to be 'formally diagnosed as autistic' (n = 144)[3] or 'non-autistic' (n = 132). Of these 276 participants, 39 reported never having experienced periods (n = 11) or did not answer all relevant questions (n = 28) and were excluded from the dataset. Subsequent analyses were therefore based on complete responses from 237 participants (autistic: n = 123; non-autistic: n = 114), ranging in age from 16 to over 60 years (see Table 55.1 for participant details).

The survey took approximately 10 min to complete and was hosted by SurveyMonkey between February and August 2016. All participants gave informed consent to take part prior to participation. Ethical approval for this study was granted by the Research Ethics Committee at UCL Institute of Education, University College London (REC 874).

DATA ANALYSIS

Descriptive results for the initial, closed questions are presented first. Next, we analysed the qualitative responses across the two open-ended questions using thematic analysis (Braun and Clarke 2006). We used an inductive ("bottom up") approach, providing descriptive overviews of the key features of the semantic content of the data within an essentialist framework. We independently familiarised ourselves with the data, and met several times to agree the initial codes, review the results, resolve discrepancies and decide on final themes and subthemes.

Table 55.1 Background information for respondents to the online survey for each (autistic, non-autistic) group

	Autistic (n = 123)	Non-autistic (n = 114)
Age range (in years)		
16–18	9	6
19–25	33	26
26–31	33	20
32–45	36	44
46–59	12	13
60+	0	5
Gender[a]		
Female (including transgender women)	83	96
Male (including transgender men)	7	0
Non-binary	26	15
Other	6	3
Prefer not to say	1	0
Also identified as a:		
Parent	14	18
Professional	14	16
Sibling	13	5

[a]Gender categories were identified in direct consultation with the autistic community
Credit: Robyn Steward, Laura Crane, Eilish Mairi Roy, Anna Remington, and Elizabeth Pellicano

Table 55.2 Participants' responses to the question, "How did you first learn about periods?"

	Autistic (n = 123)	Non-autistic (n = 114)
The internet	3	3
Friends	13	26
Parents	70	66
School	55	46
Doctor or other medical professional	1	5
I don't know	6	6
Other	19	25

The total numbers exceed the number of participants in each group because participants could endorse more than one category. Most of the responses in the 'other' category related to printed material (books, magazines or leaflets), "my sister" or "when it happened"
Credit: Robyn Steward, Laura Crane, Eilish Mairi Roy, Anna Remington, and Elizabeth Pellicano

RESULTS

Quantitative Results

The majority of participants first obtained information about periods through either their parents or through school (see Table 55.2), with some finding out about menstruation from friends and printed material, and, for a handful of people, from medical professionals or the internet. This pattern was similar across groups, with the exception of fewer autistic women discovering information from friends than non-autistic women.

Qualitative Results

Figure 55.1 displays themes and subthemes identified from the analysis. For the sake of brevity, we report the themes below collapsed across both groups (subthemes are italicised). When attributing quotes, 'A' refers to autistic respondents, 'NA' to non-autistic respondents. Participant numbers are included to illustrate the breadth of responses.

It's a Normal and Natural Part of Life
Respondents spoke of the taboo associated with having periods. They emphasised the need to reassure young people that periods are "not dirty" (A23) or "something to fear or be ashamed of" (NA36) or "embarrassed by" (NA65); rather, they are a "healthy" (NA96), "natural" (A6) and "normal part of growing up" (NA68). They also felt that it was important for everyone—even those who have not experienced periods—to understand this, so that they "understand what their mum/sister/etc. is going through" (A71). Both autistic and non-autistic respondents further reported that young people need to understand "how their body works and why" (NA94), including biological

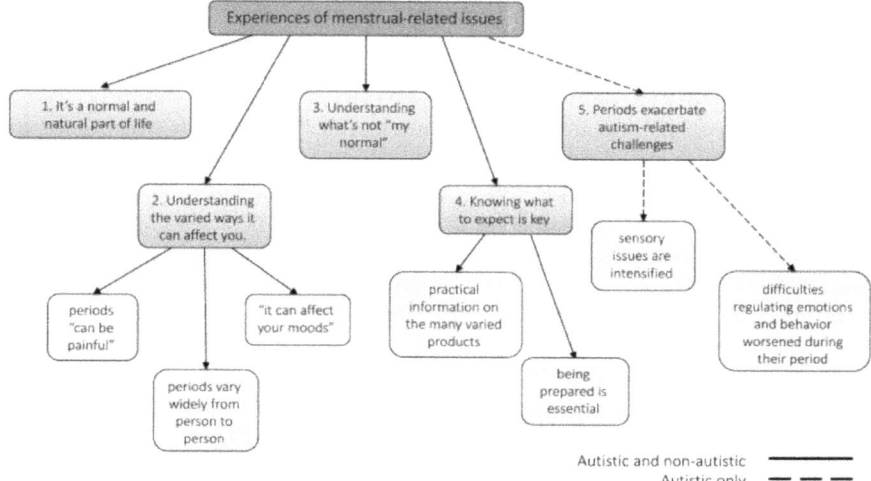

Fig. 55.1 Respondents' experiences of menstrual-related issues: themes and sub-themes (Credit: Robyn Steward, Laura Crane, Eilish Mairi Roy, Anna Remington, and Elizabeth Pellicano)

information about "how the menstrual cycle and organs work, and the anatomy of the vagina/vulva/ uterus etc." (NA59), and "why [periods] happen" (A57). Autistic respondents in particular wanted "more detail about this" (A73).

Understanding the Varied Ways it Can Affect You
Respondents highlighted the need for young people to know "what happens before their first period" (A8) so that it "doesn't come as a terrifying surprise" (NA16). In particular, they should understand that the cramps that come with *periods "can be painful"* (NA48), and that there can be other physical symptoms, too, including bloating (water retention), "tender breasts" (A22), "skin changes" (A110) or acne, changes to eating habits, constipation, and "the way it affects your weight" (NA60). Several autistic respondents noted specifically that "it was helpful to know beforehand that I wasn't dying" (A7) from the bleeding and/or pain.

Respondents also cited the many ways that *"it can affect your moods"* (NA60)—"before and possibly after your period, not just during" (A29)— and that these changes "are normal" (A66). Some respondents noted the importance of providing an explanation of these "mood swings" (A78) and "the reason one is acting in a particular way" (NA52). Indeed, one autistic participant described, "about once a month, I get anxious and melancholic for no reason. This mood lasts for a day or two, at which point my period arrives, and my 'normal' mood resumes. Understanding that about myself makes that melancholy a lot more bear-able/manageable and helps my partner understand my behavior/mood" (A52).

Both autistic and non-autistic participants emphasised the need to improve knowledge on "how long it lasts, how often it happens, how heavy/light it will be" (A84) but, critically, that *periods vary widely from person to person.* Indeed, they felt that knowing that "everyone gets them differently" (A85) would have been helpful for them to understand "that what's normal for me is not what's normal for everyone else" (NA41). One autistic respondent summed it up: "Some people get them heavy, some people get them light, not everybody gets them regularly, not everybody gets them exactly 28 days apart, some people get pains, some people get moody, sometimes medical conditions make them irregular, some people get them for more than a week at a time, some people get them for only a couple of days" (A85).

Understanding What's Not "My Normal"
Respondents felt that it was important to "pay attention to what normal is for me, and to know how to react if my normal changes" (NA41). They spoke of the need for young people to "know how to track their cycles effectively so that they can be aware of how their body is … and be prepared for any irregularities that might occur" (A22), including the amount of pain, the amount of blood and the frequency between periods. They stressed the need to know how to distinguish between, for example, "normal period cramps and unusually painful periods" (NA60) to identify "how much is acceptable before you need to worry" (NA71) and "what could be a warning sign of a medical problem" (A65).

Knowing What to Expect is Key
Respondents suggested that young people needed *practical information on the many varied products* to use, including a "chance to familiarise themselves with a variety of products before they start" (A50), information on "how to choose menstrual hygiene products" (NA19), where to get them, and "disposal methods" (A28). They stressed that introductory information should be accessible, "breaking down initial concepts" (A15) with a "step-by-step 'this is how you deal' instructions and tips" (A42), particularly for young autistic people. They also wanted young people to know about the different strategies available to relieve pain (including medication), and how to deal with mood swings and other symptoms or issues like "acne, cramps, bloating, nausea" (NA35).

Respondents also noted that *being prepared is essential.* This preparation could include a "script for what to say to a nurse or parent when you start your first one or are unprepared" (A50), knowing "what to do at school, if you are out in a public place, if you are on holiday" (NA113), and having an "action plan for mishaps (stains, stains in public places, forgot supplies, etc.)" (NA50). They also wanted to reassure them that "it's ok to tell your teachers if you need more time for the bathroom" (NA38) and that it was important to identify "who to ask questions about periods" (NA13) and who to turn to for support, especially "if you might feel there is a problem" (A104).

Periods Exacerbate Autism-Related Challenges
Although some participants were unsure whether their menstrual experiences were related to being autistic ("I have only ever been an autistic person having a period!"; A80), many autistic participants felt that their "symptoms worsen dramatically" (A47), often making "life much more difficult to manage during periods" (A93). One participant stated: "autism does play a role. It can become much more overwhelming and harder to maintain control of the things that already take a lot of effort for us to keep on top of, during a period" (A13). Participants highlighted sensory and self-regulation difficulties in particular.

Respondents felt that *sensory issues are intensified* during menstruation, as described above. For the most part, these related to pre-existing hypersensitivities becoming "extra sensitive during my period" (A107), such that "everything is magnified when it's that time of the month" (A78). Participants described "being sensitive to the smell of the blood" (A17), "finding my skin and body more sensitive in general" (A43), being "more sensitive and reactive to noise, touch and visual stimuli" (A99), and "struggling the most with the physical pain from cramps" (A22). Pain could be particularly difficult to bear: "When it's at its worst, I find myself unable to focus well because all I can focus on is the ache and the sense of where in my body that pain lives" (A44). Participants also reported how these hypersensitivities and the "sensory overload [that] happens far more frequently just before and during a period" (A103) further exacerbated other autistic experiences, including "dealing with unrelated problems caused by my autism (harder to filter noise etc.)" (A28) and being "more prone to self-injurious behavior" (A94).

Participants also highlighted how *difficulties regulating emotions and behavior worsened during their period*. Some noted that "executive dysfunction gets worse when I have cramps" (A98), which "made dealing with periods difficult—keeping clean and changing pads" (A74). They also highlighted difficulties "recognising and managing my emotions, which is amplified just before and during my periods" (A45) and even "an inability to describe my emotions while experiencing PMS" (A41). One participant explained: "understanding my own emotions has always been difficult for me so any mood swings made life even more difficult" (A17). They also reported that exaggerated difficulties often led to "heightened anxiety" (A99) and, most commonly, meltdowns: "I have more meltdowns, and worse meltdowns, just before my period" (A30). One participant also noted that their epilepsy was affected, with increased seizures during menstruation. These cyclical symptoms were often so severe that participants sought (usually hormonal) medication to manage them.

Discussion

This preliminary study directly elicited, for the first time, autistic people's views and experiences on menarche and menstruation. Notably, autistic and non-autistic respondents cited many similar issues, including needing to understand what was "normal" in terms of menstrual cycle length, amount and duration of flow, the often-dramatic effects that menstruation can have on mood, behavior and bodily symptoms and, importantly, what was normal for them. The importance of educating young people and their parents in what to expect at menarche has been emphasised previously (American Academy of Pediatrics 2006) but rarely so for young autistic people, who may be at increased risk for serious premenstrual-related disorders (Obaydi and Puri 2008). The current study suggests that, consistent with the albeit-limited research (Burke et al. 2010; Hamilton et al. 2011; Lee 2004; Obaydi and Puri 2008), autistic people's menstrual experiences are in some ways distinct from those of non-autistic people, placing extra strain on what can be already-challenging lives.

Indeed, for the autistic participants sampled here, menstruation was seen as a particularly difficult and distressing event (akin to women with other developmental disabilities; Ditchfield and Burns 2004), during which their preexisting challenges—especially regarding sensory hyper-sensitivies and difficulties with regulating emotions and behavior—become exacerbated before, during, and after menses. The amplification of pre-existing autistic features has been reported previously (e.g., Hamilton et al. 2011; Lee 2004) but no study has sought to understand the impact of menstruation on the individuals themselves. Those sampled here described overwhelmingly negative experiences, especially exaggerated sensory issues and intensified executive and emotion-regulation problems, which had often-serious consequences, including "shutdown", withdrawal, and heightened anxiety—and therefore reduced participation in work, social and community life. Understanding the prevalence of premenstrual-related symptoms in individuals across the autism spectrum, the causes of such symptoms (which may be related to higher levels of hormonal fluctuations; Obaydi and Puri 2008), and their associated impact (including the potential treatment side-effects of cyclical changes) is critical for further research.

One important step to mitigate potential problems following menarche is to increase knowledge of menstrual-related issues in young people and their parents (see American Academy of Pediatrics 2006), particularly in the form of accessible, step-by-step guides and strategies for how to deal with pain and mood changes in particular (see Steward 2019). This is especially important for young autistic people, who may be less likely to gather information about sexual topics from informal social settings (with peers—as evidenced in Table 55.2), whose parents might be reticent to discuss puberty and sexual health and may begin these conversations later (Pownall et al. 2012; Sedgewick et al. 2018; Cridland et al. 2014, for discussion), and whose clinicians may fail to notice (or prioritize) any link between menstruation and mood- or behavior-related features (Kaminer et al. 1988). The current absence of this knowledge rather worryingly means that the particularly severe symptoms reported by some girls and women may be going unrecognised by clinicians and therefore not treated appropriately.

Given the nature of the current methodology, it was possible neither to confirm, for self-declared autistic respondents, where they lie on the autism spectrum nor to ensure that we did not oversample those with particularly problematic menstrual experiences. Notwithstanding, these preliminary findings serve to stress the importance of these issues for autistic people and call for greater attention on women's health issues across the lifespan, including systematic investigations on the causes, correlates, and consequences of menstruation (particularly with regard to mental health) for autistic young people and adults—from their own perspectives and the perspectives of supportive others (parents, teachers).[4]

Author Contributions RS and AR designed the study; RS implemented the study; RS, ER, LC and EP analysed and interpreted the data; EP drafted the manuscript; all authors contributed to, and approved, the final manuscript.

Funding This study was funded by grants from Pears Foundation (RS) and a Philip Leverhulme Prize 2015 awarded to EP from the Leverhulme Trust.

NOTES

1. "Life is Much More Difficult to Manage During Periods": Autistic Experiences of Menstruation by Robyn Steward, Laura Crane, Eilish Mairi Roy, Anna Remington, Elizabeth was first published in 2018 in *Journal of Autism and Developmental Disorders*, 48 (12): 4287–4292. Reprinted with permission. [OA CC-BY 4.0].
2. Two further questions were asked, relating to growing up more generally and about strategies for helping school-age children to access the toilet during their period. These questions were not analyzed in this current study, but were designed for use by the first author, to inform the content of a forthcoming book (Steward 2019).
3. Note that some individuals reported that they either were going through the process of a clinical diagnosis (n = 32) or were self-diagnosed as autistic (n = 88). These participants were excluded from all analyses.
4. We are grateful to all of our participants for taking part in this study.

REFERENCES

American Academy of Pediatrics. (2006). Menstruation in girls and adolescents: Using the menstrual cycle as a vital sign. *Pediatrics, 118*, 2245–2250.
American Psychiatric Association. (2000). *Diagnostic and statistical manual of mental disorders* (revised 4th ed.). Washington, DC: American Psychiatric Association.

American Psychiatric Association. (2013). *Diagnostic and statistical manual of mental disorders* (revised 5th ed.). Washington, DC: American Psychiatric Association.

Braun, V., & Clarke, V. (2006). Using thematic analysis in psychology. *Qualitative Research in Psychology, 3*, 77–101.

Burke, L. M., Kalpakjian, C. Z., Smith, Y. R., & Quint, E. H. (2010). Gynecologic issues with Down Syndrome, autism, and cerebral palsy. *Journal of Pediatric and Adolescent Gynecology, 23*, 11–15.

Burrows, A., & Johnson, S. (2005). Girls' experiences of menarche and menstruation. *Journal of Reproductive and Infant Psychology, 23*, 235–249.

Cridland, E. K., Jones, S. C., Caputi, P., & Magee, C. A. (2014). Being a girl in a boys' world: Investigating the experiences of girls with autism spectrum disorders during adolescence. *Journal of Autism and Developmental Disorders, 44*, 1261–1274.

Ditchfield, H., & Burns, J. (2004). Understanding our bodies, understanding ourselves: The menstrual cycle, mental health and woman with learning disabilities. *Tizard Learning Disability Review, 9*, 24–32.

Hamilton, A., Marshal, M. P., & Murray, P. J. (2011). Autism spectrum disorders and menstruation. *Journal of Adoelscent Health, 49*, 443–445.

Jeffery, E., Kayani, S., & Garden, A. (2013). Management of menstrual problems in adolescents with learning and physical disabilities. *The Obstetrician and Gynaecologist, 15*, 106–112.

Kaminer, Y., Feinstein, C., Barrett, R. P., Tylenda, B., & Hole, W. (1988). Menstrually related mood disorder in developmentally disabled adolescents: Review and current status. *Child Psychiatry and Human Development, 18*, 239–249.

Lee, D. O. (2004). Menstrually related self-injurious behavior in adolescents with autism. *Journal of the American Academy of Child and Adolescent Psychiatry, 43*, 1193.

Loomes, R., Hill, L., & Mandy, W. (2017). What is the male-to-female ratio in autism spectrum disorder? A systematic review and meta-analysis. *Journal of the American Academy of Child & Adolescent Psychiatry, 56*, 466–474.

Mason, L., & Cunningham, C. (2008). An exploration of issues around menstruation for women with Down Syndrome and their carers. *Journal of Applied Research in Intellectual Disabilities, 21*, 257–267.

Obaydi, H., & Puri, B. K. (2008). Prevalence of premenstrual syndrome in autism: A prospective observer-rated study. *The Journal of International Medical Research, 36*, 268–272.

Pownall, J. D., Jahoda, A., & Hastings, R. P. (2012). Sexuality and sex education of adolescents with intellectual disability: Mothers' attitudes, experiences and support needs. *Intellectual and Developmental Disabilities, 50*, 140–154.

Rodgers, J., & Lipscombe, J. (2005). The nature and extent of help given to women with intellectual disabilities to manage menstruation. *Journal of Intellectual and Developmental Disability, 30*, 45–52.

Rodgers, J., Lipscombe, J., & Santer, M. (2006). Menstrual problems experienced by women with learning disabilities. *Journal of Applied Research in Intellectual Disabilities, 19*, 364–373.

Sedgewick, F., Hill, V., & Pellicano, E. (2018). Parent perspectives on autistic girls' friendships and their futures. *Autism & Developmental Language Impairments*. https://doi.org/10.1177/2396941518794497.

Steward, R. (2019). *The autism friendly guide to periods*. London: Jessica Kingsley Publishers.

CHAPTER 56

Not a "Real" Period?: Social and Material Constructions of Menstruation

Katie Ann Hasson

Menstruation, as a "natural" bodily process, seems self-evident; this allows it to work as an enduring and central marker of difference in essentialist conceptions of sex and gender.[1] Considered to be a nearly universal female experience,[2] it functions as a literal and symbolic marker of sex and sexuality, fertility, age, and health. Challenging essentialism, feminist scholarship has analyzed the varying social construction of menstruation in different places and times (Bobel 2010; Buckley and Gottlieb 1988; Delaney, Lupton, and Toth 1988; Fingerson 2006; Freidenfelds 2009; Lee and Sasser-Coen 1996; Martin [1987] 2001). This research has focused most often on representations and individual experiences of menstruation, mirroring the analytic split between sex as biological and gender as cultural and social, which often leaves the biological outside the range of feminists' critical examination (Fausto-Sterling 2005; Roberts 2007). The cultural and scientific categories with which we categorize and study menstruation are important to how we understand, experience, and intervene in it. Drawing on analyses of the material aspects of hormones and the menstrual cycle, recent feminist scholarship has brought necessary attention to the ways biology interacts with and is dynamically shaped by social and cultural environments (Fausto-Sterling 2000; Roberts 2007). Overall, feminist scholars have paid more attention to the cultural meanings and norms that govern menstruation than to what menstruation *is* or how it is *defined*. In doing so, what counts as menstruation

© The Author(s) 2020
C. Bobel et al. (eds.), *The Palgrave Handbook of Critical Menstruation Studies*, https://doi.org/10.1007/978-981-15-0614-7_56

is taken for granted, and the complex choreography that can alternately shift or stabilize definitions of menstruation is overlooked.

Despite seeming self-evident, menstruation is not always easily defined. When technologies alter whether or when menstruation occurs, they often raise definitional questions, and menstruation's close links to reproduction raise the stakes. We can see this, for example, when asking whether medications to "bring on blocked menses" act on menstruation or pregnancy. Scholars have questioned whether the bleeding produced by these medications should be considered menstruation, miscarriage, or abortion (Luker 1984; van de Walle and Renne 2001). It took significant, coordinated effort by nineteenth-century physicians and politicians in the United States to shift the understanding of their use; what had been generally accepted as the restoration of healthy menstruation was redefined as illegal abortion (Luker 1984; Reagan 1997). Similarly, the feminist self-help health technique of "menstrual extraction,"[3] developed by U.S. women's health activists in the 1970s, purposely occupied an ambiguous position between regulating menstruation and regulating reproduction (Mamo and Fosket 2009; Murphy 2012). Historical cases like these provide examples of how definitions of menstruation have been negotiated and institutionalized in conjunction with technologies that materially alter bleeding.

Although it is rarely discussed in this way, hormonal birth control is a prime site for the elision of regulating reproduction and regulating menstruation. When oral contraceptives were first approved by the Food and Drug Administration (FDA) and introduced in the United States, they benefited from the ambiguity of the effects exogenous hormones have on bleeding and conception. Enovid, the first hormonal birth control pill, was FDA approved in 1957 not as a contraceptive but rather as a medication for treating menstrual symptoms; it was accompanied by a warning that it had the side effect of stopping ovulation (Watkins 1998). As a treatment for menstrual symptoms, Enovid could pass through FDA scrutiny and enter the market, sidestepping (some) moral debates on birth control while maintaining the possibility of its off-label use as contraception. It was not until 1960 that Enovid's producer requested and received FDA approval for its use as a contraceptive (Junod and Marks 2002; Watkins 1998). The initial developers of the birth control pill knew that the hormones in the pill suppressed both ovulation and menstruation (Marks 2001). To emphasize the "naturalness" of the pill and increase its acceptability (to users, pharmaceutical executives, and religious officials), they designed a regimen with a pill-free break that produced bleeding similar to a menstrual period (Marks 2001; Watkins 1998). Menstrual suppression was thus always a possible use for the pill, but deliberate decisions by developers obscured these effects on bleeding.

The introduction of menstrual suppression birth control pills—oral contraceptives such as Seasonale, Seasonique, and Lybrel, which are taken in an extended regimen to produce fewer periods per year—made the ways that oral contraceptives regulate bleeding highly visible. Menstrual suppression birth control pills are nearly identical to existing hormonal birth control, changing only the regimen of how many pills with active hormones are taken before a "break" from hormones that allows bleeding to occur (Watkins 2012). This change incorporates previously known but little-mentioned ways that all hormonal birth control changes bleeding and widely practiced but "unofficial" or "off-label" uses, recreating oral contraceptives as menstrual suppression in the process. Available since 2003 in the United States, menstrual suppression birth control has attracted a great deal of attention; however, the focus on *suppressing* menstruation as a novel feature of birth control has precluded attention to both the continuities present in *regulating* menstruation and the way menstrual suppression includes a *redefinition* of menstruation itself. To make birth control into "menstrual suppression," monthly bleeding—once relied on to emphasize the naturalness of hormonal birth control—has been redefined as a side effect with no relation to "natural" menstruation. How were these understandings and definitions of menstruation changed following the introduction of menstrual suppression birth control? What are the implications of these changes for gendered embodiment?

The redefinition of menstruation, including how it is understood in relation to birth control hormones, is central to understanding menstrual suppression birth control and its effects on gendered embodiment. I argue that, following the introduction of menstrual suppression, menstruation itself was redefined, proliferating new categories of bleeding across clinical, regulatory, and popular discourse. New definitions and terminology for menstruation explicitly aimed to change women's understandings of menstruation and normalize menstrual suppression. The case of menstrual suppression birth control pills provides an opportunity to study the work of redefining a biological process understood as quintessentially natural and deeply significant for gendered embodiment. Further, this case provides a challenge to gender scholarship to consider both the social and material construction of gendered bodies. What material aspects of menstruation—what interactions of bodies and hormones, what timing and sources of bleeding—are drawn on to define and redefine menstruation and menstrual suppression?

In what follows, I first present a brief discussion of the literature on gendered bodies and technologies, particularly as they relate to menstruation and menstrual technologies. After introducing my data and methods, I turn to an examination of how menstruation was redefined in two key sites: the first is the creation and institutionalization of new scientific categories describing bleeding, and the second is the promotion of new frameworks for understanding bleeding through marketing websites directed at potential users. The first site shows how previously known distinctions related to bleeding were made newly salient as clinical categories. The second site, marketing websites, shows how changed definitions of bleeding were conveyed to women. I conclude by exploring how studying both the social and material construction of menstruation enhances our understanding of gendered embodiment.

GENDERED BODIES, TECHNOLOGY, AND MENSTRUATION

There is a rich body of feminist research on gender, bodies, and technologies. In this section, I bring together feminist literature on the social and material construction of gendered bodies with technology studies literature emphasizing how technologies, inscribed with gendered ideologies, are actively used in the performance of gender. Together, this work explains why promoters of menstrual suppression would seek to change beliefs that monthly menstruation is natural and necessary (Mamo and Fosket 2009). Past research has shown that when gender identities or behaviors inscribed in technologies do not match dominant gender norms, developers can attempt to change those norms (Oudshoorn 2003a). The introduction of menstrual suppression prompted not only a challenge to norms of gendered embodiment, but also a redefinition of menstruation itself.

Feminist studies of bodies and embodiment have provided key tools for this analysis by showing the ways in which gender norms and ideologies shape how individuals display and "do" gender using their bodies (Butler 1990; West and Zimmerman 1987); how sexed and gendered bodies are perceived (Friedman 2013); and the appearance, capacities, and biology of bodies themselves (Fausto-Sterling 2005; Young 1990). These scholars theorize the body as neither purely biological/natural nor purely cultural/ social, but rather as disrupting these boundaries (Fausto-Sterling 2005; Grosz 1994; Haraway 1991). Technologies, especially those (like medications) that change the body invisibly from within, make clear the simultaneously social and material construction of bodies and the dynamic nature of biology (Mamo and Fosket 2009; Roberts 2007). Recent feminist theories of materiality go even further, showing that bodies take shape and shift in relation to both the material conditions that surround them and the conceptual tools that shape our understanding of and interventions into biology (Frost 2014; Pitts-Taylor 2016; Roberts 2007; Wilson 2010).

Because menstruation is so closely associated with biological femaleness, technologies that change menstruation make visible the construction of gendered bodies and challenge the equation of gender and biological sex (Mamo and Fosket 2009). Menstruation has long served as a central aspect of essentialized, biological understandings of sex difference (at least since the nineteenth century) (Laqueur 1992; Martin [1987] 2001), serving as a marker of biological sex and thus a site of the production of gender on the body. Since the early twentieth century, the hormonal model of sex has dominated medical understandings of menstruation (Oudshoorn 1994). Under the hormonal model, monthly cycles define both female biology and sexual difference. "Unexamined assumptions that normal equals periodically regular" (Martin 1999, 104) meant that any disruption to or absence of regular cycles was considered a pathological deviation from the female norm and in need of medical intervention. Menopause has thus been considered doubly problematic, with medical literature equating the lack of menstrual cycles to both maleness and a breakdown of communication among bodily systems (Martin [1987] 2001; Roberts 2007).

One of the first uses of exogenous sex hormones for women was to correct menstrual irregularity (Oudshoorn 1994). Later, when oral contraceptives were first introduced, they were designed to include regular bleeding, a decision that most accounts explain was intended to make the pill more acceptable (to women, the church, and pharmaceutical executives) (Marks 2001; Watkins 1998). The idea that monthly bleeding would make oral contraceptives more acceptable depends on the medical and popular belief that regular menstrual cycles define "normal" women's bodies. By building an approximation of menstruation into hormonal contraceptives, developers simultaneously made birth control seem more natural *and* reinforced the idea that monthly periods should be the norm by physically producing them in women's bodies (Oudshoorn 1994).

Menstruation is just one of many examples that feminist scholars have drawn on to show how ideologies of gender work through technologies to shape bodies and subjectivities (Balsamo 1995; De Lauretis 1987; Mamo and Fosket 2009) and how individuals use technologies to perform gender (Oudshoorn and Pinch 2003b). "Technologies of gendered bodies" construct the body both materially and discursively to bring them in line with ideologies of gender (Balsamo 1995; Mamo and Fishman 2001). One way this happens is through the process of configuration, in which developers "configure" or "script" gendered users as they anticipate—and, ultimately, produce—the knowledge, identities, and behaviors ideal users will bring to their interactions with technologies (Akrich 1992; Oudshoorn and Pinch 2003a; Woolgar 1991). Gendered identities and behaviors can be inscribed in technologies in ways that "invite or inhibit specific performances of gender identities and relations," but users must enact these performances through active use; in doing so, they can accept, reject, or modify gendered scripts (Oudshoorn and Pinch 2003b, 10). Attention to technology is a necessary component of theorizing how individuals do gender and how bodies materialize sex/gender (Oudshoorn 2003b; Roberts 2007).

Users are more likely to take up gendered scripts if they align with dominant gender norms and culturally prevalent embodiments (Oudshoorn 2003b; Roberts 2007, 155). When new technologies challenge hegemonic gender identities, the configuration process may be used to shift gender norms (Oudshoorn 2003b). In the case of menstrual suppression birth control, transforming the body to eliminate menstruation challenges several aspects of dominant understandings of menstruation and women's bodies. Medical and popular understandings identify regular monthly bleeding as a defining feature of the menstrual cycle and of female bodies; normalizing menstrual suppression, therefore, meant shifting prominent beliefs about menstruation across multiple discourses.

One way to shift popular beliefs is through advertising campaigns. Advertisements for menstrual products significantly shaped popular discourse and understandings of menstruation across the twentieth century. Emerging scientific understandings of menstruation and new public discourses of the body coincided with the mass-production of disposable personal products and the advertising intended to popularize them (Freidenfelds 2009). Early ads for pads and tampons sought to instill in women new "modern" understandings of the body and menstruation, emphasizing hygienic practices and secrecy. They worked to persuade women that they did not need to restrict their activities during menstruation—particularly with the assistance of commercially produced menstrual products that would allow them to conceal evidence of menstruation (Freidenfelds 2009; Vostral 2008). Later advertisements emphasized freedom, empowerment, and continuous productivity (Vostral 2008). Other popular culture depictions have furthered ideas of menstruation as a source of shame and embarrassment, requiring vigilant self-surveillance to hide all signs of bleeding (Kissling 2006).

These cultural understandings and expectations of menstruation affect women's embodied experiences (Martin [1987] 2001). They are critical components of culturally prevalent embodiments, particular to a certain place and time (Lock 1993). As shown by Freidenfelds (2009), U.S. women's understandings and experiences of menstruation changed throughout the twentieth century as they embodied the shift to modernity through the use of new menstrual technologies. These products facilitated (and were required by) new norms of embodiment, although their use was taken up unevenly in ways that reflected and exacerbated stratification by class, race, and ethnicity. Menstrual technologies and the discourse of "the modern period" also initiated the "body project" of menstrual surveillance and management (Brumberg 1998; Freidenfelds 2009). As women increased their activity outside the home during menstruation amid prevailing norms insisting that bleeding be hidden, women became responsible for an increasing range of self-monitoring and body-management tasks. Stigma and secrecy meant that for many women menarche and menstruation were characterized by shame and embarrassment (Lee and Sasser-Coen 1996), even as the body project of managing menstruation could also provide a site of agency, pride, and resistance (Bobel 2010; Fingerson 2006).

While feminists hold a range of positions on menstruation and menstrual technology, a key focus of many feminist engagements has been valuing women's experiential ways of knowing and diminishing stigma (Aengst and Layne 2010; Bobel 2010). In addition to critiquing scientific representations of menstruation, feminist researchers have worked to generate new ones (Dan and Lewis 1992; Fausto-Sterling 2000; Martin 1991). Women's health activists have challenged regulatory stances on hormonal birth control and tampon regulation, innovated new ways of producing knowledge about women's bodies through self-help health, and developed new technologies of menstrual management such as menstrual extraction (Bobel 2010; Morgen 2002). Feminist challenges to dominant understandings of menstruation have emphasized valuing monthly menstruation as a sign of good health, challenging shame and secrecy, and highlighting how the profit-driven menstrual product industry risks women's health (Bobel 2010).

Despite their significant differences, feminist challenges share with mainstream medical and popular understandings the belief that regular, monthly bleeding defines healthy menstruation. The introduction of menstrual suppression birth control challenged this belief, which cyclic hormonal birth control had played a significant role in reproducing and reinforcing. By transforming the body to eliminate both visible and experienced signs of menstrual cycles, menstrual suppression challenges cyclicity as a defining feature of female bodies. If medical technologies must align with culturally prevalent embodiments (Roberts 2007) and gender norms and identities (Oudshoorn 2003b), how was menstrual suppression birth control introduced and promoted? I argue that definitions of menstruation were reworked across clinical, regulatory, and popular discourse. While the taken-for-granted naturalness of menstruation had allowed the periodic bleeding built into the Pill to naturalize hormonal contraception, the task of normalizing menstrual *suppression* required redefining menstruation to exclude some forms of bleeding. This explicit renegotiation of the definition of menstruation provides a clear view of the social construction of menstruation, as well as an opportunity to explore how its social construction matters to what menstruation is and its relation to gendered embodiment.

METHODS

For this research, I analyzed two sets of texts through which new definitions of bleeding were institutionalized and disseminated. This analysis is situated in a larger project examining medical, advertising, and regulatory discourses surrounding menstrual suppression pills and practices more broadly, as well as how prescribers and women engage these technologies. In this article, I focus specifically on changing definitions of menstruation in regulatory discussions and advertising for menstrual suppression pills. The first part of this article demonstrates the creation and institutionalization of new *clinical* definitions of menstrual bleeding, drawing on two sources: transcripts from the January 23-24, 2007, meeting of the FDA's Reproductive Health Drug Advisory Committee, and two articles published in the January 2007 issue of the journal *Contraception* that discussed terminology and standards for measuring bleeding in clinical trials of combined hormonal contraceptives (CHCs) (Mishell et al. 2007a, 2007b). The second examines new definitions and terminology for menstruation used on websites promoting menstrual suppression.

First, I focused on a public meeting of the FDA's Reproductive Health Drugs Advisory Committee that convened to discuss the testing and approval of hormonal contraceptives. This meeting stands as the only public regulatory discussion of extended regimen birth control. The Advisory Committee convened to discuss clinical trial procedures and regulatory guidelines for the approval of new hormonal contraceptives, including issues pertaining specifically to extended regimen birth control. FDA advisory committee meetings are convened periodically to provide an additional level of review for the approval of new products, to evaluate safety data and reconsider previously approved drugs, or to address general questions about the drug approval process (as in this case) (Institute of Medicine et al. 1992; Leiter and White 2015). Advisory committees supplement the FDA's in-house expertise and provide additional scientific legitimacy to FDA decisions that are "too political" to be decided internally (Jasanoff 1994). The FDA determines in advance a specific set of questions the committee will address. At the January 2007 meeting, the committee was made up of 20 members, mainly academic physicians (primarily specialists in obstetrics/gynecology), statisticians, and epidemiologists, along with one representative each for consumers, patients, and the pharmaceutical industry. Eight FDA employees also participated in the meetings (but did not vote). The topics on the meeting agenda included clinical trial design issues, contraceptive efficacy and risk/benefit analysis, cycle control, extended-dosing regimens, and product labeling.

The meeting agenda, transcripts, and materials are available on the FDA website.[4] I downloaded the meeting transcripts and coded and analyzed them using Atlas.ti. Through a recursive process of open coding, I identified categories and themes in the transcripts, such as defining bleeding, patient responsibility, physician counseling, and providing information. My analysis here focuses on definitions of menstruation and bleeding, which appeared primarily in discussions of "cycle control" (i.e., the ability of hormonal contraception to produce bleeding at scheduled times and not at other times) and how to define, measure, and analyze bleeding data in clinical trials.

In the agenda and questions posed to the advisory committee, the FDA specifically asked for guidance on new guidelines proposed by a group of clinical trial researchers for defining, measuring, and analyzing bleeding in clinical trials of hormonal contraceptives. For this reason, I have included in my analysis the two journal articles proposing these guidelines. Published in the January 2007 issue of *Contraception*, the articles examined bleeding definitions and measures used in previous clinical trials for CHCs and proposed guidelines for future trials (Mishell et al. 2007a, 2007b). I analyzed these articles with a particular focus on their discussion of why new guidelines were necessary, the importance of studying bleeding related to CHCs, and their specific proposals for defining, measuring, and analyzing bleeding in clinical trials.

Second, to assess how definitions of menstruation were conveyed to potential users of menstrual suppression birth control, I analyzed websites promoting menstrual suppression pills and practices. As individuals increasingly turn to the Internet for health information, pharmaceutical companies have featured websites as a central component of their direct-to-consumer marketing

(Ebeling 2011). These include websites that advertise specific products, as well as those that are part of "disease education campaigns" meant to raise awareness of particular diseases while "branding" their association with a particular medication (Dumit 2012). To examine the meanings circulated to users, I studied the websites of the available brands of menstrual suppression pills and those promoting the practice of menstrual suppression more generally. I draw here on analysis of five websites accessed in January and February of 2009. Three were websites for the brands of extended-cycle and continuous-use oral contraceptives available at that time: Seasonique, Seasonale, and Lybrel. The fourth was fewerperiods.com, an unbranded website promoting the use of hormonal contraceptives (of all kinds) to suppress menstruation. Although it did not promote any particular brand or product, the website was produced by DuraMed/Barr Laboratories, Inc., makers of Seasonale and Seasonique, and the company logo could be found at the bottom of the page.[5] Finally, I included materials from the website of the Association of Reproductive Health Professionals,[6] including the interactive tool, "Menstrual Suppression: What It Is, and How to Do It," and the fact sheet, "Health Matters: Understanding Menstrual Suppression." Both were developed using unrestricted grants from both Barr and Wyeth.[7]

These websites represent the web-based marketing for all of the available brands of menstrual suppression pills at the time of analysis, plus two prominent sites explicitly promoting the practice of menstrual suppression that purported to provide independent, authoritative, medically sanctioned information. While this analysis does not include all online discussions of menstrual suppression, it does provide a sample of sites that were highly visible and that were presented as authoritative information about menstrual suppression (vs. blogs or forums). In analyzing the text and images on the website, I examined how menstruation was portrayed, how the websites described what menstrual suppression birth control is and what it does, and how the websites constructed potential or ideal users of menstrual suppression birth control.

Redefining Menstruation

Scientific Categories and Institutionalization

In January 2007, members of the FDA's Reproductive Health Drug Advisory Committee met to discuss a range of issues affecting clinical trials, FDA approval, and postapproval follow-up for new hormonal contraceptives.[8] Cycle control was one of seven main topics on the meeting agenda. Two points were provided to frame the discussion: (1) New birth control pills approved in recent years (including one under FDA review at the time of the meeting) were taken in an extended cycle that postponed bleeding and promised women fewer periods; however, users complained of frequent and unpredictable bleeding. (2) Changes in the hormonal makeup of contraceptives, especially lower doses of estrogen, made breakthrough bleeding more common. The FDA posed two questions related to cycle control. The first was how researchers should assess bleeding patterns and cycle control in clinical trials—specifically, whether the FDA should adopt in their clinical trial guidelines a recently published proposal for defining and measuring bleeding. The second question was whether the FDA ought to ensure that extended regimen pills actually produced fewer overall days of bleeding. To prepare the committee for discussing the first question, Dr. James Trussell, one of the authors, presented the research and proposed guidelines to his fellow committee members. Because they were at the center of the advisory committee's discussion, I analyze the proposed guidelines in detail before returning to the committee meeting.

The guidelines were developed at the 2005 Hormonal Contraceptives Trial Methodology Consensus Conference (Mishell et al. 2007a, 2007b). Consensus conferences such as this one are often convened to develop and agree on new diagnoses, assessments of medical technologies, or guidelines for clinical decision making (Fishman 2004). This particular conference was funded by an unrestricted educational grant from Duramed Pharmaceuticals, which was at that time in the midst of clinical trials for Seasonique, a follow-up product to Seasonale. Two articles, coauthored by several clinical trial researchers, were published to convey the results of the conference. Of the seven coauthors, six disclosed financial and other ties to Duramed and Wyeth, the two companies with menstrual suppression birth control pills on the market or under FDA review (Mishell et al. 2007a).[9] These ties to Duramed and Wyeth suggest that, while the guidelines do not explicitly reference extended regimen pills, the authors were likely aware that their proposals could affect clinical trials and FDA review of menstrual suppression pills under review or in development.

The first of the two articles examined the methods of evaluating bleeding used in clinical trials of CHCs approved between 1975 and 2003 (Mishell et al. 2007a). Explaining that lower hormone doses in more recent contraceptives resulted in higher rates of breakthrough bleeding and spotting, the authors argued that physicians need better data on how CHCs affect bleeding patterns to inform their prescribing decisions and patient counseling. The information available to physicians and consumers is, in some ways, limited to the information requested and reviewed by the FDA in the process of approving a new drug. Therefore, changing the information requested by regulators would also change the data collected by clinical trials and the information available to prescribing physicians and consumers. The FDA reviews drugs based on criteria of safety and efficacy; because bleeding patterns had not previously been considered relevant, the FDA had not required companies to provide information on cycle control when applying for approval of a new hormonal contraceptive. The article showed that individual regulators had at times asked for and evaluated information on bleeding patterns according to their own interest and expertise, but that the amount and format of information included in the reviews varied widely—and thus so had the information available to physicians and consumers as they decided which hormonal contraceptives to use (Mishell et al. 2007a).

The authors' review of existing clinical trials focused on the lack of consistent measures for bleeding and the shortcomings of the most commonly used terminology. Previous clinical trials and New Drug Applications for CHCs varied widely in how they defined bleeding, how they collected information about it, and how they analyzed and reported their data. Despite overall inconsistency, the most commonly used bleeding categories were those defined by the "Belsey criteria," introduced by World Health Organization researchers in the 1980s. Half of the studies used the criteria in some form (Mishell et al. 2007a, 6). The Belsey criteria established two categories: "bleeding" and "spotting." Bleeding is defined as "any bloody vaginal discharge that requires the use of such protection as pads or tampons," and "spotting" as "any bloody vaginal discharge that is not sufficient to require protection" (Belsey, Machines, and d'Arcangues 1986, 255). The main distinction between these categories is the amount of bleeding, determined by whether menstrual products would be used to manage it.

The authors critiqued the Belsey criteria for not accounting for the timing of bleeding and not distinguishing between bleeding produced by the withdrawal of contraceptive hormones and *menstruation* occurring in the absence of exogenous hormones. For these reasons, the authors argued that the Belsey criteria were not well suited to describing bleeding in studies of CHCs, despite their frequent use. They stated:

> While valuable for providing definitions and global descriptions of bleeding patterns associated with various methods of contraception, these criteria are not . . . . particularly useful for the reporting of cyclic bleeding in women using CHCs. . . . [T]he unmodified criteria do not differentiate bleeding occurring during active hormone therapy from that occurring during placebo interval, making the determination of the incidence of unscheduled bleeding/spotting difficult. (Mishell et al. 2007a, 9)

To address these shortcomings, the authors put forth in a second article new guidelines for defining, measuring, and reporting bleeding patterns (Mishell et al. 2007b). They proposed that researchers track the *timing* of bleeding, particularly with respect to whether it occurs during "placebo intervals" or while taking active hormones. The new guidelines stressed that researchers concerned with bleeding patterns related to hormonal contraception must distinguish *bleeding* from *menstruation*, and provided new terminology for doing so:

> The use of traditional terminology ("periods" or "menses") should be abandoned with regards to CHCs and replaced by the use of "scheduled bleeding"

or "withdrawal bleeding." E.g., any bleeding or spotting that occurs during hormone-free intervals regardless of the duration of the regimen. "Scheduled bleeding" emphasizes to the woman that her bleeding with hormonal methods is not the same as menstruation. (Mishell et al. 2007b, 13)

The authors proposed replacing the "traditional terminology" of periods and menses with new terminology that "emphasizes *to the woman*" the differences between hormonally produced bleeding and menstruation. In accordance with pill developers' intentions, many pill users recognize the familiar experience of monthly bleeding as a period, regardless of its physiological differences from menstruation. In contrast, the proposed terminology emphasizes bleeding as a side effect of birth control hormones that is "not the same as menstruation." While the published recommendations are directed to clinical researchers, here the authors specifically focus on changing how women think about their bleeding patterns.

Since before the introduction of the first hormonal contraceptive, researchers had known the distinction between bleeding produced by the withdrawal of birth control hormones and menstruation in the absence of synthetic hormones (Marks 2001). Specifying in the proposed guidelines that the terminology would apply to regimens of any duration suggests that this distinction is particularly important in the context of extended regimen contraceptives. The introduction of birth control pills intended for menstrual suppression provided the conditions in which this distinction became newly salient, prompting efforts to convince researchers of its significance.

At this point, my analysis returns to the FDA, where these new guidelines and categories structured the advisory committee's discussion of cycle control. In setting the agenda for the advisory committee meeting, the FDA directly asked the committee to evaluate the proposed guidelines and make a recommendation about whether the FDA should adopt them. Following a presentation of the guidelines, the committee's discussion focused primarily on the proposed terminology. Many committee members concurred that new terminology would communicate to women that withdrawal bleeding is not menstruation. Committee chair Lockwood, for example, praised the recommended language for emphasizing the differences between bleeding on and off hormonal contraceptives, stating that the new vocabulary "moves away from mixing metaphors with both physiologic and pharmacological processes" (U.S. FDA transcript, January 24, 2007, 15-16). Committee member Johnson favored the new standards because "both physicians and patients need to know what we mean when we say bleeding and spotting" (U.S. FDA transcript, January 24, 2007, 15).

However, not everyone at the meeting agreed that the new language would be easier to understand. One senior FDA official asked the committee to revisit the question of how to convey cycle control information to consumers. Scott Monroe, acting director of the FDA Division of Reproductive and Urologic Products, asked whether moving away from familiar language for describing bleeding might cause more, rather than less, confusion for consumers. Specifically, he asked how to describe bleeding patterns produced by extended regimen pills:

> When we think in terms of a traditional monthly cycle, it is fairly easy to conceptualize things [bleeding patterns]. . . . But then when you are talking about longer intervals . . . let's say it is an 84/7 [extended regimen] or a continuous [regimen] . . . then you have to start doing all this mental sort of arithmetic if you are trying to go back and relate it to a more traditional pill. We wondered if . . . everybody can instantly do the mental mathematics . . . ? I don't think your average person thinks in terms of numbers of anticipated bleeding days over a year, the numbers of withdrawals, and so on. (U.S. FDA transcript, January 24, 2007, 42-43)

Monroe highlighted that shifting to the language of withdrawals or anticipated bleeding days, instead of "periods," might make it more difficult for women to understand and compare extended and traditional pill regimens and their effects on bleeding.

The advisory committee voted to approve the proposed guidelines. In doing so, they institutionalized new categories for different types of bleeding. Because FDA guidelines dictate the information that future clinical trials will collect, they can determine what information the FDA can provide to physicians and consumers. Adopting these guidelines meant that potential users would also need to adopt—or at least under-stand—these categories to make sense of the information available about hormonal contraceptives.

The proposal and discussion of new guidelines for measuring and evaluating bleeding demonstrates the work required to make the known distinction between menstruation and withdrawal bleeding salient for clinical researchers, regulators, and physicians. Further, the discussion of terminology highlights that both researchers and regulators were explicitly concerned with changing how women think about bleeding. Establishing and institutionalizing new terminology and clinical categories of bleeding facilitated the normalization of menstrual suppression by changing the object on which menstrual suppression birth control acts: not menstruation, but "scheduled" or "withdrawal" bleeding.

Not a "Real" Period

Prior to the initial consensus conference and ensuing FDA discussions, pharmaceutical advertising campaigns had already been working to change popular understandings of menstruation as they introduced menstrual suppression birth control (Mamo and Fosket 2009). On their websites and in print campaigns, the companies introduced potential users to menstrual suppression through images of an idealized lifestyle achievable with menstrual suppression and information that preempted questions about what is normal, natural, and safe. They presented detailed explanations of the menstrual cycle and introduced women to the "pill period"—a new term for the "scheduled" bleeding that occurs when women are taking cyclic hormonal birth control. In this section, I focus on these explanations of menstruation and the distinctions made between different kinds of periods.[10] Although the websites continued to use the language of "periods," they stressed the same distinction between bleeding on and off the pill that was highlighted in the clinical guidelines. Promoting a new understanding of the experience of menstruation, this distinction brought attention to the ways that *all* hormonal birth control alters menstruation. In this way, pharmaceutical companies pioneered efforts to change how women understand menstruation that researchers and regulators took up later.

The language of "pill periods" is specifically tailored to the context of menstrual suppression; it is therefore important to examine closely how the websites construct these *kinds* of periods. The work of redefining menstruation figured prominently on the websites: all five devoted significant space to describing how the menstrual cycle works with and without the effects of hormonal birth control. The descriptions were structured around the distinction between bleeding on and off the pill, divided into three categories: the "regular" or "basic" menstrual cycle, the cycle when taking hormonal contraceptives, and the cycle when using hormonal birth control to suppress menstruation.[11] See Figure 56.1 for a representative example from the Lybrel website.

As described in Figure 56.1, the "regular monthly period" occurs when hormones "naturally rise and fall." In contrast, the "traditional birth control period," or "pill period," results from the withdrawal of hormones during the placebo week. The new intermediary category of the "pill period" is built on the assumption of a fundamental difference between bleeding that occurs "naturally" and bleeding that occurs when taking hormonal contraceptives. The "pill period" is explained as a *side effect* of the pill itself. Finally, the website states that with Lybrel women "do not have regular menstrual periods or 'pill periods,'" because the pills provide continuous levels of hormones. Distinguishing between types of bleeding takes on new importance with the existence of menstrual suppression technology: "pill periods" are suppressed, not "regular periods." The purpose of introducing the "pill period," then, is to promote its suppression. As an intermediary between the "natural" menstrual cycle and menstrual suppression, the "pill period" normalizes menstrual suppression and repositions some forms of monthly bleeding as artificial, rather than natural.

The idea that women "do not have regular menstrual periods" seems relatively clear in the case of Lybrel, which is taken continuously to suppress both "regular" menstruation and "pill periods." However, the websites for Seasonale and Seasonique make similar claims, even though they are designed to produce four bleeding periods per year. How do these websites define menstruation in a way that distinguishes the periodic bleeding they produce

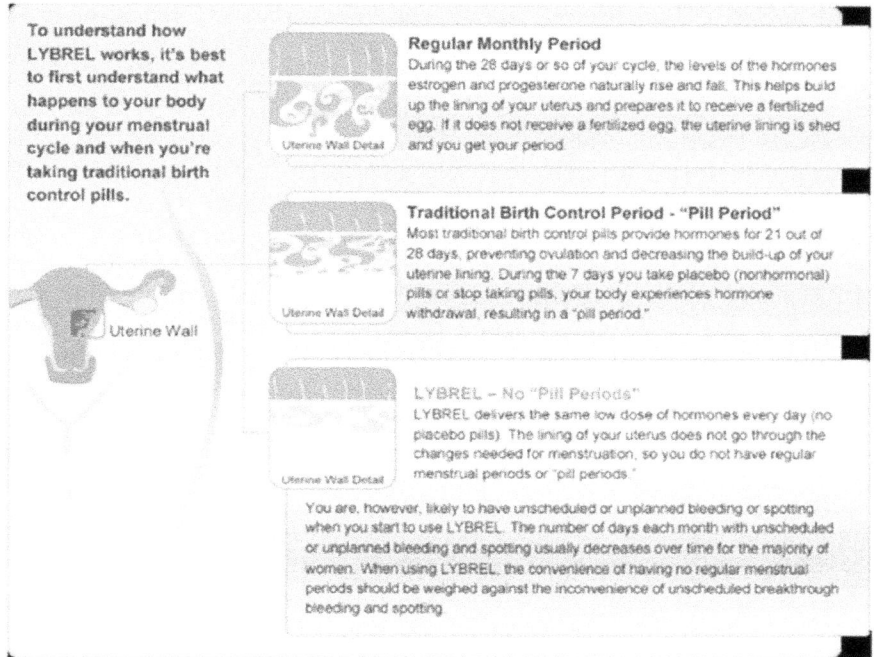

To understand how LYBREL works, it's best to first understand what happens to your body during your menstrual cycle and when you're taking traditional birth control pills.

Uterine Wall

Regular Monthly Period
During the 28 days or so of your cycle, the levels of the hormones estrogen and progesterone naturally rise and fall. This helps build up the lining of your uterus and prepares it to receive a fertilized egg. If it does not receive a fertilized egg, the uterine lining is shed and you get your period.

Uterine Wall Detail

Traditional Birth Control Period - "Pill Period"
Most traditional birth control pills provide hormones for 21 out of 28 days, preventing ovulation and decreasing the build-up of your uterine lining. During the 7 days you take placebo (nonhormonal) pills or stop taking pills, your body experiences hormone withdrawal, resulting in a "pill period."

Uterine Wall Detail

LYBREL – No "Pill Periods"
LYBREL delivers the same low dose of hormones every day (no placebo pills). The lining of your uterus does not go through the changes needed for menstruation, so you do not have regular menstrual periods or "pill periods."

You are, however, likely to have unscheduled or unplanned bleeding or spotting when you start to use LYBREL. The number of days each month with unscheduled or unplanned bleeding and spotting usually decreases over time for the majority of women. When using LYBREL, the convenience of having no regular menstrual periods should be weighed against the inconvenience of unscheduled breakthrough bleeding and spotting.

Uterine Wall Detail

Fig. 56.1 "How Lybrel Works" from www.lybrel.com. (*Source* https://www.lybrel.com/works/. Credit: Wyeth Pharmaceuticals)

from "regular menstrual periods"? As seen in this quote from seasonique.com, one approach is to deemphasize menstruation as the key event of the menstrual cycle:

> When you take a birth control pill, you don't have a menstrual cycle—your body doesn't prepare for pregnancy because you don't ovulate. As a result, your uterine lining doesn't build up, so there's no need to shed it. This is the reason you don't need to bleed every month when you take the Pill.[12]

To say that women taking birth control pills "don't have a menstrual cycle," the websites identify ovulation as the defining feature of the menstrual cycle. Because all hormonal birth control suppresses ovulation, by focusing on ovulation they suggest that *all* birth control pills are already suppressing the menstrual cycle. The fewerperiods.com website makes this point explicitly, stating that the "pill period" is "the reason having fewer periods is possible when you use hormonal birth control." In other words, *all* hormonal birth control pills suppress "real" periods, but "traditional" pills produce unnecessary bleeding as a side effect and disguise it as a "pill period." These statements assume that potential users already use hormonal birth control and will accept menstrual suppression once they learn that they are already suppressing their menstrual cycles and inducing a "fake" pill period.

In this way the websites work to revise—and perhaps even delegitimize—women's knowledge and experiences of menstruation. The "Fewer Periods" website addresses how hormonal contraceptive users *experience* menstruation when it says, "The light, short bleeding that you experience on the Pill isn't a real menstrual period—it's actually a 'Pill period,' which is due to the withdrawal of hormones in your active pills."[13] Women taking hormonal birth control who experience regular monthly bleeding often understand that bleeding to be a menstrual period (as developers intended), and some might find this distinction irrelevant if the bodily sensations and effects are the same. Menstrual suppression websites encourage users to question this embodied experience of monthly bleeding, reinterpreting it as a "pill period." Redefining menstrual cycles around fluctuating hormone levels and ovulation shifts the defining event from something women feel and see—bleeding—to something (most) women experience less directly.

The "pill period" narrative revises existing knowledge about menstruation and provides a new interpretation for current and past experiences of bleeding. It encourages pill users to reinterpret all previous experiences of bleeding while on the pill, applying a new understanding that these were never "real periods." Like the clinical guidelines, the websites focus on changing how women think about menstruation, emphasizing that the bleeding that occurs when taking hormonal birth control is not actually menstruation, and thereby introducing new categories of bleeding. While this information is technically more accurate, it is important to ask why it appears only in conjunction with discussions of menstrual suppression.

Menstruation as Multiple

What work does this proliferation of bleeding categories do to undermine the seemingly obvious naturalness of menstruation? The revision of clinical, regulatory, and popular understandings of menstruation reversed the efforts of hormonal contraceptive developers, who built in regular monthly bleeding intending that it would be perceived as a period and that the assumed naturalness of menstruation would extend to birth control itself. To the extent that the experience of withdrawal bleeding has been considered a "period," normalizing menstrual suppression birth control necessitates *denaturalizing* bleeding by breaking the automatic association of monthly bleeding with "natural" menstruation. To redefine menstruation, researchers, regulators, and advertisers identified and highlighted certain material differences in bleeding—the presence or absence of "naturally" fluctuating hormones or ovulation, for example—to create new categories of bleeding.

Proliferating types of bleeding reveals as many and changeable what seemed to be a single natural phenomenon. Introducing the idea that many kinds of bleeding have been and continue to be misrecognized as menstruation made visible "the coexistence of multiple entities that go by the same name" (Mol 2002, 151). Menstruation was revealed as "multiple," "a single [bodily process] that in practice appears to be more than one— without being fragmented into many" (Mol 2002, 151). This multiplicity had been made coherent and stable through the socially meaningful experience of bleeding once per month and the taken-for-granted naturalness of menstruation. As Mol notes, when conflicting multiples of the body become apparent, "one reality wins" (2002, 55). Coherence could be (re) achieved by instituting hierarchy—"real" menstruation as defined by "natural" hormonal fluctuations was placed above the newer categories of "scheduled" or "withdrawal" bleeding and "pill periods," which would no longer be considered menstruation.

In this case, the proliferation of bleeding categories was re-hierarchized and used to narrow the definition of menstruation and support its suppression. However, there is opportunity for gender scholars in recognizing menstruation as multiple by taking seriously the material differences in bleeding that were mobilized to redefine menstruation. The hormone levels produced by hormonal contraceptives and the effects on the uterine lining of their abrupt withdrawal during a placebo week do differ from a menstrual cycle unaffected by external hormones. In what ways does it matter—and to whom—that withdrawal bleeding is different from menstruation? Examining "multiple" menstruation as revealed in this process of its redefinition allows feminist scholars to theorize how this particular "meeting of the conceptual and the biological has material effects" (Pitts-Taylor 2016, 42).

Conclusion

Cultural associations of menstruation with biological sex, nature, reproduction, and femininity make it an overdetermined site for the examination of gendered embodiment. Considering technologies that intervene in gendered bodies provides a vantage point from which to reconsider aspects of embodiment taken for granted as fixed, as they are socially and materially reconstructed. While we often take for granted what menstruation is, this case reveals coordinated effort across multiple realms to redefine menstruation by disaggregating the many types of bleeding that have been understood as menstruation.

This article has shown how categories of bleeding developed for use in clinical trials and institutionalized by the FDA aligned with online marketing campaigns for menstrual suppression by emphasizing the distinction between menstruation and withdrawal bleeding. This distinction challenges a longstanding equation of menstruation with both nature and biological sex that has been central to the relationship between birth control hormones and bleeding. Almost 50 years after the development of oral contraceptives, with the introduction of menstrual suppression birth control, pharmaceutical companies worked to undo the acceptance of all monthly bleeding as menstruation. They disclosed that oral contraceptives produced a "pill period" that was not a "real" period, and therefore unnecessary. Advertising campaigns promoted this new definition for reinterpreting the familiar experience of monthly bleeding. Scientists and regulators also took aim at the misrecognition as menstruation of multiple forms of bleeding. They proposed using the category of "scheduled" or "withdrawal" bleeding to distinguish it from menstruation.

The redefinition of menstruation undoubtedly provides another example of the social construction of the body and gender, and a further rejection of biological essentialism. The explicit re-negotiation of biological facts across a wide range of institutions—scientific research, medicine, pharmaceutical industry, federal regulators—offers further evidence that even the most irreducibly biological processes have been and continue to be subject to reinscription with new meanings. And yet, stopping here would again shy away from grappling with biology, with what is going on materially in the body. What response can feminist scholars and advocates give to the claim that "pill periods" are not "real" periods and are therefore unnecessary? Without theoretical tools to address material differences in bleeding, some may be left to call on something like a "natural" period and its importance to women's biology to critique or reject menstrual suppression.

In past encounters with these "interactions between the social and the biological that are not, or not only, linguistic or discursive" (Frost 2014, 316), feminist scholars have missed chances to analyze how "the meeting of the conceptual and the biological has material effects" (Pitts-Taylor 2016, 42). Feminist science studies scholars continue to argue that the dynamic multiplicity of biology undermines and explodes essentialist categories and biological determinism (Fausto-Sterling 2005; Frost 2014; Pitts-Taylor 2016; Roberts 2007; Wilson 2010). Taking seriously the material differences highlighted here in the process of redefining menstruation allows feminist scholars to theorize menstruation—and gendered embodiment—as multiple, highlighting the variation and dynamism of biology in its interactions with technology (Pitts-Taylor 2016).

The axiom of menstrual activism that "not all women menstruate, and not all menstruators are women" does not need to be taken as a rejection of biology and its relationship to gendered embodiment. It can be a call to investigate the interrelated social *and* material construction of menstruation and gender, exploring and highlighting "the capacity of biological substance to forge complex alliances and diverse forms" (Wilson 2010, 197). Complex arrangements of organs, tissues, hormones—produced in the body or taken in from outside—generate embodied experiences of regular, irregular, or absent bleeding. These occur in the context of gender binaries and cultural norms that demand management and concealment of bleeding, along with menstrual technologies that facilitate this. We can analyze these "complex alliances" that produce women that do or do not menstruate, menstruators who may or may not be women. Taking for granted what menstruation is and how it is defined obscures what can be learned by remaining open to menstruation's multiplicity and its relationship to gendered embodiment. Sociologists of gender and embodiment can enhance and strengthen our analysis by theoretically and empirically engaging the material aspects of biology as always in relation and interaction with their representations in specific social and historical contexts.[14]

Notes

1. "Not a Real Period?": Social and Material Constructions of Menstruation by Katie Ann Hasson was first published in 2016 in *Gender & Society* 30 (6): 958–983. Reprinted with permission. No further reproduction or distribution of the material is allowed without permission from the publisher.
2. Menstrual activists challenge this belief, pointing out that not all women menstruate and not all who menstruate are women, and use the term "menstruator" rather than women (Bobel 2010).
3. Menstrual extraction is a technique developed by feminist self-help health activists for removing the contents of the uterus at the (expected) start of a woman's period (Murphy 2012).
4. http://www.fda.gov/ohrms/dockets/ac/cder07.htm#rhdac.
5. The fewerperiods.com website is now defunct and the URL redirects to seasonique.com. DuraMed Pharmaceuticals and its parent company, Barr Laboratories, have been acquired by Teva Women's Health, Inc.
6. ARHP was founded as "the education arm of Planned Parenthood Federation of America (PPFA) and incorporated as an independent organization in 1972" (http://www.arhp.org/about-us/about-arhp).
7. These websites were part of Duramed's advertising strategy for Seasonique (Saul 2007).
8. The discussion included oral contraceptives as well as transdermal and vaginal rings, but excluded injectable contraceptives (DepoProvera) and implants (Implanon, etc.).
9. The only author who did not report financial conflicts of interest was James Trussell. These connections do not necessarily indicate sinister intent. Clinical trial researchers are structurally positioned between drug developers, regulators, and prescribers (Fishman 2004). However, financial ties to pharmaceutical companies have been shown to affect study outcomes (Sismondo 2008).
10. Elsewhere I explore how depictions of menstrual suppression's ideal users and lifestyles convey classed and racialized meanings that reflect and reinforce stratified access to reproductive technologies.
11. In contrast, websites for cyclic birth control pills (e.g., Ortho TriCyclen Lo and LoEstrin) did not present detailed information about the menstrual cycle or discuss "pill periods" and how they differ from "regular" periods. Cyclic pills have little to gain from making a distinction between different *kinds* of periods.
12. http://seasonique.com/Consumer/BodyAndPeriod/FewerPeriodsPossible.aspx [emphasis added].

13. http://fewerperiods.com/FewerPeriodsPossible/Default.aspx.
14. I would like to thank Rene Almeling, Kate Darling, Dawn Dow, Hannah Landecker, Martine Lappe, Sarah Macdonald, Theresa MacPhail, Silvia Pasquetti, Leslie Salzinger, Jade Sasser, Rachel Washburn, Raka Ray and the members of her gender working group, and the USC Sociology Social Organization reading group for their feedback on earlier versions of this paper, with special thanks to Jo Reger and the anonymous reviewers from Gender & Society for their generous comments. This work has also benefited from feedback provided after presentations to the Body and Embodiment Section at the 2013 ASA Annual Meeting and the Center for Feminist Research at USC.

References

Aengst, Jennifer, and Linda Layne. 2010. The need to bleed? A feminist technology assessment of menstrual-suppressing birth control pills. In *Feminist technology*, edited by Linda L. Layne, Sharra L. Vostral, and Kate Boyer. Urbana: University of Illinois Press.

Akrich, Madeleine. 1992. The de-scription of technological objects. In *Shaping technology/building society*, edited by Wiebe E. Bijker and John Law. Cambridge, MA: MIT Press.

Association of Reproductive Health Professionals. 2006. Menstrual suppression: What it is and how to do it. http://www.arhp.org/Publications-and-Resources/Patient-Resources/Interactive-Tools/menstrual-suppression.

Association of Reproductive Health Professionals. 2008. Health matters: Understanding menstrual suppression. http://www.arhp.org/Publications-and-Resources/Patient-Resources/fact-sheets/Understanding-Menstrual-Sup-pression.

Balsamo, Anne. 1995. *Technologies of the gendered body: Reading cyborg women*. Durham, NC: Duke University Press.

Belsey, Elizabeth M., David Machines, and Catherine d'Arcangues. 1986. The analysis of vaginal bleeding patterns induced by fertility regulating methods. *Contraception* 34 (3): 253-60.

Bobel, Chris. 2010. *New blood: Third-wave feminism and the politics of menstruation*. New Brunswick, NJ: Rutgers University Press.

Brumberg, Joan Jacobs. 1998. *The body project: An intimate history of American girls*. New York: Random House.

Buckley, Thomas C. T., and Alma Gottlieb. 1988. *Blood magic: The anthropology of menstruation*. Berkeley: University of California Press.

Butler, Judith. 1990. *Gender trouble: Feminism and the subversion of identity*. New York: Routledge.

Dan, Alice J., and Linda L. Lewis. 1992. *Menstrual health in women's lives*. Urbana: University of Illinois Press.

De Lauretis, Teresa. 1987. *Technologies of gender: Essays on theory, film, and fiction*. Bloomington: Indiana University Press.

Delaney, Janice, Mary Jane Lupton, and Emily Toth. 1988. *The curse: A cultural history of menstruation*. Urbana: University of Illinois Press.

Dumit, Joseph. 2012. *Drugs for life: How pharmaceutical companies define our health*. Durham, NC: Duke University Press.

Ebeling, Mary. 2011. "Get with the program!": Pharmaceutical marketing, symptom checklists and self-diagnosis. *Social Science & Medicine* 73 (6): 825-32.

Fausto-Sterling, Anne. 2000. *Sexing the body: Gender politics and the construction of sexuality*. New York: Basic Books.

Fausto-Sterling, Anne. 2005. The bare bones of sex, Part 1: Sex and gender. *Signs* 30 (2): 1491-1527.

Fewer Periods. 2009. http://fewerperiods.com.

Fingerson, Laura. 2006. *Girls in power: Gender, body, and menstruation in adolescence*. Albany: State University of New York Press.

Fishman, Jennifer R. 2004. Manufacturing desire: The commodification of female sexual dysfunction. *Social Studies of Science* 34 (2): 187-218.

Freidenfelds, Lara. 2009. *The modern period: Menstruation in twentieth-century America*. Baltimore, MD: Johns Hopkins University Press.

Friedman, Asia. 2013. *Blind to sameness: Sexpectations and the social construction of male and female bodies*. Chicago: University of Chicago Press.

Frost, Samantha. 2014. Re-considering the turn to biology in feminist theory. *Feminist Theory* 15 (3): 307-26.

Grosz, Elizabeth A. 1994. *Volatile bodies: Toward a corporeal feminism*. Bloomington: Indiana University Press.

Haraway, Donna Jeanne. 1991. *Simians, cyborgs, and women: The reinvention of nature*. New York: Routledge.

Institute of Medicine, Richard A. Rettig, Laurence E. Earley, and Richard A. Merrill. 1992. Food and Drug Administration advisory committees. Washington, DC: National Academy Press.

Jasanoff, Sheila. 1994. *The fifth branch: Science advisers as policymakers*. Cambridge, MA: Harvard University Press.

Junod, Suzanne White, and Lara Marks. 2002. Women's trials: The approval of the first oral contraceptive pill in the United States and Great Britain. *Journal of the History of Medicine and Allied Sciences* 57 (2): 117-60.

Kissling, Elizabeth Arveda. 2006. *Capitalizing on the curse: The business of menstruation*. Boulder, CO: Lynne Rienner.

Laqueur, Thomas Walter. 1992. *Making sex: Body and gender from the Greeks to Freud*. Cambridge, MA: Harvard University Press.

Lee, Janet, and Jennifer Sasser-Coen. 1996. *Blood stories: Menarche and the politics of the female body in contemporary U.S. Society*. New York: Rout-ledge.

Leiter, Valerie, and Shelley K. White. 2015. Enmeshed in controversy: Claims about the risks of vaginal mesh devices. *Health, Risk & Society* 17 (1): 64-80.

Lock, Margaret M. 1993. *Encounters with aging: Mythologies of menopause in Japan and North America*. Berkeley: University of California Press.

Luker, Kristin. 1984. *Abortion and the politics of motherhood*. Berkeley: University of California Press.

Lybrel. 2009. http://lybrel.com.

Mamo, Laura, and Jennifer R. Fishman. 2001. Potency in all the right places: Viagra as a technology of the gendered body. *Body & Society* 7 (4): 13-35.

Mamo, Laura, and Jennifer Ruth Fosket. 2009. Scripting the body: Pharmaceuticals and the (re)making of menstruation. *Signs: Journal of Women in Culture and Society* 34 (4): 925-49.

Marks, Lara. 2001. *Sexual chemistry: A history of the contraceptive pill*. New Haven, CT: Yale University Press.

Martin, Emily. 1991. The egg and the sperm: How science has constructed a romance based on stereotyped male and female roles. *Signs: Journal of Women and Culture in Society* 16:485-501.

Martin, Emily. 1999. The woman in the flexible body. In *Revisioning women, health and healing: Feminist, cultural, and technoscience perspectives*, edited by Adele Clarke and Virginia L. Olesen. New York: Routledge.

Martin, Emily. (1987) 2001. *The woman in the body: A cultural analysis of reproduction*. Boston: Beacon Press.

Mishell, Daniel R., John Guillebaud, Carolyn Westhoff, Anita L. Nelson, Andrew M. Kaunitz, James Trussell, and Ann Jeanette Davis. 2007a. Combined hormonal contraceptive trials: Variable data collection and bleeding assessment methodologies influence study outcomes and physician perception. *Contraception* 75 (1): 4-10.

Mishell, Daniel R., John Guillebaud, Carolyn Westhoff, Anita L. Nelson, Andrew M. Kaunitz, James Trussell, and Ann Jeanette Davis. 2007b. Recommendations for standardization of data collection and analysis of bleeding in combined hormone contraceptive trials. *Contraception* 75 (1): 11-15.

Mol, Annemarie. 2002. *The body multiple: Ontology in medical practice*. Durham, NC: Duke University Press.

Morgen, Sandra. 2002. *Into our own hands: The women's health movement in the United States, 1969-1990*. New Brunswick, NJ: Rutgers University Press.

Murphy, Michelle. 2012. *Seizing the means of reproduction: Entanglements of feminism, health, and technoscience*. Durham, NC: Duke University Press.

Oudshoorn, Nelly. 1994. *Beyond the natural body: An archeology of sex hormones*. London: Routledge.

Oudshoorn, Nelly. 2003a. Clinical trials as a cultural niche in which to configure the gender identities of users: The case of male contraceptive development. In *How users matter: The co-construction of users and technologies*, edited by Nelly Oudshoorn and T. J. Pinch. Cambridge: MIT Press.

Oudshoorn, Nelly. 2003b. *The male pill: A biography of a technology in the making*. Durham, NC: Duke University Press.

Oudshoorn, Nelly, and Trevor J. Pinch. 2003a. *How users matter: The co-construction of users and technologies*. Cambridge: MIT Press.

Oudshoorn, Nelly, and Trevor J. Pinch. 2003b. Introduction: How users and non-users matter. In *How users matter: The co-construction of users and technologies*, edited by Nelly Oudshoorn and T. J. Pinch. Cambridge: MIT Press.

Pitts-Taylor, Victoria. 2016. *The brain's body: Neuroscience and corporeal politics*. Durham, NC: Duke University Press.

Reagan, Leslie J. 1997. *When abortion was a crime: Women, medicine, and law in the United States, 1867-1973*. Berkeley: University of California Press.

Roberts, Celia. 2007. *Messengers of sex: Hormones, biomedicine, and feminism*. New York: Cambridge University Press.

Saul, Stephanie. 2007. Pill that eliminates the period gets mixed reviews. *The New York Times*, 20 April.

Seasonale. 2009. http://seasonale.com.

Seasonique. 2009. http://seasonique.com.

Sismondo, Sergio. 2008. How pharmaceutical industry funding affects trial outcomes: Causal structures and responses. *Social Science & Medicine* 66 (9): 1909-14.

United States Food and Drug Administration. 2007. Department of Health and Human Services, Food and Drug Administration, Center for Drug Evaluation and Research, Advisory Committee for Reproductive Health Drugs, Volume I (transcript). January 23.

United States Food and Drug Administration. 2007. Department of Health and Human Services, Food and Drug Administration, Center for Drug Evaluation and Research, Advisory Committee for Reproductive Health Drugs, Volume II (transcript). January 24.

van de Walle, Etienne, and Elisha P. Renne. 2001. *Regulating menstruation: Beliefs, practices, interpretations.* Chicago: University of Chicago Press.

Vostral, Sharra L. 2008. *Under wraps: A history of menstrual hygiene technology.* Lanham, MD: Lexington Books.

Watkins, Elizabeth Siegel. 1998. *On the pill: A social history of oral contraceptives, 1950-1970.* Baltimore, MD: Johns Hopkins University Press.

Watkins, Elizabeth Siegel. 2012. How the pill became a lifestyle drug: The pharmaceutical industry and birth control in the United States since 1960. *American Journal of Public Health* 102 (8): 1462-72.

West, Candace, and Don H. Zimmerman. 1987. Doing gender. *Gender & Society* 1 (2): 125-51.

Wilson, Elizabeth A. 2010. Underbelly. *Differences* 21 (1): 194-208.

Woolgar, Steve. 1991. Configuring the user: The case of usability trials. In *A sociology of monsters: Essays on power, technology and domination,* edited by John Law. New York: Routledge.

Young, Iris Marion. 1990. *Throwing like a girl and other essays in feminist philosophy and social theory.* Bloomington: Indiana University Press.

CHAPTER 57

Painting Blood: Visualizing Menstrual Blood in Art

Ruth Green-Cole

INTRODUCTION

The notion of 'gendered blood' (Lupton 1993, 3) is a concept that struggles against patriarchal traditions, which traditionally denigrate and suppress images of menstruation, while European and American art valorize women's bodies as vehicles for male scopic desire. Biologically, blood may not be gendered, but society's values transform it into female, dirty, discharge, abject, shame. Moreover, in Jewish, Christian, and Islamic traditions, menstrual blood is associated with immorality or accursedness due to Eve's temptation of Adam. While Christ's blood is elevated in the Eucharist, as is the blood of heroes lost on the battlefield, 'women with issue' are humiliated or are unspeakable. Earlier art used blood, especially depictions of Christ's blood, in controlled ways to show concepts of humanity, sacrifice, and transcendent spiritual values. Gendered blood in contemporary art is contested and has become an image or medium appropriated from traditional masculinity (and the censorship concerning feminine blood understood as abject) for different subject positions and identities not dealt with in earlier art.

Patriarchal values have kept menstruation as a subject in the margins of art and art history until it emerged in feminist art in the 1970s. Contemporary works discussed in this essay challenge what is acceptable and ask questions about what should and should not be visible. Many menstrual works mean to shock, in true avant-garde fashion, but not needlessly or nihilistically. They are critical interrogations of aesthetic authority and decorum, challenging ignorant and popular beliefs about art's functions, as much as they create visibility and space within art and the public domain for contested values associated with blood. These works may not be easy to look at or decent

© The Author(s) 2020
C. Bobel et al. (eds.), *The Palgrave Handbook of Critical Menstruation Studies*, https://doi.org/10.1007/978-981-15-0614-7_57

and palatable, as they refuse the requirement that art can only be about traditional prettiness and chaste femininity. Artworks that deal with menstruation in many different ways are important because they work against negative stereotypes and actively revalue gendered blood, showing it in a positive, defiant, or ambiguous light.

Drawing on a feminist reading of Maurice Merleau-Ponty's phenomenology in conversation with work by gender theorist Judith Butler provides a framework to understand gendered performativity. Butler explains that gender becomes constructed through the 'sedimentation' of 'performative acts,' which gradually and through iteration and mutual recognition collectively assemble gender identity. To read menstruation as 'gendered blood' is to recognize the ritualization of difference through cleansing practices, psychoanalytic classification, and other constructs that ultimately affect the position of women. Consequently, visualizing menstruation challenges and exposes ingrained psychological behaviors and categories, because it challenges the valorization in many cultures of the classical, male, sealed body. The 'leaky body' (Irigaray 1985) associated with the weaker sex and its vulnerability, lack of self-control, and natural biological state closer to base materiality is set up in contradistinction. Feminist artists have been important in challenging these kinds of representations of women and creating a space within art and art history for women's voices. Still, Butler challenges gender dualism by claiming that acting aggressively toward patriarchy can reinforce the very notion that feminists are trying to eradicate. Butler elaborates "[Y]et, in this effort to combat the invisibility of woman as a category, feminists run the risk of rendering visible a category which may or may not be representative of the concrete lives of women" (Butler 1988, 523).

In another direction, feminist artists and philosophers such as Luce Irigaray have attempted to be more specific about female phenomenology and to provide positive accounts of such phenomenology, rather than theorizing how gender essentialism is established through social norms. As Irigaray has observed, "[f]luids are implicitly associated with femininity, maternity, menstruation and the body. Fluids are subordinated to that which is concrete and solid" (Irigaray 1985, 113). The continuous concealment of the reproductive body from society (for example, the practice of menstrual etiquette, homebound pregnant women, and the sexualized breast) has created the false assumption that a leaky body is unnatural, and only when it is concealed is the body acting normally. Therefore, the body is seen as something that should be controlled to enable a prescribed representation of self-identity. Menstruation, in particular, has been codified as something uncontrollable that needs to be organized, managed, and contained.[1]

This chapter is developed in two parts: the first briefly traces menstruation in art history, and the second examines visual interpretation and phenomenological approaches to paintings in the contemporary setting that deal directly with this theme. The impetus for retracing representations of menstruation through art history is to show how historical and modern art has conceptualized and depicted menstrual blood as a way to discuss how the contemporary artworks examined in this essay are innovative and transgressive. Furthermore, this essay is explicitly centered on how contemporary art challenges preconceived views of menstruation, through an examination of the semiotics of menstrual blood. 'Painting blood' as a medium is discussed in parallel to 'Blood as paint' and is similar to the canonical concept of 'paint as flesh,' as thematized by art historians. For example, artists such as Jusepe de Ribera, Rembrandt, Willem de Kooning, and Lucian Freud all create fleshly paintings with their brushwork. I also address contemporary artists who considered the application of paint to be symbolic of menstrual processes and a comment on masculine painting traditions. Blood performances also provide artistic contexts for how blood has featured as a part of art and public space; I would like to use these performances to situate the relatively under-researched and more particular concept of 'gendered blood' as explored by various contemporary artists.

Controlling, elite, and dominating systems of decorum and aesthetics can be countered by reevaluating the abject, transgressive, and marginalized, which are valorized in contemporary art. The primary intention of this essay is to illustrate how art can transvalue menstruation, thereby disturbing or affecting gender essentialism, and create new and meaningful narratives within visual culture.

MENSTRUATION IN EARLIER ART HISTORY

Tracing the representation or visualization of menstruation through art history has not been without difficulty as it is not often explicitly depicted.[2] Nevertheless, menstruation has made its presence felt in painting through indirect and transgressive ways. Modern paintings such as Paul Gauguin's 'grotesque exotic' *Parau na te Varua ino (Words of the Devil)*, an oil on canvas painted in 1892, shows a young Tahitian woman standing naked amidst a tropical garden in front of a masked she-devil. Her posture is that of a woman fallen from grace as she covers her genitals with a white cloth, a symbol of purity. At the feet of the disgraced Eve is a brilliant pink field, encircled by vine roots that are reminiscent of writhing serpents, "leading us to the banks of vermillion flowers that reinforce associations of blood— perhaps menstrual, perhaps that of a passage that has just occurred from virginity into the status of a 'fallen' Eve" (Childs 2003, 178). The modernist artists represented female sexuality from a patriarchal perspective, as a disembodied *other* that is either weak or writhing in pain. Another example is Marc Chagall's 1910 oil on canvas *Birth,* which appears to be one of the earliest paintings of a birth scene that openly portrays uterine blood. Gauguin and Chagall's contemporaries Paula Moderson-Becker and Frida Kahlo challenged these male heterosexual images of women.

While Kahlo never explicitly worked with menstruation as a theme, the image of blood is prevalent throughout her oeuvre, perhaps because she was exposed to many scenes of blood and gore in her life. Kahlo introduces us to the depiction of uterine blood through works such as *Henry Ford Hospital* and *My Birth (Mi Nacimiento)*, both painted in 1932. Each of these works includes the visual depiction of blood underneath a birthing mother. Painting blood in her personal narratives gives these paintings an emotional clarity and impact that perhaps they would not otherwise have. The black and white lithograph *Frida and the Miscarriage (El Aborto, 1932)* illustrates a desire to investigate the meaning of her second unsuccessful pregnancy in two years. The lithograph has been divided vertically down the middle to create alternate possibilities, the reality of miscarriage and fantasy future of successful pregnancy. While the painting of blood in *Henry Ford Hospital* may signify pain and the suffering of miscarriage, the illustration of blood in *Frida and the Miscarriage (El Aborto)* shows a coming to terms with that suffering, a sign of rebirth and growth.

This theme of birth and rebirth is curiously explored in the later painting of the same year, *My Birth*, as it confronts the spectator with a frontal presentation of labor. On the surface, the *ex-voto* painting is an allegorical homage to Our Lady of Sorrows and the loss of her child Jesus. The blood represents loss, and the deity image of Our Lady of Sorrows represents Kahlo's own sorrow. Thus, the painted blood on white sheets partly constructs an iconography of sorrow and loss. The ambiguity is in the title and lack of inscription. *My Birth* implies that someone in this work is the artist. Is Kahlo the figure giving birth or the baby being born? Or, is it a double maternal failure: the death of Kahlo's own mother two months after her miscarriage? The infant appears to have Kahlo's iconic monobrow; perhaps it is a self-rebirth? The intended meaning of the loss of Kahlo's mother and child could mean that all she can now give birth to is endless self-portraits. Yet it is an *ex-voto*, one that gives thanks for survival. Kahlo's paintings and their vivid depiction of uterine blood bear witness to her pain and suffering.

SEVENTIES FEMINISM AND ART

The feminist art that developed in the late 1960s, through the 'second-wave' of feminism in the United States and England, was heralded by a long history of feminist activism. There were two distinct approaches at the time. New York City had a substantially fixed gallery and museum system, and women artists such as Nancy Spero and Louise Bourgeois worked mainly for equal representation in art institutions. This was in contrast to California, where women artists such as Judy Chicago and Miriam Schapiro concentrated on creating a different, separate space for women's art, rather than fighting an entrenched order of male-dominated artists, gallerists, and critics. Feminist artists often combined aspects from various art movements, including conceptual art, body art, performance art, and video art into works that exhibited a message about women's lived experience and the need for gender equality, often in very physical and visceral ways. Performance art in particular, because it is created on a very intimate level and often with no distinction or separation between the artist and artwork, became difficult to ignore or consign to the margins.

During the early 1960s, some years before the growth of an organized feminist movement in the visual arts, Fluxus artists Shigeko Kubota and Carolee Schneemann positioned their bodies in, and as, their art. They employed the modernist language of abstract expressionism, but in contrast to its patriarchal expectations. These artists were "interested in debunking or overthrowing modernism because of its supposedly reactionary desire to ensure the artist's presence" and its dismissal of body art projects. The female artist's body became a gesturing, expressive body, a mode of projecting nonconformity, suffering, activism, and excess, "as a way of laying claim to 'being' itself" (Jones and Warr 2012, 21). As Kubota said in her *Video Poem* (1968–76), "I, a woman, feel, 'I Bleed, therefore I am'" (Stiles and Selz 2012, 504)."

Kubota's *Vagina Painting* (1965), Schneemann's *Interior Scroll* (1975), and Ana Mendieta's *Body Tracks* (1974) (as well as subsequent homages by Schneemann and Nancy Spero to the aforementioned *Body Tracks*) show a spatial and metaphorical use of paint as gendered blood through female expression. In a subjugation of Abstract Expressionism, artists like Kubota used an intermingling of painting and performance to demonstrate that women have a voice within art making and to demarcate space for women as (re)action against the dominant ideologies surrounding painting. Artworks such as Kubota's *Vagina Painting* and Schneemann's *Meat Joy* (1964) destabilized the conventions of art history and criticism, paving the way for widespread feminist art making.

Judy Chicago's depictions of menstruation, *Red Flag* (1971) and *Menstruation Bathroom* (1971–2), have become some of the most documented artworks visualizing menstruation in popular culture. Similarly, Schneemann's *Interior Scroll*—the canonical performance of vulvic space—has now become the ultimate anticlassical-sealed body presentation, referencing menstruation without the semiotic of blood. There were, however, other important blood prints from the seventies, such as Schneemann's *BLOODWORK DIARY* (1972) and Judy Clark's *Menstruation* (1973), as well as occasions of menstrual performance. Catherine Elwes' *Menstruation I* and *Menstruation II* (1979) were performances of horizontal blood inscriptions created through processes of bleeding, drawing, and writing. Through performance, bloodletting, or painterly gestures incorporating a trace of the body, we can also include Marina Abramović, Gina Pane, Ana Mendieta, and later Orlan. These artists all inscribed their blood as gendered, ultimately demarcating a space for contemporary visualizations of menstruation.

Subjects such as sexual violence, menstruation, childbirth, media images, and abortion were high on the agenda of American feminist artists working in the 1970s. Judy Chicago determined, "I just know that I wish to base my art-making entirely on female reality - using objects, items, references that relate to beauty parlors, makeup, menstruation, victimization" (Levin 2007, 179). Chicago refers to menstruation as one of the many female 'issues' that is worthy of individual attention. Chicago's representation of a used tampon in her photographic lithograph, *Red Flag* (1971), is typical of the American feminist objective in the 1970s. She elaborates,

> I wanted to validate overt female subject matter in the art community and chose to do so by making "Red Flag" as a handmade litho[graph], which is a 'high art' process, usually confined to much more neutralized subject matter. By using such overt content in the form, I was attempting to introduce a new level of permission for woman artists. It really worked. (Delaney, Lupton, and Toth 1976, 275)

Chicago's lithograph *Red Flag* initially conjured frightful visions of a bloodied penis, which is typical of Freud's 'phantasy' springing to life. Those who consider the image more carefully would note the silhouette of a vulva and blood-soaked string as indicative that the image shows Chicago in the act of tampon removal, rather than a partially castrated penis. Chicago commented that such confusion was "a testament to the damage done to our perceptual powers by the absence of female reality" (Bobel 2010, 47; Rosewarne 2012, 168). The practice of 'menstrual denial' was exactly what Chicago was trying to abolish. In hindsight, the notion that the lithograph is something other than what it depicted, a blood-soaked tampon, seems strange. However, it seems logical to posit that this might have been one of the earliest images that depicted this routine action, one that women know intimately, yet which is visually absent in many Euro–American cultures (Delaney, Lupton, and Toth 1976, 275).

The significance for *Red Flag*, according to American feminist art historian Arlene Raven, was that "the title, hidden language for menstruation and revolution, would also be hidden information to most men and even some women. Nevertheless, the directness and strength of the visual image elicit immediate, powerful reactions" (Levin 2007, 183). These reactions were a case in point for what Chris Bobel discusses as 'menstrual activism' (2010). The purpose of such 'abjectifying' projects was to "rectify some glaring omissions in art" (Rosewarne 2012, 168) and to bring menstruation out of the "great medicine cabinet of American culture" (Rosewarne 2012, 164). The reaction evoked by *Red Flag* transferred to Chicago's next project, *Menstruation Bathroom*, for the collaborative installation *Womanhouse* (1972) with feminist artists Miriam Schapiro and Faith Wilding, among others.[3]

While the emphasis in the 1970s was on constructing gender equality, the concern now tends to be avoiding gender essentialism so that women can think of themselves outside of patriarchally determined dichotomies of same/other, male/female, and subject/object. It is worth considering whether the subject of menstruation in art is situated within these two phases of feminism differently. In the seventies, it was important to draw attention to (and be proud of) issues and topics, such as menstruation, that had been swept under the rug. Looking forward, has anything changed substantially? Do artists dealing with menstruation within contemporary art have something new to say, or are they still struggling to get patriarchal societies to own up to their prejudices surrounding menstruation? What contribution will visualizing menstruation make to social change? The following section discusses a range of painters and printmakers who utilize menstruation or menstrual blood as a medium to transgress boundaries and challenge normative assumptions about the menstrual cycle.

PAINTING BLOOD

Paint has the capacity to evoke the body's viscera, both internally and externally. Curator Mark Scala proclaims that paint becomes a 'human paste,' as it allows painters to capture the human qualities of the body, revealing something of the subject's psyche and also of their own (Frist Art Museum 2009). 'Paint as flesh' has a long tradition that reaches back to the Renaissance. Referring to the Spanish artist Jusepe de Ribera (1591–1652), art historians have written about his "marks and lines in wet paint creating topography of human flesh as a mirror of the human spirit" (Held and Posner 1971, 85). Many modern and contemporary artists fall along the art historical continuum of 'paint as flesh.' We see this in Bacon's visceral rendering of meat and flesh, as well as in Jenny Saville's truncated portraits of butchered meat and her plastic surgery-inspired portraits, with paint so thick that the skin appears almost translucent. All this fleshiness can be achieved with paint. Flesh and paint are configured within a unified semiotics in painting, and can be extended to include 'blood as paint,' as seen in Rembrandt's *Rape of Europa* (1632), which enacts blood as paint within Europa's dress. Paint as blood and blood as paint can be employed metaphorically and exchanged or creatively elided.

Kubota's Fluxus performance, *Vagina Painting*, mentioned above, took place right after she moved from Japan to New York. It was performed at the Perpetual Fluxus festival in 1965. Kubota informed her audience that she would be using her vagina to paint. In fact, the paintbrush was attached to her underwear but was made to look as though it was directly inserted in the vagina. Upon dipping the paintbrush into a pail of red paint meant to 'invoke menstrual blood' (Butler 2007, 256), Kubota crouched over large rolls of white paper spread out on the floor and pushed down against the paper to make marks. She repeated this gestural, painterly process several times, as the brush dried regularly, in order to cover the roll of white paper. Painting in a horizontal direction above the painting surface is a gesture to the 'eastern calligraphic tradition' (Yoshimoto 2005, 178), as the individual movements painted with the brush created red script-like marks on a crisp white background.

Art Historian Kristine Stiles asserts that Kubota's performance "[r]edefined Action Painting according to the codes of female anatomy" (Stiles 1993, 82), insofar as Jackson Pollock can be considered to have 'masturbated' and 'ejaculated' paint upon a horizontal un-stretched canvas. *Vagina Painting* makes a comment on the tradition of masculine Action Painting, by trespassing on patriarchal aesthetics. Stiles adds that "[t]he direct reference to menstrual cycles seems to compare the procreation/creation continuum lodged in the interiority of women with the temporal cycles of change and growth she experienced in her own art and life after moving from Japan to the United States" (Stiles 1993, 82). Her performance piece made both female and male viewers feel a sense of discomfort. The female spectators voiced concerns about Kubota's paintbrush 'coming out of her vagina' and empathetically discussed how uncomfortable she must have been. Male viewers were faced with symbolic references to menstruation through the use of red paint. Kubota draws a parallel "between human procreation and artistic creation using the vagina as the artistic medium" (Yoshimoto 2005, 179). Not only does Kubota redefine an artistic process, but she also creates within a space of otherness on the margins of Abstract Expressionism. Kubota's spatial art practice blurs the boundaries of masculinist approaches to artmaking. *Vagina Painting* is a cyclic expression of women's difference that plays upon the semiotic between vagina as sexual organ and menstruation. This work builds on Irigaray's 'counter-penis' argument, an action toward the increasingly difficult situation of making work in a patriarchal paradigm.

The iconography of blood on white fabric recalls an old trope that has become synonymous with 'menstrual shame.' Referring back to Kahlo and Kubota, nothing stands out more than the sight of a bloodstain on white cloth. Nonetheless, menstrual symbolism is something that Irigaray (1980, 71) celebrates:

> I love you: body shared, undivided. Neither you nor I severed. There is no need for blood split between us. No need for a wound to remind us that blood exists. It flows within us, from us. It is familiar, close. You are quite red, and still so white. Both at once. You don't lose your candor as you become ardent. You are pure because you have stayed close to the blood. Because we are both white and red, we give birth to all the colors; pinks, browns, blonds, greens, blues . . . For this whiteness is no sham, it is neither dead blood nor black blood. Sham is black: it absorbs everything, closes up and tries to come alive, but in vein . . . The whiteness of this red appropriates nothing. It gives back as much as it receives, in luminous mutuality.

Irigaray uses 'red' and 'white' to critique color associations of passion and purity and uses them to create a locus where 'opposites' may coexist at once, but in a new way. The combination of red blood on a white background constitutes a reference to menstruation as it signifies menstrual blood that cannot be contained.

Gendered Blood in Contemporary Art

Sarah Maple is a provocative contemporary British painter, photographer, and performance artist of Iranian and English descent, who uses the symbolic power of blood on white fabric to create a painting that transgresses social and cultural borders. Maple, in her self-portrait *Menstruate with Pride* (2010), brazenly confronts the taboo of menstruation. The three-paneled painting fits within a broader theme of women and shame, where she uses herself as the subject of humiliation to circumvent patriarchal values in society and, in particular, her Islamic religion. By employing the triptych convention on a two-dimensional support, she underscores the connection to traditional religious altars. Because of her challenge to traditional notions of religion, Maple has been the subject of death threats and protests about her art, and her exhibitions have been vandalized. Maple asserts that her work is a product of a mixed religious upbringing and is not intended to insult, but rather to provoke thought. Curators have continued to exhibit her work despite the "attack on freedom of expression" (BBC 2008). Maple, in an artist statement, narrates the purpose of her art in the third person, "Maple explores the phallocentric regions of pop-culturally transgressive and transcendental phenomena. Her endeavours strike fear at the status quo by erecting an almost architecturally [sic] notion of symbolic terror thus rendering the seemingly un-flinched established order limp with its own theoretically paradoxical flaccidity. She splices binarisms in a comely, fearsomely and castrative manner." This "feminist quasi-conceptualisation" (Maple 2010) makes a statement about how women are perceived and presented from a unique lived experience of dual heritage, which can often expose contradictions.

Maple's painting portrays a semicircle of people enclosing her as she looks out of the canvas toward the spectator, daring us to make a judgment. The onlookers in the painting are of mixed gender and ethnicity, although there is not a diverse representation of age. One child is present with her eyes shielded by her mother's hand, which is an effort to protect the young girl from seeing the inevitable. The group blocks Maple's path, giving her nowhere to go; they lean away in horror. The focus of fear and ridicule is the menstrual stain upon her white dress. Her body transgresses social norms. Maple's personal space is crowded and therefore compromised. Despite the outburst of disgust that surrounds her, the artist raises her fist to salute the viewer, presenting the symbol of resistance and unity, the opposite of a white flag of surrender. Maple challenges the collective belief that menstrual blood is revolting as socially constructed. Women are taught to be ashamed when blood leaks on to one's clothing; Maple is saying that this normative perspective is prescribed and unjust. Instead, women should stand proud and menstruate (Fig. 57.1).

Reproduction is represented by the womb-like lamppost in the background, hovering above Maple's head like a female deity, symbolic of a uterus. Maple has taken care to paint a variety of emotions. In the female subjects, expression ranges from humiliation to empathy, objectification, and menophobia, while the male subjects show predominantly terror and disgust. For the group, Maple ceases to be a person with feelings as she is subjected to this public display of judgment; even the monkey is laughing. As in a history painting, Maple includes a self-portrait in assistance; on the right and to the rear of the group, she is wearing the Muslim *hijab*. Even though her mouth is covered she is smiling with her eyes. Maple's duplication comes from an earlier work, the provocative c-type print, *Sharia Law the Third* (2009), a photographic self-portrait holding a blue book inscribed in gold foil, "Sharia The Islamic Law." The joke is not on the artist but on the crowd surrounding her.

Fig. 57.1   Sarah Maple, Menstruate with Pride, oil on canvas, 275 × 275 cm 2011 (Credit: Sarah Maple)

Maple is a menstrual activist making a humorous comedy about the universal prejudice of 'embarrassing' leakiness.

Maple's paintings draw inspiration from Kahlo, who also regularly used the self as a subject to depict personal, societal, and cultural injustices.[4] However, her application of paint is akin to that of muralist Diego Rivera, firmly placing her painting in conversation with political and protest murals. With a twist of pop realism, she employs 'tongue-in-cheek' humor to balance the political protest in a bid to make visible the ridiculous nature of taboo. To give an example, Maple's signature is suggestive of blood-dripped letters from a horror movie, as well as tagging or graffiti art, which links back to the suburban scene of this painting. *Menstruate with Pride* is conventional in the sense that it adheres to the rules of painting. Maple opens the space of the painting through a one-point perspective to draw the viewer into the scene. She also uses foreshortening effectively to give the characters dynamism, and so the audience senses their abjection in association with menstrual blood.

Many societies associate menstruation with a time of emotional instability, discomfort, and annoyance, resulting in a general consensus that both women and men loathe menstrual blood. Interestingly, creative women such as Louise Bourgeois have referred to menstruation as being "a blessing" and the "best, most creative time," which aligns female fertility with artistic creation and generation (MacMillian 2008, 75; Coxon 2010, 110). Although none of Bourgeois' works directly reference menstruation, except perhaps one anatomical drawing, her paintings depicting sexuality, motherhood, and birth use the color red in a way that suggests menstrual blood. For Bourgeois, red conveys a multitude of meanings:

> Red is the colour of blood
> Red is the colour of paint
> Red is the colour of violence
> Red is the colour of danger
> Red is the colour of shame
> Red is the colour of jealousy
> Red is the colour of grudges
> Red is the colour of blame. (Bernadac 1998, 40; Coxon 2010, 22)

Bourgeois' menstrual iconography is extensive, seen vividly in many works from the 2008 'Nature Study' exhibition on display at the Edinburgh Royal Botanic Gardens. Her blood-red gouache drawings are complemented by a selection of botanical illustrations that provide a dialogue concerning nature, sensuality, and reproduction. In works like *Birth* (2008), Bourgeois fluidly depicts the separation of mother and child. Both figures are painted blood red, composed with a viewpoint that suggests a doctor's examining gaze over splayed legs and a baby descending into this world. In *The Couple* (2007), the red wispy lines appear as parallel bloodstains linked together by an umbilicus. The work is watery and bloody, abstract yet figurative; they are bloodstains personified. We search for some kind of facial recognition within these fluid forms and see two figures emerge standing opposite each other, male and female, their sexual organs reaching toward each other.

'Paint as blood' becomes literal when artists begin to use human blood as a medium. As early as 1973, Judy Clark created a test strip of menstrual bloodstains. She often uses "paint as a stain or sense of seepage" as it allows her to "bring out aspects of human experience that can be hidden or unrecognized" (Clark 2014). Clark's work on paper is laid out on a formal grid, each square holding a circular, organic stain. Each bloodstain has layers, which show that the stain has dried at different rates. This work on paper was produced for a show called *Issues* at Garage Art Ltd, London, where it accompanied other works that used alternative substances from the home including dust, a collection of human hair, and a collection of used plasters, complete with old skin particles, blood, and pus.

Jess Von der Ahe uses blood as a medium for painting and drawing. She mixes it with resin for a body of painterly drawings she calls *drolleries*, also known as 'grotesques.' These hybrid forms are usually a combination of animals, or of humans and animals. Drolleries are found in the margins of illuminated manuscripts featuring the words of God, and usually have a thematic connection with the text. They were popular around 1250 and look similar to creatures found in Hieronymus Bosch's painting *The Garden of Earthly Delights* (1503–4). Von der Ahe's complex droll, *Adam and Eve (After Lucas Cranach the Elder)* (2005), is a composite work of naked human figures all converging to create the form and detail of Adam and Eve. The erotically positioned figures come together as a mass of bodies in an orgy within the internal frame of the first woman and man. Von der Ahe's drawing simply titled *9-5-05* (2005) draws parallels with Bourgeois' *The Couple* (discussed above). The difference is that in Von der Ahe's version the woman is part wheel and the male figure is holding a flower.

Menstrual blood is a potent symbol of inner fluid that brings forth life, and therefore the image is a powerful claim for women's importance to the human race. Using this fluid as a medium in painting and printmaking makes this potency more literal. Perhaps the ultimate transgression is the use of menstrual blood as medium, as it is a step further than representation. Blood as a substance for painting complements the notion of painting as autobiography or personal expression. Blood, human essence, is thus impregnated into works by Von der Ahe and Clark. The use of actual blood on paper implies blood flowing out of control. The bloodstains in Clark's work are clinical; while they appear organic, they are controlled by the formation of the grid. Von der Ahe's stained drawings, on the other hand, are more intuitive; they are surreal figures morphing from autonomous drips of menstrual blood. They are blood drawings of pure psychic automatism.

Conclusion

Menstrual blood, which is gendered blood, has moved from the margins into the public realm of art to challenge dominant ideologies, prejudices, and fears. As a result, menstrual blood and themes of menstruation are now visible within contemporary art and are being used productively to question gender essentialism. It is important to keep in mind that there is not one right or correct way to understand what menstruation, menstrual blood, and taboo mean. Nevertheless, it is essential that artists should be allowed to analyze and expose how traditional structures of power can make people feel vulnerable and ashamed. In very diverse and personal ways, the artists I have discussed in this essay share this common aim. Some artists, such as Chicago, Abramović, and Mendieta, use their art to critique how menstrual blood is made into gendered blood and associated with a stereotype of the weaker and inferior sex. This biological determinism remains unchallenged if assigned to the margins of the non-visible. Thus, their art functions as a social critique of gender essentialism joined to biological essentialism. Some artists, such as Kubota, Maple, and Von der Ahe, go further and celebrate menstruation in their art as a way to turn stigma and shame into transgressive and creative acts. This causes some conflict, the kind characterized by the different traditions of feminism indebted to Butler and Irigaray. While Butler vehemently opposed gender essentialism and saw gender as a social and cultural construct assigning negative traits to women, Irigaray was content to accept some of these traits and celebrate them as a 'female imaginary,' a way of transgressing against patriarchy. It could be argued that this tension continues to exist in the multiplicity of approaches to menstruation in art. It is crucial that such discourse be aired in public, made visible, and divested of any historical or habitual residues of shame or deference to models of decorum and femininity.

This chapter analyzed how artists have used paint to signify or stand in for blood as a challenge to the decorum of modernist formalism, which conveniently erased women's issues. It has made the important contribution of situating menstruation within the discourse of art history. While works of art made with menstrual blood and about menstruation have been created by feminist artists as early as the mid-sixties, this women's 'issue' has never been exclusively researched outside of larger projects on abjection, performance, or feminism. This research and the contemporary artworks that it engages with have the potential to put a marker in the sand, to say no longer will women hide in the menstrual closet, no longer will women be made to feel ashamed by human biological processes.

Notes

1. The contraceptive pill is a way that women can control menstruation in the same sense that waxing is a way to control unwanted body hair. It allows women to emulate this notion of containment of the classical sealed body.

2. For an in-depth study of menstruation in early modern England see, Sara Read, *Menstruation and the Female Body in Early Modern England* (2013).
3. For a list of collaborator names, visit the *Womanhouse* Official Website, http://womanhouse.refugia.net/.
4. This can be seen in Maple's painting, *If I loved you it was because of your hair. Now that you are without hair, I don't love you anymore* (2010), oil and acrylic on canvas (170 × 190 cm). This painting is a portrait of Britney Spears in homage to Kahlo in her *Self Portrait with Cropped Hair* of 1940, oil on canvas (40 × 27.9 cm).

REFERENCES

BBC. 2008. "Muslim Artist Gets Death Threats." Last Modified 31 October 2008. Accessed January 25, 2014. http://news.bbc.co.uk/2/hi/uk_news/england/london/7701168.stm.

Bernadac, Marie-Laurre. 1998. *Louise Bourgeois: Recent Work*. Exhibition Catalogue. London: CAPC Musee d'art Contemporain de Bordeaux/Serpentine Gallery.

Bobel, Chris. 2010. *New Blood Third-Wave Feminism and the Politics of Menstruation*. New Brunswick, NJ: Rutgers University Press.

Butler, Cornelia H., et al., 2007. *WACK!: Art and the Feminist Revolution*. Los Angeles: Museum of Contemporary Art; Cambridge, MA.

Butler, Judith. 1988. "Performative Acts and Gender Constitution: An Essay in Phenomenology and Feminist Theory." *Theatre Journal* 40 (4): 519–31.

Childs, Elizabeth C. 2003. "Eden's Other: Gauguin and the Ethnographic Grotesque." In *Modern Art and the Grotesque*, edited by Frances S. Connelly, 175–92. New York: Cambridge University Press.

Clark, Judy. 2014. "Gallery Statement." Accessed January 21, 2014. http://www.judyclarkartist.co.uk/gallery.php?page=gallery.

Codreanu, Florina. 2009. "Dynamics of Blood in Frida Kahlo's Creation." *Bulletin of the Transilvania University of Brasov* 2 (51), Series IV: Philology and Cultural Studies: 249–56.

Coxon, Ann. 2010. *Louise Bourgeois*. London: Tate Publishing.

Creed, Barbara. 1993. "Medusa's Head: The Vagina Dentata." In *The Monstrous-Feminine: Film, Feminism and Psychoanalysis*, edited by Barbara Creed, 105–121. London: Routledge.

Delaney, Janice, Mary Jane Lupton, and Emily Toth. 1976. *The Curse: A Cultural History of Menstruation*. New York: EP Dutton and Co. Inc.

Frist Art Museum. 2009. "Paint Made Flesh Exhibition at the Frist Center." Filmed 31 March 2009 at Frist Art Museum, Video, 05:12.

Held, Julius S., and Donald Posner. 1971. *Seventeenth and Eighteenth Century Art, Baroque Painting, Sculpture and Architecture*. New York: Prentice Hall/H.N. Abrams.

Irigaray, Luce. 1980. "When Our Lips Speak Together." Translated by Carolyn Burke. *Signs* 6 (1), Women: Sex and Sexuality, Part 2 (Autumn): 69–79.

———. 1985. *This Sex Which Is Not One*. Ithaca New York: Cornell University Press.

Jones, Amelia, and Tracey Warr. 2012. *The Artist's Body*. London; New York, NY: Phaidon.

Levin, Gail. 2007. *Becoming Judy Chicago: A Biography of the Artist*. 1st ed. New York: Harmony Books.

Longhurst, Robyn. 2000. *Bodies: Exploring Fluid Boundaries*. New York: Routledge.

Lupton, Mary Jane. 1993. *Menstruation and Psychoanalysis*. Urbana: University of Illinois Press.

MacMillian, Duncan. 2008. "Blood Ties." *Modern Painters*, May 2008.

Maple, Sarah. 2010. "Artist Statement 2010–2012." Accessed July 3, 2013. http://www.sarahmaple.com/mixedmedia.htm.

Read, Sara. 2013. *Menstruation and the Female Body in Early Modern England*. Basingstoke: Palgrave Macmillan.

Rosewarne, Lauren. 2012. *Periods in Pop Culture: Menstruation in Film and Television*. Lanham, MD: Lexington Books.

Stiles, Kristine. 1993. "Between Water and Stone, Fluxus Performance: A Metaphysics of Acts," In *In the Spirit of Fluxus: Published on the Occasion of the Exhibition*, edited by Elizabeth Armstrong et al., 64–99. Minneapolis: Walker Art Center.

Stiles, Kristine, and Peter Selz. 2012. *Theories and Documents of Contemporary Art: A Sourcebook of Artists' Writings*. Berkeley, CA: University of California Press.

Yoshimoto, Midori. 2005. *Into Performance: Japanese Women Artists in New York*. New Brunswick, NJ: Rutgers University Press.

CHAPTER 58

To Widen the Cycle: Artists Engage the Menstrual Cycle and Reproductive Justice

Curated and Edited by Jen Lewis

In order to break the social and cultural stigma around the menstruating body, we must expand the conversation, or widen the cycle. We must begin to see all the ways that the menstrual cycle touches lives. We must acknowledge that menstruation is a human right that deserves a public dialogue, not simply the shame and silence that have been handed down from generation to generation. One way to do this is through the power of art.

In 2015 at the Society for Menstrual Cycle Research biennial conference in Boston, USA, I curated *Widening the Cycle*—a diverse collection of work by 36 artists from ten countries that situated menstruation within the reproductive justice framework. The aim of the exhibit was to make the invisible visible and push the boundaries of what we know about the menstrual cycle. This chapter titled "To Widen the Cycle: Artists Engage the Menstrual Cycle and Reproductive Justice" presents a selection of works and artists' statements from that historic exhibit. (For more information, visit www.wideningthecycle.com.) It builds on the Feminist and Body Art Movements that started in the 1960s. Whether directly or indirectly influenced, the artists in this collection continue a dialogue started by Judy Chicago, Carolee Schneemann, and Ana Mendieta. Their work reflects an intersectional approach to feminism, a rise in radical self-acceptance, a resurgence of the DIY spirit popularized with the Riot Grrl movement of the 1990s, and the power of social media to facilitate a revolution. The art selected for this Handbook and the wider collection from which it comes threads together global voices to raise consciousness about menstruation and reproductive justice through feminist art.

Aesthetically, the pieces that follow vary greatly, but they are united by the desire to reframe the social narrative surrounding menstruation and reclaim

C. Bobel et al. (eds.), *The Palgrave Handbook of Critical Menstruation Studies*, https://doi.org/10.1007/978-981-15-0614-7_58

this natural process as something positive rather than negative—erasing "the curse" mentality. Perhaps the most important takeaway from "To Widen the Cycle" is recognizing that this stigmatization is about gendered blood. For example, Ingrid Goldbloom Bloch's piece "Feminine Protection" illustrates the disparity between blood spilled in violence and blood shed during the menstrual cycle. The former is widely accepted and a common staple of popular entertainment, while the latter is reviled despite its connection to life and creation. Illuminating our monthly blood inserts menstruation into the broader gender equality discussion, empowers us to neutralize stigma, normalizes our bodies, and revolutionizes the way society sees bodies that menstruate.

ARTIST'S STATEMENT

Danielle Boodoo-Fortuné

O green god, first mother
She who made the water,
lay your hands on me
shake the firmament
of my body,
deliver me from doubt. (Fig. 58.1)

ARTIST'S STATEMENT

Gabriella Boros

In Hebrew, *niddah* means a menstruating woman, literally, "one who is excluded." According to Jewish law, a man is forbidden to maintain sexual relations with his wife during and after her menses, since anything she touches is impure and can cause impurity to others. In this project, the concepts betray a negativity that is inherent in the Talmudic view of women's cycles (written in 2 AD), as well as my ambivalence to the bodily process.

The Curses illustrates the negative physical manifestations that women may have during menstrual periods. *The Curses*, named after the Victorian term for menstruation, are depicted in elegant single linear fashion with a traditional female medium: embroidery. By abstracting the figures somewhat, I create a beautiful image for painful or embarrassing side effects. In addition, by styling the images on banners, I reverse the usual proud representation of a family coat of arms to a shameful or hard issue. At the bottom of the banners are *bdikah* cloths, used by Orthodox Jewish women to check for purity in the seven days following menses. I have painted them with abstractions rather than the light stains of menstrual blood to give these negative side effects a positive, colorful secondary focus (Fig. 58.2).

Fig. 58.1 *The Burden of Bearing.* Medium: watercolor, acrylic, collage, and ink (Credit: Danielle Boodoo-Fortunè 2013)

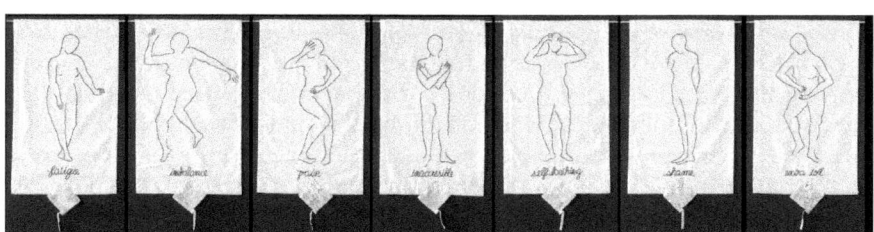

Fig. 58.2 From left to right, the text on the banners reads: fatigue, imbalance, pain, inaccessible, self loathing, shame, extra toil (Credit: Gabriella Boros 2012)

Fig. 58.3 *The Lost Ones.* Nichole Speciale (Credit: Nichole Speciale 2014)

ARTIST'S STATEMENT

Nichole Speciale

The Lost Ones is a group of nine small hoops embroidered with white embroidery thread that has been dyed with menstrual blood. The project seeks to bring attention to both the gendered practice of embroidery and the shaming of the female body. Thread has for many decades been associated with feminine home craft and disregarded as a legitimate art-making medium. In its simplest interpretation, thread is a continuous line. Thread, like the womb, is absorptive. It reflects its environment; it creates new forms from its own tissue. *The Lost Ones* connects the womb to this gendered material, allowing the thread to act as a conduit for the womb's unused material. On a more social level, the project works to expose the artist's own female body. It reveals an otherwise obscured material that is lessened to the status of excrement, even though it is a material responsible for the creation of life. The depicted fetus shapes serve as reminders of the potential held in the menses (Fig. 58.3).

Fig. 58.4 *Threaded Together*. Johanna Falzone. Thread, toilet. Completed in 2013. Originally installed at the Howard Johnson Motel in St. Augustine, FL (Credit: Johanna Falzone 2013)

Artist's Statement

Johanna Falzone

Threaded Together is a site-specific installation work originally on view at a Howard Johnson Motel. The work symbolizes a common thread women have through menstruation. Each pad and tampon is stitched to show how different women may feel about this cycle. Images range from love and reproduction to squiggle blobs, each representing these impressions of what menstruation symbolizes. The installation is placed with a toilet because this is where women change tampons and pads. The toilet is also where fortunate or unfortunate reproductive events, such as miscarriage or using a pregnancy test, may take place. Pregnancy and miscarriage can be both positive and negative outcomes for a woman. No matter how a woman feels about reproduction, this is a cycle women have in common. Menstruation is why women must empathize and celebrate with one another; despite what one's personal beliefs may be, we must respect each woman's choice (Figs. 58.4 and 58.5).

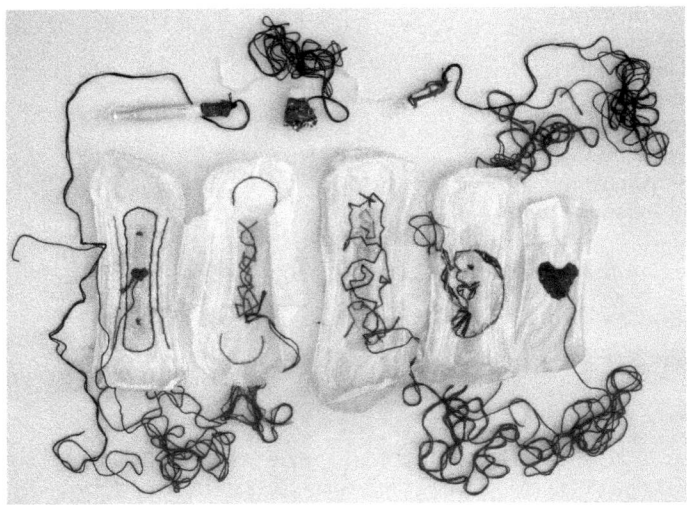

Fig. 58.5 *Threaded Together.* Johanna Falzone. Thread, tampons, maxi-pads. Completed in 2013. Originally installed at the Howard Johnson Motel in St. Augustine, FL (Credit: Johanna Falzone 2013)

Artist's Statement

Phoebe Man

In the southern part of China, there is a custom of eating red eggs to celebrate a baby's birth. I combined the red eggs with sanitary napkins to stress the relationship between menstruation and women's reproductive abilities. Arranging these two materials into blossoming flowers shows my positive attitude toward the naturalness of these bodily functions.

The mirror invites the audience to come closer, in order to be with the works and celebrate the bodily functions together. However, in my own experience, when viewers came closer to my work and realized what the materials were, their facial expressions changed greatly. Some even said, "disgusting!" Why did people's attitudes toward the work change so fast?

The monthly period is still a stigma in Chinese society. It is regarded as unlucky and dirty. Why? This is something I want to question through my work.

The work *My Mirror* is inspired by my anxiety around periods when I was a student. I wanted to enjoy sexual pleasure but did not want to get married and have babies. It was a nightmare when my period came late. These experiences made me want to raise the issue and generate more discussion (Fig. 58.6 and 58.7).

Fig. 58.6 *My Mirror* by Phoebe Chin Ying Man. Sanitary napkins, egg shells, and a mirror, 55 cm × 55 cm × 4 cm (Credit: Phoebe Chin Ying Man 2014)

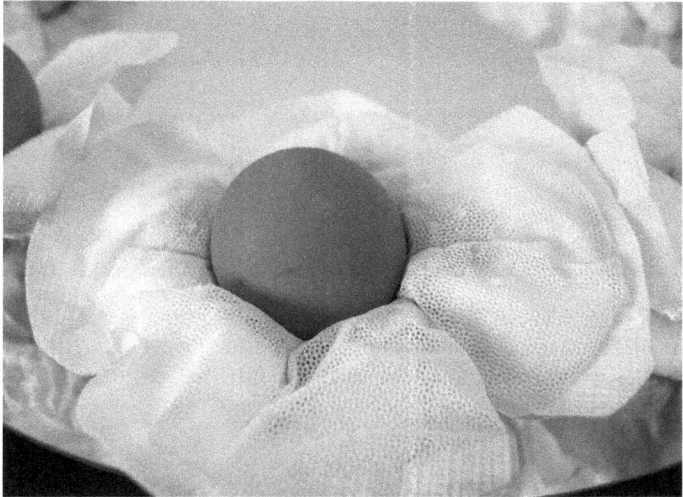

Fig. 58.7 *My Mirror* (detail) by Phoebe Chin Ying Man. Sanitary napkins, egg shells, and a mirror, 55 cm × 55 cm × 4 cm (Credit: Phoebe Chin Ying Man 2014)

Artist's Statement

Ingrid Goldbloom Bloch

Stop the Flow of Violence. PERIOD. is part of a series that transforms recycled and upcycled materials into something entirely different from their intended purpose with the goal of stimulating conversations around polarizing topics. *Stop the Flow of Violence. PERIOD.* speaks for itself. Through a play on words and the use of a gun made out of surprisingly unconventional materials, I hope to draw viewers into conversations around the politics of gun control/violence and the social taboos surrounding menstruation.

I strive to produce art that is both humorous and surprising. I find beauty in common objects, roadside debris, and cast-offs and hope others will too, as they stop and examine how something was made and transformed into something unexpected (Fig. 58.8).

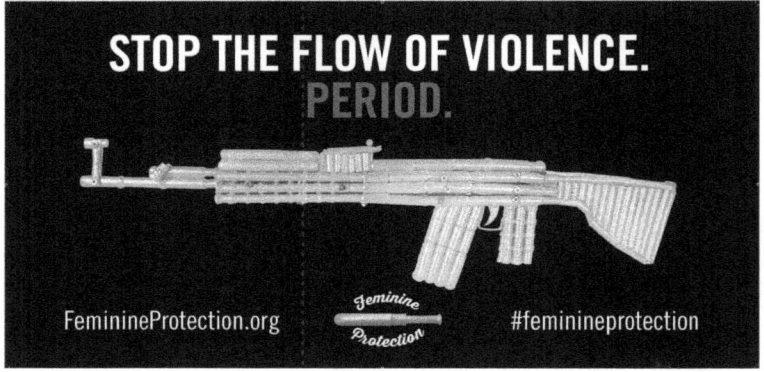

Fig. 58.8 *Stop the Flow of Violence. Period.* From Feminine Protection? series. Ingrid Goldbloom Bloch. Materials: Plastic tampon applicators, woven together in the shape of an AK-47 (Credit: Ingrid Goldbloom Bloch 2015. Photography: Deb Dutcher. Graphic Design: Cheryl Robock)

CHAPTER 59

The Modern Way to Menstruate in Latin America: Consolidation and Fractures in the Twenty-First Century

Eugenia Tarzibachi

INTRODUCTION: THE MODERN (AND AMERICAN) WAY TO MENSTRUATE AS A FEMININE DISCIPLINARY PRACTICE

The so-called "feminine care" or "Femcare" industry that designed and commercialized disposable pads and tampons during the twentieth century changed the experience of the menstrual body in many (but not all) countries of the world. A new disciplinary practice,[1] which Lara Freidenfelds has called "the modern way to menstruate," took hold as the dominant form of femininity[2] among bio-women[3] (2009). It implied more than just a new way to *do* something about menstrual blood every month; these products conceal it efficiently and get rid of the associated burden of the menstrual experience in a more pragmatic way. The disposability of the products became a metaphor for the disposability of the menstrual "crisis"—that is, exposed, public menstruation risked by a barbaric, uncontrollable version of women's bodies. This new way to menstruate also implied a new way to *think* and *talk*, spreading the knowledge accumulated by the modern bio-medical discourse about menstruation (Freidenfelds 2009; Brumberg 1997).

One critical component of the modern way to menstruate was the replacement of homemade menstrual management products made of reusable pieces of towels, cloth, and/or cotton, which were portrayed by the Femcare industry as outdated. The new "feminine protection" products were marketed as a modern way to manage the hygienic "problem" of women's bodies. The products were meant to liberate bio-women from a natural aspect of their biology that had been framed as a monthly problem. From a subjective perspective, women felt liberated by the pads and tampons that helped them

C. Bobel et al. (eds.), *The Palgrave Handbook of Critical Menstruation Studies*, https://doi.org/10.1007/978-981-15-0614-7_59

pass as non-menstruators, but the industry perpetuated and capitalized on the shame and disgust that continued to be associated with their menstrual bodies, albeit "under wraps" (Vostral 2008; Kissling 2006). Menstrual stigma seemed to disappear when the experience of the "menstrual crisis" was significantly reduced, but the shame women felt about their menstrual bodies was perpetuated subtly. The corporeal ideal of a (masculine) non-menstrual body was never problematized; on the contrary, when the menstrual body could be concealed in a more efficient way the ideal was reproduced and continued to nurture menstrual stigma. Therefore, the idea that women were liberated by pads and tampons is paradoxical, since they were now compelled to conceal the menstrual blood through standardized technologies and, in doing so, became somewhat alienated from their periods.

From a Latin-American perspective, this modern way to menstruate was also American, since most of the companies that opened and consolidated the "Femcare" market in Latin America originated in the United States: Johnson & Johnson, Kimberly-Clark, and Procter & Gamble (all now transnational conglomerates). My in-depth study of the experiences of Argentinean women who lived in big cities like Buenos Aires when these products were first available shows that Argentineans who used these products felt that they were entering modernity. For them, it represented not only social status in general, but also alignment with the most developed societies of the world.

In the countries of Las Americas, the proliferation and dissemination of disposable menstrual management technologies occurred at different times, depending on the markets' geography, social class, race, and age. Beyond these particulars, the overall impact of this process can be seen in the global, multimillion-dollar market consolidated by these companies. The major Femcare companies continue to expand, testing new rhetoric that in some cases deeply challenges menstrual stigma. Simultaneously, they are facing competition from new menstrual management technologies, such as reusable products or contraceptives that suppress menstrual bleeding (Hasson 2016; Sanabria 2016).

This chapter, based on some of the main findings of my doctoral and postdoctoral research conducted from 2009 to 2018, describes and analyzes how the dissemination and proliferation of disposable pads and tampons unfolded in the region, and how this process reinscribed traditional narratives about gender. I will focus on a counterpoint between the United States and Argentina as paradigmatic of what happened in the rest of Latin America (Tarzibachi 2017). I will also summarize some of the current initiatives that purport to break down menstrual stigma, as well as look at "new" discourses about other menstrual management technologies that are negotiating their own position alongside the disposable pads and tampons that remain the dominant way to menstruate.

THE EARLY FEMCARE INDUSTRY IN LAS AMERICAS: ORIGINS AND DISSEMINATION OF PADS AND TAMPONS IN THE UNITED STATES AND ARGENTINA

The first disposable pad in the United States—Lyster Towels, by Johnson & Johnson—actually appeared at the end of the nineteenth century, but it was a commercial failure. It was not until after World War I that Kimberly-Clark found a way to successfully monetize women's menstruation. Numerous factors contributed to their success, including growing industrialization, a deep menstrual taboo, and a desire to "civilize" bodies that was nourished by the social hygiene movement. But above all, the invention of feminine pads was driven by economic benefit, a way to capitalize on an invention from World War I (Vostral 2008). The history of the pad's emergence is an example of how women's reproductive bodies were placed at the service of maximizing corporate profits and how a product created for wartime was adapted for civilian use in the postwar period.

Ever since the American Civil War and emancipation, cotton prices had risen steadily. In 1915, Kimberly-Clark began to manufacture cellucotton—a cheaper, more absorbent substitute for cotton—to sell to hospitals as surgical dressings (Heinrich and Batchelor 2004; Sahlberg 2011). In 1917, when the United States entered World War I, the company began selling cellucotton to the North American Navy and the Red Cross; it had contracts to provide some 375 tons to each and opened new factories to produce these immense quantities. Then the signing of the armistice in November 1918 canceled the contracts and left the factories idle (Heinrich and Batchelor 2004, 44). The company decided to optimize both assets by reconfiguring the product and creating feminine hygiene pads (Sahlberg 2011, 34). According to Heinrich and Batchelor, the potential waste was perhaps "the only factor that could persuade the businessmen of those decades to think about menstrual hygiene and precipitated the search for alternative uses for the cellulose product that led to the introduction of Kotex" (2004, 2). Kimberly-Clark's institutional narrative says that the idea came from The American Fund for French Wounded, which "received letters from the nurses of the North American Navy mentioning having improvised hygienic towels with cellucotton surgical gauze" (48).

The company's aversion to loss was so motivating that it confronted social taboos, albeit through a marketing strategy of extreme discretion. In 1920 the company created a subsidiary, Cellucotton Products Company, to launch the Kotex brand, thereby shielding the Kimberly-Clark name from association with menstruation. Ads did not include the words "sanitary" or "sanitary towel," so women simply had to ask retailers for the neutral-sounding "Kotex," a brand name that suggested cellucotton, cotton, and a "cotton-like-texture" (Heinrich and Batchelor 2004, 48). In 1921, the first Kotex ad was published in *Ladies' Home Journal*. This first ad did not refer to feminine hygiene; rather, it was linked to patriotism, war, and science, to con-note the technological protection of wounded (vulnerable) bodies. Following the introduction of Kotex, Johnson & Johnson quickly launched Modess to compete with Kimberly-Clark and many other minor brands.

Tampons were introduced in the United States in the 1930s, about a dec-ade after pads. While there are records of many brands of tampons—mostly with applicators—Tampax dominated the market. The applicators may have removed some concerns women had with tampons, such as touching their genitals and protecting the hymen. Acceptance was also enhanced by ad cam-paigns that normalized menstruation and encouraged the use of tampons to facilitate women's work outside the home during World War II. The wide-spread acceptance of tampon use in the United States might also be attrib-uted to a pragmatic culture and strong feminist movement.

In Argentina, the first generation of disposable pads was launched by the end of the 1930s; tampons arrived in the late 1960s through the 1970s, but they have not achieved the same level of acceptance as in the United States. Modess from Johnson & Johnson was the main brand of the first generation of pads. Kotex arrived in Argentina around 2010, almost a century after its appearance in the American market. Johnson & Johnson's o.b. tampons were launched almost simultaneously in the United States and Argentina (around 1975); they are the generic name of tampons in Argentina, although today there are many other brands in the local market. While pads were advertised in Argentina just one decade after the American market, tampons lagged by three to four decades. Both cultural and political factors contributed to this lag. One reason for this is that the tampon required women to be willing to touch their own genitalia and have some knowledge about it. In this highly conservative society, where Catholic influence on the culture and the inter-ruption of a dictatorship slowed the development of feminist movements, these requirements were often missing.

MENSTRUAL BODIES, MANAGEMENT, AND MEANINGS: THE SOCIOCULTURAL COMPONENT OF THE TRANSITION FROM REUSABLE RAGS TO MANUFACTURED AND DISPOSABLE PRODUCTS

This section provides a comparison between existing research, which details when and under what circumstances women adopted changing menstrual technologies in the United States, with my own research on this transi-tion in Argentina. I interviewed women from the city and province of Buenos Aires who were young in the 1970s—those who lived the tran-sition between the reusable and disposable products.[4] For the most part, American and Argentinian women adopted disposable pads first and tam-pons later—and only some accepted tampons. This happened differently in each country and depended largely on women's social class. Also, while the use of pads was embraced without controversy in the United States (around the 1950s) and in Argentina (around the 1960s and 1970s in urban sectors), the same was not true for tampons.

Reusable Cloth

In the United States, studies indicate that, before the appearance of disposable pads, most women used homemade cloths that they boiled and reused (Freidenfelds 2009). In the nineteenth century, they made their own pads, folding and sewing cotton, gauze, or rags that they fastened with hooks to their underwear. Some wore petticoats and rubber aprons to avoid stains on clothing (Vostral 2008, 243). In Argentina, the women interviewed indicated that at the time of their menarche (sometime between 1950 and the beginning of 1970) they, like their mothers, used leftover cuts of clothes, sheets, T-shirts, and towels, usually white. Some mentioned that those "rags" or "cloths" could also be purchased already made in the pharmacy and then washed and reused. They washed them by hand with cold water and white soap until the blood stain was gone. Some women from low-income sectors went into much more detail about the washing. A few mentioned that they boiled the cloth if the stain did not come out. Noemí (55 years old), who grew up in a town in Santiago del Estero and came to Buenos Aires in the mid-1970s, said:

> I remember that my mother gave us rags. We could not throw the rags [away]. We had to wash them and hang them and you can't imagine the shame we felt . . . . We first took the blood out of the rags with cold water, then put them in white soap, rinsed them, and finally put them under the sun. And they turned whiteeeee. And if the blood did not come out, if the rags were stained, they were boiled in a pot [over a fire] that we always had at that time . . . exclusively for that. Like the shoes, when we washed them, we put chalk on them.

Many women mentioned that they tried to avoid using bleach, since it irritated the vulva and if you used it, "you went all bald." They were also emphatic about the need to hide the materials, especially from men, both during the laundering process and while they were being worn. For a better grip, some wore girdles, hooks, or special belts.

Disposable Pads

By between 1940 and 1950, women in the United States overwhelmingly used disposable pads. By that time, using rags had become a symbol of poverty and disposable pads became a necessity, even for those who did not yet belong to the middle class. Work by Lara Freidenfelds (2009) shows that, for poor and working-class white Americans, African Americans, and daughters of immigrants, disposable pads constituted a way of becoming Americanized and belonging to the middle class to which they aspired. Among adolescents of working-class families, especially immigrants, disposable pads were a way of assimilating, of leaving behind their mothers' Old World culture and becoming modern Americans (Brumberg 1997). Brumberg's research shows that in the 1930s and 1940s, the daughters of immigrants still used "homemade protectors," and in the Italian community, for example, mothers did not encourage their daughters to change the cloths frequently, not only because of the cost but because a bloody pad meant fertility (1997, 44). Learning about menstruation:

> was less private and more social in the working-class, because poor girls had more opportunities to mix with women of different ages in work, the kitchen, or in common bathrooms . . . . For the daughters of immigrants, it was extremely important to be 'scientific' and 'up to date' on issues related to menstruation and appearance. Girls born in America . . . understood, before their mothers or grandmothers, that there was a North American way of menstruating, that it required participation in the broader consumer society. (Brumberg 1997, 45)

Among the Argentinian middle-class women interviewed—mostly daughters of European immigrants who lived all their lives in the province of Buenos Aires or the federal capital—the adoption of disposable pads occurred throughout the 1960s; as in America, it was a way of belonging to the middle class (Cosse 2014). As young women, they had to question the menstrual management practices that their mothers had taught them, and they came to regard women who still washed "rags" as "barbaric" or poor. This was reinforced by Johnson & Johnson's educational material, "Learn to be a woman," which circulated in schools in the Federal Capital in the late 1960s. The material concluded with a clear description of what it took to be a "modern woman": "You will not really be a modern girl if you do not use MODESS for your intimate protection. FREE YOURSELF AND LIVE."

In the Argentinian interviews, discussion of the move from washable to disposable products often prompted laughter as well as pejorative value judgments regarding the practices of their mothers. It also exposed a dimension of social snobbery. The middle-class women remembered Modess as their first disposable pads and used the term generically; by contrast, none of the women from lower-income sectors recalled their first brand. Those from lower-income sectors—almost all of whom were born in the northern provinces or in countries bordering Argentina and later migrated to Buenos Aires or the federal capital—associated the use of disposable pads with their relocation; this was where they became aware of the products, realized these pads were necessary to become "urbanized," and found them economically accessible. They all had to work, and disposable pads simplified that. However, this consumption was an extravagance for some, even long after the 1970s. They said that the pads "that were thrown" were more hygienic, but they used them only to leave their homes because they were "thrown silver." Carina

(65 years old), for example, came from Paraguay in 1974 and started using disposable pads when she arrived in Buenos Aires, but because they were expensive she alternated with the cloths—a habit she maintained until menopause. This alternating pattern was also seen with disposable baby diapers in the 1980s: Moms used them only to take the baby to the doctor. Otherwise, they were viewed as an extravagance.

Tampons

Tampons were first used in the United States in the 1940s by women who worked outside the home; ballerinas and athletes were early adopters (Vostral 2008). By the time use became widespread, between 1960 and 1970, most versions included applicators (Freidenfelds 2009). With the sexual revolution, tampons were especially attractive to young American women. Many of the women interviewed by Freidenfelds, born between the late 1930s and 1950s, also first tried them in the 1960s.

In the United States, there was extensive medical discussion of tampons. Beginning with their introduction in the late 1930s, there was concern in the medical community and among women about tampons' effectiveness, safety, and sexual implications. Concerns about sexuality centered on the potential rupture of the hymen, autoeroticism, and promiscuity—all based on the assumption that the tampon imitated heterosexual intercourse (Vostral 2008; Freidenfelds 2009). Kimberly-Clark educational material from as early as 1952 tried to dispel these concerns, stating that the tampon was smaller than the opening to the hymen and therefore safe for even the youngest menstruator. They also attempted to educate women about the safety of submerging in water. The issue of tampon use prompting masturbation was simply ignored (Freidenfelds 2009, 175). By the 1960s tampons were less controversial and ads and educational materials encouraged adolescents to use tampons. Medical discussion dissipated, until it resumed in the late 1970s and early 1980s following the deaths of women from Toxic Shock Syndrome (TSS) (see Vostral, in this volume).

Brumberg (1997) examines the sexual implications of tampon use, taking into account the significance of the hymen as a symbol of purity, aptitude for marriage, class membership, moral standard, and object of family control. These concepts were already losing strength as gynecological exams became part of women's health care, and these exams were closely correlated with using tampons. Industry materials undertook work that could be called the "de-sexualization of the tampon," disseminating knowledge about the female genital tract, including the role of the clitoris and how tampons bypassed it completely (Brumberg 1997). Likewise, doctors steered mothers to more fact-based information about their daughters' bodies.

There were some factors of tampon adoption particular to the daughters of immigrants, African Americans, and the poorest white women, who had to cross borders of race, class, culture, and religion to test and use it (Freidenfelds 2009). In this sense, tampons were a culminating expression of the modern management of the menstrual body, an ideal that originated from and reflected the values of white, urban, and educated women of the American middle class of the early twentieth century (Freidenfelds 2009).

In Argentina, the 1960s and 1970s were a time that could be referred to as "authoritarian modernization." The strong conservative (Catholic) culture of those years, crossed by successive military coups, left a very narrow place for women. Some conservative women even judged a single woman's decency by whether or not she used tampons. While the tampon was a symbol of the modern woman who embraced new innovations, most of the Argentinian women I interviewed stated or implied that they didn't use tampons until after they were married, had their first sexual relationship, or had children. The preservation of virginity—which the tampon endangered in their imaginations—was a powerful social value in relation to the sexual morality that governed not only the social order but women themselves. It is important to note that, unlike in the United States, gynecologists in Argentina did not use speculums when examining young girls who had not had their first sexual relationship. According to medical articles published in Argentina (for example, Siemaszcko et al. 1980), this was true even in the 1980s, when doctors used ultrasound scans instead.

It was within this context that tampons began to gain acceptance among young, educated, middle-class women who saw what Isabella Cosse (2009) called the emergent prototype of the "liberated young woman." Most middle-class women and adolescents in the 1970s used tampons at least sometimes—to play sports, wear the tight-fitting jeans that had come into style, and especially to swim and sunbathe during the summer (Manzano 2009, 2010). Although the issue of "internal use" remained an obstacle, this was mostly true among women from lower-income sectors, many of whom never even tried tampons. The boundaries of social class were significant: certainly economics was a large consideration, but symbolism factored in as well. Among these women, lack of knowledge about their non-visible genitalia made the tampon seem foreign and threatening.

When disposable pads and tampons became popular, they had two fundamental impacts for bio-women. First, they offered modernity—a new, practical, and hygienic way for bio-women to manage menstrual fluids. For the poorest white American women, African Americans, and immigrants, they represented a way to access the American way of life. In Argentina, low-income women accessed these products years later than the middle class, both for economic and symbolic reasons. Middle-class Argentine women, on the other hand, saw pads and tampons as a way of aligning with the prototype of modern women, specifically from countries such as the United States. The industry tapped into this aspiration by marketing disposable products as the antithesis of cultural and economic backwardness. Second, these products enabled women to mask and discard their menstrual blood, which also distanced them from their bodies. This led to some alienation of the body itself, even as it greatly eased menstrual management and the tension associated with hiding something seen as culturally disgusting.

So, if the disposable products were associated with women's liberation, what were women liberated from? In short, the technologies of "feminine protection" liberated women from a menstrual body that symbolized something shameful and uncontrollable—a body that could leave them out of the ideal of social acceptability in the public world. However, they did not liberate themselves from the social meanings that, in much more subtle forms, continued to signify their menstrual bodies as something wretched and deplorable.

Public Discourse on Menstruation: Advertising and Educational Materials

This section uses ads and educational materials from the United States and Argentina between 1920 and 1980 to summarize two main concepts that the "Femcare" industry conveyed transnationally about menstrual bodies. The ads are particularly important, as they comprised the main public discourse about menstrual bodies[5] and had to resonate with the then-current cultural meanings in each social context.

During the twentieth century, the Femcare industry promoted both positive and negative conceptions of the menstrual body. On the one hand, menstruation continued to be portrayed as a construct of an impolite femininity, and disposable products added a more efficient way to deny the body. This reinforced traditional gender narratives of the fictional feminine ideal and what was necessary to be desirable. Advertisements represented women as being able to "fix" their biological "problem" and be productive 24/7 in the public domain, with all traces of the menstrual body (the look, smell, and noise of both the blood and the products) safely hidden. On the other hand, materials continued to present the menstrual body as something positive: a symbol of femininity (becoming a woman); a fertile body, ready to become pregnant (a mother) from menarche on. The negative conceptions appeared mainly in advertisements; the more positive ones were in educational materials targeted toward teenagers.

As mentioned previously, the first publicity campaign for Kotex pads used the discourse of patriotism to launch its brand to the US market (Vostral 2008; Sahlberg 2011). That first ad featured a nurse taking care of wounded American soldiers on the French battlefield with the line, "To save men's lives, science discovered Kotex." Without ever explicitly mentioning menstruation, the ad suggested the pads as protectors of vulnerable menstruating women, and connected that to the strong, patriotic soldiers who were wounded (Fig. 59.1).

Fig. 59.1 Ladies Home Journal, 1921 (Credit: Copyright Kimberly-Clark Worldwide, Inc. Reprinted with permission)

From this foundational image, we can begin to reconstruct a story in which protection, hygiene, and liberation formed a semantic tripod throughout the twentieth century to advertise pads and tampons. Within this context, we find two "periods"—that is, period as in time, but also a metaphor that gave a social meaning to these technologies and, collaterally, reproduced feelings about the menstrual body. A first period, from the 1920s until around the 1950s, relied more heavily on concepts of hygiene and female protection; it extolled the hygienist discourse of the beginning of the century, mainly referring to a menstruating body as being simultaneously vulnerable and dangerous, as well as dirty. A second period, starting in the 1960s, aligned with the discourse of female liberation, freeing women from the internal enemy of menstruation in the form of "a mercantile feminism" (Goldman 1992). Like the earlier period, this second period highlighted the idea of a menstruating body as being vulnerable/dangerous, but this era also capitalized on the concept of menstruation as oppressive.

Educational materials focused on the idea of "becoming a young lady" through menarche. That involved (1) introducing a gender pedagogy of different bodily practices; (2) promoting the idea that femininity demanded that menstruation be concealed; and (3) reinforcing the ideal that maternity was every woman's correct and natural destiny. This mainstream explanation of menstruation as "the preparation to become a mother" never questioned whether a bio-woman wanted to become one and, as Emily Martin ([1987] 2001) showed, reproduced gender stereotypes in their explanation of how procreation works (a passive egg that is conquered by an active sperm). Elsewhere, I discuss in more detail how this process unfolded in the United States and Argentina (Tarzibachi 2017). Many of these concepts are still embodied in descriptions that essentialize womanhood through the biological processes of menstruation and reproduction.

GLOBAL FLUXES, IN NUMBERS: THE FEMCARE INDUSTRY IN LAS AMERICAS TODAY

Today, the Femcare or "sanitary protection" sector is second only to the diaper market in the global market for disposable hygienic products (Euromonitor International 2016a). In 2018, its global performance was around $95 billion USD, with the United States and China being the two largest markets (Euromonitor International 2016a). This sector is expected to continue expanding, mainly due to the growth of emerging economies and "population changes and social and cultural shifts that influence women's lifestyles" (Euromonitor International 2018). However, the industry is already considering some threats to its market: reusable products, the increasing use of contraceptives that suppress menstrual bleeding, and rising global rates of obesity (which can change or stop periods) (Euromonitor International 2018).

Procter & Gamble Co., Kimberly-Clark Corp., and Johnson & Johnson Inc., have developed the biggest multinational operations in this sector (Bobel 2010; Euromonitor International 2016a). Disposable pads are the main product sold globally, with nearly 10 times the total sales of tampons, and pads with wings lead in terms of total sales and growth (Euromonitor

International 2018). In 2015, global sales by product were distributed as follows: pads, $22.23 billion; tampons, $2.84 billion; and daily protectors $4.45 billion (Euromonitor International 2016a). Tampon consumption varies widely. It is significantly lower in Latin America, the Middle East, Africa, and the Asian Pacific compared to North America, Australasia, and Western Europe (Euromonitor International 2016a).

Several factors explain this disparity. Given the higher cost of tampons compared to pads, they are easier to market in high-income countries. However, other societal factors seem to contribute to this uneven pattern of consumption. As discussed in earlier sections, there are often significant concerns related to sexuality, including loss of virginity, encouraging masturbation or promiscuity, and lack of knowledge about the body. Therefore, in cultures that place high symbolic value on virginity, closely regulate women's sexuality, and/or lack comprehensive sex education, there may be significant resistance to tampon use. Euromonitor International (2016a) identified several of these factors as barriers to tampon consumption at the global level. Additionally, women with a higher education—especially those who remember the 1970s and 1980s—are concerned about TSS, which made the news in the United States at about the same time that tampons began advertising in Latin America. More recently, the famous case of model Lauren Wasser, who had both legs amputated because of TSS, has renewed those fears (Bever 2017).

The following table summarizes the current state of the Femcare market in Las Americas. For each of the nine reference countries in that region, it illustrates the companies that dominate the market, product sales, projected growth, and the type of product most consumed (Table 59.1).

United States

Pads and tampons are consumed evenly (Euromonitor International 2016b). Of tampons sold in the United States, those with an applicator are by far the biggest sellers. Tampax continues to dominate. This is completely different from the situation in Argentina, where, since the launch of tampons in the late 1970s, its most economical version—without an applicator—has led sales.

Argentina

Given the economic crisis in Argentina, the volume of menstrual management product sales has decreased, especially among lower-income consumers (Euromonitor International 2016c). Disposable pads continue to be the most-consumed menstrual management product in Argentina; tampon usage was never significant.

Table 59.1 The Femcare market in Las Americas

Country	Sales per Year (USD)	Leading companies in the local market	Most-used products	Growth projection in the local market
United States	$3.1 billion	Procter & Gamble Kimberly-Clark	Even consumption between pads and tampons with applicator	Mature, almost "saturated." Growth of reusable products, organic tampons, and internet retailing
Argentina	$300 million	Johnson & Johnson Procter & Gamble Kimberly-Clark	Pads, with digital tampons (no applicator) in lower proportion	In expansion
Brazil	$823 million	Johnson & Johnson Procter & Gamble do Brazil SA Kimberly-Clark do Brazil Industria e Comercio de Productos de Higiene Ltda	Pads, with digital tampons in lower proportion (although the tampon with applicator is growing)	In expansion
México	$330 million	SCA México Kimberly-Clark Procter & Gamble	Pads, with tampons with applicator in lower proportion	In expansion
Perú	$187 million	Productos Sancela Perú SA Kimberly-Clark del Perú SA Johnson & Johnson del Perú SA	Pads, with tampons in lower proportion	In expansion
Colombia	$185 million	Productos Familia Sancela SA Johnson & Johnson de Colombia SA Colombiana Kimberly- Colpapel SA	Pads, with tampons in lower proportion	In expansion
Chile	$95 million	Productos Familia Chile Procter & Gamble Chile	Pads, with tampons in lower proportion (but surpasses average consumption in LATAM region)	Mature but still in expansion
Uruguay	$39 million	Kimberly-Clark Uruguay	Pads, with tampons in lower proportion	In expansion
Bolivia	$27 million	Kimberly-Clark Bolivia	Pads, with tampons in lower proportion	In expansion

Source 2016 and 2018 Euromonitor International reports
Credit: Eugenia Tarzibachi

Brazil

"Hygiene protection" product sales are higher in Brazil than in other Latin American countries, but remain significantly lower than in the United States and Europe (Euromonitor International 2016g, 1). The frequency of use differs widely between the upper-middle socioeconomic sectors (which resemble North America and Europe) and low-income sectors (which are far below the national average). The most-consumed product in Brazil is the pad. Tampons are two or three times more expensive than a standard pad, and marketing strategies for tampons focus on women of higher socioeconomic strata. The digital tampon is dominant, although those with applicators are slowly gaining, with a 30% market share. Global companies have an 85% share of the main market. Given the economic crisis in Brazil, annual growth of the local market has slowed, although some growth is still expected.

Mexico

Products are very segmented by social strata in Mexico (Euromonitor International 2016d). There are still rural communities where disposable products are not available, and women use rags that they wash and reuse. Marketing targets low-income women to increase per capita consumption. Digital tampons are no longer consumed; they are considered difficult to use and less hygienic, and they face more resistance than tampons with an applicator.

Peru

Sales in Peru registered a record growth of 9% in 2015, due to reduced poverty and increased purchasing power of the population (Euromonitor International 2016f). Pads are the best-selling product of the category, but the penetration of pads is only 70% in urban areas and 30% in rural areas. The tampon represents only 1% of Femcare sales, and 65% of those sold are tampons with applicators. Barriers to tampon use include fears about possible infections and the pull of traditional knowledge of Peruvian women, although there is greater openness among younger generations, particularly among 22- to 40-year olds with middle to high income (Euromonitor International 2016f, 1).

Colombia

National companies, such as Produsa SA with its Ellas brand, have small, regional distribution in the eastern part of the country, while multinationals have nationwide distribution (Euromonitor International 2016e). There are no specified cultural or religious factors that discourage the consumption of Femcare products; however, the tampon is still rejected.

Chile

Sales are expected to hit 70 billion Chilean pesos in 2020 (Euromonitor International 2016h). Tampon use is low, as in the rest of Latin America, although it exceeds the average for that region. Sales are mainly seasonal, with peaks in the summer. The pad is the most-used product, and consumers indicate that the quality is markedly superior to pads in the rest of Latin America.

Uruguay

Tampons are not widely accepted in Uruguay. The women who use them are those with higher educational levels and greater purchasing power, and they prefer digital tampons to those with applicators. Women from lower income and educational backgrounds refer to powerful myths about tampons, and many consider them dangerous. Cotton and fabrics are still used by low-income women, especially those in rural areas. This is so despite the fact that low-cost brands are offered by several companies (Euromonitor International 2016j).

Bolivia

The pad is the most-consumed product in Bolivia, particularly the standard pad with wings. According to market research, there are no religious or traditional barriers to the tampon, but women prefer pads because of their lower cost and because they are unaware of the tampon's benefits (Euromonitor International 2016i, 2). A tampon with an applicator wasn't introduced in Bolivia until 2015. Cotton and rags are still used in remote rural areas, although expanded penetration and lower-cost products are reducing these traditional approaches. However, in some rural areas the intense taboo surrounding menstruation increases women's comfort with disposable products.

CHALLENGING THE LEGITIMATE MEANINGS OF MENSTRUATION: MENSTRUAL MANAGEMENT AND THE MENSTRUAL TABOO IN THE TWENTY-FIRST CENTURY

Today, the Femcare industry fosters very different discussions of the menstrual body than were prevalent during the twentieth century. Much of this is due to criticism the industry has received for practices—like showing blue liquids as a substitute for blood—that reproduce the menstrual taboo. "Female empowerment" is the new signifier, although it is sometimes used very conservatively, in ways that do not significantly challenge existing menstrual taboos. For example, the well-known campaign "Like a girl," features a politically palatable message of empowerment, but without going as far as showing menstrual blood. Perhaps the most revolutionary piece that has come out has been Libresse brand's *Blood Normal*[6] (2018). This ad earned the Glass Lion for Change Grand Prix on the final night of the Cannes Lions International Festival of Creativity. Johnson & Johnson is now wading into these waters in Argentina and other Latin American markets by showing a red liquid for the first time in the history of its advertising there. The commercial *Siempre Libre, Siempre Juntas*[7] (2018) is an example.

Further, the Femcare industry is no longer the dominant or only source of public discussion of menstruation, particularly in countries where the industry did not make inroads in the past century. A different type of intervention has been trending in these countries: efforts by international organizations focused on menstrual hygiene management to advocate for access to menstrual management products. The menstrual discourses promoted by menstrual hygiene efforts have been carefully reviewed by Bobel (2019). Another central actor is the pharmaceutical industry, which offers contraceptive methods that suppress menstruation (Sanabria 2016). Discussion of these technologies tends to portray menstruation as a useless physiological process that may even be dangerous for a woman's health. Competing discourses arise from small companies that produce menstrual cups, panties with pads, or reusable pads. These tend to encourage environmental sustainability and promote positive connotations that can reconnect women with the experience of menstruation—for example, Mother Earth, the different phases of the moon, the cyclic nature of menstruation. However, while some of these reusable technologies have new designs, many tend to evoke those used by the grandmothers of the Argentinian women interviewed, who regarded this form of menstrual management as culturally backward. Other small and little-known producers offer still more innovations, such as the vibrating tampon (a tampon with a vibrator inside that can be activated by women when they experience menstrual cramps).

These examples illustrate the varied approaches to the menstrual body taken by Femcare producers and others in the world of menstrual management. Some promise to protect bio-women from their own blood with disposable pads and tampons; others want women to connect with the benefits of menstruation as a way to promote social causes, such as environmental care. And others seek to suppress the menstrual cycle altogether, splitting a powerful cultural association between femininity and menstruation.

Menstrual activism is also challenging the meanings attached to menstruation. Artists and activists have transmuted the blood that for centuries was mere waste into products that can be, for many, disgusting: Art and facial masks made with menstrual blood are just two examples. The taboo of menstruation is being denounced even in high-visibility institutions such as the Oscar Awards, which in 2019 gave the Best Documentary (Short Subject) award to *Period. End of sentence*. And a new menstrual emoji brings representations of menstruation further into the mainstream. Yet research shows that the stigma of menstruation, which many think is a thing of the past, continues to unfold. Studies show how menstruation and menstrual products generate health risks, private costs absorbed by women, and environmental waste. The issue of sustainability is a double-edged sword: it can be used to individually blame menstruators for harming the planet, without challenging the cultural norms that require them to conceal their menstruation. Finally, research can reveal the multiple areas in which the topic of menstruation has been neglected, particularly in economic, environmental, gender, and reproductive and sexual health agendas. This is especially true for policies that address sexual education for children and adolescents.

These different discourses, and the ways they are antagonistic to each other, demonstrate how the meaning of the menstrual body is currently in dispute. Economic and other political interests are in conflict to define and control the symbolic power of menstruation. Some of these battles are driven by economic interests that have little concern for the dignity of menstruating people, while others aim to offer menstruators other modes of existence than dejection and shame. As more interests come into the arena—the Femcare industry, NGOs, activists, environmentalists—the result is that they often operate on the menstrual body as a battlefield.

Notes

1. Michael Foucault defined a disciplinary practice as a way of subjecting the body to a process of constant surveillance and examination that enables a continuous and pervasive control of individual behavior. In this case, it can be applied to the social construction of femininity (Foucault 1977, 138–39). Perpetual surveillance is internalized by individuals to produce the kind of self-awareness in relation to a social ideal of normality that defines the modern subject.
2. Femininity, here, is understood as a performance (Butler [1993] 2002), an artifice (Bartky 1990).
3. I follow Paul Preciado (2008) in using the term bio-women to avoid essentializing the category of women in a way that equates it with what our culture considers the female human biological body. Similarly, "menstrual bodies" is used as a neutral term that is meant to include menstruators with diverse gender identities.
4. My research in Argentina focuses on the generation that lived through the seventies as young women. This decade saw the introduction of the tampon, the last new "feminine protection" technology of the century, at the same time that disposable pads and disposable diapers became widely available. These effective, disposable products alleviated some of the physical and emotional work that was culturally assigned to women (for example, changing and laundering cloths for reuse and the emotional labor of masking menstruation). While they eased the burden of the tasks imposed on bio-women and were associated with women's liberation, they did not affect the dominant gendered narratives around what women should do and feel in order for their bodies to be considered socially acceptable.
5. You can find further discussion of the advertisements in each country, with a focus on their contextual particularities, in Chapter 3 of my book *Cosa de Mujeres: Menstruacion, Genero y Poder* (Tarzibachi 2017) and my conference at the Library of Congress of the United States in 2016 (https://www.loc.gov/today/cyberlc/feature_wdesc.php?rec=7642).

6. https://www.youtube.com/watch?v=lm8vCCBaeQw.
7. https://www.jnjarg.com/siempre-libre/siempre-juntas.

References

Bartky, Sandra. 1990. *Femininity and Domination: Studies in the Phenomenology of Oppression*. New York: Routledge.

Bever, Lindsey. 2017. "This Model Lost Her Leg Because of Toxic Shock Syndrome: Here's What She Wants You to Know." *The Washington Post*, December 19, 2017. https://www.washingtonpost.com/news/to-your-health/wp/2017/12/19/this-model-lost-a-leg-because-of-an-infection-from-tampons-heres-what-she-wants-you-to-know/?noredirect=on.

Bobel, Chris. 2010. *New Blood: Third Wave Feminism and the Politics of Menstruation*. London: Rutgers University Press.

———. 2019. *The Managed Body: Developing Girls and Menstrual Health in the Global South*. Boston, MA: Palgrave Macmillan.

Brumberg, Joan J. 1997. *The Body Project: An Intimate History of American Girls*. New York: Random House.

———. 1999. "'Something Happens to Girls.' Menarche and the Emergence of the Modern American Hygiene Imperative." In *Women and Health in America*, edited by Judith Leavitt. Wisconsin: University of Wisconsin Press.

Butler, Judith. [1993] 2002. *Cuerpos que importan. Sobre los límites discursivos del sexo*. Buenos Aires: Paidós.

Cosse, Isabella. 2009. "Los nuevos prototipos femeninos de los años 60 y 70: de la mujer doméstica a la joven liberada" En AA.VV. al cuidado de Andrea Andujar, Débora D'Antonio, Fernanda Gil Lozano, Karim Grammático y María Laura Rosa, *De minifaldas, militancias y revoluciones. Exploraciones sobre los 70 en Argentina*. Buenos Aires, Ediciones Luxemburgo.

———. 2014. "Mafalda: Middle Class, Everyday Life, and Politics in Argentina, 1964–1973." *Hispanic American Historical Review* 94 (1): 35–77.

Euromonitor International. 2016a. "Sanitary Protection: Evolving Category in the Changing World of Womanhood." June 2016.

———. 2016b. "Sanitary Protection in the US," June 2016.

———. 2016c. "Sanitary Protection in Argentina," March 2016.

———. 2016d. "Sanitary Protection in México," April 2016.

———. 2016e. "Sanitary Protection in Colombia," March 2016.

———. 2016f. "Sanitary Protection in Perú," March 2016.

———. 2016g. "Sanitary Protection in Brazil," April 2016.

———. 2016h. "Sanitary Protection in Chile," March 2016.

———. 2016i. "Sanitary Protection in Bolivia," March 2016.

———. 2016j. "Sanitary Protection in Uruguay." April 2016.

———. 2018. "STRATEGY BRIEFING." Feminine Care Industry Global Outlook.

Foucault, Michel. 1977. *Discipline and Punish: The Birth of the Prison*. Translated by A. Sheridan. Harmondsworth: Peregrine.

Freidenfelds, Lara. 2009. *The Modern Period: Menstruation in Twentieth Century America*. Baltimore: John Hopkins University Press.

Goldman, R. 1992. *Reading Ads Socially*. NY: Routledge.

Hasson, Katie Ann. 2016. "Not a 'Real' Period? Social and Material Constructions of Menstruation." *Gender & Society* 30 (6): 958–83. https://doi.org/10.1177/0891243216672662.

Heinrich, T., and B. Batchelor. 2004. *Kotex, Kleenex, Huggies: Kimberly-Clark and the Consumer Revolution in American Business.* Columbus, OH: Ohio State University.

Kissling, Elizabeth. 2006. *Capitalizing on the Curse: The Business of Menstruation.* London: Lynne Reinner Publishers.

Manzano, V. 2009. "The Blue Jean Generation: Youth, Gender, and Sexuality in Buenos Aires, 1958–1975." *Journal of Social History* 42 (3): 657–76.

———. 2010. "Juventud y modernización sociocultural en Argentina." *Desarrollo Económico* 50 (199): 363–90.

Martin, Emily. [1987] 2001. *The Woman in the Body: A Cultural Analysis of Reproduction.* Boston: Beacon Press.

Preciado, B. 2008. *Testo Yonqui.* Madrid: Espasa-Calpe.

Sahlberg, J. 2011. "Kotex: An Early 20th Century Demonstration of Media Campaigns Addressing Stigma." *The Yale Historical Review* I (5): 33–43.

Sanabria, Emilia. 2016. *Plastic Bodies: Sex Hormones and Menstrual Suppression in Brazil.* Durham: Duke University Press.

Siemaszcko, K., J. M. Mendez Ribas, B. Warman, N. Pereyra Pachecho, N. Macagno, and R. Nicholson. 1980. "Anticoncepción en adolescencia (I)." *Obstetricia y Ginecología Latino Americana* 38, Año 38, julio agosto: 221–29.

Tarzibachi, Eugenia. 2017. *Cosa de Mujeres. Menstruación, Genero y Poder.* Buenos Aires: Sudamericana.

Vostral, Sharra. 2008. *Under Wraps: A History of Menstrual Hygiene Technology.* Plymouth: Lexington Books.

———. 2010. "Tampons. Re-scripting Technologies as Feminist." In *Feminist Technology*, edited by Linda Layne, Sharra Vostral, and Kate Boyer. Chicago: University of Illinois Press.

———. 2011. "Rely and Toxic Shock Syndrome: A Technological Health Crisis." *Yale Journal of Biology and Medicine* 84: 447–59.

CHAPTER 60

Challenging the Menstruation Taboo One Sale at a Time: The Role of Social Entrepreneurs in the Period Revolution

Maria Carmen Punzi and Mirjam Werner

INTRODUCTION

What does entrepreneurship have to do with the menstruation taboo? What is the connection between innovation in menstrual products and advancement in gender equality? In the last few years, there has been a noticeable increase in attention toward menstruation (Radnor 2017). In 2015 alone, menstruation surpassed 167 mentions in the top five national news outlets, tripling from the numbers in the 2011 to 2014 period (Weiss-Wolf 2019). Actors spanning multiple roles and industries have stepped up to change how society talks about and relates to menstruation. As menstruation gains traction as a social issue, a number of astonishing problems are gaining increased visibility: lack of access to basic products to manage menstruation, especially—but not only—in the Global South (House, Mahon, and Cavill 2013); practices and traditions that threaten women's security and social inclusion during menstruating days (McMahon et al. 2011); and lack of transparency in the menstrual products industry (Kounang 2015), which has no specific requirements to disclose the ingredients in their products (Rabin 2017).

As controversial cultural and market practices continue to be exposed, the picture is becoming crystal clear: menstruation needs to be looked at in its complexity and has the potential to reveal new dimensions of discourses of gender equality (Schechtman 2015), good health (Sommer et al. 2015), access to education, and human rights (Thomas 2007; Human Rights Watch 2017). Unfortunately, milestone documents on global gender equality, like the European Commission 2016–2020 Strategic Engagement for Gender Equality (2015), the Sustainable Development Goals (2015), and the Global

© The Author(s) 2020
C. Bobel et al. (eds.), *The Palgrave Handbook of Critical Menstruation Studies*, https://doi.org/10.1007/978-981-15-0614-7_60

Strategy for Women's, Children's and Adolescents' Health 2016–2030 (2015), represent missed opportunities to recognize how the menstruation taboo hinders gender equality, sustainable consumption, and a healthy life for all.

While institutions seem not to have grasped the urgency of the grand challenge of menstruation, activists throughout the globe have started using their platforms and voices to denounce the taboo that connects women confined to menstrual huts in Nepal (Bowman 2018); young girls skipping school days in the UK (Marsh 2017; Brooks 2018; George 2017); women's feelings of shame about their bodies; and under-regulation of the Femcare industry. Street marches and protests against the "tampon tax" started taking place, with campaigners calling for free menstrual products for girls in need, in the workplace, and in schools (Radnor 2017). Menstrual activism seems to offer a new angle to those fighting for gender equality, one that is inclusive and connects to common struggles. As pointed out by the *Newsweek* article that crowned 2015 as "the year of period," people have started "talking about gender equality, feminism, and social change through women's periods" (Jones 2016).

Building on interviews with 35 social entrepreneurs, communication with current and former employees of mainstream menstrual product companies, and participant observation of menstrual activists, this chapter aims to analyze the role of social entrepreneurs in this so-called 'period revolution.' Exploring the market strategies and social media messaging of social enterprises, along with the ways other activists in the menstrual equity movement question or support their work, this study analyzes the opportunities and potential pitfalls presented by social enterprises' involvement in the movement.

SOCIAL ENTERPRISES AND THE MARKET FOR MENSTRUAL PRODUCTS

Given that menstrual experiences are mediated by their use, it is crucial to critically look at the market for menstrual products—expected to be worth $42.7 billion by 2022 (Allied Market Research 2016). As Bobel (2010) points out, researchers and practitioners have not inquired into the real reasons why women hate their period more than other bodily processes and what role consumerism plays in that hate. For nearly a century, brands have reinforced narratives of secrecy and embarrassment, promising solutions that would save women from revealing to anyone that they were menstruating (Kissling 1996; Jackson and Falmagne 2013). Advertisements created needs and insecurities in consumers, in order to then suggest the purchase of a certain product which would solve them (Park 1996; Karzai 2010). The message was clear: the best sanitary product is the one that keeps women safe, doesn't leak, and hides their "condition" from the outer world (Newman 2010; Malefyt and McCabe 2016; Vostral 2008). Given that menstrual experiences are profoundly affected by how women choose to manage their cycle, the brands selling menstrual products have a powerful role in shaping the way society feels about periods. It is important, therefore, to look at the players that are creating and marketing the products millions of women and girls use every day to manage their period.

Currently, 80% of the Femcare industry is dominated by four main brands: Procter & Gamble (introduced Always in 1983); Johnson & Johnson/ Edgewell (developed o.b. in 1950); SCA (introduced Libresse in the 1940s); and Kimberly Clark (launched Kotex in 1920). While disposable pads and tampons have been the biggest players in the market for a long time, they haven't been the only ones. Brands selling sustainable options like menstrual cups and washable pads have been around since long before the 2015 wave of the period revolution and before they were tied to concepts like "social enterprises" and "B-corps." Before the turn of the twenty-first century, however, reusable products' brands tended to emphasize the environmental benefits of their product rather than their social impact, mainly because menstrual equity was not seen as an issue affecting the Global North. The companies that did sell cups and reusable pads—mainly in health food stores—were not positioning themselves as direct competitors or alternatives to the most established brands. Nor did they seek to spread awareness about "period poverty," a recently coined term (Weiss-Wolf 2019) that recasts menstruation as a rights-based issue (Bobel 2019).

Interviews with representatives of the reusable brands that have been on the market the longest reveal that their original target groups were mostly 'eco-consumers.' An interviewee from DivaCup—in business since 2002—reported:

> When we had first started marketing the [cup] there was a lot of resistance from mainstream channels because it was a niche product at the time. [. . .] We worked through natural product channels, so a lot of health food stores and online retailers whose values lined up with what our product could offer.

A co-founder of one of the best-known menstrual cups reflected:

> A menstrual cup is for everyone, not just for some health freaks or chemist freak—that's the reputation the menstrual cup had especially back in 2012-13. Most women thought it was only for these people who go to organic shops and never eat anything that is not 100% organic. Hippies kind of product.

An interviewee from GladRags, another long-standing reusable brand in the field, testifies:

> We were founded in 1993 and in the beginning [. . .] we were marketed mostly to the very eco-friendly people who were really concerned about environmentalism. Now I think we've definitely shifted away from talking just about environmental benefits [of reusable products] because that's a little more obvious to people.

In the past few years, new companies have emerged, often selling similar products but with different motivations. These "social enterprises" are organizations whose entrepreneurial activity has an embedded social purpose (Austin, Stevenson, and Wei–Skillern 2006). Much has been written about the term "social entrepreneurship" and there is not a common definition that scholars agree on. This is partly due to its recent appearance in the academic literature and partly due to the difficulty of measuring and putting a strict label on what "social" means. While "social enterprise" has been the most popular label among scholars, similar concepts have been introduced. Hockerts (2006), for example, calls "social purpose business ventures" those for-profit businesses whose main purpose of existence is to create external social benefits; Dorado (2006) renames "social entrepreneurial ventures" those ventures that blend business principles and social goals in new and creative ways, while other scholars have referred to them as "double bottom line organizations" (Dees 1998).

The core idea is that if commercial businesses tend to focus on—or even create—new needs, social businesses aim to serve basic, enduring needs more effectively by means of innovative approaches. The social entrepreneur recognizes a social need, demand, or market failure before starting the enterprise, while a commercial one looks for a large or growing total market (Austin, Stevenson, and Wei–Skillern 2006). These hybrid organizations often recognize and pursue opportunities to serve the social mission (Dees 1998), using a "business-like" and innovative approach to fulfill it. The emerging social innovation is seen as a business opportunity and turned into a commercial for-profit aimed at generating both social and economic value, a process that Emerson calls 'blended value' creation (2003).

In the last five or six years, organizations in the menstrual health space have started experimenting with this model (Ratcliff 2017), "using the power of business to fight for gender equality and provide products, education and jobs," as stated on the website of CORA, one of the best-known social enterprises in the space. They sell innovative products with the aim of changing the conversation about the menstrual cycle (Douglas 2017). By offering organic cotton pads and tampons or reusable options, these enterprises strive to change the Femcare industry—so far, loosely regulated and dominated by few disposable brands—and improve women's experience of menstruation.

The social enterprises working in and with the Global South mostly offer reusable options like menstrual cups or washable pads because they are inexpensive, long-term solutions (for example, RubyCup, EcoFemme, AfriPads). They can sell them at competitive prices thanks to subsidies and profits from sales in the Global North. The social enterprises don't stop at product distribution: they organize educational sessions and workshops that often feature local facilitators, creating safe spaces for users to learn about the menstrual cycle alongside the new product. Some of these initiatives also involve local women in the production process, building livelihoods and creating jobs. The objective is to help women and girls overcome the belief that their body is dirty while understanding that periods are normal and shouldn't exclude them from society (McMahon et al. 2011; Boosey and Wilson 2014). On the other hand, the businesses that focus their sales on the Global North countries strive to become lifestyle brands, making the menstrual experience comfortable through high-quality products (for example, FLEX, AuntFlow, DivaCup). By offering customized delivery services, customer support, and opportunities to ask questions online and return products, as well as ways to become part of a broader community, these social entrepreneurs have positioned their products as a means for women to take care of themselves.

Simultaneously, these enterprises are shedding light on something that the Global North has never considered its problem: "menstrual equity" (Weiss-Wolf 2019). Like the founder of the British social enterprise No More Taboo explains, people don't think of the Global North when thinking of menstruation as a social justice issue:

> In the UK, [when asked about the menstruation taboo] people are like: "don't be silly, that just happens in Africa." Actually, when you start digging down and finding out, it's more subtle, [for example] people think you cannot swim on your period at all and all the things about menstrual blood being dirty, people still do believe that in the UK.

This attitude summed up how the lack of conversation and discussion about periods has resulted in the menstrual needs of homeless or underprivileged women being ignored in wealthier countries. A number of media outlets have recently called attention to how low-income families in the UK, especially those with more than one daughter, struggle to find the money for menstrual products (Marsh 2017; PLAN International UK 2018). In more than one case, data showed that girls miss school days and experience great embarrassment, discomfort, and isolation because they're unable to properly take care of their period.

Understanding the danger of dismissing menstrual equity as a "problem of the poor," some of the social enterprises make a conscious effort to overcome the wealth divide between beneficiaries and customers by inspiring a sense of empathy and connection among women worldwide. The language used on the companies' websites reflects this aim:

> When you buy a Ruby Cup, you aren't just investing in a worry-free period for yourself, you're also sharing the same benefit with someone in need. (Ruby Cup 2020)

> The cloth pad is good for your body, for the earth and connects you with others as your purchase gives dignified livelihood to rural women who stitch pads and enables access to product to poor girls. (EcoFemme, n.d.)

> People helping people. Period. (Aunt Flow 2020)

The social entrepreneurs are cautious of working on a clear-cut division of wealthy versus poor or of assuming, generalizing, and misunderstanding practices that need to be fully contextualized in order to be correctly addressed. Religious and cultural understandings of menstruation are different for each community and geographical location; they have changed in a number of places and are in the process of being reassessed in others. This means that working with menstrual products—wherever in the world—requires sensitivity and a willingness to receive and integrate feedback from users, no matter their income or cultural beliefs.

SKEPTICISM AND CRITIQUES

The social enterprise model has been both contested and praised, in other fields as much as in menstrual health. Supporters highlight its potential benefits: empowering girls and women with education and access to more financially and environmentally sustainable solutions, as well as offering freedom from the short-term funding and bureaucracy nonprofit organizations often struggle with. Social entrepreneurship offers the agility of a startup, the ability to hear first-hand accounts from product users through social media and customer engagement, and the opportunity to combine product innovation with social impact and financial sustainability. Nonetheless, there is a certain degree of skepticism around the rapid growth of these hybrid organizations: Is there a need for social enterprises? Are they free-riders or are they genuinely contributing to the period revolution? Critics include activists, scholars, and mainstream Femcare brands.

Activists committed to the period revolution raise the question of how genuine the work of the social entrepreneurs is. Looking at the companies' efforts to make their social projects an appealing point of distinction from other brands, journalist Miller voices activists' concerns, asking: "how authentic is your mission if you're trying to sell stuff?" (2016). The fact that "giving back" can be used as a selling point for these brands raises some concerns. "Does it matter which business is most noble? Not really and there's something inherently icky about branding a natural biological process that actually has nothing to do with self-expression as a capital F feminist movement," continues Miller (2016).

Can a business genuinely take up an activist message like the one of the period revolution? How honest can participation in a social movement be when there is an economic exchange between the company and customers? Some activists remark that a number of brands claiming to be steering the period revolution are simply revisiting product concepts that have been on the market for decades. From their point of view, making organic cotton tampons and period underwear sound like the invention of the century is deceiving and unfair to those companies that started selling those products a long time ago. In the words of Jennifer Weiss-Wolf, "is branding menstruation – even if packaged in all the popular vernacular of feminism, autonomy and girl power – a genuine, viable form of activism?" (2019, 163). The lawyer and tampon tax expert stresses the hard balance between advancing a social agenda and running a successful company. Considering the agenda and messaging of the well-established menstrual product brands in the past decades, it is not surprising that activists worry about companies jumping on the menstrual equity bandwagon, using a social change message as a marketing strategy. Market mechanisms have been considered incompatible with a social agenda and the Femcare industry has done little to challenge that.

To differentiate themselves from mainstream pad and tampon brands, the social enterprises have depicted their own products as sustainable, transparent, and innovative alternatives. However, mainstream brands and activists alike raise doubts about the actual environmental and social impact of offering organic cotton products in the Western market or introducing reusable solutions in underprivileged communities in the Global South. From an ecological standpoint, the incumbents argue that it's still uncertain whether organic cotton is a more environmentally friendly option than traditional cotton. Cotton crops are generally subject to highly chemical processes that have an impact not only on the final products and therefore on consumers, but on farmers as well. Furthermore, while some organic cotton pads and tampons are marketed as biodegradable, for the decomposition process to take place they would need to be disposed of in a composting facility, which is often not possible. An R&D expert working for one of the main traditional brands of pads and tampons explained:

> "*I'm making an organic and biodegradable napkin,*" [they say] but there is no such thing. There are a lot of people claiming that, [but] if you understand the technology, they are not making a biodegradable napkin, they are making a napkin that might biodegrade if you put it in a composting facility. But since there are very few of them and since the product contains pathogens. It's never going to end there. (2019)

Despite the skepticism, the Femcare industry has started reacting to the momentum around sustainable, environmentally friendly products. In October 2018, Procter & Gamble launched the first Tampax menstrual cup. A few months before that, they had acquired the startup *This is L.*, a manufacturer of organic pads and tampons; the transaction was rumored to be worth around $100 million (Magistretti 2019). In April 2019, Edgewell launched an organic cotton version of the best-seller o.b. tampons, available with or without a plant-based applicator. While this may not be a direct result of the work of social entrepreneurs, traditional brands are undeniably trying to adjust their messaging and product offers to a fast-changing market.

Additionally, critics point out a major weakness in the social enterprise model: scalability and access to the resources needed to expand beyond localized projects. A former employee of one of the leading multinationals said, "I think it's sweet that these social enterprises think they're the only ones trying to do something about it, when their budget is ridiculous compared to the one of multinationals, which means that so is their impact." An R&D expert from another major brand commented that, in his view, it is only possible to tackle the hygiene or the waste issue when providing access to menstrual products in disadvantaged communities (personal communication 2017). Disposable solutions are safer in terms of hygiene, because they don't require proper washing and drying and can be either thrown away or burned once used. However, they represent a significant challenge in terms of trash, considering that in many developing countries there is little—if any—waste infrastructure available. Reusable products, however, come with the risk of girls not washing, drying, and storing them properly, potentially raising the risk of infections.

On the social side, both activists and scholars have expressed caution and concerns when assessing the impact of (only) distributing or selling products. Sinu Joseph, a local Indian activist working in menstrual health education, wrote on her website "Mythri Speaks":

> When I started the Menstrual Hygiene awareness initiative 3 years ago [2010], I assumed the answer to menstrual hygiene was distributing Sanitary Napkins. So, we got donors and napkins and gave them for free in rural government schools. And then, we ran out of donors. And then, a small voice in my head asked me why I never bothered to ask the girls if they did need Sanitary Napkins and what they would do if I (or others like me) stopped supplying these. (Joseph 2013)

Scholar and expert Chris Bobel has also problematized the tendency to demonize the use of cloth or rags for menstrual management, which is sometimes described as an unhygienic and primitive practice, in order to justify the introduction of new products (2018).

When asked about the potential shortcomings of the solutions they offer, the social enterprises prove to be aware of them, but also convinced that they're doing their part in contributing to a positive shift in the industry. The social enterprises marketing organic disposables underline how they're changing the industry from the inside, offering a category of products for which there will always be demand:

> No matter what product you're using, there is always going to be waste associated with it [. . .]. Don't get me wrong, I love the idea of reusables. If we look at other things like cloth and nappies, I totally support that, but there is always going to be people who are going to be buying [tampons] and we are just trying to create the best product for that waste.

Organizations working in the Global South, on the other hand, claim to do their best in being mindful of the context they work in:

We were skeptical and getting more and more uncomfortable with the growing number of organizations that were sort of jumping on this bandwagon of a horror scenario of girls in India, the need to fix it [. . .] which was probably well intentioned but a bit alarming because it's not contextualized. I think we have to be really careful when we come from another culture not to superimpose our own assumptions or projections of what's good or bad, sanitary or unsanitary.

Contrasting Institutional Logics

The concept of institutional logics can be helpful for understanding why social entrepreneurs face such skepticism when entering the menstrual health space. Institutional logics are defined as "taken-for-granted social prescriptions that represent shared understandings of what constitutes legitimate goals and how they may be pursued" (Battilana and Dorado 2010, 1420). While the menstrual activists and the big mainstream brands stand on opposite sides of the spectrum when it comes to approaching periods, they adhere to the same institutional logic: businesses are supposed to offer customers products, not engage in social change, and activists are supposed to advocate for political change, not promote products. This clear-cut division of tasks and responsibilities among societal stakeholders belongs to a traditional vision of the world that sees for-profit organizations and the market as sharply separated from the third sector and social impact. However, the institutional complexity that characterizes the twenty-first century world calls for collaborative and innovative responses. Hybrid organizations like social enterprises seem particularly adept at finding solutions as they are able to tap into a multiplicity of institutional elements, discourses, and tools in order to address these multifaceted issues in innovative ways (Battilana and Lee 2014; Tracey, Phillips, and Jarvis 2011). Combining competing institutional logics (Pache and Santos 2013), social enterprises apply business strategies to social change and vice versa.

The work and rationale of the social enterprises need to be understood and contextualized in relation to mainstream brands and not-for-profit organizations dedicated to women's health and sanitation. The social entrepreneurs in the menstrual health space see themselves as activists and social innovators while still caring for the success of their business operations. They are challenging the traditional way to do activism by maintaining that, while presenting the urgency of the issue is essential, it is potentially paralyzing if people are not offered means to act on it. By presenting a variety of innovative products, the social enterprises aim to empower women and men to take action in their daily lives and contribute to making change. While these are the objectives that the social entrepreneurs set out for themselves, for other stakeholders working in the space the question remains: does menstruation really need social entrepreneurship? Is there room among the propellers of the period revolution for such unconventional activists? To understand whether that's the case, it is worthwhile to look at what the approaches of stakeholders working in the realm of menstrual health have been so far.

The branding and advertising of companies that have dominated the menstrual products space for decades have contributed to the secretive tones and the culture of shame surrounding menstruation (Kissling 1996; Vostral 2008; Freidenfelds 2009). It could be argued that this lack of conversation benefitted big companies by preventing competition and discouraging consumers from questioning the ingredients in the products or asking for more choices. In terms of their philanthropic efforts, the big players in the industry mainly engage in sporadic in-kind donations in developing countries. The issue with this kind of initiative is that they often do not take into account that free products discount the value of the items, harm local markets, and make communities dependent on products they won't be able to get access to in the long-run (Wydick, Katz, and Janet 2014). From a business point of view, in terms of product innovation and for-profit logics, social enterprises strive to offer improved options, challenging the rules and mechanisms of the Femcare industry. For them, innovating also means getting to know their target audience: who they are talking and selling to, how they experience menstruation, and how that can be improved. They acknowledge the importance of customer support, offering continuous help and integrating feedback. Further, they have been able to create a strong social media following and online communities where people share experiences, doubts, and tips.

When it comes to the work of non-governmental organizations (NGOs), most programs targeting girls and women of reproductive age fall into either the Water Sanitation and Hygiene (WASH) or the Sexual and Reproductive Health and Rights (SRHR) areas. Menstrual health, located at the intersection of the two, has been mainly tackled as a collateral effect of other programs and not as an area to be addressed independently. Menstrual Hygiene Day, for example, was instituted by the German NGO WASH United, heavily focused on water and sanitation programs. Other long-standing NGOs working in family planning and sexual and reproductive health have mainly looked at menstruation through the lens of contraceptive methods and reproductive disorders. Essentially, the nonprofit world has not focused on menstrual health until recently, and even when it did, it has limited its programming to the Global South, due to donor requirements and budget availability. This represents a missed opportunity to help people in the Global North recognize that the menstruation taboo is part of their reality and beliefs too. Given that nonprofit organizations receive funding through individual and corporate donations, government support, and grants, some of their biggest struggles include short-term project design, bureaucratic processes, and dependence on short-term funding (Goggins Gregory and Howard 2009). One of the cornerstones of the social entrepreneurs' mission is to modernize those traditional philanthropic procedures; their model goes beyond simple product distribution, favoring long-lasting and holistic solutions that are supported by education, conversations, and engagement with local communities. While their holistic approach aims to improve livelihoods by involving women in the production of the products, engaging men and communities, and introducing the importance of body literacy, in practice this is not always successful.

REFRAMING MENSTRUATION: FROM SOCIAL MEDIA TO BODY POSITIVITY

While menstruation has gained considerable attention on global media outlets and social networks in the last couple of years, the topic still remains entrenched in discomfort and silence in most places in the world. A survey conducted by PLAN International UK in July 2017 with 1000 UK girls and young women revealed that 48% of them felt embarrassed by their period, with the figure rising to 56% among 14-year-olds (PLAN International UK 2018). A survey of 1500 U.S. women commissioned by period-proof underwear brand THINX revealed that nearly 60% of women feel embarrassed when they menstruate (Siebert 2018). In the same sample, 71% of the women declared that they had hidden a pad or tampon from view on their way to the bathroom. Based on statistics like these from recent surveys and their own understanding of an opportunity for social change, the social entrepreneurs seek to start conversations, partnerships, and collaborations with nonprofit organizations, politicians, and artists working on all aspects of menstrual, reproductive, and women's health. Their mission is to change the narrative around menstruation by means of an innovative product that is likely to catch the eye of the consumer and spark a conversation.

What tools and strategies do social entrepreneurs use in their quest to change the way society perceives, talks about, and relates to menstruation? Firstly, social entrepreneurs use empowering language, shifting from the notion of female hygiene to menstrual health:

> We started using more language like "caring for your cycle" and self-care. [Other brands use the term] feminine hygiene care but we really felt that speaking more about care for the body and the whole self, that would give women this understanding that 'Oh, rather than this being a disgusting experience, I can see this as being caring for my body.'

Their messaging is meant to inspire body positivity and encourage women to become knowledgeable about their menstrual cycle and overall health. Examples of this can be found in their social media profiles or websites, which feature statements like:

> We believe that periods are precious, that sweat is therapy, and that feeling 100% protected during both is the stuff of dreams. We're down with our discharge, even prouder of our pregnancy shows, runaway bladders are welcome here too. (Modibodi 2020)

> We believe it's time that all women lived totally, unapologetically free. Free from judgement. Free from self-doubt. And free to be yourself. That's why every intimates product that we design (and redesign and redesign) is made with one goal in mind: to make you feel more comfortable in your own skin. (Knix, n.d.)

They consciously drop the term hygiene, using health to convey a holistic sense of well-being beyond just cleanliness. While hygiene can be considered a personal matter, often tied to a responsibility for the woman to be attractive and socially accepted, health is a global concern. The social entrepreneurs want to bring awareness to menstruation as a social issue, not just a female problem.

Secondly, social entrepreneurs aim to inspire empathy and build global connections through the purchase of menstrual products. While a working woman living in a European city with a good salary and social security does not face the same menstrual challenges as a girl in rural East Africa, buying a menstrual cup or a pair of period underwear can make the former aware of the challenge that the latter faces monthly. This way, menstruation can be profoundly felt as a shared human experience. The founder of CORA, a U.S.-based organic cotton tampons and pads subscription, reflects on this:

> [You're] giving women that perspective and helping them to have empathy for another woman who is going through this same experience but maybe having [. . .] much more negative consequences. [These women start seeing their period] as an opportunity to give back and to help another girl and actually perceive it as a force for good.

The chance of connecting does not end with the purchase: the social entrepreneurs have created online communities (for example, Blood Milk; THINX Periodical) and safe spaces where women share their stories of heavy and painful periods, doubts about health, and accounts of years being unsatisfied with the menstrual products available. Women have responded by connecting, getting personal, and stepping up in online and offline platform to share their experiences. These communities offer taboo-free spaces to find answers and access educational resources.

Thirdly, social entrepreneurs have invested in comfortable and high-quality products to encourage consumers to choose reusable or organic options over the inexpensive and practical ones that have always seemed to work fine. These include subscription boxes, reusable options that last up to a couple of years, and lux-feel products with stylish packaging. "I get a lot of customers commenting to me and taking pictures of the products on their shelves saying 'I am proud to have them on display now' whereas in the past they would have been tucked in the back cupboard or in the drawer," recalls the founder of Tsuno. The social entrepreneurs have steered away from the environmental argument predominantly used by reusable brands in the past and describe their menstrual products as comfortable, cost-effective, and easy to use, with less focus on eco-friendliness. The founder of easy., a subscription box that delivers organic cotton tampons to your door, has decided to include pampering products in the box:

> I think people are loving that, the fact that our packaging comes with a fair trade chocolate bar and a little quote, I think it does make it into something that when you see your [tampon subscription] box you're excited, which is a total spin on what you usually feel when you get your period, "augh, here it is, again." So it's a way to make it into somewhat of a positive experience.

Finally, the social entrepreneurs have started speaking of themselves as innovators. Slogans like "We believe Lunette is the future of period care" (Lunette, n.d.), "Welcome to the New Period" (OrganiCup, n.d.), or "The future of menstrual care. To a better period" (LENA 2020) show how these brands are positioning themselves as disruptors and modernizers. From one generation to another, girls learn how to manage their period and keep it a discreet, personal matter. More often than not, they start their menstrual cycle journey by adopting the same product or brand their mother or elder female figure uses. This fact, combined with the secrecy that characterizes the menstrual discourse, results in little if any demand for innovation in product quality, performance, and brand attractiveness. To contrast this trend, social entrepreneurs consciously emphasize the importance of continuously improving menstrual product choices and performance for women all over the world. They see a link between stigmatization of menstruation and a lack of innovation in products. Because menstruation often comes with physical and psychological discomfort, women internalize the idea that they are destined to have a painful experience for the rest of their fertile years. This is reinforced by the negative language surrounding menstruation, resulting in a lack of effort toward improving menstrual experiences.

By stressing innovation, the social entrepreneurs demonstrate that menstruation is worth an investment of time, technology, and money. The co-founder of YONI, a Dutch social enterprise selling organic cotton pads, tampons, and panty liners, notes:

> I don't think anything very inspirational happens in terms of what most tampons and pads packaging looks like [. . .] and so we felt like there was a very big opportunity to start approaching the subject and the product itself in a different way, that's been part of our company from the start.

While not all of these enterprises are offering a radically new product, they show a willingness to modernize the industry: they invest in marketing strategies, provocative messaging, and revolutionary packaging, while working on optimizing the quality of the product and the brand positioning. They embrace the for-profit model, making the case that financial sustainability can be a strong enabler for social innovation.

Conclusion

This chapter has looked at social enterprises in the menstrual health space, analyzing their characteristics and how they are changing discourses of menstruation, and comparing them to other menstrual activists, philanthropic organizations, and mainstream Femcare brands. There is growing evidence that social entrepreneurs are affecting the organizations already working in this space. In the last year, some of the most established menstrual product brands have started advertising their philanthropic efforts and launching campaigns to inspire confidence in young girls or address 'period poverty': Always—Procter & Gamble's brand—has committed to #EndPeriodPoverty in the UK and North America by donating menstrual products to girls and women in need (Always, n.d.-a, -b). In 2017, for the first time in the history of traditional brands' advertisements, BodyForm debuted an ad featuring red liquid to represent menstrual blood, instead of the ever-present blue period (BodyForm 2017). Moreover, most tampons and pads brands have added a section to their websites that addresses concerns about Toxic Shock Syndrome (TSS) and ingredients in their products, even though tampon boxes in the United States have had TSS info since the 1980s, suggesting that consumers increasingly search for this kind of information online. Even though information and explanations stay quite vague, this change can be interpreted as a sign of increased customer awareness and demand for transparency. The most surprising change is surely the launch of the Tampax menstrual cup (Procter & Gamble), on the market since October 2018. Simultaneously, non-governmental and nonprofit organizations are gravitating toward menstrual health, often looking into setting up social enterprise-like models in the countries they work in, like in the case of Simavi and Days for Girls.

Selling menstrual products helps social entrepreneurs address the societal discomfort around menstruation, uncovering the lack of understanding of women's bodies and fertility as well as society's fear of female imperfection (that is, being "leaky" or "dirty"). Instead of condemning practices, habits, and beliefs from other cultures, social enterprises make use of open dialogue and context sensitivity to establish a connection with women both in the Global South and Global North countries, ditching hierarchical and ideological models in favor of user-centered and innovative products. Where educational policies initiated by governments seem to slowly be adapting to current debates, social enterprises have an incomparable agility to communicate with consumers and improve their knowledge of menstrual issues through product experience and shame-free conversations on user-friendly platforms.

The contribution of the social enterprises to the period revolution may challenge the idea that the market and its mechanisms can only do harm. While advertisements and packaging were once thought of as the way for companies to create unnecessary needs and insecurities in consumers that could only be solved by purchasing their product, social entrepreneurs now use them to convey positive and educational messages. The question remains about what is going to happen to these social enterprises in the long run, considering the fast pace they are growing at. Will they reach untapped markets? Will they keep innovating products indefinitely? Will they outgrow the mainstream brands or be acquired by them? Could their social mission eventually be outweighed by business concerns? One thing is for sure: the period revolution is on, and the social entrepreneurs aspire to take the driver seat.

References

Allied Market Research. 2016. "Feminine Hygiene Products Market by Type and Distribution Channel—Global Opportunity Analysis and Industry Forecast, 2015–2022." https://www.alliedmarketresearch.com/feminine-hygiene-market.

Always. n.d.-a. "ALWAYS Commits to #EndPeriodPoverty in the UK." Accessed April 29, 2020. https://www.always.co.uk/en-gb/about-us/endperiodpoverty.

Always. n.d.-b. "Join Always to End Period Poverty!" Accessed April 29, 2020. https://always.com/en-us/about-us/end-period-poverty.

Aunt Flow. 2020. "Our Mission." Accessed April 29, 2020. https://www.goauntflow.com/pages/our-mission.

Austin, James, Howard Stevenson, and Jane Wei–Skillern. 2006. "Social and Commercial Entrepreneurship: Same, Different, or Both?" *Entrepreneurship Theory and Practice* 30, no. 1 (January): 1–22. https://doi.org/10.1111/j.1540-6520.2006.00107.x.

Battilana, Julie, and Matthew Lee. 2014. "Advancing Research on Hybrid Organizing—Insights from the Study of Social Enterprises." *Academy of Management Annals* 8 (1): 397–441. https://doi.org/10.5465/19416520.2014.893615.

Battilana, Julie, and Silvia Dorado. 2010. "Building Sustainable Hybrid Organizations: The Case of Commercial Microfinance Organizations." *The Academy of Management Journal* 53 (6): 1419–40. http://www.jstor.org/stable/29780265.

Bobel, Chris. 2010. *New Blood: Third-Wave Feminism and the Politics of Menstruation.* New Brunswick: Rutgers University Press.

Bobel, Chris. 2019. *The Managed Body: Developing Girls and Menstrual Health in the Global South.* Boston, MA: Palgrave Macmillan.

BodyForm. 2017. "Blood Normal." Video, 0:20. Posted [October 2017]. https://www.youtube.com/watch?v=QdW6IRsuXaQ.

Boosey, Robyn, and Emily Wilson. 2014. "A Vicious Cycle of Silence: What Are the Implications of the Menstruation Taboo for the Fulfilment of Women and Girls' Human Rights and to What Extent Is the Menstruation Taboo Addressed by International Human Rights Law and Human Rights Bodies?" *Sheffield School of Health and Related Research (ScHARR)*, ScHARR Report Series No. 29.

Bowman, Verity. 2018. "Woman in Nepal Dies after Being Exiled to Outdoor Hut during Her Period." *The Guardian*, January 12, 2018. https://www.theguardian.com/global-development/2018/jan/12/woman-nepal-dies-exiled-outdoor-hut-period-menstruation.

Brooks, Libby. 2018. "Period Poverty: Scotland Poll Shows Women Go to Desperate Lengths." *The Guardian*, February 5, 2018. https://www.theguardian.com/society/2018/feb/05/period-poverty-scotland-poll-shows-women-go-to-desperate-lengths.

Dees, J. Gregory. 1998. "The Meaning of Social Entrepreneurship." Kauffman Foundation and Stanford University. https://centers.fuqua.duke.edu/case/wp-content/uploads/sites/7/2015/03/Article_Dees_MeaningofSocialEntrepreneurship_2001.pdf.

Dorado, Silvia. 2006. "Social Entrepreneurial Ventures: Different Values so Different Process of Creation, No?" *Journal of Developmental Entrepreneurship* 11 (4): 319–43. https://doi.org/10.1142/s1084946706000453.

Douglas, Lucy. 2017. "'We're Trying to Normalize the Conversation': The UK Firms Rethinking Tampons." *The Guardian*, November 23, 2017. https://www.theguardian.com/small-business-network/2017/nov/23/were-trying-to-normalise-the-conversation-the-uk-firms-rethinking-tampons.

Ecofemme. n.d. "EcoFemme • Join the Cloth Pad Revolution!" Accessed April 29, 2020. https://ecofemme.org/.

Emerson, Jed. 2003. "The Blended Value Proposition: Integrating Social and Financial Returns." *California Management Review* 45 (4): 35–51. https://doi.org/10.2307/41166187.

European Commission. 2015. "Strategic Engagement for Gender Equality 2016–2019." Accessed March 29, 2018. http://ec.europa.eu/antitrafficking/sites/antitrafficking/files/strategic_engagement_for_gender_equality_en.pdf.

Freidenfelds, Lara. 2009. *The Modern Period: Menstruation in Twentieth-Century America*. Baltimore, MD: Johns Hopkins University Press.

George, Amika. 2017. "The Shame of Period Poverty Is Keeping British Girls Out of School: Let's Break the Silence." *The Guardian*, December 19, 2017. https://www.theguardian.com/commentisfree/2017/dec/19/british-girls-period-poverty-menstruation-sanitary-products.

Goggins Gregory, Ann, and Don Howard. 2009. "The Nonprofit Starvation Cycle." *Stanford Social Innovation Review* 7 (4): 49–53.

Hockerts, Kai. 2006. "Entrepreneurial Opportunity in Social Purpose Business Ventures." In *Social Entrepreneurship*, edited by J. Mair, J. Robinson, and K. Hockerts, 142–54. Basingstoke: Palgrave Macmillan. https://doi.org/10.1057/9780230625655_10.

House, Sarah, Thérèse Mahon, and Sue Cavill. 2013. "Menstrual Hygiene Matters: A Resource for Improving Menstrual Hygiene Around the World." *Reproductive Health Matters* 21 (41): 257–59. https://doi.org/10.1016/s0968-8080(13)41712-3.

Human Rights Watch. 2017. "Understanding Menstrual Hygiene Management & Human Rights." Accessed March 29, 2018. https://www.hrw.org/sites/default/files/news_attachments/mhm_practitioner_guide_web.pdf.

Jackson, Theresa E., and Rachel Joffe Falmagne. 2013. "Women Wearing White: Discourses of Menstruation and the Experience of Menarche." *Feminism & Psychology* 23 (3): 379–98. https://doi.org/10.1177/0959353512473812.

Jones, Abigail. 2016. "The Fight to End Period Shaming Is Going Mainstream." *Newsweek*, April 20, 2016. http://www.newsweek.com/2016/04/29/womens-periods-menstruation-tampons-pads-449833.html.

Joseph, Sinu. 2013. "Menstrual Hygiene Is Not about Sanitary Napkins." *Mythri Speaks*. Accessed April 25, 2019. https://mythrispeaks.wordpress.com/2013/04/08/menstrual-hygiene-is-not-about-sanitary-napkins-2/.

Karzai, Suraya. 2010. "A Bloody Business: How the Feminine Hygiene Industry Sells Taboos." *The York Scholar* 7 (1): 24–32.

Kissling, Elizabeth Arveda. 1996. "Bleeding Out Loud: Communication about Menstruation." *Feminism & Psychology* 6 (4): 481–504. https://doi.org/10.1177/0959353596064002.

Knix. n.d. "About Us." Accessed April 29, 2020. https://knix.com/pages/about-us.

Kounang, Nadia. 2015. "What's in Your Pad or Tampon?" *CNN*, November 13, 2015. https://edition.cnn.com/2015/11/13/health/whats-in-your-pad-or-tampon/index.html.

LENA. 2020. "Why LENA." Accessed April 29, 2020. https://lenacup.com/pages/what-is-a-menstrual-cup.

Lunette. n.d. "About Us." Accessed April 29, 2020. https://www.lunette.com/pages/about-us.

Magistretti, Bérénice. 2019. "FemBeat: P&G Acquires Organic Period Care Startup This Is L." *Forbes*, February 6, 2019. https://www.forbes.com/sites/berenicemagistretti/2019/02/06/fembeat-pg-acquires-organic-period-care-startup-this-is-l/#193bdd2e7bb6.

Malefyt, Timothy, and Maryann McCabe. 2016. "Women's Bodies, Menstruation and Marketing 'Protection': Interpreting a Paradox of Gendered Discourses in Consumer Practices and Advertising Campaigns." *Consumption Markets & Culture* 19 (6): 555–75. https://doi.org/10.1080/10253866.2015.1095741.

Marsh, Sarah. 2017. "Girls from Poorer Families in England Struggle to Afford Sanitary Protection." *The Guardian*, March 17, 2017. https://www.theguardian.com/society/2017/mar/17/girls-from-poorer-families-in-england-struggle-to-afford-sanitary-protection.

Mcmahon, Shannon A., Peter J. Winch, Bethany A. Caruso, Alfredo F. Obure, Emily A. Ogutu, Imelda A. Ochari, and Richard D. Rheingans. 2011. "The Girl with Her Period Is the One to Hang Her Head Reflections on Menstrual Management among Schoolgirls in Rural Kenya." *BMC International Health and Human Rights* 11 (1). https://doi.org/10.1186/1472-698x-11-7.

Miller, Jennifer. 2016. "A New Crop of Companies Want to Make Your Period Empowering." *Bloomberg*, May 18, 2016. https://www.bloomberg.com/news/articles/2016-05-18/a-new-crop-of-companies-want-to-make-your-period-empowering.

Modibodi. 2020. "Our Story." Accessed April 29, 2020. https://eu.modibodi.com/pages/our-story.

Newman, Andrew Adam. 2010. "Rebelling Against the Commonly Evasive Feminine Care Ad." *The New York Times*, March 15, 2010. https://www.nytimes.com/2010/03/16/business/media/16adco.html.

OrganiCup. n.d. "OrganiCup Menstrual Cup: Easier, Healthier and Greener Periods." Accessed April 29, 2020. https://www.organicup.com/.

Pache, Anne-Claire, and Filipe Santos. 2013. "Inside the Hybrid Organization: Selective Coupling as a Response to Competing Institutional Logics." *Academy of Management Journal* 56 (4): 972–1001. https://doi.org/10.5465/amj.2011.0405.

Park, Shelley M. 1996. "From Sanitation to Liberation?: The Modern and Postmodern Marketing of Menstrual Products." *The Journal of Popular Culture* 30 (2): 149–68. https://doi.org/10.1111/j.0022-3840.1996.00149.x.

PLAN International UK. 2018. "Break the Barriers: Girls' Experiences of Menstruation in the UK." https://plan-uk.org/file/plan-uk-break-the-barriers-report-032018pdf/download?token=Fs-HYP3v.

Rabin, Roni Caryn. 2017. "Period Activists Want Tampon Makers to Disclose Ingredients." *The New York Times*, May 24, 2017. https://www.nytimes.com/2017/05/24/well/live/period-activists-want-tampon-makers-to-disclose-ingredients.html.

Radnor, Abigail. 2017. "We're Having a Menstrual Liberation': How Periods Got Woke." *The Guardian*, November 11, 2017. https://www.theguardian.com/society/2017/nov/11/periods-menstruation-liberation-women-activists-abigail-radnor.

Ratcliff, Rebecca. 2017. "Tampons That Care: Helping Girls Across the World to End 'Shame of Periods'." *The Guardian*, May 28, 2017. https://www.theguardian.com/global-development/2017/may/27/one-for-one-firms-target-stigma-menstruation.

Ruby Cup. 2020. "How we Give." Accessed April 29, 2020. https://rubycup.com/pages/how-we-give.

Schechtman, Lisa. 2015. "Why Tackling the Stigma Around Menstruation Is Key to Gender Equality." *World Economic Forum*, May 29, 2015. https://www.weforum.org/agenda/2015/05/why-tackling-the-stigma-around-menstruation-is-key-to-gender-equality/.

Siebert, Valerie. 2018. "Nearly Half of Women have Experienced 'Period Shaming.'" *New York Post*, January 3, 2018. https://nypost.com/2018/01/03/nearly-half-of-women-have-experienced-period-shaming.

Sommer, Marni, Jennifer S. Hirsh, Constance Nathanson, Richard G. Parker. 2015. "Comfortably, Safely, and Without Shame: Defining Menstrual Hygiene Management as a Public Health Issue." *American Journal of Public Health* 105 (7): 1302–11. https://doi.org/10.2105/ajph.2014.302525.

Thomas, Erika. 2007. "Menstruation Discrimination: The Menstrual Taboo as a Rhetorical Function of Discourse in the National and International Advances of Women's Rights."*Contemporary Argumentation and Debate* 28: 65–90.

Tracey, Paul, Nelson Phillips, and Owen Jarvis. 2011. "Bridging Institutional Entrepreneurship and the Creation of New Organizational Forms: A Multilevel Model." *Organization Science* 22 (1): 60–80. https://doi.org/10.1287/orsc.1090.0522.

United Nations. 2015. "Transforming Our World: The 2030 Agenda for Sustainable Development." Accessed March 29, 2018. https://sustainabledevelopment.un.org/post2015/transformingourworld.

Vostral, Sharra Louise. 2008. *Under Wraps: A History of Menstrual Hygiene Technology*. Lanham, MD: Lexington Books.

Weiss-Wolf, Jennifer. 2019. *Periods Gone Public: Taking a Stand for Menstrual Equity*. New York: Arcade Publishing.

World Health Organization. 2015. "Global Strategy for Women's, Children's and Adolescents' Health 2016–2030." Accessed March 29, 2018. http://www.who.int/life-course/publications/global-strategy-2016-2030/en/.

Wydick, Bruce, Elizabeth Katz, and Brendan Janet. 2014. "Do In-Kind Transfers Damage Local Markets? The Case of TOMS Shoe Donations in El Salvador." *Journal of Development Effectiveness* 6 (3): 249–267. http://dx.doi.org/10.1080/19439342.2014.919012.

Ziv, Stav. 2017. "Periods Are Normal, Says First U.K. Commercial to Use Red Liquid on a Pad." *Newsweek*, October 19, 2017. http://www.newsweek.com/periods-are-normal-says-first-uk-commercial-use-red-liquid-pad-688699.

CHAPTER 61

Transnational Engagements: Smashing the Last Taboo—Caring Corporations in Conversation

Edited by Milena Bacalja Perianes

As Maria Carmen Punzi and Mirjam Werner explore in Chapter 60, social enterprises and businesses are increasingly active in the areas of menstrual health and menstrual product distribution. Motivated by social missions but with for-profit models, these companies reflect a new way of doing business where positive social change and financial profit are not mutually exclusive. These hybrid ventures and commercial businesses have emerged as key actors addressing the diverse needs of women and girls, menstruators and those who don't menstruate, in a range of communities. Their recent proliferation, particularly in the menstrual health "space," raises a range of questions and avenues for analysis regarding their approach and contributions to the sector as a whole.

This dialogue features reflections by those working in the private sector whose products and services address a broad range of menstrual health needs in countries around the world. They discuss how they understand their work, their contribution to communities, and their role in a larger movement for menstrual health equity for all. What are their goals? How do they balance the demands of making profits and promoting social change, especially where these goals may be in tension? How do they understand their mission differently than the traditional Femcare corporations? How do they measure their success beyond profits?

While many praise the rise of private sector actors within this space for their innovative approaches, as Punzi and Werner point out, they also face critiques, which can come from both mainstream Femcare corporations and menstrual activists. The Femcare industry questions the sustainability and scalability of the social enterprise approach and products, while menstrual health advocates question whether social enterprises and commercial

C. Bobel et al. (eds.), *The Palgrave Handbook of Critical Menstruation Studies*, https://doi.org/10.1007/978-981-15-0614-7_61

businesses are truly committed to social change. Is their aim to change the discourse around menstruation or are they co-opting feminist politics for a marketing agenda? Proponents of the private sector's increased role argue that their approach allows them to out-innovate the big corporations, focusing locally and listening to their customers to provide products that fit their diverse needs. At the same time, their marketing provides a platform to change the discourse around menstruation and their for-profit model allows them to sustain this work. Unlike NGOs, companies are not restricted by the politics of foreign aid and financing and can more readily and iteratively respond to social problems. But the question will always remain, what are the consequences of addressing social problems through private sector approaches?

The three companies represented below focus on improved menstrual health outcomes for their customers (and women more broadly) as central components of their business models. Within each of their contexts, menstrual health education and advocacy are fundamental to not only the acquisition of new customers, but also to a broader movement around health and gender equality. With social enterprises, start-ups, and large corporations all capitalizing on the market and social opportunity of women's bodies, we have to ask: is this a good or bad trend? Does reframing consumption and consumer choice as activism actually undermine the effort needed to achieve gender equality? Or are we witnessing the rise of a new form of activism that works to achieve social change and improve the lives of women, girls, and their communities by means of business innovation?

These are tricky questions and the dialogue below tackles them head-on. Each of the contributors has different understandings and answers to these complex questions, but all are committed to providing girls and women with positive, shame-free experiences of their bodies and periods. Together, they tell a story of how corporations can care, with a focus on how designing products that meet the needs of underserved and neglected populations is a valuable market and social opportunity. Each is thoughtful about their roles and responsibilities in this hybrid space between traditional Femtech corporations and the broad range of activist and nonprofit approaches.

Mana Care: Ethically Sourced and Environmentally Friendly Solutions to Menstrual Health in the Pacific

Authors: Isabella Rasch and Angelica Salele
Location: Samoa

Mana Care Products is a Samoa-based social enterprise that provides women and girls with affordable, safe, and environmentally friendly menstrual products.

Why Femcare/Femtech?

Women and girls (especially those in developing parts of the world) have specific health needs that have not been met due to a number of reasons:

negative cultural norms, menstrual taboos, and broader systemic gender equality issues. Increasingly, it has become more and more relevant to talk about Femcare or develop Femtech products, because slowly but surely societies across the world are beginning to allow for safer places and conversations where women's and girls' needs can be brought into the light. These conversations would have been nearly impossible to have ten or fifteen years ago, because people and society were not ready to tackle these issues. At Mana Care, we believe people from diverse backgrounds are ready to engage with taboo topics and tackle menstrual health because there is greater awareness of its impact on women's and girls' lives. We choose to build Mana Care because we believe women and girls have suffered in silence for long enough.

The idea to build a company focused on sustainable menstrual products in Samoa was driven by an environmental perspective on women's and girls' health needs. Our desire was to create a product that did minimal harm to the environment. From this point of view, we began researching and engaging in conversations to understand both the environmental and socioeconomic impact of menstrual health in Samoa. We found that access, or more accurately a lack of access, was a clear barrier to menstrual health, with women and girls in Samoa and the Pacific commonly using cloths and rags to manage their periods. By understanding this context, we sought to create an improved product that would meet their needs.

What do you do differently compared to others in the space?

Our product is not only environmentally friendly but is aesthetically attractive to consumers. We use a variety of beautiful and fashionable prints to create pads that our customers purchase to help manage their menstrual cycle. Furthermore, we have built a business model with considerable thought to our brand and marketing. Our main target market is high-income consumers—women who are interested in products because of their aesthetic appeal, and as an environmentally friendly alternative to traditional products (single-use plastics). As part of our branding, we promote the idea that by supporting Mana Care, a social enterprise, customers are supporting their local community. Our main product (the reusable menstrual pad) is also hand made with love by local Samoan women, which makes us unique in this industry. With limited innovation around menstrual health in the Pacific region, Mana Care works to address the needs of women and girls in this regard. We also ensure vulnerable women and girls are able to access the products at a lower cost or through a donation model. At Mana Care, we fundamentally believe in creating a high quality, locally produced, and environmentally friendly product that makes women and girls feel more comfortable and prettier while on their period—and we think it works!

What marketing/communications choices does your company make to stand out from its competitors?

Our emphasis in marketing Mana Care and our products more broadly focuses on the environment and the "plastic-free" movement. We also have just launched a "pad for pad" marketing campaign, which enables us and our buyers to make an additional purchase for lower income women. This is supported by the purchase of limited-edition pads, custom made upon request, from high-end fabric stores.

What role do companies have in breaking menstrual taboos and the silence around women's health?

Awareness-raising is an important part that companies can play in order to help break down menstrual taboos and encourage more open conversations around women's health. As young women, particularly Pacific island women, we understand the challenges and struggles that come with menstruating and how it can be especially hard and confusing when you first get your period. We want to end the confusion, shame, and misinformation perpetuated by negative taboos and cultural norms, in order to allow women and girls to talk more openly about their individual needs. By reshaping the way women think about their bodies and natural processes like menstruation, companies have the power to increase the quality of life for many people, not just women and girls, because menstruation affects men's livelihoods as well. We actively contribute to this process in Samoa and the Pacific region more broadly.

Can companies manage social and financial interests within this space?

We believe companies can do both by showing that there is an unaddressed need, identifying the needs of different market segments, and working to address them. The full social and financial interests within this space are still yet to be fully discovered because menstrual health management is a relatively unexplored area, especially in the developing world. We believe there are major technologies or inventions which could completely change the way women and girls manage their periods. We know there are certainly many that already do, but we are even more hopeful to see what is yet to be discovered.

How do you see new businesses and social enterprises challenging the big Femcare companies?

Buyers and consumers are becoming increasingly aware and cautious about where and how they spend their money. There is evidence to show that more individuals now want to invest or do business with smaller companies that have a social purpose and make positive changes to communities. There is a clear trend toward smaller, locally owned businesses. We foresee a massive change coming to the big Femcare companies because single-use plastic is unsustainable. As more and more consumers demand environmentally friendly products, they will be pressured to pivot or change their products. How long it will take is the real question, and whether they will recognize the need to do so quickly, because the patterns of change we are seeing today indicate that women and girls are going to start switching from the traditional products that are available on shelves today.

What are the biggest challenges facing social enterprises and businesses in the menstrual health/Femtech space?

Menstrual taboos remain a key barrier to improved menstrual health. While menstrual health poses a new opportunity, the negative cultural practices are still strong and taboos continue to exist. The initial investment or purchase price of a pack of reusable menstrual health products is also another issue, because while they are a good investment in the long term, the initial purchase is still very expensive compared to the cost of a month's worth of single-use/disposable products. For some consumers, particularly low-income households, the initial price can be a deterring factor and discourage a good portion of one of our market segments.

MAS Holdings: Innovating in Femtech from the Global South

Author: Ginny Mendis
Location: Sri Lanka

MAS Holdings is one of Sri Lanka's largest apparel tech enterprises, comprising an ecosystem of design offices and apparel-manufacturing facilities with an integrated supply chain. This operation is powered by over 98,000 employees spread across 16 countries. MAS is noted for its emphasis on ethical & sustainable work environments and is the creator of UK-based Become menopause clothing. MAS also developed patent-pending technology for underwear that can absorb body fluids, which is currently available online and in retail stores.

Why Femcare/Femtech?

At MAS Holdings, we see Femtech as a large market opportunity because women's needs have remained mostly underserved and undervalued until now. We strongly believe in creating innovative, functional, wellness-oriented products and solutions across the Femtech space, drawing from our readily available apparel technology and over 30 years of expertise in the apparel and textile industry. As a largely female-powered enterprise, our auxiliary objective is to continue our efforts in women's empowerment and utilize our proficiency to give the female consumer more control over their health, their bodies, and their lifestyles. Another benefit of the rise of Femtech is the greater awareness of women's health issues, which is yet another reason for our enthused efforts in the space.

What do you do differently compared to others in the space?

Seventy percent of our workforce are women, and women drive our Innovation Team. Our strategic goal is to create solutions that better address women's unmet health needs, so our team continuously consults women at every stage of product development. Our core set of values encourages the discovery of each individual's potential. MAS has imbued a culture of fearlessness and new thinking across the company, which has naturally led us to break conventional boundaries in product and solution development.

With women being both the creators and the beneficiaries of our unique products, we also take action on women's empowerment as a core business strategy. Our program, which supports the skills and career development of women in our workforce, has made an unparalleled impact on female employees, the garment industry, and Sri Lanka generally in the last 15 years.

What marketing/communications choices does your company make to stand out from its competitors?
We are big believers in authentic marketing. With 'Become,' our menopause apparel brand, we wanted to build a narrative that menopausal women feel a part of and that spoke directly to their experiences. Generally speaking, apparel brands tend to focus on younger demographics and older women's needs are regularly neglected. It is not often you see older voices or bodies reflected in marketing campaigns, let alone those which are not airbrushed to perfection. The women in our marketing and communications materials are "real menopausal women." They are those who use our product and reflect diverse bodies and experiences. We are especially focused on building a community around menopause, much like that which has emerged around menstrual products. We have a Facebook group which has thousands of women sharing their stories. It is not just about selling a product, but also creating spaces for information-sharing and education.

What role do companies have in breaking menstrual taboos and the silence around women's health?
We have worked in the women's health space and across different issues for quite some time and have seen the role that companies can play in reinforcing stigma around "taboo" issues, but also the potential to break it. At MAS, we think companies have a responsibility to set a higher standard for how they choose to design and market products. Companies have a duty not only to solve the physical health issues or limitations but also to make an effort to normalize and engage with the emotional and social factors that impact women's health. For us, we think cocreating products with our users and making room for women's voices and real experiences is critical to this.

Can companies manage social and financial interests within this space?
For us, this is critical to everything we do. We are deeply focused on corporate responsibility because we believe being a good corporate citizen means actually supporting the communities in which we work. We consciously invest resources into our women's empowerment programs, as well as the sustainability and environmental impact of our factories. This is both financially and socially prudent for us.

How do you see new businesses and social enterprises challenging the big Femcare companies?
As a diversified conglomerate, MAS Holdings has already embraced the entrepreneurial spirit of start-ups. Our Twinnery team is focused on

innovation and design for those not served by current products. We think of ourselves as responsive to current trends as a result. There is more than enough room for a variety of small and large actors to create new products and services for women in this space. According to a recent market analysis, Femtech is emerging as the next big disruptor in the global healthcare market with a USD 50 billion market potential by 2025.

There are an incredible number of untapped markets and needs across women's health. While menstrual products have seen a resurgence in innovation and popularity, many other issues such as menopause or incontinence are still taboo. Whether big or small, these issues pose incredible potential for market and social opportunities for everyone.

What are the biggest challenges facing social enterprises and businesses in the menstrual health/Femtech space?

Whether an enterprise or a company, we all need to make sure we are designing with our users. This seems obvious but is not always the case. Consumer demands, be it in the Femtech space or any other, can be distinctly identified and disruptive technologies—including big data, artificial intelligence, machine learning, and interactive digital applications—can be used to tailor solutions. The challenge lies in our desire to utilize these technologies to explore far greater potential in the space. For decades, healthcare products and solutions were designed and developed with minimum attention paid to the physiological differences between men and women. Today we see a shift in the industry to better serve women beyond products and services focused on maternal health and childcare. There is a greater need for differentiated products and the challenge is to keep up with the rapid pace of technological advancement and innovation, and apply it fittingly to meet our consumer demands.

Ruby Cup: Bridging the Gap between Customers and Beneficiaries

Author: Alfred Muli
Location: Kenya

Ruby Cup is a social menstrual health business that produces and sells menstrual cups around the world. Our mission is to provide sustainable menstrual health solutions to all menstruators regardless of their income. We strive to dismantle taboos surrounding menstruation by delivering sound education on reproductive health and menstrual care so everyone can live their periods with dignity, free of shame. We work in and distribute across 32 countries around the world. Our model is based on a 'Buy One, Give One' concept: for every cup that is purchased, another cup is donated to a girl or woman in need, mainly in Africa and South East Asia. We do this through strategic partnerships and with our colleague organizations based on the ground who implement menstrual health-focused programming.

Why Femcare/Femtech?

Initially this space was totally focused on hygiene. However, a great deal of evidence and data demonstrate the link between menstrual health and gender equality. This links back to the overall well-being of a woman or a girl. I think that is why Femcare and Femtech are gaining in popularity because it links back to so many other sectors and parts of women's and girls' lives. So if you are interested in education, menstrual health is relevant. If you are interested in water and sanitation, menstrual health is also there. If your focus is sustainability and environmental management, menstrual health is also important in this regard. But at the end of the day it is a business opportunity as well. It is a big money business and a good social opportunity, so that has brought a lot of innovation to the space. It is not just the menstrual cup, but period panties, for instance, that are examples of good innovation.

It is important not to underestimate the amount of global advocacy that has brought menstrual health to global attention and invigorated the Femcare/Femtech space. This championing by diverse groups, including feminists, practitioners, and activists, has been key in bringing all these players into this space and putting menstrual health on the agenda.

What do you do differently compared to others in the space?

Ruby Cup's donation model is unique in that we provide education as well as a product. This has proven critical to our success. One of the challenges with product-based interventions—and let us be honest here, most people are just focused on the product—is the need for more than simply distribution of the disposable or reusable sanitary pad or the cup. We believe education is very important and our menstrual health workshops are critical to reaching our target audience.

We are well known for our country-based, national programming. We have worked hard to develop materials and educational workshops that are tailored to the needs of women and girls on the ground. Our approach of working collaboratively with partners is critical to our success. These workshops are run by our ambassadors, who are trained by us and are Ruby Cup users themselves. It is important to us that our trainings are conducted by someone who not only uses the cup but understands the challenges of first-time use. Ensuring that our trainers are also users helps with converting users and building authentic relationships with women and girls.

We also look at how we can sustain support beyond the initial training. This is rather unique! For instance, if you are a new user, you can call the trainer and say, "I don't know how to use my cup" or "Today I'm having trouble with my cup," and they continue to offer ongoing support and guidance. Our training and support is not just around menstrual health, but also puberty and social taboos.

What marketing/communications choices does your company make to stand out from its competitors?

At Ruby Cup, we let the community members tell their stories from their own experiences. So the stories do not come from us, they come from the girls and the women in the community. We are incredibly proud of the partners we work with, because they then become the drivers of our brand and business. We understand the benefit we have provided in the community in terms of the practical impact on women's and girls' lives. Choice is critical to our approach. We do not force any products on any users. Instead we ensure they have all the information they need to make the best decisions for themselves. Fundamentally, we want girls to have a choice. That is important to our communication and branding.

What role do companies have in breaking menstrual taboos and the silence around women's health?

Addressing menstrual taboo is a big part of the responsibility of being in this space. When you produce products to help women and girls feel comfortable, you cannot make assumptions. Companies must listen to their users and understand what the users want, in order to produce products that the users want and need. We cannot just distribute a product and forget the rest. As companies we have to invest in understanding the reasoning behind women's decisions, especially when dealing with a marginalized issue that is stigmatized. We have to work on some aspects of tackling taboos to do so.

This should not just be left to the small organizations. We all need to contribute. If you leave this all to the grassroots organizations, you will never be able to market or position your product. In the end, you will be marketing a myth, because myths are so prevalent around menstruation. You need marketing that is positive and recognizes that menstruation is a biological process and not anything to be ashamed or afraid of.

Can companies manage social and financial interests within this space?

There are a lot of social interests which are critical to business within the menstrual space. At Ruby Cup, we think companies have a responsibility to be more inclusive. They must also invest in research and development to create safe products which meet high standards for quality. We believe that every company should continually invest in, listen to, and learn from their users in order to develop eco-friendly, high-quality, sustainable menstrual products for all. It is doable for a business. It is what we do.

How do you see new businesses and social enterprises challenging the big Femcare companies?

The scale of the company is not what matters most. What is important is the participation and trust of users. While smaller companies will not have the financial resources of the big multinationals that can just flood the market with a particular product, their impact is just as important. The changes we make in a few lives are rewarding and long lasting. That is something large companies can learn from smaller ones.

Large companies have to continue investing in research and development to ensure innovation. While they have the resources, they are not always the first to release or market innovative products. The dynamics are changing in this regard. Look at Procter and Gamble, who just released a menstrual cup! They are learning from smaller social businesses like ours that there is a need and market for this type of product. Different options are important but so is having diverse companies and actors talking about these issues. It helps normalize it.

What are the biggest challenges facing social enterprises and businesses in the menstrual health/Femtech space?
There are a wide variety of challenges facing local businesses. Regulation remains a huge challenge across this space, especially around sanitary pads and menstrual cups. For social enterprises this can be a barrier to market access. Secondly, funding and investment at the early grant stage is an ongoing stress factor for any small enterprise or business. It can be difficult to convince investors and donors of the market opportunity that exists. As much of our work is focused on Africa, the affordability of our product in a small or emerging market can be a particular challenge. In Kenya, for example, the costs of raw or basic products are still very expensive and this affects the pricing. Social enterprises and businesses have to think carefully about how to price their products when working in these kinds of contexts.

PART VI

Menstruation as Narrative

CHAPTER 62

Introduction: Menstruation as Narrative

Elizabeth Arveda Kissling

Personal stories, urban legends, literature, media representations, and other kinds of narratives provide means of sharing information about menstruation, including what women and other menstruators should and should not do during their periods. For instance, no book has had more impact upon pubescent North American girls than Judy Blume's 1970 *Are You There God? It's Me, Margaret*. Girls growing up in the 1970s and onward, in a cultural milieu where they were encouraged to silence their questions and hush their bodies, had a protagonist with whom to identify and empathize. While *Margaret* has been updated to reflect changes in menstrual products, little has changed with respect to norms of menstrual silence.

In conceptualizing menstruation as narrative, we recognize story as a fundamental element of human culture. Any situation is a story, "a humanly constructed set of meanings that make sense out of phenomena" (Gadow 1994, 306). Every culture throughout history has used narratives to process, share, and remember life experience (Gottschall 2012; Hsu 2008). That includes experiences and knowledge of menstruation, menarche, and menopause. Stories—whether the personal narratives of menarche instruction passed from mother to daughter or sister to sister or the scripted, sterile, wooden speeches of menstrual education films of the mid-twentieth century—are more than language. Indeed, they are not merely words or symbols (Reinsborough and Canning 2010). The words, heeded and performed, become actions. Stories do more than describe the world—stories *make* the world. They are more than mere discursive structures; it is only through new narratives that extant cultural narratives can be reframed and altered. Activists and advertisers know this (Ganz 2001; Couto 1993; Kissling 2018); so do politicians and filmmakers (Day 2011).

Stories take many different forms. Although many Westerners conceptualize narratives as linear—with a beginning, middle, end, and point—that

© The Author(s) 2020
C. Bobel et al. (eds.), *The Palgrave Handbook of Critical Menstruation Studies*, https://doi.org/10.1007/978-981-15-0614-7_62

structure is by no means universal. For example, there is the kernel story, a brief, unstructured capsule version or reference to a story identified in (Kalčik's 1975) study of women's personal experience narratives; and the interactive call-and-response practice common in African–American oratory, musical, and literary traditions (Asante 1998; Callahan 2001). And no general narrative can speak for everyone, especially when the goal is resisting and revising dominant narratives, such as the limited biomedical narratives (Gadow 1994) of the menstrual cycle. It is unsurprising, therefore, that research and exploration of menstruation narratives takes many forms and occurs both across and outside many academic disciplines.

The variety of approaches, methods, linguistic styles, national origins, and phases of menstrual life cycle in this section indicates the great breadth and depth of menstrual narratives and narrative analysis. Some stories are told through traditional sociological qualitative methods, collecting stories from interviews with community members and analyzing the stories with interpretive feminist frameworks informed by the participants' views. Among these are stories shared in private—for example, those revealed to Rothschild and Piya (Chapter 66) that were passed across generations of women in Nepal, and those collected by Singh and Sivakami (Chapter 70) that were shared among women in India. Other stories move between intimate partners, as those seen in Fahs' research (Chapter 69) who were discussing and/or practicing period sex.

Sharing stories through media is another narrative form; some are shared on screens in schools, seen in Ghanoui's analysis of twentieth-century U.S. menstruation education films (Chapter 67), while others are shared globally, like the menstrual stories of young Danish YouTubers presented in Andreasen's discussion of monstrous menstruators (Chapter 65). These media narratives are examined through critical lenses, considering historical contexts, ideology, and performance.

Some stories challenge the biomedical menstruation narratives that exclude and distort—for example, the growing awareness among activists, academics, and, increasingly, the general public that 'not only women menstruate and not all women menstruate,' which is articulated in Rydström's analysis of trans menstruation (Chapter 68). What some call 'queering menstruation' is gaining traction in various social, commercial, educational, and political spaces, as Guilló-Arakistain's challenge to menstrual normativity asserts (Chapter 63). While these two pieces incorporate ethnographic material, they also draw upon the language and methods of critical theory to construct new stories and definitions of menstruators. Bobel and Fahs (Chapter 71) also use a feminist, critical approach in their overview of menstrual activism in the United States, building a political history to ground their call for action.

Still other stories are told by voices that provide an unexpected perspective. Berkley Conner combines feminism and rhetorical analysis to show how period stories can be used as political tools in the online 'Periods for Pence' campaign (Chapter 64). The varied personal experience narratives collected by Perianes (Chapter 72) remind us that family stories can serve as instruction, warning, and even entertainment.

We hope the menstruation narratives collected here—and their analyses— educate and entertain you, and in some cases warn you or inspire you.

REFERENCES

Asante, Molefi K. 1998. *The Afrocentric Idea*, Revised and Expanded. Philadelphia: Temple University Press.

Callahan, John. 2001. *In the African-American Grain: Call-and-Response in Twentieth-Century Black Fiction*. Urbana: University of Illinois Press.

Couto, Richard A. 1993. "Narrative, Free Space, and Political Leadership in Social Movements." *The Journal of Politics* 55 (1): 57–79.

Day, Amber. 2011. *Satire and Dissent*. Bloomington: Indiana University Press.

Gadow, Sally. 1994. "Whose Body? Whose Story? The Question about Narrative in Women's Health Care." *Soundings* 77 (3–4): 295–307.

Ganz, Marshall. 2001. "The Power of Story in Social Movements." Paper presented at Annual Meeting of the American Sociological Association, Anaheim, CA.

Gottschall, Jonathan. 2012. *The Storytelling Animal: How Stories Make Us Human*. Boston, MA: Houghton-Mifflin.

Hsu, Jeremy. 2008. "The Secrets of Storytelling: Our Love for Telling Tales Reveals the Workings of the Mind." *Scientific American Mind* 19 (4). Retrieved from EBSCO Academic Search Complete.

Kalčik, Susan. 1975. "'… Like Ann's Gynecologist or the Time I Was Almost Raped': Personal Narratives in Women's Rap Groups." *The Journal of American Folklore* 88 (347): 3–11.

Kissling, Elizabeth Arveda. 2018. *From a Whisper to a Shout: Abortion Activism and Social Media*. London: Repeater Books.

Reinsborough, Patrick, and Doyle Canning. 2010. *Re:Imagining Change*. Oakland, CA: PM Press.

CHAPTER 63

Challenging Menstrual Normativity: Nonessentialist Body Politics and Feminist Epistemologies of Health

Miren Guilló-Arakistain

In the last decade initiatives related to menstruation have multiplied and diversified to include workshops on the experience of menstruation and alternative management of menstrual bleeding; visual and performance art, production of independent zines; art exhibitions, street actions, websites and social media; conferences, and scholarly research.[1] Among such activities, the way the cycle is experienced is primarily analyzed collectively; for example, different alternatives are offered for blood management (such as the menstrual cup, sponge tampons, reusable cloth pads, et cetera). These activities represent a changing *politics of menstruation*.[2] In addition, I use the term "Alternative Cultures and Politics of Menstruation" to refer to these political actions that go beyond the hegemonic view of menstruation to create new and alternative meanings and images. As a broad and dynamic term, this concept incorporates various initiatives of differing intensity and features,

This chapter is based on research carried out for my doctoral thesis, centered on an analysis of menstrual politics, gender, and corporeality (University of the Basque Country, UPV/EHU). Special thanks to Mari Luz Esteban, Margaret Bullen, and Elizabeth A. Kissling for their helpful contributions. Prior to beginning my academic research in 2009, and during the course of it, I participated in a collective in which menstruation was addressed, and I have led and facilitated workshops on the topic in different formats for many years. For this reason, body itineraries, observing events related to the cycle, and the autoethnographic perspective have been fundamental methodological strategies. The names of all research participants have been changed to protect their confidentiality.

C. Bobel et al. (eds.), *The Palgrave Handbook of Critical Menstruation Studies*, https://doi.org/10.1007/978-981-15-0614-7_63

focusing on collective politics as well as individual initiatives (although the latter distinction is rarely clear). Politics, in its Arendtian sense as a realm of human fulfillment, is a collective action that engenders power; or put simply, it is those processes oriented toward the achievement of group objectives. Moreover, I understand culture as a complex network of dynamic meanings and relationships, as well as changes fostered by human action. Therefore, (for example, feminist) individual and collective agency figures centrally in these alternative cultures and politics of menstruation.

In doing so, discourses are produced which question and re-signify medical and pathological approaches to the western biomedical vision of menstruation.[3] This resignification occurs in many different ways and in diverse ideological–political fields: academe, social movements (particularly feminist, environmental, and health movements), the alternative artistic scene, alternative medicine, spiritualties, and what could be termed feminist biomedicine. These politics are connected, in turn, to specific ways of understanding bodies, care, the (self) management of health (particularly reproductive health), social, and gender relations, and ways in which biomedical conceptions of bodies, based on a rigid dimorphic vision of sex and gender, are laid open to debate.[4] However, as it is not easy to break with biomedical ideologies, it is important to analyze critically both the cracks and continuities emerging in these body politics. This will enable us to reflect on corporeal and identity diversity as well as inclusivity exploring each manifestation in more detail. In this way, the more we sharpen the edge of these alternative politics, the more they will fissure the hegemonic ideology of menstruation.

In this chapter, I aim to examine more closely the ways in which alternative politics of menstruation are challenging the paradigm of sexual dimorphism and heteronormativity. To do so, I will first refer to the contemporary western biomedical definition of menstruation, addressing the ideology of *menstrual normativity*. I will then focus on the potential continuity of this biomedical ideology within these politics, observing which ideas are reproduced, but, above all, how breaks and processes of dissidence and subversion emerge in activism. To this end, I will examine the experience of some non-menstruating cisgender women, and finish by summarizing the challenges inherent in understanding menstruation as a more dynamic and complex process, with the goal of contributing to a more flexible approach to our body image and gender frameworks. This will allow for further exploration of alternative corporeal politics that are mindful of body and identity diversity.

THE IDEOLOGY OF MENSTRUAL NORMATIVITY

Menstruation is a highly productive space for analyzing the social body (Douglas 1966; Scheper-Hughes and Lock 1987), while the menstruating body also helps us understand how all bodies are shaped culturally. The traditional western vision of the menstrual cycle is linked to dirtiness, alterity, and taboo (De Miguel 1971; Martin 1987; Laws 1990; Houppert 1999; Esteban 2001; Šribar 2004; Kissling 2006; Ortiz Gómez 2006; Valls-Llobet 2009; Stein and Kim 2009; Bobel 2010; Chrisler 2008; Johnston-Robledo and Chrisler 2011; Guilló 2014; Fahs 2016; Irusta 2018) and, while a more positive approach is currently adopted, medical discourses—both scientific and educational—continue to emphasize the most pathological aspects.[5] In addition to being a source of social stigma, in some societies menstruation is the central axis of the western biological construction of the female body—construed as an undeniable fact of those bodies—and is used to categorize male and female bodies.[6] In the current biomedical definition of menstruation, which appears in discourses and a variety of contemporary gynecological manuals, the menstrual cycle always appears within the biomedical apparatus as part of the female: it occurs in all of them, and *only* in them. These definitions barely recognize the complexity of biological materiality. However, this process is highly variable: there are many women who, for whatever reasons, do not menstruate, while some transgender men and other people with masculine gender identities do. This is something that is rarely included in medical manuals, yet is reproduced in the popular imagination.

For a woman to produce blood she must be at a specific phase of her life cycle, since ovulation does not occur any earlier or later in her life. This stage is marked by the ages of her first period and her last bleeding cycles. These tend to appear in the medical literature as objective, concise, and static stages, with little mention of the enormous variability that results from geographical, temporal, social, and/or health factors. For example, technological advances, improvements in public health, diet, and urban development systems influence biological processes such as the age of menarche[7] (Sau 1980; Parera et al. 1997; Esteban 2001). Apart from the increasing availability of food and the recent decline in the incidence of infectious disease, there are additional influences on the falling age of puberty including environmental exposures to endocrine-disrupting chemicals that alter the age of menarche (Steingraber 2007). Furthermore, the cycle itself and ovulations vary widely: a variability laden with cultural, personal, and social meanings, and dependent on time and place.

In medical discourse, the biological explanation for the cycle is presented as a hierarchy, with hormonal processes being regulated by the hypothalamus, rather than a set of interacting processes, as Emily Martin (1987) pointed out. More technically, the hypophysis (pituitary) directs this process. Following stimulation by the hypothalamus, it begins to produce the follicle-stimulating hormone (FSH) and the luteinizing hormone (LH). This stimulates the ovaries to secrete estrogens and other so-called sexual hormones. Furthermore, this explanation of the cycle is based on the logic of the non-fertile ovum. Put simply, the cycle is explained as a failed process from the moment the ovum is unfertilized (no pregnancy), marking the logical end of the cycle (Martin 1987). So, unlike male bodies, both the cycle and the female body are defined exclusively in terms of their reproductive function and in relation to a hormonal explanatory model (Esteban 2001). In turn, these processes are explained by assuming the supposed neutrality and objective nature of hormones, although we know that materiality is already infused with fixed ideas about sexual differences. It is understood that all bodies— whether female or male—respond to certain hormones, and anything that is perceived to deviate from this assumption is pathologized. Nonetheless, we are clearly faced with a paradox. On the one hand, hormones are the hard fact of bodies and those that supposedly do not correspond to a specific body, which go beyond the norm, are pathologized. On the other hand, the reality is that all humans (male, female, and others) present great variability in the type and quantity of hormones (Oudshoorn 1994; Fausto-Sterling 2000). For example, in the case of hormones that impact the menstruation process, what is termed the (LH) is present in all bodies but it tends to be emphasized much more in the female body.[8] While hormones may be seen as 'the truth of bodies,' they are, rather, part of the medical plasticity of those bodies, given that they are a basic primary material destined to *repair* and *clarify* the expectations of the sexual dimorphism paradigm. They serve to clarify any variety or material diversity,[9] and ultimately, they function as biopolitical control mechanisms and as a means of managing bodies.

Another characteristic of this biomedical logic of menstruation is that it is based on a fragmented view that is related to, among other things, the separation of medical knowledge into specialties (Lock 1993; Oudshoorn 1994; Perdiguero and Comelles 2000). This division into specialties—for example, endocrinology and gynecology or dermatology and gynecology—hinders a more holistic or systematic explanation of the cycle in particular, and of bodily process in general. In this way, the menstrual cycle is defined exclusively on the basis of its reproductive function, and therefore it is not understood in broader or more complex ways nor in relation to other physiological processes. Little attention is paid to what are termed peripheral or systemic effects such as its influence on metabolism, the osseous or vascular system, the skin, or mucosae (Valls-Llobet 2009). In fact, the menstrual cycle is an important indicator of general health and a vital tool for self-awareness, and that is exactly why it is essential to bear in mind its complexity.

These are some biomedical ideological principles of menstruation that enable us to understand how this ideology of menstrual normativity is constructed within a binary normativity and a sexual dimorphism paradigm, which barely acknowledges the diversity of biological materiality, social experience, or their interrelationship. Furthermore, it is understood as a universal physiological process, with little mention of the sociocultural processes that influence it and is thus an ethnocentric view. Indeed, everything that falls outside this menstrual norm is explained via anomalous language and pathologized so that the margins become uncertainty spaces, and the ambivalent becomes utterly hopeless, abject. (For further discussion of menstrunormativity, see Persdotter, Chapter 29.)

Nevertheless, not all women menstruate. Divesting the idea of menstruation as the defining concept for the female body was one of the conclusions of my Master's thesis (2009). I conducted research based on the corporeal itineraries of people who were interpreted and experienced as women, and who were supposedly of reproductive age. I found that many of them experienced no bleeding at all for different reasons. This led me to look more deeply into and question the idea of menstruation as a concept that defines the female body and to think about the cycle as something that occurs in more (and also fewer) bodies than we may think; that something is related to female social subordination.

As stated above, to menstruate, women must be at a particular point in their lives, and ovulations are also influenced by social, geographic, and temporal factors. However, there are also multiple cases where women do not have periods: women who are pregnant or breastfeeding,[10] those that have experienced hysterectomy, transgender women, some intersex women,[11] women who are taking hormonal contraceptives to suppress their cycles, women with anovulatory androgen excess, or certain athletes or others who lose weight very quickly. There are various examples refuting menstruation as the only 'universal truth' of the female body. If we consider the number and variety of women who do not menstruate and the transgender men and genderqueer individuals who do, we perceive the cycle as something that appears only within the life cycles of certain bodies.

Thus, very distinct characteristics reflect the diversity, variability, and dynamic nature of both biological and social experience. Even when attempting to comprehend this from a more positive view, greater value is attributed to this particular body process than to other bodily activities that also affect our lives. It is not a question of denying the importance of the cycle for health, since it clearly affects all organs and bodily systems, and any alteration of the cycle may be an indicator of one's health status (Valls-Llobet 2009, 133; Barranco et al. 2016). Less still is my intention to make menstruation even more invisible than it already is. Yet, while we consider it a health indicator and fundamental for analyzing the differential morbidity of women, if we consider the menstrual cycle to be the single fact, the irreducible truth, we reproduce a reductionist perspective of corporeal diversity and complexity, which then hinders a dynamic and relational understanding of corporeal processes and gender. First, women who do not menstruate perceive this as a deficiency in their body, and those that do experience it, but who do not feel female[12] or who do not want to get pregnant, experience it as a problem. Second, one of the main negative consequences is that the female body is reduced to the reproductive arena, thereby corroborating the socially established role for women.

Therefore, despite certain changes, the biomedical definition of menstruation continues to be crucial for defining and normativizing the female body in a certain way, claiming that all female bodies experience menstruation, thereby distinguishing them from male bodies, a (biological) difference that, following Esteban (2001), is transformed into social inequality, given that biological differentiation becomes a justification for the differential and hierarchical organization of work, knowledge, and spaces (for example, in relation to caregiving and child raising). Furthermore, the normativization of menstruation has not contributed to increasing research of the cycle itself or women's health in general, but rather contributed to its invisibilization.

Alternative Politics of Menstruation and the Theory of Sexual Dimorphism: Complexity in a Field Study

Having outlined the premise of the menstrual normativity ideology and its consequences for the way people experience their bodies and identities, I now turn to the consequences of this biomedical ideology in the management of alternative cultures and politics of menstruation: continuities, discontinuities, and dissidence. While alternative narratives are critical discourses which question the tendency to medicalize or pathologize menstruation, they adopt different approaches to gender, (political) consumption, and other ideologies (Kissling 2006; Bobel 2010; Guilló 2014). Beneath the alternative discourses of the reproductive body as a political body, there clearly exists defined but diverging ideas of the implications of *being a woman* and specific ways of understanding *femininity*, *sexual difference*, and *body*, which comprise a heterogeneous feminist spectrum of equally specific and heterogeneous gender practices. Consequently, while new menstrual imaginaries are created and corporeal empowerment occurs in those corporeal politics, there is sometimes continuity in emphasizing the biological dimension of menstruation.

This complexity can be illustrated by different experiences. Maite is a 31-year-old woman who studied fine arts. She has lived in several European cities, but when I interviewed her, she was living in a farmhouse in the small fishing village where she was born. She combined her studies and artistic work with a job waiting tables and activism in different social movements. Studying arts has been for her a scenario for corporeal exploration and experimentation, which, in its own way, is also linked to her menstrual experience. My interest in interviewing her arose from the fact that she had created an artistic installation about blood. When interviewed, Maite was worried because for reasons unknown to her doctors, she does not bleed, and she believes this impedes her body from cleansing itself:

> It's just that having a period is not just having a period, it implies cleansing, and I'm not cleansing myself. I see it as the cleansing of a process that your body creates, all that has to be cleansed. I define it like that because it seems dirty to me. As a woman, I am a complete woman, I have my breasts, I'm a woman in body, there's no doubt about that, but I lack something. A flaw in a female body. It's not a drama, but it is an anomaly. What is worse, nobody has explained it scientifically to me and that makes it even more difficult for me. I experience it like a dark cave. And I often feel I'm dirty, that I want to clean up what's in there.

Maite has never experienced any bleeding without the help of hormone therapy. She has sought answers in the biomedical field and in alternative and complementary medicines such as homeopathy, but without satisfactory answers, which has frustrated her. Her menstrual experience spills over into her general health, her lifestyle, and her relationships with other people. At the same time, watching her health has been represented as a process of bodily empowerment. For Maite, art has been a way of "connecting with her body and a way of connecting with nature"; it has also been a means of channeling the questions that hound her and her quest for her sense of herself in her body. For her, being a woman without menstruation is not only a health-related concern; she feels *lacking* in her femininity. In different parts of my fieldwork, this woman-menstruator schema recurs. However, in recent years, I've seen this idea increasingly questioned.

Along with the lack of blood, Maite mentions that her body is not cleansing itself as it should: she feels her body is 'dirty.' As I have observed in some menstrual initiatives, menstruation is resignified as a cleansing process, a health process, which forms part of common lore. Although the intention is to give it a positive meaning, sometimes there is no discontinuity with the menstruation-dirtiness paradigm (Esteban 2001).

Luna's experience, although very different, also shows us this material diversity of menstrual experience. She works in various social initiatives she feels passionate about, and is currently working in a cultural association.[13] Throughout the course of her life she has adopted a holistic view of health. When I met Luna, she was mentoring a work group on menstruation and developing a project about the cycle and balanced nutrition habits. Though she lives in a postindustrial city, she has plenty of contact with nature. She has been through some tough experiences, but has long been working on her emotional health and she thinks she is now reaping what she once sowed. She does not feel physically ill and her cycles are not painful. However, since she reached menarche, she has experienced several episodes in which she has not menstruated for months. The first time she did not menstruate was when she first went on a macrobiotic diet. It was a vital moment in which she needed to find herself, and during those months of consuming a stricter diet, she did not bleed: "They told me that, somehow, my body was regaining strength." The second time without periods was when she was in jail (due to her political activity) something, she says, that is very common in women prisoners. (For further discussion of menstruation among women prisoners, see Bozelko, Chapter 5, and Roberts, Chapter 6.)

> I went some months without menstruating, but it was not what most worried me, my body being as vulnerable as it was. At that moment, the last thing I was worried about was that. At the end of the day, for me the cycle is a mirror, it is a tool that helps me to know how I am.

This same example shows how the menstrual cycle is a clear indicator of well-being. Moreover, in this case, we are talking about cases of stress and trauma that involve much bodywork. This example also reveals that irregularity is a common and vital feature of many people's bodies. As Luna says, it is a 'mirror,' but I would add that it is a mirror of the human body's immense variability.

Nonetheless, in some of the workshops and other encounters in which I have participated as part of my ethnography, this idea of bleeding as something which makes us women—this woman=menstruator schema—has been repeated, although more recently I have found it challenged. This is often common in workshops related to topics where health and spirituality converge, the so-called holistic and therapeutic field (Heelas and Woodhead 2005; Cornejo and Blázquez 2013). (For more about gendered perspectives in the holistic and spiritual field, also see Fedele and Knibbe 2013; Guilló 2018.) In these contexts, essentialist discourses occasionally emerge or others relate menstruation to spirituality involving the search for a more felt, corporeal, sensitive, sentient, and conscious female specificity. Thus, the cycle may be defined not only as a cleansing process, but also as creativity, as internal regeneration and as part of 'female spirituality.' For this reason, the narratives that are created in these therapeutic–spiritual contexts offer alternative readings of women's bodies, with emphasis on emotions and different ways of experiencing corporeality, highlighting the importance of self-care. Ultimately, scenarios of well-being and processes of agency are created which imply a critical perspective toward patriarchal and hierarchical values. Hence, it is vital to acknowledge the empowering circumstances they reveal and the complexity and variety of these discourses. Moreover, we sometimes see that the female body is totally in opposition to the male body (or related to dirtiness, when menstruation is presented as a cleansing process). Female bodies are thus shown to be cyclical, hormonal, and circular bodies that are totally influenced by the menstrual cycle as opposed to other bodily or life cycles. It must be noted that the discourses and practices in these arguments are dynamic and complex and gender ideologies are not unique and homogenous. The further one examines and understands the different experiences, the more diversity one sees, but it is important to keep a critical perspective.

DISSIDENCE, RESISTANCE, AND THE REAPPROPRIATION OF THE ABJECT BODY

As I have shown, experiences of menstruation vary enormously and are not always key to identity. Returning to Maite's experience, as we have seen, she feels there is something missing in her female corporeality and has conducted artistic work using blood as a means of reappropriating the abject body, attempting to create fissures and cracks in that female identity. One of her works was an installation in which she played with liquids such as milk and blood, using copious streams of blood to explore, question, and interrogate by means of an abject and monstrous corporeality. In the case of Luna, although she has always had an interest in health, it was after leaving jail that she incorporated health research and macrobiotics as a life view. Nonetheless, since being very young, she had been interested in the menstrual cycle and alternative menstrual care products, even when these resources were not well known. She grasps the phases of the cycle with gusto and in each phase sees a new opportunity to explore her needs. She has undertaken several projects about the cycle individually and co-lead a project to analyze it collectively.

In Ljubljana, Slovenia, where I spent a research stay in 2014, I got to know the *Vstajniške socialne delavke* collective, a feminist group which uses blood in its activities, and in which the women themselves problematize the rigidity of the sexual dimorphism paradigm. These activities involve the reappropriation of the body-space, enabling its transgression from the margins. The abject thus becomes a space of resistance. In this way, the menstruating body becomes a subversive body, questioning the biomedical paradigm of sexual dimorphism. Several artists have used menstrual blood and products to challenge menstrual silence and shame, frequently as a mechanism to challenge the binary paradigm.

Some trans men that I have interviewed also reflect on the consequences of this rigid biomedical ideology. For example, Jon, who lives on a small rural inland, tells me how at one point in his life his male corporeal identity was not compatible with his menstrual cycle. He remembers wanting to be a boy from early childhood: "But then, around the age of 12, I went through a phase of 'I'm a girl', that I had to be a girl. When I first had periods, it was like a relief. At least it made me feel like the others, right? Because in the end, I was fucked up not feeling like them. That gave me a sense of equality."

However, that feeling soon changed. Jon has a very dynamic way of conceiving bodies and sexes and is very critical of the biomedical model, through both his participation in environmentally related social movements and as a result of his own medical, gynecological, and social experiences. He points out that, if it were a question of context, if his social world had not been so rigid, if it were not for the pain suffered, he would have no problem in being a man and menstruating: "At the end of the day, it would be just one more part of my body; and well, if I'm not disgusted by my body, well, fine, right?" (For more about the menstrual experiences of trans menstruators, see Rydström, Chapter 68, and Frank and Dellaria, Chapter 7.)

The corporeal and life experiences of Maite, Luna, Jon, and other people I interviewed—defined in hegemonic terms as pathological or imperfect bodies that do not meet the expectations of the gender system—in fact demonstrate the rigidity of Western gender models and the violence that occurs in the health field. These experiences question both the rigid, dichotomous biomedical perspective and the resulting corporeal fiction. Moreover, their trajectories reveal the agency of non-submissive bodies, and the different strategies for managing their experiences in different settings. **I believe that if menstruation itself was not so negatively interpreted, people's experiences would be more positive, not just in the case of people whose corporeal identity does not fit with their periods, but for all people who do (or do**

not) menstruate. It is vital to prioritize social and medical research of the cycle to improve health processes and life conditions, and above all the most painful or problematic scenarios (such as dysmenorrhea or endometriosis). But it is important that such research is undertaken with more attention to the diversity of the corporeal materiality and identities.

QUEERING MENSTRUATION: PROPOSALS
FOR NONESSENTIALIST BODY POLITICS

Refusing to assume who does and does not menstruate is one way of challenging the rigid gender binary that perpetuates privilege and oppression. (Society for Menstrual Cycle Research 2011, 1)

The rich research surrounding the Western construction of bodies (Foucault 1977; Lock 1993; Grosz 1994; Laquer 1990; Fausto-Sterling 2000; Butler 2004; Esteban 2004) provides clues for reviewing and rethinking the corporeal politics that are implemented in different contexts. In addition, in feminist practices of menstruation, experimentation is one important kind of knowledge and, via debates and analysis, collective knowledge emerges, which increasingly refines both insight and practice. Following the thread of the theoretical ideas and ethnographic cases presented throughout this chapter, I consider it necessary to undertake critical readings which demonstrate the consequences of biologistic frameworks that reproduce the hegemonic medical vision of sexual difference, thereby contributing to the perpetuation of a biological and social difference between women and men, whether in the form of essentialist, ethnocentric, or universalist discourses. This critical gaze also contributes to a more flexible view of our corporeal and gender frameworks, thus giving rise to alternative corporeal theories and politics based on corporal and identity diversity.

It is necessary to dispense with the idea of menstruation that forges a very specific gender and social identity, with its normative and reductionist take on the menstrual experience (and more generally the human experience) in such a way that not all bodies, not all people, fit into that approach. We need to understand menstruation as more than an identity process, as one more (health and) corporeal process which may serve as a space of resistance, creativity, dissidence, and subversion. 'Deficient bodies,' 'abject bodies,' and 'erroneous bodies' all become, within these body politics, scenarios of resistance, bodies of influence for a more profound social transformation.

Another important challenge would be to consider distinct gender biases. In this chapter, I have explored a medical gender bias which distinguished men and women as different and opposed. This bias has reduced 'women's health' to their sexual and reproductive health, disregarding the gender perspective of illnesses common to both sexes, and through the stereotypes produced, the health sciences have *invented* specific disorders and syndromes for women, attributing them to internal-individual causes, *normal* processes or emotional discomforts which, in reality, are the products of social inequalities and gender norms, pathologized and medicalized (García-Dauder and Pérez Sedeño 2017, 198). As previously stated, exacerbating the differences and interpreting them as biological and innate, naturalizes gender inequalities and transforms them into fixed notions (Hare-Mustin and Marecek 1994; Esteban 2001). However, considering the different gender biases is a fundamental challenge. As well as the bias that exaggerates the differences, we should also consider the bias that ignores them: the typical androcentric bias which takes all things male as normal and obvious or devalues all things female, without addressing differential morbidity or specific symptoms of women's illnesses (Hare-Mustin and Marecek 1994; García-Dauder and Pérez Sedeño 2017, 13).

Therefore, in corporeal epistemologies and politics, it is important to bear in mind these reductionist tendencies specific to processes of constructing an androcentric pattern. According to García-Dauder and Pérez Sedeño (2017, 205–6), "their handling will depend on the necessities of each context, on understanding the differences in their historicity and on addressing other possible important variables which run through the experience and subjectivity of gender." Consequently, differential morbidity should be a research priority, and it is critical to address subjective experience or the meaning of women's symptoms in their social context. In this way, we could deal with the material diversity of bodies, and bodily and identity processes which do not respond to dualism, and produce a more reflexive and critical feminist understanding that challenges biological determinism and heteronormative ideology. This implies a continuous revision of the educational contexts in which the cycle is examined, within the sphere of what is termed the Femcare industry, healthcare professionals, feminist technologies, workshops, and the discourses and practices created within such alternative knowledge. Indeed, I believe that these alternative cultures and politics of menstruation are critical and transformative of the contemporary world: of the economic model, the gender model, and the biomedical paradigm. It is vital to be reflexive in relation to corporeality, which narratives are produced, and the places of enunciating these resignifications—not just in terms of sexual dimorphism—but also with respect to any characteristics which demonstrate corporal or identity diversity in very different contexts, in order to continue constructing corporal epistemologies and politics that pay more attention to diversity in all its forms.

NOTES

1. See, for example, Ruth Green Cole, Chapter 57, and Jen Lewis, Chapter 58, for examples of menstrual art; Lise Ulrik Andreasen, Chapter 65, for discussion of menstruation and social media; Berkley Conner, Chapter 64, for discussion of social media used to conduct menstrual activism.
2. By *politics of menstruation* I refer to the processes of giving meaning to and governing menstruation and reproductive health (in general), carried out both by different institutions (medical, religious, economic, media, and so on) as well as by collectives and individual subjects (Guilló 2013).

3. The Western hegemonic perspective of menstruation can be a problematic concept: on the one hand, because the concept Western implies certain geopolitical characteristics within power relations. On the other hand, it is difficult to pinpoint a hegemonic view, because "there is no single or hegemonic view or model of menstruation in a particular society. Rather, there is a range of views and experiences that menstruation may produce across the social divides that structure women's lives" (Gottlieb 2002, 286). Despite these difficulties, I use this concept conscious of the very nuances and variability that I want to draw attention to.

4. As we know, feminist criticism has challenged in different ways the dual logic of positivist science (Fox-Keller 1985; Harding 1986; Haraway 1988), calling into question different assumptions such as the debates on nature/culture, subjectivity/objectivity, and discriminatory hierarchical binomials (Fox-Keller 1985). Analyzing the paradigm of sexual dimorphism is an important task in order to comprehend how biomedical bodies are understood; but also, social and political bodies. The contemporary western perspective of menstruation is constructed within this paradigm.

5. Beyond the western traditional view on menstruation, anthropological studies have shown that the symbolism attached to it show a great deal of variation across diverse cultural settings. Furthermore, not all menstrual prohibitions are equally disadvantageous; some women reaffirm their agency in certain menstrual practices (Buckley and Gottlieb 1988; Gottlieb 2002).

6. I will use these concepts as social categories taking into account the difficulties implied at the same time as I endeavor to expose the rigid gender binary ideology. In selected cases, I will clarify the gender spectrum, including cisgender, transgender, et cetera. (See Rydström, Chapter 68, for further discussion of trans menstruation.)

7. For example, in Spain a hundred years ago the average age of the first cycles was 15 years old, and the last bleedings at 35 (Sau 1980; Esteban 2001).

8. For example, the luteinizing hormone produces testosterone, and the follicle-stimulating hormone also makes possible the production of sperm. However, references to these two hormones are more typical in gynecological manuals about women.

9. Beyond the processes of sex change, for example, in fieldwork many people have explained that they have been prescribed endocrinological treatments for very different reasons: acne, hormonal irregularities, menstrual irregularities, for having 'too much testosterone', et cetera.

10. If fertilization has taken place in the uterus, there will be no ova in that period. The case of lactation is much more variable: in the biomedical discourse, menstrual cycles tend to occur once lactation has finished, but this does not happen automatically.

11. I use *intersex* broadly here, although I am aware it is a concept that many people do not identify with and that the intersexual variations are very different from one another. For example, women with Androgenic Insensitivity Syndrome do not menstruate, while some women with Congenital Superrenal Hyperplasia may. The same happens with some men. Studies show that some women experience this as a devaluation of femininity (see Gregori 2017).

12. For example, menstruation can also affect the sense of masculinity or of lack of safety for transgender men. On attitudes toward and experiences of menstruation in the masculine of center and transgender communities, see Chrisler et al. 2016; Rydström, Chapter 68 in this book).

13. Both Maite and Luna are socially active, and, like the majority of my interviewees, they are participating in these corporeal resignifications. In the Basque society which both are a part of, there is a long tradition of associationism and social movements.

REFERENCES

Barranco, E., I. Jiménez-Díaz, L. M. Iribarne-Durán, O. Ocón, E. Salamanca, M. F. Fernández, and N. Olea. 2016. "Determination of Personal Care Products—Benzophenones and Parabens—In Human Menstrual Blood." *Journal of Chromatography B* 1035: 57–66.

Bobel, Chris. 2010. *New Blood: Third-Wave Feminism and the Politics of Menstruation*. New Brunswick, NJ: Rutgers University Press.

Buckley, Thomas, and Gottlieb, Alma. 1988. "A Critical Appraisal of Theories of Menstrual Symbolism." In *Blood Magic. The Anthropology of Menstruation*, edited by Thomas Buckley and Alma Gottlieb, 3–50. Oakland: University of California Press.

Butler, Judith. 2004. *Undoing Gender*. New York: Routledge.

Chrisler, Joan. 2008. "Fear of Losing Control: Power, Perfectionism, and the Psychology of Women." *Psychology of Women Quarterly* 32: 1–12.

Chrisler, Joan C., Jennifer A. Gorman, Jen Manion, Michael Murgo, Angela Barney, Alexis Adams-Clark, Jessica R. Newton, and Meaghan McGrath. 2016. "Queer Periods: Attitudes toward and Experiences with Menstruation in the Masculine of Center and Transgender Community." *Culture Health, & Sexuality* 18 (11): 1238–50.

Cornejo, Mónica, and Maribel Blázquez. 2013. "La Convergencia de Salud y Espiritualidad en la Sociedad Postsecular. Las Terapias Alternativas y La Constitución del Ambiente Holístico." *Revista de Antropología Experimental* 13 (2): 11–30.

Douglas, Mary. 1966. *Purity and Danger: An Analysis of Concepts of Pollution and Taboo*. Oxford: Routledge.

Esteban, Mari Luz. 2001. *Re-producción del Cuerpo Femenino. Discursos y Prácticas acerca de la Salud*. San Sebastian: Gakoa.

———. 2004. *Antropología del Cuerpo, Género, Itinerarios Corporales, Identidad y Cambio*. Barcelona: Bellaterra.

Fahs, Breanne. 2016. *Out for Blood: Essays on Menstruation and Resistance*. Albany: State University of New York Press.

Fausto-Sterling, Anne. 2000. *Sexing the Body: Gender Politics and the Construction of Sexuality*. New York: Basic Books.

Fedele, Ana, and Kim Knibbe, eds. 2013. *Gender and Power in Contemporary Spirituality: Ethnographic Approaches*. New York: Routledge.

Foucault, Michel. 1977. *Histoire de la Sexualité I. La Volonté de Savoir*. Paris: Gallimard.

Fox-Keller, Evelyn. 1985. *Reflections on Gender and Science*. New Haven: Yale University Press.

García-Dauder, S., and Eulalia Pérez Sedeño. 2017. *Las Mentiras Científicas sobre las Mujeres*. Madrid: Catarata.

Gottlieb, Alma. 2002. "Afterword." *Ethnology* 41 (4): 381–90.

Gregori, Nuria. 2017. "Encuentros y Des-encuentros en torno a las Intersexualidades/DSD: Narrativas, Procesos y Emergencias." PhD diss., Universitat de València.

Grosz, Elisabeth. 1994. *Volatile Bodies: Toward a Corporeal Feminism*. Bloomington: Indiana University Press.

Guilló Arakistain, Miren. 2009. "Hilekoa: Diskurtsoak, Praktikak eta Tentsioak. Esanahi Berrien Bila Hausnartzen." Thesis for Master's in Feminist and Gender Studies, University of the Basque Country UPV/EHU.

———. 2013. "La Incorporación de la Investigación: Políticas de la Menstruación y Cuerpos (Re)productivos." In *La sensibilidad: Potencia y Resistencias, Revista Nómadas*, edited by Manuel Roberto Escobar and Nina Alejandra Cabra, no. 39, 233–45. Bogotá: Universidad Central.

———. 2014. "Mujeres Jóvenes y Menstruación: Contracultura y Resignificacion del Ciclo Menstrual en el País Vasco." In *Jóvenes, Desigualdades y Salud. Vulnerabilidad y Políticas Públicas*, edited by Oriol Romaní and Linda Casadó, no. 13, 143–65. Tarragona: URV.

———. 2018. "Habitar lo Imponderable. Espiritualidades Contemporáneas y Lecturas Alternativas de la Menstruación." In *Etnografías Feministas. Una Mirada al Siglo XXI desde la Antropología Vasca*, edited by Mari Luz Esteban and Jone M. Hernández García. Barcelona: Bellaterra.

Haraway, Donna. 1988. "Situated Knowledges: The Science Question in Feminism and the Privilege of Partial Perspective." *Feminist Studies* 14 (3): 575–99.

Harding, Sandra. 1986. *The Science Question in Feminism*. New York: Cornell University Press.

Hare-Mustin, Rachel T., and Marecek, Jeanne. 1994. *Marcar la diferencia: psicología y construcción de los sexos*. Barcelona: Herder.

Heelas, Paul, and Linda Woodhead. 2005. *The Spiritual Revolution: Why Religion Is Giving Way to Spirituality*. Oxford: Blackwell.

Houppert, Karen. 1999. *The Curse: Confronting the Last Unmentionable Taboo: Menstruation*. New York: Farrar, Straus and Giroux.

Irusta, Erika. 2018. *Yo Menstrúo: un manifiesto*. Barcelona: Catedral.

Johnston-Robledo, Ingrid, and Joan C. Chrisler. 2011. "The Menstrual Mark: Menstruation as Social Stigma." *Sex Roles* 68 (1–2): 9–18.

Kissling, Elizabeth Arveda. 2006. *Capitalizing on the Curse: The Business of Menstruation*. Colorado and London: Rienner.

Laquer, Thomas. 1990. *Making Sex: Body and Gender from the Greeks to Freud*. Cambridge, MA and London: Harvard University Press.

Laws, Sophie. 1990. *Issues of Blood: The Politics of Menstruation*. London: Macmillan.

Lock, Margaret. 1993. "Cultivating the Body: Anthropology and Epistemologies of Bodily Practice and Knowledge." *Annual Review of Anthropology* 22: 133–55.

Martin, Emily. 1987. *The Woman in the Body: A Cultural Analysis of Reproduction*. Boston: Beacon Press.

De Miguel, Jesús. 1971. *El Mito de la Inmaculada Concepción*. Barcelona: Anagrama.

Ortiz Gómez, Teresa. 2006. *Medicina, Historia y Género, 130 años de investigación feminista*. Oviedo: KRK.

Oudshoorn, Nelly. 1994. *Beyond the Natural Body: An Archeology of Sex Hormones*. London: Routledge.

Parera, Nuria, J. Penella, and J. M. Carrera. 1997. "Menarquia, Avance Secular y Datos Antropométricos." In *Progresos de Obstetricia y Ginecología*, 40 (1): 30–37.

Perdiguero, Enrique, and Josep Comelles. 2000. *Medicina y Cultura. Estudios entre la Antropología y la Medicina*. Barcelona: Bellaterra.

Sau, Victoria. 1980. "Mito y Realidad del Fenómeno de la Menstruación." In *Comportamientos Sexuales*, edited by Josep Farré, Manuel Valdés, and Eudald Maideu. 316–28. Barcelona: Editorial Fontanella.

Scheper-Hughes, Nancy, and Margaret Lock. 1987. "The Mindful Body: A Prolegomenon to Future Work in Medical Anthropology." *Medical Anthropology Quarterly* 1 (1): 6–41.

Society for Menstrual Cycle Research. 2011. "The Menstrual Cycle: A Feminist Lifespan Perspective." (Fact sheet) Sociologists for Women and Society. https://socwomen.org/wp-content/uploads/2018/03/fact_4-2011-menstruation.pdf.

Šribar, Renata. 2004. *O menstruaciji: telo v diskurzu, diskurz v telesu*. Ljubljana: Delta.

Stein, Elissa, and Susan Kim. 2009. *Flow: The Cultural Story of Menstruation*. New York: St. Martin's Griffin.

Steingraber, Sandra. 2007. *The Falling Age of Puberty in U.S. Girls: What We Know, What We Need to Know*. San Francisco: Breast Cancer Fund.

Valls-Llobet, Carme. 2009. *Mujeres, Salud y Poder*. Madrid: Cátedra.

CHAPTER 64

Menstrual Trolls: The Collective Rhetoric of Periods for Pence

Berkley D. Conner

After two months of amending, a 37–13 Senate vote, and a 60–40 House vote in favor, on March 24, 2016, Indiana Governor Mike Pence signed House Enrolled Act 1337 into law. Authored by Indiana Representative Casey Cox and co-authored by Indiana Representatives Peggy Mayfield, Ronald Bacon, and Chris Judy, the law served to mitigate the fear that women often seek abortions after they have learned of a fetus' genetic makeup; would-be parents are either unhappy with the race or sex of the fetus; or that they worry that the future child may be born with genetic anomalies. As such, HEA 1337 sought to strengthen the severity of Indiana's existing abortion laws in three major ways: first, by prohibiting abortion on the basis of race, sex, or diagnosis of disability; second, by enhancing informed consent provisions by adding to the list of information that an abortion provider must impart to anyone seeking an abortion; and, third, by adding a requirement that the remains of any miscarriage or abortion be cremated or buried (Tuttle 2016). While the law was met with much criticism by Hoosiers, HEA 1337's measure regarding interment generated the most outrage.

Because fertilized eggs can be released and expelled during menstruation, any period could be a technical miscarriage, making menstrual fluid subject to HEA 1337s interment clause. The thought of being legally required to bury her menstrual blood compelled Carmel, Indiana, resident Laura Shanley to call Mike Pence's office to offer details about her period, feigning concern that she would be prosecuted under the new law. When asked about her call, Shanley told the *Indianapolis Star*, "If the governor is this interested in what's going on in my body, I might as well call and tell him" (Rudavskey 2016, para. 2). What started as an angry phone call quickly became an organized effort to protest HEA 1337. Shanley started a Facebook page and, shortly after, a Twitter account using the moniker 'Periods for Pence,' where

© The Author(s) 2020
C. Bobel et al. (eds.), *The Palgrave Handbook of Critical Menstruation Studies*, https://doi.org/10.1007/978-981-15-0614-7_64

she transcribed her phone conversation and invited other Indiana menstruators to follow suit. From March 2016 through July 2016 (when Pence was announced as Donald Trump's vice-presidential running mate) the members of Periods for Pence made calls straight to Pence's office in the interest of sharing information about their individual menstrual cycles.[1] Shanley transcribed these phone conversations on Facebook and Twitter for followers to read, like, and leave commentary.

While it may be easy to dismiss Periods for Pence as another instance of internet 'slacktivism,' the goal of the campaign was to mobilize menstruators to persuade lawmakers to make legislative change using their periods. The calls made on behalf of Periods for Pence are funny, sarcastic, biting, and copious, which earned the campaign national recognition. *Cosmopolitan* reported that in April 2016, Pence's office was forced to shut down their phone lines because it was receiving so many calls about periods (Gupta 2016), *New York Magazine* claimed that Periods for Pence's brand of "menstrual trolling" was "the best new kind of trolling" (Rinkunas 2016, para. 5), and *Triple Pundit* argued the campaign was "the perfect example of spurring change at the policy level" (Mazzoni 2016, para. 6). To date, the campaign's Facebook page boasts nearly 100,000 'likes' while the Twitter account is followed by nearly 6500 users.

Because it is such a variable phenomenon, mobilizing around menstruation is difficult (Weiss-Wolf 2017, xviii–xix). However, in this chapter, I explore how the activists of Periods for Pence relied on diverse experiences with menstruation to protest anti-abortion legislation. Using Tasha N. Dubriwny's (2005) theory of collective rhetoric, I argue that the activists of Periods for Pence engage in acts of narrative sharing, humor, and symbolic reversal to: (1) craft a cohesive narrative of the many experiences of menstruation, and (2) draw on logics of menstruation to rhetorically re-moralize abortion as necessary. Finally, I will demonstrate that Periods for Pence successfully employs a second-wave feminist strategy, consciousness-raising, in a third-wave feminist context. Thus, I also draw on Chris Bobel's notion of *radical menstruation* (2010, 110) to argue that Periods for Pence can help scholars rethink traditional conceptualizations of static 'waves' of feminism and feminist rhetorical theorizing.

I examine the Periods for Pence Facebook page and Twitter account between the dates of March 28, 2016 (the day the accounts were created) and April 21, 2016, the day of the Rally for Women's Rights at the Indiana Statehouse. The posts between these dates are most consistently transcriptions of phone calls made to Pence's office. After the rally, the accounts mainly became used for sharing news articles related to Pence and reproductive health. I analyze 20 Facebook and Twitter posts, comments, and replies from 19 different rhetors that best represent the discourse of Periods for Pence. These posts are created by people participating in the campaign, including Laura Shanley, either as callers or supporters seeking advice about making calls.

Menstruation, Feminism, and Activism

Feminist thinkers in sociology, anthropology, and women's studies have taken up the question of menstruating bodies largely by contributing to research about the Western, cultural menstrual taboo. This foundational work importantly links the taboo to a history of systemic sexism that identifies bodily fluid as both feminine and dirty. Elizabeth Grosz (1994) argues that, centuries ago, bodily fluids came to be indicative of physiological and social differences between men and women.[2] Marking the woman's body as "a body which leaks, which bleeds, which is at the mercy of hormonal and reproductive functions" (Grosz 1994, 204) allows for easier subjugation of both the blood and of the person who bleeds. Indeed, menstrual blood and its bleeders are given meaning based on the ideals of male, heterosexual desire. In other words, menstrual blood is dirty and so are menstruating women. Mary Douglas (2002) asserts that this point of view birthed the menstruation taboo, which persists because of society's association of menstrual blood with a bodily type of pollution. Often, the notion of pollution is mobilized against menstruators to justify policing the menstruating body. Janet Lee and Jennifer Sasser-Coen (1996) explain that once a woman feels shame related to the dirtiness of her body, a symbolic resemblance exists between the limits of the body and danger to the community, freeing society to intervene and manage the body for the good of the community. In this way, sexism becomes a public safety measure.

Centralizing menstruation in conversations about sexism and taboo opened doors for rhetorical scholars to pay closer attention to how taboo materializes as marginalizing legislation. Although rhetorical conversations do not typically focus on menstrual blood, rhetorical literature has focused on blood in various contexts, cementing the necessity of understanding blood discourse. For example, some rhetorical blood scholarship interrogates questions of biopolitics that are often linked to toxicity. Jeffrey Bennett (2015) argues that the FDA's ban of blood donations from gay men is couched in a rhetoric of contagion that bolsters the political and social othering of these men. For women, toxicity extends beyond menstrual blood to often include reproductive organs writ large. Mollie Murphy (2017), in her exploration of the link between reproductive and environmental justice, contends that the toxicity of the womb is highly stigmatized and isolated to the private sphere where medical issues related to childbirth and abortion are less likely to be acknowledged. While these scholars highlight the importance and materiality of social constructions of blood, others are interested in how bleeders utilize these negative constructions in order to recast blood in a positive way. Elizabeth Dickinson, Karen Foss, and Charlotte Kr{\o}lkke (2017) explore postpartum placenta consumption and burial and reveal that these placenta practices operate rhetorically within logics of disgust. Specifically, mothers who engage in placenta practices rearticulate what 'disgust' means in the context of Western medicine in order to reshape public perception of placenta practices. Importantly, rhetoricians who study blood are interested not only in how certain blood becomes weaponized against its bleeders, but also how these bleeders may disrupt this process.

Rhetorical work that does engage menstruation often focuses on how menstrual blood is policed through the deployment of lifestyle feminism. This typically happens at the corporate level in service of capitalism. Elizabeth Kissling (2013) and Carly Woods (2013) have both explored how oral contraceptives are marketed to highlight their potential as menstrual suppressors. This decision is often made in the name of postfeminism or choice feminism; pharmaceutical companies capitalize on the menstruation taboo in order to sell the idea that women are happier, healthier, and more able when they are free of their periods. Woods contends that these advertisements succeed in marketing an illusion of 'choice' that offers women a false sense of control over their bodies and lives, while, at the same time, reinstating ideologies of the patriarchy. For both Kissling and Woods, menstruation is an important battleground on which feminists are sparring over matters of agency.

While multidisciplinary feminist thought and rhetorical scholarship engaging menstruation and blood are critical to understanding how humans have come to know menstruation, putting these fields in conversation with each other offers academics interesting ways to think through menstruation's current political moment. Performance studies scholar Shauna MacDonald (2007) models an avenue through which to link interdisciplinary feminist work that traces the lineage of menstrual oppression, and rhetorical scholarship that considers how blood is rhetorically constructed. MacDonald thinks about bodily fluids, specifically menstrual blood, as existing in a liminal space when they have exited the body. In a liminal space, matter is not subject to social structures—it is, if only temporarily, unknowable. Fluids such as menstrual blood may be engendered to the feminine, but an important characteristic of fluid is that it is able to dissolve dichotomous boundaries by literally seeping through them. This notion of liminality suggests rhetorical critics could consider how (and what is at stake when) menstrual blood is deployed by bleeders. If it is true that society has come to know menstruators as dirty, unruly, incapable, and subordinate, MacDonald suggests that the mere sight of blood via a leak (or possibly even an intentional act such as free-bleeding) puts the power of social construction back into the hands of the bleeder because fluid has effectively transcended the bounded control of the body. Because menstruation is a deeply individual experience, in order for menstrual blood to be deployed in the interest of large-scale social movements, a cooperative discourse of menstruation must be able to form. In the following section, I offer a methodological strategy that can help scholars interested in menstrual blood's liminality think about the potential of menstrual blood in the arena of activism.

A Collective Understanding of Menstruation

I believe MacDonald's (2007) understanding of the liminality of menstrual blood has the potential to shape how scholars might think about menstruation and activism, but at the same time I recognize Weiss-Wolf's (2017) concern that it is hard to unify around issues of menstruation because every experience with menstruation is unique. As a solution, I offer that Tasha Dubriwny's (2005) theory of collective rhetoric may be a fruitful way to understand how menstrual blood becomes collectively understood and mobilized by menstruators, even when their individual experiences with menstruation are drastically different.

Dubriwny (2005) analyzes how Redstockings, a second-wave radical feminist collective, employed consciousness-raising as a means of collectively reframing public opinion about abortion in 1969. Her theory details the way "collective rhetoric emerges through strategies that enable the collaborative creation and validation of worldviews through the articulation, or the strategic linking, of individual experiences" (396). As these individual experiences amass, oppressed groups are able to create new meaning from the articulation and validation of shared, lived experiences through three rhetorical processes: personal narrative, use of humor, and symbolic reversal. Burke (1984) refers to this overall process as *perspective by incongruity*, and Dubriwny argues that feminist rhetoricians (Demo 2000; Foss 1979) have adopted Burke's strategy for their own purposes. What Burke calls "perspective by incongruity," Dubriwny terms "atom cracking," or the "strategic juxtaposition of incongruent ideals, values, practices, and symbols that not only call into question gender ideologies but also re-moralize them" (398). For Dubriwny, atom cracking is the ultimate goal of collective rhetoric because once an ideology is re-moralized, oppressed groups develop solidarity, which can bring about change on social and political levels.

Mobilizing Around Menstruation

Because a theory of collective rhetoric privileges how individual narratives come together to form a cohesive narrative in service of larger political goals, I see the theory as being useful for attuning to how individual narratives about menstruation can coalesce for activist purposes. Thus, using collective rhetoric as my theoretical framework, in this section I analyze the Periods for Pence posts to illuminate how the members of the campaign unified around the invocation of menstruation to protest anti-abortion legislation and sensibilities.[3] The activists do this by creating a shared narrative of menstruation, using humor, and employing symbolic reversal. Their activism, I argue, re-moralizes abortion by couching the conversation in logics of menstrual health.

Sharing a Shedding Narrative

Initially, and perhaps most importantly, Periods for Pence is built on a network of activists who share two important characteristics: they menstruate and they demand the right to have an abortion. Inevitably, then, the campaign provides these activists with a cause they can rally around and a platform on which they can support each other. Dubriwny (2005) argues that engaging in co-construction of narratives by sharing stories that build on one another helps "create a community by allowing participants to affirm and validate each other's experiences by sharing parts of their own experiences" (p. 405). In the case of Periods for Pence, shared narrative is most typically seen in the form of comments to original Periods for Pence posts in which members engage with material from the original post and proceed to share their similar experiences. In this tweet thread, several members share both their experiences with menstruation as well as their concerns about their menstrual cycles in light of HEA 1337:

@PeriodsforPols: Contact @GovPenceIN to report your periods in response to HEA1337! Because it IS his business, now! (March 30, 2016)

|**@PriestessofAres**: @periodsforpence @GovPenceIN Ah yes . . . Would he like to know in gory detail what every girl goes through? Graphic detail? (March 30, 2016)

|**@danapokie**: @PriestessOfAres @periodsforpence @GovPenceIN Some symptoms of pre menopause are quite fascinating. Soaked thru 5 Playtex Ultras in 1 hr! (March 30, 2016)

@SpoolandThimble: @periodsforpence @GovPenceIN if I remove my tampon & there's a big clot, should I have that interred? What if I'm in a med facility restroom? (March 30, 2016)

@ma2therisa: @periodsforpence @GovPenceIN I'm PMS'ing this week, just thought that you'd like to know. I'll be tweeting you again next week. (March 30, 2016)

The narrative becomes even more colorful when posters use vivid detail to describe what it feels and looks like to be a menstruating person. These three separate posts from Facebook and Twitter highlight the aesthetic of periods in a way that builds on the narrative from the examples above:

Periods for Pence: Me: My period is three days late, but I'm starting to get the white mucus discharge. I thought the gov should know since his hand is in my uterus.

Man: Ummm, please *stifles giggles* contact your physician. (April 11, 2016)

@kaedlen: @periodsforpence @GovPenceIN Started my PERIOD 3 days early over here in HK ... Should I worry Gov? OMGS THERE'S SO MUCH BLOOD! (April 7, 2016)

@LoopsOFury: @periodsforpence @GovPenceIN It happened in the middle of a doozy of a period–8 days! I could feel it sort of … slither out. (April 5, 2016)

In these examples, several different posters construct a narrative of menstruation's complications. Their tweets build off of each other until users can see a complete story detailing multiple experiences with menstruation. From symptoms to menstrual product economy to the feeling of blood slithering out of the body, these menstruators craft a vision of living as a menstruator.

What is interesting about narrative sharing in Periods for Pence is that the social media platform makes it possible for these members to create a cohesive narrative not only about their menstrual cycles, but about their involvement in the campaign. Laura Shanley tells *Cosmopolitan* that once she started transcribing her calls on the campaign's Facebook page, "people just started calling as well" (Gupta 2016, para. 6). Writing the content of the calls on a public platform allows the group's mission to be visibly supported and reproduced via social media functions such as liking, commenting, replying, sharing, and retweeting. Often, Periods for Pence supporters will see an original post and then either voice their desire to join in, or simply join in right away. For example:

Periods for Pence: Me: Good morning. I just wanted to call and let the good Governor know that I am still not pregnant, since he seems to be so worried about women's reproductive rights.
Irritated lady on the other end of the phone: And can I get your name, please?
Me: Sure, it's Not Pregnant Laura. (March 29, 2016)
Anjali Mirmira: My latest call:
Me: hello, I was wondering if you could tell me where Gov. Pence went to medical school?
Rep: Excuse me? He's hasn't from what I am aware of
Me: Then why does he feel like he can be the healthcare provider of every woman in Indiana and know nothing about women's healthcare?
Rep: Ma'am, do you have a constructive question? Because we've been getting these calls too often
Me: Sure. Where's the closest planned parenthood from Zionsville?
Rep: *hangs up.* (March 29, 2016)
|**Shawn Fisher:** Wow! I wonder if Gov. Pence considers my tubal ligation and years later hysterectomy to be bad choices? Perhaps I should call? Thoughts? *hmmm.* (March 29, 2016)
Carl Lampman: Call and ask. (March 29, 2016)
Kathy Marquis Waugh: Please someone post his number I so want to call, thank you. (March 29, 2016)

In this Facebook comment thread, members of Periods for Pence see the original post reflecting a phone call made by Shanley and proceed to describe their own phone calls, ask if they should call regarding certain issues related to their own cycles, voice their individual interest in participating, and support the activists in the campaign. In this way, the campaign functions to also foster a cohesive narrative about activism. The generative meaning that is produced and reproduced fosters a doors-open policy with regard to discussing personal anatomy and physiology, but also casts a wider net of interest in the campaign's cause.

Dubriwny's conception of shared narrative exists in the Periods for Pence campaign in that the commenters and responders collectively build a community that supports and solidifies the group's intentions, messages, and experiences. The activists discuss bleed-throughs, clots, and general issues with their menses—building up a narrative of the complications that arise from a bodily function they cannot easily control. At the same time, the group builds cohesion around their experiences calling Pence's office.

Bloody Hilarious

What is so appealing about Periods for Pence, even for those who would not typically enjoy casual conversation about menstrual blood, is the hilarity with which the members describe their menses. In fact, Gupta (2016) praises the campaign as being perhaps the most hilarious way to contact your representatives. Dubriwny (2005) argues that the use of humor in feminist protest, particularly when considering it through the lens of collective rhetoric, is strategic and essential. She mentions that the type of humor used for consciousness-raising centers on self-denigration and common understanding; it often speaks to the larger narrative of women's struggles to achieve bodily autonomy. Dubriwny's understanding of humor speaks to Kalčik's (1975) valuation of humor as a strategic rhetorical tool. She argues that strategic humor often initially appears to be devaluing the individual who employs its use, typically through plays on stereotypes. What makes humor based on stereotype so effective in protest is that it usually carries implicit critiques of the stereotypes it appears to propagate. For example, Dubriwny explains that Redstockings activists relied on humor that made them appear as if they did not know anything about their bodies. In reality, having knowledge about their bodies (as opposed to the men who were legislating them), was crucial to the success of the women's rights movement. To make themselves appear as if they had no knowledge of their own bodies, then, worked to reveal the dangers of not knowing about women's bodies. Content generated by the members of Periods for Pence mostly serves to humorously point out the logical flaws of HEA 1337 by highlighting the body. For example, Twitter user @lynneebrown highlights the shallow logic used in the crafting of HEA 1337 by noting that it fails to account for menstruators who have had surgical procedures and menstruators who may not biologically reproduce:

@periodsforpence @GovPenceIN I've heard you're interested in women's reproductive health. FYI I've had a hysterectomy. Plus lesbian. (April 1, 2016)

Facebook user Rick Matlock plays on the stereotype of menstruating people as cranky and needing comfort as a way to feign interest in reporting his wife's reproductive health to Pence and inquiring about the extent to which Pence would remain financially interested in all facets of menstruators' reproductive healthcare:

> My wife seems to be doing fine, just thought I would check in since the State has taken the time to be concerned with her . . . Her irritability level was 8. Her chocolate intake level was 9. Her snippy comments were a level 10 . . . Just contact me if you need more info . . . Thank You for your concern . . . BTW, do you reimburse for OB/GYN visits? Do you cover the parking garage fees as well? (April 6, 2016)

Pence's failure to understand the intricacies of the lack of menstrual blood present in a menstruator's biology is also frequently used for comedic effect. An anonymous contributor writes:

> I'm 60 and stopped having my periods 10 years ago. Will I be fined for such activity? (April 6, 2016)

Another anonymous Facebook contributor feigns concern about faux medical issues to highlight Pence's lack of foresight in signing HEA 1337, as well as to call attention to Pence's history of marginalizing minority citizens:

> So I knew a girl 30 years ago who had an abortion. Do you think Pence would want me to stone her the next time I see her? And since I'm gay, should I also stone myself afterwards? So many questions and not enough answers. (April 2, 2016)

While each example may not be laugh-out-loud funny, the brilliance of Periods for Pence's humor lies in the work it assigns to its audience. Chvasta (2006) argues that an enthymematic strategy such as the one employed by the Periods for Pence activists highlighted above allows audience members to find value in the protestors' cause for themselves, which increases the overall efficacy of the movement. Essentially, this strategy saves Periods for Pence protestors the burden of arguing with those who cannot and will not support their cause. Rather, their efforts are focused on those members of the audience who can see the humor in the logical gaps they have created and understand that those gaps are indicative of more serious problems embedded in the law itself as well as the law's inevitable repercussions.

A Reversal in Red

In order for Periods for Pence to have any sort of political practicality, the campaign needed to convince their audience to believe in the serious dangers of HEA 1337. By engaging in both the spinning of cohesive narratives about reproductive health issues and the formation of enthymematic humor, the members of Periods for Pence position themselves closer to establishing an overall support system for their ideology. Before political success can be achieved, however, Dubriwny (2005) argues that its members must re-moralize abortion for their audience. In other words, Periods for Pence must convince their audience that abortion is not only a necessity but a right. The campaign attempts this by engaging in what Dubriwny calls *symbolic reversal*, or the re-moralizing of a contentious ideology. Regarding Periods for Pence, efforts to enact symbolic reversal are targeted toward the linking of menstruation and abortion. The campaign's argument hinges on juxtaposing the unpredictability of menstruation, as well as a myriad of logical flaws, with compliance to HEA 1337. For example:

> **@Sasha827**: @GovPenceIN @periodsforpence Started my cycle today. When will you be by to check my used pads for HEA1337 compliance so I know to be home? (April 5, 2016)

As HEA 1337 suggests, any menstrual expulsion is subject to its jurisdiction. The goal of Periods for Pence, then, is to reject this notion. To do so is to acknowledge the link between menstruation and pregnancy. Periods for Pence does this using several techniques, all of which are humorous, and all of which indicate that regulation of the fetus inevitably involves the regulation of the menstrual cycle—a near impossible feat given menstruation's irregularity. For example, one of Laura Shanley's early calls to Pence's office is simply a report that she could not be pregnant given that her period had come and gone. This particular call highlights the impracticality of claiming to be concerned with women's reproductive health without understanding that pregnancy and menstruation are linked:

> Them: Good Morning, Governor Pence's office
> Me: Good Morning. I just wanted to inform the Governor that things seem to be drying up today. No babies seem to be up in there. Okay?
> Them: (Sounding strangely horrified and chipper at the same time) Ma'am, can we have your name?
> Me: Sure. It's Sue.
> Them: And your last name?
> Me: Magina. That's M-A-G-I-N-A. It rhymes with –
> Them: I've got it. (March 31, 2016)

In a follow-up call, Shanley details just how easily a pregnancy can go wrong against a mother's wishes, often resulting in expelled blood:

Me: Hello, this is Sue Magina again. I just hit a pothole on I-70. It was a doozy! I'm worried it might have shaken something around up in there, and I wanted to make sure that was addressed in this new abortion law. I knew Governor Pence would be worried. Thanks. (April 1, 2016)

And perhaps the best example to demonstrate Periods for Pence's use of symbolic reversal is Shanley's very first phone call to Pence's office, used as a previous example as well. She lets the staff know that she still is not pregnant (likely due to the arrival of her period), and effectively translates the entire point of Periods for Pence:

Me: Good morning. I just wanted to call and let the good Governor know that I am still not pregnant, since he seems to be so worried about women's reproductive rights.
Irritated lady on the other end of the phone: And can I get your name, please?
Me: Sure, it's Not Pregnant Laura. (March 29, 2016)

While Periods for Pence does support the notion that abortion is a moral, legal right, the campaign, as exemplified above, mostly concerns itself with circumstances in which a terminated pregnancy lies outside a menstruator's impetus of control. In Periods for Pence's eyes, HEA 1337 is Mike Pence's way of criminalizing abortions that were had via informed decision making, but the law also punishes those who did *not* make a choice to terminate a pregnancy. In order to convince those with political power that the law is not sound enough to remain, Periods for Pence had to get their audience to understand that HEA 1337 had ramifications beyond those that Mike Pence sought to implement.

On Feminism, Fluids, and Fluidity

The explicit detail with which the activists of Periods for Pence protest HEA 1337 spins a singular narrative of menstruation while celebrating differences in the phenomenon. In the interest of this campaign, talking about menstruation diminishes the boundaries that varied experiences with periods might build between menstruators. Through their posts, menstruation is cast as complicated, dangerous, funny, annoying, painful, messy, and even banal. The support for these experiences by other members of group indicate that menstruation can, overall, be thought of as unpredictable and part of living. Importantly, this is how Periods for Pence rhetorically links menstruation discourse with abortion discourse. In much the same way that menstruation is unable to be controlled, the circumstances of pregnancy, and often, the termination of that pregnancy, are unable to be controlled.

If it is true that bodily fluid can dissolve boundaries, we might think about how menstrual blood is implicated in this notion. What differences exist between different displays of menstrual blood in protest? Is an accidental menstrual leak different from a purposeful display of menstrual blood? If so, how? Can menstrual fluid dissolve racial boundaries? How can we theorize the menstruating bodies of people who are transgender or gender non-binary?[4] These questions can be answered if rhetorical scholars pay attention to how blood is deployed by bleeders. Periods for Pence helps us start to think about this because even though no actual blood is used, it is absolutely invoked in great detail. What is the performative potential of talking about menstrual blood? This case study suggests that deploying Kissling's (1996) notion of the communication taboo for political purposes is useful. In the wake of Periods for Pence, Planned Parenthood of Indiana and Kentucky planned to request a preliminary injunction to block Pence's restrictions. In June of 2016, U.S. District Court judge Tanya Walton Pratt granted Planned Parenthood of Indiana and Kentucky's injunction, writing that HEA 1337 "directly [contravenes] the principle established in *Roe v. Wade*" (Pratt 2016, 2). In accordance with the injunction, HEA 1337 has not yet taken effect, although Indiana state attorneys remain open to filing an appeal. If Periods for Pence demonstrates anything, it is that scholars can and should turn their attention to bleeding (whether physically or verbally) and its potential to create policy change.

Historically, academics, public intellectuals, and activists alike have, likely for organizational purposes, distinguished between waves of feminism. But, analysis of Periods for Pence reveals that it is plausible that acts of protest in the name of feminism may not fall neatly into one wave or another. Rather, the work of the members of Periods for Pence illustrates that feminist activism has always been, and remains, fluid. Rights to reproductive health care and services remain under constant threat. As my analysis demonstrates, the same strategies used to advocate for abortion rights in 1969 are being used to advocate for abortion rights in 2017. Consciousness-raising, collective rhetoric, and embodiment strategies span decades. The strategies are not new, they are just being carried out in new ways. Periods for Pence highlights the craftiness of activists; they find ways to use the same strategies in spaces that are popular in the current time.

As such, I suggest that feminist thinkers should embrace the *fluidity* of feminism. Chris Bobel (2010) urges feminist scholars to continue to do work that aids feminism in delicately acknowledging its past while crafting its future. Rhetorical attention paid to menstrual blood can help feminist scholars trace a lineage of the use of blood to advocate for feminism. Where has our blood been, and where is it going? If Mike Pence learned anything during his tenure as Indiana Governor, it is that his phone lines were not prepared for a state full of menstruating people scorned. As he and like-minded politicians serve Americans from the White House at least until 2021, it is certain that the future is red. Perhaps only time will tell if blood will be shed in effect.

NOTES

1. I will refer to those who participate in Periods for Pence activism as either 'members,' 'activists,' or 'posters' since I cannot be certain of their pronouns or gender identities.
2. Much of this work refers to menstruating people as 'women' and 'females.' While I understand that this choice is essentializing, I will include the language these scholars use in order to trace the lineage of this scholarship.
3. I will typically indicate who made each post by prefacing the content with the poster's name or Twitter handle in a fashion similar to dialogue in a script. Vertical lines indicate a post was made in response to the post above it. When Pence became Trump's vice-presidential running mate, the name of the campaign was changed from Periods for Pence to Periods for Politicians, and as such the tweets from this account are indicated by @PeriodsforPols. All typos present appear in the original posts.
4. See SE Frank and Jac Dellaria, Chapter 7, Miren Guilló-Arakistain, Chapter 63, and Klara Rydström, Chapter 68, for discussion of these questions.

REFERENCES

Bennett, Jeffrey A. 2015. *Banning Queer Blood: Rhetorics of Citizenship, Contagion, and Resistance*. Tuscaloosa: University of Alabama Press.

Bobel, Chris. 2010. *New Blood: Third-Wave Feminism and the Politics of Menstruation*. New Brunswick, NJ: Rutgers University Press.

Burke, Kenneth. 1984. *Permanence and Change: An Anatomy of Purpose*. 3rd ed., With a new afterword. Berkeley: University of California Press.

Chvasta, Marcyrose. 2006. "Anger, Irony, and Protest: Confronting the Issue of Efficacy, Again." *Text & Performance Quarterly* 26 (1): 5–16. https://doi. org/10.1080/10462930500382278.

Demo, Anne Teresa. 2000. "'The Guerilla Girl's Comic Politics of Subversion.'" *Women & Language* 23 (2): 63.

Dickinson, Elizabeth, Karen Foss, and Charlotte Kroløkke. 2017. "Empowering Disgust: Redefining Alternative Postpartum Placenta Practices." *Women's Studies in Communication* 40 (1): 111–28. https://doi.org/10.1080/07491409.2016.124 7400.

Douglas, Mary. 2002. *Purity and Danger: An Analysis of Concept of Pollution and Taboo*. Routledge Classics. London and New York: Routledge.

Dubriwny, Tasha N. 2005. "Consciousness-Raising as Collective Rhetoric: The Articulation of Experience in the Redstockings' Abortion Speak-Out of 1969." *Quarterly Journal of Speech* 91 (4): 395–422.

Foss, Sonja K. 1979. "Feminism Confront Catholicism: A Study of the Use of Perspective by Incongruity." *Bulletin: Women's Studies in Communication* 3 (1): 7–15.

Grosz, E. A. 1994. *Volatile Bodies: Toward a Corporeal Feminism*. Theories of Representation and Difference. Bloomington: Indiana University Press.

Gupta, Prachi. 2016. "Why a Woman Called Mike Pence's Office Every Day to Talk about Her Period." *Cosmopolitan*, November 2, 2016. http://www.cosmopolitan. com/politics/a8061278/periods-for-pence-sue-magina/.

Kalčik, Susan. 1975. "'… Like Ann's Gynecologist or the Time I Was Almost Raped': Personal Narratives in Women's Rap Groups." *The Journal of American Folklore* 88 (347): 3–11. https://doi.org/10.2307/539181.

Kissling, Elizabeth Arveda. 1996. "'That's Just a Basic Teen-Age Rule': Girls' Linguistic Strategies for Managing the Menstrual Communication Taboo." *Journal of Applied Communication Research* 24 (4): 292–309.

———. 2013. "Pills, Periods, and Postfeminism." *Feminist Media Studies* 13 (3): 490–504. https://doi.org/10.1080/14680777.2012.712373.

Lee, Janet, and Jennifer Sasser-Coen. 1996. *Blood Stories: Menarche and the Politics of the Female Body in Contemporary U.S. Society.* New York: Routledge.

MacDonald, Shauna M. 2007. "Leaky Performances: The Transformative Potential of Menstrual Leaks." *Women's Studies in Communication* 30 (3): 340–57.

Mazzoni, Mary. 2016. "'Periods for Pence' Is Social Media Gold." *Triple Pundit*, April 15, 2016. http://www.triplepundit.com/2016/04/periods-for-pence-social-media-gold/.

Murphy, Mollie K. 2017. "What's in the World Is in the Womb: Converging Environmental and Reproductive Justice through Synecdoche." *Women's Studies in Communication* 40 (2): 155–171.

Periods for Politicians/Periods for Pence. *Periods for Politicians/Periods for Pence* [Facebook Post]. March 29, 2016. https://www.facebook.com/REALP4P/.

Periods for Pols [PeriodsforPols]. *Periods for Pols* [Tweets]. March 30, 2016. https://twitter.com/periodsforpols?lang=en.

Periods for Politicians/Periods for Pence. *Periods for Politicians/Periods for Pence* [Facebook Post]. March 31, 2016. https://www.facebook.com/REALP4P.

Periods for Politicians/Periods for Pence. *Periods for Politicians/Periods for Pence* [Facebook Post]. April 1, 2016. https://www.facebook.com/REALP4P.

Periods for Politicians/Periods for Pence. *Periods for Politicians/Periods for Pence* [Facebook Post]. April 2, 2016. https://www.facebook.com/REALP4P.

Periods for Politicians/Periods for Pence. *Periods for Politicians/Periods for Pence* [Facebook Post]. April 6, 2016. https://www.facebook.com/REALP4P.

Periods for Politicians/Periods for Pence. *Periods for Politicians/Periods for Pence* [Facebook Post]. April 11, 2016. https://www.facebook.com/REALP4P/.

Pratt, Tonya Walton. 2016. "Planned Parenthood of Indiana and Kentucky, INC., Dr. Marshall Levine, M.D. v. Commissioner, Indiana State Department of Health in his official capacity, et al.

Rinkunas, Susan. 2016. "Indiana Women Are Calling the Governor about Their Periods." *The Cut*, April 1, 2016. https://www.thecut.com/2016/04/indiana-women-calling-governor-mike-pence-about-their-periods.html.

Rudavskey, Shari. 2016. "Carmel Mom Is the Previously Unidentified Woman behind Periods for Pence." *Indianapolis Star*, November 2, 2016. https://www.indystar.com/story/news/2016/11/02/woman-behind-periods-pence/93186838/.

Tuttle, Ian. 2016. "Mike Pence Abortion Law Should Be Taken Seriously." *National Review*, April 8, 2016. http://www.nationalreview.com/article/433879/mike-pence-abortion-law-should-be-taken-seriously.

Weiss-Wolf, Jennifer. 2017. *Periods Gone Public: Making a Stand for Menstrual Equity.* New York: Arcade Publishing.

Woods, Carly S. 2013. "Repunctuated Feminism: Marketing Menstrual Suppression Through the Rhetoric of Choice." *Women's Studies in Communication* 36 (3): 267–87. https://doi.org/10.1080/07491409.2013.829791.

CHAPTER 65

Menstruation Mediated: Monstrous Emergences of Menstruation and Menstruators on YouTube

Lise Ulrik Andreasen

INTRODUCTION: MENSTRUATION ON YOUTUBE

YouTuber Kristine Sloth's video *Tea-time with . . .* from 2014 was one of the first videos made by a Danish YouTuber to address menstruation. Since then YouTuber Julia Sofia has been another pioneer in talking about menstruation on YouTube. Menstruation has now become almost an imperative theme for any self-respecting Danish YouTuber. This means that most menstruating YouTubers have at least one video about menstruation and the popularity of the theme has resulted in some of the most popular YouTubers doing commercial collaborations with major menstrual product companies. Videos made by teenage YouTubers for teenage viewers have become increasingly popular in Denmark, where 99% of all Danish teens between 15 and 18 years use YouTube (Statistics Denmark 2017). Some Danish YouTubers have hundreds of thousands of subscribers and YouTube is now the online platform where most files are shared (Balleys 2017).

Media and youth scholar Claire Balleys observes that YouTube "provides a brand new world for expressing, experimenting with and negotiating adolescence" (Balleys 2017, 227). As menarche and menstruation are central events for many going through adolescence, YouTube also provides a space for re-negotiations and new emergences of menstruation and menstruating subjectivities. The two YouTubers in the center of this chapter, Julia Sofia and Kristine Sloth, are not a part of the global group of feminist activists and scholars who through the last decade have been successfully making menstruation visible in the mainstream (Bobel 2010). When Danish teen YouTubers choose to talk about menstruation, they speak in the confessional, intimate

© The Author(s) 2020
C. Bobel et al. (eds.), *The Palgrave Handbook of Critical Menstruation Studies*, https://doi.org/10.1007/978-981-15-0614-7_65

style of the teen YouTube genre, where YouTubers take their own life experiences as vantage point for the conversation:

> APOLOGIZE (sic) in advance to the delicate souls who cannot handle the honest talk about menstruation. That is the way it is on my channel. Story of being woman. Here is my morning routine! (Kristine Sloth 2015, "My morning routine")

> To be honest, menstruation is something that gets to all of us, and instead of just going around whining about what a freaking downer it is - because it is, then we can do somethings to help ourselves and make it all a little easier, so we don't have to worry all of the time. Because in that way, we are winning and not the devil menstruation. Does it make sense? (Julia Sofia 2016, "Beat your Menstruation")

This genre represents new forms of intimacy made possible by the distance of the internet (Andreassen et al. 2017) and this enables the YouTubers to connect with their viewers. As Balleys puts it, "[b]y making themselves the subject and expressing themselves publicly, teenagers are conducting an exercise in social and identity positioning, leveraging intimacy to tell their stories and triggering recognition and identification in their online audience" (Balleys 2017, 227). When watching teen YouTube videos about menstruation, some of the menstrual negotiations stick out from how menstruation has emerged in the past. The move away from more traditional sources of menstrual knowledge, like youth magazines, classrooms (for example, see Ghanoui, this volume, Chapter 67), or the school nurse's office, and into the hands of teens with digital knowledge, means that menstruation is re-negotiated by teen menstruators themselves, which makes new emergences possible.

Where menstruation in traditional, mainstream discourses has been portrayed through narratives of happiness, cleanliness, and control (Kissling 2006; Ussher 2006), some YouTubers now deal with menstruation through affective and, I argue, subversive strategies where humor and disgust intersect. They do this by engaging with what media scholar Maria Parsons refers to as the *menstruous—monstrous* (2009).

IN-BETWEEN AND ELSEWHERE

As uterine blood that has left the body is considered in-between and an object of otherness, so is the monster. Throughout time, writers and scholars have explored cultural ideas connecting bodies that bear a "seeping, leaking bleeding womb" (Ussher 2006, 1) to the monstrous (Braidotti 2000; Cohen 1996; Haraway 1992; Kristeva 1982; Parsons 2009; Ussher 2006). Representative of uncontrollable nature, moisture, pain, and weakness, the idea of the womb-bearing subject as monstrous can be seen as pointing out limited and undesirable subject positions for menstruators: positions of suppression, shame, and internalized hate toward one's body (Ussher 2006), which can be understood as effects of an archaic patriarchal and sexist culture.

However, considering these new emergences of menstruation on YouTube, I will argue that menstrual connections to the monstrous might work as affirmative and immanent critiques (Butler 2002; Foucault 1997). Exploring relationships of power is central to feminist critique (Butler 2002), and thus an affirmative critique of menstrual norms will aim at identifying and exploring normative breaking points and possible turns toward a promise of a menstrual everyday utopia which offers new, multiple, and inclusive ways and spaces of doing menstruation and being/becoming menstruator. In the words of feminist philosopher Donna Haraway, this utopia could be an undefined place of feminist science fiction and promise, simply called *Elsewhere* (Haraway 1992).

Through my initial skeptical viewing, I first considered the menstruation-themed YouTube videos as merely another way of capitalizing on adolescent menstruators' fears of losing control over their bodies (Kissling 2006). Some of the videos are most likely exactly that, but through a careful and open reading (Staunæs 2007), which turned my attention toward the complexity and messiness of how menstruation emerges in the videos, they seemed to hold potential for something else.

The Promise of Monsters

In the horror movie *Carrie* (1970), a bullied teenager gains special powers when entering menarche (Parsons 2009). Something similar happens in the Canadian horror movie *Ginger Snaps* (2000), when a teen gets bitten by a werewolf on the night of her menarche. The title character in *Buffy the Vampire Slayer* (1992) is a high school student with special vampire-slaying powers, who experiences PMS and cramping as a warning when a vampire is around. There are numerous accounts in popular culture and cultural history where the supernatural is connected to menstruation (Parsons 2009; Ussher 2006). While these stories can be read as merely reinforcing stereotypes of menstruators as freaks of nature, I would argue that menstruous–monstrous narratives hold imaginative and emancipatory potential. In a normative and restrictive matrix where menstruators are expected to manage their body through concealment, medicine, and technology, maybe the menstrual monster we meet in tales like these, who is unintelligible and moves across the lines of menstrual norms, can be viewed as a positive figure of transgression. With the words of Henriksen et al., "What kind of critical and imaginative work, does the monster as a guide make (im)possible?" (Henriksen et al. 2017, 4).

Haraway argues that because transformation happens through difference, the monster can be a promising figure (Haraway 1992). In a similar way monster scholar Jeffrey Jerome Cohen proposes the monster can act as a guide for imaginative thinking to access alternative spaces:

The monster is difference made flesh, come to dwell among us. In its function as dialectical Other or third-term supplement, the monster is an incorporation of the Outside, the Beyond—of all those loci that are rhetorically placed as distant and distinct but originate Within. Any kind of alterity can be inscribed across (constructed through) the monstrous body, but for the most part monstrous difference tends to be cultural, political, racial, economic, sexual. (Cohen 1996, 7)

The monster as difference is a suitable inhabitant roaming the liminal locus of adolescence, where for womb-bearing subjects, menstruation and the coming of menarche in itself de-centralizes and places the subject in a position of otherness, by shaking it with affects such as pain, shame, alienation, not-knowingness, and discomfort (Oinas 2001; Pendergast 1992; Rembeck 2008; Ussher 2006). The expectation of menarche is the expectation of the crossing of a border-to-come and to menstruate is to become (an)Other.

To follow the monster means looking to process (Haraway 1992) and to let go of, what Butler refers to as *proper objects* (Butler 1994). Feminist philosopher Rosi Braidotti proposes, "the monster is a process without a stable object. It makes knowledge happen by circulating sometimes as the irrational non-object" (Braidotti 1994, 300). Engaging with the monstrous thus directs my attention toward the moving and affective entanglements through which menstruation is co-constructed on teen YouTube videos. Understanding affects as local, cultural, and relational forces (Ahmed 2004), affect becomes a context through which we can read the monster: "The monster is born only at this metaphoric crossroads, as an embodiment of a certain cultural moment—of a time, a feeling, and a place. The monster's body quite literally incorporates fear, desire, anxiety, and fantasy . . . " (Cohen 1996, 4).

With this in mind, I will take the monster by its slimy hand, to look for signs of YouTubers performing what I will refer to as *monstrous menstruation* and *menstrual trouble*. Inspired by Butler (1990), I see trouble as small agential cracks in normative scripts of how to do menstruation. Social psychologist Dorthe Staunæs gives a helpful account of trouble as "an antagonistic force, which goes against the force, which is currently dominating. It is a challenging force up against the force, which is currently working towards making something the norm. Trouble is the tremble and the tension, which worries, threats or promise to subvert or destroy already sediment systems" (Staunæs 2007, 263). In this case, menstrual trouble can be read as resistance toward narrow menstrual norms.

Multiple Agents

The monster is *embodiment* and *difference made flesh* (Cohen 1996) and when navigating the field of menstruation, the presence of materiality is insistent, in the form of blood, bodies, and menstrual products. This materiality, together with affective intensities and the technological context of YouTube, means that, in addition to menstruating subjects, there are endless and multiple agents in play. This calls for an analytical perspective able to capture process, mess, and complication.

Thinking menstruation as situated in culture, time, and space (Hasson 2016; Oinas 2001) and co-produced by multiple agents (Barad 2003) resonates with findings in my own empirical material (Andreasen, forthcoming), where young menstruators experience their menstruating bodies as not being separate from discourse (Oinas 2001). It enables me to engage with empirical material in a way that will better understand complexity by conceptualizing emergences of menstruation as the moving sum of an entangled and messy bundle. This mess gives meaning to what we know as menstruation, or as feminist philosopher Karen Barad more accurately explains it, "It is through specific agential intra-actions that the boundaries and properties of the 'components' of phenomena become determinate and that particular embodied concepts become meaningful" (Barad 2003, 16).

Thinking along the lines of Barad's post-humanist re-development of feminist philosopher Judith Butler's theory of performativity (Butler 1990), I can move *beyond* thinking menstruation, bodies, and subjectivity as either-or, but as always already all (Barad 2003). By doing a strategic de-centering of the menstruating subject, I attempt to do an inclusive move, which also invites us to problematize and think beyond rigid categories. Menstruation is not purely subjective or objective, biological or social, natural or cultural, but is co-produced in encounters of discourse, affect, matter, technology, space, and time. The menstruating body is situated, and with its flow of blood and its cramping uterus, it carries agency in itself. Thus, I understand *relational* as the always already moving connected processes and intra-actions that entangle, co-construct, and materialize menstruation, menstruating bodies, and menstruating subjects.

Becoming and Emerging on YouTube

To get a better understanding of how menstruation and menstruating subjects emerge and become (Kember and Zylinska 2012) through monstrous performances in this specific context of YouTube, I look to media scholars Sarah Kember and Joanna Zylinska, who suggest a movement away from the exploration of *media* as an object, and instead encourage focusing on *mediation*, "a key trope for understanding and articulating our being in and becoming with the technological world, our emergence and ways of intra-acting with it as well as the acts and processes of temporarily stabilizing the world into media, agents, relations, and networks" (Kember and Zylinska 2012, xv). The menstruating subject is thus not separate from the digital space of YouTube, but exists within and with it. In this sense, here menstruation and menstruating subjects come into being with YouTube and the dichotomy of real and virtual life is rendered useless.

Following Kember and Zylinska I use the Deleuzian terminology of becoming and emerging to reflect my focus on the relationality and process that is menstruation. I use the terms broadly, to also cover how the phenomenon of menstruation is continuously re-constructed as part of the process that points out lines of flight for possible menstruating subjectivities. It is worth noticing that becoming can only be a move toward otherness, since the norm already "is." This means becoming can only be a move away from normative positions (Parsons 2009). Menstruation can be seen as an impetus for such a movement away from white hetero-cis-gender hegemony (which already "is") and toward otherness. As Parsons writes, "Menarche and menstruation are the initiators of the line of flight for 'becoming'" (Parsons 2009, 38), hence becoming-women. The terms of becoming and emergence are meant to serve as affirmative images of menstruation and subjectivities, as always moving, always in the making and as always Other.

Cuts

YouTube videos are also always moving, and when cutting and drawing lines in images, sound, and theory, which grabs and freezes images, the researcher acts as a co-constructing agent. Writing about moving audio-visual images, one is doomed to miss important dimensions. My cuts in the material are marked by (. . .), to further transparency by indicating missing parts of the videos. The YouTubers also make cuts that matter (Barad 2007). Together with fast voices, bodies, music, spaces, clothes, make up, camera angles, zooms, and text applied in editing programs, multiple agents are involved in the mediation of menstruation and all of this matters for how menstruation emerge. In the case of the YouTubers' editing, a style of extensive and abrupt cutting produces high-energy videos, where menstruation emerges through intensity.[1] The way the videos are produced and edited thus decides how and what we know about menstruation.

The Smoothie Challenge

Hey friends, and welcome to this video! Today we are going to do a mega nice video, because this video is made in collaboration with Libreeeeeeesse![2]

Julia Sofia's hands appear in the frame, each holding and shaking a pink package of menstruation pads. Julia Sofia's younger sister Ida is in the video as well and she explains that they are going to do a *smoothie challenge*.

Julia Sofia and Ida are both white, middle-class teenagers. They are well dressed in casual good quality clothes, accessorized with trendy and relatively expensive jewelry. Their hair is healthy-looking and their makeup follows certain codes for Danish middle-class everyday makeup, dictating "not too much."

Doing different kinds of challenges is a well-known and popular feature within the YouTube genre. Often it involves the YouTuber being challenged to consume something considered disgusting. Julia Sofia explains that she and her sister will take turns answering questions about menstruation, and choose red ingredients for each other's smoothies. At the end of the quiz, they are going to taste the smoothies and test how well the menstruation pads can absorb the red liquids.

(. . .)

The frame zooms out, and we see Julia Sofia and Ida sitting on a white couch behind a coffee table, where there is a spread of red foods and liquids, which Julia Sofia identifies as follows:

> There is some red soda, and some cranberry juice. Then there is some pomegranate seeds, and some pickled beetroot, some raspberries, a bit of ketchup. We have some raspberry marmalade and finally we have some Siracha, which is stinking up the whole room.

Julia Sofia then explains how they are going to blend it all together and Ida adds, "You get a sense of the red theme right? Menstruaaation!!!", while she does a V-sign with her hand and both sisters laugh. On the wall behind the couch are two posters with the inspirational quotes, "Your body loves you, love it back" and "Be your own kind of beautiful."

Our Little Juices

As the video moves along, each girl answers questions about menstruation. Wrong answers are penalized by a sister choosing a red ingredient for the other to put in their blenders.

When it comes to the moment of blending, an out of place event occurs. Julia Sofia's sister Ida blends her ingredients to a bright red liquid, while anxiously uttering, "I am seriously scared." She then screws the lid of the bottle and both sisters smell the liquid with disgust: "Ooooh it smells baaad!" Laughing while making a face of repugnance, Julia Sofia shakes her blender bottle while stating, "I'm scared to shake this, because it contains soda." Then she places the container on the blender and blends the ingredients to a soft red liquid. After blending, Julia Sofia, now displaying a skeptical expression, screws the lid off and the red liquid squirts out of the bottle, as Julia Sofia laughingly screams, "Oh no Ida!".

The video then CUTS to slow-motion, and replays the incident with distorted sound. The container that was to serve as a border between the red matter and the outside has failed. As the red liquid moves beyond the borders of the civilizing bottle, it becomes out of place. The red liquid escapes from the bottle, accompanied by Julia Sofia's slowed and distorted monster-like face and voice underlining the disorderliness of menstruation. The slow temporality of the clip enables us to see the CO_2 induced force behind the mixed red ingredients, as the liquid squirts out of the bottle. The guttural and slow "ooooooooh noooooooooo . . . " of Julia Sofia, is not the sound of a human, but the deep roar of a monster. Julia Sofia with her slow moving lips and distorted face with mediation becomes a monstrous menstruator, with squirting liquids out of place. By the agency of technology, time is done differently, which alters affectivities, matter, space, sound, and thus the emergence of menstruation.

The monstrous performance continues when the sisters drink the red smoothies resembling menstrual blood. The smoothie makes Ida drool, and she holds her hands under her chin, to catch the escaping fluid, and we again see a display of disgust entangled with laughter. In this way menstruation emerges through connections with affective displays of disgust (*Ooooh it smells bad*), fear (*I am scared*), and laughter. Except for their reaction to the nontraditional combination of ingredients, the sisters' performance can be read as connected to a long tradition of positioning the menstruating body in a place of fear and disgust. However, as Julia Sofia and Ida each take a sip of their smoothies, and drink what mimics menstrual blood, cracks in the foundation of normative menstrual behavior appear. As the sisters drink, they perform outside proper performances of middle-class femininity, by breaking norms for how to engage with menstrual blood (Kristeva 1982; Ussher 2006).

The two sisters move on to the final part of the video, where they test how well the menstrual pads will absorb the red liquid. As each girl tears open a pink and white wrapper, Julia Sofia says, "This is my favorite part," to which Ida replies, "Yes, just the sound." CUT. "So what we are testing now is whether these pads can absorb our little juices." As Ida spreads a teaspoon of red liquid across the pad, which has been stuck to the coffee table, Julia Sofia exclaims, "Yes thank you, Yes thank you!" Her sister replies, "Oh yes!" Both sisters laugh, and Julia Sofia says, "OK, that's actually a little more intense than I expected it to be." CUT.

This section is edited with cross clips from Julia Sofia's smart phone, which she uses for close-ups of the pads. The close-ups allow the viewer to see how lumpy and foamy the liquids are. Ida says, "Julia, there is a little too much foam on yours." She continues, "Well, you cannot get around there being some bits of pomegranate in this," while she swings the pad from side to side, which makes Julia burst out into a high pitch laugh. CUT. The video concludes with Julia Sofia exclaiming: "We just did this!" Both sisters are holding the pads with the smoothie surface facing the camera. This time they are holding it with two fingers, as if using a pair of tweezers.

When Julia Sofia and Ida construct an assemblage of affects, objects, sounds, and words which makes menstruation emerge in an entanglement of materiality and affects like disgust, fear, and humor and as menstruous–monstrous (Parsons 2009), it could lead to the conclusion that they are reinforcing the very stigma they aim to break. Nevertheless, I would argue that the dichotomous division of stigma or not-stigma is too simplistic here. Instead, I want to propose the idea that the connection to disgust in the case of Julia Sofia's smoothie challenge is what enables menstruation to emerge at all. The YouTubers play with disgust can actually be seen as an act of subversion and a way of gaining power over what menstruation can be. The fact that the red, lumpy, and foaming liquid is in the very center of the video, and that it is not only produced, but also drunk and then applied to menstruation pads, can be considered menstrual trouble. It stands in contrast to traditional mediated emergences of menstrual blood as a clear blue liquid. When the sisters drink the lumpy red liquid, which they connect to their own menstrual liquids by naming it "our little juices," they consume what is considered unclean, abject, and out of place matter (Kristeva 1982; Ussher 2006) and what Julia Sofia herself in another video refers to as "the devil menstruation." By doing this, they make menstrual trouble, which according to Julia Sofia feels "a little intense"—an intensity catalyzed by the performance of a transgressive and subversive act.

My Morning Routine

YouTuber Kristine Sloth is, like Julia Sofia, a young, white, middle-class cis-woman. Her video *My Morning Routine* starts with soft music, and we hear the sarcastically excited voice-over of Kristine as we see her rolling around her bed. Kristine is wearing a sweater and no trousers, her hair is greasy and messy and she is not wearing makeup. Her room is also messy, her laptop is thrown on her bed and things are scattered all over. CUT. We see Kristine sitting on the toilet, where she picks up a packet of menstrual pads. It is empty. She throws it on the floor, and then picks another one up as the voice-over continues.

> Yes, I am looking for pads in this scene, because I have my menstruation these days. And yes, it's the worst thing I can think of. You all have been real lucky for this morning routine, because you have won me on all the best times of the month, but eeeh, I always use night pads because I bleed worse than a hippo, so yes, these are the biggest diapers on the market, for the woman with menstruation, these are from Always, they are really fucking great, because they are also scented by the way, yes but I'm really fucking tired, yes I'm really pretty in this shot.

We see Kristine yawning, still sitting on the toilet. CUT. Kristine is now sitting by her computer, looking at her phone, when she opens her mouth widely and bends over slowly, as we hear the voice-over: "Right here I'm complaining about my stomach cramps, because I have my menstruation as I told you, and I'm drinking a sip of coffee from yesterday." CUT. We see Kristine go back to bed, where she takes her laptop and starts watching the screen.

> I'm just lying here, with my delicious double-chin enjoying life. I would just like to say, and defend all other people who cannot be bothered going to the fitness center and eat a lot of healthy foods every morning, that this is completely normal and that I also do it, so I'm really sympathizing with you, because I

really don't have the energy to do anything, especially not when I'm menstruating, but yes good night again!

Kristine falls backwards into her bed and closes her eyes. CUT. Kristine appears again, this time talking directly into the camera, about her motives for doing her morning routine in what she describes as a "realistic" style, as opposed to the perfect morning routines she has seen others post on YouTube.

Making Menstrual Trouble

The menstruating body might experience anger and pain, it might be tired, and it might be bloated and can in itself be considered trouble. As sociologist Elina Oinas puts it, the woman's body is a trouble-making body (Oinas 2001, 58). In this sense, everything we see in Kristine's video is trouble. Kristine appears unclean and the fact that she shows herself sitting on the toilet, while menstruating, troubles normative performances of femininity and menstruating. Through humorous realism, Kristine uses this messy and monstrous performance to propose a way to do menstruator differently. Her sarcastic, energetic, and fresh-sounding voice in the start stands in contrast to her tired looking body, which ignites humor. All the way through the video, she is performing what I would call *the monstrous menstruating clown*.

By talking about how heavy her flow is, and comparing herself to a very large animal (*I bleed worse than a hippo*), Kristine not only breaks with menstrual norms of concealment, but also positions herself as non-human and Other. A hippo, monstrous by its mere size (Petersen 2017), places her within the realm of nature, which is beyond control. I would argue that what Kristine is doing, when she performs troublesome monstrous menstruator, can be read as a critique of neo-liberal body ideals, where the individual is responsible for having, building, and keeping a body which appears and functions in accordance with contemporary ideals of strength, thinness, and ability (Eriksen 2017; Fritsch 2015).

Oinas points out that

(. . .) the Nordic girl is marked by paradoxical opportunities and pressures that require both individual strength and social relationality in ways that should be carefully scrutinized rather than sweepingly celebrated. Contemporary girls and young women are expected to embody the autonomous and neo-liberal subject ideal. (Oinas 2017, 179)

Kristine herself refers to the pressure of eating healthy and going to the gym, and she directly protests against this by going back to bed after having her day-old coffee, while her body is kicking back (Barad 2007) in flows of blood and waves of cramps. As feminist scholar Sarah Ahmed would put it, Kristine opposes the "duty of being happy" (Ahmed 2010, 7), by how she does menstruator differently.

As mentioned above, Parsons suggests a reading of the monster as a promising and transgressive figure, by arguing that . . . the 'leaky' body or its, so-called menstruous-monstrosity is potentially positive, multiple and transgressive and that the 'real' horror in these texts instead resides within the 'Frankensteinian' agencies of family, state, and religion, and in the methods they employ to both create and destroy the 'monster' (Parsons 2009, 192).

So the *real* horror might lie within contemporary neo-liberal ideals, where menstruators are expected to do anything bleeding and Kristine's negotiation of menstruation through performing monstrous menstruator, can be read as an immanent critique, which points beyond the edges of normative ideas of menstruation.

Kristine's critique takes form through performing a constant menstruating monster, where Julia Sofia performs a happy and clean menstruator, which only turns monstrous at times. However, what is important to notice about both videos is the constant vibrations of humor, which makes doing menstruation differently possible. Both YouTubers take an already established truth of menstruation as disgusting and do something different with it. They work with humor as an affective strategy of subversion. The impetuses for laughter are the affective intensities of the forbidden, the disgusting, and the scary: the monstrous. When the disgusting and scary monster meets humor, it resembles a clown, and the menstruating clown can touch upon the untouchable and become the matter of intense fun. In this way, the monster might awaken through disgust and fear, but it is legitimized through humor. This affective entanglement of fun, fear, and disgust thus becomes the chisel that makes cracks and carves out spaces for menstruation to emerge differently.

IMAGINATIVE RE-EMERGENCES OF MENSTRUATION

Can the troubling menstrual praxis of the YouTubers be seen as imaginative? And can the videos offer a glimpse of an open-ended menstrual utopia? Perhaps a feminist menstrual sci-fi Elsewhere?

Braidotti argues that "[w]e need to learn to think of the anomalous, the monstrously different not as a sign of pejoration but as the unfolding of virtual possibilities that point to positive alternatives for us all" (Braidotti 2000, 172). If we follow this idea, then the monstrous emergences of menstruation, offered by the young menstruating YouTubers, might point to such positive alternatives or merely some alternatives for "some." Some, because there is an important problem, which needs to be addressed and thought about further.

Both Julia Sofia and Kristine Sloth embody the normative ideals that they are also disturbing and they cannot be seen as representative for all menstruators. From their positions of privilege, they can afford to do menstruation differently without experiencing serious sanctions. The same videos made by less privileged menstruators would result in different outcomes. Menstruation is not a fixed or universal phenomenon and it is important to notice the context in which menstruation is situated. Menstruation is local and multiple (Hasson 2016). Menstruation is embodied, it is gendered, it is classed, it is racialized, it is geo-political (Oinas 2017) and thus emerges in various forms and shapes, dependent on the various entanglements through which it becomes intelligible. Menstruation in the context of these Danish white middle-class YouTubers is thus not to be taken as a universal account of a socio-material phenomenon, but as a reading of a localized and situated praxis. It is a praxis that might hold emancipatory potential for some.

The entanglement of whiteness, middle-class, able bodies, technology, and affectivity here act together and allow menstruation to be visible. Through humor, at times generated by editing and use of voice-over, the menstruous monster becomes laughable and maybe even desirable. This carves out a space for the menstrual monster to stay with us and to act as a transgressive figure, which points its monstrous finger toward Elsewhere, a way of doing menstruation differently. By embracing the monstrous, like Kristine Sloth and Julia Sofia do, I read their videos as imaginative work through which they make room for re-negotiations and make different emergences of menstruation possible. No matter possible ulterior motivations like commercial earnings, I will argue that these new openings can be seen as affirmative and as imaginative possibilities. As Haraway puts it, "It's not a happy ending we need, but a non-ending" (Haraway 1992, 110).

NOTES

1. To get a fuller experience, I highly recommend watching the videos in question, in relation to reading this chapter. Links for the videos are provided in the list of references.
2. Libresse is internationally known as Bodyform.

REFERENCES

Andreasen, Lise Ulrik. forthcoming. *Menstruating Subjects and Everyday Life*. PhD thesis.

Andreassen, Rikke, M. N. Petersen, K. Harrison, and T. Raun, eds. 2017. *Mediated Intimacies: Connectivities, Relationalities and Proximities*. London: Routledge.

Ahmed, Sarah. 2004. *The Cultural Politics of Emotions*. Edinburg: Edinburg University Press.

———. 2010. *The Promise of Happiness*. Durham: Duke University Press.

Balleys, Claire. 2017. "Teen Boys on YouTube: Representations of Gender and Intimacy." In *Mediated Intimacies: Connectivities, Relationalities and Proximities*, edited by Rikke Andreassen, Michael Nebeling Petersen, Katherine Harrison, & Tobias Raun. London: Routledge.

Barad, Karen. 2003. "Posthumanist Performativity: Toward an Understanding of How Matter Comes to Matter." *Signs* 28, no. 3 (Spring): 801–31.

———. 2007. *Meeting the Universe Halfway: Quantum Physics and the Entanglement of Matter and Meaning*. Durham: Duke University Press.

Bobel, Chris. 2010. *New Blood: Third-Wave Feminism and the Politics of Menstruation*. New Brunswick: Rutgers University Press.

Braidotti, Rosi. 1994. *Nomadic Subjects: Embodiment and Sexual Difference in Contemporary Feminist Theory*. New York: Columbia Press.

———. 2000. "Teratologies." In *Deleuze and Feminist Theory*, edited by Ian Buchanon and Claire Colebrook. Edinburgh: Edinburgh University Press.

Butler, Judith. 1990. *Gender Trouble: Feminism and the Subversion of Identity*. New York: Routledge.

———. 1994. "Against Proper Objects." *Differences: A Journal of Feminist Cultural Studies* 6 (2–3): 1–26.

———. 2002. "What Is Critique? An Essay on Foucault's Virtue." In *The Political: Readings in Continental Philosophy*, edited by David Ingram. London: Basil Blackwell.

Cohen, Jeffrey Jerome. 1996. "Monster Culture (Seven Theses)." In *Monster Theory: Reading Culture*, edited by Jeffrey Jerome Cohen. Minneapolis: University of Minnesota Press.

Eriksen, Camilla Bruun. 2017. "It's Not Over Until the Fat Lady Loses Weight - Det er ikke slut, før den fede dame taber sig." En form – og normkritik af tv-programmet Min krop til skræk og advarsel. I: *Tidsskrift for Kjønnsforskning*. Årgang 41 (2): 106–23.

Foucault, Michel. 1997. "What Is Critique?" In *The Politics of Truth*, edited by Sylvère Lotringer and Lysa Hochroth. Translated into English by Lysa Hochroth. New York: Semiotext(e).

Fritsch, Kelly. 2015. "Desiring Disability Differently: Neoliberalism, Heterotopic Imagination and Intracorporeal Reconfigurations." *Foucault Studies* 1 (19): 43–66.

Haraway, Donna. 1992. "The Promise of Monsters: A Regenerative Politics for in/ Appropriated Others." In *Cultural Studies*, edited by Lawrence Grossberg, Cary Nelson, and Paula A. Treichler. New York: Routledge.

Hasson, Katie Ann. 2016. "Not a Real Period: Social and Material Constructions of Menstruation." *Gender and Society* 30 (6): 958–83.

Henriksen, L., M. Bülow, and E. Kvistad. 2017. "Monstrous Encounters: Feminist Theory and the Monstrous." *Kvinder, Køn & Forskning* 26 (2–3). https://doi. org/10.7146/kkf.v26i2-3.100801.

Kember, Sarah, and Joanna Zylinska. 2012. *Life After New Media: Mediation as a Vital Process*. Cambridge: MIT Press.

Kissling, Elizabeth Arveda. 2006. *Capitalizing on the Curse: The Business of Menstruation*. Boulder, CO: Lynne Rienner Publishers.

Kristeva, Julie. 1982. *Powers of Horror: An Essay of Abjection*. New York: Columbia University Press.

Oinas, Elina. 2001. *Making Sense of the Teenage Body: Sociological Perspectives on Girls, Changing Bodies and Knowledge*. Åbo: Åbo Akademi University Press.

———. 2017. "The Girl and the Feminist State? Subjectification Projects in the Nordic Welfare State." In *Nordic Girlhoods: New Perspectives and Outlooks*, edited by Bodil Forsmark, Heta Mulari, and Myry Voipio Formark. London: Palgrave Macmillan.

Parsons, Maria. 2009. *The Menstruous-Monstrous: Female Blood in Horror*. Doctoral thesis, School of English, University of Dublin, Trinity College. Accessed date November 12, 2017, Parsons TCD THESIS 9484 The Menstruous Monstrous.pdf.

Pendergast, Shirley. 1992. *This Is the Time to Grow Up: Girls Experiences of Menstruation in School*. London: The Family Planing Association.

Petersen, Daniel Otto Jack. 2017. "At the Mountains of Monstrosity: Reading Ontology in a Fjord." *Women, Gender and Research* 2: 3–17.

Rembeck, Gun. 2008. *The Winding Road to Womanhood: Adolescents' Attitudes Towards Menstruation, Womanhood and Sexual Health—Observational and Interventional Studies*. Gothenburg, Sweden: University of Gothenburg, Sahlgrenska Academy.

Statistics Denmark. 2017. "It-Anvendelse I Befolkningen." Accessed April 20, 2018. https://www.dst.dk/Site/Dst/Udgivelser/nyt/GetPdf.aspx?cid=24235.

Staunæs, Dorthe. 2007. "Subversive analysestrategier - eller governmentality med kjole, fjerboa og Sari." *Magtballader*, edited by Jette Kofoed and Dorthe Staunæs. Copenhagen: Danmarks pædagogiske Universitetsforlag.

Ussher, Jane Mary. 2006. *Managing the Monstrous Feminine: Regulating the Reproductive Body*. London and New York: Routledge.

YouTube Videos (Authors English Translations)

Sloth, Kristine. 2014. *Tetid med Kristine #1!* (Tea-Time with Kristine #1!). Accessed August 5, 2017. https://www.youtube.com/watch?v=CoS0len51X0.

———. 2015. *Min morgenrutine!* (My Morning Routine!). Accessed August 5, 2017. https://www.youtube.com/watch?v=jaifDrQ7H80.

Sofia, Julia. 2016. *<3 Beat din menstruation // Libresse<3* (<3 Beat Your Menstruation // Libresse<3). Accessed August 5, 2017. https://www.youtube.com/watch?v=HQUwbxZEpec.

———. 2017. *THE SMOOTHIE CHALLENGE I Julia Sofia<3*. Accessed August 5, 2017. https://www.youtube.com/watch?v=8_r2jrPp4bg.

CHAPTER 66

Rituals, Taboos, and Seclusion: Life Stories of Women Navigating Culture and Pushing for Change in Nepal

Jennifer Rothchild and Priti Shrestha Piya

INTRODUCTION

For the last few years, a steady stream of news stories has emerged from Nepal, detailing the segregation of menstruating women. This traditional practice of living in menstrual huts (*chaupadi*)—remains most widespread in Western rural areas but is also practiced in other parts of the country, despite the Nepali Supreme Court ban in 2005 and numerous physical, emotional, and mental dangers associated with *chaupadi*. Our research team of Nepali and American scholars collected the life histories of 84 women in Nepal over a period of 16 months starting in June of 2016. These women ranged in age from 17 to 61 years old. The study area included the middle or hill areas of Nepal and the ethnic groups who reside primarily in the hill area, including high caste women and *Dalit* (so-called untouchable) women. We contend that women's life histories are a meaningful location for studying menstrual health and hygiene. We ground our analysis in the specific sociocultural context of Nepali women themselves and their particular lived experiences.

How might micro-level examples such as individual women in Nepal and their life stories illuminate social structures and macro-level social change? We assert that researching gender and reproductive health at the micro-level reveals important dynamics about gender formation, the perpetuation of power, and the resistance to gendered constructions, which then better equips us to understand and develop more effective ways to support women and adolescent girls to empower themselves and enhance their menstrual health and hygiene. These findings can be placed in conversation with similar

© The Author(s) 2020
C. Bobel et al. (eds.), *The Palgrave Handbook of Critical Menstruation Studies*, https://doi.org/10.1007/978-981-15-0614-7_66

studies from other locations to find where the ideas converge and separate, and provide a more holistic view of menstrual practices and politics.

SETTING THE CONTEXT: NEPAL

Nepal is a multiethnic, multilingual, and multicultural country with a population of about 30 million. A majority of people are Hindu (81.3% of the total population), followed by Buddhist (9%), Islam (4.4%), and other religions (5.2%) (Central Bureau of Statistics 2011). Nepal has one of the lowest-ranked levels of human development by UNDP (United Nations Development Program) standards: the UNDP ranked Nepal 144th in their Human Development Index and 115th in their Gender Inequality Index out of 188 countries (UNDP Human Development Report 2016). Families and communities reflect strict and consistent gender inequality in Nepal as these institutions are founded upon cultural and religious beliefs that create and reinforce Nepal's gendered social order. Amidst gender disparity and other discriminatory social norms, Nepali women and girls face a complex set of challenges related to puberty and sexuality. Early marriage, early sexual activity, and early childbearing are common, culturally entrenched practices. Child marriage (bride below 18 years of age) is still prevalent in some parts of Nepal, although it is legally prohibited. The Central Bureau of Statistics (2015) estimated the adolescent fertility rate as 71 per 1000 women aged 15–19 years with wide differences in urban and rural settings (33 in urban and 80 in rural), and 16% of women aged 20–24 years have had a live birth before the age of 18.

In the context of Nepal, where the average age of menarche is 13.5 years old, adolescent women lack consistent access to education on sexual, reproductive, and menstrual health. Many Nepali women lack access to hygienic menstrual materials and disposal options, access to a private place to change menstrual cloths or pads, and clean water to wash their hands, bodies, and (if used) reusable products. Women are left to manage their periods in ineffective, uncomfortable, and unhygienic ways, including using bark, leaves, and dirty rags (WaterAid 2009, 2015). The dearth of affordable hygienic products and facilities is often compounded by cultural attitudes that view menstruation as shameful or dirty. As a result, many women and adolescent girls are excluded from fully participating in social and cultural life, including religious activities. Despite the great need, there have been very few studies on menstrual health and hygiene in Nepal.

LISTENING TO WOMEN'S VOICES

Our goal in this chapter is to illuminate previously overlooked dynamics within the social constructions of gender as they play out around menstruation among the understudied population of women living in Nepal.

The focus on these understudied women themselves and their individual experiences is an important way of expanding our epistemology and consideration of so-called legitimate knowledge vis-á-vis Dorothy Smith's (1987) 'everyday world' and Patricia Hill Collins' (1990) 'everyday knowledge.' We, therefore, ground our analysis in the everyday practices of 'knowers' (Smith 1987, 183–92). Each woman's story in this research project is significant in its own right. Drawing from these life history narratives, we argue that looking closely at the micro-level illuminates what is possible at the macro-level.

BLEEDING IN NEPAL: REINFORCING TRADITION BUT PUSHING FOR CHANGE

Women's identities are constructed and intricately connected to the social construction of motherhood and reproductive viability. Specifically, who a woman is (and who she identifies as and how she is identified by others) is a complex web of the socially constructed concepts of womanhood, reproduction, and motherhood. An adolescent girl's menarche (first menstruation) is a social marker of a girl's entry into womanhood. The subjective experience of menarche is not only important for understanding adolescents' first-hand viewpoint, but also important from a societal standpoint for understanding how menarche is treated as an opportunity to reinforce gender expectations and roles.

Experiencing Menarche (First Menstruation)

The average age of menarche varied by age of the participants in this study. On average, women above the age of 40 first menstruated when they were 15 years old or older, whereas women between the ages of 26 and 40 had their first period after the age of 14. Younger participants first menstruated, on average, around 13 years old. Thus, younger women reported getting their periods sooner than their older counterparts (see Fig. 66.1). We found that most women were not well informed about mensturation at the time of their first period, leaving them scared, confused, and ashamed when they first saw blood.

For example, Sumangala Khadka (40, Sindhupalchowk) who had her period at 14, shared, "I had thought that it was a leech but the blood was a lot more . . . After seeing the blood and checking my body properly for leeches and not finding them, I was then sure that I had my period." And Rekha Sunuwar (20, Kathmandu) told us, "I cried during my first time . . . I remember when I was nearing that age my mom started to control me, saying don't go out and far . . . I was out and playing . . . during my first time . . . I didn't know how to use [a] pad . . . ."

For the women in our study, their shared experiences of menarche can best be characterized by a lack of preparation that exacerbated feelings of

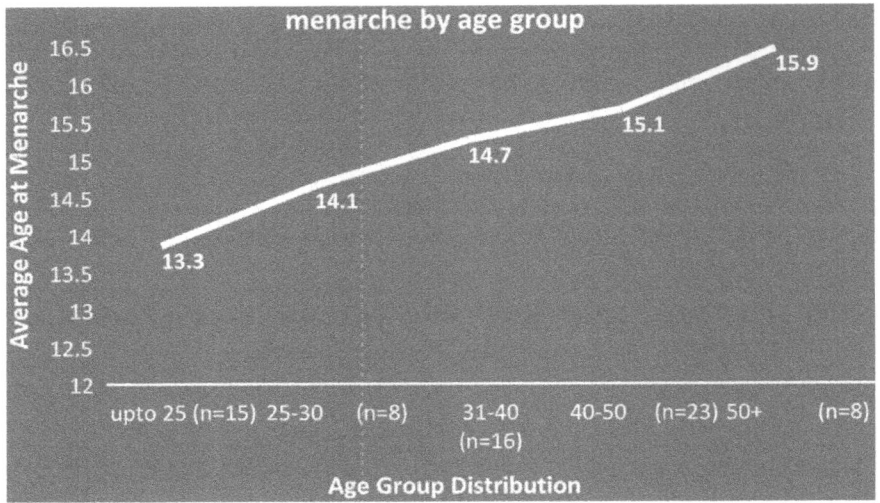

Fig. 66.1 Menarche by age group (Credit: Jennifer Rothchild and Priti Shrestha Piya)

fear, shame, and discomfort. Rather than harness their initial curiosity about menstruation that might be cultivated into a healthy approach for this natural process, their experiences were often surprise, confusion, and isolation.

Menstruation Education

Most of the women in the study said they had not received any education on menstruation. This finding is reflected in a 2012 study by the Ministry of Health [Nepal], New Era, and ICF International Inc. that found almost a quarter of teenage girls had 'no idea' what their menarche was prior to its onset. Siru Shrestha (50, Gorkha), who dropped out of school after fifth grade and had her first period at 16, said, "I thought some insect bit me when I saw blood going down my legs . . . Mother was away . . . I told a woman living in (renting) our house, (and) my sister scolded me . . . I cried terribly. I didn't eat anything." Likewise, Gita Silwal (44, Kathmandu) shared,

> I didn't know what menstruation was. My relatives used to tease . . . that I will experience something new when I reach the age of 13 or 14. I used to feel scared and think how does it feel to have these kinds of experiences? I had it at 12 years . . . I cried a lot . . . because I could not understand what is happening in my body . . . we feel shame.

Women also complained about the lack of information from their mothers. Nayana (46, Kathmandu) stated,

She (her mother) didn't give any information. She just said that when you reach to certain age, you'd have menstrual bleeding. But I didn't know that it happens every month and almost for lifetime. But I knew during menstruation we are prohibited to touch certain things and should also be clean . . . I only knew behavioral term, but had no idea about biological aspect.

Prior to the day of her first menstruation, Rekha Sunuwar's (20, Kathmandu) mother made no mention of it, and instead, talked about how one day she would "grow up . . . and will get married and so on." Rekha recalled that during her first period, "I used the pad in a wrong way . . . the sticky part was placed upside . . . (It) makes me laugh a lot (now) . . . (after periods) I feel I have become mature . . . (I realized) I should not go here and there (roam around) . . . ."

Some women narrators had greater access to information. They cited sources like friends, teachers, or health workers, who explained menstruation to them and told them not to feel anxious about it. Tripti Shrestha (45, Gorkha) remembered,

It would have been awkward at home but since I was in the hostel with girls of same age, going through similar things as me, it wasn't as uncomfortable . . . Sometimes I felt bad but friends used to tell me that it was just the rule of nature.

Similarly, Saloni Rajbhandari (25, Nuwakot) shared, "our hostel in-charge made me understand that it was (a) normal and natural process." It would seem that adolescent girls with access to education and accurate information about menstruation were not only more likely to feel more confident about their changing bodies but also more likely to maintain menstrual health and hygiene.

MANAGING MENSTRUAL HEALTH AND HYGIENE IN THIS CONTEXT

Conversely, for many women in our study, not having open and informative conversations about menstruation at an early age and not having emotional support from others directly affected how these women managed their menstruation. As we will discuss later in this chapter, the importance placed on segregation and isolation of menstruating women far outweighed the importance of healthy and proper menstrual health management (MHM). As a result, MHM was rarely discussed and remained largely unpracticed.

Managing Menstrual Bleeding

Many older women (40+) in our study confessed that there was no practice of wearing underwear or undergarments in their youth. In fact, most of these women above 40 years shared that there was no orientation on wearing anything. Tripti Shrestha (45, Gorkha) said, "There was no system of wearing a pad. They would just let it (the blood) flow." And Kripa Achami (59, Gorkha) recalled,

> No, it (pad) was not there. There was even no practice of (wearing a) petticoat. I used to wrap up only Fariya (saree) . . . we used to wrap up Fariya and roll it inside like that . . . and after that got wet . . . I washed it . . . it would dry up and another day I used.

These women, as well as other older women in our study, said they eventually learned from others to use *taalo*, a piece of cloth such as old sarees or bed sheets to manage bleeding. Older women expressed a preference for using *taalo*. Further, because disposable pads were often too expensive and/or were not readily available in the markets, most of the women in our study used *taalo*. Younger women, especially from urban areas, with more income and more education, were more likely to use menstrual pads than older and rural women.

Managing Menstruation at School

Poor sanitation facilities and a lack of adequate water supply further exacerbate poor menstrual hygiene among women. Only 28% of public schools in Nepal have separate facilities with toilets for girls (UNICEF 2015). Likewise, a lack of adequate water supply or locks on toilet doors to maintain privacy negatively affect adolescent girls' attendance and focus on studies. Additionally, there is a dearth of menstrual products available and little to no disposal pits for soiled pads in school facilities. Thus, during menstruation, many adolescent girls do not attend school (WaterAid 2009, 2015). In some districts, parents do not allow their daughters to go to school during menstruation; instead, parents insist daughters take rest at home. All of these factors contribute to high rates of absenteeism for adolescent girls. Samjhana Sharma (45, Gorkha) shared, "We could not go to school when we were having our periods. The first time, it was 22 days that I was hidden (secluded). I didn't go to school. Second time, it was for 11 days. Third time for seven. When this continued, I didn't go to school altogether." And Sabina Thapa Magar (25, Dolakha) recalled, "It was after the five days of my menstruation. So, I didn't use the cloth at that time. When the sir (teacher) entered in the class it started bleeding . . . So I ran from the class to house . . . The boys in the class were shameless and they used to tease the girls knowingly in such days."

Other women also reported facing embarrassing situations in school. These issues regarding managing menstruation at home and at school underscore the role of harmful norms in shaping MHM. While adolescent women might prefer pads, and many women of all ages understand the value for improved MHM, all women in our study were caught in a complex web of attitudes and beliefs that place women's menstruation in a negative light.

Rather than understanding menstruation as a meaningful and positive step toward adulthood, most women in our study remembered battling shame and embarrassment during this adolescent phase of life.

BELIEFS, TABOOS, AND RITUALS ASSOCIATED WITH MENSTRUATION

The sociocultural beliefs about and perceptions of menstruation have been constructed by multiple factors having to do with cultural beliefs about women's fertility and lack of purity. We found that strong beliefs about menstruating women as impure were present across all castes and ethnic groups in the lives of the women in our study, yet each caste and ethnic group maintained its own customs about recognizing and influencing MHM for women and girls.

Most women and adolescent girls experience restricted mobility and participation in normal activities during menstruation and were forced to observe traditional norms and practices of isolation and segregation, despite physical discomfort and/or lack of resources. Family members and communities perpetuated these sociocultural beliefs and taboos (Fig. 66.2).

Women reported that during menstruation they were forbidden from touching plants, manual grain grinding machines (*dhiki/jaato*), water sources, or places where food was stored. Touching anything while menstruating, left it polluted—too dirty for anyone else to use. A menstruating woman should be careful to not let even a drop of water fall from her mouth while drinking, as that drop could then pollute the ground. Women interviewees talked about how at the end of their menstruation, they were instructed to bathe, as well as wash the items (utensils, clothes, bed sheets, towels, et cetera) that they used separately from everyone else during the fourth day of their period.

Examples of Taboos Reported by Our Participants

"We were not allowed to touch copper." - Garima Jirel (46, Dolakha)

"We had to eat in the same plate for five days. After fifth day's bathing, cow's urine is used to purify us." - Shanti Khati (30, Dolakha)

"For four days of period, I was not allowed to consume salt.- Sita Bohora (45, Gorkha)

"We could not even spill a drop of water while drinking." – Champadevi Sapkota (33, Nuwakot)

"They didn't let us dry our clothes at other's land. They didn't let us step there saying that the road would die." – Shanta Lamichanne (45, Sindhupalchowk)

Fig. 66.2 Common taboos reported in Nepal (Credit: Jennifer Rothchild and Priti Shretstha Piya)

Women also shared how community members seemed intent on restricting the mobility of adolescent girls after menarche; the belief was that an adolescent girl should no longer roam around freely as she now could become pregnant. Barsha Khatri (55, Dolakha) said, "People used to tell that we must not go to others' (place) after menstruation . . . If she goes, she may get pregnant and so on . . . (After menstruating) I used to think like that, now I must do marriage and search a good man to marry." Women also reported that as adolescents, they were discouraged from maintaining friendships with boys. Concerns like these reflect fears that a premarital pregnancy would negatively affect a family's reputation and a young woman's standing in the community. As a mechanism of social control, family members instilled their daughters with shame and fear.

Menstruation Beliefs and Rituals Among High Caste Hindus

While menstruation is a normal physiological process, Hindus consider women impure, untouchable, and undesirable during menstruation (Ueda 2012). Dominant Hindu practices are based on the belief that when women menstruate, impure blood leaves the body, and the body becomes impure. The woman's impurity forbids her from practicing religious and other sacred activities. Societal pressures to maintain menstrual restrictions become even more potent when menstruating women internalize these beliefs and begin to practice self-exclusion. Believing the mythical interpretations of menstruation and internalizing the concept of having an 'impure body,' menstruating girls and women become fearful that the gods might bring misfortunes onto them or their families. Menstrual problems such as abdominal pain and cramping are interpreted as punishment for not abiding by the restrictions. And because the menstruating woman is removed from religious rites, it becomes easier for larger society to disregard menstruation as a natural process, and instead, blame the menstruating woman for any unfortunate events that might occur, for example, a family member becoming ill or a landslide that destroys a home.

Almost all the high caste Hindu women in the study shared that the first reaction of the person (mostly sisters, mothers, or female relatives) who they told after seeing blood was to hide them from the male members of their immediate family. As Ramita Dahal (33, Dolakha) explained, "(I) told my mother . . . Then, my mother scolded me for looking at my brother's face and (she) said, 'Do you want to eat (kill) your one and only brother looking at his face while having period? . . . In our Chhetri caste, a girl should not look the face of her brother while having her first period so I was taken to another place to hide."

This practice of hiding often left the women feeling afraid, ashamed, guilty, and confused about the possibility of something terrible happening to their male family members if they did not comply with this practice. As illustrated by Sarita Neupane (32, Nuwakot), "Mother used to explain about these things . . . if this (period) happens then you shouldn't look at father or brothers . . . I felt like crying . . ."

What is notable in these shared stories was how menstruation was framed as a concern for men or boys. This framing (or reframing) implies that menstruation becomes an area deserving of interest and focus <u>only when it has a negative impact for males</u>.

To protect their male family members, menstruating women are hidden, that is, put into seclusion. This practice of seclusion is often more strictly observed among high caste Hindu women and typically means living away from their own home and keeping a distance from kitchens, prayer rooms, and temples. One of the most extreme forms of menstrual seclusion practice in Nepal is *chaupadi*, which despite being first outlawed in 2005, forces menstruating girls and women to live outside their homes in a *chaupadi* shed (or animal shed) for four to seven days. *Chaupadi* comes from the belief Hindu scriptures dictate secretions associated with menstruation and childbirth to be religiously impure, deeming women untouchable, and prohibiting menstruating women and girls from inhabiting public space, socializing with others, and sharing food and water sources (Ranabhat et al. 2015). Although beliefs and practices are gradually changing, even today, in many parts of the country, women and girls either are forced to spend three to four days outside of their homes, often in sheds, or in a separate room or area while they are menstruating. The practice is more deeply rooted in and widely practiced among certain Hindu communities living in Far and Mid-Western regions of Nepal. However, the practice persists throughout the country (Central Bureau of Statistics 2015). For example, from our study, Archana Rana, now 35, was hidden in a cowshed in a suburb of Kathmandu during her first menstruation. She recalled, "There was a cow shed on the ground floor of the house . . . so I was there in the house but inside an in-built cow shed . . . I used to go out early in the morning and bathe and come back, and my food and all necessities would be brought to the cowshed."

This practice can increase risks of infections, including reproductive tract and genital infections, due to poor hygiene linked to lack of access to clean water (Singh et al. 2001). Women are also forbidden from consuming dairy products, meat, and other nutritious foods, for fear they will spoil them, and must survive on a diet of dry foods, such as rice. As a result, anemia and emaciation among women are two times higher in *chaupadi*-practicing areas (Ministry of Health [Nepal], New Era, and ICF International Inc. 2012). Despite these health hazards, women in our study frequently talked about living in cowsheds during menarche and in subsequent menstruations. Sumangala Khadka (40, Sindhupalchowk) shared, "During periods we either slept in the shed with cattle or in the veranda outside the house."

Beyond the physical, the effects of *chaupadi* are extremely dehumanizing and stressful. Psychological problems linked to stigmatization and isolation, including substance abuse (for example, smoking and alcohol consumption), are often a consequence of the practice, while isolation leaves women vulnerable to rape and animal attacks (Ranabhat et al. 2015; Robinson 2015). Rubina Mishra (43, Nuwakot) recalled, "I felt bad . . . I had to sit on the same place and couldn't go out even in the afternoon."

Despite the psychological and physical tolls on girls and women, the Nepali government has not made *chaupadi* a priority: The Nepali Supreme Court outlawed the practice in 2005, but it took more than a decade for the Nepali government to enact a law to criminalize the custom.[1] To date, the law has yet to come into effect. Government officials argue that prosecuting violators will be difficult.

Instead, the patriarchal system reinforces prescribed gender roles. The seclusion practices remind women that they must police themselves once they start menstruating, that young women become 'polluted' and 'impure,' and that they must conduct themselves accordingly, including secluding themselves from men and boys. Many women interviewees shared that they were told repeatedly that the time had come when they could no longer wander about as they once had as children.

Menstruation Among Newars

The Newars of Kathmandu and Bhaktapur have a culture of *baarha* or *gufa rakhne* (with 'gufa' literally meaning 'cave') ritual that is similar to the hiding or seclusion practices of other high caste Hindus, but this ritual differs in the sense that it is more of a celebration of adolescent girls coming of age and a traditional way to prepare girls for adulthood and menstruation. Siru Shrestha (50, Gorkha) shared, "In the Newar community, when a girl reaches (about) ten years old, she is taken to gufa room. She has to remain in this gufa room for twelve days."

During those twelve days, the girl is visited by female relatives who teach her how to dress and apply makeup. Although Newars are more accepting of menstruation as a rite of passage in comparison to other high caste Hindus, the *gufa* ritual also serves to reinforce gender norms and gendered expectations as girls learn how to present themselves like adult women.

Tripti Shrestha (45, Gorkha) explained that, "In Newar households, (After that Gufa ritual we) . . . avoid going to the Puja (worship) room and avoid touching holy materials (during subsequent menstruations)." The *gufa* room is intended to hide the menstruating girl from all men and the sun and is typically a designated room in the girl's own home. Thus, despite a greater social acceptance of the onset of menarche, Newars still consider menstruation a pollution of adolescent girls' bodies that needs to be hidden.

Prevalence of Rituals Among Non-Hindus and Low Castes

Among non-Hindus, *Janajati* groups, and low castes and *Dalits*, the restrictions and seclusion regarding menstruation are typically not practiced, moderately practiced, or rapidly changing. These groups tend to be more socially lenient about menstruation compared to their Hindu counterparts and often regard it as simply a natural process. Yet, as noted by M. N. Srinivas (1952), lower-caste groups have historically sought upward mobility in the caste hierarchy by emulating the rituals and practices of the upper or dominant castes, in a process he called "sanskritization." Perhaps this is why many non-Hindu groups throughout Nepal are observed as practicing menstruating rituals like Hindus.

Nirmala Pariyar (23, Kathmandu) shared in that Dalit community, "We stay at another's house for seven days. After the seventh day, mother would come to take me back home with new clothes to change. Before that, I should wash myself, then I would be purified (sprinkled) with gold water (water dipped with gold) and cow urine."[2] Goma Rai (23, Nuwakot) noted, "We have restriction for five days . . . We (can) stay in home, we don't go near the stove, (but) we don't separate the utensils with which we eat." Likewise, Shradha Sunuwar (29, Dolakha), shared that, "My mother told me not to cook food for father and brothers . . . but not to hide. In Chhetri and Brahmin (families), they hide but we are Matawali (a lower caste) people like Rai and Sunuwar (and) people don't hide."

While some non-Hindu women in our study practiced restrictions in moderation under the influence of Hindus, other women (especially Gurung, Jirel, and Sherpa women) thought of menstruation as a natural phenomenon and focused instead on staying clean. Notably, women from these particular ethnic groups have historically enjoyed social standing on par with men as compared to Hindu groups. Kusum Tamang, aged 59, from Nuwakot explained, "In our Tamang culture we don't have to abstain . . . I know these Chhetri and Brahmin (women) abstain during menstruation. I don't believe in these things." Samjhana Jirel (27, Dolakha) shared her experience, "When I told to my mother, she told (me) you must keep it clean and take bath." *Janajati* women also shared that they had to continue working in the field and house as they normally do without any rest during menstruation.

Listening to the stories of women who are members of families and groups in which menstrual restrictions are not enforced and the reasons why they refrain from these cultural practices that are otherwise so common in rural areas of Nepal is significant, as these individuals could provide the key insights to designing more effective interventions to improve MHM nationally.

Navigating Culture and Pushing for Change

As women shared with us their life stories and talked about menstruation practices in the past as well as in the present, many believed that social practices are rapidly changing. Sometimes, practices were altered for practical reasons or logistical purposes; for example, requiring menstruating women to take a bath in the early morning every day while menstruating was just not possible for those who had to go to work or college. Some women, including high caste Hindus, reported making changes in the number of days of hiding to accommodate exam dates or school attendance. Srijhana Karki (30, Kathmandu) shared that her mother-in-law and grandmother-in-law had always been strict about her following the practice of not working the four days of her period, as prescribed by Hindu tradition. Yet, at the time when a relative passed away and death rituals needed to carried out and the workload became too much, Srijhana noted that her in-laws then decided it would be all right if she were only restricted three days. After that, she could go back to her domestic responsibilities. Importantly, Srijhana herself noted that cultural traditions, historically held steadfast, were being interpreted and implemented according to convenience.

Women and adolescent girls seem to be increasingly inquisitive, putting menstrual restrictions to the test and subsequently altering practices based on what they then realize. For example, Nayana (46, Kathmandu) shared, "There was also a prohibition during menstruation to not touch the plant. . . . so I always wondered what will happen if I touched . . . one day during menstruation I did touch a plant but nothing happened. So I do touch a plant during menstruation but other (religious) practices I do follow them."

Women like Nayana noted restrictions being lessened and modified in significant ways, but she and others were still careful to maintain their religious purity and not sin by worshipping during menstruation. It would appear that women are trying to gain control of their bodies, their activities, and themselves, while also seeking a balance that does not interfere with their core beliefs.

Rekha (20, Kathmandu) noted, " . . . We used to follow the house rule like not touching this and that and not coming in the kitchen . . . but now, we don't follow this." Education and increasing awareness have led to changing seclusion practices at her home. Rekha went on to explain,

> I used to feel, why I should not touch here and there? . . . And why I am kept in this room? . . . Even during religious ceremonies, I didn't go to places . . . I didn't go myself into the kitchen . . . Now my mom has become a little more liberal . . . entering kitchen and other places . . . All I know is I should be clean . . . my mom also became aware by going to nearby health places and also she heard from others too.

And Champadevi Sapkota (33, Nuwakot) recalled,

> They said . . . we could not touch food too. One day I was very hungry and nobody was home . . . I was feeling so hungry . . . so I couldn't stop myself from taking out that food. I was also very curious about that. I thought what would happen even if I touched. But nothing happened . . . since I was a child

I told them that I took the food from there and ate it but nothing happened. They said that it is a sin and I will have to face problem later. Then I told them I have already touched . . . where is the sin? What does the sin mean?

In the sharing of their life stories, women, both young and old, raised important questions about menstrual restrictions, and often connected these ideas with the ability to challenge existing patterns of discrimination and exclusion.

Conclusions and Implications

In Nepal, there is clearly a dire need for evidence-based, effective approaches to improve MHM. Policy planners and program implementers need a deeper understanding of the issues that Nepali women face today. Our life history interviews reveal that the knowledge and attitudes of family members determine and shape adolescent girls' future actions and beliefs about menstruation and how menstrual health should be maintained. Sociocultural beliefs also transmit important messages about female roles in society more broadly. Specifically, after menarche, adolescent girls' menstruation become increasingly subject to sanction and physical separation enforced by patriarchal attitudes regarding men's and women's roles.

Further, the restrictive attitudes and misinformation about menstruation carried over into other areas of women's lives: education, family relations, and self-perceptions about their role and potential in society. Deep feelings of shame, coupled with poor MHM, meant that women's potential to contribute to their existing environment and situations were greatly diminished. While we observed variation in families' beliefs, taboos, and restrictive practices, many women in our study felt compelled to follow physical seclusion. These sociocultural practices instill shame and fear in menstruating women and serve as a control mechanism for policing gender norms and women's sexuality. The Nepali government could institute change at the macro-level and has officially prohibited *chaupadi*, but seems reluctant to criticize sociocultural practices couched in Hindu beliefs common among its citizens.

What emerges strongly from women's life stories is a scenario in which entire communities (and the country as a whole) become complicit in maintaining a patriarchal structure in which controlling adolescent girls and women is seen as essential to maintaining societal health. We find that such discriminatory beliefs, attitudes, and social norms around menstruation form a complex web of control that limits adolescent girls' and women's independence, and seeks to strengthen the position of men in society by placing the burden of responsibility for managing menstruation on women themselves, but without giving them authority and freedom to manage it responsibly and in a healthy manner. Therefore, women and their individual stories should be at the forefront of MHM research. And on the basis of the findings, effective policies and programs must be contextualized, designed, and implemented in order to create awareness and change people's understandings of and practices surrounding menstruation not only in the context of Nepal, but elsewhere as well.

Micro-level stories illuminate what is possible at the broader societal level. Putting women at the center of our efforts gives way to a better understanding of individuals' circumstances, limitations, daily struggles, motivations for change, hopes, dreams, and fears. This offers a holistic picture not just of MHM or even reproductive health, but of a perspective that could be the key to unlocking better health practices for all people.

NOTES

1. According to the law enacted in August 2017 and purported to come into effect in 2018, anyone forcing a woman to follow the custom will face a three-month jail sentence or a fine (of approximately $30 USD) or both. "The bill criminalizing *chhaupadi* became law as the House endorsed two of the five bills formulated to replace decades-old *Muluki Ain* (General Code), in an attempt to modernize the country's legal system" ("New Law Criminalises Chhaupadi Custom" 2017).
2. Cow and cow urine are considered pure and sacred in Hindu practice.

REFERENCES

Central Bureau of Statistics. 2011. *National Population and Housing Census 2011, National Report*. Kathmandu, Nepal: Central Bureau of Statistics.

———. 2015. *Nepal Multiple Indicator Cluster Survey 2014, Final Report*. Kathmandu, Nepal: Central Bureau of Statistics and UNICEF Nepal.

Collins, Patricia Hill. 1990. *Black Feminist Thought: Knowledge, Consciousness, and the Politics of Empowerment*. Boston: Unwin Hyman.

Ministry of Health [Nepal], New ERA, and ICF International Inc. 2012. *Nepal Demographic and Health Survey 2011*. Kathmandu: Ministry of Health [Nepal], New ERA, and ICF International Inc. (Calerton, MD).

"New Law Criminalises Chhaupadi Custom." 2017. *Kathmandu Post*, August 10.

Ranabhat, C., C. B. Kim, E. H. Choi, A. Aryal, M. B. Park, and Y. Ah Doh. 2015. "Chhaupadi Culture and Reproductive Health of Women in Nepal." *AsiaPacific Journal of Public Health* 27 (7): 785–95.

Robinson, Hannah. 2015. "Chhaupadi: The Affliction of Menses in Nepal." *International Journal of Women's Dermatology* 1 (4): 193–194. https://doi.org/10.1016/j.ijwd.2015.10.002.

Singh, M. M., R. Devi, S. Garg, and M. Mehra. 2001. "Effectiveness of Syndromic Approach in Management of Reproductive Tract Infections in Women. *Indian Journal of Medical Sciences* 55 (4): 209–14.

Smith, Dorothy. 1987. *The Everyday World as Problematic: A Feminist Sociology*. Toronto: University of Toronto Press.

Srinivas, M. N. 1952. *Religion and Society among the Coorgs of South India*. Oxford: Clarendon Press.

Ueda, Misaki Akasaka. 2012. "Creating New Social Norms for Changing the Harmful Practice for the Advancement of Adolescent Girls in Nepal." Unpublished paper.

UNICEF. 2015. *ONE WASH Annual Report 2015, WASH Programme.* UNICEF.

United Nations Development Programme (UNDP). 2016. *Human Development Report 2016.* New York: United Nations.

WaterAid. 2009. *Is Menstrual Hygiene and Management an Issue for Adolescent School Girls? A Comparative Study of Four Schools in Different Settings of Nepal.* Kathmandu, Nepal: WaterAid.

———. 2015. *Formative Research on Menstrual Hygiene Management in Udaypur and Sindhuli Districts of Nepal: Final Report.* Kathmandu, Nepal: WaterAid.

CHAPTER 67

From Home to School: Menstrual Education Films of the 1950s

Saniya Lee Ghanoui

By the 1940s sex hygiene films had become one of the most common forms of sex education, because they constituted a popular form of 'edutainment'—a portmanteau of 'education' and 'entertainment.' While Walt Disney did not use the term until the 1950s, he had popularized the concept through his numerous propaganda films produced during World War II (Disney 1954). Sex education was a popular category of edutainment films, including menstrual education films. Menstrual education films came from two categories of filmmakers, the hygiene product companies that sponsored the films or the non-profit organizations that created them for public health information (Vostral 2011). Due in no small part to the success of the first U.S. menstrual education film, *The Story of Menstruation* (1946), schools became the go-to place for menstrual education and film became the sought-after tool. No longer did the home serve as the sole place of instruction, the school overtook such necessities, and this meant that schools could purchase films for instruction, something that pleased the filmmakers and menstrual product companies.

While *The Story of Menstruation* remains one of the most well-known menstrual education films, the subsequent decades saw a surge of menstrual instruction motion pictures that had a lasting impact.[1] Scholarship has neglected menstrual education films of the 1950s, instead focusing on *The Story of Menstruation* or the new documentary-form menstrual education films of the 1960s which saw the introduction of color, advanced technology and graphics, and more racially diverse casts.[2]

Walt Disney Productions released *The Story of Menstruation* (1946) in association with International Cellucotton Products Company, a branch of the Kimberly-Clark Corporation and makers of Kotex. It took some time

© The Author(s) 2020
C. Bobel et al. (eds.), *The Palgrave Handbook of Critical Menstruation Studies*, https://doi.org/10.1007/978-981-15-0614-7_67

for teachers to warm up to *The Story of Menstruation*, as the teachers expressed apprehension due to the commercially sponsored nature of the film. To combat this, Kimberly-Clark created a teaching manual and menstrual education program that educators could use in combination with the film in classrooms (Kennard 1989, 117). The use of *The Story of Menstruation* set the standard for how other menstrual education films incorporated teaching, commercialism, and health into the classroom, and as a result most of the literature on the topic covers that film. Absent from the menstrual education film literature is a discussion of the 1950s and how menstrual hygiene films of that era established their own approach to the topic.

In this essay I examine two menstrual hygiene films of the 1950s, *Molly Grows Up* (1953) and *As Boys Grow* (1957), to investigate how the films portrayed the menstrual cycle and how educational literature received the films. I argue that the films became so popular because they eased the teaching responsibilities of school instructors without undermining the authority figure of teachers. In other words, the films could supplement traditional menstrual education in the school while the teacher would serve the role of facilitator. I show that this occurred in these two films as a plot device, and in the real life use of the menstrual hygiene films. I contend that menstrual education films of the 1950s constructed the home as the starting place for menstrual education, but portrayed the school as the final authority. This teacher as authority figure laid the blueprint for subsequent menstrual education films of the 1960s and 1970s.

To show this I conduct a contextual analysis of the two films before turning to the widely read *Educational Screen* magazine to explore how it covered the films. *Molly Grows Up* is the second most well-known menstrual education film in the United States, after *The Story of Menstruation*. Medical Arts Productions produced *Molly Grows Up*, and the company had a reputation for its straight-forward presentation of sex education material (Smith 1999). Medical Arts Productions produced numerous films throughout the 1950s specifically made for junior high and high school students. The film *As Boys Grow* may seem an odd addition to the group since its main demographic was pubescent and pre-pubescent boys. In my research, I have yet to find any indication that the film was, at any time upon initial release, viewed by girls. However, I include the film here for two key reasons. First, the same production company, Medical Arts Production, produced both *As Boys Grow* and *Molly Grows Up*; there are similarities between the two films in both content and style. Second, despite the male audience, the period and the menstrual cycle take up a quarter of the time in the film, four minutes out of the 16-minute film. I explore why this film dealt with menstruation so deeply if only boys saw it, while *Molly Grows Up* only pays scant attention to the male body and male puberty.

There is another difference between the two films that make them of interest to study. *Molly Grows Up* is a menstrual hygiene film while *As Boys Grow* is a sex hygiene film. While menstrual education is very much a part of sex education, it does not constitute a comprehensive instruction curriculum. As in the case with *The Story of Menstruation* and, as I will show, *Molly Grows Up*, the films only educate on the menstrual cycle and do not address other aspects of sexual education including sexual anatomy, reproductive health, sexual activity, or birth control and contraception. Part of this reason was due to the historical context, as sex education films of the time did not cover all of these topics. *As Boys Grow*, however, does include discussion of both sex and menstrual education, and the differences between it and *Molly Grows Up* warrant this examination.

As stated above, *The Story of Menstruation* and *Molly Grows Up* were the two most well-known menstrual education films of the twentieth century. Both were used for decades after production, even when the variety of sex education films increased in the 1970s and beyond. Personal Products Corporation sponsored the film alongside Medical Arts Productions; Personal Products Corporation actively incorporated itself into menstrual education and created films that were meant to replace their older counterparts. *Molly Grows Up* was the first film for Personal Products Corporation that focused on menstruation. The company would have continued success with *It's Wonderful Being a Girl* (1968), *Naturally . . . A Girl* (1973), and *Growing Up On Broadway* (1984) (Martin 1997).

Menstrual Performance in *Molly Grows Up*

Molly Grows Up tells the story of pre-adolescent Molly who is jealous of her older sister and close friends who have already started menstruating. Molly finally gets her first period and goes home to tell her mother about it. That evening she asks her mother questions about menstruation, while at school the next day the school nurse gives a lecture to the girls in Molly's class about menstruation and their growing bodies. Opening with a three-ring binder that displays the cast list, the black-and-white *Molly Grows Up* establishes itself as distinct from its famed Disney predecessor, thanks to this opening shot. The relation between education about the menstrual cycle at school and home is constructed as a necessity; a young girl's education may start in one place but it must carry over to the other. The binder sits on a table with a checkered tablecloth, conjuring an image of a pre-adolescent student sitting at the kitchen table doing her homework. Written in the notebook is the cast list and, as a hand comes into turn the page, the audience sees the crew, consultants, writer, and director of the film—each text on a new page. Thus, both students and audience learn that the film they are about to see is meant to be informational and has a knowledgeable authority over the students. Indeed, the film had four consultants—two medical doctors and two PhDs—giving a gravitas to *Molly Grows Up* that did not occur in *The Story of Menstruation*. Furthermore, such a list of medical professionals grants the film a claim to the scientific, tying it to objectives of social and moral standards and communicating that the film will not be indecent or crass. The medical professionals listed function as moral specialists by linking the film not to sex, but to medicine.

In addition, to reinforce the notion that audiences' comfort is of utmost importance, the film creates the idea that menstruation is a girls-only and positive experience. Medical Arts Productions, and its sponsor Modess of the Personal Products Corporation, brought in consultants to address the medical components of the film and gave them opening credits.[3] While the director, Charles Larrance, and writer/consultant, Donald M. Hatfield, are revealed to be men, three out of four of the consultants' genders—those who are meant to know and discuss the menstrual cycle—are kept unknown and only listed by their initials. Including consultants and medical professionals in the making of the film was a revolutionary idea, given medicine's history with sex hygiene films.

By the 1920s the medical profession did its best to distance itself from sex education films. This was due to the image that medical professionals crafted for themselves—an image that projected professional, educated doctors who rejected the popular sex hygiene films of the era because of their lack of medical foundations. In the decades preceding World War II, the medical community tightened its qualifications on who could practice medicine, who could qualify as a physician, and how medical knowledge was disseminated. Medical historian George Rosen noted that in the interwar period "scientific and clinical knowledge and expertise were used to establish criteria with which to circumscribe areas of specialization, to validate the competence of those who wished to practice in them, and to exclude self-styled specialists." Rosen went on to state that the institutionalization of medicine "fostered an increasingly hierarchical structure of medical practice, involving not just physicians but also growing numbers of auxiliary paramedical personnel" (Rosen 1983, 115). Medical professionals rarely contributed to early sex education films. As such, the consultants on *Molly Grows Up* counter established social conventions regarding medicine's input to hygiene education films while also lending expert knowledge. This created a film that pushed against any notion of illicit or even pornographic material.

Molly Grows Up presents itself as an educational television drama with characters and a plotline. In his discussion on the forms of television, Raymond Williams notes that education by seeing is enacted as a means of "experience[ing] a process rather than being taught 'about' it" (Williams 1975, 74). Even though *Molly Grows Up* contains the stereotypical diagrams of the reproductive system, the audience experiences it in the same manner as the character of Molly, a student in a classroom. *Molly Grows Up* operates as a response to negative feedback concerning educational films—particularly a lack of attention regarding communication to audience members (students) and a rehashing of already established teaching tools. In other words, even by the mid-1930s, educators were already jaded by educational films that did not contribute to the curriculum in ways that not only assisted teachers, but enhanced learning as well. Early social and sex hygiene films did little more than disseminate information, and educators insisted that films used in the classroom should have definite "bearing upon the broader objectives of education. Furthermore, [the films] must present those aspects of subject matter which cannot be equally well presented by the use of other mediums of communication" (Arnspiger 1936, 147). *Molly Grows Up* no longer served to feed spongy spectators menstrual information; rather, we, as audience members, are to relate to Molly and her menstrual misgivings. We are even instructed by the authority figure, the school nurse Miss Jensen, to meet Molly and her family so we can understand her frustration and share in her excited anticipation as she awaits her first period. As a result, the audience's awareness of performance is part of the makeup of the film: the performance of the actors, the performance of menstrual envy in that Molly is jealous she does not have her period yet, and the performance of body maintenance (Fig. 67.1).

At the beginning of the film Molly is heartbroken that she has yet to get her period and goes to great lengths to present herself as a mature woman: she discusses fashion with her neighbors, she plays with and wears her older sister's stylish hat, and she takes a Modess sanitary napkin from her sister's room and turns it over and over in her hand musing about the joys of her not-yet-started period. The period is, undeniably, a natural function but the act and construction of a menstrual identity is far from organic. Molly's interaction with her period, her desire to start her period, is a site of what Judith Butler calls a "dissonant and denaturalized performance that reveals

Fig. 67.1 *Molly Grows Up* Jensen image: "Miss Jensen teaches the students on the reproductive cycle" (Credit: Medical Arts Productions for Personal Products Corp. [Modess] 1953)

Fig. 67.2 *Molly Grows Up* Molly image: "Molly plays with her older sister's sanitary pad" (Credit: Medical Arts Productions for Personal Products Corp. [Modess] 1953)

the performative status of the natural itself" (Butler 1999, 186). As a form of menstrual performance, Molly is mimetically a woman—her failure is exhibited by her inability to bleed at a time when she thinks she is supposed to bleed, when society tells her she is supposed to bleed. Molly's own father acknowledges that a time has occurred for her to start bleeding, by indicating that she is no longer a baby and thus her performance as a woman is dependent upon her ability to menstruate (Fig. 67.2).

Male Perspective of the Period

With the success of *Molly Grows Up*, Medical Arts Production followed up with *As Boys Grow*, a film aimed specifically at adolescent boys. With a similar form to *Molly Grows Up*, this black-and-white tale uses actors, again set in school, and an authority figure to serve as educator about puberty and sexual experiences. The film's authority figure is high school gym coach Gene Douglas (played by the father from *Molly Grows Up*, Joe Miksak), who conducts a lesson that is guided by the freshman track team's questions on masturbation, ejaculation, nocturnal emissions, and menstruation. As with Nurse Jensen, the coach breaks the fourth wall in an attempt to relate to the audience members by means of introduction.

In the two films we, as audience members, are included in the discussion and talks by first being introduced to the characters via the authority figures. Jensen compels us to get to know Molly better saying that "I think you might like to know her, too," while Douglas tell us that "maybe you'd like to meet some of them," and subsequently introduces us to the members of the track team. Breaking the fourth wall, while often unexpected, can result in a positive connection between program and viewer (Auter and Davis 1991). Doing so creates a viewer interest involvement that counters the traditionally passive expectations of audience members. More so, in the two films discussed here the fourth wall is not simply broken to address the audience but to invite them to participate in the following program. Both films want an active audience that participates in the education and learning of puberty and the menstrual cycle.

In addition to audience participation, *As Boys Grows* constructs the same needed notion of authenticity, and a list of medical and professionals are credited at the onset of the film; two names worked on both *Molly Grows Up* and *As Boys Grow*. In the latter, W. M. Mitchell served as a psychological consultant, along with two other medical consultants, and D. M. Hatfield produced the short film. Noteworthy is that in *Molly Grows Up* the four respective doctors are simply billed as consultants without any distinction made between their roles. However, in *As Boys Grow*, the psychological consultant is billed separately from the medical consultants, indicating that the filmmakers wanted two points of views, one psychological and one medical, when producing the film and the roles were so great that a distinction was made. This raises the question, then, why was a separate billing of professionals deemed not needed on *Molly Grows Up* but was necessary on *As Boys Grow*?

As mentioned, Medical Arts Productions had a reputation for its frank, open, and blunt discussions of sexual education during the 1950s, and *As Boys Grow* is no exception. However, while *Molly Grows Up* concerns itself only with education on the female body—not once is the male body mentioned nor is there any discussion about conception—*As Boys Grow* challenges both those norms. Initially dancing around the issue of sexual intercourse when discussing the male sex organs—only mentioning that sperm is a "guy's part of the baby" or that an egg must be fertilized by a sperm cell to produce a fetus—it is when the coach discusses the female reproductive system that sexual intercourse is brought into the discussion. Using a diagram of the cross section of a uterus and vagina, he explains how sexual intercourse works and concludes with the piece of advice that going out with girls is fun, and sex is not the only reason for socializing. I contend that because the film discusses both sex-specific reproductive systems and sexual intercourse, more caution was taken in the writing and framing of the explanation, caution that needed both medical and psychological professionals.

While in *Molly Grows Up* the girls in the film function as one unit to ask questions and get answers about the menstrual cycle, how the complete cycle works, and the emotional and physical impact the period will have on them, the main concern of *As Boys Grow* is to teach sexual education. Menstruation is thus explained only so far as its impact on men—specifically how menstruation plays a part in fertilization. The track coach, using a cardboard diagram, explains how an egg travels through the fallopian tubes and into the uterus, and he then explains a woman's readiness for pregnancy. Coverage of the menstrual cycle is incomplete, but the significance of this presentation can be understood, especially in its coverage of sexual intercourse as related to the period.

Where *As Boys Grow* falters is in its overt support for the traditional gender roles as related to menstruation. *As Boys Grow* frames the period as natural but an event that only women must experience and something that often incites acts of anger. The four boys in the film wear workout clothing in a gymnasium, playing off male-desired athleticism, and discuss menstruation with numerous attempts to one-up each other with questions and comments about the female reproductive system.[4] As Butler cites, "culture so readily punishes or marginalizes those who fail to perform the illusion of gender essentialism," the film creates boys who reinforce their own masculinity and its role in our binary gendered society by laughing at the experience of women (Butler 1988, 528). When an image of a fertilized egg appears on screen, an egg with only the tail end of the sperm cell visible, one boy comments with a hint of laughter that it is a "pollywog," or tadpole. The coach does correct him, noting that it is a fertilized egg, only after answering with a smile and small chuckle. By joking about the images of the female reproductive system the boys in the film reassure the male-dominated audience that the subject of menstruation and female anatomy is in some way more humorous than that of male anatomy. As such, the male gender performance in the film is accomplished by degrading the process of fertilization in females and celebrating the boys for acting in a manner that strengthens the "illusion of gender essentialism."

EDUCATIONAL SCREEN MAGAZINE

Educational Screen started in January 1922 and called itself "the independent magazine devoted to the new influence in national education," that new influence, of course, being the motion picture (*Educational Screen* 1922, 3). *Educational Screen* was popular with both educators and filmmakers who wanted to reach schools. Because of its vast influence, the journal was a pinnacle for reviews, commentary, advertisements, and essays on education film, including sex education. Pulling any mention of *Molly Grows Up* and *As Boys Grow* from *Educational Screen* from the 1950s I have found two categories of reference: advertisements and commentary. The advertisements tell us that teachers were the target audience for these films since they could benefit the most. Further, the advertisements acknowledge other nurses and educators who have already watched the film, and give their cited opinions. The commentary sections are like reviews, as the contributors analyze the film and then give critical analysis on it.

Educational Screen featured the film *Molly Grows Up* prominently in its magazine, including both advertisements and commentary.[5] The first mention of the film occurred in February 1954, which contained an advertisement for *Molly Grows Up* calling it a "NEW motion picture on Normal Menstruation." The emphasis on both the newness and the normalcy of the film indicated two larger problems the magazine subtly addressed. First, 'NEW' signaled the immediacy of the research in the film, and the desire for educators to use the latest tools in their teaching. As it had been nearly eight years since *The Story of Menstruation*, advertisers quickly latched on to the selling point that educators could use *Molly Grows Up* as the latest and greatest film. Second, the term 'Normal Menstruation' has two meanings, the normalcy of the film and the normalcy of the cycle. One obstacle that filmmakers had to face included concerns that audiences would construe films on menstrual education or sex hygiene as immoral. Filmmakers paid close attention to the language used in the film and keywords like normal served to recognize the moral and wide acceptance of the film (Heinrich and Batchelor 2004). The advertisement noted that the film is for 9- to 15-year-old girls, again assigning an age demographic to the film to preemptively combat any negative criticism that the film could somehow damage the minds of young girls. This implied that boys did not need to see the film, thus keeping any talk about menstruation out of the hands of boys. The advertisement ended with information on how to preview, rent, or purchase the film from Medical Arts Productions, appealing to educators through different means of obtaining the film. That exact same advertisement appeared again in the four subsequent issues of the magazine: March 1954, April 1954, May 1954, and the Summer 1954 (*Educational Screen*, February 1954, 72).

In the same issue, under the listing of new motion pictures for rent or purchase, the magazine listed *Molly Grows Up* and described it as "the story of a 13-year-old as she experiences her first menstrual period and learns something of its significance." *Education Screen* identified every film by its production company and mailing address in order to obtain the film. The magazine did not shy away from addressing the finances of *Molly Grows Up*. Most other films were attached only to their production companies, however the magazine described *Molly Grows Up* as "financed in substantial part by Personal Products Corporation." The transparency of this statement functioned as a disclaimer, noting that a commercial product is directly involved in the making of the film (*Educational Screen*, February 1954, 77).

Two months later, the magazine published a summary and analysis of the film in a large write-up. The unsigned article commented that when Molly gets her first period and rushes home to tell her mother, Molly's mother is calm, shows no sign of embarrassment or "emotional disturbance," and affectionately asks her daughter how she feels and promises to talk further with her after dinner. Most pointedly both the film and the *Educational Screen* article spent time addressing why society should not use language such as 'the curse' and that girls and women can treat the menstrual cycle not as a nuisance. The anonymous reviewer called the film "a calm, forthright, faithful, and sympathetic portrayal of a girl's experiences connected with the growing-up process." The article's intended audiences concerned both educators and parents. The author noted that teachers would find the film useful in explaining menstruation while allaying any fears the adolescent girls may have while, at the same time, parents would like the film for its helpful "ways of handling their daughter's questions about menstruation and developing desirable attitudes toward the whole process of growing up." Most strikingly, the author employed educators and parents to discuss the film after viewing, even by having group discussions. The film does not end the talk on menstruation but rather it begins it. Whereas many sex education films of decades before functioned as independent educational entities and did not foster talk, as seen by *Molly Grows Up*'s reception the film intended its audience to talk about it afterwards. I do not want to lose sight of the fact that since Personal Products Corporation sponsored *Molly Grows Up*, discussion of the film also meant discussion of the products used during menstruation. The *Educational Screen* article claimed that the "incidental references to the sponsor's products do not seem to distract from the value of the film," but the sponsor's subtle presence allowed for further discussion afterwards, since the film did not feel like a sponsor's advertisement (*Educational Screen*, April 1954, 152–53).

By the autumn of 1954, *Educational Screen* had advertised *Molly Grows Up* numerous times throughout the year, and in September the advertising message began to shift. The film was no longer the 'new' must-have motion picture; rather it morphed into an even more valuable educational tool that supported educators because the film had the luxury of time to establish itself as a legitimate motion picture for school use. That month an advertisement directed squarely at teachers played into the beginning of the school semester: "Preview your print of *Molly Grows Up* before classroom use this fall!" (*Educational Screen*, September 1954, 258). The following month a different advertisement pleaded with its audience to "preview before classroom use" and noted that *Molly Grows Up* contained a companion filmstrip, useful since schools often had a projector that could be used for the filmstrip (*Educational Screen*, October 1954, 336). These two advertisements worked off the positive reputation *Molly Grows Up* had crafted by this point. The magazine reiterated this idea in November 1954 by adding a disclaimer to one of its *Molly Grows Up* advertisements that an unnamed "School Nurse" called it "superior to anything else I have used" (*Educational Screen*, November 1954, 382).

The gravitas afforded to *Molly Grows Up* signified its importance, particularly in comparison to the advertisement for *As Boys Grow*. During the 1950s, there is not one advertisement for *As Boys Grow* in *Educational Screen*. Medical Arts Productions, and by extension the Personal Products Corporation, achieved a consistent advertising campaign with *Molly Grows Up* aimed at educators in the magazine, while the same production company did nothing for *As Boys Grow*. *Educational Screen* featured advertisements for *Molly Grows Up* throughout the late 1950s.

Molly Grows Up and *As Boys Grow* both exist within the spaces that pre-pubescent boys and girls would be intimately familiar with: the home and the school. *The Story of Menstruation* achieved its educational value through its narrator who communicated without the need of a plotline and characters. As a result, *The Story of Menstruation* does not take place in one specific location. Filmmakers of 1950s menstrual hygiene films turned away from such a structure and, instead, placed characters within spaces that provided opportunity for educational moments. These home-to-school moments reflect the way educators felt menstruation should be taught. Particularly, young students' menstrual education may happen at school, but it must first start at home. By including medical professionals in the making of each film, Medical Arts Productions granted the films a claim to scientific objectivity, thus binding it to social and moral standards. Doing so preemptively decreased any potential controversy around the films' subject matter. Menstrual education films evolved greatly from *The Story of Menstruation* to those of the 1950s, which approached the subject through new narration and served to assist school instructors with their teaching responsibilities. Even if educators still believed the home to be the starting place for menstrual education, they were keenly aware that such instruction was not occurring. With menstrual education films teachers could have the equipment to educate students on their bodies with the authority afforded to them by the films.

Notes

1. During the 1950s films dealing with the menstrual cycle were either called 'menstrual education films,' 'menstrual instruction films,' or 'menstrual hygiene films.' I use all three terms here, interchangeably, because that is how educators and filmmakers used them during the 1950s.
2. For more on *The Story of Menstruation* see Bob Cruz, Jr. 2011. "Paging Dr. Disney: Health Education Films, 1922–1973," in *Learning from Mickey, Donald and Walt: Essays on Disney's Edutainment Films*, edited by A. Bowdoin Van Riper, 127–44. Jefferson, NC: McFarland & Company, Inc.; Thomas Heinrich and Bob Batchelor. 2004. *Kotex, Kleenex, Huggies: Kimberly-Clark and Consumer Revolution in American Business*. Columbus: Ohio State University Press; Sharra Vostral. 2008. *Under Wraps: A History of Menstrual Hygiene Technology*. Lanham, MD: Lexington Books.
3. The consultants are listed in the following order as W. M. Mitchell, PhD; D. C. Harrington, M. D.; E. M. Marsh, M. D.; and D. M. Hatfield, PhD.
4. The film sounds as if the filmmakers shot it in a gymnasium without any sound equipment. The boys shout questions/answers that are muffled and, at times, hard to understand. Even though Medical Arts Productions created *Molly Grows Up* and *As Boys Grow*, the latter does not appear to have had the same financial backing as its predecessor.
5. I recognize that *Educational Screen* did not create the specific advertisements for *Molly Grows Up*, but the magazine's willingness to run the advertisements coupled with its educator demographic makes the magazine a ripe place for analysis into how it advertised *Molly Grows Up* and its respective commentary.

REFERENCES

Arnspiger, V. C. 1936. "The Educational Talking Picture." *Journal of Educational Sociology* 10, no. 3 (November): 143–50.

As Boys Grow. Directed by George Watson. Medical Arts Productions, 1957.

Auter, Philip J., and Donald M. Davis. 1991. "When Characters Speak Directly to Viewers: Breaking the Fourth Wall in Television." *Journalism Quarterly* 68 (1/2): 165–71.

Butler, Judith. 1988. "Performative Acts and Gender Constitution: As Essay in Phenomenology and Feminist Theory." *Theatre Journal* 40 (4): 519–31.

———. 1999. *Gender Trouble: Feminism and the Subversion of Identity*. New York: Routledge.

Disney, Walt. 1954. "Educational Values in Factual Nature Pictures." *Educational Horizons* 33 (2): 82–84.

Educational Screen, January 1922, 1, no. 1.

Educational Screen, April 1954, 33, no. 4.

Educational Screen, February 1954, 33, no. 2.

Educational Screen, November 1954, 33, no. 9.

Educational Screen, October 1954, 33, no. 8.

Educational Screen, September 1954, 33, no. 7.

Heinrich, Thomas, and Bob Batchelor. 2004. *Kotex, Kleenex, Huggies: Kimberly-Clark and Consumer Revolution in American Business*. Columbus: Ohio State University Press.

Kennard, Margot Elizabeth. 1989. "The Corporation in the Classroom: The Struggles Over Meanings of Menstrual Education in Sponsored Films, 1947–1983." PhD, University of Wisconsin-Madison.

Martin, Michelle H. 1997. "Periods, Parody, and Polyphony: Fifty Years of Menstrual Education through Fiction and Film." *Children's Literature Association Quarterly* 22 (1): 21–29.

Molly Grows Up. Directed by Charles Larrance. Medical Arts Productions for Personal Products Corp., 1953.

Rosen, George. 1983. *The Structure of American Medical Practice: 1875–1941*. Philadelphia: University of Pennsylvania Press.

Smith, Ken. 1999. *Mental Hygiene: Classroom Films: 1945–1970*. New York: Blast Books.

Vostral, Sharra L. 2011. "Advice to Adolescents: Menstrual Health and Menstrual Education Films, 1946–1982." In *Gender, Health, and Popular Culture: Historical Perspectives*, edited by Cheryl Krasnick Warsh, 47–64. Waterloo, ON, Canada: Wilfried Laurier University Press.

Williams, Raymond. 1975. *Television: Technology and Cultural Form*. New York: Schocken.

CHAPTER 68

Degendering Menstruation: Making Trans Menstruators Matter

Klara Rydström

INTRODUCTION

As scholars and activists we have the responsibility to continually evaluate the work we do. Within the field of critical menstruation studies, we must pay attention to our depictions of menstruation and menstruators, and the knowledge we produce in the pursuit to de-stigmatize menstruation. The recognition that not all women menstruate and not all who menstruate are women illustrates such an awareness. Yet, even though we know that menstruators are of various gender identities, there are few scholarly pieces with a focus on menstruators other than cis women. This chapter offers a theoretical intervention in the cisgendering of menstruation, informed by my own empirical work (Rydström 2018) and that of others (Bobel 2010; Chrisler et al. 2016; Fahs 2016; Berg 2017).

Within the framework of my MA thesis, I conducted qualitative interviews with nine Swedish trans people—self-defining as trans man (3), trans boy (1), trans guy (1), nonbinary and trans person (2), nonbinary and man (1), and trans person (1)—about their experiences with menstruation. All participants were assigned-female-at-birth and either had an ongoing menstrual cycle and bleedings, or they had menstruated in the past. Based on the understanding of menstruation as often being cisgendered, I also attempted to understand what the participants believed must occur for a possible degendering of menstruation to happen (Rydström 2018).

Building on these findings, this chapter asserts that degendering menstruation is necessary for us to deal with menstrual shame, taboos, and stigmas in a gender-inclusive manner. Drawing on theoretical insights provided by post-constructionist theory[1] (Lykke 2010) and transgender studies (Stone 2006; Stryker 1994; Nordmarken 2014; Preciado 2013), I suggest

C. Bobel et al. (eds.), *The Palgrave Handbook of Critical Menstruation Studies*, https://doi.org/10.1007/978-981-15-0614-7_68

that a recognition of the multiplicity of menstrual realities among different menstruators, and non-menstruators, is key therein. Thus, the chapter elucidates various experiences with menstruation among trans people and casts light on areas wherein cis menstruators are made the normative menstruator and trans menstruators the Other.

The 'Nature' of Menstruation

The cisgendering of menstruation is closely linked to the idea of nature as an ontological fact. Nature as given comes with the belief that certain bodies are closer to this naturalness than others; women are then commonly designated as the weaker sex because of their closeness to nature (Jackson and Falmagne 2013, 380). Women in general (Braidotti 1997, 59–80) and menstruating women in particular (Chrisler 2010, 204; Chrisler and Caplan 2012, 275, 283–84) are seen as monstrous because of their assumed lack of control over their unruly bodies. In a similar manner, references to nature are applied to build the argument that trans people hold monstrous bodies (Stryker 1994; Nordmarken 2014; Barad 2015). While cis women are perceived as monsters because of their supposed closeness to nature, the monstrosity of trans people derives from their assumed unnaturalness.

So, what is nature? The assumption that people necessarily belong to one of two gender categories—men and women—essentially linked with one of two sexed bodies—male and female—is based on a faulty ontology (Preciado 2013, 101–3). Following the post-constructionist theory, nature is inseparable from culture and the two are constituted in conjunction with one another; "[n]ature is neither a passive surface awaiting the mark of culture nor the end product of cultural performances" (Barad 2007, 183). Consequently, reality is not made up of discernible 'things,' given by nature or culture, but of material-discursive phenomena (Barad 2007, 139–40). As things, bodies, too, are phenomena and "it is through specific intra-actions[2] that phenomena come to matter—in both senses of the word" (Barad 2007, 140). In other words, no body is more 'natural' than another, but all bodies are materially discursively constructed.

Within cisnormative contexts, cis bodies and trans bodies come to matter differently. Cisnormativity refers to the system of norms structuring societies according to the belief of sex/gender as binary where everyone is assumed to identify with the sex assigned to them at birth—an ideology and practice that divides all humans into the categories male/man or female/woman. Those who do not correspond to these structures are considered deviant (Nord, Bremer, and Alm 2016, 4–7). Cis bodies appear as fixed and given by nature—as compared to trans bodies defined by their expected physical modification and, therefore, unnaturalness (Finn Enke 2012, 6, 12).

Intra-actions determine what we perceive as a body and, specifically what we perceive as a *menstruating* body. Our identities, our bodies and regulative norm systems, intra-act and make menstruation appear as an experience natural to cis-female bodies, whereas it is considered unnatural to trans people who are seen as deviant/monstrous/Other. Thus, trans menstruators per se, are not only challenging, but actually *materializing* menstruation as other than ciswoman phenomena.

The potential for a degendering of menstruation is likewise found in the multiplicity of it as phenomena. By showing that no menstruating body is more natural than another, we can counter the Othering of trans menstruators based on the conception of unnaturalness (Stryker 2013, 149; Barad 2015, 412–13). In my opinion, it is our responsibility as scholars and activists to do so; as part of the acknowledgment that there are no naturally pre-existing boundaries of phenomena (Barad 2007, 139–40), I agree with post-constructionist scholars arguing that we are all accountable for the kind of phenomena we materialize (Barad 2003, 827; 2007, 90–91; Haraway 2016, 7). Within the field of critical menstruation studies, our depictions of menstruation and menstruators, and the knowledge we produce to de-stigmatize and de-taboo menstruation, directly affect menstruation as phenomena—not only in discursive but also material terms.

One way to advance a more inclusive discourse and practice is to bring in voices of trans menstruators. This requires non-trans identified scholars and activists to decenter their own voices to make space for trans voices, though I assert that cisgender scholars and activists can make meaningful contributions. For one, scholars like me can offer critiques of menstruation as cisgendered phenomena, rather than represent 'truths' about trans menstruators. We can also change the way we talk about menstruation, including using the term menstruators, 'menstrual products' instead of 'feminine hygiene products' (Quint 2016) and rename 'female healthcare' centers with more gender-neutral terms (Rydström 2018, 74). After all, while these shifts can make *trans* menstruators matter, a degendering of menstruation will serve all of us, irrespective of identity and experience.

Trans Experiences of Menstruating

Menarche is commonly recognized as the moment where one goes from being a girl to becoming a woman (Hawkey et al. 2017; Jackson and Falmagne 2013; Rembeck 2008). Previous research has shown that menarche is perceived by some cis menstruators in positive terms, whereas others respond with more mixed feelings or resist the claims of womanhood (Jackson and Falmagne 2013, 382, 386–88). Equivalently, and not very surprisingly, trans menstruators' encounters with the first period vary as well. Some of my research participants stated that they remember it as a positive thing, whereas others recall how it marked them with a gender identity they do not identify with, especially through other people's new-found way of approaching them as women (Rydström 2018, 42–43; similar findings are made by Devor 1997 and Rubin 2003, cited in Bobel 2010, 161). While not all trans menstruators experience their menarche in negative terms, it seems fair to state that it is an event marked by cisnormativity.

Experiences of menstruating later in life vary among menstruators as well. Some do not suffer from their periods in direct relation to their gender identity. Others do, as they disidentify with the body as a whole and/or with certain body parts such as the genitalia or the uterus, or with the bodily function of menstruation. This suffering is sometimes related to gender dysphoria (Rydström 2018, 44–49). Notably, how gender dysphoria is experienced and felt is individual. Sade Kondelin (2017) has found that it is explained by many as "becoming conscious of one's embodiment [. . .] as gendered in a way that does not match one's sense of self" (26). As shown by participants in my study, for some menstruators that consciousness has to do with menstruation being cis-coded and, hence, mismatching with their trans-coded being. In practice, this means that some choose to stay at home when menstruating, either because the dysphoric feelings are too strong or to avoid any encounter that could trigger dysphoria (Rydström 2018, 44–45).

Not everyone experiences dysphoria in relation to their menstruation. Given that gender dysphoria is not only an embodied experience but also a diagnosis regulated by the World Health Organization's International Classification of Diseases (ICD)[3] (World Health Organization 2018a), it is important to reckon with the diversity of experiences. Apart from enabling trans people to get access to gender-confirming treatments, the diagnosis contributes to the pathologization of their lives (Butler 2004, 4–5, 76). Previous studies indicate that some trans menstruators do not experience dysphoria but feel that they are seen by others as "not trans enough" (Berg 2017, 35; Rydström 2018, 49). Therefore, future studies should explore how transnormativity[4] (Johnson 2016) shapes the experiences of trans menstruators.

Among those who do suffer from their menstruation, there are various ways to deal with their periods. In one study, trans menstruators are more positive toward menstrual suppression than cis menstruators (Chrisler et al. 2016, 1244). Aligned with such findings, testosterone treatments are a method adopted by some trans menstruators to get rid of unwanted bleeding. Preventing the menstrual period is not necessarily the main reason for using testosterone, but it can be one among several desired outcomes. Moreover, testosterone can help alleviate premenstrual symptoms and, as such, improve general well-being. For others, a side effect of testosterone is that it actually worsens well-being. One of my participants explained that healthcare personnel had proposed that he could adjust the dose, that is, lower it to not interfere with the mental health, while keeping it high enough to prevent his periods. Still, he told me, dealing with testosterone injections and keeping track of hormone levels is a more demanding alternative than menstruating (Rydström 2018, 52–53). Others choose to be on testosterone, while still keeping menstruation intact (Chrisler et al. 2016, 1241). I agree with Chrisler et al. (2016, 1241) who emphasize that research is needed to find out more about testosterone usage and menstruation.

What we do know is that not all who desire to suppress their menstruation are given access to testosterone. In Sweden, gaining access to testosterone requires assignment of the diagnosis "gender identity disorder, unspecified" (Transformering 2018). In other words, testosterone is not an alternative for menstruators outside the diagnostic framework. Moreover, nonbinary trans people who have been in touch with the trans-specific healthcare sector in Sweden have experienced their trans identity as discredited, as they do not fit into the binary idea of sex/gender (SOU 2017:92, 614–22). Similarly, my own research found that having a nonbinary gender identity can result in worries and anxiety for those who wish to get help from the healthcare sector to suppress their periods (Rydström 2018, 53). Here, it becomes evident that testosterone exists within a larger biopolitical framework wherein some bodies are considered more desirable—or more needy—than others (Preciado 2013), resulting in the regulated use of testosterone.

A common strategy to deal with unwanted menstruation is to choose menstrual products for their ability to conceal the periods. Some menstruators who feel aversion toward the genitalia, or toward penetration of the genitalia, prefer period-proof underwear (Chrisler et al. 2016, 1247) or pads, while others who find it distressing to feel to blood, and/or experiences pads as bulky, choose tampons as they keep the period less visible. The menstrual cup is used for the same reasons, or because tampons are physically uncomfortable (Rydström 2018, 54–55). Menstrual cups are sometimes preferred because the cup can stay within the body for a relatively long period of time once it is inserted and, thereby, does not require as many changes as tampons or pads (Chrisler et al. 2016, 1246; Fahs 2016, 82). Considering the long life span of a cup, it likewise minimizes the times one has to purchase menstrual products. Some trans menstruators find this beneficial as purchasing menstrual products in itself can be mentally demanding (Berg 2017, 25).

Menstrual blood, as opposed to venous blood, is commonly seen as differentiating women from men, a fluid seen as making the female body weak (Bramwell 2001, 89–90). It is abject, meaning it is "neither subject nor object" (Kristeva 1982, 1) and perceived with loathing and shame. Consequently, menstrual products are depicted as saving or protecting the body from blood and, as such, advertisements and commercials contribute to the objectification of women (Rosengarten 2000, 92–94, 99). Yet, for trans menstruators suffering from their periods, menstrual blood can be more than a bodily fluid charged with societal shame; the blood may be a reminder of a body, a body part, or a bodily function they do not identify with (Rydström 2018, 57–58). Thus, it must be acknowledged that for some menstruators, menstrual products can be a rescue, just because the right product keeps the unwanted blood at a distance.

Other trans menstruators deal with unwanted periods by subverting the meaning of menstruation. For example, the blood can be visualized as making oneself manlier (Fahs 2016, 83–84). One of my research participants explained that putting on a specific pair of briefs helped him look upon his bleedings as a superpower making him special, as compared to other men. Another participant understood menstruation as a sign of bodily health (Rydström 2018, 51–52). Moreover, some of the people I interviewed indicated that they had become comfortable in their bodies over time and, in doing so, improved their perception of their periods (Rydström 2018, 46–47). Such subversive efforts can be seen as abjectification, that is, processes wherein the abject is reclaimed and reidentified by the abject-holders themselves. For some, abjectification is a strategy of resistance toward the system which oppresses them, for others it is simply a question of survival (Wasshede 2017, 35, 48). But for some trans menstruators, the only way out is to simply ignore and mentally repress unwanted periods (Rydström 2018, 52). Each of these examples makes clear that menstruation as phenomena are not fixed but vary among menstruators and across time.

Menstruators, Not Menstruating Women

The term 'menstruator' likely comes as no surprise for any critical menstruation scholar. Chris Bobel's use is prominent, with her book *New Blood: Third-Wave Feminism and the Politics of Menstruation* (2010) underlining how the term "expresses solidarity with women who do not menstruate, transgender men who do, and intersexual and genderqueer individuals" (12). Here, it is clear that a move beyond the perception of menstruation as a female experience not only functions to make our work inclusive of trans and intersex people, but the menstruator term likewise substantiates theorizing around the fact that not all women menstruate, for example, trans women, postmenopausal women, pregnant women, and those experiencing amenorrhea. I would like to stretch it further to state that the menstruator term is a foundational part of our terminology; it captures the critical engagement driving the field of menstruation studies.

Referring to people who menstruate as 'menstruators' does not come without criticism. Related to a more general debate within feminism(s), the term reifies the question: is it possible to make resistance toward patriarchal structures while leaving behind the term 'woman?' Here, the division between those aligning with sexual difference theory—stressing the necessity of the categorical label 'women'—versus those aligning with gender theory—recognizing gender as a construct—is made explicit (Bobel 2010, 155–56). Nevertheless, taking into consideration that "transsexual lives are lived, hence livable" (Scheman, quoted in Stone 2009), the perceived issue at stake becomes a nonissue. Menstruators *are* of a variety of gender identities (far beyond those who identify as trans) and, hence, menstruation cannot be equated singularly with cis/womanhood. To argue otherwise would be to ascribe menstruation to a biologically essentialist idea of corporeality.

As a development of the menstruator term, we can talk about cis menstruators and trans menstruators, the latter as expressed through some menstrual activism (Toni the Tampon 2017) as well as in children's books (Cairney 2017). While the cis menstruator term is not as recognized, I would like to emphasize its usefulness. The two terms allow us to explore what it means to be a menstruator within cisnormative contexts. It is important to note, however, the vast diversity of menstrual experiences across the gender spectrum. There is no uniform menstrual reality for trans or cis menstruators. Thus, the two menstruator terms should be considered conceptual tools applicable to critical explorations of menstruation as cisnormative phenomena; they should not be applied to make assumptions about menstrual experiences based on gender identity.

Cis Menstruators as the Normative Menstruator, Trans Menstruators as the Other

Taken together, the cis menstruator and trans menstruator terms facilitate analyses of menstruation as phenomena affecting different menstruators in different ways. This terminology means that we should not focus exclusively on the supposed Other but, rather, recognize the power relations and structures underlying menstruation. Josefin Persdotter's concept of menstrunormativity, (Chapter 29, this volume) makes visible the processes through which certain menstruators (and menstruations) are seen as the 'right' ones, whereas others are considered 'wrong.' She further states that menstrunormativity could be thought of as a conceptual sibling to cisnormativity and heteronormativity (Persdotter 2020). Heteronormativity builds on the assumption of people being heterosexual and, hence, desiring people of 'the opposite' sex/gender. Affecting not only non-heterosexual people, heteronormativity impinges upon societies as a whole (Herz and Johansson 2015, 1012). Considered together, menstrunormativity, cisnormativity, and heteronormativity intra-act, and are mutually reinforcing, in the Othering of trans menstruators within various areas related to menstruation.

One such area is menstrual activism. Even though the existence of trans menstruators has been raised within activism by trans people themselves—for example, Cass Bliss, who does prominent work spreading awareness through social media—trans menstruators are not represented within menstrual movements in general. As briefly mentioned, Bobel (2010) points out that for many activists, the question of trans inclusion in menstrual activism comes down to the same question as is present in many other feminist movements, namely: "[w]hat should feminists do about the category 'woman'? [. . .] work within it or destroy it?" (156). If we abolish the category 'woman' and thus, detach menstruation from gender, can we still claim that menstruators are oppressed because the meaning of menstruation is shaped by the devaluation of women qua women? I believe that I have already answered this question. Because some trans people do menstruate, they must be recognized within any menstrual activist agenda. The truth lies at the surface of the body. Our menstrual discourse must include all bodies that menstruate, as well as all non-menstruating bodies that are affected by menstrual norms, regardless of gender identity. It might just be that simple.

Cisnormativity and menstrunormativity as norm systems are both detectable within menstrual activism. Menstruation as phenomena are surrounded by stigma; menstrual blood is commonly depicted as something that should be kept within the body and, thereby, menstruators are encouraged to conceal their periods (Johnston-Robledo and Chrisler 2013, 11). Menstruating women can feel united through this shared experience of periods as a private matter that should be publicly undetectable (Brantelind, Nilvér, and Alehagen 2014, 611). One of my participants expressed that the menstrual concealment results in 'female' rooms becoming a safe space to discuss menstruation, as that is where one is free from the stigmatizing public jargon. Correspondingly, another participant emphasized that, in general, it is only cis women who participate within menstrual activism. In their own experience, doing menstrual activism has resulted in disapproval from other activists, just because they are not a woman (Rydström 2018, 59, 63). Gender norms and menstrual norms intra-act within menstrual activism, resulting in it being represented by, and representing, cis menstruators.

Public bathrooms constitute a second area wherein trans menstruators are Othered. Public bathroom access is commonly debated when discussing the living conditions of trans people in the Global North (Halberstam 2018, 133–35). Bathroom access is, of course, an especially acute issue for trans and nonbinary menstruators, per se. Chrisler et al. (2016) report that more than 60% of their 150 participants, all belonging to the masculine of center and transgender community, feel unsafe and/or uncomfortable when using a "men's room" during their periods (1246). One of the informants in Breanne Fahs' essay "The Menstruating Male Body" (2016) raises the same issue when declaring a fear of changing tampons in public bathrooms as it potentially will "out" him as trans (82). Evidently, the issue is deeper than a dispute over access to bathrooms, it is a question of trans people's health and safety (Schuster, Reisner, and Onorato 2016). Given that menstruation requires regular visits to bathrooms, it is imperative to include trans menstruators in future discussions on the topic.

Just as within menstrual activism, cis norms and menstrual norms intra-act in the Othering of trans menstruators within the area of public bathrooms. The lack of menstrual infrastructure within the 'men's room,' such as basins and bins to wash or toss menstrual products, is not only a practical issue affecting the individual menstruator. It likewise materializes the normative assumption that menstruators use the 'women's room.' As a result, trans menstruators who prefer to use the 'men's room' are challenging the binary and risk being seen as abject (Cavanagh 2013, 433–36). An anecdote told by one of the trans menstruators I interviewed, Mika, indicates that the issue is even more complex. Mika once visited a 'men's room' on his period where the menstrual infrastructure actually was in place. Yet, he did not dare to use the bin as he was afraid it would not be emptied, and he feared that his bloody menstrual product would eventually start to smell (Rydström 2018, 61). Generally speaking, menstrual blood in itself is perceived as dirty/unclean/impure (Bramwell 2001, 90–93) and it seems perceived as even more abject when the menstruator is non-conformant to the idea of menstruation as a female experience.

Menstrual products represent the third area wherein trans menstruators are being Othered. Scholars have critically examined the representation of menstrual products as mutually constitutive of menstrual stigma and taboo; pads and tampons are depicted as keeping the menstrual blood at a distance (Erchull 2013), protecting menstruators from leakage, and keeping the body in control (Rosengarten 2000, 92–96). The blue liquid in advertisements, supposedly representing menstrual blood, is a well-known example of such discourses[5] (Merskin 1999, 955). To quote Persdotter (2020): "[m]enstrunormativity is built into menstrual technologies" (p. 357) and, as explained below, the business of menstrual products materializes cis menstruators as the normative menstruator.

The Othering of trans menstruators in relation to menstrual products is present on various levels. Firstly, the framing of products reveals who is counted as a menstruator. The term 'feminine hygiene products' is commonly applied to label pads/tampons/cups/period underwear/et cetera, while some activists use 'menstrual products' as a gender-neutral alternative (Quint 2016). As confirmed by my own research, some trans menstruators feel invisible when looking for information about menstrual products, just because many companies target girls and women. Relatedly, the packaging of menstrual products—feminine-coded colors and patterns—can cause discomfort to people who do not identify with femininity (Rydström 2018, 59–60). The design of the products themselves, specifically the design of pads, is likewise part of the Othering. Pads are predominantly designed to fit to panties and, hence, are difficult to use for menstruators who prefer briefs or boxers (Rydström 2018, 60; The Period Prince 2018). The company Pyramid Seven, with their "boxer briefs for periods, not for gender" (Pyramid Seven, n.d.), offers an alternative. However, they are an exception within a generally cisnormative industry.

The healthcare sector constitutes the final area I note wherein trans menstruators are Othered. It has been shown that some menstruators fear healthcare personnel's lack of knowledge and transphobic reception (Chrisler et al. 2016, 1239, 1247; Rydström 2018, 62–63) and, at a closer look, this can be intertwined with context-specific concerns. Taking Sweden as an example, the gendering of personal identity numbers[6] is a general issue affecting trans people in various societal instances (SOU 2017:92, 173) and testimonies of trans menstruators indicate that the numbers result in health-care workers having preconceived ideas about patients before they enter the room. This is expressed in transphobic and heterosexist treatments (Rydström 2018, 62–63). Another issue applicable to the Swedish context is the naming of clinics focused on sexual and reproductive issues; the label *kvinnokliniker* ("women's clinics") not only affects trans people seeking pregnancy health services (SOU 2017:92, 724) but it likewise makes some trans people hesitant to contact the clinics with menstrual-related issues (Rydström 2018, 63, 71). Cisnormativity and heteronormativity clearly play an important role in the Othering of trans menstruators within the Swedish healthcare sector, and I hypothesize that this complex of problems is applicable to other geographical contexts as well.

When cis and trans menstruators are seen as binary categories given by nature, the intra-actions of cisnormativity, heteronormativity, and menstrunormativity make cis women the normative menstruator, a standard against which trans menstruators (as in *non-cis* menstruators) are deviant. In the contexts of menstrual activism, cisgendered public bathrooms, the design and marketing of menstrual products, and the failure of the healthcare sector to be trans inclusive, menstruation as cis-female phenomena are reinscribed. In other words, trans menstruators are being Othered as their existence challenges prevalent perceptions of menstruation.

THE INTERSECTION OF CRITICAL MENSTRUATION STUDIES AND TRANSGENDER THEORY

Recognizing critical menstruation studies as an interdisciplinary field, I suggest that the theoretical framework of transgender studies is useful to make our work more inclusive. Sandy Stone (2006), one of the first scholars of transgender studies, criticizes the ways in which trans people have often been theorized by the medical establishment as unnaturally embodied. Constrained by such narratives, they adapt and align with the norms by which they are subordinated and admit their Otherness. Thus, she stresses, by drawing on their embodiment and experiences, trans people can redefine their existence and efficiently disrupt conventional ideas of gender (230–31). Along this line, trans scholars have associated trans bodies with monstrous bodies; trans people have been perceived as nonhuman freaks by non-trans scholars—Janice Raymond with her book *The Transsexual Empire* (1979) is probably the most well-known of these—but this monstrosity has likewise been reclaimed by some trans people themselves. The concept of monstrosity, and transgender theory in general, questions the idea of nature, a nature which trans people are perceived to violate. It is described by trans scholars as birthing a productive rage toward the system of norms which Others them (Stryker 1994; Nordmarken 2014).

There are similarities between transgender theory and previous explorations of menstruation. Here, I am aligning with the post-constructionist (Lykke 2010) strand of critical menstruation scholars, such as Jessica Shipman Gunson (2016) and Katie Ann Hasson (2016), who demonstrate that menstruation as phenomena are material-discursive; menstruation (or its lack) is perceived and experienced in a variety of ways and, hence, there is nothing singular in its being. Rather, menstruation is multiple. For instance, some menstruators differentiate menstruation—as 'natural'—from menstrual suppression—as 'unnatural.' Others consider suppressing bleeding just as natural as bleeding, and some do not even think in terms of naturalness versus unnaturalness. Thus, it would be inadequate to argue that something is intrinsically natural/unnatural when it comes to menstruation because the conceptualizations vary (Shipman Gunson 2016, 317–20). Similar to how transgender scholars challenge prevalent perceptions of nature, we can confront the idea of menstruation as a natural female experience by exploring the plurality of menstruation.

While pointing out the usefulness of transgender theory for critical menstruation studies, I also acknowledge my own social location. I am not trans. Historically, trans people have been denied the space and authority to conduct research and build theory about their lives (Stone 2006, 229–30). Today, this discrepancy persists. Similar to Bobel (2010), I have been struggling with these thoughts. Initially, when conducting research for *New Blood*, she intended to interview trans people. Yet, aiming to be a responsible trans ally, she decided not to carry forward her quest in response to the feedback she received from trans identified people she consulted regarding her project (161–63). I received different feedback and made a different choice (Rydström 2018). I still do not know if it was a proper one.

However, my main point is to bring forward the argument that we—as scholars and activists—can make our work inclusive of a variety of menstruators, and non-menstruators, by questioning continuously the nature of menstruation. Here, Jacob Hale's advice to non-trans identified writers to "[f]ocus on: what does looking at transsexuals, transsexuality, transsexualism, or transsexual _____ tell you about *yourself*, *not* what does it tell you about trans" (Hale, quoted in Stone 2009) provides productive guidance. What does looking at trans experiences with menstruation tell us about menstruation per se? By looking at trans and nonbinary people's experiences with menstruation, or the experiences of menstruators belonging to other gender categories, the multiplicity of menstruation is articulated. This tells us that there is no given relationship between menstruation and cis/womanhood—an observation that efficiently facilitates the degendering of menstruation as phenomena.

NOTES

1. Nina Lykke (2010) coined the term "post-constructionism" to label the group of heterogeneous theories moving beyond biological essentialism and social constructionism. Different terms have been applied by other scholars to refer to the same kind of theories and theorists, for example, "material feminisms" (Alaimo and Hekman 2008).
2. "Intra-action" refers to relational movements between human as well as nonhuman agents. Different from "interaction"—which occurs between ontologically independent entities—such agents are intertwined and do not exist outside of one another (Barad 2007, 139–40).
3. At present, gender dysphoria is labeled as a mental health condition. With the new edition, the ICD 11, it will be placed under the umbrella term "sexual health." These changes will be formally presented in 2019 and set into motion in 2022 (WHO 2018b).
4. Transnormativity refers to the system of norms structuring the lives of trans people through privileging certain narratives. The medicalization narrative is one such stipulation, assuming that all trans people wish to reshape their bodies. Trans people who do not desire a medical transition hence break with the normative idea of what makes a trans person (Johnson 2016, 465–66, 468–69).
5. The Body Form commercial series "Blood Normal," making visible a red liquid, suggests a change may be on its way (Libresse 2016).
6. In Sweden, one is either assigned a female personal identity number, where the second to last number is even, or a male number, where the second to last number is uneven (SOU 2017:92, 173).

REFERENCES

Alaimo, Stacy, and Susan Hekman, eds. 2008. *Material Feminisms*. Bloomington and Indianapolis: Indiana University Press.

Barad, Karen. 2003. "Posthumanist Performativity: Toward an Understanding of How Matter Comes to Matter." *Signs: Journal of Women in Culture and Society* 38 (3): 801–31.

———. 2007. *Meeting the Universe Halfway: Quantum Physics and the Entanglement of Matter and Meaning*. Durham and London: Duke University Press.

———. 2015. "Transmaterialities: Trans*/Matter/Realities and Queer Political Imaginings." *GLQ: A Journal of Lesbian and Gay Studies* 21 (2–3): 387–422. https://doi.org/10.1215/10642684-2843239.

Berg, Beatrice. 2017. "Transgendering Menstruation: en kvalitativ studie av uppfattningar kring menstruation i relation till transidentitet i Sverige." BA thesis, University of Gothenburg.

Bobel, Chris. 2010. *New Blood: Third-Wave Feminism and the Politics of Menstruation*. New Brunswick, NJ and London: Rutgers University Press.

Braidotti, Rosi. 1997. "Mothers, Monsters, and Machines." In *Writing on the Body: Female Embodiment and Feminist Theory*, edited by Katie Conboy, Nadia Medina, and Sarah Stanbury, 59–80. New York: Columbia University Press.

Bramwell, Ros. 2001. "Blood and Milk: Constructions of Female Bodily Fluids in Western Society." *Women and Health* 34 (4): 85–96. https://doi.org/10.1300/J013v34n04_06.

Brantelind, Ida Emelie, Helena Nilvér, and Siw Alehagen. 2014. "Menstruation during a Lifespan: A Qualitative Study of Women's Experiences." *Health Care for Women International* 35 (6): 600–16. https://doi.org/10.1080/07399332.2013.868465.

Butler, Judith. 2004. *Undoing Gender.* New York and Abington: Routledge.

Cairney, Gemma. 2017. *Open: A Toolkit for How Magic and Messed Up Life Can Be.* Basingstoke: Pan Macmillan.

Cavanagh, Sheila. 2013. "Touching Gender: Abjection and the Hygienic Imagination." In *The Transgender Studies Reader 2*, edited by Susan Stryker and Aren Z. Aizura, 426–42. New York and London: Routledge.

Chrisler, Joan C. 2010. "Leaks, Lumps, and Lines: Stigma and Women's Bodies." *Psychology of Woman Quarterly* 35 (2): 202–14. https://doi.org/10.1177/0361684310397698.

Chrisler, Joan C., Jennifer A. Gorman, Jen Manion, Michael Murgo, Angela Barney, Alexis Adams-Clark, Jessica R. Newton, and Meaghan McGrath. 2016. "Queer Periods: Attitudes toward and Experiences with Menstruation in the Masculine of Centre and Transgender Community." *Culture, Health & Sexuality* 18 (11): 1238–50. https://doi.org/10.1080/13691058.2016.1182645.

Chrisler, Joan C., and Paula Caplan. 2012. "The Strange Case of Dr. Jekyll and Ms. Hyde: How PMS Became a Cultural Phenomenon and a Psychiatric Disorder." *Annual Review of Sex Research* 13 (1): 274–306.

Erchull, Mindy J. 2013. "Distancing through Objectification? Depictions of Women's Bodies in Menstrual Product Advertisements." *Sex Roles* 68 (1–2): 32–40. https://doi.org/10.1007/s11199-011-0004-7.

Fahs, Breanne. 2016. *Out for Blood: Essays on Menstruation and Resistance.* Albany: State University of New York Press.

Finn Enke, Ann, ed. 2012. *Transfeminist Perspectives in and Beyond Transgender and Gender Studies.* Philadelphia: Temple University Press.

Halberstam, Jack. 2018. *Trans*: A Quick and Quirky Account of Gender Variability.* Oakland: University of California Press.

Haraway, Donna J. 2016. *Manifestly Haraway.* Minneapolis: University of Minnesota Press. ProQuest Ebook Central.

Hasson, Katie Ann. 2016. "Not a "Real" Period? Social and Material Constructions of Menstruation." *Gender & Society* 30 (6): 958–83. https://doi.org/10.1177/0891243216672662.

Hawkey, Alexandra J., Jane M. Ussher, Janette Perz, and Christine Metusela. 2017. "Experiences and Constructions of Menarche and Menstruation among Migrant and Refugee Women." *Qualitative Health Research* 27 (10): 1473–90. https://doi.org/10.1177/1049732316672639.

Herz, Marcus, and Thomas Johansson. 2015. "The Normativity of the Concept of Heteronormativity." *Journal of Homosexuality* 62 (8): 1009–20. https://doi.org/10.1080/00918369.2015.1021631.

Jackson, Theresa E., and Rachel Joffe Falmagne. 2013. "Women Wearing White: Discourses of Menstruation and the Experience of Menarche." *Feminism & Psychology* 23 (3): 379–98. https://doi.org/10.1177/0959353512473812.

Johnson, Austin H. 2016. "Transnormativity: A New Concept and Its Validation through Documentary Film about Transgender Men*." *Sociological Inquiry* 86 (4): 465–91.

Johnston-Robledo, Ingrid, and Joan C. Chrisler. 2013. "The Menstrual Mark: Menstruation as Social Stigma." *Sex Roles* 68 (1–2): 9–18. https://doi. org/10.1007/s11199-011-0052-z.

Kondelin, Sade. 2017. "'If I Could Touch It, It Would Be Something Sticky and Cold': Gender Dysphoria as an Embodied Experience." *Lambda Nordica* 22 (1): 15–37.

Kristeva, Julia. 1982. *Powers of Horror: An Essay on Abjection.* New York: Columbia University Press.

Libresse Sverige. "Libresse Blood." Youtube Video, 26 April, 2016. Accessed March 17, 2018. https://youtu.be/sL2xE5kkL_Q.

Lykke, Nina. 2010. "The Timeliness of Post-Constructionism." *NORA—Nordic Journal of Feminist and Gender Research* 18 (2): 131–36. https://doi.org/10.1080/08038741003757760.

Merskin, Debra. 1999. "Adolescence, Advertising, and the Ideology of Menstruation." *Sex Roles* 40 (11–12): 941–57.

Nord, Iwo, Signe Bremer, and Erika Alm. 2016. "Redaktionsord: Cisnormativitet och feminism." *Tidskrift för Genusvetenskap* 37 (4): 2–14.

Nordmarken, Sonny. 2014. "Becoming Ever More Monstrous: Feeling Transgender In-Betweenness." *Qualitative Inquiry* 20 (1): 37–50. https://doi. org/10.1177/1077800413508531.

Persdotter, Josefin. 2020. "Introducing Menstrunormativity: Towards a Complex Understanding of 'Menstrual Monsterings'." In *The Palgrave Handbook of Critical Menstruation Studies,* edited by Chris Bobel, Inga Winkler, Breanne Fahs, Katie Ann Hasson, Elizabeth Kissling, and Tomi-Ann Roberts. Palgrave MacMillan.

Preciado, Paul. 2013. *Testo Junkie: Sex, Drugs, and Biopolitics in the Pharmacopornographic Era.* New York City: Feminist Press.

Pyramid Seven. n.d. "Pyramid Seven." Accessed June 7, 2018. https://www.pyramid-seven.com.

Quint, Chella. 2016. "Queeriods." Accessed May 31, 2018. https://periodpositive. files.wordpress.com/2016/02/queeriods-lgbt-stem-poster-feb-2016.pdf.

Raymond, Janice. 1979. *The Transsexual Empire: The Making of the She-Male.* Boston: Beacon Press.

Rembeck, Gun. 2008. "The Winding Road to Womanhood: Adolescents' Attitudes towards Menstruation, Womanhood and Sexual Health—Observational and Interventional Studies." PhD diss., University of Gothenburg.

Rosengarten, Marsha. 2000. "Thinking Menstrual Blood." *Australian Feminist Studies* 15 (31): 91–101.

Rydström, Klara. 2018. "Trans Menstruators Matter: A Qualitative Research Exploring Swedish Trans Experiences with Menstruation." MA thesis, University of Oviedo and University of Hull.

Schuster, Mark A., Sari L. Reisner, and Sarah E. Onorato. 2016. "Beyond Bathrooms—Meeting the Health Needs of Transgender People." *New England Journal of Medicine* 375 (2): 101–3.

Shipman Gunson, Jessica. 2016. "Nature, Menstrual Suppression, and the Value of Material Feminism." *Health Sociology Review* 25 (3): 312–25. https://doi.org/10.1023/A:1018881206965.

SOU 2017:92. *Transpersoner i Sverige: Förslag för stärkt ställning och bättre levnadsvillkor.*

Stone, Sandy. 2006. "The Empire Strikes Back: A Posttransexual Manifesto." In *The Transgender Studies Reader*, edited by Susan Stryker and Stephen Whittle, 221–35. New York and London: Routledge.

———. 2009. "Suggested Rules for Non-Transsexuals Writing about Transsexuals, Transsexuality, Transsexualism or Trans____." Accessed November 16, 2018. https://sandystone.com/hale.rules.html.

Stryker, Susan. 1994. "My Words to Victor Frankenstein above the Village of Chamounix: Performing Transgender Rage." *GLQ* 1 (3): 237–54. https://doi.org/10.1215/10642684-1-3-237.

———. 2013. "Trans Health Is Queer (and Queer Health Isn't Normal)." *Lambda Nordica* 18 (3–4): 147–65.

The Period Prince. 2018. "Rainbow Reviews: GladRags." Accessed June 7, 2018. https://www.bleedingwhiletrans.com/single-post/2018/05/21/Rainbow-Reviews-GladRags.

Toni the Tampon (@tonithetampon). 2017. "The Christian Post Calls My Move to Include Trans and Queer Menstruators 'Child Abuse'." Instagram Photo, March 10, 2017. https://www.instagram.com/tonithetampon/.

Transformering. 2018. "Underlivskirurgi." Accessed May 23, 2018. http://www.transformering.se/vardhalsa/hormoner.

Wasshede, Cathrin. 2017. "Queer Hate and Dirt Rhetoric: An Ambivalent Resistance Strategy." *Journal of Resistance Studies* 3 (1): 29–61.

World Health Organization. 2018a. "Classifications." Accessed April 28, 2018. http://www.who.int/classifications/icd/icdonlineversions/en/.

———. 2018b. "WHO Releases New International Classification of Diseases (ICD 11)." Accessed June 25, 2018. http://www.who.int/news-room/detail/18-06-2018-who-releases-new-international-classification-of-diseases-(icd-11).

CHAPTER 69

Sex During Menstruation: Race, Sexual Identity, and Women's Accounts of Pleasure and Disgust

Breanne Fahs

In tandem with cultural taboos about menstruation, women have historically reported ambivalent relationships between their body images, sexualities and their menstrual cycles.[1] From the historic menstrual hut to the recent invention of Premenstrual Dysphoric Disorder (PMDD) to the pervasive social norm encouraging women to hide menstruation from others (Delaney et al., 1988), menstruation has been labeled disgusting, socially deviant and, at times, pathological. Similarly, contemporary women often feel shameful about their bodies in the context of sex, citing body image problems, sexual 'dysfunction', and concerns about feeling normal as key sexual problems (Kleinplatz, 2001; Nobre and Pinto-Gouveia, 2008; Plante, 2006; Tiefer, 2001). Women routinely engage in a variety of normative body practices to manage their 'disgusting' bodies (Roberts and Goldenberg, 2007): hiding menstruation, shaving (Tiggemann and Lewis, 2004), wearing makeup, controlling weight, concealing body odors and grooming extensively.

While some research has addressed women's feelings, emotions and reactions to menstruation – particularly negativity toward menstruation – surprisingly little research has explored how these cultural interpretations of menstruation have influenced women's *sexual* lives. This study extends existing research by qualitatively examining women's experiences and perceptions of menstruating *while* having sex (either with male or female partners). Why are women's

C. Bobel et al. (eds.), *The Palgrave Handbook of Critical Menstruation Studies*, https://doi.org/10.1007/978-981-15-0614-7_69

responses polarized, with some embracing menstrual sex and others rejecting it? How might such attitudes relate to partner choice, body image and social identities such as race and sexual identity? This study draws upon qualitative accounts of menstrual sex to add to the growing literature about sexuality, bodies and menstruation by highlighting narratives of pleasure and disgust.

The context within which women menstruate affects not only the discourse surrounding menstruation but also *how women themselves experience menstruation* (Mansfield and Stubbs, 2007). Historically, menstruation was considered taboo and in need of management (Delaney et al., 1988), as menstrual blood indicated disease, social taboos about decency and shame, and spiritual corruption (Read, 2008). Recent constructions of menstruation in the west carry these same negative cultural connotations (Delaney et al., 1988). Common western cultural constructions treat menstruation as failed reproduction (Kerkham, 2003), particularly because the state constructs women's bodies as reproduction machines (Martin, 2001). Yet, even though negativity permeates cultural and patriarchal attitudes toward menstruation, some cultures (e.g. some African tribes) celebrate it or even attempt to simulate it as a powerful, revered practice (Brain, 1988).

Contemporary western women are barraged with portrayals of menstruation that warn of its so-called disgusting, disabling and tainting qualities. When selling menstrual products, advertisers portray the female body as unclean and unfeminine to market panty liners, pads, and tampons (Berg and Coutts, 1994; Kissling, 2006). Oral contraceptives (e.g. Seasonale) feature menstrual suppression as an asset of taking birth control pills (Johnston-Robledo et al., 2006), thereby amplifying women's interest in suppressing menstruation (Johnston-Robledo et al., 2003; Rose et al., 2008). The invention of premenstrual syndrome (PMS) and PMDD – and their accompanying drug therapies – also promote menstruation as a pharmaceutically managed illness (Endicott et al., 1999). Such portrayals depict women's bodies as debilitated by their monthly cycles, while men's bodies remain 'normal' (Chrisler et al., 2006; Cosgrove and Caplan, 2004; Nicolson, 1995; Offman and Kleinplatz, 2004). Films regularly portray menstruation as revolting (e.g. *Superbad* (Mottola, 2007)), using menstruation to depict horror and disgust (Briefel, 2005; Kissling, 2002). These portrayals construct women's bodies as dirty, disgusting, and in need of sanitizing, deodorizing, medicating, managing, exfoliating and denuding (Kissling, 2006).

Feminists and other activists have responded to negative portrayals of menstruation by promoting – often through media campaigns, consciousness-raising, educational campaigns and attacks on mainstream representations of menstruation - affirming views of menstruation within institutions such as education, the health industry, media and families (Kissling, 2006). Menstrual activists advocate that women *and* men cultivate a critical consciousness about menstruation that facilitates positive views of women and their bodies (Bobel, 2008) and have specifically focused their efforts on stopping PMDD from entering the *Diagnostic and Statistical Manual of*

Mental Disorders (DSM) (Offman and Kleinplatz, 2004), de-medicalizing menstruation as 'disease' (Koekse, 1983), challenging the toxicity and dangers of commercial menstrual products (Bobel, 2006), embracing menstruation as power and political protest (Aretxaga, 1995), critically examining menstrual suppression products (Johnston-Robledo et al., 2006), and embracing more positive, self-loving messages about women's bodies and menstrual cycles (Kissling, 2006; Stubbs and Costos, 2004). Some argue that cultivating positive attitudes about menstruation unites women, particularly mothers and daughters (Stubbs and Costos, 2004), and may lead to better acceptance of 'imperfect' bodies (Roberts and Waters, 2004).

Despite these efforts, educational settings overwhelmingly promote negative messages about menstruation, as girls learn to associate menstruation with fear, embarrassment, disgust and revulsion (Erchull et al., 2002; Havens and Swenson, 1989). Alarmingly, a recent analysis of menstrual education pamphlets between 1932 and 1997 revealed little differences over time, as *all* pamphlets promoted secrecy about menstruation (Erchull et al., 2002). Girls in school settings have expressed a strong desire to hide menstruation from others, believing menstruation is an *illness* rather than a natural occurrence (Burrows and Johnson, 2005) – a phenomenon that persists up through menopause (McKinley et al., 2008).

Given this onslaught of negative messages, imagery and socialization about menstruation, women and girls describe negative attitudes toward menstruation and their bodies. For instance, while nearly 43 percent of girls disliked their bodies in one study, they *especially* disliked their menstruating bodies (Rembeck et al., 2006). Adult women associated menstruation with uneasiness, pain and maturity (Amann-Gainotti, 1994). Negative attitudes about menstruation persist throughout the lifespan, though some are buffered more against such negativity. Younger girls (Rembeck et al., 2006), those who self-objectify less often (Roberts and Waters, 2004), those with less body shame, with more sexual experience (Schooler et al., 2005), who described breastfeeding as positive (Johnston- Robledo et al., 2007), who have experienced less name-calling from others and who communicated with their mothers about menstruation (Rembeck et al., 2006) described less disgust toward menstruation than others.

Although US culture socializes women to feel negatively toward menstruation, some variation occurs in who embraces such notions. Only one study has examined differences between lesbian and heterosexual women's menstrual attitudes, finding no differences between heterosexual and lesbian-identified women in Mexico (Zepeda and Eugenia, 1999). That said, lesbian women often suffer from internalized homophobia that manifests itself as body shame and dissatisfaction (Pitman, 1999), even while lesbian identity may buffer against full-blown body hatred (Owens et al., 2002).

With regard to gender and menstruation, men described menstruation more negatively than women, saying that menstruation debilitated women, dramatically affected their mood (Brooks-Gunn and Ruble, 1986), and negatively affected their academic performance (Walker, 1992). Men also expressed uneasiness about women's disclosure of menstruation, even while publicly ascribing women's mood shifts and hostility to menstruation (Christensen and Oei, 1990; Laws, 1992). Such studies raise questions about the role of partner attitudes toward menstruation in (heterosexual) sexual encounters, given this expressed negativity.[2]

While no studies have addressed racial differences in menstrual attitudes, research has explored racial differences among women for body image, self-objectification and attitudes toward fatness. In US culture, women of color are compared to white women as the 'idealized' standard, engendering women of color to dislike their natural bodies, ethnic features (e.g. 'kinky' hair, almond-shaped eyes), and culturally distinct body practices (Chapkis, 1986; Hill-Collins, 2005). That said, communities of color more often affirm larger bodies, which helps to partially buffer women of color from body shame about weight (Kronenfeld et al., 2010; Miller et al., 2000). Though underdeveloped as a literature, one study addressed race, body image and sexuality, finding that African-American women masturbated far less often, and (contradicting aforementioned studies) had worse body image when compared to white women (Shulman and Horne, 2003).

Some cultural and ethnic differences exist in attitudes toward menstruation, though most cultures describe menstruation negatively. US and Mexican respondents considered menstruating women irritable and moody, and had almost universally negative attitudes toward them, particularly among younger men (Marván et al., 2005, 2008). Similarly, traditional and indigenous Chinese girls described menstruation as a negative event (Yeung et al., 2005). Indian women rated menstruation as more natural and less distressing than US women (Hoerster et al., 2003), indicating that disgust and shame do not universally define women's menstrual experiences.

Despite this sparse literature on connections between women's attitudes about menstruation, body image and sexuality across social identity lines, one recent study directly addressed college women's attitudes toward sex while menstruating (Allen and Goldberg, 2009) (though notably did not collect information about race). This study found that less than half of women engaged in menstrual sex, and that these women were typically younger, in committed relationships, more mature and less disgusted by menstrual sex than others; further, over 80% of women reported polarized feelings about menstrual sex (e.g. one-third would *never* do it, while roughly one-third engaged in it regularly and without restrictions) (Allen and Goldberg, 2009).

Other studies have found links between menstruation, sexuality and body image, though women were not directly asked about menstrual sex attitudes. Comfort with menstruation correlated to comfort with sexuality, and engaging in menstrual sex was associated with arousal to romantic activities and less disgust toward one's body (Rempel and Baumgartner, 2003). Women who engaged in menstrual sex also had more partner support than those who avoided menstrual sex (Hensel et al., 2007), again suggesting that partner attitudes influence women's menstrual sex attitudes. Also, women who espoused general comfort about menstruation reported more body comfort, sexual assertiveness and sexual experience, but less sexual risk-taking (Schooler et al., 2005).

Across age/racial groups, women have sex far more often while *not* menstruating than while menstruating (Hensel et al., 2004), though no research has examined possible reasons for this. Some report ovulating women felt more sexual arousal and sexual attractiveness than menstruating women (Jarvis and McCabe, 1991; Röder et al., 2009), though researchers disagree about whether social and cultural reasons (e.g. cultural disgust toward menstruation, negative socialization) drive these findings (Leiblum, 2002), or whether evolutionary and hormonal causes (Mass et al., 2009) are responsible.

Given that the normative climate surrounding menstruation often dictates negative feelings of disgust, shame and embarrassment for menstruating women, and because negativity toward menstruation correlates with poorer body image, I explored the way women interpreted and rationalized engaging in or avoiding sex during menstruation. I also examined whether embracing menstrual sex might link to masturbation attitudes. Further, because women of color face pressures to conform to white norms of body image and sexuality, and because partner choice affects body image attitudes, I explored how social identities related to race and sexuality corresponded with attitudes toward menstrual sex.

Method

This study utilized two waves of qualitative data collection from a sample of 40 women recruited over the span of three years (2005–07). Half the participants (wave 1) were interviewed in a medium-sized Midwestern US city (a 'college town'); the other half (wave 2) in a large metropolitan Southwestern US city. Participants in wave 1 were recruited through local entertainment and arts listings ($N = 20$) distributed free to the local community. Participants in wave 2 were recruited through local entertainment and arts listings ($N = 12$) distributed free to the local community as well as the volunteers section on the local online section of Craigslist ($N = 8$).

Participants were screened only for their gender, racial/ethnic background, sexual identity and age; no other pre-screening questions were asked. A purposive sample was selected: sexual minority women and racial/ethnic minority women were intentionally oversampled; a diverse range of ages was represented (see Tables 69.1 and 69.2). As advertised in the recruitment materials (and approved by the institutional ethics review board), participants were compensated US$20.00 for participating. All participants consented to have their interviews audiotaped and interviews were fully transcribed using an orthographic style. Identifying data was removed and each participant received a pseudonym to ensure anonymity. Information reported during the

Table 69.1 Participant demographic information

Name	Age	Race	Sexual identity
Margaret	54	White	Heterosexual
Aya	25	Chinese-American	Heterosexual
Ophelia	56	White	Lesbian
Susan	45	White	Heterosexual
Fiona	24	White	Bisexual
Mitra	29	Indian-American	Heterosexual
Geena	51	White	Heterosexual
Courtney	23	White	Heterosexual
Kate	25	White	Heterosexual
Leigh	21	White	Lesbian
Nora	23	White	Bisexual
Maria	21	Mexican-American	Lesbian
Priya	23	Indian-American	Heterosexual
Dorothy	19	Chinese-American	Heterosexual
Julie	23	White	Bisexual
Charlene	44	Native American	Bisexual
Melanie	21	White	Heterosexual
Pam	42	African-American	Heterosexual
Anita	46	Mexican-American	Heterosexual
Lori	33	White	Bisexual
Esther	44	African-American	Heterosexual
Janet	21	White	Lesbian
Brynn	25	White	Lesbian
Shonda	44	African-American	Heterosexual
Dawn	43	White	Bisexual
Bonnie	51	White	Heterosexual
Marilyn	54	White	Heterosexual
Lucy	25	Mexican-American/Filipina	Bisexual
Diana	50	White	Lesbian
Charlotte	34	African-American/Indian-American	Heterosexual
Ruth	46	White	Heterosexual
Jasmine	37	Mexican-American	Lesbian
Sally	20	White	Heterosexual
Edie	19	White	Heterosexual
Jill	32	White	Heterosexual
Niko	23	Asian-American	Heterosexual
Ciara	24	White	Bisexual
Sonja	24	White	Heterosexual
Emily	22	White	Bisexual
Carol	38	White	Bisexual

Note Demographics are self-reported. In the case of sexual identity, their reported behavior, attitudes and attraction do not always align with their self-reported sexual identities

interviews indicated a range of socioeconomic and educational backgrounds, employment histories, and parental and relationship statuses. No significant demographic differences appeared between wave 1 and wave 2, except that wave 1 participants more often mentioned that they had college or advanced degrees than wave 2 participants.

Table 69.2   Summary table: Participant demographic information

Demographic	Percentage	No. of participants
Age		
18–31	52	21
32–45	28	11
46–59	20	8
Race		
White	65	26
Women of color	35	14
AfricanAfrican-American	7.5	3
Chicana/Latina	7.5	3
Indian-American	5	2
Asian-American	7.5	3
Native American	2.5	1
Biracial	5	2
Sexual identity		
Heterosexual	58	23
Bisexual	24	10
Lesbian	18	7
Total	100	40

I interviewed participants using a semi-structured interview protocol that lasted for approximately 1.5 to 2 hours, where participants responded to 30 questions about their sexual histories, sexual practices, and feelings and attitudes about sexuality. Several addressed issues relevant to this study. For example, women were asked, 'Many women report that their feelings about their own bodies greatly affect their experience of sex. How do you feel your body image affects your sexual experiences?', with the follow up question: 'How do you feel about having sex while menstruating?' These 30 questions were scripted, but served to open up other conversations and dialogue about related topics, as follow-up questions were free-flowing and conversational. The original questions served as 'sensitizing concepts' that allowed previous research to lay the groundwork for topics and themes to look for (Charmaz, 2006).

Responses were analyzed qualitatively using thematic analysis. This type of analysis was considered the most effective and useful because it allowed for groupings of responses based on women's attitudes and feelings (e.g. acceptance of, neutrality about or disgust toward sex while menstruating, along with accompanying sub-themes). This method of analysis also supported an examination of the intersection between sex and menstruation as well as other components of women's sexual lives (e.g. masturbation). To conduct the analysis, I familiarized myself with the data by reading all of the transcripts thoroughly, and then identified patterns for common interpretations posed by participants. In doing so, I reviewed lines, sentences and paragraphs of the transcripts, looking for patterns in their ways of discussing menstrual sex (Braun and Clarke, 2006). I selected and generated themes through the process of identifying logical links and overlaps between participants. After creating these themes, I compared them to previous themes in order to identify similarities, differences and general patterns. This type of thematic analysis relied upon a data-driven inductive approach (Boyatzis, 1998) in which themes were generated prior to interpretation (Boyatzis, 1998). As such, initial themes were identified, codes were applied and then connected back to the themes, and these themes were then corroborated and legitimized, per Fereday and Muir-Cochrane's (2006) method of inductive thematic analysis.

Throughout this process, I relied upon a critical realist framework that contextualized themes and patterns found in the transcripts. While sorting and naming themes required some level of interpretation, I generally did not delve into covert, implicit, or subtextual meanings in the transcript. Women's stories of menstrual sex and masturbation were grouped thematically without undermining their explicit narratives. This study took a contextualist approach as its epistemological framework, as women's narratives were examined discursively for their treatment of gender, sexuality and the body.

RESULTS

Overall, 25 women described negative reactions to sex while menstruating, two described neutral reactions and 13 women described positive reactions. Responses did not overlap: no-one identified mixed feelings about sex while menstruating, although women's responses *did* sometimes overlap between sub-themes (e.g. reporting several negative reactions), and interesting intersections between other sexual practices (masturbation) and identities (race, sexual identity) were noted.

Negative responses to sex during menstruation

Negative responses cohered around four sub-themes: (1) women's physical discomfort and physical labor; (2) overt partner discomfort; (3) negative self-perception; and (4) emotional labor to manage partner's judgments and feelings. At times, categories overlapped.

Women's physical discomfort and physical labor. Nine women (Bonnie, Aya, Geena, Margaret, Kate, Niko, Sally, Carol and Ruth) mentioned their physical discomfort with menstrual sex, citing logistic problems of cleaning a messy bed, along with the general sentiment that menstruation was difficult to manage. Women took responsibility for the *physical labor* of cleaning and managing menses, as gendered norms dictate women's management of household cleanliness as 'their job' (Brines, 1994). Bonnie avoided menstrual sex because of cleaning responsibilities: 'It's more my discomfort that my partner's discomfort. I think the messiness of it just gets on things. It's hard to get out of the sheets, the clothes.' Aya also expressed concern about cleaning the mess, linking blood to general dirtiness:

> I wouldn't have sex on my period because I'd rather not dirty the sheets. I'm sure there are guys that don't mind. Even if they wanted to, I don't think that I'd want to. I think blood is dirty. It just seems like such a hassle. Maybe in the shower, but that's the only place that would make sense to me.

Geena provided a particularly vivid example of how physical labor of cleaning deterred her, but also how self-consciousness and emotional labor to manage her partner's judgment fueled disinterest in menstrual sex:

> I avoid it, but it's not because I'm not comfortable with it. It's because it makes such a damn mess. You have to get dark-colored towels. It's the pain of cleaning it up and not leaving stains on stuff is why I avoid it, and there's part of me that's always thought that if I'm in the thick of throwing clots, men are not really going to want to do this. They may act all brave and like it's no big deal, but I don't really trust that. It's probably just my own projection, not wanting to deal with the mess.

Overt partner discomfort. Six women (Sally, Pam, Priya, Mitra, Charlene and Sonja) noted that their partners, particularly men, expressed overt discomfort and disgust around sex during menstruation. For example, Sally talked about negative reactions to blood and her boyfriend's aversion: 'My boyfriend didn't like blood, so he would get totally grossed out and we wouldn't have sex for the whole week I was on my period.' Other women, such as Pam, described partners having negative feelings about menses, such as her husband who found it 'gross': 'My husband is uncomfortable about the idea of penetrating me with blood in there. He prefers not to do it. He thinks my period is gross.' Charlene also noted the 'gross' quality of it: 'He thinks it's just messy. And kinda gross.' The concept of 'grossness' appeared frequently in women's talk about menstruation and sex during menstruation, perhaps revealing the strength of their own, and others', disgust reactions toward menstrual sex. The concept of women's bodies as inherently disgusting appeared particularly strongly in these partner aversion stories.

Perhaps testifying to the centrality of partner (dis)affirmation in women's body image and assessment of menstruation, Priya noted that her partner's overt discomfort fed into her own negative feelings about menstruation:

> I was with someone and we were going to have sex on my period – this was someone I had been dating for a while, so I was really comfortable – and he was just like, 'I don't want to watch'. I think he was more traumatized than I was and we didn't end up doing it. He was just like, 'I don't want to see blood. I'm going to feel like I'm killing you'. I feel I have so many negative associations with my own period, just having a period on my own is hard so the last thing I need to be doing is doing it with someone else… I've always felt odd because my old roommate told me about how her boyfriend would always go down on her during her period and I just felt like that was something he'd never do.

Negative self-perception. Fifteen women (Kate, Sonja, Dorothy, Ruth, Jasmine, Edie, Margaret, Aya, Leigh, Priya, Anita, Charlotte, Pam, Courtney and Geena) described negative self-perceptions of body shame and discomfort as key reasons why they avoided menstrual sex. Kate noted that menstruating negatively affected her self-image: 'When I actually have my period I don't usually feel like having sex so much. I feel physically not great, and while I really like to be hugged and to feel physically close, sex feels different. I don't like my body at that time.' Notably, her desire for physical closeness yet no sex suggests that body shame may manifest itself more strongly in *sexual* contexts compared with physical comfort. Sonja also felt this way, noting that her body shame negatively affected her sexual desire: 'It makes me feel a little bit more uncomfortable about my body. It's kind of grosser. I just feel uncomfortable. I have had sex before on my period, but I always feel gross.' Again, the lack of analysis around where the 'gross' assessments came from (e.g. societal pressures) suggests a link between physical disgust/ aversion and emotions, as women failed to critically examine why they felt 'gross'.

Other women linked menstruation to poor physical well-being, perhaps to manage negative feelings from self and others. Dorothy complained about her menstrual symptoms as a deterrent to sex: 'I'm probably bloated and having cramps and just unhappy already. It just seems like more for the guy, and there's less they can do with your body. It's gross.' Ruth disliked the feel of the blood while having sex, saying: 'It was uncomfortable. The Ph balance, the blood, had a rougher feel than natural lubrication.'

Some women articulated common cultural assumptions about menstruation as frankly unattractive, using a strong language of disgust and repulsion. Jasmine stated direct aversion to the idea of menstrual sex: 'No, no, no. It just doesn't look pretty. I may be horny, but that's just gross.' Edie echoed this, but unlike most others, reflected on the social location of these personal views: 'My period smells and I don't have a light period. I've been brainwashed somewhere along the lines to think that it's unnatural I guess.' Margaret said, simply: 'It's disgusting.'

Emotional labor to manage partner's judgment. Six women (Mitra, Charlotte, Courtney, Geena, Priya and Ophelia) discussed the emotional labor aspects of managing their partner's judgments about menstrual sex. Mitra worried about her partner's disgust, suggesting that her emotional management of his aversion loomed larger than her own disgust:

> My partner isn't fine with it. He's not that into it. I think a lot of men are like that. He's kind of a neat freak so the idea of seeing blood on his penis freaks him out a little bit. It just seems like I would have to put towels down and it just feels like a big thing to worry about – his feelings and this mess.

Charlotte observed that, based on her past experiences, men disliked menstrual sex: 'I've tried it with guys and they didn't like it. They're not used to it and they avoid it with me.' Interestingly, Courtney expressly pathologized men who enjoyed menstrual sex as she considered her desire to manage their emotional responses to sex: 'Guys don't think it's hot when women are bleeding. If they do, they're creepy... I try to always shower and stay clean before sex so the guy will enjoy sex.' In summary, women's emotional and physical labor, self-perception and partner perceptions collectively imply that individual, cultural and relational issues all affect women's subjective experiences. These responses point to the centrality of women's internalized body shame narratives as well as both *anticipatory* and *actual* responses from partners (particularly men). Women with negative responses often expressed strong, definitive, emotional reactions about women's bodies as 'gross' or repulsive (Fausto-Sterling, 2000). Similarly, even when women did not harbor these views internally, they *anticipated* that others would feel negatively, and thus internalized a need to 'manage' anti-menstrual attitudes by avoiding sex. This collectively reveals the many layers of shame and negativity women faced when confronting menstrual sex, as tensions between physical and emotional labor, and between partner and self, appeared.

Neutral responses to sex during menstruation

Two women (Ciara and Emily) described neutral responses to sex while menstruating, noting that it was 'no big deal'. Ciara talked about menstrual sex with her husband in a casual and nonchalant way: 'It wasn't a bad experience. It was kind of weird, but there's nothing bad about it.' Emily also felt this way, saying: 'It's fine. I mean, it makes a lot of laundry [laughs] but I don't think it's a big deal.' These responses reveal that, while many women internalized a strong reaction (negative or positive) to menstrual sex, others did not. While these neutral reactions appeared rarely, their occurrence suggests that some women may resist negative characterizations of their body simply by constructing menstruation as 'no big deal'.

Positive responses to sex during menstruation

Positive responses to sex while menstruating were categorized into two themes: (1) Physical and emotional pleasure from sex while menstruating; and (2) Rebellion against anti-menstrual attitudes.

Physical and emotional pleasure. Ten women (Susan, Janet, Dawn, Diana, Maria, Lucy, Marilyn, Julie, Esther and Brynn) described menstrual sex as physically and/or emotionally pleasurable. Susan talked about enjoying sex while menstruating despite its messiness: 'I accept it. I'll say, "If you don't care about it, I don't care about it." Let's put down a towel. The sex on my period probably feels the best of almost any time of the month, so it doesn't bother me.' Janet had overwhelmingly positive feelings, noting that menstrual sex conveyed love and acceptance from her partner:

> The first time it happened it was like, 'Oh my God', a totally different experience. It was better than anything I've ever known. In your mind you're like, 'That's a bloody mess!' In my heart it was like, damn, she doesn't really care, she really loves me, 'cause she doesn't care if I'm bleeding or not, and I don't care if she is.

This suggests that sex while menstruating can represent a combination of physical pleasure and emotional pleasure, perhaps because of the *coupled* subversion of anti-menstrual attitudes. In other words, satisfaction with menstrual sex may reflect feelings of acceptance, validation, warmth and love, all of which typically fall under *emotional* satisfaction from sex. This supports other research that has shown that women often conflate physical and emotional satisfaction when reporting on general sexual satisfaction (McClelland, 2010).

Women also described having intense orgasms during sex while menstruating, such as Dawn: 'I have no problem with sex while menstruating. I actually have some of the most intense orgasms right at the beginning of my period and right at the end of my period. Of course I'm always happy I'm not pregnant!' Diana also embraced the pleasurable aspects of sex while menstruating, saying: 'I'm perfectly fine with it if it's not heavy... It can be really arousing.' This differed from earlier negative responses in which women identified feeling 'horny' yet avoided sex because of negative social penalties (e.g. Jasmine).

Some women also described increased bodily responsiveness while menstruating, noting that physical changes felt erotic. These women situated their desire for menstrual sex as *bodily* rather than about partners, and described sensual details about the encounters. Marilyn described her body as 'more moist, just so much more moist, so it's pleasurable'. Julie also felt more sexual desire while menstruating: 'It feels really good. It's a time when I often desire it, particularly with my partners that are men, and it feels really good to them too. It's really warm and good. It's messy and I like that. I think it's fun.' Esther similarly noted: 'Right before my period, I'm at a peak when I really feel a strong desire for sex. My cramps are usually so bad the first couple of days, so there's no way I can think about it, but after that point, it really doesn't bother me. It's fun.' Brynn described her body's physical changes as arousing, and admitted to seeking out sex while menstruating: 'I would have to say that sex while menstruating is awkwardly a bit more pleasurable because my senses are a little more heightened, my feelings are more on edge. It's more pleasurable. It's just a totally different experience.'

Rebellion against anti-menstrual attitudes. Finally, three women (Nora, Fiona and Lori), notably all feminist-identified, constructed menstrual sex as a way to rebel against anti-menstrual attitudes by embracing their natural bodies and affirming their womanhood. Nora discussed her progression from reluctance to fully embracing menstrual sex as a wonderfully messy occurrence:

At first I was really kind of amazed, held back by it. I remember learning very early that you could have sex on your period, and I was blown away by that and kind of grossed out at the same time, but then after I started having sex on my own, I enjoyed it. Usually it can help with my cramps... [My first boy-friend] allowed me to finger- paint on his chest with my menstrual blood. I really enjoyed that as a great expression. This one guy recently asked me to put a towel down and I was like, 'Whatever.' I've definitely made total messes and there's some satisfaction in that to be honest. I've definitely messed up white towels and white comforters and totally made things messy, and for whatever reason I don't have any regret or shame about that. I'm just like, 'Oh, that's nice.'

Fiona described menstrual sex as rejecting negative stereotypes about women's bodies: 'I don't like to think about my body as smelly or dirty. Having sex on my period is a way to prove that my boyfriend doesn't think of me this way either.' Lori believed that menstrual sex affirmed her body image and created a shared rebellion between her and her partner: 'Most of the time girls are told that their bodies are gross and disgusting, and I always feel like having sex while I'm bleeding is a way to go against that, like my boyfriend and I are staging a private little rebellion. I feel like an animal or something, kind of like when I gave birth and I felt just this life force coming out of me.'

Connections to masturbation experiences

To expand the discussion of menstrual sex attitudes into a larger conversation about body shame, attitudes about masturbation were also examined. Interestingly, positive attitudes about menstrual sex also closely aligned with positive attitudes toward masturbation, as all but two of the 13 women describing menstrual sex positively had positive experiences with masturbation as well. Similar findings did not appear for those with negative attitudes toward menstrual sex, as their masturbation attitudes were more inconsistent. Specifically, of the 25 women with negative feelings toward menstrual sex, eight did not enjoy masturbating. That positive feelings toward menstrual sex appeared often in tandem with positive attitudes toward masturbation suggests that women who affirmed menstruation and experienced less body shame about menstrual sex may also have felt more positively about their bodies in general.

To highlight some positive responses, consider Dawn's views about enjoyment of masturbation and her framing of this as body affirmation: 'I masturbate about once per week. It makes me feel in control, like I have a deeper connection with my body.' Emily also happily expressed shameless enjoyment of masturbation along with positive experiences with menstrual sex:

I love it! I have been masturbating for about as long as I can remember. In middle school, I must have tripled my parent's water bill just by playing around in the shower [laughs]. I masturbate daily, unless I get interrupted at it or something. I'm very unapologetic about masturbation. I'm like, 'Can a girl masturbate in peace please?'

Lucy also noted the importance of shared communities around sexual openness, as talking about masturbation (something women often avoid, see Jones, 2002) helped her confidence and positivity toward masturbation:

> I have had really good experiences masturbating and I think it's taught me what I like, and I've been able to have better sex because of it. I masturbate probably three or four times a week.. . I'm kind of a hippie, so all of my friends that are hippies are also very, very open with masturbation too.

In contrast, negativity toward menstrual sex often aligned with negativity toward masturbation, though this was much more inconsistent. Charlotte described self-doubt and anxiety about masturbation:

> I don't mean to sound like a prude, but I've never done that. It just seems weird that everyone's like, 'It's normal. You can get a toy or get a guy to get a toy.' I tried once and it felt weird, just so awkward. I felt like, 'What are you doing? You're a weirdo.'... I've talked to people who say it's a piece of cake, but I just have never had the pleasure to experience it like that.

Interestingly, Charlotte situated her lack of interest in masturbation as clearly freakish or abnormal, which did *not* appear in women's discussion of menstrual sex; perhaps women more clearly normalized masturbation as 'something everyone does', while menstrual sex may still lack a social comparison dimension. Nonetheless, clear similarities between *liking* masturbation and liking menstrual sex suggest that more body affirmation and less body shame led to positive feelings about these dimensions of sexuality.

Intersections of identities and menstrual sex experiences. In addition to connections between masturbation and menstrual sex attitudes, social identities also factored centrally in women's accounts of sex while menstruating. Race and sexual identity differences appeared, as a mere three out of 13 positive responses toward menstrual sex were provided by women of color, indicating that white women in this sample felt more positively toward menstrual sex than did women of color. Notably, women who embraced positive feelings about menstrual sex were more often white and bisexual or lesbian. Only three of the heterosexual-identified respondents described menstrual sex as positive; the remaining 22 had negative feelings. Both women who described neutral feelings identified as white and bisexual. More women of color and heterosexual women (including heterosexual women of color specifically) described sex while menstruating as a negative experience. These results collectively raise questions about the relationship between social identities and menstrual sex attitudes, discussed further in the next section.

DISCUSSION

Collectively, these results raise questions about *how* women learn to embrace sex while menstruating, *who* they embrace it with and *why* they embrace it. Equally interesting questions are raised about how, why and with whom some women reject menstrual sex. Though I initially did not prioritize examining the relationship between social identities and menstrual sex attitudes, these findings suggest that both race and sexual identity may relate to women's attitudes about sex while menstruating. Those describing positive experiences with menstrual sex overwhelmingly included white, lesbian or bisexual-identified women, indicating that women of color and heterosexual women far more often reported negative experiences with sex while menstruating. Additionally, those espousing positive experiences with menstrual sex more often described positive experiences with masturbation compared to those with negative menstrual sex attitudes. Further, intersections between race and sexuality were noted, as more women of color identified as heterosexual, while more white women identified as bisexual or lesbian, indicating that many of the results speak to the experiences of heterosexual women of color and bisexual/lesbian white women.

Such results tap into existing research on racial differences in body image and sexual practices, as white women receive more cultural and social validation for their bodies (albeit within narrowly prescribed limits of thinness and beauty), and they may benefit from a culture that values and celebrates white bodies (Molloy and Herzberger, 1998; Patton, 2010; Stephens and Phillips, 2005). There are many possible reasons why white women more often expressed positive feelings toward menstrual sex and masturbation. The oversexualization of women of color, particularly in pop media (Sengupta, 2006), combined with the pervasive influence of whiteness as a primary criterion of beauty (Hill-Collins, 2005), may influence women's relationship to menstruation and other body norms (e.g. shaving, dress, hairstyles, etc.). Within communities of color, a taboo against sex while menstruating may exist more prominently than it does for whites, likely because menstrual taboos exist more prominently within communities of color (Bobel, 2010). As Bobel argued in a recent interview, 'Menstrual activism is risky business for all, and especially for women of color, whose bodies have been denigrated throughout history. Taking on the menstrual taboo can make others see you as nasty, gross, improper... and if you're already struggling to be accepted and taken seriously, then why go "there"?' (Nack, 2010).

One could compellingly argue that partner aversion also played a major role in these findings, as most women who discussed partner aversion were women of color (eight out of 12). Or, negative menstrual sex attitudes may stem from both women and their partners engaging in a cyclical dance of aversion. Clearly, women battle not only their own internalized body shame about menstruation and masturbation, but also their *partner's* communication of this body shame as well (Bobel, 2010); women of color may face steeper battles in affirming both masturbation and menstruation. These findings are particularly interesting in light of studies that show African-American women reporting lower body dissatisfaction about weight (Kronenfeld et al., 2010; Miller et al., 2000; Sabik et al., 2010), indicating a contrast between attitudes about different body 'taboos'. These discrepancies also suggest that conceptualizations of body image and body shame may have overvalued emphasis on weight and thinness and undervalued other social norms about bodies (Davis et al., 2010). Nonetheless, this study suggests that women of color and white women differently experienced sex during menstruation, warranting further research about why women of color are not more buffered against negative menstrual attitudes.

Additionally, the sexual identity findings point to some troubling aspects of heterosexuality and the possibly liberatory aspects of lesbianism and bisexuality, as lesbian and bisexual women overwhelmingly embraced menstrual sex compared to heterosexual women. Notably, negative feelings about menstrual sex did not exclusively correlate with the choice of male partners instead of female partners, as bisexual women having sex with men overwhelmingly had *positive* experiences with sex while menstruating – and mostly they spoke of sex with *male* partners. Rather, heterosexual *identity* seemed more related to shame about menstrual sex and negative experiences with masturbation. Perhaps queer identity makes women less likely to seek out the male gaze (Walker, 1995), or perhaps queer women have considered alternatives to male partnerships and thus have chosen different types of men. For example, one study found that queer women more often viewed their sexual identity as involving a conscious, effortful process than did heterosexual women (Konik and Stewart, 2004), implying that bisexual women may choose men consciously and less as a social mandate. As such, if bisexual and lesbian identity link with positive and affirming self-attitudes, future research could interrogate the strengths and limitations of this finding, particularly as they relate to intersections between race and sexuality. For example, perhaps heterosexual women of color might use negative menstrual sex attitudes to deny men sexual access to their bodies, indicating a subversion of the 'sexually available' woman of color cultural narrative (Hurtado, 1989).

Looking more broadly, these findings suggest that women may resist body shame by choosing affirmative partners. Janet's description of menstrual sex as loving and affirming and Lori's conceptualization of her mutual private rebellion suggest this possibility. Similarly, those managing partner aversion show that partner choice can *negatively* impact women's menstrual sex attitudes. Heterosexual women often prioritize their partners' pleasure over their own when assessing sexual satisfaction (Nicolson and Burr, 2003; Traeen and Skogerbo, 2009), so *avoiding* menstrual sex may also satisfy their partners' needs or wants if their partners dislike menstrual sex. Women's attitudes may change from partner to partner as well, so longitudinal data could prove interesting. Clearly, the *gender* of women's partners matters, as women's female partners more often embraced menstrual sex than did women's male partners. That said, bisexual women often selected *male* partners who appeared to eroticize menstruation. More research about mutually affirming body processes in lesbian partnerships could be useful, particularly since lesbian relationships were found to be more satisfying, egalitarian, empathic and better at conflict resolution compared to heterosexual relationships (Ussher and Perz, 2008), as would research about *which men* support women's body affirmations.

This study adds to existing literatures on sexuality, menstruation and masturbation in several ways. First, few studies have directly examined women's qualitative *feelings about* menstrual sex (Allen and Goldberg, 2009, is the notable exception, though they utilized a heterosexual college sample and did not interrogate race). Research studies have typically neglected material about women's thought processes, feelings, emotions and rationalizations, preferring instead to study frequencies and behavior. Qualitative methods allow a deeper, more textured, interrogation of these aspects of women's lives. Second, while this study nuances women's overwhelmingly negative attitudes toward menstrual sex – layering together their own disgust, emotional labor and partner aversion – *many women* successfully subverted anti-menstruation attitudes by seeing menstruation as positive and erotic. This study offers qualitative insight about the frameworks within which women rebel against social norms that require them to keep menstruation hidden and secretive. Women's body affirmations occur often, and in spite of, negative normative social pressures. Whether this represents *feminist consciousness* remains an open question, though the groundwork for feminist beliefs appeared in several responses (e.g. Edie's suggestion that she had been 'brainwashed'); certainly, feminist consciousness has predicted better body image and comfort with bodily processes (Murnen and Smolak, 2009; Peterson et al., 2006), and all three 'rebellion' comments in this study were made by feminist women.

Additionally, this study helps connect the dots between attitudes toward menstrual sex and other aspects of body affirmation such as enjoyment of masturbation; women may subvert norms of body shame in *multiple spheres* of their lives. Though limited in scope and detail, a connection between menstrual sex and masturbation appeared. Future research could determine what other social attitudes might attach to these positive or affirming views (e.g. refusing to fake orgasms, rejecting shaving norms, embracing natural or home births, rejecting psychotropic medications, affirming fatness, and so on). These connections may expand ideas about bodies and sexualities. Women who feel less shame about menstruation may have more comfort with sexuality in general (Schooler et al., 2005), while masturbation experiences may fuel better body image (thus addressing a notable gap in existing literatures – see Coleman, 2002). Women in this study could, and often did, resist negative characterizations of their bodies as disgusting, abhorrent, dirty and shameful by instead creating new narratives of affirmation and support. More research is needed to determine *which women* better resist cultural constructions of women's bodies as disgusting, and *how they go about* creating these alternative narratives. Knowing about the relationship between social identities, body image and narratives of shame can nuance our understanding of how different women are differently affected by these social norms. Further, it can support relevant social, psychological and public policies about body image (e.g. eating disorder treatments, educational interventions, psychotherapy treatments, medical interventions and body activism).

In light of this, perhaps other social identities may also contribute to attitudes toward menstrual sex, as socioeconomic class, geographic location and educational background likely affect women's experiences with their bodies. Clearly, educational experiences matter, because women's studies classes foster greater feminist consciousness among female students (Henderson-King and Stewart, 1999; Stake and Rose, 1994), and feminist consciousness is directly associated with better body image (Murnen and Smolak, 2009; Peterson et al., 2006). Also, social class affects women's perceptions of their bodies (Fahs and Delgado, in press; McLaren and Kuh, 2004), indicating it likely also affects menstrual attitudes. Similarly, life experiences surely affect women's relationships with their bodies, as sexual assault (Campbell and Soeken, 1999), parental divorce (Billingham and Abrahams, 1998), and pregnancy and post-pregnancy (Clark et al., 2009) predict poorer body image. Future research could examine whether sexual assault, abortion, divorce, pregnancy and post-pregnancy affect women's ideas about menstrual sex.

Several limitations are noted. First, though a framework for some patterns was established – including patterns surrounding race, sexual identity and masturbation – quantitative research could differently address these connections, perhaps situating women along a continuum for each dimension. Though this study examined race and included many women of color from diverse backgrounds, future qualitative work could more directly discuss women's ideas about how race and sexual identity relate to their sexuality, body image, and ideas about menstrual sex. Also, this sample may have over-selected for those more interested in sexuality who had less shame about sexuality. Notably, women sometimes avoid sex research as a privacy invasion (Boynton, 2003), and may narrate their sexual experiences in relation to the wishes and needs of their (male) partners (McClelland, 2010).

In sum, while this study suggests that most women felt negatively toward menstrual sex – thus confirming common cultural assumptions about menstruation as dirty and shameful – many women resisted such characterizations by instead eroticizing menstruation and, in some cases, affirming their bodies in other ways (e.g. masturbation). Women's partnerships with women and also (some) men may serve as important sites for such resistance. Further, the finding that heterosexual *identity* more than heterosexual *activity* seemed related to negative attitudes toward menstrual sex presents a curious interpretation of gender, sexuality and identity. Much work remains in facilitating women's acceptance and embracing of their bodies, particularly around common cultural taboos such as menstruation. Research must continue to examine – both qualitatively and quantitatively – the interplay between pleasure, sexuality, emotions, identities and social justice.[3]

NOTES

1. "Sex during Menstruation: Race, Sexual Identity, and Women's Qualitative Accounts of Pleasure and Disgust" by Breanne Fahs was first published in 2011 in *Feminism & Psychology* 21 (2): 155–78. Reprinted with permission. No further reproduction or distribution of the material is allowed without permission from the publisher.
2. In response to men's negative attitudes, Gloria Steinem (1986) has sarcastically suggested that, if men could menstruate, congress would subsidize menstrual pads, men would brag about their flow, and the military would require menstruation to enlist.
3. Special thanks to Jennifer Bertagni, Kathleen Courter, Judith Sipes, and Eric Swank for their help in preparing this manuscript.

REFERENCES

Allen KR and Goldberg AE (2009) Sexual activity during menstruation: A qualitative study. *Journal of Sex Research* 6(1): 1–11.

Amann-Gainotti M (1994) Adolescent girls' internal body image. *International Journal of Adolescent Medicine and Health* 7(1): 73–86.

Aretxaga B (1995) Symbolic and over determination and gender in Northern Ireland ethnic violence. *Ethos* 23(2): 123–148.

Berg DH and Coutts LB (1994) The extended curse: Being a woman every day. *Health Care for Women International* 15(1): 11–22.

Billingham R and Abrahams T (1998) Parental divorce, body dissatisfaction, and physical attractiveness ratings of self and others among college women. *College Student Journal* 32(1): 148–1052.

Bobel C (2006) 'Our revolution has style': Contemporary menstrual product activists 'doing feminism' in the third wave. *Sex Roles* 54(5–6): 331–345.

Bobel C (2008) From convenience to hazard: A short story of the emergence of the menstrual activism movement, 1971–1992. *Health Care for Women International* 29(7): 738–754.

Bobel C (2010) *New Blood: Third Wave Feminism and the Politics of Menstruation*. Camden, NJ: Rutgers University Press.

Boyatzis R (1998) *Transforming Qualitative Information: Thematic Analysis and Code Development*. Thousand Oaks, CA: Sage.

Boynton PM (2003) 'I'm just a girl who can't say no'? Women, consent, and sex research. *Journal of Sex & Marital Therapy* 29(1): 23–32.

Brain JL (1988) Male menstruation in history and anthropology. *The Journal of Psychohistory* 15(3): 311–323.

Braun V and Clarke V (2006) Using thematic analysis in psychology. *Qualitative Research in Psychology* 3(1): 77–101.

Briefel A (2005) Monster pains: Masochism, menstruation, and identification in the horror film. *Film Quarterly* 58(3): 17–27.

Brines J (1994) Economic dependency, gender, and the division of labor at home. *The American Journal of Sociology* 100(3): 652–688.

Brooks-Gunn J and Ruble DN (1986) Men's and women's attitudes and beliefs about the menstrual cycle. *Sex Roles* 14(5–6): 287–299.

Burrows A and Johnson S (2005) Girls' experiences of menarche and menstruation. *Journal of Reproductive and Infant Psychology* 23(3): 235–249.

Campbell JC and Soeken KL (1999) Forced sex and intimate partner violence: Effects on women's risk and women's health. *Violence Against Women* 5(9): 1017–1035.

Charmaz C (2006) *Constructing Grounded Theory*. London: Sage.

Chapkis W (1986) *Beauty Secrets: Women and the Politics of Appearance*. Boston, MA: South End Press.

Chrisler JC, Rose JG, Dutch SE, Sklarsky KG and Grant MC (2006) The PMS illusion: Social cognition maintains social construction. *Sex Roles* 54(5–6): 371–376.

Christensen AP and Oei TP (1990) Men's perception of premenstrual changes on the premenstrual assessment form. *Psychological Reports* 66(2): 615–619.

Clark A, Skouteris H, Wertheim EH, Paxton SJ and Milgrom J (2009) My baby body: A qualitative insight into women's body-related experiences and mood during pregnancy and the postpartum. *Journal of Reproductive and Infant Psychology* 27(4): 330–345.

Coleman E (2002) Masturbation as a means of achieving sexual health. *Journal of Psychology and Human Sexuality* 14(2–3): 5–16.

Cosgrove L and Caplan PJ (2004) *Bias in Psychiatric Diagnosis: A Project of the Association for Women in Psychology*. Lanham, MD: Jason Aronson Press.

Davis DS, Sbrocco T, Odoms-Young A and Smith DM (2010) Attractiveness in African American and Caucasian women: Is beauty in the eyes of the observer? *Eating Behaviors* 11(1): 25–32.

Delaney J, Lupton MJ and Toth E (1988) *The Curse: A Cultural History f Menstruation*. Chicago: University of Illinois Press.

Endicott J, Amsterdam J, Eriksson E, Frank E, Freeman E, Hirschfeld R, Ling F, Parry B, Pearlstein T, Rosenbaum J, Rubinow D, Schmidt P, Severino S, Steiner M, Stewart DE and Thys-Jacobs S (1999) Is premenstrual dysphoric disorder a distinct clinical entity? *Journal of Women's Health and Gender-Based Medicine* 8(5): 663–679.

Erchull MJ, Chrisler JC, Gorman JA and Johnston-Robledo I (2002) Education and advertising: A content analysis of commercially produced booklets about menstruation. *Journal of Early Adolescence* 22(4): 455–474.

Fahs B and Delgado D (in press) The specter of excess: Race, class, and gender in women's body hair narratives. In: Bobel C and Kwan S (eds) *Embodied resistance: Breaking the rules, challenging the norms*. Nashville: Vanderbilt University Press.

Fausto-Sterling A (2000) *Sexing the Body: Gender Politics and the Construction of Sexuality*. New York: Basic Books.

Fereday J and Muir-Cochrane E (2006) Demonstrating rigor using thematic analysis: A hybrid approach of inductive and deductive coding and theme development. *International Journal of Qualitative Methods* 5(1): 1–11.

Havens BB and Swenson I (1989) A content analysis of educational media about menstruation. *Adolescence* 24(6): 901–907.

Henderson-King D and Stewart AJ (1999) Educational experiences and shifts in group consciousness: Studying women. *Personality and Social Psychology Bulletin* 25(3): 390–399.

Hensel DJ, Fortenberry JD and Orr DP (2007) Situational and relational factors associated with coitus during vaginal bleeding among adolescent women. *Journal of Sex Research* 44(3): 269–277.

Hensel DJ, Fortenberry JD, Harezlak J, Anderson JG and Orr DP (2004) A daily diary analysis of vaginal bleeding and coitus among adolescent women. *Journal of Adolescent Health* 34(5): 391–394.

Hill-Collins P (2005) *Black Sexual Politics: African Americans, Gender, and the New Racism.* New York: Routledge.

Hoerster KD, Chrisler JC and Rose JG (2003) Attitudes toward and experience with menstruation in the US and India. *Women and Health* 38(3): 77–95.

Hurtado A (1989) Relating to privilege: Seduction and rejection in the subordination of white women and women of color. *Signs* 14(4): 833–855.

Jarvis TJ and McCabe MP (1991) Women's experience of the menstrual cycle. *Journal of Psychosomatic Research* 35(6): 651–660.

Johnston-Robledo I, Barnack J and Wares S (2006) 'Kiss your period good-bye': Menstrual suppression in the popular press. *Sex Roles* 54(5–6): 353–360.

Johnston-Robledo I, Ball M, Lauta K and Zekoll A (2003) To bleed or not to bleed: Young women's attitudes toward menstrual suppression. *Women & Health* 38(3): 59–75.

Johnston-Robledo I, Sheffield K, Voigt J and Wilcox-Constantine J (2007) Reproductive shame: Self-objectification and young women's attitudes toward their reproductive functioning. *Women and Health* 46(1): 25–39.

Jones R (2002) 'That's very rude, I shouldn't be telling you that': Older women talking about sex. *Narrative Inquiry* 12(1): 121–143.

Kerkham P (2003) 'Menstruation – The gap in the text? *Psychoanalytic Psychotherapy* 17(4): 279–299.

Kissling EA (2002) On the rag on screen: Menarche in film and television. *Sex Roles* 46(1–2): 5–12.

Kissling EA (2006) *Capitalizing on the curse: The business of menstruation.* Boulder, CO: Lynne Rienner.

Kleinplatz P (2001) *New directions in sex therapy: Innovations and alternatives.* New York: Routledge.

Koekse R (1983) *Lifting the curse of menstruation: Toward a feminist perspective on the menstrual cycle.* Pittsburg, PA: Haworth Press.

Konik J and Stewart AJ (2004) Sexual identity development in the context of compulsory heterosexuality. *Journal of Personality* 72(4): 815–844.

Kronenfeld LW, Reba-Harrelson L, Von Holle A, Reyes ML and Bulik CM (2010) Ethnic and racial differences in body size perception and satisfaction. *Body Image* 7(2): 131–136.

Laws S (1992) 'It's just the monthlies, she'll get over it': Menstrual problems and men's attitudes. *Journal of Reproductive and Infant Psychology* 10(2): 117–128.

Leiblum SR (2002) Reconsidering gender differences in sexual desire: An update. *Sexual and Relationship Therapy* 17(1): 57–68.

Mansfield PK and Stubbs ML (2007) The menstrual cycle: Feminist research from the Society for Menstrual Cycle Research. *Women & Health* 46(1): 1–5.

Martin E (2001) *The woman in the body: A cultural analysis of reproduction.* Boston, MA: Beacon Press.

Marva´ n ML, Corte´ s-Iniestra S and Gonza´ lez R (2005) Beliefs about and attitudes toward menstruation among young and middle-aged Mexicans. *Sex Roles* 53(3–4): 273–279.

Marva´ n MI, Islas M, Vela L, Chrisler JC and Warren EA (2008) Stereotypes of women in different stages of their reproductive life: Data from Mexico and the United States. *Health Care for Women International* 29(7): 673–687.

Mass R, Ho¨ lldorfer M, Moll B, Bauer R and Wolf K (2009) Why we haven't died out yet: Changes in women's mimic reactions to visual erotic stimuli during their menstrual cycles. *Hormones and Behavior* 55(2): 267–271.

McClelland SI (2010) Intimate justice: A critical analysis of sexual satisfaction. *Social and Personality Psychology Compass* 4(9): 663–680.

McKinley NM, Lyon LA, McKinley AF, Nita M and Lyon LA (2008) Menopausal attitudes, objectified body consciousness, aging anxiety, and body esteem: European American women's body experiences in midlife. *Body Images* 5(4): 375–380.

McLaren L and Kuh D (2004) Women's body dissatisfaction, social class, and social mobility. *Social Science and Medicine* 58(9): 1575–1584.

Miller KJ, Gleaves DH, Hirsch TG, Green BA, Snow AC and Corbett CC (2000) Comparisons of body image dimensions by race/ethnicity and gender in a university population. *International Journal of Eating Disorders* 27(3): 310–316.

Molloy BL and Herzberger SD (1998) Body image and self-esteem: A comparison of African-American and Caucasian women. *Sex Roles* 38(7–8): 631–643.

Mottola G (2007) *Superbad.* USA: Columbia Pictures.

Murnen SK and Smolak L (2009) Are feminist women protected from body image problems? A meta-analytic review of relevant research. *Sex Roles* 60(3–4): 186–197.

Nack A (2010) Bedside manners: New blood, and a new view of menstrual activism. *Girl w/ Pen: Bridging Feminist Research and Popular Reality.* Available at: girl-wpen.com/. ?p¼1902.

Nicolson P (1995) The menstrual cycle, science and femininity: Assumptions underlying menstrual cycle research. *Social Science and Medicine* 41(6): 779–784.

Nicolson P and Burr J (2003) What is 'normal' about women's (hetero)sexual desire and orgasm? A report of an in-depth interview study. *Social Science & Medicine* 57(9): 1735–1745.

Nobre PJ and Pinto-Gouveia J (2008) Cognitive and emotional predictors of female sexual dysfunctions: Preliminary findings. *Journal of Sex & Marital Therapy* 34(4): 325–342.

Offman A and Kleinplatz PJ (2004) Does PMDD belong in the DSM? Challenging the medicalization of women's bodies. *Canadian Journal of Human Sexuality* 13(1): 17–27.

Owens LK, Hughes TL and Owens-Nicholson D (2002) The effects of sexual orientation on body image and attitudes about eating and weight. *Journal of Lesbian Studies* 7(1): 15–33.

Patton TO (2010) Hey girl, am I more than my hair? African American women and their struggles with beauty, body image, and hair. In: Moore LJ and Kosut M (eds) *The Body Reader: Essential Social and Cultural Readings.* New York: New York University Press, 349–366

Peterson RD, Tantleff-Dunn S and Bedwell JS (2006) The effects of exposure to feminist ideology on women's body image. *Body Image* 3(3): 237–246.

Pitman GE (1999) Body image, compulsory heterosexuality, and internalized homophobia. *Journal of Lesbian Studies* 3(4): 129–140.

Plante R (2006) *Sexualities in Context: A Social Perspective.* Boulder, CO: Westview Press.

Read S (2008) Thy righteousness is but a menstrual clout: Sanitary practices and prejudice in early modern England. *Early Modern Women: An Interdisciplinary Journal* 3: 1–25.

Rembeck GI, Moller MM and Gunnarsson RK (2006) Attitudes and feelings towards menstruation and womanhood in girls at menarche. *Acta Paediatrica* 95(6): 707–714.

Rempel JK and Baumgartner B (2003) The relationship between attitudes toward menstruation and sexual attitudes, desires, and behavior in women. *Archives of Sexual Behavior* 32(2): 155–163.

Roberts TA and Goldenberg JL (2007) Wrestling with nature: An existential perspective on the body and gender in self-conscious emotions. In: Tracy JL, Robins RW and Tangney JP (eds) *The Self-conscious Emotions: Theory and Research.* New York: Guilford Press, 389–406.

Roberts TA and Waters PL (2004) Self-objectification and that 'not so fresh feeling': Feminist therapeutic interventions for healthy female embodiment. *Women and Therapy* 27(3–4): 5–21.

Röder S, Brewer G and Fink B (2009) Menstrual cycle shifts in women's self-perception and motivation: A daily report method. *Personality and Individual Differences* 47(6): 616–619.

Rose JG, Chrisler JC and Couture S (2008) Young women's attitudes toward continuous use of oral contraceptives: The effect of priming positive attitudes toward menstruation on women's willingness to suppress menstruation. *Health Care for Women International* 29(7): 688–701.

Sabik NJ, Cole ER and Ward ML (2010) Are all minority women equally buffered from both negative body image? Intra-ethnic moderators of the buffering hypothesis. *Psychology of Women Quarterly* 34(2): 139–151.

Schooler D, Ward LM, Merriwether A and Caruthers AS (2005) Cycles of shame: Menstrual shame, body shame, and sexual decision-making. *Journal of Sex Research* 42(4): 324–334.

Sengupta R (2006) Reading representations of black, east Asian, and white women in magazines for adolescent girls. *Sex Roles* 54(11–12): 799–808.

Shulman JL and Horne SG (2003) The use of self-pleasure: Masturbation and body image among African Americans and women. *Psychology of Women Quarterly* 27(3): 262–269.

Stake JE and Rose S (1994) The long-term impact of women's studies on students' personal lives and political activism. *Psychology of Women Quarterly* 18(3): 403–412.

Steinem G (1986) *Outrageous Acts and Everyday Rebellions.* New York: Holt.

Stephens DP and Phillips L (2005) Integrating Black feminist though into conceptual frameworks of African American adolescent women's sexual scripting processes. *Sexualities, Evolution, and Gender* 7(1): 37–55.

Stubbs MI and Costos D (2004) Negative attitudes toward menstruation: Implications for disconnection within girls and between women. *Women and Therapy* 27(3–4): 37–54.

Tiefer L (2001) *Sex is not a Natural Act and Other Essays*. Boulder, CO: Westview Press.

Tiggemann M and Lewis C (2004) Attitudes toward women's body hair: Relationship with disgust sensitivity. *Psychology of Women Quarterly* 28(4): 381–387.

Traeen B and Skogerbo A (2009) Sex as an obligation and interpersonal communication among Norwegian heterosexual couples. *Scandinavian Journal of Psychology* 50(3): 221–229.

Ussher JM and Perz J (2008) Empathy, egalitarianism, and emotion work in the relational negotiation of PMS: The experience of women in lesbian relationships. *Feminism & Psychology* 18(1): 87–111.

Walker A (1992) Men's and women's beliefs about the influence of the menstrual cycle on academic performance: A preliminary study. *Journal of Applied Social Psychology* 22(11): 896–909.

Walker L (1995) More than just skin-deep: Fem(me)ninety and the subversion of identity. *Gender, Place, & Culture* 2(1): 71–76.

Yeung DYL, Tang CSK and Lee A (2005) Psychosocial and cultural factors influencing expectations of menarche: A study on Chinese premenarchal teenage girls. *Journal of Adolescent Research* 2(1): 118–135.

Zepeda G and Eugenia B (1999) Erotic sexuality in lesbian and heterosexual females. *Archivos Hispanoamericanos de Sexologia* 5(1): 93–115.

CHAPTER 70

Normality, Freedom, and Distress: Listening to the Menopausal Experiences of Indian Women of Haryana

Vanita Singh and M. Sivakami

INTRODUCTION

Menopause is a biological phenomenon marked by permanent cessation of menstruation. Because it is experienced differently across cultures, its meanings and management are highly contested. There are two popular and contrasting perspectives of menopause: biomedical and feminist (Hyde et al. 2010). The biomedical model constructs menopause as a disease of estrogen deficiency, while some feminists view menopause as a natural process shaped by social forces. However, both models are criticized for their failures to engage the viewpoints of individual women who experience menopause (Murtagh and Hepworth 2003). Ferguson and Parry (1998) argue that the medicalization of menopause has sidelined the voices of women in the current discourse, resulting in incomplete understanding of women's experience. Anthropologists and other social scientists, as well as epidemiologists, have highlighted the differences in the symptom experience of women from different cultures and different socioeconomic backgrounds (Beyene 1986; Flint and Samil 1990; WHO 1996; Lock 1993; Avis et al. 2001; Lock and Kaufert 2001). Although the biological changes associated with menopause are universal, menopausal experiences, including the reporting of menopausal symptoms, are largely explained by concerns about aging and the change in social roles once women attain menopause (Obermeyer 2000; Obermeyer, Ghorayeb, and Reynolds 1999; Zeserson 2001). Thus, there is a need to understand the uniqueness in the menopausal experience of women within the broader social context.

© The Author(s) 2020
C. Bobel et al. (eds.), *The Palgrave Handbook of Critical Menstruation Studies*, https://doi.org/10.1007/978-981-15-0614-7_70

CONTESTED MEANINGS OF MENOPAUSE

The field of endocrinology, which enabled the extraction of synthetic estrogen, was established in 1930, leading to an era of interest in menopause by medical practitioners and researchers (Coney 1993). Menopause became popularly understood as hormone deficiency disease leading to loss of femininity. The movement promoting hormone replacement therapy (HRT) in the 1960s to help women remain 'feminine forever' is highly criticized by feminists for its ageist and sexist agenda (Bell 1987; Coney 1993). Approaches to menopause within the medical community, however, have evolved from treating women and their bodies without reference to social context. Bell (1987) traces the history of medicalization of menopause in the United States and identifies three models—biological, psychological, and environmental—guiding medical practice. The biological model defines menopause as "a physiological process caused by cessation of ovarian function," emphasizing hormone deficiency and thus treatment with estrogen or HRT. The psychological model believes that women's personalities affect their symptom experience, so psychotherapy is the appropriate line of treatment. According to the environmental model, women's symptoms are the result of stresses and strains posed by changing social roles and responsibilities during midlife; it proposes that women change their lifestyle and habits to manage menopausal symptoms. All three models have identified the cause of distress during menopause as existing within women and thus advocate for medical intervention for every menopausal woman. Though the environmental model acknowledges the impact of external factors, the solution proposed is internal (Bell 1987). Feminists critique medical models for devaluing older woman (Bell 1987; Coney 1993), presenting menopause as a 'hormonal deficiency disease,' and representing the menopause experience as universal and definitive, all of which have led to the promotion of HRT as the "elixir of youth" (Bell 1987; Coney 1993; Klein 1992).

The use of HRT for managing menopausal symptoms became more contentious after the Women's Health Initiative (WHI) study that linked HRT use with increased risk of breast cancer, endometrial cancer, and cardiac morbidity (Hyde et al. 2010). The WHI study was one of the largest prospective studies assessing the long-term impact of HRT on women's health; however, it was stopped prematurely due to the alarming findings of increased risk of breast cancer and cardiac diseases among study participants (Hammond 2005). The media picked up the study findings, and physicians became highly cautious of using HRT for their patients (Lazar et al. 2007). However, the study soon was criticized in medical, epidemiological, and social science literature for its design, sample selection, and the results not being significant (Derry 2004; Hammond 2005; Wren 2009; Gurney et al. 2014). The feminist-medical debate on the use of HRT for managing menopause resurfaced with the failure of the WHI study. However, we believe strongly that listening to women's own voices is important to further our understanding of menopause. Hence, the focus of this current study is to bring out the uniqueness in the experiences of menopause among Indian women of Haryana and to place them within the broader social context.

The next section discusses the literature on menopause in India. We then explain the methodology adopted for this study, followed by a discussion of our findings and the three narratives that emerged. In the discussion section, we illuminate our findings using anthropological and sociological theories around life transitions and illness. We assert that the meanings of menopause are created by the complex interplay of social, cultural, and biological factors.

MENOPAUSE IN THE INDIAN CONTEXT

India, like many nations, is governed by patriarchal values where women's voices are often ignored or muted (Patel 2007; Syamala and Sivakami 2005). One area in which their experiences are silenced is the menstrual cycle. Menopause and menarche are taboos; they are rarely discussed in Indian society, reflected by the dearth of studies on menopause in the Indian context. Studies on menopause in India have mainly focused on the differences in age at menopause and factors affecting it (Kriplani and Banerjee 2005; Syamala and Sivakami 2005; Sharma et al. 2007; Dasgupta and Ray 2009; Singh and Sivakami 2014). These differences are attributed to rural/urban residence, lifestyle factors, age, environment, education status, occupation status, and reproductive history (Stanford et al. 1987; Brambilla and Mckinlay 1989; Hill 1996; Hidayet et al. 1999; Harlow and Signorello 2000; Kriplani and Banerjee 2005; Syamala and Sivakami 2005; Dasgupta and Ray 2009; Singh and Sivakami 2014; Agarwal et al. 2018). Some studies found women who report no symptoms at all, while others report many symptoms, including hot flashes, night sweats, anxiety, irritability, loss of vision, joint pains, vaginal dryness, numbness, and tingling (Singh and Arora 2005; Bernis and Reher 2007; Dasgupta and Ray 2009; Singh and Sivakami 2014). There are two plausible explanations for this disparity: first, according to Aaron, Muliyil, and Abraham (2002), women do not report menopausal symptoms, as they link them with aging rather than menopause; and second, talking about menarche and menopause is still taboo in Indian culture.

The extant studies on the perception of menopause in India indicate that menopause is seen as positive, since the end of menstrual bleeding removes the societal restrictions that come with the cultural view that menstruation is polluting (Singh and Arora 2005). The negative views about menstruation and menstruating women have appeared in the accounts of women from Haryana (Singh 2006). Singh's (2006) study in rural Haryana of practices during menstruation found that only 0.4% of the sample population was using sanitary napkins to manage menstrual bleeding because of the difficulty in accessing menstrual products. This created a source of monthly distress for these women. Moreover, studies on poor women from India (George 1994) have reported the difficulties women face in maintaining hygiene and managing menstrual bleeding in poor living conditions with limited space and no indoor toilets. Systematic review and meta-analysis of 138 studies on menstrual hygiene management among adolescent girls in India between 2000 and 2015 point out that adolescent girls experience menstruation as "shameful and uncomfortable" due to lack of water, sanitation, and hygiene facilities, insufficient puberty education, and poor hygienic practices to manage menstruation (van Eijk et al. 2016). The same study also showed the societal issues faced by Indian adolescent girls in the form of various restrictions, which adds to the negative construction of menstruation. These difficulties construct menstruation as a monthly distress for these women, who later consider postmenopause as a "freedom from monthly tension" (Singh and Sivakami 2014). Also, because menstruation prevents them from entering places that are considered sacred or pure, menopause is considered a type of liberation in Haryana (Singh and Arora 2005; Singh and Sivakami 2014; van Eijk et al. 2016).

It is important to understand the meanings of menopause as constructed by women themselves as an alternative to overreliance on the medical discourse. Some have argued that the reporting of menopausal symptoms as well as their management is based on the studies of white women (Avis et al. 2001), and our policies reflect that. The policies need to be guided by the "local biologies" (Lock and Kaufert 2001) to ensure effective management of menopause. This study, undertaken to amplify the voices of menopausal women from lower socioeconomic strata, was conducted in the North Indian state of Haryana. Haryana is known for its gender discrimination practices, which are reflected in the state's sex ratio—that is, the ratio of males to females in a population. Sex ratio is a powerful indicator of social health in any society (Patel 2007), and Haryana's is the lowest in India.

Methods and Participants

In 2012, we conducted in-depth qualitative interviews with 28 postmenopausal women from the lower socioeconomic strata in the state of Haryana. We recorded information related to women's socioeconomic status, educational and working status, parity, breastfeeding practices, lifestyle and eating habits, and menopause status, using a structured interview schedule as part of a larger study focusing on sociocultural differences in the age at menopause and symptom experience (see Singh and Sivakami 2014). A subset of women who identified themselves as postmenopausal and who agreed to share their menopause experience in detail were part of the qualitative narratives that form the sample participants for this study. We use pseudonyms throughout the paper to maintain privacy and confidentiality. The interviews lasted for 45–60 minutes, and the first author made a detailed summary of the interviews. The first author is familiar with the local language, and thus women discussed their experience using their own words, in their own space and time.

The women interviewed were between 45 and 60 years old. They identified themselves as postmenopausal if their menstruation had stopped more than a year ago. The mean age at menopause was 47 years for urban women and 44 years for rural women. Of the 28 postmenopausal women, three women reported having had hysterectomies and thus had surgical menopause, while the rest had conventional menopause. In terms of education, 17 of the women were illiterate, 5 had completed education up to primary level, and only 6 women were educated up to senior secondary level. Four women were widowed and the rest were currently married. Almost all had children of marriageable age or were already married. Many of them were grandmothers. Six of them were currently working as unskilled labor and the rest did not work outside the home.

After establishing a rapport with the women, we asked them to discuss their menopausal experience, and the topics ranged from the changes in their menstrual cycle, changes in body, managing menopause, family and work life, social support, peer comparisons, and local beliefs about menopause and its management. Their experiences ranged from having had no symptoms to suffering from a multitude of symptoms. The gamut of experiences suggested the role of social context in the uniqueness of each woman's experience. The narrative accounts of each woman were crafted from the information obtained during the interviews. We conducted narrative analysis to obtain rich insights into women's experience of menopause and identified themes and subthemes to gauge the significance of sociocultural context in the experience of menopause; we then linked them to theoretical models (Ryan and Bernard 2003). Three distinct but co-occurring narratives, discussed below, emerged from the detailed accounts of North Indian women from lower socioeconomic strata. We also read the narrative accounts repeatedly to identify five key themes grounded in the data: bodily changes or complaints, support system, visit to health provider, local beliefs about menopause and its management, and attitude toward menopause. The themes were then compared and contrasted across all interviews using constant comparison method (Corbin and Strauss 1990). We drew on sociological and anthropological theories of illness and transition to discuss the differences and similarities across narratives.

Ultimately, the three distinct but co-occurring narratives that emerged were: menopause as a normal life transition, menopause as distress because it's taboo, and menopause as freedom from monthly distress and societal restrictions. We found that although these narratives co-occur as women reflect back or anticipate the future, one narrative dominated. We draw insights from anthropological and sociological theories of menopause, illness, and transition states (Kaufert 1982; Hyden 1997; Ballard, Kuh, and Wadsworth 2001) to illuminate the differences and similarities in narratives. We also compare and contrast the different narratives based on the five identified key themes.

FINDINGS

Menopause as a Normal Life Transition

> I do not have any issue . . . all of a sudden it stopped and I was happy. (Kamla, 49, rural resident)

Accounts of menopause as a normal life transition showed that women tend to normalize their symptoms by peer comparisons and by acknowledging the universality of menopause. Women used phrases like "every woman's issue" and "it's like childbirth" to normalize the experience.

> Every woman has to go through this, women are made to suffer, it's like child-birth you know. (Meena, 56, rural resident)

> It's over now. . . . . I had to suffer from heavy bleeding for six months which every woman has to suffer. . . . I have heard that menopause happens this way only. (Jyoti, 47, urban resident)

Many women reported having had no symptoms and abrupt cessation of menstruation; for these women, menopause was a natural transition. They considered menopause symptom-free; they mentioned joint pains or head-aches but associated those with aging rather than menopause (Singh and Sivakami 2014). Similar results have been reported by other Indian studies as well, indicating that Indian women report fewer symptoms, as they frequently link menopausal symptoms with symptoms of aging (Aaron, Muliyil, and Abraham 2002).

In contrast to Western studies, where women are influenced by biomedical discourse and visit health providers to make sense of their experience and confirm their menopausal status (see Hyde et al. 2010), our study depends on "vernacular health theories": using popular beliefs and perceptions in local communities to understand illness or a life transition (Goldstein 2000) and to make sense of their bodily changes. These women depend on the affirmation of their community, as Kamla (49, rural resident) offers: "I discussed with other women in my community and they told me, it's menopause (*mahi-na-bandh*) for you. . . . . " The conception of menopause as a normal life transition also comes from the accounts of women who expressed the insignificance of menopause in their lives once their family was complete. Many women stated that they did not need the menstrual cycle past age 40, as menstruation is equated with reproduction, especially among illiterate women. As one woman said, "Why you need this? . . . Once your family is complete, you don't need it" (Gyani, 58, rural resident).

Biomedical discourse also influenced women's perception of early menopause. Few who went to health providers mentioned their fear of getting uterine cancer. Most of the women acknowledged that they suffered from unmanageable, heavy, and painful bleeding during the menopause phase; however, they call it normal. They believe that if the 'bad blood' were to remain inside the body, they may develop cancer. This suggests the power of vernacular health theories in normalizing distress, which is crafted by the medical discourse of menopause (Goldstein 2000). Menopause became insignificant for those women who were busy managing the chaos of life, most notably poor women, whose struggle every day to ensure that their families get fed takes priority over reflection on their own health issues. Reena (58, urban resident), a domestic servant who did not recall her menopause experience, reflects this view: "I don't remember how it stopped . . . it's been 10 years now. . . . I think I had no complaints. . . . At that time I was busy doing my job [domestic servant]. I used to work in 10 houses and was busy whole day." Seema (53, rural resident) is another woman who failed to recall her menopause experience: "I never had time to think about menopause and all. . . . I don't even remember exactly when I stopped menstruating. . . . I was too busy managing my household chores and managing field work [agricultural fields]."

Beena (49, rural resident) has a daughter of marriageable age, which in their culture is as soon as they turn 18. Beena is more concerned about the marriage of her daughter than her menopausal status. The latter has no bearing on her identity, while the former shapes her identity as a mother. She said, "It's a woman's issue which she has to bear. . . . Now, it's over but I have lot of other tensions. . . . I have to arrange for my daughter's wedding, she is 26 now. . . . It's getting late." Here, menopause, as a midlife transition is overshadowed by other events, and it becomes a routine and normal midlife transition, consistent with Ballard, Kuh, and Wadsworth (2001) finding that other life transitions, such as changing relationships with one's partner or children or changes in work life, often overshadow the experience of menopause.

Most women in our study reported that menopause was an insignificant life event that either passed without any symptoms or without any symptoms that they remember. Those who did have unwanted symptoms normalized them by ascribing them universality, as reflected through peer comparisons and phrases such as "every woman has to suffer this" and "women are made to suffer."

Menopause as Distress

In this second thematic narrative, distress emerged mainly from somatic changes (heavy and painful bleeding, sleepless nights, irritability, anxiety, mood swings, and frequent headaches), often exacerbated by negative life events and an inability to share their experience with anyone. Rajni (52, urban resident) recalls her struggle with painful and heavy menstruation during her perimenopause phase: "I suffered during menopause for 2 years. . . . It was heavy and painful bleeding which kept me awake . . . [I] used to feel very hot, couldn't sleep, had frequent headaches . . . was feeling depressed. . . . . During that time my husband died . . . It was a tough time."

In Rajni's account, there is no mention of depression or other body changes emerging from menopause. Rather, she reports that perimenopause became tough at the point where her husband died. That is, it was the death of her husband and her inability to share her distress with anyone that made perimenopause a difficult time for her. Sarla (51, rural resident), who is currently postmenopausal, recalls suffering and difficulties during perimenopause, particularly managing menstrual bleeding and maintaining hygiene in the presence of a male family member (her son). In her culture, it is unacceptable to discuss issues such as menarche and menopause with male members, as she mentions,

> It's a suffering. . . . I suffered heavy and painful bleeding for nine years. . . . I used to bleed continuously for 15–20 days, was exhausted. . . . I had sleepless nights due to fear of spoiling the bed-sheet. . . . I used to be so tired . . . had no energy to stand and work . . . wished I had someone to understand my situation and help with household chores. . . . My daughters were married by that time and I can't discuss all this with my son. . . . Whenever I used to lie down in bed, he used to ask me what has happened to you . . . and I had no answer. . . . Now I am free from all such tensions.

Sarla's wish for someone to understand her situation points to a lack of emotional and social support in her life. Though she mentions a number of symptoms—sleeplessness, exertion, dizziness—she does not intend to seek medical intervention. Rather, she expressed the need to be understood. In Sarla's account, the narratives of freedom and distress coexist. Initially, her difficulties managing menopause and its associated symptoms in the presence of male members in her house made her feel distressed about menopause; however, once her period stopped completely, she felt liberated from monthly tensions. In her culture, menstruation or menopause is a women's issue and should not be discussed with men.

We found that women who lacked a support system were more likely to feel the distress of menopause. For Sita (54, urban resident), the postmenopausal phase became very difficult because she had no one with whom to share her grief over the death of her husband. Her complaint about God highlights her helplessness and distress. She complains of frequent headaches and joint pains:

> For two years, I suffered from heavy bleeding and painful menstruation. . . . It was difficult, you know very difficult . . . it was painful. . . . Now I am not menstruating, but I have all sorts of pain, headache I get every day, my vision has diminished, all my joints ache. . . . My husband died three months back

[tears roll down her eyes] . . . my daughter-in-law is not good . . . my son is busy with his own life. . . . [There is no one] whom to tell my problem . . . even God doesn't listen to my prayers.

Similarly, Geeta (59, rural resident) remembers her menopause experience and calls it a "suffering." She mentions her frequent fights with her husband and even her son. She expresses the need to be understood in the following excerpt:

During menopause, I suffered a lot. . . . I had very heavy bleeding . . . which was unmanageable. . . . I tried many home-remedies . . . but nothing worked. . . . I was unable to sleep, fully exhausted . . . didn't wanted to talk to anyone, became anxious, used to get irritated over small things . . . had frequent fights with my husband . . . and even with my son. . . . They don't understand what a woman is going through . . . and you can't explain them. . . . It was a difficult time.

On the other side, Gita (52, rural resident) boasts about her daughter-in-law and mentions that she is enjoying this carefree life as a grandmother:

There was very heavy bleeding before I stopped menstruating . . . but that's normal . . . every woman has to go through this. . . . My daughter-in-law is very nice . . . she managed all the household chores at that time . . . she helped me a lot. . . . Now she works and I look after her kids . . . now I am free and enjoying life.

Menopause as Freedom

The third thematic narrative carried two meanings of freedom: freedom from the stressful management of menstruation and freedom from societal restrictions. Most women acknowledged that managing heavy and painful menstrual bleeding is difficult at an older age, thus they are happy when menstruation ends. The management is further complicated by the taboos attached to menstruation and menopause, which leads them to suffer in silence. Gita (52, rural resident) mentions, "Three to four years back it stopped. . . . It's good that it stops when you become old." Women expressed their difficulties in managing the pain and chaos of menstruation every month and thus embraced menopause as entry into a carefree phase of life. Women are happy in their new roles (mother-in-law or grandmother), which they would have found difficult with menstruation. As Gita adds, "It's good it stopped before I became grandmother as it's difficult to manage menstruation in old age."

For some women, the narrative of distress and freedom co-occurred. When they reflect back on their experience during perimenopause, they call it "suffering" but they ended with phrases like "I am free now," "I am enjoying now," and "it's over now." For these women, the narrative of freedom has emerged from the narrative of distress. For instance, Nirali, (53, urban resident) mentions, "two to three months I was bleeding heavily and then it stopped. . . . It was difficult to manage . . . it was annoying, now it's over."

For most of the rural and some urban women, ending menstruation liberated them from the societal restrictions on entering sacred places and participating in sacred rituals like wedding *pheras*. Rekha (48, rural resident) recalls that she was barred from entering wedding *mandap* of her daughter. She mentions, "Even during my daughter's wedding I was menstruating and thus could not perform many rituals." Seema (47, rural resident) earns a living by organizing *kirtans* (spiritual gatherings). In her community, it is not acceptable that she attends a *kirtan* while she is menstruating, as menstruation is considered to be polluting. She is happy that she has reached menopause, as now she can plan and organize spiritual gatherings any time: "I am happy that it stopped. . . . I run a *mandli* (group) for performing spiritual gatherings (*kirtan*). . . . It was so difficult to manage when I was menstruating. . . . I cannot plan a *kirtan* when I am menstruating. . . . Now, I am free."

All of these women are happy that they have attained menopause. They can freely go to temple, plan outings, and lead a carefree life. Women were more concerned about their new roles as grandmothers, and they felt liberated from monthly distress; by contrast, Western studies report that women are more concerned about their fertility status (Nosek, Kennedy, and Gudmundsdottir 2012). Some sociological and anthropological studies have reported that Indian women enjoy greater control over resources when they enter postmenopause and when they become mothers-in-law (Kaufert 1982; Inhorn 2006; Patel 2007). Also, Indian women enjoy higher social status in post-reproductive years due to freedom from the so-called 'polluting' menstruation and power dynamics (Vatuk 1998, 289–306; Aaron, Muliyil, and Abraham 2002).

DISCUSSION

The narratives of normality, distress, and freedom emerged from the accounts of North Indian women from the lower socioeconomic strata. Each narrative surfaced from the complex interaction of biological changes, changes in family and work life, social status changes, and cultural beliefs. Though the narratives were distinct, they co-occurred in the accounts of many women.

The narratives of normality and freedom suggest the positive attitude of these women toward menopause. Our findings are reflected in earlier studies from India that examined perceptions of menopause (Aaron, Muliyil, and Abraham 2002; Singh and Arora 2005). The studies reported that Indian women embrace menopause, as they are free from the societal restrictions and the 'polluting' effects of menstruation. The narrative of normality was influenced by cultural perceptions about menopause, shaped either by their personal experience or that of their community members (Aaron, Muliyil, and

Abraham 2002). It has been found that societies that value fertility, youth, and sexual attractiveness view menopause negatively (Kaufert 1982; Khademi and Cooke 2003; Hall et al. 2016), while societies in which menopause is considered to be socially liberating embrace it (Aaron, Muliyil, and Abraham 2002; Singh and Arora 2005; Syamala and Sivakami 2005). In these settings, postmenopausal women enjoy greater self-esteem (Hall et al. 2016).

Menopausal normality also emerged from the insignificance of menopause in the lives of many women who were busy managing other chaos of life. Ballard, Kuh, and Wadsworth (2001) report similar findings from their study of menopause experience in social context. The authors argue that social events compete with menopause for attention, and significant life events often overshadow the menopause experience. We found that life events such as the death of a husband, the birth of a grandson, or the marriage of a son or daughter overshadowed the menopause experience. These major life events that reshaped family structures rendered menopause insignificant.

The distress of menopause reported by studies from the West is different from the narrative of distress that has emerged in this study. For Western women, the distress primarily emerges from the anticipation of aging, losing fertility, and becoming less attractive (de Salis et al. 2018; Nosek, Kennedy, and Gudmundsdottir 2012); their accounts seem to be influenced more by the biomedical perspective (Nosek, Kennedy, and Gudmundsdottir 2012). Conversely, in our study, distress stemmed mainly from difficulty in managing heavy and painful bleeding, further exacerbated by taboos attached to menopause. The grief and distress expressed by one of the rural women in our study was palpable, as she described being barred from her daughter's wedding *mandap* because she was menstruating. The distress in the lives of poor women from Haryana seems to stem from the complex interaction of patriarchy, gender, and poverty.

The narrative of distress as well as freedom has its roots in the patriarchal structure of Indian society—broadly defined as the domination of women by men—as illuminated in the review of ethnographic works by Inhorn (2006). Inhorn asserts that patriarchy has a demoting effect on women's health through both the "micropatriarchy" in a doctor–patient relationship and the "macropatriarchy" in the family structure, in which men exert domination over females of the house. Patriarchy is also seen as women being discriminated against and/or abused by their husbands. There is also an age dimension, as older women exert control over younger women and girls in the household, sometimes tormenting them for being infertile or not doing household chores (Inhorn 2006).

The three thematic narratives are interlinked. Narratives of normality and freedom dominated, and for most women freedom emerged from the narrative of distress. The narrative of distress, rooted in patriarchal values and practices, co-occurred with normality and freedom for many women. Women cannot share their menopausal experiences with men in the house because a woman is considered polluting while menstruating. Even the accounts of normality reflect the workings of patriarchy in the sense that women are supposed to bear the pain and show themselves as culturally competent. The narrative of freedom departs from normative and oppressive power structures, however, in the way that it challenges the dominant negative image of menopause found in medical discourses that cast menopause as disease in need of treatment.

As these narratives have emerged from the accounts of postmenopausal women, we must note that the phase of menopause per se may be influential in shaping the meaning of menopause. We may expect different findings if premenopausal and perimenopausal women were also part of this narrative analysis. Furthermore, the findings are based on the accounts of women from low socioeconomic strata and a single state in India. Future research may look into women of more diverse backgrounds.

In conclusion, this study identified a spectrum of menopausal experiences of low-income Indian women, whose voices are rarely heard. When we listen to women's own stories, located in the social context, we can capture what menopause actually means to women. In this case, the result found a complex interplay of social, cultural, and biological factors. From here, we can develop strategies of support that enable healthy and empowered aging that meets women's diverse needs.

REFERENCES

Aaron, Rita, Jayaprakash Muliyil, and Sulochana Abraham. 2002. "Medico-Social Dimensions of Menopause: A Cross-Sectional Study from Rural South India." *National Medical Journal of India* 15 (1): 14–17.

Agarwal, Anil K., Nirmala Kiron, Rajesh Gupta, Aditi Sengar, and Preeti Gupta. 2018. "A Study of Assessment of Menopausal Symptoms and Coping Strategies among Middle Age Women of North Central India." *International Journal of Community Medicine and Public Health* 5 (10): 4470–77.

Avis, Nancy E., Rebecca Stellato, Sybil Crawford, Joyce Bromberger, Patricia Ganz, Virginia Cain, and Marjorie Kagawa-Singer. 2001. "Is There a Menopausal Syndrome? Menopausal Status and Symptoms across Racial/Ethnic Groups." *Social Science and Medicine* 52 (3): 345–56.

Ballard, Karen D., Diana J. Kuh, and Michael E. J. Wadsworth. 2001. "The Role of the Menopause in Women's Experiences of the 'Change of Life'." *Sociology of Health & Illness* 23 (4): 397–424.

Bell, Susan E. 1987. "Changing Ideas: The Medicalization of Menopause." *Social Science & Medicine* 24 (6): 535–42.

Bernis, Cristina, and David Sven Reher. 2007. "Environmental Contexts of Menopause in Spain: Comparative Results from Recent Research." *Menopause* 14 (4): 777–87.

Beyene, Yewoubdar. 1986. "Cultural Significance and Physiological Manifestations of Menopause a Biocultural Analysis." *Culture, Medicine and Psychiatry* 10 (1): 47–71.

Brambilla, Donald J., and Sonja M. McKinlay. 1989. "A Prospective Study of Factors Affecting Age at Menopause." *Journal of Clinical Epidemiology* 42 (11): 1031–39.

Coney, Sandra. 1993. *The Menopause Industry: A Guide to Medicine's 'discovery' of the Mid-Life Woman*. Australia: Spinifex Press.

Corbin, Juliet M., and Anselm Strauss. 1990. "Grounded Theory Research: Procedures, Canons, and Evaluative Criteria." *Qualitative Sociology* 13 (1): 3–21.

Dasgupta, Doyel, and Subha Ray. 2009. "Menopausal Problems among Rural and Urban Women from Eastern India." *Journal of Social, Behavioral, and Health Sciences* 3 (1): 2.

Derry, Paula S. 2004. "Hormones, Menopause, and Heart Disease: Making Sense of the Women's Health Initiative." *Women's Health Issues* 14 (6): 212–19.

van Eijk, Anna Maria, M. Sivakami, Mamita Bora Thakkar, Ashley Bauman, Kayla F. Laserson, Susanne Coates, and Penelope A. Phillips-Howard. 2016. "Menstrual Hygiene Management among Adolescent Girls in India: A Systematic Review and Meta-Analysis." *BMJ Open* 6 (3): e010290.

Ferguson, Susan J., and Carla Parry. 1998. "Rewriting Menopause: Challenging the Medical Paradigm to Reflect Menopausal Women's Experiences." *Frontiers: A Journal of Women Studies* 19 (1): 20–41.

Flint, Marcha, and Ratna Suprapti Samil. 1990. "Cultural and Subcultural Meanings of the Menopause." *Annals of the New York Academy of Sciences* 592 (1): 134–47.

George, Annie. 1994. "It Happens to Us: Menstruation as Perceived by Poor Women in Bombay." In *Listening to Women Talk about Their Health: Issues and Evidence from India*, edited by Joel Gittelsohn, Margaret E. Bentley, Pertti J. Pelto, Moni Nag, Saroj Pachauri, Abigail D. Harrison, and Laura T. Landman, 168–83. New Delhi, India: Har-Anand Publications.

Goldstein, Diane E. 2000. "'When Ovaries Retire': Contrasting Women's Experiences with Feminist and Medical Models of Menopause." *Health* 4 (3): 309–23.

Gurney, Elizabeth P., Margaret J. Nachtigall, Lila E. Nachtigall, and Frederick Naftolin. 2014. "The Women's Health Initiative Trial and Related Studies: 10 Years Later: A Clinician's View." *The Journal of Steroid Biochemistry and Molecular Biology* 142: 4–11.

Hall, Lisa, Lynn Clark Callister, Judith A. Berry, and Geraldine Matsumura. 2016. "Meanings of Menopause." *Journal of Holistic Nursing* 25 (2): 106–18.

Hammond, Charles B. 2005. "The Women's Health Initiative Study: Perspectives and Implications for Clinical Practice." *Reviews in Endocrine & Metabolic Disorders* 6 (2): 93–99.

Harlow, Bernard L., and Lisa B. Signorello. 2000. "Factors Associated with Early Menopause." *Maturitas* 35 (1): 3–9.

Hidayet, N. M., S. A. Sharaf, S. R. Aref, T. A. Tawfik, and I. I. Moubarak. 1999. "Correlates of Age at Natural Menopause: A Community-Based Study in Alexandria." *EMHJ-Eastern Mediterranean Health Journal* 5 (2): 307–19.

Hill, Kenneth. 1996. "The Demography of Menopause." *Maturitas* 23 (2): 113–27.

Hyde, Abbey, Jean Nee, Etaoine Howlett, Jonathan Drennan, and Michelle Butler. 2010. "Menopause Narratives: The Interplay of Women's Embodied Experiences with Biomedical Discourses." *Qualitative Health Research* 20 (6): 805–15.

Hydén, Lars-Christer. 1997. "Illness and Narrative." *Sociology of Health and Illness* 19 (1): 48–69.

Inhorn, Marcia C. 2006. "Defining Women's Health: A Dozen Messages from More Than 150 Ethnographies." *Medical Anthropology Quarterly* 20 (3): 345–78.

Kaufert, Patricia A. 1982. "Myth and the Menopause." *Sociology of Health & Illness* 4 (2): 141–66.

Khademi, S., and M. S. Cooke. 2003. "Comparing the Attitudes of Urban and Rural Iranian Women toward Menopause." *Maturitas* 46 (2): 113–21.

Klein, Renate. 1992. "The Unethics of Hormone Replacement Therapy*." *Bioethics News* 11 (3): 24–37.

Kriplani, Alka, and Kaberi Banerjee. 2005. "An Overview of Age of Onset of Menopause in Northern India." *Maturitas* 52 (3–4): 199–204.

Lazar, Felipe, Jr., Lucia Costa-Paiva, Sirlei S. Morais, Adriana Orcesi Pedro, and Aarão Mendes Pinto-Neto. 2007. "The Attitude of Gynecologists in São Paulo, Brazil 3 Years after the Women's Health Initiative Study." *Maturitas* 56 (2): 129–41.

Lock, Margaret. 1993. "The Politics of Mid-Life and Menopause: Ideologies for the Second Sex in North America and Japan." *Knowledge, Power, and Practice: The Anthropology of Medicine and Every Day Life*: 330–36.

Lock, Margaret, and Patricia Kaufert. 2001. "Menopause, Local Biologies, and Cultures of Aging." *American Journal of Human Biology* 13 (4): 494–504.

Murtagh, Madeleine J., and Julie Hepworth. 2003. "Feminist Ethics and Menopause: Autonomy and Decision-Making in Primary Medical Care." *Social Science & Medicine* 56 (8): 1643–52.

Nosek, Marcianna, Holly Powell Kennedy, and Maria Gudmundsdottir. 2012. "Distress during the Menopause Transition: A Rich Contextual Analysis of Midlife Women's Narratives." *Sage Open* 2 (3). https://doi.org/10.1177/2158244012455178.

Obermeyer, Carla Makhlouf. 2000. "Menopause across Cultures: A Review of the Evidence." *Menopause* 7 (3): 184–92.

Obermeyer, Carla Makhlouf, Françoise Ghorayeb, and Robert Reynolds. 1999. "Symptom Reporting Around the Menopause in Beirut, Lebanon." *Maturitas* 33 (3): 249–58.

Patel, Tulsi. 2007. "The Mindset Behind Eliminating the Female Fetus." In *Sex-Selective Abortion in India: Gender, Society and New Reproductive Technologies*, 135–74. Thousand Oaks, CA: Sage.

Ryan, Gery W., and H. Russell Bernard. 2003. "Techniques to Identify Themes." *Field Methods* 15 (1): 85.

de Salis, Isabel, Amanda Owen-Smith, Jenny L. Donovan, and Debbie A. Lawlor. 2018. "Experiencing Menopause in the UK: The Interrelated Narratives of Normality, Distress, and Transformation." *Journal of Women and Aging* 30 (6): 520–40.

Sharma, Sudhaa, Vishal R. Tandon, and Annil Mahajan. 2007. "Menopausal Symptoms in Urban Women." *Alcohol* 4: 3–41.

Singh, Amarjeet. 2006. "Place of Menstruation in the Reproductive Lives of Women of Rural North India." *Indian Journal of Community Medicine* 31 (1): 10.

Singh, Amarjeet, and Arvinder Kaur Arora. 2005. "Profile of Menopausal Women in Rural North India." *Climacteric* 8 (2): 177–84.

Singh, Vanita, and M. Sivakami. 2014. "Menopause: Midlife Experiences of Low Socio-Economic Strata Women in Haryana." *Sociological Bulletin* 63 (2): 263–86.

Stanford, Janet L., Patricia Hartge, Louise A. Brinton, Robert N. Hoover, and Ronald Brookmeyer. 1987. "Factors Influencing the Age at Natural Menopause." *Journal of Chronic Diseases* 40 (11): 995–1002.

Syamala, T. S., and M. Sivakami. 2005. "Menopause: An Emerging Issue in India." *Economic and Political Weekly* 40: 4923–30.

Vatuk, Sylvia. 1998. "The Indian Woman in Later Life: Some Social and Cultural Considerations." In *Women's Health in India: Risk and Vulnerability*, edited by Monica Das Gupta, Lincoln C. Chen, and T. N. Krishnan, 289–306. New Delhi, India: Oxford India Paperbacks.

World Health Organization. 1996. "Research on the Menopause in the 1990s: Report of a WHO Scientific Group." Geneva: Switzerland. World Health Organization. https://apps.who.int/iris/handle/10665/41841.

Wren, Barry G. 2009. "The Benefits of Oestrogen Following Menopause: Why Hormone Replacement Therapy Should Be Offered to Postmenopausal Women." *Medical Journal of Australia* 190 (6): 321–25.

Zeserson, Jan Morgan. 2001. "How Japanese Women Talk about Hot Flushes: Implications for Menopause Research." *Medical Anthropology Quarterly* 15 (2): 189–205.

CHAPTER 71

The Messy Politics of Menstrual Activism

Chris Bobel and Breanne Fahs

WHAT WE TALK ABOUT WHEN WE TALK ABOUT MENSTRUATION[1]

When we pay attention to menstrual health and its potential to inspire political resistance, we tap into a complex and enduring project of loosening the social control of women's bodies. Menstrual activism works to move embodiment from object to subject status—to see the body not as trivial or unimportant, but as something foundational, urgent, and politically relevant. When we take seriously the (menstruating) body, we link up with others who engage in critical embodiment work, from human trafficking to eating disorders to sexual assault. This is why #menstruationmatters (https://twitter.com/hashtag/menstruationmatters) really should be a rallying call for *everyone* who cares about social justice and gender equality. Menstruation unites the personal and the political, the intimate and the public, the minutiae and the bigger stories about the body. *It IS about so much more than blood.*

It may not be obvious at first but those working to improve menstrual health, whether using humor, poetry, empirical research, school curricula, or promoting a better menstrual absorbent, must counter the internalization of destructive messages about womanhood including notions of bodies as messy, unruly things that need to be tidied up, medicated, plucked, smoothed, and trimmed. "Managing" menstrual cycles evokes the range of activities and practices that women do to "manage" other parts of their bodies, including grooming body hair, making fashion choices, hiding breastfeeding, losing weight, and more. In this essay, we argue that feminists must challenge generations of silence and shame

© The Author(s) 2020
C. Bobel et al. (eds.), *The Palgrave Handbook of Critical Menstruation Studies*, https://doi.org/10.1007/978-981-15-0614-7_71

that obstruct quality menstrual health education. We must also promote a culture of curiosity and informed decision-making about caring for our bodies. Finally, we must counter the assumption that menstruation matters *only* to menstruators.

When we pull back and see menstrual health in context, we can see what is really at stake in menstrual activist work. Because a challenge to the menstrual status quo is itself a critique of gender norms about embodiment, it productively leads us to ask some tough questions about what we take for granted. What can we learn about our cultural value systems when we consider enduring menstrual restrictions? What can we learn when we consider the popularity of skin lightening creams or steroid abuse among teens? Who benefits from these values-in-practice? Who suffers?

Certainly, menstruation is personal, but feminists have long understood that the personal *is* political, that is, while we may experience something—a monthly period, an act of intimate partner violence, an unplanned pregnancy—the way we respond to these events and the support, or lack of support, we can access is the consequence of something far bigger than ourselves. As Carol Hanisch, an American feminist, wrote in 1969: ". . . personal problems are political problems. There are no personal solutions at this time. There is only collective action for a collective solution" (http://www.carol-hanisch.org/CHwritings/PIP.html). Hanisch, here, calls for remedies to the injustices of women's lives that compel us all. That's what menstrual activism is—a mobilizing effort that challenges menstrual taboos and insists that menstruators have the support they need to live healthy happy lives, throughout their cycles and throughout their lives. When we pay attention to menstruation, we work toward a world that is safer and more just, a world where everyone is supported in whatever body they inhabit.

Taking seriously the call for collective action, this chapter first describes a brief history of menstrual activism alongside its more recent iterations in both policy and radical social activism. This is followed by an analysis of menstrual humor, menstrual art, and menstrual activism today. We then turn toward the hazards and possibilities of doing menstrual activist work, including politics of menstrual language and the trivializations and hostilities that can plague this work, followed by a politically charged outline for the future of menstrual activism.

A Movement Dawns, and Finally Gets Brighter

In April 2016, Newsweek ran a feature on menstrual activism. The cover featured an unwrapped tampon against a deep red background. The words, large, bold and in contrasting white, read: "There Will Be Blood. Get Over It. Period Stigma is hurting the economy, schools and the environment. But the crimson tide is turning." When a mainstream, high-circulation news organization ran a feature like this, it signaled that something had shifted in the urgency around menstrual culture. Despite what many assume, menstrual activism is *not* new. It has been eating at the edges of body-based shame and stigma for decades. According to Bobel (2007, 2010), "feminist spiritualist menstrual activists" broke ground in the late 1960s with their reframing of menstruation as a source of power and sisterhood. They refused the taken-for-granted assumption that menstruation was merely a nuisance, a "curse," and offered a conceptual reframing through art, including filmmaking, music, poetry, and ritual. They built upon a second wave cultural feminist sensibility that embraced rather than obscured sexual differences.

Concurrently, the women's health movement emerged, cultivating a robust resistance to the androcentric and often patronizing medical establishment. These activists spurred a healthy scrutiny of many tacit practices, especially those around reproductive health care, which led to attention to menstrual and menopausal health care. Animated by a need for bodily sovereignty and the tools and resources to make informed choices regarding providers, diagnoses and treatments, these activists joined consumer rights advocates when thousands of women developed Toxic Shock Syndrome (TSS) (and 38 died) as a result of a new super absorbent, fully synthetic tampon call Rely ("It Even Absorbs the Worry" was its tagline). Rely's makers, Procter & Gamble, pulled the tampon from the shelves because it was an aggressive incubator for the potentially lethal bacterial strain staphylococcus aureus. This tragedy led activists to work with the US government to better regulate the industry, leading to mandated TSS warnings in tampon packages and standardized absorbency ratings to help consumers choose the appropriate tampon to meet their needs.

Some environmentalists also became menstrual activists, adding to the critique of conventional care by bringing to light the polluting effects of single use menstrual care products and promoting greener alternatives such as organic tampons and pads, reusable cloth pads, cups, and sponges. Further, as third-wave feminism took shape, it found alignment with Punk and anarchism's anti-capitalism and Do It Yourself ethos (Leblanc 1999; Marcus and McKay 2010). This intersection became a site for the emergence of what Bobel terms the "radical menstruation" wing of the movement (Bobel 2010). Alternative products were championed, as well as free bleeding, as were zine making, early blogging and menstrual health education. Creative actions such as tampon art and tampon "send backs," in which activists return menstrual products to their manufacturers, defined menstrual activism through the turn of the century.

Another hallmark of the radical menstruation activist approach was its disinterest in structural reform. While earlier activists worked tirelessly with both government and industry to protect consumers, radical menstruation activists took a more cultural approach, instigating attitudinal change through visual art, performance, and humor. While feminist spiritualists promoted personal transformation through a celebration of menstruation, radical menstruation focused on building a more inclusive movement and "undoing gender," in the words of theorist Judith Butler (1990). Some began to use the word 'menstruator,' a term that embraced anybody that menstruated, heretofore assumed to be women. This linguistic move splits the biological (menstruation) from the socio sociocultural (the social construction of gender).

Today's menstrual activism is difficult to categorize, revealing its dynamism and its responsiveness to an ever-globalizing world where boundaries around identities are shifting. Accordingly, scholarship on menstrual activism is emerging and complicating definitions and categorizations (see Barkardóttir 2016; Fahs 2016; Kafai 2019; Tarzibachi 2017). Persdotter (2013), for example, situating her inquiry in western Europe, refers not to menstrual activism, but instead to the "menstrual Countermovement": "the mass of actions, and agents that purposefully work towards challenging the repressive mainstream menstrual discourse of shame and silence" (13). She argues for a more nuanced and diverse categorization that makes room for consumer-oriented change makers, such as those who produce and sell alternative menstrual products, ritualists, and others whose work is at once activist and income generating. With this historical overview in mind, we turn now to selected activist actions, moving between past and present. Our aim is to illustrate both important continuities with and key departures from menstrual activism's little known history.

Everything Old Is New Again

Laughing While Bleeding: Using Humor to Break the Silence

In 1978, Gloria Steinem penned her now-iconic satire "If Men Could Menstruate." The essay, what she dubbed "a political fantasy," first appeared in *Ms.* and has been widely and regularly republished and excerpted since. Cleverly using satiric humor, Steinem guides the reader through a thought experiment led by the question: "So what would happen if suddenly, magically, men could menstruate and women could not?" She goes on: "Clearly, menstruation would become an enviable, worthy, masculine event. Men would brag about how long and how much. Young boys would talk about it as the envied beginning of manhood. Gifts, religious ceremonies, family dinners, and stag parties would mark the day" (Steinem 1978). A few years later, "If Men Could Menstruate" was republished in a collection of feminist humor titled "Pulling Our Own Strings" (Kaufman and Blakely 1980), which included several menstrual-themed pieces in the book's lead section titled "Periodic Hysteria." The book opens with an introduction explaining what makes humor *feminist*: at once an acknowledgment of sexist oppression and a vision of change. The book includes an argument for feminist menstrual humor in which "the attitude is that menses is not to be hidden (as shameful) but to be joked about (as normal) or even celebrated (as naturally female) . . . the expression of such humor attacks the unhealthy and oppressing idea cultivated for thousands of years that women's bodies are foul" (Kaufman 1980, 14).

This claim similarly animates contemporary menstrual activist humor. The Crimson Wave is a stand-up comedy duo of Natalie Norman and Jess Beaulieu from Toronto, Canada that describes their feminist podcast as being "about periods/vaginas/Beyonce's vagina where guests tell hilarious stories, anecdotes, and theories about their lovely menses." In this same vein, UK-based solo comedian, educator and activist Chella Quint performs her gender-inclusive show aimed at menstrual stigma through playful deconstruction of vintage menstrual product ads. In her shows, she introduces her tongue-in-cheek "product": Stains.™ The use of the bright red stain—as earrings, as cufflinks, as felt stains ready to be pinned to any part of the body—deploys the visual to challenge menstrual invisibility. The use of the visual has been a mainstay of menstrual activists since the dawn of the movement.

The Visual Is Political: Menstrual Art as Change Agent

In 1971, Judy Chicago, feminist art pathbreaker, created a shocking photolithograph: a self-portrait of Chicago, legs spread, withdrawing a tampon from her vagina. Titled "Red Flag," the piece, reports Chicago, represented something so profoundly absent from our visual landscape that many people assumed the object was a penis.

The following year, Chicago further explored menstrual realities when she created "Menstruation Bathroom" in her multiroom installation collaboration with Judith Schapiro, "Womanhouse." In Menstruation Bathroom, numerous tampons and pads (both single use and reusable cloth), used and unused, are strewn throughout the room. They sit on shelves, piled in the trash bin, on the floor and hanging on the clothesline. Here, one encounters the scope and scale of menstrual care across many cycles, even years. Interestingly, the blood in this piece is relatively contained—only a few splatters appear on the floor. The lid to the toilet is closed.

Some forty years later, self-described 'menstrual designer' Jen Lewis and collaborator and photographer Rob Lewis lifted the lid in their work and began showing portraits of menstrual blood moving through water. As a menstrual cup user, Lewis grew fascinated by the designs she observed as she emptied her menstrual cup in the toilet. Her work has been featured in mainstream and fringe media and widely circulated across social media. In 2015, she curated the largest art show featuring menstrual art: Widening the Cycle: A Menstrual Cycle and Reproductive Art Show. Framing menstrual art as part of the reproductive justice movement, the show displayed the work of 38 artists from 10 countries "to disrupt the current cultural narrative and replace it with one that reflects the real thoughts, emotions and experiences of menstruators" (http://www.wideningthecycle.com/). Soon after, "Our Bodies Our Blood," "an art project about creating a safe space to express thoughts on menstruation, cultural shame and our stories," went on view in Nova Scotia. In 2017, there were two menstrually focused group art shows—"Period" in Miami, Florida (http://rojasrubensteenprojects.com/new-page/) and "The Crimson Wave: Art About and With Menstrual Blood" in Bangalore, India (http://www.boondh.co/the-crimson-wave.html). These shows are evidence that menstrual artists form a crucial part of the battle to make menstruation visible and culturally relevant.

2015 was the same year *Cosmopolitan* dubbed "the year the period went public" evinced by an illustrated timeline cataloging "The 8 Greatest Menstrual Moments of 2015." The 12-month history included Instagram's much-maligned decision to twice remove a photo of artist and *New York Times* bestselling poet Rupi Kaur lying in bed wearing period-stained pants. The photo quickly went viral. *Cosmo*'s timeline also included a reference to Kiran Gandhi's decision to free bleed while running the London Marathon. Initially, Gandhi chose to go without menstrual absorbent for ease and comfort, but as she ran, she began to consider the political implications of her act, transforming her personal decision into a moving protest, a type of circumstantial performance piece (Gandhi 2015). Gandhi has continued her menstrual activism, with a focus on menstrual product access (https://madamegandhi.blog/hi/) and legislative action. Her turn to legislative action represents a return to the roots of menstrual activism, when collaborations between feminist health and consumer rights advocates pressured the government to provide more careful oversight of the industry. While menstrual activists of the late twentieth and early twenty-first centuries showed disinterest, springing from distrust of lawmakers and corporate players, the newer crop of social change has doubled down on making change through legal reform.

Menstrual Activists (Go Back) to Washington

In some menstrual products there are trace amounts of dioxins, which are potentially carcinogenic. Recent studies have found phthalates and furanes as well (Gloaguen 2017), but the contents of menstrual products, classified in the US as a class of medical device, are not made available to the public. Since 1997, Congresswoman Carolyn B. Maloney has brought a bill before Congress nine times, and each time it languished in committee; she called the bill the Tampon Safety and Research Act. Her most recent version is the Robin Danielson Feminine Hygiene Product Safety Act of 2015, named after a woman who died in 1988 of Toxic Shock Syndrome. Maloney proposes the National Institute of Health fund research on the synthetics and chemicals used in menstrual products and test for reproductive problems as well as common health conditions "to close this research gap" (O'Hara 2015). In addition, the FDA would check manufacturers' data and make this information available to the public (O'Hara 2015). Maloney's bill is now joined by US Rep. Grace Meng's introduction of the Accurate Labeling of Menstrual Products Act of 2016, which calls for menstrual product companies to disclose the ingredients of their products.

As menstrual activism gains force, some legislators are linking issues of safety to broader issues of access and affordability. In New York City, Mayor Bill de Blasio signed legislation to provide public school grades 6–12, homeless shelters, and correctional facilities with free pads and tampons. US Rep.

Grace Meng proposed a second menstrual-related bill: the five-part Menstrual Equity Act of 2017, the product of a recent spate of interest and energy around the cost of menstrual products. The fact that many cities, states, and countries tax menstrual products, commonly referred to as the "tampon tax," has become a lightning rod for those frustrated by gender inequality. Considering that items such as men's shaving cream, Viagra, candy, and soda do not carry this added tax, it is easy to regard the extra cost as a form of discrimination against females qua females. For menstruators, menstrual supplies are as essential as toilet paper.

No doubt, the reach of social media and an evolving public attitude toward menstruation facilitated high profile efforts such as Jennifer Weiss-Wolf's joint campaign with *Cosmopolitan* magazine; "Stop Taxing our Periods! Period." To date, it has garnered over 68,000 signatures. Weiss-Wolf coined the phrase 'menstrual equity,' to frame "what it means to consider the ability to manage menstruation in the context of full democratic and civic participation" (Jones 2016a). Menstruation, she claims "transcends all the other things about women's bodies that make us targets for the right, and this one doesn't" (Arriaga 2017). This may explain bipartisan support for removing the 'tampon tax' in a climate where other reproductive issues garner little Republican support.

Access and Affordability: Today's Menstrual Activism's Rallying Cry

In a marked departure from their predecessors who, for the most part, had not yet taken stock of the particular realities of various populations, today's activists are keen to address the menstrual needs of the marginalized. For low-income, homeless, and incarcerated menstruators, removing a tax or testing for potential health risks is less a priority than gaining access to menstrual products. A 2011 Feeding America Survey places menstrual products among the top eight basic essentials; however, they are not covered by food stamps. Many low-income menstruators cannot afford the products they need—even without taxes in certain places. The cost is estimated at approximately $2500 over the course of a lifetime (Meng 2017).

Homeless menstruators face not only a lack of access to menstrual products but also to clean spaces in which to care for their menstruating bodies. A handful of campaigns and organizations are distributing menstrual care products to low-income and homeless menstruators. #Happy Period and Period: the Menstrual Movement are two such organizations, both newcomers to the movement. Many of the new activists are under 30. Their age may explain a growing sensitivity to the needs of menstruators in schools, including not only the cost of products but also access to supplies in school bathrooms.

Universities have also provided free menstrual products to students so that all students can focus on their education despite their income or level of preparedness–or gender identity. For example, a student-led initiative at Brown

University put free pads and tampons in women's, men's and gender-inclusive bathrooms around campus. Other schools—such as Emory, University of Arizona, Reed College, University of Minnesota-Twin Cities, and the University of Wisconsin at Madison—have followed suit with similar programs.

Another profoundly marginalized population—incarcerated menstruators—has become a focus of some activists. In prisons and jails, menstrual products are often rationed, restricted, traded, or used by guards in power games. The lack of menstrual supplies among inmates affects not only their hygiene but also their self-esteem. Chandra Bozelko, a former inmate, said, "To ask a macho guard for a tampon is humiliating. But it's more than that: it's an acknowledgement of the fact that, ultimately, the prison controls your cleanliness, your health and your feelings of self-esteem" (Ronan 2015). Not only is there a dearth of materials, but also the ones provided are small, poor quality and lack adhesive. Some incarcerated menstruators are forced to wear these for multiple days, which can cause bacterial or fungal infections, or lead to bleeding through clothes.

In April 2017, Colorado passed an amendment agreeing to spend $40,000 to provide tampons for female inmates. Senators Elizabeth Warren and Cory Booker introduced the Dignity for Incarcerated Women Act on July 11, 2017; the legislation proposes distributing free quality pads and tampons to inmates. Soon thereafter, and ostensibly to avoid a legislative mandate, the US Department of Justice issued an Operations Memorandum in August 2017 that "ensures that female inmates have access to a range of feminine hygiene products related to menstruation" (https://www.bop.gov/policy/om/001_2017.pdf). Because the mandate does not apply to state and municipal facilities, the fight continues. As Bozelko (2017) forcefully writes:

> Talking openly about menstruation can't be restricted to hygiene, as important as it is. The appeal to public health sidles up to the issue by making it seem like it's everyone's problem. That just gets us halfway there. The truth is that it shouldn't be anyone's problem because your period really isn't a predicament. It's proof of life.

Bozelko's observation resonates. While early menstrual activism of the past did attend to products, the focus was safety and promoting alternative, more environmentally friendly cups and cloth pads in a context of creative resistance to menstrual invisibility. Today, the product focus dominates the activist landscape, one centered on product access and affordability. While these are worthy projects indeed, they are narrowly focused on what Sharra Vostral (2011) calls "technologies of passing," that enable menstruators to "pass" as non-menstruators in order to comply with the cultural norms—keep it hidden, keep it quiet. As such, products serve merely to accommodate rather than resist the menstrual mandate of shame, silence, and secrecy. In this way, contemporary menstrual activism has dulled its radical edge through a neoliberal engagement with menstrual *management*. When menstrual activism's main focus devolves to a preoccupation with "something to bleed on," it betrays its feminist roots of challenging the misogynist framing of the polluted and disgusting menstrual body.

HAZARDS AND POSSIBILITIES OF THE WORK

Identity Politics

Menstrual activism is uniquely positioned to push feminists and other critically minded scholars and activists toward thinking about the complexities of gender and the body. That said, the hazards of doing justice to the menstrual experience—particularly with regard to gender identity—have been particularly messy in recent years. Menstruation is, at once, deeply gendered and coded as women's experience and also expansive and transgressive in its gender politics. Many women menstruate but not *all* women menstruate. Some women, such as women who have gone through menopause, those who have had hysterectomies, pregnant women, young women, women with severe eating disorders, women suppressing their periods, women on certain kinds of birth control, and some competitive athletes do not menstruate; this then complicates simplistic notions that *all women menstruate.*

Similarly, *some* men do menstruate. Trans men often still menstruate, particularly if they are not on testosterone hormone therapies. Further, trans men who menstruate often report distress and body dysmorphia during their menstrual periods, while other trans men associate menstruation more positively (Fahs 2016; Reading 2014). Holly Devor (1999) found that trans men reported intensely negative emotions about menstruation, with 51% saying they felt emotional discomfort. Trans men and transmasculine people's menstrual experiences are under-researched, in part because menstrual bleeding is not typically associated with cultural ideals about masculinity. This can lead not only to trans men feeling distress about their periods but also, as one recent study found, avoiding restrooms and working toward menstrual suppression (Chrisler et al. 2016).

Endometriosis— where tissue that normally lines the uterus grows outside of the uterus and causes pain—also complicates the 'who menstruates?' question. Cara Jones (2016b) has argued that not all bodies with endometriosis are female and that endometriosis has been found in infants, postmenopausal bodies, post-hysterectomy bodies, trans men, and cisgender men. At the same time, trans women and transfeminine people's experiences who undergo hormone replacement therapy often report "menstrual" symptoms and pain such as soreness, swelling, nausea, cramping, dizziness, migraines, muscle fatigue, joint pain, bloating, depression, and mood changes on a cyclical basis; researchers have largely ignored this (Riedel 2016).

The move to recognize the breadth of the menstrual experience has appeared more vividly in recent years. While little research has examined non-binary gender identity and menstruation, some activist and "artivist" work has started to make room for these experiences; for example, Cass Clemmer (https://www.tonithetampon.com/) developed the character "Toni the Tampon" for a coloring book about menstruation in order to broaden cultural ideas about who menstruates. In July 2017, Clemmer posted a photo on Facebook that showed Clemmer with obvious menstrual blood on their pants; the photo went viral and inspired a series of news articles and commentaries (Dupere 2017).

The move from gendered language to non-gendered language creates a conversation that, on the one hand, expands notions of "who menstruates" and, on the other hand, erases some of the ways menstruation is coded as a cisgender female experience. Some menstrual activists argue that non-gendered language of "menstruators" should be used in tandem with "women" and "girls" as a way to both broaden the language of menstruation while still marking menstruation as feminized and menstrual negativity as grounded in misogyny (Przybylo and Fahs 2018). How to best challenge the frank sexism of much menstrual discourse remains an open question that menstrual activists are continuing to take up. Expanding the existing circle of menstruators better represents the wide swath of people affected by menstrual cycle changes. Thus, "Menstrual bleeding in this sense is complex: it is both highly gendered and not attached as a material reality to only one gender" (Przybylo and Fahs 2018).

Additionally, menstrual activism has also grown savvier in recent years in its approach to race and class diversity, particularly as menstrual activists recognize that different groups of menstruators have different needs. There has been a major push toward thinking about menstrual health as more of a *global* issue rather than a solely Western issue; this means that menstrual activists may simultaneously work within the contexts of the US and the Western world and in the Global South. In the last several years, a development subsector referred to as "Menstrual Hygiene Management," has rapidly proliferated. Using human rights (Boosey and Wilson 2014; WASH United and Human Rights Watch 2017; Winkler and Roaf 2014) and public health frames (Sommer et al. 2015), MHM advocates seek to challenge menstrual shame, silence, and stigma through initiatives that provide girls and women access to menstrual care products, improved infrastructure (access to toilets, water, and soap) and puberty education.

Policy efforts are sometimes engaged as well, such as persuading governments to provide menstrual care resources—including menstrual products and menstrual health education—in government schools (Bobel 2015). For example, Menstrual Hygiene Day now has a global platform with 350 events in 54 countries as of 2017, including educational events in schools, community rallies, concerts to raise awareness about menstrual health needs, advocacy workshops with governments, and product donations. India was the most active of all nations with 67 separate events. Online media coverage has been abundant, with pieces in *Huffington Post, The Guardian,*

El Pais, Metro, and *Glamour*, just as digital campaigning has been abundant and successful. All key development partners working in the field of menstrual health management participated in menstrual hygiene day (for example, UNICEF, WaterAid, WSSCC, Global Citizen, USAID, PLAN International, and PATH). These efforts are part of a larger global trend of centering development efforts on girls, a move that has garnered some feminist critique as overly instrumentalist and only superficially focused on girls' most acute needs (Hayhurst 2011; Koffman and Gill 2013).

As discussed earlier, social class issues have also been foregrounded in the recent policy work of menstrual activists, particularly as menstrual activists recognize the importance of addressing underserved populations like homeless and incarcerated menstruators. Initiatives to give homeless menstruators access to menstrual products have begun, as have policies to make high-quality tampons and pads available to those in prison. Beyond this, the push toward reusable products as a *class-based* issue has also occurred; rather than framing reusable products only as an environmentally friendly choice, menstrual activists have encouraged women and other menstruators to see these products as ways to cut ties with corporate control over periods *and* save money while doing so (Edwards 2015; Mok 2004).

Corporate and Media Appropriation

As with any resistance movement, one of the dangers of pushing for progressive social change is that the work is often swiftly appropriated, distorted, and/or used for unintended purposes. For example, for years, menstrual activists critiqued the use of 'blue liquid' to signify menstrual blood, noting that it distorted the ordinariness of the menstrual experience by erasing blood. This has recently been taken up in an advertisement for Kotex that makes fun of the blue liquid commercials in order to *sell* disposable tampons and pads (https://www.youtube.com/watch?v=lpypeLL1dAs).

Similarly, activists' efforts to promote the empowerment of women have been distorted by corporate and media entities as "girl power" packaged to sell products, particularly in menstrual advertisements. Products that purport to "empower" women, like tampon subscription services, often end up merely recreating menstrual shaming and taboo; for example, Club Monthly advertises, "Feminine products at your door, without the shame of the store" (Davis 2014). Many menstrual products have "better and better" technologies that serve to better hide or mask menstruation. Tampon companies inject scented perfumes into tampons to "deodorize" the vagina, just as disposable pad companies design products to be better absorbent and "leak proof." The language of appropriation used to sell tampons and disposable pads exemplifies one of the hazards of menstrual activist work and serves as a reminder of why menstrual activists must always remain one step ahead of such corporate and media appropriation.

Hostilities and Trivialization

Menstrual activists have also faced hostilities from right-wing internet trolls, bloggers, and journalists for the work they do, resulting in painful clashes about the value of making menstruation more public. Given that menstrual activist work is often in the public eye, such hostilities have intensified in recent years. The hostile climate of the Trump presidency and his policies and practices of xenophobia, racism, sexism, and classism have only worsened these attacks. Clemmer grappled with a host of negative reactions to their work on gender-inclusive menstrual education and their "Toni the Tampon" character (Clemmer 2017). Their work has been met with harsh criticisms and hateful, vitriolic rhetoric from some conservatives (we avoid replicating that here in order to not further and reproduce hate speech) and even some doctors. We, too, have dealt with such hostilities. Breanne Fahs's recent book cover for *Out for Blood*—which features a realistic depiction of menstrual blood running down a woman's leg—started a tweetstorm online in early 2017 after some right-wing bloggers found it "disgusting"; later in 2017, Fahs also watched

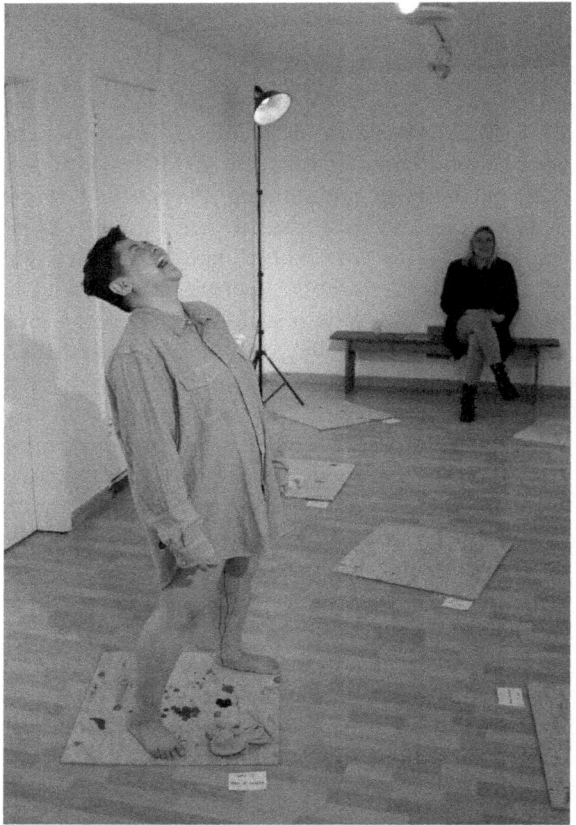

Fig. 71.1 Day 5 of "Sloughing," a 28-day performance by Raegan Truax. Pictured: Thao P. Nguyen (performing) and Raegan Truax (artist) at Royal NoneSuch Gallery in Oakland, CA. www.raegantruax. com/sloughing (Credit: Jeremiah Barber 2017)

conservative internet trolls move freely from mocking and trivializing her work on fatness narratives to mocking and trivializing her work on menstruation.

Menstrual activists also face numerous other obstacles related to the trivialization of their work, particularly as many activists work to get policies changed, funding secured, or research published. Those working within the academy face critiques that their work is not on a "serious" subject, or that it is of little academic value (Fahs et al. 2018). The taboo of menstruation, or beliefs that people should not discuss menstruation publicly, have negatively impacted public conversations about governmental policy initiatives and media coverage for menstrual activist work (Bobel 2007, 2010). Recent debates concerning menstrual leave in the workplace reveal the enduring perception of menstruation as a minor matter that should be quietly managed. This view often stems from a liberal feminist ideology that fails to engage the complexity of menstrual experience as at once biological *and* sociocultural. As long as menstruation is portrayed as trivial, silly, gross, or unimportant, much of the important work on menstruation continues to get sidelined (Fig. 71.1).

THE FUTURE OF MENSTRUAL ACTIVISM

After reviewing some of the major accomplishments of menstrual activism, and some of the recent hazards and challenges of doing menstrual activist work, we turn an eye toward the future of menstrual activism. Ultimately, we argue that menstrual activism has radical potential to deeply unsettle many assumptions about gender, bodies, political activism, embodied resistance, and feminist coalition-building. Indeed, menstrual activism has at its core a vision of radical politics that we find exciting, timely, and relevant—one that digs deep into the root structures of inequalities to expose some of the fundamentally problematic aspects of misogyny, sexism, racism, homophobia, and classism. That said, menstrual activism also has the potential to shun its political roots and move toward assimilationist or liberal politics; such a move could have many problematic consequences—for example, without radical energies menstrual activism is more vulnerable to appropriation in ways that serve corporate and/or pharmaceutical interests, just as it could become reduced to "liking your body" or "girl power."

We assert that menstrual activism has the potential to fight back against the forces that silence and shame women's bodies. In these regressive and conservative times, menstrual activism has even more relevance and importance, but if it succumbs to its more assimilationist/liberal impulses, it will diminish some of its major successes and fracture some of its major inroads as a global powerhouse. As such, we end this chapter with a call to action for the vision of (radical) menstrual activism we find most exciting and impactful:

- Menstrual activism must continue to prioritize underserved populations and engage in activism that prioritizes these groups' needs. A major thread of its work is to keep a close eye on menstruators who are overlooked and ignored. This may include menstruators who are: incarcerated, disabled, trans/non-binary, homeless, low-income, HIV-positive, refugees, pre-menarche, indigenous, in the military, outside of Western contexts, rural, and who have undergone forced sterilization.
- Menstrual activists must continue to link arms with other progressive social movements and groups, forging alliances that will be productive and co-constructed. This may include groups that prioritize: anti-xenophobia, anti-racism, sustainability, homelessness, LGBT justice, transnational feminism, disability rights, immigrant rights, anti-femicide, sexual violence prevention, education, reproductive justice, abortion rights and the social class struggle.
- The notion of what menstrual activism is, and the work these activists do, must move from a primary focus on menstrual products and instead emphasize a much wider array of menstrual activist priorities including mental health, global feminisms, cultural critiques, humor and mockery and educational changes as means to promote menstrual literacy and fight stigma.
- Menstrual activism must prioritize work that speaks to the immediate, urgent issues of the day, just as they must work on long-range projects with long-term impacts. This means that menstrual activists have a special responsibility to see their work as connected to the political climate of their time. Some ways that menstrual activists could directly combat the existing political climate of repression, xenophobia, misogyny, and classism include: flash mob workshops, tweeting (#periodsarenotaninsult), free bleeding in public as a form of protest, making snarky or satirical menstrual ads, menstrual performance art (see Raegan Truax's 2017 piece called *Sloughing* for an example), engaging in open dialogue about menstruation, turning toward old-fashioned consciousness-raising, educational disruptions, art and storytelling, menstrual humor and clowning, and remembering that we can use our everyday bodies as a form of political protest.
- Menstrual activism must resist letting go of its radical impulses; it must always push to become *more* radical. This means thinking deeply about the root structures of menstrual negativity and taboo, and working to link the inequalities that surround menstruation to deeper stories about power and identity. For example, menstrual activists could work to link menstrual activist work to conscious capitalism, pushing back against co-optation of menstrual activist work, confronting and changing menstrual narratives, working on connecting other kinds of policies (for example, breastfeeding, menstrual leave) to existing activist work, continuing to engage in internal critiques of the work ("productive unsettling"—see Bloomfield 2015), engaging in menstrual stunts, challenging men's attitudes about menstruation, connecting policy change to art and cultural change, and using humor widely and wisely.

- Ultimately, menstrual activism must resist appropriation by remembering to eschew respectability, commodification, and neoliberalism in favor of other tactics and ideologies: risky actions, unexpected coalitions, non-self-defeating self-criticism, sabotage, cross-contamination between scholarship/art/activism, and an insistence on strengthening connective tissue between menstrual activists and their allies.[2]

NOTES

1. "The Messy Politics of Menstrual Activism" by Chris Bobel and Breanne Fahs was first published in 2018. In Reger, J. (Ed), *Nevertheless, They Persisted: Feminisms and Continued Resistance in the U.S. Women's Movement*. New York, NY: Routledge, 151–169. No further reproduction or distribution of the material is allowed without permission from the publisher.
2. Special thanks to Eric Swank, Jessica Brofsky, Kimberly Koerth, Ayanna Shambe, and the Feminist Research on Gender and Sexuality Group for their contributions to this manuscript. We also thank the participants at our June 2017 Society for Menstrual Cycle Research workshop on menstrual activism for their thoughtful ideas about the future of menstrual activism.

REFERENCES

Arriaga, Alex. 2017. "Tampons in Men's Rooms? It's Just a Small Part of 'Menstrual Equity,' Campus Activists Say." *The Chronicle of Higher Education*. http://www.chronicle.com/article/Tampons-in-Men-s-Rooms-/240091.

Barkardóttir, Freyja Jónudóttir. 2016. "Hope of Failure: Subverting Disgust, Shame and the Abject in Feminist Performances with Menstrual Blood." Masters Thesis, Central European University.

Bloomfield, Mandy. 2015. "Unsettling Sustainability: The Poetics of Discomfort." *Green Letters* 19 (1): 21–35.

Bobel, Chris. 2007. "'I'm Not an Activist, Though I've Done a Lot of It': Doing Activism, Being Activist and the 'Perfect Standard' in a Contemporary Movement." *Social Movement Studies* 6 (2): 147–59.

———. 2010. *New Blood: Third-Wave Feminism and the Politics of Menstruation*. New Brunswick, NJ: Rutgers University Press.

———. 2015. "Finding Problems with Simple Solutions: Ethnographic Insights into Development Campaigns to Support Menstruating Girls." National Women's Studies Association Annual Conference, Milwaukee, WI.

Boosey, Robin, and Emily Wilson. 2014. "A Vicious Cycle of Silence." Research Report. ScHARR Report Series (29). School of Health and Related Research, University of Sheffield.

Bozelko, Chandra. 2017. "Federal Prisoners Now Get Free Pads and Tampons. Our Battle Isn't Over Yet." *The Lily*. https://thelily.com/federal-prisoners-now-get-free-pads-and-tampons-our-battle-isnt-over-yet-93d4278b1e5.

Butler, Judith. 1990. *Gender Trouble: Feminism and the Subversion of Identity*. New York: Routledge.

Chrisler, Joan C., Jennifer A. Gorman, Jen Manion, Michael Murgo, Angela Barney, Alexis Adams-Clark, Jessica R. Newton, and Meaghan McGrath. 2016. "Queer Periods: Attitudes toward Experiences with Menstruation in the Masculine of Centre and Transgender Community." *Culture, Health & Sexuality* 18 (11): 1238–50.

Clemmer, Cass. 2017. Toni the Tampon. Presented at the *Society for Menstrual Cycle Research Conference*, Kennesaw, GA, June 24.

Davis, Lisa Selin. 2014. "In 2014, Startups Discovered Women Get Their Periods—Every Month." *QZ*. https://qz.com/310086/in-2014-startups-discovered-women-get-their-periods-every-month/.

Devor, Holly. 1999. *FTM: Female-to-Male Transsexuals in Society*. Bloomington: Indiana University Press.

Dupere, Katie. 2017. "Trans Artist Destroys Period Stigma with One Seriously Bold Facebook Post." *Mashable*. http://mashable.com/2017/07/22/transgender-periods-menstruation-facebook/#wsxYbPkmbkqm.

Edwards, Charlotte. 2015. "Ladies, We Found a Way to Help You Save $100 a Year. Period." *The Penny Hoarder*. https://www.thepennyhoarder.com/smart-money/reusable-menstrual-supplies/.

Fahs, Breanne. 2016. *Out for Blood: Essays on Menstruation and Resistance*. Albany, NY: State University of New York Press.

Fahs, Breanne, Rebecca F. Plante, and Sara I. McClelland. 2018. "Working at the Crossroads of Pleasure and Danger: Feminist Perspectives on Doing Critical Sexuality Studies." *Sexualities* 21 (4): 503–519.

Gandhi, Kiran. 2015. "Sisterhood, Blood and Boobs at the London Marathon 2015." https://madamegandhi.blog/2015/04/26/sisterhood-blood-and-boobs-at-the-london-marathon-2015/.

Gloaguen, Audrey. 2017. "Tampon: Our Closest Enemy." Documentary Film Dreamway Productions.

Hayhurst, Lyndsay L. 2011. "Corporatising Sport, Gender, and Development: Postcolonial IR Feminisms, Transnational Private Governance, and Global Corporate Social Engagement." *Third World Quarterly* 32 (3): 223–41.

Jones, Abigail. 2016a. "Periods, Policy and Politics: Menstrual Equity Is the New Thing." *Newsweek*, May 8, 2017. https://www.newsweek.com/periods-policy-and-politics-menstrual-equity-new-thing-596027.

Jones, Cara. 2016b. "The Pain of Endo Existence: Toward a Feminist Disability Studies Reading of Endometriosis." *Hypatia* 31 (3): 554–71.

Kafai, Shayda. 2019. "Blood as Resistance: Photography as Contemporary Menstrual Activism." In *Body Battlegrounds: Transgressions, Tensions, Transformations*, edited by Chris Bobel and Kwan Samantha. Nashville, TN: Vanderbilt University Press.

Kaufman, Gloria. 1980. "Introduction." In *Pulling Our Own Strings: Feminist Humor & Satire*, edited by Gloria J. Kaufman and Mary Kay Blakely, 13–16. Bloomington, IN: Indiana University Press.

Kaufman, Gloria J., and Mary Kay Blakely. 1980. *Pulling Our Own Strings: Feminist Humor & Satire*. Bloomington, IN: Indiana University Press.

Koffman, Ofra, and Rosalind Gill. 2013. "'The Revolution Will Be Led by a 12-Year-Old Girl': Girl Power and Global Biopolitics." *Feminist Review* 105 (1): 83–102.

Leblanc, Lauraine. 1999. *Pretty in Punk: Girls' Gender Resistance in a Boys' Subculture.* New Brunswick, NJ: Rutgers University Press.

Marcus, Sara, and McKay, Julie. 2010. *Girls to the Front: The True Story of the Riot Grrrl Revolution.* New York: Harper Perennial.

Meng, Grace. 2017. "Meng Renews Effort to Make Menstrual Hygiene Products More Accessible and Affordable to Women." *Media Center.* https://meng.house.gov/media-center/press-releases/meng-renews-effort-to-make-menstrual-hygiene-products-more-accessible.

Mok, Kimberley. 2004. "7 Powerful Reasons Why You Should Switch to Reusable Menstrual Products." *Tree Hugger.* https://www.treehugger.com/health/reasons-why-you-should-switch-to-reusable-menstrual-products.html.

O'Hara, Mary E. 2015. "'Robin Danielson Act' Would Mandate Independent Testing on Tampon Safety." *Rewire.* https://rewire.news/article/2015/04/21/robin-danielson-act-mandate-independent-testing-tampon-safety/.

Persdotter, Josefin. 2013. "Countering the Menstrual Mainstream: A Study of the European Menstrual Countermovement." Masters Thesis, Götesborg Universitet.

Przybylo, Ela, and Breanne Fahs. 2018. "Feels and Flows: On the Realness of Menstrual Pain and Cripping Menstrual Chronicity." *Feminist Formations* 30 (1): 206–29.

Reading, Wiley. 2014. "My Period and Me: A Trans Guy's Guide to Menstruation." *Everyday Feminism.* http://everydayfeminism.com/2014/11/trans-guys-guide-menstruation/.

Riedel, Sam. 2016. "Yes, Trans Women Can Get Period Symptoms." *The Establishment.* https://theestablishment.co/yes-trans-women-can-get-period-symptoms-e43a43979e8c.

Ronan, Alex. 2015. "Menstruation Can Become Humiliation in Prisons." *The Cut,* June 16. https://www.thecut.com/2015/06/menstruation-can-become-humiliation-in-prisons.html.

Sommer, Marni, Jennifer S. Hirsch, Constance Nathanson, and Richard G. Parker. 2015. "Comfortably, Safely, and Without Shame: Defining Menstrual Hygiene Management as a Public Health Issue." *American Journal of Public Health* 105 (7): 1302–11.

Steinem, Gloria. 1978. "If Men Could Menstruate." *Ms Magazine,* October, 110.

Tarzibachi, Eugenia. 2017. *Women's Things: Menstruation, Gender and Power.* Argentina: Penguin Random House.

Vostral, Sharra L. 2011. *Under Wraps: A History of Menstrual Hygiene Technology.* Lanham, MD: Lexington Books.

WASH United and Human Rights Watch. 2017. "Understanding Menstrual Hygiene Management and Human Rights." *Human Rights Watch.* https://www.hrw.org/sites/default/files/supporting_resources/mhm_practitioner_guide_web.pdf.

Winkler, Inga, and Virginia Roaf. 2014. "Taking the Bloody Linen Out of the Closet: Menstrual Hygiene as a Priority for Achieving Gender Equality." *Cardozo Journal of Law and Gender* 21 (1): 1–37.

CHAPTER 72

Transnational Engagements: Women's Experiences of Menopause

Edited by Milena Bacalja Perianes and Elizabeth Arveda Kissling

BACKGROUND

The questions that form the basis of these discussions were inspired by a conversation about menopause that Elizabeth had many years ago with her closest aunt, who was near her fiftieth birthday at the time. Aunt Patricia is the youngest of five sisters (including Elizabeth's mother) and had sought information and advice to prepare for menopause from her older sisters and was frustrated by their responses. 'They all say that it was nothing or they don't remember!' she told Elizabeth in disbelief. 'How is that possible?' Even knowing that experiences of menopause are highly variable, Elizabeth agreed that it seemed unlikely that all four women had completely uneventful transitions, with none of the common signs they had both read about: irregular or heavier periods, hot flashes, night sweats, headaches, mood changes, insomnia, dry skin, or weight gain. Nevertheless, Patricia—and Elizabeth, ten years later—was left to her own resourcefulness to navigate menopause.

As chief innovation and research officer of the international organization Menstrual Health Hub, Milena managed connections with NGO directors, health practitioners, and other community leaders who facilitated conversations with participants in four nations to explore questions about menopause knowledge and experiences. We are extremely grateful for their assistance and insight and especially appreciative of the women and men who spoke with them about menopause. As readers will soon learn, this was not an easy task for many of the participants or questioners.

The first discussion comes from a transcript Shardi Nahavandi sent us of a conversation about menopause between her and her Persian mother (57). Her mother grew up in a traditional, rather conservative family in Iran. She now lives in London, and Shardi grew up between London and Tehran.

© The Author(s) 2020
C. Bobel et al. (eds.), *The Palgrave Handbook of Critical Menstruation Studies*, https://doi.org/10.1007/978-981-15-0614-7_72

Shardi reported that talking to her mother about menopause and her health was an emotional and awkward conversation, and not because it was in Farsi and she translated it into English for our benefit.

> I found the difficulty not in the physical language barrier, but the ability for her to express herself, and me to really understand her experiences. If I am honest, my mum was more open than me. I almost didn't want to hear what she had to say because it was painful to think about her in pain and distress. I don't know how to help her and in our culture it's not appropriate for a daughter to talk to her mother about these experiences. I've thought about it a lot since we spoke.

Our second source is a conversation Swetha Sridhar recorded between three Indian women from one family: a grandmother (76), her daughter (55), and her granddaughter, Swetha (27), the interviewer. Her grandmother grew up in Kerala, a small state on the western coast of India. She experienced menarche and menopause in the same house. Her daughter, Swetha's mother, traveled the world, finally landing on the opposite coast of India, in the city of Chennai, where she dealt with both the onset of menopause and widowhood at the same time.

A third source comes from Ursula Maschette Santos, who works in health promotion for women and girls. She conducted a group interview with three women (48–57) in various phases of their menopause transition from her community in São Paulo, Brazil.

Finally, Jennifer Poole conducted a focus group discussion with 17 participants of the NGO Medical Services Pacific in Fiji. The 11 women and 6 men represent the two largest ethnic groups of Fiji: native Fijians or Melanesians (Itaukei) and Indo-Fijians, who have Indian descent.

Meanings of Menopause

Our Persian participant succinctly summarized her view of menopause in Iran: 'It is very negative, no one likes to talk about it and everyone wants to pretend such a thing doesn't exist.' Itaukei women of Fiji did not necessarily view menopause as positive or negative within their community, just as something that happened: 'Nothing else changes.' Itaukei women, however, acknowledged that menopause is a taboo subject, stating it is only discussed between close female relatives and that men were not included in discussions of women's health. One woman said shame is associated with women's health in general, and 'no one wants to talk about it.' Another said menopause is not spoken about at all: 'You only learn about menstruation, and then menopause as you go through it.' Women in Iran and Fiji sound just like Elizabeth's American aunties!

Jennifer Poole told us that the community members she spoke with stated they did not know how menopause was understood in the community because 'no one ever spoke about it publicly.' She found this especially salient for Indo-Fijian women, whose cultural norms continue to support silence around menstrual and vaginal health. When asked whether a woman's position in society changes as a result of menopause or if it changes the way they see themselves, a 55-year-old woman from India responded, 'It's not publicized. At least not in our community. How would anyone know? Unlike when you get your period first, there is some little bit of celebration or whatever, when you stop your period, there is no announcement or public gathering.'

More specifically, women of several nations and communities discussed menopause in terms of the end of reproductive life, as a sign of aging, or both:

It means women are old! (Fiji)

Menopause means old age! (Laughs.) I don't think there are any positive or negative connotations as such. Probably because we don't talk about it at all. But when I was reading about it, I realized that there are many associated things that can make life tiresome, like hot flashes and dry skin. I've only heard older women talk about this, but not menopause itself. (India)

Women have very little value here. After menopause, well, this makes us much less valuable. When you are young, women are seen as reproductive machines and then with menopause you completely lose that one function that society associates with you. We lose the little value we had. (Iran)

It is very sad [that there are] stigmas and taboos around menopause in our society. With the menopause we start to feel old, like a grandma. It gets worse because the children you have, your sons, expect you to look after your grandchildren and not to take care of yourself, your body. (Brazil)

We as women do not usually have a good position in society, our position is to be mothers, and with menopause it doesn't get any better, even if your children are already adults. I have already heard the phrase, "Come on your phase as a young lady has gone, do not use this short skirt, you have to behave yourself in this way, you are an old lady now." The push back women receive happens in all different stages of life, however, it does not stop with the onset of menopause. Society keeps pushing women to behave in certain ways, saying how she should dress up, behave and live. (Brazil)

As the final quotation shows, for some women reaching the age of menopause is accompanied by warnings or advice of how a proper older woman should behave: 'Be quiet, not sexy,' as one Brazilian respondent summarized. Even though their reproductive years are past, women in Brazil still experience the pressure to look and act like a good mother. One woman in Fiji felt her value in society was diminished, as she was still working outside the home after menopause, which is unusual in Fiji. She was teased by her sisters-in-law about what she wore to work and about 'still trying to find a man,' which was very upsetting to her. Other women in the Fiji focus group spoke of not wanting to be viewed as 'old grandmas' with no sex drive, saying, 'Women still want to be sexy and noticed, even if their sex drive is down.' At the same time, there was significant concern in the group regarding sexual obligations in heterosexual relationships, and the ways 'men don't understand why we don't have sexual desire.' This created some friction in relationships, with some women feeling unable to talk with their partners about sex and menopausal changes: 'Relationships can change between husband and wife because of decreased sexual desire. Some men don't listen, think women are making excuses.'

Women everywhere also praised the many benefits of menopause. The end of reproductive capacity represents sexual freedom for some; one Brazilian woman said she is now free to 'explore my sexuality without the pressure and fear of getting pregnant.' To some women, menopause means the onset of their sexual liberation, where they feel freer to explore their sexuality in many different ways. 'Sexuality means freedom during my menopause,' said one of the women from Brazil. 'I feel freer to have a sexual relationship with a person and this opens other types of freedom. What was once was a barrier to me, now no longer is. I am more open now and much more open-minded.' Another Brazilian mentioned that her 60-year-old friends carry condoms in their bags, something unthinkable some years ago.

Menopause brings other freedoms as well. Indo-Fijian women stated there were various religious taboos around menstruation that impacted both menstruation and menopause: 'A Muslim woman cannot pray while bleeding, for example!' Postmenopausal women do not face this restriction. One woman from India said, 'In many ways, it gives us freedom, where otherwise you may not be allowed in public spaces, you can now do what you like.' Another Indian woman said, 'A lot of us felt relief—a freedom from the cultural norms that governed our period, the lack of infrastructure and technology needed to cope with our period hygienically, and just the quiet stigma of it. It felt like it was all lifted.' Sometimes the relief was expressed in concrete and specific terms: One Itaukei woman in Fiji said it had taken pressure off her family, as 'pads are expensive and need to be prepared for in time of menses.' She said she normally stayed home to do household duties, so it was challenging for her when she had her menses to go out and buy pads. Brazilian women also spoke of freedom from menstrual discomfort, such as cramps, and the expense of menstrual products. Another woman in Fiji said, 'I can swim with my friends and never have to worry about my monthly friend.'

These internal positive changes are not necessarily perceived externally in these women's communities. Indo-Fijian and Itaukei women say that after menopause, women are seen as old and no longer as valuable in society: 'We are just there to advise but not to participate.' Many in this group worried specifically about their sexual drive and desirability; one said, 'I cannot play my role as wife in satisfying my husband's needs.' However, the men in the group stated that women's status and participation in society does not change. This discrepancy could be due to a lack of understanding around menopause. Ironically, none of the male participants expressed a view of menopause as negative or positive; rather, they seemed to see it as a natural part of life. A woman in India, now long past menopause, remembers, 'I had a colleague who had a hysterectomy and therefore hit menopause very young. I remember her husband publicly mocking her, and asking what kind of a woman she was, if she didn't even have her period.'

LEARNING ABOUT MENOPAUSE

Most of our information came from overhearing conversations among the elder women in the house, rather than a focused conversation with our mothers or our doctors. We heard and saw what was happening at home and assumed that it was natural and would happen to us. We didn't make a fuss about these things as you tend to today; menses happens, menopause happens, life goes on for a woman. What's there to talk about? (India)

The quote above from an Indian woman in her mid-70s could easily be Elizabeth's mid-70s mother in the United States or this grandmother in Iran:

My parents never talked about it as they were very conservative and you never speak about this topic, or other similar ones. My mother is still alive but I have no idea when exactly she experienced menopause. There is no education around this in Iran. When it happens you have no idea what is happening to you or how you should tackle it. I am going through it now and my husband has no idea what I'm going through. Although I tell him he just can't grasp the concept or have any understanding towards it.

Although Brazilian women interviewed for this project have learned about menopause through informal conversations among friends and observing the experiences of their close relatives, they, too, say there is 'silence' and 'loneliness' around the topic and they do not feel comfortable talking about it:

We learned about menopause in a very bad way, because we did not even learn from our mother, as their generation also do not know much about it. Even though women are more open today, women generally do not like to say that they are entering on menopause. (Brazil)

I used to hear that our period was going to stop coming at some point around a certain age, however, no one taught me how to handle it. So, I went to search on my own. Only when the symptoms started coming, and my period began to get irregular, did I start searching more about this subject. (Brazil)

Within the Fiji focus group, women reported learning about menopause through different channels. Many Itaukei women had learned about menopause through female family members and female friends. While three of the women (two Itaukei and one Indo-Fijian) had learned about it through medical professionals, this was largely due to broader experiences with health professionals regarding their reproductive health. Two women had total hysterectomies and understood their reproductive health more comprehensively compared to other women in the group. Of the 11 female participants, one woman of Indian descent had never learned about menopause and required further education on when and how it begins. Most of the women stated that menopause wasn't widely spoken about and was for women to discuss only. None of the woman had ever spoken about it with male relatives or friends. The older men in the group (46–65) understood that menopause is 'when the bleeding stops' and only knew about it because of their wives' experiences. However, younger men (40–45) stated they had been taught about menopause through secondary education and female relatives. The younger men displayed greater understanding of menopause-related signs and symptoms compared to their elders.

It is worth noting that the most well-informed in the group had developed knowledge through reproductive health crises. A woman from India also reported hysterectomy as a source of knowledge about menopause: 'About menopause?! Hmm. Probably from the elders in the family, from when they went through menopause. Just this sense that it [periods] would stop, after a certain age.

Continuing Conversations

Part of the difficulty and challenge of menopausal changes is experiencing and managing them in isolation. Several women in the Fiji group stated that having a supportive husband/partner helped them through menopause, but the interviews and conversations shared from all four nations revealed a great deal of silence around menopause. A woman from Iran currently in perimenopause states, 'I feel absolutely horrified for the journey I have ahead of me. Also I am away from my children so it makes it even more difficult to cope with. On top of that I am caring for my 80-year-old mother, which makes it even more difficult to get any support for myself.' Her British-Iranian daughter says,

> When I speak to my Persian friends about our mothers and menopause, they all say the same thing: "She has gone crazy!" I know we shouldn't talk like that but it's hard to grasp. At least now I know. We are talking more about it. I am trying to be more patient. I also suggested we go see a menopause specialist in London. I am helping her to navigate the health services to help her through it all. The crazy thing is she doesn't talk to any of her friends about it.... You get education from women on these sorts of things—women's bodies and health. . . . This is a woman's issue, and we aren't supporting or talking to each other.

Despite the epithet 'crazy,' this quote reflects the transformation in attitude that is possible when people begin to talk about openly about menopause. A single conversation sparked a change in this daughter's response to her mother's menopausal moods and behavior.

FEELING MENOPAUSE

> I am going through menopause at the moment. I have become very emotional and I have no control around my emotions—good or bad. Also, on top of that there is this feeling that I am no longer a proper woman. [Iran]

Like this woman from Iran, several women in the Fiji group expressed particular concern regarding the psychological symptoms associated with menopause. Stress and forgetfulness were common, and most worrying: 'Sometimes I just get fed up of doing everything. I am beginning to forget about where I leave things.' One woman linked her depression to the exhaustion she was feeling. All of the men in the group saw mood swings as a sign and symptom in their female partners. Women reported high levels of irritability and a low tolerance toward family members, with one Itaukei woman explaining how difficult this change has been for her. Her children and grandchildren say, 'you angry old woman,' and 'every time you always angry.' This hurt her deeply but she didn't know it was due to menopause, so she was unable to explain her frustration to her family members.

Most of the women interviewed experienced other physical changes as their periods ended: sweating, hot flashes, fatigue, irregular and often heavy periods, migraines, weight fluctuations, and dry skin. Although some expressed these changes were annoying or uncomfortable, they understood that menopause is not itself a disease, and most of the changes are not illnesses. But one of the Brazilian women discussed the difficulty of managing these changes, saying that some women think 'they are ill once they start their menopause process, they think they will die.' The irregular and often heavier periods were cited as particularly troublesome:

> I go to the gym, so I use a menstrual cup, but some of my friends find it strange. Due to the lack of information and the right products to deal with heavy flows, some women over this period isolate themselves as they don't want to risk having problems with leakage and so forth. (Brazil)

Two Brazilian participants mentioned having experienced negative physical aspects of menopause, as well as psychological (depression, anxiety) and emotional (PMS, irritability) symptoms. One of the three reported no negative changes and referred to menopause as a very light and uneventful time in her life.

A woman from India reported struggling with physical and emotional changes: 'Physically, I had a lot of symptoms—hot flashes, dryness of skin, tiredness, pigmentation. I would cry at the drop of a hat, but also wild mood swings. . . . I think I also felt helpless at things in general then. So, I used this to cry a lot.' When asked if she had learned any cultural practices to ease her discomfort, she said, 'Actually my life coach suggested Evening Primrose tablets. Later my gynecologist prescribed some other tablet, which would help to cope with the fluctuating hormones. But nothing cultural.' Her mother, 21 years older at 76, chimed in:

> We used a lot of Ayurveda to ease our symptoms. It was easy for us. Most of the medicines could be made from herbs in our kitchen gardens, so we could put them together, without telling anyone else, to help ease pain or nausea, or ease hot flashes. I remember telling your sister to have some when she had her hysterectomy. (India)

Many women in the Fiji focus group stated that they felt uncomfortable with their aging bodies as they started managing changes but, with time, have come to accept these changes. How women in Fiji come to understand and respond to menopause appears to be linked with their social status and ethnic background. While Indo-Fijian women frequently sought medical advice for their symptoms, Itaukei women had no routine way of responding to menopausal symptoms. The mother from Iran reported also that women in Iran have few options to manage the physical and emotional changes of menopause, so

> [t]hey just internalize it. They don't really do much about it. The doctors don't know much about it so there isn't much to do really. The doctors in the UK have more knowledge and can help, but in Iran we struggle. On a daily basis, well, you pray that it will be a normal day without any unpredicted episodes.

Physical changes, as well as emotional and psychological changes, can impact one's relationships with others. All of the women in the Fiji group experienced fatigue, for instance, affecting their family and work lives. Some worried about the reduction in sexual desire they were feeling, even though they felt good about their periods ending. One of the participants from Brazil mentioned that vaginal dryness was really disturbing her life and significantly affected her sexual life:

> With vaginal dryness, it's very painful having sex. As a married woman for many years, I can say that not all husbands are concerned about giving pleasure to their wives in bed, and they [do] not care if their wives are wet enough to have penetra[tive sex]. Many of my friends are embarrassed to buy lubricants because their husbands have prejudice if their wives use such "heretic" products.

As with other physical changes, there is great variation here, too. An Indian grandmother in her seventies said, 'Again it depends. I had a hysterectomy.

And I could still, you know, be satisfied. It was not difficult. I don't know why people think of them as impossible.'

CHANGING MENOPAUSE EXPERIENCES

[W]e need much more education before it happens. I also think we need more empathy for what women go through. It's hard to not understand the changes you go through. There is so much stigma and taboo around menopause in society. It's like menopause is disgusting and whatever happens to women during this time doesn't matter. (Iran)

One of the last questions asked in these informal interviews was some version of, 'What else is important to understand about women's experiences of menopause?' The women from Brazil urged the inclusion of information about menopause along with education about menarche and an end to the medicalization of menopause.

Commonly during the menopause women lose the vaginal muscle tone and they might become incontinent. This kind of information should have been taught to us [when] we were a child. Our vagina is like a muscle like any other, so we could have exercised them to prevent the loss of vaginal tone. By exercising that muscle certainly it would have been more firm during menopause. For us to have a better relationship with menopause, we would need to change our first experience with menstruation, talking more about our body changes and stop sending the message that women should only be mothers.

They also emphasized that everyone's experience is different and that every menstruator should have access to what she needs, although these Brazilian women prefer to see more alternative approaches to help manage menopause.

The Fiji participants all expressed the need for more information about menopause in schools and in the community. One participant said, 'What we need is menopausal preparedness, like they do at schools with menstruation. . . . It is easier for young people to get information. Not for women in villages who don't have phones.' They want to see a deeper understanding of what women go through and why. Younger men in the group believe greater knowledge of menopause would help them argue less and communicate more effectively with their female partners. Men in the group expressed interest in attending information sessions on menopause.

CONCLUSION

Although we point out common themes and many similarities in these discussions of menopause experiences, we caution readers against generalizing from these stories. None of these women or men represent their countries or even necessarily their specific communities. Their words represent only their own individual experiences. Nevertheless, their stories are important and worth sharing.

These stories reiterate to us the importance of making menopause culturally visible. This means representation in the broadest sense, including literary and media representation that tells women's stories of menopause and menopausal women's stories. It also calls for greater openness about menopause in all aspects of life. Our Iranian respondent said, 'Women don't really show the transition to menopause. Our society is not welcoming towards this.' Let's make it welcome, with more talk and more stories. Elizabeth is going to start by sending this essay to Aunt Patricia.

INDEX

© The Editor(s) (if applicable) and The Author(s) 2020
C. Bobel et al. (eds.), *The Palgrave Handbook of Critical Menstruation Studies*, https://doi.org/10.1007/978-981-15-0614-7

GPSR Compliance
The European Union's (EU) General Product Safety Regulation (GPSR) is a set
of rules that requires consumer products to be safe and our obligations to
ensure this.

If you have any concerns about our products, you can contact us on

ProductSafety@springernature.com

In case Publisher is established outside the EU, the EU authorized
representative is:

Springer Nature Customer Service Center GmbH
Europaplatz 3
69115 Heidelberg, Germany

2 04